Feature Papers in Eng 2023

Feature Papers in Eng 2023

Editor

Antonio Gil Bravo

MDPI

Basel • Beijing • Wuhan • Barcelona • Belgrade • Novi Sad • Cluj • Manchester

Editor
Antonio Gil Bravo
Universidad Publica de
Navarra
Pamplona
Spain

Editorial Office
MDPI
St. Alban-Anlage 66
4052 Basel, Switzerland

This is a reprint of articles from the Special Issue published online in the open access journal *Eng* (ISSN 2673-4117) (available at: https://www.mdpi.com/journal/eng/special_issues/1E75I822B4).

For citation purposes, cite each article independently as indicated on the article page online and as indicated below:

Lastname, A.A.; Lastname, B.B. Article Title. *Journal Name* **Year**, *Volume Number*, Page Range.

ISBN 978-3-7258-0871-7 (Hbk)
ISBN 978-3-7258-0872-4 (PDF)
doi.org/10.3390/books978-3-7258-0872-4

Contents

eng

MDPI

Editorial

Special Issue: Feature Papers in *Eng* 2023

Antonio Gil Bravo

INAMAT^2, Science Department, Public University of Navarra, Building Los Acebos, Campus of Arrosadia, E-31006 Pamplona, Spain; andoni@unavarra.es

1. Introduction

The aim of this third *Eng* Special Issue is to collect experimental and theoretical re-search relating to engineering science and technology. The topics included in *Eng* are as follows: electrical, electronic, and information engineering; chemical and materials engineering; energy engineering; mechanical and automotive engineering; industrial and manufacturing engineering; civil and structural engineering; aerospace engineering; biomedical engineering; geotechnical engineering and engineering geology; and ocean and environmental engineering. Many of these topics have been selected with the idea of contributing to the circular economy and sustainable development. Therefore, these aspects are being addressed from various points of view and have the support of the field of engineering and its applications. The following editorial presents a representative selection of these articles published in our journal in 2023.

Legislative requirements and the principles of the circular economy and sustainable development make waste valorization the best strategy for its management. The biodegradable fraction of industrial waste is a sustainable source of biomass, optimizing its management through energy recovery, reducing the amount of waste to be managed (and its economic costs), minimizing the environmental impact and health risks, and reducing the high dependence on industries on primary sources and fossil fuels [1–7]. Although traditional sources of biomass, such as wood, crops, agricultural and forestry residues, and food and municipal wastes, are renewable, sustainable, and profitable, they compete with food, and their energy processes release waste into the environment. On the other hand, there is also a non-biodegradable fraction of industrial waste. If its composition is taken into account, in many cases it is considered an inorganic waste. Here, its valorization is not as standardized as the biodegradable fraction, since the energy aspect is rarely taken into account, and it is the possible applications that give rise to the interest in valorizing this material, rather than depositing it in landfills [8].

Other topics discussed in this Special Issue are as follows:

- Amazon natural fibers for application in engineering composites and sustainable actions [9–12].
- Rheological behavior of modern cementitious materials [13–15].
- Vibration monitoring techniques for predictive maintenance of rotating machinery [16,17].
- Integrating multi-criteria decision-making methods with sustainable engineering [18,19].

These topics allow for greater discussion among potential readers. For more information, please see the Contributions.

2. Overview of the Published Articles

This Special Issue contains 34 papers, including eight reviews, published by several authors interested in cutting-edge developments in the field of engineering. The authors are from 25 countries, including Australia, Canada, Bosnia and Herzegovina, Brazil, Egypt, France, Germany, Greece, Hong Kong, Hungary, India, Italy, Japan, Mexico, Montenegro, Portugal, Russia, Serbia, Sweden, Taiwan, The Netherlands, United Arab Emirates, United Kingdom, USA, and Yemen.

Citation: Gil Bravo, A. Special Issue: Feature Papers in *Eng* 2023. *Eng* **2024**, 5, 562–565. https://doi.org/10.3390/eng5020031

Received: 28 March 2024
Revised: 29 March 2024
Accepted: 29 March 2024
Published: 3 April 2024

3. Conclusions

The articles published in this Special Issue present important advancements in the field of this journal. I would like to express my sincere gratitude to all the authors, who have contributed to this Special Issue, and I would also like to thank the managing editors and reviewers who contributed by improving the papers. I hope that the included articles are interesting and inspiring for readers, especially young scholars who are eager to learn about recent advances and contribute future research to the field.

Acknowledgments: The author is grateful for financial support from the Spanish Ministry of Science and Innovation (MCIN/AEI/10.13039/501100011033) through project PID2020-112656RB-C21.

Conflicts of Interest: The author declares no conflicts of interest.

List of Contributions

1. Kara, P.; Ognjanovic, I.; Maindorfer, I.; Mantas, J.; Wippelhauser, A.; Šendelj, R.; Laković, L.; Roganović, M.; Reich, C.; Simon, A.; et al. The Present and Future of a Digital Montenegro: Analysis of C-ITS, Agriculture, and Healthcare. *Eng* **2023**, *4*, 341–366. https://doi.org/10.3390/eng4010021.
2. Junio, R.; da Silveira, P.; Neuba, L.; Monteiro, S.; Nascimento, L. Development and Applications of 3D Printing-Processed Auxetic Structures for High-Velocity Impact Protection: A Review. *Eng* **2023**, *4*, 903–940. https://doi.org/10.3390/eng4010054.
3. Elshaboury, N.; Mohammed Abdelkader, E.; Al-Sakkaf, A.; Zayed, T. A Critical Review and Bibliometric Analysis on Applications of Ground Penetrating Radar in Science Based on Web of Science Database. *Eng* **2023**, *4*, 984–1008. https://doi.org/10.3390/eng4010059.
4. Sharafeldin, M.; Ksaibati, K.; Gerow, K. Frequency Analysis of Equivalent Property-Damage-Only (EPDO) Crashes at Intersections. *Eng* **2023**, *4*, 1116–1126. https://doi.org/10.3390/eng4020064.
5. Hasan, A.; Hasan, U.; AlJassmi, H.; Whyte, A. Transit Behaviour and Sociodemographic Interrelation: Enhancing Urban Public-Transport Solutions. *Eng* **2023**, *4*, 1144–1155. https://doi.org/10.3390/eng4020066.
6. Comuzzi, L.; Tumedei, M.; Di Pietro, N.; Romasco, T.; Montesani, L.; Piattelli, A.; Covani, U. Are Implant Threads Important for Implant Stability? An In Vitro Study Using Low-Density Polyurethane Sheets. *Eng* **2023**, *4*, 1167–1178. https://doi.org/10.3390/eng4020068.
7. Pontarolli, R.; Bigheti, J.; de Sá, L.; Godoy, E. Microservice-Oriented Architecture for Industry 4.0. Eng **2023**, *4*, 1179–1197. https://doi.org/10.3390/eng4020069.
8. Johnson, J. Rapid Prediction of Leaf Water Content in Eucalypt Leaves Using a Handheld NIRS Instrument. *Eng* **2023**, *4*, 1198–1209. https://doi.org/10.3390/eng4020070.
9. Kakinuma, T.; Hisada, M. A Numerical Study on the Response of a Very Large Floating Airport to Airplane Movement. *Eng* **2023**, *4*, 1236–1264. https://doi.org/10.3390/eng4020073.
10. Flamarion, M.; Castro, E.; Ribeiro-Jr, R. Pressure Anomalies Beneath Solitary Waves with Constant Vorticity. *Eng* **2023**, *4*, 1306–1319. https://doi.org/10.3390/eng4020076.
11. Shyam Kumar, P.; Ramasamy, M.; Varadan, V. A Comparison of Personalized and Generalized LSTM Neural Networks for Deriving VCG from 12-Lead ECG. *Eng* **2023**, *4*, 1337–1355. https://doi.org/10.3390/eng4020078.
12. Shan, F.; He, X.; Xu, H.; Armaghani, D.; Sheng, D. Applications of Machine Learning in Mechanised Tunnel Construction: A Systematic Review. *Eng* **2023**, *4*, 1516–1535. https://doi.org/10.3390/eng4020087.
13. Štilić, A.; Puška, A. Integrating Multi-Criteria Decision-Making Methods with Sustainable Engineering: A Comprehensive Review of Current Practices. *Eng* **2023**, *4*, 1536–1549. https://doi.org/10.3390/eng4020088.
14. Nilimaa, J.; Zhaka, V. An Overview of Smart Materials and Technologies for Concrete Construction in Cold Weather. *Eng* **2023**, *4*, 1550–1580. https://doi.org/10.3390/eng4020089.
15. Rosta, S.; Gáspár, L. Skid Resistance of Asphalt Pavements. *Eng* **2023**, *4*, 1597–1615. https://doi.org/10.3390/eng4020091.
16. García-Pintos, C.; Aldana-Murillo, N.; Ovalle-Magallanes, E.; Martínez, E. A Deep Learning-Based Visual Map Generation for Mobile Robot Navigation. *Eng* **2023**, *4*, 1616–1634. https://doi.org/10.3390/eng4020092.

17. Lee, J.; Tayfur, H.; Hamza, M.; Alzahrani, Y.; Diaconeasa, M. A Limited-Scope Probabilistic Risk Assessment Study to Risk-Inform the Design of a Fuel Storage System for Spent Pebble-Filled Dry Casks. *Eng* **2023**, *4*, 1655–1683. https://doi.org/10.3390/eng4020094.
18. Fiorin, F.; Sartori, L.; Méndez, M.; Ferreira, C.; França, M.; Krueger, E. The Learning Curve of People with Complete Spinal Cord Injury Using a NESs-FESs Interface in the Sitting Position: Pilot Study. *Eng* **2023**, *4*, 1711–1722. https://doi.org/10.3390/eng4020097.
19. Romanssini, M.; de Aguirre, P.; Compassi-Severo, L.; Girardi, A. A Review on Vibration Monitoring Techniques for Predictive Maintenance of Rotating Machinery. *Eng* **2023**, *4*, 1797–1817. https://doi.org/10.3390/eng4030102.
20. El Bitouri, Y. Rheological Behavior of Cement Paste: A Phenomenological State of the Art. *Eng* **2023**, *4*, 1891–1904. https://doi.org/10.3390/eng4030107.
21. da Silva, F.; de Castro, P. Redesign of a Failed Hoisting Shaft of a Vertical Transfer Device. *Eng* **2023**, *4*, 1981–2002. https://doi.org/10.3390/eng4030112.
22. Hasan, U.; AlJassmi, H.; Hasan, A. Microsimulation Modelling and Scenario Analysis of a Congested Abu Dhabi Highway. *Eng* **2023**, *4*, 2003–2014. https://doi.org/10.3390/eng4030113.
23. Chliveros, G.; Kontomaris, S.; Letsios, A. Automatic Identification of Corrosion in Marine Vessels Using Decision-Tree Imaging Hierarchies. *Eng* **2023**, *4*, 2090–2099. https://doi.org/10.3390/eng4030118.
24. Sosnin, M.; Berestenko, E.; Mounassar, E.; Shorstkii, I. Cold Plasma Technology for Tomato Processing By-Product Valorization: The Case of Tomato Peeling and Peel Drying. *Eng* **2023**, *4*, 2167–2177. https://doi.org/10.3390/eng4030124.
25. Durden, S.; Vilnrotter, V.; Shaffer, S. Assessment of Techniques for Detection of Transient Radio-Frequency Interference (RFI) Signals: A Case Study of a Transient in Radar Test Data. *Eng* **2023**, *4*, 2191–2203. https://doi.org/10.3390/eng4030126.
26. Provatidis, C. Power Transmission Using Circular Elements Bounded by Given Central Angle in Rolling Contact. *Eng* **2023**, *4*, 2309–2335. https://doi.org/10.3390/eng4030132.
27. Oliveira, C.; Lima, T. Setup Time Reduction of an Automotive Parts Assembly Line Using Lean Tools and Quality Tools. *Eng* **2023**, *4*, 2352–2362. https://doi.org/10.3390/eng4030134.
28. Gonzalez, Y.; Prati, R. Comparative Study of Musical Timbral Variations: Crescendo and Vibrato Using FFT-Acoustic Descriptor. *Eng* **2023**, *4*, 2468–2482. https://doi.org/10.3390/eng4030140.
29. Schmitz, M.; Menz, F.; Grunau, R.; Mandischer, N.; Hüsing, M.; Corves, B. Robot Cooking—Transferring Observations into a Planning Language: An Automated Approach in the Field of Cooking. *Eng* **2023**, *4*, 2514–2524. https://doi.org/10.3390/eng4040143.
30. Franco, B.; Alves Ribeiro, J.; Sánchez-Arce, I. Development of an Airbag Geometry Specific for Autonomous Vehicles. *Eng* **2023**, *4*, 2553–2570. https://doi.org/10.3390/eng4040146.
31. Rodrigues, E.; Santos, D.; Buiochi, F. Comparison of Three Linear Digital Filters Applied to Improve the Quality of the Measured Acoustic Field. *Eng* **2023**, *4*, 2582–2600. https://doi.org/10.3390/eng4040148.
32. Koury Costa, G. Modelling Air Flow through Pneumatic Valves: A Brief Review with an Experimental Case Study. *Eng* **2023**, *4*, 2601–2614. https://doi.org/10.3390/eng4040149.
33. Patle, B.; Patel, B.; Jha, A.; Kashyap, S. Self-Directed Mobile Robot Navigation Based on Functional Firefly Algorithm (FFA). *Eng* **2023**, *4*, 2656–2681. https://doi.org/10.3390/eng4040152.
34. da Silveira, P.; Cardoso, B.; Marchi, B.; Monteiro, S. Amazon Natural Fibers for Application in Engineering Composites and Sustainable Actions: A Review. *Eng* **2024**, *5*, 133–179. https://doi.org/10.3390/eng5010009.

References

1. Castaldi, M.M.; van Deventer, J.; Lavoie, J.M.; Legrand, J.; Nzihou, A.; Pontikes, Y.; Py, X.; Vandecasteele, C.; Vasudevan, P.T.; Verstraete, W. Progress and Prospects in the Field of Biomass and Waste to Energy and Added-Value Materials. *Waste Biomass Valor.* **2017**, *8*, 1875–1884. [CrossRef]
2. Qazi, W.A.; Abushammala, M.F.M.; Azam, M.-H.; Younes, M.K. Waste-to Energy Technologies: A Literature Review. *J. Solid Waste Technol. Manag.* **2018**, *44*, 387–409. [CrossRef]
3. Bolis, V.; Capón-García, E.; Roca-Puigros, M.; Gazzola, A.; Hungerbühler, K. Opimal Design and Management of Industrial Waste-to-Energy Systems. *Ind. Eng. Chem. Res.* **2019**, *58*, 3735–3753. [CrossRef]
4. Seljak, V.; Buffi, M.; Valera-Medina, A.; Chong, C.T.; Chiaramonti, D.; Katrašnik, T. Bioliquids and their Use in Power Generation—A Technology Review. *Renew Sustain. Energy Rev.* **2020**, *129*, 109930. [CrossRef]

5. Bora, A.P.; Gupta, D.P.; Durbha, K.S. Sewage Sludge to Bio-Fuel: A Review on the Sustainable Approach of Transforming Sewage Waste to Alternative Fuel. *Fuel* **2020**, *259*, 116262. [CrossRef]
6. Gil, A. Current insights into Lignocellulose Related Waste Valorization. *Chem. Eng. J. Adv.* **2021**, *8*, 100186. [CrossRef]
7. Gil, A. Challenges on Waste-to-Energy for the Valorization of Industrial Wastes: Electricity, Heat and Cold, Bioliquids and Biofuels. *Environ. Nanotechnol. Monit. Manag.* **2022**, *17*, 100615. [CrossRef]
8. Gil, A.; Korili, S.A. Management and Valorization of Aluminum Saline Slags: Current Status and Future Trends. *Chem. Eng. J.* **2016**, *289*, 7484. [CrossRef]
9. Kieling, A.C.; de Macedo Neto, J.C.; Del Pino, G.G.; da Silva Barboza, R.; Diáz, F.R.V.; Rivera, J.L.V.; Fernández, M.V.; Ketterer, C.G.; Ortega, A.G.; Abarzúa, R.I. Development of an Epoxy Matrix Hybrid Composite with Astrocaryum Aculeatum (Tucumã) Endocarp and Kaolin from the Amazonas State in Brazil. *Polymers* **2023**, *15*, 2532. [CrossRef] [PubMed]
10. Neto, J.S.; de Queiroz, H.F.; Aguiar, R.A.; Banea, M.D. A Review on the Thermal Characterisation of Natural and Hybrid Fiber Composites. *Polymers* **2021**, *13*, 4425. [CrossRef] [PubMed]
11. Kieling, A.C.; Santana, G.P.; Santos, M.C.D.; Neto, J.C.D.M.; Pino, G.G.D.; Santos, M.D.D.; Duvoisin Jr, S.; Panzera, T.H. Wood-Plastic Composite Based on Recycled Polypropylene and Amazonian Tucumã (*Astrocaryum aculeatum*) Endocarp Waste. *Fibers Polym.* **2021**, *22*, 2834–2845. [CrossRef]
12. Neves, P.D.; Cabral, M.R.; Santos, V.; Mafra, M.R.P.; Junior, H.S. Technical Assessment of Leaf Fibers from Curaua: An Amazonian Bioresource. *J. Nat. Fibers* **2022**, *19*, 5900–5909. [CrossRef]
13. Nicolas, R.; Richard, B.; Nicolas, D.; Irina, I.; Temitope, K.J.; Dirk, L.; Viktor, M.; Romain, M.; Arnaud, P.; Ursula, P.; et al. Assessing the Fresh Properties of Printable Cement-Based Materials: High Potential Tests for Quality Control. *Cem. Concr. Res.* **2022**, *158*, 106836. [CrossRef]
14. Biricik, Ö.; Mardani, A. Parameters Affecting Thixotropic behavior of Self Compacting Concrete and 3D Printable Concrete; a State-of-the-Art Review. *Constr. Build. Mater.* **2022**, *339*, 127688. [CrossRef]
15. Li, R.; Lei, L.; Plank, J. Influence of PCE Superplasticizers on the Fresh Properties of Low Carbon Cements Containing Calcined Clays: A Comparative Study of Calcined Clays from Three Different Sources. *Cem. Concr. Compos.* **2023**, *139*, 105072. [CrossRef]
16. Lee, S.B.; Stone, G.C.; Antonino-Daviu, J.; Gyftakis, K.N.; Strangas, E.G.; Maussion, P.; Platero, C.A. Condition Monitoring of Industrial Electric Machines: State of the Art and Future Challenges. *IEEE Ind. Electron. Mag.* **2020**, *14*, 158–167. [CrossRef]
17. Shukla, K.; Nefti-Meziani, S.; Davis, S. A Heuristic Approach on Predictive Maintenance Techniques: Limitations and Scope. *Adv. Mech. Eng.* **2022**, *14*, 16878132221101009. [CrossRef]
18. Stojcic, M.; Zavadskas, E.K.; Pamučar, D.; Stević, Ž.; Mardani, A. Application of MCDM Methods in Sustainability Engineering: A Literature Review 2008–2018. *Symmetry* **2019**, *11*, 350. [CrossRef]
19. Banasik, A.; Bloemhof-Ruwaard, J.M.; Kanellopoulos, A.; Claassen, G.; Van Der Vorst, J.G. Multi-Criteria Decision Making Approaches for Green Supply Chains: A Review. *Flex. Serv. Manuf. J.* **2018**, *30*, 366–396. [CrossRef]

eng

Article

Self-Directed Mobile Robot Navigation Based on Functional Firefly Algorithm (FFA)

Bhumeshwar K. Patle [1], Brijesh Patel [2,*], Alok Jha [3] and Sunil Kumar Kashyap [4]

1 Department of Mechanical Engineering, MIT Art, Design and Technology University, Pune 412201, India; balu_patle@rediffmail.com
2 Department of Mechanical Engineering, National Taiwan University of Science and Technology, Taipei 10607, Taiwan
3 Department of Mechanical Engineering, Shri Ramdeobaba College of Engineering and Management, Nagpur 440013, India; alokjha.30@gmail.com
4 Department of Mathematics, MATS University, Raipur 492001, India; drsunilkk@matsuniversity.ac.in
* Correspondence: aero.brijesh@gmail.com

Abstract: This paper proposes an optimized mobile robot navigation strategy using a functional firefly algorithm (FFA) and choice function. This approach has two key advantages: first, the linear objective function performs efficiently with the single degree and finite-order polynomial time operation, and second, the cartesian constraint performs compactly with the chosen degree of freedom on the finite interval. This functional approach optimizes the size of operational parameters in context with key size, operation time, and a finite range of verification. The choice function achieves parameter order (size) reduction. The attraction characteristic of fireflies is represented by the choice function for optimizing the choice between low and high intensities of fireflies. In 2D and 3D environments, the proposed robot navigation performs well in an uncertain environment with static and dynamic obstacles. This efficiency includes the robot's speed as determined by the choice function's minimum path lengths. The collision-free path is achieved by the non-void family of non-void sets. The obtained results are optimal in terms of path length and navigational time. The proposed controller is also compared with the other existing controllers, and it is observed that the FFA gives the shortest path in less time for the same environmental condition.

Keywords: mobile robot navigation; firefly algorithm; choice function; path planning; obstacle avoidance

Citation: Patle, B.K.; Patel, B.; Jha, A.; Kashyap, S.K. Self-Directed Mobile Robot Navigation Based on Functional Firefly Algorithm (FFA). *Eng* **2023**, *4*, 2656–2681. https://doi.org/10.3390/eng4040152

Academic Editor: Antonio Gil Bravo

Received: 9 September 2023
Revised: 15 October 2023
Accepted: 16 October 2023
Published: 18 October 2023

1. Introduction

With the growing demand for autonomous systems in household work, industry, entertainment, medical care, transportation, and especially national security, mobile robots play an important role and are heavily utilized. These mobile robots are used as UAVs, marine robots, or ground robots to perform critical tasks, especially where human interaction is impossible [1]. While performing the task in an unstructured environment, autonomous navigation is the major challenge for any mobile robot, as it involves many non-linear constraints. To enable the means for autonomous navigation, the mobile robot should be equipped with a power control unit, sensory mechanisms, and an intelligent path-planning algorithm [2]. The success of any autonomous system completely depends on selecting and implementing effective path planners. Therefore, path planners must be able to determine the best-fit parameters among all possibilities to generate an optimal path by avoiding obstacles. Autonomous navigation is not limited to a single objective function; hence, an artificially intelligent computational approach is required to deal with the multi-objective problem of fulfilling the goal of effective navigation in a complex, unstructured environment [3]. In mobile robot navigation (MRN) [4], the major challenges observed are path planning in the presence of static and dynamic obstacles, path planning in the presence of dynamic goals, and navigation in the presence of multiple robots. These challenges become very complex when counting over-optimization (obtaining the shortest path in the minimum navigational time). Additionally, they require significant computing efforts for seemingly simpler problems, including comparison with other intelligent approaches.

This work presents the application of the FA along with the CF as a multifunctional approach to static and dynamic conditions. Real-time robots encounter challenges in real-world scenarios, including noisy sensor data, perceptual uncertainties, dynamic environmental complexities, computational efficiency constraints, limited model generalization, and the critical need for safety and collision avoidance. To address these issues comprehensively, strategies such as sensor fusion for data refinement, adaptive obstacle avoidance, efficient algorithm design, machine learning for environment adaptation, and robust safety protocols must be implemented to enhance the robot's performance and reliability in dynamic, uncertain environments. Achieving the goal of navigation requires the quick computation of efficient paths, which is presented here by introducing first the CF and then the FA. The key advantage of adopting the CF in MRN is an extension of the option for selecting the optimum path by the characteristics of the non-void family and non-void set. Another noteworthy advantage is to present the optional optimal paths into the classes by FL. The obstacles are classified into non-void sets, allowing the robot to make a quick decision and become more efficient. The choice of probability and the choice of optimality are the two major characteristics of the choice function. These characteristics are applied to improve the firefly algorithm for mobile robot navigation. The chaos of fireflies is transformed into the index set of the non-void family in this paper. The obstacle avoidance function comprises the distributed probability, and the distance–time function comprises the fuzzy logic-based index set. Thus, MRN is studied and applied using mobile robot navigation, multiple mobile robots, chaotic target seeking, multiple targets seeking, chaotic obstacle position, topological spaces, and the firefly's micro- and macro-attraction. The simulation and real-time results are provided for validation in an uncertain environment, and the obtained results are optimal compared to other navigational controllers. As per the author's belief and knowledge, very little work has been published on the path planning of mobile robots in an unstructured environment in the presence of static and dynamic (obstacle and goal) conditions using FA.

This paper is organized as follows. The introduction to the literature review is presented in Section 2. Section 3 presents the proposed functional firefly algorithm with its mathematical analysis. In Section 4, the simulation and experimental results are presented and compared. The conclusion and the future scope are presented in Section 5.

2. Review of Literature

From the review of the available literature on intelligent path-planning techniques such as cell decomposition [5], fuzzy logic [6], neural network [7], particle swarm optimization algorithm [8], ant colony algorithm [9], bacterial foraging optimization [10], harmony search algorithm [11], cuckoo search algorithm [12], and dragonfly algorithm [13], it is clear that the applications of metaheuristic algorithms for solving mobile robot navigational problems are growing rapidly compared to heuristic algorithms due to their high-performance capabilities. The "randomization" and "local search" features of the metaheuristic algorithm are critical. Randomization provides a good way to move away from local search to search on a global scale, and therefore, the metaheuristic algorithm is intended to be suitable for global optimization.

In 2008, Yang [14] proposed the firefly algorithm based on the behavior of fireflies for solving various optimization problems in engineering. This firefly algorithm holds two main characteristics of fireflies, i.e., flashing patterns and biological behavior. However, this firefly algorithm follows the three fundamental principles under the two characteristics as defined below:

1. Fireflies are unisex, but their attraction is based on intensity rather than gender;
2. The attraction is proportional to brightness, from lesser brightness to greater brightness;
3. The brightness interacts with the landscape of the objective function.

These three rules are necessary and sufficient for applying the FA in various behavioral applications. The generalization of these rules is possible because of the specific requirements and applications, i.e., 3D navigation of mobile robots, target-seeking applications, chaotic obstacle positions, topological spaces, etc.

The basic formula of attractiveness interacts with the intensity of light presented as follows:

$$\beta = \beta_0 e^{-\gamma r^2} \tag{1}$$

where β is the variation of the attractiveness, γ is the light absorption coefficient, and β_0 is the attractiveness at $\gamma = 0$. Here, the choice of the axiom is applied to the existing firefly algorithm. Next, the movement of a firefly (*i*) attracts a firefly (*j*) due to greater brightness, and it is presented as follows:

$$x_i^{t+1} = x_i^t + \beta_0 e^{-\gamma r_{ij}^2}\left(x_j^t - x_i^t\right) + \alpha_t \epsilon_i^t \tag{2}$$

where α and ϵ are the randomized parameters, although ϵ is a vector of random numbers defined over the Gaussian number as uniform distribution.

FA has been studied and implemented to solve various optimization problems in engineering and science. The fault detection in robots [15], economic emission dispatched problem [16], reliability–redundancy optimization [17], mixed variable structural optimization problem [18], cooperative networking problem [19], combinatorial optimization problem [20], learning from demonstration problem [21], and the dynamic environment problem [22] are a few of them. The FA has shown great performance and created a good impact in the category of the population-based algorithm. The FA has the ability to solve multi-model optimization and extremely non-linear problems excellently. It has a better convergence speed for finding a global solution in a complex environment and starts iteration processes without a good initial solution. As a result, FA and CF are chosen here to investigate the environment for mobile robot navigation, which includes dynamic obstacles and dynamic goals. Many researchers use the FA to solve mobile robot navigation problems. However, very few of the papers meet the requirements of the navigation. The navigation of the mobile robot using FA in the presence of a static environment was demonstrated by Liu et al. [23] and Hidalgo-Paniagua et al. [24]. However, in both approaches, the navigation strategies were presented in the simulation environment in the presence of the static obstacle. The navigation task is difficult in the presence of a moving obstacle, so Brand et al. [25] and Patle et al. [26] worked to develop the FA for a dynamic environment. In their work, the environment with single and multiple moving obstacles was tested for a single mobile robot system, respectively. The analysis of a multi-mobile robot system in the presence of multiple obstacles was demonstrated by Kim et al. [27], but the results of the navigation were limited to the simulation environment only and not to the real environment. The study of single and multiple mobile robot systems over real-time experiments was carried out by Patle et al. [28,29]. His work shows multiple mobile robots navigating multiple obstacles. The application of FA is not only limited to ground robots but also includes the navigation of aerial and underwater robots. Wang et al. [30] developed an FA-based path-planning strategy for the aerial robot. It outperformed other approaches in avoiding hazardous areas in a complex, crowded environment and reducing fuel costs. Similarly, the FA-based underwater path planning strategy was examined by Sutantyon et al. [31]. Their work primarily focused on the scheduling strategy of swarm robots to avoid interface and jamming in underwater conditioning using the principles of FA. In other work, they also presented underwater navigation in a partially known environment using a leavy-flight–firefly-based approach. To explore and enhance the performance of the mobile robot system, the FA has been introduced into a few techniques such as Q-learning [32], invasive weed optimization [33], radial basis function neural network [34], pareto-based optimization algorithm [35], and many more. Table 1 gives an overview of the related work using the firefly algorithm.

Table 1. Comparison of Related Work for the Firefly Algorithm.

Reference No.	Single Robot/Multi-Robot/Aerial Robot/Underwater Robot	Simulation	Experimental	Static/Dynamic Obstacle	Hybrid Technique Used
[20]	Single	Y	N	Static	N
[21]	Single	Y	N	Static	N
[22]	Single	Y	N	Static and dynamic	N
[23]	Single	Y	N	Static and dynamic	N
[24]	Multi-robot	Y	N	Dynamic	N
[25]	Multi-robot	Y	Y	Dynamic	Y
[26]	Multi-robot	Y	Y	Dynamic	Y
[27]	Aerial robot	Y	N	Dynamic	N
[28]	Underwater	Y	N	Dynamic	N

3. Proposed Functional Firefly Algorithm

Various challenges exist in the real environment; some of them are the localization of position in the environment, the determination of a goal and an organized path towards it, obstacle avoidance mechanisms, and the generation of an optimal path in a minimal amount of time. Figure 1a shows that no obstacle is present in the environment between the robot and the goal position; hence, the robot reaches the goal position by using Euclidian distance, which is optimal. But in Figure 1b, when the robot moves from its initial position to the goal position, it detects the obstacle and stops. The obstacle avoidance mechanism activates, and then, the robot avoids the obstacle, as shown in Figure 1c. While avoiding the obstacles, it may produce several paths up to the goal position, while achieving the shortest path from the robot's initial position to the robot's goal position is the proposed study's primary goal. The proposed algorithm addresses the issues mentioned above of robot navigation over the choice function. Any mobile robot navigation is based on the likelihood of selecting the best path, which is a function of distance and time. Therefore, probability plays an important role in executing the navigation of mobile robots. As a non-void family, CF comprises a set of probabilistic choices. Here, the axiom of choice generalizes to FA. The fireflies are defined over a finite set, and the set of fireflies with a distributed probability then generates the function of choice. The non-void family of fireflies with a distributed probability comprises the classification. The proposed FFA holds several advancements in context to theory and application. Flashing pattern feasibility is defined on the finite interval to optimize the trajectory. The basic firefly algorithm is based on a variation of the attractiveness of fireflies. The variable β is attractiveness. β_0 exists if attractiveness is defined at the distance $\gamma = 0$.

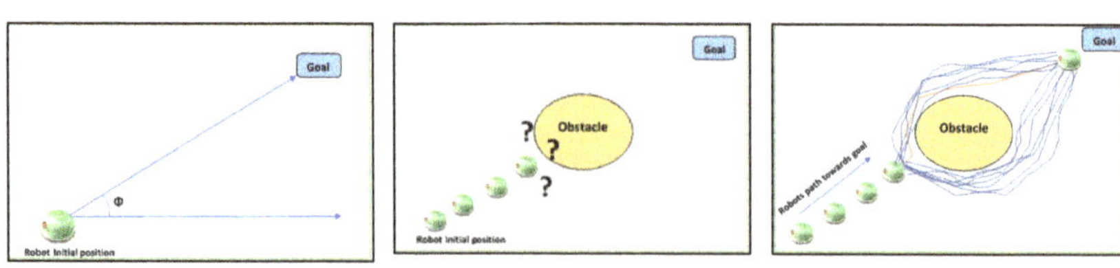

(a) (b) (c)

Figure 1. Navigational challenges for a mobile robot in (**a**) exploring the environment for the shortest distance, (**b**) obstacle avoidance, and (**c**) selection of an optimal path.

The key idea of this paper is to define attractiveness at the neighborhood of the distance of not $\gamma = 0$ exactly but very close to zero. This approach is called calculus. The neighborhood of β_0 is defined by an interval $(\beta_0 - \delta,\ \beta_0 + \delta)$, where δ is the small, positive real number. $\beta_0 - \delta$ is the point at the left-hand side from $\gamma = 0$, and $\beta_0 + \delta$ is the point at the right-hand side from $\gamma = 0$. The reason for applying this concept is to achieve optimization. This is an optimal input that gives the optimal output correspondingly. For example, if β_0 is 2, and δ is 0.0001, then the neighborhood will be $(2 - 0.0001,\ 2 + 0.0001)$ or $(1.9998,\ 2.0001)$. Thus, this is very close to 2 but not exactly equal to 2, and hence, the resultant output β will be approaching the corresponding value. Hence, Equation (3) holds the limit (L) represented by $|\beta - L| < \varepsilon$, where ε is a small, real positive number. Hence, Equation (3) is represented by a limit as follows:

$$lim_{\gamma \to 0} \beta = L \tag{3}$$

The attractiveness is measured from the left-hand side (from $\beta_0 - \delta$ to β_0) and denoted by β_e for attractiveness at the point $\beta_0 - \delta$, represented as follows:

$$lim_{\gamma \to 0^-} \beta^- = \beta_e e^{-\gamma r^2} \tag{4}$$

Similarly, the attractiveness is measured from the right-hand side (from $\beta_0 + \delta$ to β_0) and denoted by β_l for attractiveness at the point $\beta_0 + \delta$, represented as follows:

$$lim_{\gamma \to 0^+} \beta^+ = \beta_l e^{-\gamma r^2} \tag{5}$$

If $lim_{\gamma \to 0^-} \beta = lim_{\gamma \to 0^+} \beta$, then Equation (1) exists, which will be unique.

The source of brightness and its area are also formulated in the proposed FFA. It depends on the size of the firefly, although the difference is minor but measurable. The source of brightness (S) is defined as a function of β_0, β_e, and β_l, which is defined as follows:

$$S = f(\beta_0, \beta_e, \beta_l) \tag{6}$$

The distance between two fireflies that are attracted to each other is also reviewed as an advantage of the FFA. A firefly attracts those whose distance is less than other fireflies, although brightness is the same. The deviation in the distance is δ, although the intensity remains the same, presented as follows:

$$\text{Left-Hand Deviation}: \ \beta = \beta_e e^{-\gamma \delta^2} \tag{7}$$

$$\text{Right-Hand Deviation}: \ \beta = \beta_l e^{-\gamma \delta^2} \tag{8}$$

The position of the firefly is measured over its central tendency in the proposed FFA. This lies in the displacements, i.e., up, down, diagonal, horizontal, etc., although brightness is the same. The deviation in the position of the fireflies exists when the following is true:

$$\beta_e e^{-\gamma \delta^2} \neq \beta_l e^{-\gamma \delta^2} \tag{9}$$

The discrete approach of attractiveness is studied in this paper. The proposed FFA establishes robot navigation based on the natural conjugation of fireflies. The probability of attraction is only by brightness, but its discrete and continuous distributions are also generalized in the proposed FFA. The choice function plays a crucial role in executing the idea of a functional firefly algorithm for robot navigation. This approach achieves optimum navigation. The application of the choice function receives the dynamic decision. The choice function is a mathematical rule applied as the association of the elements of the two non-empty sets such that each element of the first set has the unique image of the element of the second set. Notable is the self-map mechanism, where the pre-image and image are identical. This identity characteristic establishes a distinct path for navigation. The self-image approach is associated with the firefly position, and its pre-image selects the unique and optimal path. The mathematical definition of the choice function is presented as follows:

Choice Function: Let Λ be a non-void set and $\lambda \in \Lambda$. f is said to be a choice function if $f(\lambda) \in \lambda, \forall \lambda \in \Lambda$. Its application to robot navigation is presented as follows: Λ is the non-empty finite set of the position of fireflies, which is coordinated in three dimensions. The brighter firefly is denoted by $\lambda \in \Lambda$, and the movement is defined by the displacement from the less-bright firefly as the pre-image or the domain as the first set to the image as the co-domain of the second set. This displacement is referred to as the range $f(\lambda) \in \lambda$ of the function, which is the real number obtained by the choice rule.

Axiom of Choice Function: There exists a choice function (*f*) for each non-void family of non-void sets if $\{X_\lambda : \lambda \in \Lambda\}$ is a family of sets such that $\Lambda \neq \phi$ and x_λ for each $\lambda \in \Lambda$; then, there exists f on Λ such that $f(\lambda) \in X_\lambda$ for each $\lambda \in \Lambda$.

Cartesian-Choice Operator: Let $\{X_\lambda : \lambda \in \Lambda\}$ be an arbitrary collection of sets induced by Λ. Then, the cartesian product of this collection is the set of all mapping.

$$X : \Lambda \to \lambda = \{X_\lambda : \lambda \in \Lambda\} : X(\lambda) \in X_\lambda \tag{10}$$

For all, $\lambda \in \Lambda$, and it is denoted by the following:

$$\Lambda\{X_\lambda : \lambda \in \Lambda\} \text{ or by; } x\{X_\lambda : \lambda \in \Lambda\} \tag{11}$$

The set X_λ is called the λth coordinate set of the product. It is used as a symbol X_λ for the image $X(\lambda)$ of X under the mapping X. Here, Λ is an index set, and $\lambda \in N$ is the set of natural numbers.

Then,

$$X_\lambda = \{x : x \in N, x \text{ is the multiple of } N\}$$

Hence,

$$X_1 = \{1, 2, 3, \ldots\}$$

$$X_2 = \{2, 4, 6, \ldots\ldots\ldots\ldots\ldots\ldots\}$$

$$X_n = \{n, 2n, 3n, \ldots\}$$

Here, f is the function of choice on Λ such that

$$f(\lambda) \in X_\lambda;\ \lambda \in \Lambda$$

Thus, the axiom of the choice function is generalized to the resultant formula of firefly for the attractiveness over the initial position, defined as follows:

$$x_i^{t+1} : \Lambda \to U\left\{x_\beta : \beta \in \Lambda\right\} : x(\lambda) \in x_\lambda \tag{12}$$

where Λ is the index set of existing properties of the FA, and U is the proposed transformation over the seven characteristics of fireflies. Hence, the resultant formula of attraction over the axiom of choice is presented as follows:

$$x_i^{t+1} = x_i^t + \frac{\Lambda}{U}[\left(\beta_0 e^{-\gamma r_{ij}^2}\right)\left(x_j^t - x_i^t\right)] + \alpha_t \epsilon_i^t \tag{13}$$

Here, Λ is an index set.

$\lambda \in N$ (the set of natural numbers)

Then,

$$x_\lambda = \{x : x \in N,\ x \text{ is the multiple of } N\}$$

The robot navigation architecture is presented in Figure 2. In an uncertain environment, the finite set of obstacles is $O = \{o_1, \ldots, o_n\}$, the finite set of fireflies of lower and higher intensity of light is $X = \left\{x_1^l, x_1^h, \ldots, x_n^l, x_n^h\right\}$, and the robots' initial position is $R(x_1^l)$ and the goal position $G(x_n^h)$. The robot (less-bright firefly) $R(x_1^l)$ is attracted towards the brighter firefly (x_1^h) with obstacle o_1 avoidance by the optimal choice function $f(\lambda_1); \lambda_1 \in \Lambda$. Similarly, $R(x_1^l)$ follows the preceding rule and reaches the goal $G(x_n^h)$ by the optimal choice function navigation: $C : R(x_1^l) \overset{f(\lambda_n)}{\rightarrow} G(x_n^h)$.

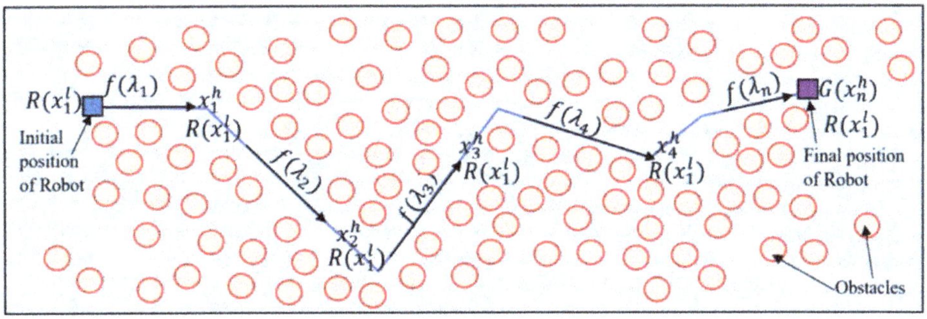

Figure 2. Functional firefly navigation space.

The robot navigation follows the attraction behavior of fireflies as the difference $(I^H - I^L \geq 0)$ of the intensity of light of fireflies by $(I^H > I^L)$, where I^H and I^L are high intensity and low intensity, respectively. Thus, the objectives of optimal navigation, i.e., the shortest path in the minimal time and collision-free path, are achieved by the CF simultaneously. Let the non-void family of the non-void set be $X = \left\{\left\{x_1^l, x_1^h\right\}, \ldots, \left\{x_n^l, x_n^h\right\}\right\}$; then, the choice function is $f(x) = \left\{\left\{x_1^l\right\}, \left\{x_1^l, x_2^h\right\}, \left\{x_2^l, x_3^h\right\}, \ldots, \left\{x_n^h\right\}\right.$. Thus, the firefly-choice function-based robot navigation is presented as FFA: $f : R \rightarrow G$.

The set of lower-intensity fireflies = $X_1 = X_\lambda^l = \left\{x_1^l, \ldots, x_n^l\right\}$;

The set of higher-intensity fireflies = $X_2 = X_\lambda^h = \left\{x_1^h, \ldots, x_n^h\right\}$.

Then, the set of functional fireflies is defined by the choice function as follows:

$$C = \{C_l, C_h\} = \left\{\left\{x_{n-1}^l, x_n^l\right\}, \left\{x_{n-1}^h, x_n^h\right\}\right\} = \{X_\lambda : \lambda \in \Lambda\} \quad (14)$$

Here, f is a function defined on Λ such that $f(\lambda) \in X_\lambda$ for each $\lambda \in \Lambda$. This choice function is applied to the above-defined sets of fireflies and modified again as follows:

$$f : \lambda \rightarrow \lambda \text{ Or, } f : X_\lambda \rightarrow X_\lambda \text{ Or, } f : X_1 \rightarrow X_2 \text{Or, } X_2 = f(X_1) \quad (15)$$

Next, the objective function is formulated. The decision variables are as given: The choice-function-based navigational path function is $f(\lambda_1), \ldots, f(\lambda_n)$; respective time is $t_1, \ldots, t_n$; and the navigation rate is $\frac{x_n^h - x_n^l}{t_n}$.

Hence, the objective function is given:

$$min\, f(\lambda) = \frac{\left(x_1^h - x_1^l\right)}{t_1} + \ldots + \frac{\left(x_n^h - x_n^l\right)}{t_1}, f(\lambda) \in X_\lambda, \lambda \in \Lambda \quad (16)$$

This is subject to $C_{m1}\frac{1}{t_1} + \ldots + C_{mn}\frac{1}{t_2} = f(\lambda_n)$ and $t_1, \ldots t_n \geq 0$. The navigation function table is presented as follows in Table 2.

Table 2. Navigation Functions.

SN.	Navigation Direction	Representations
1	Left Move	$\beta_0 + \delta$
2	Straight Move	$\beta_0 = \delta$
3	Right Move	$\beta_0 - \delta$
4	Up Move	β_0 / δ
5	Down Move	δ / β_0
6	Constant	β_0
7	Left Curve Move	$(\beta_0 + \delta)^n; n \geq 2$
8	Right Curve Move	$(\beta_0 - \delta)^n; n \geq 2$

Hence, by the Cartesian product of the X_1 and X_2, we find the move according to the position of obstacles. There are finite options to choose the move, as defined by the choice function.

$$f : X_1 \times X_2 = \left\{ \left(x_1^l, x_1^h \right), \ldots, \left(x_n^l, x_n^h \right) \right\} \tag{17}$$

The distance vector is defined for controlling the obstacle position. Let the set of the position point of the obstacle be $\{D_1, \ldots, D_n\}$. Each point has the vector by the characteristic of this set of finite sums of intervals that hold the probabilistic decision of optimal path. The probability of optimal path is $\{p(D_1), \ldots, p(D_n)\}$. The domain of probability for the rule (choice function) to decide the optimal path is represented by the following matrix:

$$D = \begin{bmatrix} d_{11} & . & . & . & . \\ . & . & . & . & . \\ . & . & . & . & . \\ . & . & . & . & . \\ . & . & . & . & d_{nn} \end{bmatrix}$$

Thus, the rule (choice function) is set by the co-domain of the probable vectors as $\{(d_{11}, \ldots, d_{1n}), \ldots, (d_{91}, \ldots, d_{9n})\}$. The permutation of the vectors of the co-domain for generating the range as a rule by the following vector spaces is illustrated in Tables 3–5.

Table 3. Vector Space-Based Navigational Decision.

Obstacle Position	Distance Classification	Choice of Axiom of (II)	Metric Decision Vector
	$D_1 \ldots D_N$	$A(D_1), \ldots A(D_N)$	$C_i\ [A(D_1), A(D_N)]$
LOD	$d_{11} \ldots d_{1n}$	$A(d_{11}), \ldots A(d_{1n})$	$[A(d_{11}), \ldots, A(d_{1n})]$
FOD	$d_{21} \ldots d_{2n}$	$A(d_{21}), \ldots A(d_{2n})$	$[A(d_{21}), \ldots, A(d_{2n})]$
ROD	$d_{31} \ldots d_{3n}$	$A(d_{31}), \ldots A(d_{3n})$	$[A(d_{31}), \ldots, A(d_{3n})]$
UOD	$d_{41} \ldots d_{4n}$	$A(d_{41}), \ldots A(d_{4n})$	$[A(d_{41}), \ldots, A(d_{4n})]$
DOD	$d_{51} \ldots d_{5n}$	$A(d_{51}), \ldots A(d_{5n})$	$[A(d_{51}), \ldots, A(d_{5n})]$
ULDOD	$d_{61} \ldots d_{6n}$	$A(d_{61}), \ldots A(d_{6n})$	$[A(d_{61}), \ldots, A(d_{6n})]$
URDOD	$d_{71} \ldots d_{7n}$	$A(d_{71}), \ldots A(d_{7n})$	$[A(d_{71}), \ldots, A(d_{7n})]$
DLDOD	$d_{81} \ldots d_{8n}$	$A(d_{81}), \ldots A(d_{8n})$	$[A(d_{81}), \ldots, A(d_{8n})]$
DRDOD	$d_{91} \ldots d_{9n}$	$A(d_{91}), \ldots A(d_{9n})$	$[A(d_{91}), \ldots, A(d_{9n})]$

where LOD, left obstacle distance; FOD, front obstacle distance; ROD, right obstacle distance, UOD, up obstacle distance; DOD, down obstacle distance; ULDOD, up left obstacle distance; URDOD, up right obstacle distance; DLDOD, down left obstacle distance; DRDOD, down right obstacle distance.

Hence, the turning or transformation or direction function of the mobile robot is given:

$$T = f\{X_D, X_S; D, \lambda \in \Lambda\} \tag{18}$$

The conditional rule for controlling the MRN is given as follows in Tables 6 and 7.

Table 4. Euclidean Metric Grading for Optimal Navigation.

Obstacle Position	Distance Classification $(D_1),\ldots(D_N)$	Choice of Axiom of (II) $[A(D_1),\ldots A(D_N)]$	Distance Decision Matrix $C_i\ [A(D_1),\ldots,A(D_N)]$
LOD		$[A(d_{11}),\ldots,A(d_{1n})]$	
FOD	VVN-VN-N-F-VF-VVF	$[A(d_{21}),\ldots,A(d_{2n})]$	
ROD	VVN—Very Very Near	$[A(d_{31}),\ldots,A(d_{3n})]$	
UOD	VN—Very Near	$[A(d_{41}),\ldots,A(d_{4n})]$	$\begin{bmatrix} A(d_{11}) & . & . \\ . & . & . \\ . & . & A(d_{9n}) \end{bmatrix}$
DOD	N—Near	$[A(d_{51}),\ldots,A(d_{5n})]$	
ULDOD	F—Far	$[A(d_{61}),\ldots,A(d_{6n})]$	
URDOD	VF—Very Far	$[A(d_{71}),\ldots,A(d_{7n})]$	
DLDOD	VVF—Very Very Far	$[A(d_{81}),\ldots,A(d_{8n})]$	
DRDOD		$[A(d_{91}),\ldots,A(d_{9n})]$	

Table 5. Speed Grading Rule for Fast Navigation.

Obstacle Position	Speed Classification $(S_1),\ldots,(S_N)$	Choice of Axiom of (II) $A(S_1),\ldots,A(S_N)$	Speed Decision Matrix $C_i\ [A(S_1),\ldots,A(S_N)]$
LOD		$A(S_{11})\ldots A(S_{1n})$	
FOD	VVF-VF-F-S-VS-VVS	$A(S_{21})\ldots A(S_{2n})$	
ROD	VVF—Very Very Fast	$A(S_{31})\ldots A(S_{3n})$	
UOD	VF—Very Fast	$A(S_{41})\ldots A(S_{4n})$	$\begin{bmatrix} A(S_{11}) & . & . \\ . & . & . \\ . & . & A(S_{9n}) \end{bmatrix}$
DOD	F—Fast	$A(S_{51})\ldots A(S_{5n})$	
ULDOD	S—Slow	$A(S_{61})\ldots A(S_{6n})$	
URDOD	VS—Very Slow	$A(S_{71})\ldots A(S_{7n})$	
DLDOD	VVS—Very Very Slow	$A(S_{81})\ldots A(S_{8n})$	
DRDOD		$A(S_{91})\ldots A(S_{9n})$	

Table 6. Linear If-Then Rule.

If	Then
$LOD = \frac{(LOD)_i}{HA}; FOD = \frac{(FOD)_i}{HA}; ROD = \frac{(ROD)_i}{HA}; HA = \frac{(HA)_i}{(LD)(RD)}; UOD = \frac{(UOD)_i}{HA};$ $ULDOD = \frac{(ULDOD)_i}{HA}; URDOD = \frac{(URDOD)_i}{HA}; DOD = \frac{(DOD)_i}{HA};$ $DLDOD = \frac{(DLDOD)_i}{HA}; DRDD = \frac{(DRDOD)_i}{HA}$	$\frac{HV}{HV_{ijkl}} \cdot \frac{VV}{VV_{ijkl}}$

where HV, horizontal velocity; VV, vertical velocity.

Table 7. Non-Linear If-Then Rule.

If	Then
$\frac{(LOD)_i}{LOD} \cdot \frac{(FOD)_i}{FOD} \cdot \frac{(ROD)_i}{ROD} \cdot \frac{(HA)_i}{HA} \cdot \frac{(UOD)_i}{UOD} \cdot \frac{(ULDOD)_i}{ULDOD} \cdot \frac{(URDOD)_i}{URDOD} \cdot \frac{(DOD)_i}{DOD} \cdot \frac{(DLDOD)_i}{DLDOD}$ $\frac{LOD}{(LOD)_i} \cdot \frac{FOD}{(FOD)_i} \cdot \frac{ROD}{(ROD)_i} \cdot \frac{HA}{(HA)_i} \cdot \frac{UOD}{(UOD)_i} \cdot \frac{ULDOD}{(ULDOD)_i} \cdot \frac{URDOD}{(URDOD)_i} \cdot \frac{DOD}{(DOD)_i} \cdot \frac{DLDOD}{(DLDOD)_i}$	$\frac{HV}{HV_{ijkl}} \cdot \frac{VV}{VV_{ijkl}}$

Thus, the compact rule for the effective navigation of robots is defined as follows:

$$W_{ijkl} = Dis_{ijkl}\left[\left(\frac{X_{LOD}}{LOD_i}\right)\left(\frac{X_{FOD}}{FOD_i}\right)\left(\frac{X_{ROD}}{ROD_i}\right)\left(\frac{X_{HA}}{HA_i}\right)\left(\frac{X_{UOD}}{UOD_i}\right)\left(\frac{X_{ULDOD}}{ULDOD_i}\right)\left(\frac{X_{URDOD}}{URDOD_i}\right)\left(\frac{X_{DOD}}{DOD_i}\right)\left(\frac{X_{DLDOD}}{DLDOD_i}\right)\right] \tag{19}$$

Similarly, the velocity function is presented:

$$(Vel)^X LV_{ijkl} = \frac{W_{ijkl}}{(vel_{LV})X_{LV_{ijkl}}};\ (Vel)^X RV_{ijkl} = \frac{W_{ijkl}}{(vel_{RV})X_{RV_{ijkl}}};$$

$$(Vel)^X HV_{ijkl} = \frac{W_{ijkl}}{(vel_{VV})X_{HV_{ijkl}}} \text{and} (Vel)^X VV_{ijkl} = \frac{W_{ijkl}}{(vel_{VV})X_{VV_{ijkl}}}$$

Hence, $LV = \left(\frac{\sum (Vel) \frac{X_{LV}}{(Vel)X_{LV}}}{\sum (Vel)X_{LV}}\right)$; $RV = \left(\frac{\sum (Vel) \frac{X_{RV}}{(Vel)X_{RV}}}{\sum (Vel)X_{RV}}\right)$; $HV = \left(\frac{\sum (Vel) \frac{X_{HV}}{(Vel)X_{HV}}}{\sum (Vel)X_{HV}}\right)$; and $VV = \left(\frac{\sum (Vel) \frac{X_{VV}}{(Vel)X_{VV}}}{\sum (Vel)X_{VV}}\right)$.

The continuous representation of the above is given as follows:

$$LV = \frac{\int (Vel) \frac{X_{LV}}{(Vel)X_{LV}} d(Vel)}{\int (Vel)X_{LV} d(Vel)};\ RV = \frac{\int (Vel) \frac{X_{RV}}{(Vel)X_{RV}} d(Vel)}{\int (Vel)X_{RV} d(Vel)};\ HV = \frac{\int (Vel) \frac{X_{HV}}{(Vel)X_{HV}} d(Vel)}{\int (Vel)X_{HV} d(Vel)};\ \text{and } VV = \frac{\int (Vel) \frac{X_{VV}}{(Vel)X_{VV}} d(Vel)}{\int (Vel)X_{VV} d(Vel)}.$$

The pseudocode for Functional Firefly Algorithm 1 is described as:

Algorithm 1 Fuctional Firefly (FFA).

#Function FunctionalFireflyAlgorithm():
Initialize robot navigation
Initialize the population of fireflies $X_\lambda = \{x_1, \ldots, x_n\}$.
Initialize the objective function $f(\lambda) \in X_\lambda; \lambda \in \Lambda$.
Initialize light intensity of fireflies (I)
Initialize absorption coefficient (γ)
Initialize the distance between two fireflies (r)
Vary attractiveness $(e^{-\gamma r})$.
Classify fireflies into two groups based on intensity:
$X_\lambda^l = \{x_1^l, \ldots, x_n^l\}$, *l: less intensity.*
$X_\lambda^h = \{x_1^h, \ldots, x_n^h\}$, *h: high intensity.*
$t = 0$
$max_{iterations} = max(X_\lambda)$
While $t < maxX_\lambda$:
 # Update lowintensity fireflies
For i in $X_\lambda{}^l$:
If $C_l \overset{f_1(\lambda)}{\rightarrow} C_h$;
$x_1^l \rightarrow x_1^h \left(x_1^l < x_1^h\right)$;
If $C_l \overset{f_2(\lambda)}{\rightarrow} C_h$:
$x_2^l \rightarrow x_2^h \left(x_2^l < x_2^h\right)$;
If $C_l \overset{f_3(\lambda)}{\rightarrow} C_h$;
$x_3^l \rightarrow x_3^h \left(x_3^l < x_3^h\right)$;
If $C_l \overset{f_4(\lambda)}{\rightarrow} C_h$;
$x_4^l \rightarrow x_4^h \left(x_4^l < x_4^h\right)$;
If $C_l \overset{f_5(\lambda)}{\rightarrow} C_h$;
$x_{35}^l \rightarrow x_5^h \left(x_5^l < x_5^h\right)$;
 # Calculate the objective function
$f(\lambda) = f_1(\lambda) + f_2(\lambda) + f_3(\lambda) + f_4(\lambda) + f_5(\lambda)$

 # Update fireflies based on the objective function
$R \overset{f(\lambda)}{\rightarrow} G$.
$t = t + 1$
 # Optimized $f(\lambda)$
Optimized function $f(\lambda)$
End robot navigation

4. Simulation and Experimental Result Analysis

4.1. Mobile Robot Navigation Simulation Results

To demonstrate the effectiveness of our developed approach across diverse environmental conditions, we conducted numerous trials in both static and dynamic environments featuring various obstacles. Simulation analysis was carried out using MATLAB R2021a software, providing the flexibility to customize the environment with different obstacle positions, robot placements, and goals. In static environments, obstacle positions remained fixed, allowing adjustments only in the initial robot and goal positions. Conversely, dynamic environments featured variable obstacle and goal positions. Our program was designed to accommodate a variable number of robots and goals. Figures 3 and 4 demonstrate the navigation of a single mobile robot in simulated environments with static configurations, presented in both 3D and 2D formats. During navigation, the robot prioritizes path safety, avoiding obstacles by maintaining a safe distance. We also demonstrated multi-robot navigation strategies in a complex environment with four robots (Figure 5a–d). Each robot had a distinct starting position and a predefined common goal. The paths created by individual robots were uniquely color-coded, illustrating our approach's efficiency in finding optimal collision-free paths even in dynamic environments. Figure 6 presents an environment with two moving obstacles (green and pink) and a fixed goal. The robot autonomously identifies approaching obstacles and adjusts its position to maintain a safe distance. Figure 7 illustrates mobile robot navigation when the goal itself is in motion. Our approach consistently generates optimal pathways in both cases, effectively addressing uncertainties in dynamic environments. These results highlight the robustness of our proposed approach.

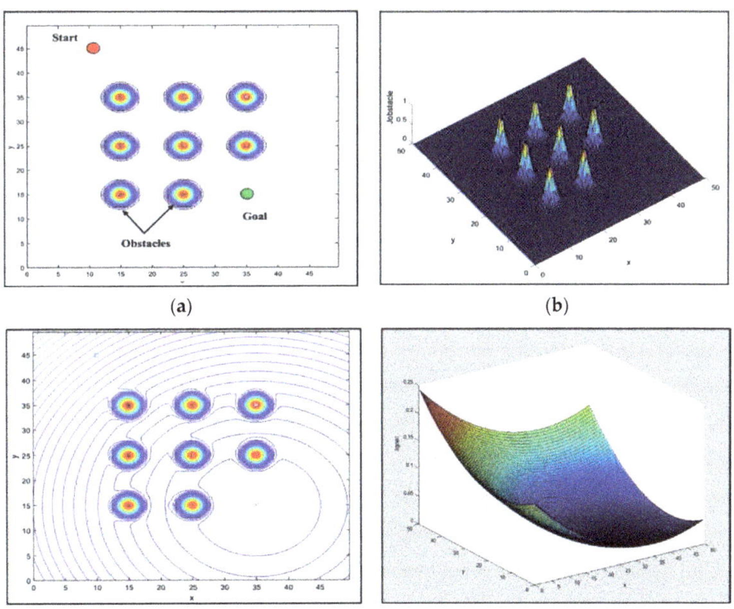

(**a**) (**b**)

(**c**) (**d**)

Figure 3. *Cont.*

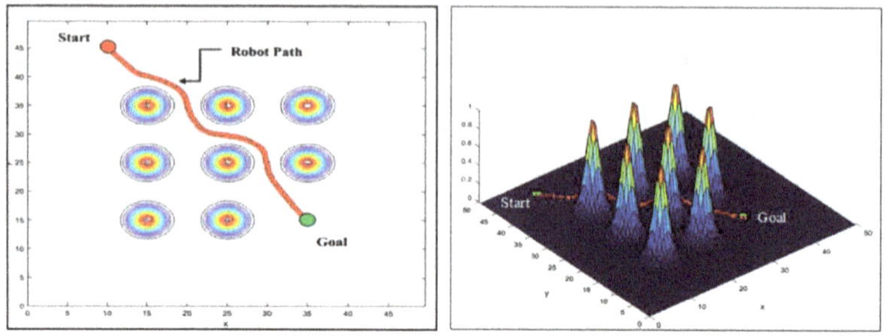

(e) (f)

Figure 3. Mobile robot navigation in a static environment. (**a**) Two-dimensional representation of the environment. (**b**) Obstacle function. (**c**) Total cost function. (**d**) Goal function. (**e**) Navigation in 2D environment. (**f**) Navigation in 3D.

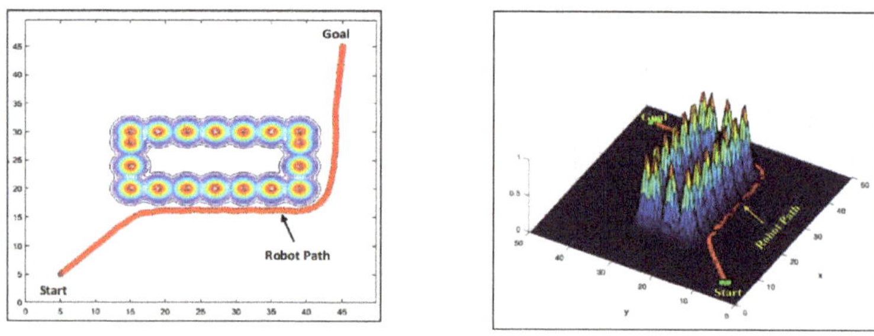

(**a**) (**b**)

Figure 4. Mobile robot navigation in static environment. (**a**) Robot navigation in 2D environment. (**b**) Robot navigation in 3D.

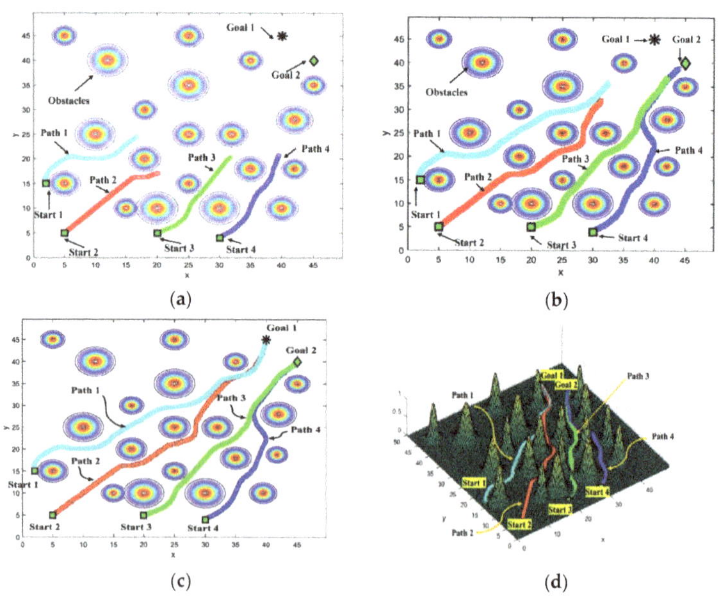

Figure 5. Demonstration of multiple mobile robot navigation in a static environment, illustrated in (**a**–**d**).

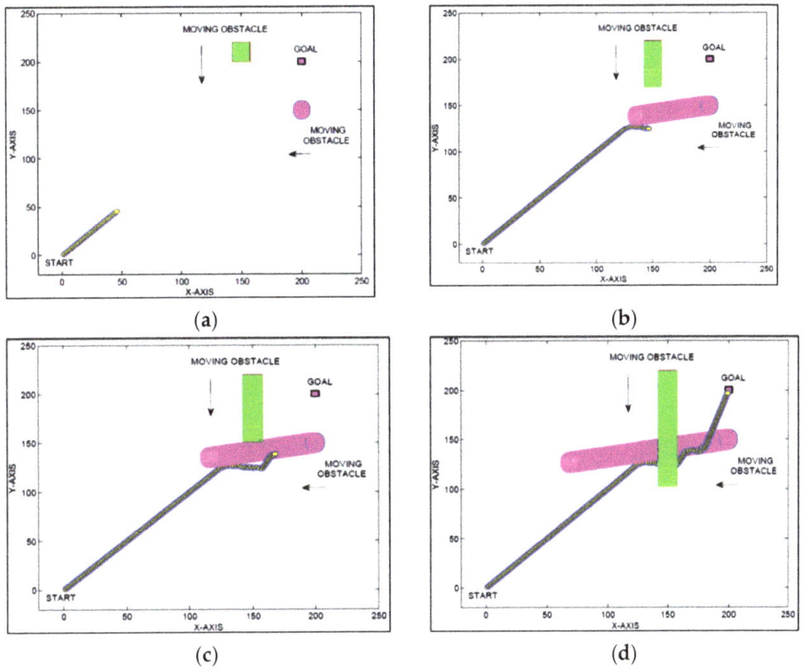

Figure 6. Demonstration of mobile robot navigation in a dynamic environment with two moving obstacles: illustrated in (**a**–**d**).

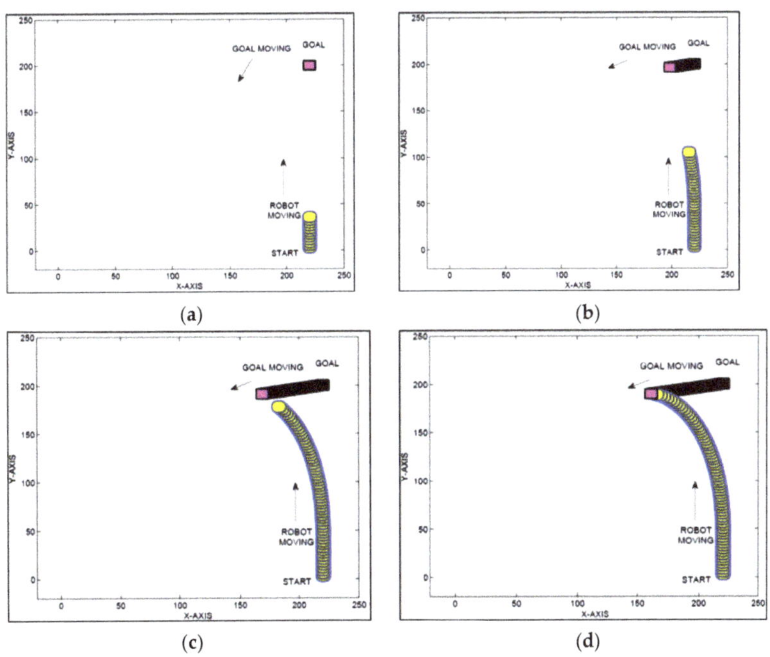

Figure 7. Demonstration of mobile robot navigation in a dynamic environment with moving goal: illustrated in (**a**–**d**).

4.2. Mobile Robot Navigation Experimental Results

To evaluate the effectiveness of the developed approach in real-time scenarios, we employed two distinct types of robots: an in-house developed robot (Figure 8) and the Khepera-II robot (Figure 9). Detailed specifications for these robots can be found in the Appendix A, specifically outlined in Table A1 for the in-house developed robot and Table A2 for the Khepera-II robot. As illustrated in Figures 10–12, we established a consistent experimental setup to assess the approach's performance for single- and multiple-robot systems. This setup confirms that the approach can generate optimal paths comparable to those achieved in simulation, as demonstrated in Figures 3–5.

Figure 8. In-house developed a robot.

Figure 9. Khepera-II robot.

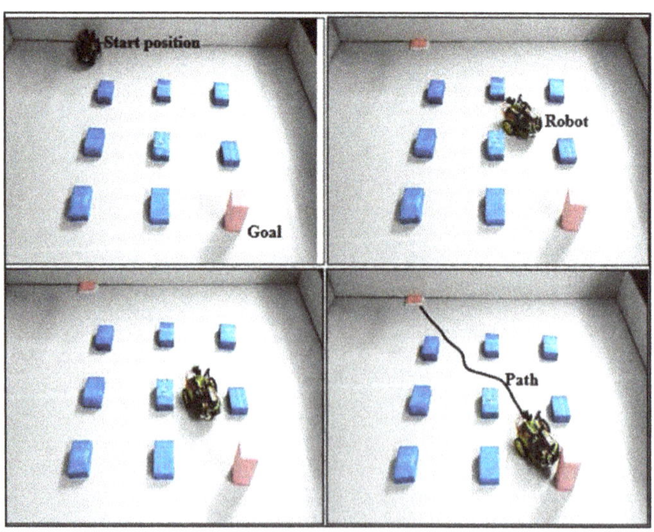

Figure 10. Mobile robot navigation in a real environment (Scenario-1).

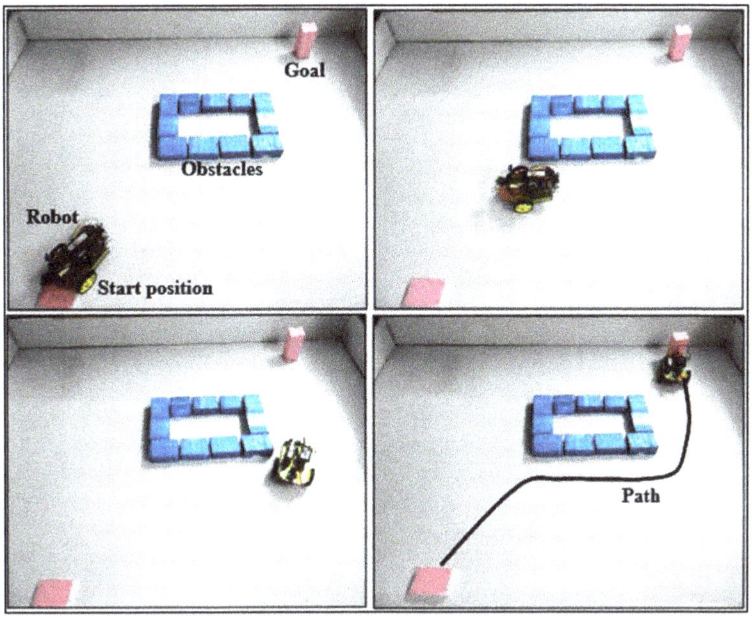

Figure 11. Mobile robot navigation in a real environment (Scenario-2).

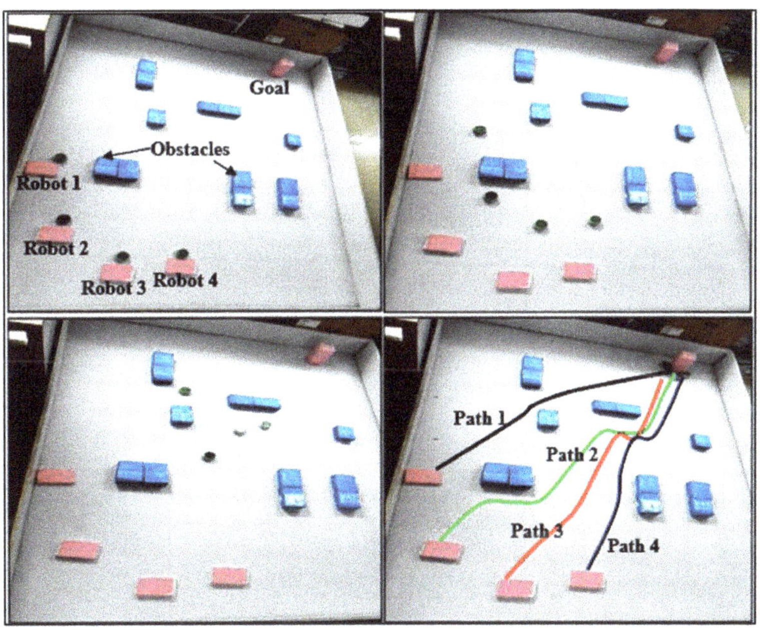

Figure 12. Mobile robot navigation in a real environment in the presence of multiple robots.

Both robots are three-wheeled, with the rear two wheels being active and the front one passive. The rear wheels can move independently to achieve the desired navigation angle. The path was traced on the platform using a pencil during the robot's movement between points. Extensive experimentation involving more than twenty trials for each environment was conducted with a maximum velocity of 0.06 m/s. The approach's robustness was further tested with various obstacle shapes and sizes. The results affirm the approach's suitability for real-time applications, showcasing minimal deviation when compared to the simulation environment. The experimental results demonstrate the robot's successful obstacle avoidance and timely attainment of its goal.

4.3. Comparative Analysis of Experimental vs. Simulation

On comparison of real-time results with simulation results, it seems that the path developed by the proposed FFA controller safely avoids static and dynamic obstacles. The observed path length and navigational time tabulated in Tables 8–11 confirm the appropriateness of the developed approach, as the percentage of deviation between real-time results and simulation results is less than 4.5%. Hence, the goal of navigation was achieved successfully in an unknown environment. The developed approach deals with the single mobile robot system that effectively and efficiently handles the multiple robot systems for crowded environments. The velocity profile for the left and right wheels in meters per second is shown in Figure 13. For a detailed analysis of path length and navigational time for single- and multiple-robot navigation systems, 20 trials and 5 trials were undertaken, respectively.

Table 8. Path length comparison for a single mobile robot system.

Sl. No.	Experimental Path Length (cm)	Simulation Path Length (cm)	% of Error
Scenario-1	133.61 (Figure 8)	127.69 (Figure 3)	4.43
Scenario-2	261.97 (Figure 9)	250.95 (Figure 4)	4.20

Table 9. Navigational time comparison for a single mobile robot system.

Sl. No.	Experimental Time during MRN (s)	Simulation Time during MRN (s)	% of Deviation
Scenario-1	15 (Figure 8)	14.2 (Figure 3)	4.40
Scenario-2	27.6 (Figure 9)	26.4 (Figure 4)	4.34

Table 10. Path length comparison for multiple mobile robot systems.

Sl. No.		Experimental Time during MRN (s) (Figure 10)	Simulation Time during MRN (s) (Figure 5)	% of Deviation
Scenario-3	Robot 1	133.12	127.5	4.22
	Robot 2	147.59	141.31	4.25
	Robot 3	163.11	156.30	4.17
	Robot 4	122.76	117.71	4.11

Table 11. Navigational time comparison for multiple mobile robot systems.

Sl. No.		Experimental Time during MRN (s) (Figure 10)	Simulation Time during MRN (s) (Figure 5)	% of Deviation
Scenario-3	Robot 1	13.60	13.02	4.26
	Robot 2	15.01	14.36	4.33
	Robot 3	16.48	15.88	4.24
	Robot 4	13.31	12.76	4.13

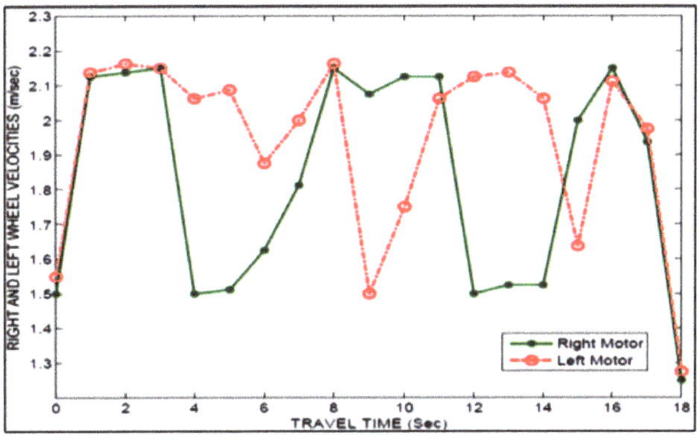

Figure 13. Left wheel velocity versus right wheel velocity profile in m/s.

4.4. Proposed FFA Controller versus Another Intelligent Controller over the Same Environmental Setup

To prove the effectiveness of the developed controller, it is necessary to check its performance with other artificially intelligent controllers over the same environmental condition; i.e., the number of obstacles, its position, and the robot position must be the same. The performance parameters considered are the path length and navigational time. For comparison, fuzzy logic (FL), particle swarm optimization (PSO), and genetic algorithm (GA) were considered. The optimized path was selected for each controller after performing more than 20 trials. While analyzing, all the positions of robots and obstacles in a static environment are shown in Figures 14 and 15, whereas Figures 16 and 17 deal with the navigation in a dynamic environment with a moving obstacle system and dynamic goal system, respectively. The movement of obstacle and goal in a dynamic environment are shown in Figures 18 and 19. The simultaneous comparison of path length and navigational time is tabulated in Tables 12–15, and from the data, it is clear that the path produced by using the proposed FFA controller in all terrain is short, and the required navigational time was greatly decreased as compared to FL, PSO, and GA.

Table 12. Path length and navigational time comparison for a single-robot system (Figures 16 and 20).

Sl. No.	Name of Controllers	Simulation Path Length (cm)	Simulation Time (s)	Real-Time Path Length (cm)	Real-Time (s)
1	FL	291.06	30.93	307.23	32.65
2	PSO	278.12	29.55	297.528	31.62
3	GA	270.03	28.69	287.82	30.58
4	**FFA**	**261.97**	**27.84**	**281.35**	**29.90**

Table 13. Path length and navigational time comparison for multiple-robot system (Figures 17 and 21).

Sl. No.	Name of Controllers		Simulation Path Length (cm)	Simulation Time (s)	Real-Time Path Length (cm)	Real-Time (s)
1	FL	Robot-1	245.78	26.12	252.2	26.802
		Robot-2	213.44	22.68	226.38	24.05
2	PSO	Robot-1	239.08	25.40	247.08	26.25
		Robot-2	210.50	22.37	216.67	23.02
3	GA	Robot-1	229.61	24.40	232.84	24.74
		Robot-2	207.21	22.02	210.21	22.34
4	**FFA**	**Robot-1**	**223.14**	**23.71**	**230.38**	**24.48**
		Robot-2	**205.97**	**21.89**	**208.74**	**22.18**

Table 14. Path length and navigational time comparison in the presence of moving obstacles (Figure 18).

Sl. No.	Name of Controllers	Simulation Path Length (cm)	Simulation Time (s)
1	FL	244.55	26
2	PSO	237.69	25.62
3	GA	232.84	24.74
4	**FFA**	**228.76**	**23.87**

Table 15. Path length and navigational time comparison in moving goal situations (Figure 19).

Sl. No.	Name of Controllers	Simulation Path Length (cm)	Simulation Time (s)
1	FL	161.70	17.18
2	PSO	153.61	16.32
3	GA	142.90	15.18
4	**FFA**	**139.29**	**14.80**

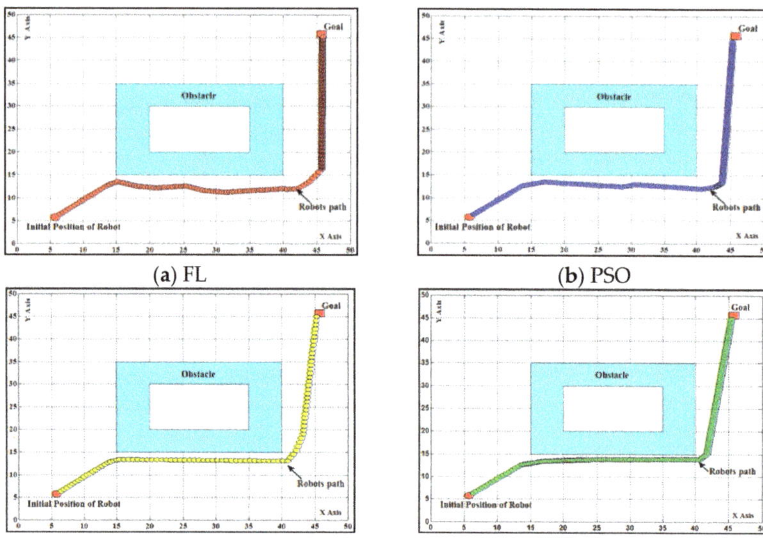

(**a**) FL (**b**) PSO (**c**) GA (**d**) FFA

Figure 14. Mobile robot (single) navigation in static simulation environment by various intelligent approaches.

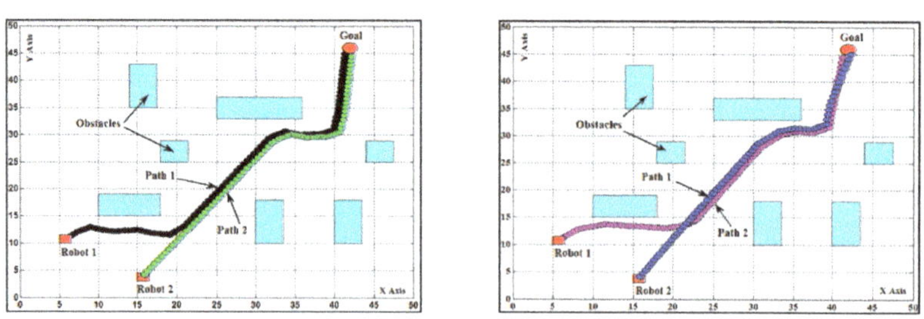

(**a**) FL (**b**) PSO

Figure 15. *Cont.*

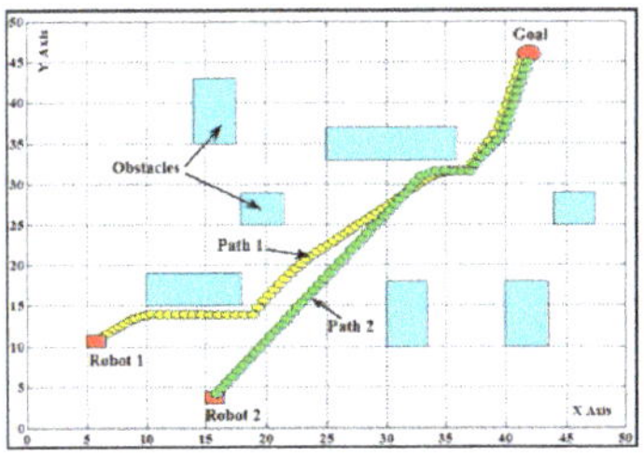

(**c**) GA

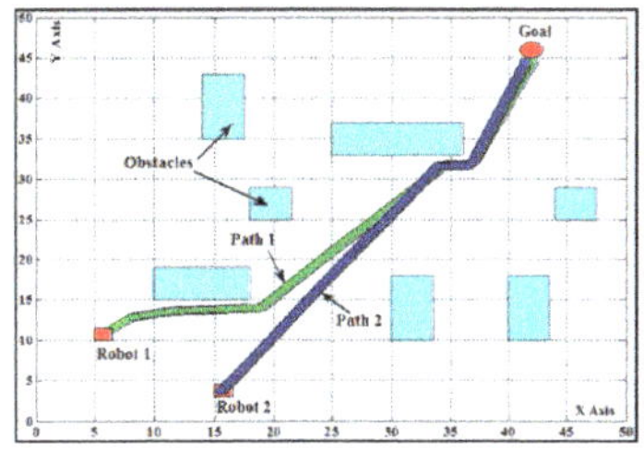

(**d**) FFA

Figure 15. Multiple mobile robot navigation in a static simulation environment by various intelligent approaches.

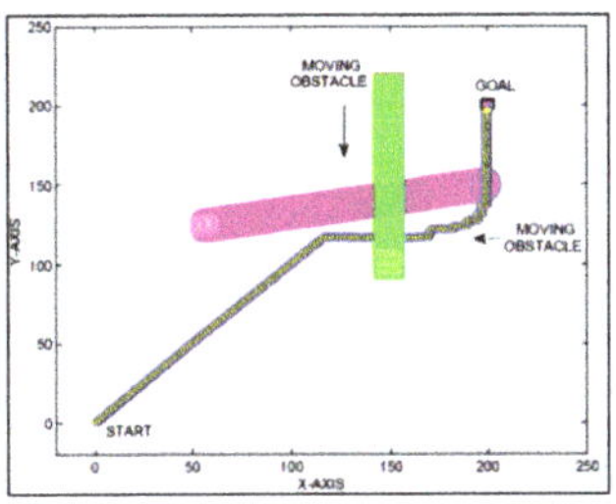

(**a**) FL

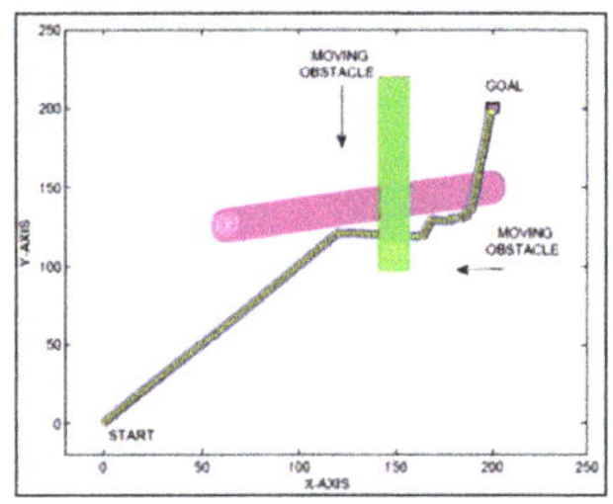

(**b**) PSO

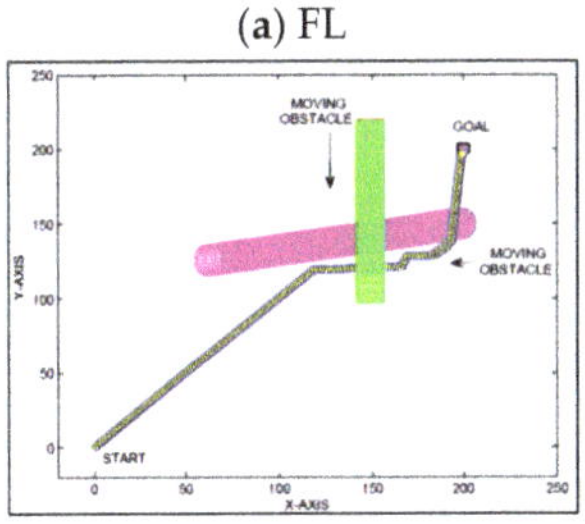

(**c**) GA

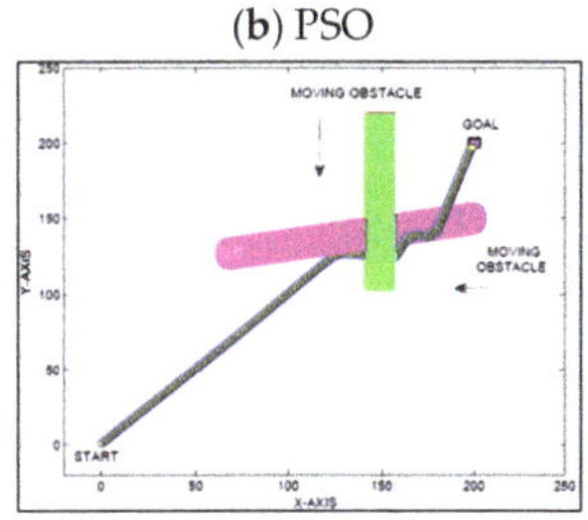

(**d**) FFA

Figure 16. Navigation in a dynamic obstacle environment by various intelligent approaches.

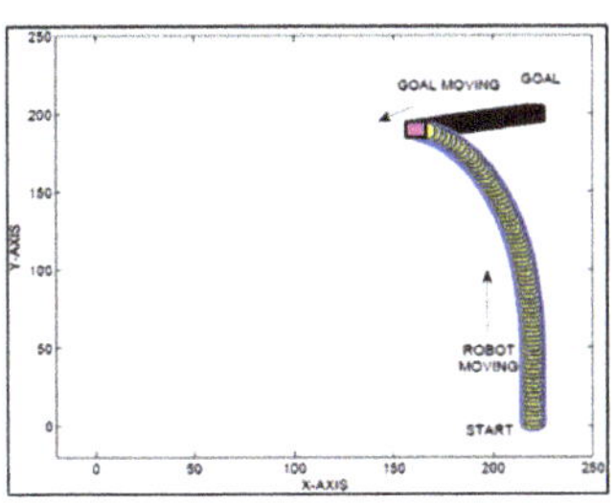

(**a**) FL

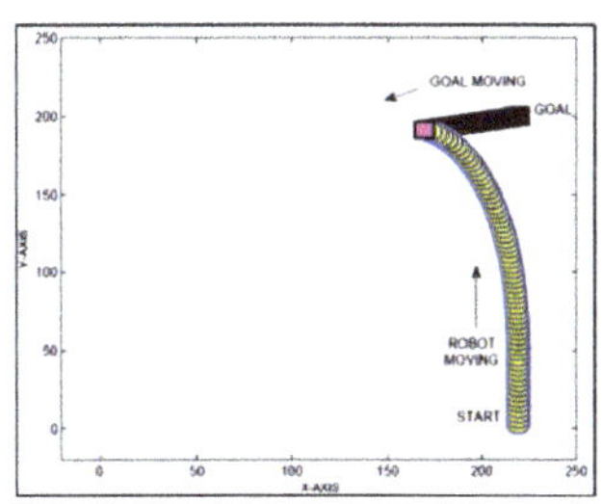

(**b**) PSO

Figure 17. *Cont.*

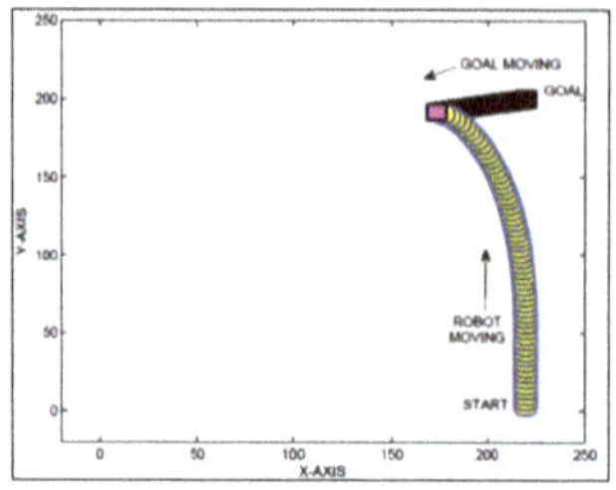

(**c**) GA

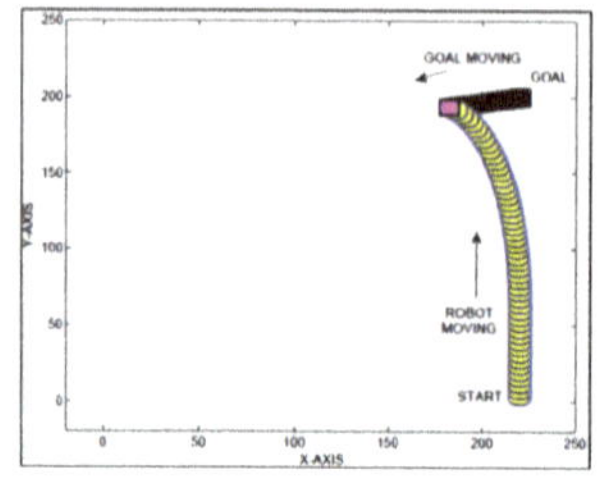

(**d**) FFA

Figure 17. Navigation in a dynamic goal environment by various intelligent approaches.

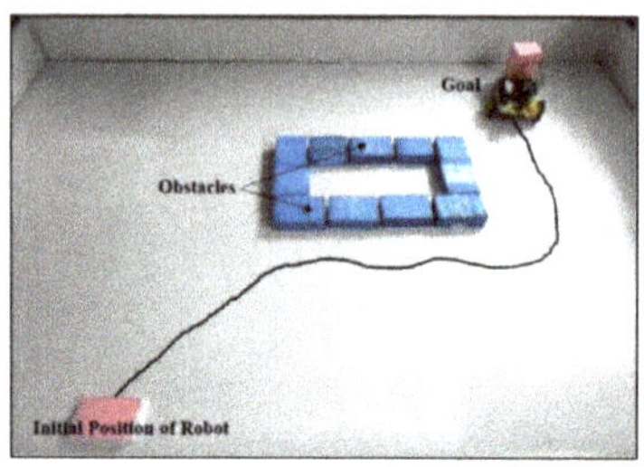

(**a**) FL

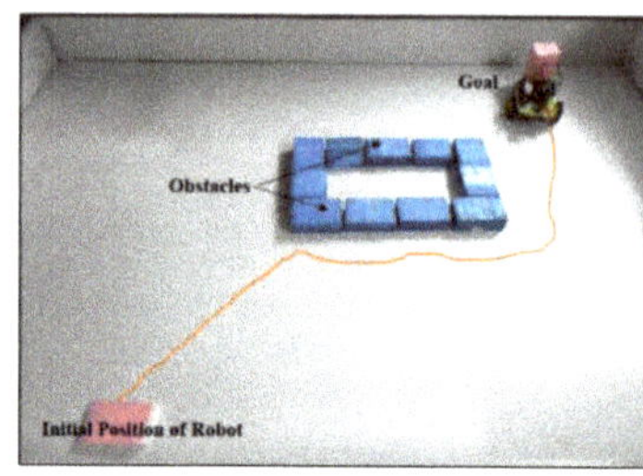

(**b**) PSO

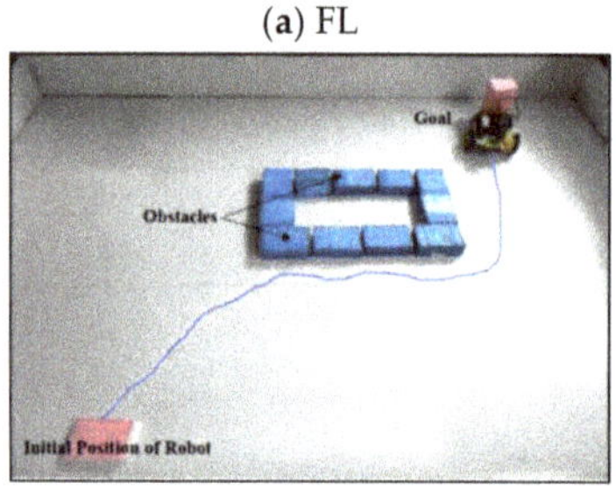

(**c**) GA

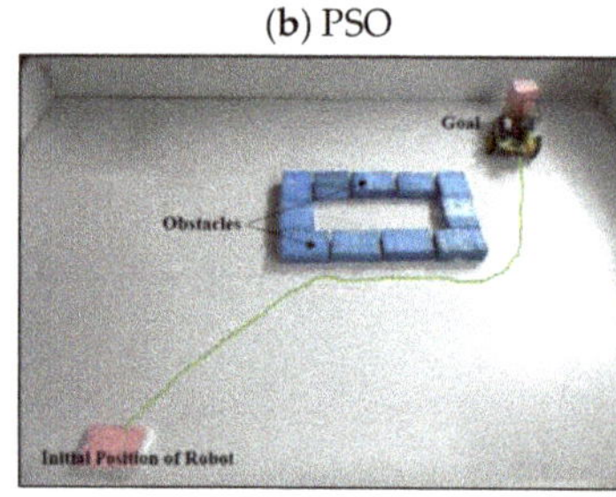

(**d**) FFA

Figure 18. Mobile robot (single) navigation in the static, real-time environment by various intelligent approaches.

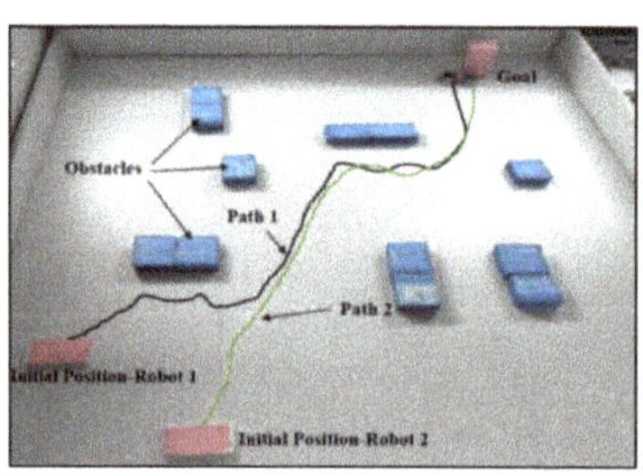

(**a**) FL

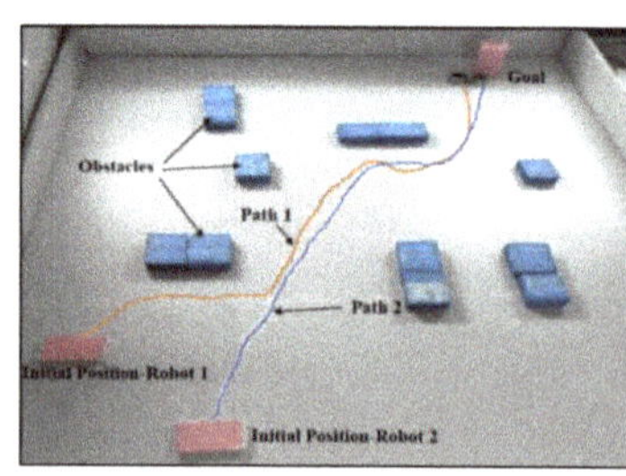

(**b**) PSO

Figure 19. *Cont.*

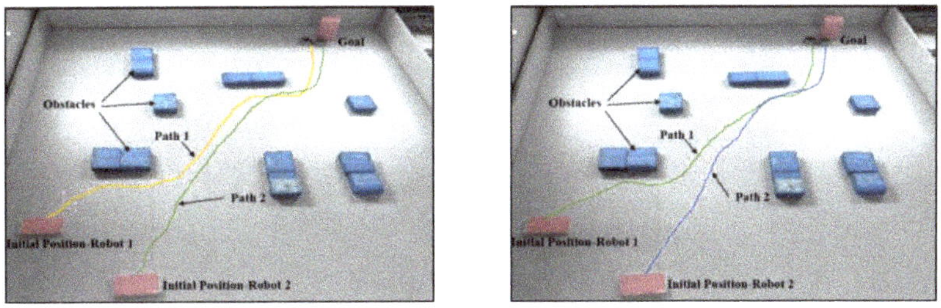

(**c**) GA (**d**) FFA

Figure 19. Multiple mobile robot navigation in the static, real-time environment by various intelligent approaches.

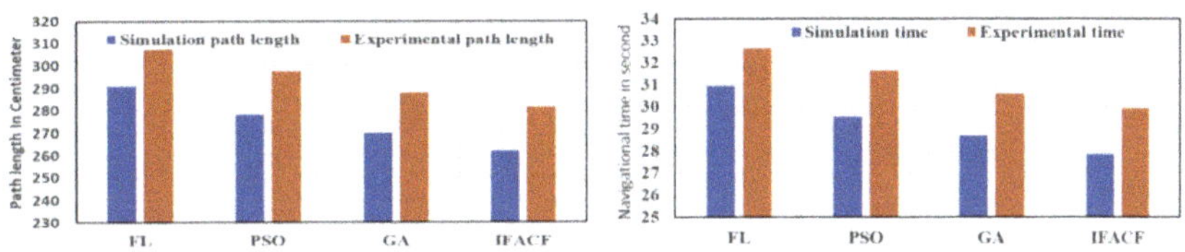

(**a**) (**b**)

Figure 20. Moving goal environment: Comparison between FL, PSO, GA, and FFA. (**a**) Path length and (**b**) navigational time.

4.5. Proposed FFA Controller versus Published Work

In this section, the work developed by the other researchers, i.e., Singh et al. [36], Montiel et al. [37], Zheng et al. [38], and Orozco-Rosas et al. [39], is considered for comparison with the proposed FFA controller. The comparison was carried out in the simulation environment only, and the performance was evaluated on the basis of path length. In [36], the navigation of robots was presented using the neural network (NN) in a complex, crowded environment where the obstacles had a rectangular shape (Figure 21a). The application of artificial potential field (APF) for the navigation of robots in a static environment was developed by [37] in the presence of three circular obstacles in a static environment (Figure 21c). Similarly, the navigation based on the Elman neural network (ENN) training technique in the presence of three rectangular obstacles (Figure 21e) and the parallel bacterial potential field algorithm (PBPFA) over a set of circular obstacles (Figure 21g) were presented by [38] and [39], respectively. Figure 21 demonstrates that the proposed FFA controller generates a smoother and shorter path than the respective AI controllers. Table 16 and Figure 22 show that there is a huge gap between the path developed by the proposed controller and other controllers, and the percentage of path length saved by using a proposed controller reached a maximum of 35.38% and a minimum of 5.7%.

Table 16. FFA versus other AI controllers.

S. N.	Start Point	Goal Point	Path Length (cm) by Other AI Controllers	Path Length (cm) by FFA Controller	% of Path Length Saved by FFA
Scenario-4	(4,8)	(90,89)	7.9 (Figure 21a)	6.6 (Figure 21b)	16.45
Scenario-5	(5,9)	(7,1)	6.5 (Figure 21c)	4.2 (Figure 21d)	35.38
Scenario-6	(1.7,1.5)	(16.9,16)	5.2 (Figure 21e)	4.9 (Figure 21f)	5.7
Scenario-7	(5,9)	(5,1)	7.6 (Figure 21g)	5.7 (Figure 21h)	25

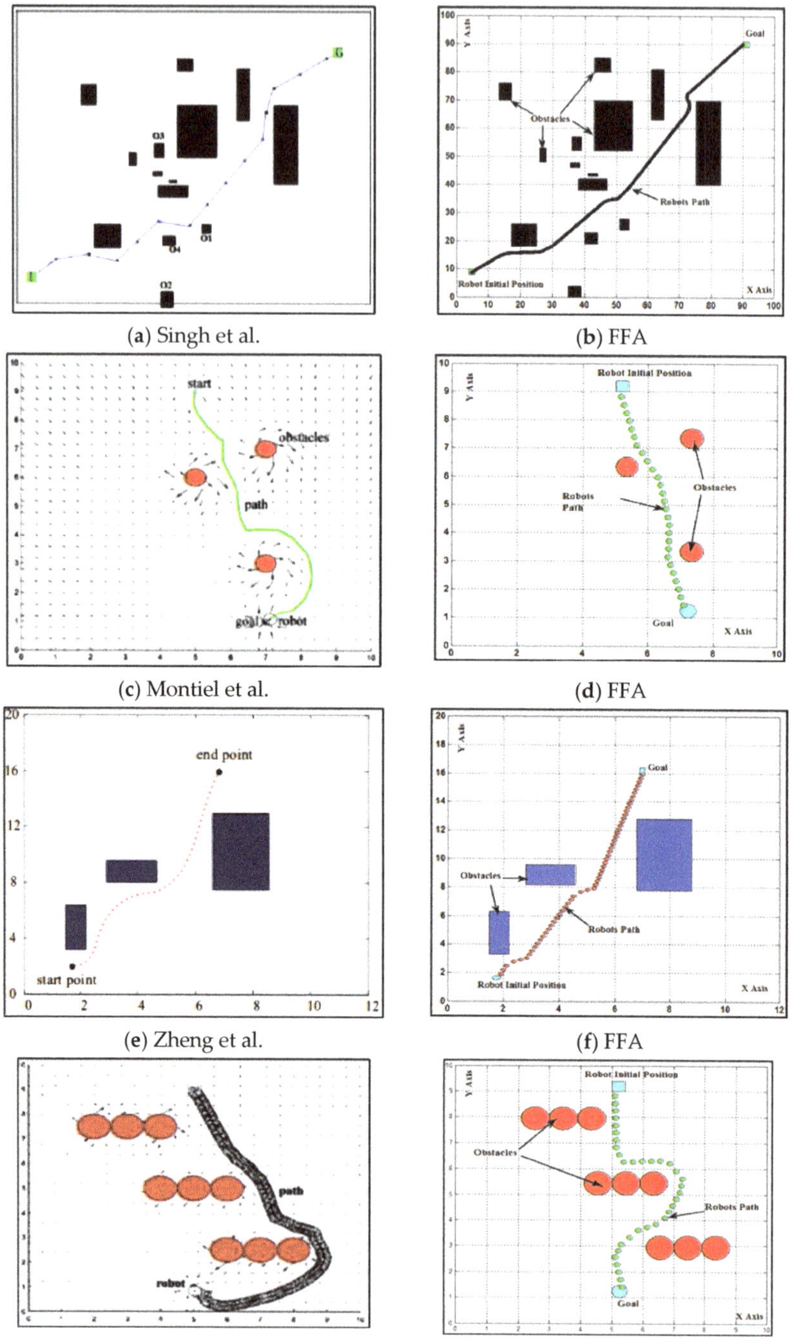

(**a**) Singh et al. (**b**) FFA

(**c**) Montiel et al. (**d**) FFA

(**e**) Zheng et al. (**f**) FFA

(**g**) Orozco-Rosas et al. (**h**) FFA.

Figure 21. Other published work versus the proposed FFA controller [36–39].

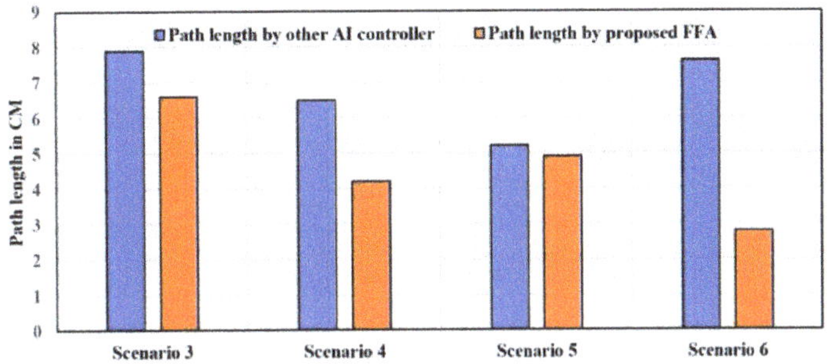

Figure 22. Path length comparison between another AI controller and FFA.

5. Conclusions

The proposed functional firefly algorithm provides the smallest size (order) parameters for optimizing the mobile robot navigation. The FFA is based on the reduction order technique from n-interval to (n-m) interval by applying the choice function. The axiom of choice function reduces the order of domain, co-domain, and range of FFA to minimize the navigation. FFA is presented with a choice function for the navigation of a mobile robot in a completely unknown environment in the presence of a static obstacle, a moving obstacle, and a dynamic goal. The choice function identifies these limitations by the non-void family of the fireflies as a set. The movement of the fireflies is classified by fuzzy, and the probability function sets the optimization of the path. The developed controller achieved the goals of navigation, i.e., obstacle avoidance and path optimization in static and dynamic environments for single and multiple mobile robot systems. The obtained results show that the robot provides a smoother trajectory with a shorter path in less navigational time. The observed percentage of deviation between simulation and real-time results is less than 4.5%, and it yielded the optimal path length with minimum navigational time when compared to FL, PSO, and GA over similar environmental conditions. In comparison with the other researcher's work on NN, APF, ENN, and PBPFA, it saves a maximum path length of 36% and a minimum of 5%. In the future, the developed approach will aim to apply real-time navigation for on-road traffic conditions (real dynamic situations). The proposed controller can be tested by hybridizing with newly developed intelligent algorithms for developing a new path planner. It can be implemented in the development of an autonomous robot in an uncertain environment. The proposed approach can also be implemented to navigate aerial vehicles and underwater robots.

Author Contributions: Conceptualization, B.K.P. and A.J.; methodology, B.K.P. and S.K.K.; software, B.K.P. and B.P.; formal analysis, B.P. and B.K.P.; investigation, B.K.P. and B.P.; writing—original draft preparation, B.K.P. and A.J.; writing—review and editing, B.P.; visualization, S.K.K. and B.P. All authors have read and agreed to the published version of the manuscript.

Funding: This research received no external funding.

Institutional Review Board Statement: Not applicable.

Informed Consent Statement: Not applicable.

Data Availability Statement: The data presented in this study are available on request from the corresponding author. The data are not publicly available due to privacy.

Conflicts of Interest: The authors declare no conflict of interest.

Appendix A

Table A1. Specification of the in-house robot used in the experiment.

	Elements	Technical Specification
1	Processor	ATmega2560 (Arduino Mega 2560, Arduino UNO, Olatus Systems, Guwahati, India)
2	RAM	8 KB, EEROM-4 KB
3	Flash	256 KB (8 KB used for boot loader)
4	Motors	2-DC gear motors with incremental encoders
5	Distance sensors	(a) Infrared sensors with up to 150 cm range (b) Ultrasonic sensors with up to 400 cm range
6	Speed	Max: 0.47 m/s, Min: 0.03 m/s
7	Power	Power adapter or Rechargeable NiMH Battery (2000 mAh)
8	Communication	USB connection to the computer
9	Size	Length: 25 cm, Width: 19 cm, Height: 12 cm
10	Weight	Approx. 1100 g
11	Payload	Approx. 4000 g
12	Remote control Software via USB cable	C/C++17 ® (on PC, MAC OS 12) MATLAB R2021a ® (on PC, MAC OS 12, Linux)

Table A2. Specification of the Khepera-II robot used in the experiment.

	Elements	Technical Specification
1	Processor	Motorola 68331 CPU, 25 MHz
2	RAM	512 KB
3	Flash	512 KB
4	Motors	2-DC brushed Servo motors with incremental encoders
5	Sensors	8 Infrared proximity and ambient light sensors with up to 100 mm range
6	Speed	Max: 0.5 m/s, Min: 0.02 m/s
7	Power	Power adapter or Rechargeable NiMH Batteries
8	Communication	Standard Serial Port, up to 115 KB/S
9	Size	Diameter: 70 mm, Height: 30 mm
10	Weight	Approx. 80 g
11	Payload	Approx. 250 g
12	Remote control software via tether or radio	LabVIEW® (on PC, MAC OS 12) using RS232 MATLAB® (on PC, MAC OS 12, Linux) using RS232 Sys Quake® (on PC, MAC OS 12, Linux) using RS232 Freeware Any other software capable of RS232 communication

References

1. Patle, B.K.; Patel, B.; Jha, A. Rule-Based Fuzzy Decision Path Planning Approach for Mobile Robot. In Proceedings of the 2018 Fourth International Conference on Computing Communication Control and Automation (ICCUBEA), Pune, India, 16–18 August 2018; pp. 1–7.
2. Dubey, V.; Patel, B.; Barde, S. Path Optimization and Obstacle Avoidance using Gradient Method with Potential Fields for Mobile Robot. In Proceedings of the 2023 International Conference on Sustainable Computing and Smart Systems (ICSCSS), Coimbatore, India, 14–16 June 2023; IEEE: Piscataway, NJ, USA, 2023; pp. 1358–1364.
3. Patle, B.K. Intelligent Navigational Strategies for Multiple Wheeled Mobile Robots Using Artificial Hybrid Methodologies. Ph.D. Thesis, National Institute of Technology, Rourkela, India, 2016.
4. Patle, B.K.; Pandey, A.; Parhi DR, K.; Jagadeesh, A. A review: On path planning strategies for navigation of mobile robot. *Def. Technol.* **2019**, *15*, 582–606. [CrossRef]
5. Kanoon, Z.E.; Al-Araji, A.S.; Abdullah, M.N. Enhancement of Cell Decomposition Path-Planning Algorithm for Autonomous Mobile Robot Based on an Intelligent Hybrid Optimization Method. *Int. J. Intell. Eng. Syst.* **2022**, *15*, 161–175. [CrossRef]
6. Wang, D.; Chen, S.; Zhang, Y.; Liu, L. Path planning of mobile robot in dynamic environment: Fuzzy artificial potential field and extensible neural network. *Artif. Life Robot.* **2021**, *26*, 129–139. [CrossRef]
7. Wang, G.; Zhou, J. Dynamic robot path planning system using neural network. *J. Intell. Fuzzy Syst.* **2021**, *40*, 3055–3063. [CrossRef]
8. Chen, Z.; Wu, H.; Chen, Y.; Cheng, L.; Zhang, B. Patrol robot path planning in nuclear power plant using an interval multi-objective particle swarm optimization algorithm. *Appl. Soft Comput.* **2022**, *116*, 108192. [CrossRef]

9. Hou, W.; Xiong, Z.; Wang, C.; Chen, H. Enhanced ant colony algorithm with communication mechanism for mobile robot path planning. *Robot. Auton. Syst.* **2022**, *148*, 103949. [CrossRef]
10. Muni, M.K.; Parhi, D.R.; Kumar, P.B. Improved motion planning of humanoid robots using bacterial foraging optimization. *Robotica* **2021**, *39*, 123–136. [CrossRef]
11. Quan, Y.; Wei, W.; Ouyang, H.; Lan, X. New Harmony Search Algorithm for Mobile Robot Path Planning. In *Advances in Guidance, Navigation and Control*; Springer: Singapore, 2022; pp. 3463–3473.
12. Singh, S.; Sharma, K.; Doriya, R. Minimizing Computation Time for Robot Path Planning Using Improvised Cuckoo Search Algorithm. In *Advances in Electrical and Computer Technologies*; Springer: Singapore, 2021; pp. 199–209.
13. Syama, R.; Mala, C. A Multi-objective Optimal Trajectory Planning for Autonomous Vehicles Using Dragonfly Algorithm. In Proceedings of the 3rd EAI International Conference on Big Data Innovation for Sustainable Cognitive Computing, Online, 18–19 December 2020; Springer: Cham, Switzerland, 2022; pp. 21–36.
14. Yang, X.S. *Nature-Inspired Metaheuristic Algorithm*; Luniver Press: London, UK, 2008.
15. Christensen, A.L.; O'Grady, R.; Dorigo, M. Synchronization and fault detection in autonomous robots. In Proceedings of the 2008 IEEE/RSJ International Conference on Intelligent Robots and Systems, Nice, France, 22–26 September 2008; pp. 4139–4140.
16. Liaquat, S.; Fakhar, M.S.; Kashif SA, R.; Saleem, O. Statistical Analysis of Accelerated PSO, Firefly and Enhanced Firefly for Economic Dispatch Problem. In Proceedings of the 2021 6th International Conference on Renewable Energy: Generation and Applications (ICREGA), Al Ain, United Arab Emirates, 2–4 February 2021; IEEE: Piscataway, NJ, USA, 2021; pp. 106–111.
17. Nalluri, M.R.; Kannan, K.; Roy, D.S. A Novel Discrete Firefly Algorithm for Constrained Multi-Objective Software Reliability Assessment of Digital Relay. In *Machine Vision Inspection Systems, Volume 2: Machine Learning-Based Approaches*; Wiley: Hoboken, NJ, USA, 2021; pp. 287–321.
18. Cheng, Z.; Song, H.; Chang, T.; Wang, J. An improved mixed-coded hybrid firefly algorithm for the mixed-discrete SSCGR problem. *Expert Syst. Appl.* **2022**, *188*, 116050. [CrossRef]
19. Xing, H.X.; Wu, H.; Chen, Y.; Wang, K. A cooperative interference resource allocation method based on improved firefly algorithm. *Def. Technol.* **2021**, *17*, 1352–1360. [CrossRef]
20. Sivaranjani, P.; Senthil Kumar, A. Hybrid Particle Swarm Optimization-Firefly algorithm (HPSOFF) for combinatorial optimization of non-slicing VLSI floorplanning. *J. Intell. Fuzzy Syst.* **2017**, *32*, 661–669. [CrossRef]
21. Dash, S.; Thulasiram, R.; Thulasiraman, P. Modified firefly algorithm with chaos theory for feature selection: A predictive model for medical data. *Int. J. Swarm Intell. Res.* **2019**, *10*, 1–20. [CrossRef]
22. Feng, Y.; Wang, G.G.; Wang, L. Solving randomized time-varying knapsack problems by a novel global firefly algorithm. *Eng. Comput.* **2018**, *34*, 621–635. [CrossRef]
23. Liu, C.; Gao, Z.; Zhao, W. A new path planning method based on firefly algorithm. In Proceedings of the 2012 Fifth International Joint Conference on Computational Sciences and Optimization, Harbin, China, 23–26 June 2012; IEEE: Piscataway, NJ, USA, 2012.
24. Hidalgo-Paniagua, A.; Vega-Rodríguez, M.A.; Ferruz, J.; Pavón, N. Solving the multi-objective path planning problem in mobile robotics with a firefly-based approach. *Soft Comput.* **2017**, *21*, 949–964. [CrossRef]
25. Michael, B.; Yu, X.-H. Autonomous robot path optimization using firefly algorithm. In Proceedings of the 2013 International Conference on Machine Learning and Cybernetics, Tianjin, China, 14–17 July 2013; IEEE: Piscataway, NJ, USA, 2013; Volume 3.
26. Patle, B.K.; Pandey, A.; Jagadeesh, A.; Parhi, D.R. Path planning in uncertain environment by using firefly algorithm. *Def. Technol.* **2018**, *14*, 691–701. [CrossRef]
27. Kim, H.C.; Kim, J.S.; Ji, Y.K.; Park, J.H. Path planning of swarm mobile robots using firefly algorithm. *J. Inst. Control. Robot. Syst.* **2013**, *19*, 435–441. [CrossRef]
28. Patle, B.K.; Parhi, D.R.; Jagadeesh, A.; Kashyap, S.K. On firefly algorithm: Optimization and application in mobile robot navigation. *World J. Eng.* **2017**, *14*, 65–76. [CrossRef]
29. Patle, B.K.; Parhi, D.; Jagadeesh, A.; Sahu, O.P. Real Time Navigation Approach for Mobile Robot. *J. Comput.* **2017**, *12*, 135–142. [CrossRef]
30. Wang, G.; Guo, L.; Hong, D.; Duan, H.; Liu, L.; Wang, H. A modified firefly algorithm for UCAV path planning. *Int. J. Hybrid Inf. Technol.* **2012**, *5*, 123–144.
31. Sutantyo, D.; Levi, P. Decentralized underwater multi robot communication using bio-inspired approaches. *Artif. Life Robot.* **2015**, *20*, 152–158. [CrossRef]
32. Sadhu, A.K.; Konar, A.; Bhattacharjee, T.; Das, S. Synergism of Firefly Algorithm and Q-Learning for Robot Arm Path Planning. *Swarm Evol. Comput.* **2018**, *43*, 50–68. [CrossRef]
33. Panda, M.R.; Dutta, S.; Pradhan, S. Hybridizing invasive weed optimization with firefly algorithm for multi-robot motion planning. *Arab. J. Sci. Eng.* **2018**, *43*, 4029–4039. [CrossRef]
34. Huang, H.C.; Lin, S.K. A Hybrid Metaheuristic Embedded System for Intelligent Vehicles Using Hypermutated Firefly Algorithm Optimized Radial Basis Function Neural Network. *IEEE Trans. Ind. Inform.* **2018**, *15*, 1062–1069. [CrossRef]
35. Duan, P.; Sang, H.; Li, J.; Han, Y.; Sun, Q. Solving multi-objective path planning for service robot by a pareto-based optimization algorithm. In Proceedings of the 2018 Chinese Control and Decision Conference (CCDC), Shenyang, China, 9–11 June 2018; IEEE: Piscataway, NJ, USA, 2018.
36. Singh, N.H.; Thongam, K. Neural network-based approaches for mobile robot navigation in static and moving obstacles environments. *Intell. Serv. Robot.* **2019**, *12*, 55–67. [CrossRef]

37. Montiel, O.; Sepúlveda, R.; Orozco-Rosas, U. Optimal path planning generation for mobile robots using parallel evolutionary artificial potential field. *J. Intell. Robot. Syst.* **2015**, *79*, 237–257. [CrossRef]
38. Zheng, W.; Wang, H.; Zhang, Z.; Lu, X. Hybrid position/virtual-force control for obstacle avoidance of wheeled robots using Elman neural network training technique. *Int. J. Adv. Robot. Syst.* **2017**, *14*, 1729881417710460. [CrossRef]
39. Orozco-Rosas, U.; Montiel, O.; Sepúlveda, R. Parallel Bacterial Potential Field Algorithm for Path Planning in Mobile Robots: A GPU Implementation. In *Fuzzy Logic Augmentation of Neural and Optimization Algorithms: Theoretical Aspects and Real Applications*; Springer: Cham, Switzerland, 2018; pp. 207–222.

eng

Article

Modelling Air Flow through Pneumatic Valves: A Brief Review with an Experimental Case Study

Gustavo Koury Costa

Federal Institute of Science and Technology of Pernambuco, Recife CEP 50740-545, PE, Brazil; gustavokoury@recife.ifpe.edu.br

Abstract: Compressible flow models are commonly used for describing air flow through pneumatic valves. Because of the difficulties in predicting viscous losses, these models ultimately rely on experimental determination of coefficients. Different equations have been proposed for different fluid speeds, having the sonic fluid velocity as a reference mark. However, one might question whether a much simpler approach, where the fluid is considered as incompressible, would still give good results within the typical range of industrial applications. Moreover, practically all models presuppose that the valve output pressure decreases in time, as in a discharge process. This paper reviews some representative one-dimensional compressible flow models and discusses the appropriateness of using equations based solely on discharging flows. Two experimental circuits, where an air reservoir is pressurized and, subsequently, decompressed, are used for comparison between different flow models. It is shown that a simpler set of equations still produces acceptable results for practical pneumatic applications.

Keywords: pneumatic circuits; orifice flow; fluid power; air flow

Citation: Koury Costa, G. Modelling Air Flow through Pneumatic Valves: A Brief Review with an Experimental Case Study. *Eng* **2023**, *4*, 2601–2614. https://doi.org/10.3390/eng4040149

Academic Editor: Antonio Gil Bravo

Received: 5 September 2023
Revised: 28 September 2023
Accepted: 12 October 2023
Published: 16 October 2023

1. Introduction

Pneumatic valves are inherently complex in their internal geometry. The simplest spool-valve is three-dimensional by nature, which demands the complete solution of the Navier–Stokes plus the energy–balance equations to determine the pressure field within. Even in the simplest case scenario, where compressibility effects are small, still the Navier–Stokes equations, consisting of four non-linear partial, second order differential equations, are the basis for modelling the air flow through the valve. Since, to this day, no analytical solution exists for those complex equations, one needs to rely on numerical methods for modelling every other valve.

A simple approach to the problem (of modelling pneumatic valves) is to assume the flow as unidimensional, or, likewise, treat the variables involved in an averaged way. Therefore, we can simply refer to the input and output pressures with no concern about the pressure field within the valve itself. Such an approach is classical in engineering and was made possible through the Reynolds Transport Theorem. Yet, difficulties are not reduced and the price to be paid is the introduction of model constants to be experimentally determined. One model in particular has been generally adopted, even by in the ISO 6358 Standards [1], as will be described in this paper. Yet, its accuracy has been questioned [2], besides the fact that it requires the experimental determination of two constants. Our aim in this paper is to discuss the accuracy of the ISO 6358 model compared with other proposed equations.

2. Orifice Flow Theory

Air flows through valves and orifices are traditionally modeled after an initial application of the mass, momentum and energy balances over a small region in space, as illustrated in Figure 1, where a convergent–divergent nozzle is represented. As air flows from point 1 to point 2, there may or may not be a significant change in density. If the density does not change much, we may treat the flow as incompressible. However, this is not the general practice adopted in pneumatic valve design.

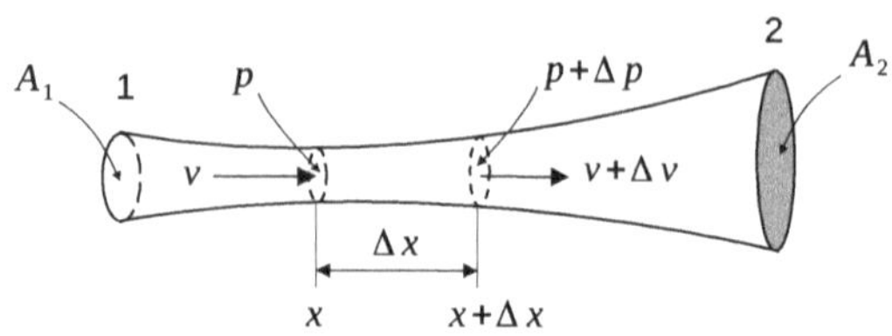

Figure 1. Unidimensional flow.

In what follows, we analyze the basis of compressible and incompressible flow modelling for the unidimensional flow represented in Figure 1 and further apply the resulting equations to the typical situation where flow is discharged and charged from and into an air tank. We begin with the incompressible flow approach.

It can be shown [3] that as long as the fluid speed remains below $0.3M$ (M is the Mach number), incompressible flow models can be applied and, in particular, the so-called "orifice equation" can be used; that is [4]

$$Q = \left\{ \sqrt{\frac{2A_2^2}{\rho\left[1-\left(\frac{A_2}{A_1}\right)^2\right]}} \right\} \sqrt{p_1 - p_2} = K\sqrt{p_1 - p_2} \quad (1)$$

where Q is the volumetric flow; p_1 and p_2 are the pressures at points 1 and 2; A_1 and A_2 are the cross-sectional areas at points 1 and 2 (see Figure 1) and ρ is the density of the fluid. Note that K is constant in Equation (1).

The mass flow, $\dot{m}$, can be obtained from (1) through the equation $\dot{m} = \rho Q$. Since we are considering that ρ remains constant in space and in time, Equation (1) can be simplified to

$$\dot{m} = C\sqrt{p_1 - p_2} \quad (2)$$

where $C = \rho K$. Here, the coefficient C can be experimentally adjusted to include pressure losses.

If fluid compressibility is considered, Equation (2) cannot be used. Still, an analytical expression can be developed if we assume that the flow is inviscid, the process is adiabatic and the flow regime has reached the steady-state. Considering Figure 1 again, it can be shown that a momentum and a mass balance between points x and $x + \Delta x$ produces the following equation [3]

$$\frac{dp}{\rho} + d\left(\frac{v^2}{2}\right) = 0 \quad (3)$$

where ρ, p and v are the density, pressure and velocity, respectively. As Δx, in Figure 1, becomes infinitesimal ($\Delta x \to dx$, $\Delta v \to dv$ and $\Delta p \to dp$).

The assumption of an adiabatic flow, for which $p/\rho^\gamma = C_\gamma$ (C_γ is a constant and γ is the ratio of specific heat capacities) allows us to integrate Equation (3) between points 1 and 2, resulting in

$$\frac{1}{2}\left(v_2^2 - v_1^2\right) = C_\gamma^{\left(\frac{1}{\gamma}\right)}\left(\frac{\gamma}{\gamma-1}\right)\left[p_1^{\left(\frac{\gamma-1}{\gamma}\right)} - p_2^{\left(\frac{\gamma-1}{\gamma}\right)}\right] \quad (4)$$

Equation (4) can be written in a more meaningful form if we write $C_\gamma^{1/\gamma} = p_1^{1/\gamma}/\rho_1$. After some mathematical work, the following equation is obtained

$$v_2^2 - v_1^2 = 2RT_1\left(\frac{\gamma}{\gamma-1}\right)\left[1-\left(\frac{p_2}{p_1}\right)^{\left(\frac{\gamma-1}{\gamma}\right)}\right] \tag{5}$$

Where $R = 287$ J/kgK is the gas constant for air and T_1 is the absolute temperature (K) inside the tank.

Of particular interest are the situations where air contained in a tank (a) discharges into the atmosphere through an orifice (or a valve), as illustrated in Figure 2a,b, is pressurized by an incoming flow, as shown in Figure 2b. Considering that air flows from point 1 to point 2, in Equation (4), we have that

- $v_1 = 0$, $v_2 = v_0$, $p_1 = p$ and $p_2 = p_0$ for the tank discharge (Figure 2a);
- $v_1 = v_0$, $v_2 = 0$, $p_1 = p_0$ and $p_2 = p$ for the tank pressurization (Figure 2b).

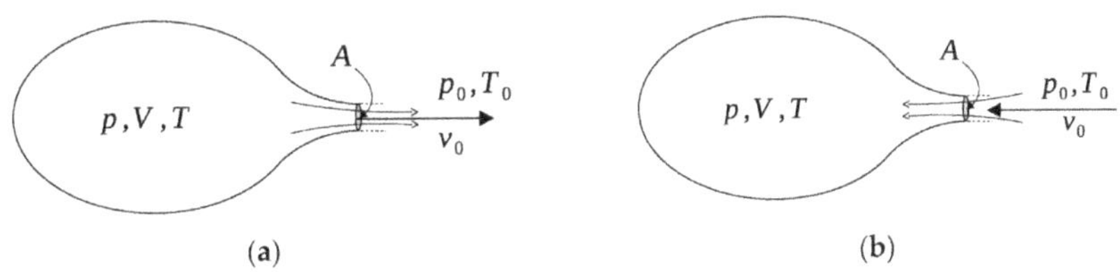

Figure 2. (**a**) Air tank discharge; (**b**) Air tank pressurization.

Substituting the values at points 1 and 2 for each case represented in Figure 2 into Equation (5), and considering the speed of sound at the tank nozzle, $c_0 = \sqrt{\gamma R T_0}$, and the corresponding Mach number, $M_0 = v_0/c_0$, it is possible to arrive at the following equations for the pressure ratio $r_p = p_0/p_1$

$$\begin{cases} r_p = \left[1-\left(\frac{T_0}{T}\right)\left(\frac{\gamma-1}{2}\right)M_0^2\right]^{\frac{\gamma}{\gamma-1}} & \text{Discharge} \\ r_p = \left[\left(\frac{\gamma-1}{2}\right)M_0^2+1\right]^{-\left(\frac{\gamma}{\gamma-1}\right)} & \text{Charge} \end{cases} \tag{6}$$

Remarks

1. The critical pressure ratio, b, is defined as the ratio, r_p, for which sonic speed is attained at the nozzle. Its value is not the same for charging and discharging, being dependent on the ratio between nozzle and tank temperatures, T_0/T, at discharge (but not on charge). Although this fact was recognized long ago [5], it has been ignored in most of the references on the topic (see, for example, references [6–8]), where in both charge and discharge, b is obtained from the second equation in (6) by making $M_0 = 1$; that is

$$b = \left(\frac{\gamma+1}{2}\right)^{-\left(\frac{\gamma}{\gamma-1}\right)} = \left(\frac{2}{\gamma+1}\right)^{\frac{\gamma}{\gamma-1}} \tag{7}$$

2. In spite of the fact that Equation (6) makes no reference to the nozzle geometry, the Mach number at the nozzle, M_0, is directly related to the cross-sectional area, A, and the way it changes along the x axis (Figure 1). For instance, if pressure decreases along the flow direction, it can be shown that, for subsonic flows, the fluid velocity increases as A is reduced. An opposite effect exists when the flow becomes supersonic. At sonic speed, A reaches a minimum value [9];
3. Considering a hypothetical scenario where the pressure tank discharges into the absolute vacuum ($p_0 = 0$ and $r_p = 0$), the first equation in (6) fails to represent the flow. To see that, we note that for a perfect gas, $p_0 = 0$ implies in $T_0 = 0$. As a result,

the first equation in (6) would result in $0 = 1$, which is incongruent. The second equation in (6) also fails to represent the flow at $r_p = 0$. In addition, liquefaction of the gaseous components of air would not allow for using the perfect-gas hypothesis at very low temperatures. These considerations pose a limitation on the use of Equation (6) for adiabatic flows through orifices.

The velocity and the mass flow at the nozzle, v_0 and $\dot{m}_0$, can be obtained from Equation (5) for both situations shown in Figure 2. Therefore, v_0 is given by

$$\begin{cases} v_0 = \sqrt{2RT}\sqrt{\left(\frac{\gamma}{\gamma-1}\right)\left[1-\left(\frac{p_0}{p}\right)^{\left(\frac{\gamma-1}{\gamma}\right)}\right]} & \text{Discharge} \\ v_0 = \sqrt{2RT_0}\sqrt{\left(\frac{\gamma}{\gamma-1}\right)\left[\left(\frac{p_0}{p}\right)^{-\left(\frac{\gamma-1}{\gamma}\right)}-1\right]} & \text{Charge} \end{cases} \tag{8}$$

To obtain the mass flow, we make $v_0 = \dot{m}/(\rho_0 A)$

$$\begin{cases} \dot{m} = \rho_0 A\sqrt{2RT}\sqrt{\left(\frac{\gamma}{\gamma-1}\right)\left[1-\left(\frac{p_0}{p}\right)^{\left(\frac{\gamma-1}{\gamma}\right)}\right]} & \text{Discharge} \\ \dot{m} = \rho_0 A\sqrt{2RT_0}\sqrt{\left(\frac{\gamma}{\gamma-1}\right)\left[\left(\frac{p_0}{p}\right)^{-\left(\frac{\gamma-1}{\gamma}\right)}-1\right]} & \text{Charge} \end{cases} \tag{9}$$

The *discharge* equation for $\dot{m}$ has been extensively used in pneumatic circuits to simulate flow through valves and is usually represented in a different way, following the work of St. Venant and Wantzel in 1839 [6]. To modify the discharge equation, we first write $p/\rho^{\gamma} = p_0/\rho_0^{\gamma}$. Then, we obtain ρ_0 from this last expression and substitute it into the first equation in (9). Finally, we write $\rho = p/(RT)$ and, after some algebraic work, arrive at the following expression for $\dot{m}$

$$\dot{m} = pA\Psi\sqrt{\frac{2}{RT}} \tag{10}$$

where Ψ, known as the "flow function", is given by

$$\Psi = \sqrt{\left(\frac{\gamma}{\gamma-1}\right)\left[\left(\frac{p_0}{p}\right)^{\frac{2}{\gamma}}-\left(\frac{p_0}{p}\right)^{\frac{\gamma+1}{\gamma}}\right]} \tag{11}$$

The following expression is obtained for $d\Psi/dr_p = 0$ (remember that $r_p = p_0/p$)

$$\left(\frac{2}{\gamma}\right){r_p}^{\left(\frac{2-\gamma}{\gamma}\right)} - \left(\frac{\gamma+1}{\gamma}\right){r_p}^{\left(\frac{1}{\gamma}\right)} = 0 \tag{12}$$

Equation (12) can be solved for r_p, which yields the maximum value of the flow function, Ψ, in Equation (11). The solution of Equation (12) is

$$r_p = b = \left(\frac{2}{\gamma+1}\right)^{\frac{\gamma}{\gamma-1}} \tag{13}$$

which is the critical pressure ratio, already obtained in Equation (7) for the case where the air tank is being charged. Substituting b, given by Equation (13), into the first equation in (8), we obtain the maximum value of the air velocity when the tank is being discharged, $v_0 = v_{max}$, as

$$v_{max} = c\sqrt{\frac{2}{\gamma+1}} \tag{14}$$

where $c = \sqrt{\gamma RT}$ is the speed of sound at temperature T. Note that v_{max} is close to sonic speed (for $\gamma = 1.4$, $v_{max} = 0.913c$).

The ratio $b = p_0/p$ corresponding to the maximum value of Ψ has been used as a divider in compressible flow discharge. Flows where $0 < r_p < b$ are termed as *critical flows* while flows where $b < r_p < 1$ are denominated *subcritical flows* [10]. Figure 3 illustrates this division in a graphical manner. Observe that the function $\Psi(r_p)$ resembles half of an ellipse at the interval $b < r_p < 1$. Figure 3 shows the ellipse as a dashed curve and its corresponding equation.

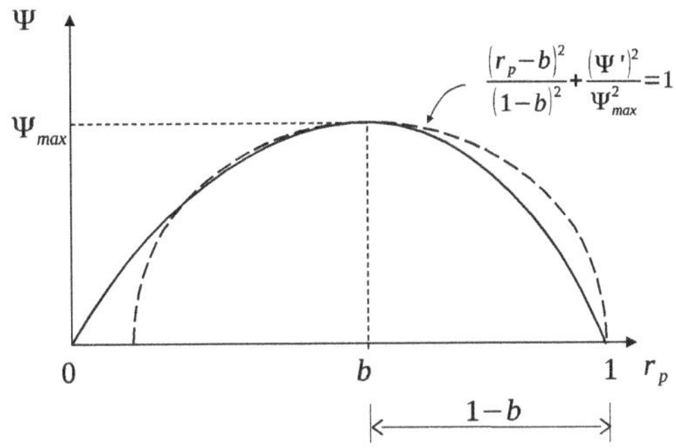

Figure 3. Flow function variation with the pressure ratio, r_p, and elliptical approximation.

We may write the ellipse equation in Figure 3 as $\Psi'(r_p)$, as follows

$$\Psi' = \Psi_{max}\sqrt{1 - \left(\frac{r_p - b}{1 - b}\right)^2} \tag{15}$$

where Ψ_{max} is obtained by substituting r_p, given by Equation (13) into Equation (11), resulting in the following expression

$$\Psi_{max} = \left(\frac{2}{\gamma + 1}\right)^{\frac{1}{\gamma - 1}} \sqrt{\frac{\gamma}{\gamma + 1}} \tag{16}$$

The theory presented so far led us to two equations that describe the steady-state behavior of incompressible and compressible air flows as they discharge through a nozzle; that is, Equations (2) and (10), plus the elliptical flow function approximation (15). Although pressure losses due to viscosity effects and the influence of nozzle geometry can be factored into the coefficient C, in Equation (2), the compressible flow model, given by Equation (10), assumes that the flow is inviscid and makes no particular reference to the nozzle geometry. One simple adjustment aiming to correct the mass flow to include these effects consists in adding a discharge coefficient, C_d, to Equation (10), as follows

$$\dot{m} = C_d p A \Psi \sqrt{\frac{2}{RT}} \tag{17}$$

Like the fluid velocity, v_0, the mass flow, $\dot{m}$, has a maximum at $r_p = b$ in the case where the upstream pressure, p, is kept constant. This is very important, because in the situation depicted in Figure 2a, where no flow enters the air tank, such assumption is not correct. The fact that $\dot{m}$ achieves a maximum when p is constant can be concluded by a simple observation of Equation (17), where all the terms on the right-hand side, except the flow function Ψ, become constant, so that $\dot{m}$ has the same behaviour as Ψ (see Figure 3).

When the upstream pressure, p, is not constant, as is the case in Figure 2a, we can write $p = p_0/r_p$ in Equation (17) so that the mass flow can then be written as $\dot{m}(r_p)$, as follows

$$\dot{m} = K_T\left(\frac{\Psi}{r_p}\right) \tag{18}$$

where K_T is given by

$$K_T = C_d p_0 A\sqrt{\frac{2}{RT}} \tag{19}$$

Assuming that K_T is constant, the mass flow, $\dot{m}$, can be plotted against the pressure ratio, r_p, for a vessel discharging into the atmosphere, as in Figure 2a. Here, we have assumed the typical industrial range for the pressure ratio, $0.1 \leq r_p \leq 1$. Figure 4 shows the curve $\dot{m}(r_p)$.

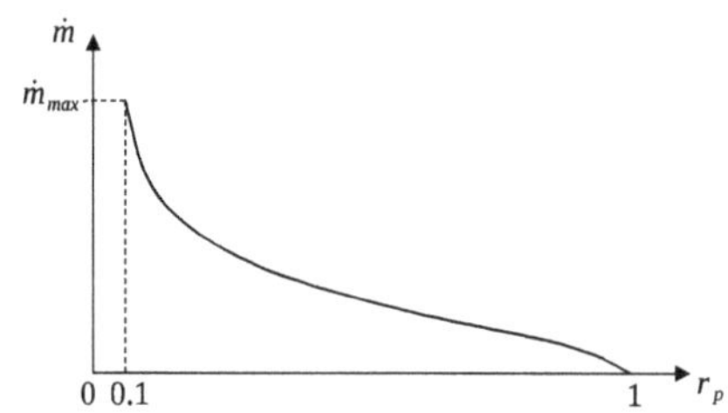

Figure 4. Evolution of the mass flow, $\dot{m}$, with the pressure ratio, r_p, for a variable upstream pressure, p.

The problem in using Equation (17) in the whole range of the pressure ratio was addressed by Lord Rayleigh in 1916 [4], who experimentally proved the non-validity of Equation (18) for the interval $0 \leq r_p \leq 0.25$, where the mass flow would, in fact, be constant for a constant value of p. His results contradicted a previous statement made by Osborne Reynolds, in 1885. According to Hartshorn [11], Reynolds wrote that the mass flow would remain constant within the interval $0 \leq r_p \leq 0.53$. Hartshorn concluded that neither Rayleigh nor Reynolds were 100% correct, for experiments showed that the interval limit where the mass flow is constant depends on the type of nozzle that is used. According to his experiments, the mass flow would become constant whenever $0 \leq r_p \leq L$, with L varying from 0.2 to 0.8, depending on the shape of the discharge nozzle. Figure 5 has been based on the experimental curves published by Hartshorn and illustrates the experimental results that he obtained.

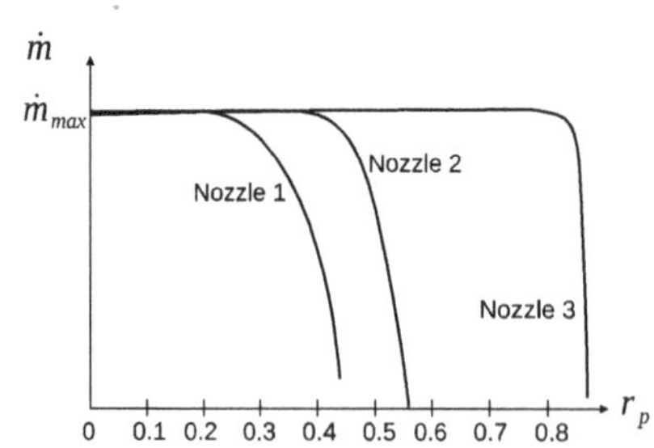

Figure 5. Mass flow in a discharging air tank for three different nozzle shapes.

In summary, the following equations can be used for adiabatic discharges through orifices and valves *only for the case where p is constant*

$$\begin{cases} \dot{m} = C_d p A \Psi\sqrt{\frac{2}{RT}} & \text{when} \quad L < r_p \leq 1 \\ \dot{m} = \dot{m}_{max} & \text{when} \quad 0 \leq r_p \leq L \end{cases} \tag{20}$$

Attempts to find the experimental curves for $\dot{m}$ have followed ever since. For instance, in 1955, Druett obtained experimental curves for divergent conic-shaped orifices [12]. However, a somewhat common practice is to assume $L = 0.528$, corresponding to the critical pressure ratio, b, for $\gamma = 1.4$ (see, for example, [2,13–21]).

3. The ISO 6358 Equations

As is usually the case, the need to find a relation between mass flow and pressure has led to attempts to standardize the mathematical model for orifice and valve flows in pneumatic circuits. One known standard that has been extensively used is the ISO 6358, where the following equations are recommended for modelling pneumatic valve flows [6]

$$\begin{cases} \dot{m} = \left(C\rho_r p\sqrt{\frac{T_r}{T}}\right)\sqrt{1-\left(\frac{r_p-L}{1-L}\right)^2} & \text{when} \quad L < r_p \leq 1 \\ \dot{m} = C\rho_r p\sqrt{\frac{T_r}{T}} & \text{when} \quad 0 \leq r_p \leq L \end{cases} \tag{21}$$

where ρ_r and T_r are the density and absolute temperature measured at a pre-established rated condition, defined by ISO as $T_r = 293.15$ K, $p_r = 100$ kPa, $\rho_r = 1.185$ kg/m^3 for a relative humidity of 65% and a corresponding gas constant $R = 288$ J/kg·K. The coefficient C (m^3/Pa·s) is denominated "sonic conductance" and, together with the "sonic pressure ratio", L, must be experimentally obtained.

The ISO standard uses the elliptical approximation, Ψ', given by Equation (15), as can be seen in second radicand of the first equation in (21). In fact, the first equation in (21) can be obtained by first substituting Ψ with Ψ', in Equation (17), as follows

$$\dot{m} = pC_d\sqrt{\frac{1}{RT}}\sqrt{1-\left(\frac{r_p-L}{1-L}\right)^2} \tag{22}$$

We can then use the perfect gas equation at a given reference state, $p_r = \rho_r R T_r$, to eliminate the constant R from Equation (22)

$$\dot{m} = \left(C\rho_r p\sqrt{\frac{T_r}{T}}\right)\sqrt{1-\left(\frac{r_p-L}{1-L}\right)^2} \tag{23}$$

where

$$C = AC_d\Psi_{max}\sqrt{\frac{2}{p_r\rho_r}} \tag{24}$$

Equation (21) has been widely used to this date for modelling pneumatic valve flows. The usual procedure, however, is to skip the experimental determination of the sonic pressure ratio, L, and use $L = 0.528$, which, as we have seen in the previous section, is not a safe assumption (Figure 5). On the other hand, the ISO 6358 Standard is highly demanding, requiring the experimental determination of two coefficients (L and C). The difficulties involved in determining the exact nature of the gas compression/expansion, being generally polytropic can move γ away from the adiabatic index ($\gamma = 1.4$), thus changing the value of b in Equation (13). For an assumed range $1 < \gamma < 1.7$, we would have $0.48 < b < 0.605$ [22]. In addition, as mentioned in the previous section, values of L as low as 0.2 were experimentally obtained. In fact, due to the difficulty in obtaining L, it is tempting to attribute a random value to L, perhaps based on some intuitive criteria, as can be concluded by reading some of the several works published on pneumatic circuits.

One question to be asked here is whether a simpler model such as the one given by Equation (2) could be used for modelling mass flows through valves. Bobrow and McDonell [2], for example, stated that "…nearly all previous results on pneumatic control incorrectly assume (14) is true", where (14) is a reference to the ISO equations in their paper. They obtained better results with the following equations, used in the context of a proportional valve connected to a cylinder

$$\begin{cases} \dot{m} = k_1\sqrt{p - p_0} & \text{Charging} \\ \dot{m} = k_2(p - p_0) & \text{Discharging} \end{cases} \tag{25}$$

where the coefficients k_1 and k_2 depend on the electric current that is applied to the proportional valve, according to the original paper [2].

We would like to investigate whether Equation (25) can still be used to model air flow through valves and orifices when k_1 and k_2 are constants. In fact, the first equation in (25) is, precisely, the incompressible flow model (2), with $C = k_1$. The linear relation indicated by the second equation, on the other hand, characterizes a laminar flow [4]. If these equations are able to represent flows through actual valves, the difficulties involved in determining L in Equation (21) are bypassed. We deal with this matter in the following section.

4. Numerical Modelling and Simulation

Consider the pressurization and subsequent decompression of an air tank, as shown in Figure 6. The schematics shown in the figure include elements that will be referred to, when we describe the test rig used in the experimental procedure.

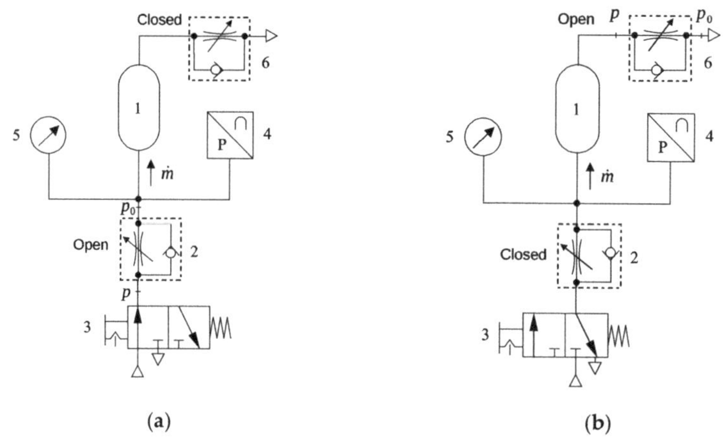

Figure 6. (**a**) Pressurization and (**b**) decompression of an air tank: 1—air tank; 2,6—unidirectional flow–control valves; 3—directional 3/2 valve; 4—pressure transducer; 5—pressure gauge.

To pressurize the tank 1, in Figure 6a, the unidirectional flow–control valve 2 can be adjusted between totally closed and totally open. The manually activated directional valve 3 provides a constant pressure, p, at the input of valve 2 while the tank pressure, p_0, changes from 0 to 0.6 MPa (gauge). The pressure evolution in time can be visually seen with the help of gauge 5 and measured in a timely manner by transducer 4, which sends data to an analog–digital converter connected to a computer. Depressurization (Figure 6b) is performed by inverting the position of valve 2 and closing it with the air tank pressure at 0.6 MPa, while air escapes to the atmosphere through valve 6. Note that p, at the valve 6 input, is variable this time, while p_0 remains constant.

Mass conservation can be applied to both situations in Figure 6. Considering that the air tank volume is V and that it remains at a constant temperature, T, the following equation is obtained

$$\dot{m} + \left(\frac{V}{RT}\right)\frac{dp}{dt} = 0 \tag{26}$$

where $\dot{m}$ is negative when flowing into the air tank, as in Figure 6a, and positive otherwise (Figure 6b). If we write $C_0 = (RT)/V$, Equation (26) can be written in a more convenient form as

$$\frac{dp}{dt} = -C_0\dot{m} \tag{27}$$

To circumvent the complexities of the analytical solutions for (27), when different expressions for $\dot{m}(p)$ are used, we have solved this equation numerically. Table 1 summarizes the equations for $\dot{m}(p)$ we have seen so far, for the situation where the air tank discharges into the atmosphere. In the table, we have grouped different factors as constants C_1 through C_3. Since the great majority of scientific and engineering works make use of only one model regardless of the direction of the air flow, only the equations corresponding to the tank discharge are being considered in Table 1. Incidentally, both Equation (25) are represented, given that the first equation in (25) corresponds exactly to the incompressible flow model (2).

Table 1. Summary of the different expressions for the air mass flow.

Equation Label	Equation
Equation (2)	$\dot{m} = C_1\sqrt{p - p_0}$
Equation (21)	$\begin{cases} \dot{m} = C_2 p\sqrt{1 - \left(\frac{r_p - L}{1-L}\right)^2} & \text{when} \quad L < r_p \leq 1 \\ \dot{m} = C_2 p & \text{when} \quad 0 \leq r_p \leq L \end{cases}$
Equation (25)	$\dot{m} = C_3(p - p_0)$

Each equation in Table 1 demands the knowledge of a constant, which must be determined through experimental data fitting. An easy way to make any of these models describe the actual air flow, however, is to change these constants into coefficients. For instance, considering a general coefficient C_k, we might think of a polynomial expansion of degree N, $C_k = \sum_n a_n r_p^n$ $(n = 0 \ldots N)$, as proposed in [13] for the ISO Equation (21). We do not favor this approach, given that, virtually, any model in Table 1 can be adjusted to fit experimental results by correctly choosing the new constants of the coefficient C_k. Moreover, once the polynomial equation for C_k is introduced back into its corresponding mathematical model, we end up with a totally strange equation, for which no physical background can be traced.

In Table 1, the constants C_1, C_2 and C_3 are given by:

$$\begin{cases} C_1 = \sqrt{\dfrac{2A_2^2}{\rho\left[1 - \left(\frac{A_2}{A_1}\right)^2\right]}} \\ C_2 = C\rho_r p\sqrt{\frac{T_r}{T}} \\ C_3 = k_2 \end{cases} \tag{28}$$

The combination of the equations in Table 1 and the differential Equation (27) results in three mathematical models for the pressure inside the air tank in Figure 6, as listed in Table 2.

Table 2. Mathematical models for the pressure inside the air tank.

Model	Equation
Model 1	$\frac{dp}{dt} = -C_0C_1\sqrt{p - p_0}$
Model 2	$\begin{cases} \frac{dp}{dt} = -C_0C_2 p\sqrt{1 - \left(\frac{r_p - L}{1-L}\right)^2} & \text{when} \quad L < r_p \leq 1 \\ \frac{dp}{dt} = -C_0C_2 p & \text{when} \quad 0 \leq r_p \leq L \end{cases}$
Model 3	$\frac{dp}{dt} = -C_0C_3(p - p_0)$

We have solved each equation in Table 2 for $p(t)$, adjusting constants C_1 through C_3 so that $p(0) = 0$ MPa and $p(12) = 0.7$ MPa, simulating the pressurization of the air tank in Figure 6a. In a reverse order, we also simulated the decompression of the air tank for $p(0) = 0.7$ MPa and $p(30) = 0$ MPa. The time setting for the simulation has been based on experimental data to be introduced in the following section. Considering an air tank volume $V = 0.4$ L and assuming that the air temperature is kept at $T = 293$ K, the constant C_0 can be calculated as

$$C_0 = \frac{RT}{V} = \frac{287 \frac{\text{J}}{\text{kg}\cdot\text{K}} \times 293\ \text{K}}{(0.4 \times 10^{-3})\text{m}^3} = \left(2.1 \times 10^8\right) \frac{\text{J}}{\text{kg}\cdot\text{m}^3} \quad (29)$$

We have assumed $L = 0.528$ and $p_0 = 0.1$ MPa. Simulation results, obtained by using the 4th order Runge–Kutta method to solve the equations in Table 2, are shown in Figure 7. Figure 7a,b show the evolution in time of the pressure inside the air tank. In both cases, there is a closer match between results given by models 1 and 2. This is due to the much higher mass flow at the beginning when the linear model 3 is used, as can be seen in Figure 7c,d. Note the constant mass flow during $0 \leq t \leq 0.528$ at pressurization when model 2 is used. This is justified by the fact that model 2 is based on the theoretical–experimental approach explained in the previous section where the mass flow becomes constant when the pressure ratio drops below a given value of r_p (see Figure 5).

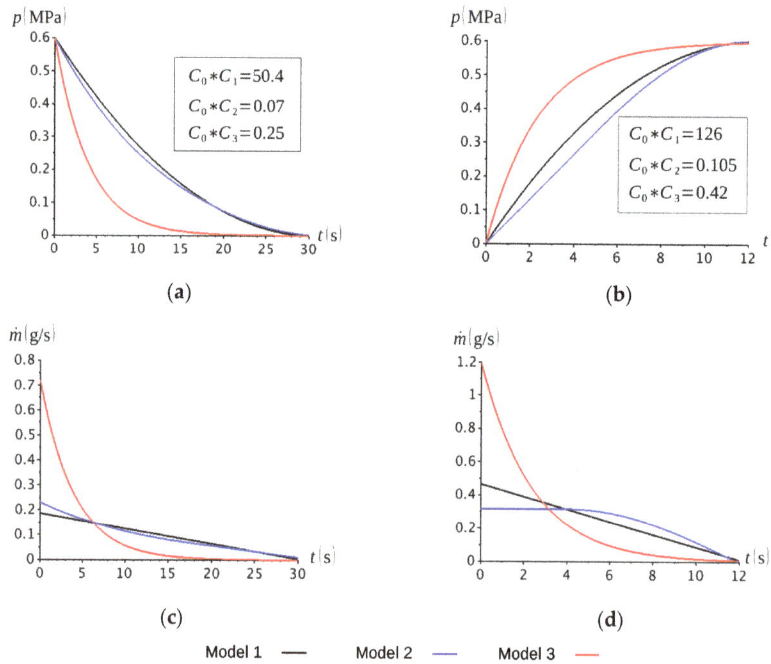

Figure 7. Gauge pressure inside the air tank during (**a**) decompression and (**b**) pressurization. Corresponding mass flows (**c**) escaping from the tank during decompression and (**d**) entering the tank during pressurization. Constants used for decompression and pressurization are shown in (**a**) and (**b**), respectively.

It is very important to observe the values of the constants C_1, C_2 and C_3 change significantly between decompression and pressurization. As a matter of fact, they also change with the flow–control valve adjustment, as will be seen in the following section, where we present experimental results that will help us better compare the models in Table 2.

5. Comparison with Experimental Results

Figure 8 shows the pneumatic test rig used to obtain the curve $p(t)$ for the two situations displayed in Figure 6. The technical specifications for each component are given in Table 3. All components are interconnected through plastic tubing whose external/internal diameters are 4 mm and 2.5 mm, respectively.

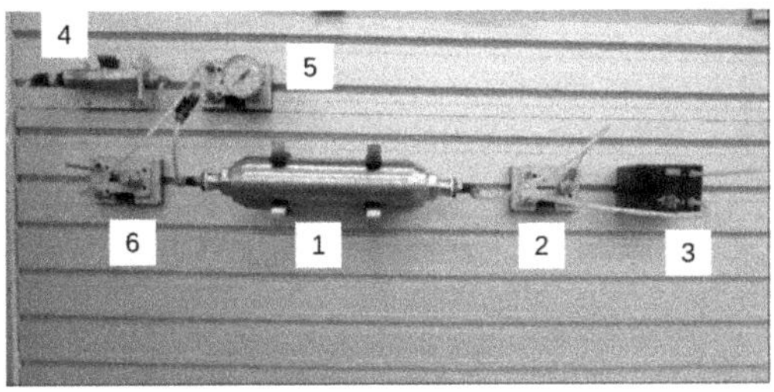

Figure 8. Test rig (pneumatic parts are numbered according to the schematics in Figure 6): 1—Air reservoir; 2, 6—Flow-control valves; 3—Directional 3/2 valve; 4—Pressure transmitter; 5—Pressure gauge.

Table 3. Technical specifications for the test rig components.

Component Number	Commercial Model
1	Air reservoir FESTO. Model CRVZS-0,4 160234 (0.4 liters)
2	Flow–control valve FESTO. Model 152881
3	Festo 3/2 valve with selector switch. Model 152863
4	Pressure transmitter FESTO. Model SDET-22T-D10-G14-U-M12 (1% accuracy)
5	Pressure gauge FESTO. Model 152865
6	FESTO 152881

A comparison between experimental results and the theoretical curves is shown in Figure 9 for two opening states of the flow–control valves. First, the valve was approximately 5% open. The curves corresponding to the decompression and pressurization of the air tank for this first setting are shown in Figure 9a,b, respectively. The second set of experimental data was obtained for a 100% opening of the flow control valves, for which corresponding curves are shown in Figure 9c,d, for decompression and pressurization, respectively. The curves were fitted so that they would match experimental data at $p = 0.6$ MPa and $p = 0$ MPa (gauge).

At a first glance, we conclude that the third model produces a far better approximation to experimental data when the air tank is discharging into the atmosphere (Figure 9a,c). On the other hand, both the incompressible flow model 1 and the ISO model 2 are better when the tank is being charged (Figure 9b,d). It is, therefore, reasonable, to prefer the one-constant modelling Equation (25) instead of the two-constant ISO Equation (21).

The reason why the linear dependence between flow and pressure provides a better approximation during discharge is a challenging one. As we already mentioned, the assumption of a laminar flow on discharge would help to explain this fact. We believe that a more detailed analysis would require the numerical solution of the Navier–Stokes equations. This is certainly one possibility for further developments to be pursued in a future work.

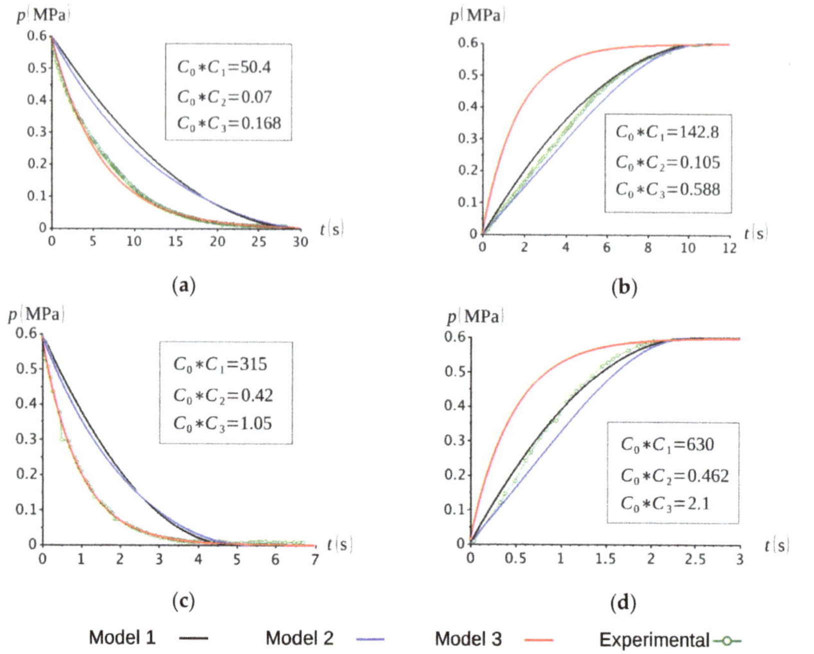

Figure 9. Gauge pressure in the air tank: (**a**,**b**)—decompression and pressurization at 5% opening of the flow–control valves; (**c**,**d**)—decompression and pressurization at 100% opening of the flow–control valves. The corresponding constants in Table 2 are displayed for each set of curves.

One might wonder whether the choice for L in model 2 would be the reason why the ISO model does not provide the best fit during tank discharge. To verify this, we simulated the same case shown in Figure 9c using the two extreme values of L, 0 and 1. The results are presented in Figure 10. Even with these changes we could not obtain a better fitting, which suggests that Equation (21) is not the best option for the case where the tank is depressurized. Moreover, the simpler incompressible model 1 also provides a good approximation for the case where the tank is being charged, with only one constant to be adjusted. We thus come to the same conclusion found in reference [2] and agree that Equation (25) best fits our experimental data.

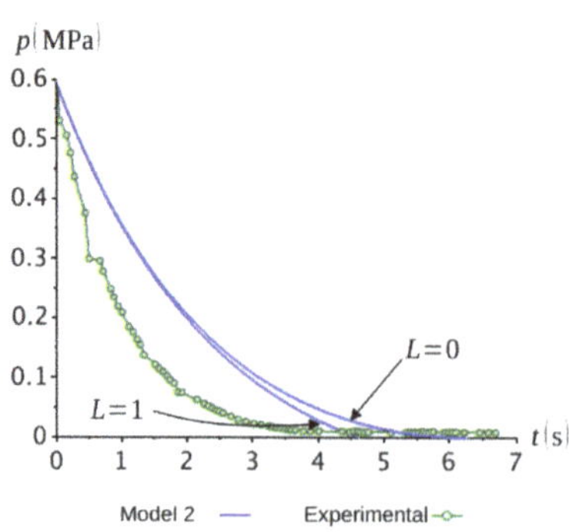

Figure 10. Tank decompression for 100% opening of the flow–control valve. Comparison with model 2 using two different values of L.

6. Conclusions

In this paper, we reviewed the most common models for pneumatic valve flow, beginning at the very foundations of the currently accepted models. We showed that equations are not the same for discharging and charging vessels and discussed the application of classical compressible-flow equations in real-life pneumatic applications, where we could see that the simple assumption of fluid incompressibility can be safely applied for the usual pressure range in industrial applications when the flow direction is towards pressurization of the receiving pneumatic element, which, in our case, was an air tank. The similarity between the results obtained using compressible and incompressible models, in this case, explains why we do not find any publication, except from the one in reference [2], where incompressible flow equations are applied. Incidentally, a numerical flow analysis carried out for a spool valve has recently shown that the internal valve flow is predominantly incompressible [23]. This fact may certainly be considered in a future analysis of the problem.

Last, it is interesting that an even simpler relation provided a more accurate match with experimental values when the flow direction was directed towards depressurizing the air tank. An actual explanation for this behavior is not in the scope of this paper and can also be an inspiring topic for future work. In the end, we have concluded that it is safe to use a simple model where only one constant needs to be experimentally determined, instead of the more complex ISO 6358 model, where two different constants are required.

Funding: This research received no external funding.

Data Availability Statement: The data presented in this study are available on request from the corresponding author.

Conflicts of Interest: The author declares no conflict of interest.

References

1. *ISO 6358-1: 2013*; Pneumatic Fluid Power–Determination of Flow Rate Characteristics of Components Compressible Fluids–Part 1: General Rules and Test Methods for Steady-State Flow. ISO: Geneva, Switzerland, 2013. Available online: https://www.iso.org/standard/56612.html (accessed on 13 October 2023).
2. Bobrow, J.E.; McDonell, W. Modeling, identification, and control of a pneumatically actuated, force controllable robot. *IEEE Trans. Robot. Autom.* **1998**, *14*, 732–742. [CrossRef]
3. Pritchard, P.J.; Mitchell, J.W. *Fox and McDonald's Introduction to Fluid Mechanics*, 8th ed.; John Wiley & Sons: New York, NY, USA, 2011; pp. 675–680.
4. Costa, G.K.; Sepehri, N. *Hydrostatic Transmissions and Actuators: Operation, Modelling and Applications*; John Wiley & Sons: Chichester, UK, 2015; pp. 81, 86.
5. Lord Rayleigh, O.M.F.R.S. On the discharge of gases under high pressures. *Lond. Edinb. Dublin Philos. Mag. J. Sci.* **1916**, *32*, 177–187. [CrossRef]
6. Beater, P. *Pneumatic Drives: System Design, Modelling and Control*; Springer: Berlin, Germany, 2007; pp. 29–31.
7. Sanville, F.E. A New Method of Specifying the Flow Capacity of Pneumatic Fluid POWER Valves. In Proceedings of the Second Fluid Power Symposium, Guildford, UK, 4–7 January 1971.
8. Pasieka, L. The applicability of the mass-flow-model according to ISO 6358 with the parameter critical conductance c and critical pressure ratio b for gases in high-pressure range up to 300 bar. In Proceedings of the 12th International Fluid Power Conference (12. IFK), Dresden, Germany, 12–14 October 2020.
9. Zucker, R.D.; Biblarz, O. *Fundamentals of Gas Dynamics*, 2nd ed.; John Wiley & Sons. Inc.: Hoboken, NJ, USA, 2002; pp. 119–120.
10. Perry, J.A., Jr. Critical Flow Through Sharp-Edged Orifices. *Trans. Am. Soc. Mech. Eng.* **1949**, *71*, 757–763. [CrossRef]
11. Hartshorn, L. The Discharge of Gases under High Pressures. *Phil. Mag.* **1916**, *32*, 178–188.
12. Druett, H.A. The Construction of Critical Orifices Working with Small Pressure Differences and Their Use in Controlling Airflow. *Br. J. Ind. Med.* **1955**, *12*, 65–70. [CrossRef] [PubMed]
13. Pugi, L.; Malvezzi, M.; Allotta, B.; Banchi, L.; Presciani, P. A parametric library for the simulation of a Union Internationale des Chemins de Fer (UIC) pneumatic braking system. *Proc. Inst. Mech. Eng. Part F J. Rail Rapid Transit* **2004**, *218*, 117–132. [CrossRef]
14. Rahman, R.A.; He, L.; Sepehri, N. Design and experimental study of a dynamical adaptive backstepping–sliding mode control scheme for position tracking and regulating of a low-cost pneumatic cylinder. *Int. J. Robust Nonlinear Control.* **2016**, *26*, 853–875. [CrossRef]

15. Pawananont, K.; Leephakpreeda, T. Sequential Control of Multichannel On–Off Valves for Linear Flow Characteristics Via Averaging Pulse Width Modulation without Flow Meter: An Application for Pneumatic Valves. *J. Dyn. Syst. Meas. Control.* **2019**, *141*, 011007. [CrossRef]
16. Mandali, A.; Dong, L. Modeling and Cascade Control of a Pneumatic Positioning System. *J. Dyn. Syst. Meas. Control.* **2022**, *144*, 061004. [CrossRef]
17. Gibson, T.J.; Barth, E.J. Design, Model, and Experimental Validation of a Pneumatic Boost Converter. *J. Dyn. Syst. Meas. Control.* **2019**, *141*, 011004. [CrossRef]
18. Richer, E.; Hurmuzlu, Y. A High Performance Pneumatic Force Actuator System: Part I—Nonlinear Mathematical Model. *J. Dyn. Syst. Meas. Control.* **2000**, *122*, 416–425. [CrossRef]
19. Jiang, Z.; Xiong, W.; Du, H.; Wang, Z.; Wang, L. Energy-saving methods in pneumatic actuator stroke using compressed air. *J. Eng.* **2021**, *2021*, 241–251. [CrossRef]
20. Lin, Z.; Xie, Q.; Qian, Q.; Zhang, T.; Zhang, J.; Zhuang, J.; Wang, W. A Real-Time Realization Method for the Pneumatic Positioning System of the Industrial Automated Production Line Using Low-Cost On–Off Valves. *Actuators* **2021**, *10*, 260. [CrossRef]
21. Zhang, L.; Yan, Y.; Zhu, Q.; Zhao, G.; Feng, D.; Wu, J. A Pneumatic Control Method for Commercial Vehicle Electronic Brake System Based on EPV Module. *Actuators* **2022**, *11*, 316. [CrossRef]
22. Urata, E.; Youn, C.; Kagawa, T. Approximate expressions for characteristics of gas-flow through orifices. In Proceedings of the 9th JFPS International Symposium on Fluid Power, Matsue, Japan, 28–31 October 2014.
23. Blasiak, S.; Laski, P.A.; Takosoglu, J.E. Rapid Prototyping of Pneumatic Directional Control Valves. *Polymers* **2021**, *13*, 1458. [CrossRef] [PubMed]

eng

MDPI

Article

Comparison of Three Linear Digital Filters Applied to Improve the Quality of the Measured Acoustic Field

Estevão Patricio Rodrigues [1], Djalma Simões dos Santos [2] and Flávio Buiochi [1,*]

1 Department of Mechatronic Engineering, Engineering School at the University of São Paulo, São Paulo 05508-010, Brazil; estevaopatricio@usp.br
2 Laboratory of Medical Imaging, Department of Imaging Physics, Delft University of Technology, 2628 CJ Delft, The Netherlands; d.simoesdossantos@tudelft.nl
* Correspondence: fbuiochi@usp.br

Abstract: The measurement of acoustic fields generated by ultrasonic transducers is important for determining the focal length, lateral resolution, and amplitudes of the lateral and grating lobes. The acoustic field is commonly characterized by a set of scans using a needle hydrophone. The output of the hydrophone can be connected to an analog filter to enhance the signal. However, the analog filter might not be sufficient to avoid the noises that distort the signals. Alternatively, linear digital filters can be advantageous to improving the acoustic-field characterization. In this work, three filters were investigated: moving average (MA), band-pass Hamming window (HW), and band-pass Blackman window (BW). The filters were implemented and evaluated in terms of the root-mean-square error (RMSE) of the measured sound field, which was filtered, in relation to the simulated acoustic field (gold standard). As a compromise between effective filtering and signal non-distortion, a method to model the MA kernel length was proposed. All the filters reduced the noise of the measured acoustic field. The HW and the BW filters were more effective (RMSE = 4.01%) than the MA filter (RMSE = 4.28%). In spite of the small quantitative difference, acoustic field comparisons showed qualitative improvements.

Keywords: acoustic field characterization; transducers; ultrasound; digital filters; window filters

Citation: Rodrigues, E.P.; Santos, D.S.d.; Buiochi, F. Comparison of Three Linear Digital Filters Applied to Improve the Quality of the Measured Acoustic Field. *Eng* **2023**, *4*, 2582–2600. https://doi.org/10.3390/eng4040148

Academic Editor: Antonio Gil Bravo

Received: 30 August 2023
Revised: 3 October 2023
Accepted: 10 October 2023
Published: 16 October 2023

1. Introduction

The experimental determination of the acoustic field generated by a transducer is a very important process in the design of ultrasound transducers [1,2], as well as for their maintenance and calibration. The measurement of the acoustic field is useful to obtain various characteristics of the beam, such as the focal length, beam width, and directivity pattern.

The acoustic field is normally acquired underwater using a needle hydrophone, which detects the ultrasound pulses emitted by the transducer in a region of interest using an xyz-positioning stage [1–3]. A waveform (i.e., a time-varying signal) associated with each grid point is obtained. The signals obtained can be represented by peak values, peak-to-peak values, average values, root-mean-square (RMS) values, or envelope peak values. Then, the generated acoustic field can be plotted in terms of the chosen values.

The needle hydrophone can be made of a piezoelectric polymer, polyvinylidene fluoride (PVDF), which has a very large sensitivity and bandwidth compared to ceramics [4,5]. In [6], a sensor for measuring acoustic waves from a high-intensity focused ultrasound (HIFU) transducer also used a PVDF polymer. While a broadband hydrophone is suitable for reliably sampling the pulse emitted by the transducer, it is also susceptible to capturing unwanted frequencies that arise from factors associated with the source and measurement setup [4,7]. These factors, which can distort the hydrophone signal and cause variation in the measured acoustic pressure, include fluctuations in electrical impedance and drive voltage, changes in water temperature (especially during long scans), alignment and positioning errors, reflections from the tank or water surface, poor water quality, electrical noise, and environmental vibrations [7–9].

In general, a needle hydrophone is used to detect weak acoustic signals, which have a very low amplitude. Due to this, the output of the hydrophone is commonly connected to a pre-amplifier to provide a gain stage to amplify the measured signal [4]. In addition, an analog low-pass filter can be combined with the hydrophone output to avoid aliasing and suppress the potential amplification of high-frequency noise. Although the low-pass filter can improve the performance of the measured acoustic signal, further improvement might be achieved by applying additional filtering beyond the low-pass filter [10]. As an alternative, digital filters can be applied to filter the signals after they are sampled and digitized [11].

In the literature, many studies have reported on the use of digital filters for different applications in acoustics. A digital filter was applied [12] to reduce the noise interference in radio-frequency (RF) signals, used in ultrasound elastography images. A filter was developed for medical ultrasound images, to reduce the speckle noises while maintaining the edges of the human tissue [13].

A band-pass finite impulse response (FIR) filter was used [14] to enhance the visualization of the audible sound field in real time using the Schlieren technique, such that the noise was reduced by removing unwanted frequency components. The sinusoidal sound fields of frequency 10 kHz and 15 kHz were obtained using a band-pass filter with the order parameter equal to 200 (for a Nth order FIR filter, there will be N + 1 coefficients [15]).

Second- and fourth-order moving average (MA) filters were proposed [16] to be implemented in an Arduino-based acquisition system to filter temperature and ultrasound echo signals. The higher-order filter improved the results, producing smoother signals. Another filter [17], a moving average hybrid FIR filter, which linearly combines MA and hybrid median filters, reduced noise and improved the edges of 2D ultrasound images.

Two low-pass filters, one with a Blackman window and the other with a flat top window, were designed in MATLAB Simulink [18]. The filters were 34th order and the normalized cutoff frequency was 0.2. The roll-off of the Blackman window was faster than the flat top window, which provided greater attenuation in the stopband. In another study [19], 4th- to 34th-order low-pass filters with Blackman Nuttall and Welch windowing were analyzed to remove white noise from electrocardiogram (ECG) signals. The Blackman Nuttall filter was more effective than the Welch filter, as it provided less distortion in the signal.

A band-pass Hamming-window filter at a central frequency of 2.25 MHz and bandwidth of 2.48 MHz (110%) was used to characterize an electroacoustic hydrophone developed in [20]. In [21], a band-pass Hamming window was used to filter the ultrasound signals from an underground mine detector system. The filter lower and upper cutoff frequencies were 39.9 kHz and 40.1 kHz, respectively, close to the operating frequency of 40 kHz. The transition band from each cutoff frequency to the stopband was 5 kHz and the sampling frequency was 212 kHz.

In [22], three types of window functions used for designing FIR low-pass filters were tested: the Kaiser window, Dolph–Chebyshev window, and Hamming window. In terms of frequency selectivity, the Hanning window played a better role than the others. To date, however, it seems that no systematic studies have been conducted that quantitatively compare the performance of different types of digital filters employed to filter hydrophone signals during the characterization of acoustic fields.

This work aims to test linear digital filters that can be easily implemented to improve the acoustic field characterization by reducing the unwanted frequency components from the hydrophone measurements. Three filters were evaluated: the moving average filter, band-pass Hamming window, and band-pass Blackman window. The measured acoustic field was filtered with these three filters and compared to the simulated acoustic field, taken as the gold standard.

2. Theory

Digital filters are very important for many kinds of devices such as smartphones, radios, wireless data transfer, multimedia devices, and so on [23]. They are essential for processing digital signals, being useful to separate signals that have been mixed and to re-store signals that have been distorted [24]. For example, digital filters allow obtaining a baby's electrocardiogram still in the womb, separating the baby's heartbeat signal from the mother's signals, such as her breathing and heartbeat [24].

Although analog filters can perform the same functions as digital filters, they are very expensive, especially at high frequencies, as they are implemented with hardware components such as operational amplifiers, capacitors, inductors, and resistors. Furthermore, digital filters have superior performances than those obtained with analog filters [23,24]

Moreover, digital filters can be implemented through software [23], allowing them to change their performance easily, for example, by increasing or decreasing the order and changing the cutoff frequencies as needed.

However, an analog low-pass filter is required before the signal digitalization (analog-to-digital conversion) to avoid the undersampling effect, also known as aliasing. When aliasing occurs, the high frequencies are digitalized as low frequencies, distorting the information. This can be avoided by adding an analog low-pass filter with a cutoff frequency (f_c) that is lower than half of the sampling frequency (f_s). This is known as the Nyquist criterion, which says that f_s must be at least twice as high as the maximum frequency of the signal to be sampled [15]. This analog low-pass filter is known as an anti-aliasing filter. Next, the linear digital filters used in this work are introduced.

2.1. Moving Average Filter

The moving average (MA) filter operates straight in the time domain and it is the most common filter in digital signals processing (DSP) due to its ease of implementation [24]. The MA is efficient at decreasing random noise and smoothing the signal. However, MA is not suitable for separating signal components in the frequency domain. Although the MA acts as a low-pass filter, its roll-off (i.e., the transition from the cutoff frequency to stopband frequency) is large. For a given digitized temporal signal $x[\ ]$, the MA-filtered signal $y[\ ]$ is defined as [24]

$$y[i] = \frac{1}{M} \sum_{j=0}^{M-1} x[i+j]\,, \tag{1}$$

where M is the number of samples used in the moving average. In short, Equation (1) means that each signal sample is replaced by the average of M adjacent samples, including the sample being filtered. For example, supposing a signal in the discrete time domain being operated by the MA with $M = 3$, the value of the signal at index 80 will be $y[80] = (x[80] + x[81] + x[82])/3$.

The MA filter can be implemented as a convolution using a simple filter kernel (the kernel of the filter is its impulsive response, $h[\ ]$). For example, the kernel of an MA with $M = 3$ is $h[\ ] = \{1/3, 1/3, 1/3\}$. The MA is a sum of the convolution of the signal with a rectangular pulse (kernel), resulting in an area equal to one. Formally, the kernel convolution sum $h[\ ]$ of length M, with the input signal $x[\ ]$, with j going from 0 to $M - 1$, results in the output signal $y[\ ]$, given as [24]:

$$y[i] = \sum_{j=0}^{M-1} h[j]x[i-j]\,. \tag{2}$$

2.2. Windowed-Sinc Filter

The windowed-sinc filter (also called window filter) is implemented by convoluting its kernel with the signal in the discrete time domain. However, the window filter kernel is modeled for different purposes, such as selecting and removing the amplitudes at specific frequencies, making it possible to remove noise and extract only the pertinent information from the signal. Thus, the window filter is suitable to be used as a low-pass, high-pass, band-pass, or band-reject filter (the latter is also called a notch filter [24]).

To understand how a window digital filter is obtained, one must start from the ideal filter response in the frequency domain. The kernel of the ideal frequency-domain filter is a rectangular function (Figure 1a), and its time-domain equivalent is a sinc function (Figure 1b), which is obtained by the inverse Fourier transform of the rectangular function. The idea is to give unity gain to the components of interest in the signal, and to give zero gain to the other components. This is accomplished by multiplying the kernel by the signal in the frequency domain, or convolving them in the discrete time domain, as seen in (2).

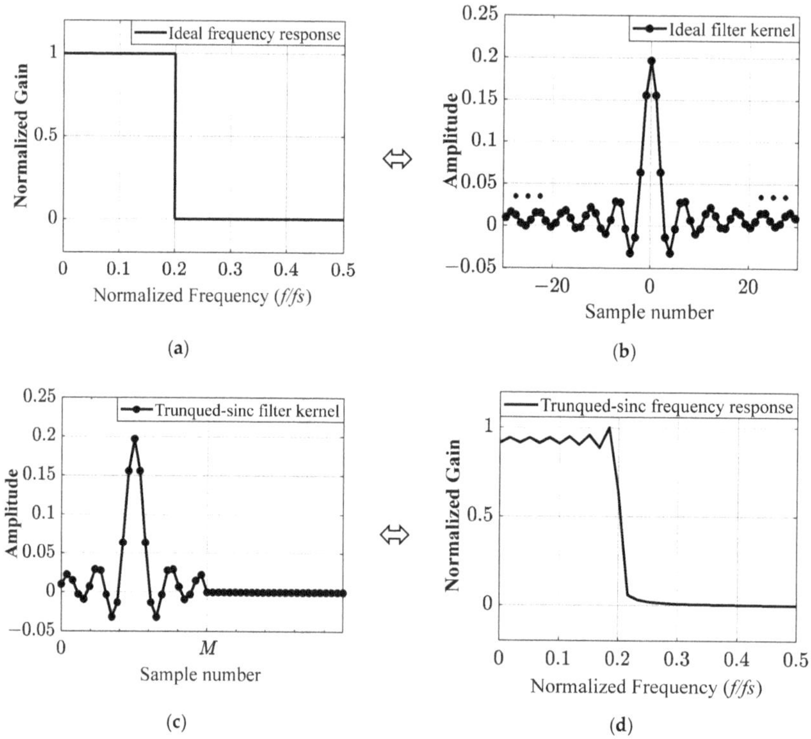

Figure 1. Windowed-sinc low-pass filter with a cutoff frequency of $0.2\,f_S$: (**a**) frequency response of the ideal filter kernel, and (**b**) its respective infinite sinc function; (**c**) kernel used from a sinc function truncated with $M + 1$ samples, and (**d**) its low-pass and overshoot near the cutoff frequency.

For sinc kernel to be implemented in a computer, that is, in a discrete domain, the sinc function must be limited (truncated) symmetrically around the main lobe, with $M + 1$ points (M must be an even number, and the sum of one is due to the central symmetry point), and all samples outside this range must be set to zero [24]. Also, the entire sequence must be shifted to the right so that the kernel can "run" from zero to M, resulting in a causal function (Figure 1c). However, the abrupt discontinuities at the ends of the truncated sinc result in unwanted ringing at the band-pass and overshoot around the cutoff frequency (Figure 1d). This distortion phenomenon is known as the Gibbs effect [15,24].

A very efficient method to reduce the distortions in the frequency domain is to multiply the truncated sinc by a weighting function, known as the apodization window (hence the name of the windowed-sinc filter). Thus, the apodization window makes the ends of the sinc function gradually decrease to zero.

There is a variety of apodization functions, but the Hamming and Blackman windows are particularly advantageous as they have higher stopband attenuation than others. While the stopbands of the rectangular (windowless), Bartlett (triangular), and Hanning (raised cosine) windows are −21 dB, −25 dB, and −44 dB, respectively, the Hamming and the Blackman windows are −53 dB and −74 dB, respectively. On the other hand, the roll-off of the Blackman window is 3 times larger than the roll-off of the rectangular window and 1.5 times larger than that of the Hamming window. Moreover, the roll-off rates are the same Hanning and Bartlett windows [15]. Therefore, for a given kernel of length $M + 1$, the choice of an apodization window is a trade-off between stopband and roll-off.

Although the stopband is dependent on the window type, the roll-off can be adjusted by increasing or decreasing M: the larger the M, the faster the roll-off, and vice versa [24]. This means that a Blackman window could achieve the Hamming roll-off while maintaining the Blackman stopband, but the filtering computing time also increases due to the convolution operation (2).

2.3. Band-Pass Filter with Hamming Window and Blackman Window

The Hamming window (HW) and the Blackman window (BW) filters were used in this work because they have larger stopbands than the other window filters mentioned here. The process for designing the two filters is very similar, so they were designed together in this section. A band-pass filter is suitable to improve the characterization of the acoustic field, as the transducer bandwidth has been previously characterized using the pulse-echo measurement. By fitting the frequency band of the band-pass filter within the bandwidth of the transducer, the pulse is separated from unwanted noise in the sampled window measured with the hydrophone.

In the design of a filter, the lower and upper cutoff frequencies (respectively, f_1 and f_2) must be normalized by the sampling frequency, f_s, and must be a value between 0 and 0.5 [24] (considering the Nyquist frequency).

The value of M must be an even positive integer. It can be approximated by [24]

$$M = \frac{4}{BW}, \tag{3}$$

where BW is the frequency bandwidth of the roll-off, which is also normalized by the sampling frequency f_s. Equation (3) is a trade-off between the roll-off speed and computation efficiency: the larger the M, the faster the roll-off (because the transition bandwidth is shorter) and the longer the computation time is, and vice versa.

Afterward, the kernel of the windowed-sinc Hamming filter was calculated by [24]

$$h[i] = k\left[\frac{\sin\left(2\pi f_2\left(i - \frac{M}{2}\right)\right)}{i - \frac{M}{2}} - \frac{\sin\left(2\pi f_1\left(i - \frac{M}{2}\right)\right)}{i - \frac{M}{2}}\right]\left[0.54 - 0.46\cos\left(\frac{2\pi i}{M}\right)\right]. \tag{4}$$

In (4), the arguments of the sinc functions, which are inside the left brackets, are subtracted by $-M/2$ to shift them to the right (see Figure 1c). In order to avoid division by zero when $i = M/2$, then the equation $h[M/2] = k\ (2\pi f_2 - 2\pi f_1)$ is used to replace Equation (4). The k is constant for a unity gain filter. In practice, k is disregarded while the kernel is being computed and then all samples are normalized as needed [24]. This can be conducted by normalizing the kernel FFT, and the kernel filter is obtained by applying the inverse of the FFT.

The Blackman window filter is obtained by replacing the expression inside the right brackets of (4), which is a Hamming window function, with the Blackman window function, as shown here [24]:

$$h[i] = k\left[\frac{\sin\left(2\pi f_2\left(i - \frac{M}{2}\right)\right)}{i - \frac{M}{2}} - \frac{\sin\left(2\pi f_1\left(i - \frac{M}{2}\right)\right)}{i - \frac{M}{2}}\right]\left[0.42 - 0.5\cos\left(\frac{2\pi i}{M}\right) + 0.08\cos\left(\frac{4\pi i}{M}\right)\right]. \tag{5}$$

3. Materials and Methods

In this section, the acoustic field measurement and simulation processes, as well as the design of the filters and performance evaluation, are presented.

3.1. Quantitative Analysis

The pure qualitative analysis of the results regarding waveforms and acoustic fields can lead to misinterpretations, as the perception of the results is subjective [13]. Thus, to obtain accurate comparisons, the filter performance was numerically calculated by using the root mean square error (RMSE) equation [25]:

$$RMSE = \sqrt{\frac{1}{N}\sum_{i=1}^{N}(AF_{SIM}[i] - AF[i])^2}, \quad (6)$$

where AF_{SIM} are the simulated values (gold-standard reference), AF are the measured values (not filtered, and filtered), i is an index of the sample of a pulse or a given point (x, z, y = 0) located at the acoustic field, and N is the number of samples or points of the respective measurement.

To quantify the point-to-point absolute error between the filtered and the gold-standard acoustic fields, the absolute error (AE) was used as a metric. This index of quality allowed comparisons of the acoustic fields by using images of errors. An error image is an image that attributes a color to a given $AE[i]$, which was calculated by [26]

$$AE[i] = |AF_{SIM}[i] - AF[i]| \quad (7)$$

3.2. Acoustic Field Measurement

The acoustic field from a 2D phased-array ultrasound transducer (Figure 2) was measured with the self-made needle hydrophone described in [5]. Because the hydrophone was not calibrated, the measurements were normalized by the maximum amplitude of the field. The phased-array transducer used in this work is the same transducer used in [27] to make 3D acoustic images of objects immersed in water. The transducer consists of 16 squared elements with sides equal to 5 mm distributed in a 4 × 4 matrix, which can emit and receive ultrasound pulses individually. The center frequency of the transducer is 480 kHz and its −6 dB bandwidth is 50% [27]. Both the transducer and the hydrophone were built in our ultrasound laboratory.

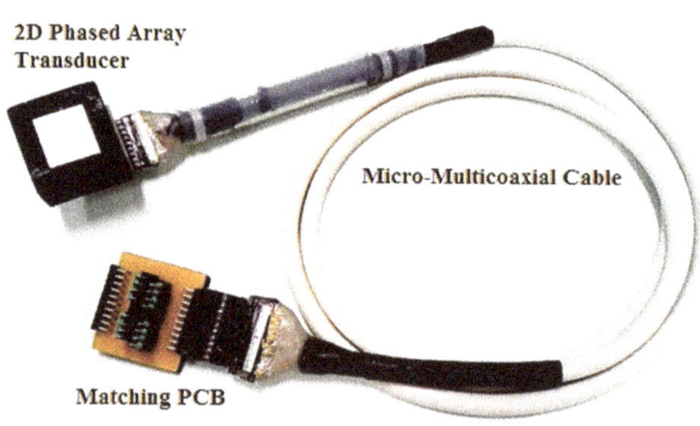

Figure 2. Two-dimensional phased-array transducer.

The transducer was excited by an ultrasonic pulser/receiver (5077PR, Panametrics-NDT, Waltham, MA, USA), which was adjusted to a 500 kHz squared pulse with an amplitude of 100 V, 10 dB gain, and a 10 MHz low-pass filter. The acoustic field generated by the transducer was measured in a water tank (Figure 3), with a needle hydrophone held by an automatic scanning system, sweeping the xz-plane at y = 0.

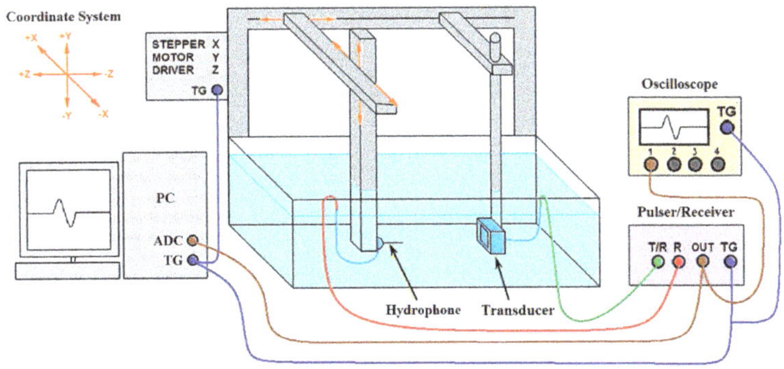

Figure 3. Signal acquisition system to measure the acoustic field.

As the pulser/receiver has only one T/R channel; all the transducer elements were connected in parallel, emitting at the same time. Thus, the measured acoustic field was equivalent to that of a single-element square transducer with a 20.6 mm side.

The ultrasonic pulse emitted by the transducer reached the hydrophone at an x, z point in the field. Consequently, the hydrophone generated an electrical pulse that was received by the R channel of the pulser/receiver, which filtered the signal with a 10 MHz analog low-pass filter. This frequency was 20 times greater than the central frequency of the transducer.

The digital oscilloscope (MSO8104A Infiniium, Agilent Technologies, Englewood, CO, USA) was only used for the transducer–hydrophone alignment process, which consisted of moving the hydrophone in front of the transducer surface. This was necessary to define the position of the center of the transducer and to adjust the scanning system.

Next, the pulser/receiver sent the pre-filtered signal for its digitization, which was performed on an 8-bit analog-to-digital converter (ADC) board that had a maximum sample rate of 100 MS/s (NI PCI-5112, National Instruments, Austin, TX, USA). In order to reduce the amount of data for processing, the signals were sampled at 25 MS/s. The digitization frequency was 2.5 times higher than the analog low-pass filter. Therefore, the Nyquist criterion was observed to avoid undersampling.

After digitalizing and storing the signal in the PC, it sent a synchronized trigger pulse to the tank driver to move the hydrophone with a step of 1 mm ($\lambda/3$) in z; when the z-axis was fully swept, the hydrophone was moved with a step of 1 mm in x and the path of z was reversed. The process was repeated until the entire plane -50 mm $\leq x \leq 50$ mm by 5 mm $\leq z \leq 150$ mm was sampled.

Before the acoustic field calculation, the DC components in the signals were removed by subtracting the mean. Then, the acoustic field was obtained in MATLAB (The MathWorks, Inc., Natick, MA, USA) by assigning to each point of the mesh the maximum pulse amplitude at its respective point. Finally, the measured acoustic field was normalized.

3.3. Acoustic Field Simulation

The steps taken to simulate the acoustic fields in MATLAB are presented in Figure 4. First, a pulse emitted by the transducer, called $vn(t)$, was obtained using the modeling of the KLM and ABCD matrices (Figure 4a–c) [28–30]. Then, the pressure wave $p(\vec{r}_Q,t)$, at a spatial point Q in the field, was calculated with the rigid-plane piston model, where $\vec{r}_Q$ is the position vector of the point Q (Figure 4d–f) [25,31,32]. This linear model describes a piezoelectric transducer as a piston in which the face particles vibrate in a phase, with velocity $vn(t)$, normal to the transducer face. The acoustic field at a position Q is the maximum amplitude of $p(\vec{r}_Q,t)$. The model used in the simulation was accurate for the purpose of the experiment, considering that the medium for measuring the acoustic field is water, which meets the boundary conditions of the model, because it is homogeneous and isotropic.

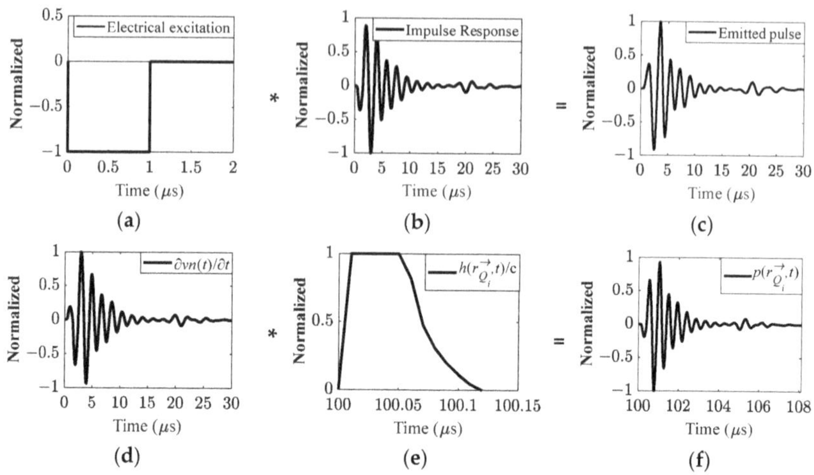

Figure 4. The acoustic-field simulation processing: (**a**) a simulated electrical excitation signal is convolved with (**b**) the impulse response of the transducer, resulting in (**c**) the pulse being emitted by the transducer, represented as $vn(t)$; (**d**) then, the derivative of $vn(t)$ is convolved with (**e**) the velocity potential impulse response of the rigid piston $h(\vec{r}_{Q_i},t)$ observed at a point Q_i. The result is multiplied by the water density ρ, resulting in (**f**) the pressure wave $p(\vec{r}_{Q_i},t)$. The maximum of the absolute pressure wave is the acoustic field at $\vec{r}_{Q_i}$. All signals in this figure were normalized exclusively for better visualization.

The developed algorithm took into account all parts of the transducer presented in Figure 2, such as piezocomposite elements, matching layer, backing layer, and also the micro-multicoaxial cable and series inductors, used to match the electrical impedance of the transducer with that of the pulser/receiver.

In addition, a squared negative electrical pulse (Figure 4a), similar to that generated by the pulser/receiver, was used for transducer excitation. It was possible to simulate a pulse with the same characteristics as the real transducer, a central frequency f_o of 480 kHz and 50% bandwidth (Figure 4c), making the simulation of the acoustic field more accurate.

Moreover, parameters such as the grid size and its discretization, as well as the sampling frequency used to simulate the acoustic field, were equal to those used in the measurements. The ultrasound propagation velocity in water was c = 1500 m/s. The noise was disregarded in the pulses. Figure 5 shows (a) the acoustic field simulated at the xz-plane, (b) its axial beam profile, and (c) its lateral beam profile. The acoustic pressure along the acoustic axis (z-axis) is useful for determining the focal distance where the amplitude is maximum. The lateral beam profile at the focal distance is useful for determining the lateral resolution, given as the full-width at the half-maximum (FWHM).

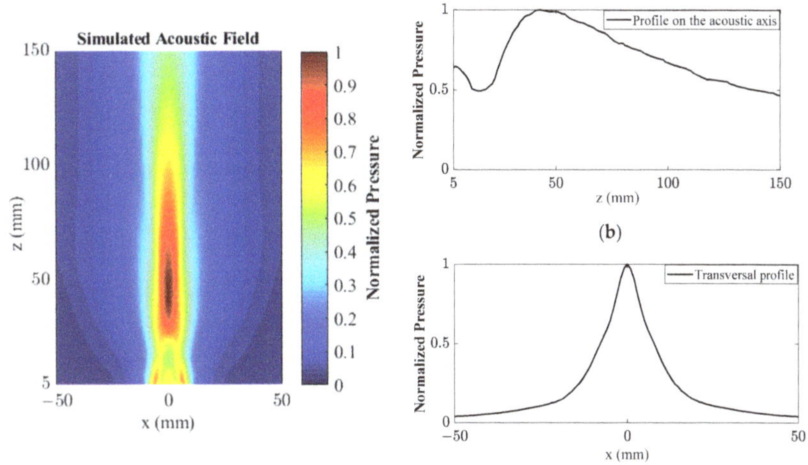

(a) (b) (c)

Figure 5. Simulated results: (**a**) acoustic field, (**b**) beam profile on the acoustic axis, and (**c**) lateral beam profile at the focal distance.

3.4. Design of the Moving Average Filter

The length of the MA kernel, *M*, was varied, and the MA filter was applied to the noiseless simulated pulse (Figure 4) to analyze how it was distorted as a function of *M*. This was carried out by calculating the RMSE (6) of the filtered pulse with a given *M*, relative to the simulated pulse.

The method was started by applying the MA filter to the simulated pulse, with *M* from 1 (no filter) to 100 (approximately twice the number of samples of a one-cycle sinusoid at the central frequency of the transducer, f_o, such that $M = 2f_{s/}f_o$) in steps of 1 (for better visualization, only $M = \{1, 10, 50, 70\}$ pulses are shown in Figure 6). Then, the RMSE was calculated for each filtered pulse (Figure 7). The simulated unfiltered pulse was the reference to calculate each RMSE(*M*).

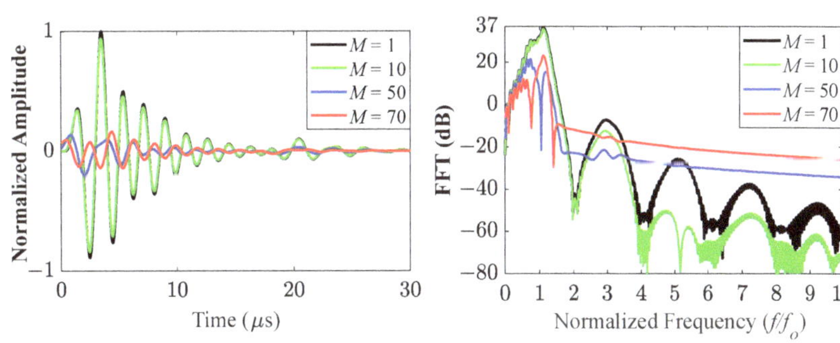

(a) (b)

Figure 6. (**a**) A noiseless simulated pulse (black line) was MA-filtered with different kernel lengths, *M*. When *M* = 10, the filtered pulse was similar to the original pulse, as the RMSE was only 2% (green line). (**b**) The spectral frequency shows that even with only *M* = 10, the MA filter acted as a low-pass filter, and the greater the *M*, the more deteriorated the pulse was.

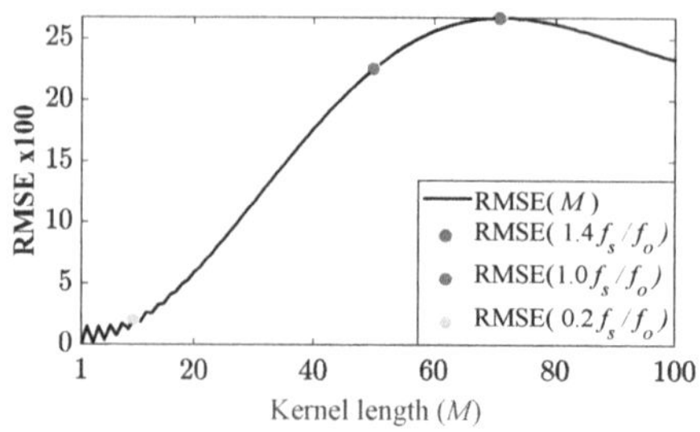

Figure 7. The RMSE of the simulated unfiltered pulse in relation to the filtered pulse with different MA kernel lengths shows the RMSE increased, i.e., the pulse was distorted when M increased.

This method was useful for choosing an M that would not significantly deteriorate a filtered pulse, as shown in Figures 6a and 7. The larger the kernel of the MA filter, the more uncharacterized the pulse was.

In turn, a shorter kernel could preserve the waveform while attenuating high frequencies, as shown in Figure 6a,b, respectively. This suggests that, although the MA filter acts as low-pass filter, care must be taken to choose a kernel length so as not to mischaracterize the pulse. After testing different kernel lengths, it was found that $M = 0.2\,f_{s/}f_o$ was a suitable kernel length for the MA filter, as the RMSE when $M = 10$ was only 2%.

3.5. Design of the Hamming Window and Blackman Window Band-Pass Filters

While the MA kernel filter was designed with a simulated pulse, the window filters were modeled using a measured signal. The pulse was measured at x = 25 mm, z = 100 mm, far off the acoustic axis, in the far field, and normalized by the highest of all measured pulses in the acoustic field.

This point was chosen because it was possible to locate the pulse emitted by the transducer (see Figure 8a, between 70 µs and 80 µs) as well as unwanted noise and oscillations. The oscillations that are evident below 70 µs have a frequency lower than that of the pulse emitted by the transducer. Averaging the measurements can indeed remove random noise. However, it was not possible to average the results because it would considerably increase the time for acquisition. The signal acquisition was automatic, i.e., the signals were continuously acquired while the hydrophone swept the field.

Applying the FFT to the signal, frequency components outside the transducer operating range ($0.75\,f_o$ to $1.25\,f_o$) were identified with magnitudes comparable to and even greater than those of the transducer operating, as noted at $0.12\,f_o$ and $0.2\,f_o$ in Figure 8b. The amplitude of the high-frequency noise was slightly lower than that of the pulse, but the duration of the low-frequency noise was longer than that of the pulse. As a consequence, the energy of the noise and oscillations was comparable to that of the pulse, making it impossible to identify its effective range in the frequency domain presented in Figure 8b.

In order to attenuate unwanted frequencies and, at the same time, not to mischaracterize the measured pulse of the transducer, the low and high cutoff frequencies of the band-pass filter were $f_1 = 0.25\,f_o$ (120 kHz) and $f_2 = 1.50\,f_o$ (720 kHz), respectively.

The transition band between the cutoff frequency and the stopband was chosen as $BW = 0.10\,f_o$ (48 kHz), resulting in $M = 2084$ (3) (considering that BW must be normalized by f_s, and M is an even number). The transducer operating band was, thus, kept within the band-pass filter band.

After calculating the kernel size, the kernels of the Hamming window filter and of the Blackman window filter were calculated using Equations (4) and (5), respectively.

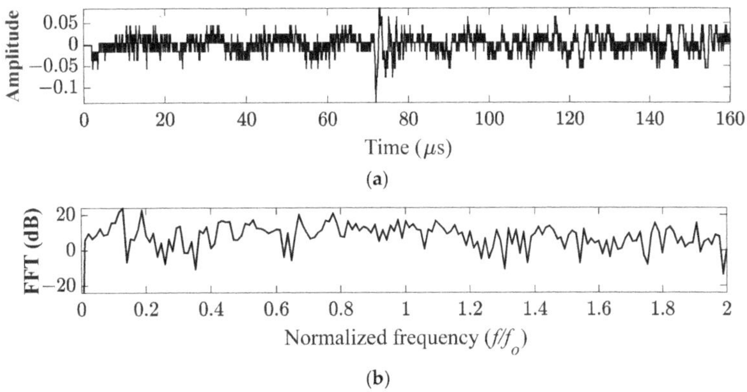

Figure 8. (**a**) Noisy pulse acquired by the ADC board (amplitude normalized by the maximum amplitude of the field). Pulse measured in the far-field region at x = 25 mm, z = 100 mm, located between 70 μs and 80 μs in the sampled window, showing unwanted oscillations at a frequency lower than that of the measured pulse. (**b**) Frequency components of the sampled signal. The normalized frequency signal shows that the highest amplitudes occurred at 0.12 f_o and at 0.2 f_o (below the transducer bandwidth). These undesired frequencies could be filtered such that the high amplitudes that remained were within the bandwidth of the transducer.

Although the kernels look very similar (Figure 9a), the Blackman kernel truncation was smoother than the Hamming kernel truncation (Figure 9b), which can reduce ringing in the band pass range and reduce overshoot near the cutoff frequencies.

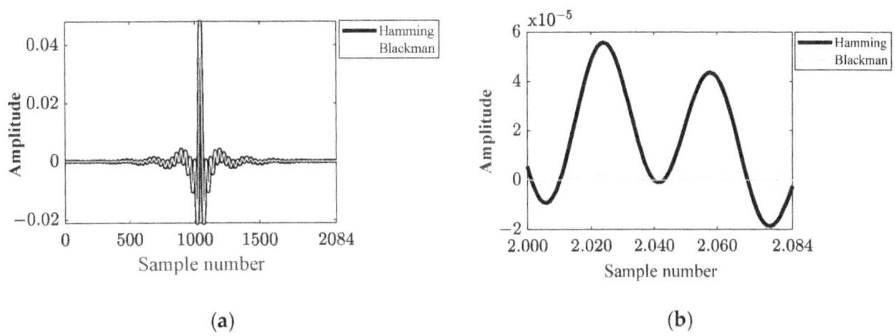

Figure 9. (**a**) The Hamming window and the Blackman window kernels look very similar, (**b**) but the apodization at the end of the Blackman window was smoother than that of the Hamming window (**b**).

The frequency response of the band-pass filters showed that the Blackman window filter increased attenuation out of the band-pass range more than the Hamming filter does (Figure 10a). The Blackman window stopband magnitude was lower than that of the Hamming window, and the roll-off from both filters was similar (Figure 10b).

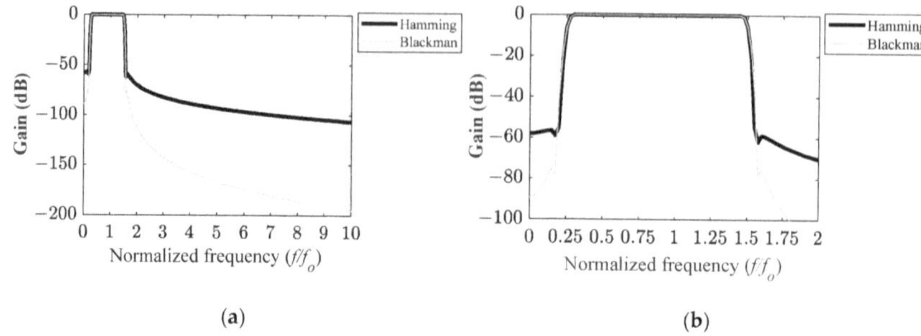

Figure 10. (**a**) The FFTs of the Hamming window and Blackman window band-pass filters show that the attenuation out of the band-pass region using the Blackman window was higher than that obtained with the Hamming window. (**b**) The roll-off was similar with both filters.

4. Results

To evaluate the performance of the filters, a pulse was measured at x = 50 mm, z = 150 mm in the plane y = 0, normalized by the maximum amplitude of the field. The location chosen to acquire the signal is critical because its amplitude is greatly attenuated off the acoustic axis and in the far field, making the pulse and noise magnitudes comparable (Figure 11a). Thus, a noise could be computed as a pulse, generating an error in the determination of the acoustic field. Although the signal was limited by the 8-bit acquisition board, it was able to show that the filters actually work.

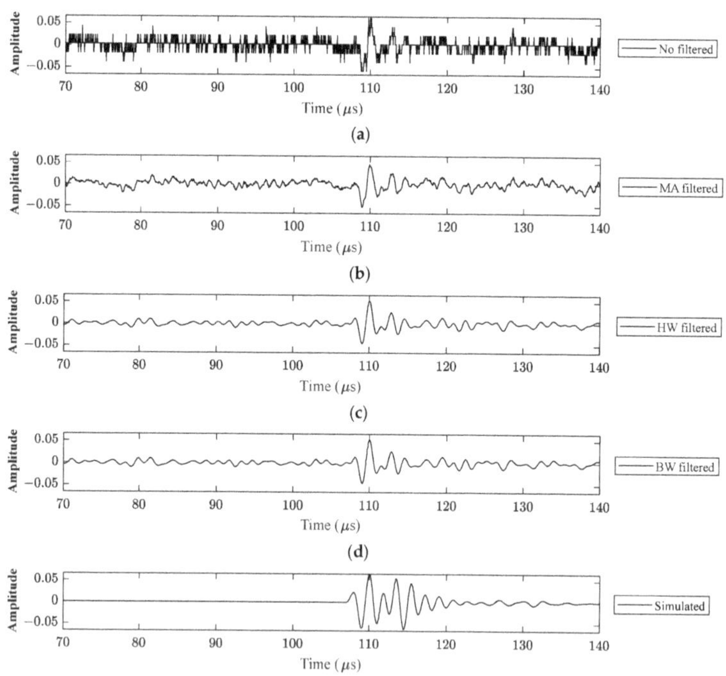

Figure 11. For the point x = 50 mm and z = 150 mm, the signals presented refer to (**a**) measured signal—unfiltered; (**b**) measured signal—filtered with MA $M = 10$; (**c**) measured signal—filtered with HW $M = 2084$; (**d**) measured signal—filtered with BW $M = 2084$; and (**e**) simulated noiseless signal.

Although the MA filter with $M = 10$ smoothed the signal, unwanted distortions remained, with significant amplitudes around 80 μs, and from 117 μs (Figure 11b). The same MA filter applied to the simulated pulse of Figure 6a made the attenuation significant from $2f_o$ on. Thus, it is reasonable to consider that the signal was smoothed when the high frequencies were filtered, and that the remaining disturbances were at frequencies in the operating range of the transducer and below that.

In turn, the HW (Figure 11c) and the BW (Figure 11d) window filters with $M = 2084$ were more effective than the MA filter, as they make it easier to identify the pulse around 110 μs. The frequency response of the filter kernels modeled in Figure 10 showed that frequencies below $0.25\,f_o$ and above $1.50\,f_o$ were greatly attenuated; one may hence assume that unwanted distortions remain at frequencies around the operating band of the transducer.

However, comparing the pulses filtered with HW (Figure 11c) and BW (Figure 11d) with the simulated noiseless pulse (gold standard—Figure 11e), some noise was removed. This suggests that unwanted frequency components are also included in the operating band of the transducer, and that the linear filters in this work cannot remove them.

The filtering of high-amplitude noise was evaluated by adding a synthetic noise, simulated in MATLAB, to the measured signal presented in Figure 11a. The synthetic noise was the result of the sum of a 50 kHz continuous wave and 200 random pulses (single-sinusoidal cycles of 10 MHz), both with twice the maximum amplitude of the measured signal.

The 50 kHz continuous wave was low-frequency noise and the 10 MHz random pulses were high-frequency noise. The unfiltered measured signal (raw) is shown in Figure 12a, and the same signal with synthetic noise (synthetic), before and after MA, HW and BW filtering, are shown in Figure 12b–d, respectively.

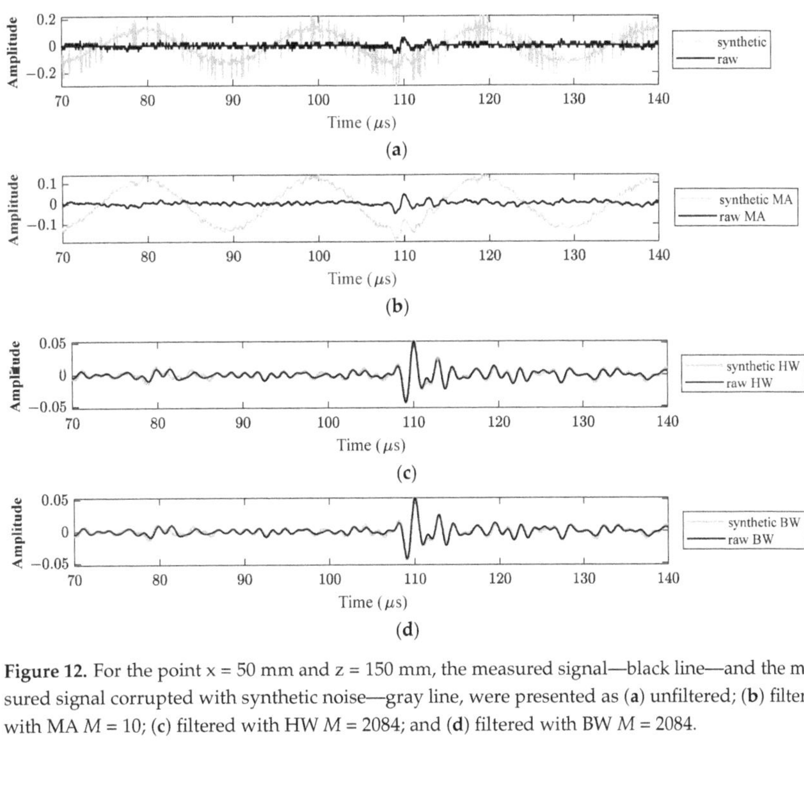

Figure 12. For the point x = 50 mm and z = 150 mm, the measured signal—black line—and the measured signal corrupted with synthetic noise—gray line, were presented as (**a**) unfiltered; (**b**) filtered with MA $M = 10$; (**c**) filtered with HW $M = 2084$; and (**d**) filtered with BW $M = 2084$.

As in Figure 11b, the MA filter in Figure 12b did not remove the low-frequency noise from the signal with the synthetic noise, but it did smooth out the high-frequency noise. On the other hand, the HW filter (Figure 12c) and the BW filter (Figure 12d) were very effective in removing low- and high-frequency noise.

The measured acoustic field (Figure 13a) was filtered by applying the filters MA with $M = 10$ (Figure 13b), HW with $M = 2084$ (Figure 13c) and BW with $M = 2084$ (Figure 13d). The filter kernel was applied in each sampled signal in the mesh, and the pressure at a given x, z point was the maximum of the absolute value over the sampled signal. All values were normalized by their respective filtered acoustic field.

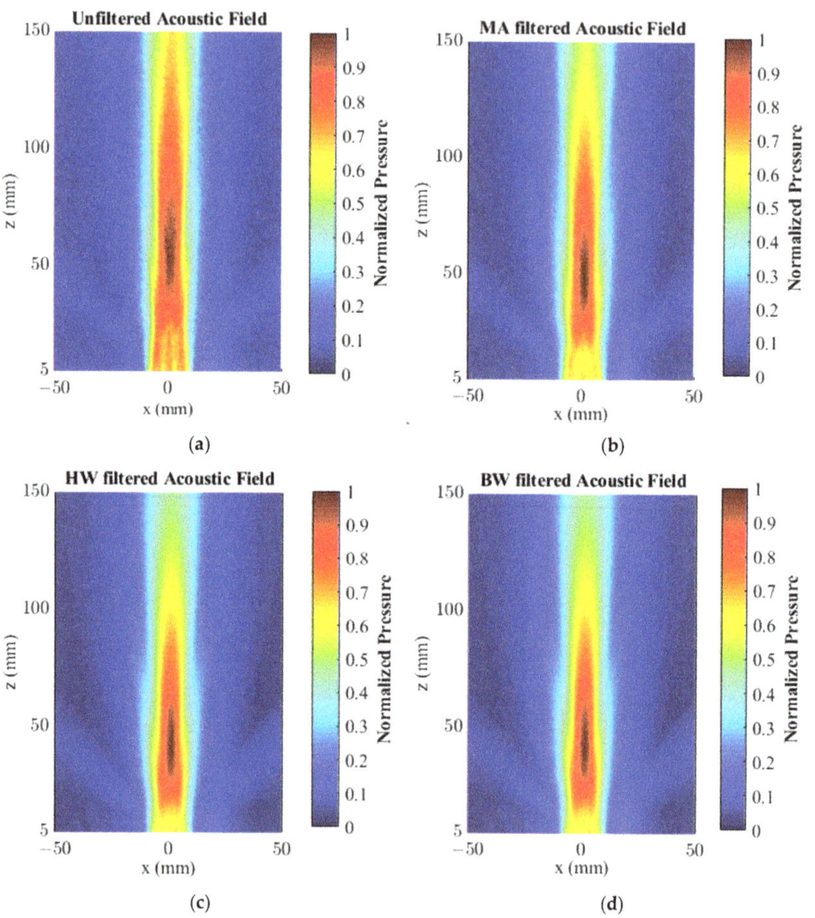

Figure 13. Measured acoustic field: (**a**) unfiltered, (**b**) MA with $M = 10$, (**c**) HW with $M = 2084$, and (**d**) BW with $M = 2084$.

The RMSEs of the measured/filtered acoustic field in relation to the simulated one were 6.34% for the unfiltered, 4.28% for the MA-filtered, and 4.01% for the HW- and BW-filtered acoustic fields. Although the window filters reduced the RMSE by only around 0.27% relative to the MA filter, there was a qualitative improvement when comparing the filtered acoustic fields of Figure 13 with the simulated noiseless one in Figure 5a.

For better visualizing the measured acoustic-field error as well as the performance of the filters, images of the absolute error of the measured and filtered acoustic fields in relation to the noise-free simulated acoustic field are given in Figure 14. By calculating the absolute error $AE[i]$ for each point (7), this figure illustrates how different the normalized pressure at each point was in respect to the gold standard. If there were no errors, the image would be all blue.

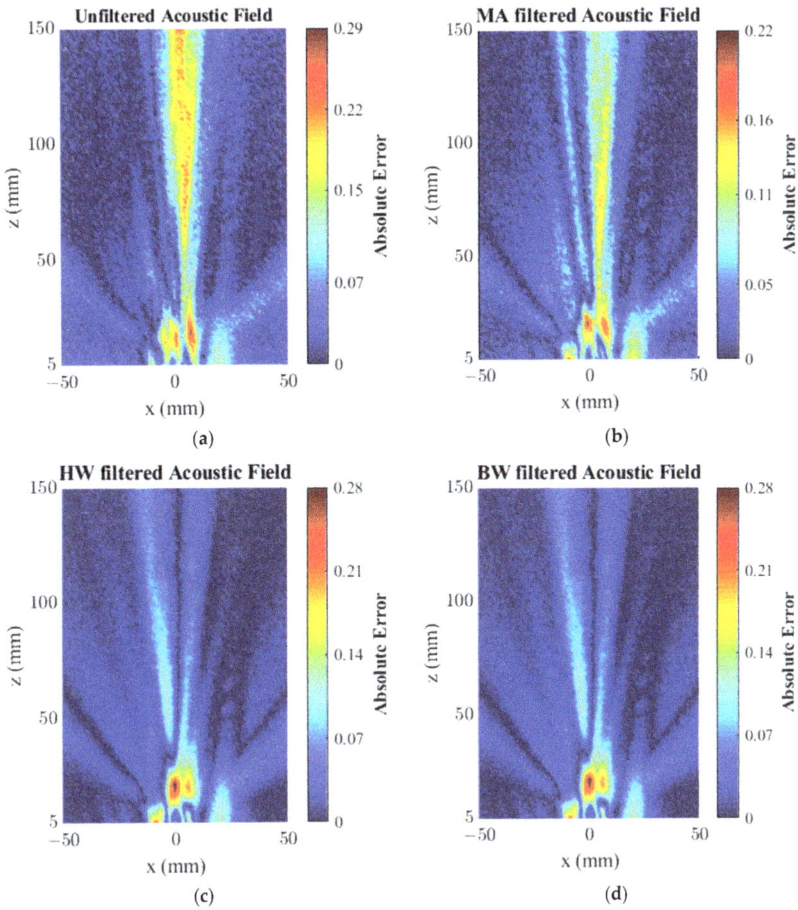

Figure 14. Absolute error of the measured acoustic field to the simulated acoustic field: (**a**) unfiltered; (**b**) MA-filtered with $M = 10$; (**c**) HW-filtered with $M = 2084$; (**d**) BW-filtered with $M = 2084$.

Although all the images presented large-amplitude errors in the near field ($z < 43$ mm), all filters improved the measured acoustic field, being more evident along the acoustic axis ($x = 0$), where the pressure was higher than in other directions (keeping in mind that all array elements were pulsed at the same time, therefore, without beam deflection). Both the unfiltered (Figure 14a) and the MA-filtered (Figure 14b) acoustic field present a granular aspect, while the acoustic fields filtered with the window filters, HW and BW, present smoother images (Figure 14c,d, respectively). This can be explained by the fact that HW and BW filters were capable of reducing the frequency components below the operating band of the transducer, as shown in Figures 8 and 11c,d. The filters HW and BW showed equivalent error images because their RMSE values were equal to 4.01%.

The focal length is the distance from the transducer up to the point where the maximum amplitude occurs, and from which the acoustic amplitude decays monotonically. The axial beam profile showed that the focal length of the measured acoustic field (Figure 15—black line) was further away than the focal length of the simulated acoustic fields (Figure 15—

yellow line). The focal length of the simulated acoustic field was 43 mm. The focal lengths of the measured signal were 61 mm for the unfiltered (41.86% further than simulated), 48 mm for MA-filtered (11.63% further than simulated), and 40 mm for the HW- and BW-filtered (6.98% closer than simulated) acoustic fields.

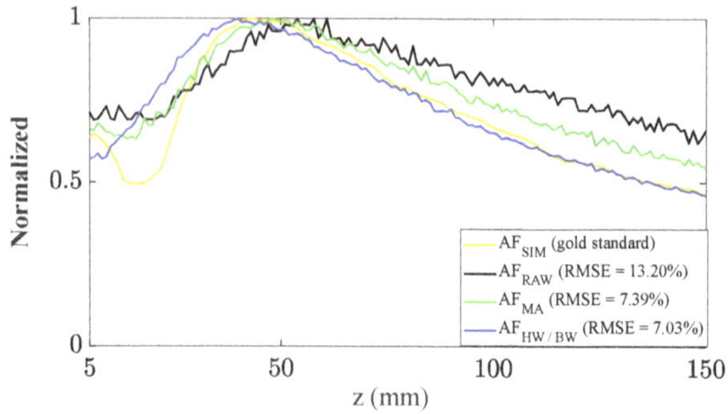

Figure 15. Acoustic pressure along the acoustic axis: noiseless simulation—yellow line; unfiltered measured—black line; MA-filtered—green line; HW- and BW-filtered—blue line.

Although this effect could be explained by a measurement error, Figure 15 shows that the focal distance was reduced when a filter was applied to the noised measured acoustic field. Taking into account that the focal distance is frequency-dependent (for a given geometry, the higher the frequency, the longer the focal length) and that the filters applied were low-pass and band-pass filters, the reduction in the focal length can be explained by the attenuation of the high-frequency components of the measured pulses.

In fact, the window filters HW and BW decreased the focal length more than the MA filter (Figure 15—blue line, and Figure 15—green line, respectively), as their higher cut-off frequency was $1.5f_o$, while the MA filter attenuated frequencies higher than $2f_o$ (see Figure 6b). In conclusion, the high-frequency noises contributed to an increase the focal length of the measured acoustic field. The RMSE of the normalized pressure profile on the acoustic axis (x = 0 mm, y = 0 mm) were as follows: the RMSE of the unfiltered field = 13.20%, the RMSE of the MA-filtered field = 7.39%, and the RMSE of the HW-filtered and BW-filtered fields = 7.03%.

The lateral beam profiles were obtained from their respective acoustic fields (see Figure 13) at their focal points (Figure 16). All the filters improved the results, and the RMSE of the MA filter and of the window filters were equivalent. The filtering performance of the MA, HW and BW filters, summarized in Table 1, shows that all filters improved the acoustic field characterization.

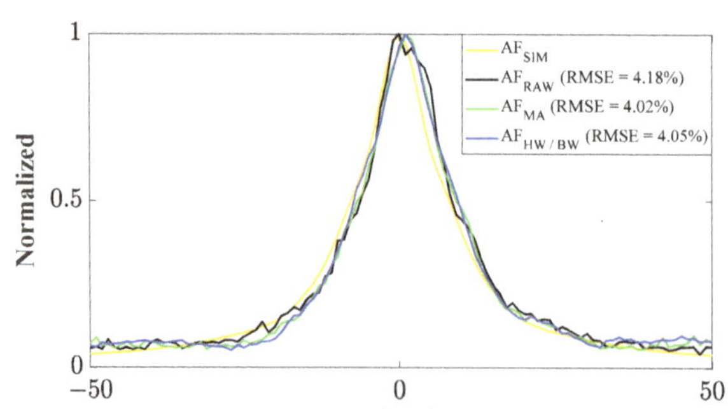

Figure 16. Lateral beam profile at the focal region: noiseless simulation—yellow line; unfiltered measured—black line; MA-filtered—green line; and HW- and BW-filtered—blue line.

Table 1. Summary of the main results: RMSE of the measured/filtered acoustic pressure over the entire field, along the z-axis and transversal at focus; and error in focal length relative to simulated values.

Filter	Kernel Length	RMSE (%)			Error (mm)
		Acoustic field	Along z-axis	Transversal at focus	Focal length
Unfiltered	-	6.34	13.20	4.18	18
MA	10	4.28	7.39	4.02	5
HW	2085	4.01	7.03	4.05	3
BW	2085	4.01	7.03	4.05	3

5. Conclusions

Linear digital filters are extensively used for many purposes, such as electronic devices, medical images, RADAR, and signal processing. This work presented an effective, low-cost, and useful method to design and implement linear digital filters that improve acoustic-field characterization, reducing noises and unwanted distortions. Once the kernel is designed, filtering is performed through a convolution of the Kernel and the ultrasound signal in the time domain, with no complex mathematical calculations.

The MA filter with a kernel of only 10 samples acted as a low-pass filter, which attenuated frequency components from twice the transducer central frequency. To model the kernel length of the MA filter, a criterion of $M = 0.2 f_s/f_o$ (wherein f_s is the sampling frequency and f_o is the central transducer frequency) was established to avoid unwanted distortion of the pulse. As a result, the RMSE of the measured acoustic field to the simulated acoustic field reduced from 6.34% (unfiltered) to 4.28% (MA-filtered).

The HW and BW filters were more effective than the MA filter. However, these filters are more difficult to model because there is a compromise between the computational load and the roll-off band to determine the kernel length. The longer the kernel, the wider the band of transition from the cut-off frequency to the stopband frequency. The HW and BW were more effective because their stopband attenuations are much higher than that of the MA. Furthermore, the HW and the BW make the implementation of the band-pass filter feasible using cutoff frequencies around the frequency band of the transducer, thus excluding frequencies that cause unwanted distortions. Although the BW has a stopband attenuation higher than that of the HW (−74 dB versus −53 dB, respectively), and for a given kernel length $M + 1$ the HW roll-off is shorter than that of the BW, their performances were equivalent (both achieved an acoustic-field RMSE = 4.01%). The better performance of window filters in comparison to the MA filter was obtained at the cost of computational load, as the kernel length of HW and BW has $M + 1 = 2085$.

Although the RMSE quality index for the acoustic field showed that the window filter was only 0.27% better than the MA filter, the acoustic fields filtered with HW and BW were visibly much closer to the simulated noiseless acoustic field, used as a reference of quality (see Figures 5a and 13). Furthermore, the results from the window filters can be improved and customized for each transducer being characterized by adjusting the cut off frequencies and the roll-off. Thus, window filters are suitable for improving the characterization of acoustic fields generated by ultrasound transducers.

In the future, we intend to apply other filters to characterize acoustic fields, such as nonlinear filters and adaptive filters, which can change the weighting coefficients according to the local statistics. These filters will be compared to the window filters presented herein.

Author Contributions: Conceptualization, E.P.R. and F.B.; methodology, E.P.R. and F.B; software, E.P.R. and F.B.; validation, E.P.R., D.S.d.S. and F.B.; formal analysis, E.P.R., D.S.d.S. and F.B.; investigation, E.P.R., D.S.d.S. and F.B.; resources, F.B.; data curation, E.P.R. and F.B.; writing—original draft preparation, E.P.R.; writing—review and editing, E.P.R., D.S.d.S. and F.B.; visualization, E.P.R.,

D.S.d.S. and F.B.; supervision, F.B.; project administration, F.B.; funding acquisition, F.B. All authors have read and agreed to the published version of the manuscript.

Funding: This research was funded by the São Paulo Research Foundation (FAPESP, grant #2017/13094-4), and the Coordination for the Improvement of Higher Education Personnel (CNPq, process #140510/2021-4).

Institutional Review Board Statement: Not applicable.

Informed Consent Statement: Not applicable.

Data Availability Statement: Not applicable.

Conflicts of Interest: The authors declare no conflict of interest.

References

1. Burgess, M.T.; Konofagou, E.E. Fast qualitative two-dimensional mapping of ultrasound fields with acoustic cavitation-enhanced ultrasound imaging. *J. Acoust. Soc. Am.* **2019**, *146*, EL158–EL164. [CrossRef]
2. Yu, Y.; Shen, G.; Zhou, Y.; Bai, J.; Chen, Y. Quantitative assessment of acoustic intensity in the focused ultrasound field using hydrophone and infrared imaging. *Ultrasound Med. Biol.* **2013**, *39*, 2021–2033. [CrossRef]
3. ASTM. E1065-08 Standard Guide for Evaluating Characteristics of Ultrasonic Search Units. *Am. Stand. Test. Mater.* **2008**, *99*, 23.
4. Saheban, H.; Kordrostami, Z. Hydrophones, fundamental features, design considerations, and various structures: A review. *Sens. Actuators A Phys.* **2021**, *329*, 112790. [CrossRef]
5. Boechat, F.M.B.; Rodrigues, E.P.; Oliveira, T.F.D.; Buiochi, F. Development of a PVDF needle-type hydrophone for measuring ultrasonic fields. In Proceedings of the 2018 13th IEEE International Conference on Industry Applications (INDUSCON), Sao Paulo, Brazil, 12–14 November 2018; pp. 1004–1007. [CrossRef]
6. Aoxuan, X.; Yonggang, C.; Yuebing, W.; Huifeng, Z. Design and fabrication of novel sensor for high pressure measurement of acoustic wave. In Proceedings of the 2019 14th IEEE International Conference on Electronic Measurement & Instruments (ICEMI), Changsha, China, 1–3 November 2019; pp. 1664–1669. [CrossRef]
7. Martin, E.; Treeby, B. Investigation of the repeatability and reproducibility of hydrophone measurements of medical ultrasound fields. *J. Acoust. Soc. Am.* **2019**, *145*, 1270–1282. [CrossRef]
8. Haller, J.; Wilkens, V. Short and longtime stability of therapeutic ultrasound reference sources for dosimetry and exposimetry purposes. In *AIP Conference Proceedings*; AIP Publishing: Melville, NY, USA, 2017; Volume 1821. [CrossRef]
9. Sapozhnikov, O.A.; Tsysar, S.A.; Khokhlova, V.A.; Kreider, W. Acoustic holography as a metrological tool for characterizing medical ultrasound sources and fields. *J. Acoust. Soc. Am.* **2015**, *138*, 1515–1532. [CrossRef]
10. Wear, K.; Liu, Y.; Gammell, P.; Maruvada, S.; Harris, G. Correction for frequency-dependent hydrophone response to nonlinear pressure waves using complex deconvolution and rarefactional filtering: Application with fiber optic hydrophones. *IEEE Trans. Ultrason. Ferroelectr. Freq. Control* **2015**, *62*, 152–164. [CrossRef] [PubMed]
11. Eichstädt, S.; Wilkens, V. Evaluation of uncertainty for regularized deconvolution: A case study in hydrophone measurements. *J. Acoust. Soc. Am.* **2017**, *141*, 4155–4167. [CrossRef]
12. Song, Y.; Liu, J. An improved adaptive weighted median filter algorithm. In *Journal of Physics: Conference Series*; IOP Publishing: Bristol, UK, 2019; Volume 1187. [CrossRef]
13. Cardoso, F.M.; Matsumoto, M.M.S.; Furuie, S.S. Edge-Preserving Speckle Texture Removal by Interference-Based Speckle Filtering Followed by Anisotropic Diffusion. *Ultrasound Med. Biol.* **2012**, *38*, 1414–1428. [CrossRef]
14. Chitanont, N.; Yaginuma, K.; Yatabe, K.; Oikawa, Y. Visualization of sound field by means of Schlieren method with spatio-temporal filtering. In Proceedings of the 2015 IEEE International Conference on Acoustics, Speech and Signal Processing (ICASSP), South Brisbane, QLD, Australia, 19–24 April 2015; IEEE: Piscataway, NJ, USA, 2015; pp. 509–513. [CrossRef]
15. Schlichthärle, D. Digital Filters: Basics and Design. *Meas. Sci. Technol.* **2001**, *12*, 2211. [CrossRef]
16. Purnamasari, D.N.; Wibisono, K.A.; Sukri, H. Digital Moving Average Filter Application for Echo Signals and Temperature. In *E3S Web of Conferences*; EDP Sciences: Les Ulis, France, 2021; Volume 328. [CrossRef]
17. Morales-Mendoza, L.J.; Shmaliy, Y.; Ibarra-Manzano, O.G.; Arceo-Miquel, L.J.; Montiel-Rodriguez, M. Moving average Hybrid FIR Filter in Ultrasound Image Processing. In Proceedings of the 18th International Conference on Electronics, Communications and Computers (Conielecomp 2008), Puebla, Mexico, 3–5 March 2008; pp. 160–164. [CrossRef]
18. Shehu, N.M.; Gidado, A.S.; Wudil, Y.S.; Gora, U.A.; Student, M. Performance Analysis of FIR Low Pass Filter Design using Blackman and Flat Top Window Methods. *Int. J. Eng. Sci. Comput.* **2016**, *6*, 3659–3665.
19. Goel, S.; Kaur, G.; Tomar, P. Performance analysis of Welch and Blackman Nuttall window for noise reduction of ECG. In Proceedings of the 2015 International Conference on Signal Processing, Computing and Control (ISPCC), Waknaghat, India, 24–26 September 2015; pp. 87–91. [CrossRef]
20. Wang, Z.; Ingram, P.; Greenlee, C.L.; Olafsson, R.; Norwood, R.A.; Witte, R.S. Design considerations and performance of MEMS acoustoelectric ultrasound detectors. *IEEE Trans. Ultrason. Ferroelectr. Freq. Control* **2013**, *60*, 1906–1916. [CrossRef]

21. Jordaan, J.P.; Kruger, C.P.; Silva, B.J.; Hancke, G.P. An ultrasonic-based localization system for underground mines. In Proceedings of the 2017 IEEE 15th International Conference on Industrial Informatics (INDIN), Emden, Germany, 24–26 July 2017; pp. 141–146. [CrossRef]
22. Rakshit, H.; Ullah, M.A. A comparative study on window functions for designing efficient FIR filter. In Proceedings of the 2014 9th International Forum on Strategic Technology (IFOST), Cox's Bazar, Bangladesh, 21–23 October 2014; pp. 91–96. [CrossRef]
23. Dwivedi, A.K.; Ghosh, S.; Londhe, N.D. Review and Analysis of Evolutionary Optimization-Based Techniques for FIR Filter Design. *Circuits Syst. Signal Process.* **2018**, *37*, 4409–4430. [CrossRef]
24. Smith, S.W. *The Scientist and Engineer's Guide to Digital Signal Processing*; California Technical Pub: San Diego, CA, USA, 1997.
25. Buiochi, F.; Martínez, O.; Gómez-Ullate, L.; Montera de Espinosa, F. A Computational Method to Calculate the Longitudinal Wave Evolution Caused by Interfaces Between Isotropic Media. *IEEE Trans. Ultrason. Ferroelectr. Freq. Control* **2004**, *51*, 181–192. [CrossRef]
26. Rabinovich, S.G. *Measurement Errors and Uncertainties: Theory and Practice*; Springer Science & Business Media: Berlin, Germany, 2005. [CrossRef]
27. Rodrigues, E.P.; Oliveira, T.F.; Matuda, M.Y.; Buiochi, F. Development of a 2-d array ultrasonic transducer for 3-d imaging of objects immersed in water. *Sensors* **2021**, *21*, 3501. [CrossRef] [PubMed]
28. Oakley, C.G. Calculation of ultrasonic transducer signal-to-noise ratios using the KLM model. *IEEE Trans. Ultrason. Ferroelectr. Freq. Control* **1997**, *44*, 1018–1026. [CrossRef]
29. Krimholtz, R.; Leedom, D.A.; Matthaei, G.L. New equivalent circuits for elementary piezoelectric transducers. *Electron. Lett.* **1970**, *6*, 398–399. [CrossRef]
30. Szabo, T.L. *Diagnostic Ultrasound Imaging: Inside Out*; Academic Press: Cambridge, MA, USA, 2004.
31. Stepanishen, P.R. Transient radiation from pistons in an infinite planar baffle. *J. Acoust. Soc. Am.* **1971**, *49*, 1629–1638. [CrossRef]
32. Emeterio, S.J.L.; Ullate, L.G. Diffraction impulse response of rectangular transducers. *J. Acoust. Soc. Am.* **1992**, *92*, 651–662. [CrossRef]

eng

MDPI

Article

Development of an Airbag Geometry Specific for Autonomous Vehicles

Bartolomeu Franco [1,2], José Manuel Alves Ribeiro [1,3] and Isidro de Jesús Sánchez-Arce [1,4,*]

1 Departamento de Engenharia Mecânica, Escola de Ciências e Tecnologia, Universidade de Trás-os-Montes e Alto Douro (UTAD), Quinta de Prados, 5000-801 Vila Real, Portugal; bartolomeu.franco@zf.com (B.F.); jmar@utad.pt (J.M.A.R.)

2 ZF Friedrichshafen AG Portugal, Parque Industrial da Gemeira, Rua da Barreira 1237, 4990-645 Ponte de Lima, Portugal

3 Centre for the Research and Technology of Agro-Environmental and Biological Sciences (CITAB), Quinta de Prados, 5000-801 Vila Real, Portugal

4 Laboratório Associado de Energia, Transportes e Aeronáutica (LAETA-INEGI), Rua Dr. Roberto Frias 400, Polo da FEUP, 4200-465 Porto, Portugal

* Correspondence: isidroarce@utad.pt

Abstract: Airbags are important safety devices in modern vehicles. However, their effectiveness is linked to the occupants being seated in standard positions. Although autonomous vehicles are less accident-prone, they are also equipped with airbags, similar to any other vehicle. Additionally, autonomous vehicles allow for occupants seated in non-standard positions, so in the case of a collision, the airbags' effectiveness decreases. In this work, an airbag design suitable for both assisted and autonomous driving conditions is proposed, the driver's airbag being the object of interest. Airbag geometry, threads, seam strength, and seam geometries were selected following Design of Experiments (DoE) methodologies and a series of experimental tests. Moreover, an adaptive system based on sewn tethers allows the airbag to adapt to the driving mode, which is proposed and validated. Finally, all the findings were experimentally tested on two different geometries. The results were satisfactory as the deployed airbag shape and dimensions were as expected, indicating that this airbag design is capable of protecting the driver of a vehicle capable of autonomous driving.

Keywords: airbag systems; autonomous driving; seam geometry; seam strength; airbag geometry; Design of Experiments (DoE)

Citation: Franco, B.; Alves Ribeiro, J.M.; Sánchez-Arce, I.d.J. Development of an Airbag Geometry Specific for Autonomous Vehicles. *Eng* **2023**, *4*, 2553–2570. https://doi.org/10.3390/eng4040146

Academic Editor: Antonio Gil Bravo

Received: 10 August 2023
Revised: 24 September 2023
Accepted: 9 October 2023
Published: 11 October 2023

1. Introduction

Airbags are important safety devices in the case of vehicle collision, having contributed to saving numerous lives [1,2]. The airbag system is composed of several components, the most important being the inflatable cushions (the airbag itself), impact sensors, ignition system, propellant, mounting hardware, and the moulded covers Nayak et al. [3]. From this list, it can be seen that the inflatable cushions, the ignition system, and the propellant are the most critical components to ensure the correct deployment of the airbag. Airbags could be installed in different parts of the vehicle, the most common are as follows: (a) driver's airbag; (b) front passenger airbag; and (c) side airbags, which includes curtain and thorax (seat) airbags. The driver and passenger airbags are mandatory for all US vehicles produced after 1998 [4].

Vehicle safety regulations directly impact the design and testing of new airbags [4] because the required tests dictate parameters such as the deployment time. For example, in 1997 in the U.S., the regulations changed the testing method from a barrier to a sledge, both at 48 km/h. This change led to using different airbag designs on vehicle models after 1998. Braver et al. [4] analysed crash and fatalities data for models from 1997 and models from 1998 to 1999. In this case, vehicles of the same make, models, and platforms were analysed (a pool of 171 makes and models). In fact, the change in airbag design (and regulation) led to a 7% reduction in fatalities and airbag-induced injuries despite the yearly increase in travelled distance. In addition, the death risk in a frontal crash for children (0–4 years old) as front seat passengers was reduced up to 65% [5]. On the other hand, the airbags developed after this regulatory change also possess features such as dual-stage inflators and a series of sensors to identify the presence and size of the occupant [5], so these features may have also contributed to the fatality and injury reduction observed. The advances in technology also bring updates in the regulations, resulting in further airbag improvements, which are accompanied by further sensors [6]; therefore, impacting positively the mortality reduction with respect to the previous airbag generations [6].

The aforementioned studies are based on the standard driving position. However, alterations to this position, e.g., being farther away from the knee bolster or highly reclined seat can lead to a phenomenon known as 'submarining' where the occupant slides down the seat belt, resulting in further injuries. Small occupants are more likely to experience submarining [7,8]. These non-standard seating positions are more likely to occur in front seat passengers, but with the advent of autonomous driving these may also occur on both front seat occupants [8,9], emphasising their relevance for airbag designs aimed for such a driving mode. Altering the location of the passenger airbag seems to reduce the risk of submarining front seat passengers [10]. These studies were mostly focused on passenger submarining; however, in the case of autonomous driving, the passenger in the driver's seat may also be affected by this phenomenon, so it is necessary that the restraint systems (seat belt and airbags) are located in a suitable position in order to offer protection while in autonomous driving mode.

Nowadays, vehicles capable of autonomous driving are available commercially, e.g., Tesla®, which may be followed by other manufacturers [11]. Although the concept of autonomous driving offers advantages over manual conduction, e.g., for elderly drivers [12], the capability of such vehicles for the two driving modes requires unique developments to ensure road users' safety. Furthermore, it is thought that vehicles in autonomous driving are less prone to accidents because their algorithms follow the road regulations [11]. There are several studies focused on the digital safety of this type of vehicle [11–13]; however, little has been reported about occupant safety within these vehicles in the event of an accident, perhaps, because their design must comply with all the vehicle safety standards.

Passenger kinematics during a crash event were also modelled for autonomous driving. The procedure is similar to that already employed for conventional vehicles, but the occupants would be seated in non-standard postures, therefore increasing the risk of injury. Diez et al. [14] studied the kinematics of an autonomous vehicle driver during a crash event, i.e., a lateral impact. Five seating positions were considered: (1) standard NCAP, (2) standard NCAP (New Car Assessment Program) hands not on the wheel, (3) work, (4) leisure, (5) relax or sleep. The study included the lateral airbags (curtains) but was focused on the forces imposed on the driver during the event, concluding that current safety standards are not sufficient for this type of vehicle. Furthermore, the vehicle while under autonomous driving may decelerate, accelerate, or change direction suddenly as a response to the traffic dangers, increasing the risk of sudden airbag deployment when the occupants' heads may be closer to the airbag deployment areas [15]. Consequently, the control modules should take into account the occupants' position in the vehicle prior to airbag deployment [15].

Regarding airbag systems for vehicles capable of autonomous driving, an inflator for an airbag up to 50 L has been developed [16,17]. Numerical and experimental approaches were followed to arrive at the final design, which allows stopping inflation after the airbag contacts the passenger [16]. However, no information about the design of the actual airbag has been found in the literature. Regarding airbags to protect the vehicle's occupants, designs such as the 'Life Cell' of Autoliv® offer mostly lateral protection [17], although no other designs offering frontal protection to the driver of an autonomous vehicle have been found in the literature.

Regarding the airbag design procedure for autonomous vehicles, no information has been found yet, but assuming that it is similar to that used for conventional vehicles, such recommendations could be followed for designing an airbag for an autonomous vehicle. During the development of an airbag static deployment, impact from a head-like object is tested; then, it is installed in a vehicle and crash tested using dummies [18]. However, the iterative nature of the design process requires time and resources for each design iteration. Alternatively, Computer Aided Engineering (CAE) offers the possibility of reducing the development time, for example, for analyses considering impacts on an inflated airbag. On the other hand, Computational Fluid Dynamics (CFD) provide information about inflation dynamics as airbags are filled with a gas. These two approaches were combined and employed for the design of a curtain-type airbag [18], in which the design was tested computationally, first for stresses and then using CFD, prior to its physical construction and testing. The CFD model led to identifying an area of probable airbag bursting during deployment; the inflation time was also non-satisfactory. Following a series of design iterations and computational models the final design was obtained. The physical airbag prototype was produced and tested, validating the computational models. In addition, the design and analysis workflow employed techniques such as the Design of Experiments (DoE) and numerical optimisation to arrive at the final prototype [18]. The use of computational models, despite their limitations, together with DoE and numerical optimisations have proven to be valuable and cost-effective tools to reduce the development time while arriving at a successful and safe airbag design. Recently, machine learning techniques and CAD were combined and used for the preliminary design of a curtain-type airbag [19]. Parametric CAD models were used to train the algorithm. Once the training process was completed, the preliminary design time for an airbag was significantly reduced [19].

As described above, non-standard seating positions compromise the effectiveness of conventional airbag designs aimed at protecting the driver. Although various airbag designs have been proposed in the literature for autonomous vehicles [17,20], they are not intended for frontal protection or for providing protection in both driving modes. Furthermore, the airbag designs proposed in the literature vary in purpose, the curtain type being mainly those described with more detail [18,21], but none regarding the driver's protection in the case of a frontal crash.

This work aims to propose and develop a driver's airbag for vehicles capable of autonomous driving. Therefore, the proposed airbag geometry should offer protection to the driver whether the vehicle is in autonomous driving mode or in assisted driving mode. The airbag's geometry is chosen from six geometries already developed using known selection criteria methodologies. Moreover, the necessary airbag volume depends on the driving mode selected, so a tether-based restraint system was proposed and developed using DoE methodologies. The development of this airbag was mostly done through experimental work because the study was performed within a company producing airbags for several car makers, i.e., ZF Friedrichshafen Portugal. Therefore, the know-how that already existed was expanded through the use of experimental and analytical work. The final geometry and design was tested experimentally using internal standards for such purpose, validating the design.

2. Materials and Methods

2.1. Airbag Geometries

The airbag geometry influences the level of protection it provides [18]. Therefore, it was necessary to investigate what geometry is more suitable for both assisted and autonomous driving modes. Various geometries were constructed and then analysed regarding how they fulfil the objectives. As a starting point, the geometries previously studied by Esteves [22] were analysed:

Double: Composed of a standard airbag at the back (steering wheel side) and a smaller airbag located at the front (passenger side) (Figure 1a). When the vehicle is driven in assisted mode, the standard airbag deploys. When driven in autonomous mode, both airbags deploy.

Double inverted: Geometrically opposite to the previous one (Figure 1b); nevertheless, the deployment is equal, i.e., the standard size airbag deploys when the vehicle is driven in assisted mode, and the whole airbag deploys when driven in autonomous mode.

Double chamber: The airbag has two internal chambers non-visible from the exterior (Figure 1c). One chamber is deployed in assisted driving mode. In autonomous driving, both chambers are inflated.

Triple chamber: Geometrically bellows-shaped (Figure 1d). Similarly to the previous cases, part of the airbag is deployed during assisted driving mode while the full airbag is deployed during autonomous driving. However, it is necessary to use constraining structures to control the airbag size.

Cylinder: Offers a concept similar to the previous one while being geometrically simpler (Figure 1e). Therefore, it requires constraining structures to control the volume and extension of the airbag depending on the driving mode.

Pillow: This concept is similar to the double chamber geometry. However, the connection between the two airbags is located upwards instead of centred (Figure 1f). The accessory bag has a rectangular shape. The displaced geometry aims to protect the driver even if not properly seated.

The geometry selection was based on their theoretical capability of offering protection during autonomous driving. The geometries were tested experimentally by inflating them with compressed air, allowing to identify their advantages and disadvantages (Figure 1). The airbags employed in these tests were made of PA 470 dTex fabric coated with 25 g of silicone. Following the testing, five parameters were evaluated for all the geometries:

Cost: The manufacturer's production costs, lower costs were preferred;

Reach: Extension of the deployed airbag, i.e., distance from the steering wheel to the airbag's front panel;

Volume: How much volume does the folded airbag occupy?

Possibility of adjustment: How easy is it to modify the airbag's dimensions to suit a specific application case?

Adaptive systems: How easy is it to implement a system to alter the airbag reach to suit two driving modes: assisted and autonomous?

A multiple-criteria decision analysis (MCDA) was performed with these parameters [23], being graded as excellent (10), good (5), average (3), below average (3), and poor (1); the number inside the parenthesis indicate the points given in each case. The punctuation attributed to each airbag geometry and parameter were selected from the company's previous experiences. Subsequently, the two geometries with the higher punctuation are selected for further study.

2.2. *Adaptive Systems*

The proposed solution to control the driver's airbag volume for both driving conditions consists of a sewed internal strap that limits the deployed airbag's reach (height), also known as 'tethers' (Figure 2); the tethers must be strong enough to support this inflating pressure. Nevertheless, when in autonomous driving mode and when the driver's seat is away from the steering wheel, the airbag is subjected to a higher inflating pressure, breaking the seams in the tethers and allowing for maximum airbag reach. Therefore, the strength of those tethers must be predictable and with lower variability. The strength of the tethers would then depend on the strength of the threads conforming the seams, the seam geometries, and the fabric material employed [24,25]. Two values of inflating pressure were used in this proposal, one per driving mode. It is worth noting that the inflating system is out of the scope of this work. Here, the airbag's reach for assisted driving was considered to be 300 mm and 400 mm for autonomous driving. These values are internal design requirements while the other dimensions are confidential. Similarly, it is expected that the seams can support 2000 N, which is another internal design requirement.

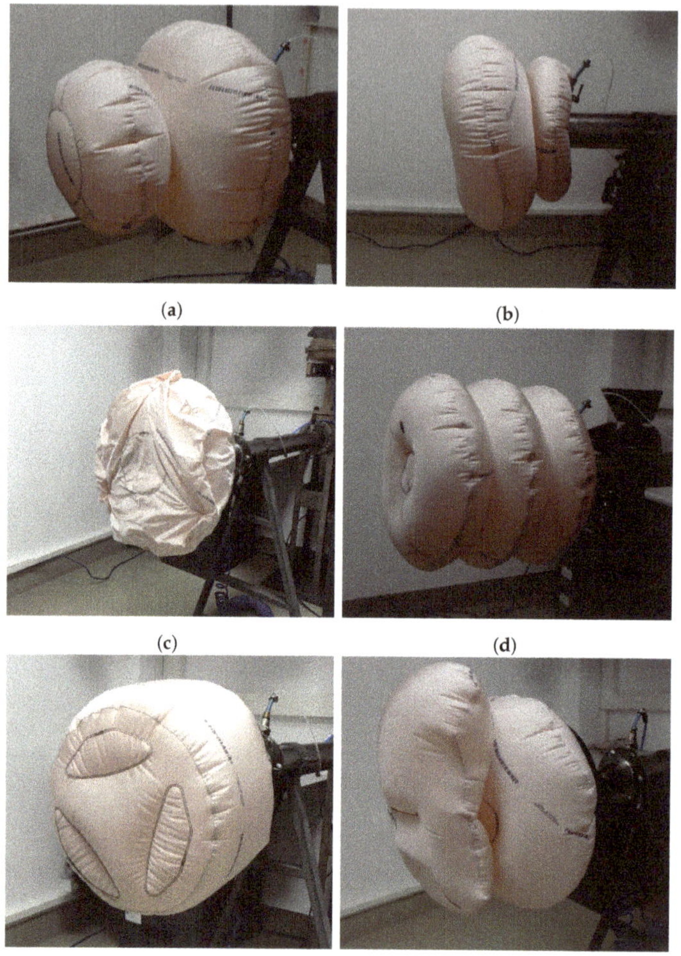

(a) (b) (c) (d) (e) (f)

Figure 1. Tested airbag geometries. (**a**) Double. (**b**) Double inverted. (**c**) Double chamber. (**d**) Triple chamber (bellows). (**e**) Cylinder. (**f**) Pillow.

Currently, the company employs six different thread types (Table 1) in the manufacture of airbags. Three of these are placed on the upper part of the seam (threads 70, 78, and 79) while the remaining are placed in the lower one (threads 69, 71, and 88). However, seam strength not only depends on the thread type but on other parameters such as fabric, cord treatment, seam threads, seam geometry, curtain orientation, and seam tension [24,25]. Therefore, the effect of these six parameters was further studied to identify the most suitable combination to be used in the tethers.

Table 1. Thread types, strengths, and other data.

Thread Type (Nm)	Tensile Strength	Colour	Reference
120/1	165.5 ± 3.5	Orange	9200-0178
60/2	33 ± 5	Black	9200-0078
40/3	50 ± 4	Brown	9200-0070
20/3	106 ± 9	Red	9200-0069
17/3	135 ± 20	Gold	9200-0217
13/3	160 ± 17	Green	9200-0145

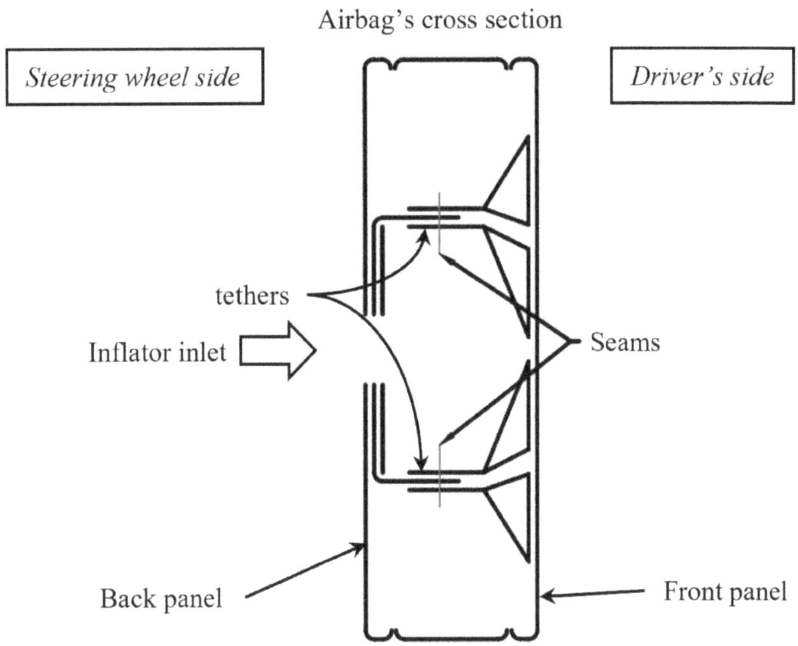

Figure 2. Cross section schematic of the proposed tether system to control the airbag's reach. The seams, in red, will break with the inflating pressure in autonomous driving.

Starting with the six parameters influencing seam strength, a Design of Experiments (DoE) was performed to identify the parameters or the combination of them that influence the most the seam strength; this analysis was done within the Minitab 17 Statistical Software (Minitab Inc., Philadelphia, PA, USA, 2010). After completing the DoE, the parameters identified were (1) the seam's threads, (2) the seam's geometry, and (3) tension in the seam's threads, as shown in Figure 3. In consequence, the influence of these parameters was investigated further, as described in the following sections.

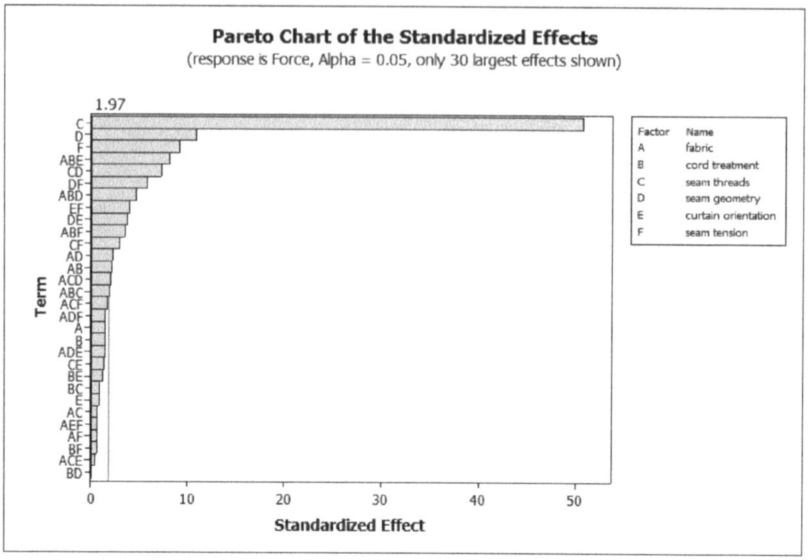

Figure 3. Identification of the parameters with larger influence in the seam strength.

2.2.1. Seam Threads

Seam strength depends on the strength of the threads used in the seam [25]. As mentioned above, three thread types are used for the upper seam while three different ones are used for the lower seam, resulting in six combinations. Then, 29 specimens of each seam combination (174 specimens in total) were tested experimentally following the standard ISO2062. After completion of the tests, a single-factor analysis of variance (ANOVA) was performed to evaluate the effects of these thread combinations.

2.2.2. Seam Geometry

The seam geometry was evaluated on a 700 dTex fabric coated with silicone (Figure 4). The seams were created with nylon threads, the upper was a Tex 30 Anefil Nylon® while the lower was a Tex 138. The mechanical tests were performed using a Universal Testing Machine (UTM) INSTRON with a 10 kN load cell. All the tests were performed at 200 mm/s and had a 5 kN preload; moreover, the gauge length for all the specimens was 200 mm [26]. An example of the specimens is shown in Figure 4.

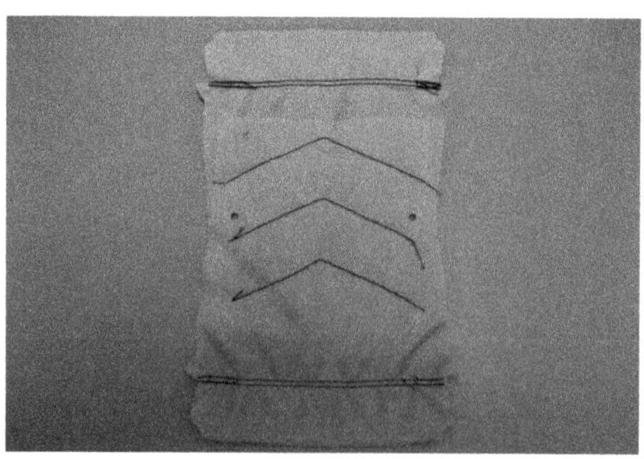

Figure 4. Example of a specimen for testing the seam geometry.

Eight seam geometries were tested, such as: (1) squared "U", (2) semicircle, (3) "U", (4) "O", (5) skinny squared "U", (6) eye, (7) "V", and (8) curved "V"; all of them with 10 stitches except for the skinny squared "U" with 11. Subsequently, variations of the chosen seam geometry were studied to identify if the variations contribute to strength improvement.

2.3. Case Study

Airbags with geometries, thread combinations, and seam geometries selected in the previous steps were manufactured and tested. The airbag testing was done in two stages: first, using a high-pressure bag tester (HPBT); second, using a steering wheel assembly and an inflator module. All the tests were performed in ZF's experimental facilities in Portugal.

The HPBT was developed by ZF and consists of a compressed air reservoir with a capacity of 10,000 L. The compressed air is stored at 10 bar. The output of this reservoir is connected to the laboratory where the airbags are tested (Figure 5). Furthermore, image-based measurement techniques are employed in the HPBT laboratory, allowing to determine the airbags' dimensions during and after the inflation tests, for which a photographic camera was used (Canon EOS 800D+18-55MM F/4-5.6 IS STM). The UTM mentioned above is located within the same facilities.

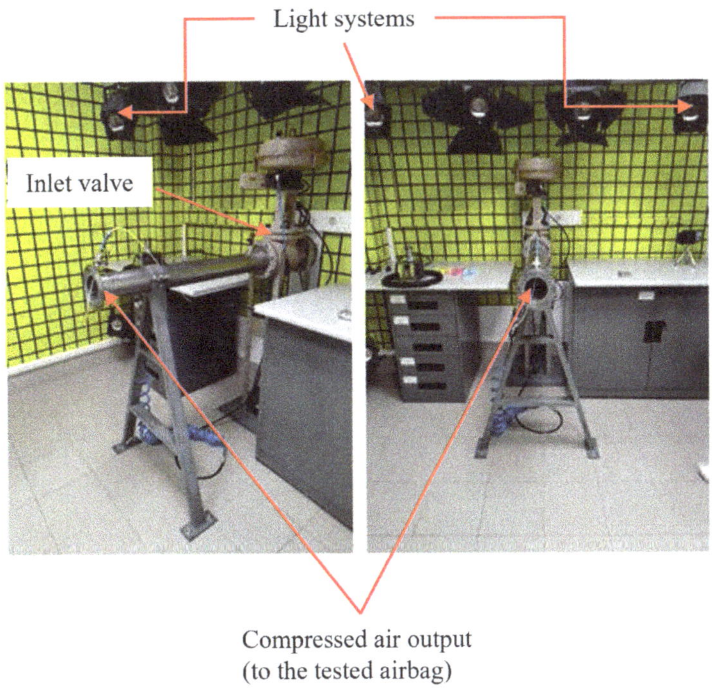

Figure 5. Laboratory facilities for the static testing of airbags. The grids on the walls are used for image-based measurements.

3. Results

The summarised methodology followed in this work is represented in Figure 6. The results of each step indicated there are described in the following subsections.

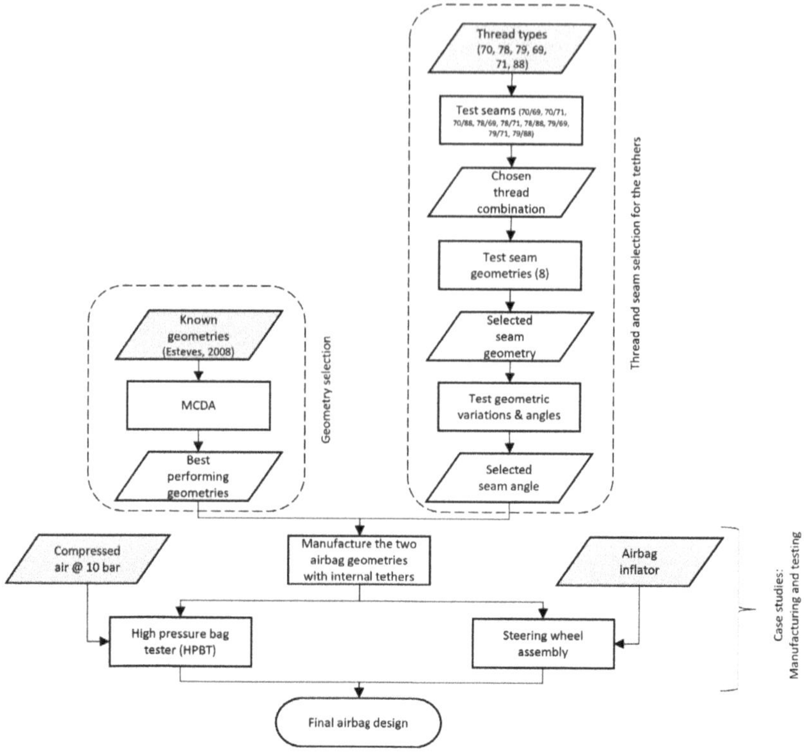

Figure 6. Schematic representation of the methodology followed in this work.

3.1. Airbag Geometry

Six airbag geometries were analysed using an MCDA and five parameters were considered: (1) manufacture cost, (2) airbag reach, (3) airbag's volume when stored, (4) adjustment easiness, and (5) suitability for adaptive systems; these parameters were described in Section 2.1. The MCDA is shown in Table 2. Two geometries obtained the highest scores, the bellows and the cylinder, Figure 1d,e. Therefore, these two geometries will be the subject of further study within this work.

Table 2. Decision matrix for the tested airbag geometries: (I) cost, (II) reach, (III) volume when stored, (IV) ease of adjustment, and (V) adaptive systems.

	Parameter and Value					
	I	**II**	**III**	**IV**	**V**	
Airbag Geometry	**10**	**7**	**7**	**5**	**5**	**Total**
Cylinder	5	5	5	5	5	170
Double	5	5	3	5	3	146
Double inverted	5	5	3	5	3	146
Triple	1	3	1	3	1	58
Pillow	1	3	3	3	1	46
Bellows	5	5	5	7	7	190

3.2. Adaptive Systems

3.2.1. Seam Threads

Seam strength was tested for each of the 6 thread combinations described in Section 2.2; also, 29 specimens per seam type were tested, giving a total of 174 specimens. Seam strength was found to be related to the thread combinations; then, the results were grouped by the type of thread used on the upper seam, as shown in Figure 7. It is worth noting that for the thread combinations 78/69 and 79/69 only 27 specimens were valid (Figure 7).

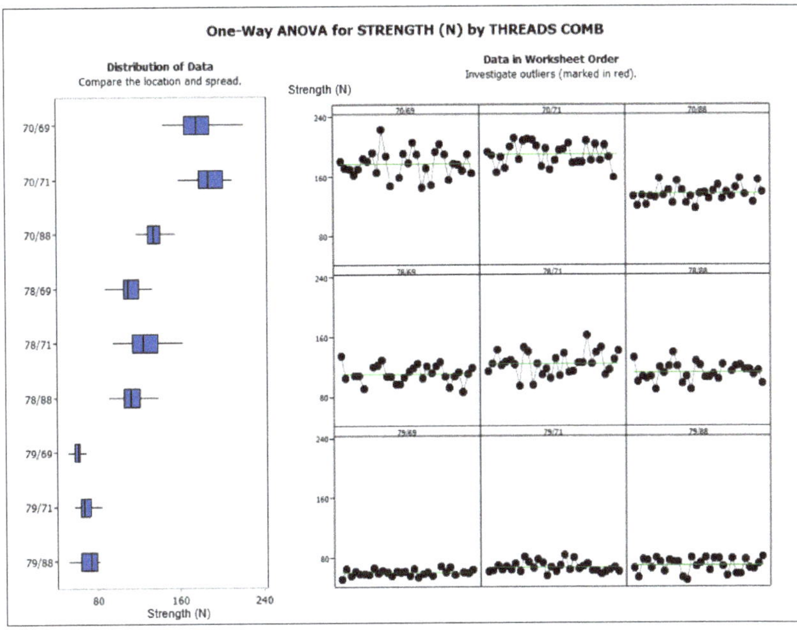

Figure 7. Analysis of the seam strength due to the different thread types and sizes. The results are grouped by upper seam thread type.

The thread with reference 69 was the strongest of those employed in the lower seam, so its use ensured that thread failure could only occur in the upper seam. The seams where the thread with reference 70 was placed on the upper seam were found to be the strongest of the set, mainly the 70/69 combination; however, such a combination also had a large dispersion. Continuing with the seams with the thread with reference 78 in the upper seam, these showed similar strength to those described above, although slightly lower. Conversely, these seams showed lower dispersion. Finally, the seams with the upper seam sewn with thread with reference 79 showed the lowest strength. Subsequently, the three seams groups 70/69, 78/69, and 79/69 were analysed to select one for the next steps of the project. The strength data of the associated seam were grouped, as shown in Figure 8.

Upon inspection of the strength data obtained from the experimental tests, shown in Figures 7 and 8, it was decided to use the seam with the thread combination 78/69 because it showed an intermediate strength ($S = 111.3$ N) of the lot with acceptable variability (Figure 8). The seam with the thread combination 79/69 showed the lowest variability but also the lowest strength; hence, its rejection.

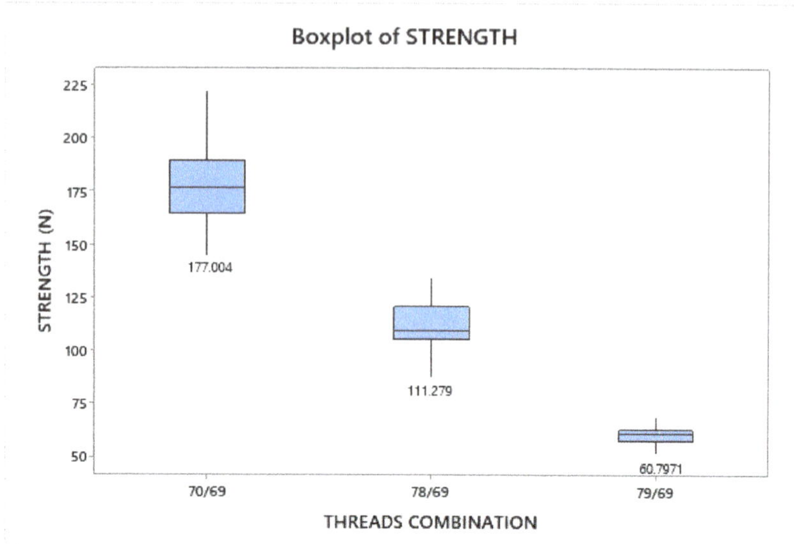

Figure 8. Seam strength of those with the thread using reference 69, in the lower part of the seam.

3.2.2. Seam Geometry

Eight seam geometric variations were tested to identify the most suitable one to be applied to the tethers (Section 2.2.2). Among the eight seam geometries tested, the squared "U" shape showed the highest strength (173.3 N); however, this geometry also presents one of the highest variabilities in the results ($CV = 6.6\%$). Conversely, the curved "V" shape showed the lowest strength (67.14 N) and high variability ($CV = 17.7\%$), as shown in Figure 9.

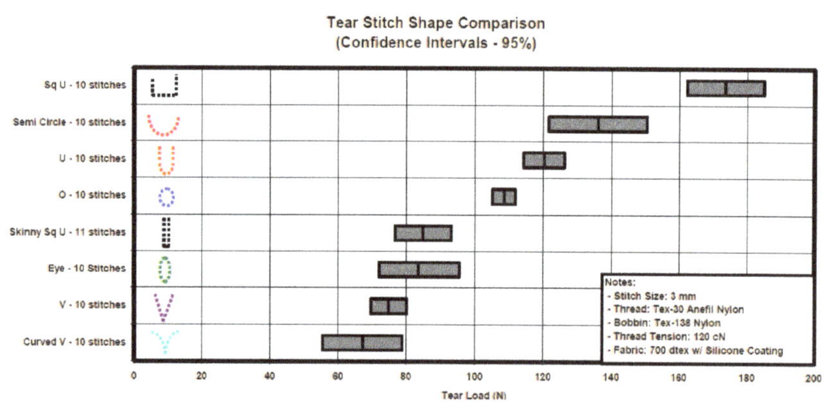

Figure 9. Comparison of stitch shape and strength.

On the other hand, the seams with "O" and "V" shapes had the lower variability, with CVs of 3.5% and 7%, respectively. The "O" shape was found to be stronger (108.57 N) than the "V" shape (74.76 N). Despite the lower strength of the latter two configurations with respect to the squared "U" stitch, they were chosen for further study because of their lower CVs. Seam strength can also be influenced by other parameters such as temperature and humidity; however, these were not considered at this stage.

Although the "V" shaped seam had the lowest strength of the chosen two geometries, it is the easiest to sew. Consequently, it was decided to study the effect that its geometric variations have on seam strength, so five additional geometries were tested: arrow, heart, rombo, teardrop, and "V" shape; each geometry was tested with two thread combinations 78/4 AS and 78/88. The testing procedure was the same as described before.

The geometric variation of the "V-shaped" seam had an effect on seam strength, as shown in Figure 10. However, the strength difference had statistical significance; furthermore, the results presented high variability, which was not desired. Among the analysed geometries, the standard "V-shaped" seam showed the lowest variability; hence, it was selected for further analyses.

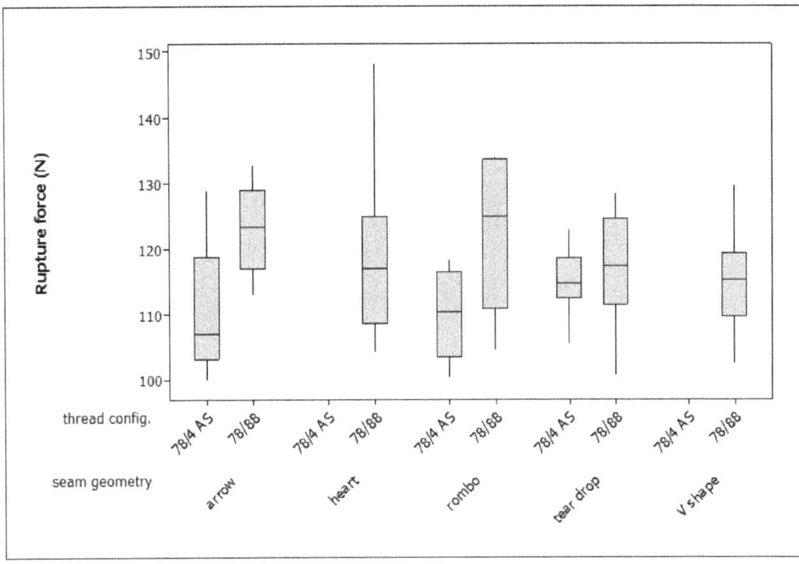

Figure 10. Strength of the "V-shaped" seam geometric variations.

Although the seam shape is defined, the angle of the "V" can also be varied, so this parameter was further studied. The "V" angle was varied from 0° to 180° in 10° increments while keeping the fabric, thread type, and seam density constant. Subsequently, 30 specimens of each were manufactured with a polyamide fabric (350 dTex), the upper seam thread was done with a 20/3 Nm thread while the lower with a 60/2 Nm one. The "V" angle influenced the seam's strength, which was found to increase with the "V" angle, as shown in Figure 11, noting that values for a few low angles were omitted (Figure 11).

Upon further inspection, the strength values obtained with the 30°, 45°, and 60° were statistically similar, so only the 30° was used. Similarly, the 90°, 100°, 110°, and 120° showed such pattern; again, the 90° was used. Then, the 130°case was employed; this being similar to the 140°. The cases above 140° were not considered for further analyses due to their high dispersion. Among the three cases chosen, the 90° presented the lowest dispersion, so it was selected for the "V" shape seam, as shown in Figure 12.

In summary, seam strength depends on thread combination and seam geometry. The aforementioned tests indicate that the thread combination 78/69, the V-shaped seam, and a "V" angle of 90° have low variability. Therefore, these parameters were chosen for a proof-of-concept through a case study.

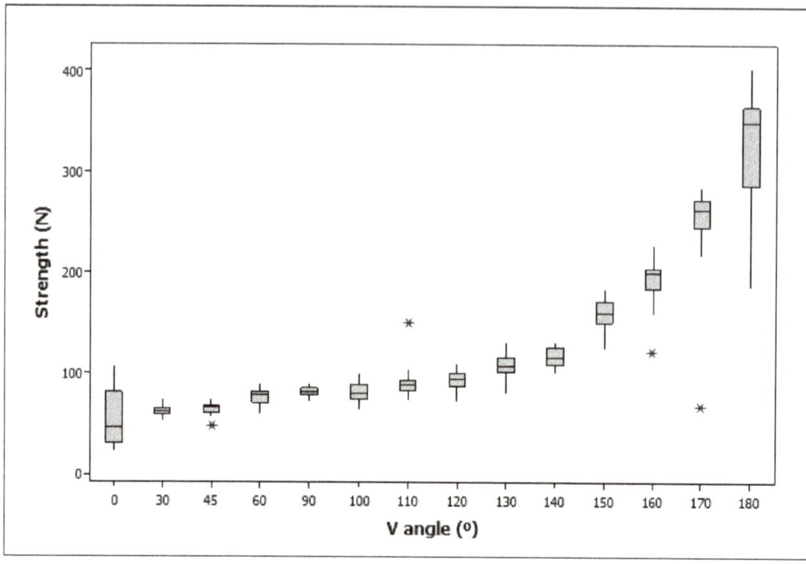

Figure 11. Seam strength variations due to the seam angle.

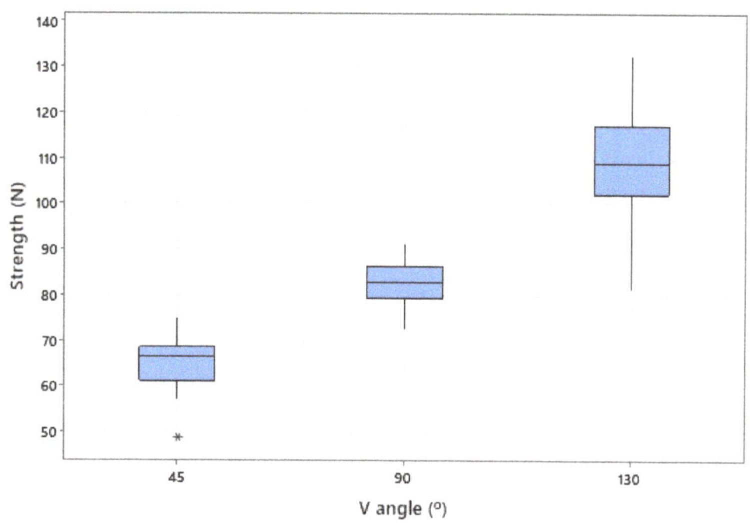

Figure 12. Strength vs. "V" angle from the selected configurations.

3.3. Case Study

The airbag is mainly composed of three main components: (1) the front panel, (2) the back panel, and (3) two tethers, the latter connecting the front and back panels (Figure 2). Moreover, in this study, two airbag geometries were employed: (1) cylindrical shape and (2) bellows shape, which were selected in Section 3.1. Then, airbags were manufactured in-house and tested to evaluate their performance under assisted and autonomous driving. For these tests, the airbags had similar dimensions to allow for comparisons. The material employed for airbag manufacture was also PA 470 dTex fabric coated 25 g of silicone (same as in Section 2.1).

The tests showed that once the tethers break, the increase in airbag extension (reach) was 120 mm and 130 mm for the cylindrical and bellows-shaped geometries, respectively, hence increasing the airbag volume. For example, a test of the bellows-shaped geometry is shown in Figure 13. Therefore, this increase in extension allows to protect the driver even when he/she is farther away from the steering wheel, as is the case of autonomous driving.

Afterwards, static tests at ambient temperature were performed to the airbags with the aim of evaluating their structural integrity, inflation time, positioning with respect to the steering wheel, and the performance of the shape control through seams.

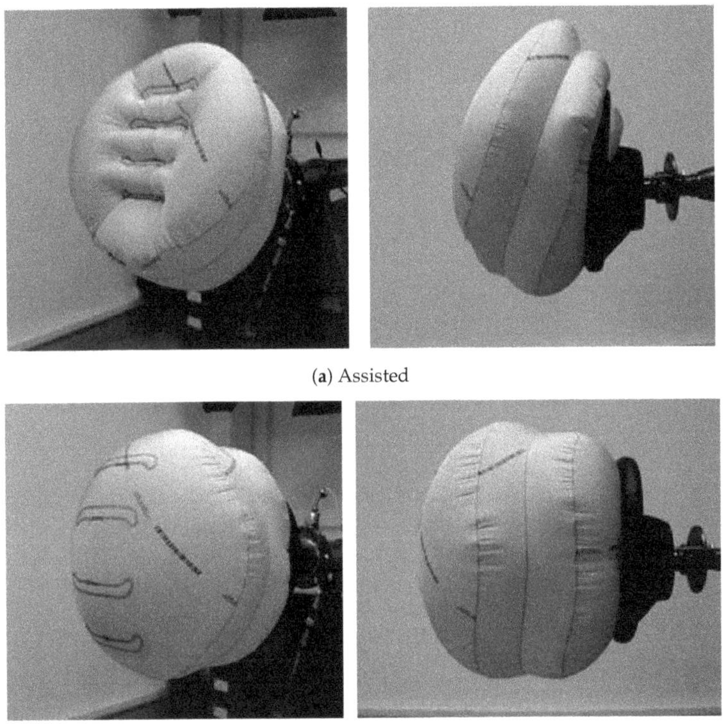

(**a**) Assisted

(**b**) Autonomous

Figure 13. Bellows-shaped airbag in both driving modes, oblique and lateral views shown for comparison. The recesses or islands shown in (**a**) correspond to the tethers' attachment to the airbag's front panel; tear of the tether seams can be observed in (**b**) because the recesses disappeared.

The recorded inflation time was 30 ms for both assisted and autonomous driving. The use of tethers to control the airbag volume and shape in both driving modes performed as expected, as shown in Figure 14 for the cylinder-shaped airbag. Therefore, the tests confirmed that the chosen geometries, materials, and shape control via the seams fulfil the safety requirements of both assisted and autonomous driving.

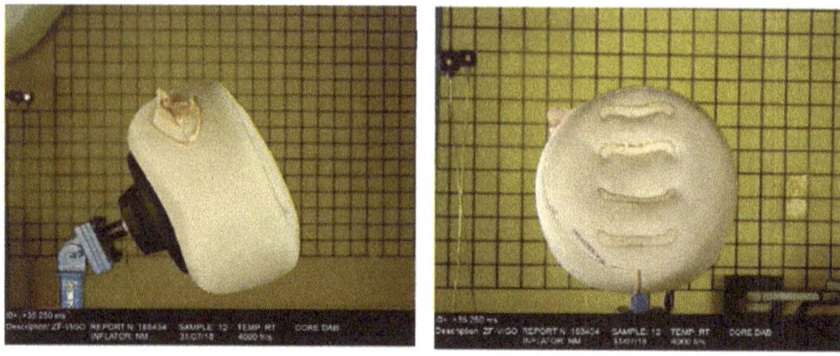

(**a**) Assisted

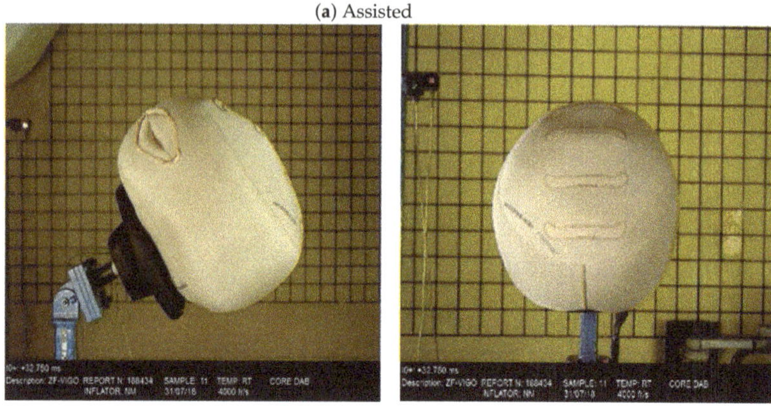

(**b**) Autonomous

Figure 14. Cylinder-shaped airbag in both driving modes, oblique and lateral views shown for comparison. The recesses or islands shown in (**a**) correspond to the tethers' attachment to the airbag's front panel; tear of the tether seams can be observed in (**b**) because the recesses disappeared.

4. Discussion

This work aimed to design an airbag capable of protecting the driver of an autonomous vehicle regardless of the driving mode, which was achieved through an experimental methodology. Experimental methodologies had been employed for developing airbags [18], although, in recent years, the use of CAE systems to perform design tasks is widespread. For example, curtain airbags had been developed and optimised through CAE and CFD methodologies [18,21] or using CAD and machine learning methodologies [19]. However, such computational approaches also require corresponding models, dimensions, properties, etc., of the vehicle or vehicles where the airbag is going to be implemented, which often is confidential; for that reason, generic airbag models are often used in the literature. On the other hand, the airbag presented here was developed within a company producing airbags for several car manufacturers; hence, the proposed airbag complies with internal and international standards, although in the future computational studies could be performed, as in Refs. [7,9,10].

Although airbag designs for autonomous vehicles had been reported in the literature, for example, the 'Life Cell' of Autoliv® [17,20], those designs are aimed to protect the occupants of the vehicle, offering mostly lateral protection. On the contrary, the airbag design developed here offers frontal protection to the vehicle's driver, so these designs are not comparable. Regarding the design process, the geometry of conventional airbags was studied to identify which could perform in both driving scenarios. Esteves [22] studied several airbag geometries for the same company; therefore, those geometries were the starting point for this project. Then, airbag volume could be controlled through tethers, which were not reported in the literature, but textiles and threads were [24–27]. Furthermore, the materials employed in this work are actual airbag materials and correspond to some of those described by Nayak et al. [3].

The design process of this airbag system was divided into individual tasks, i.e., geometry, seams, seam strength and geometry, and case studies, allowing to verify each step and component prior to introducing them to the whole assembly. In addition, the DoE methodology employed is common practice in many experimental studies and plays an important role in airbag development [15,18,21]. Consequently, the developed airbag design followed the usual route for this type of product.

The tests performed on the proposed geometries, confirm that they perform in accordance with the design requirements. The in-house pressure testing allowed to test the strength of the whole assembly, while its mounting within a steering wheel and using an actual airbag inflator validated its deployment performance. The dimensions of the deployed airbags were determined with image-based techniques, which are commonly employed for these purposes [16,17]. Although the developed airbag geometry was validated experimentally using internal and international standards, these tests were mostly about inflation performance for both driving modes. Therefore, tests with head-shaped impactors would further validate the designs. In addition, computational models would contribute to a deeper understanding of the airbag behaviour during inflation and impact.

5. Conclusions

Although the airbag designs already employed in vehicles can be used in autonomous ones, it is necessary to adapt their design so they properly protect the occupants. However, it has been observed that in vehicles able to drive in both autonomous and assisted modes, the occupant in the driver seat may change seating posture while in autonomous driving, therefore reducing the airbag effectiveness in the case of a collision.

In this work, an airbag geometry to be mounted in the steering wheel and offering protection whether the vehicle is in autonomous driving mode or not, was developed using an experimental methodology. The airbag volume is controlled by tethers sewn internally, which will tear with the inflation pressure. The experimental validation confirmed the expected airbag behaviour for both driving modes. The concept of controlling the airbag volume using tethers was found to be successful, so it can be applied to other airbag geometries such as knee bolsters. Furthermore, to the authors' best knowledge, this is the first time this type of airbag solution has been reported. Although the geometry was validated experimentally, there is an opportunity for future computational studies regarding this geometry and its performance within the vehicle.

Tether strength was found to be influenced by thread type and its strength, the combination of threads in the upper and lower seam locations, and seam geometry. A "V-shaped" geometry forming a 90° angle was chosen for the tether seam because it showed low dispersion while offering adequate strength (90 N) and being easy to sew. The lower variability of seam strength allowed for a predictable behaviour of the tethers, and in consequence, of the whole airbag.

In summary, the geometry proposed here offers protection to the occupants regardless of the driving mode by controlling the number of chambers inflated, which is controlled electronically. Furthermore, the strength and geometry during and after deployment are controlled through the stitching patterns and thread types as well as their location within the airbag. In consequence, the proposed geometry protects the vehicle occupants regardless of the driving mode in which the vehicle is at the moment of a frontal collision.

Author Contributions: Conceptualisation, B.F. and J.M.A.R.; methodology, B.F. and J.M.A.R.; resources, B.F.; formal analysis, B.F., J.M.A.R. and I.d.J.S.-A.; writing the original draft, B.F., J.M.A.R. and I.d.J.S.-A.; supervision, J.M.A.R. All authors have read and agreed to the published version of the manuscript.

Funding: This research received no external funding.

Institutional Review Board Statement: Not applicable.

Informed Consent Statement: Not applicable.

Data Availability Statement: The data is confidential because of the industrial nature of this project.

Acknowledgments: The authors thank ZF Friedrichshafen Portugal for all the support to complete this project.

Conflicts of Interest: The authors declare no conflict of interest.

Abbreviations

The following abbreviations are used in this manuscript:

ANOVA	Analysis of variance
CAE	Computer Aided Engineering
CFD	Computational Fluid Dynamics
CV	Coefficient of variation
DoE	Design of Experiments
HPBT	High pressure bag test
MCDA	Multiple-criteria decision analysis
NCAP	New Car Assessment Program
Nm	Metric count (unit), number of hanks of 1000 m/kg
UTM	Universal testing machine

References

1. Starnes, M. *Trends in Non-Fatal Traffic Injuries: 1996–2005;* Techreport DOT HS 810 944; National Highway Traffic Safety Administration: Washington, DC, USA, 2008.
2. Glassbrenner, D. *Lives Saved Calculations for Seat Belts and Frontal Air Bags;* Techreport DOT HS 811 206; National Highway Traffic Safety Administration: Washington, DC, USA, 2009.
3. Nayak, R.; Padhye, R.; Sinnappoo, K.; Arnold, L.; Behera, B.K. Airbags. *Text. Prog.* **2013**, *45*, 209–301. [CrossRef]
4. Braver, E.R.; Kyrychenko, S.Y.; Ferguson, S.A. Driver Mortality in Frontal Crashes: Comparison of Newer and Older Airbag Designs. *Traffic Inj. Prev.* **2005**, *6*, 24–30. [CrossRef] [PubMed]
5. Braver, E.R.; Scerbo, M.; Kufera, J.A.; Alexander, M.T.; Volpini, K.; Lloyd, J.P. Deaths among Drivers and Right-Front Passengers in Frontal Collisions: Redesigned Air Bags Relative to First-Generation Air Bags. *Traffic Inj. Prev.* **2008**, *9*, 48–58. [CrossRef] [PubMed]
6. Braver, E.R.; Shardell, M.; Teoh, E.R. How Have Changes in Air Bag Designs Affected Frontal Crash Mortality? *Ann. Epidemiol.* **2010**, *20*, 499–510. [CrossRef] [PubMed]
7. Thorbole, C.K. Dangers of Seatback Recline in a Moving Vehicle: How Seatback Recline Increases the Injury Severity and Shifts Injury Pattern.In Proceedings of the ASME International Mechanical Engineering Congress and Exposition, Houston, TX, USA, 13–19 November 2015. [CrossRef]
8. McMurry, T.L.; Poplin, G.S.; Shaw, G.; Panzer, M.B. Crash safety concerns for out-of-position occupant postures: A look toward safety in highly automated vehicles. *Traffic Inj. Prev.* **2018**, *19*, 582–587. [CrossRef] [PubMed]
9. Rawska, K.; Gepner, B.; Kulkarni, S.; Chastain, K.; Zhu, J.; Richardson, R.; Perez-Rapela, D.; Forman, J.; Kerrigan, J.R. Submarining sensitivity across varied anthropometry in an autonomous driving system environment. *Traffic Inj. Prev.* **2019**, *20*, S123–S127. [CrossRef] [PubMed]
10. Rawska, K.; Gepner, B.; Kerrigan, J.R. Effect of various restraint configurations on submarining occurrence across varied seat configurations in autonomous driving system environment. *Traffic Inj. Prev.* **2021**, *22*, S128–S133. [CrossRef] [PubMed]
11. Fagnant, D.J.; Kockelman, K. Preparing a nation for autonomous vehicles: Opportunities, barriers and policy recommendations. *TRansportation Res. Part A Policy Pract.* **2015**, *77*, 167–181. [CrossRef]
12. Rhiu, I.; Kwon, S.; Bahn, S.; Yun, M.H.; Yu, W. Research Issues in Smart Vehicles and Elderly Drivers: A Literature Review. *Int. J. Hum.-Comput. Interact.* **2015**, *31*, 635–666. [CrossRef]
13. Cui, J.; Liew, L.S.; Sabaliauskaite, G.; Zhou, F. A review on safety failures, security attacks, and available countermeasures for autonomous vehicles. *Ad Hoc Netw.* **2019**, *90*, 101823. [CrossRef]
14. Diez, M.; Abajo, J.; de Prada, J.V.; Negro, A.; Fernández, M.T. Sitting posture influence in autonomous vehicles for the evaluation of occupant safety in side impact. *Saf. Sci.* **2023**, *159*, 106002. [CrossRef]
15. Diederich, A.; Bastien, C.; Blundell, M. The prediction of autonomous vehicle occupants' pre-crash motion during emergency braking scenarios. *Proc. Inst. Mech. Eng. Part D J. Automob. Eng.* **2023**, 095440702311532. [CrossRef]

16. Breed, D.; Yurchenko, N.; Vynogradskyy, P.; Kuzmenko, K.; Zhang, S.; Li, B. The analysis and experimental development of aspirated airbags for conventional and autonomous vehicles. In Proceedings of the 26th International Technical Conference on the Enhanced Safety of Vehicles, Eindhoven, The Netherlands, 10–13 June 2019; pp. 122–131.
17. Yurchenko, N.F.; Breed, D.S.; Zhang, S. Design of the Airbag Inflation System Applicable to Conventional and Autonomous Vehicles. *Automot. Innov.* **2021**, *4*, 390–399. [CrossRef]
18. Zhang, H.; Ma, D.; Raman, S.V. CAE-Based Side Curtain Airbag Design. *SAE Trans.* **2004**, *113*, 488–494.
19. Rad, M.A.; Salomonsson, K.; Cenanovic, M.; Balague, H.; Raudberget, D.; Stolt, R. Correlation-based feature extraction from computer-aided design, case study on curtain airbags design. *Comput. Ind.* **2022**, *138*, 103634. [CrossRef]
20. Parvez, M.; Rahman, M.; Samykano, M.; Ali, M.Y. Current advances in fabric-based airbag material selection, design and challenges for adoption in futuristic automobile applications. *Mater. Today Proc.* **2023**. [CrossRef]
21. Yun, Y.W.; Choi, J.S.; Park, G.J. Optimization of an automobile curtain airbag using the design of experiments. *Proc. Inst. Mech. Eng. Part D J. Automob. Eng.* **2014**, *228*, 370–380. [CrossRef]
22. Esteves, L.P.d.S. Airbag e a sua concepção: Redução de custos em sacos airbag condutor: Estudo de novos conceitos de sacos. Master's Thesis, Facultade de Engenharia da Universidade do Porto, Porto, Portugal, 2008.
23. Dieter, G.E.; Schmidt, L.C. *Engineering Design*, 4th ed.; McGraw-Hill: New York, NY, USA, 2009.
24. Chowdhary, U.; Poynor, D. Impact of stitch density on seam strength, seam elongation, and seam efficiency. *Int. J. Consum. Stud.* **2006**, *30*, 561–568. [CrossRef]
25. Ünal, B.Z. The prediction of seam strength of denim fabrics with mathematical equations. *J. Text. Inst.* **2012**, *103*, 744–751. [CrossRef]
26. *ISO13935-1:2014*; Textiles-Seam Tensile Properties of Fabrics and Madeup Textile Articles. Part 1: Determination of Maximum Force to Seam Rupture Using the Strip Method. ISO: Geneva, Switzerland, 2014.
27. Schwarz, I.G.; Kovačević, S.; Kos, I. Physical–mechanical properties of automotive textile materials. *J. Ind. Text.* **2014**, *45*, 323–337. [CrossRef]

eng

MDPI

Article

Robot Cooking—Transferring Observations into a Planning Language: An Automated Approach in the Field of Cooking

Markus Schmitz *, Florian Menz, Ruben Grunau, Nils Mandischer, Mathias Hüsing and Burkhard Corves

Institute of Mechanism Theory, Machine Dynamics and Robotics, RWTH Aachen University, 52062 Aachen, Germany; mandischer@igmr.rwth-aachen.de (N.M.); huesing@igmr.rwth-aachen.de (M.H.); corves@igmr.rwth-aachen.de (B.C.)

* Correspondence: schmitzm@igmr.rwth-aachen.de

Abstract: The recognition of human activities from video sequences and their transformation into a machine-readable form is a challenging task, which is the subject of many studies. The goal of this project is to develop an automated method for analyzing, identifying and processing motion capture data into a planning language. This is performed in a cooking scenario by recording the pose of the acting hand. First, predefined side actions are detected in the dataset using classification. The remaining frames are then clustered into main actions. Using this information, the known initial positions and virtual object tracking, a machine-readable planning domain definition language (PDDL) is generated.

Keywords: human activity recognition; learning from demonstration; PDDL; robot cooking; task planning; motion tracking

Citation: Schmitz, M.; Menz, F.; Grunau, R.; Mandischer, N.; Hüsing, M.; Corves, B. Robot Cooking—Transferring Observations into a Planning Language: An Automated Approach in the Field of Cooking. *Eng* **2023**, *4*, 2514–2524. https://doi.org/10.3390/eng4040143

Academic Editor: Antonio Gil Bravo

Received: 3 August 2023
Revised: 21 September 2023
Accepted: 22 September 2023
Published: 7 October 2023

1. Introduction

The automation of cooking processes is currently still subject to some limitations. A vision of extracting recipe and process data using only tracked motion data, making this information available to the robot in machine-readable form, and thus enabling automated cooking by a robot or even in a human-robot collaboration, is presented in this paper. In a cooperation between human and machine, it is essential that the machine can correctly interpret and categorize the actions of the human. This is especially relevant in a situation where the human is supposed to teach the machine specific work processes. In passive observation, a method of learning from demonstration (LfD), the machine learns only by observing the human performing a task [1]. One important aspect of this method is to enable non-experts to teach robots or automated systems specific work processes. Traditionally, robot programming has been a complex and technical task, requiring specialized knowledge. However, with LfD approaches, even individuals without programming expertise can instruct machines effectively [2].

The increasing demand for automation in cooking processes is evident both in the rising sales of multipurpose kitchen appliances (e.g., Vorwerk's Thermomix [3]) and in the expected growth of the global market for smart kitchens [4]. Although multipurpose kitchen appliances offer some degree of automation, there is still significant room for improvement. Integrating robot arms into the cooking process could enhance the level of automation and streamline various tasks. While there have been some concepts and prototypes exploring the use of robot arms in cooking tasks [5,6], none have yet made it to the mainstream commercial market. This indicates that there are still challenges and complexities, such as the high price and the necessity of preprocessing ingredients, to overcome before such systems become widely available and accessible to the general public [7].

Several studies show that, in principle, robotic hardware is already capable of providing all the necessary functions for household and cooking processes [8,9]. The more challenging part is designing and enabling the grippers to grasp tools and selecting ingredients of the cooking process of varying texture, shape and solidity [10,11]. Another important aspect is safety when using robots in open workspaces in the household environment. Dangers arise particularly from collisions between humans and robots [12] and especially when tools such as knives are used in the context of the cooking scenario [13]. In addition to safety, ethical considerations such as acceptance must also be taken into account. From today's point of view, people would rather let a robot serve food or do the cleaning than let it cook food for them [14]. Cognition has an important role in the field of cooking processes and includes the recognition of cooking states, ingredients, and the composition of dishes obtained from motion data [15]. However, the use of sensing technology is expensive and the development of robust algorithms is challenging to realize the recognition of cooking states, ingredients, and the overall composition of cooked dishes, as well as the learning of subjective taste preferences [16].

Motion planning is a challenging task, since collision-free motion has to be achieved in a complex 3D environment, while also taking into account required contact with objects, including controlled intrusion (e.g., when cutting) [13,17]. In addition to motion planning, there must be efficient task planning that can divide the given recipe into individual tasks and orchestrate their execution [18,19]. Ideally, the tasks can be divided between different actors, such as multiple robots and/or humans, to prepare a tasty dish together [8].

In this context, the primary objective of this paper is to observe and analyze human activities in the domain of cooking processes. To achieve this goal, an automated method is developed to enable LfD through passive observation of human hand posture. By combining this information with the known objects and their positions, a machine-readable planning domain definition language (PDDL) is created. PDDL is a formal language, commonly used in the field of artificial intelligence and robotics, for representing and describing planning problems and actions [20]. It provides a structured and unambiguous way to model the actions, consisting of preconditions and effects, involved in a task, making it an ideal choice for translating human actions into machine-readable instructions. The PDDL representation will facilitate the machine's understanding and interpretation of human actions, thereby allowing it to effectively learn and perform cooking tasks and even collaborate with a human actor.

2. Concept

The challenge to be solved in cooking scenarios is that the actions and their effects and preconditions are largely unknown and vary greatly depending on the recipe. First, a cooking process is recorded. This is carried out using passive markers and a motion capture system. The question arises of how the recorded motion data can be analyzed. Motion data contains the position and orientation of the hand for each frame. These can naturally be supplemented with descriptive statistics, describing, for instance, positions, orientations, velocities, and accelerations within a time window. The basic idea and challenge for this research is exemplified in Figure 1: The progression from the recording of the cooking process to the execution of the resulting plan by robots.

To create this plan, the executed actions have to be identified from the motion data. A distinction is made between two different types of actions: "processing" and "moving/changing position". While there can be an infinite number and variety of actions related to processing, there are only a few actions (e.g., pick, move, place) that change an object's position. It is also essential to closely examine the effects and preconditions of each action type. The processing actions can have very different effects and variations, whereas the change in position is the central effect of the "moving/changing position" action. Conversely, it is necessary to uniquely identify position-changing effects as they contribute to causality and enable process interpretation based on motion data. In this way, the emerging plan can subsequently be decomposed into a PDDL (planning domain definition language).

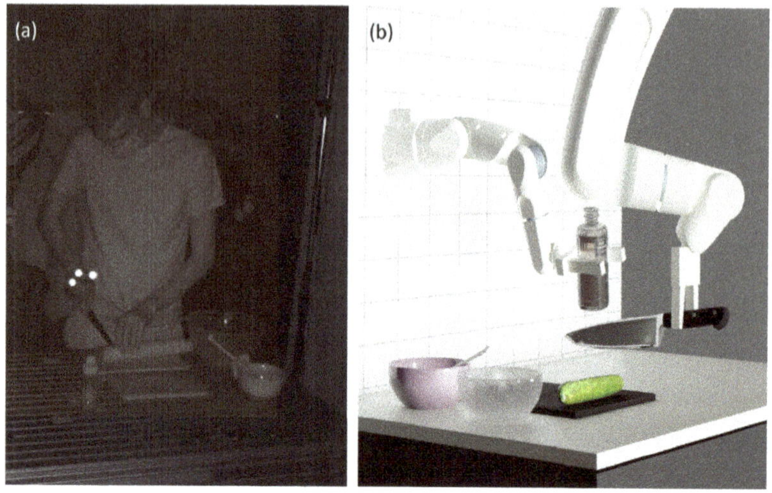

Figure 1. (**a**) The recording of the cooking process for this project; (**b**) a simplified conceptual rendering of the execution of the resulting plan.

These requirements align with the selection criteria of supervised and unsupervised learning. Classification is suitable when there is a finite number of classes. However, this requires training data that behave similarly for different individuals performing actions like pick, place, and move, for example. For the case of an unknown number of actions and a still unknown structure, clustering is a method of unsupervised learning to discover data patterns and correlations in the motion data.

After the motion data has been analyzed, it has to be transferred into a PDDL. In the context of our study, the machine-readable PDDL generated from the observed human activities and their combination with known positions of objects will serve as the bridge between the human demonstrator and the robot or automated system. The PDDL representation allows the machine to interpret and understand the sequence of actions performed by the human during the cooking process. By leveraging this formal representation, the machine can efficiently plan and execute the same task or collaborate with the human actor seamlessly. An overview of the complete process, including recording, classification, clustering, and transfer into PDDL, which are subject of the research in this paper, is provided by Figure 2. The goal of our approach is to provide a general solution, independent of the specific actors involved in the execution of the resulting plan. Therefore, the execution aspect is excluded from the scope of this paper.

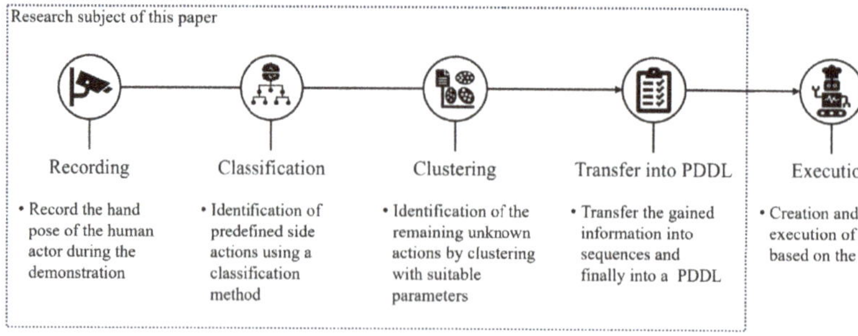

Figure 2. Schematic representation of the concept.

3. Recording

First, the cooking process is recorded by a motion capture system consisting of seven PrimeX13 cameras, sourced from OptiTrack (Corvallis, OR, USA). The schematic representation of the workspace is given in Figure 3. During the recording, the cook wears a glove with three passive markers placed on the back of the hand in a defined fixed position relative to each other. The position of each marker is determined by triangulation. These three points provide enough information to derive a pose. Specifically, the back of the hand is considered as a rigid body with 6 degrees of freedom. The recording was captured at 120 frames per second.

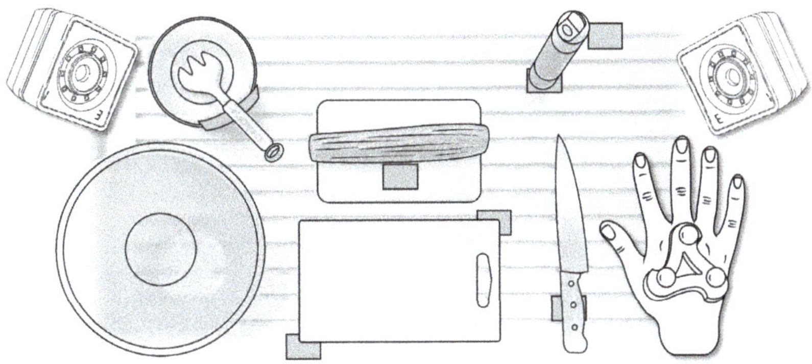

Figure 3. Schematic view of the workspace. Cameras (**top left** and **right**), hand with passive markers (**bottom right**) and used tools and ingredients.

In addition to the continuous recording of the trajectory, all objects in the workspace are identified once before the cooking process, as well as their initial positions. In the productive scenario, this is to be automated by an additional camera and image recognition. As previously mentioned, all recorded data are basically poses with time stamps. Other variables such as velocity, acceleration and angle in relation to the tabletop were derived from this. A look into this raw data already reveals some characteristics which are to be worked out as illustrated in Figure 4.

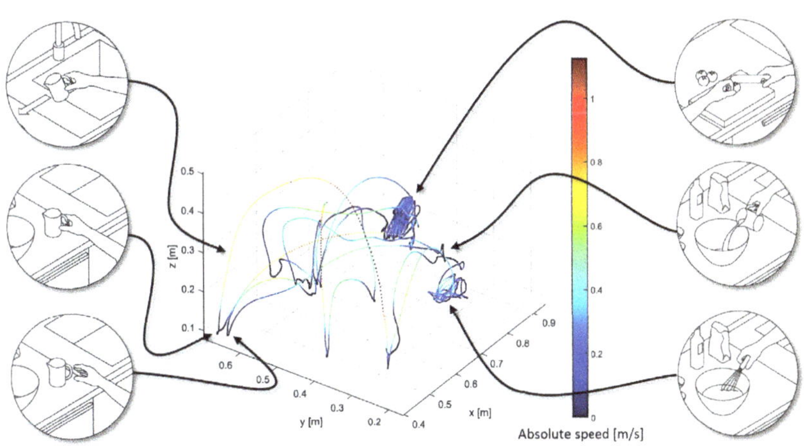

Figure 4. Illustration of the absolute speed within the space using Matlab R2021b.

4. Classification

When analyzing this cooking process, it is noticeable that some short actions occur repeatedly, which are called side actions in the following. These actions are part of the distinct actions "moving/changing position" that are important to track the positions of objects, as presented in the concept section. These are:

- Pick
- Move
- Place

Assuming that these actions are very consistent in their execution and occur in every cooking process, a classifier is trained to find these actions in the dataset. This ensures the generality of the procedure, since no main actions are anticipated, which are very individual. To train the classifier, a separate dataset is first recorded. In the training dataset, an action is assigned manually to each frame, with a numerical value assigned to it. There is a wide range of classification methods, from simple decision trees to weighted decision forests and neural networks. For our purposes, the RUS Boosted Trees method was chosen based on some preliminary research. RUS Boosted Trees is an algorithm that combines random under-sampling (RUS) with Boosted Trees to address the issue of imbalanced datasets. By applying RUS, the algorithm removes samples from majority classes randomly, thereby achieving a balanced class distribution. This approach improves the training process, mitigating runtime and storage problems [21].

A large number of parameters are available for the classification. Since these are partially dependent on each other and under the assumption that more parameters are not equivalent to a better classification, a parameter analysis is performed. With the classification method and parameters now selected, the classifier is trained and applied to the original dataset. Assuming that an action takes significantly longer than a single frame (recording at 120 frames per second), the classification results are smoothed. This is achieved by performing a check for each frame using a sliding window. Within this window, the number of times each action occurs is counted. Then the most frequent action within the window is selected for this frame. Classification provides a basic structure to the dataset, allowing the analysis of the remaining actions to continue with reduced complexity.

5. Clustering

The remaining frames are then to be categorized. This is the aim of the clustering. In this step, only those frames are considered which have not already been assigned by the classification, as shown in Figure 5.

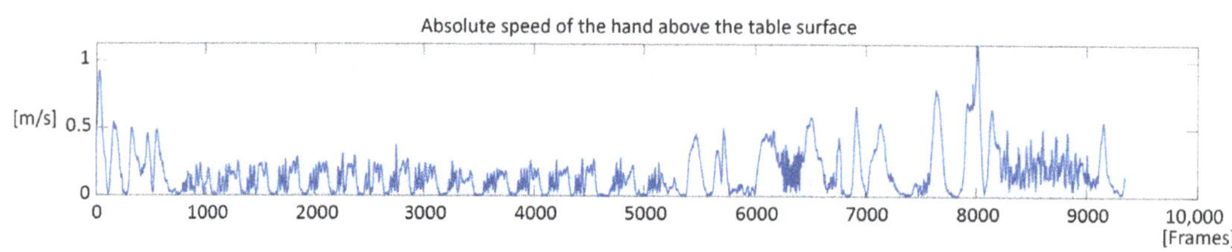

Figure 5. Graph of the absolute speed of the hand above the table surface within the cooking process. The frames with a white background have already been identified by the classification.

In contrast to classification, there is no search for a defined number of predefined classes. This approach makes it possible to analyze the cooking process despite a high degree of variation in the execution. Without this dynamic approach, there would always be limitations—a classification with its training dataset can only in a limited way represent the diversity of a cooking process in a non-standardized environment. The downside is that, in contrast to classification, the results are only clusters, without any description. A stirring action, for example, is only recognized as "cluster2". The approach to this problem will be considered later on.

The selection of suitable parameters for clustering includes the absolute speed of the hand above the table surface and the angular changes of the hand. In contrast to the absolute speed in space, the values of the absolute speed above the table surface reveal significant variations between actions, for example, low velocities during cutting, medium velocities during stirring, and high velocities during tilting. With respect to angle, the variation in hand rotation is considered, not the absolute values. Here, the angle changes show low values for cutting, larger values for tilting, and very high values for stirring. Before the actual clustering analysis, the dataset goes through pr-processing where the standard deviation of the angle is calculated for each frame within a defined window width and then smoothed. Using both parameters simultaneously allows for clearer discrimination of values and leads to improved clustering results (see Figure 6) compared to using a single parameter. The characteristics of other parameters are less significant and are not taken into account in the analysis.

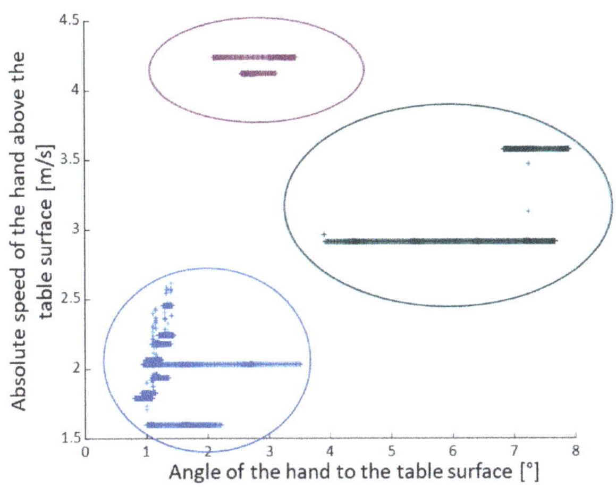

Figure 6. Three found clusters identified by analyzing the absolute speed of the hand above the table surface and the angle of the hand to the table surface. Tilt is marked as purple, stir is green and slice is blue.

In order to obtain a clustering result that is as accurate as possible, the different clustering approaches are evaluated in terms of their quality. For this purpose, the training dataset is used, where the actions have already been assigned to the frames manually. An intuitive approach could be to determine the percentage of frames correctly assigned by clustering. However, this would lead to an overvaluing of long-lasting actions. In addition, it should also be evaluated how accurately all existing individual actions were recognized. Clustering alone does not provide any information on which actions are covered by a specific cluster. Therefore, each individual cluster had to be evaluated with respect to each individual action. The combination of the two evaluation approaches presented below turned out to be the most suitable solution to establish these associations and to evaluate how accurately the clusters represent different actions:

- Determining the percentage of frames in the cluster that belong to the specific action. To do this, the number of frames associated with this action within the cluster is divided by the total number of frames in the cluster.
  For example: "90 percent of the cluster belongs to the slice action".

$$score(cluster, action) = \frac{nFrames(cluster, action)}{nFrames(cluster)} \quad (1)$$

However, this alone is not meaningful because a high score does not necessarily mean that the majority of the action was detected. Therefore, a second approach was used.
- Determining which percentage of a specific action is covered by this cluster. To do this, the number of frames associated with that action within the cluster is divided by the total number of frames associated with that action within the entire recording. For example: "70 percent of the slice frames are present in this cluster".

$$score(cluster, action) = \frac{nFrames(cluster, action)}{nFrames(recording, action)} \quad (2)$$

Combining these two approaches by multiplying Formulas (1) and (2), a result is obtained that answers the following questions: How consistent and complete is the cluster regarding a specific action?

$$score(cluster, action) = \frac{nFrames(cluster, action)^2}{nFrames(recording, action) * nFrames(cluster)} \quad (3)$$

An algorithm then uses these ratings to determine the most reasonable combinations of cluster and action. The corresponding scores are combined into an arithmetic mean, which is the overall score of the clustering. This automated evaluation makes it possible to quantitatively compare the approaches found. The parameters thus found are then used to apply clustering to the main dataset. In our study, a unique fingerprint is found for each action, consisting of the orientation of the back of the hand and its absolute speed of the hand above the table surface. The raw data of these two parameters are processed in such a way that the characteristics lead to the formation of the required clusters.

6. Transfer into PDDL

The motion data and the combined classification and clustering results, as well as the locations, are used to create sequences that summarize specific sections of the frames. The sequences, which can be assigned to actions known from the classification, are directly named accordingly. Location information is added for each sequence—both the location at the beginning and at the end of the action. Numeric values are transferred directly to the known locations, assuming the location with the least distance. Objects are divided into different subtypes. For example, a container like a bowl can contain ingredients. Objects such as knives and spoons are assigned to the tools subtype, while cuke and dressing are considered ingredients. The positions and states of the objects at the beginning of the scenario are known.

The schedule, which consists of the individual sequences, is now analyzed. For each action, it is recorded which object is at the active position. A "pick" action causes this object to be in the hand afterwards. Only one object can be in the hand, but, if this is a container, it can contain another object. For example, the dressing can be in the bottle. A "move" action with an object in hand changes the status of that object to be "on hand". A "place" action sets the new position of the object. Thus, virtual object tracking is implemented. If an unknown action is performed, all available information is used to describe the action. The objects at the active position, the object in the hand, and the active position itself are used. For example, it could be:

Cuke is processed by knife on cuttingboard.

The analyzed schedule is transferred to PDDL, where the domain name and requirements are defined statically. Then the types are defined following the structure in Figure 7, representing the different objects, actors, locations and actions. The subtypes of objects mentioned before are defined as movables. Furthermore, under the type hand, the possible actors, robot and human, are defined. The specific locations are listed in the subtype location. All unknown actions detected in the analysis of the schedule are summarized under treatment.

Then, the predicates are statically defined to describe the relationships and states of the objects and actors. Five predicates are used:

- *on* describes the location of a moveable or hand
- *in* describes that an ingredient is in a container
- *is* describes the treatment of an ingredient
- *holding* describes which movable a hand is holding
- *hand-empty* describes that the hand of an actor is empty (or not)

These predicates are already used in object tracking. The information is prepared during object tracking in such a way that the predicates can be read out after each sequence. This allows the start and end situations to be read directly from the predicates in the PDDL. Furthermore, this allows the observation of changes in the predicates in the case of unknown actions. The known standard actions are also defined statically.

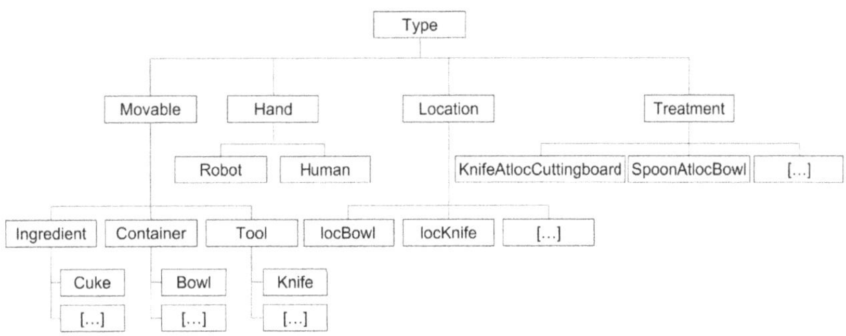

Figure 7. The structure of the types in the PDDL.

Unknown actions are now generated. Here a loop runs through all sequences of the analyzed schedule, whereby the distinction between *ToolOnHand*, *IngredientOnHand* and *ContainerOnHand* provides the general structure. This allows unknown actions to be captured and described. The problem and the associated domain are defined statically to define the specific task. The objects are named concretely and both the objects and the involved actors (hands) are specified.

The initial state is defined, with the hands empty and the objects at their initial locations. The target state is also defined to capture the intended result of the actions and, thus, describe the goal of the task. Some examples of the resulting PDDL and the plan are given in Figure 8.

(a)
```
(:action pick
    :parameters (?hand - hand ?mov - moveable ?loc - location )
    :precondition (and
        (on ?hand ?loc)
        (on ?mov ?loc)
        (hand-empty ?hand)
    )
    :effect (and
        (not (on ?mov ?loc))
        (holding ?hand ?mov)
        (not (hand-empty ?hand))
    )
)
```

(b)
```
(:action KnifeAtlocCuttingboard
    :parameters (?hand - hand ?locCuttingboard - locCuttingboard
                 ?cuttingboard - cuttingboard ?knife - knife
                 ?ingredient1 - ingredient ?ingredient2 - ingredient
                 ?ingredient3 - ingredient ?KnifeAtlocCuttingboard -
                 KnifeAtlocCuttingboard )
    :precondition (and
        (on ?hand ?locCuttingboard)
        (on ?cuttingboard ?locCuttingboard)
        (holding ?hand ?knife)
        (on ?ingredient1 ?locCuttingboard)
        (on ?ingredient2 ?locCuttingboard)
        (on ?ingredient3 ?locCuttingboard)
    )
    :effect (and
        (not(on ?ingredient1 ?locCuttingboard))
        (in ?ingredient1 ?cuttingboard)
        (not(on ?ingredient2 ?locCuttingboard))
        (in ?ingredient2 ?cuttingboard)
        (not(on ?ingredient3 ?locCuttingboard))
        (in ?ingredient3 ?cuttingboard)
        (is ?ingredient1 ?KnifeAtlocCuttingboard)
        (is ?ingredient2 ?KnifeAtlocCuttingboard)
        (is ?ingredient3 ?KnifeAtlocCuttingboard)
    )
)
```

(c)
```
(:goal
    (and
        (on cuttingboard1 locKnife1)
        (on bowl1 locBowl1)
        (on bottle1 locBottle1)
        (on knife1 locKnife1)
        (on spoon1 locSpoon1)
        (in cuke1 bowl1)
        (in dressing1 bowl1)
        (is cuke1 KnifeAtlocCuttingboard1)
        (is cuke1 CuttingboardAtlocBowl1)
        (is cuke1 SpoonAtlocBowl1)
        (is dressing1 SpoonAtlocBowl1)
        (is dressing1 BottleAtlocBowl1)
    )
)
```

(d)
```
001: (move robot2 lochuman1 loccuttingboard1)
002: (move robot1 locrobot1 locbottle1)
003: (pick robot1 bottle1 locbottle1)
004: (move robot1 locbottle1 locbowl1)
005: (bottleatlocbowl robot1 locbowl1 bowl1 bottle1 dressing1
         dressing1 dressing1 bottleatlocbowl1)
006: (move robot1 locbowl1 locbottle1)
007: (drop robot1 bottle1 locbottle1)
008: (move robot1 locbottle1 locbowl1)
009: (move robot2 loccuttingboard1 loccuke1)
010: (pick robot2 cuke1 loccuke1)
011: (move robot2 loccuke1 loccuttingboard1)
012: (drop robot2 cuke1 loccuttingboard1)
013: (move robot2 loccuttingboard1 locknife1)
014: (pick robot2 knife1 locknife1)
015: (move robot2 locknife1 loccuttingboard1)
016: (knifeatloccuttingboard robot2 loccuttingboard1
         cuttingboard1 knife1 cuke1 cuke1 cuke1
         knifeatloccuttingboard1)
017: (move robot2 loccuttingboard1 locknife1)
018: (drop robot2 knife1 locknife1)
```

Figure 8. Snippets of the resulting PDDL: (**a**) The side task pick; (**b**) the unknown task cut; (**c**) The described goal state; (**d**) Extract from the final plan.

7. Conclusions

The presented approach stands out in the landscape of LfD methods due to its extensive flexibility in converting human-guided demonstrations into executable task plans using a PDDL, making it suitable as a versatile solution that can be used for various applications.

After the three side tasks are classified once, the presented approach enables one-shot robot teaching. In this way, our approach addresses a common limitation of many existing LfD methods, which is their struggle to adapt to new application areas. In addition, the transfer to a PDDL offers great agility in the creation of task schedules depending on the robotic or human actors involved.

While we highlight the potential of our approach, it is worth mentioning that this research is likely to face some challenges and limitations. Variability in human actions, changes in kitchen layouts, and handling unforeseen situations are some of the issues that require further research and development. These challenges are not unique to our approach but represent inherent difficulties in the automation of complex tasks.

As the focus of this paper is limited to the creation of task planning based on LfD, its execution requires overcoming corresponding challenges, such as mechanical implementation, robot programming, and trajectory planning. Solving these challenges requires future research.

Moreover, as household robotics advance, we must consider the ethical implications of increased automation and collaboration with machines in daily life. Ethical considerations should include privacy, safety, and the impact on human employment and society. Addressing these concerns, including transparency and accountability, should be integrated into the development and deployment of such technologies.

In conclusion, the proposed method of using passive observation to teach machines through LfD in the context of the cooking process shows great promise in enhancing automation and making it accessible to non-experts. As technology advances and more research is conducted in this area, we can expect significant strides towards achieving seamless human–machine cooperation in various real-world applications.

Author Contributions: Conceptualization, M.S., F.M., R.G. and N.M.; methodology, M.S., F.M. and R.G.; software , M.S., F.M. and R.G.; validation, M.S., F.M. and R.G.; formal analysis, M.S., F.M. and R.G.; investigation, M.S., F.M. and R.G.; resources, M.S., F.M. and R.G.; data curation, M.S., F.M. and

R.G.; writing—original draft preparation, M.S., F.M. and R.G.; writing—review and editing, M.H. and B.C.; visualization, M.S., F.M. and R.G.; supervision, M.H. and B.C.; project administration, M.S. All authors have read and agreed to the published version of the manuscript.

Funding: This research received no external funding.

Institutional Review Board Statement: Not applicable.

Informed Consent Statement: Not applicable.

Data Availability Statement: Not applicable.

Conflicts of Interest: The authors declare no conflict of interest.

Abbreviations

The following abbreviations are used in this manuscript:

PDDL	Planning domain definition language
LfD	Learning from demonstration
RUS	Random under-sampling

References

1. Ravichandar, H.; Polydoros, A.; Chernova, S.; Billard, A., Recent Advances in Robot Learning from Demonstration. *Annu. Rev. Control Robot. Auton. Syst* **2020**, *3*, 297–330. [CrossRef]
2. Wake, N.; Arakawa, R.; Yanokura, I.; Kiyokawa, T.; Sasabuchi, K.; Takamatsu, J.; Ikeuchi, K. A Learning-from-Observation Framework: One-Shot Robot Teaching for Grasp-Manipulation-Release Household Operations. In Proceedings of the 2021 IEEE/SICE International Symposium on System Integration (SII), Iwaki, Fukushima, Japan, 11–14 January 2021; pp. 461–466.
3. Infografik: Thermomix-Geschäft Brummt. Available online: https://de.statista.com/infografik/9513/umsatz-von-vorwerk-im-geschaeftsbereich-thermomix/ (accessed on 25 July 2023).
4. Smart Kitchen Market Revenue Worldwide from 2017 to 2027. Available online: https://www.statista.com/statistics/1015395/worldwide-smart-kitchen-market-revenue/ (accessed on 1 September 2023).
5. Samsung Bot Chef. Available online: https://news.samsung.com/global/tag/samsung-bot-chef (accessed on 25 July 2023).
6. Moley Robotics. Available online: https://www.moley.com/ (accessed on 25 July 2023).
7. Jiang, A.Z.; Zhou M. Design of Affordable Self-learning Home Cooking Robots. In Proceedings of the 2022 IEEE International Conference on Networking, Sensing and Control (ICNSC), Shanghai, China, 15–18 December 2022; pp. 1–6.
8. Ilic, S.; Hughes, J. Complexity and Similarity Metrics for Unsupervised Identification of Programming Methods for Robot Cooking Tasks. In *Human-Friendly Robotics 2022, Proceedings of the HFR: 15th International Workshop on Human-Friendly Robotics, Delft, The Netherlands, 22–23 September 2022*; Springer Proceedings in Advanced Robotics; Springer: Berlin/Heidelberg, Germany, 2022; Volume 26.
9. Bao, S.; Luo, B. Design of a fully automatic intelligent cooking robot. *J. Phys. Conf. Ser.* **2021**, *1986*, 012101.
10. Yang, Y.; Zhu, H.; Liu, J.; Li, Y.; Zhou, J.; Ren, T.; Ren, Y. An Untethered Soft Robotic Gripper with Adjustable Grasping Modes and Force Feedback. In Proceedings of the IEEE International Conference on Robotics and Biomimetics (ROBIO), Jinghong, China, 5–9 December 2022; pp. 1–6.
11. Lee, Y.C.; Lim, S.J.; Hwang, S.W.; Han, C.S. Development of the robot gripper for a Home Service Robot. In Proceedings of the ICCAS-SICE, Fukuoka, Japan, 18–21 August 2009; pp. 1551–1556.
12. Paez-Granados, D.; Billard, A. Crash test-based assessment of injury risks for adults and children when colliding with personal mobility devices and service robots. *Sci. Rep.* **2022**, *12*, 5285. [CrossRef] [PubMed]
13. Mitsioni, I.; Karayiannidis Y.; Kragic, D. Modelling and Learning Dynamics for Robotic Food-Cutting. In Proceedings of the IEEE 17th International Conference on Automation Science and Engineering (CASE), Lyon, France, 23–27 August 2021; pp. 1194–1200.
14. Ivanov, S.; Webster C. Restaurants and robots: public preferences for robot food and beverage services. *J. Tour. Future* **2022**, *9*, 229–239. [CrossRef]
15. Malmaud, J.; Huang, J.; Rathod, V.; Johnston, N.; Rabinovich, A.; Murphy, K. What's Cookin'? Interpreting Cooking Videos using Text, Speech and Vision. In Proceedings of the 2015 Conference of the North American Chapter of the Association for Computational Linguistics: Human Language Technologies, Denver, CO, USA, 31 May–5 June 2015; pp. 143–152.
16. Fedorov, F.S.; Yaqin, A.; Krasnikov, D.V.; Kondrashov, V.A.; Ovchinnikov, G.; Kostyukevich, Y.; Osipenko, S.; Nasibulin, A.G. Detecting cooking state of grilled chicken by electronic nose and computer vision techniques. *Food Chem.* **2021**, *345*, 128747. [CrossRef] [PubMed]
17. Miao, R.; Jia, Q.; Sun, F. Long-term robot manipulation task planning with scene graph and semantic knowledge. *Robot. Intell. Autom.* **2023**, *43*, 12–22. [CrossRef]
18. Bollini, M.; Tellex, S.; Thompson, T.; Roy, N.; Rus, D. Interpreting and Executing Recipes with a Cooking Robot. In *Experimental Robotics*; Springer Tracts in Advanced Robotics; Springer: Berlin/Heidelberg, Germany, 2013; Volume 88, pp. 481–495.

19. Jeon, J.; Jung, H.R.; Yumbla, F.; Luong, T.A.; Moon, H. Primitive Action Based Combined Task and Motion Planning for the Service Robot. *Front. Robot. AI* **2022**, *9*, 713470. [CrossRef] [PubMed]
20. Fox, M.; Long, D. PDDL2.1: An Extension to PDDL for Expressing Temporal Planning Domains. *J. Artif. Intell. Res.* **2003**, *20*, 61–124. [CrossRef]
21. Tanha, J.; Abdi, Y.; Samadi, N.; Razzaghi, N.; Asadpour, M. Boosting methods for multi-class imbalanced data classification: an experimental review. *J. Big Data* **2020**, *7*, 70. [CrossRef]

eng

Article

Comparative Study of Musical Timbral Variations: Crescendo and Vibrato Using FFT-Acoustic Descriptor

Yubiry Gonzalez * and Ronaldo C. Prati

Center of Mathematics, Computer Science and Cognition, Federal University of ABC, Av. Dos Estados, 5001, Santo André 09210-580, Brazil; ronaldo.prati@ufabc.edu.br
* Correspondence: yubiry.gonzalez.17@gmail.com

Abstract: A quantitative evaluation of the musical timbre and its variations is important for the analysis of audio recordings and computer-aided music composition. Using the FFT acoustic descriptors and their representation in an abstract timbral space, variations in a sample of monophonic sounds of chordophones (violin, cello) and aerophones (trumpet, transverse flute, and clarinet) sounds are analyzed. It is concluded that the FFT acoustic descriptors allow us to distinguish the timbral variations in the musical dynamics, including crescendo and vibrato. Furthermore, using the Random Forest algorithm, it is shown that the FFT-Acoustic provides a statistically significant classification to distinguish musical instruments, families of instruments, and dynamics. We observed an improvement in the FFT-Acoustic descriptors when classifying pitch compared to some timbral features of Librosa.

Keywords: FFT-acoustic descriptor; timbral variations; Random Forest algorithm; musical acoustics

Citation: Gonzalez, Y.; Prati, R.C. Comparative Study of Musical Timbral Variations: Crescendo and Vibrato Using FFT-Acoustic Descriptor. *Eng* **2023**, *4*, 2468–2482. https://doi.org/10.3390/eng4030140

Academic Editor: Antonio Gil Bravo

Received: 1 August 2023
Revised: 14 September 2023
Accepted: 15 September 2023
Published: 21 September 2023

1. Introduction

Various techniques are used in musical composition to add expressiveness to the performance; the most common being those generated by subtle variations in dynamics and pitch. In musical instruments, the intensity variations in the impulse (tension in the strings or air pressure) and/or variations in the frequency of the pulsation produce secondary waves of sounds that propagate through the musical instrument: in the boxes and resonance tubes of chordophones and aerophones, respectively. When sound propagates in the resonant cavities of musical instruments, reflection, diffraction, and interference phenomena take place, which generally produce secondary sound waves, which overlap the fundamental frequency of the natural vibration mode (characteristic of each musical sound). Therefore, there will be slight timbre variations between two musical instruments of the same type that were manufactured differently (between two violins or between two flutes, etc.). Such timbre variations are due to changes in the envelope of the wave that forms the musical sound.

The most common variation in dynamics in music is the crescendo or gradual increase in the intensity of the sound, that is, a transitional dynamic nuance [1,2]. From an acoustic point of view, a crescendo occurs in aerophones when the musician gradually increases the amount of air blown into the instrument, thereby increasing the amplitude of the sound waves that are produced. The intensity of the sound produced depends on the amount of air entering the instrument and the pressure exerted by the musician's lips and tongue. As the musician increases the intensity of the musical note, they can change the pressure exerted by the lips and tongue to maintain the desired tonal quality. Similarly, in chordophones, the crescendo is produced when the musician gradually increases the pressure exerted on the strings of the instrument, which increases the amplitude of the sound waves that are produced. The intensity of the produced sound will depend on the pressure exerted on the strings and the position and speed of the musician's hand on the fingerboard and frets. When a musician uses the crescendo technique on bowed string instruments, the musician gradually increases the pressure exerted by the bow on the strings, increasing the amplitude of the sound waves produced. The crescendo technique can also affect the tonal quality of the produced sound. As the player increases the intensity of the note, the musician can slightly change the position of their hand on the fingerboard to maintain the desired tonal quality.

In addition, in acoustic terms, vibrato [1,3] occurs when the player oscillates the frequency of the played note by a small amount compared to its fundamental frequency. The frequency of the musical note produced in aerophones depends on the length of the tube and the tension of the musician's lips. When the musician uses the vibrato technique, she or he modulates the tension of the lips and the speed of the blowing air, which alters the frequency of the note. Vibrato on aerophones produces a series of additional formants and harmonics that overlap the fundamental note. These secondary transverse waves can be stronger or weaker depending on the speed and amplitude of the vibrato, and can contribute to the tonal quality and harmonic richness of the sound. In chordophones, the sound frequency depends on the length, tension, and mass of the strings, as well as the way they are played. When the musician uses the vibrato technique, they slightly move their finger up and down the string in the fingerboards, which alters the effective length of the string and, therefore, the frequency of the note. Consequently, a series of additional formants and harmonics are produced that are superimposed on the fundamental note as a result of the interaction of the string pulse with the resonant box. These harmonics can be stronger or weaker depending on the speed and amplitude of the vibrato and can contribute to the tonal quality and harmonic richness of the produced sound. In addition, vibrato can also affect the intensity and duration of the note, since the movement of the finger on the string can influence how much energy is transmitted to the string and how it is released.

On the other hand, the main timbral characteristics of the digital audio records must be somehow inscribed within the FFT through the succession of the pairs of amplitude and frequencies that comprise the sinusoidal components and that enable the recording and subsequent reproduction of musical sound. The collection of amplitude and frequency pairs in the FFT represents the intensities and tonal components of the audio recordings. Consequently, the timbre characteristics of digitized musical audio, which allow for discrimination between musical sounds, octaves, instruments, and dynamics, must be contained in some way in the FFT [4,5]. Several representations of timbre descriptors can be computationally derived from statistical spectrum analysis (FFT). As many of them are derivatives or combinations of others and, in general, are correlated among themselves [6,7], we adopt the dimensionless acoustic descriptors proposed in [4,8] to describe the timbral variations in the playing techniques associated with the only existing magnitudes in the FFT: amplitudes (crescendo) and frequencies (vibrato).

The objective of this paper is to use acoustic descriptors to compare timbre variations in a sample of monophonic audio recordings, corresponding to the aerophones clarinet, transverse flute, and trumpet, as well as the chordophones violin and violoncello. We will describe the methodology in the next section. Then, the results and a brief discussion are shown in Section 3, covering the comparison of this family of instruments (Section 3.1) by musical dynamics (Section 3.2) according to timbre variations in amplitude or crescendo (Section 3.3) and timbre variations in frequencies or vibrato (Section 3.4). The accuracy of the FFT acoustic descriptors is then compared with other timbral coefficients of statistical features through the Random Forest machine learning algorithm (Section 4), and the conclusions are provided in the last section.

2. Databases and General Formalism

We used the Good-Sounds dataset [9], which contained monophonic recordings of single notes with different timbral characteristics (in mezzo-forte musical dynamics: *mf*, crescendo, and vibrato modes). Only the fourth-octave musical was used, C4, C#4, D4, D#4, E4, F4, F#4, G4, G#4, A4, A#4, and B4, in the musical scale of equal temperament, the most typical in western music. The selection of musical instruments corresponded to the aerophones clarinet, transverse flute, and trumpet, and the chordophones violin and violoncello. The Tynisol database [5,10] in dynamic pianissimo (*pp*) and fortissimo (*ff*) was also used as a comparison reference for these musical instruments. The dynamic messoforte (*mf*) corresponds to a sound intensity in the order of 10^{-5} W m^{-2}. The *ff* and *pp* dynamics are equivalent to average intensities of the order of 100 times higher and lower, respectively. Also, the Tynisol database is used in Section 4 for the Random Forest algorithm of automatic classifications and is compared with other timbral features.

For each audio record, we obtained the FFT spectrum normalized by the ratio of the greatest amplitude of each spectrum. Note that all the audio records are monophonic and have the same duration (5 s), so the complete FFT was performed with a constant window function (unit step). Noise in the spectrum was also reduced by considering only amplitudes greater than 10% of the maximum amplitude. Then, each monophonic audio record was digitized by FFT as a discrete, finite, and countable collection of pairs of numbers that represent the relative amplitudes and frequencies, in Hertz, of the spectral components and the fundamental frequency (f_0).

Digital audio records store a set of pairs of numbers that represent the frequencies and amplitudes of the FFT of the corresponding monophonic sound. Then, all the relevant timbre information must be contained in that register. Thus, the musical Timbre can be defined operationally as the set of amplitudes and frequencies that accompany the fundamental frequency in the FFT of the audio recordings.

To describe the timbre in each FFT spectrum, we used the fundamental frequency (f_0) and its amplitude (a_0), plus a set of six dimensionless magnitudes denominated timbral coefficients [4,5,8]: "Affinity" A, "Sharpness" S, "Mean Affinity" MA, "Mean Contrast" MC, "Harmonicity" H, and "Monotony" M. The A and S timbral coefficients provide a measure of the frequency and relative amplitude of the fundamental signal with respect to the FFT spectrum. The coefficient H is a measure of the quantity and quality of the harmonics present in a spectral distribution. The coefficient M describes the average increase–decrease in the spectrum envelope. The MA and MC coefficients provide a measure of the mean frequency and mean amplitude of the spectral distribution, respectively (see Table 1). Figure 1 shows the timbral coefficients as a function of musical sounds and frequencies for the instruments selected from the Goodsound database, fourth octave, and mezzo-forte.

Table 1. Timbral coefficients associated with the FFT of monophonic musical sounds.

Coefficient	Operational Definition	Description
(A) Affinity	$A \equiv \frac{1}{f_0} \frac{\sum_{i=1}^{N} a_i f_i}{\sum_{i=1}^{N} a_i}$	Relative measurement of the centroid with respect to the fundamental frequency
(S) Sharpness	$S \equiv \frac{a_0}{\sum_{i=1}^{N} a_i}$	Relative measure of the amplitude of the fundamental frequency
(MA) Medium affinity	$MA \equiv \frac{\sum_{i=1}^{N} \left\lvert f_i - \frac{\sum_{i=1}^{N} a_i f_i}{\sum_{i=1}^{N} a_i} \right\rvert}{N f_0}$	Average deviation of the partial frequencies from the average frequency
(MC) Medium Contrast	$MC \equiv \frac{\sum_{j=1}^{N} \lvert a_0 - a_j \rvert}{N}$	Mean deviation of the partial amplitudes from the amplitude of the fundamental frequency
(H) Harmonicity	$H \equiv \sum_{i=1}^{N} \left(\frac{f_j}{f_0} - \left[\frac{f_j}{f_0} \right] \right)$	Average value of the harmony of the partial frequencies
(M) Monotony	$M \equiv \frac{f_0}{N} \sum_{j=1}^{N} \left(\frac{a_{j+1} - a_j}{f_{j+1} - f_j} \right)$	Deviation from regularity in the distribution of amplitudes with respect to frequencies

See [8] for detail.

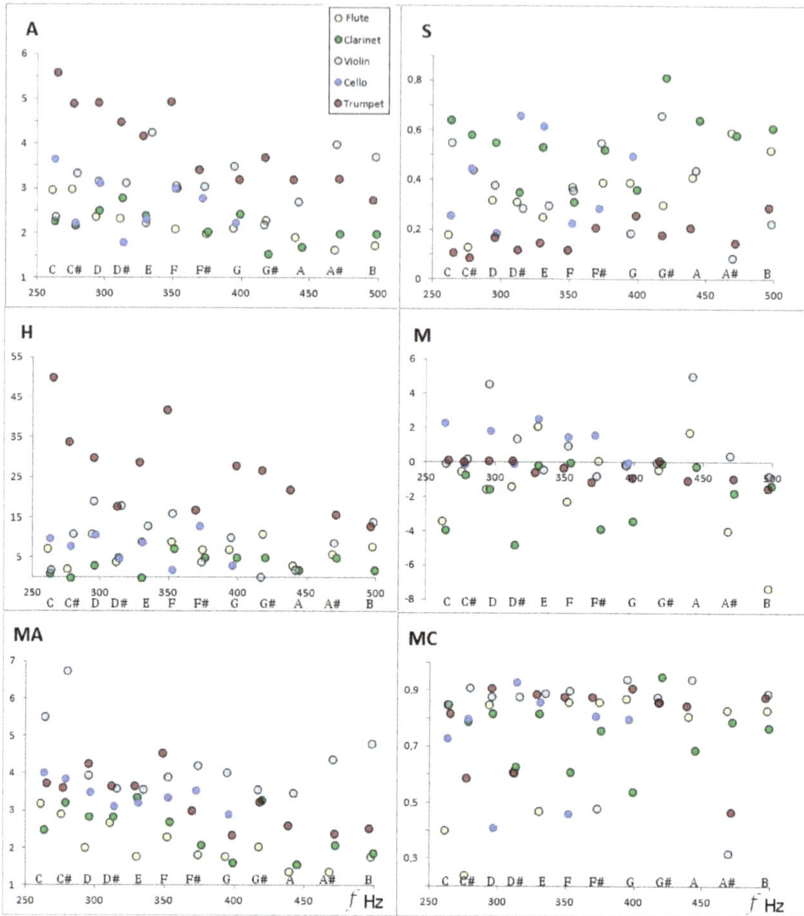

Figure 1. Representations in the frequency of the FFT–acoustic descriptors (timbral coefficients) for the Goodsound dataset.

Then, each FFT spectrum can be represented by a mean 7-tuple (f_0, A, S, H, M, MA, MC) in an abstract configurational space. These 7-tuples that characterize the amplitude-frequency distribution present in each FFT spectrum provide a morphism between the frequency space and the seven-dimensional vector space. This 7-space can be called timbral space, since the musical timbre consists precisely of the set of spatial frequencies (formants and harmonics) that accompany each musical sound produced by a certain musical instrument, a certain dynamic, and the set of techniques of the performing musician. Note that the 7-tuples are real numbers and admit the definition of a module or Euclidean norm along with equivalence relations; therefore, they formally constitute a Moduli space, represented by a geometrical place that parametrizes the family of related algebraic objects [11].

3. Euclidean Metric in Timbral Space

The timbral variations in the same musical sound due to the considered instrument (Section 3.1), the musical dynamics (Section 3.2), and the musical performance techniques used by the player, crescendo (Section 3.3) and vibrato (Section 3.4), are shown below through the Euclidean distance between the characteristic vectors of each FFT of the audio record, classified by musical sound (among the 12 possible in the fourth octave of the tempered scale).

All audio registers form a 7-tuple $(f_0, A, S, H, M, MA, MC)$ in abstract timbral space. So, any two audio records (subscripts i and j) can be timbrically compared via Euclidean distance as:

$$d_{ij} \equiv \sqrt{(f_{0i} - f_{0j})^2 + \Delta A_{ij}{}^2 + \Delta S_{ij}{}^2 + \Delta H_{ij}{}^2 + \Delta M_{ij}{}^2 + \Delta MA_{ij}{}^2 + \Delta MC_{ij}{}^2} \quad (1)$$

where Δ represents the algebraic difference between timbral coefficients.

Although this distance is mainly governed by the difference in frequency values, it should be noted that, for the purposes of comparing musical timbres, it is assumed that the audios being compared have the same tone, that is, they correspond to the same musical sounds. Therefore, in practice, the difference in the fundamental frequencies of the real audios (first sum of the radical in Equation (1)) is small, of the order of tens. In general, it is evident that the timbral similarity cannot be provided only by the distance; the orientación with respect to the axes of the 7-dimensional space is also required. Its spatial location in the abstract space of seven dimensions is important. However, for sufficiently small distance values (i.e., fundamental frequencies very close) between the position of two audios in Timbral space, timbral proximity criteria can be established.

3.1. Instruments

Figure 2 shows the standardized distances between monophonic audio recordings of instruments grouped by musical sounds. We observe that the registers are separated by notes, and the distance is a function of the tempered-scale sequence. The difference between the tables is due to the specific values of the timbral coefficients, as shown in Figure 1. Each musical sound corresponding to an instrument occupies a single point of timbre space.

The distances between different instruments, grouped by musical sounds, are illustrated with various examples in Figure 3. Note that for the same musical notes, the distances are smaller between instruments of the same type: flute and clarinet, both wooden aerophones (panel a). It is greater between aerophones and chordophones (panel b), between the chordophone and the wooden aerophone (panel c) and between the metal aerophone and the wooden aerophone (panel d).

On the other hand, the results show that some sounds seem close to each other, although they were from different musical instruments with different classifications, for example, the B4 sound. Figure 4 shows the FFTs for that sound. Notice the decrease in pulses, as well as the number and position of the partial frequencies. It cannot be affirmed that there is timbre similarity only because of the distance, since what defines the timbre is the vector and not only its module, and although the distance is equivalent between violin–trumpet and clarinet–trumpet, the sounds of these three instruments are in different regions of the timbre space (different clusters). To have timbral similarity, the sounds must be in the same cluster or region of the timbral space and must also be close to each other [4]. This is equivalent to saying that they must be from audio recordings of the same instrument or type of instrument, and also have a distance that is less than the distance between adjacent musical sounds.

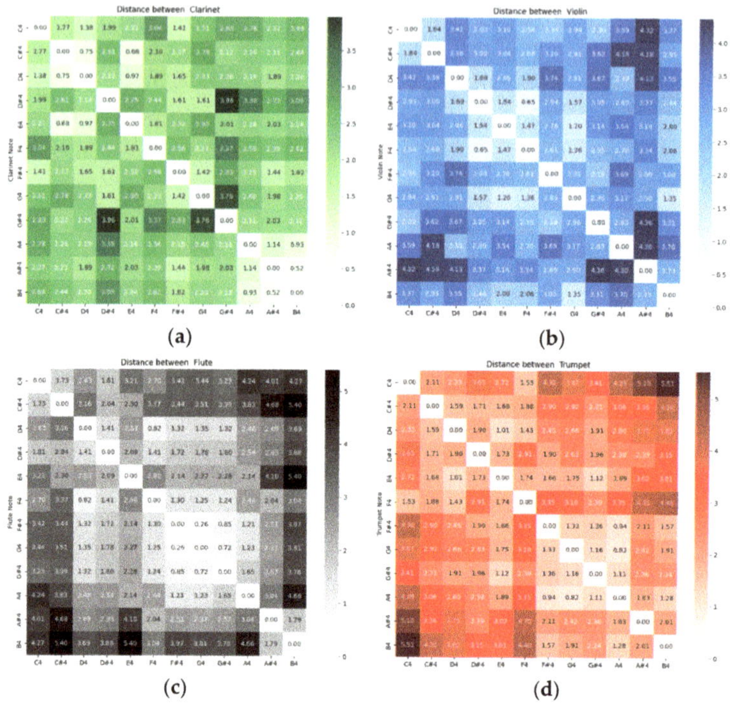

Figure 2. Matrix of Euclidean distances (statistically normalized by converting each value into its typical score) between musical sounds of the clusters that make up the proper subspace of each musical instrument: (**a**) Clarinet (**b**) Violin (**c**) Flute and (**d**) Trumpet.

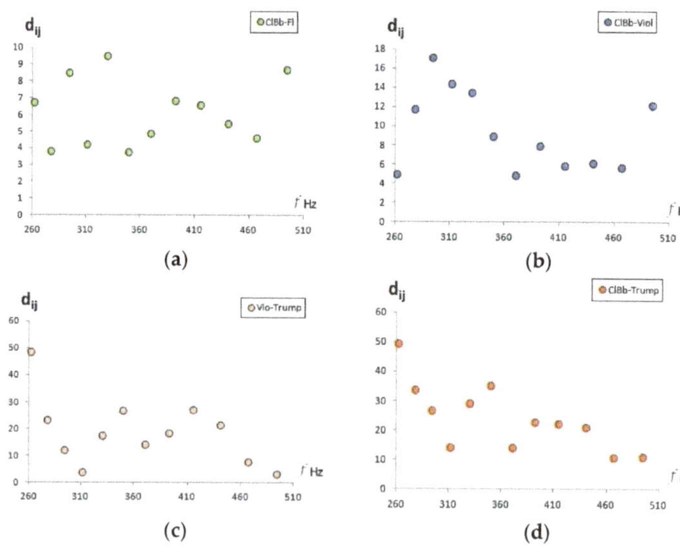

Figure 3. Comparison of Euclidean distances between musical sounds of different instruments: (**a**) clarinet–flute; (**b**) clarinet–violin; (**c**) violin–trumpet; (**d**) clarinet–trumpet.

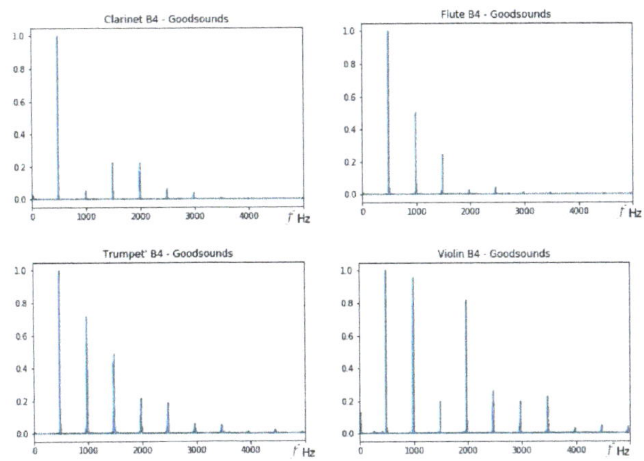

Figure 4. Fourier Transforms of the B4 Goodsound for different instruments. See the text for details.

3.2. Musical Dynamics

Given a musical sound and an instrument, the variations in the intensity of the performance (musical dynamics) should produce timbrally similar sounds, and consequently, their timbral representation should be close to the mezzo-forte sound. Indeed, that is what is observed in Figure 5 for the sounds in the Goodsound database compared to the Tinysol database records for different dynamics. Note that the minimum distances are always equal musical sounds and are less than 15.6, which is the minimum separation between two different musical notes of the tempered scale (between C4 and C#4), and therefore is also less than any other pair of sounds (in the fourth octave).

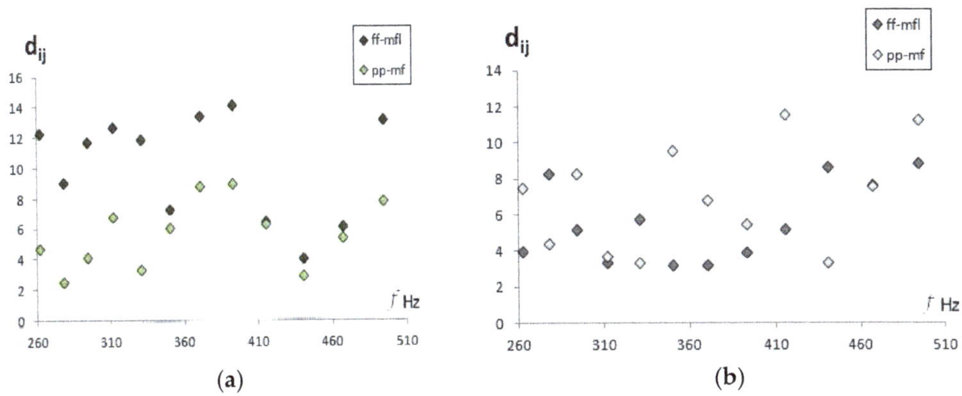

Figure 5. Euclidean distances between musical sounds from mezzo-forte Goodsounds and their cluster dynamics using the Tynisol dataset with the proper subspace of each musical instrument: (**a**) clarinet; (**b**) flute.

Timbral variations due to musical dynamics are shown by the increase in formants and harmonics in the FFT as we increase the intensity. Thus, the envelope of the FFT spectrum must be more extended, and the average value of the amplitudes changes. Hence, the acoustic descriptor of medium contrast, timbral coefficient MC, must vary in all musical sounds for the same instrument, as shown in Figure 6 for clarinet and flute.

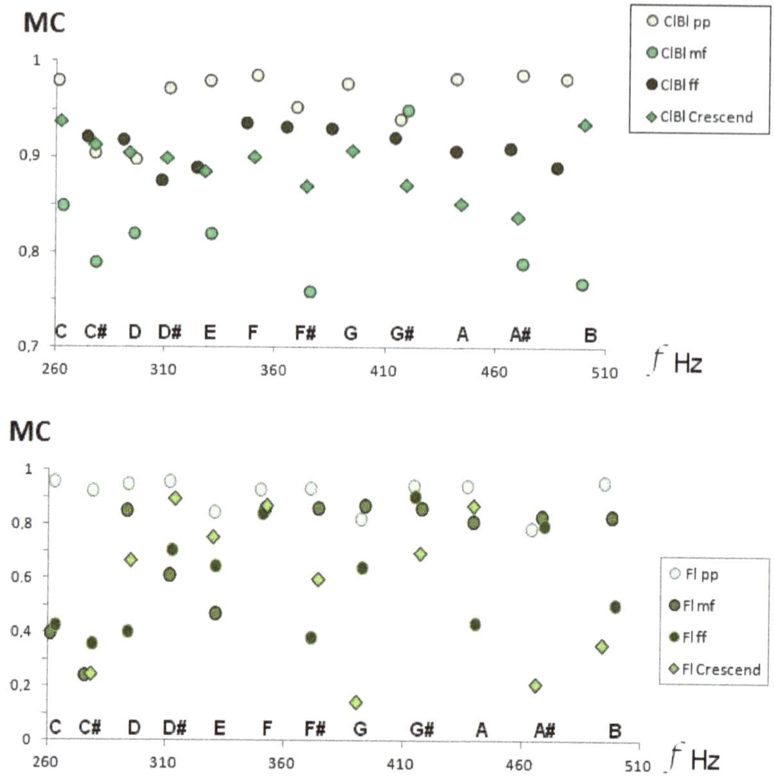

Figure 6. Variations in the MC timbre coefficient as a function of musical dynamics: clarinet, upper panel and flute, lower panel. Also, note the variation in M when performing the crescendo technique (Section 3.3).

3.3. Crescendo

The crescendo is an instrumental performance technique that consists of the gradual variation in the musical dynamics. Consequently, the timbre effect with respect to timbre in the mezzo-forte audio recordings should be similar. For the flute and the clarinet, we can see in Figure 6 how a decrease in the Mean Contrast (MC) occurs when we compare the dynamics of the pianissimo and mezzo-forte, also observing that the behavior of the crescendo effect decreases in the clarinet when we advance the frequencies. Figure 7 shows the same effect for the other instruments in the sample, so we can conjecture that, in general, the crescendo modifies the timbre coefficient of MC by incorporating more secondary frequencies in all instruments.

The right panel of Figure 7 shows the values of the timbre coefficient M in the crescendo technique with respect to mezzo-forte audio recordings for both aerophones and chordophones. We notice that the timbral variation in the crescendo reduces the monotony value, which is a timbre coefficient that quantifies the envelope in the FFT. A decrease in the absolute value of monotony implies that the envelope softens, that is, that the average value of the variations in amplitude with respect to the fundamental frequency decreases.

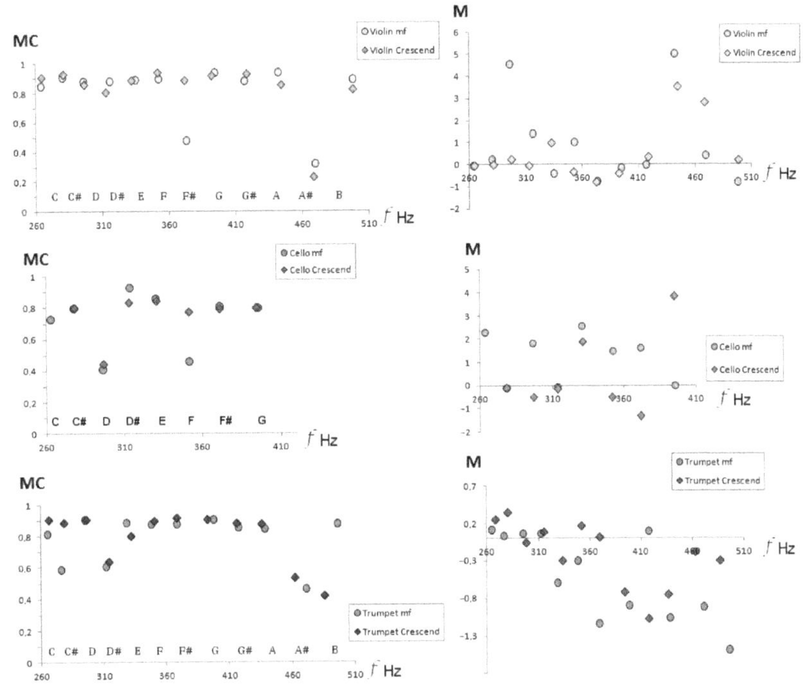

Figure 7. Medium contrast (**left panel**) and monotony (**right panel**) timbral coefficients, in the Goodsound database audios of violin (**top**), cello (**center**) and trumpet (**bottom**).

The audio recordings made with the crescendo technique must, similar to the dynamic musical variations, be close to the corresponding sounds in mezzo-forte. To illustrate this proximity, the Euclidean distances between each crescendo sound are shown in Figure 8. Note again that all distances are less than 15.6 (separation between C4 and C#4 sounds).

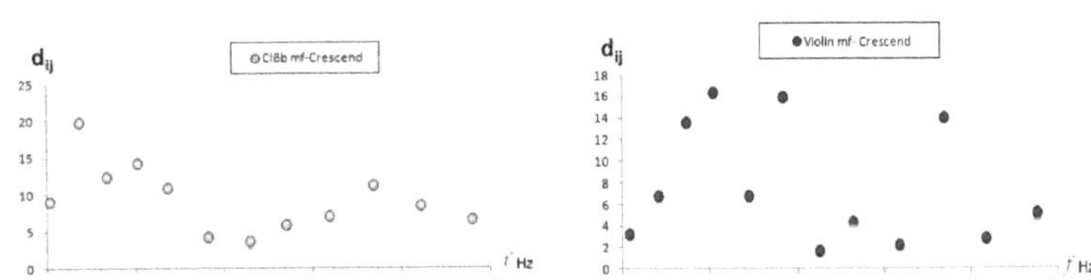

(**a**) (**b**)

Figure 8. Euclidean distances between the musical sounds of the crescendo and mezzo-forte Goodsound audio records: (**a**) clarinet; (**b**) violin.

The crescendo technique increases the average intensity of the sounds; this implies that the formants and harmonics increase in intensity and, therefore, the value of the timbral coefficient of Affinity (A), Mean Affinity (MA), and Harmonicity (H) increases with respect to the values in mezzo-forte dynamics, as observed in the FFT of the audio recordings of Figure 9 for the aerophones.

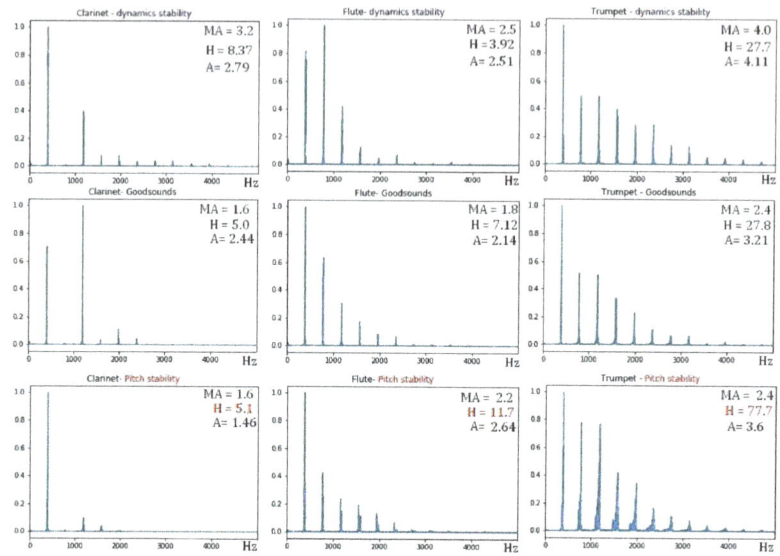

Figure 9. FFTs G4 sound of clarinet (**Left Column**), flute (**Central Column**) and trumpet (**Right Column**); normal register mezzo-forte (**middle row**), with crescendo technique (**upper row**) and vibrato (lower row). The values of the timbral coefficients of the Mean Affinity (MA), Harmonicity (H), and Affinity (A) are highlighted.

3.4. Vibrato

During vibrato, there is a slight variation in the fundamental frequency of the corresponding musical sound. Consequently, secondary frequencies that accompany the fundamental must appear; then, the Affinity (A) and Mean Affinity (MA) coefficients must change since they explicitly depend on the frequency values of the audio recording. Figure 9 compares the Mean Affinity values with the Goodsound mezzo-forte records. Although the change in the value of MA is uniform with respect to the musical sounds of the fourth octave, it is not the same for all instruments. Vibrato increases the MA value on the cello and decreases it on the clarinet and violin. Similarly, Figure 10 shows that vibrato also modifies monotony, as expected, because an increase in partial frequencies leads to a change in the envelope of the FFT spectrum.

The details of why some instruments increase the average of the partial frequencies (MA) and others decrease them are related to the geometry of the chordophone resonance box. The acoustics of chordophones are especially complicated because the wave generated by the vibration of the strings propagates in the air as a transversal wave, but in the soundbox, this pulsation originates transversal and longitudinal waves in the solid of the resonant cavity in addition to the transversal sound waves inside the air chamber. Therefore, it is beyond the objectives of this communication to elucidate this issue.

Also, since the variations in the frequency of the vibrato are less than the variation between adjacent musical notes in the tempered scale, it would be expected that the vibrato audio recordings would occur at relatively close distances to the Goodsounds mezzo-forte recordings. Figure 11 shows a clarinet that behaves in the described manner, but in the case of the violin, greater distances appear in some sounds. This could be due to an incorrect musical performance of the vibrato or due to the effects of the violin sound box. Unlike the cello, the violin is more diverse in its musical performance of vibrato, due to the addition of the bow to the tension placed on the string by hand and due to the influence of the jaw resting on the body of the violin, which can modify the vibration modes of the formants.

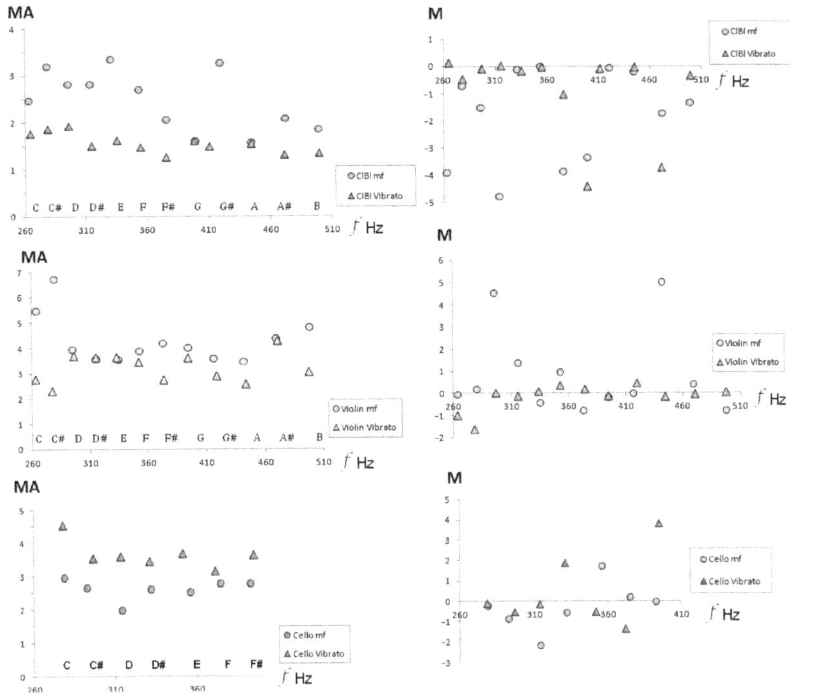

Figure 10. Medium Affinity (**left panel**) and monotony (**right panel**) timbral coefficients, in the Goodsound database audios of clarinet (**top**), violin (**center**), and cello (**bottom**).

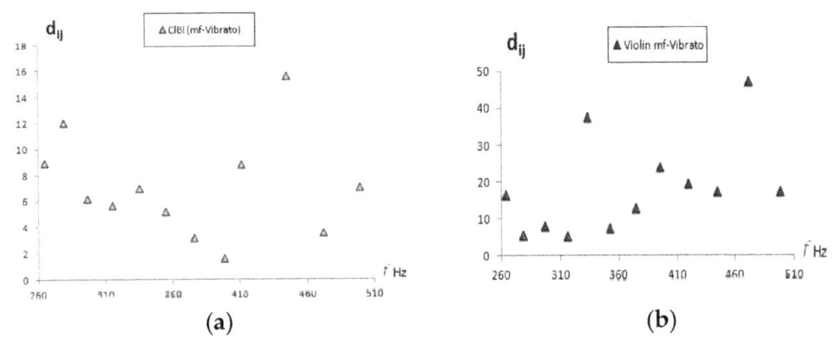

Figure 11. Euclidean distances between the musical sounds of the vibrato and mezzo-forte Goodsound audio recordings: (**a**) clarinet; (**b**) violin.

It is understandable that the oscillations in the main frequency in the vibrato increase the coefficient H since the partial frequencies that are generated will not be harmonic (greater H, less harmonicity), as can be seen in the lower panel of Figures 9 and 12. Figure 12 also shows that vibrato decreases the value of the Mean Affinity (MA) for chordophones.

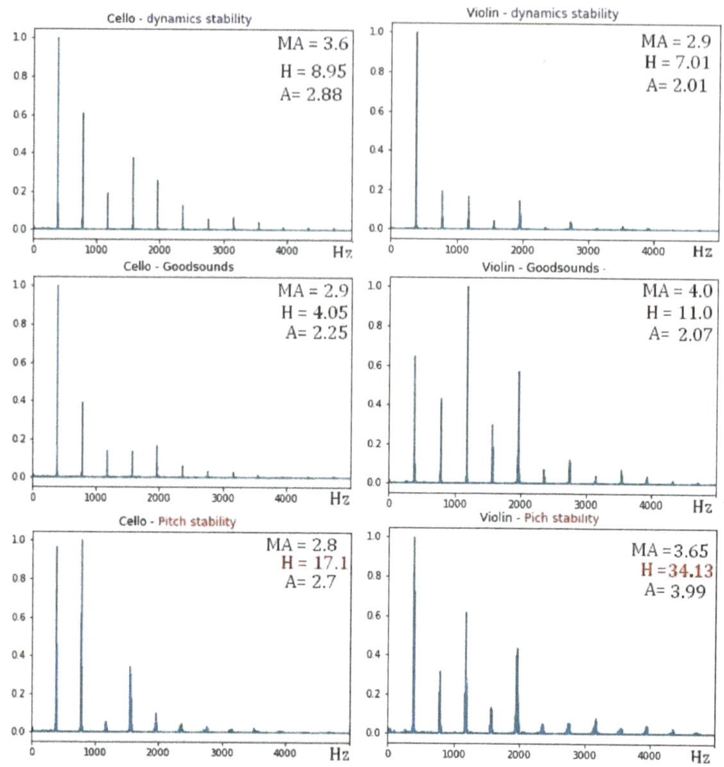

Figure 12. FFTs of G4 sound: cello (**left column**) and violin (**right column**); normal register mezzoforte (**middle row**), with crescendo technique (**upper row**) and vibrato (**lower row**). The values of the timbral coefficients of Mean Affinity (MA), Harmonicity (H), and Affinity (A) are highlighted.

In the upper panel of Figure 12, it is observed that H does not always increase in the chordophones. This may be due to the interaction with the resonance box of the instrument, since the vibrations of some harmonics can cause destructive interference with the generated formants due to the geometry of the musical instrument being considered. However, the change due to musical dynamics is evidenced by the increase in Mean Affinity, even for chordophones. The variation in the Affinity is not conclusive, since, in this technique, as in vibrato, the musical performer can, according to their discretion and personal taste, modify the fundamental frequency during the performance of the crescendo; if they do this, the information will not be recorded in the Goodsound datasets.

Vibrato not only causes variations in frequency, but also oscillates the timbre of the sound, that is, causes the greater or lesser prevalence of one component or another. This oscillation in the sound quality caused by the vibrato violin is a characteristic feature of this instrument. The acoustic explanation of this feature of the violin lies in the properties of its sound box, which responds differently to very close frequency components. Finally, all instruments allow for the performer to make their own vibrato, and this resource is a very important part of characterizing the sound.

We have seen that the timbral coefficients allow for characterization of the timbral variations; however, it is worth asking how these acoustically motivated descriptors compare with other descriptors of the FFT based on statistical distributions. This is discussed in the next section.

4. Automatic Classification of Musical Timbres

The problems of classification can be resolved using supervised learning. These classification algorithms have been used in music style recognition problems through music feature extraction [12], musical instrument classification problems [13], and the use of an intelligent system for piano timbre recognition [14], among other techniques. We are going to compare the classification capacities of the timbral coefficients proposed by González and Prati [8] with some timbral features extracted using Librosa: Chroma stft, spectral contrast, spectral flatness, poly features, spectral centroid, spectral rolloff, and spectral bandwidth [15].

For this, we use the TinySol database through the MIRDATA library [16], which offers a standardization to work with audio attributes more efficiently. After defining the meta-attributes, we explore timbral classification capabilities by considering certain variations, such as instruments (violin, cello, transverse flute, clarinet, and trumpet), dynamics (pianissimo, mezzo-forte, and fortissimo), musical notes (considering the entire range of each instrument) and instrument families (chordophones, wooden aerophones, and metal aerophones).

We evaluated some classification algorithms, such as Random Forest (RF), Support Vector Classifier (SVC), K-Nearest Neighbor (KNN), and logistic regression, and we observed better statistical behavior in terms of classification for our subject of study with the Random Forest algorithm; this behavior occurs in benchmark tests [17]. This is a conjoint learning method that combines multiple decision trees to create a more robust and accurate predictive model [18].

We used the data split provided in the MIRDATA library, which divides the data into five folds. We applied a 5-fold cross-validation, where, in each iteration, one fold is used for testing and the remaining folds are used for training. The process is repeated five times, using a different test split each time. Using the Random Forest algorithm, we computed the mean accuracy using the timbral coefficients and the LibRosa features. Table 2 presents the results.

Table 2. Comparative results of the Random Forest classification algorithm (mean accuracy ± Standard Deviation) for category recognition: musical instrument, musical dynamics, musical note, and musical instrument families.

	Instrument	Dynamics	Pitch	Family
Timbral Coefficients [8]	0.78 ± 0.02	0.63 ± 0.038	0.65 ± 0.046	0.92 ± 0.017
Timbral features (Librosa)	0.89 ± 0.029	0.97 ± 0.011	0.22 ± 0.014	0.91 ± 0.018
Test T (*p*-value)	0.0000209	0.0000136	0.000115	0.0185

To statistically compare the results, we use a paired T-test for each possible class. The last row of Table 1 shows the *p*-value of the test. Statistically significant differences were observed for the timbral coefficients when compared with Librosa in the classification by musical notes (pitch); this may be because the musical timbre, as an acoustic characteristic, is a frequency-independent property of the musical timbre. On the other hand, if we consider a significance interval of 99%, we can see that the timbral coefficients behave well when classifying instruments and families of instruments, and are better for the classification according to dynamics with respect to timbral features (Librosa).

5. Conclusions

Timbral variations in monophonic musical sounds can be characterized from an FFT analysis of audio recordings. More particularly, due to the techniques of musical performances of variations in amplitude (crescendo) and frequency (vibrato), these timbre variations differ between instruments according to their acoustic characteristics.

The acoustic FFT descriptors proposed by Gonzalez and Prati [4,8] provide a representation of the characteristic timbral space of each audio recording. Its position in the timbral space [4] and the Euclidean distance between the registers allow for us to distinguish the timbral variations due to the family of instruments, the musical dynamics, and the variations in the execution technique. The latter can modify the envelope of the FFT and consequently change the values of monotonicity (M) and harmonicity (H). The crescendo modifies the Mean Contrast (MC) coefficient and the vibrato modifies the Affinity (A).

The Random Forest technique applied to evaluate the accuracy of the proposed classification shows statistically significant results for the FFT-Acoustic descriptors and timbral features of Librosa when classifying instruments, dynamics, and families of instruments, observing a better classification by pitch in the FFT-Acoustic descriptors when comparing them with Librosa features. It is important to perceive that Librosa does not discriminate between the dynamic variations in crescendo and vibrato, while the FFT-Acoustic descriptors do allow for them to be discriminated.

Author Contributions: Conceptualization, Y.G. and R.C.P.; methodology, Y.G. and R.C.P.; software, Y.G. and R.C.P.; validation, Y.G. and R.C.P.; formal analysis, Y.G. and R.C.P.; investigation, Y.G. and R.C.P.; resources, Y.G. and R.C.P.; data curation, Y.G. and R.C.P.; writing—original draft preparation, Y.G.; writing—review and editing, Y.G. and R.C.P.; visualization, Y.G.; supervision, R.C.P.; project administration, Y.G. and R.C.P.; funding acquisition, Y.G. and R.C.P. All authors have read and agreed to the published version of the manuscript.

Funding: This study was financed in part by the Coordenação de Aperfeiçoamento de Pessoal de nível Superior—Brasil (CAPES)—Finance Code 001.

Institutional Review Board Statement: Not applicable.

Informed Consent Statement: Not applicable.

Data Availability Statement: The sounds used in this work are available at the following link: https://zenodo.org/record/3685367#.XnFp5i2h1IU%22, accessed on 15 May 2023.

Conflicts of Interest: The authors declare no conflict of interest.

References

1. Randel, D.M. *The Harvard Dictionary of Music*; Harvard University Press: Cambridge, MA, USA, 2003; p. 224.
2. Gough, C. Musical acoustics. In *Springer Handbook of Acoustics*; Springer: New York, NY, USA, 2014; pp. 567–701.
3. Almeida, A.; Schubert, E.; Wolfe, J. Timbre Vibrato Perception and Description. *Music Percept.* **2021**, *38*, 282–292. [CrossRef]
4. Gonzalez, Y.; Prati, R.C. Similarity of musical timbres using FFT-acoustic descriptor analysis and machine learning. *Eng* **2023**, *4*, 555–568. [CrossRef]
5. Gonzalez, Y.; Prati, R. Acoustic Analysis of Musical Timbre of Wooden Aerophones. *Rom. J. Acoust. Vib.* **2023**, *19*, 134–142.
6. McAdams, S. The perceptual representation of timbre. In *Timbre: Acoustics, Perception, and Cognition*; Springer: Cham, Switzerland, 2019; pp. 23–57.
7. Peeters, G.; Giordano, B.L.; Susini, P.; Misdariis, N.; McAdams, S. The timbre toolbox: Extracting audio descriptors from musical signals. *JASA J. Acoust. Soc. Am.* **2011**, *130*, 2902–2916. [CrossRef] [PubMed]
8. Gonzalez, Y.; Prati, R.C. Acoustic descriptors for characterization of musical timbre using the Fast Fourier Transform. *Electronics* **2022**, *11*, 1405. [CrossRef]
9. Romaní Picas, O.; Parra-Rodriguez, H.; Dabiri, D.; Tokuda, H.; Hariya, W.; Oishi, K.; Serra, X. A real-time system for measuring sound goodness in instrumental sounds. In Proceedings of the 138th Audio Engineering Society Convention, AES 2015, Warsaw, Poland, 7–10 May 2015; pp. 1106–1111.
10. Carmine, E.; Ghisi, D.; Lostanlen, V.; Lévy, F.; Fineberg, J.; Maresz, Y. TinySOL: An Audio Dataset of Isolated Musical Notes. Zenodo 2020. Available online: https://zenodo.org/record/3632193#.Y-QrSnbMLIU (accessed on 15 May 2022).
11. Kollár, J. Moduli of varieties of general type. In *Handbook of Moduli*; Vol. II, 131–157, Adv. Lect. Math.(ALM); Int. Press: Somerville, MA, USA, 2013; p. 25.
12. Zhang, K. Music style classification algorithm based on music feature extraction and deep neural network. *Wirel. Commun. Mob. Comput.* **2021**, *2021*, 9298654. [CrossRef]
13. Chakraborty, S.S.; Parekh, R. Improved musical instrument classification using cepstral coefficients and neural networks. In *Methodologies and Application Issues of Contemporary Computing Framework*; Springer: Singapore, 2018; pp. 123–138.
14. Lu, Y.; Chu, C. A Novel Piano Arrangement Timbre Intelligent Recognition System Using Multilabel Classification Technology and KNN Algorithm. *Comput. Intell. Neurosci.* **2022**, *2022*, 2205936. [CrossRef] [PubMed]

15. McFee, B.; Raffel, C.; Liang, D.; Ellis, D.P.; McVicar, M.; Battenberg, E.; Nieto, O. librosa: Audio and music signal analysis in python. In Proceedings of the 14th python in science conference 2015, Austin, TX, USA, 6–12 July 2015; pp. 18–25.
16. Bittner, R.M.; Fuentes, M.; Rubinstein, D.; Jansson, A.; Choi, K.; Kell, T. Mirdata: Software for Reproducible Usage of Datasets. In Proceedings of the 20th International Society for Music Information Retrieval (ISMIR) Conference, Delft, The Netherlands, 4–8 November 2019.
17. Fernández-Delgado, M.; Cernadas, E.; Barro, S.; Amorim, D. Do we need hundreds of classifiers to solve real world classification problems? *J. Mach. Learn. Res.* **2014**, *15*, 3133–3181.
18. Michalski, R.S.; Carbonell, J.G.; Mitchell, T.M. *Machine Learning: An Artificial Intelligence Approach*; Springer: Berlin/Heidelberg, Germany, 2013.

eng

MDPI

Article

Setup Time Reduction of an Automotive Parts Assembly Line Using Lean Tools and Quality Tools

Cátia Oliveira [1] and Tânia M. Lima [1,2,*]

[1] Department of Electromechanical Engineering, University of Beira Interior, Rua Marquês de D'Ávila e Bolama, 6201-001 Covilhã, Portugal; meneses.oliveira@ubi.pt

[2] C-MAST-Center for Mechanical and Aerospace Science and Technologies, Rua Marquês de D'Ávila e Bolama, 6201-001 Covilhã, Portugal

* Correspondence: tmlima@ubi.pt

Abstract: The business world is becoming more competitive. Therefore, it is crucial to increase the flexibility of production by decreasing the time used in the processes of preparing the production lines for new items' production, reducing changeover and setup times. This paper presents a case study where the main goal is to reduce the setup time of welding robots. Single Minute Exchange of Die (SMED) was implemented, using other tools such as the Spaghetti Diagram, ERCS Analysis (Eliminate, Rearrange, Combine, Simplify), Gemba Walk, Standardized Work, Flowcharts, and Pareto Diagram. The setup time decreased by 36% in the welding robots studied, decreasing the motions by 43% during the changeover process and reducing the time from the categories: "transportation", "main", "other", and "waiting". In addition to SMED implementation, this study offers an integrated study of several Lean tools and Quality tools to achieve the maximum reduction of changeover and setup times.

Keywords: Lean manufacturing; SMED; automotive industry; ERCS analysis; Spaghetti Diagram; Gemba Walk

Citation: Oliveira, C.; Lima, T.M. Setup Time Reduction of an Automotive Parts Assembly Line Using Lean Tools and Quality Tools. *Eng* **2023**, *4*, 2352–2362. https://doi.org/10.3390/eng4030134

Academic Editor: Antonio Gil Bravo

Received: 31 July 2023
Revised: 10 September 2023
Accepted: 11 September 2023
Published: 13 September 2023

1. Introduction

Competitiveness in the industry is increasingly intense, especially in the automotive sector. Companies need to reduce costs, optimize production, and need to achieve products with higher quality to become more competitive [1,2]. To reach these goals, companies adopt the Lean manufacturing philosophy, which focuses on eliminating waste [3].

Nowadays, customers value products with higher quality, lower cost, and higher variety. Therefore, processes must become more flexible and efficient to meet the demand of product diversification and smaller batch production. It is mandatory to differentiate and thrive in an increasingly competitive and saturated market [4].

Production time and the diversity of items have become a new critical factor for the profitability of today's companies [5]. When the item to be fabricated changes, it is necessary to prepare the production line for the new fabrication. It is important to perform adjustments to equipment, change tools, and prepare raw materials. These are activities that do not add value to the final product. However, they are extremely important for ensuring their proper manufacturing [4]. The faster the changeover process is, the greater capacity there will be to respond to market needs, and the more flexible a company will be. This means that the number of changeovers can increase and it is possible to offer more variety of products and batch sizes [5].

The case study presented in this paper was made in a factory in the automotive sector. The factory has three similar production lines A, B, and C, where two of them spent about 40 min doing the changeover process and the other one was around 90 min. The production line chosen was the production line B; it was the one with a higher changeover time. The main goal of this paper is to reduce the welding robot's setup time of the production line chosen.

This paper starts with a literature review of the Lean Manufacturing Philosophy and Quality Tools. This provides background knowledge to define the problem presented and to find a solution to solve it. Then, the case study and the methodology used to solve the problem are presented just as the results and the discussion. Finally, a conclusion is made, where all the limitations of this study as future proposals can be read.

The study in this paper was not just an implementation of SMED. All the tools used served to underpin SMED and make the results even smaller. Tools such as the Spaghetti Diagram, Gemba Walk, Flowcharts, ERCS analysis, and Pareto diagrams led to improved efficiency and a swifter implementation of the SMED methodology.

Lean Manufacturing Philosophy aims to maximize the value from the customers' point of view by offering items with higher quality. To achieve this goal, it is necessary to eliminate waste as much as possible, contributing to the increase of the processes' flexibility and efficiency [6,7]. Therefore, this philosophy has the objective of reducing costs (maintaining the products' high quality), eliminating waste, and enhancing customers' satisfaction [8].

The seven wastes of Lean are [7–9]:

- Overproduction: It occurs when the offer (quantity of products manufactured) is higher than the demand by customers. Overproduction leads to waste like production costs for goods that are not in demand, time, space used for storage, and transportation costs;
- Waiting: It is when an operation is stopped waiting for the conclusion of the previous ones. It also occurs when operators wait for a machine to finish its job, wait for orders, or wait for tools;
- Transportation: It happens when materials and tools are moved from one site to another, with no need. This type of activity does not add value to the final product and generates costs;
- Over-processing: This occurs when offering products comes with more characteristics than customers' requirements, and when there are more operations in a product fabrication than necessary;
- Motion: It occurs when some equipment or people are in motion without making operations. It includes motions, such as walking, looking for tools or information, and reaching and stacking parts or tools. There should be plans in action in every workplace to eliminate unnecessary movement;
- Inventory: It occurs when an excess of stock is not used for production, including raw materials or intermediate products. It can lead to longer delivery times, obsolescence of materials, transportation and storage costs, and damaged goods;
- Defects: It happens when products do not have the characteristics required by customers. These problems result in internal quality issues and cause wasted handling, time, and effort;

Below are presented all the Lean tools that seem to fit with the objective in the available time:

- SMED: It decreases the time used to prepare the production line and equipment to produce a new product, reducing setup and changeover times, and contributing to a quick and efficient change [6,10]. Setup time is the time of preparing machines or tools, and is also the time between the previous compliance item's part fabrication and the next compliance item part [11]. Changeover is all the activities of a production line preparation, and is the time between the previous compliance product and the next compliance product [6,11].

  The time spent performing the changeover is considered waste according to the Lean philosophy because changeover does not add value for the customer. Its elimination brings numerous advantages, such as stock reduction, increased production capacity, elimination of setup errors, improved quality, reduced production time, reduced production costs, and simplified use of tools [11].

- Gemba Walk: It occurs when someone goes to the shop floor to watch what is going on. It is essential to go to the local site where everything is made, watch and take notes about the process, and talk with people. By visiting the shop floor, you can find crucial information for eliminating problems such as cycle times, waiting times, stocks, and rework. It enables management to understand their employees' daily challenges, allowing leaders to have two different points of view: the management view and the operational view [12–14].
- Eliminate, Rearrange, Combine, and Simplify (ERCS) Analysis: It is used to analyze the processes and consists of eliminating all the non-value-added activities, rearranging the operations made, combining operations that can be conducted together, and simplifying all the tasks as much as possible to simplify processes [15,16]. The ERCS acronym is explained below [15,16]:
    - E: It is the elimination of all non-value-added activities;
    - C: It is the combination of two or more operations;
    - R: It is the reorganization of the processes sequence;
    - S: It is the simplification of operations, becoming easier to perform.
- Standardized Work: It is a set of working instructions and sequences of all the operations that establish a uniformization of all activities performed [17,18]. This documentation defines the optimal way of carrying out tasks and leads to increased quality levels, reduced variability, reduced injuries and strain, standardized takt-time, and it can also be a starting point for continuous process improvement activities [18,19]. The main objectives of Standardized Work are individual responsibility, experiential learning, and discipline in execution [17].
- Spaghetti Diagram: The representation of all motions in a workplace, including people motion, materials, or tools transportation [20,21]. The representation in the layout allows the identification of the process inefficiency so that unnecessary motions can be identified and eliminated, reducing or eliminating motion waste. It is also a tool used in the proposal of representations for improvement related to movements, such as reorganizing the layout or eliminating motions [20,21].

Quality can be defined as the level of customer satisfaction, depending on their requirements [22]. Quality tools are applied to obtain improvements in productive processes and quality control. These tools can support the analysis of non-compliance and contribute to defining some actions to be implemented and eliminate the problem's routes [23,24]. According to Djekic and Tomasevic [25], Karou Ishikawa has presented seven quality tools:

- Flowchart: It is a visual representation of all steps from a work process, leading to an easier process understanding [22,24].
- Pareto Chart: It is a graphic that represents the occurrence of events and it is used to categorize and analyze operational performance, challenges, situations, and causes. The Pareto principle holds that 80% of defects are determined by 20% of causes [25,26].
- Check Sheet: It is a tool used to collect and record data so that they can be further analyzed. These sheets are presented in simple columns and rows to be easily interpreted by everyone [27,28].
- Histogram: It is a bar chart that translates the shape of the data distribution [28].
- Control Chart: It represents the position of a sample relative to the mean [27].
- Fishbone Diagram: It shows the causes of a given problem, defining corrective actions to undertake and resources to invest [29].
- Scatter Diagram: It shows a potential relationship between some values employing a graph containing all the values [27].

To achieve the best solution to the main objective of the case study presented in this paper, other case studies performed in the automotive industry were reviewed. In the literature, SMED is usually used to minimize the time to prepare the production lines. Some cases also use the Pareto Chart to verify the most critical operation during changeover time.

Analyzing the case studies, it could be observed that conventional SMED is performed in five steps [11,30–32]:

1. Data collection (video recording for analysis if possible);
2. Classification of internal and external activities;
3. Conversion of internal activities into external ones;
4. Simplification of setup tasks;
5. Analysis of results.

The cases that only used task reorganization achieved improvements by 30%, so if this value was accomplished in this study, it would be a success [11,30]. It was also noticed that Quality and Lean Tools could boost the decrease in setup time. In the same case study, the use of conventional SMED reached a 50% improvement, and using other lean tools reached 62% [33].

This study is further than a SMED implementation to reduce setup time. It was developed involving several Lean and Quality tools to decrease the setup time as much as possible. This is a relevant study because companies must minimize their production line preparation time to become more competitive and survive in the market.

2. Materials and Methods

This study was developed in a factory that produces automobile parts formed by several welding robots and its focus is to reduce the setup time of two of them. When a new product is ordered, all production lines need to do their preparation as fast as possible. The factory has three similar production lines: A, B, and C. A and C spent around 40 min doing their changeover, while B wasted around 90 min. The production line presented in this paper is line B. It was chosen because it was the one with the highest changeover time.

The changeover process in the welding robots consists of exchanging the welding tool used to fabricate the previous product for a tool that will be used to produce the next product. To exchange from one tool to another, it is necessary to unscrew the screws that fix the tool used to fabricate; clean the tool; transport the previous tool to a specific spot; transport the next tool to the welding robot; and screw the screws.

All the steps performed to achieve the main goal of this case study are described as follows:

1. Go to the shop floor to watch all the processes: Communicate with people, and do a flowchart about the process. As Gemba Walk indicates, it is important to be where everything is made. This step will help to understand the production line requirements and what needs to be changed when a new product has to be fabricated;
2. Data collection by video recording: Watch the video and list all the activities and their times and draw the Spaghetti Diagram;
3. Split the tasks mentioned into four categories: The categories are transportation, waiting, main, and other. Transportation and waiting are two of the seven wastes considered by Lean, as explained before. The main tasks are the ones where it is crucial to change the tool when it is necessary to fabricate other different products (it is screw and unscrew the screws). "Other" is a category to include all activities that do not fit with any of the categories mentioned before, such as cleaning activities or tool adjustments needed before screwing the screws. This is also applied to the four categories presented in Figures 6 and 7;
4. Classify every task as internal or external: Internal activities are the ones made while the setup process is counting and external activities are the ones made before or after the setup/changeover process [5,32];
5. Convert internal tasks into external and do Analysis ERCS: Convert internal tasks as much as possible to reduce the time spent in the changeover. At the same time, it is important to classify every activity into eliminate, rearrange, combine, and simplify, taking into account the explanation made before;

6. Do a Pareto Chart: To observe the most critical operation;
7. Define an action plan: Where every step presented before has to be considered as well as all the perceptions felt while the setup was being performed. In this case, it was noticed that the worker did not know where the materials used to change the tool were and there were a lot of motions and transports to pick up all the materials. The responsibilities of each worker were not defined and there was no transportation tools flow (racks used during the manufacturing process are in the production line, and there was no space to move the tools). Therefore, the action plan must include: a list of tasks for the worker, a changeover trolley/rack with all the necessary materials to perform the setup, and a reorganization of the production line before setup starts.
8. Repeat the setup and collect new data: Repeat steps 2, 3, 4.
9. Analysis of the results. In this step, a dashboard was made to have a better and easier understanding of all the improvements made. This dashboard is shown in Appendix A.

3. Results

In this chapter, all the results obtained in this study are presented. The Spaghetti Diagram, Internal and External Activities, Process Category, and Setup Time will be shown.

3.1. Spaghetti Diagram

Figure 1 shows all the motions without materials/tools (red) and transports of materials/tools (blue) performed in the first setup. To compare all the movements completed in the second setup in Figure 2, by observing the two figures, it can be proved that all the motions and transports have been reduced from 54 motions to 31 and organized. It was verified a 43% decrease in all motions. In Figures 1 and 2, black boxes represent the place where the changeover material is. It is shown that the material used in the changeover was placed all together, and due to this change, it is possible to pick up the changeover material before the setup process starts. Therefore, this has contributed to the motion reduction.

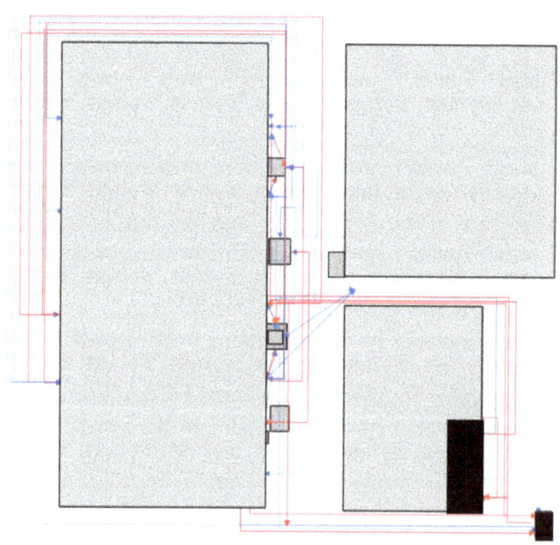

Figure 1. Spaghetti Diagram in the first setup.

3.2. Internal vs. External Activities

Figure 3 presents the number of internal and external activities comparing the first setup with the second setup. Internal tasks are those executed during the changeover period and external tasks are executed outside the changeover period [5,32]. The internal tasks converted into external ones are all related to the preparation of every material used in the setup process before the process starts; this is the preparation of cleaning material and the keys used to screw and unscrew. Unscrewing or screwing the screws, and transporting and cleaning the tools, are internal tasks that are impossible to convert into external tasks because it is impossible to perform these activities while the robots are running. Figures 4 and 5 show the percentage of internal and external tasks in the first setup and the second setup, respectively. The number of external activities has increased from 1 to 3 tasks, increasing their importance from 1% to 6%. The number of internal activities has decreased from 93 to 44, decreasing their significance from 99% to 94%.

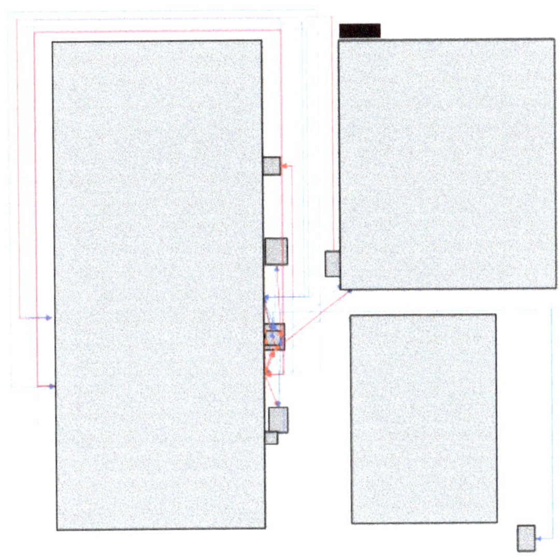

Figure 2. Spaghetti Diagram in the second setup.

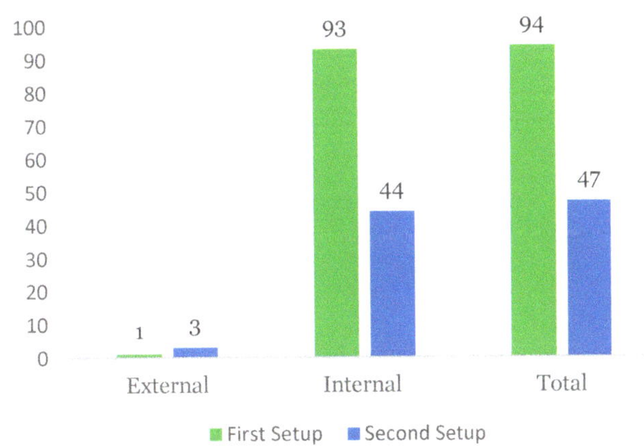

Figure 3. Number of internal and external activities in the two setup processes.

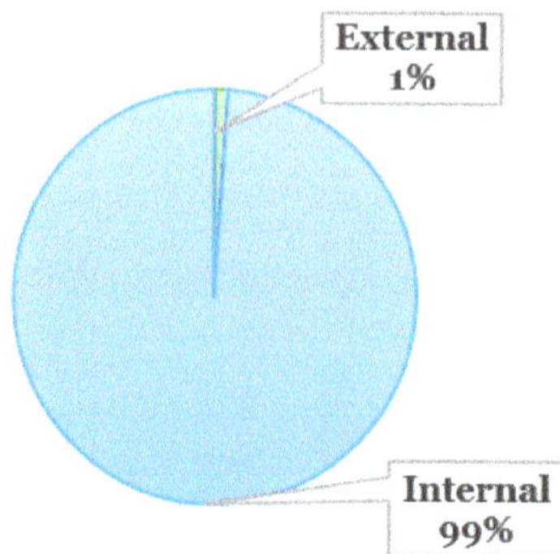

Figure 4. Percentage of internal and external activities in the first setup.

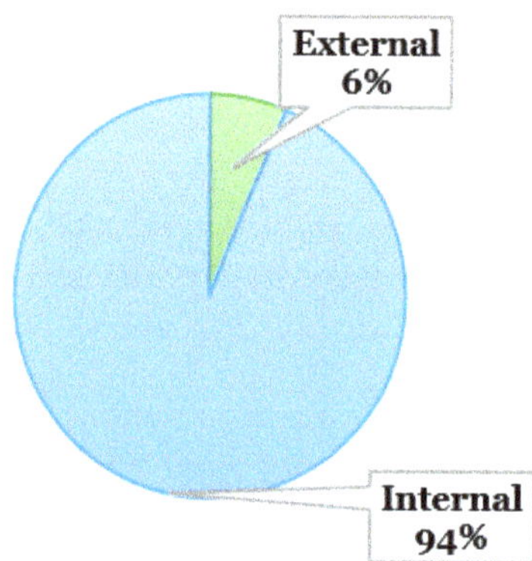

Figure 5. Percentage of internal and external activities in the second setup.

3.3. Process Category and Setup Time

Figure 6 presents the time difference, in seconds, spent in setup by the four categories selected: transportation, main, other, and waiting. Figure 7 shows the difference in percentage in each correspondent category.

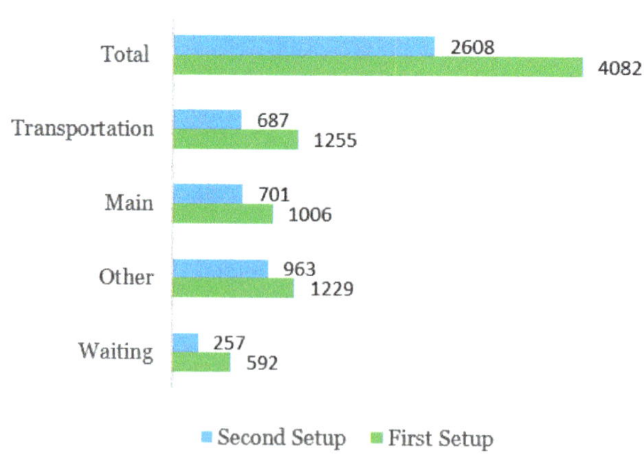

Figure 6. Spent time by each category in seconds.

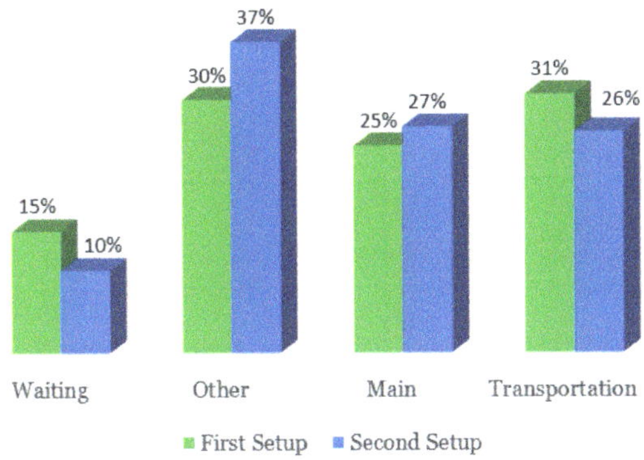

Figure 7. Process category in percentage.

The total time of SMED in pre-implementation was 4082 s and in post-implementation was 2608 s. It can be observed that every time spent in each category has reduced. The bigger reduction was in transportation and it can be due to the reduction of motions and transportation shown in Spaghetti Diagrams above. Transportation and waiting have decreased their weight in the total setup time, at the same time the main and other categories have increased.

3.4. Standardized Work with Takt-Time

Takt-time is the frequency with which the product or part of the product is required by the customer, relating the time available to the customer demand. It is the ratio of available time to consumer demand [3].

Standardized work was implemented by applying a list of tasks, such as: taking the wrench to loosen the screws; loosening all the screws counterclockwise; adjusting the new tool to the robot; and tightening all the screws.

The fact that the setup time has been reduced means that the time available for manufacturing has increased, allowing more products to be manufactured while maintaining the same takt-time. The 24-min reduction in setup time means that 24 more parts can be manufactured without altering the takt-time. In other words, it is possible to satisfy an increase in customer demand.

4. Discussion

The first setup lasted 4082 s and the second setup 2608 s, reducing 1474 s from one setup to another. In other words, the first setup was completed in 1 h 8 min and two seconds, and the second one spent 43 min and 28 s. There was a reduction from 24 min and 34 s. The reduction of setup time by 36% was due to all the steps being performed and explained above. These results were achieved only with task and space reorganization, without any monetary investment.

There were recorded 94 tasks performed by the operator before the SMED implementation. After implementation, a total of 47 tasks were recorded, and there was a reduction of tasks performed by 50%. In pre-implementation, there was only one internal task and 93 external tasks and in post-implementation, there were three external tasks and 44 internal tasks. In terms of percentage, it can be seen that there was an increase in the weight of external tasks from 1% to 6%. Although the increase in external tasks was not very strong in quantity, it represented 5% more of the total tasks than before.

The importance of the main activities was demonstrated, and the "other" category increased its weight, in the total setup time. As can be seen in Figure 7, the main activities increased their importance in the total setup time from 25% to 27% and the "other" activities from 30% to 37%.

The main activities are all the tasks that allow the preparation of a machine for the production of the new reference, so the weight increase of this activity in the total setup time is quite positive. It means that this activity started to have more representativeness in the total setup time. The activities classified as "other" are also important because they assist the main activities and, by assuming greater importance in the total time, indicate that there was an improvement associated with the implementation of SMED. The increases in the importance of these categories in the total setup time represent a positive balance, as it is highlighted that the indispensable activities are increasing their importance in the setup. However, it would be even more beneficial for a higher setup time reduction if the activities of the "other" category reduced their weight in the setup time, and the activities of the main category increased their importance even more.

5. Conclusions

This SMED implementation was a success. The reduction reached had a good value, taking into account that the measures implemented were only about the reorganization of tasks and the workplace. In the analysis of the case studies, it was verified that the application of SMED using only task reorganization measures returns results of around 30%. Therefore, it can be concluded that the reduction of 36% of the robots was successful, having reached the expected values. The conversion of internal tasks to external tasks was not significant to achieving these improvement values, although it also contributed.

The main activities and "others" have increased their relevance in the setup time, while the others reduced it. This represents an improvement in the process since they are essential activities in the preparation of the machines for a new reference and are the ones that occupy more of the setup time. However, it is important that the activities of the category "others" are reduced so that the main activities occupy even more time of the total time.

Therefore, the implementation of SMED led to numerous benefits through the reduction of changeover time, such as cost reduction and increased process efficiency. It can then be proven that SMED was successfully implemented. These results demonstrate the importance of implementing Lean tools and methodologies in companies to increase their competitiveness in the market.

The limitations of this study are related to data collection. Data were only collected once after implementation, and to obtain results more reliably, it would be better to repeat the setup process and do this analysis again. The available time to perform this study was also seen as a limiting factor since it was not possible to make a monetary investment to boost the time decrease.

In the future, it would be interesting if this methodology was implemented in other factories of other sectors to optimize and increase their efficiency. To improve the analysis conducted in this case, it would be valuable to do a financial analysis.

Author Contributions: Conceptualization, C.O. and T.M.L.; methodology, C.O. and T.M.L.; validation, T.M.L.; formal analysis, T.M.L.; investigation, C.O.; resources, C.O. and T.M.L.; data curation, C.O.; writing—original draft preparation, C.O.; writing—review and editing, C.O and T.M.L.; visualization, C.O. and T.M.L.; supervision, T.M.L. All authors have read and agreed to the published version of the manuscript.

Funding: This research received no external funding.

Institutional Review Board Statement: Not applicable.

Informed Consent Statement: Not applicable.

Data Availability Statement: Not applicable.

Acknowledgments: This work was supported in part by Fundação para a Ciência e Tecnologia (FCT) and C-MAST (Centre for Mechanical and Aerospace Science and Technologies), under project UIDB/00151/2020.

Conflicts of Interest: The authors declare no conflict of interest.

Appendix A

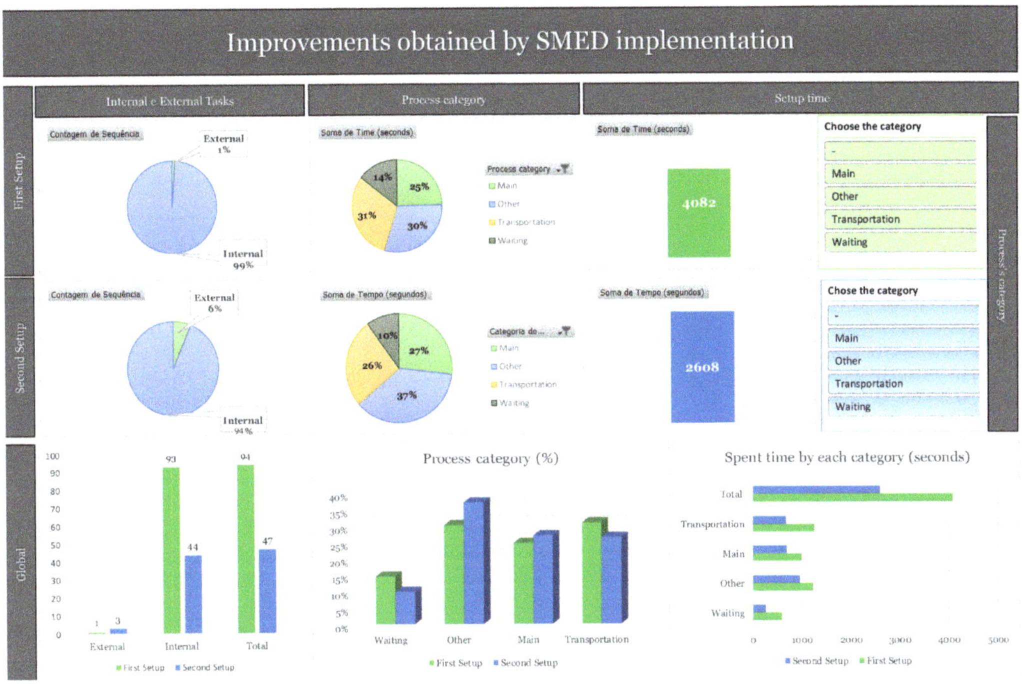

Figure A1. Results Dashboard.

References

1. Pinto, G.F.L.; Silva, F.J.G.; Campilho, R.D.S.G.; Casais, R.B.; Fernandes, A.J.; Baptista, A. Continuous improvement in maintenance: A case study in the automotive industry involving Lean tools. *Procedia Manuf.* **2019**, *38*, 1582–1591. [CrossRef]
2. Rosa, C.; Silva, F.J.G.; Ferreira, L.P.; Pereira, T.; Gouveia, R. Establishing Standard Methodologies to Improve the Production Rate of Assembly Lines Used for Low Added-Value Products. *Procedia Manuf.* **2018**, *17*, 555–562. [CrossRef]
3. Tiwari, K.V.; Sharma, S.K. The Impact of Productivity Improvement Approach Using Lean Tools in an Automotive Industry. *Process Integr. Optim. Sustain.* **2022**, *6*, 1117–1131. [CrossRef]
4. Toki, G.F.I.; Ahmed, T.; Hossain, M.E.; Alave, R.K.K.; Faruk, M.O.; Mia, R.; Islam, S.R. Single Minute Exchange of Die (SMED). A sustainable and well-timed approach for Bangladeshi garments industry. *Clean. Eng. Technol.* **2023**, *12*, 100592. [CrossRef]
5. Godina, R.; Pimentel, C.; Silva, F.J.G.; Matias, J.C.O. A Structural Literature Review of the Single Minute Exchange of Die: The Latest Trends. *Procedia Manuf.* **2018**, *17*, 783–790. [CrossRef]
6. Ghobadian, A.; Talavera, I.; Bhattacharya, A.; Kumar, V.; Garza-Reyes, J.A.; O'Regan, N. Examining legitimatization of additive manufacturing in the interplay between innovation, lean manufacturing and sustainability. *Int. J. Prod. Econ.* **2020**, *219*, 457–468. [CrossRef]
7. Velmurugan, V.; Karthik, S.; Thanikaikarasan, S. Investigation and implementation of new methods in machine tool production using lean manufacturing system. *Mater. Today Proc.* **2020**, *33*, 3080–3084. [CrossRef]
8. Kumar, N.; Hasan, S.S.; Srivastava, K.; Akhtar, R.; Yadav, R.K.; Choubey, V.K. Lean manufacturing techniques and its implementation: A review. *Mater. Today Proc.* **2022**, *64*, 1188–1192. [CrossRef]
9. Liker, J.K. *The Toyota Way: 14 Management Principles from the World's Greatest Manufacture*; McGraw-Hill: New York, NY, USA, 2004.
10. Karam, A.A.; Liviu, M.; Cristina, V.; Radu, H. The contribution of lean manufacturing tools to changeover time decrease in the pharmaceutical industry. A SMED project. *Procedia Manuf.* **2018**, *22*, 886–892. [CrossRef]
11. Sugarindra, M.; Ikhwan, M.; Suryoputro, M.R. Single Minute Exchange of Dies as the Solution on Setup Processes Optimization by Decreasing Changeover Time, A Case Study in Automotive Part Industry. *IOP Conf. Ser. Mater. Sci. Eng.* **2019**, *598*, 012026. [CrossRef]

12. Hoefsmit, P.C.; Schretlen, S.; Does, R.J.M.M.; Verouden, N.J.; Zandbergen, H.R. Quality and process improvement of the multidisciplinary Heart Team meeting using Lean Six Sigma. *BMJ Open Qual.* **2023**, *12*, e002050. [CrossRef] [PubMed]
13. Taggart, M.; Willis, C.; Hanahoe, J. Not seeing the wood for the trees—A gemba walk through a timber framed housing development. In Proceedings of the 27th Annual Conference of the International Group for Lean Construction (IGLC), Dublin, Ireland, 3–5 July 2019.
14. Reynders, P.; Kumar, M.; Found, P. 'Lean on me': An integrative literature review on the middle management role in lean. *Total Qual. Manag. Bus. Excell.* **2022**, *33*, 318–354. [CrossRef]
15. Saravanan, R.; Srinivasa Rao, M.S.; Sunkara, N.; Malyadri, T. Six sigma's ECRS technique to down cost and time of manufacturing-An experimental investigation. *AIP Conf. Proc.* **2020**, *2283*, 020070. [CrossRef]
16. Waiyanet, P.; Khongkaew, P.; Bunvachcharapai, S. Implementation of validate invoice and packing list document process by Microsoft Access a case study of ABC logistics company. In Proceedings of the 2018 5th International Conference on Business and Industrial Research: Smart Technology for Next Generation of Information, Engineering, Business and Social Science, ICBIR, Bangkok, Thailand, 17–18 May 2018. [CrossRef]
17. Frédéric, R.; Florian, M.; Laurent, J.; Forget, P.; Pellerin, R.; Samir, L. Lean 4.0: Typology of scenarios and case studies to characterize Industry 4.0 autonomy model. *IFAC-PapersOnLine* **2022**, *55*, 2073–2078. [CrossRef]
18. Marinelli, M.; Ali Deshmukh, A.; Janardhanan, M.; Nielsen, I. Lean manufacturing and industry 4.0 combinative application: Practices and perceived benefits. *IFAC-PapersOnLine* **2021**, *54*, 288–293. [CrossRef]
19. Palmqvist, A.; Vikingsson, E.; Li, D.; Fast-Berglund, Å.; Lund, N. Concepts for digitalization of assembly instructions for short takt times. *Procedia CIRP* **2021**, *97*, 154–159. [CrossRef]
20. Balaji, M.; Dinesh, S.N.; Raja, S.; Subbiah, R.; Manoj Kumar, P. Lead time reduction and process enhancement for a low volume product. *Mater. Today Proc.* **2022**, *62*, 1722–1728. [CrossRef]
21. Guzel, D.; Asiabi, A.S. Increasing Productivity of Furniture Factory with Lean Manufacturing Techniques (Case Study). *Tehnicki Glasnik.* **2022**, *16*, 82–92. [CrossRef]
22. Djekic, I.; Tomasevic, I. Tools in Improving Quality Assurance and Food Control. In *Food Control and Biosecurity*; Academic Press: Cambridge, MA, USA, 2018; pp. 63–104. [CrossRef]
23. Kharub, M.; Limon, S.; Sharma, R.K. The application of quality tools in effective implementation of HACCP: An empirical study of food and pharmaceutical industries. *Int. J. Qual. Reliab. Manag.* **2018**, *35*, 1920–1940. [CrossRef]
24. Realyvásquez-Vargas, A.; Arredondo-Soto, K.C.; Carrillo-Gutiérrez, T.; Ravelo, G. Applying the Plan-Do-Check-Act (PDCA) cycle to reduce the defects in the manufacturing industry. A case study. *Appl. Sci.* **2018**, *8*, 2181. [CrossRef]
25. Solanki, A.B.; Sonigra, S.S.; Vajpayee, V. Implementation of quality tools and effective strategies to boost production market standards for forged crankshafts: A case study of forging industry. *Mater. Today Proc.* **2021**, *47*, 5970–5976. [CrossRef]
26. Silva, F.L.; Fushita, Â.T.; Cunha-Santino, M.B.; Bianchini, I. Adopting basic quality tools and landscape analysis for applied limnology: An approach for freshwater reservoir management. *Sustain. Water Resour. Manag.* **2022**, *8*, 65. [CrossRef]
27. Barsalou, M. Determining which of the classic seven quality tools are in the quality practitioner's RCA tool kit. *Cogent Eng.* **2023**, *10*, 2199516. [CrossRef]
28. Siraj, I.; Bharti, P.S. Embedding Quality in Extrusion-Based Additive Manufacturing Technologies. *J. Mater. Eng. Perform.* **2022**, *31*, 5100–5117. [CrossRef]
29. Leone, F.; Viviani, F.; Grech, E.; Villa, R. Wire sweep phenomenon overview by Ishikawa diagram. In Proceedings of the 24th Electronics Packaging Technology Conference, EPTC, Singapore, 7–9 December 2022. [CrossRef]
30. Sahin, R.; Kologlu, A. A Case Study on Reducing Setup Time Using SMED on a Turning Line. *Gazi Univ. J. Sci.* **2022**, *35*, 60–71. [CrossRef]
31. Basri, A.Q.; Mohamed, N.M.Z.N.; Yasir, K.A.S.H.M.; Fazi, H.M.; Fudzin, A.F. The validation of productivity on the changeover activity at the automotive stamping press line by comparing the embedded SMED framework versus SMED approach: A witness simulation case study. *IOP Conf. Ser. Mater. Sci. Eng. Inst. Phys. Publ.* **2019**, *469*, 012005. [CrossRef]
32. Martins, M.; Godina, R.; Pimentel, C.; Silva, F.J.G.; Matias, J.C.O. A Practical Study of the Application of SMED to Electron-beam Machining in Automotive Industry. *Procedia Manuf.* **2018**, *17*, 647–654. [CrossRef]
33. Vieira, T.; Sá, J.C.; Lopes, M.P.; Santos, G.; Félix, M.J.; Ferreira, L.P.; Silva, F.J.G.; Pereira, M.T. Optimization of the cold profiling process through SMED. *Procedia Manuf.* **2019**, *38*, 892–899. [CrossRef]

eng

MDPI

Article

Power Transmission Using Circular Elements Bounded by Given Central Angle in Rolling Contact

Christopher G. Provatidis

School of Mechanical Engineering, National Technical University of Athens, 9, Iroon Polytechniou Str., 15780 Athens, Greece; cprovat@central.ntua.gr or cprovat@gmail.com

Abstract: In the present study, the concept of utilizing two circular cam-track disks, of the same central angle, in combination with one circular roller is presented. The roller is restrained to move within a vertical groove, and at the same time it rotates with rolling-contact on both cam tracks. When the upper cam is fully travelled by the roller, the same occurs with the lower one, despite their different lengths. Therefore, during the rolling contact, the two cams always sweep the same central angle. The aforementioned configuration of the two circular arcs may be considered as a unit cell, which can be repeated an even number of times, and when folded forms a closed circular groove between two cam-track disks. For better understanding, a manufactured prototype and 3D CAD-models have been developed. The operation of this setup as a gearless automotive differential is demonstrated by performing two bench experiments, which are then explained by a simplified mechanical model. The latter focuses on the implementation of the principle of the inclined plane, in which an upper limit of the inclination angle is imposed in accordance with the coefficient of friction at the friction disks. Previous patents on gearless differentials are discussed and other possible applications in mechanical engineering are outlined.

Keywords: cam-track disks; differential mechanism; gearless transmission; kinematics; rolling-contact

1. Introduction

Power transmission is usually performed using gears; the current research is mainly concerned with the modification of tooth profiles and the use of alternative materials in order to increase the lifetime, radiated noise, etc. (for example, see [1–3]). One of the disadvantages of using gears is the friction that appears at the contact point between the teeth of the conjugate bodies due to the sliding velocity [4], as well as backlash, which in turn give rise to other undesired dynamic phenomena (i.e., rattling) [5]. Furthermore, conventional planetary systems based on gears, such as mechanical differential gear devices, deliver very little (if any at all) torque when one of the steering wheels loses traction (i.e., "spins out" on loose substrates such as snow, mud, sand or gravel). In addition, gearboxes are widely used in industrial and military applications, for example in helicopters, where many crashes have occurred due to ruptured gears [6].

All these reasons have motivated researchers and inventors to look for better solutions and/or alternative power-transmission means without gears (gearless), some of which have also been adopted by the industry (see review in [7]). For particular spherical cams applied to robotic devices and automotive differentials, the reader is referred to [8,9]. In general, there is a great interest in replacing the gear-boxes in several drives at present [10–13].

As reported by others [7], historical evidence for the existence of an early differential device includes the Antikythera mechanism that contained a differential gear, which is interpreted as a mechanism with two degrees of freedom [14]. The conventional automobile differential was invented in 1827 by a Frenchman named Onésiphore Pecqueur. It was used first on steam-driven vehicles and was a well-known device when internal-combustion

Citation: Provatidis, C.G. Power Transmission Using Circular Elements Bounded by Given Central Angle in Rolling Contact. *Eng* **2023**, *4*, 2309–2335. https://doi.org/10.3390/eng4030132

Academic Editor: Antonio Gil Bravo

Received: 19 July 2023
Revised: 6 September 2023
Accepted: 8 September 2023
Published: 11 September 2023

engines appeared at the end of the 19th century. If we restrict the discussion to automotive differential devices, an important patent describing a construction wherein all the working parts are encased in such a way that all parts may be disconnected from the driving-wheels, and removed from the casing for the purpose of repair or readjustment, was granted in 1906 to Mooers [15], while the first patent regarding a gearless device (using sliding blocks) was granted in 1918 to Patch [16]. Among several others, the idea of introducing rolling elements was proposed within the years 1920–1933 by Ford, who used conical cams [17], while a superior idea using wavy cam-track disks was developed in 1943 by Beucher [18] and continued by others (Randall [19], Altman [20], Tsiriggakis [21,22], etc.). The common characteristics of the latter concepts [17–22], developed within the period 1943–1985, are the following:

(i) The usual side gears, which are attached to the shafts of the steering wheels, are replaced by conical or wavy cam-track disks;
(ii) The usual spiders (gears orbiting around the abovementioned side gears), which are also attached to the ring gear (crown wheel), are replaced by sliding elements, which interfere (slide or roll) with the cam tracks and also slide in several patented ways within a cage fixed to the aforementioned crown wheel.

The question that arises is whether the abovementioned sliding elements are fully sliding (in both their contact with the cam tracks and their support in the cage) or may operate by rolling contact with the opposite cam tracks. If the latter (i.e., the rolling contact) is possible, then we could generalize the simple concept of the "rolling-element bearing" by using wavy bearing rings in differential devices by modifying the cylindrical or conical races that are mass-produced at present. Then, all known technology in the rolling bearings industry could possibly be transferred to the automotive differential mechanism.

A recent study has shown that sinusoidal- and circular-shaped cam-track disks can achieve rolling contact, while a finite element analysis has indicated that service life in the order of 200,000 km is feasible [23]. Despite this fact, since the commercial software that was used, RecurDyn® V9R3, is based on spline representation, there are doubts as to whether the circular track was accurately represented. Moreover, the large-scale finite element model could not efficiently reveal the parameters that influence the stress concentration.

Therefore, the aim of this paper is:

(i) To explain the process followed to conceive the innovative idea.
(ii) To conduct a thorough theoretical study on the kinematics regarding rolling contact to certify previous findings.
(iii) To perform an elementary force analysis and compare this with the previous finite element analysis.
(iv) To use the abovementioned elementary force analysis to reveal and roughly optimize the most critical dimensional parameters that affect the magnitude of the maximum stress affecting the fatigue life.
(v) To use the elementary force analysis and reveal the essence of the differential mechanism.
(vi) To provide additional evidence for a better understanding of the operation of this differential mechanism.

The structure of this paper is as follows. Section 2 presents the development of the new concept. Section 3 is the theoretical study of kinematics, which proves the ideal rolling contact of the rollers between the circular segments associated with the meshed cam-track disks. Section 4 is a continuation of Section 3, and shows the way in which the circular segments can be repeated in so that they eventually form a closed track of 360 degrees. Section 5 is an enhancement of the elementary kinematical study of Section 3, now presenting the parametric equations of all the meshed moving bodies and their centrodes. Section 6 is the implementation of the new kinematical concept to the design of automotive differentials based on circular arcs. Section 7 is concerned with the force analysis of a simplified mechanical model. Section 8 presents results obtained from two bench experiments and manual computations regarding the mechanics of the gearless differential device based on circular arcs. Section 9 discusses the disadvantages of three previous patents on gearless automotive differentials, as well as details regarding the proposed design. Section 10 summarizes the conclusions. Appendix A explains the principle of the inclined plane.

2. Development of the New Concept

Power transmission is usually performed using gears in several configurations, one of which concerns two racks and a pinion. When the two racks are equally displaced so that the upper rack moves to the right (by distance $+\delta$ at velocity V) and the lower rack to the left (by distance $-\delta$ at velocity $-V$), the pinion rotates in the clockwise direction while its center remains at rest (as shown in Figure 1). Regarding the induced normal and shear (friction) forces at the contact points between the teeth in the pinion and the racks, the well-known sliding velocity appears on the teeth surface (see [4,24]).

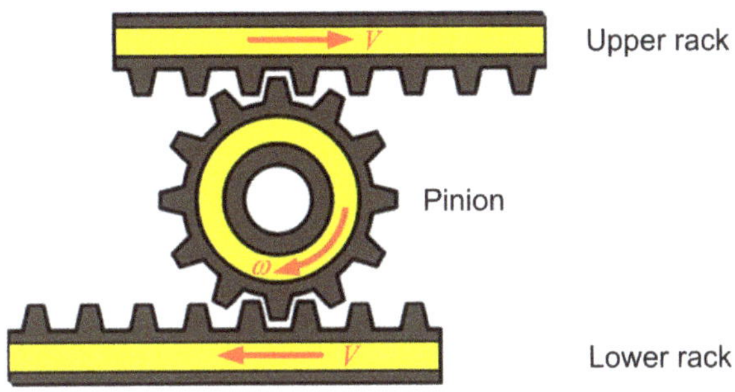

Figure 1. Double rack and a pinion.

Similar kinematics could be achieved when the pinion of Figure 1 is replaced by a circular disk or a sphere which covers its pitch circle, thus becoming a roller, while the profiles of the racks become straight plates, similar to those in the usual rolling bearings [25], as shown in Figure 2. However, then we need to impose sufficient normal pressure (i.e., a set of distributed forces, F) to allow for the development of proper friction forces ($T_f \leq \mu F$, where μ is the friction coefficient) to transmit the available power through the contact points between the roller and the plates. The kinematics of a very similar case with a set of two identical spheres in rolling contact with two planes (so-called 'cage-plane motion') has been studied by Freudenstein and Soylemez [26].

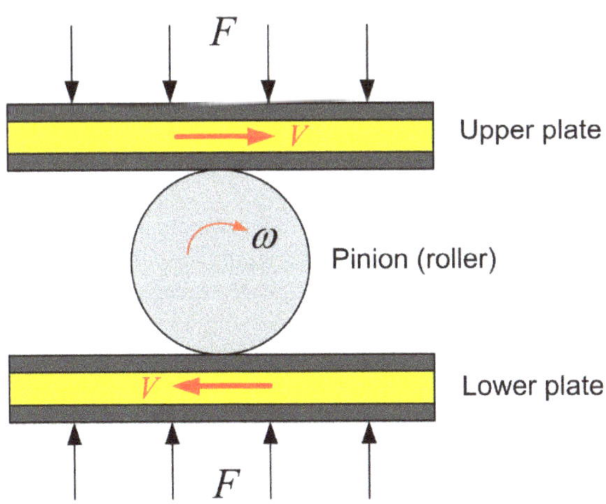

Figure 2. Pre-stressed plates and a roller.

3. Theoretical Study of Profiled Cams in Power Transmission

Now, the question is whether there is a standard shape to replace the plates in Figure 2, which ensures rolling contact, and thus has a high fatigue life, and does not require pre-compression by the illustrated distributed forces F.

The general problem of the kinematic synthesis of conjugate profiles has been treated extensively (e.g., [27–30] and papers therein). Applications are restricted to overrunning clutches [31], gearless reducers [32] and small gear-boxes [33]. Older patents discuss asymmetric [18–20] or symmetric [21,22] gearless differentials using curved cam track disks (in the place of the aforementioned plates). Mostly, the sinusoidal shape has been studied [34]. Moreover, undulating face gears, which combine wavy tracks with geared surfaces, appeared a few years ago [35].

In this paper, the central issue is how to replace the abovementioned straight and conical plates with profiled ones; thus, the force F shown in Figure 2 is no longer needed.

It will be shown that two circular cams in conjunction with a circular roller, as schematically shown in Figure 3, provide a working solution. After the forthcoming definitions, a new theorem of kinematics will be formulated and its proof will be derived. We start with an elementary proof based on Euclidian geometry and kinematics, and then we apply the concept of fixed and moving centrodes (polodes or polhodes) [36,37].

Definition 1. *Following Figure 3, let us consider two concentric circular cam tracks ($\widehat{DAE}$ and $\widehat{FBG}$) of the same central angle ($\alpha_c = \widehat{DOE}$, in degrees), which initially have the same vertical bisecting line ($v = OACB$). Based on the mean radius $OH = OC = OI = R_m$, the radius of the upper cam will be ($OA = R_m - r$), while the radius of the lower cam will be ($OB = R_m + r$). Clearly, the gap between these two cam track disks is covered by a circular roller of center C and radius $CA = CB = r$, which has the freedom to move along the vertical bisecting line (v) toward the y-direction and to rotate about its oscillating center. Technical solutions will be discussed in Section 5. Starting from this configuration, when the upper cam track is forced to translate horizontally at a velocity $+V = \delta/t$ to the right, we consider that the lower cam track will translate horizontally at a velocity $-V = -\delta/t$ to the left. After a certain amount of time t, the aforementioned opposite velocity components ($\pm V = \delta/t$) will lead to equal and opposite horizontal cam displacements $\pm\delta$, as shown in Figure 3.*

One may observe that, in the initial position of the cams shown in Figure 3, the contact points are (A, B) and the corresponding tangent lines between the roller and the cams are horizontal lines. Later, we shall see that there is always a specific relationship between these two tangent lines.

Theorem 1. *Given two cam tracks and a roller, according to the above definitions, for corresponding horizontal displacements where $\pm\delta$ perpendicularly to the bisecting line of the common central angle $\left(\widehat{FDO}, \widehat{OEG}\right)$ (as illustrated in Figures 3–5), we shall show that:*

1. *The center of mass of the roller (point C) moves along the vertical guide (v) in such a way that rolling contact appears at two points (A, B), which, after the displacement $\pm\delta$, occur (A_1, B_1) between the roller and the two meshed cam tracks, as shown in Figure 4 (for the sake of clear visualization, point B_1 is shown later in Figure 5). Obviously, the points (A_1 and B_1 of Figure 4) belong to the upper and lower cams, respectively.*
2. *The slopes ($\lambda_1 = tan\alpha_1, \lambda_2 = tan\alpha_2$) of the two tangents at the abovementioned contact points (A_1, B_1) are equal in measure and opposite in sign (i.e., $\lambda_1 = -\lambda_2$, and $|\alpha_1| = |\alpha_2|$).*
3. *The abovementioned two contact points (A_1, B_1) are always symmetric with respect to the variable horizontal line passing through the center of the roller.*
4. *In Figure 3, the initial position of the contact points (A, B) is the middle of the arcs $\widehat{DE}$ and $\widehat{FG}$. After the displacement $\pm\delta$, the new contact points will be (A_1, B_1), as shown in Figures 4 and 5, and form the angles: $\widehat{AOA_1} = \alpha_1 < 0$ and $B\hat{O}B_1 = \alpha_2 > 0$. The points*

(A_1, B_1) split the arcs in the same ratio measured from the displaced ends D and G; that is, $\widehat{D_1A_1}/\widehat{DE} = \widehat{G_1B_1}/\widehat{FG}$. Always, we have $|\alpha_1| = |\alpha_2| \leq \alpha_c/2$.

5. *Points (A and B) constitute the first, whereas the extreme points (D and G) shown in Figure 3 constitute the last contact pair in the mesh between the two cams and the roller.*

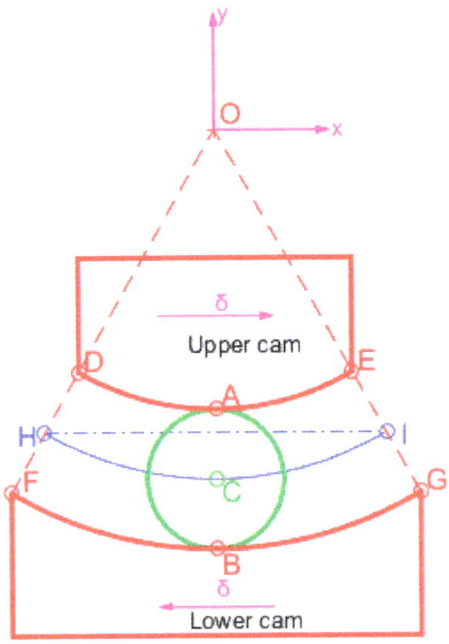

Figure 3. Cam tracks and a following roller.

Proof of Theorem 1. First, let us assume that the rotating axis of the roller (perpendicular to the plane of the image shown in Figure 3) is pushed from the bottom to the top by a spring; thus, when the upper cam track translates horizontally to the right at a velocity (so the curve DE is displaced to D_1E_1 by δ, and the center moves similarly, from O to O_1, as illustrated in Figure 4), the roller will be in rolling-contact. Therefore, point A_1 will possess two velocity components, i.e., the tangential ωr due to the instantaneous angular velocity ω of roller's rotation about the center C_1 and the vertical component V_b due to sliding along the vertical guide (v), as shown in Figure 4. Since the vector sum equals $+V$, for the vertical direction, we can obtain:

$$V_b = V \tan \alpha_1, \tag{1}$$

with

$$\sin \alpha_1 = \delta / R_m. \tag{2}$$

Velocity compatibility in the horizontal direction can be denoted as $V = \omega r \cos \alpha_1$, with ω denoting the abovementioned instantaneous angular velocity of the roller, whence:

$$\omega = \frac{V}{r \cos \alpha_1}. \tag{3}$$

Similarly, if we focus on the lower cam track considering a spring to push the roller downwards (Figure 5), and the curve $\hat{FG}$ is displaced to $F_1\hat{G}_1$ by δ to the left, the condition of rolling-contact in the vertical direction will lead to:

$$V_b = V \tan \alpha_2, \tag{4}$$

with

$$\sin \alpha_2 = \delta / R_m. \tag{5}$$

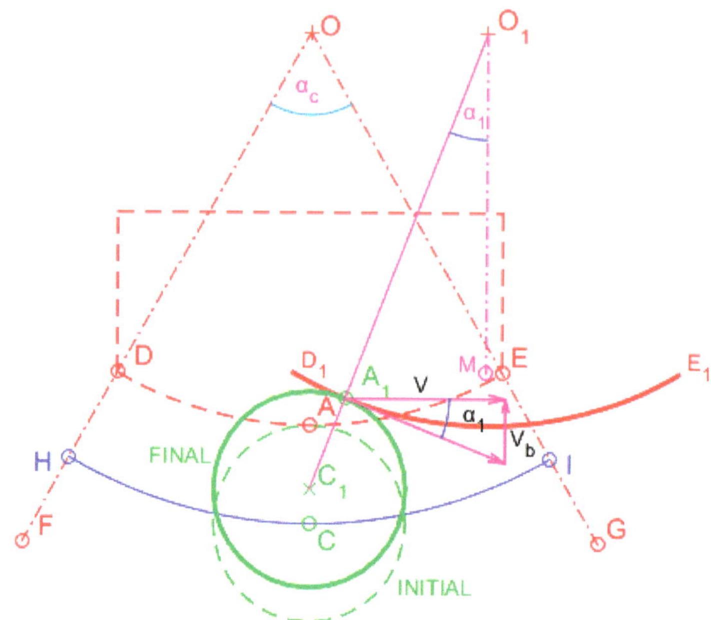

Figure 4. Contact between the roller and the upper cam-track disk.

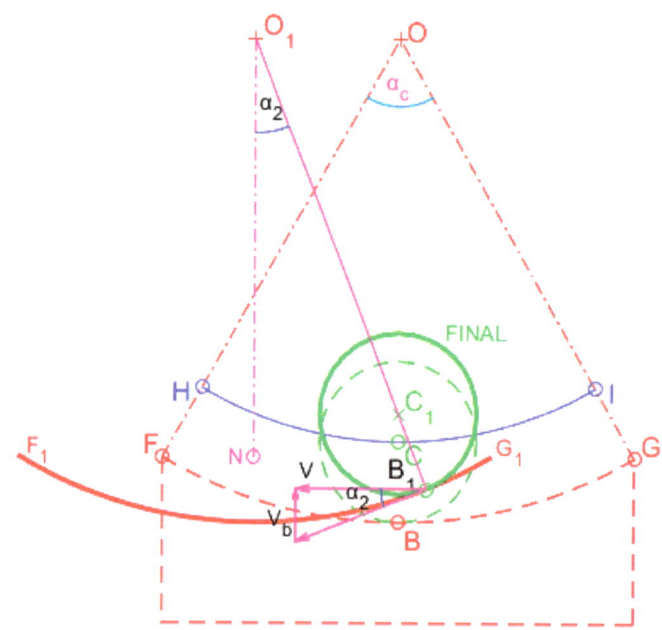

Figure 5. Contact between the roller and the lower cam-track disk.

Velocity compatibility in the horizontal direction means that $-V = -\omega r \cos \alpha_2$, whence:

$$\omega = \frac{V}{r \cos \alpha_2}. \tag{6}$$

Since Equations (2) and (5) share the same right-hand side, i.e., $\sin \alpha_1 = \sin \alpha_2 = \delta / R_m$, we have $\alpha_1 = \alpha_2$. At the same time, Equations (1) and (4) provide a unique value for the vertical velocity V_b of the roller. Finally, since $\cos \alpha_1 = \cos \alpha_2$, Equations (3) and (6) determine a unique angular velocity ω, which fulfills the conditions of rolling contact (velocity compatibility). In other words, we have shown that there is always a unique downward velocity V_b of the center C of the roller, and a unique angular velocity ω in the condition of rolling-contact with the two cams, i.e., the upper and the lower one. This completes the proof of part 1 and part 2 of the Theorem.

Let us now deal with the position of the two contact points of the roller, i.e., at point A_1 on the upper and point B_1 on the lower cam track disk. We recall that, generally, when two circles are in contact, their unique contact point is collinear to the two centers, i.e., it belongs to the unique line, which is determined by the two centers of the circles. This trivial theorem of Euclidian Geometry is applied twice: once for the upper cam, where we deduce

$$\frac{x_{A_1}}{\delta} = \frac{r}{R_m}, \tag{7}$$

and once more for the lower cam, whence

$$\frac{x_{B_1}}{\delta} = \frac{r}{R_m}. \tag{8}$$

Comparing the above Equations (7) and (8), we obtain:

$$x_{A_1} = x_{B_1}. \tag{9}$$

Therefore, the normal projections of both contact points on the horizontal plane coincide (this completes the proof of part 3 of the Theorem), while the slopes of the tangents are equal in measure ($|\alpha_1| = |\alpha_2|$) but of the opposite sign (i.e., $\alpha_1 = -\alpha_2$).

At the displaced position shown in Figure 4, we see that the angle formed by the middle point M of the straight segment D_1E_1, as well as the displaced center O_1 and the displaced point A_1, equals $\widehat{MO_1A_1} = \alpha_1$. In other words, the circular arc D_1A_1 corresponds to a central angle equal to $\left(\frac{\alpha_c}{2} - |\alpha_1|\right)$. Similarly, Figure 5 shows that the circular arc B_1G_1 corresponds to a central angle equal to $\left(\frac{\alpha_c}{2} - |\alpha_2|\right)$. Since $|\alpha_1| = |\alpha_2|$, we deduce that the circular arcs D_1A_1 and B_1G_1 correspond to equal central angles; thus, they split the total arcs to which they belong (DE and FG, respectively) to the same ratio: $\frac{\widehat{DA_i}}{\widehat{DE}} = \frac{\widehat{GB_i}}{\widehat{FG}}$. This completes the proof of part 4 of the Theorem.

Part 5 of the Theorem is a corollary of Part 4, with $|\alpha_1| = |\alpha_2| = \frac{\alpha_c}{2}$, and this completes the proof of the Theorem. □

Interestingly, if we combine Figure 4 with Figure 5, then Figure 6 shows the simultaneous contacts points (A_1, B_1) which are found to the right of the vertical sliding guide (v). Moreover, one may observe the wedge formed by the displaced curves D_1E_1 and F_1G_1. In this situation, the initial position of the three meshed surfaces are illustrated in dashed lines, the roller (of center C_1) is pulled upwards in compression and rotates in the clockwise direction. Both the normal forces at (A_1, B_1) pass through the displaced center C_1. Therefore, if power is transmitted to the vertical guide (v), it further flows to the cams by rolling contact.

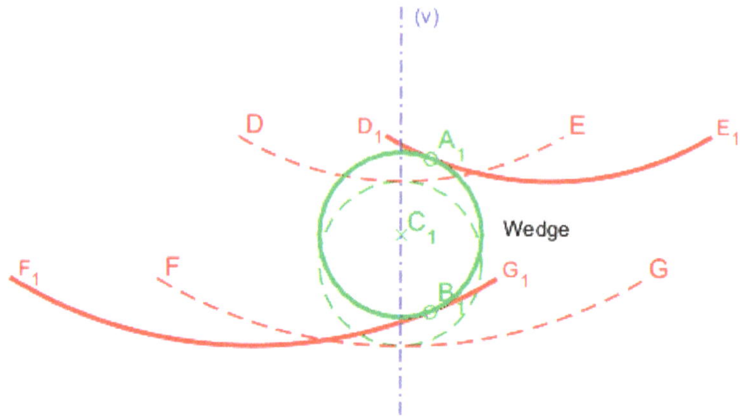

Figure 6. A wedge made by the roller and the two cam-track disks ($\alpha < \alpha_c/2$).

4. Repetition of the Initial (Reference) Circular Segment

From the above Theorem 1, we recall that when the horizontal displacement OO_2 of the center in the lower cam-track disk takes its maximum value, $\delta_{max} = R_m \sin(\alpha_c/2)$, the initial points (D, G) are then meshed at $(D_2,\ G_2)$, with the points O_2, D_2, C_2 being collinear, as shown in Figure 7.

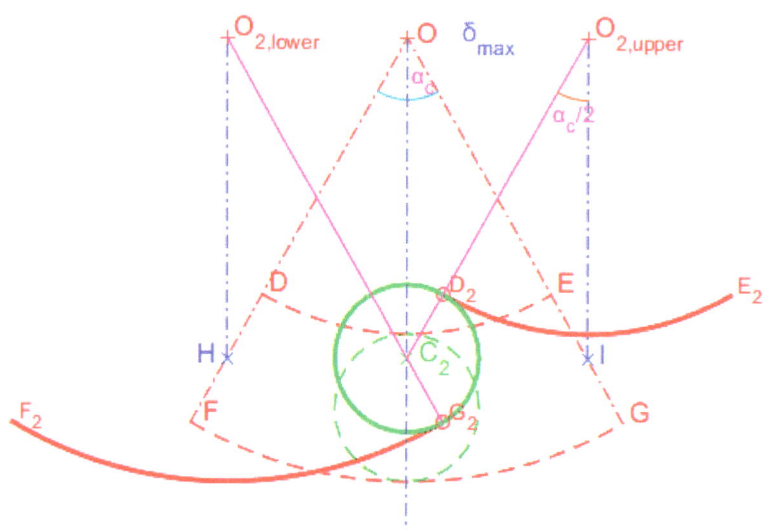

Figure 7. Extreme arrangement with rolling contact ($\alpha = \alpha_c/2$).

Therefore, to continue the rolling contact between the two cam surfaces and the roller, we can merely reverse the curvature of the circular arcs, as indicated in Figure 8. Note that each of the four rollers is restrained to move in the horizontal direction due to vertical grooves (guides), which do not affect the rolling-contact between the rollers and the tracks. Therefore, in this setup, the centers of mass of the rollers move only in the vertical direction.

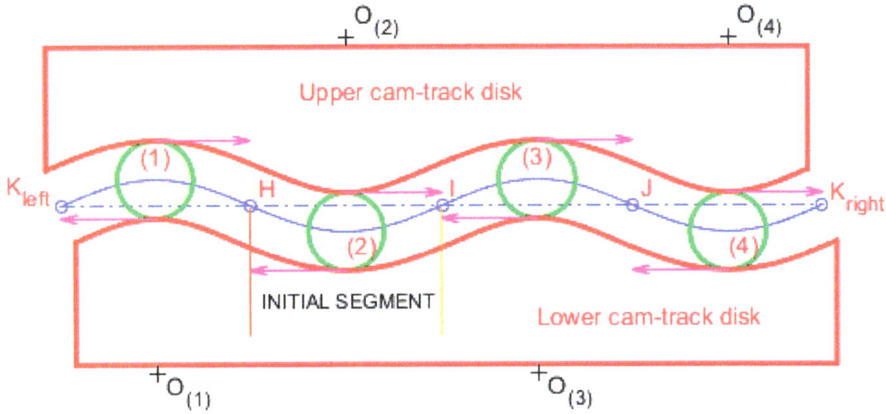

Figure 8. Extension of the rolling contact to the left and right of the initial segment.

5. Parametric Equations

For the sake of completeness, we also present parametric equations in terms of the past time t for all three bodies in motion.

We select the axis origin of the coordinate system to occur at the initial position of the center O of the two circular arcs (see, Figures 3–5). Starting from point O (at time $t = 0$) and terminating at the extreme point $O_{2,upper}$ (with $\delta_{max} = \left|\overrightarrow{OO_{upper}}\right|$, shown in Figure 7) at time $t_{max} = \delta_{max}/V$, we have the following equations for the trajectory of the center of mass of the roller, as well as for the instantaneous position of the two contact points.

5.1. Center of Mass C of the Roller

By construction, the center of mass C of the roller moves along a vertical slide guide, which is here assumed to be fixed in space (note that in an open differential, the sliding guide is fixed to the usually geared retainer (cage)). Therefore, by definition, the center C does not move toward the x-direction. As the upper cam moves to the right, it leaves space for the caged point C to cover the gap moving upward. Therefore, the equations of motion of point C are:

$$\begin{cases} x_C(t) = 0 \\ y_C(t) = -R_m\sqrt{1-(V/R_m)^2t^2} \end{cases}. \tag{10}$$

5.2. Contact Point A_1 between Upper Cam and Roller

$$\begin{cases} x_{A_1}(t) = (rV/R_m)t \\ y_{A_1}(t) = (-R_m+r)\sqrt{1-(V/R_m)^2t^2} \end{cases}. \tag{11}$$

During the time period in which the cam tracks ($D\hat{A}E$ and $F\hat{B}G$) are meshed, the application of Equation (11) to the locus of contact point A_1 results in the cyan-colored line shown in Figure 9.

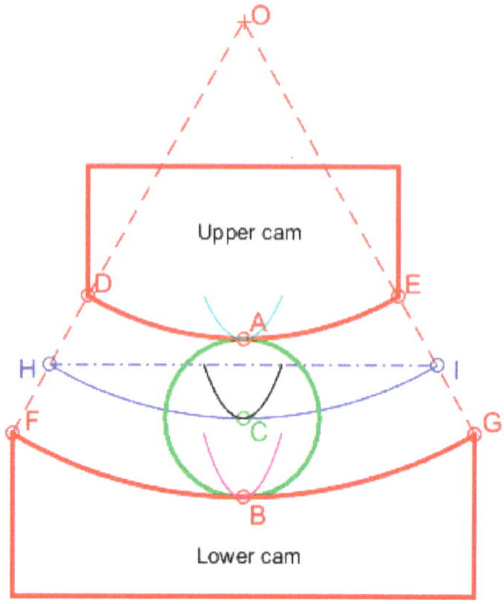

Figure 9. The three centrodes associated with the contact points (A, B) and the center of mass C of the roller.

5.3. Contact Point B_1 between Lower Cam and Roller

$$\begin{cases} x_{B_1}(t) = (rV/R_m)t \\ y_{B_1}(t) = -(R_m + r)\sqrt{1 - (V/R_m)^2 t^2} \end{cases}. \tag{12}$$

During the time period in which the cam tracks ($D\hat{A}E$ and $F\hat{B}G$) are meshed, the application of Equation (12) to the locus of the contact point B_1 results in the magenta-colored line shown in Figure 9.

From Equations (10)–(12), we can obtain the following relationships:

$$\begin{cases} x_{A_1}(t) = x_{B_1}(t) \\ y_{A_1}(t) = y_C(t) + r\sqrt{1 - (V/R_m)^2 t^2} \\ y_{B_1}(t) = y_C(t) - r\sqrt{1 - (V/R_m)^2 t^2} \end{cases}. \tag{13}$$

Equation (13) reflects the proposed Theorem 1, according to which the contact points (A_1, B_1) have the same x-coordinate, while they lie at the same distance from a line passing through the moving center of mass C and are parallel to the horizontal x-axis.

5.4. Locus of Instantaneous Pole (Centrode)

With respect to the fixed space system Oxy, the pole of the roller lies along a horizontal line passing through the instantaneous center of mass of the roller, and thus has the following parametric equations:

$$\begin{cases} x_p(t) = r \sin \alpha_1 = r(V/R_m)t \\ y_p(t) = y_O - R_m \cos \alpha_1 = -R_m\sqrt{1 - (V/R_m)^2 t^2} \end{cases}. \tag{14}$$

Taking the origin of the Cartesian co-ordinate system at the initial point O (see Figure 3), a trivial manipulation of Equation (14) leads to

$$(x_p/r)^2 + (y_p/R_m)^2 = 1. \tag{15}$$

Equation (15) depicts that the locus of the instantaneous pole of the roller (centrode), for as far as the circular arcs $\widehat{DE}$ and $\widehat{FG}$ are meshed with the roller, moves along a part of an ellipse centered at O with horizontal and vertical semi-axes equal to r and R_m, respectively. This is shown by the black-coloured curve in Figure 9, which is obviously tangent with the circular arc $\widehat{HI}$ at initial point C.

6. Implementation of the New Concept to Automotive Differentials

6.1. General Remark

The advantage of the rolling contact between the rollers and the two cams is compensated by the friction induced at the vertical sliding guides. From the other point of view, the technology required in the proposed concept is closer to the rolling bearings than the gears.

6.2. Application to Automotive Differentials

One of the possible applications of the proposed curves is the gearless differential, of which a full study, including sinusoidal curves as well as stress (finite element) and fatigue analyses, may be found in [23].

In the case of the gearless differential, the extreme points K_{left} and K_{right} (shown in Figure 8) are coincident, as in Figure 10. Clearly, the outer roller of Figure 10 represents the sketch element (1) of Figure 8, while the developable line $K_{left} \div K_{right}$ becomes a whole circumference.

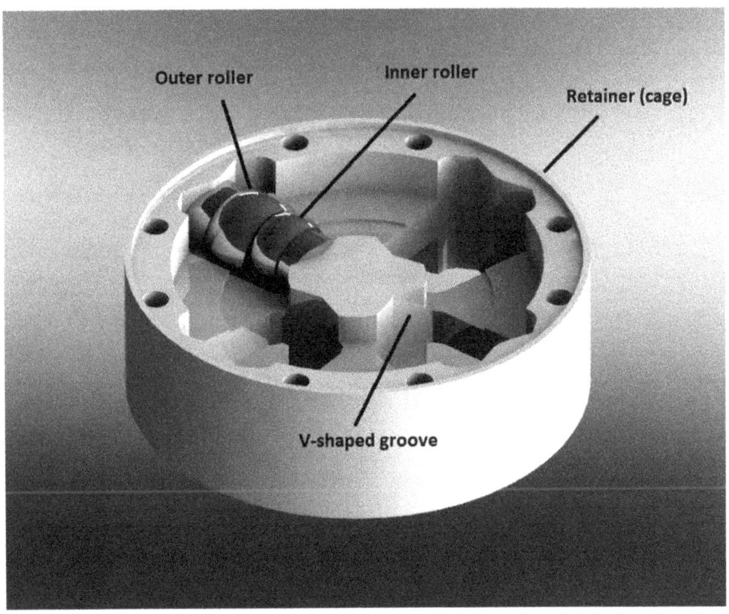

Figure 10. Retainer (cage) with one (out of the four pairs) of outer and inner rollers within vertical grooves.

The reason for using two rollers per vertical groove is due that when all the rollers hit the top of the cam profiles, uncertainty will arise; thus, the mechanism will work intermittently. Therefore, it is necessary to install two parallel rows of rollers at the distance of one row relative to the other, as shown in Figure 11. One may observe that, between the maxima of the outer ring and the maxima of the inner ring, there is a phase-difference of exactly 45° degrees.

Figure 11. Outer and inner grooves at some distance in one manufactured cam-track disk.

Clearly, the rollers are encaged in the retainer and can move on V-shaped vertical grooves by sliding (Figure 10). The rolling contact with which this paper deals refers to the contact of the two rolling members (shown in Figure 10) with the two cam-track disks (the lower one is shown in Figure 11), one upper and one lower, forming the differential mechanism. If we perform a horizontal cut in the middle, and the retainer is left out to increase our visibility, in Figure 12 we obtain the 3D shape of an entire cam-track disk, with two rows at a phase difference of 45° degrees, four vertical grooves and four pairs of rolling members (one pair per vertical groove). An almost horizontal cut of the rolling members was performed to increase our understanding of the relative motion between the rolling members and the vertical grooves.

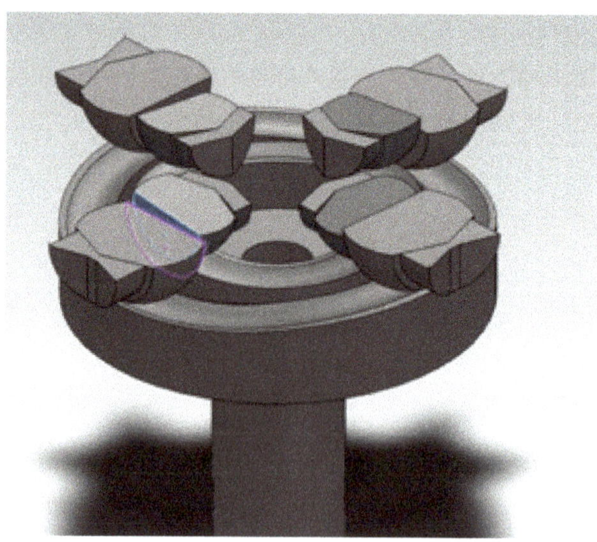

Figure 12. A complete cam-track disk with all the rolling members (the blue-colored area shows the internal contact between outer and inner rollers).

In the design of two rows shown above, we avoid the case where all rollers are found at the extreme points (maxima and minima), thus ensuring at least one working roller per track under pressure, as illustrated in Figure 6. It is also worth mentioning that both cam-track disks, upper and lower, are identical in shape and size. thus minimizing manufacturing and storage costs.

The advantage of the proposed circular cams is the supposed self-regulation of the differential device to variable road conditions without additional means (i.e., those used in locking and limited slip ones). Note that if the central angle α_c is relatively small with respect to the coefficient of friction between the outer support of the cam-track disks in the housing (for design details, see [21]), i.e., when $\tan(\alpha_c/2) < \mu$, the wedge shown in Figure 6 obeys the rule of the inclined plane, a self-blocking performance has been noticed [23]. In other words, the entire mechanism operates as an equivalent one-dimensional inclined plane in the circumferential direction.

The upper cam-track disk of the abovementioned graphs is actually connected to the (say) left driving wheel of the automobile vehicle. Similarly, the lower cam-track disk is connected to the right driving wheel.

7. Force Analysis in a Simplified Model

7.1. Mechanical Model

Figure 13a represents any of the four active rollers operating between the two cam-track disks. Under certain conditions, the totality of the aforementioned four active rollers constitutes an equivalent roller. Thus, Figure 13a shows that the circumferential force F_u, which is transmitted to the roller through the vertical grooves of the retainer, is cancelled by two contact forces F_n; these should be almost vertical to the inclined plane of the averaged angle $\overline{\alpha}_m$ (as the rolling friction is very small). Of course, the frictional force along the vertical groove at which the rollers slide changes the situation; thus, the normal forces F_n are not exactly equal to one another. Again, the truth is that the circumferential force F_u is cancelled by the sum of the two normal forces F_n plus the unknown frictional force, a matter which will be discussed later.

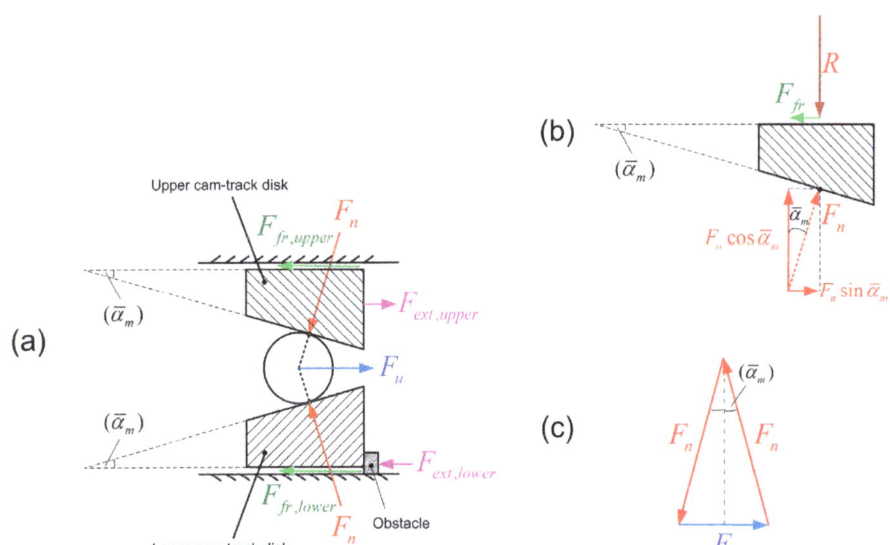

Figure 13. (**a**) One-dimensional analog of the entire differential mechanism as well as force equilibrium of (**b**) the upper cam-track disk and (**c**) the roller (ignoring the friction at grooves).

Below we present an elementary mechanical analysis for the determination of forces and contact pressures.

In an automotive differential, the engine power is transmitted from the geared crown to the attached retainer (cage), and then through V-shaped supporting grooves to the rollers. Thus, circumferential forces, F_u^{out} and F_u^{in} are induced in the outer and the inner tracks, respectively. The identification of the working elements is as follows. Only those elements of which the horizontal components of the outward unit vector are directed toward the circumferential velocity of the crown (as shown in Figure 14) are actually working and transmitting power. As previously mentioned, this occurs for only half of them in each track (i.e., two in the outer track and another two in the inner track), which can be easily identified in Figure 15, as denoted by arrows ($\rightarrow$).

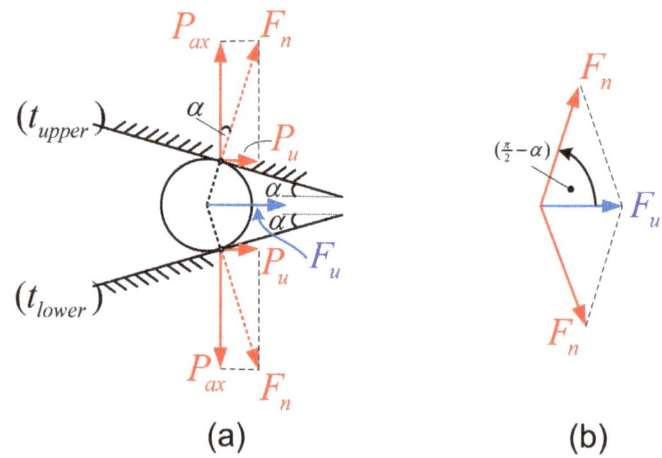

Figure 14. Analysis of circumferential force F_u (**a**) at a wedge (**b**) in two components F_n normal to the upper and lower track. The projection of F_n is the axial force F_{ax}.

In each track, the serial numbering of the rollers ranges from 1 to 4, as shown in Figure 8. Rollers in contact but in different tracks have the same serial number. Since, at each time instance, only half of the rollers push each track, either No. (1,3) or No. (2,4), at a phase difference of 180 degrees (in anti-diametrical positions), we can safely consider that the exerted moment in the outer track is due to one pair of forces (for example, those exerted on No. (1,3), shown in Figure 15f–i), thus forming a moment of force couple $F_u^{out} D_{cam}^{out}$. Similarly, another pair of forces in the inner track with a corresponding moment of force couple $F_u^{in} D_{cam}^{in}$ is exerted in the inner track. Therefore, the total moment $M_{t,crown}$, which is transmitted from the geared crown to the cage of the differential device, will be:

$$M_{t,crown} = F_u^{out} D_{cam}^{out} + F_u^{in} D_{cam}^{in} \quad (16)$$

Assuming that the friction between the rollers and the cam tracks is negligible (rolling-contact), and also neglecting the friction at the V-shaped grooves, each of the above circumferential forces F_u is analyzed into two equal force components F_n of equal size, both normal to the surface of the surrounding tracks. Then, each of the F_n forces is further analyzed into one axial to the cam-track disk (P_{ax}) and another circumferential/radial component (P_u) according to Figure 16a.

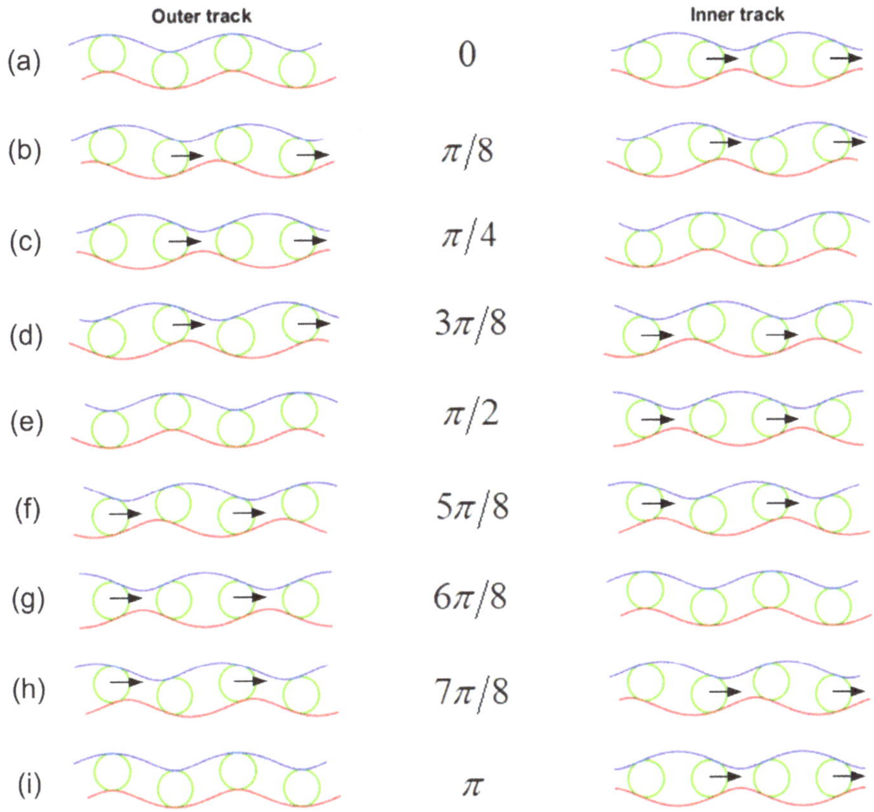

Figure 15. Several phases of the cam tracks, for relative rotation θ every $\pi/8$, for the outer track and the inner track (uniform scale). (**a**) $\theta = 0$. (**b**) $\theta = \pi/8$. (**c**) $\theta = 2\pi/8$. (**d**) $\theta = 3\pi/8$. (**e**) $\theta = 4\pi/8$. (**f**) $\theta = 5\pi/8$. (**g**) $\theta = 6\pi/8$. (**h**) $\theta = 7\pi/8$. (**i**) $\theta = \pi$.

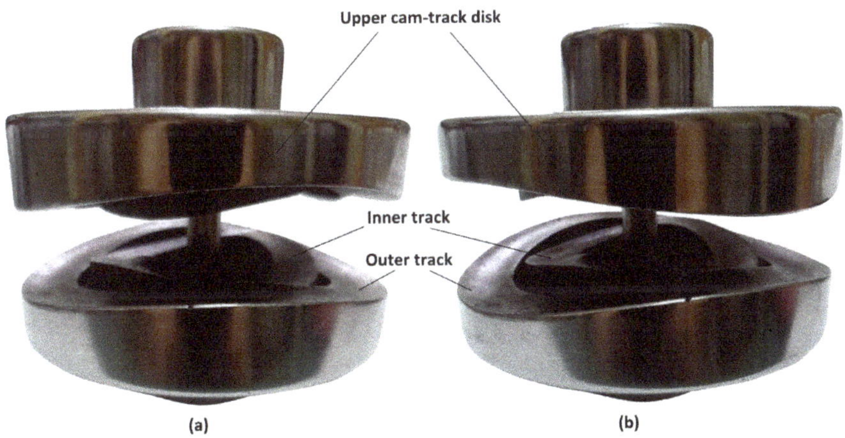

Figure 16. Cam-track disks with outer tracks (**a**) in parallel state and (**b**) maximum wedge angle.

Ignoring the dynamic effect as well as the significant frictional force at the V-shaped grooves, the static equilibrium of the roller overestimates the force components as follows:

$$\text{Outer track}: \; P_u^{out} = \frac{F_u^{out}}{2}, \; P_{ax}^{out} = \frac{F_u^{out}}{2\tan\alpha_{out}} \tag{17}$$

and

$$\text{Inner track}: \; P_u^{in} = \frac{F_u^{in}}{2}, \; P_{ax}^{in} = \frac{F_u^{in}}{2\tan\alpha_{in}} \tag{18}$$

Although the system is not statically determined (redundant) and the relationship between F_u^{out} and F_u^{in} is a matter of elasticity, to obtain a closed-form analytical solution we further assume rigid-body conditions; thus, the circumferential forces are proportional to their distance from the center:

$$\frac{F_u^{in}}{F_u^{out}} = \frac{D_{cam}^{in}}{D_{cam}^{out}} = \xi \tag{19}$$

Then, solving in F_u^{in} from Equation (19), and substituting into Equation (16), we obtain:

$$M_{t,crown} = F_u^{out} D_{cam}^{out}\left(1 + \xi^2\right) \tag{20}$$

It should become clear that only half of the moment given by Equation (20) is transmitted to each cam-track disk. Since each cam-track disk undertakes the moment:

$$M_{t,cam} = P_u^{out} D_{cam}^{out} + P_u^{in} D_{cam}^{in}, \tag{21}$$

the total transmitted moment will be:

$$M_{t,crown} = 2M_{t,cam} = 2\left(P_u^{out} D_{cam}^{out} + P_u^{in} D_{cam}^{in}\right) \tag{22}$$

By virtue of Equation (20), Equation (22) becomes:

$$M_{t,crown} = F_u^{out} D_{cam}^{out}\left(1 + \xi^2\right) \tag{23}$$

Therefore, if the transmitted moment $M_{t,crown}$ is known, the circumferential force in the outer track is given by:

$$F_u^{out} = \frac{M_{t,crown}}{D_{cam}^{out}(1 + \xi^2)} \tag{24}$$

Combining Equation (19) with Equation (24), we can obtain the circumferential force in the inner track:

$$F_u^{in} = \frac{\xi M_{t,crown}}{D_{cam}^{out}(1 + \xi^2)} \tag{25}$$

7.2. Normal Forces

Based on the above circumferential forces, F_u^{out} and F_u^{in}, we can calculate the normal forces in both tracks.

Therefore, the normal force in the outer track is given by:

$$F_n^{out} = \frac{F_u^{out}}{2\sin\alpha_{out}} = \frac{M_{t,crown}}{2D_{cam}^{out}(1 + \xi^2)\sin\alpha_{out}} \tag{26}$$

Also, the normal force in the inner track is given by:

$$F_n^{in} = \frac{F_u^{in}}{2\sin\alpha_{in}} = \frac{\xi F_u^{out}}{2\sin\alpha_{in}} = \frac{\xi M_{t,crown}}{2D_{cam}^{out}(1 + \xi^2)\sin\alpha_{in}} \tag{27}$$

Each active spherical roller presses, by the normal force F_n (either F_n^{out} or F_n^{in}), the corresponding track, which has two curvatures in space. According to the generalized Hertz theory, which refers to the contact of two ellipsoidal surfaces [38], at the contact point of each track, the equivalent curvatures in the two perpendicular x- and y-directions are given by:

$$\frac{1}{R_{ex}} = \frac{-R_{1x} + R_{2x}}{R_{1x}R_{2x}}, \quad \frac{1}{R_{ey}} = \frac{-R_{1y} + R_{2y}}{R_{1y}R_{2y}}, \tag{28}$$

and the radius R_e of an equivalent sphere is:

$$R_e = \sqrt{R_{ex}R_{ey}}, \tag{29}$$

Based on the normal contact force F_n and the abovementioned equivalent radius R_e, the maximum contact pressure becomes:

$$p_0 = \left[\frac{6F_n(E^*)^2}{\pi^3(R_e)^2}\right]^{\frac{1}{3}}. \tag{30}$$

In Equation (30), F_n refers to either (F_n^{out}, F_n^{in}), R_e to either (R_e^{out}, R_e^{in}), while the equivalent elastic modulus E^* is determined by:

$$\frac{1}{E^*} = \frac{1-\nu_1^2}{E_1} + \frac{1-\nu_2^2}{E_2}, \tag{31}$$

where (E_1, ν_1) and (E_2, ν_2) are the elastic moduli and the Poisson's ratios of the rollers and the tracks, respectively.

7.3. Axial Forces

The abovementioned moment $M_{t,crown}$ is equally transmitted from the rolling members to both the cam-track disks by the two working normal forces F_{ax}^{out} and F_{ax}^{in} (the same per track for each cam-track disk, directed to the wheels) of total magnitude

$$F_{ax,tot} = 2(F_{ax}^{out} + F_{ax}^{in}). \tag{32}$$

The above sum includes two equal forces per track (which shows the involved factor "2" in Equation (32)) and is cancelled by the friction on the friction disk.

It should become clear that only half of the moment given by Equation (23) is transmitted to each cam-track disk. Actually, each cam-track disk undertakes the moment given by Equation (21).

Combining Equations (22) and (23), we have:

$$M_{t,cam} = \frac{1}{2}F_u^{out}D_{cam}^{out}(1+\xi^2) \tag{33}$$

Since the friction should not exceed the maximum static value, we may set a desired threshold for the maximum moment at each cam for which full blocking is ensured; thus, we can write:

$$M_{t,cam} \leq \underbrace{\mu P_{ax,tot}}_{Friction}\frac{D_R}{2}, \tag{34}$$

where μ is the coefficient of friction and D_R is the equivalent diameter of friction. From standard machine elements and design books, we know that the latter variable depends on the assumptions imposed in clutch theory, either uniform-pressure or uniform-wear conditions. Following the uniform-pressure condition, we have (see, [39,40]):

$$D_R = \frac{2}{3}\frac{d_{out}^3 - d_{in}^3}{d_{out}^2 - d_{in}^2}, \tag{35}$$

where d_{out} and d_{in} is the outer and the inner diameters of the friction disk, respectively.

Substituting Equation (17) to Equation (19) into Equation (32), we can see that each cam-track disk undertakes the total axial force:

$$F_{ax,tot} = F_u^{out}\left(\frac{1}{\tan\alpha_{out}} + \frac{\xi}{\tan\alpha_{in}}\right), \tag{36}$$

Then, substituting Equations (33) and (36) into Equation (34), after a reduction in the common factor F_u^{out}, we obtain:

$$\left(\frac{1}{\tan\alpha_{out}} + \frac{\xi}{\tan\alpha_{in}}\right) \geq \frac{D_{cam}^{out}(1+\xi^2)}{\mu D_R}, \tag{37}$$

In order to make further analysis easy, we assume that the angles are small and that similarity conditions between the outer and inner track are assumed; thus, according to [23], the sum of the two contact angles is constant:

$$\alpha_{out} + \alpha_{in} = \alpha_m, \tag{38}$$

where

$$\alpha_m = \alpha_{central}/2 \tag{39}$$

Under these circumstances, the function at the left part of Equation (37) may be approximated in terms of only α_{out}, as follows:

$$f(\alpha_{out}) = \left(\frac{1}{\alpha_{out}} + \frac{\xi}{\alpha_m - \alpha_{out}}\right), \tag{40}$$

Since the right-hand side of Equation (37) is a constant, we are seeking a condition that will ensure that this equation will be valid even for the minimum value of the left-hand side. Actually, the function $f(\alpha_{out})$ may obtain close-to-infinite values when $\alpha_{in} \to \alpha_m$; thus, Equation (37) will be fulfilled for small values of α_{out} (where $\alpha_{in} \to \alpha_m$), but it is easy to see that it also possesses a minimum value. Equating the first derivative of $f(\alpha_{out})$ to zero, the only acceptable solution for the optimality condition (less than α_m) is:

$$(\alpha_{out})_{opt} = \frac{\alpha_m}{1+\sqrt{\xi}} \tag{41}$$

Thus, for the particular value of the outer contact angle given by Equation (41) and the associated inner contact angle given by Equation (38), i.e., $\alpha_{in} = \alpha_m - \alpha_{out}$, Equation (37) finally implies the following critical threshold:

$$\alpha_m \leq \alpha_{m,CR} = \frac{D_R}{D_{cam}^{out}}\frac{\left(1+\sqrt{\xi}\right)^2}{(1+\xi^2)}\mu. \tag{42}$$

In conclusion, Equation (30) is the most important formula regarding the maximum induced stress between the rollers and the cam-track disks, while Equation (42) is a good estimation of the maximum inclination angle, which equals half the central angle of the meshed circular arcs.

Typical data of the model are given in Table 1 and will be used in Section 8.

Table 1. Parameters of the model.

Parameter	Value
Input torque (applied to the crown attached to the cage/retainer)	260 Nm
Number of rolling elements at each cam track	4
Phase-difference between outer and inner cam track	45°
Diameter of outer cam track: D_{cam}^{out}	85 mm
Diameter of inner cam track: D_{cam}^{in}	50 mm
Radius of rolling element in the outer track: r_{out}	15 mm
Radius of rolling element in the inner track: r_{in}	12 mm
Outer diameter of friction disk: d_{out}	95 mm
Inner diameter of friction disk: d_{in}	45 mm
Coefficient of friction: μ	0.066
Inclination angle (half the central angle) of outer track: α_{out}	15°
Inclination angle (half the central angle) of inner track: α_{in}	27.25°

8. Results

The results concern (a) experiments and (b), with manual calculations based on previous sections.

8.1. Experiments

Using a prototype that was manufactured according to [21], two simple experiments were conducted as follows:

Bench experiment No.1: We put the right shaft (i.e., the extension of the lower cam-track disk) into a mechanical clamp, while the other shaft (extension of the upper cam-track disk) is left free. Then, a torque is progressively exerted on the retainer (between the two shafts). It was observed that the rotation of the retainer is *impossible* even if a very high torque is applied. Nevertheless, by applying even a small torque to the left shaft, the latter can easily rotate twice as fast as the retainer.

Bench experiment No.2: We carefully put the retainer into a mechanical clamp. We observe that it is impossible to turn only one shaft, regardless of the magnitude of the applied torque. If, however, a small torque is applied to the other shaft in the opposite direction, then both shafts rotate (in opposite directions) at the same angular velocity.

The above two experiments are very enlightening and support the theory of the previous Section 7.3.

8.2. Rough Estimation of the Inclination

According to Table 1, the following data were adopted.

Diameters of cam-track disks: $D_{cam}^{in} = 50$ mm, $D_{cam}^{out} = 85$ mm.

Rolling members: $r_{out} = 15$ mm, $r_{in} = 12$ mm.

Friction disks: $d_{out} = 95$ mm, $d_{in} = 45$ mm, and $\mu = 0.066$.

Substituting the above figures in Equation (35), the equivalent diameter D_R in the friction disks is estimated as:

$$D_R = \frac{2}{3}\frac{d_{out}^3 - d_{in}^3}{d_{out}^2 - d_{in}^2} = \frac{2}{3} \times \frac{95^3 - 45^3}{95^2 - 45^2} = 72.9762 \text{ mm}. \quad (43)$$

We assume that the force-to-force and diameter-to-diameter ratios are equal to each other, as in the pivoted connections; thus, Equation (19) implies:

$$\xi = \frac{D_{cam}^{in}}{D_{cam}^{out}} = \frac{50}{85} \cong 0.5882 \quad (44)$$

Substituting the above numerical values of Equations (43) and (44) into Equation (42), we obtain:

$$\alpha_m \leq \left[\frac{72.9762}{85} \times \frac{\left(1+\sqrt{0.5882}\right)^2}{\left(1+(0.5882)^2\right)} \times 0.066 \right] \times \frac{180}{\pi} \cong 7.53^\circ \tag{45}$$

Therefore, using the numerical result given by Equation (45), for similar tracks, the maximum allowable central angle will be twice the value shown by Equation (45); thus, its upper limit will be:

$$\alpha_{central} \leq 15.1^\circ \tag{46}$$

8.3. Calculated Stresses

The basic parameters of the model were obtained according to Table 1. The stresses were calculated according to Equation (30), in which the normal force F_n depends only on the standard inclination angles of the two tracks and the standard input torque (all found in Table 1). In contrast, the radius R_e of the equivalent sphere in the denominator of Equation (30) is a very crucial design parameter, which (by virtue of Equation (28)) highly depends on the double curvature of each track.

Based on the above data, in conjunction with a typical input torque $M_{t,crown} = 260$ Nm applied to the crown attached on the cage, for the above inclination angle we apply Equations (26) and (27) to derive the normal forces on the outer (F_n^{out}) and the inner track (F_n^{in}), respectively.

Furthermore, regarding the derivation of a large value for R_e, after many trials, which all resulted in high stress values, acceptable stresses were obtained for at least the following design details (these reference values are close to those of the prototype, for which a finite element analysis has been documented in [23]):

$$\begin{cases} \text{Roller in the outer track}: R_{1x}^{out} = R_{1y}^{out} = r_{out} = 15\text{ mm} = 0.015\text{ m} \\ \text{Roller in the inner track}: R_{1x}^{in} = R_{1y}^{in} = r_{in} = 12\text{ mm} = 0.012\text{ m} \\ \text{Outer track}: R_{2x}^{out} = 0.220\text{ m and } R_{2y}^{out} = 1.1 r_{out} \\ \text{Inner track}: R_{2x}^{in} = 0.058\text{ m and } R_{2y}^{in} = 1.1 r_{in} \end{cases} \tag{47}$$

The substitution of Equation (47) into Equations (28)–(30) leads to contact pressures equal to 1.6 MPa and 1.2 GPa for the outer and the inner track, respectively.

It is worth mentioning that, as also shown in Table 2, an accurate three-dimensional finite element analysis (FEA) for the same dimensional parameters and input torque resulted in smaller values, i.e., 1.3 GPa and 0.8 GPa, respectively [23]. The deviation between the above simplified mechanical model and FEA is mainly attributed to the existence of frictional forces at the V-shaped groves; thus, the assumed synthesis of forces in Figure 14b is not absolutely correct because—eventually—the friction cannot be ignored.

Table 2. Calculated stresses (in GPa).

Simplified Model		Finite Element Analysis (Ref. [23])	
Outer Track	**Inner Track**	**Outer Track**	**Inner Track**
1.6	1.2	1.3	0.8

Despite the above differences in the two models, the simplified mechanical model was of major significance, because Equations (28)–(30) could quickly consider and elucidate the influence of the curvatures in the tracks. Clearly, for each candidate combination of detailed parameters determined by Equations (28)–(30), a corresponding FEM model has to be developed afterwards.

Note that when the design includes spherical rollers, the first two equalities of Equation (47) remain untouched because both radii of the roller are equal to the radius of the corresponding sphere. In contrast, we shall show that the induced stresses are sensitive to the chosen curvatures on each track. Between the many combinations, let us preserve the same inclination angles as previously used (according to Table 1), and then let us standardize the radius $R_{2x}^{out} = 0.22$ m for the outer track, and the radius $R_{2x}^{in} = 0.058$ m for the inner track. Considering Equation (30) from the simplified model of Section 7.2, Table 3 shows the effect of the chosen radii R_{2y}^{out} and R_{2y}^{in} in the outer and the inner track, respectively (note that equal factors were imposed for r_{in} and r_{out}).

Table 3. Calculated stresses (in GPa) for varying curvatures of the tracks.

		Simplified Model: Equation (30)	
Outer Track	Inner Track	Outer Track	Inner Track
$R_{2y}^{out} = 1.10r_{out}$	$R_{2y}^{in} = 1.10r_{in}$	1.57	1.20
$R_{2y}^{out} = 1.09r_{out}$	$R_{2y}^{in} = 1.09r_{in}$	1.52	1.16
$R_{2y}^{out} = 1.08r_{out}$	$R_{2y}^{in} = 1.08r_{in}$	1.46	1.12
$R_{2y}^{out} = 1.07r_{out}$	$R_{2y}^{in} = 1.07r_{in}$	1.40	1.07
$R_{2y}^{out} = 1.06r_{out}$	$R_{2y}^{in} = 1.06r_{in}$	1.34	1.02
$R_{2y}^{out} = 1.05r_{out}$	$R_{2y}^{in} = 1.05r_{in}$	1.26	0.96
$R_{2y}^{out} = 1.04r_{out}$	$R_{2y}^{in} = 1.04r_{in}$	1.18	0.90

Another result for this sub-section refers to the sensitivity of the inclination angles. As a reference, we can consider the initial case (according to Table 1) in which the inclination angles are $\alpha_{out} = 15°$ and $\alpha_{in} = 27.25°$, respectively. Then, by imposing a uniform ±10% change around the reference values, Equation (30) leads to the results shown in Table 4 (change of about 3%).

Table 4. Calculated stresses (in GPa) for varying inclination angles of the tracks.

		Simplified Model: Equation (30)	
Outer Track	Inner Track	Outer Track	Inner Track
$\alpha_{out} = 1.1 \times 15°$	$\alpha_{in} = 1.1 \times 27.25°$	1.5211	1.1611
$\alpha_{out} = 1.0 \times 15°$	$\alpha_{in} = 1.0 \times 27.25°$	1.5690	1.1953
$\alpha_{out} = 0.9 \times 15°$	$\alpha_{in} = 0.9 \times 27.25°$	1.6239	1.2351

As a last note, a uniform variation of R_{2x}^{out} and R_{2x}^{in} by ±10% influences the contact stresses by less than ±1%.

8.4. Interpretation of the Experiments

The contents of Section 7.3 and the numerical results of Section 8.3 suggest that the main design philosophy of this gearless differential is concerned with an adequately small central angle α_c, so that the averaged inclination angle fulfills the condition $\tan \overline{\alpha}_m < \mu$, where μ is the coefficient of friction between the outer 'side boundary' of the cam-track disks and the housing. The conditions of Experiment No.1 are illustrated in Figure 13a. The equilibrium of the upper cam-track disk is shown in Figure 13b, in which the normal force F_n is analyzed into two components, which are cancelled by the frictional force F_{fr} and the normal reaction R. According to the well-known rule of the inclined plane (for the sake of completeness, as repeated in Appendix A), for such a small angle $\overline{\alpha}_m$, no matter the size of the normal force F_n, the static friction F_{fr} does not exceed its maximum static value; thus, no sliding occurs in this position.

Moreover, if we exert a force $F_{ext,upper}$ (in the same direction as the F_u) on the upper cam-track disk to overcome the side friction $F_{fr,upper}$ in the housing (see, Figure 13a), due to this net force, the horizontal motion of this part becomes possible. If the right driving wheel of a vehicle is blocked, it is possible to operate the differential mechanism even by exerting a small force on the free left driving wheel, a fact that justifies the abovementioned blocking behaviour of this mechanism.

8.5. Typical Screenshots during the Operation of the Gearless Mechanism

The initial position of the two tracks ($\theta = 0°$) is that shown in Figure 8. It is also repeated in Figure 15a for the outer and the inner track (parallel in the form of circumferences/rows separated by a distance). In Figure 15i one may observe that the same picture, as the top (Figure 15a), appears after half a turn of each cam-track disk (rotational angle $\theta = 180°$). In general, when the upper cam-track disk (left shaft) rotates by θ, the lower one (right shaft) will rotate by $-\theta$; thus, their phase-difference will be $\Delta\theta = 2\theta$. Therefore, the angle that appears in the middle column of Figure 15 refers to the aforementioned rotational angle θ of each cam-track disk (shaft). Obviously, within each cam-track disk (upper or lower), the outer and inner tracks (rows) rotate by the same amount as a rigid body. In more detail, from Figure 15, we can obtain the following results:

- At the angle $\theta = 0°$, all four rollers of the outer track hit the tops of the cam profiles (see also Figure 8), while those of the inner track do not.
- At the angle θ =45°, all four rollers of the inner track hit the tops of the cam profiles, while those of the outer track do not.
- At the angle θ =90°, all four rollers of the outer track hit the tops of the cam profiles, while those of the inner track do not.
- At the angle θ =135°, all four rollers of the inner track hit the tops of the cam profiles, while those of the outer track do not.
- At the angle $\theta = 180°$, all four rollers of the outer track hit the tops of the cam profiles (see also Figure 8), while those of the inner track do not.

From the above discussion, it is clear that every half a turn of each shaft ($\theta = 180°$), we take exactly the same picture of the eight rollers. This happens because, by construction (according to Figure 8), we can obtain two complete periods ($360°/2 = 180°$).

Furthermore, in Figure 15b, which corresponds to $\theta = 22.5°$ (i.e., each shaft has performed 1/16 of a full revolution), one may observe that two rollers in the outer track and another two in the inner track press the profiled cams (in the direction of the arrows), thus transmitting power. Interestingly, as the outer track rotates by $\theta = 22.5°$, it sweeps an angle equal to 1/4 of the central angle $\alpha_{out} = 15°$, and at the same time the inner track rotates in the same direction by the same amount ($\Delta\theta = 22.5°$), now sweeping an angle equal to 1/4 of the central angle $\alpha_{in} = 27.25°$.

Concentrating on Figure 15, one may observe that, at ($\theta = \pi/8,\ 3\pi/8,\ 5\pi/8,\ 7\pi/8$), four out of the eight rollers, depicted by arrows ($\rightarrow$), transmit power:

- Two out of the four rollers in the outer tracks are active, pushing them in the circumferential direction (thus transferring part of the power).
- Two out of the four rollers in the inner tracks are active, pushing them in the circumferential direction (thus transferring the rest part of the power).

In contrast, in the remaining five cases ($\theta = 0,\ \pi/4,\ \pi/2,\ 3\pi/4,\ \pi$), only two rollers are active. In other words, there at least two rollers are engaged in a wedge formed at the contact points with the cam-tracks.

9. Discussion

9.1. Other Patents on Gearless Differentials

Clearly, although repeated (sinusoidal like) curves have been used for many years [18–22], to date, the property of fully rolling contact cam track surfaces has neither been revealed nor studied. Readers who have experience in evaluating patents may refer to the originals [18–20] and could find the following comments to be useful:

In Beucher's patent Nr. 741,812 (granted in 1943) [18], there is an outer and an inner groove. In each groove, the curves of the cam-track disks are repeated cylindrical arcs of interrupted shape with sharp edges; thus, high-contact stresses are induced and fatigue phenomena are anticipated. There is substantial sliding between the sliding plate-like elements (in pairs) and the cam track disks. This mechanism works as a clutch, with the result of a low coefficient of efficiency. The wear is high due to friction and high temperature, and thus its operation is very problematic. Power transmission is circumferential. The shape of driving elements is different than that proposed (rectangular with smooth sliding surfaces). The inventor himself states that his invention is for the back shaft only.

In Randall's patent No. 2,651,214 (applied in 1950, granted in 1953) [19], there is again an outer and an inner groove. The curves of cam-track disks have sharp edges; thus, high-stress concentration is anticipated. Power transmission is circumferential. During differentialization, high friction and a high temperature develop, which wears rolling members at points that then rolls them on the driven cam-track disks, leading to the destruction of rolling. Rolling members have a frusto-conical shape. This works as a clutch.

In Altmann's patent No. 2,967,438 (filled in 1958, granted in 1961) [20], there is only one groove, in contrast to the abovementioned patents [18,19]. The curves in cam-track disks are sinusoids, but this fact does not result in the full rolling of driving elements without sliding. In more detail, the curves are unsymmetrical with each other (one cam track has 5 maxima and 5 minima, whereas the other has 6 maxima and 6 minima), thus leading to unequal inclination angles. The asymmetry between two driven cam-track disks results in unequal action on the wheels. Equalizing means are provided at both driven sides.

9.2. Overall Advantages of the Proposed Concept

From the above discussion in around Figure 13, it becomes obvious that the proposed concept is against the operation of the open differential. Clearly, experiments reveal that when one of the drive wheels meets mud or snow, even a minor resistance to the other drive wheel offers motion. However, this is not a limited slip differential in the classical sense, since no additional means exist, except for two bronze disks in the outer part of the cam-track disks (for construction details, see [21,22]). Clearly, the particular shape of the cam tracks, which leads to an interchangeable shape for both of them, can be efficiently manufactured through CNC machining centers, thus reducing overall costs. This shape ensures rolling contact and a kind of 'self-regulation', meaning that the power transmission obtained through this differential is somehow adjusted to road conditions.

In the particular case of helicopters in which the gearbox suffers from possible fractured gear teeth, it is obvious that the adoption of the proposed concept, in conjunction with a small inclination angle α, would not lead to full destruction because the small angle and the associated static friction (inclination rule) would prevent this.

9.3. Design Aspects of the Present Concept

In contrast to the above relevant patents [18–20], the use of the proposed circular curves (indicated in Figure 3) allows for the transmission of power through a radial arrangement of the rolling members. The support of the rolling members is achieved by sliding grooves (guides) into the retainer, but this does not affect the rolling contact between them and the cam-track disks. Rolling members are either in couples or independent. Rolling members are drum-shaped (spherical, conical, etc.), i.e., of an axisymmetric form. The same concept works with any even number on each curve: 2, 4, 6, 8, and so on. Then, the total number of rolling members is 4, 8, 12, 16, or any other multiple of 4. Rolling members are arranged in two concentric circumferences (one outer and one inner). The particular case of $n = 4$ rollers per circumference is shown in Figure 16: (a) for the initial state of parallel tracks and (b) after a rotation of the upper cam-track disk by 90° degrees in the counter-clockwise direction. Note that when the outer curves are parallel (Figure 16a), the inner ones form the maximum wedge angle. In contrast, when the outer curves form the maximum wedge angle (Figure 16b), the inner ones become parallel. In addition, details of the prototype at a regular size, including the rollers as well as stress and fatigue analysis, may also be found in [23]. Although this has been previously noted, it is instructive to repeat that, in cases where the prototype of Figure 16 is used as a gearless automotive differential, the upper cam-track disk is firmly linked to the left half-shaft of the drive wheel, while the lower one is linked to the right half-shaft.

By neglecting the friction at the V-shaped grooves, in the simplified model of Section 7.2, we used Equation (17) to determine the total axial force P_{ax} directed toward the half-shafts and pressed the friction disk on the housing of the gearbox. For a given coefficient of friction μ, the obvious condition is that the maximum torque transmitted between a friction disk and the housing of the gearbox should not exceed the value imposed by the maximum static friction. This inequality was used to analytically determine the maximum allowable average inclination angle for the whole mechanism. This model somehow suffers due to the hard assumption of frictionless operation at the V-shaped grooves. Of course, since the relative velocities $\dot{u}$ are analytically known everywhere in the power-train, one could introduce an assumption for the friction (e.g., $F_f = -c\dot{u}$), but this issue is beyond the scope of this paper.

9.4. Other Applications

Although the motivation for developing the concept of the two conjugate wavy cam-track disks was primarily the differential gear (useful for tractors as well as for military and passenger vehicles), the theory covers a lot of other power transmission applications in mechanical engineering. For example, the same concept could be applied to replace gearboxes in several drives, such as wind-turbines, mills, conveyors, mixers, pitch controls in aeronautical engineering, high-safety transmissions for helicopters, multi-step marine-type gearboxes, robotic devices such as pitch–roll wrists, electric cars, etc.

The wavy form of the cam tracks may also be useful for other applications where controlled oscillations are needed, such as the replacement of the camshaft (obvious) and/or the crankshaft (giving it a circular shape) in (single- or double-stroke) internal combustion engines, as well as for piston pumps and compressors, among others.

As a last note, if we wish to categorize the proposed new concept, it might be considered a practical contribution to power transmission through kinematic contact (in German, Formschluss [41]).

10. Conclusions

The findings of this paper suggest that:

(1) Ideal (pure) rolling (without sliding) of a roller on the two surfaces of cams (profiled plates) is achieved when the center of mass of the roller moves along a circular arc bounded by a given central angle.
(2) If this design concept is applied to an automotive differential for an even number of repetitions, it leads to two identical cam-track disks (i.e., symmetrical differential), thereby saving manufacturing and storage costs.
(3) If the central angle is small, self-regulation and blocking are achieved through the principle of the inclined plane.
(4) The proper selection of particular dimensional parameters, such as double curvature, leads to mechanical stresses, which are within the usual allowable limits.
(5) A weakness of this study is that elastic deformation of the cam-track disks has not been considered.

Funding: This research received no external funding.

Institutional Review Board Statement: Not applicable.

Informed Consent Statement: Not applicable.

Data Availability Statement: Upon request.

Conflicts of Interest: The author declares no conflict of interest.

Nomenclature

Quantity	Explanation
$D^{out}_{cam} = 2R^{out}_{cam}, D^{in}_{cam} = R^{in}_{cam}$	Diameter and radius of cam-track disks (outer and inner).
ξ	Diameter ratio $D^{in}_{cam} / D^{out}_{cam}$.
r_{out}, r_{in}	Radius of rollers (outer and inner track).
d_{out}, d_{in}	Diameters of friction disks (outer and inner).
D_R	Equivalent diameter of frictions disks (with uniform pressure).
μ	Coefficient of friction on friction disks.
$\alpha^{out}_{central}, \alpha^{in}_{central}$	Central angle (outer and inner track).
$\alpha_{out}, \alpha_{in}$	Contact angle (outer and inner track).
$M_{t,crown}$	Torque transmitted from the geared crown to the totality of two cam-track disks.
$M_{t,cam}$	Torque transmitted to each cam-track disk.
F^{out}_u, F^{in}_u	Circumferential force at each active roller (outer and inner track).
F^{out}_n, F^{in}_n	Normal force on cam-track disk transmitted by active roller (outer and inner track).
P^{out}_u, P^{in}_u	Circumferential force on cam-track disk transmitted by active roller (outer and inner track).
$P^{out}_{ax}, P^{in}_{ax}$	Axial force on cam-track disk transmitted by active roller (outer and inner track).

Appendix A

We consider a particular groove (external or internal). The corresponding reaction force R (see Figure 13b) cancels the horizontal projection of the normal force F_n; thus,

$$R = F_n \cos\alpha \tag{A1}$$

Also, the friction T cancels the vertical component of force F_n; thus,

$$T = F_n \sin\alpha \tag{A2}$$

Dividing (A1) and (A2) by parts, we receive:

$$T = R \tan\alpha \tag{A3}$$

Considering that no sliding occurs, the static friction is smaller than the maximum one:

$$T \leq \mu R \tag{A4}$$

From (A3) and (A4), one can obtain the well-known inequality:

$$\tan\alpha \leq \mu \tag{A5}$$

In other words, when the inclination angle of a cam surface is smaller than a critical limit, there is no sliding between the planar surface of the corresponding cam-track disk and the surrounding housing (for a possible practical implementation, see [18]); thus, it works as a 'blocking' differential mechanism. Otherwise, if $\tan\alpha > \mu$, it works as a usual differential mechanism.

References

1. Bahk, C.-J.; Parker, R.G. Analytical investigation of tooth profile modification effects on planetary gear dynamics. *Mech. Mach. Theory* **2013**, *70*, 298–319. [CrossRef]
2. Huai, C.; Zhao, Y. Meshing theory and tooth profile geometry of toroidal surface enveloping conical worm drive. *Mech. Mach. Theory* **2019**, *134*, 476–498. [CrossRef]
3. Ghosh, S.S.; Chakraborty, G. On optimal tooth profile modification for reduction of vibration and noise in spur gear pairs. *Mech. Mach. Theory* **2016**, *105*, 145–163. [CrossRef]
4. Litvin, F.L.; Fuentes, A. *Gear Geometry and Applied Theory*, 2nd ed.; Cambridge University Press: Cambridge, UK, 2004.
5. Sakaridis, E.; Spitas, V.; Spitas, C. Non-linear modeling of gear drive dynamics incorporating intermittent tooth contact analysis and tooth eigenvibrations. *Mech. Mach. Theory* **2019**, *136*, 307–333. [CrossRef]
6. Liang, X.; Zuo, M.J.; Feng, Z. Dynamic modeling of gearbox faults: A review. *Mech. Syst. Signal Process.* **2018**, *98*, 852–876. [CrossRef]
7. Mihailidis, A.; Nerantzis, I. Recent Developments in Automotive Differential Design. In *Power Transmissions, Mechanisms and Machine Science*; Dobre, G., Ed.; Springer Science + Business Media: Dordrecht, The Netherlands, 2013; Volume 13, pp. 125–140. [CrossRef]
8. Bai, S.; Angeles, J. The design of spherical multilobe-cam mechanisms. *Proc. Inst. Mech. Eng. Part C J. Mech. Eng. Sci.* **2009**, *223*, 473–482. [CrossRef]
9. Chaudhary, M.; Angeles, J.; Morozov, A. Design of a Spherical Cam-Roller Mechanism for an Automotive Differential. *Trans. Can. Soc. Mech. Eng.* **2016**, *40*, 243–252. [CrossRef]
10. Butterfield, S.; Smith, J.; Petch, D.; Sullivan, B.; Smith, P.; Pierce, K. *Advanced Gearless Drivetrain-Phase I Technical Report*; National Renewable Energy Laboratory: Golden, CO, USA, 2012. [CrossRef]
11. Andersen, S.B. Design and Optimization of Gearless Drives Using Multi-Physics Approach. Ph.D. Thesis, Technical University of Denmark (DTU), Lyngby, Denmark, 2012.
12. Chau, K.; Lee, C. Development of Reliable Gearless Motors for Electric Vehicles. In Proceedings of the 2017 IEEE International Magnetics Conference (INTERMAG), Dublin, Ireland, 24–28 April 2017. [CrossRef]
13. Sheu, K.-B.; Chien, C.-W.; Chiou, S.-T.; Lai, T.-S. Kinetostatic analysis of a roller drive. *Mech. Mach. Theory* **2004**, *39*, 819–837. [CrossRef]
14. de Price, D.S.J. Gears from the Greeks: The Antikythera mechanism—A calendar computer from ca. 80 BC. *Trans. Am. Philos. Soc. New Ser.* **1974**, *64*, 1–70. [CrossRef]
15. Mooers, L.P. Power Transmitting Mechanism for Automobiles. USA Patent No. 832991, 9 October 1906.
16. Patch, H.M. Differential Power Transmission Device. USA Patent No. 1283283, 29 October 1918.
17. Ford, A.F. Gearless Differential. US Patent No. 1897555, 14 February 1933.
18. Beucher, J.J. Ausgleichsgetriebe Besonders für Kraftfahrzeuge. Deutsches Patentschrift No. 741812, 12 March 1943.
19. Randall, R.R. Gearless Differential. Patent No. 2651214, 8 September 1953.
20. Altmann, W.E. Self-locking Differential Gear. USA Patent No. 296738, 10 January 1961.
21. Tsiriggakis, T. Differential Gear. USA Patent No. 4509388, 9 April 1985.
22. Tsiriggakis, T. Differential Getriebe. Europaisches Patentamt No. 0066122/B1, 15 May 1985.
23. Provatidis, C.G.; Kalligeros, C.; Spitas, V. Design considerations and simulation of a rolling-contact gearless automotive differential. *Proc. Inst. Mech. Eng. Part C J. Mech. Eng. Sci.* **2023**. [CrossRef]
24. Dudley, D.W. *Gear Handbook: The Design, Manufacture, and Application of Gears*; McGraw-Hill: New York, NY, USA, 1962.
25. Niemann, G.; Winter, H.; Höhn, B.-R.; Stahl, K. *Maschienenelemente 1: Konstruktion und Berechnung von Verbindungen, Lagern, Wellen*; Springer: Berlin/Heidelberg, Germany, 2019.
26. Freudenstein, F.; Soylemez, E. On the motion of spheres between two parallel planes. *ASME J. Eng. Ind.* **1975**, *97*, 294–302. [CrossRef]
27. Koetsier, T. Mechanism and Machine Science: Its History and Its Identity. In *International Symposium on History of Machines and Mechanisms-Proceedings HMM 2000*; Ceccarelli, M., Ed.; Springer Science + Business Media: Dordrecht, The Netherlands, 2000; pp. 5–24.
28. Kozhevnikov, S.N.; Antonyuk, E.Y. Synthesis of a Cam-Differential Mechanism with Periodic Dwell of the Output Link. *Mech. Mach. Theory* **1974**, *9*, 219–229. [CrossRef]
29. Ghosh, A.; Yadav, R.P. Synthesis of cam-follower systems with rolling contact. *Mech. Mach. Theory* **1983**, *18*, 49–56. [CrossRef]
30. Kerle, H.; Corves, B.; Hüsing, M. *Getriebetechnik*; Springer: Wiesbaden, Germany, 2015.
31. Nagler, N.; Lohrengel, A. Improved design criterion for frictionally engaged contacts in overrunning clutches. *Forsch. Ing.* **2021**, *85*, 1053–1063. [CrossRef]
32. Terada, H. The Development of gearless reducers with rolling balls. *J. Mech. Sci. Technol.* **2010**, *24*, 189–195. [CrossRef]
33. Ding, J.; Wei, Y.; Liu, L.; Geng, T. Novel design method for constrained conjugate contact curves of line gears based on rolling-up of pitch pattern. *J. Mech. Sci. Technol.* **2022**, *36*, 4535–4548. [CrossRef]
34. Lehman, M. Sinoiden Getriebe. *Konstruktion* **1981**, *33*, 109–112.
35. Liu, D.; Ren, T.; Jin, X. Geometrical model and tooth analysis of undulating face gear. *Mech. Mach. Theory* **2015**, *86*, 140–155. [CrossRef]

36. Uicker, J.E.; Pennock, G.R.; Shigley, J.E. *Theory of Machines and Mechanisms*, 5th ed.; Oxford University Press: New York, NY, USA, 2017; pp. 164–166.
37. Bottema, O.; Roth, B. *Theoretical Kinematics;* Dover: Mineola, NY, USA, 1990; pp. 21+211–212.
38. Wang, Q.J.; Zhu, D. Hertz Theory: Contact of Ellipsoidal Surfaces. In *Encyclopedia of Tribology*, 1st ed.; Wang, Q.J., Chung, Y.-W., Eds.; Springer: Berlin/Heidelberg, Germany, 2013; p. 1650.
39. Norton, R.L. *Machine Design: An integrated Approach*, 3rd ed.; Pearson International Edition: Upper Saddle River, NJ, USA, 2006; p. 887.
40. Spotts, M.F.; Shaup, T.E. Design of Machine Elements. Prentice-Hall: Upper Saddle River, NJ, USA, 1998; p. 337.
41. Roth, K. *Konstruieren mit Konstruktions-Katalogen*; Springer: Berlin/Heidelberg, Germany, 1982.

eng MDPI

Article

Assessment of Techniques for Detection of Transient Radio-Frequency Interference (RFI) Signals: A Case Study of a Transient in Radar Test Data

Stephen L. Durden *, Victor A. Vilnrotter and Scott J. Shaffer

Jet Propulsion Laboratory, California Institute of Technology, Pasadena, CA 91109, USA
* Correspondence: sdurden@jpl.nasa.gov

Abstract: The authors present a case study of the investigation of a transient signal that appeared in the testing of a radar receiver. The characteristics of the test conditions and data are first discussed. The authors then proceed to outline the methods for detecting and analyzing transients in the data. For this, they consider several methods based on modern signal processing and evaluate their utility. The initial method used for identifying transients is based on computer vision techniques, specifically, thresholding spectrograms into binary images, morphological processing, and object boundary extraction. The authors also consider deep learning methods and methods related to optimal statistical detection. For the latter approach, since the transient in this case was chirp-like, the method of maximum likelihood is used to estimate its parameters. Each approach is evaluated, followed by a discussion of how the results could be extended to analysis and detection of other types of transient radio-frequency interference (RFI). The authors find that computer vision, deep learning, and statistical detection methods are all useful. However, each is best used at different stages of the investigation when a transient appears in data. Computer vision is particularly useful when little is known about the transient, while traditional statistically optimal detection can be quite accurate once the structure of the transient is known and its parameters estimated.

Keywords: radar; interference; transient signals; computer vision; machine learning; deep learning; maximum likelihood estimation; matched filter; detection theory

Citation: Durden, S.L.; Vilnrotter, V.A.; Shaffer, S.J. Assessment of Techniques for Detection of Transient Radio-Frequency Interference (RFI) Signals: A Case Study of a Transient in Radar Test Data. *Eng* **2023**, *4*, 2191–2203. https://doi.org/10.3390/eng4030126

Academic Editor: Antonio Gil Bravo

Received: 28 June 2023
Revised: 12 August 2023
Accepted: 14 August 2023
Published: 21 August 2023

1. Introduction

This work describes an investigation of transient interference observed in radar testing. In this introductory section, we provide a general description of the investigation. We then follow with a summary of the different state-of-the-art for approaches and applications of transient signal identification.

1.1. Investigation Overview

When operating radar or radio receivers, it is not unusual to see interfering signals in addition to expected signals and ubiquitous thermal noise. For such interfering signals, typical questions center around the rate of occurrence, signal characteristics, the effects on the desired operation of the system, and whether interference can be reliably detected and possibly removed. When faced with the problem of detecting and characterizing unwanted transients in data, the analyst finds that a number of methods are available; however, picking which one or ones to use can be challenging. The next subsection provides a high-level summary and review of some of the available methods. To review and compare methods, we present a case study of a particular radio-frequency interference (RFI) signal occasionally seen in the noise-only testing of a radar receiver. This receiver is part of a spaceborne radar in development, with a launch planned in the near future. After surveying previous work on transient detection, we provide details about the radar receiver, test conditions, and initial observations of the transient in Section 2. Additionally, we provide the transient's characteristics, estimated manually using a standard spectrogram. In Section 3, several techniques for automatically detecting the presence of the transients in test datasets are discussed in some detail. Section 4 describes the results of applying these techniques to our data, with conclusions provided in Section 5.

1.2. Overview of State-of-the-Art

Applications of transient signal detection include radar, microwave radiometry, radio astronomy, underwater acoustics, and RF fingerprinting [1–6]. Methods have included traditional signal processing and, more recently, the application of machine learning techniques [6]. We summarize a number of methods here, grouping them by method rather than by application. The methods with perhaps the longest history have their foundations in statistical decision theory. A standard reference within the electrical engineering community is that of Van Trees [7]. This volume presents the classical detection theory of communications or radar signals in noise, including matched filtering for the detection of known signals. The paper by Friedlander and Porat [8] applies statistical analyses to linear transformations of signals, including common time-frequency representations. Such representations have also been used in some of the more recent machine learning procedures described below. A relatively recent review of statistical detection methods applied to transient problems is found in [9]. Following this review, the reference describes a new method based on sequential probability ratio tests. Another relatively recent work on statistical detection is described in [10]. The application used here is radar, and the approach is based on a generalized likelihood ratio test (GLRT) for detecting a Gaussian signal in Gaussian noise. As described in Section 3, GLRT denotes a test in which unknown parameters of the signal are estimated prior to applying the likelihood ratio test. A related work [11] addresses this problem using statistical tests other than GLRT. The overview of RF fingerprinting in [5] primarily covers statistical techniques that are often used in this application; however, the problems there tend to be more related to a change in the received signal than a finite signal in noise. This distinction is further described in [9]. As such, the methods in [5] are related to but are not necessarily directly applicable to RFI transient detection.

Many of the recent methods for transient detection use machine-learning approaches and are applied to areas like drone signature recognition and radio astronomy. The latter also uses statistical detection theory but is often generalized for typical radio astronomy measurements, such as using sensor arrays. The technique described in [6] examines signals emitted by drones. Such signals can be used to identify intruding or otherwise unwanted drones. In [6], received data are first transformed, e.g., wavelet, and then used as an input to a hierarchical classifier. In this application, as also noted for the RF fingerprinting applications in [5], the transient is usually a change in signal rather than one of finite duration (temporary) interference. Nevertheless, the problem here is related to RFI detection, especially when viewed using machine learning. This is also true of [12], which is again applied to drones and RF fingerprinting; it uses a neural network that is applied directly to the received complex data. The problem described in [13] is similar to that being investigated here, namely RFI in radar data. The approach there uses time-frequency images of the data as input to a neural network classifier, which can detect the RFI. An alternative machine learning approach is a support vector machine, which is applied in [14] to recognize transients from features also extracted from a time-frequency representation. Reference [15] describes the application of two types of neural networks for RFI detection. There, RFI is not only detected but classified as to its likely source. Both [14] and [15] describe astronomical applications, as does [16], which uses a neural network to directly classify time-frequency images.

This brief overview of methods, especially those discussed in recent publications, provides a background for the specific techniques that we use and compare. These are described in Section 3, following the discussion of the radar data and transient overview in the next section.

2. Data

The data used here were recorded by the radar. The radar receiver has a bandwidth of 80 MHz and operates with a carrier near 1.257 GHz (L-band). The received L-band data are filtered and sampled at 240 MHz and then are digitally down-converted to a 96 MHz complex sample rate. The Nyquist sampling criterion is satisfied because the complex sampling results in a single-sideband signal that fits unambiguously inside the 0–96 MHz spectrum. The data recorded within each interpulse period are referred to as a range line; each range line has 105,152 complex samples at a 96 MHz sample rate (roughly 1 ms in length). Each dataset used here contains up to about one hundred thousand range lines, corresponding to nearly a minute of the radar receiver operation, based on 0.5 ms between range lines. All eight datasets were acquired indoors in the noise-only mode with no transmitted waveform. The overall noise level depended primarily on the environment (noisy laboratory versus test chamber); the inherent gain and noise level of the receivers were stable.

The left panel in Figure 1 shows an example of a typical, noise-only range line. The very small amplitudes near 300 and 850 μs are due to receiver blanking during radar pulsing, although pulsing was not happening in this case. The statistics of these data showed approximate Gaussian behavior of the in-phase and quadrature (I/Q, or real and imaginary) parts of the data. This indicated relatively pure thermal noise, as would be expected for a high-quality receiver with no signal inputs. Additionally, as shown in Figure 1 (right panel) a spectrogram for this noise range line is presented. Although some harmonics or spurious tones (vertical lines) are present, these are low and not are expected to affect the radar measurements. Figure 2, left, shows a different range line that appears similar to that in Figure 1 but also contains an unexpected transient signal starting around time 900 μs. However, this level iss nearly the same as the noise; therefore, the transient is almost invisible in the time domain. The corresponding spectrogram for this range line, on the right, shows the transient much more clearly, highlighted by an ellipse. The transient appears to have a roughly linear frequency increase with time, making it approximately a linear, frequency-modulated (LFM) chirp, and increasing in frequency from 83 MHz to about 85 MHz as time increases from roughly 900 to 1000 μs.

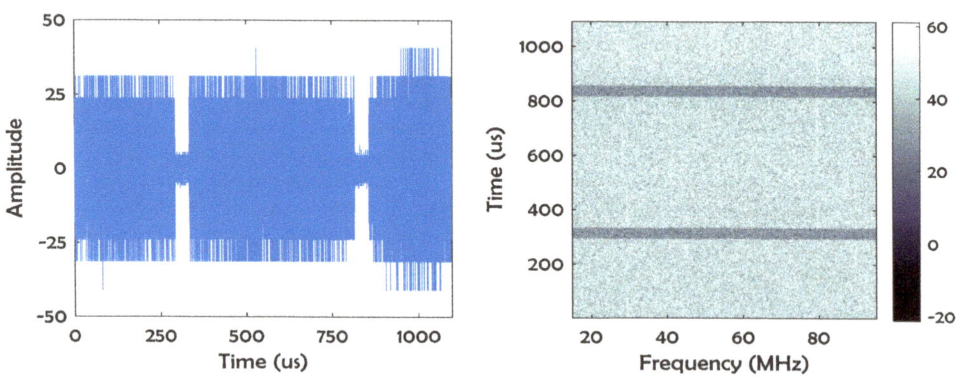

Figure 1. Left: Time domain plot of a typical range line with no transients. **Right**: Spectrogram of this same range line, plotting power in dB as a function of time on the vertical axis and frequency on the horizontal axis.

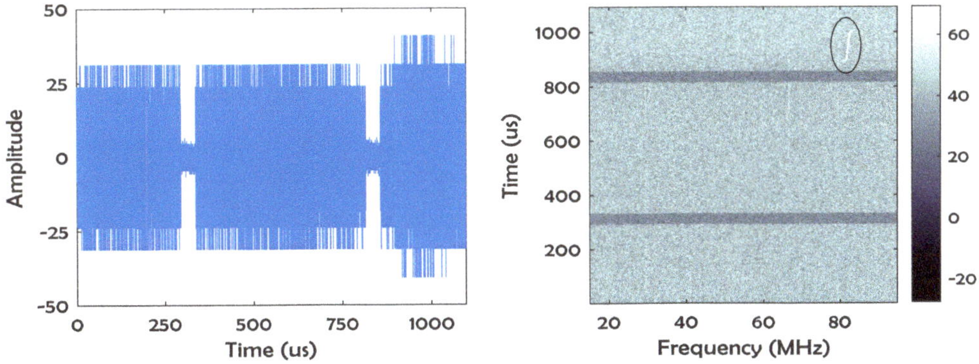

Figure 2. Same as Figure 1, except for a different range line, containing a highlighted transient.

3. Data Analysis Methodology

3.1. Computer Vision

A priority when investigating a newly noticed transient is the determination of how often it occurs in the data (i.e., rate of occurrence), followed by the determination of the characteristics of the transients. In light of a lack of visibility in the transient in the time domain (Figure 2, left), spectrograms were used (see right panels in Figures 1 and 2). These were computed using the Short-Time Fourier Transform, which computed the Discrete Fourier Transform of time segments in the data. As a first step, a simple video, or movie, of all the spectrograms in a dataset allowed the manual detection of some of the transients. For automated detection, several approaches were possible, as summarized in Section 1.2. One approach used early in our investigation included computer (or machine) vision [17,18], which could be applied to spectrograms. This method was considered because it can find arbitrary objects in images without being sensitive to the exact time or frequency within the spectrogram. Since computer vision algorithms extract features from images and use these features for decisions, we consider this a relative of the machine learning techniques mentioned in Section 1.2. Although the feature extraction approaches used in [6,14] might not be classical computer vision, the overall approach of feature extraction and decision-making is analogous to that considered here. While we are not aware of a specific application of computer vision to RFI transient detection, we note in [19], for example, the application of computer vision to the detection of anomalous signals or events for network security.

Because the noise is visible even in normal spectrograms (Figure 1, right), an estimated thermal noise level is used to set a threshold. All pixels below the threshold are set to zero, while all above are set to one, yielding a binary image. In some cases, there are linear features above the threshold, as shown in the binary images. These are found by summing the binary image both horizontally and vertically; the presence of a line results in a larger-than-average sum for a problematic row or column, allowing it to be removed. Following this, we use morphological image processing functions [17,20] to remove isolated pixels above the threshold. This processing results in images that are relatively clean, except for the transients, when present. Figure 3 shows the application of these steps to the spectrogram in Figure 2 on the right. Once we have a cleaned binary spectrogram, we apply a computer vision routine (called bwboundaries), which locates objects in a binary scene [20]. It returns information on the size, shape, and location of each object found. From the object dimensions and the spectrogram pixel sizes, we perform a somewhat crude estimate of the starting time and frequency, the bandwidth, and the duration of each transient. For the example transient in Figures 2 and 3, the function returned a set of boundary indices that allowed the code to calculate a duration of 88 μs and a bandwidth of 1.87 MHz. The code saves this information, along with the object's peak power, range line number, and range sample number for each detection. This allows an examination of the statistics of the transient characteristics over the dataset, which is discussed in Section 4. In the remainder of this paper, we denote this approach as a computer vision algorithm or CVA.

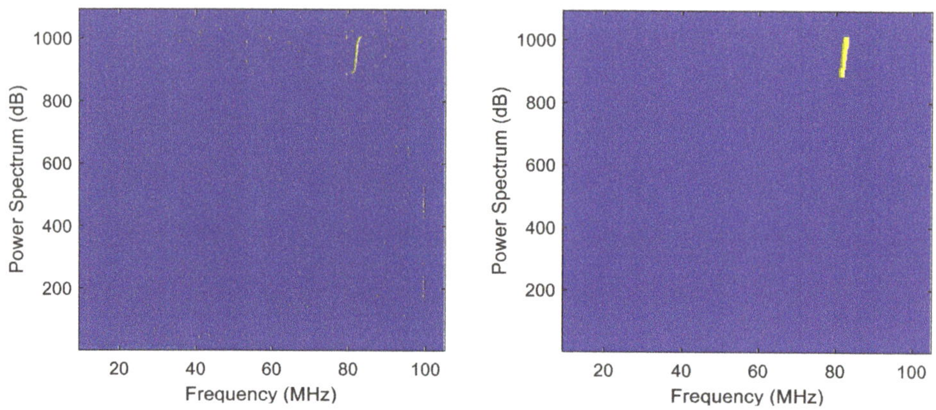

Figure 3. Example of spectrogram processing steps prior to computer vision. On the **left** of the spectrogram, the linear power domain has been thresholded to eliminate noise; blue pixels are 0 and yellow pixels are 1. On the **right**, the thresholded image has been further processed with line removal filtering and morphological image processing. The transient here corresponds to that in Figure 2.

3.2. *Convolutional Neural Network*

Another method considered was the convolutional neural network (CNN) [21]. This approach is already noted in Section 1.2 as being used for RFI detection in radar data [13]. To avoid conducting pre-processing, we let the CNN operate directly on the spectrograms [13,16]. This was reasonable since CNNs excel at extracting features from images and using them for classification. While one could train deep CNNs from scratch, such training typically requires hundreds of thousands of images, which were not available in this case. Instead, we used the approach of transfer learning. As described in [16,21–23], the general idea of transfer training is to start with a pre-trained deep network and adapt it for a new task by replacing the last hidden layer and the output layer and re-training on a much smaller, task-specific dataset. The studies described in [16,22] demonstrate RFI detection using CNNs and transfer learning, with the input data in [22] based on continuous wavelet transform, which is analogous to the spectrograms used here. For the problem of transient RFI detection in our radar noise data, we also started with a pre-trained CNN, in this case, SqueezeNet [21,23]. SqueezeNet is a CNN with 18 convolutional layers that are designed to classify color images of size $227 \times 227 \times 3$ into 1000 different categories. It was trained on over 1 million input images; image categories included common scenes and objects, such as various animals, pencils, keyboards, and coffee mugs. While these categories are very different from spectrograms with RFI, the method can work because much of a deep network, starting with the input layers, is good at extracting features. The retraining of output layers changed the way the network used the extracted features for the new problem. For the Matlab implementation of SqueezeNet, we replaced the last convolutional layer with a new convolutional layer, setting the number of filters in the layer to match the number of classes, which was two in this case, corresponding to either RFI or noise. The output classification layer was also replaced. Prior to training, the learning rate for the new convolutional layer was set to a much larger number than the rate for the original SqueezeNet layers. When training was started, these different rates caused only the new layer to be trained. This transfer learning technique allowed the deep CNN to be re-trained in only a few minutes on a typical laptop.

3.3. Statistical Decision Theory

Lastly, we considered methods that are traditionally considered statistical detection methods within the signal-processing community. The approach here is similar to the statistical approaches surveyed in Section 1.2. The basic strategy is to make the choice that optimized some error criterion. As described in [7,24–26], the general strategy computes a likelihood ratio and compares this with a threshold. Using the notation of [26], the likelihood ratio is the ratio of the conditional probability of the data with the signal present $f_1(\tilde{x})$ compared to the conditional probability of the data when the signal is not present $f_0(\tilde{x})$, where $\tilde{x}$ is the observed data, in our case, a noisy range line that either has or does not have a transient signal in it. In the simplest model, the signal to be detected is fully deterministic. The optimal detection method is the matched filter (MF), in which a deterministic signal is cross-correlated with the data [7,24–26]. The optimal detection approach for the case of a set of unknown parameters ψ is also discussed in [7,24–26]. This problem becomes a computation of a so-called generalized likelihood ratio, which is the ratio of the maximum of $f_1(\tilde{x};\psi)$ over the possible ψ when the chirp is present to the likelihood function $f_0\left(\tilde{x}\right)$ with no chirp. The GLRT is discussed in Section 1.2 in relation to [10] and was needed in our case since we did not have a priori knowledge of the chirp starting frequency and rate. We implemented this case by correlating with a chirp whose parameters were estimated with the maximum likelihood estimation (MLE). In [27], it is shown that the MLE of chirp parameters could be found by maximizing the peak of the cross-correlation of a model chirp with the data. The search was carried out over the set ψ, here consisting of the starting frequency f_0 and the chirp rate α; these parameters were chosen to maximize:

$$X(f,\alpha) = \left|\sum x_i\, e^{j2\pi\left(f_0 t_i + \frac{\alpha}{2} t_i^2\right)}\right| \quad (1)$$

where j is the square root of -1, $t_i = (i-1)\Delta T$, ΔT is the sampling interval, and the complex samples x_i represent the observed data. The exponential was complex conjugated to correspond to the usual definition of cross-correlation. While matched filtering was also used in the power domain (e.g., in image processing), the use of complex data was preferred so that the output SNR could be boosted by the effect of coherent integration. Hence, the above approach was preferred on theoretical grounds to the matched filtering of the spectrogram by a time-frequency representation of the chirp power. Details of MLE for chirp parameters are provided in Appendix A.

4. Results

4.1. Generation of Training and Validation Data

Following the initial notice of our transient in a few spectrograms, our investigation turned to the methods described above in Section 3. To provide quantitative results for these methods, we created subsets for training and validation by arbitrarily selecting range lines within the various full datasets noted in Section 2. The subsets needed to be large enough that sufficient data could be used but small enough that the classifications could be verified by a human viewing of their contents. Our procedure used the results of the CVA to create 3000 spectrograms, half with transients and half without. All 3000 spectrograms were manually inspected to check for correct identification (requiring several seconds per image). Incorrect classifications were noted for CVA accuracy estimates and then replaced with new spectrograms; hence, the final set of 3000 contained the desired half-and-half mix. This set was used for the training and validation of the CNN, as described in Section 4.3 below. In this case, 1020 of the 3000 spectrograms were used for independent validation. In the generation of the spectrograms, the transients were shifted to other, randomly selected frequencies to keep the CNN from memorizing cases with one frequency. The validation set had 510 spectrograms with transients and 510 without, which were verified manually. The CVA output was also used to identify a set of complex data (range lines) for the testing of the matched filtering method (described in Section 4.4). As with the CNN training data, the range lines were manually inspected to verify the assessment and the presence/absence of transients. Hence, the complex validation set had 510 range lines with transients and 510 without, matching the CNN validation set.

4.2. CVA Results

As noted in Section 3.1, for automated detection and characterization, CVA is a good first choice because it can detect fairly arbitrary objects, requiring minimal assumptions about the data. Additionally, it provides not only detections but the basic characteristics of the objects it finds. Figure 4 shows example histograms from a selected dataset with transients. There is no averaging in the values shown; each value of a characteristic corresponds to one transient. Duration (time from transient start to stop) tends to be near 100 μs, while bandwidths are mostly less than a few MHz. The transients seemed to occur at several different frequencies, while the starting time within a range line (denoted as fast time) could be anywhere from 0 to nearly 1 ms. Figure 4c shows the occurrence of transients detected using CVA in slow time (proportional to the pulse number); they tended to occur in clusters separated by roughly 12 s. As all timing on the radar is much faster than 12 s, this result suggests a source outside the radar. Zooming into this plot, which is not shown, the fast time looks mostly random, but we could see evidence of systematic changes in fast time location between the adjacent pulses. This varying and generally random fast-time behavior (Figure 4d) suggested that the transients were not synchronized with radar pulsing and were likely external. The hypothesis that the transients are external is further supported by their non-occurrence in subsequent testing in a different environment.

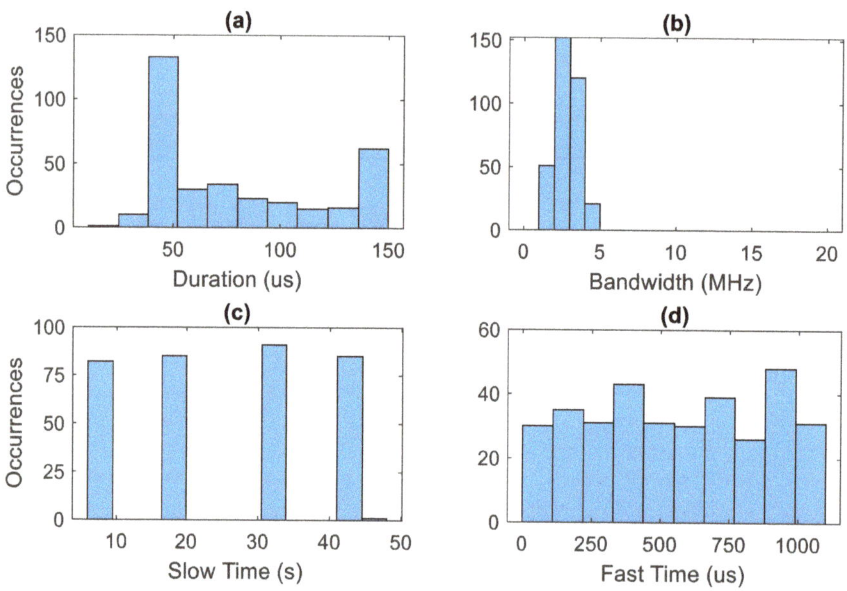

Figure 4. Results of applying the computer vision algorithm (CVA) (**a**) Transient duration, (**b**) Bandwidth, (**c**) Slow time (time location relative to start of dataset), and (**d**) Fast time (time location relative to the start of the receive window).

To quantitatively evaluate the accuracy of the CVA, we refer back to Section 4.1 and the generation of the spectrograms. For the 3000 spectrograms initially selected by CVA, a manual inspection found 10 false alarms, corresponding to about 4 false alarms in the test, or validation, data set. This result is reflected in the value of 1016 out of 1020 spectrograms that were correctly identified by CVA (99.6%), as shown in Table 1. These results for CVA support our claim that it can successfully detect transient events and can provide useful characteristics; however, it relies, to some extent, on knowing how to set the threshold to separate transients from noise. This and some additional parameters of the algorithm, including the size of the erosion and dilatation filters, are best set by experimenting with CVA performance.

Table 1. Comparison of RFI detection methods.

Method	Est. Run Time (s)	Validation Data	Correct
Computer Vision Algorithm (CVA)	40	1020 spectrograms	1016
Convolutional Neural Net (CNN)	45	1020 spectrograms	1007
Matched filter (MF)	25	1020 range lines	990

4.3. CNN Results

The next method evaluated was the deep learning of spectrograms with a CNN. Since CNNs operate directly on the spectrogram of each range line, there were no explicit thresholds to be set. However, unlike CVA, this method requires accurately trained data and so is typically not a good first choice for the analysis of a newly discovered transient signal. Indeed, as noted in Section 4.1, both the CNN and the statistical detection algorithms rely on the prior use of CVA for the preliminary characterization of the transients and to detect a sufficient number of them so that training data that can be developed. Figure 5 shows the results of re-training with the SqueezeNet network, which was conducted with the transfer learning approach. The final accuracy after 1320 training iterations was 98.7% on the independent validation data, where accuracy is the percentage of correct classifications for both signals present and absent. Hence, CNN detects about the same number of transients as that detected by CVA (Table 1). CNN is also relatively robust since the validation data consist of transients at differing chirp rates and time-frequency locations. Indeed, it is highly likely that the validation data contained transients with frequencies not represented in the training data, indicating generalization by the CNN.

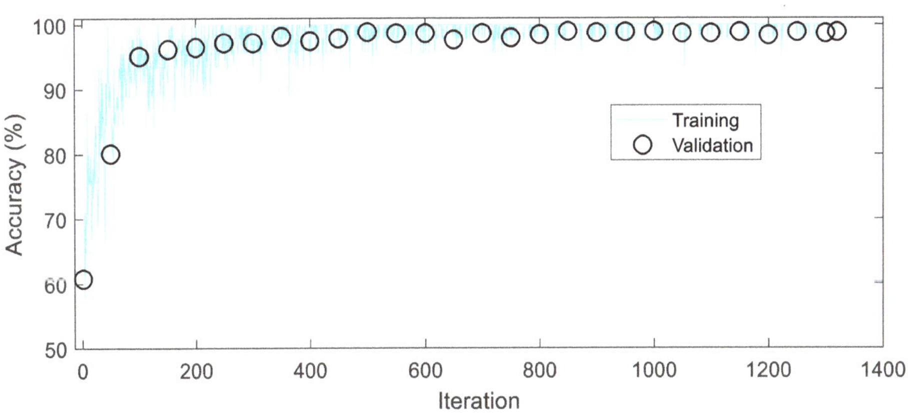

Figure 5. Training of CNN for spectrograms with and without transients. The black circles are for the testing of the network on the 1020-spectrogram validation data, which was not used in training.

4.4. MF Results

This section reports on the statistical detection of transients, implemented here as matched filtering. We note that this matched filter, or MF, processing is identical to that in pulse compression radars, in which the transmitted waveform is correlated with the received signal. MFs using observed data and an ideal chirp are shown in Figure 6. The observed chirp was extracted from noisy range line data (e.g., Figure 2, left) using a filtering technique that we developed, retaining the chirp while removing most of the noise. The observed chirp has significant amplitude modulation, unlike the ideal chirp. Furthermore, the amplitude modulation could vary between observations. We tested both filters by applying them to the 1020 lines of complex validation data. To use the observed chirp, its frequency was shifted to match transients in the validation data set. We found that with a well-selected threshold, the detection rate was 97%, with either the ideal or observed chirps used as an MF (Table 1). However, estimating chirp parameters with MLE and then generating a corresponding ideal chirp was somewhat simpler than trying to extract and modify an observed pulse from another dataset. Figure 7 illustrates the performance of a matched filter on example range lines.

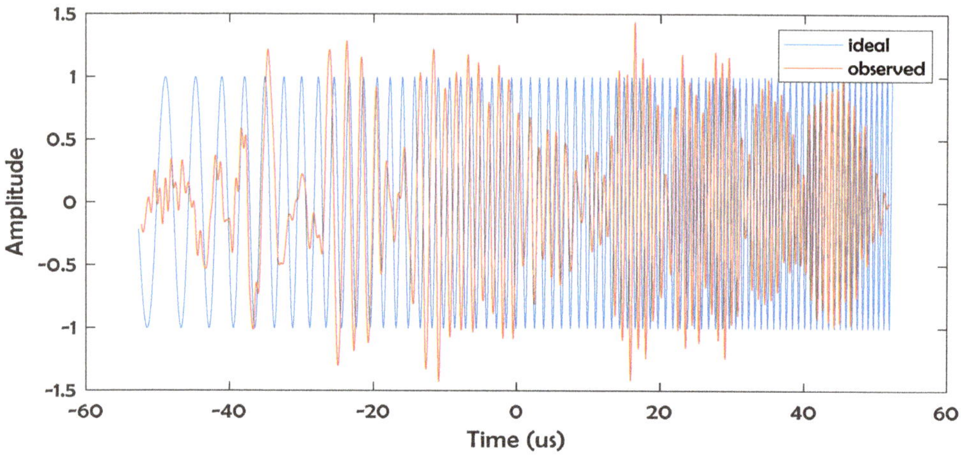

Figure 6. Transient signal extracted by filtering a single range line, and ideal chirp using chirp parameters estimated from the data.

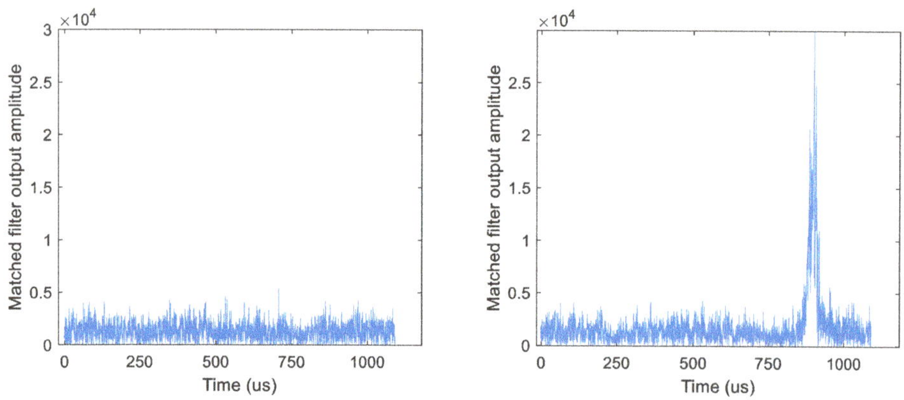

Figure 7. Left: Noise-only data after filtering with a matched filter. **Right**: Data containing a transient after matched filtering.

An inspection of the incorrectly classified range lines verified that all had transients that were missed. To determine the reason for each miss, we looked at the power spectrum of the corresponding range line. We found that all apparent misses did, in fact, contain a transient but with different start frequencies or chirp rates than those assumed in the matched filter. While it would be feasible to apply a combination of MLE and MF on each range line, the time required to search the entire range line would be significant, likely resulting in slower run times than CVA. If the frequencies and rates of all the transients were somehow known, a simpler algorithm could apply a set of matched filters to the data, with the set being based on all the known transient signals. The test statistic, in this case, would be the maximum of all the outputs on a given range line; based on the above tests, such an algorithm should detect all the transients. The missed detections here highlight MF's lack of ability to generalize.

5. Conclusions

We have reported on a case study of a specific type of transient observed in noise-only radar testing. The purpose of this work was to compare several techniques that could be used in such a situation. While our particular transient was approximately a linear frequency-modulated chirp, we believe that the approach here could be generalized to many kinds of radio-frequency interference occurring in test data. Once the transients were noted in our data, we began an investigation with computer vision techniques to detect unknown "objects" within spectrograms or their equivalent. While we did need to estimate the level of background noise in the spectrograms for thresholding, the algorithms were able to clean up the thresholded spectrograms and found objects with relatively little adjustments and a priori knowledge. An exception is morphological processing, which did require some changes based on the experiment to better keep the objects of interest while removing noise. We found the computer vision approach to be extremely useful, providing most of the information needed for characterizing the transients in our data, as well as allowing the creation of training and validation data for other methods. Because it does not depend on the transient being a linear frequency-modulated chirp, we believe that computer vision can be a useful first step for investigating a variety of transients.

The convolutional neural network, trained via transfer learning, proved to be accurate and had the advantage of working directly from spectrograms without the need for thresholds or other adjustable parameters. However, without first using the computer vision algorithms, identifying sufficient training data for the neural network was very difficult. Furthermore, the training data had to be rather carefully generated, making sure to cover the basic ways in which the transient could occur in a spectrogram. As with the computer vision approach, the neural network can be applied to many types of transients. The last method evaluated was statistical detection using matched filtering. The matched filter applied to the complex range lines was quite sensitive (Figure 7). However, this approach required a reference function that matched the transient interference signal of interest. Consequently, one needed to estimate the transient parameters and then generate the matched filter. This process can be applied to many types of transient signals. Once the transient has been characterized, more general models, such as in [28,29], could be used, with the maximum likelihood estimation of model parameters. However, the MF approach is best used only if it is already known that the transients are all identical or have only a few sets of parameters. The experience here confirms that all the methods can be very accurate but differ significantly in their need for a priori information and their ability to generalize. While computer vision recognizes fairly arbitrary objects, CNNs are likely more specific but can generalize. Matched filters only use their given parameters and do not generalize well to transients with significantly different parameters. To summarize, the following approach for investigating arbitrary transients is suggested: (1) A computer vision for obtaining general characteristics of objects in spectrograms or other time-frequency images. Once training data can be identified, a neural network approach (2) using transfer learning could potentially achieve high accuracy with no assumptions beyond the training data. Finally, classical matched filtering methods (3) could potentially provide a very high detection accuracy; however, the transient properties must be well-known to allow for the construction of an accurate matched filter or filters.

Author Contributions: Conceptualization, S.L.D. and S.J.S.; Methodology, S.L.D. and V.A.V.; Software, S.L.D. and V.A.V.; Validation, S.L.D.; Investigation, S.L.D. and V.A.V.; Writing—original draft, S.L.D.; Writing—review & editing, V.A.V. and S.J.S.; Supervision, S.J.S. All authors have read and agreed to the published version of the manuscript.

Funding: This research was funded by the US National Aeronautics and Space Administration (80NM0018D0004).

Institutional Review Board Statement: Not applicable.

Informed Consent Statement: Not applicable.

Data Availability Statement: The data used here are archived along with all the test data from this program. The full set of data is not publicly available; however, inquiries about the data can be made to the corresponding author.

Acknowledgments: The research described here was performed by the Jet Propulsion Laboratory, California Institute of Technology, under a contract with the US National Aeronautics and Space Administration (80NM0018D0004). N. Niamsuwan of JPL initiated this work, starting with noticing the transients in a few spectrograms and creating videos. He also provided guidance and encouragement along the way.

Conflicts of Interest: The authors declare no conflict of interest.

Appendix A

In this appendix, we briefly summarize the theory underlying the maximum likelihood estimation (MLE). This method is covered in detail in numerous sources, including [7,26]. For a general transient signal (vector) $\tilde{s}$, the observed data is $\tilde{r} = \tilde{s} + \tilde{n}$, where $\tilde{n}$ is Gaussian white noise. Then, let the subscript i be the time index of each signal component; hence the element of $\tilde{r}$ for time t is r_i with $t = (i-1)\Delta$ and Δ is the spacing between samples. Since the noise is complex Gaussian and white, with total variance σ_n^2, the joint probability density function (PDF) for the noise signal can be written as:

$$p(\tilde{n}) = \left(\pi\sigma_n^2\right)^{-N}\prod_{i=1}^{N} \exp(-|n_i|^2/\sigma_n^2) \quad \text{(A1)}$$

Letting ψ represent a set of parameters characterizing the signal $\tilde{s}$, the conditional PDF of the "signal plus noise" vector becomes:

$$p(\tilde{r}|\psi) = \left(\pi\sigma_n^2\right)^{-N}\prod_{i=1}^{N} \exp(-|r_i - s_i|^2/\sigma_n^2) \quad \text{(A2)}$$

Here, ψ denotes a parameter vector, as opposed to the accent over the signal symbols, indicating data vectors. The maximum likelihood (ML) method assumes that the best estimate of the unknown parameters are those values that maximize the probability of the observed data. Specifically, these parameters were chosen to simultaneously maximize the conditional joint PDF, or, equivalently, its natural logarithm, which is known as the log-likelihood function [7]. This was derived from (A2) by expanding $|\tilde{r}_i - \tilde{s}_i|^2$ and then taking the natural logarithm:

$$\Lambda(\tilde{r}|\psi) \equiv \ln[p(\tilde{r}|\psi)] = -N\ln(\pi\sigma_n^2) + \tfrac{2}{\sigma_n^2}\mathrm{Re}\left(\textstyle\sum_{i=1}^{N} r_i s_i^*\right) - \tfrac{1}{\sigma_n^2}\textstyle\sum_{i=1}^{N}|s_i|^2 - \tfrac{1}{\sigma_n^2}\textstyle\sum_{i=1}^{N}|r_i|^2 \quad \text{(A3)}$$

Equation (A3) provides the likelihood function for a general signal $\tilde{s}$ with unknown parameters ψ.

For the specific case of a chirp, $\widetilde{s}$ has the form $A\exp(j\theta)\widetilde{s}_0(\tau, f_0, \alpha)$, where τ is the time delay, f_0 is the start frequency, and α is the chirp rate. Substituting this form for $\widetilde{s}$ into (A3) yields:

$$\Lambda(\widetilde{r}|\psi) = -N\ln(\pi\sigma_n^2) + \frac{2A}{\sigma_n^2}\mathrm{Re}\left\{\exp(j\theta)\sum_{i=1}^{N} r_i s_{0,i}^*]\right\} - \frac{NA^2}{\sigma_n^2} - \frac{1}{\sigma_n^2}\sum_{i=1}^{N}|r_i|^2 \quad \text{(A4)}$$

The amplitude A and phase θ are nuisance parameters that are of no interest, but must be estimated to derive the max-likelihood estimates of the desired parameters τ, f_0, and α. Ignoring terms that do not explicitly contain the parameters of interest (and cannot contribute to the maximization), we obtained the following simplified version of (A4):

$$\Lambda(\widetilde{r}|\psi) = \mathrm{Re}\left\{\exp(j\theta)\sum_{i=1}^{N} r_i s_{0,i}^*\right\} \quad \text{(A5)}$$

For any complex number z, the expression $\mathrm{Re}\{z\exp(j\theta)\}$ was maximized with respect to θ when we let $\theta = \arg(\widetilde{z})$, resulting in a value of $|\widetilde{z}|$ for the expression. Letting $z = \sum_{i=1}^{N} r_i s_{0,i}^*$ in (A5) and carrying out the maximization yielded the ML estimate of phase, $\hat{\theta}$, at any value of the delay:

$$\hat{\theta} = \arctan\left[\frac{\mathrm{Im}\left(\sum_{i=1}^{N} r_i s_{0,i}^*\right)}{\mathrm{Re}\left(\sum_{i=1}^{N} r_i s_{0,i}^*]\right)}\right] \quad \text{(A6)}$$

Substituting this estimate into the simplified log-likelihood function $\Lambda(\widetilde{r}|\psi)$ maximized it with respect to θ for any values of τ, f_0, and α, yielding $|\widetilde{z}|$; hence we can write:

$$\underset{\theta}{\mathrm{argmax}}\ \Lambda\left(\widetilde{r}\middle|\psi\right) = \underset{\theta}{\mathrm{argmax}}\ \mathrm{Re}\left\{\exp(j\theta)\sum_{i=1}^{N} \widetilde{r}_i \widetilde{s}_{0,i}^*\right\} = \left|\sum_{i=1}^{N} r_i s_{0,i}^*\right| \quad \text{(A7)}$$

From (A8), joint estimates of τ, f_0, and α can now be expressed as:

$$(\hat{\tau}, \hat{f}_0, \hat{\alpha}) = \underset{\tau, f_0, \alpha}{\mathrm{argmax}}\Lambda_0(\widetilde{r}|\psi) = \underset{\tau, f_0, \alpha}{\mathrm{argmax}}\left|\sum_{i=0}^{N-1} r_i s_{0,i}^*(\tau, f_0, \alpha)]\right| \quad \text{(A8)}$$

Numerical optimization is normally used to find the estimates provided by (A8). When τ is known (e.g., after using computer vision to isolate a set of transients), it can be fixed, so that the maximization is only over f_0 and α, as conducted in (1) in the main text.

References

1. Zhou, H.; Wen, B.; Wu, S. Dense radio frequency interference suppression in HF radars. *IEEE Signal Process. Lett.* **2005**, *12*, 361–364. [CrossRef]
2. Misra, S.; Kristensen, S.S.; Sobjaerg, S.S.; Skou, N. CoSMOS: Performance of kurtosis algorithm for radio frequency interference detection and mitigation. In Proceedings of the IGARSS, Barcelona, Spain, 23–28 July 2007; pp. 2714–2717.
3. Leshem, A.; van der Veen, A.-J.; Deprettere, E. Detection and blanking of GSM interference in radio-astronomical observations. In Proceedings of the 2nd IEEE Workshop on Signal Processing Advances in Wireless Communications, Annapolis, MD, USA, 9–12 May 1999; pp. 374–377.
4. Yang, D.; Xiao, D.; Zhang, L. The parameters estimation and the feature extraction of underwater transient signal. In Proceedings of the IEEE International Conference on Signal Processing, Communications and Computing (ICSPCC), Xi'an, China, 14–16 September 2011; pp. 1–4.
5. Soltanieh, N.; Norouzi, Y.; Yang, Y.; Chandra Karmakar, N. A review of radio frequency fingerprinting techniques. *IEEE J. Radio Freq. Identif.* **2020**, *4*, 222–233. [CrossRef]
6. Medaiyese, O.O.; Ezuma, M.; Lauf, A.P.; Adeniran, A.A. Hierarchical learning framework for UAV detection and identification. *IEEE J. Radio Freq. Identif.* **2022**, *6*, 176–188. [CrossRef]
7. Van Trees, H.L. *Detection, Estimation, and Modulation Theory Part I: Detection, Estimation, and Linear Modulation Theory*; Wiley: New York, NY, USA, 2001.

8. Friedlander, B.; Porat, B. Performance analysis of transient detectors based on a class of linear data transforms. *IEEE Trans. Inf. Theory* **1992**, *38*, 665–673. [CrossRef]
9. Guepie, B.K.; Fillatre, L.; Nikiforov, I. Detecting a Suddenly Arriving Dynamic Profile of Finite Duration. *IEEE Trans. Inf. Theory* **2017**, *63*, 3039–3052. [CrossRef]
10. Besson, O.; Coluccia, A.; Chaumette, E.; Ricci, G.; Vincent, F. Generalized likelihood ratio test for detection of Gaussian rank-one signals in Gaussian noise with unknown statistics. *IEEE Trans. Signal Process.* **2017**, *65*, 1082–1092. [CrossRef]
11. Besson, O. Adaptive detection of Gaussian rank-one signals using adaptively whitened data and Rao, gradient and Durbin tests. *IEEE Signal Process. Lett.* **2023**, *30*, 399–402. [CrossRef]
12. Yang, J.; Gu, H.; Hu, C.; Zhang, X.; Gui, G.; Gacanin, H. Deep complex-valued convolutional neural network for drone recognition based on RF fingerprinting. *Drones* **2022**, *6*, 374. [CrossRef]
13. Itschner, S.; Li, X. Radio frequency interference (RFI) detection in instrumentation radar systems: A deep learning approach. In Proceedings of the IEEE Radar Conference, Boston, MA, USA, 22–26 April 2019; pp. 1–5.
14. Jiang, M.; Cui, B.; Yu, Y.-F.; Cao, Z. DM-Free curvelet based denoising for astronomical single pulse detection. *IEEE Access* **2019**, *7*, 107389–107399. [CrossRef]
15. Czech, D.; Mishra, A.; Inggs, M. A CNN and LSTM-based approach to classifying transient radio frequency interference. *Astron. Comput.* **2018**, *25*, 52–57. [CrossRef]
16. Agarwal, D.; Aggarwal, K.; Burke-Spolaor, S.; Lorimer, D.R.; Garver-Daniels, N. FETCH: A deep-learning based classifier for fast transient classification. *MNRAS* **2020**, *497*, 1661–1674. [CrossRef]
17. Shapiro, L.G.; Stockman, G.C. *Computer Vision*; Prentice-Hall: Englewood Cliffs, NJ, USA, 2001.
18. Davies, E.R. *Machine Vision: Theory, Algorithms, Practicalities*; Morgan-Kaufmann: Amsterdam, The Netherlands, 2005.
19. Zhao, J.; Masood, R.; Seneviratne, S. A review of computer vision methods in network security. *IEEE Commun. Surv. Tutor.* **2021**, *23*, 1838–1878. [CrossRef]
20. MathWorks. *Matlab Image Processing Toolbox User's Guide*; The Mathworks: Natick, MA, USA, 2022.
21. Khan, S.; Rahmani, H.; Shah, S.A.A.; Bennamoun, M. *A Guide to Convolutional Neural Networks for Computer Vision*; Springer: Cham, Switzerland, 2018.
22. Ujan, S.; Navidi, N.; Landry, R., Jr. An efficient radio frequency interference (RFI) recognition and characterization using end-to-end transfer learning. *Appl. Sci.* **2020**, *10*, 6885. [CrossRef]
23. Beale, M.H.; Hagan, M.T.; Demuth, H.B. *Matlab Deep Learning Toolbox User's Guide*; The Mathworks: Natick, MA, USA, 2022.
24. Garth, L.M.; Poor, H.V. Detection of non-Gaussian signals: A paradigm for modern statistical signal processing. *Proc. IEEE* **1994**, *82*, 1061–1095. [CrossRef]
25. Kailath, T.; Poor, H.V. Detection of stochastic processes. *IEEE Trans. Inf. Theory* **1998**, *44*, 2230–2259. [CrossRef]
26. Abraham, D.A. *Underwater Acoustic Signal Processing*; Springer: Cham, Switzerland, 2019.
27. Abatzoglou, T.J. Fast maximum likelihood joint estimation of frequency and frequency rate. *IEEE Trans. Aerosp. Electron. Syst.* **1986**, *AES-22*, 708–715. [CrossRef]
28. Boyer, R.; Abed-Meraim, K. Damped and delayed sinusoidal model for transient signals. *IEEE Trans. Signal Process.* **2005**, *53*, 1720–1730. [CrossRef]
29. Golden, S.; Friedlander, B. Maximum likelihood estimation, analysis, and applications of exponential polynomial signals. *IEEE Trans. Signal Process.* **1999**, *47*, 1493–1501. [CrossRef]

eng

MDPI

Article

Cold Plasma Technology for Tomato Processing By-Product Valorization: The Case of Tomato Peeling and Peel Drying

Maxim Sosnin, Egor Berestenko, Emad H. A. Mounassar and Ivan Shorstkii *

Advanced Technologies and New Materials Laboratory, Kuban State Technological University, 2, Moskovskaya Str., Krasnodar 350072, Russia
* Correspondence: i-shorstky@mail.ru

Abstract: The tomato processing industry is focused on product yield maximization, keeping energy costs and waste effluents to a minimum while maintaining high product quality. In our study, cold atmospheric plasma (CAP) pretreatment enhanced tomato processing to facilitate peelability, a specific peeling process, and enhance peel drying. Peeling force analysis determined that CAP pretreatment of whole tomatoes improved peelability under the conditions used. The specific peeling force after CAP treatment decreased by more than three times. It was observed that cold atmospheric plasma pretreatment reduced the duration of infrared drying of tomato peels by 18.2%. Along with that, a positive effect on the reduction of the specific energy consumption of peel drying was shown for CAP-pretreated tomato peels. The obtained data show that the technology of cold atmospheric plasma pretreatment, in particular, when processing whole tomatoes and tomato peels, has a promising application in industry, as it can significantly reduce the specific energy consumption for peeling and drying procedures.

Keywords: cold plasma; process engineering; drying; by-product; tomato peel; energy efficiency

Citation: Sosnin, M.; Berestenko, E.; Mounassar, E.H.A.; Shorstkii, I. Cold Plasma Technology for Tomato Processing By-Product Valorization: The Case of Tomato Peeling and Peel Drying. *Eng* **2023**, *4*, 2167–2177. https://doi.org/10.3390/eng4030124

Academic Editor: Antonio Gil Bravo

Received: 20 July 2023
Revised: 11 August 2023
Accepted: 14 August 2023
Published: 15 August 2023

1. Introduction

The use of advanced electrical technologies in food processing engineering is a global trend for sustainable development [1,2]. The possibility of reducing the costs of implementing food processes to obtain safe products allows for facilitating the industrial development of such technologies. At the same time, the effectiveness of the suggested emerging technologies must be considered for each specific task.

The removal of the integumentary tissue (skin) from plant materials is an important preparation stage of industrial tomato processing. This operation is in high demand for tomato paste production as well as jams, fruit and vegetable purees [3,4]. The degree of by-product valorization depends on the efficiency and degree of purification of the integumentary tissue of the tomato fruits [5,6]. The cell membrane is a barrier for the transfer of intracellular components (moisture, juice and other dissolved substances) from the plant structure during industrial processing. According to research data, the energy costs for the peeling procedure of tomato fruits can reach 15% of the total energy costs of processing [5]. Progressive climate change, as well as increased control by regulatory authorities on the level of carbon emissions, contribute to the active search for alternative treatment methods to minimize energy costs. In this regard, the use of modern and effective methods of preparing tomato fruits for the peeling procedure, as well as further processing analysis, is of great theoretical and practical interest.

The valorization of tomato processing by-product is of great interest as an alternative pectin source [3,7]. Grassino [7] noted that fresh tomato pomace contains about 32% protein and 30% carbohydrates, making it a valuable by-product. Several authors have noted that the content of pectin substances in the integumentary tissue of tomatoes can reach 25% [8].

The effective peeling is hindered by the strong connection of the mesocarp with the integument of tomato tissue [6]. To weaken this connection, various methods of tomato preparation are used in the industry: physical (thermal), steam-thermal, mechanical, chemical and a combination of suggested methods [9–12]. The currently available methods have both advantages and disadvantages. For example, when processing via the thermal method, the phenomenon of local overheating of the structure of tomato fruits occurs, which decreases the overall quality characteristics. The mechanical method of tomato preparation in the form of an incision in the integumentary tissue improves the efficiency of skin removal. However, mechanical preparation can be used only in combination with thermal or steam-thermal methods.

In this paper, an emerging pretreatment method of tomato fruit using cold atmospheric plasma (CAP) to enhance peelability and peel drying is considered. Cold plasma treatment for fruit and vegetable processing is a novel and prospective method that has appeared in recent years. The mechanism of CAP technology is based on the use of short electrical high-voltage plasma discharges, which cause the electroporation of cell membranes in the air gap [13–15]. Depending on the specific energy consumption, CAP technology can significantly enhance the mass transfer of intracellular compounds [16] and lead to microbial inactivation [17]. Recently, Bao et al. [18] developed the application of cold plasma pretreatment for improving phenolic extractability from tomato pomace. Extracts from plasma-treated tomato pomace had a higher total phenolic content and antioxidant capacity. An overview of the cold plasma preprocessing for the extraction of bioactive compounds was reported by Bezerra et al. [19] and Du et al. [15].

From a thermodynamic point of view, the resulting intense mass transfer in plant materials subjected to electroporation is caused by the formation of a large number of micropores spontaneously and randomly located on the surface of the material along the lines of electric field strength [20]. CAP pretreatment positively affects the dynamics of mass transfer in plant materials due to changes in volumetric porosity. In comparison with other electroporation-based methods, such as pulsed electric field (PEF), CAP treatment can be applied in the air gap. Andrew et al. [21] reported a significant peeling enhancement using pulsed electric field pretreatment for tomatoes. It was found that PEF treatment (0.5–1.5 kV/cm, 0–8000 pulses, 15 μs pulse width) reduced the work required for peel detachment by up to 72.3%. Currently, CAP technology is actively used in various areas of the food industry to improve technological processes, for example, in the production of juices, drying and extraction of plant materials [22–24]. At the same time, the CAP technology is already carried out in a continuous mode of transportation on a scraper conveyor [25].

The purpose of this work is to study the effect of cold plasma treatment on the efficiency of tomato peeling and tomato peel drying, with an assessment of specific energy consumption.

2. Materials and Methods

2.1. Materials

An industrial tomato processing technology based on a local manufacturer (Krasnodar, Russia) with cold plasma application was used as the engineering object in this study. The experimental scheme is shown in Figure 1. Two major processes were chosen for the application of cold plasma technology: tomato skin removal and tomato skin drying.

To compare the effect of CAP technology, the following processing protocols were considered and analyzed: protocol A, whole tomato + cold plasma treatment + peeling; protocol B, whole tomato + peeling; protocol C, tomato peel + CAP treatment + infrared drying; protocol D, tomato peel + infrared drying. In our experiments, we used tomatoes of the Aurora variety as the object of research. The tomatoes were obtained from a local manufacturer (Krasnodar, Russia).

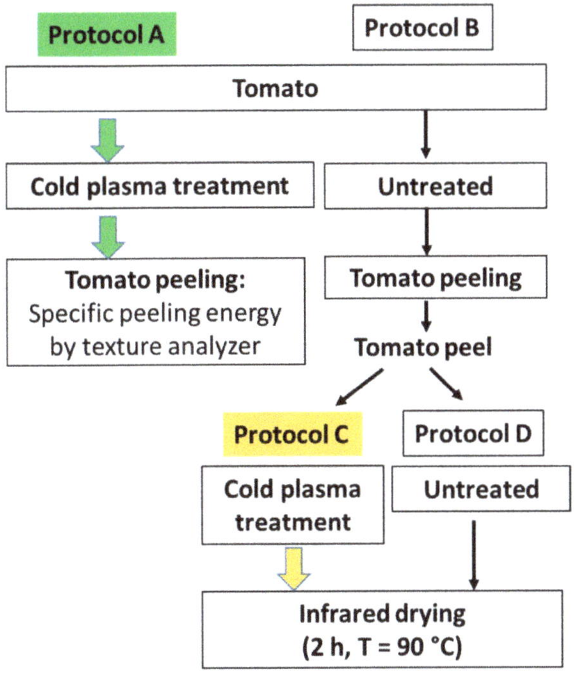

Figure 1. Experimental scheme of tomato processing with cold plasma implementation.

2.2. Cold Plasma Treatment

The principle of operation of the experimental setup in accordance with cold plasma technology [13] is shown in Figure 2a. To generate cold atmospheric plasma in the air gap, a "point-plate" type electrode configuration was used, which included a stainless steel plate as a grounded electrode, and a point steel electrode with a diameter of 1 mm as a high-voltage electrode in a dielectric holder. The gap between the electrodes was set to 70 (for whole tomatoes) and 15 mm (for tomato peels). The processing chamber had a square shape (length 20 cm) made of dielectric material. The grounded electrode in the dielectric holder was mounted on a positional platform with two stepper motors to provide movement along the X-Y axis. The trajectory of the cold plasma treatment nozzle was set using the authors' intelligent object recognition system.

Cold atmospheric plasma (CAP) treatment was performed using a high voltage power supply system (Matsusada AMPS 20B20, Matsusada Precision, Otsu, Japan) in combination with an Agilent functional generator (Agilent 33220A, Agilent Technologies, Santa Clara, CA, USA) [13]. In all experiments, the pulse duration and the frequency of the plasma discharge were set to 50 microseconds and 100 Hz, respectively. The selected electrical parameters made it possible to precisely control the processing parameters for tomato and tomato peel treatments. Each pulse supplied a voltage of up to 20 kV. In addition, positive pulses with an electric field strength of 2.5 and 8 kV/cm were used for whole tomato and tomato peel, subsequently. The average specific energy consumption of CAP treatment for all experiments was 1.7 kJ/kg at 6000 discharges. The total CAP treatment time was 3 s per cm^2. The temperature difference between the pretreated tomato samples and the control sample was less than 2 °C.

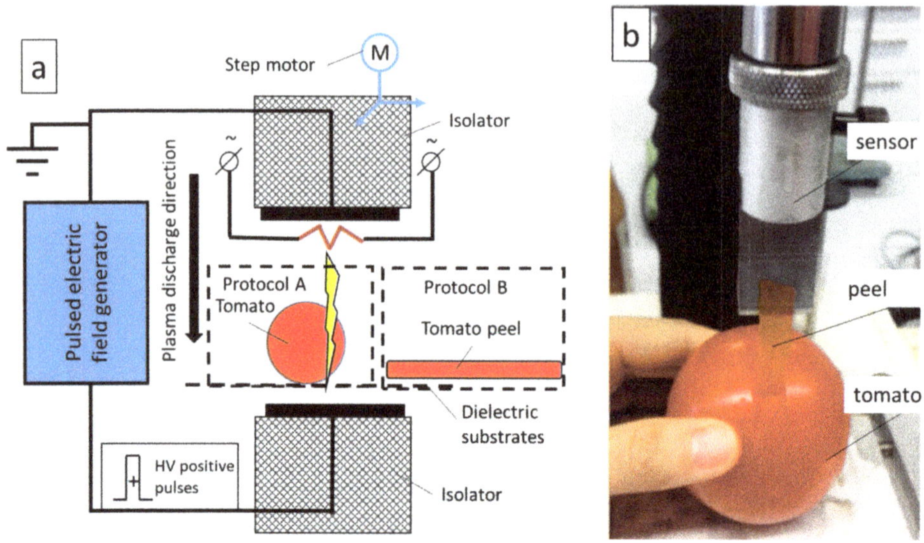

Figure 2. CAP treatment experimental scheme (**a**) and tomato peeling procedure based on texture analyzer system (**b**).

2.3. Skin Removal

In our study, the tomato peeling was performed mechanically using a texture analyzer (CT3-4500 texture analyzer, Brookfield Ametek, Middleborough, MA, USA). The selected precut area on the tomato skin was removed, and the specific peeling force was calculated according to the methodology described in [26]. Figure 2b demonstrates the peeling procedure.

The specific peeling force (F_p) was calculated from the peeling energy and the area of the peeled skin as follows:

$$F_p = \frac{E_p}{S} \quad (1)$$

where E_p—peeling energy from texture analyzer, N; S—area of the peeled skin sample, m^2.

To determine the area of the peeled skin, the sample was photographed on a whiteboard with a tripod positioned at a fixed distance. Afterwards, the pictures were analyzed using ImageJ (v 1.52e), a Java-based image-processing program.

Consequently, the area under the force–displacement curve was interpreted as the peeling energy required for tearing off the tomato peel. The same methodology was described in [27].

2.4. Peel Drying

Tomato peels were dried in a single layer. Drying experiments were carried out in an infrared dryer with IR lamps (Ballu BIH-l-0.3, Hong Kong, China). An average radiative heat flux of 0.11 W/m^2 with a radiation wavelength of 3 microns was applied. The operating parameters were selected based on the data in the literature regarding the depth of IR radiation penetration into apple tissue (about 10 mm at a wavelength of 3 microns) [28]. During the experiment with IR drying, the moisture loss of tomato peel samples was measured at intervals of 10 min. IR drying was conducted until the final moisture content in the samples reached ~0.06 g/100 g (dry matter). The value of the final moisture content was chosen based on the conditions for further storage of tomato peels. The kinetic curves and drying rates were calculated according to a well-known methodology [29].

2.5. Extraction and Quantification of Total Carotenoids and Total Phenolic Compounds

To study the quality effects of CAP treatment, two intracellular compounds were analyzed as follows: carotenoids and polyphenols from tomato peels. Carotenoids were extracted from dried tomato peels using acetone as solvent. Briefly, 6–7 g of CAP-treated (protocol C) and untreated (protocol D) samples were mixed with 2 mL of acetone and 0.01 g of magnesium carbonate. The extracts were centrifuged for 5–10 min at 3000 rpm (AWTech MPW—260 RH centrifuge, Poland) to separate the supernatant. Three mL of petroleum ether was added to the supernatant to separate the upper layer of the organic phase. The organic phase was transferred to a centrifuge tube with 2 g of predried sodium sulfate.

The content of total carotenoids was estimated spectrophotometrically. The absorbance of extracts was measured at 450 nm. The content of total carotenoids was calculated using the following equation:

$$\text{Totalcarotenoids} = \text{A}\cdot 4.00\cdot\frac{\text{V}_1}{\text{V}_2} \quad (2)$$

where A is the optical density of the solution of extracted substances; 4.00 is an indicator equal to the ratio of the mass concentration (in milligrams per cubic decimeter) of a β-carotene solution in petroleum ether to its optical density at a wavelength of 450 nm and an optical path length of 10 mm; V_1 is the volume of the test sample of the product; V_2 is the volume of the extract in ether.

The concentration of total polyphenols (mg of gallic acid/kg of tomato product) in the tomato peels was determined according to the method described in [21]. The extraction of tomato waste was carried out with a suitable liquid-to-solid ratio (10:1), allowing the maintenance of a homogenous solid–liquid extraction. Ten grams of tomato peels were mixed with 50% ethanol solution and were agitated for 20 min. Total phenolic compounds were quantified using the Folin–Ciocalteu method [30].

2.6. Energy Aspects of CAP Treatment and Drying

The peeling-specific energy consumption (PSEC) was calculated as follows [13]:

$$W_{PSEC} = \frac{W_{peeling} + W_{CAP} + W_{TE}}{M_{tomato}} \quad (3)$$

where $W_{peeling}$—total energy consumption of the peeling procedure, kW/h; W_{CAP}— total energy consumption of cold plasma pretreatment, kW/h; W_{TE}—energy consumption of thermionic emission source, kW/h; M_{tomato}—the weight of tomatoes, kg.

The values of $W_{peeling}$ were calculated as the area under the force–displacement curve, which was interpreted as the peeling energy required for tearing off the tomato peel. The values of W_{TE} were obtained using a wattmeter (IC-M207D wattmeter, Cartool, Ningbo, China) according to the experimental scheme (Figure 2). The energy consumption of the cold plasma treatment W_{CAP} was calculated based on the volt-ampere characteristics as follows:

$$W_{CAP} = n\cdot\int U(t)\cdot I(t)\cdot dt \quad (4)$$

where n—number of electrical discharges; $U(t)$—instantaneous voltage on the electrodes, V; $I(t)$—discharge current passing through the sample, A.

The drying-specific energy consumption (DSEC) was calculated as follows:

$$W_{DSEC} = \frac{W_{IR} + W_{CAP} + W_{TE}}{M_{peel}} \quad (5)$$

where W_{IR}—total energy consumption of the infrared drying procedure, kW/h; M_{peel}—the weight of the tomato peel, kg.

The values of W_{IR} were obtained using a wattmeter.

2.7. Statistical Analysis

All measurements of the above-mentioned characteristics were performed in at least five replicates. Statistical evaluation of the physical properties was performed via ANOVA, using SigmaPlot (Version 14) with the least significant difference (LSD) at $p < 0.05$. Also, a pair-wise Tukey's test was used to find significant differences between treatments using α 0.05.

3. Results and Discussion

3.1. Improvement in Peeling of Whole Tomato

In our experiments, the untreated and CAP-treated samples were compared. When analyzing the cut of tomato fruits for protocol A, the mesocarp color changed. The white streak between the integument and the endocarp was modified, which indicates a change in the internal mass transfer processes. The same effect was observed for the pulsed electric field treatment of the whole kiwifruit [26].

From the authors' point of view, this phenomenon of peelability is described as follows: moisture starts to migrate from the endocarp region to the zone between the mesocarp and the integumentary tissue via the additionally formed pores in the internal structure of the tomato fruit (Figure 3). In this case, the process of internal mass transfer is carried out due to the emerging turgor intracellular pressure. The resulting layer of liquid, due to its hydrodynamic force, contributes to the effective removal of the integumentary tissue. A similar effect was observed by Andreou et al. [21] in a study on the use of a pulsed electric field when the tomato was peeling.

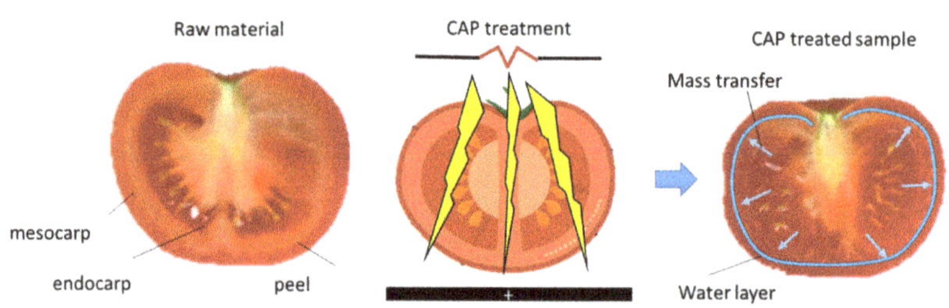

Figure 3. Water migration mechanism in tomatoes caused by cold plasma treatment.

The dynamics of tomato peel removal are presented in Figure 4. For each protocol, the maximum specific peeling force was defined from the graph and is summarized in Table 1. Based on the experimental results, CAP treatment input appeared to significantly influence the tomato peeling procedure. As shown in Figure 4, the peeling procedure of the untreated sample (protocol B) was aborted due to the breakage of the tomato skin. CAP treatment significantly enhanced the peelability of tomato fruits.

Table 1. Comparison of peeling energy between CAP-treated tomatoes (protocol A) and fresh untreated tomatoes (protocol B).

Sample	Specific Peeling Force, N/cm	Peelability, cm^2/g	Energy Consumption of Peeling Procedure $W_{peeling}$, W/kg
Protocol A	18.3 ± 3.35	0.2 ± 0.03	0.065 ± 0.009
Protocol B	52.8 ± 1.12	0.18 ± 0.03	0.170 ± 0.012

The results of our study demonstrated that CAP treatment of whole tomatoes (protocols A and B) produced less specific peeling force as well as energy consumption during the peeling procedure (Table 1). The suggested CAP treatment can be potentially used as an alternative to traditional technologies such as chemical and thermal treatment.

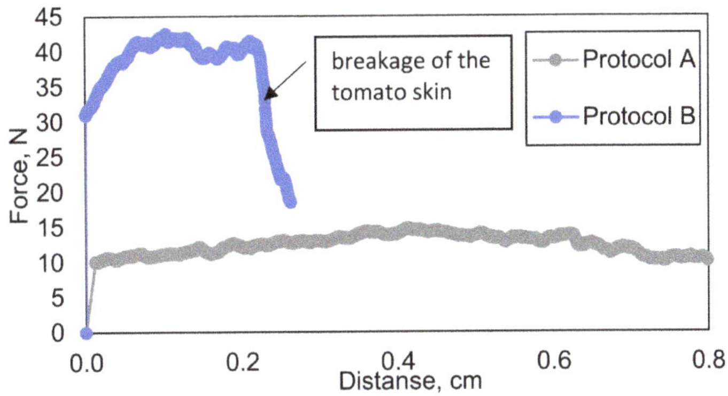

Figure 4. A specific peeling force graph of instrumental tomato peeling analysis for protocols A and B.

Khudyakov et al. [31] compared the efficacy of peeling with pulsed electric field pretreatment (E = 1 kV/cm; specific energy 1, 5, and 10 kJ/kg) for the tomato peeling procedure. It was found that tomato peeling was more effective at 1 kJ/kg. The specific force of mechanical peel removal decreased by 10%. The same PEF technology applied for a tomato peeling procedure resulted in a significant decrease ($p \leq 0.05$) of up to 43% in the force required for mechanical peeling. In comparison with PEF technology, CAP treatment requires less energy. Currently, such methods (PEF and CAP) are still in their developmental stages, requiring optimization and pilot tests before they can be considered for commercialization.

3.2. Improvement in Tomato Peel Drying

Figure 5 shows the kinetics and drying rate of the tomato peels in a thin layer. CAP treatment improved the drying kinetics of tomato peels by forming electrically induced channels with a tree-like structure, which reduced the resistance to moisture transfer during the drying process. A similar result was obtained by the authors Zhang et al., who treated chili peppers with cold plasma and found that the drying time after pretreatment was significantly reduced [32]. A similar explanation for the formation of micro-holes was presented by Zhou et al. when processing wolfberries using CAP treatment [33]. In our experiments, the drying time decreased by 18.2% for tomato peels. The total drying time of the CAP-treated (protocol C) and control (protocol D) samples was 119 and 143 min subsequently. The period of decrease in the diffusion rate, which varied between the control and pretreated samples, was the dominant physical mechanism of the IR drying method [14]. In comparison with control samples (protocol D), where the drying rate reached its peak only after ~10 min, the drying rate of CAP-pretreated tomato peels reached its peak at the very beginning of the drying process. This behavior of the drying rate curve showed that the rate of moisture evaporation from the surface of the sample pretreated by CAP was higher than that which occurred from the inside.

Thus, it can be assumed that with the small-scale production of tomato juice of about 1500 tons of raw materials per day [34], the technology of CAP pretreatment can potentially increase peel-processing capacity by about 10–12%. This can potentially provide additional income, which increases both the rate of return and the return on investment in CAP equipment. Currently, the technologies of CAP pretreatment developed by our research group show great prospects, especially in providing textural and organoleptic characteristics of dried tomato peels. This, in particular, occurs as a result of the use of a stabilized form of electric field distribution.

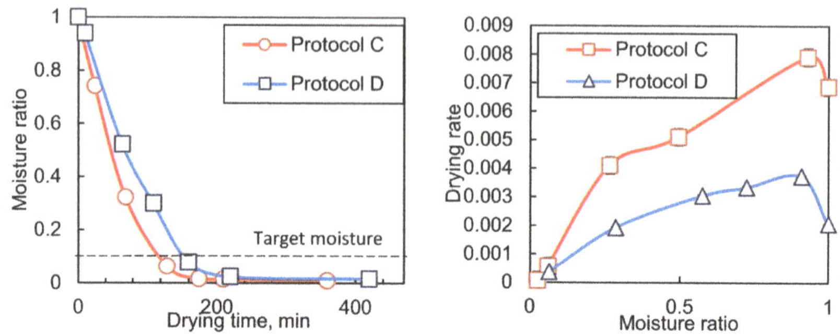

Figure 5. Drying kinetic and drying rate curves of tomato peels for protocols C and D.

3.3. Improvement in Quality Characteristics

Table 2 shows the influence of CAP treatment for protocols C and D on the extraction of high-added-value compounds from tomato peels. Both quality parameters, the total carotenoids and total phenolic compounds, resulted in higher concentrations of all intracellular compounds studied. The results demonstrate the same behavior as discussed by several authors using the pulsed electric field [21] and cold plasma [18]. The electroporation effect might enhance the extraction process due to the better release of these compounds. Andreou et al. [21] reported that PEF treatment doubled the total phenolic compound extraction yield compared to the control. From the authors' point of view, such results might be obtained using electroporation technologies prior to the extraction process. In our experiments, the tomato peels were treated before the drying procedure. The induced pores were used to enhance the drying procedure (Figure 5).

Table 2. Comparison of the drying specific energy consumption of CAP-treated tomato peels (protocol C) and untreated tomatoes (protocol D).

Sample	Total Carotenoids (mg Carotenoids/100 g Tomato Peels)	Total Phenolic Compounds (mg of Gallic Acid/kg Tomato Peels)	The Drying Specific Energy Consumption W_{DSEC}, kW/kg
Protocol C	14.29 ± 3.18	24.43 ± 3.18	3.7 ± 0.21
Protocol D	18.78 ± 2.76	29.51 ± 3.08	4.5 ± 0.33

3.4. Improvement in Energy Consumption

The oscillogram of the current and voltage of the cold plasma discharge is shown in Figure 6. The oscillogram for protocols A and C demonstrates a similar behavior and value due to the high moisture content of the whole tomato as well as tomato peel. The effect of CAP pretreatment on the specific energy consumption of W_{DSEC} for protocols C and D is shown in Table 2. Using Equation (3), an oscillogram of the current and voltage, the value of W_{CAP} was calculated. The total specific energy consumption of W_{PSEC} for CAP-pretreated samples was lower than that of the control samples ($p < 0.05$).

In general, the values of W_{PSEC} for CAP pretreatment depend on the characteristics of the plasma discharge. Since the values of W_{CAP} and W_{TE} are significantly lower than those of W_{IR}, the energy costs for the pretreatment of CAP can be neglected. Potentially, the efficiency of CAP pretreatment can lead to a significant reduction in the energy consumption of the drying procedure of tomato peels. Finally, it was found that the drying-specific energy consumption for CAP-treated samples was 17% lower than that of the control samples.

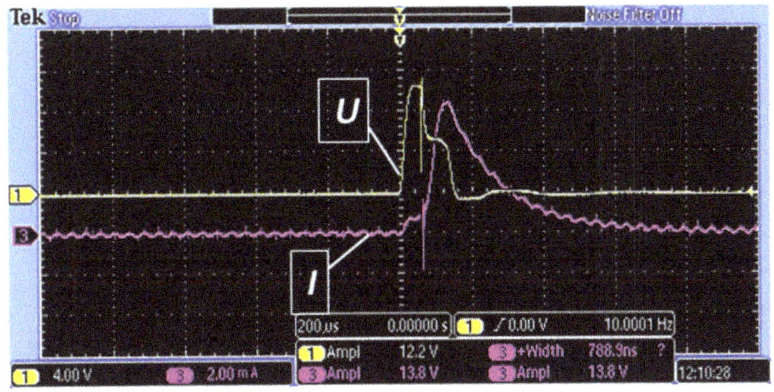

Figure 6. CAP treatment oscillogram of current and discharge.

Santos et al. [35] reported the efficiency of PEF for mango peel drying at E = 4.5 kV cm^{-1} and drying temperature at 70 °C, where the maximum reduction in drying time reached 67%. In comparison with CAP technology, PEF treatment demonstrated a higher effect on peel drying; however, further analysis should be performed.

4. Conclusions

The results obtained in this research showed that cold atmospheric plasma treatment could be applied as a useful tool in the tomato processing industry, leading to decreased energy consumption and increased productivity. Generally, cold atmospheric plasma pretreatment enhanced the peeling process, including less specific peeling force consumption. In the case of peeling, enhanced peelability can be explained via the migration of water from the mesocarp region under the tomato skin as a result of electroporation. This led to a pressure difference across the tomato skin, reducing the surface resistance and facilitating its removal. Cold atmospheric plasma can easily replace the existing mechanical peeling processes, leading to energy requirements.

Taking into consideration that CAP treatment has low energy requirements (~1.7 kJ/kg of raw material), it could be an economically viable approach for tomato processing industries. CAP treatment could be applied to tomato peel resulting from the tomato processing industry in order to reduce the drying time. The drying-specific energy consumption of the CAP-treated samples was 17% lower than that of the control samples. These could be natural alternatives to high productivity for further tomato peel processing, such as the peel extraction of valuable compounds. In the case of tomato by-product valorization, the efficacy of CAP pretreatment was implemented on tomato waste. CAP treatment could be applied to tomato waste (peel) resulting from the tomato processing industry in order to obtain high extraction yields of carotenoids and phenolic compounds. Future tendencies are directed towards the use of eco-friendly processing technologies to avoid excessive waste generation and promote the circular economy.

Author Contributions: Conceptualization, I.S. and E.H.A.M.; methodology, M.S.; software, E.B.; validation, M.S. and E.B.; resources, I.S.; data curation, E.H.A.M.; writing—original draft preparation, I.S., M.S. and E.H.A.M.; writing—review and editing, I.S.; visualization, M.S.; supervision, I.S.; project administration, I.S.; funding acquisition, I.S. All authors have read and agreed to the published version of the manuscript.

Funding: This research was carried out with the financial support of the Kuban Science Foundation and FSBEI HE "KubSTU" under the framework of scientific project No. MFI-P-20.1/40.

Institutional Review Board Statement: Not applicable.

Informed Consent Statement: Not applicable.

Data Availability Statement: The data that support the finding of this study are available from the corresponding author upon reasonable request.

Acknowledgments: The authors gratefully acknowledge Claudia Siemer for the useful comments.

Conflicts of Interest: The authors declare no conflict of interest.

References

1. Eslami, E.; Carpentieri, S.; Pataro, G.; Ferrari, G. A Comprehensive Overview of Tomato Processing By-Product Valorization by Conventional Methods versus Emerging Technologies. *Foods* **2023**, *12*, 166. [CrossRef] [PubMed]
2. Alam, M.F.B.; Tushar, S.R.; Zaman, S.M.; Gonzalez, E.D.R.S.; Bari, A.B.M.M.; Karmaker, C.L. Analysis of the drivers of Agriculture 4.0 implementation in the emerging economies: Implications towards sustainability and food security. *Green Technol. Sustain.* **2023**, *1*, 100021. [CrossRef]
3. Gharibzahedi, S.M.T.; Smith, B.; Guo, Y. Pectin extraction from common fig skin by different methods: The physicochemical, rheological, functional, and structural evaluations. *Int. J. Biol. Macromol.* **2019**, *136*, 275–283. [CrossRef] [PubMed]
4. Nawaz, A.; Ali, S.W.; Irshad, S.; Irshad, F.; Ahmed, A.; Sharmeen, Z.; Khan, I. Effect of peeling and unpeeling on yield, chemical structure, morphology and pasting properties of starch extracted from three diverse potato cultivars of Pakistan. *Int. J. Food Sci. Technol.* **2020**, *55*, 2344–2351. [CrossRef]
5. Wu, X.; Yu, L.; Pehrsson, P.R. Are processed tomato products as nutritious as fresh tomatoes? Scoping review on the effects of industrial processing on nutrients and bioactive compounds in tomatoes. *Adv. Nutr.* **2022**, *13*, 138–151. [CrossRef]
6. Lu, Z.; Wang, J.; Gao, R.; Ye, F.; Zhao, G. Sustainable valorisation of tomato pomace: A comprehensive review. *Trends Food Sci. Technol.* **2019**, *86*, 172–187. [CrossRef]
7. Grassino, A.N.; Brnčić, M.; Vikić-Topić, D.; Roca, S.; Dent, M.; Brnčić, S.R. Ultrasound assisted extraction and characterization of pectin from tomato waste. *Food Chem.* **2016**, *198*, 93–100. [CrossRef]
8. Sengar, A.S.; Rawson, A.; Muthiah, M.; Kalakandan, S.K. Comparison of different ultrasound assisted extraction techniques for pectin from tomato processing waste. *Ultrason. Sonochem.* **2020**, *61*, 104812. [CrossRef]
9. Vidyarthi, S.K.; El Mashad, H.M.; Khir, R.; Zhang, R.; Tiwari, R.; Pan, Z. Quasi-static mechanical properties of tomato peels produced from catalytic infrared and lye peeling. *J. Food Eng.* **2019**, *254*, 10–16. [CrossRef]
10. Koch, Y.; Witt, J.; Lammerskitten, A.; Siemer, C.; Toepfl, S. The influence of Pulsed Electric Fields (PEF) on the peeling ability of different fruits and vegetables. *J. Food Eng.* **2022**, *322*, 110938. [CrossRef]
11. Gavahian, M.; Sastry, S.K. Ohmic-assisted peeling of fruits: Understanding the mechanisms involved, effective parameters, and prospective applications in the food industry. *Trends Food Sci. Technol.* **2020**, *106*, 345–354. [CrossRef]
12. Grassino, A.N.; Pedisić, S.; Dragović-Uzelac, V.; Karlović, S.; Ježek, D.; Bosiljkov, T. Insight into high-hydrostatic pressure extraction of polyphenols from tomato peel waste. *Plant Foods Hum. Nutr.* **2020**, *75*, 427–433. [CrossRef]
13. Shorstkii, I. Application of cold filamentary microplasma pretreatment assisted by thermionic emission for potato drying. *Innov. Food Sci. Emerg. Technol.* **2020**, *66*, 102540. [CrossRef]
14. Bassey, E.J.; Cheng, J.H.; Sun, D.W. Novel nonthermal and thermal pretreatments for enhancing drying performance and improving quality of fruits and vegetables. *Trends Food Sci. Technol.* **2021**, *112*, 137–148. [CrossRef]
15. Du, Y.; Yang, F.; Yu, H.; Xie, Y.; Yao, W. Improving food drying performance by cold plasma pretreatment: A systematic review. *Compr. Rev. Food Sci. Food Saf.* **2022**, *21*, 4402–4421. [CrossRef]
16. Shishir, M.R.I.; Karim, N.; Bao, T.; Gowd, V.; Ding, T.; Sun, C.; Chen, W. Cold plasma pretreatment–A novel approach to improve the hot air drying characteristics, kinetic parameters, and nutritional attributes of shiitake mushroom. *Dry. Technol.* **2020**, *38*, 2134–2150. [CrossRef]
17. Chizoba Ekezie, F.-G.; Sun, D.-W.; Cheng, J.-H. A review on recent advances in cold plasma technology for the food industry: Current applications and future trends. *Trends Food Sci. Technol.* **2017**, *69*, 46–58. [CrossRef]
18. Bao, Y.; Reddivari, L.; Huang, J.-Y. Development of cold plasma pretreatment for improving phenolics extractability from tomato pomace. *Innov. Food Sci. Emerg. Technol.* **2020**, *65*, 102445. [CrossRef]
19. De Araújo Bezerra, J.; Lamarão, C.V.; Sanches, E.A.; Rodrigues, S.; Fernandes, F.A.N.; Ramos, G.L.P.A.; Esmerino, E.A.; Cruz, A.G.; Campelo, P.H. Cold plasma as a pre-treatment for processing improvement in food: A review. *Food Res. Int.* **2023**, *167*, 112663. [CrossRef] [PubMed]
20. Shorstkii, I.; Koshevoi, E. Drying Technology Assisted by Nonthermal Pulsed Filamentary Microplasma Treatment: Theory and Practice. *ChemEngineering* **2019**, *3*, 91. [CrossRef]
21. Andreou, V.; Dimopoulos, G.; Dermesonlouoglou, E.; Taoukis, P. Application of pulsed electric fields to improve product yield and waste valorization in industrial tomato processing. *J. Food Eng.* **2020**, *270*, 109778. [CrossRef]
22. Tappi, S.; Berardinelli, A.; Ragni, L.; Dalla Rosa, M.; Guarnieri, A.; Rocculi, P. Atmospheric gas plasma treatment of fresh-cut apples. *Innov. Food Sci. Emerg. Technol.* **2014**, *21*, 114–122. [CrossRef]
23. Chaple, S.; Sarangapani, C.; Jones, J.; Carey, E.; Causeret, L.; Genson, A.; Duffy, B.; Bourke, P. Effect of atmospheric cold plasma on the functional properties of whole wheat (*Triticum aestivum* L.) grain and wheat flour. *Innov. Food Sci. Emerg. Technol.* **2020**, *66*, 102529. [CrossRef]

24. Bao, T.; Hao, X.; Shishir, M.R.I.; Karim, N.; Chen, W. Cold plasma: An emerging pretreatment technology for the drying of jujube slices. *Food Chem.* **2021**, *337*, 127783. [CrossRef] [PubMed]
25. Sakudo, A.; Misawa, T.; Yagyu, Y. Equipment design for cold plasma disinfection of food products. In *Advances in Cold Plasma Applications for Food Safety and Preservation*; Elsevier: Amsterdam, The Netherlands, 2020; pp. 289–307.
26. Shorstkii, I.; Stuehmeier-Niehe, C.; Sosnin, M.; Mounassar, E.H.A.; Comiotto-Alles, M.; Siemer, C.; Toepfl, S. Pulsed Electric Field Treatment Application to Improve Product Yield and Efficiency of Bioactive Compounds through Extraction from Peels in Kiwifruit Processing. *J. Food Process. Preserv.* **2023**, *2023*, 8172255. [CrossRef]
27. Li, X.; Pan, Z.; Atungulu, G.G.; Zheng, X.; Wood, D.; Delwiche, M.; McHugh, T.H. Peeling of tomatoes using novel infrared radiation heating technology. *Innov. Food Sci. Emerg. Technol.* **2014**, *21*, 123–130. [CrossRef]
28. Krishnamurthy, K.; Khurana, H.K.; Soojin, J.; Irudayaraj, J.; Demirci, A. An overview on ultrafiltration in food processing. *Compr. Rev. Food Sci. Food Saf.* **2008**, *7*, 2–13. [CrossRef]
29. Liu, Z.; Song, Y.; Guo, Y.; Wang, H.; Wu, Z. Influence of Pulsed Electric Field Pretreatment on Vacuum Freeze-dried Apples and Process Parameter Optimization. *Adv. J. Food Sci. Technol.* **2017**, *13*, 224–235. [CrossRef]
30. Velioglu, Y.S.; Mazza, G.; Gao, L.; Oomah, B.D. Antioxidant Activity and Total Phenolics in Selected Fruits, Vegetables, and Grain Products. *J. Agric. Food Chem.* **1998**, *46*, 4113–4117. [CrossRef]
31. Khudyakov, D.; Sosnin, M.; Munassar, E.M.A.; Techakanon, C.; Siemer, C.; Toepfl, S.; Shorstkii, I. Pulsed Electric Field Processing as an Effective Tomato Peeling Method. *Food Process. Tech. Technol.* **2022**, *1748*, 189–198. [CrossRef]
32. Zhang, X.; Zhong, C.; Mujumdar, A.S.; Yang, X. Full title: Cold plasma pretreatment enhances drying kinetics and quality attributes of chili pepper (*Capsicum annuum* L.). *J. Food Eng.* **2019**, *241*, 51–57. [CrossRef]
33. Zhou, Y.H.; Vidyarthi, S.K.; Zhong, C.S.; Zheng, Z.A.; An, Y.; Wang, J.; Wei, Q.; Xiao, H.W. Cold plasma enhances drying and color, rehydration ratio and polyphenols of wolfberry via microstructure and ultrastructure alteration. *Lwt* **2020**, *134*, 110173. [CrossRef]
34. Koulouris, A.; Kotelida, I. Simulation-based reactive scheduling in tomato processing plant with raw material uncertainty. In *21st European Symposium on Computer Aided Process Engineering*; Pistikopoulos, E.N., Georgiadis, M.C., Kokossis, A.C., Eds.; Computer Aided Chemical Engineering; Elsevier: Amsterdam, The Netherlands, 2011; Volume 29, pp. 1020–1024.
35. Santos, N.C.; Almeida, R.L.J.; de Oliveira Brito, A.C.; de Alcantara Silva, V.M.; Albuquerque, J.C.; Saraiva, M.M.T.; Santos, R.M.S.; de Sousa, F.M.; de Alcantara Ribeiro, V.H.; de Oliveira Carvalho, R.; et al. Effect of pulse electric field (PEF) intensity combined with drying temperature on mass transfer, functional properties, and in vitro digestibility of dehydrated mango peels. *J. Food Meas. Charact.* **2023**. [CrossRef]

eng

MDPI

Article

Automatic Identification of Corrosion in Marine Vessels Using Decision-Tree Imaging Hierarchies

Georgios Chliveros [1,*], Stylianos V. Kontomaris [1] and Apostolos Letsios [2]

[1] Department of Engineering, Metropolitan College, Marousi Campus, 15125 Athens, Greece; skontomaris@mitropolitiko.edu.gr
[2] Maritime Academy, Metropolitan College, Piraeus Campus, 18535 Pireas, Greece
* Correspondence: gchliveros@mitropolitiko.edu.gr

Abstract: We propose an unsupervised method for eigen tree hierarchies and quantisation group association for segmentation of corrosion in marine vessel hull inspection via camera images. Our unsupervised approach produces image segments that are examined to decide on defect recognition. The method generates a binary decision tree, which, by means of bottom-up pruning, is revised, and dominant leaf nodes predict the areas of interest. Our method is compared with other techniques, and the results indicate that it achieves better performance for true- vs. false-positive area against ideal (ground truth) coverage.

Keywords: corrosion detection; image segmentation; entropy pruning; industrial inspection

Citation: Chliveros, G.; Kontomaris, S.V.; Letsios, A. Automatic Identification of Corrosion in Marine Vessels Using Decision-Tree Imaging Hierarchies. *Eng* **2023**, *4*, 2090–2099. https://doi.org/10.3390/eng4030118

Academic Editor: Antonio Gil Bravo

Received: 26 June 2023
Revised: 13 July 2023
Accepted: 18 July 2023
Published: 26 July 2023

1. Introduction

Corrosion on marine merchant vessels typically presents itself as rust on surfaces, but when treated, it depends primarily on the area (and the extent thereof) it occupies. In cases of extensive oxidation of steel surfaces on the hull, the vessel is 'dry-docked' in order for rust to be removed, most commonly by means of water blasting. The most widespread method for protecting against oxidation is that of cathodic protection with sacrificial electrodes and antifouling pigment paint. Corrosion areas have to be measured using a combination of ultrasound instrumentation (e.g., ultrasound images [1]) and visual RGB images, operated by class surveyors/human operators. Even if protective coatings are used, all merchant marine vessels must complete an inspection of the hull in a dry dock at least three times in a five-year period, with intermediate surveys being performed within 36 months. This regulation [2] relates to dry-dock visual marine hull inspection (oil tankers and bulk carriers) requirements due to the Convention for Safety of Life At Sea (SOLAS).

The automation of such a laborious measurement and visual inspection task would save not only person hours but also the time required for a vessel to remain in dry dock. As already mentioned, dry-dock inspection is performed by means of both ultrasonic [1,3] and optical (RGB) imaging. The problem of corrosion surveying via visual inspection in the optical space (RGB images) and its automation can be seen as an extension of semantic segmentation in images and the association of pixels with a specific class/label (corrosion) as a means to recognize (interpret) specific items in a visual scene (vessel hull). Recent advances in deep learning models have spawned renewed interest in the domain of machine vision for industrial inspection and diagnosis of defects, inclusive of cracks and corrosion detection.

Visual inspection of corrosion has been investigated using classic techniques, such as multilevel thresholding [4], k-means clustering [5], and histogram-based clustering [6], as well as deep learning models, as is the case with convolutional neural networks [7,8] and self-organising maps [9]. The availability of such methods in corrosion detection is investigated as part of the general defect issue in industrial inspection, also known as pitting and uniform corrosion. Corrosion is defined as isolated (corroded) area units on a structure's surface that are difficult to both detect and predict, with a diverse geometrical shape. It is difficult to postulate prior knowledge on the basis of a generalised geometry and morphology in terms of visual inspection and image processing, although several segmentation algorithms have been applied to the problem and assessed [10–12]. Corrosion image processing techniques typically refer to:

- Chroma patterns of corroded (e.g., rusted) areas that present a foreground qualitatively ranging from yellow to red (e.g., use of histograms in [13–15]);
- Texture and/or roughness of surface areas that increase with respect to the level of corrosion (e.g., the fractal dimension index used in [16,17]);
- Features selected from preconditioned datasets of training examples that are fed to convolution and feed-forward neural networks [18–23].

Using specific colour (chroma space) patterns as a priori corrosion knowledge for the 'foreground' scene, i.e., the collection of pixels that should be separated to investigate corrosion, can be instigated. However, in marine ship hulls, this is not an aid, since most of the lower ship hull is of a red background colour, making it difficult to extract the foreground. Texture can be taken as a means to fine tune in order to filter out artifacts, such as shading or crack-like formations that generate irregularities in terms of neighbouring pixels. However, these form compensating measures that do not factor in changes in light and other environmental conditions, leading to said visual artifacts not being taken into account.

Feature descriptors constructed from texture and colour spaces have been applied as input to discriminant functions [14,15], and decision trees [24] have been used for corrosion classifiers. Similar feature descriptors from datasets of training examples have also been used as input for deep learning approaches (e.g., convolution and feed-forward neural networks (CNNs/FFNNs)) [18–23]. These techniques seem to achieve substantial performance with respect to detection/classification, with accuracy in the range of 75 to 90% of the total area versus the ground truth. However, these methods are supervised in nature, heavily depend on the size of the training datasets and how well the data-driven approach is structured, and are prone to raising many false positives [23]. Additionally, in the absence of widely available corrosion datasets, the complexity of the geometry/morphology of corrosion areas [23,25–27] makes it difficult to generate semisynthetic training datasets, as is the case in many other areas (e.g., in agriculture to determine a specific plant species [28]). As a result, in the domain of image corrosion detection, several challenges still remain:

- Decreasing recognition accuracy in the presence of increased crack-like texture, leading to increased local illumination variability;
- High false-positive detection rates in the presence of high 'colour' similarity between groups of pixels, i.e., a decreasing 'coverage' percentage of correctly segmented regions of interest, with fine tuning required to enhance specificity;
- Lack of specificity from available datasets with accompanying annotations and data size, which is especially important for supervised/deep learning techniques.
- Lack of freely available deep learning models for corrosion detection.

We introduced an unsupervised approach in corrosion detection via pruned decision tree hierarchies over raw (RGB) image inputs and interrogated binary splits for classification of corrosion areas over significant clusters. Our method effectively produces substantially increased defect area coverage (in comparison to the ground truth and other methods), requiring none of the aforementioned assumptions or fine tuning to elicit a priori scene knowledge. Furthermore, our pruning method (although it may not improve recognition accuracy) substantially increases the recognition of the true-positive (TP) area of interest, with smaller false-positive (FP) areas. Therefore, the application of our decision criteria leads to better discrimination of true-positive versus false-positive segmented areas.

2. Materials and Methods

In the absence of publicly available data, we devised the dataset used in this work [29] by collecting images of the external hull over two different merchant vessels, i.e., oil tankers. The marine vessel inspections correspond to a ship being under 'dry-dock' maintenance conditions. The images have been collected in using two cameras to device high and low pixel resolution examples: high pixel resolution 3799 × 2256 (72 dpi, at 24 bit depth), and low pixel resolution 1920 × 1080 pixels (96 dpi at 24 bit depth). For performance testing, we have maintained images labeled by an expert human operator to be used as ground truth. The dataset incorporates several artifacts due to environmental conditions (e.g., changing lighting conditions, surface artifacts) and objects in front of the hull (e.g., maintenance ladders) not associated with the marine vessel surface.

As a result, the dataset contains: (a) high-resolution RGB images of 3799 × 2256 pixels, (b) low-resolution RGB images of 1920 × 1080 pixels, alongside (c) labelled (ground truth) images that have RGB triplet values for each pixel that correspond to corrosion and zero values elsewhere. The ground truth images are used only for performance evaluation among different methods in Section 3. We present a revised segmentation decision tree methodology for corrosion applied in this dataset. The decision criteria of this method are based jointly on information entropy and eigen values of nodes, as opposed to the more typical decision tree pruning criteria based on node information gain (e.g., in [30]).

2.1. Hierarchical Decision Tree Method

We have implemented an unsupervised methodology via pruned decision tree hierarchies over raw image inputs and interrogated binary splits for classification of corrosion areas over significant clusters. This method is based on eigen features collected from image areas, and tree-splits are performed based on the extent of similarity of new nodes. That is to say, each node of the tree represents a subset of pixels (cluster) in the original image, as serialised image pixels x_ℓ (with $\ell = i \times j \times 3$) belonging to a cluster C_n. The children of any node partition the members of the parent node into two new sets (clusters). The algorithm constrains the partitioning of the binary-tree based on a set of pixel colours corresponding to a node. The set of image pixels corresponding to nodes $2n, 2n+1$ is denoted by C_{2n}, C_{2n+1} and splits under certain criteria. The tree grows until either a node (or set of nodes) reaches a dominant colour, or until a predefined number of nodes is reached.

An eigen tree decomposition [31] selects a candidate node and, under certain conditions, produces a split into leaf nodes. In our case, this manifests as a binary split of two leaf nodes, whereas two quantisation levels (Q_{2n}, Q_{2n+1}) are estimated and each member of a cluster is associated with that of the closest quantisation level. The mean intensity value of each colour channel is the histogram point with the least variance in the eigen space, leading to a specific quantisation level Q_n [32]. The quantization level of each colour channel, and for each node, is defined as $Q_n = M_n/N_n$, where $N(n)$ is the number of clusters for the group of pixels indices. In effect, we revise the tree generation algorithm to

- Perform calculations: utilise the smallest eigen value over current node branch candidates;
- Decide binary split: nodes of a local cluster are split by using the largest eigen value over current iteration clusters.

These methodological insertions introduce a more natural generation of nodes/clusters and reduce the computations otherwise required over all nodes (as was the case, for example, in [18,22]). We introduce a formal representation of decision trees with binary splits that is such that we select an appropriate hyperplane $h(\mathbf{w})$ that hierarchically separates data into clusters in a sequence. Inasmuch, current iteration clusters are not re-evaluated in their totality but only in the sequence of a current branch. For a binary decision tree, this means that the average of square distances of all data points (image pixel quantisation levels) from a hyperplane h_n sequentially generates leaf nodes of new clusters C_{2n}, C_{2n+1} [31]. The parameters of the new nodes C_{2n}, C_{2n+1} in the n-th leaf with hyperplane h_n and parameters vector $\mathbf{w}_n$ can be estimated by finding the minimum and maximum values, as per Equation (1). The optimisation problem of Equation (1) can be solved using the generalised eigenvalue problem [33].

$$\begin{aligned} \mathbf{w}_{2n} &= \arg\max_{w \neq 0} \frac{\mathbf{w}_n^{\mathrm{T}} \mathbf{R}_{2n+1} \mathbf{w}_n}{\mathbf{w}_n^{\mathrm{T}} \mathbf{R}_{2n} \mathbf{w}_n} \\ \mathbf{w}_{2n+1} &= \arg\min_{w \neq 0} \frac{\mathbf{w}_n^{\mathrm{T}} \mathbf{R}_{2n+1} \mathbf{w}_n}{\mathbf{w}_n^{\mathrm{T}} \mathbf{R}_{2n} \mathbf{w}_n} \end{aligned} \tag{1}$$

In contrast with standard decision tree approaches using chroma quantisation, we use the eigen vector that corresponds to the smallest eigen value magnitude. Ergo, a binary split decision at each parent node is taken within the tree decomposition algorithm based on either $\max\{\lambda_n\}$ or $\min\{\lambda_n\}$, since the two hyperplanes generate two new child nodes. In effect, we perform calculations by utilising the smallest eigen value, but for node binary splits of a local cluster into two new clusters, we use the largest eigen value (Equation (3)) over all (current iteration) clusters. The order and direction in which the node is split are determined by selecting that eigenvector e_n which corresponds to the largest eigenvalue λ_n stored from all previous node eigenvalues of covariance. However, in this step, a normalised covariance form is used $\hat{\mathbf{R}}_n$, whereas each node covariance is calculated by:

$$\hat{\mathbf{R}}_n = \mathbf{R}_n - \frac{1}{N_n} \mathbf{M}_n \mathbf{M}_n^{\mathrm{T}} \tag{2}$$

In effect, a node split is determined by its eigenvector being that of the largest eigenvalue over all previous nodes, which in turn determines the pixel indices in cluster C_n that will be assigned into the new clusters C_{2n}, C_{2n+1}. The binary split for node image indices ℓ associated in cluster $C_n \rightarrow \{C_{2n}, C_{2n+1}\}$ is performed using the schema:

$$\begin{aligned} C_{2n} &= \{\ell \in C_n : e_n^T x_\ell \geq e_n^T Q_n\} \\ C_{2n+1} &= \{\ell \in C_n : e_n^T x_\ell < e_n^T Q_n\} \end{aligned} \tag{3}$$

The procedure has been summarised in pseudo-code as per Algorithm 1.

Algorithm 1 Generate decision tree

```
1: Set image as root node
2: Calculate R_1, M_1, N_1 and w_0, λ_0
3: for all n nodes (cluster) do
4:     Find leaf n that λ_n is max
5:     Form node n by Equation (3)
6:     Calculate R_n, M_n, N_n, w_n for new nodes where λ_n is min
7: end for
8: for each node do
9:     Find leaf where w_n λ_n is max
10:    Form new nodes 2n, 2n + 1 by Equation (3)
11:    Calculate {R_2n, M_2n, N_2n, w_2n} and {R_2n+1, M_2n+1, N_2n+1, w_2n+1}
12: end for
```

2.2. Pruning and Defect Prediction

In addition to our revised tree generation process, we introduce pruning based on entropy calculation (bottom-up setup). Given the preserved intermediate (parent) C_n nodes or edges of the tree, we calculate the entropy vector H corresponding to each node's entropy. The information entropy is calculated by means of

$$H(x_\ell) = -\sum_n p_n(x_\ell) \log_2(p_n(x_\ell)) \tag{4}$$

where $H(x_\ell)$ is the entropy considered as a random variable of x_ℓ, and the probability $p_n(\cdot)$ is that of outcome x_ℓ occurring, with $(n, \ell$ representing all possible outcomes). The probability density p_n is calculated by approximating over the channel-level histograms, where histogram bins represent possible states.

We use the cluster information entropy as a measure of information content; i.e., an interpretation of the uncertainty of the parent node. The corresponding states of quantisation levels that an individual pixel can adopt are evaluated and it is determined whether the image information of a specific node is sufficient for the node to be pruned, or not. The procedure is such that, initially, the corresponding leaf nodes C_{2n}, C_{2n+1} of some intermediate node C_n are identified. Based on entropy value comparison, a decision is made for selected leaf nodes C_{2n+k}, C_{2n+k+1} with C_{n+k} as follows:

- Suppress the leaf nodes, if $\{H_{2n+k}, H_{2n+k+1}\} \leq H_{n+k}$ and move upward to $n+k-1$ node set;
- Preserve the leaf nodes, if $\{H_{2n+k}, H_{2n+k+1}\} \geq H_{n+k}$ and proceed to a neighbouring branch $n+k$ node set.

In order to predict the leaf node that best captures the corroded regions within the input frame, decision criteria have been implemented. The procedure requires iteration over the leaf nodes C_n, C_{n+k}, where k is the pruning invariant depth, identifying the nodes that correspond to the maximum eigenvalue λ_n and maximum entropy H_n. In the event that $\max\{\lambda_n\}$, $\max\{H_n\}$ point to different leaf nodes, the node of $\max\{H_n\}$ is eliminated from the candidate pool. Conversely, if both max values refer to the same node, we assume said node to be the maximum of all entropy values. Since the predicted leaf node contains only pixel indices, it reconstructs a predicted frame based on the preserved cluster indices.

3. Results

We compare our method with standard deep-learning image segmentation methods, with performance results illustrated in Figure 1. Considering the deep-learning techniques, we have applied a supervised learning fully Convolutional Neural Network implementation known as U-Net [34], using the library, framework and models available at SegmentationModels github (last accessed on 18 July 2023), and an unsupervised learning Self Organising Maps algorithm [35] using the library, framework and examples available at MiniSOM github (last accessed on 18 July 2023). In general, and for the unsupervised case, it has been noted that decision trees with pruning are of complexity $\mathcal{O}(n \log n)$, whilst SOM is of complexity $\mathcal{O}(n^2)$; hence, decision trees can, in principle, achieve better computational times than SOMs. The CNN method is supervised, and thus its complexity is linear, since it depends on the input vector size (n) and the total number of layer nodes in the supervised network.

The CNN (unet) implementation was trained and tested under a 60:40 split of the dataset using the 'seresnet34' backbone with a sigmoid activation function and the Adam optimiser, with binary loss cross-entropy function enabled for training, whereas the learning rate is 10×10^{-3} for 50 epochs. We used the U-Net neural network which performs semantic segmentation that more closely matched our revised method. As a result, a comparison between U-Net and other techniques in this application domain allows for better comparison between true-positive/positive and false-positive/negative areas. The self-organising map (miniSOM implementation) was tested in such a way as to mimic the maximum number of allowed clusters in the Eigen tree method (i.e., orthogonal lattice topology of 3×2, 4×2, 5×2).

The dataset we used is a set of two image folder collections [29]. The collected raw data were gathered from the hull areas that are likely (but not necessarily) deemed to be problematic. To produce the labelled images (ground truth) that were later used for methods' performance evaluation, a trained human inspector highlighted the regions of interest by manually labeling areas identified as corroded. It is important to note that areas in the image manually annotated by human inspectors are labelled as regions of interest characterised by rust. However, this includes areas that are deemed to be corroded and could produce rust on the surface of the hull in the near future.

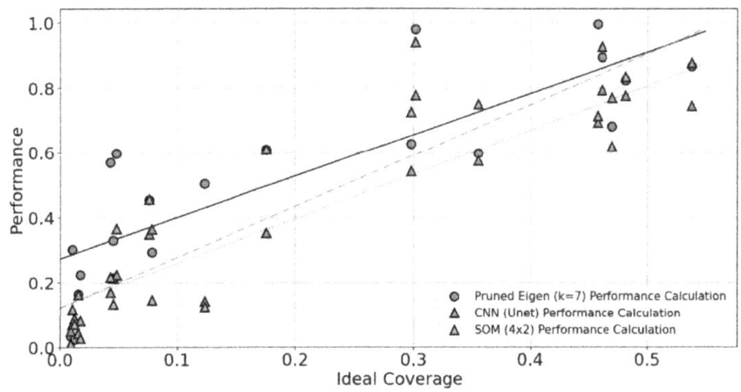

Figure 1. Performance metric comparison between CNN (unet-seresnet34), SOM (4×2 ortho-lattice), and Pruned Eigen tree (tree depth from seven to six nodes). The corresponding lines illustrate a linear fit over the methods' performance.

To assess the performance of corrosion detection, the pixel coordinates of corrosion in an image under investigation are found through comparison to the ground truth annotated images. This is performed by applying on the input image, the labelled image mask on the dominant cluster result. The pixels defined as 'True Positive' pixels (TP) of a cluster are those that match the labelled mask, and 'False Positive' pixels (FP) are those that do not fall within the labelled mask.

We define as 'Recognition Accuracy' the percentage of correctly classified images as containing some corroded area. In the case of our method, this means an image whose predicted dominant cluster contains some arbitrary percentage of captured corroded region above a threshold of 10% of true positive pixels. For example, Figure 2 illustrates the case of a confusion matrix for tree nodes $k = 7$. It is evident that recognition accuracy remains the same at 94% for pruned and unpruned Eigen tree. Similarly produced confusion matrices for $k = 5, 10$ result in an accuracy of 93% and 95%, respectively, (as they appear in Table 1). As will become evident later, recognition accuracy does not necessarily depend on our pruning method, but rather on the implemented decision criteria as evidenced in Table 1. We note that there is no direct representation of true coverage versus ground truth in this metric.

We refer to 'Ideal Coverage' as the number of labelled pixels over total image pixels, i.e., percentage of maximum coverage that could be achieved by any given method. This aids the comparison to Ideal Coverage as a reference value. The number of TP pixels over the total number of image pixels is the True 'Area Coverage' (TAC), with the number of FP pixels over total number of image pixels being False 'Area Coverage' (FAC). The closer TAC is to Ideal Coverage, the higher the expected accuracy in terms of capturing defective pixels. However, FAC can be any given percentage outside of the label region, and as such should be used to 'penalise' a method's performance measure. The 'Performance' metric is defined as per the metric:

$$\text{Performance} = \frac{TAC}{TAC + FAC} \tag{5}$$

The metric of Equation (5) is an indication of the algorithms' performance, since FAC is not related to ideal coverage and it penalises the method behaviour. The CNN and SOM versus Pruned Eigen tree method comparison is summarised in Figure 1, with example results provided visually in Figure 3.

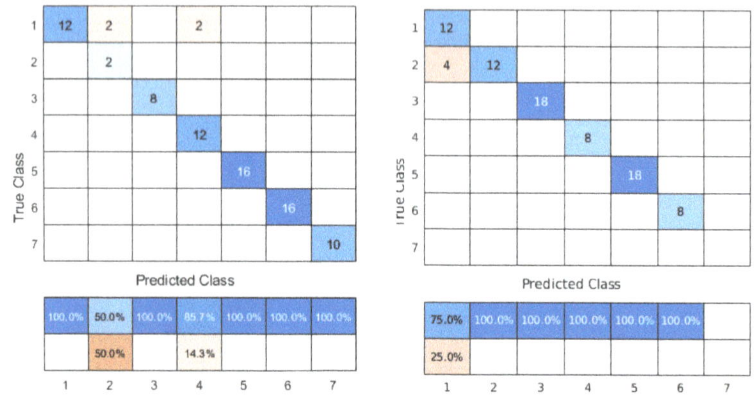

(**a**) Full Tree (unpruned) with seven nodes (**b**) Pruned node tree from seven to six nodes

Figure 2. Pruned versus unpruned eigen tree confusion matrices. A light to dark blue colour map has been applied for correctly classified images, and a light to dark orange colour map for erroneously classified images.

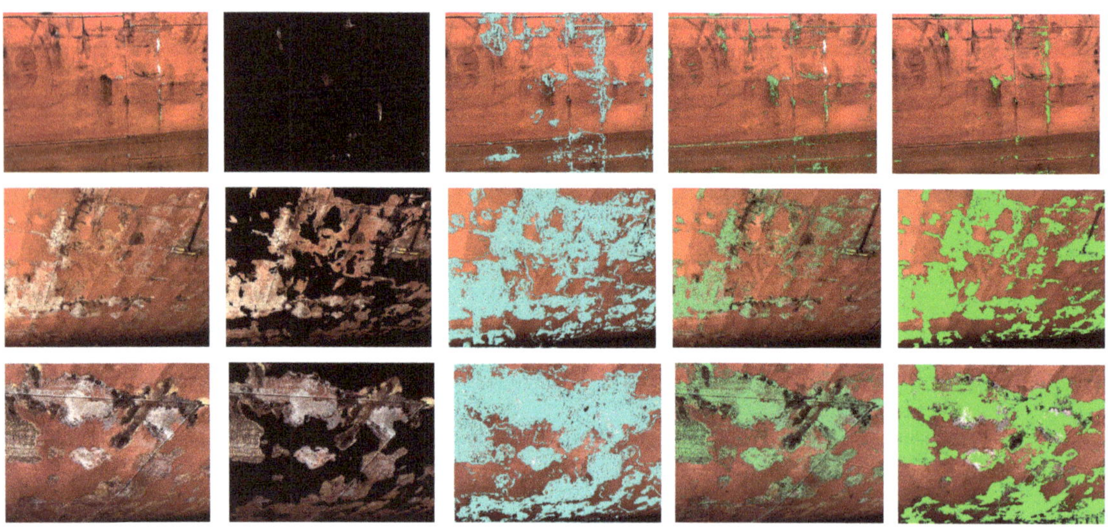

(**a**) Input Image (**b**) Ground Truth (**c**) CNN (**d**) SOM (**e**) Pruned Tree

Figure 3. Representative examples of the application of methods in Figure 1 (CNN, SOM, Pruned Eigen) on our compiled dataset.

Table 1. Comparison to other methods; the Area Coverage metric reported value TP relates to 'true positive' pixels and FP to 'false positive' pixels. All reported values are mean values over the entire dataset.

		Eigen Tree	Pruned Tree	SOM	CNN
Recognition Accuracy (%)	k = 5	95%	94%	94%	
	k = 7	94%	94%	95%	96%
	k = 10	93%	93%	94%	
Area Coverage TP, FP (%)	k = 5	9.7%, 10%	15.8%, 13.1%	10%, 16%	
	k = 7	7%, 4%	13.4%, 5.9%	9%, 10%	15%, 20%
	k = 10	6%, 3.5%	12.7%, 5%	8%, 9%	
Performance (%)	k = 5	50%	52%	44%	
	k = 7	55%	56%	47%	44%
	k = 10	54%	54%	48%	

4. Discussion

We provided evidence that our pruned Eigen tree exhibits better performance than CNN and SOM architectures, by means of Figures 1 and 3. We further substantiate this by means of accumulated results in Table 1, whereas the mean values of the defined metrics over the totality of the dataset instances are reported. It is evident that the implemented CNN architecture (inspired by U-Net) outperforms other techniques under investigation in recognition accuracy. However, and at least for our dataset and annotations, CNN and SOM do not outperform the Eigen tree methods (pruned/unpruned), with respect to Area Coverage ratios and, thus, performance. In fact, on average, the CNN architecture has a better true positive (TP) coverage at the expense of increased false positive (FP) coverage, whilst the remainder of the reported methods seem to be more balanced in the sense of TP/FP area coverage ratios.

Furthermore in Table 1 the pruned Eigen tree has a significantly higher TP coverage without sacrificing FP coverage, as opposed to the unpruned Eigen tree. The same applies for the CNN and SOM methods, although the number of clusters maximum depth seems to have a minor effect on dominant cluster selection, and thus average performance. Conversely, the pruned Eigen tree has a better TP to FP area coverage, although on aggregate, performances between pruned and unpruned versions of the tree seem to be comparable. For example, in Table 1 and for our method (Eigen Tree/Pruned Tree) at $k = 5$, it is evident that the true positive (TP) area detected is higher than the false positive (FP); i.e., 15.8 versus 13.1%, leading to a performance of 52%. The performance of 52% is higher than the 44% achieved in SOM (4×2 ortho-lattice) and CNN methods, since both of these methods have an FP % higher than the TP % (i.e., in SOM the FP to TP area coverage of 16 versus 10%, and for CNN 20 versus 15%).

We note that the applied CNN model has an increased false-positive area coverage (20% in Table 1) which is higher than that of pruned eigen tree (for tree depth of seven nodes, FP area 5.9%) and SOM (FP area 10%). This leads to the inferior (as defined) CNN performance, similar to SOM implementations, but to a much smaller extent. This should not come as a surprise since, as previously mentioned (see Section 1), neural networks, and similar deep learning techniques, seem to suffer from crack-like pixel groups and changing light conditions, both of which are extensively present in our dataset. For the pruned eigen tree, this can be said to lead to significantly better performance than CNN and SOM, particularly in the case of low ideal coverage. However, it should be noted that SOM seems to outperform CNN and be very close to the pruned Eigen tree in terms of performance at high coverage (Figure 1).

We recognise that sample size in the used dataset is an issue for CNN training. It remains to be seen whether increasing the data size would lead to better CNN and possibly SOM performance; albeit that the TP to FP ratio examined herein, and the identified CNN, SOM vs. pruned Eigen tree trends, does not seem to support the data size issue. At present, we can postulate that at least for this dataset, the pruned Eigen tree leads to better overall area coverage and performance.

5. Conclusions

We have presented a revised eigen tree decomposition method alongside a pruning methodology/dominant node selection criteria. We have tested and validated these alongside other methods in a domain-specific dataset. The pruning method does not improve recognition accuracy, but substantially increases true positive (TP) area coverage with smaller false positive (FP) areas. The decision criteria, in conjunction with the pruning method, lead to much better FP to TP ratios.

Our method seems to outperform similar unsupervised segmentation (clustering) techniques, which is even more evident in the pruned Eigen tree case. Furthermore, the pruned Eigen Tree method achieves comparable recognition accuracy and better average area coverage in the domain of corrosion detection to neural network architectures; namely the supervised CNN (U-Net-like) model, and the unsupervised SOM model. This is particularly prominent when low ideal coverage images are expected and high TP area coverage is to be desired.

Future work is required to establish whether a data size increase for CNN model training would lead to better performance. We shall further investigate this route by taking a data-driven model and examining the dataset equal distribution of raw and annotated images in predefined low, medium, and high ideal coverage cases.

Author Contributions: Conceptualization, G.C. and A.L.; methodology, G.C.; software, G.C.; validation, S.V.K. and A.L.; formal analysis, G.C. and A.L.; investigation, S.V.K. and G.C.; data curation, G.C. and A.L. All authors have read and agreed to the published version of the manuscript.

Funding: This research received no external funding.

Institutional Review Board Statement: Not applicable; data acquired are standard camera images that do not include any human or vessel identification features or markings.

Informed Consent Statement: Informed consent was obtained from human operators involved in the study at the data collection stage.

Data Availability Statement: The raw data from which results are presented in this study are open and freely available at [29].

Acknowledgments: We acknowledge collaboration and help from: A. Andreous for aiding with the final version English language editing, I. Tzanetatos for the software testing and development; and K. Kamzelis for material and arrangements on data collection.

Conflicts of Interest: The authors declare no conflict of interest.

References

1. Liu, C.; Li, M.; Chen, P.; Zhang, C. Method to detect the bolt thread defect in the aerospace vehicle with ultrasonic image and its implementation. *Anti-Corros. Methods Mater.* **2023**, *70*, 197–201. [CrossRef]
2. International Maritime Organization. International Convention for the Safety of Life at Sea. Regulation 3–6: Access to and within Spaces in, and Forward of, the Cargo Area of Oil Tankers and Bulk Carriers. Available online: https://www.refworld.org/docid/46920bf32.html (accessed on 1 May 2023).
3. Gao, T.; Sun, H.; Hong, Y.; Qing, X. Hidden corrosion detection using laser ultrasonic guided waves with multi-frequency local wavenumber estimation. *Ultrasonics* **2020**, *108*, 106182. [CrossRef]
4. Merzban, M.H.; Elbayoumi, M. Efficient solution of Otsu multilevel image thresholding: A comparative study. *Expert Syst. Appl.* **2019**, *116*, 299–309. [CrossRef]
5. Dhanachandra, N.; Manglem, K.; Chanu, Y. Image Segmentation Using K -means Clustering Algorithm and Subtractive Clustering Algorithm. *Procedia Comput. Sci.* **2015**, *54*, 764–771. [CrossRef]

6. Yang, W.; Cai, L.; Wu, F. Image segmentation based on gray level and local relative entropy two dimensional histogram. *PLoS ONE* **2020**, *15*, e0229651. [CrossRef]
7. Forkan, A.R.M.; Kang, Y.B.; Jayaraman, P.P.; Liao, K.; Kaul, R.; Morgan, G.; Ranjan, R.; Sinha, S. CorrDetector: A Framework for Structural Corrosion Detection from Drone Images using Ensemble Deep Learning. *arXiv* **2021**, arXiv:2102.04686.
8. Wang, H.; Wang, Y.; Zhang, Q.; Xiang, S.; Pan, C. Gated Convolutional Neural Network for Semantic Segmentation in High-Resolution Images. *Remote Sens.* **2017**, *9*, 446. [CrossRef]
9. Calabrese, L.; Galeano, M.; Proverbio, E.; Pietro, D.D.; Donato, A. Topological neural network of combined AE and EN signals for assessment of SCC damage. *Nondestruct. Test. Eval.* **2020**, *35*, 98–119. [CrossRef]
10. Kapsalas, P.; Zervakis, M.; Maravelaki-Kalaitzaki, P. Evaluation of image segmentation approaches for non-destructive detection and quantification of corrosion damage on stonework. *Corros. Sci.* **2007**, *49*, 4415–4442. [CrossRef]
11. Zhu, Y.; Wang, L.; Behnamian, Y.; Song, S.; Wang, R.; Gao, Z.; Hu, W.; Xia, D.H. Metal pitting corrosion characterized by scanning acoustic microscopy and binary image processing. *Corros. Sci.* **2020**, *170*, 108685. [CrossRef]
12. Das, A.; Ichi, E.; Dorafshan, S. Image-Based Corrosion Detection in Ancillary Structures. *Infrastructures* **2023**, *8*, 2412–3811. [CrossRef]
13. Bonnin-Pascual, F.; Ortiz, A. Corrosion Detection for Automated Visual Inspection. *Dev. Corros. Prot.* **2014**, 619–632.
14. Ghanta, S.; Karp, T.; Lee, S. Wavelet Domain Detection of Rust in Steel Bridge Images. In Proceedings of the IEEE International Conference on Acoustics, Speech and Signal Processing, Prague, Czech Republic, 22–27 May 2011; pp. 1033–1036.
15. Khayatazad, M.; De Pue, L.; De Waele, W. Detection of corrosion on steel structures using automated image processing. *Dev. Built Environ.* **2020**, *3*, 100022. [CrossRef]
16. Xu, S.; Weng, Y. A new approach to estimate fractal dimensions of corrosion images. *Pattern Recognit. Lett.* **2006**, *27*, 1942–1947. [CrossRef]
17. Ping, A.Q.; Xiaoming, J.; Wang, T. Corrosion Damage Assessment of AerMet100 Steel Based on Image Analysis. *IOP Ser. Mater. Sci. Eng.* **2018**, *394*, 52–66.
18. Ortiz, A.; Bonnin-Pascual, F.; Garcia-Fidalgo, E.; Company-Corcoles, J. Vision-Based Corrosion Detection Assisted by a Micro-Aerial Vehicle in a Vessel Inspection Application. *Sensors* **2016**, *16*, 2118. [CrossRef]
19. Ma, Y.; Yang, Y.; Yao, Y.; Li, S.; Zhao, X. Image-based corrosion recognition for ship steel structures. In *Smart Structures and NDE for Industry 4.0*; Meyendorf, N.G., Clingman, D.J., Eds.; SPIE: Bellingham, WA, USA , 2018; Volume 10602, pp. 143–149.
20. Atha, D.J.; Jahanshahi, M.R. Evaluation of deep learning approaches based on convolutional neural networks for corrosion detection. *Struct. Health Monit.* **2018**, *17*, 1110–1128. [CrossRef]
21. Liu, L.; Tan, E.; Cai, Z.Q.; Yin, X.J.; Zhen, Y. CNN-based Automatic Coating Inspection System. *Adv. Sci. Technol. Eng. Syst. J.* **2018**, *3*, 469–478. [CrossRef]
22. Wang, J.; Fu, P.; Gao, R. Machine vision intelligence for product defect inspection based on deep learning. *J. Manuf. Syst.* **2019**, *51*, 52–60. [CrossRef]
23. Nash, W.; Zheng, L.; Birbilis, N. Deep learning corrosion detection with confidence. *NPJ Mater. Degrad.* **2022**, *6*, 26. [CrossRef]
24. Aijazi, A.K.; Malaterre, L.; Tazir, M.L.; Trassoudaine, L.; Checchin, P. Detecting and Analyzing Corrosion Spots on the Hull of Large Marine Vessels Using Colored 3D LIDAR Point Clouds. *ISPRS Ann. Photogramm. Remote Sens. Spat. Inf. Sci.* **2016**, *III3*, 153–160. [CrossRef]
25. Nash, W.; Drummond, T.; Birbilis, N. A review of deep learning in the study of materials degradation. *NPJ Mater. Degrad.* **2018**, *37*, 37. [CrossRef]
26. Nash, W.; Drummond, T.; Birbilis, N. Deep Learning AI for corrosion detection. In *NACE CORROSION*; NACE: Bethlehem, PA, USA , 2019.
27. Coelho, L.B.; Zhang, D.; Van Ingelgem, Y.; Steckelmacher, D.; Nowé, A.; Terryn, H. Reviewing machine learning of corrosion prediction in a data-oriented perspective. *NPJ Mater. Degrad.* **2022**, *6*, 8. [CrossRef]
28. Güldenring, R.; Nalpantidis, L. Self-supervised contrastive learning on agricultural images. *Comput. Electron. Agric.* **2021**, *191*, 106510. [CrossRef]
29. Chliveros, G.; Tzanetatos, I.; Kamzelis, K. MaVeCoDD Dataset: Marine Vessel Hull Corrosion in Dry-Dock Images. *Mendeley Data* **2021**.
30. Hoang, N.; Nguyen, Q. A novel method for asphalt pavement crack classification based on image processing and machine learning. *Eng. Comput.* **2019**, *35*, 487–498. [CrossRef]
31. Sander, T.; Sander, J. Tree decomposition by eigenvectors. *Linear Algebra Its Appl.* **2009**, *430*, 133–144. [CrossRef]
32. Orchard, M.T.; Bouman, C.A. Color Quantization of Images. *IEEE Trans. Signal Process.* **1991**, *39*, 2677–2690. [CrossRef]
33. Manwani, N.; Sastry, P.S. Geometric Decision Tree. *IEEE Trans. Syst. Man, Cybern. Part B (Cybern.)* **2012**, *42*, 181–192. [CrossRef]
34. Shelhamer, E.; Long, J.; Darrell, T. Fully Convolutional Networks for Semantic Segmentation. *IEEE Trans. Pattern Anal. Mach. Intell.* **2017**, *39*, 640–651. [CrossRef]
35. Kohonen, T. Essentials of the self-organizing map. *Neural Netw.* **2013**, *37*, 52–65. [CrossRef] [PubMed]

eng

MDPI

Article

Microsimulation Modelling and Scenario Analysis of a Congested Abu Dhabi Highway

Umair Hasan [1,*], Hamad AlJassmi [1,2] and Aisha Hasan [1]

1 Emirates Center for Mobility Research, UAE University, Abu Dhabi 999041, United Arab Emirates
2 Department of Civil & Environmental Engineering, UAE University, Abu Dhabi 999041, United Arab Emirates
* Correspondence: umair.hasan@uaeu.ac.ae

Abstract: Today's roadways are subject to traffic congestion, the deterioration of surface-assets (often due to the overreliance on private vehicle traffic), increasing vehicle-operation and fuel costs, and pollutant emissions. In Abu Dhabi, private car traffic forms the major share on urban highways, as the infrastructure was built to a high quality and the public transport network needs expansion, resulting in traffic congestion on major highways. These issues are arguably addressable by appropriate decisions at the planning stage. Microsimulation modeling of driving behavior in Abu Dhabi is presented for empirical assessment of traffic management scenarios. This paper presents a technique for developing, calibrating, validating, and the scenario analysis of a detailed VISSIM-based microsimulation model of a 3.5 km section of a 5-lane divided highway in Abu Dhabi. Traffic-count data collected from two sources, i.e., the local transport department (year 2007) and municipality (2007 and 2015–2016) were used. Gaps in traffic-counts between ramps and the highway mainline were noted, which is a common occurrence in real-world data situations. A composite dataset for a representative week in 2015 was constructed, and the model was calibrated and validated with a 15% (<100 vehicles per hour) margin of error. Scenario analysis of a potential public bus transport service operating at 15 min headway and 40% capacity was assessed against the base case, for a 2015–2020 projected period. The results showed a significant capacity enhancement and improvement in the traffic flow. A reduction in the variation between vehicle travel times was observed for the bus-based scenario, as less bottlenecking and congestion were noted for automobiles in the mainline segments. The developed model could be used for further scenario analyses, to find optimized traffic management strategies over the highway's lifecycle, whereas it could also be used for similar evaluations of other major roads in Abu Dhabi post-calibration.

Keywords: microsimulation; public transport; highways; travel time; mobility management

Citation: Hasan, U.; AlJassmi, H.; Hasan, A. Microsimulation Modelling and Scenario Analysis of a Congested Abu Dhabi Highway. *Eng* **2023**, *4*, 2003–2014. https://doi.org/10.3390/eng4030113

Academic Editors: Antonio Gil Bravo and Sanjay Nimbalkar

Received: 24 April 2023
Revised: 16 June 2023
Accepted: 15 July 2023
Published: 17 July 2023

1. Introduction

Transport policymakers are often required to implement service provision by prioritizing route design, supply–demand balance, mode-shift uncertainties, and generic passenger attitudes, while working within the confines of social, political, economic, administrative, and environmental thresholds [1,2]. Alawadi [3] noted that transparent and inclusive transport planning policy is required for multicultural and climate-sensitive cities, where different population groups need to be included within the decision-making process, particularly regarding the argument between linear and scattered development of urban designs. Traffic control-related strategies are important for achieving long-term goals, since they affect fleet-management systems and enhance urban traffic flow. For example, focusing on traffic routing conditions, eco-driving and eco-routing both aim to cut pollutant generation and energy use through traffic congestion reduction, while optimizing different aspects [4].

Other studies have argued that the car-centric urban design of cities can promote unsustainable mode choices, such as overreliance on private cars for urban mobility, instead of public transport or micro-mobility options [5]. Hence, transport agencies gauge three policymaking issues: the optimal grouping of variables predicting passenger travel patterns; the magnitude of probable temporal divergence trends; and whether this divergence in travel behavior could be directed towards a sustainably optimal option through control of variable combinations. Passenger satisfaction with the existing public transport service, alongside public-expected service quality attributes, may aid decision-makers in increasing public transport user uptake [6,7], towards the ultimate lifecycle environmental asset management goal of a low carbon city.

In Abu Dhabi, the transport infrastructure was built to the highest international standards, but it is car-centric [8], which resulted in approximately 80% of the passengers opting for private and shared car transport in most regions of Abu Dhabi [9]. The majority of the population in Abu Dhabi, similar to other Gulf Cooperation Countries (GCC), are male expatriates from South Asian and Middle Eastern countries as a full-time workforce, with a monthly income profile at an average of USD 1000 [10]. The transport plans of the Abu Dhabi transport authorities and policy-makers target shifting the mode choice of these expatriate resident population groups from private cars to more sustainable transit modes, such as public transport [11], since it may be more complicated to trigger a shift in the mobility choices of the so-called wealthier population groups, due to social status and income dynamics [12].

It is critical to project the expected benefits in terms of travel time savings, reductions in congestion, etc., that would potentially result from adoption of public transport on major highways. Microsimulation modeling can be used to understand a high-resolution per vehicle change in the traffic flow profile, but this requires proper model development for the studied region's driving behavior [13]. Contrary to macroscopic models that use aggregated quantities to describe the traffic flow and are easier to calibrate, microscopic or microsimulation models are more complex to calibrate but give more details about the traffic flow profile once the traffic flow characteristics of a region's car-following behavior have been properly modelled.

This research aimed to develop, calibrate, and validate a VISSIM-based microsimulation model, to reveal the car-following behavior of Abu Dhabi city, as a representative of the GCC and Middle Eastern drivers, where such models are not highly used for research and traffic management policy evaluations. However, the majority of sustainable road research in the region has focused on alternate material research [14] and traffic safety behavioral studies from a psychological response perspective [15,16], with minimal attention to actually modeling the driver behavior. The model was then applied to analyze the efficacy of extending public transport services to a major highway in the City of Abu Dhabi, using in-field data and traffic projection models, which can be used to perform scenario modeling of not only traffic management strategies, as demonstrated here, but traffic safety research when considering the impact of flow management policies on changes in aggressive driving behavior.

2. Background

Traffic congestion due to excessive private car usage by urban residents is a serious concern, in terms of the overall sustainability of road transportation systems, while also affecting the journey experience and travel times [17]. Built-up regions and open space geometries have a profound influence on the morphologies of cities, traffic mobility, and congestion, due to the topological links between such spaces, spatial identities, and the concentrations of people, goods, and services. Street metrics characterize the urban typologies and affect the flow of traffic in a city [18]. This indirectly affects the mode choice and traffic flow pattern of residents and, as such, city design should acknowledge these metrices and their eventual impact on traffic and congestion pocket formation. According to Gao et al. [19], in locations with higher traffic densities, repeated automobile deceleration–acceleration cycles result in excessive fuel consumption. Likewise, the journey time of any major expressway is dependent upon vehicle-fleet speeds, deceleration–acceleration, flow rates, and traffic densities. The traffic volume rises each year, corresponding to a sharp decline in engine speeds and vehicle velocities, as the saturation flow rate for the highway is reached, as is also evident in the transport literature [19].

Public-transport-based solutions are recognized as alleviating traffic congestion, user time delays, road deteriorations, and for reducing costs and the energy and pollutant burden in major cities [20]. Public transportation projects, in any form, are aimed at assuaging congestion from overcrowded road networks. For example, studies [21,22] have noted the adverse effect of accessibility, journey time, network coverage, and on-board crowding, etc., on public intention to use public transport. Similarly, at a rudimentary service level, public bus transport can either serve as a feeder network to the largescale BRT or LRT mobility services or act as an early proof-of-concept in a crowded metropolitan area, before the overhaul of a transit network from car-centric to more sustainable transit modes. To that end, the travel habits and mode choices of existing transit users may be directed in favor of sustainable public transit services by traffic management policies that cater better to user needs and improve their journey attributes. A study by Hensher [23] on bus operators in Australia investigated the perception of buses on the basis of service quality and performance indices, and a methodology for cost per kilometer and commuter perception of quality of service by operators was established. Lavery et al. [24] found travel satisfaction motivated passengers towards a pro-environmentalist (e.g., public bus service) approach.

As stated earlier, simulating the per vehicle change in the traffic flow profile can show immediate changes as a result of different traffic management strategies. This work argues that, in addition to this, traffic growth models can be used to project the benefits over extended periods of time, particularly when relieving daily traffic congestion is a concern. It should be noted that none of these benefits can be assessed without properly creating, calibrating, and validating a dedicated microsimulation model. Although, some macroscopic models exist for the urban planning purposes in Abu Dhabi and the GCC region at large [16,25], these are largely concerned with personal mobility or alternate vehicle type evaluations and do not model the general driving behavior of the city's residents on a major urban road, which is required to compare different traffic management strategies at a micro level.

On a global scale, many studies have incorporated variables such as vehicle acceleration–deceleration profiles, congestion wave speed, jam densities, and free-flow speeds to create and calibrate vehicle simulation models [26]. Others [27,28] have utilized a microsimulation calibration approach to present accurate car-following driving behavior and traffic flow profile models using software such as VISSIM and AIMSUN; however, these approaches were within a limited demographic context. The case study Abu Dhabi region has a much more diverse population, with a majority of expatriates and residents from over 200 nationalities, which makes it critical to create a dedicated and well-calibrated microsimulation model for simulating its daily traffic and evaluating future traffic management strategies. It is the largest emirate, as well as the federal capital of the United Arab Emirates, with a population of over 1.8 million urban residents, a population density of 1900 person per km^2, and estimated land-size of 972 km^2. Approximately 85% of the population are the expatriate workforce, and while this ratio may be higher or lower in other GCC countries, the distribution is similar, with expatriates still forming the largest share of the population [29,30].

Generally, developing transit policies to achieve a mode shift in favor of public transport is a broad concept. The psychometric properties of passenger satisfaction is often complicated by various underlying and interdependent factors, such as travel bias, public transit quality attributes (journey time, transit stations, accessibility etc.), travel cost, ride quality, and service frequency [31,32]. This is covered in a separate study under the umbrella of this project. Nonetheless, that study [33] yielded interesting observations, providing a foundation for this exploration of the impact of implementing a traffic management strategy in the study region for improving public (bus service) transit ridership based on public requirements. The project deals with altering travel habits, user perception, reducing congestion, and alleviating the pressure of car traffic from major highways in the city of Abu Dhabi. The findings of this work could help towards meeting the energy consumption and pollutant decrement goals of local authorities at a micro level, while also producing real-time improvements in the journey of passengers on a case study highway.

3. Methodology

This project aimed to utilize the power of VISSIM microsimulation software for modelling traffic flow behaviors, delays, and queue formations for a case study road section located near the main modern shopping areas, business districts, and offices in the city. It begins at Sheikh Zayed Bridge (both a major bridge and a tourist attraction connecting the island of Abu Dhabi to the mainland) and winds its way around Abu Dhabi's eastern edge, until it meets Corniche Road (a 8 km road along the Abu Dhabi beach, parks, tourist, and recreational facilities). An existing four-lane road was extended and repaired to finish the project in 2009.

The Abu Dhabi transport department and municipality originally measured traffic-counts as vehicle volumes at continuous 15 min intervals in November 2007 for both inbound and outbound traffic, considering all vehicle types (passenger cars, vans and coaches, minibuses, light, and heavy trucks). However, the data were only gathered for the traffic stations located before the HW3 and HW5 traffic stations (location is off frame in Figure 1) and lacked data for all the traffic-count stations displayed in Figure 1. Subsequent traffic-counts were gathered by the Abu Dhabi Municipality between July 2014 and July 2016 for inbound traffic but only for ADM4 (location is off frame in Figure 1), which was located right before the HW3 traffic station on the mainline freeway. Nonetheless, the traffic surveys revealed that the peak traffic periods in Abu Dhabi are

- Morning-peak: 07:00 a.m.–08:00 a.m.
- Afternoon-peak: 14:00 p.m.–15:00 p.m.
- Evening-peak: 19:00–20:00
- A median traffic-count on 27 April 2015 was also gathered by the Abu Dhabi Municipality, covering all the traffic-count stations displayed in Figure 1 for the peak 7 a.m. time. Although lacking counts for other time periods, this count created a comprehensive dataset of the traffic in- and outbounds on the case study road, as recorded by all of the traffic stations. Figure 1 illustrates the traffic-count station locations. This random distribution of traffic-count days and locations is a common occurrence in real-world data situations, as noted by Gomes et al. [34]. Therefore, the first stage was to gather 24 h traffic-counts for a representative single week (inclusive of both weekdays and weekends). An initial review of both *"Year 2007"* and *"Years 2015–2016"* traffic-counts exhibited a general agreement between the traffic flow behavior observed from both sources.

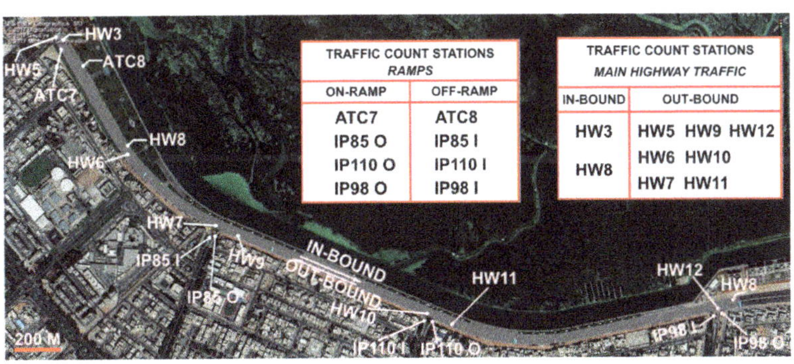

Figure 1. Case study section of Sheikh Zayed bin Sultan Street and traffic-count station locations.

This study had 2015 set as its base year, as this was the latest common traffic-count year between both inbound and outbound traffic volumes. Initially, the traffic-count for the stations HW5 and HW3 was estimated and the missing data from the Abu Dhabi Municipality for outbound traffic were estimated by separately calculating the traffic level ratios between the inbound and outbound traffic from the "*Year 2007*" counts for weekdays and weekends and multiplying by the inbound traffic volumes from "*Year 2015*", to generate the approximate outbound traffic volumes for the traffic stations located before HW5 and HW3, as shown in Equation (1) for 2015.

$$\underbrace{\rho_o(k)_\lambda^m}_{base\ year\ out-bound\ traffic\ count} = \underbrace{\rho_i(k)_\lambda^m}_{base\ year\ in-bound\ traffic\ count} \times \underbrace{\frac{\rho_o(k)_{\alpha\lambda}^m}{\rho_i(k)_{\alpha\lambda}^m}}_{2007\ \ traffic\ count} \tag{1}$$

where "$\rho_o(k)\ _\lambda{}^m$" and "$\rho_i(k)\ _\lambda{}^m$" are traffic-counts at outbound and inbound station "k" on mainline "m" for any hour "λ", respectively. The outbound traffic volume for the peak at 7 a.m. on 27 April 2015 traffic was then used to estimate the traffic volumes for all of the traffic-counting stations displayed in Figure 1 using Equation (2) below, where "$\rho_o(k)\ _{\bar{\lambda}}{}^m$" is the peak 7 a.m. traffic for any traffic station "k".

$$\rho_o(k)_\lambda^m = \underbrace{\rho_o(k-1)_\lambda^m \times \frac{\rho_o(k)_{\bar{\lambda}}^m}{\rho_o(k-1)_{\bar{\lambda}}^m}}_{mainline\, traffic} - \underbrace{\rho_o(k-1)_\lambda^{\bar{\bar{r}}} \times \frac{\rho_o(k)_{\bar{\lambda}}^{\bar{\bar{r}}}}{\rho_o(k-1)_{\bar{\lambda}}^{\bar{\bar{r}}}}}_{off-ramp\, traffic} + \underbrace{\rho_o(k-1)_\lambda^r \times \frac{\rho_o(k)_{\bar{\lambda}}^r}{\rho_o(k-1)_{\bar{\lambda}}^r}}_{on-ramp\, traffic} \tag{2}$$

After the traffic volumes for all the traffic-counting stations were computed for two representative weeks (for calibration and validation), *week 1* traffic data were then used to model OD matrices for the simulation of the current traffic situation in the case study area.

Developing a VISSIM Microsimulation Model

The studied 3.5 km dual-carriageway section was modelled in VISSIM with a 3.65 m single lane's width apart from the shoulder/exit-lanes, with a 3 m width in each direction. Traffic profiles, comprising the fleet composition, vehicle types, and distribution of 2D/3D models, are based on previous literature and the provision of preliminary data by the Abu Dhabi Municipality and the Abu Dhabi Department of Transport.

The lane behavior and speed profile were developed based upon the field observations gathered by the authors on several site visits to the case study location and local laws in the City of Abu Dhabi. Minibuses, vans and coaches, and HGVs were barred from entering the two extreme-left high-speed lanes, and driving behavior was modelled as per the right-hand side driving rule. The speed profile of the different vehicles was based on the minimum and maximum on-site posted speed limits for the case study road.

4. Results and Analysis

4.1. Current Traffic Situation

Figure 2 presents the current traffic situation on the studied case study highway section. The workday commencement resulted in 7–8 a.m. workdays peaks, corresponding to morning rush hours. This is especially relevant for passenger car traffic of small- (8700 units) and regular-sized cars (7200 units). The 2–3 p.m. time window corresponds to lunch break hours, once more exhibiting a higher traffic density and spike in car traffic. Similarly, the work day end traffic of 7–8 p.m. again showed this traffic pattern.

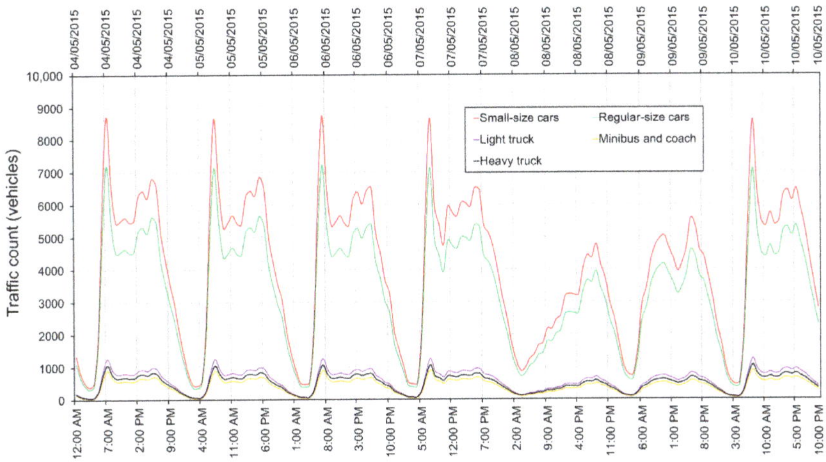

Figure 2. Current daily traffic situation on the studied road (base year 2015).

The period from 8th to 9th May 2015 corresponds to a weekend, and thus the trend for peak hour periods was representative of a conventional non-working day. The traffic peak period trends were hence indicative of typical non-working day flow patterns; for example, the resident travel behaviors for leisure reasons may have been responsible for showing the 5–7 p.m. passenger car traffic peak. The results exhibit the predominant contribution of cars to the overall traffic situation, as well as advantage of potentially introducing a public transport service as an alternate traffic management strategy.

4.2. Calibration of the VISSIM Microsimulation Model

The model was primarily run with the default car-following and driving behavior parameters provided in the VISSIM settings, based on the Wiedemann 99 Freeway (free lane selection) model. Traffic-counts for *week 1* from the representative weekday traffic dataset calculated above were utilized to provide the OD matrix. The model run results at per-hour resolution for the entire week are graphed alongside the field-based vehicle flow profiles (vehicles/hour) in Figure 3. The first simulation run resulted in several errors, with >2000 vehicles becoming lost vehicles and disappearing after waiting more than 60 s to change lanes. In general, the simulation model failed to achieve the GEH cut-off criteria of under 5 (85% cases).

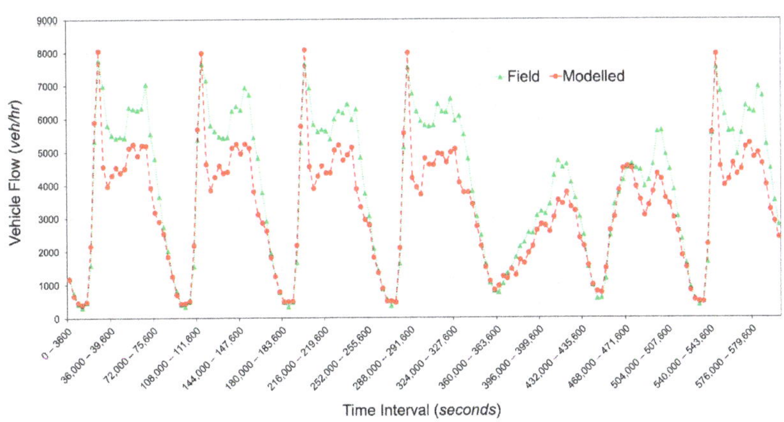

Figure 3. Traffic volume profile from model run with default VISSIM driving behavior parameters.

As a consequence of the simulation output from the model being constructed based on default parameters differentiating from the field data, especially for the peak period traffic volumes, the driving behavior was modified and the model was run several times. Figure 4 shows the simulation results for a calibrated model constructed after several rigorous trial-and-error runs, to develop a microsimulation model replicating the field-based measurements as closely as possible. Even though the simulated traffic volumes varied from the field data for some peak hours, the variation in traffic volume was negligible and within a 15% error (<100 vehicles per hour).

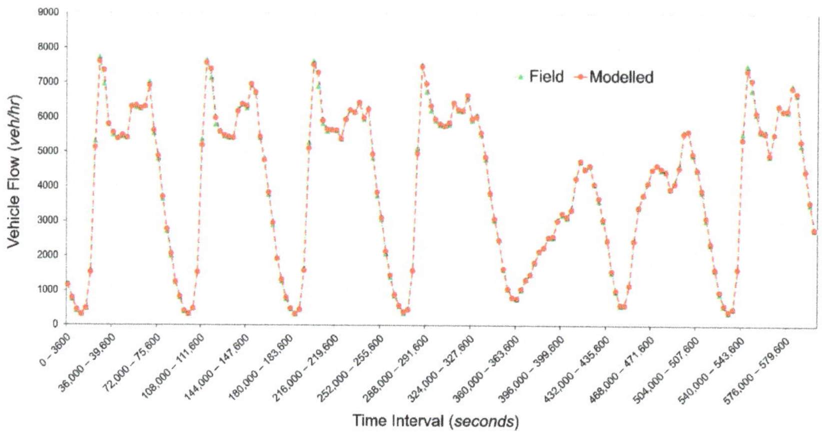

Figure 4. Traffic volume profile from model run with calibrated VISSIM driving behavior parameters.

4.3. Validation of the VISSIM Microsimulation Model

The calibrated base case model was then validated using the *week 2* traffic-count data applied to the constructed microsimulation model. Figure 5 illustrates the evaluation of the field-based and modelled traffic-volume profiles. It may be deduced from the results that, apart from some minor errors, the simulated traffic-counts successfully replicated the field traffic conditions, as the error was within the required bounds.

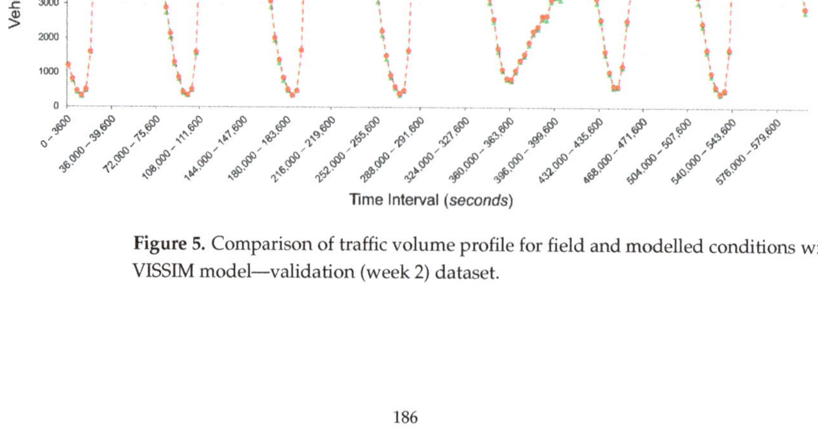

Figure 5. Comparison of traffic volume profile for field and modelled conditions with the calibrated VISSIM model—validation (week 2) dataset.

4.4. Public Transportation Scenario Analysis

According to 2015 base year measurements, the studied section serves over 9500 peak-hour vehicles in each direction. Growth in traffic and changes in vehicle driving variables were simulated here using high-resolution microsimulation models. The goal of constructing a traffic microsimulation model was to compare scenarios assessing the lifecycle impacts from implementing traffic management strategies. This was modelled and simulated for the years 2015–2020 as

1. **Do nothing:** The existing lane configuration, no public transport on the studied section. Utilizing secondary growth models from transport research, the traffic was assumed to increase by 6% annually.
2. **Traffic Management Scenario:** Assuming a hypothetical bus service operated on the studied section, mode-share switches to match the mode-share in other parts of Abu Dhabi, wherein public bus transport is currently operational (Table 1).

Table 1. Mode share of the current passenger journey profile.

Transport Mode	Base Case	Traffic Management Scenario
Small cars	56.5%	45.2%
Regular cars	43.5%	34.8%
Public transport (bus)	0%	20%

The simulation results for the 2015 base case confirmed the initial findings of the field observations, with regular traffic jams occurring in the peak hour traffic periods. The vehicle queue lengths were also recorded for the traffic levels across different years, as shown in Figures 6 and 7. These figures show that the queue lengths experienced by vehicles travelling across the case study area increased for the same time durations. Once the traffic flow profiles had been generated for every scenario and lifecycle year from 2015 to 2020, the vehicle travel times, queue length, and vehicle flow were tabulated against the lifecycle year.

In the traffic management scenario, the current lane configuration was kept; however, public transport was provided, in the form of a 50-seater public bus operating at a 15 min headway and 40% capacity. Similarly to the base case, this model aimed to be run for a representative week every subsequent year by escalating the traffic level by 6%. The mode shift of passenger vehicles was based upon the "Surface Transport Master Plan 2009" of the Abu Dhabi Department of Transport, as described in Table 1. A total of eight bus-stops at approximately 500 m distance were created for the case study highway mainline links, with passengers permitted to board and alight from right side, and a dwell time of $\pm$20 s was set after several subsequent runs, to find the options with the least errors, as well as to replicate the field settings given by Abu Dhabi Department of Transport.

The microsimulation for the traffic management scenario was run for the years 2015–2020. Random simulation instances were examined, to observe the traffic flow behavior of the simulated scenario, to ensure that the model flow followed real-world behavior. It should also be noted that, except for initial runs, no pedestrians were simulated, as this is beyond the current study's scope and may require further work on VisWalk for a more accurate and realistic pedestrian simulation. Figure 8 shows a comparison of the vehicle network performance between the base case and the implemented traffic management scenario [9]. A significant disparity in the network performance was observed for the base case scenario, where public transport is not provided for the case study mainline network. In this case, the journey time experienced by vehicles escalated near peak hours (7 a.m.–8 a.m.), due to the high vehicle-volume entering the network.

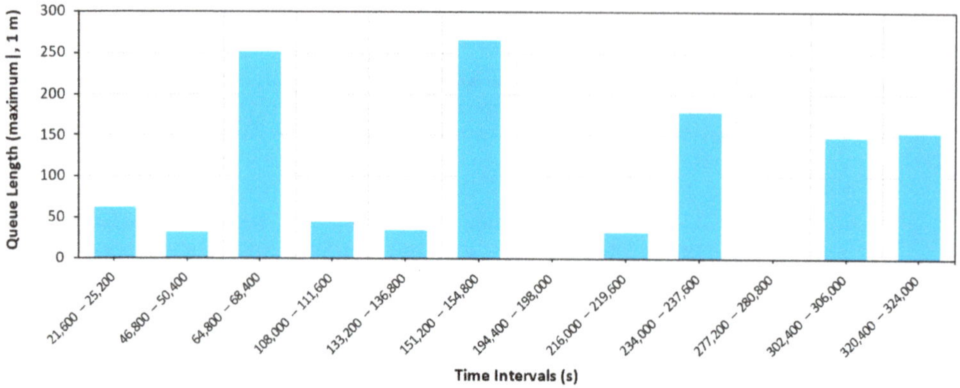

Figure 6. Queue length for in-bound traffic—base case year 2015.

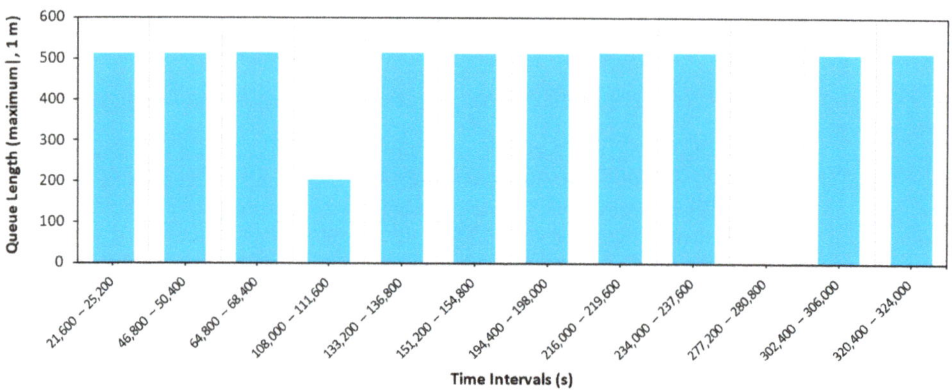

Figure 7. Queue length for in-bound traffic—base case year 2020.

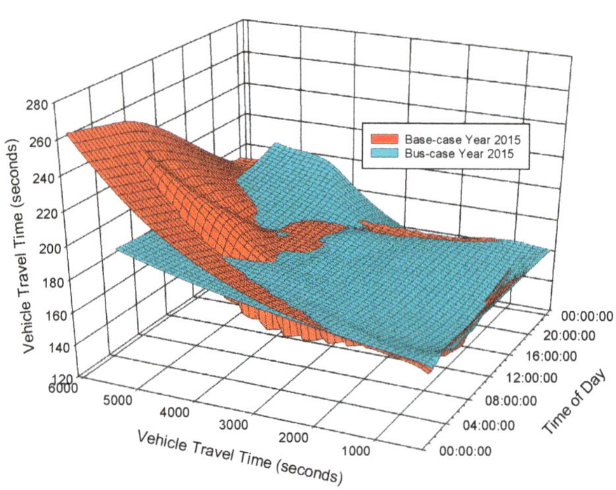

Figure 8. Comparison of vehicle network performance for the base case and traffic management scenario (public bus-case) for a representative weekday.

The morning peak hours saw the biggest increase in journey time and vehicles volume, likely brought by commuters going to work. The travel time decreased after that, until the afternoon peak hours (2 p.m.–3 p.m.), when it tended to increase again and decline before reaching the evening peak hours (7 p.m.–8 p.m.) (as our earlier study [1,35] predicted). A reduction in the variation between the vehicle travel times was observed for the bus-based scenario, as less bottlenecking and congestions was noted for automobiles on the mainline segments, thus lessening the day-wide journey time fluctuations.

5. Conclusions and Future Work

This study aimed to present the efficiency of microsimulation models for empirically illustrating the car-following driving behavior of urban residents on a major highway, within the context of Abu Dhabi and the GCC region. For this purpose, a traffic-count dataset was gathered from local municipal and transport departments, in a series of data collection attempts for a 3.5 km 5-lane divided highway road segment from the Abu Dhabi city network experiencing a peak hour traffic of more than 9000 vehicles. Following data curation and analysis, a comprehensive daily traffic dataset for a representative week (constituting of all vehicle types) was deduced, despite the time gaps between various on- and off-ramp and mainline traffic-counting stations.

The findings suggest that proposed calibrations of the car-following variables (standstill distance, headway time, following variation, following thresholds, speed dependency of oscillation, and acceleration–deceleration parameters) of the Wiedmann 99 model were capable of accurately simulating the real-world driving behavior for the studied highway, with under 5% error noted across the different simulated periods, throughout the weekdays or weekends. The calibrated model was then used to simulate the traffic flow profile for another week, and the traffic data were compared against the actual field-collected data. It was found that the model successfully simulated a whole week's traffic flow profile within acceptable bounds of error. The microsimulation model calibration parameters developed in this study could be used by the local authorities to evaluate and simulate the traffic management scenarios in the studied region.

The success of any transit solution in meeting a city's traffic management goals depends on user adoptability and is gaged accordingly. Since increasing congestion on urban highways due to car traffic is a growing concern in many developing metropolitan areas and enhancing or developing public transport is a common strategy, alternatives aiming to divert passenger traffic from private vehicles to public transport should be backed by strong data assessment, to project their expected benefits. Currently, the comparison of traffic management strategies in Abu Dhabi typically involves identification and development of an initial plan to meet the need, feasibility, and economic viability of the different planning alternatives and, finally, the implementation of a selected system. The microsimulation model developed in this study is presented as a tool for investigating the long-term (2015–2020) benefits of implementing a public transport bus service on the studied highway section, in terms of the reduction in congestion, queue lengths, and journey times. The findings indicate that, if transportation is not improved, there will be significant traffic jams on the studied highway, as no public transport currently exists. The proposed bus service was found to immediately ease the traffic flow situation, by reducing queue length formation and minimizing fluctuations in the travel time between peak and off-peak hours throughout the day, and more significantly on weekdays, where the highest congestions occur.

Multiple research directions can be adopted from this observation. For example, future works could investigate the relative influence of financial and spatial constraints on travel behavior, while strictly considering mobility as a service in a consumer market. This may result in a more realistic research outcome, where vehicle routing, operational and fare policies, transport systems, and alternate on-demand services are evaluated while considering the whole transportation system and how these factors influence mobility at smaller to larger scales. User mobility, benefits to commuters and the environment, and the logical integration of the economic and environmental factors spanning the transportation system's elements can then be balanced. Marketing the transportation infrastructure system as a compact package may also provide an added level of comparability during the decision-making process. Additionally, the performance of new technologies such as autonomous vehicles, which offer better car-following and effective platooning behavior and thereby enhance the traffic capacity of a road, could also be evaluated using microsimulation models for long-term models, instead of the short-term or daily flow projection models that are currently presented in the transport literature.

Author Contributions: Conceptualization, U.H. and H.A.; Methodology, U.H., H.A. and A.H.; Software, U.H.; Validation, U.H.; Formal analysis, U.H., H.A. and A.H.; Investigation, A.H.; Resources, H.A.; Data curation, H.A.; Writing—original draft, U.H., H.A. and A.H.; Writing—review & editing, U.H.; Supervision, U.H. and H.A.; Project administration, U.H. and H.A.; Funding acquisition, U.H. All authors have read and agreed to the published version of the manuscript.

Funding: This research project is supported by an Australian Government Research Training Program (RTP) scholarship.

Institutional Review Board Statement: Not applicable.

Informed Consent Statement: Not applicable.

Data Availability Statement: Data will be provided upon request.

Conflicts of Interest: The authors declare no conflict of interest.

References

1. Hasan, U.; Whyte, A.; Al Jassmi, H. Public bus transport service satisfaction: Understanding its value to urban passengers towards improved uptake. *Trans. Transp. Sci.* **2021**, *12*, 25–37. [CrossRef]
2. Kussl, S.; Wald, A. Smart Mobility and its Implications for Road Infrastructure Provision: A Systematic Literature Review. *Sustainability* **2023**, *15*, 210. [CrossRef]
3. Alawadi, K. Rethinking Dubai's urbanism: Generating sustainable form-based urban design strategies for an integrated neighborhood. *Cities* **2017**, *60*, 353–366. [CrossRef]
4. Sheikh, M.S.; Peng, Y. A comprehensive review on traffic control modeling for obtaining sustainable objectives in a freeway traffic environment. *J. Adv. Transp.* **2022**, *2022*, 1012206. [CrossRef]
5. Ding, C.; Wang, D.; Liu, C.; Zhang, Y.; Yang, J. Exploring the influence of built environment on travel mode choice considering the mediating effects of car ownership and travel distance. *Transp. Res. Part A Policy Pract.* **2017**, *100*, 65–80. [CrossRef]
6. de Luca, S. Public engagement in strategic transportation planning: An analytic hierarchy process based approach. *Transp. Policy* **2014**, *33*, 110–124. [CrossRef]
7. Currie, G.; Delbosc, A. Spiral Plot Analysis of Variation in Perceptions of Urban Public Transport Performance between International Cities. *Transp. Res. Rec. J. Transp. Res. Board* **2015**, *2538*, 54–64. [CrossRef]
8. Qamhaieh, A.; Chakravarty, S. Drive-through cities: Cars, labor, and exaggerated automobilities in Abu Dhabi. *Mobilities* **2020**, *15*, 792–809. [CrossRef]
9. Hasan, U.; Whyte, A.; Al-Jassmi, H. A life-cycle decision-making framework to assess the need for autonomous mobility. *Transp. Res. Procedia* **2019**, *42*, 32–43. [CrossRef]
10. Hasan, A.; Hasan, U.; AlJassmi, H.; Whyte, A. Transit behaviour and sociodemographic interrelation: Enhancing urban public-transport solutions. *Eng* **2023**, *4*, 1144–1155. [CrossRef]
11. Dhabi, D.A. *Surface Transport Master Plan—A Vision for Connecting Abu Dhabi*; Department of Transport: Abu Dhabi, United Arab Emirates, 2009; p. 171.
12. Balsa-Barreiro, J.; Menendez, M.; Morales, A.J. Scale, context, and heterogeneity: The complexity of the social space. *Sci. Rep.* **2022**, *12*, 9037. [CrossRef] [PubMed]
13. Mahmud, S.S.; Ferreira, L.; Hoque, S.; Tavassoli, A. Micro-simulation modelling for traffic safety: A review and potential application to heterogeneous traffic environment. *IATSS Res.* **2019**, *43*, 27–36. [CrossRef]
14. Hasan, U. Experimental Study on Bentonite Stabilisation Using Construction Waste and Slag. Master's Thesis, Department of Civil Engineering, Curtin University, Perth, WA, Australia, 2015; p. 181.
15. Bener, A.; Verjee, M.; Dafeeah, E.E.; Yousafzai, M.T.; Mari, S.; Hassib, A.; Al-Khatib, H.; Choi, M.K.; Nema, N.; Özkan, T.; et al. A Cross "Ethnical" Comparison of the Driver Behaviour Questionnaire (DBQ) in an Economically Fast Developing Country: DBQ in Arab Gulf and South Asian countries. *Glob. J. Health Sci.* **2013**, *5*, 165. [CrossRef] [PubMed]
16. Tulimat, M.; Obaid, O.; Obaid, L.; Hamad, K. Quantifying impacts of connected and automated vehicles on traffic operation of a diamond interchange in UAE using micro-simulation. In Proceedings of the 14th International Conference on Developments in eSystems Engineering (DeSE), Sharjah, United Arab Emirates, 7–10 December 2021.

17. Gray, D.; Laing, R.; Docherty, I. Delivering lower carbon urban transport choices: European ambition meets the reality of institutional (mis)alignment. *Environ. Plan. A Econ. Space* **2016**, *49*, 226–242. [CrossRef]
18. Hermosilla, T.; Palomar-Vázquez, J.; Balaguer-Beser, Á.; Balsa-Barreiro, J.; Ruiz, L.A. Using street based metrics to characterize urban typologies. *Comput. Environ. Urban Syst.* **2014**, *44*, 68–79. [CrossRef]
19. Hasan, U.; Whyte, A.; AlJassmi, H. A microsimulation modelling approach to quantify environmental footprint of autonomous buses. *Sustainability* **2022**, *14*, 15657. [CrossRef]
20. Abenoza, R.F.; Cats, O.; Susilo, Y.O. Travel satisfaction with public transport: Determinants, user classes, regional disparities and their evolution. *Transp. Res. Part A Policy Pract.* **2017**, *95*, 64–84. [CrossRef]
21. Brons, M.; Givoni, M.; Rietveld, P. Access to railway stations and its potential in increasing rail use. *Transp. Res. Part A Policy Pract.* **2009**, *43*, 136–149. [CrossRef]
22. Woldeamanuel, M.G.; Cyganski, R. Factors affecting travellers's satisfaction with accessibility to public transportation. In *Proceedings of the European Transport Conference 2011, Glasgow, UK, 10–12 October 2011*; Association for European Transport: Warwickshire, UK, 2011.
23. Hensher, D.A. The Relationship between Bus Contract Costs, User Perceived Service Quality and Performance Assessment. *Int. J. Sustain. Transp.* **2014**, *8*, 5–27. [CrossRef]
24. Lavery, T.; Páez, A.; Kanaroglou, P. Driving out of choices: An investigation of transport modality in a university sample. *Transp. Res. Part A Policy Pract.* **2013**, *57*, 37–46. [CrossRef]
25. Schweizer, J.; Danesi, A.; Rupi, F.; Traversi, E. Comparison of static vehicle flow assignment methods and microsimulations for a personal rapid transit network. *J. Adv. Transp.* **2012**, *46*, 340–350. [CrossRef]
26. Peng, G.; Yang, S.; Zhao, H. The difference of drivers' anticipation behaviors in a new macro model of traffic flow and numerical simulation. *Phys. Lett. A* **2018**, *382*, 2595–2597. [CrossRef]
27. Haq, M.T.; Farid, A.; Ksaibati, K. Estimating passing sight distances for overtaking truck platoons–Calibration and validation using VISSIM. *Int. J. Transp. Sci. Technol.* **2022**, *11*, 255–267. [CrossRef]
28. Acuto, F.; Coelho, M.C.; Fernandes, P.; Giuffrè, T.; Macioszek, E.; Granà, A. Assessing the environmental performances of urban roundabouts using the VSP methodology and AIMSUN. *Energies* **2022**, *15*, 1371. [CrossRef]
29. Hasan, A.; Hasan, U.; Whyte, A.; Al Jassmi, H. Lifecycle analysis of recycled asphalt pavements: Case study scenario analyses of an urban highway section. *CivilEng* **2022**, *3*, 242–262. [CrossRef]
30. SCAD. *Statistical Yearbook of Abu Dhabi*; Statistics Centre—Abu Dhabi (SCAD): Abu Dhabi, United Arab Emirates, 2020; p. 272.
31. Ponte, C.; Melo, H.P.M.; Caminha, C.; Andrade, J.S.; Furtado, V. Traveling heterogeneity in public transportation. *EPJ Data Sci.* **2018**, *7*, 42. [CrossRef]
32. Kuhnimhof, T.; Chlond, B.; von der Ruhren, S. Users of Transport Modes and Multimodal Travel Behavior Steps Toward Understanding Travelers' Options and Choices. *Transp. Res. Rec. J. Transp. Res. Board* **2006**, *1985*, 40–48. [CrossRef]
33. Hasan, U. Development of a Multi-Criteria Decision-Making Framework for Sustainable Road Transport Systems: Integrating Stakeholder-Cost-Environment-Energy Lifecycle Impacts. Ph.D. Thesis, School of Civil and Mechanical Engineering, Curtin University, Perth, WA, Australia, 2019.
34. Gomes, G.; May, A.; Horowitz, R. Congested Freeway Microsimulation Model Using VISSIM. *Transp. Res. Rec. J. Transp. Res. Board* **2004**, *1876*, 71–81. [CrossRef]
35. Hasan, U.; Whyte, A.; Al Jassmi, H. Life-cycle asset management in residential developments building on transport system critical attributes via a data-mining algorithm. *Buildings* **2018**, *9*, 1. [CrossRef]

eng

MDPI

Article

Redesign of a Failed Hoisting Shaft of a Vertical Transfer Device

Filipe Alexandre Couto da Silva [1] and Paulo M. S. T. de Castro [2,*]

[1] Departamento de Engenharia Mecânica, Instituto Superior de Engenharia do Porto—ISEP, Instituto Politécnico do Porto, Rua Dr. António Bernardino de Almeida, 4249-015 Porto, Portugal; fcs@isep.ipp.pt
[2] Faculdade de Engenharia, Universidade do Porto, Rua Dr. Roberto Frias, 4200-465 Porto, Portugal
* Correspondence: ptcastro@fe.up.pt

Abstract: The redesign of a failed hoisting shaft belonging to a 10 m stroke vertical transfer device (VTD) is presented. Firstly, the operation of the VTD is thoroughly analysed, the variation of loads and moments along the operating cycle is characterised, and transients such as emergency stop loads are calculated. The selection of safety factors and duty cycle factors was followed by the shaft sizing. After an initial rough sizing, the high-cycle fatigue (HCF) design for cyclic bending moments was performed, first considering constant torque and then considering cyclic torque. The number of bending and torsion cycles performed by the hoisting shaft over 10 years was shown to exceed 10^6, and an infinite life design is mandatory. The analyses showed that the initial shaft diameter was insufficient, thus justifying the failures observed before the present redesign. A classical fatigue model combining torsional shear stresses with bending stresses was used to take into account reversed torsional loading and ensure infinite fatigue life. This work highlights the need to thoroughly understand a machine's operating cycle so that the wrong premises for fatigue design calculations are not assumed.

Keywords: fatigue; machine elements; redesign; shafts; VTD

Citation: da Silva, F.A.C.; de Castro, P.M.S.T. Redesign of a Failed Hoisting Shaft of a Vertical Transfer Device. *Eng* **2023**, *4*, 1981–2002. https://doi.org/10.3390/eng4030112

Academic Editor: Antonio Gil Bravo

Received: 13 June 2023
Revised: 10 July 2023
Accepted: 11 July 2023
Published: 14 July 2023

1. Introduction

The present case study concerns a vertical transfer device (VTD) hoisting shaft failure and redesign. The shaft steel is 34CrNiMo6. The size of the machine is suggested by the hoisting stroke—10 m. The machine operation began in 2017, and until 2019, malfunctions of the upper sprocket assembly were reported by the customer. During that period, the hoisting shaft fractured and was replaced by another equal in size and material. Sometime after this replacement, the customer noticed an abnormal play between the shaft and the sprocket on the key area. At the beginning of 2019, it was decided to investigate the root cause of this abnormal behaviour. At the end of 2019, the upper sprocket assembly was replaced by a redesigned one according to the conclusions of the present document, and no more problems were found since this intervention.

The service loads acting upon hoisting shafts typically lead to a number of cycles in excess of 10^7, implying fatigue design for infinite life under rotating bending and torsional loading.

The need for a thorough understanding of the machine operating cycle, so that wrong premises are not assumed for the fatigue design calculations, is emphasised in this paper.

Stress-based high-cycle fatigue (HCF) considerations are used throughout the work, and a classical approach to fatigue was adopted. Harris and Jur recall in [1] that 'the long-taught classical methodology is useful and accurate as both a design and an analysis tool'. The classical fatigue methodology is presented in many references, e.g., Childs [2,3], Beswarick [4,5] or D'Angelo [6], and is widely used in industry, for design as well as for failure analyses, where interpretation of failure causes and redesign of failed parts are objectives to be pursued [7].

Milela [8] or Lee et al. [9] present comprehensive overviews of fatigue, and discuss research on biaxial fatigue as experienced in situations of combined bending and torsion moments, typical of shafts. A survey of recent trends of multiaxial fatigue is given by Anes et al. [10]. Although shaft fatigue is the object of continued research efforts, in the present work, the classical approach commonly used in industry was used for the redesign of a failed shaft subjected to high-cycle fatigue (HCF). Early presentations of the subject are found in, e.g., d'Isa [11], Hall et al. [12] or Spotts [13].

The paper is organized as follows: the VTD operation mode is thoroughly analysed and modelled in Section 2, as a starting point for the hoisting shaft redesign. After a presentation of the VTD mode of operation, the radial load, bending and torsional moments variations along the operating cycle are evaluated, and transients as emergency stop loads are characterized, leading to the inputs for the design calculations and selection of safety and duty cycle factors, as discussed in Section 3.

Using the static failure criterion, a first sizing for the peak load is presented in Section 4. Then, using the normal operation loads calculated in Section 5, a fatigue design using different approaches is presented in Sections 6–8.

The areas with keyseats are particularly sensitive in fatigue calculation. The use of steels with higher tensile strength does not proportionally increases the fatigue allowable in keyseat areas.

Given the organization of the work, relevant references are introduced and discussed throughout the text, as needed. Due to the large number of variables considered, their definitions are given in the Nomenclature of the document, and/or when called for in the text.

2. Detailed Analysis of the Vertical Transfer Device Operation Cycle

Vertical transfer devices are under the scope of the standard EN 619:2002+A1:2010 [14] (EN 619:2002+A1:2010 has a new edition in 2022, not yet harmonised; only the harmonised standards can be used to demonstrate that products comply with the relevant European legislation), where the machine safety requirements are defined. According to this standard, a VTD is a device with raising and lowering movements of more than 200 mm in the path of conveyors, in which unit loads (in logistics terminology, a unit load corresponds to the pallet plus the handled goods) can be transferred from one defined level to one or more defined levels by a carrying element. The hoisting system of the VTD analysed is composed of a driving motorized sprocket on the bottom and a driven sprocket on the top. The suspension consists of one chain that is anchored to the hoisting carriage on one side and a counterweight on the other side; refer to Figure 1 for more details. One conveyor is assembled on the hoisting carriage, and it is responsible for the transfer of the unit load to the grounded conveyors on each transfer level.

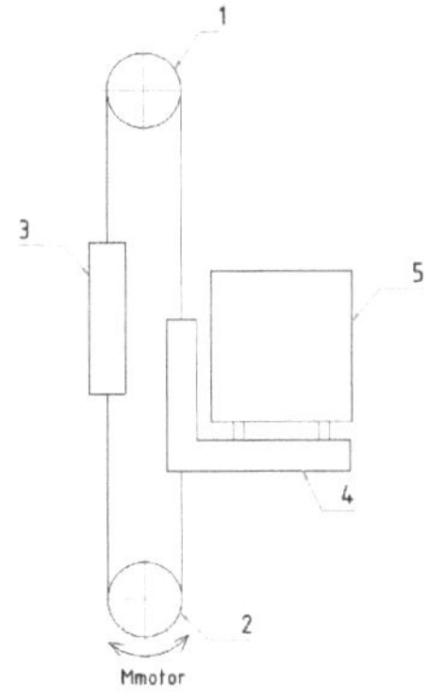

Figure 1. VTD working scheme: 1—upper chain sprocket (driven); 2—bottom chain sprocket (driving); 3—counterweight; 4—hoisting carriage with load handling device; 5—unit load.

In this VTD, the hoisting stroke is 10 m, and the operating cycle consists of the steps described in Figure 2, repeated 24 h/day. Figure 3 presents the evolution of the VTD carriage hoisting speed (m/s) with time (s).

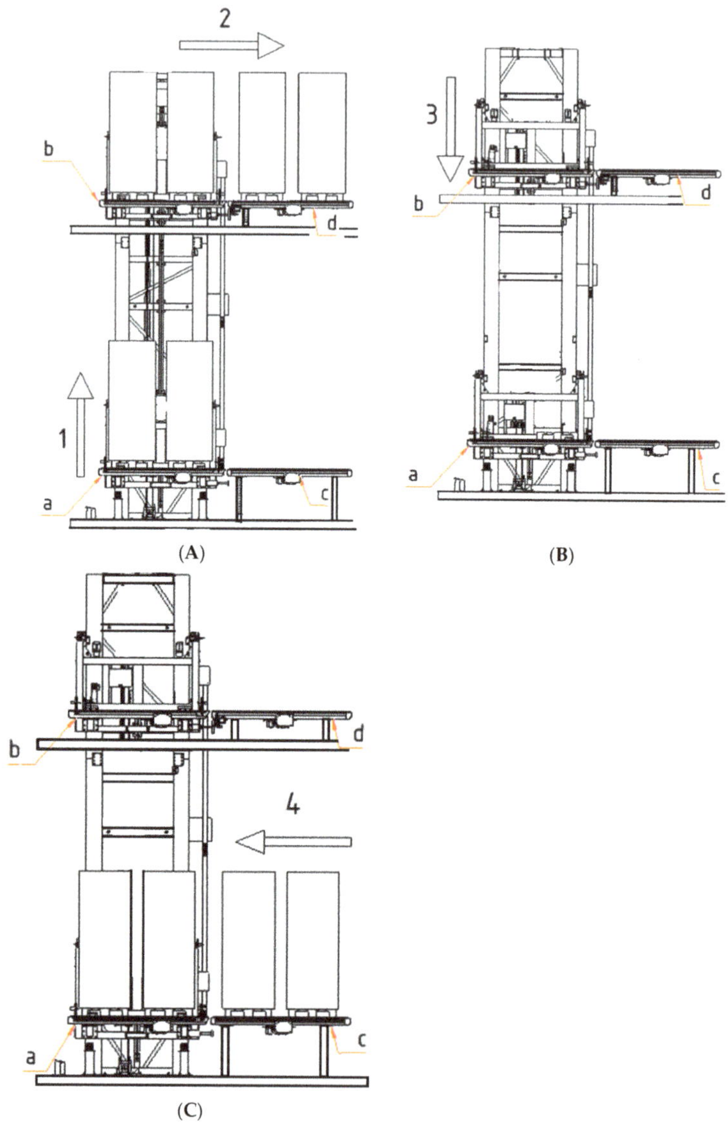

Figure 2. VTD operating cycle. (**A**) Movements 1 and 2. Hoisting and load transfer on the upper level. (**B**) Movement 3. Downward movement from the upper level to level zero. (**C**) Movement 4. Transfer of the unit load from the grounded conveyor to the conveyor mounted on the hoisting carriage: a—conveyor mounted on the VTD hoisting carriage—level zero; b—conveyor mounted on the VTD hoisting carriage—upper level; c—grounded conveyor on level zero; d—grounded conveyor on the upper level.

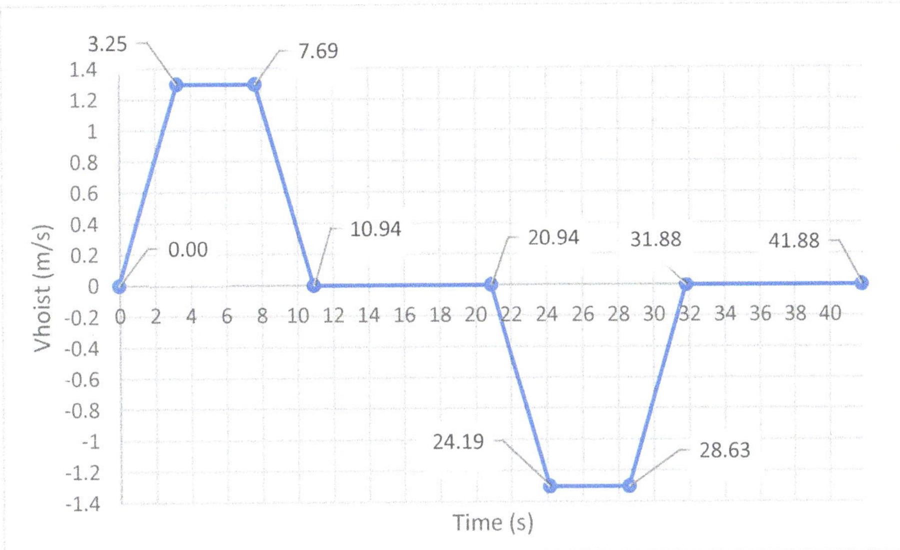

Figure 3. VTD carriage hoisting speed (m/s) versus time (s) chart.

The operating cycle steps are:

1—Upward movement of the hoisting carriage with load on the conveyor, Figure 2A. According to Figure 3, from 0 s up to 3.25 s, the hoisting carriage accelerates, reaching the speed of 1.3 m/s. Between 3.25 s and 7.69 s, the hoisting carriage will move upward at a constant speed, followed by a 3.25 s deceleration.

2—Transfer of the load from the conveyor mounted on the hoisting carriage, to the grounded conveyor on the upper level, Figure 2B. This transfer movement will take 10 s. During this time, the hoisting carriage is stopped.

3—Downward movement of the hosting carriage from the upper level to level 0, Figure 2B. This movement will start at 20.94 s and is composed of an acceleration of 3.25 s, followed by a constant speed movement of 4.44 s, and finally, a 3.25 s deceleration.

4—The load will be transferred from the grounded conveyor at level zero, to the conveyor mounted on the hoisting carriage, which is stopped. This transfer will take 10 s, Figure 2C.

Speeds and acceleration values in these machines are not prescribed. Instead, the design of the device, of its components and safety devices must be rooted in those values: the EN 619:2002+A1:2010 standard [14] does not define maximum speeds and accelerations for the VTD, but states that the safety related components must be selected according to the effective nominal speed and acceleration of the machine.

The upper sprocket assembly, which is the object of the present case study, is shown in Figure 4. It is composed of one shaft, two plummer blocks with roller bearings and one sprocket for the hoisting chain. Note that the failed shaft diameter was 50.0 mm.

Table 1. VTD upper drive assembly data before redesign.

Parameter	Value	Unit
v_{hoist} Hoisting carriage hoisting speed	1.3	m/s
a_{hoist} Hoisting carriage hoisting acceleration	0.4	m/s^2
D_p Sprocket pitch diameter	0.21304	m
a Position of the load on the shaft length	0.11	m
b Position of the load on the shaft length	0.11	m
d_i Initial shaft diameter, before redesign	0.05	m
c Initial sprocket length, before redesign	0.091	m
l Initial shaft length between supports	0.22	m
$m_{hoistcar}$ Hoisting carriage mass	600	kg
m_{load} Unit load mass	1600	kg
m_{lhd} Conveyor mass	350	kg
m_{count} Counterweight mass	1600	kg
m_{chain} Chain mass	300	kg
N Shaft rotating speed	12.15 (116)	rad/s (RPM)
η_l Load efficiency	0.9	

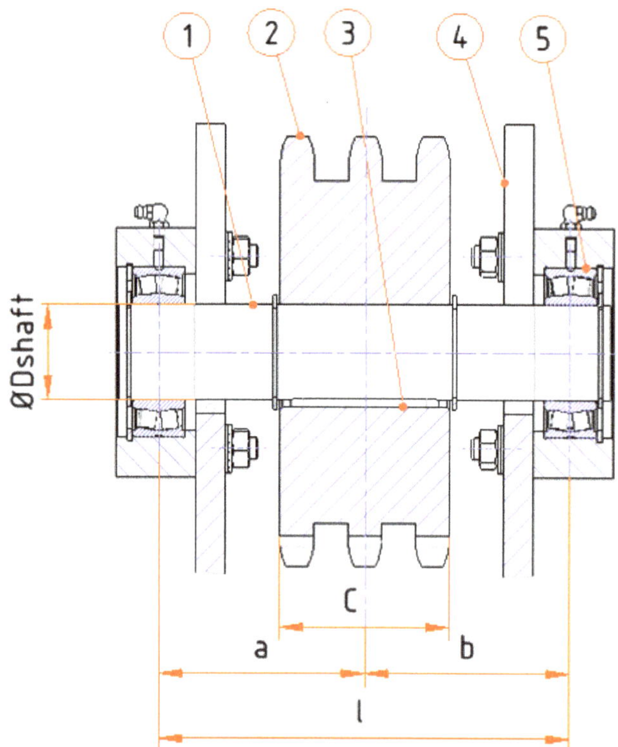

Figure 4. Components of the upper sprocket assembly: 1—upper sprocket shaft; 2—upper chain sprocket; 3—key; 4—upper sprocket assembly support; 5—roller bearings. See Table 1 for nomenclature.

The failure motivated the need for component redesign. As discussed in the following sections, a thorough understanding of the machine's operating mode is necessary for the selection of inputs to calculate the redesigned shaft diameter.

3. Assumptions for Calculation and Safety Factors

This section presents the selection of data to be used as inputs when calculating the shaft diameter and selecting the safety factors. Table 1 compiles the data for the case study system.

The load efficiency—η_l = 0.9—is the value advised by [15] for the efficiency of sprocket-chain hoisting systems, as some amount of the gearmotor output torque is not used to hoist the load, but to compensate friction losses, for instance.

The data for the gearmotor were retrieved from [15,16] and are compiled in Table 2.

Table 2. Gearmotor data for calculation; source: [16].

Parameter	Value	Unit
n_m Motor rotating speed	247.14 (2360)	rad/s (RPM)
J_M Motor moment of inertia	0.0381	kg·m^2
η_{motor} Motor efficiency	0.7833	-
i_t Total gearbox gear ratio	20.25	-
$\eta_{gearbox}$ Gearbox efficiency	0.96	-
a_{emerg} Gearmotor emergency deceleration, with brake	−4.829	m/s^2
M_{amax} Maximum nominal gearmotor torque	1500	N·m
M_{aemerg} Gearmotor torque in the event of an emergency deceleration with brake	1929	N·m

Unless there are problems associated with bearings, which is unlikely if these are properly selected and fitted according to the applicable tolerances, in these constructions, the likely cause of shaft failure is fatigue. Corrosion is excluded, given the shaft surface protection by phosphating surface treatment, and the machine's permanent location inside the warehouse.

A first check of the bearing life showed that the bearings were correctly chosen, and accordingly, bearing calculation and selection will be outside the scope of this paper. Since the failure was not related to the bearings, it was decided to further investigate the shaft design with a view to redesign. Given the level of responsibility involved, several calculation methods were used, and the results compared, to decrease the risk level.

Firstly, it was necessary to carefully define the assumptions for the calculation. The bending and torsion moments and the radial load to be considered for the design calculations were evaluated. Figures 5–7 present the evolution of these loads along the VTD operating cycle. For conciseness, the calculations behind Figures 5–7, based on [15,16], are not fully presented. Nevertheless, Section 5 includes the calculation of the maximum loads per operating cycle (i.e., the maximum values of the diagrams).

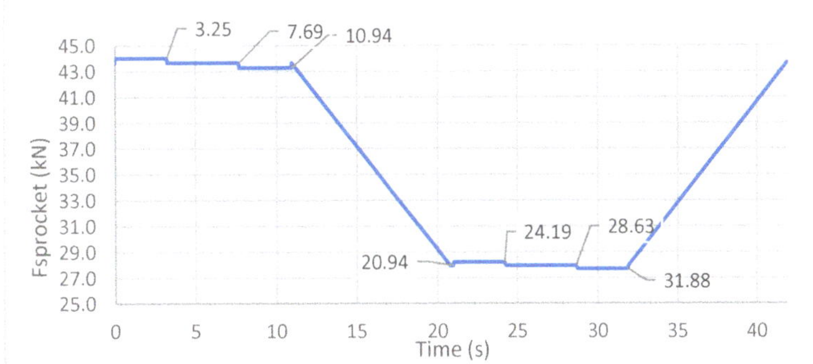

Figure 5. Radial load on the VTD upper sprocket (kN) vs. time (s).

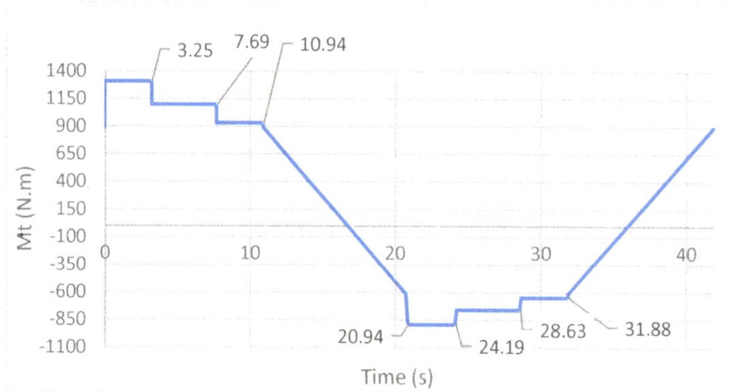

Figure 6. Torsional moment on the VTD hoisting shaft vs. time (s).

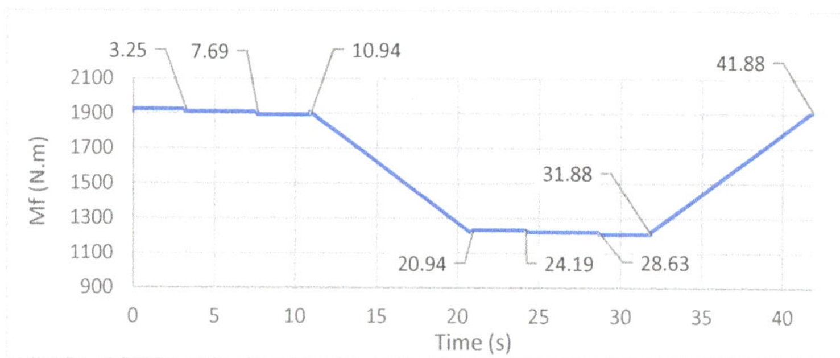

Figure 7. Bending moment on the VTD hoisting shaft vs. time (s).

The maximum vertical load on the sprocket, the maximum torsion moment and the maximum bending moment all occur when the hoisting carriage is accelerating upwards with its full maximum load. However, this only occurs between 0 and 3.25 s, i.e., in a small fraction of the machine's operating cycle, which takes nearly 41.9 s.

The total time with a full load on the hoisting carriage is about 50% of the overall working time. In the other 50%, the hoisting carriage is moving down without the unit load.

Shafts subjected to stresses below the yield strength but above the fatigue limit (also known as endurance strength) will most likely fail from fatigue. The fatigue limit is the operating stress at which the specimens do not fail after at least 10^6 cycles. Excluding consideration of giga-cycle fatigue, which is not relevant for this application, stresses below the fatigue limit lead to infinite life.

Milela [8] or Lee et al. [9] give overviews of a variety of fatigue topics; within that vast domain, this work concentrates on stress-based fatigue analysis and design for high-cycle fatigue (HCF), adopting the classical approach commonly used in industry. References such as [17,18] introduce classical fatigue design for HCF with a focus on machine elements.

It is crucial to understand which load values should be considered in the calculation methods described in Sections 6–8, and what the safety factors should be. Section 6 concerns the use of Niemann's approach [19,20], Section 7 concerns the use of the ANSI/ASME B106.1M:1985 standard [21], and a classical fatigue approach is used in Section 8. The methods considered give some guidance on these subjects but do not define them entirely.

Niemann, [19,20], introduces the safety factor C_v, and the overload factor C. To establish C_v, it is necessary to consider the consequences of overloads, e.g., danger of death, long interruptions of the operation and production line or ease of repair or replacement of the damaged shaft. The overload factor C is defined as:

$$C = \frac{F_{serv}}{F} \tag{1}$$

where F_{serv} is the maximum load that occurs periodically in the machine cycle, and F is the nominal load.

The ANSI/ASME B106.1M:1985 standard [21] does not directly define the values for the safety factor (SF). Nevertheless, it states that SF should be considerably higher than 1 if there are great uncertainties and the consequences of failure are serious, namely, in terms of safety and production stopping time. The same standard also introduces the duty cycle factor k_e and states that a shaft usually withstands variable amplitude loadings in service. Thus, the shaft design must consider start and stop cycles, transient overloads, vibrations and shocks, since these transients can have a significant impact on fatigue life. Usually, the values for the constant amplitude loads are known with sufficient accuracy, but the data for transient loads are not so well defined. Nevertheless, in the present case study, it was possible to determine the transient loads that occur in the acceleration and deceleration phases of the VTD operating cycle. The ANSI/ASME B106.1M:1985 standard [21] states that it is not advisable to design a rotating shaft for finite fatigue life, as it will be confirmed in the following paragraphs.

An emergency stop will cause an overload on the hoisting shaft, and such events will certainly occur during the VTD service life. It is not possible to foresee how many times an emergency stop event will occur, but it will not be a very rare event. More unusual is the event of a free-fall with actuation of the safety gear. Overloads originated by the transfer of a load greater than the nominal maximum may also occur, but in this case, the hoisting carriage's vertical movement will not be initiated. The EN 619:2002 + A1:2010 [14] states that the system incorporating the VTD should prevent this from happening, for example, by assembling load cells on the conveyors that feed the VTD.

After analysing the inputs from these methods, some important decisions need to be made. If the nominal static load on the shaft is considered, an overload factor $C > 1$ should be selected. Another option is to use the dynamic load (including the VTD acceleration), and the overload factor C is taken as 1. Recall that according to Niemann [19], Equation (1), F_{serv} is a load that repeats periodically.

It was decided to use the loads that occur during the initial 3.25 s of the operating cycle, where the hoisting carriage is accelerating upwards, in all the fatigue calculation methods. It is important to maintain the consistency of the previous decision with the safety factors and coefficients that will be used in each of the formulations, to avoid over-design. As explained previously, within the operating cycle, the VTD is half of the time with load and half of the time without load, which has a beneficial effect on fatigue life.

In [2], Childs suggests the following safety factors for shaft design:

- "1.25 to 1.5 for reliable materials under controlled conditions subjected to loads and stresses known with certainty,
- 1.5 to 2.0 for well-known materials under reasonably constant environmental conditions subjected to known loads and stresses,
- 2.0 to 2.5 for average materials subjected to known loads and stresses,
- 2.5 to 3.0 for less well-known materials under average conditions of load, stress, and environment,
- 3.0 to 4.0 for untried materials under average conditions of load, stress, and environment, and
- 3.0 to 4.0 for well-known materials under uncertain conditions of load, stress, and environment".

The material used for the shaft under study is well known and reliable, and the environmental conditions are controlled and constant. The normal operation loads and the emergency stop loads are known, however, the number of occurrences of an emergency stop is unclear. As mentioned previously, the present work will size the shaft for the maximum loads within the equipment operating cycle, which only occur during 3.25 s of the overall cycle time. This is a defensive approach; therefore, a small safety factor within the range was used—1.5. Another option might have been to consider the weighted average between the maximum loads that occur during the 50% of the time that the hoisting carriage is going upwards with maximum load and the 50% of the time that the hoisting carriage is moving downwards without load. If the last approach had been used, the safety factor would have been increased to 2.

The decision was to consider the loads occurring in the upward acceleration phase; therefore, $C = 1$, according to Niemann [19,20]. The safety factor was defined to be $C_v = 1.5$, instead of using higher safety factors, since the hoisting carriage is under load only half of the operating cycle time.

Likewise, in the ANSI/ASME B106.1M:1985 standard [21] method, the safety factor was defined as $SF = 1.5$, and the duty cycle factor was chosen to be $k_e = 1$.

The stress originated by the loads acting on the hoisting shaft must be classified. It is intuitive that the bending moment due to the radial load on the sprocket will cause an alternating stress on the rotating shaft. It is less intuitive how the torsional moment should be categorized. Niemann [20] defines a steady, an oscillating and an alternating stress. Notice that the direction of rotation of the hoisting shaft only reverses two times in each VTD operation cycle. Therefore, it would seem too defensive to consider the torsion as an alternating stress. At first sight, it would seem nearly steady. However, as discussed in the following paragraphs, the torsional load will cause an alternating stress that should be taken into account for infinite life calculation.

Another question that may arise is whether the infinite fatigue life design, corresponding to more than 10^6 cycles, is necessary. Would it be acceptable to calculate the shaft for a finite life, which would result in a smaller shaft diameter?

A shaft replacement is a complicated operation that may cause long production line downtime. Thus, it may be reasonable to require that the shaft not need to be replaced within the machine's lifespan. In this case study, 10 years were considered based on the assessment of the number of load cycles. In each complete rotation of the shaft, the hoisting carriage moves 0.67 m up. Since the VTD hoisting stroke is 10 m, the shaft will complete almost 15 rotations in the upward movement plus 15 complete rotations in the downward movement. In each 180 degrees of shaft rotation, there is one alternating bending cycle. So, 30 complete shaft rotations will correspond to 60 bending cycles. The throughput of the VTD is 85 pallets per hour, so the shaft will suffer 5134 bending cycles per hour. Since the VTDs will be operating in an automatic system that works 24 h a day, the shaft will withstand 123,186 bending cycles/day, 30,796,594 per year, and 308×10^6 cycles in 10 years. Thus, it is mandatory to calculate the shaft for infinite life. In the previous calculation, 250 working days per year were considered, according to DIN 15020:1974 [22].

Regarding the torsion stress, and applying the same rational, there will be 85 alternating torsional stress cycles per hour (two per machine up and down cycle), 4122 in each day, 1,030,585 per year and around 10×10^6 cycles in 10 years, much lower than the number of alternating stresses due to the rotational bending but still larger than the 10^6 cycles, which justify an infinite life calculation. Figure 8 shows schematically the relationship between both stresses. The graphic was simplified to allow better visualization, since 30 periods of the alternating bending stress curve would occur before the inversion of the signal of the alternating torsional load stress.

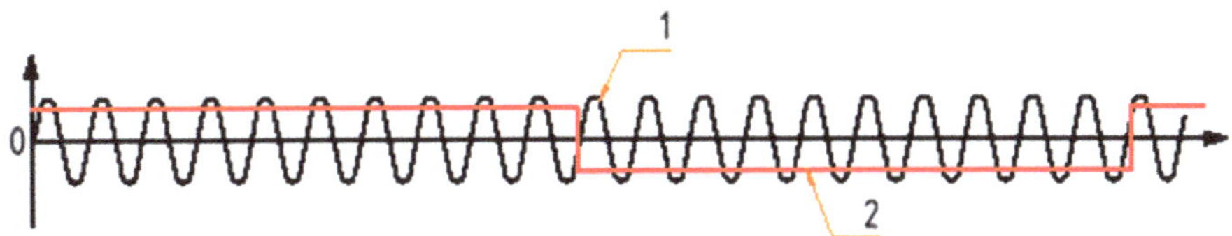

Figure 8. Schematic representation of bending and torsion stresses along the machine life. The discontinuity in the red line is a consequence of the torsion moment changing signal when the vertical movement is reversed: 1—Alternating bending stress: black line; 2—Alternating torsion stress: red line.

4. Shaft Diameter Calculation—Peak Loads

After the thorough analysis of the system and the definition of the input values for calculations, a first analysis concerns shaft behaviour under peak loads. The input data are found in Tables 1 and 2, including the parameters description. The notation for this and subsequent sections is given in Nomenclature part. The ANSI/ASME B106.1M:1985 standard [21] notes that there is not a comprehensive method to determine the impact of the peak load on the shaft fatigue life. Miner's law could be used if the occurrence of peaks could be quantified, but in the present case, the number of occurrences is unpredictable, precluding their explicit consideration in fatigue calculation.

The peak load to be considered results from the emergency stop by the gearmotor brake, which results in a torsional moment M_{aemerg} = 1929 N·m, and an acceleration of a_{emerg}= -4.829 m/s^2. The previous values were obtained using the software [16]; refer to Table 2. The radial load F_{peak}, acting on the hoisting shaft on an emergency gearmotor brake, is calculated as

$$F_{peak} = \left(m_{hoistcar} + m_{load} + m_{lhd}\right)\cdot\left(g - a_{emerg}\right) + m_{count}\cdot\left(g + a_{emerg}\right) + m_{chain}\cdot g \quad (2)$$

$$p_{peak} = \frac{F_{peak}}{c} \quad (3)$$

Figure 9 presents a model and the notation used. Recall that the failed shaft diameter was d_i = 50 mm.

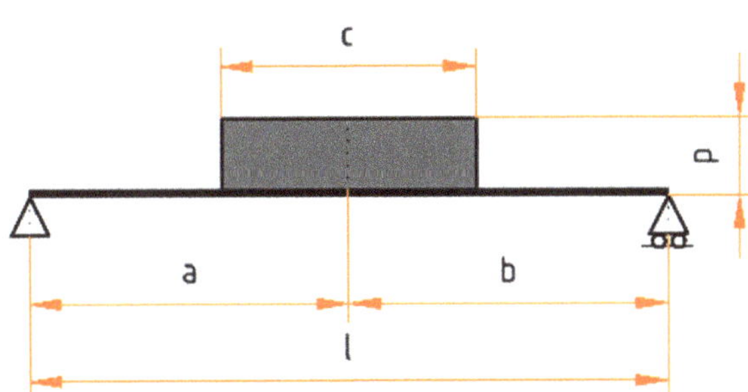

Figure 9. Load diagram.

In the shaft critical section and under the peak load, the bending moment and stress, the shear stress due to the load and to the torque, are calculated as:

$$M_{f\,peak} = \frac{p_{peak}bc}{2l}\left(2a - c + \frac{bc}{l}\right) \quad (4)$$

$$I_{d_i} = \frac{\pi d_i^4}{64} \tag{5}$$

$$I_{p_{di}} = \frac{\pi d_i^4}{32} \tag{6}$$

From Equation (4), M_{fpeak} = 2092 N·m; from (5), $I_{D_2} = 3.07 \times 10^{-7}$ m^4; and from Equation (6), $I_{p_{di}} = 6.14 \times 10^{-7}$ m^4.

$$\sigma_{peakbending} = \frac{M_{fpeak}}{I_{d_i}} \cdot \frac{d_i}{2} \tag{7}$$

$$\tau_{peaktorsion} = \frac{M_{aemerg}}{I_{p_{di}}} \cdot \frac{d_i}{2} \tag{8}$$

$$\tau_{mpeak} = \frac{4}{3} \frac{V_{peak}}{A_{di}} \tag{9}$$

$$V_{peak} = \frac{p_{peak} bc}{l} - p \cdot \left(\frac{c}{2} - a + x\right),\ a = b = \frac{l}{2} \tag{10}$$

From Equation (7), $\sigma_{peakbending} = 170$ MPa; from Equation (8), $\tau_{peaktorsion} = 79$ MPa. Replacing $x = l/2$, in Equation (10), $V_{peak} = 0$ N, in the critical middle section of the keyseat area; thus, $\tau_{mpeak} = 0$ MPa, from Equation (9). The total shear stresses will be $\tau_{peak}^{eq} = 79 + 0 = 79$ MPa.

To conclude this static strength analysis, the von Mises criterion was used:

$$\sigma_{peak} = \sqrt{\sigma_{peakbending}{}^2 + 3 \cdot \tau_{peak}^{eq}{}^2} < \frac{\sigma_{yield}}{N_{peak}} \tag{11}$$

For steel 34CrNiMo6, refer to Table 3, and considering $N_{peak} = 1.5$, Equation (11) gives $\sigma_{peak} = 218$ MPa < 533 MPa, confirming that the shaft can withstand the peak load, and implying that its failure is likely due to fatigue. Unfortunately, the broken pieces were not available, so no scanning electron microscopy could be used to possibly identify striations. The work proceeded with a fatigue analysis using several methodologies.

Table 3. Comparison of the mechanical properties of the steels.

European Standard Designation	Tensile Strength σ_r for 40 < d < 100 mm	Yield Strength- $Rp_{0.2}$ for 40 < d < 100 mm
S355 Jr	500 MPa	325 MPa
34 CrNiMo 6 (1.6582)	1000 MPa	800 MPa

5. Calculation of the Normal Operation Loads on the Hoisting Shaft

According to Section 3, the acceleration loads will be considered for the fatigue calculation, and the safety factor will be adjusted to avoid overdesign. The input data are found in Tables 1 and 2, including the parameters description. The other notation is given in Nomenclature part. The gearmotor will transmit a torque to the bottom and upper sprocket to compensate the unbalance between the total maximum suspended mass—m_{tmsm}—and the counterweight mass. The chain does a closed loop, so the gearmotor torque does not need to take into account the chain weight; refer to Figure 1.

$$F_{unbalance} = (m_{tmsm} - m_{count}) \cdot g \tag{12}$$

$$M_{static} = \frac{F_{unbalance} \cdot \frac{D_p}{2}}{\eta_l} \tag{13}$$

From (12), the $F_{unbalance}$ = 9320 N, and from (13), the gearmotor static torque—M_{static} = 1103 N·m. M_{static} is the output torque required to the gearmotor to ensure the hoisting movement at constant speed. However, the load must be accelerated and decelerated within the operating cycle. Thus, the acceleration torque—M_H—takes into account the inertia of the moving bodies, including the gearmotor and motor inertias. The equations to calculate M_H are given in [15], and some data and inputs can be found by using the calculation software [16]. Table 2 presents input data for calculation.

$$J_x = 91.2 \cdot \frac{F_{unbalance}}{g} \cdot \left(\frac{v_{hoist}}{n_m}\right)^2 \tag{14}$$

$$M_L = \frac{F_{unbalance} \cdot v_{hoist} \cdot 9.55}{n_m} \tag{15}$$

$$M_H = \frac{\left(J_M + \frac{J_X}{\eta_{motor}}\right) \cdot n_m}{9.55 \cdot t_a} + \frac{M_L}{\eta_{motor}} \tag{16}$$

From (14), $J_x = 0.02629$ kg.m^2, and from (15), $M_L = 49$ N·m. Since $v_{hoist} = 1.4$ m/s and $a_{hoist} = 0.3$ m/s^2, $t_a = 3.25$ s. With these values, from (16), the torque on the motor output shaft—$M_H = 68$ N·m. The torque in the gearmotor output shaft—M_{taccel} is calculated according to Equation (17).

$$M_{taccel} = M_H \cdot i_t \cdot \eta_{gearbox} \tag{17}$$

According to the input data from Table 2, the total gearbox transmission ratio $i_t = 20.25$, and the gearbox efficiency $\eta_{gearbox}$ = 96%, resulting in a M_{taccel} = 1323 N·m.

The radial load on the sprocket during the upward accelerating movement of the hoisting carriage with maximum load—F_{accel}—is calculated according to (18), and the bending moment is calculated by using the Equations (19) and (20), resulting in F_{accel} = 44035 N and M_{faccel} = 1927 N·m.

$$F_{accel} = (m_{hoistcar} + m_{load} + m_{lhd}) \cdot (g + a_{hoist}) + m_{count} \cdot (g - a_{hoist}) + m_{chain} \cdot g \tag{18}$$

$$p_{accel} = \frac{F_{accel}}{c} \tag{19}$$

$$M_{faccel} = \frac{p_{accel} bc}{2l}\left(2a - c + \frac{bc}{l}\right) \tag{20}$$

6. Calculation of the Shaft Diameter According to the Niemann Method

After defining the load values to be used as input for all the shaft diameter calculation methods (refer to Section 5), the first calculation was done according to Niemann [19] (see Table 17.2 of that reference). For a torsional load M_{taccel}= 1323 N·m, it defines a shaft diameter of around 80 mm, whereas before the present redesign, the diameter was only 50 mm. Even if the reference used (Table 17.2 of [19]) considers a lower strength steel, as will be seen later, this first calculation showed that the shaft design required attention.

Still in [19], a more detailed calculation method may be found (Table 17.5 of that reference). For a shaft under torsional and bending loads, Equations (21) and (22) are introduced.

$$M_{eq} = \sqrt{{M_f}^2 + \left(\frac{a_f}{2} M_t^2\right)} \tag{21}$$

$$d_N = 2.17 \cdot \sqrt[3]{\frac{M_{eq}}{\sigma_f} b_N} \tag{22}$$

For a solid shaft, $b_N = 1$, $a_f = 1$ for oscillating torsion with alternate bending, and $a_f = 1.7$ for alternating torsion and bending. As shown in Section 3, and contrary to how it may seem at first sight, the torsional load is also to be considered alternating.

The Niemann method defines a σ_{fad} = 500 kgf/cm^2 (~50 MPa), for ST50.11 steel on a hoisting shaft application. The mechanical properties of the ST50 steel could be considered equivalent to the current steel S355 Jr of NP EN 10025 + A1:1994 [23]. Currently, higher strength steels are used in this type of application. Table 3 compares the mechanical properties of the S355 Jr steel (NP EN 10025 + A1:1994, [23]) with the 34CrNiMo6 steel (EN 10083:2006 [24]).

On first thought, the fatigue strength—σ_f—in a keyway torque transmission area of a 34CrNiMo6 shaft might be expected to be approximately twice that of a ST50 one. However, as shown in Niemann [20], the use of high strength steels does not improve so much the allowable fatigue stress for a shaft under alternating bending in a keyway area—σ_{fa10}. In [20] (Figure 3.27 of that ref.), σ_{fa10} is 105 MPa for the ST50 steel, and increases to around 120 MPa for a steel with σ_r = 800 MPa, the highest value considered. Extrapolating, see the blue dotted line in Figure 10, σ_r = 1000 MPa corresponds to $\sigma_{fa10} = 125$ MPa. Notice that σ_{fa10} is the allowable fatigue strength in keyway under alternating bending for one test specimen with a diameter of 10 mm and a material with a given σ_r.

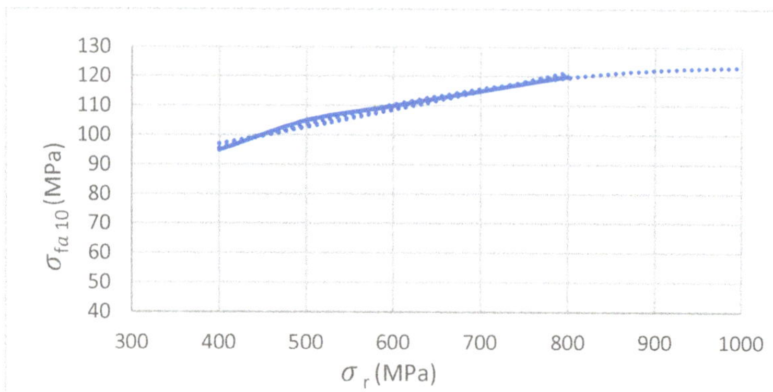

Figure 10. Allowable fatigue strength in a keyseat area σ_{fa10}, for steels with tensile strength σ_r.

To calculate the allowable fatigue strength in the critical area of the shaft, the keyway area, it is necessary to consider other factors according to:

$$\sigma_{fa} = \frac{\sigma_{fa10}.b_0}{C_v.C} \tag{23}$$

Considering the shaft diameter of 50 mm, the size factor is $b_0 = 0.7$; refer to Table 4. According to Section 3, the safety factor C_v is 1.5, and the overload factor $C = 1$. From (23), σ_{fa} = 58 MPa in the keyseat area.

Table 4. Size factor for different shaft diameters according to [20].

Shaft diameter—d_{shaft}	10	20	30	50	100
Size factor b_0	1	0.9	0.8	0.7	0.6

The results from the calculation with Equations (21) and (22) are summarized in Table 5. The first iteration resulted in a shaft diameter of 73 mm, so the b_0 value was corrected to 0.65 in the second iteration. According to [20], for oscillating torsion with alternate bending, the diameter of the shaft in the keyway area would need to be 72.5 mm considering $a_f = 1$, and 74.8 considering $a_f = 1.7$ for alternating torsion and bending. Recall that before the redesign, the shaft diameter was 50 mm only, so it looked unsatisfactory. Even considering $C_v = 1$, $C = 1$ and $a_f = 1$, the required shaft diameter would need to be higher than 50 mm, around 63 mm.

Table 5. Results from the application of Niemann's method, [20].

	Equivalent Moment M_{eq}	Resulting Shaft Diameter d_N
Alternating bending and oscillating torsion $a_f = 1$	2037 N·m	72.5 mm
Alternating bending and torsion $a_f = 1.7$	2230 N·m	74.8 mm

As already explained, the input values in the previous calculation were $M_f = M_{faccel} = 1927$ N·m and $M_t = M_{taccel} = 1323$ N·m.

7. Calculation of the Shaft Diameter According to ANSI/ASME Methodology

Given the losses incurred with the failure, and since the redesigned shaft is intended for use in several future machines, it was decided, for comparison purposes, to use yet another procedure to redesign the shaft for infinite life, following the ANSI/ASME B106.1M:1985 standard, [21]. Although now withdrawn, this standard is a commonly used guide for shaft design, as stated, e.g., by Childs, [2,3], and it continues to be included in ANSI/CEMA B105.1-2015, [25]. The notation from the standard [21] is used here.

According to [21], for steels with ultimate tensile strengths lower than 1400 MPa, such as the 34CrNiMo6 steel, in the absence of detailed testing, the approximation $S_f^* = 0.5S_u$, where S_f^* is the fatigue limit of polished unnotched test specimen in reversed bending, and S_u is the ultimate tensile strength of the steel, should provide reasonable accuracy.

According to ANSI/ASME B106.1M:1985, [21], the shaft diameter is calculated using Equation (24) based upon the von Mises criterion,

$$d = \left(\frac{32FS}{\pi}\right)^{\frac{1}{3}} \left[\sqrt{\left(\frac{M_f}{S_f}\right)^2 + \frac{3}{4}\left(\frac{T}{S_y}\right)^2}\right]^{\frac{1}{3}} \quad (24)$$

where d is the shaft diameter (m), FS is the factor of safety and S_f is the corrected endurance (fatigue) limit of the shaft in reversed bending, calculated using Equation (25), S_y is the tensile yield strength (N/m^2) of the steel, T is the static mean torque (N·m) and M_f is the reverse bending moment (N·m). Using Tresca instead of von Mises, this classical result is found, e.g., in [26] (chapter 13).

The following Equation (25) from [21] is also presented in many publications. Given its important role in the present analysis, its use is presented in some detail in the following paragraphs.

$$S_f = k_a k_b k_c k_d k_f k_e k_g S_f^* \quad (25)$$

The first correction factor in Equation (25) is the surface finish factor k_a. This factor accounts for the difference in the surface condition between the shaft under evaluation and one highly polished test specimen. As recalled in the ANSI/ASME standard, experiments have shown that the surface condition can have an important effect on the fatigue strength, since fatigue cracks are usually initiated at the surface of the shaft, where stresses are higher. For reference, the machined surface category is to be considered for shafts with surface roughness ranging from Ra 1.6 to Ra 6.3 µm.

According to [21], for the shaft under evaluation with Ra 3.2 μm, manufactured in 34CrNiMo6, with ultimate tensile strength S_u = 1000 MPa, $k_a = 0.72$, Figure 11.

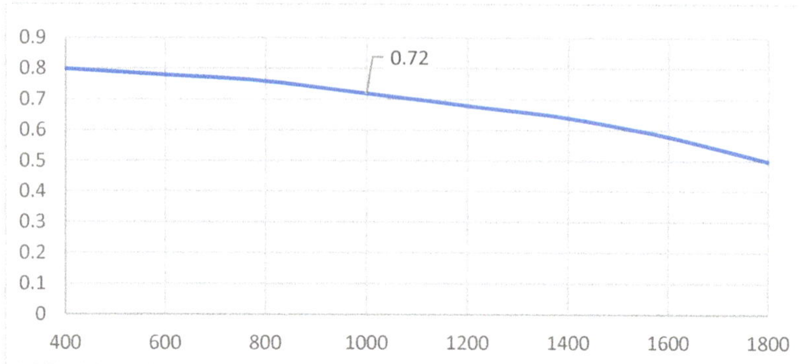

Figure 11. Surface finish factor—k_a—for a machined shaft, according to [21].

The size factor k_b quantifies the decrease of the fatigue limit with the increase of the diameter, as briefly discussed in [21], where different equations for k_b depending on the diameter range, are presented. For shafts with a diameter larger than 50 mm and lower than 254 mm, Equation (26) is to be used; thus, for the case study diameter of 50 mm, $k_b = 0.88$.

$$k_b = 1.85d^{-0.19} \tag{26}$$

According to [21], the reliability factor k_c should be taken in account, due to the variability of the mechanical properties of the specimen. In a safety risk assessment analysis, the VTD hoisting shaft would be considered critical, as its rupture would allow the hoisting carriage to free fall. The factor k_c reduces the fatigue limit, so that a smaller percentage of the population fails during the machine's life, increasing reliability. In this calculation, a reliability rate of 90% was chosen, so k_c = 0.897; refer to Table 6.

Table 6. Reliability factors k_c according to [21].

Shaft Nominal Reliability	k_c
0.5	1
0.9	0.897
0.99	0.814

The temperature factor k_d, translates the effect of the temperature on the fatigue limit. There is a trend to use VTDs in cold temperatures, down to −30 °C in deep freeze warehouses, so this factor appeared to be relevant. The ANSI/ASME standard states that for operating temperatures from −57° to 204°, the fatigue limit is not affected by temperature for most steels, so $k_d = 1$.

The duty cycle factor—k_e—was already discussed in Section 3. The value $k_e = 1$ will be considered for this calculation (refer to Section 3 for more details).

The miscellaneous effects factor—k_g—concerns different factors that may affect the fatigue life (e.g., residual stresses from the manufacturing process, corrosion, surface coating). Its value was taken as $k_g = 1$, addressing this uncertainty with the safety factor.

Failure usually occurs in a notch, keyway, shoulder or other discontinuity where the stresses are amplified. The fatigue stress concentration factor—k_f—represents the effect of the stress concentration on the fatigue limit of the shaft, according to Equation (27).

$$k_f = \frac{fatigue\ limit\ of\ the\ notched\ specimen}{fatigue\ limit\ of\ a\ specimen\ free\ of\ notches} = \frac{1}{K_f} \tag{27}$$

where K_f is the fatigue strength reduction factor and is calculated according to Equation (28). Equations (27) and (28) are used by the ANSI/ASME B106.1M:1985 standard [21], found in many references that address the fatigue damage mechanism, such as [11,18], for instance.

$$K_f = 1 + q(K_t - 1) \tag{28}$$

where q is the notch sensitivity of a given material. The notch sensitivity can be used to relate the fatigue strength reduction factor—K_f—to the theoretical stress concentration factor K_t. It is interesting to know that experience has shown that low-strength steels are less sensitive to fatigue at notches than the high-strength steels. For more details, refer to [18] (Section 1.9). Ref. [21] gives directly the value for k_f for a profiled keyway under bending stress in solid round steel shafts, according to Table 7, without the need to determine the notch sensitivity factor q, nor the theoretical stress concentration factor K_t.

Table 7. Fatigue stress concentration factor k_f, for a profiled keyway under bending stress in solid round steel shafts, according to [21].

Steel	k_f
Annealed (less than 200 BHN [27])	0.63
Quenched or drawn (more than 200 BHN [27])	0.5

The shaft material—steel 34CrNiMo6, EN 10083-3:2006 [24], has a Brinell hardness number—BHN [27]—higher than 200, so $k_f = 0.5$.

By entering all the values for the factors in Equation (25), S_f = 142 MPa. From Equation (24), considering FS = 1.5, the minimum shaft diameter would be 59.3 mm. This shows that the 50 mm diameter is not enough to ensure the fatigue strength of the shaft.

As already explained, the input values in the previous calculation were $M_f = M_{faccel} = 1927$ N·m, and $T = M_{taccel}$ = 1323 N·m.

The ANSI/ASME standard considers reversed bending and steady or nearly steady torsion. However, as seen in Section 3, in the present case, the torsion is also alternating, albeit with a lower frequency. Although the number of reversed torsion cycles is much lower than the number of reversed bending cycles, it will surpass 10^6 cycles and should be considered in the shaft design. These circumstances led to a final fatigue calculation that could fully accommodate the service conditions of the shaft, as presented in the next section (Section 8).

8. Calculation of the Shaft Diameter Taking into Account the Alternating Torsion

The ANSI/ASME B106.1M:1985 standard [21] is based on classical fatigue design considerations. It represents a particular situation of a more general treatment presented, e.g., in [26], or in [4,5,28], and other references mentioned in Section 1. In the following, the notation of [17] is adopted.

Refs. [17,21] define the fatigue limit strength—S'_n—as $S'_n = 0.5S_u$, where S_u is the ultimate tensile strength, and it is necessary to affect the fatigue limit strength with the correction factors. Table 8 defines the load type factor C_L.

Table 8. Fatigue strength correction factors C_L for each type of loading, according to [17].

	Load Type		
Fatigue Strength Correction Factor—Load Factor	**Axial**	**Bending**	**Torsion**
C_L	0.85	1	0.58

The gradient factor, also known as size factor C_G, is equivalent to the ANSI/ASME method factor k_b and is shown in Table 9.

Table 9. Gradient correction factors C_G for each type of loading, according to [17].

Fatigue Strength Correction Factor—Gradient Factor	Load Type		
	Axial	Bending	Torsion
C_G for $10 < d_{shaft} < 50$	0.7 to 0.9	0.9	0.9
C_G for $50 < d_{shaft} < 100$	0.6 to 0.8	0.8	0.8

The surface finish factor C_S, is equivalent to the ANSI/ASME method factor k_a. According to [17], $C_S = 0.72$, the same as k_a. The temperature factor is $C_T = 1$ for the case study appliance environmental temperatures.

Similar to the ANSI/ASME method, the reference [17] introduces the reliability factor C_R, with the same values shown on Table 6 of Section 7. Thus, $C_R = 0.897$.

So, the limit fatigue strength will be corrected by these factors according to Equation (29) for the rotating bending stress, and (30) for the alternate torsional load.

$$S_{nb} = S'_n C_{Lb} C_G C_S C_T C_R \tag{29}$$

$$S_{nt} = S'_n C_{Lt} C_G C_S C_T C_R \tag{30}$$

resulting in $S_{nb} = 258$ MPa for the rotational bending load, and $S_{nt} = 150$ MPa for the reversed torsional load.

As in the ANSI/ASME method, ref. [17] gives directly K_f for profiled keyways, according to Table 10.

Table 10. Fatigue stress concentration factor k_f, for a profiled keyway under bending stress in solid round steel shafts, according to [17].

Steel	Bending	Torsion
	K_{fb}	K_{ft}
Annealed (less than 200 BHN [27])	1.6	1.3
Quenched or drawn (more than 200 BHN [27])	2.0	1.6

As explained in Section 7, BHN is larger than 200, so $k_{fb} = 2$ and $k_{ft} = 1.6$.

In the previous table, it is possible to notice the fundamental tendency for the harder and stronger materials to be more notch-sensitive. This means that changing from a soft to a harder and stronger steel normally increases part fatigue strength, but the increase is not as great as might be expected because of the increased notch sensitivity. As can be seen in Table 10, the harder material with 200 BHN has a stress concentration factor—K_f—greater than the softer material.

There are different types of loads acting on the keyway area—bending and shear stress due to the radial load on the sprocket, torsional load due to the gearmotor output torque. Equivalent stresses σ^{eq} were calculated as

$$\sigma^{eq} = \sigma_m + \frac{S_y}{S'_n C_{Lb} C_G C_S C_T C_R} K_{fb} \sigma_{abending} \tag{31}$$

$$\tau^{eq} = \tau_m + \frac{\frac{S_y}{\sqrt{3}}}{S'_n C_{Lt} C_G C_S C_T C_R} K_{ft} \tau_{atorsion} \tag{32}$$

where σ_m and τ_m are the mean stresses. The bending stress has zero mean stress, $\sigma_m = 0$ (or $R = \sigma_{min}/\sigma_{max} = -1$). The alternate bending stress—$\sigma_{abending}$—results from the bending moment calculated in Section 5—$M_{faccel} = 1927$ N·m, and the shear stress—$\tau_{atorsion}$—

results from the alternating torsional load calculated in Section 5 $M_{taccel} = 1323$ N·m.

$$\sigma_{abending} = \frac{M_{faccel}}{I_{d_i}} \cdot \frac{d_i}{2} \tag{33}$$

$$\tau_{atorsion} = \frac{M_{taccel}}{I_{p_{di}}} \cdot \frac{d_i}{2} \tag{34}$$

$$\tau_m = \frac{4}{3}\frac{V_{accel}}{A_{di}} \tag{35}$$

Considering a shaft diameter of 50 mm, as in the design before the redesign, using Equations (5) and (6), $I_{d_i} = 3.07 \times 10^{-7} \text{m}^4$, and $I_{p_{di}} = 6.14 \times 10^{-7} \text{m}^4$.

From Equation (33), $\sigma_{abending} = 157$ MPa, and from (31), $\sigma^{eq} = 972$ MPa for the bending stress. For the torsional and shear stresses, from (35), $\tau_m = 0$ MPa. As seen in Section 4, $V_{accel} = 0$ N in the middle section of the shaft—$\frac{l}{2}$—that is the critical section for fatigue calculation. From (34), $\tau_{atorsion} = 54$ MPa, and (35), $\tau^{eq} = 266$ MPa.

The final validation was done by applying the Tresca criterion according to Equation (36). In the context of classical fatigue design, the use of Tresca criterion is found, e.g., in [11].

$$\tau_{mx} = \sqrt{\left(\frac{\sigma^{eq}}{2}\right)^2 + \tau^{eq2}} < \frac{S_y}{2N_s} \tag{36}$$

where N_s is the safety factor.

For a shaft diameter of 50 mm, we have from Equation (36), $\tau_{mx} = 554$ MPa, higher than the allowed 400 MPa, for $N_s = 1$; thus, far from ensuring infinite fatigue life!

It would be necessary to increase the shaft diameter to 63.9 mm to ensure infinite fatigue life, taking into account the alternating rotational bending load and alternating torsional load, with a safety factor of 1.5. Recall that this value is higher than the 59.3 mm that resulted from the ANSI/ASME method, Section 7, that did not take alternate torsion into account.

9. Conclusions

A redesign of a failed 50 mm diameter hoisting shaft was presented. The importance of making a thorough assessment of the VTD operation before the shaft redesign was emphasized, making it possible to realize that the torsional load from the gearmotor output torque should be considered alternating and taken into account as such in the fatigue redesign performed.

The peak load $\sigma_{peak} = 218$ MPa was seemingly low when compared with the fatigue limit $\sigma_{f0} = 500$ MPa (34CrNiMo6 steel). This might lead to the misleading consideration that the calculation of the shaft for fatigue was unnecessary, but that would be an expensive mistake.

In fact, the keyseat area is subjected to high stress concentration, with a serious impact on the maximum fatigue allowable stress. All the methods described in the present work indicate that the fatigue limit allowable stress in keyseat areas does not increase proportionally with the ultimate strength of the steel. This means that changing from a soft to a harder and stronger steel normally increases the part fatigue strength, but the increase is not as great as might be expected.

For the shaft material, according to a rough sizing presented by Niemann, the maximum allowable stress for infinite fatigue life in the keyseat area was 58 MPa, whereas it was 142 MPa according to ANSI/ASME, illustrating the conservativeness of the Niemann method.

The ANSI/ASME method was developed for steady torsional loads. To consider reversed torsional loading and ensure infinite fatigue life under reversed torsion and rotational bending, a classical fatigue model combining shear and torsional stresses with bending and normal stresses using equivalent stresses was performed.

Overloads, including possible misuses of the VTD, are difficult to quantify. According to ANSI/ASME and to Niemann, these transient loads need to be considered through the incorporation of safety factors that take into account the severity of a hoisting shaft rupture in terms of safety, machine damage and machine downtime.

The redesigned shaft diameter was, according to Niemann, $d = 74.8$ mm; according to the ANSI/ASME method, it was $d = 59.3$ mm; and according to the classical fatigue model, it would be $d = 63.9$ mm. Taking into account the transient overloads, finally, $d = 70$ mm was adopted for the redesigned shaft.

Author Contributions: Conceptualization, F.A.C.d.S. and P.M.S.T.d.C.; methodology, F.A.C.d.S.; investigation, F.A.C.d.S.; resources, F.A.C.d.S.; data curation, F.A.C.d.S.; writing—original draft preparation, F.A.C.d.S.; writing—review and editing, F.A.C.d.S. and P.M.S.T.d.C.; visualization, F.A.C.d.S. and P.M.S.T.d.C.; supervision, F.A.C.d.S.; project administration, F.A.C.d.S. All authors have read and agreed to the published version of the manuscript.

Funding: This research received no external funding.

Institutional Review Board Statement: Not applicable.

Informed Consent Statement: Not applicable.

Data Availability Statement: The data used in the work are presented in the text.

Acknowledgments: F.A.C.d.S. gratefully acknowledges Instituto Superior de Engenharia do Porto—ISEP, specifically, for access to technical literature relevant for researching the subject.

Conflicts of Interest: The authors declare no conflict of interest.

Nomenclature

Section 4		
A_{di}	Section area of the shaft with initial diameter	m^2
F_{peak}	Peak load on the hoisting shaft sprocket in a emergency stop	N
I_{di}	Inertia of the initial shaft diameter section	m^4
$I_{p_{di}}$	Polar inertia of the initial shaft diameter section	m^4
M_{fpeak}	Bending moment caused by the emergency stop radial load on the sprocket	N·m
M_{aemerg}	Peak emergency torque	N·m
N_{peak}	Safety factor for peak load stresses validation	---
p_{peak}	Distributed emergency radial load on the sprocket length	N/m
V_{peak}	Peak shear load on shaft, on $l/2$	N
σ_{peak}	Von Mises peak stress originated in emergency stop	MPa
$\sigma_{peakbending}$	Bending stress caused by the peak emergency radial load on the shaft critical section	MPa
σ_{yield}	Steel yield stress	MPa
τ_{mpeak}	Shear stress caused by the emergency stop radial load	MPa
τ^{eq}_{peak}	Equivalent shear stress originated in emergency stop	MPa
$\tau_{peaktorsion}$	Torsion stress caused by the peak emergency torque	MPa
Section 5		
F_{accel}	Load on the sprocket during the VTD acceleration and deceleration	N
$F_{unbalance}$	Load caused by the unbalance between m_{tmsm} and the counterweight mass	N
J_x	External moment of inertia reduced to the motor shaft	kg.m^2
M_{faccel}	Bending moment during VTD acceleration and deceleration	N

M_H	Acceleration torque	N·m
M_L	Load torque	N·m
M_{static}	Required gearmotor torque necessary to ensure the hoisting movement at constant speed	N·m
M_{taccel}	Output gearmotor torque during acceleration	N·m
p_{accel}	Distributed load on the sprocket during acceleration and deceleration	N
t_a	Acceleration time	s
m_{tmsm}	Total maximum suspended mass	kg
V_{accel}	Shear load on the shaft critical section equal to $\frac{F_{accel}}{2}$	N
η_{motor}	Motor efficiency	---
Section 6		
a_f	Factor for oscillating bending and alternating bending according to the Niemann method [20].	---
b_0	Size factor according to the Niemann method [20].	---
b_N	Factor for solid shafts or hollow shafts according to the Niemann method [20].	---
d_N	Resulting shaft diameter according to the Niemann method [20].	mm
M_{eq}	Equivalent moment for calculation of the shaft diameter according to the Niemann method [20].	N·m
M_f	Bending moment for calculation of the shaft diameter according to the Niemann method [20].	N·m
M_t	Torsion moment for calculation of the shaft diameter according to the Niemann method [20].	N·m
σ_{fa10}	Allowable fatigue strength in keyway under alternating bending for the test specimen with diameter 10 mm and a material with a given σ_r	---
σ_{fad}	Allowable fatigue strength in keyway under alternating bending for a given shaft diameter	MPa
σ_r	Ultimate tensile strength of the shaft steel	MPa
Section 7		
FS	Factor of safety according to ANSI/ASME [21]	---
K_f	Fatigue strength reduction factor according to ANSI/ASME [21]	---
K_t	Theorical stress concentration factor in bending according to ANSI/ASME [21]	---
k_a	Surface finish factor according to ANSI/ASME [21]	---
k_b	Size factor according to ANSI/ASME [21]	---
k_c	Reliability factor according to ANSI/ASME [21]	---
k_d	Temperature factor according to ANSI/ASME [21]	---
k_f	Fatigue stress concentration factor in a keyseat area under reversed bending according to ANSI/ASME [21]	---
$k_{ftorsion}$	Fatigue stress concentration factor in a keyseat area under reversed torsion according to ANSI/ASME [21]	---
k_g	Miscellaneous effects factor according to ANSI/ASME [21]	---
q	Notch sensitivity factor	---
S_f	Corrected fatigue limit of the shaft in reversed bending according to ANSI/ASME [21]	N/m^2
S_f^*	Fatigue limit of polished, unnotched test specimen in reverse bending according to ANSI/ASME [21]	N/m^2
S_{fa}	Allowable corrected fatigue limit of shaft in reversed bending according to ANSI/ASME [21]	N/m^2
S_u	Ultimate tensile strength of the shaft steel	N/m^2
S_y	Tensile yield strength of the steel	N/m^2
$S_{yctorsion}$	Allowable corrected fatigue limit of shaft in reversed (alternating) torsion	N/m^2
T	Static mean torque	N·m

Section 8		
C_G	Gradient factor	---
C_L	Load type factor	---
C_{Lb}	Load type factor for reversed bending	---
C_{Lt}	Load type factor for reversed torsion	---
C_R	Reliability factor	---
C_S	Surface finish factor	---
K_{fb}	Stress concentration factor in a keyseat area of a shaft in reversed bending	---
K_{ft}	Stress concentration factor in a keyseat area of a shaft in reversed (alternating) torsion	---
N_s	Safety factor for the conventional classical fatigue model calculation	---
S'_n	Fatigue limit strength of the steel	MPa
S_{nb}	Allowable corrected fatigue limit of shaft in reversed bending according to the conventional classical fatigue model	MPa
S_{nt}	Allowable corrected fatigue limit of shaft in reversed torsion according to the conventional classical fatigue model	MPa
S_u	Ultimate tensile strength of the steel	MPa
S_y	Tensile yield strength of the steel	MPa
$\sigma_{abending}$	Alternating bending stress	MPa
$\tau_{atorsion}$	Alternating torsion stress	MPa
σ_m	Mean normal stress	MPa
σ^{eq}	Equivalent normal stress, resulting from bending and traction loads combination	MPa
τ_m	Mean shear stress	MPa
τ^{eq}	Equivalent shear stress, resulting from torsion and shear loads combination	MPa

References

1. Harris, D.; Jur, T. Classical fatigue design techniques as a failure analysis tool. *J. Fail. Anal. Prev.* **2009**, *9*, 81–87. [CrossRef]
2. Childs, P.R.N. *Mechanical Design Engineering Handbook*, 2nd ed.; Elsevier/Butterworth-Heinemann: Oxford, UK, 2019.
3. Childs, P.R.N. *Mechanical Design: Theory and Applications*, 3rd ed.; Elsevier/Butterworth-Heinemann: Oxford, UK, 2021.
4. Beswarick, W.J. *Shaft with Fluctuating Loads. SEED—Sharing Experience in Engineering Design*; Engineering Design Procedural Guide; SEED: Hatfield, UK, 1988; guide MPT 6.2.
5. Beswarick, W.J. Shaft with fluctuating load. In *Rotary Power Transmission Design*; Hurst, K., Ed.; McGraw Hill: London, UK, 1994; pp. 142–148.
6. D'Angelo, A. *Machine Design for Technology Students. A Systems Engineering Approach*; Springer Nature: Cham, Switzerland, 2021. [CrossRef]
7. de Castro, P.M.S.T.; Fernandes, A.A. Methodologies for failure analysis: A critical survey. *Mater. Des.* **2004**, *25*, 117–123. [CrossRef]
8. Milella, P.P. *Fatigue and Corrosion in Metals*; Springer: Milan, Italy, 2013.
9. Lee, Y.L.; Pan, J.; Hathaway, R.; Barkey, M. *Fatigue Testing and Analysis. Theory and Practice*; Elsevier Butterworth–Heinemann: Oxford, UK, 2005.
10. Anes, V.; de Freitas, M.; Reis, L. The damage scale concept and the critical plane approach. *Fatigue Fract. Eng. Mater. Struct.* **2017**, *40*, 1240–1250. [CrossRef]
11. d'Isa, F.A. *Mechanics of Metals*; Addison-Wesley: Reading, MA, USA, 1968.
12. Hall, A.S.; Holowenko, A.R.; Laughlin, H.G. *Theory and Problems of Machine Design*; McGraw-Hill: New York, NY, USA, 1961.
13. Spotts, M.F. *Design of Machine Elements*, 3rd ed.; Prentice-Hall: Englewood Cliffs, NJ, USA, 1962.
14. *EN 619:2002+A1:2010*; Continuous Handling Equipment and Systems—Safety and EMC Requirements for Equipment for Mechanical Handling of Unit Loads. European Committee for Standardization: Brussels, Belgium, 2010.
15. SEW EURODRIVE. *Project Planning of Drives—Edition 10/2001*; SEW EURODRIVE: Bruchsal, Germany, 2001.
16. SEW EURODRIVE. SEW Workbench. Available online: https://www.sew-eurodrive.pt/servicos/engenharia_selecao/ferramentas_de_engenharia/ferramentas_de_engenharia.html (accessed on 4 March 2023).
17. Juvinall, R.C.; Marshek, K.M. *Fundamentals of Machine Component Design*, 6th ed.; John Wiley & Sons: Hoboken NJ, USA, 2017.
18. Pilkey, W.D.; Pilkey, D.F. *Peterson's Stress Concentration Factors*, 3rd ed.; John Wiley & Sons: Hoboken, NJ, USA, 2008.
19. Niemann, G. *Elementos de Máquinas*, 7th printing; Editora Edgard Blücher Ltda: São Paulo, Brazil, 2004; Volume II.
20. Niemann, G. *Tratado Teórico-Práctico de Elementos de Máquinas—Cálculo, Diseño e Construcción*, 2nd ed.; Editorial Labor, S.A.: Barcelona, Spain, 1973.
21. *ANSI/ASME B106.1M:1985*; Design of Transmission Shafting. ANSI/ASME: Washington, DC, USA, 1985.

22. *DIN 15020-1:1974-02*; Lift Appliances; Principles Relating to Rope Drives; Calculation and Construction. DIN: Berlin, Germany, 1974.
23. *EN 10025+A1:1994*; Produtos Laminados a Quente de Aços de Construção Não Ligados. Instituto Português da Qualidade: Caparica, Portugal, 1994.
24. *EN 10083-3:2006*; Steels for Quenching and Tempering—Part 3: Technical Delivery Conditions for Alloy Steels. European Committee for Standardization: Brussels, Belgium, 2006.
25. *ANSI/CEMA B105.1-2015*; Specifications for Welded Steel Conveyor Pulleys with Compression Type Hubs. ANSI/CEMA: Washington, DC, USA, 2015.
26. Shigley, J.E. *Mechanical Engineering Design*, 3rd ed.; McGraw-Hill Kogakusha: Tokyo, Japan, 1977.
27. The Engineering ToolBox. Brinell Hardness Number BHN. Available online: https://www.engineeringtoolbox.com/bhn-brinell-hardness-number-d_1365.html (accessed on 26 December 2022).
28. Shigley, J.E. *El Proyecto en Ingeniería Mecánica*; Ediciones del Castillo SA: Madrid, Spain, 1970.

eng MDPI

Review

Rheological Behavior of Cement Paste: A Phenomenological State of the Art

Youssef El Bitouri

LMGC, IMT Mines Ales, University Montpellier, CNRS, 30100 Ales, France; youssef.elbitouri@mines-ales.fr

Abstract: With the interest aroused by the development of modern concretes such as printable or self-compacting concretes, a better understanding of the rheological behavior, directly linked to fresh state properties, seems essential. This paper aims to provide a phenomenological description of the rheological behavior of cement paste. The first part is devoted to the most common testing procedures that can be performed to characterize the rheological properties of cement suspensions. The second one deals with the complexities of the rheological behavior of cement paste including the non-linearity of flow behavior, the viscoelasticity and yielding, and the structural build-up over time.

Keywords: cement paste; rheology; yield stress

Citation: El Bitouri, Y. Rheological Behavior of Cement Paste: A Phenomenological State of the Art. *Eng* **2023**, *4*, 1891–1904. https://doi.org/10.3390/eng4030107

Academic Editor: Antonio Gil Bravo

Received: 22 June 2023
Revised: 29 June 2023
Accepted: 30 June 2023
Published: 4 July 2023

1. Introduction

The development of modern cementitious materials, such as 3D-printable cementitious materials [1–3], low-carbon cement [4,5], or self-compacting concrete [6,7], requires a deep understanding of the rheological behavior at the fresh state. Other types of concretes including recycled aggregates or plastic fiber pose challenges with respect to their fresh state behavior [8,9]. In addition, several processes, such as pumping or casting, are strongly conditioned by the rheological behavior. For instance, the development of 3D-printable concretes based on extrusion generates several challenges that directly concern the rheology of cementitious materials. Indeed, 3D concrete formulations must be pumpable but stable in shape once placed without formwork. In addition, the sequential layers put in place by extrusion must adhere and structure themselves quickly. Finally, the geometry of the printed parts must be controlled within the appropriate geometric tolerances [3,10].

At the fresh cement paste scale, the rheological behavior is dictated by the cement particles' organization and the interparticle forces [11]. Several methods and procedures were developed to characterize this behavior and to examine the effect of influencing factors such as superplasticizers [12–14], supplementary cementitious materials (SCMs) [15,16], or formulation parameters [17]. Moreover, the rheological behavior of cement paste exhibits different complexities such as non-linearity (shear thinning/shear thickening), viscoelasticity, yielding, structural build-up at rest, and chemical evolution. In general, the rheograms of cement paste are not linear and can present a shear-thinning phenomenon leading to the decrease in viscosity with an increasing shear rate. This behavior is characteristic of flocculated suspensions [15,18] and is attenuated with the incorporation of a dispersant (superplasticizer). In fact, the flow behavior becomes linear with viscosity independent from the shear rate due to the enhanced dispersion by the addition of the superplasticizer.

Furthermore, cementitious materials cannot be described as thixotropic materials [19–23], since the evolution of rheological properties at rest is not fully reversible. At rest, a structural build-up due to flocculation and chemical evolution occurs and allows for evaluating the loss of workability during the dormant period of cement hydration. An irreversible part due to chemical hydration remains despite the application of a strong shearing which theoretically can erase the structural build-up [20]. The addition of dispersants or retarders induces a decrease in the structural build-up, and the behavior of the cement paste tends towards a thixotropic behavior [20].

It thus appears that the rheological behavior of cement paste is complex due to several factors such as the polydispersity of cement particles, their morphology, the interaction between particles, and the chemical evolution (dissolution/precipitation). In this paper, the testing procedures commonly used to characterize the rheological behavior of cement paste are presented first. Then, the rheological behavior of cement paste is described through flow behavior including steady and transient modes, as well as dynamic rheology. Finally, the structural build-up of cement paste during the dormant period of cement hydration is described.

2. Testing Procedures

Rheological measurements can be carried out using rotational rheometers with different types of geometries. It has to be kept in mind that several artifacts could occur and lead to misinterpretations of results [24]. In this section, only the most relevant testing procedures that can be performed for cement paste are presented.

2.1. Flow Tests

Flow tests are usually performed using rotational rheometers by applying an imposed shear rate or shear stress. The most relevant procedure for cement paste consists of the application of a strong preshear in order to induce a structural breakdown and ensure a same initial state for the tested samples, followed by a flow sweep (or stepped flow) with a decreasing shear rate, as shown in Figure 1. The applied time at each step must be greater than the characteristic time, which depends on the shear rate. In fact, the characteristic time is inversely proportional to the shear rate. So, for a high shear rate, the required time to reach the steady state flow is negligible, while for a very low shear rate, this time increases and is of the order of 10 to 50 s. It should be noted that most modern rheometers are equipped with steady state sensors that allow the equilibrium state to be detected at each applied shear rate.

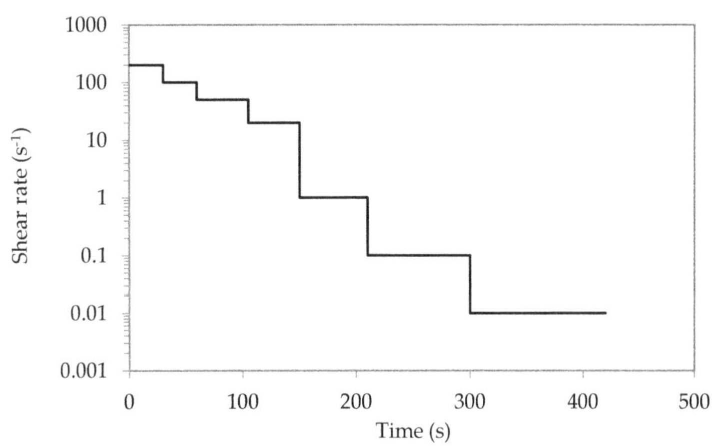

Figure 1. Flow sweep test with a decreasing shear rate.

The response given by such flow test in terms of shear stress and viscosity is presented in Figure 2. The flow curve is non-linear and displays shear-thinning behavior. In fact, the viscosity decreases with an increasing shear rate, which characterizes the breakdown of agglomerated particles. For a low shear rate, the shear stress remains almost constant, which describes the flow stoppage of freshly mixed cement paste. This shear stress is defined as the dynamic yield stress and corresponds to the minimum stress required to initiate the flow onset.

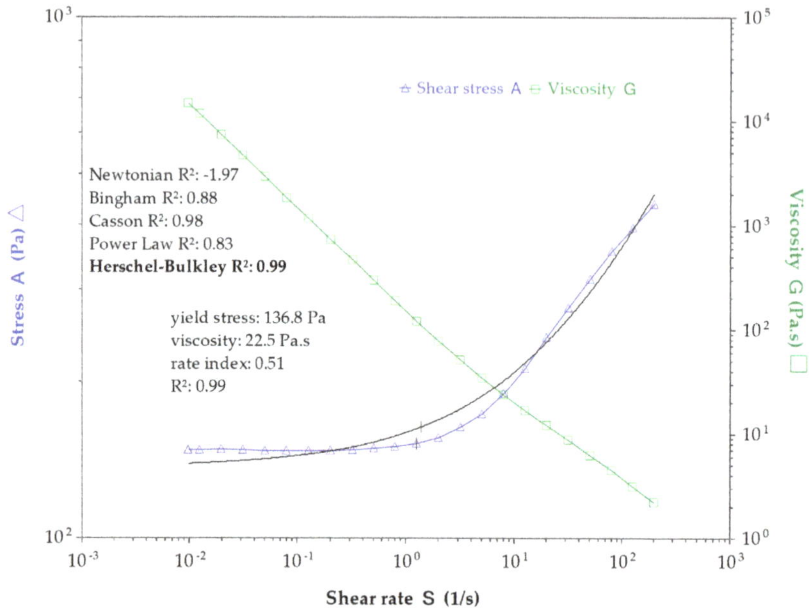

Figure 2. Shear stress and viscosity as a function of the shear rate for cement paste with a w/c ratio of 0.4 after 20 min of hydration (test performed with Rheometer AR2000Ex from TA Instruments equipped with Vane geometry).

As presented in Figure 2, several laws allow for fitting the flow curve. However, for non-Newtonian and yield stress fluid such as cement paste, the flow behavior can be described by the Casson law as well as by the Herschel–Bulkley model. This latter is widely used for cement paste behavior and is defined as follows:

$$\tau = \tau_0 + k\dot{\gamma}^n \tag{1}$$

where τ is the shear stress (Pa), τ_0 is the yield stress (Pa), $\dot{\gamma}$ is the shear rate (s^{-1}), k is the viscosity (Pa·s), and n is the rate index. k and n are the fitting parameters. A thinning behavior is represented by a rate index n inferior to 1, as shown in Figure 2. When the rate index is equal to 1, the Herschel–Bulkley law is equivalent to the Bingham law and represents a linear behavior for which the viscosity remains constant with the shear rate. A shear-thickening phenomenon is represented by a rate index superior to 1.

The Casson model can also describe a non-linear flow behavior (shear-thinning) with a yield stress term. It is defined as follows:

$$\tau^{1/2} = {\tau_0}^{1/2} + k_0^{1/2}\dot{\gamma}^{1/2} \tag{2}$$

where τ_0 is the yield stress (Pa) and k_0 is a model constant (Pa·s).

2.2. Oscillation Metods

Dynamic rheology consists of small-amplitude oscillation shear (SAOS) [17,21,25] and large-amplitude oscillation shear (LAOS) [26]. These procedures seem to be more appropriate for examining the microstructure of cement paste. The LAOS procedure appears to be relevant to investigating the evolution of non-linear viscoelastic properties at different frequencies and amplitudes represented by the Pipkin diagram [26]. The SAOS procedure is more appropriate for examining the structural build-up during the dormant period of cement hydration [21,22,27–29].

Oscillation procedures include strain sweep, frequency sweep, and time sweep tests. They consist of the measurement of the viscoelastic properties such as the storage (or elastic) modulus G′, the loss (viscous) modulus G″, critical strains, and other parameters (phase angle, oscillation stress . . .). The frequency sweep mode is generally performed to investigate the cement paste stability [30]. The time sweep procedure is appropriate for examining the structural build-up and thixotropy [19,21,22]. Strain (or stress) sweep tests can be carried out to probe the microstructure of cement paste and can give information concerning the network of cement particles and the forces acting inside this network.

The amplitude sweep can be performed by applying an increasing stress (or strain) at a frequency of 1 Hz [17,19,23]. The typical response of amplitude sweep is presented in Figure 3. It is worth noting that this kind of non-destructive test allows the microstructure of fresh cement paste to be probed [17,27,31]. As shown in Figure 3, the evolution of storage and loss moduli during an amplitude sweep follows three phases. The first phase in which the moduli remain almost constant is defined as the linear viscoelastic domain (LVED). In this domain, the microstructure of cement paste is maintained [27]. The SAOS procedure allowing the structural build-up to be characterized is commonly performed in this strain region. In addition, cement paste will behave like a solid if the storage modulus G′ is greater than the loss modulus G″. The liquid-like behavior corresponds to G′ < G″. The end of the LVED is characterized by a significant drop of the moduli. The strain associated with this decrease onset is of the order of 10^{-2}% and can be interpreted as a signature of the breakage of the links between cement particles (C-S-H links and attractive colloidal forces) [17,23,27,29,32].

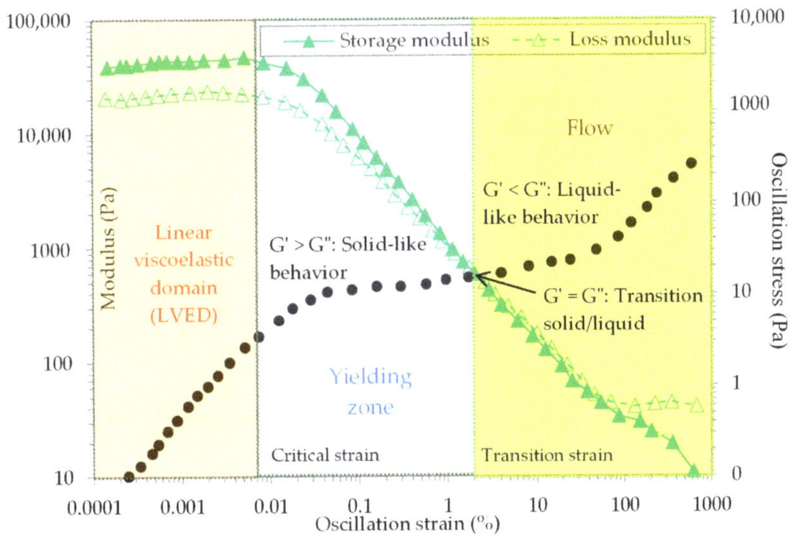

Figure 3. Example of an amplitude (strain or stress) oscillation test (test performed with Rheometer AR2000Ex from TA Instruments equipped with Vane geometry).

The second phase corresponds to the yielding zone. The oscillation stress remains almost constant in this region, and the moduli (G′ and G″) decrease significantly. In addition, there is a transition point corresponding to the intersection of the storage modulus and the loss modulus curves (G′ = G″) which defines a solid/liquid transition, i.e., the flow onset. This transition point is associated with a large strain (transition strain). The yield stress can be determined at this point.

The last phase is associated with a large strain and describes the flow behavior (liquid-like behavior G′ < G″).

2.3. Transient Mode

The transient mode includes stress growth, stress relaxation, and creep/recovery procedures. This paper focuses on the stress growth procedure widely used to measure the static yield stress.

The stress growth procedure consists of the application of a very low shear rate in the range of 10^{-3} s^{-1}–10^{-2} s^{-1}, depending on the rheometer sensitivity. In fact, the yield stress corresponds to the minimum stress required to initiate flow. Theoretically, it corresponds to the shear stress at rest, i.e., for a shear rate equal to 0. Since rheometers cannot apply a shear rate of zero, this shear rate must be as low as possible. The correct measurement of the yield stress consists of the application of a strong pre-shearing phase, followed by a sufficient resting time to allow the structure to be rebuilt [23,33–36]. Then, a very low shear rate in the range of 10^{-3} s^{-1}–10^{-2} s^{-1} is applied to the cement paste during a sufficient time to reach the steady state flow. This time has to be greater than the characteristic time of flocculation, which is of the order of 10 s [23]. Vane geometry appears to be the most suitable for such a test. However, it is important to determine beforehand the geometric constants making it possible to calculate, respectively, the shear stress and the shear rate from the applied torque and the rotational velocity. This can be accomplished using the Couette analogy method described by Aît-Kadi et al. [37].

The typical curve given by the stress growth procedure is presented in Figure 4. It can be observed that the evolution of shear stress as a function of shear strain follows at least two steps. First, the shear stress increases almost linearly with the strain until it reaches a peak followed by a plateau. The linear increase for a low strain is consistent with elasticity (solid regime) and can be represented as follows:

$$\tau = G\gamma \quad (3)$$

where τ is the shear stress (Pa), G is the instantaneous shear modulus (Pa), and γ is the shear strain.

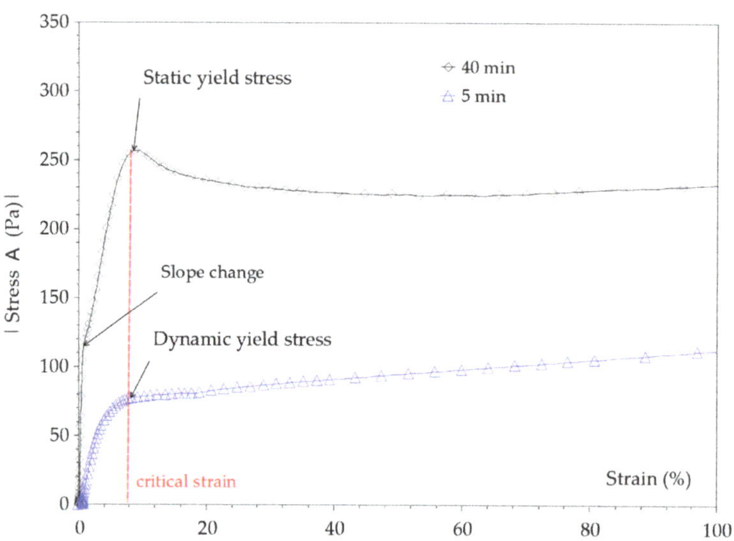

Figure 4. Shear stress as a function of shear strain during the stress growth procedure (cement paste with w/c = 0.4) performed after 5 min and 40 min of hydration with AR2000Ex equipped with Vane geometry.

The peak defines the static yield stress corresponding to the flow onset, i.e., the shear stress where the suspension just starts to flow. In cement suspensions, the yield stress originates from the network of interacting particles due to direct contacts and colloidal attractive forces [11]. This peak is not always observed. Its observation depends on several factors such as the particle properties (shape, particle size distribution), solid volume fraction, and time of hydration. In cement pastes, this peak becomes more visible with the progress of the hydration, as shown in Figure 4 [20]. Moreover, one can simply observe a plateau defining the steady state flow under a low shear rate. This plateau defines the dynamic yield stress (Figure 4). In the plateau, the shear stress is related to the shear rate through the Newton law as follows:

$$\tau = \eta\dot{\gamma} \tag{4}$$

where η is the viscosity (Pa·s), and $\dot{\gamma}$ is the shear rate (s^{-1}).

The transition point between the linear regime and the plateau which defines the yield stress can be associated with a characteristic time required to reach the steady state flow. This characteristic time is inversely proportional to the shear rate as follows:

$$t_c \sim \frac{\eta}{G} \sim \frac{\gamma_c}{\dot{\gamma}} \tag{5}$$

where γ_c is the critical strain of the order of few %. For a shear rate of $10^{-2}\ s^{-1}$ and a critical strain of about 10%, the characteristic time is of the order of 10s, which is close to the characteristic time of flocculation [23].

Furthermore, as reported by Roussel et al. [23], in the very first stages of the shearing process, the shear stress–shear strain curve can exhibit an abrupt change in the slope (Figures 4 and 5). Fourmentin et al. [38] reported the same trend. In fact, at a very low shear strain, there is a quick increase in the shear stress associated with a high instantaneous shear modulus, which could characterize a very stiff material [23,38]. This slope change is associated with a very low strain which cannot be associated with the flow onset. According to Roussel et al. [23], this low strain could be associated with the breakage of the contacts between cement particles formed by C-S-H links. The critical strain of the order of few % at the flow onset describes large structural changes in the suspension [11,23].

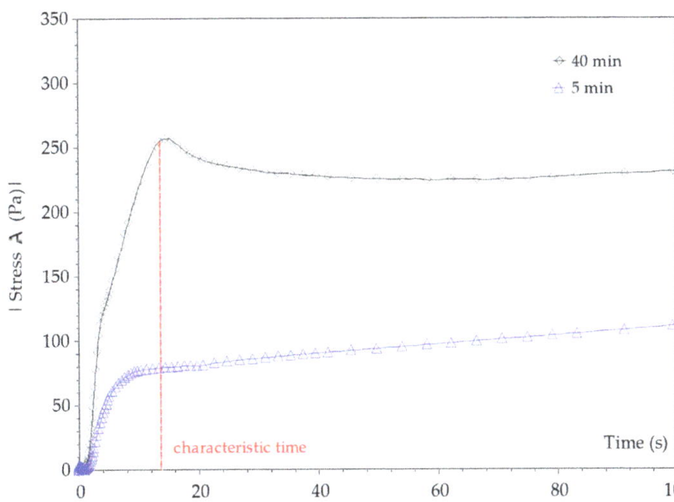

Figure 5. Evolution of shear stress as a function of time during the stress growth procedure (cement paste with w/c = 0.4) performed after 5 min and 40 min of hydration.

3. Rheological Behavior of Cement Paste

3.1. Flow Behavior

The cementitious suspensions generally display a shear-thinning behavior with a continuous decrease in viscosity with the shear rate. This behavior, characteristic of flocculated suspensions, describes the deflocculation under shearing. For simplification, the rheological behavior of cement paste is often assimilated to a Binghamian behavior under certain conditions. In addition, cement pastes (without dispersant) exhibit a yield stress that has to be overcome to initiate flow. Viscoplastic models such as the Herschel–Bulkley model or Bingham model seem to be relevant and are widely used to describe the flow behavior of cement pastes. The flow curve obtained by a stress (or shear rate) sweep is fitted with such models to determine the dynamic yield stress at a very low shear rate.

The effect of several parameters such as the properties of cement, water-to-cement ratio (w/c), admixtures, and supplementary cementitious materials (SCMs) on the flow behavior has been widely studied [15,23,32,39–45]. The effect of w/c on rheological parameters can be described by the dependence on the solid volume fraction through the Krieger–Dougherty law for viscosity [46] or yield stress [47] or by the yield stress model (YODEL) developed by Flatt and Bowen [48]. The effect of admixtures (especially superplasticizers) and SCMs has been the subject of a large number of publications.

Although it is often assumed, by simplification, that cement pastes behave like a Bingham fluid, it turns out that many experimental results reveal behaviors marked by nonlinear flow curves, especially with shear-thinning behavior. However, cement pastes can also display a shear-thickening behavior, particularly in the presence of superplasticizers and/or certain mineral additions [15,49,50]. The shear-thickening behavior refers to the abrupt or continuous increase in viscosity with the shear rate. This behavior can be described with the Herschel–Bulkley model with a rate index $n > 1$. Different assumptions have been proposed to explain the origins of this behavior.

By carrying out dynamic simulations, Bossis and Brady [51] proposed a mechanism based on particle clustering to explain the shear thickening. The cluster formation results from the lubrication forces. In a suspension with short-range repulsive interparticle forces (electro-steric effect; Brownian motion) such as cement paste with a superplasticizer, the cluster formation could be avoided, especially at low or moderate shear rates. With an increasing shear rate, the hydrodynamic forces increase until their intensity exceeds that of the repulsive forces, thus inducing the formation of particle clusters.

Another mechanism based on the order–disorder transition has been proposed to explain the shear thickening [15,52–54]. According to this theory, the shear-thickening behavior would be the consequence of the transition from a layered flow, where the particles are ordered in successive layers, to a locally disordered state, where the particles are dislodged from the layer structure. In fact, the hydrodynamic forces cause this instability, which breaks up the layered flow. This local instability induces particle jamming, which probably involves cluster formation, leading to an increase in viscosity.

In the case of cement paste, Roussel et al. suggest that these mechanisms do not seem relevant. Indeed, according to the authors, the macroscopic flow behavior of cement pastes can be characterized by two transitions: macroscopic shear thinning resulting from the transition between a colloidal and viscous regime, and shear thickening corresponding to the transition between a viscous and inertial regime. At a high shear rate, particle inertia dominates hydrodynamic effects, which may lead to shear thickening.

3.2. Linear Visco-Elastic Domain

The linear visco-elastic domain (LVED) can be determined by oscillation rheology (Section 2.2). In this domain, the storage and loss moduli remain constant. Generally, fresh cement pastes (without a superplasticizer) exhibit a solid-like behavior in this domain with a storage modulus greater than the loss modulus ($G' > G''$). The end of this domain is associated with a shear strain of the order of 10^{-2}%. This shear strain can be attributed to the breakage of the links between particles due to early hydrates nucleation [23] and/or the attractive colloidal forces [29,32]. It has to be kept in mind that this critical strain remains identical from fresh cement paste to hardened cement paste [29]. In addition, it is of the same order of magnitude as the critical strain for which cracks propagate in hardened concrete [23].

The effect of several factors (w/c ratio, SP dosage, SCMs ...) on the LVED of fresh cement paste has been investigated, but the number of studies remains relatively limited [17,19,27,29,55,56]. Recently, an interesting study dealing with the effect of the w/c ratio and superplasticizer (SP) on the viscoelastic properties of fresh cement pastes has been carried out [17]. The results show that both the storage modulus and viscoelastic stress increase with the decreasing w/c ratio, which is consistent with the fact that the stiffness of cement paste increases with the solid volume fraction. In fact, the storage modulus could follow a Krieger–Dougherty law [47].

Furthermore, it appears that the critical strain at the end of the LVED is strongly affected by the superplasticizer dosage, while the effect of the w/c ratio is less significant. Only a high w/c ratio leads to an increase in this critical strain [17]. The effect of the SP dosage thus appears to be more significant than that of the w/c ratio (Figure 6). The interpretation of this phenomenon remains complex, and some ambiguities remain in the literature. First, the critical strain at the end of the LVED is attributed to the breakage of C-S-H bridges between cement particles [17,23]. However, in a calcite suspension (chemically inert), which has granular properties close to those of cement paste and is sometimes used to mimic the fresh behavior of cement paste [57,58], the end of the LVED is characterized by a critical strain of the same order of magnitude [59].

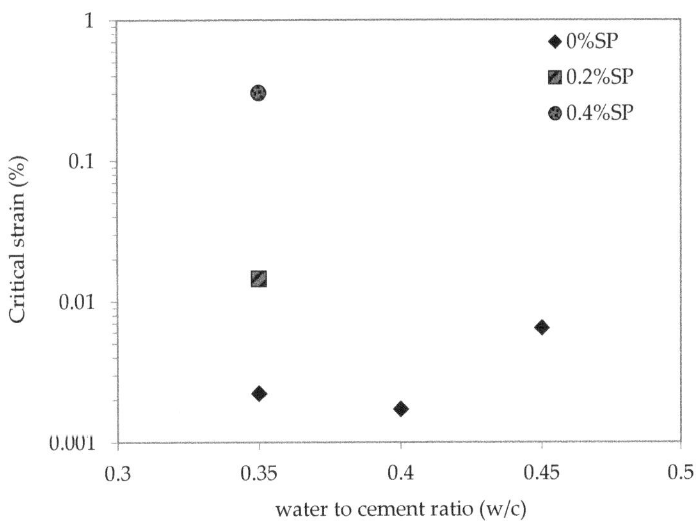

Figure 6. Effect of water-to-cement ratio (w/c) and superplasticizer dosage on the critical strain (data from [17]).

Concerning the effect of w/c on the critical strain, Jiao and De Schutter [17] explain it by the increase in the dissolution rate, leading to the increase in early hydrates formation. In addition, these early hydrates could be more fragile when w/c increases. The effect of the superplasticizer on the critical strain is explained by the entanglement of superplasticizer molecules with each other and the possible enhancement of the C-S-H bridges. According to these authors, the cohesive bonding between cement particles is improved, which increases the deformation capacity of the connections. The oscillatory shear strain required to initiate the relative movement between particles thus increases. However, some ambiguities could be pointed out. First, there is no consensus on the formation of early hydrates, which can form links between cement particles. According to Sujata and Jennings [60], alite (C_3S) particles get covered by C-S-H after 10 min of hydration, while Gauffinet et al. [61] reported 1 h after the contact between alite and lime solution. In addition, Zingg et al. [62] found that, in the presence of a superplasticizer (dispersed cement paste), the early hydrates precipitate mainly in the pore solution.

In calcite suspensions, the dependence of the critical strain on the solid volume fraction has been investigated by Liberto et al. [59]. It was found that the critical strain varies in a non-monotonic way with the solid volume fraction and displays a minimum for a volume fraction of 20% of about 10^{-2}%. The authors attributed this critical strain to the inter-floc rupture associated with a weak-link structure. The second critical strain (where $G' = G''$) of the order of few % is associated with the fragmentation of flocs.

It thus appears that the interpretation of the viscoelastic properties of fresh cement pastes is complex. Further investigations are required to deepen the comprehension of the origins of each transition in the microstructure of fresh cement paste.

3.3. Structural Build-Up

The term thixotropy reflects the fact that the rheological properties (viscosity) are time-dependent. Thixotropic materials thus become more fluid with an increasing shear time (at a constant shear rate) or more viscous when kept at rest. Therefore, the viscosity of thixotropic material gradually increases with the resting time (build-up), and when it is sheared, it must recover its initial state (break-down). Thixotropy therefore assumes that the evolution of rheological properties over time is reversible. Thus, the application of a shear makes it possible to erase the history of the structuration at rest. This reversible phenomenon is often attributed to reversible physico-chemical phenomena such as flocculation/deflocculation in colloidal suspensions [23].

The term "thixotropy" is often used for cementitious materials to describe the reversible evolution at the macroscopic scale, e.g., the maintenance of workability [35,44,63,64]. However, due to cement hydration, the initial state cannot be completely recovered. This is referred to as workability loss (or slump loss). In fact, during the low-activity period of cement hydration, also called the dormant period, chemical changes occur in the cementitious material, leading to the formation of hydrates bridges between cement particles [23]. Despite a strong shear, the history of the structuration is not fully erased. This is why the term "structural build-up" seems to be relevant to describing the time-dependence of the rheological behavior of cementitious materials. It can be noted that the addition of admixtures such as superplasticizers allows for approaching an ideal thixotropic behavior due to the retarding/dispersive effect [20].

Various testing procedures and approaches have been reported in the literature. The first procedure, called the hysteresis loop, consists of applying an increasing and decreasing shear rate. The area between the ascending and descending curve is an indicator of the thixotropy. This method can be used as a preliminary attempt to assess the thixotropy [65,66]. In fact, the hysteresis area depends not only on time but also on shear history (shear rate, test condition, step duration). The most common test consists of the measurement of the static yield stress (using the stress growth procedure) after different resting periods [20,21,67,68]. The slope of the yield stress–resting time curve (A_{thix}) is a strong indicator of the structural build-up. Another interesting method is based on time sweep measurements using small-amplitude oscillatory shear (SAOS). It consists of the determination of the evolution of both the storage and loss moduli with rest time [21,22].

The contribution of physical and chemical parts to the structural build-up is an interesting challenge. In fact, a deep understanding of the structuration of fresh cementitious materials during the dormant period is essential to controlling the placement of modern concretes such as 3D-printable concretes, low-carbon concretes, or self-compacting concretes.

Mostafa and Yahia [21] found that the physical structural changes at rest can be described by the percolation time. This latter corresponds to the time needed to form a colloidal percolated network, where the phase angle (determined by SAOS measurements) reaches its lowest and steady value. In addition, the linear increase at rest of the storage modulus within the dormant period can describe the chemical rigidification rate of a formed network. The authors suggested that these parameters can be used in combination to describe the physical and chemical structuration of cement suspensions with different water-to-cement ratios, superplasticizer dosages, and SCMs replacements.

Another interesting study, performed by Zhang et al. [22] through SAOS measurements, allowed the chemical and thixotropic part to be quantified. The authors suggested that the structural build-up is the sum of the chemical and thixotropic parts. Moreover, it appears that the thixotropic part dominates the chemical contribution until the beginning of the acceleration period of hydration, where the chemical contribution dramatically increases. The addition of a superplasticizer (SP) allows for reducing the thixotropic part, while the chemical contribution is comparable with and without SP.

Another study based on a succession of stress growth procedures (with or without the pre-shear phase) has been carried out to assess the contribution of reversible and irreversible parts of the structural build-up. It was found that the structural build-up of a chemically inert material (calcite) is completely reversible. The structural build-up in such colloidal suspension is due to flocculation induced by attractive colloidal forces. Furthermore, for cement paste without a superplasticizer, there is a significant increase in the static yield stress despite the strong pre-shear which is supposed to erase the structural build-up (irreversible part) (Figure 7). The evolution of the static yield stress with hydration time showed two stages. The first one is characterized by a linear increase in the static yield stress with time until 40 min of hydration, and the second one is characterized by a significant increase from 40 min (slope change) (Figure 7). The irreversible part (after the pre-shear phase) showed linear evolution without a slope change. This suggests that the physical part (flocculation) is not necessarily reversible in cement suspensions and/or that the chemical changes probably contain a reversible part. The addition of a superplasticizer strongly affects the structural build-up due to the retarding and dispersive effects. In the presence of a superplasticizer, the contribution of the cement hydration to the structural build-up could be neglected. The cement paste thus behaves like a thixotropic material with reversible changes. The use of a superplasticizer allows for reducing the workability loss during the dormant period [20].

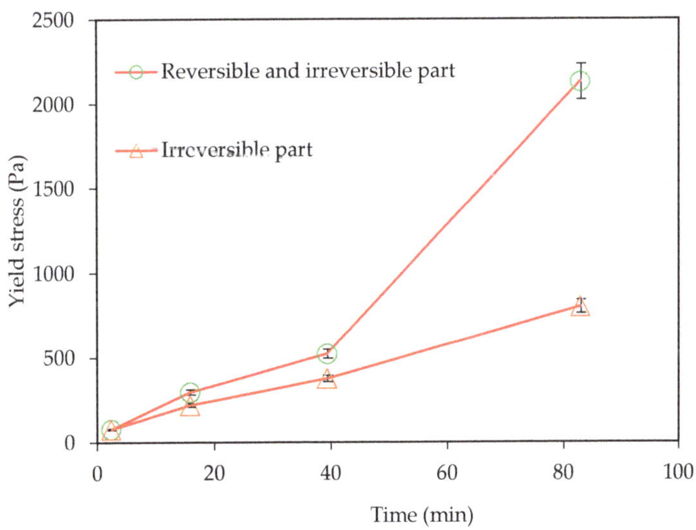

Figure 7. Evolution of the static yield stress with and without a pre-shear phase [20].

Recently, an interesting study was performed by Zhang et al. [68] to clarify and quantify the contribution of colloidal forces and C-S-H formation to the structural build-up. It was shown that the evolution of the static yield stress during the dormant period follows three stages (Figure 8). The initial stage, in the first 30 min, corresponds to the rapid linear growth of the static yield stress. In the induction period (30–60 min), the static yield stress increases slowly. Finally, the acceleration stage (>60 min) is marked by an exponential growth of the static yield stress. This evolution is almost comparable to that presented in Figure 7.

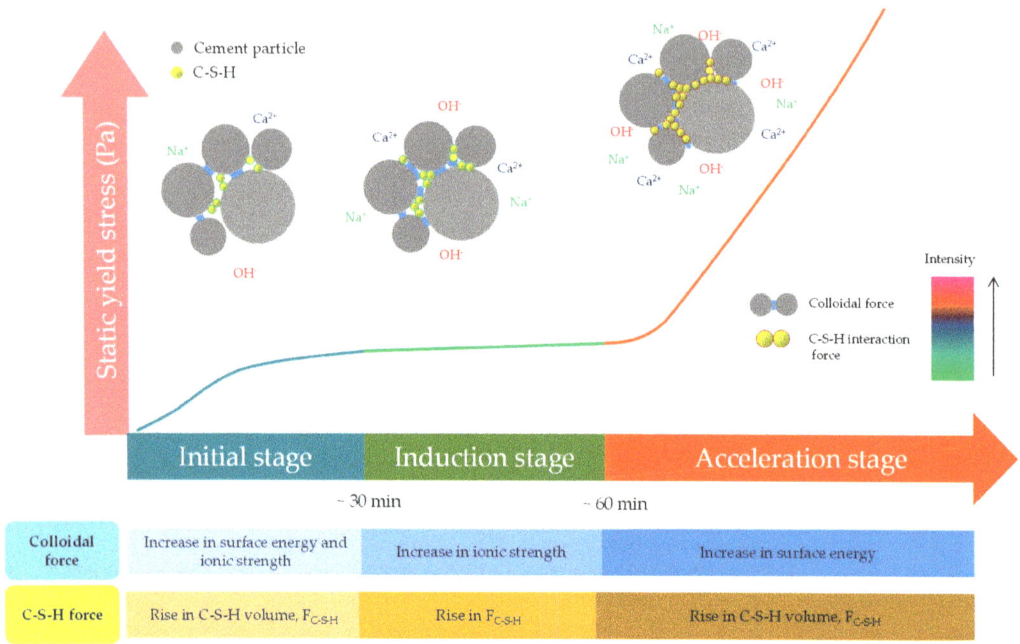

Figure 8. Schematic illustration of the static yield stress and interaction force evolution of fresh cement paste [68].

Zhang et al. [68] showed that the interparticle force, which is the combination of the colloidal force and interaction force between C-S-H particles, is the driving force for the evolution of static yield stress. The contribution of colloidal force appears to be greater in the early period (the initial and induction period), while the C-S-H force determines the evolution trend of the static yield stress since the induction period.

4. Conclusions

This paper presented a phenomenological description of the rheological behavior of cement paste. In fact, this rheological behavior displays several complexities such as non-linearity, yielding, time-dependence and chemical changes.

The most common testing procedures allowing for the characterization of this behavior have been reported. For flow behavior, the most relevant protocol consists of the application of a ramp sweep (shear rate or shear stress) so that the flow curve (shear stress as a function of the shear rate) is at steady state flow. The structural build-up can be characterized by static yield stress measurements or the SAOS protocol for various resting times.

The non-linearity of flow behavior corresponds to a shear-thinning/shear-thickening phenomenon that can be represented by the Herschel–Bulkley model. If the shear-thinning phenomenon can be related to the flocculation state, the origins of shear thickening are not clearly elucidated. Theories based on order–disorder transition, particle clustering, or viscous/inertial regime transition have been developed to explain the appearance of shear thickening.

Furthermore, oscillation rheology allows for determining the linear viscoelastic domain (LVED) in which both storage and loss moduli remain constant. The end of this linear domain is associated with a critical strain of the order of 10^{-2}%. This critical strain could be the signature of the breakage of interparticle bonds formed by early hydrates (C-S-H) and/or attractive colloidal forces, especially for cements with supplementary cementitious materials (SCMs) and alternative cements such as calcium sulfoaluminates cements or belite ye'elimite ferrite (BYF) cements.

Funding: This research received no external funding.

Conflicts of Interest: The author declares no conflict of interest.

References

1. Nicolas, R.; Richard, B.; Nicolas, D.; Irina, I.; Temitope, K.J.; Dirk, L.; Viktor, M.; Romain, M.; Arnaud, P.; Ursula, P.; et al. Assessing the fresh properties of printable cement-based materials: High potential tests for quality control. *Cem. Concr. Res.* **2022**, *158*, 106836. [CrossRef]
2. Biricik, Ö.; Mardani, A. Parameters affecting thixotropic behavior of self compacting concrete and 3D printable concrete; a state-of-the-art review. *Constr. Build. Mater.* **2022**, *339*, 127688. [CrossRef]
3. Roussel, N. Rheological requirements for printable concretes. *Cem. Concr. Res.* **2018**, *112*, 76–85. [CrossRef]
4. Li, R.; Lei, L.; Plank, J. Influence of PCE superplasticizers on the fresh properties of low carbon cements containing calcined clays: A comparative study of calcined clays from three different sources. *Cem. Concr. Compos.* **2023**, *139*, 105072. [CrossRef]
5. Ayati, B.; Newport, D.; Wong, H.; Cheeseman, C. Low-carbon cements: Potential for low-grade calcined clays to form supplementary cementitious materials. *Clean. Mater.* **2022**, *5*, 100099. [CrossRef]
6. Gaimster, R.; Dixon, N. Self-compacting concrete. In *Advanced Concrete Technology*; Japan Concrete Institute: Tokyo, Japan, 2003; pp. 1–23.
7. Brouwers, H.J.H.; Radix, H.J. Self-compacting concrete: Theoretical and experimental study. *Cem. Concr. Res.* **2005**, *35*, 2116–2136. [CrossRef]
8. Wu, H.; Hu, R.; Yang, D.; Ma, Z. Micro-macro characterizations of mortar containing construction waste fines as replacement of cement and sand: A comparative study. *Constr. Build. Mater.* **2023**, *383*, 131328. [CrossRef]
9. Duan, Z.; Deng, Q.; Liang, C.; Ma, Z.; Wu, H. Upcycling of recycled plastic fiber for sustainable cementitious composites: A critical review and new perspective. *Cem. Concr. Compos.* **2023**, *142*, 105192. [CrossRef]
10. Roussel, N.; Bessaies-Bey, H.; Kawashima, S.; Marchon, D.; Vasilic, K.; Wolfs, R. Recent advances on yield stress and elasticity of fresh cement-based materials. *Cem. Concr. Res.* **2019**, *124*, 105798. [CrossRef]
11. Roussel, N.; Lemaître, A.; Flatt, R.J.; Coussot, P. Steady state flow of cement suspensions: A micromechanical state of the art. *Cem. Concr. Res.* **2010**, *40*, 77–84. [CrossRef]
12. Hsu, K.C.; Chiu, J.J.; Da Chen, S.; Tseng, Y.C. Effect of addition time of a superplasticizer on cement adsorption and on concrete workability. *Cem. Concr. Compos.* **1999**, *21*, 425–430. [CrossRef]
13. Chiocchio, G.; Paolini, A.E. Optimum time for adding superplasticizer to Portland cement pastes. *Cem. Concr. Res.* **1985**, *15*, 901–908. [CrossRef]
14. Houst, Y.F.; Flatt, R.J.; Bowen, P.; Hofmann, H.; Mäder, U.; Widmer, J.; Sulser, U.; Bürge, T.A. Influence of Superplasticizer Adsorption on the Rheology of Cement Paste. In Proceedings of the International RILEM Conference on "The Role of Admixtures in High Performance Concrete", Monterrey, Mexico, 21–26 March 1999; pp. 387–402.
15. Cyr, M.; Legrand, C.; Mouret, M. Study of the shear thickening effect of superplasticizers on the rheological behaviour of cement pastes containing or not mineral additives. *Cem. Concr. Res.* **2000**, *30*, 1477–1483. [CrossRef]
16. Khayat, K.H.; Yahia, A.; Sayed, M. Effect of supplementary cementitious materials on rheological properties, bleeding, and strength of structural grout. *ACI Mater. J.* **2008**, *105*, 585–593.
17. Jiao, D.; De Schutter, G. Insights into the viscoelastic properties of cement paste based on SAOS technique. *Constr. Build. Mater.* **2022**, *357*, 129320. [CrossRef]
18. Legrand, C. Contribution à l'étude de la rhéologie du béton frais. *Matér. Constr.* **1972**, *5*, 379–393. [CrossRef]
19. Yuan, Q.; Zhou, D.; Khayat, K.H.; Feys, D.; Shi, C. On the measurement of evolution of structural build-up of cement paste with time by static yield stress test vs. small amplitude oscillatory shear test. *Cem. Concr. Res.* **2017**, *99*, 183–189. [CrossRef]
20. El Bitouri, Y.; Azéma, N. On the "Thixotropic" Behavior of Fresh Cement Pastes. *Eng* **2022**, *3*, 677–692. [CrossRef]
21. Mostafa, A.M.; Yahia, A. New approach to assess build-up of cement-based suspensions. *Cem. Concr. Res.* **2016**, *85*, 174–182. [CrossRef]
22. Zhang, K.; Mezhov, A.; Schmidt, W. Chemical and thixotropic contribution to the structural build-up of cementitious materials. *Constr. Build. Mater.* **2022**, *345*, 128307. [CrossRef]

23. Roussel, N.; Ovarlez, G.; Garrault, S.; Brumaud, C. The origins of thixotropy of fresh cement pastes. *Cem. Concr. Res.* **2012**, *42*, 148–157. [CrossRef]
24. Ewoldt, R.H.; Johnston, M.T.; Caretta, L.M. *Experimental Challenges of Shear Rheology: How to Avoid Bad Data BT—Complex Fluids in Biological Systems: Experiment, Theory, and Computation*; Spagnolie, S.E., Ed.; Springer: New York, NY, USA, 2015; pp. 207–241. ISBN 978-1-4939-2065-5.
25. Ukrainczyk, N.; Thiedeitz, M.; Kränkel, T.; Koenders, E.; Gehlen, C. Modeling SAOS Yield Stress of Cement Suspensions: Microstructure-Based Computational Approach. *Materials* **2020**, *13*, 2769. [CrossRef] [PubMed]
26. Conte, T.; Chaouche, M. Rheological behavior of cement pastes under Large Amplitude Oscillatory Shear. *Cem. Concr. Res.* **2016**, *89*, 332–344. [CrossRef]
27. Schultz, M.A.; Struble, L.J. Use of oscillatory shear to study flow behavior of fresh cement paste. *Cem. Concr. Res.* **1993**, *23*, 273–282. [CrossRef]
28. Kawashima, S.; Chaouche, M.; Corr, D.J.; Shah, S.P. Rate of thixotropic rebuilding of cement pastes modified with highly purified attapulgite clays. *Cem. Concr. Res.* **2013**, *53*, 112–118. [CrossRef]
29. Nachbaur, L.; Mutin, J.C.; Nonat, A.; Choplin, L. Dynamic mode rheology of cement and tricalcium silicate pastes from mixing to setting. *Cem. Concr. Res.* **2001**, *31*, 183–192. [CrossRef]
30. Kolawole, J.T.; Combrinck, R.; Boshoff, W.P. Rheo-viscoelastic behaviour of fresh cement-based materials: Cement paste, mortar and concrete. *Constr. Build. Mater.* **2020**, *248*, 118667. [CrossRef]
31. Thiedeitz, M.; Kränkel, T.; Gehlen, C. Viscoelastoplastic classification of cementitious suspensions: Transient and non-linear flow analysis in rotational and oscillatory shear flows. *Rheol. Acta* **2022**, *61*, 549–570. [CrossRef]
32. Lei, W.-G.; Struble, L.J. Microstructure and Flow Behavior of Fresh Cement Paste. *J. Am. Ceram. Soc.* **1997**, *80*, 2021–2028. [CrossRef]
33. Mahaut, F.; Mokéddem, S.; Chateau, X.; Roussel, N.; Ovarlez, G. Effect of coarse particle volume fraction on the yield stress and thixotropy of cementitious materials. *Cem. Concr. Res.* **2008**, *38*, 1276–1285. [CrossRef]
34. Perrot, A.; Lecompte, T.; Khelifi, H.; Brumaud, C.; Hot, J.; Roussel, N. Yield stress and bleeding of fresh cement pastes. *Cem. Concr. Res.* **2012**, *42*, 937–944. [CrossRef]
35. Roussel, N. Steady and transient flow behaviour of fresh cement pastes. *Cem. Concr. Res.* **2005**, *35*, 1656–1664. [CrossRef]
36. El Bitouri, Y.; Azéma, N. Contribution of turbidimetry on the characterization of cement pastes bleeding. *Adv. Cem. Res.* **2023**, *35*, 180–190. [CrossRef]
37. Aït-Kadi, A.; Marchal, P.; Choplin, L.; Chrissemant, A.S.; Bousmina, M. Quantitative analysis of mixer-type rheometers using the couette analogy. *Can. J. Chem. Eng.* **2002**, *80*, 1166–1174. [CrossRef]
38. Fourmentin, M.; Ovarlez, G.; Faure, P.; Peter, U.; Lesueur, D.; Daviller, D.; Coussot, P. Rheology of lime paste—A comparison with cement paste. *Rheol. Acta* **2015**, *54*, 647–656. [CrossRef]
39. Papo, A.; Piani, L. Flow Behavior of Fresh Portland Cement Pastes. *Part. Sci. Technol.* **2004**, *22*, 201–212. [CrossRef]
40. Ivanov, Y.P.; Roshavelov, T.T. Flow behaviour of modified cement pastes. *Cem. Concr. Res.* **1993**, *23*, 803–810. [CrossRef]
41. Colombo, A.; Geiker, M.R.; Justnes, H.; Lauten, R.A.; De Weerdt, K. On the effect of calcium lignosulfonate on the rheology and setting time of cement paste. *Cem. Concr. Res.* **2017**, *100*, 435–444. [CrossRef]
42. Plank, J.; Sakai, E.; Miao, C.W.; Yu, C.; Hong, J.X. Chemical admixtures—Chemistry, applications and their impact on concrete microstructure and durability. *Cem. Concr. Res.* **2015**, *78*, 81–99. [CrossRef]
43. Barneoud-Chapelier, A.; Le Saout, G.; Azéma, N.; El Bitouri, Y. Effect of polycarboxylate superplasticizer on hydration and properties of belite ye'elimite ferrite cement paste. *Constr. Build. Mater.* **2022**, *322*, 126483. [CrossRef]
44. Otsubo, Y.; Miyai, S.; Umeya, K. Time-dependent flow of cement pastes. *Cem. Concr. Res.* **1980**, *10*, 631–638. [CrossRef]
45. Emoto, T.; Bier, T.A. Rheological behavior as influenced by plasticizers and hydration kinetics. *Cem. Concr. Res.* **2007**, *37*, 647–654. [CrossRef]
46. Krieger, I.M.; Dougherty, T.J. A Mechanism for Non-Newtonian Flow in Suspensions of Rigid Spheres. *Trans. Soc. Rheol.* **1959**, *3*, 137–152. [CrossRef]
47. Mahaut, F.; Chateau, X.; Coussot, P.; Ovarlez, G. Yield stress and elastic modulus of suspensions of noncolloidal particles in yield stress fluids. *J. Rheol.* **2008**, *52*, 287–313. [CrossRef]
48. Flatt, R.J.; Bowen, P. Yodel: A yield stress model for suspensions. *J. Am. Ceram. Soc.* **2006**, *89*, 1244–1256. [CrossRef]
49. Odler, I.; Duckstein, U.; Becker, T. On the combined effect of water solubles lignosulfonates and carbonates on portland cement and clinker pastes 1. Physical properties. *Cem. Concr. Res.* **1978**, *8*, 469–479. [CrossRef]
50. Roy, D.M.; Asaga, K. Rheological properties of cement mixes: III. The effects of mixing procedures on viscometric properties of mixes containing superplasticizers. *Cem. Concr. Res.* **1979**, *9*, 731–739. [CrossRef]
51. Brady, J.F.; Bossis, G. The rheology of concentrated suspensions of spheres in simple shear flow by numerical simulation. *J. Fluid Mech.* **1985**, *155*, 105–129. [CrossRef]
52. Hoffman, R.L. Explanations for the cause of shear thickening in concentrated colloidal suspensions. *J. Rheol.* **1998**, *42*, 111–123. [CrossRef]
53. Ovarlez, G.; Bertrand, F.; Rodts, S. Local determination of the constitutive law of a dense suspension of noncolloidal particles through magnetic resonance imaging. *J. Rheol.* **2006**, *50*, 259–292. [CrossRef]

54. Barnes, H.A. Shear-Thickening ("Dilatancy") in Suspensions of Nonaggregating Solid Particles Dispersed in Newtonian Liquids. *J. Rheol.* **1989**, *33*, 329–366. [CrossRef]
55. Jiao, D.; Lesage, K.; Yardimci, M.Y.; El Cheikh, K.; Shi, C.; De Schutter, G. Rheological behavior of cement paste with nano-Fe_3O_4 under magnetic field: Magneto-rheological responses and conceptual calculations. *Cem. Concr. Compos.* **2021**, *120*, 104035. [CrossRef]
56. Jiao, D.; Shi, C.; De Schutter, G. Magneto-responsive structural build-up of highly flowable cementitious paste in the presence of PCE superplasticizer. *Constr. Build. Mater.* **2022**, *327*, 126925. [CrossRef]
57. Mikanovic, N.; Jolicoeur, C. Influence of superplasticizers on the rheology and stability of limestone and cement pastes. *Cem. Concr. Res.* **2008**, *38*, 907–919. [CrossRef]
58. Tramaux, A.; Azéma, N.; El Bitouri, Y.; David, G.; Negrell, C.; Poulesquen, A.; Haas, J.; Remond, S. Synthesis of phosphonated comb-like copolymers and evaluation of their dispersion efficiency on $CaCO_3$ suspensions part II: Effect of macromolecular structure and ionic strength. *Powder Technol.* **2018**, *334*, 163–172. [CrossRef]
59. Liberto, T.; Le Merrer, M.; Barentin, C.; Bellotto, M.; Colombani, J. Elasticity and yielding of calcite paste: Scaling laws in a dense colloidal suspension. *Soft Matter* **2017**, *13*, C6SM02607A. [CrossRef] [PubMed]
60. Sujata, K.; Jennings, H.M. Formation of a Protective Layer During the Hydration of Cement. *J. Am. Ceram. Soc.* **1992**, *75*, 1669–1673. [CrossRef]
61. Gauffinet, S.; Finot, É.; Lesniewska, E.; Nonat, A. Direct observation of the growth of calcium silicate hydrate on alite and silica surfaces by atomic force microscopy; [Observation directe de la croissance d'hydrosilicate de calcium sur des surfaces d'alite et de silice par microscopie a force atomique]. *C. R. l'Acad. Sci. Ser. IIa Sci. Terre Planet.* **1998**, *327*, 231–236.
62. Zingg, A.; Holzer, L.; Kaech, A.; Winnefeld, F.; Pakusch, J.; Becker, S.; Gauckler, L. The microstructure of dispersed and non-dispersed fresh cement pastes—New insight by cryo-microscopy. *Cem. Concr. Res.* **2008**, *38*, 522–529. [CrossRef]
63. Barnes, H.A. Thixotropy—A review. *J. Nonnewton. Fluid Mech.* **1997**, *70*, 1–33. [CrossRef]
64. Lapasin, R.; Longo, V.; Rajgelj, S. Thixotropic behaviour of cement pastes. *Cem. Concr. Res.* **1979**, *9*, 309–318. [CrossRef]
65. Roussel, N. A thixotropy model for fresh fluid concretes: Theory, validation and applications. *Cem. Concr. Res.* **2006**, *36*, 1797–1806. [CrossRef]
66. Petkova, V.; Samichkov, V. Some influences on the thixotropy of composite slag Portland cement suspensions with secondary industrial waste. *Constr. Build. Mater.* **2007**, *21*, 1520–1527. [CrossRef]
67. Billberg, P. Form Pressure Generated by Self-Compacting Concrete: Influence of Thixotropy and Structural Behaviour at Rest. Ph.D. Thesis, KTH Royal Institute of Technology, Stockholm, Sweden, 2006.
68. Zhang, Z.; Jia, Z.; Shi, J.; Jiang, Y.; Banthia, N.; Zhang, Y. Clarifying and quantifying the driving force for the evolution of static yield stress of cement pastes. *Cem. Concr. Res.* **2023**, *167*, 107129. [CrossRef]

eng

MDPI

Review

A Review on Vibration Monitoring Techniques for Predictive Maintenance of Rotating Machinery

Marcelo Romanssini *, Paulo César C. de Aguirre , Lucas Compassi-Severo and Alessandro G. Girardi *

Computer Architecture and Microelectronics Group, Federal University of Pampa, Alegrete 97546-550, Brazil; paulo.aguirre@unipampa.edu.br (P.C.C.d.A.); lucas.severo@unipampa.edu.br (L.C.-S.)
* Correspondence: marceloromanssini.aluno@unipampa.edu.br (M.R.); alessandro.girardi@unipampa.edu.br (A.G.G.)

Abstract: Machine failure in modern industry leads to lost production and reduced competitiveness. Maintenance costs represent between 15% and 60% of the manufacturing cost of the final product, and in heavy industry, these costs can be as high as 50% of the total production cost. Predictive maintenance is an efficient technique to avoid unexpected maintenance stops during production in industry. Vibration measurement is the main non-invasive method for locating and predicting faults in rotating machine components. This paper reviews the techniques and tools used to collect and analyze vibration data, as well as the methods used to interpret and diagnose faults in rotating machinery. The main steps of this technique are discussed, including data acquisition, data transmission, signal processing, and fault detection. Predictive maintenance through vibration analysis is a key strategy for cost reduction and a mandatory application in modern industry.

Keywords: predictive maintenance; rotating machine; vibration

Citation: Romanssini, M.; de Aguirre, P.C.C.; Compassi-Severo, L.; Girardi, A.G. A Review on Vibration Monitoring Techniques for Predictive Maintenance of Rotating Machinery. *Eng* **2023**, *4*, 1797–1817. https://doi.org/10.3390/eng4030102

Academic Editor: Antonio Gil Bravo

Received: 15 May 2023
Revised: 21 June 2023
Accepted: 23 June 2023
Published: 26 June 2023

1. Introduction

Rotating machinery is used in a variety of industries. They include equipment such as motors, pumps, fans, generators, compressors, and more. Motors and generators are essential to the operation of an industrial facility to maintain productivity, efficiency, and safety of operations. Research on the reliability of electrical machines has identified that failures can occur in all engine components [1]. Machine failures often occur due to continuous operation and various cyclic loading situations. This process leads to gradual wear of components, which increases the risk of failure [2]. This wear of machine components can be considered normal and is a consequence of machine operation. What is not normal, however, is the operation of these components under critical conditions, compromising the integrity of components in good condition and exposing the machine to total failure.

Machine failure results in production losses and increased maintenance costs. According to the literature, maintenance costs account for between 15% and 60% of the manufacturing cost of the final product, and in heavy industry, these costs can be as high as 50% of the total production cost [3]. These costs can be avoided by choosing an efficient maintenance strategy, which allows for detecting and correcting the problem in time. The main objective of the maintenance techniques strategy is to increase the availability of machines with lower maintenance costs [4].

Maintenance techniques can be basically divided into three types, breakdown maintenance, preventive maintenance, and predictive maintenance (PdM) [4,5]. Among the techniques used for equipment maintenance, PdM has proven to be the most efficient in the industrial environment. PdM is based on the analysis of data collected through monitoring or inspections [6]. The data are collected from machines to determine the health status and define the maintenance strategy. Various techniques are available for monitoring machine health, such as acoustic emission, vibration monitoring, temperature monitoring, noise monitoring, current monitoring, oil and debris monitoring, and corrosion monitoring. Each technique has its proper characteristic of application and use [2–4].

Faults can be detected by a variety of diagnostic methods. Among the various techniques used in predictive maintenance, vibration analysis has emerged as a valuable tool. By analyzing the vibration patterns of machines, it becomes possible to detect abnormalities and early signs of faults. Vibration monitoring has proven to be an effective method for locating faults in machine components [3,4,7]. Vibrations are oscillatory movements of equipment around its equilibrium position. Any change in signal amplitude or frequency indicates that machine performance is impaired [8].

Vibration analysis can be an effective tool for diagnosing faults of looseness, eccentricity, imbalance, blade defects, misalignment, defective bearings, damaged gears, and cracked or bent shafts [9,10]. As a result, this technique has emerged as a powerful and well-established PdM technique for rotating machines [11]. Compared to other PdM techniques, vibration analysis offers several advantages, such as high accuracy, sensitivity to a wide range of defect types, and it is a noninvasive and nondestructive method [12,13]. However, this method also has some disadvantages, such as the difficulty of fault detection in machines with low rotations [14], the need for continuous monitoring, and the need for reliable sensors to collect machine data.

Figure 1 shows an example of a system installed in an electrical machine located in an industry for continuously monitoring vibration. Figure 2 shows the detail of an IoT vibration monitoring system that is able to measure four points simultaneously in the same machine.

Figure 1. Example of vibration monitoring system installed in an electrical rotating machine for predictive maintenance.

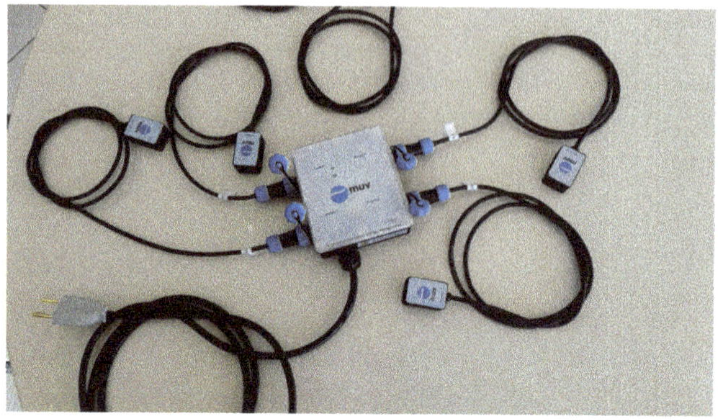

Figure 2. A vibration monitoring system composed of 4 sensors for simultaneously measuring 4 points in the same rotating machine [15].

The process of fault diagnosis in machine monitoring by vibration analysis mainly consists of four steps: data acquisition, data transmission, signal processing, and fault detection. The main steps of fault diagnosis in rotating machinery by vibration analysis are shown in Figure 3. Data acquisition can be performed using many vibration measurement devices available on the market. These devices can use different types of transducers to perform a measurement. Among the types of sensors used to acquire the vibration signal, the accelerometer is the most commonly used [16]. Signal processing consists of manipulating, filtering, digitizing, and analyzing raw data to extract meaningful information. It is a crucial aspect of vibration analysis because it allows the extraction of patterns and insights from a large amount of vibration data that would otherwise be difficult to interpret [17,18]. Fault detection is the final step of the vibration analysis process. In this stage the vibration signal is recorded in the time or frequency domain. Then, this signal is interpreted by an expert to determine the type of fault and its location [19].

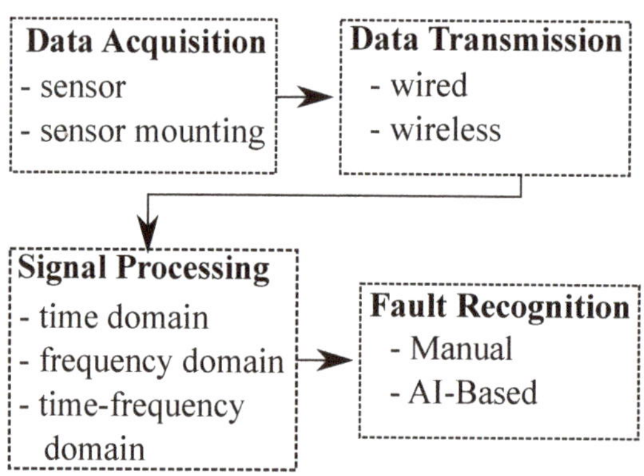

Figure 3. Main steps for fault diagnosis in rotating machinery through vibration analysis.

Evaluating life prediction through vibration analysis is challenging in terms of capturing the hidden nonlinear fault dynamics and adequately representing them with engineering characteristics. Vibration signals in rotating machinery are non-stationary, which complicates their analysis due to changing time–frequency characteristics. Bearing faults present a particular challenge because traditional methods assume only rolling behavior, while a combination of rolling and sliding causes dynamic frequency shifts. Non-stationary early vibration signals dominated by external vibrations and the presence of multiple simultaneous faults further complicate accurate fault diagnosis. Disturbances from additional vibration sources, such as bearing looseness, increase the complexity of the analysis. Overcoming these challenges requires advanced signal processing, feature extraction, and fault diagnosis algorithms capable of handling nonlinear dynamics and extracting relevant information from complex vibration signals [20].

To improve fault analysis, different types of signals can be acquired simultaneously, such as vibration signals, acoustic emissions, temperature, etc. More system information leads to a more accurate estimate of the machine's condition. Compared to the scenario where data from a single sensor are used, better predictive performance can be achieved by fusing data from multiple sensors [21].

Artificial intelligence (AI) and machine learning (ML) have been used for detecting faults. This method does not require an expert to perform the analysis and has been the subject of much research in recent years [19,22]. One issue facing the adoption of machine learning algorithms is the need for large data sets, which generally require access to machine data from multiple companies and factories. Despite the potential benefits of data sharing, this solution is not usually preferred due to the importance of data privacy in real-world industries. To address this problem, the work of [23] proposes a federated transfer learning method for machine fault diagnosis, where customer-invariant features can be extracted for diagnosis while maintaining data privacy.

Given the importance of vibration analysis for the predictive maintenance of rotating machines in order to reduce maintenance costs, as well as reduce machine downtime, this work provides an overview of some techniques and tools used to collect and analyze vibration data, as well as methods of interpreting and diagnosing faults in rotating machinery using this data. The rest of this article is organized as follows: Section 2 describes the types of sensors and the techniques for mounting the sensor on the machine that are used to collect vibration data; Section 3 discusses the main ways to transmit the acquired data, namely conventional cable transmission and wireless transmission; Section 4 presents the main techniques for processing vibration signals and methods for identifying faults in rotating machinery; and Section 5 provides concluding remarks.

2. Data Acquisition

To measure machinery vibration, a transducer or a vibration pickup is used. A transducer is a device that converts changes in mechanical quantities into changes in other physical quantities, usually an electrical signal proportional to a parameter of the experienced motion. There are three commonly used transducers for vibration measurement: displacement sensors, velocity sensors, and accelerometers [24]. Each sensor has some advantages and disadvantages, depending on the application. The type of sensor used is basically determined by the frequency range, sensitivity, and operating limits.

New approaches have been proposed, such as the use of vision data from the event-based camera [25]. However, accelerometers are most commonly used because of their greater accuracy, measurement range, ease of mounting, and cost. Moreover, it is relatively simple to numerically integrate the acceleration signal and obtain the velocity and displacement [26,27]. The next subsections discuss the main characteristics of these three types of sensors.

2.1. Displacement Transducers

Displacement transducers use capacitive, optical, or ultrasonic principles to measure vibration displacement. They are suitable for measuring vibration frequencies below 10 Hz [28]. There are several types of displacement transducers, some of which are based on variable resistance and others on induced currents. The most used for predictive maintenance in rotating machines are those based on induced currents [26]. These transducers, also called eddy current sensors or gap current sensors, are installed a short distance from the surface whose vibrations are to be measured. The eddy current sensor uses a high-frequency current in a coil inside the sensor head to generate a high-frequency magnetic field. When this magnetic field achieves a conductor in the measuring object, an eddy current is generated on the surface of the measuring object, and the impedance of the sensor coil changes. This change in impedance is proportional to the gap between the transducer and the vibrating surface [29]. The main advantage of this type of sensor over others is its application in low-frequency measurements and its great temperature stability [26]. Furthermore, it requires a simple processing circuit. This sensor can easily identify problems such as imbalance and misalignment in electrical motors or generators. On the other hand, the disadvantages are that this type of sensor is difficult to install, it is susceptible to shocks, and the calibration depends on the type of surface material [30].

2.2. *Velocity Transducer*

Velocity transducers are electromechanical sensors designed to directly measure vibratory movement. The velocity sensor is basically composed of three parts: a permanent magnet, a coil of wire, and spring supports. The schematic of the velocity sensor is shown in Figure 4. This type of sensor is based on the principle of electromagnetic induction. The movement of a coil within the magnetic field results in the generation of an induced voltage across the end wires of the coil. This voltage is produced by the transfer of energy from the magnetic field of the magnet to the wire coil. When the coil is subjected to vibration, it experiences relative movement with respect to the magnet, which leads to the induction of a voltage signal. This voltage signal is directly proportional to the speed of vibration applied to the sensor. An advantage is that this sensor does not need any external power supply for its operation. The sensitivity of the velocity is constant over a specified frequency range, usually between 10 Hz and 1000 Hz. The sensitivity decreases at low vibration frequencies, which causes inaccurate readings at vibration frequencies below 10 Hz [24,28].

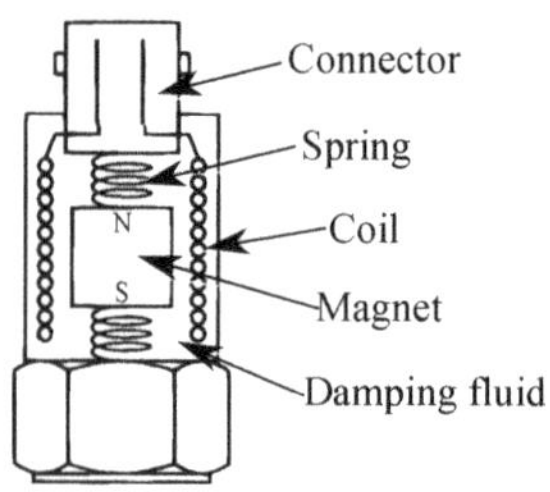

Figure 4. Velocity sensor schematic with the indication of main components.

Other advantages of velocity sensors are the ease of installation, strong signal in the mid-frequency range, and low cost when compared to piezoelectric accelerometers. The disadvantages include the relative large size, weight, variable sensitivity to input frequencies, narrow frequency response, moving parts, and sensitivity to magnetic interference [24,26].

2.3. *Accelerometers*

Accelerometers are electromechanical transducers designed for measuring linear acceleration and are the most popular transducers used for rotating machinery applications [24]. There are many types of accelerometers, however, for measuring the vibration of rotating machines, the most used are the piezoelectric and microelectromechanical system (MEMS) accelerometers. These sensors can be uniaxial—detecting acceleration in only one axis—or triaxial—when the accelerometer can identify movements in three dimensions. Compared to the uniaxial accelerometer, the triaxial accelerometer demands a larger memory capacity, resulting in a higher cost [30].

2.3.1. Piezoelectric Accelerometer

The piezoelectric accelerometer produces an electrical signal in the output proportional to the incident acceleration. The working mechanism is based on the piezoelectric effect, which converts mechanical motion to a voltage signal. When the piezoelectric crystal of the sensor is deformed by an external force (acceleration), it generates a certain potential difference between its terminals that is proportional to the force to which it is subjected [26,31].

A representation of the piezoelectric sensor and its components can be seen in Figure 5. This type of accelerometer is one of the most used transducers for measuring vibrations, as it presents the best general characteristics when compared to the other transducers. It has a wide frequency range and presents a dynamic range with good linearity. It is relatively robust and stable so its characteristics remain stable for a long period of time. Piezoelectric accelerometers have greater reliability when compared to other types of sensors, being able to operate in a frequency range of 1 Hz to 30 kHz; therefore, they are suitable for measuring high-frequency vibrations [24,27].

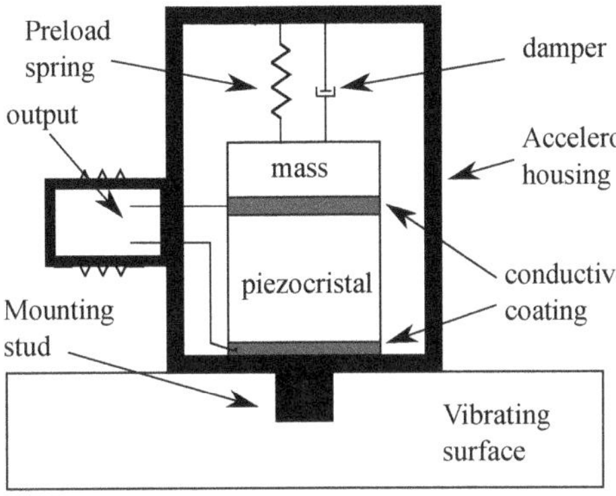

Figure 5. Schematic of a piezoelectric accelerometer.

2.3.2. MEMS Accelerometer

The rapid development of semiconductor microfabrication techniques made possible the creation of devices composed of mechanical parts with dimensions of up to a few micrometers [32]. It led to the development of micro-electro mechanical system (MEMS) accelerometers. These devices are characterized by their small size and low cost compared with the piezoelectric accelerometers [33]. As a result, MEMS accelerometers are particularly attractive for vibration monitoring in rotating structures [34].

MEMS accelerometers can be implemented based on piezoresistive or capacitive principles. Capacitive MEMS accelerometers are less sensitive to thermal excitation, which enables capacitance sensing to provide a wider operating temperature range [33]. They present three fundamental structures for their operation: the mobile test mass, the spring region, and the fixed structures or capacitive fingers. Figure 6 depicts these elements. The capacitive fingers are placed on both sides of the accelerometer. The accelerometer design allows for lateral movement of the test mass. When the sensor is at rest, the capacitance is equal on both sides of the test mass. When the device is under the effect of acceleration in a given direction, the mass moves in the opposite direction, so the capacitances formed between the fingers and the fixed structure on both sides are different. The acceleration is measured by reading the changes in the differential capacitance [35].

Most of the MEMS accelerometers available in the market are capable of measuring accelerations in three perpendicular directions simultaneously. Furthermore, MEMS accelerometers allow the easy acquisition of analog or digital signals, even with cheap microcontrollers. This can be considered the biggest advantage over traditional accelerometers, which are more accurate and reliable but require wires to transmit the collected data and still need a more robust signal conditioning circuit [34].

MEMS accelerometers have been implemented and tested for vibration measurement in a wide variety of machines, primarily because of their ease of integration into IoT systems. Rossi et al. [34] compared the use of a piezoelectric accelerometer with a MEMS accelerometer connected to a Raspberry PI microcontroller to measure vibration in a rotating machine. As a result, they found a difference of less than 5% between the data measured by the MEMS accelerometer system and the piezoelectric accelerometer. In recent years, several publications and studies that use MEMS accelerometers to measure vibration can be found. This is due to the evolution of fabrication technology that makes MEMS accelerometers more accurate, with a wide operating frequency range and at a lower cost compared to piezoelectric accelerometers.

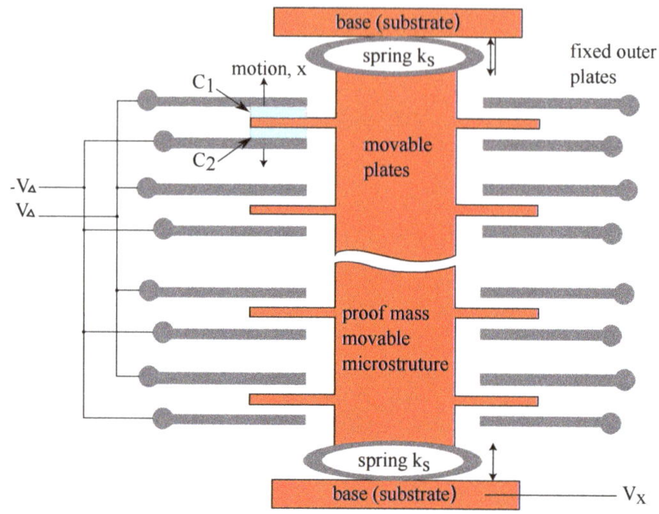

Figure 6. Diagram of an integrated MEMS accelerometer.

2.4. Sensor Mounting

Effective acquisition of vibration data is highly dependent on the proper technique of sensor mounting on the machine. In continuous or online machine condition monitoring, vibration sensors are usually mounted at a specific location on the machine [30]. The mounting methods depend largely on the sensor to be used. However, there are four main methods that can be used for both velocity and acceleration sensors: stud bolt mount, adhesive mount, magnetic mount, and unmounted [24].

In stud mounting, the sensor is screwed into a stud to attach it to the machine. This technique is extremely reliable and secure, making it ideal for permanently mounted applications. It also provides the best frequency response compared to other methods. It is important to ensure that the mounting surface is clean and free of paint to avoid irregularities that could cause incorrect readings or damage to the sensor [24].

If the machine cannot be drilled for stud mounting, adhesive mounting is a good alternative. This method involves applying epoxy, glue, or wax to the mounting surface. It is easy to apply, but the dampening effect of the adhesive reduces measurement accuracy. In addition, sensors mounted with adhesive are more difficult to remove compared to other mounting methods [24].

The magnetic mount is typically used for temporary vibration measurements with portable analyzers. It is not recommended for permanent monitoring because the sensor can be inadvertently moved and the multiple surfaces and materials of the magnet can interfere with the high-frequency vibrations [24]. This can be mitigated by using neodymium magnets, the strongest type of permanent magnets commercially available. The magnetic mount is the most flexible mounting method, as the sensor can be attached and removed countless times without damaging the device or machine. Figure 7 illustrates magnetic mounting in an electric motor.

Figure 7. Magnetic mounting of two MEMS accelerometers on an electrical motor.

Finally, the unmounted method uses a probe tip with no external mechanism. It is often used in hard-to-reach places. However, the length of the probe tip can affect measurement accuracy, with longer probes leading to greater inaccuracies [30]. It is also used in manual vibration measurements, where the probe tip is placed on the machine surface at the point of interest for a few seconds and then removed.

In addition to the four methods described, there are other techniques for mounting vibration sensors using clamps, brackets, and flexible cables. These methods provide additional flexibility for mounting sensors on rotating machinery, but may introduce harmonics into the measured signal.

Choosing the right method for mounting vibration sensors on rotating machinery is critical to obtaining accurate and reliable data. Each method has its advantages and limitations, and selection should be based on the application and the equipment to be used. Proper installation and placement of the sensors is also critical for accurate measurements.

3. Data Transmission

Permanent or long-term measurement of vibration in rotating machinery requires a reliable means of storing and transmitting measurement data. There are several ways to establish communication between measuring devices to transmit vibration data. Communication can be direct from the device/sensor to the Internet, where the data are stored for later analysis, or communication can be from sensor to sensor to the end device, which must have access to the Internet. The fourth industry revolution (Industry 4.0) is based on automation and digitalization. This includes the introduction of the Internet of Things (IoT), machine-to-machine communication, improved data transmission and communication, and condition monitoring [36–38].

With the evolution of technology and the insertion of IoT in industries, various forms of communication and data transmission are available. Among them, the most widespread are: wired, Bluetooth, Wi-Fi, and LoRa/LoraWAN [36,39–41].

Wired data transmission is a stable and secure method for connecting sensors to the monitoring system [42–44]. The main advantage of data transmission via cables is the high data transfer rates that can be achieved. Additionally, cables can transmit data over long distances and provide a high level of security. Unlike wireless communication methods, wired communication is not susceptible to interference or hacking, making it a secure and reliable choice for transmitting confidential data. However, there are some disadvantages. This method involves high costs, complicated cable installation and maintenance, and is still not scalable [36,45]. In summary, wired connections provide a stable and reliable means of data transmission that is less susceptible to interference than wireless connections. However, wired connections are less practical in terms of mobility and may require additional hardware and setup.

Bluetooth is a widely used wireless communication technology for short-distance data transmission that can be used to transmit vibration data from rotating machinery [46]. It is a simple and easy-to-use technology with low power consumption and relatively low cost [47]. However, Bluetooth has some limitations, such as limited range and the potential for interference from other wireless technologies operating on the same frequency band [47,48]. To overcome these limitations, Bluetooth Mesh technology emerged. It is a mesh networking protocol that allows large-scale networks of Bluetooth devices to be built, providing greater coverage and flexibility. Bluetooth Mesh is more reliable than traditional Bluetooth, with built-in error correction and redundancy features. However, the use of multiple devices can create security vulnerabilities [49]. Despite these advantages and disadvantages, Bluetooth and Bluetooth Mesh remain popular choices for many applications that require wireless data transmission, such as vibration monitoring of rotating machinery. Using Bluetooth Mesh in large-scale networks can provide better coverage and flexibility, while traditional Bluetooth can be a more cost-effective solution for small-scale applications with limited range.

Wi-Fi is also widely used for wireless communication in IoT applications, including vibration monitoring of rotating machinery in industry [37,50]. Wi-Fi is a widely used wireless communication technology that can transmit data over longer distances than Bluetooth [51]. Wi-Fi Mesh offers the advantage of scalability, allowing large networks of Wi-Fi devices to be built to cover larger areas and support more devices because devices can communicate with each other and create multiple paths for data transmission [52]. Wi-Fi devices can be easily connected to other Wi-Fi-enabled devices such as computers and smartphones, making it easier to access and analyze vibration data [22]. However, there are also some limitations to using Wi-Fi and Wi-Fi Mesh for vibration monitoring, such as high power consumption, possible interference, and security concerns. The use of Wi-Fi and Wi-Fi Mesh may also require additional infrastructure and installation costs depending on the size and complexity of the network [37,51].

LoRa/LoraWAN technologies are wireless communication methods used to transmit vibration data from rotating machinery [53,54]. These technologies offer long ranges, low power consumption, and high network capacity [55]. LoRa technology, developed by Semtech Corporation, is a physical layer technology, while LoRaWAN is a network protocol built on top of LoRa [56]. They are suitable for monitoring machines in remote locations, offering a large network capacity and interoperability between different devices and networks [57]. However, their relatively low data rates make them suitable for low-to-moderate data rate applications, such as vibration monitoring. Overall, LoRa/LoraWAN technologies offer reliable and cost-effective methods for wireless transmission of vibration data, especially for monitoring rotating machinery in remote and hard-to-reach locations [58].

In summary, several methods are available for transmitting the vibration data acquired from rotating machinery, ranging from wired to wireless communication technologies. A summary of the characteristics of the main communication and data transmission methods can be found in Table 1. Wired communications provide stable and reliable data transmission with high security, but can be less convenient and more expensive. Bluetooth and Wi-Fi offer wireless communication options with varying range, scalability, and potential security concerns. LoRa/LoraWAN offers long range and high network capacity, but with lower data transfer rates. The choice of communication method depends on specific application requirements, such as the distance between sensors and the monitoring system, the amount of data to be transmitted, and the level of security required. Overall, the development of IoT and digitalization has greatly expanded the options for transmitting vibration data and offers more efficient and cost-effective solutions for monitoring rotating machinery in various industries.

Table 1. Comparison between common communication methods [37,59–61].

	Wired	Bluetooth	Wi-Fi	LoRa
Frequency band	-	2.4 GHz	2.4–5 GHz	sub-GHz, 2.4 GHz
Typical range	-	10 m	100 m	3–12 km
Range on factory floor	-	≈5 m	≈25–50 m	-
Max Data rate		1 Mb/s	35 Mb/s–1 Gb/s	00.18–37.5 kbps, 31.72–253.91 kbps
Latency	Lowest	Moderate	Low	-
Throughput	High	Low	Moderate	
Scalability	Difficult	Easy	Easy	Easy
Interference susceptibility	Low	High	High	
Power consumption	-	Moderate	High	Low

4. Techniques for Signal Processing

Obtaining information through signal processing is one of the main elements for the analysis of vibration in machines. At the same time, signal processing can be considered demanding, since it aims at highlighting the features of the collected vibration signals, which are generally noisy and complex. Therefore, the data must be processed in such a way that the features of interest can be extracted.

There are several vibration signal processing methods that can be applied in monitoring rotating machinery to identify and diagnose defects or characteristic variations in the measured signal that indicate possible failures. These techniques can be divided into time domain analysis, frequency domain analysis, and time-frequency analysis. The choice of technique depends heavily on the signal to be analyzed and the characteristics of the signal to be evaluated to identify possible defects.

4.1. Time Domain Analysis

The technique of vibration analysis of rotating machinery in the time domain is the simplest analysis that can be performed. Many features such as the presence of amplitude modulation, shaft frequency components, shaft imbalance, transients, and higher frequency components can be identified visually by analyzing portions of the waveform in the function of time [62]. However, this is not sufficient to effectively detect changes in the vibration signal caused by potential faults. More sophisticated parameters and approaches should be used for time domain analysis, such as statistical parameter trends in the time domain [63]. Several statistical parameters can be defined, such as root-mean-square (RMS), peak, crest factor, and kurtosis [4]. These parameters are described hereafter.

4.1.1. Peak

The peak is the maximum value of signal $x(t)$ in the measured time interval and is defined as [62,64]:

$$Peak = \max(|x(t)|) \tag{1}$$

4.1.2. Root-Mean-Square (RMS)

Root-mean-square is related to the energy of the sampled signal, so it can contain useful information about signal construction [64,65]. This parameter is defined as:

$$RMS = \sqrt{\frac{1}{N}\sum_{i=1}^{N}(x_i)^2} \tag{2}$$

Here, N is the number of measured points and x_i is the value of the ith sensor output signal.

4.1.3. Crest Factor (CRF)

Crest factor is the ratio of peak and *RMS* value of the signal, which shows the spikiness of the vibration signal. A *CRF* near 1 represents a lower spiky signal [64,65]. Crest factor is defined as:

$$CRF = \frac{Peak}{RMS} \tag{3}$$

4.1.4. Kurtosis (KUR)

Kurtosis is the measure of the tailedness of the probability density function of a time series. This number is related to the tails of the distribution. A high kurtosis value corresponds to a greater extremity of deviations (or outliers). The kurtosis is defined as the standardized fourth moment [64,65]:

$$\mathrm{KUR} = \tilde{\mu}_4 = \frac{\mu_4}{\sigma^4}, \tag{4}$$

where μ_4 is the unstandardized central fourth moment and σ is the standard deviation.

A summary of the advantages and disadvantages of the time domain vibration analysis techniques can be seen in Table 2.

Table 2. Advantages and disadvantages of time domain methods.

Time Domain Methods	Advantages	Disadvantages
Peak	Simple technique.	Considers only the maximum value of $x(t)$ because this technique is sensitive to noise.
RMS	Easy technique, RMS values are not affected by isolated peaks in the signal.	It is not able to detect failures in the early operating stages.
Crest factor	Easy to estimate.	Reliable only in the presence of a spiky signal.
Kurtosis	High performance in detecting faults; independent of the signal amplitude.	Its effectiveness depends on the presence of significant impulsivity in the signal.

4.1.5. Application of Statistical Parameters for Vibration Analysis

The statistical parameters can be used individually or together with other parameters to analyze vibration signals to detect failures in machines. A failed machine presents an increase in the vibration peak value, and the type and severity of the failure can be evaluated based on the characteristics of the corresponding peak. The severity of failures can be evaluated by comparing features in a different derivation order. For example, a vibration signal from a machine with a bearing failure may have a peak value seven times higher than the peak value for the vibration signal collected from the same machine without a failure [66]. Peak is a simple method, but it is very susceptible to noise.

The RMS value is very useful for detecting unbalance in rotating machinery. In the time domain, the RMS value is the easiest way to identify faults in a rotating machine [67]. RMS values of a vibration signal are not affected by isolated peaks in the signal, which reduces sensitivity to incipient gear failures. This method is also not significantly affected by short bursts or low intensity vibration [30]. Therefore, the RMS method is not able to detect failures when the problem is in its early stages.

Crest factor is commonly used in rotating machinery to detect tooth breakage or failure of bearing outer rings. These faults generate pulse-like vibration signals so that the crest factor increases, which helps in detecting gear or bearing faults [67]. The introduction of a defect on any contact surface generates pulses, changing the distribution of the vibration signal and increasing the kurtosis value [68,69]. The kurtosis method does not interfere with velocity or load changes, but its effectiveness depends on the presence of significant impulsivity in the signal [70].

Figure 8 shows an example of time domain analysis applied to vibration data. The scenario is an electric motor whose vibration was measured under two different conditions: with the shaft aligned (healthy machine) and with the shaft misaligned (fault-prone machine). In the first graph (Figure 8a), the blue line represents the data collected for the machine in a healthy condition (aligned shaft), and the red line represents the data collected for the same machine but with the shaft in a misaligned state. Both lines were numerically integrated to determine the vibration velocity, as shown in Figure 8b. From this figure, the RMS and peak velocity can be calculated. The increase in RMS velocity in the misaligned shaft is clearly seen, going from 2.01 mm/s to 6.98 mm/s. The same is true for the peak velocity, which increased from 4.31 mm/s to 13.06 mm/s. These two parameters are sufficient to determine that the machine with misaligned shaft needs maintenance.

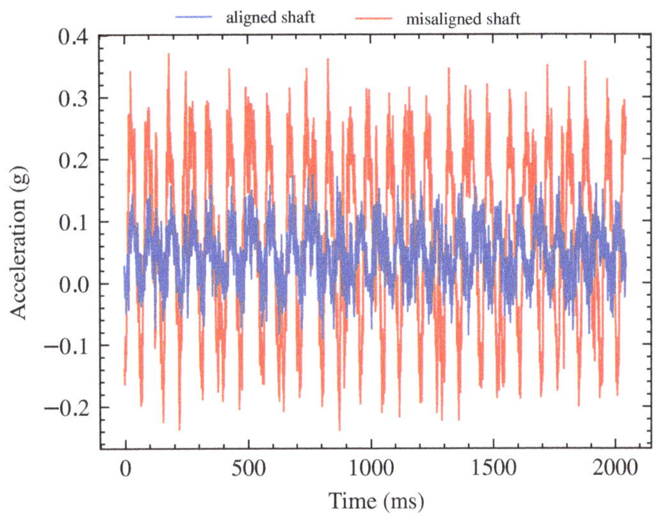

(**a**) Measured acceleration.

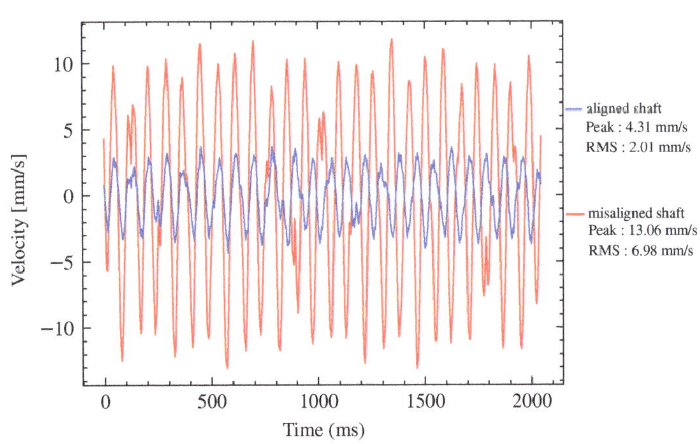

(**b**) Calculated velocity of vibration.

Figure 8. Vibration levels for a machine with aligned (blue) and misaligned (red) shaft. (**a**) Acquired acceleration data; (**b**) calculated velocity of vibration.

4.2. Frequency Domain Analysis

The characteristics of a signal in the frequency domain are often used for fault detection in rotating machinery through vibration analysis. Frequency domain analysis is a powerful tool for analyzing vibration signals in rotating machinery to diagnose faults. This method helps in identifying the frequency components present in a signal and their amplitudes [67]. Many signal features that are not visible with time domain analysis can be observed with frequency domain analysis. However, frequency analysis is not suitable for signals whose frequency varies with time [30]. The main frequency domain methods for detecting faults in rotating machinery are described below.

4.2.1. Fast Fourier Transform (FFT)

The Fast Fourier Transform is a computer algorithm that computes the discrete Fourier transform (DFT) much faster than other algorithms [28,71]. Through the FFT, it is possible to convert a signal from the time domain to the frequency domain. Using this signal represented in the frequency domain, the intensity of the different frequency components (the power spectrum) of a signal can be checked in the time domain. Vibration analysis in rotating machinery benefits from this technique because each component of the machine contributes a specific frequency component to the vibration signal. Therefore, one of the ways to detect faults is to compare the frequency components and their amplitudes to a signal from the same machine operating under perfect conditions. FFTs are used in predictive maintenance to detect various types of faults in rotating machinery, such as misalignment, imbalance, and bearing faults [72–77].

4.2.2. Cepstrum Analysis

Cepstrum analysis is the inverse Fourier transform of the logarithmic spectrum of the signal and is defined as [78]:

$$C(x(t)) = \mathcal{F}^{-1}(log(X(\omega))) \tag{5}$$

Here, $\mathcal{F}$ is the inverse of the Fourier Transform, $x(t)$ is the signal in the time domain, and $X(\omega)$ is the signal in the frequency domain. Cepstrum analysis involves analyzing the logarithm of the power spectrum to detect any periodic structure in the spectrum, such as harmonics, side bands, or echoes [28]. It is useful in detecting faults such as bearing and localized tooth faults that produce low harmonically-related frequencies. There are four types of cepstrum, with power cepstrum being the most commonly used in machine diagnostics and monitoring. Cepstrum analysis has been used in gearbox diagnosis and monitoring, detection of friction in sliding bearings, and diagnosis of faults in a universal lathe machine [79–83]. The Cepstrum analysis can be sensitive to noise present in the vibration signals. This can lead to inaccurate or distorted results, especially at lower frequencies.

4.2.3. Envelope Analysis

Envelope analysis is a technique used to separate low-frequency signals from background noise in rolling element bearings and in low-speed machine diagnostics [84]. The technique involves bandpass filtering and demodulation to extract the signal envelope, which can contain diagnostic information. Envelope analysis has the advantage of early detection of bearing problems, but determining the best frequency band for this technique is a challenge [28]. The introduction of quadratic envelope analysis solved the problem of noise components in the signal. Envelope analysis has been applied in several studies to detect faults in bearings and induction motors [85–88], but it has shown poor performance compared to other techniques, such as acoustic emission analysis [28].

4.2.4. Power Spectral Density (PSD)

Power spectral density (PSD) represents the power of a signal at different frequencies. The PSD is calculated by taking the Fourier Transform of a signal and squaring the magnitude spectrum [24]. The PSD is a powerful tool for analyzing the strength of signal fluctuations as a function of frequency. It allows the detection and measurement of oscillatory signals in time series data and indicates the frequencies at which the oscillations are strong or weak. The PSD is a graphical representation of the energy distribution of the signal over different frequencies and is commonly used for fault diagnosis in induction machines [28]. Vibration analysis using PSD offers several advantages, such as a clear frequency domain of the signal, which allows the identification of specific frequency components associated with faults or anomalies in rotating machinery. Furthermore, it enables quantitative comparisons between different signals or different operating conditions, facilitating trend analysis and condition monitoring [28,89]. However, it is important to consider some limitations of PSD analysis. Often the assumption of stationarity is made, which means that the statistical properties of the signal are assumed to be constant over time [90].

All presented vibration analysis methods in the frequency domain have advantages and disadvantages, which are summarized in Table 3.

Table 3. Summary of main advantages and disadvantages of frequency domain analysis.

Frequency Domain Analysis	Advantages	Disadvantages
Fast Fourier Transform	Easy to implement.	It is not efficient for detecting failures if the frequency and amplitude signals of the machine in normal operation are unknown.
Cepstrum Analysis	Easy technique, useful to detect harmonics, side bands, or echoes.	Sensitive to noise present in the vibration signals.
Envelope Analysis	Early detection of bearing problems.	Determining the best frequency band for this technique is a challenge.
Power Spectral Density	Clear frequency domain of the signal, which allows identification of specific frequency components associated with faults or anomalies in rotating machinery.	Specialist is required for graphical interpretation of the signal.

4.3. *Time–Frequency Domain Analysis*

In the real world, most signals are not stationary, i.e., the spectrum may change with time. In the case of vibration in machines, it can vary during operation. The vibration signal may contain different frequency components at different instants of time [28]. This variation is a problem for frequency domain analysis [30]. To overcome this challenge, time–frequency domain analysis techniques have been developed that can provide information about the time-varying frequency content of vibration signals. Time–frequency analysis allows not only the representation of the signal in three dimensions (time–frequency amplitude), but also the detection and tracking of the evolution of defects that produce weak vibration performance [78]. Conventional vibration analysis methods rely on stationary assumptions that are unsuitable for analyzing nonstationary signals. Therefore, time–frequency domain analysis methods such as the short-time Fourier transform (STFT), wavelet transform (WT), Hilbert–Huang transform (HHT), Wigner–Ville distribution (WVD), and power spectral density (PSD) are used to identify local features in the time and frequency domains [30]. These techniques are discussed in more detail below and the summary of the advantages and disadvantages can be seen in Table 4.

Table 4. Main advantages and disadvantages of time–frequency domain analysis.

Time–Frequency Domain Analysis	Advantages	Disadvantages
STFT	More efficient than conventional analysis methods in the time and frequency domain; low computational complexity.	The resolution is determined by the size of the window.
WT	Ability to detect local changes in vibration signals; improved time resolution.	Need a careful selection of the wavelet function.
WVD	High time–frequency resolution; ability to detect and locate transient events with high accuracy.	The presence of interference can make it difficult to interpret the results.
HHT	Suitable for analyzing stationary, non-stationary and transient signals; high time-frequency resolution; ability to capture transient phenomena; low computation time.	Sensitivity to noise; generation of undesirable IMFs in the low-frequency range; difficulty in separating low-frequency components.

4.3.1. Short-Time Fourier Transform (STFT)

This technique was developed to overcome the problems of FFT. It is basically an addition to the FFT's ability to analyze nonstationary or noisy signals. The STFT consists of a method that divides the nonstationary vibration signal into many small segments that can be assumed to be locally stationary, and applies the conventional FFT to these segments [78]. The STFT is defined as:

$$S_t(\omega) = \frac{1}{2\pi}\int_{-\infty}^{\infty} e^{-j\omega\tau} S(\tau)h(\tau - t)d\tau \tag{6}$$

Here, a signal $S_t(\tau)$ is obtained by multiplying the signal by a window function $h(\tau)$ centered on (τ) to produce a modified signal that emphasizes the signal around time τ. With that, the Fourier Transform reflects the frequency distribution at that time [30,78]. The main drawback of the STFT is the tradeoff between time and frequency. The resolution is determined by the size of the window. A large window gives good resolution in the frequency domain and poor resolution in the time domain and vice versa [78]. Despite this drawback, the STFT method is more efficient than conventional analysis methods in the time and frequency domains and is widely used in the analysis of vibration signals to monitor machine conditions [91–95].

4.3.2. Wavelet Transform (WT)

The Wavelet Transform is a linear transformation in which a time signal is decomposed into wavelets, i.e., local functions of time endowed with a predetermined frequency content [30]. Wavelet transforms are a powerful technique for analyzing vibration signals in rotating machinery [96,97]. By decomposing a nonstationary signal into its individual frequency components, WT can reveal time-varying features and identify transient events that may be missed by conventional Fourier transform-based methods. The wavelet scalogram provides a time–frequency representation that aids in visualization and analysis of the signal [78,98]. The advantages of using WT for vibration analysis in rotating machinery include the ability to detect local changes in vibration signals and improved time resolution. However, there are limitations to its use, including careful selection of the wavelet function and the possibility of cross terms in the wavelet scalogram [30,78]. Despite these limitations,

WT is a valuable tool that is becoming increasingly popular in industry and academia for the analysis of transient vibration signals [99].

4.3.3. Wigner–Ville Distribution (WVD)

The Wigner–Vielle distribution is based on the cross-correlation function between the signal and a time-lagged version of itself [100]. It decomposes the signal into a series of elementary waveforms, each of which has its own time and frequency characteristics. Thus, the time–frequency representation is independent of the windowing function, allowing simultaneous analysis of the signal in the time and frequency domains [78]. The advantages of WVD for vibration analysis include its high time–frequency resolution and the ability to detect and locate transient events with high accuracy [30,101]. However, WVD has some limitations, such as the presence of interference terms, which can make interpretation of the results difficult [102]. Despite its limitations, WVD is a valuable tool for analyzing nonstationary signals in rotating machinery, especially for detecting and diagnosing faults in bearings, broken rods in induction, and gears [103,104].

4.3.4. Hilbert—Huang Transform (HHT)

The Hilbert–Huang Transform is a method for analyzing stationary, non-stationary, and transient signals. It combines empirical mode decomposition (EMD) and the Hilbert transform to obtain a Hilbert spectrum that can be used for fault diagnosis in machines [30]. The HHT consists of two main steps. First, the EMD method decomposes the signal into a series of intrinsic mode functions (IMFs), which are essentially vibration components with well-defined instantaneous frequencies. Each IMF represents a specific frequency component of the signal, which allows for a more detailed analysis of the time-varying features. After obtaining the IMFs, the Hilbert transform is applied to each IMF to calculate the instantaneous frequency as a function of time. In this way, a time-varying frequency representation of the signal is obtained, which allows the detection of transient events and the analysis of frequency fluctuations [105,106].

The advantages of HHT for vibration analysis in rotating machinery include its adaptability to nonstationary and nonlinear signals, its high time-frequency resolution, its ability to capture transient phenomena, and its low computation time. It is particularly effective in identifying and analyzing fault signatures associated with bearings, gears, and other rotating components. However, the HHT has certain limitations, such as sensitivity to noise, generation of undesirable IMFs in the low-frequency range, and difficulty in separating low-frequency components [28,30].

In summary, HHT is a valuable technique for vibration analysis in rotating machinery, providing a detailed time–frequency representation of non-stationary signals and enabling the detection and diagnosis of faults and transient events. Its application in conjunction with other analysis methods can improve the understanding of vibration behavior and contribute to effective condition monitoring and maintenance strategies [107–109].

5. Conclusions

This paper provides a comprehensive review of vibration monitoring techniques for predictive maintenance of rotating machinery. We explore the main types of transducers used to acquire vibration signals, as well as the options for transmitting and analyzing data. We describe the key features and the advantages and disadvantages of each transducer. Each component, whether in acquisition, transmission, or analysis, is very important for accurate evaluation and thus for identifying potential faults in rotating machines before they become serious problems.

In summary, the field of vibration monitoring for predictive maintenance of rotating machinery is constantly evolving due to technological advances and the need for increased reliability. As highlighted in this review, the selection of the appropriate vibration monitoring technique is critical for effective machine condition assessment. Future research should focus on further refining these techniques and exploring innovative approaches, such as integrating the Internet of Things (IoT) and cloud-based platforms to enable real-time monitoring and analysis, as well as applying artificial intelligence and machine learning techniques to automatically diagnose faults.

Funding: This study was financed by the Brazilian research agency Fundação de Amparo à Pesquisa do Estado do Rio Grande do Sul (FAPERGS)—Grant 22/2551-0000841-0.

Data Availability Statement: Not applicable.

Conflicts of Interest: The authors declare no conflict of interest.

Abbreviations

The following abbreviations are used in this manuscript:

AI	Artificial Intelligence
CRF	Crest Factor
DFT	Discrete Fourier Transform
EMD	Empirical Mode Decomposition
FFT	Fast Fourier Transform
IoT	Internet of Things
HHT	Hilbert–Huang transform
IMF	Intrinsic Mode Function
KUR	Kurtosis
MEMS	Micro-Electro Mechanical System
ML	Machine Learning
PdM	Predictive Maintenance
PSD	Power Spectral Density
RMS	Root-Mean-Square
STFT	Short-Time Fourier Transform
WT	Wavelet Transform
WVD	Wigner–Ville distribution

References

1. Lee, S.B.; Stone, G.C.; Antonino-Daviu, J.; Gyftakis, K.N.; Strangas, E.G.; Maussion, P.; Platero, C.A. Condition monitoring of industrial electric machines: State of the art and future challenges. *IEEE Ind. Electron. Mag.* **2020**, *14*, 158–167. [CrossRef]
2. Madhukar, G.; Boosi, S.; Adhoni, Z.A.; Bhaskar, K.A.; Naik, A.V. A Machine Learning Based Methodology for Fault Diagnosis in Rotating Machine. In Proceedings of the 2023 IEEE International Conference on Integrated Circuits and Communication Systems (ICICACS), Raichur, India, 24–25 February 2023; pp. 1–5. [CrossRef]
3. Shukla, K.; Nefti-Meziani, S.; Davis, S. A heuristic approach on predictive maintenance techniques: Limitations and scope. *Adv. Mech. Eng.* **2022**, *14*, 16878132221101009. [CrossRef]
4. Kumar, A.; Sathujoda, P.; Bhalla, N.A. Vibration signal analysis of a rotor-bearing system through wavelet transform and empirical mode decomposition. *IOP Conf. Ser. Mater. Sci. Eng.* **2022**, *1248*, 012027. [CrossRef]
5. Efeoğlu, E.; Tuna, G. Machine Learning for Predictive Maintenance: Support Vector Machines and Different Kernel Functions. *J. Mach. Manuf. Reliab.* **2022**, *51*, 447–456. [CrossRef]
6. Villarroel, A.; Zurita, G.; Velarde, R. Development of a Low-Cost Vibration Measurement System for Industrial Applications. *Machines* **2019**, *7*, 12.
7. Umbrajkaar, A.; Krishnamoorthy, A.; Dhumale, R. Vibration analysis of shaft misalignment using machine learning approach under variable load conditions. *Shock Vib.* **2020**, *2020*, 1650270. [CrossRef]
8. Nascimento, E.G.S.; Liang, J.S.; Figueiredo, I.S.; Guarieiro, L.L.N. T4PdM: A Deep Neural Network based on the Transformer Architecture for Fault Diagnosis of Rotating Machinery. *arXiv* **2022**, arXiv:2204.03725.
9. Althubaiti, A.; Elasha, F.; Teixeira, J.A. Fault diagnosis and health management of bearings in rotating equipment based on vibration analysis—A review. *J. Vibroeng.* **2022**, *24*, 46–74. [CrossRef]
10. Tiboni, M.; Remino, C.; Bussola, R.; Amici, C. A review on vibration-based condition monitoring of rotating machinery. *Appl. Sci.* **2022**, *12*, 972.
11. Ranjan, R.; Ghosh, S.K.; Kumar, M. Fault diagnosis of journal bearing in a hydropower plant using wear debris, vibration and temperature analysis: A case study. *Proc. Inst. Mech. Eng. Part E J. Process Mech. Eng.* **2020**, *234*, 235–242. [CrossRef]
12. Li, Q.; Zhang, W.; Liu, J.; Chen, M. Fault diagnosis of wind turbine gearbox based on improved ELM algorithm and multimodal sensor fusion. *Measurement* **2021**, *180*, 109598.
13. Cheng, Y.; Lin, M.; Wu, J.; Zhu, H.; Shao, X. Intelligent fault diagnosis of rotating machinery based on continuous wavelet transform-local binary convolutional neural network. *Knowl.-Based Syst.* **2021**, *216*, 106796. [CrossRef]

14. Popescu, T.D.; Aiordachioaie, D.; Culea-Florescu, A. Basic tools for vibration analysis with applications to predictive maintenance of rotating machines: An overview. *Int. J. Adv. Manuf. Technol.* **2021**, *118*, 2883–2899. [CrossRef]
15. Muv Tecnologia. Available online: http://www.muvtecnologia.com.br (accessed on 15 May 2023).
16. Pedotti, L.A.S.; Zago, R.M.; Fruett, F. Instrument based on MEMS accelerometer for vibration and unbalance analysis in rotating machines. In Proceedings of the 2016 1st International Symposium on Instrumentation Systems, Circuits and Transducers (INSCIT), Belo Horizonte, Brazil, 29 August–3 September 2016; IEEE: Piscataway, NJ, USA, 2016; pp. 25–30.
17. Amezquita-Sanchez, J.P.; Adeli, H. Signal processing techniques for vibration-based health monitoring of smart structures. *Arch. Comput. Methods Eng.* **2016**, *23*, 1–15.
18. John Lu, Z.Q. The elements of statistical learning: Data mining, inference, and prediction. *J. R. Stat. Soc. Ser. Stat. Soc.* **2010**, *173*, 693–694. [CrossRef]
19. Zhao, Z.; Wu, J.; Li, T.; Sun, C.; Yan, R.; Chen, X. Challenges and opportunities of AI-enabled monitoring, diagnosis & prognosis: A review. *Chin. J. Mech. Eng.* **2021**, *34*, 1–29.
20. Shaheryar, A.; Yin, X.C.; Yousuf, W. robust feature extraction on vibration data under deep-learning framework: An application for fault identification in rotary machines. *Int. J. Comput. Appl.* **2017**, *167*, 37–45. [CrossRef]
21. Li, X.; Xu, Y.; Li, N.; Yang, B.; Lei, Y. Remaining Useful Life Prediction with Partial Sensor Malfunctions Using Deep Adversarial Networks. *IEEE/CAA J. Autom. Sin.* **2023**, *10*, 121–134. [CrossRef]
22. Zhang, F.; Sun, W.; Wang, H.; Xu, T. Fault diagnosis of a wind turbine gearbox based on improved variational mode algorithm and information entropy. *Entropy* **2021**, *23*, 794. [CrossRef] [PubMed]
23. Zhang, W.; Li, X. Data privacy preserving federated transfer learning in machinery fault diagnostics using prior distributions. *Struct. Health Monit.* **2022**, *21*, 1329–1344. [CrossRef]
24. Scheffer, C.; Girdhar, P. *Practical Machinery Vibration Analysis and Predictive Maintenance*; Elsevier: Amsterdam, The Netherlands, 2004.
25. Li, X.; Yu, S.; Lei, Y.; Li, N.; Yang, B. Intelligent Machinery Fault Diagnosis with Event-Based Camera. *IEEE Trans. Ind. Inform.* **2023**, 1–10. [CrossRef]
26. Correa, J.C.A.J.; Guzman, A.A.L. *Mechanical Vibrations and Condition Monitoring*; Academic Press: Cambridge, MA, USA, 2020.
27. Pedotti, L.A.D.S.; Zago, R.M.; Fruett, F. Fault diagnostics in rotary machines through spectral vibration analysis using low-cost MEMS devices. *IEEE Instrum. Meas. Mag.* **2017**, *20*, 39–44. [CrossRef]
28. Goyal, D.; Pabla, B. The vibration monitoring methods and signal processing techniques for structural health monitoring: A review. *Arch. Comput. Methods Eng.* **2016**, *23*, 585–594. [CrossRef]
29. Dowell, M.; Sylvester, G. Turbomachinery prognostics and health management via eddy current sensing: Current developments. In Proceedings of the 1999 IEEE Aerospace Conference, (Cat. No. 99TH8403), Snowmass, CO, USA, 7 March 1999; Volume 3, pp. 1–9. [CrossRef]
30. Mohd Ghazali, M.H.; Rahiman, W. Vibration analysis for machine monitoring and diagnosis: A systematic review. *Shock Vib.* **2021**, *2021*, 9469318. [CrossRef]
31. Dzhudzhev, B.; Ivancheva, V.; Kachulkova, S.; Gospodinova, E. Vibration measurement with piezoelectric transducer. In Proceedings of the XLVIII International Scientific Conference on Information, Communication and Energy Systems and Technologies ICEST 2013, Ohrid, North Macedonia, 26–29 June 2013.
32. Guru Manikandan, K.; Pannirselvam, K.; Kenned, J.J.; Suresh Kumar, C. Investigations on suitability of MEMS based accelerometer for vibration measurements. *Mater. Today Proc.* **2021**, *45*, 6183–6192. [CrossRef]
33. Varanis, M.; Silva, A.; Mereles, A.; Pederiva, R. MEMS accelerometers for mechanical vibrations analysis: A comprehensive review with applications. *J. Braz. Soc. Mech. Sci. Eng.* **2018**, *40*, 527. [CrossRef]
34. Rossi, A.; Bocchetta, G.; Botta, F.; Scorza, A. Accuracy Characterization of a MEMS Accelerometer for Vibration Monitoring in a Rotating Framework. *Appl. Sci.* **2023**, *13*, 5070. [CrossRef]
35. Tez, S.; Akin, T. Fabrication of a sandwich type three axis capacitive MEMS accelerometer. In Proceedings of the SENSORS 2013 IEEE, Baltimore, MD, USA, 3–6 November 2013; IEEE: Piscataway, NJ, USA, 2013; pp. 1–4.
36. Wu, D.; Liu, S.; Zhang, L.; Terpenny, J.; Gao, R.X.; Kurfess, T.; Guzzo, J.A. A fog computing-based framework for process monitoring and prognosis in cyber-manufacturing. *J. Manuf. Syst.* **2017**, *43*, 25–34. [CrossRef]
37. Sicard, B.; Alsadi, N.; Spachos, P.; Ziada, Y.; Gadsden, S.A. Predictive Maintenance and Condition Monitoring in Machine Tools: An IoT Approach. In Proceedings of the 2022 IEEE International IOT, Electronics and Mechatronics Conference (IEMTRONICS), Toronto, ON, Canada, 1–4 June 2022; IEEE: Piscataway, NJ, USA, 2022; pp. 1–9.
38. Swamy, S.N.; Kota, S.R. An Empirical Study on System Level Aspects of Internet of Things (IoT). *IEEE Access* **2020**, *8*, 188082–188134. [CrossRef]
39. Rizzi, M.; Ferrari, P.; Flammini, A.; Sisinni, E.; Gidlund, M. Using LoRa for industrial wireless networks. In Proceedings of the 2017 IEEE 13th International Workshop on Factory Communication Systems (WFCS), Trondheim, Norway, 31 May–2 June 2017; IEEE: Piscataway, NJ, USA, 2017; pp. 1–4.
40. Leonardi, L.; Battaglia, F.; Patti, G.; Bello, L.L. Industrial LoRa: A novel medium access strategy for LoRa in Industry 4.0 applications. In Proceedings of the IECON 2018—44th Annual Conference of the IEEE Industrial Electronics Society, Washington, DC, USA, 21–23 October 2018; IEEE: Piscataway, NJ, USA, 2018; pp. 4141–4146.
41. Cheikh, I.; Aouami, R.; Sabir, E.; Sadik, M.; Roy, S. Multi-layered energy efficiency in LoRa-WAN networks: A tutorial. *IEEE Access* **2022**, *10*, 9198–9231. [CrossRef]

42. Sinha, J.K. *Vibration Analysis, Instruments, and Signal Processing*; CRC Press: Boca Raton, FL, USA, 2014.
43. Hou, L.; Bergmann, N.W. Novel Industrial Wireless Sensor Networks for Machine Condition Monitoring and Fault Diagnosis. *IEEE Trans. Instrum. Meas.* **2012**, *61*, 2787–2798. [CrossRef]
44. Sofi, A.; Regita, J.J.; Rane, B.; Lau, H.H. Structural health monitoring using wireless smart sensor network—An overview. *Mech. Syst. Signal Process.* **2022**, *163*, 108113. [CrossRef]
45. Bal, M. Industrial applications of collaborative wireless sensor networks: A survey. In Proceedings of the 2014 IEEE 23rd International Symposium on Industrial Electronics (ISIE), Istanbul, Turkey, 1–4 June 2014; IEEE: Piscataway, NJ, USA, 2014; pp. 1463–1468.
46. Seferagić, A.; Famaey, J.; De Poorter, E.; Hoebeke, J. Survey on wireless technology trade-offs for the industrial internet of things. *Sensors* **2020**, *20*, 488. [CrossRef] [PubMed]
47. Rondón, R.; Mahmood, A.; Grimaldi, S.; Gidlund, M. Understanding the performance of bluetooth mesh: Reliability, delay, and scalability analysis. *IEEE Internet Things J.* **2019**, *7*, 2089–2101. [CrossRef]
48. Chang, K.H. Bluetooth: A viable solution for IoT? [Industry Perspectives]. *IEEE Wirel. Commun.* **2014**, *21*, 6–7. [CrossRef]
49. Martínez, C.; Eras, L.; Domínguez, F. The Smart Doorbell: A proof-of-concept Implementation of a Bluetooth Mesh Network. In Proceedings of the 2018 IEEE Third Ecuador Technical Chapters Meeting (ETCM), Cuenca, Ecuador, 15–19 October 2018; pp. 1–5. [CrossRef]
50. Jaber, A.A.; Bicker, R. Design of a Wireless Sensor Node for Vibration Monitoring of Industrial Machinery. *Int. J. Electr. Comput. Eng.* **2016**, *6*, 639–653.
51. Fellan, A.; Schellenberger, C.; Zimmermann, M.; Schotten, H.D. Enabling communication technologies for automated unmanned vehicles in Industry 4.0. In Proceedings of the 2018 International Conference on Information and Communication Technology Convergence (ICTC), Jeju, Republic of Korea, 17–19 October 2018; IEEE: Piscataway, NJ, USA, 2018; pp. 171–176.
52. Bruno, R.; Conti, M.; Gregori, E. Mesh networks: Commodity multihop ad hoc networks. *IEEE Commun. Mag.* **2005**, *43*, 123–131. [CrossRef]
53. Ikpehai, A.; Adebisi, B.; Rabie, K.M.; Anoh, K.; Ande, R.E.; Hammoudeh, M.; Gacanin, H.; Mbanaso, U.M. Low-Power Wide Area Network Technologies for Internet-of-Things: A Comparative Review. *IEEE Internet Things J.* **2019**, *6*, 2225–2240. [CrossRef]
54. Joshitha, C.; Kanakaraja, P.; Bhavani, M.D.; Raman, Y.N.V.; Sravani, T. Lorawan based cattle monitoring smart system. In Proceedings of the 2021 7th International Conference on Electrical Energy Systems (ICEES), Chennai, India, 11–13 February 2021; IEEE: Piscataway, NJ, USA, 2021; pp. 548–552.
55. Zhou, W.; Tong, Z.; Dong, Z.Y.; Wang, Y. LoRa-Hybrid: A LoRaWAN Based multihop solution for regional microgrid. In Proceedings of the 2019 IEEE 4th International Conference on Computer and Communication Systems (ICCCS), Singapore, 23–25 February 2019; IEEE: Piscataway, NJ, USA, 2019; pp. 650–654.
56. Wixted, A.J.; Kinnaird, P.; Larijani, H.; Tait, A.; Ahmadinia, A.; Strachan, N. Evaluation of LoRa and LoRaWAN for wireless sensor networks. In Proceedings of the 2016 IEEE SENSORS, Orlando, FL, USA, 30 October–3 November 2016; IEEE: Piscataway, NJ, USA, 2016; pp. 1–3.
57. Mikhaylov, K.; Moiz, A.; Pouttu, A.; Rapún, J.M.M.; Gascon, S.A. LoRa WAN for wind turbine monitoring: Prototype and practical deployment. In Proceedings of the 2018 10th International Congress on Ultra Modern Telecommunications and Control Systems and Workshops (ICUMT), Moscow, Russia, 5–9 November 2018; IEEE: Piscataway, NJ, USA, 2018; pp. 1–6.
58. Ertürk, M.A.; Aydın, M.A.; Büyükakkaşlar, M.T.; Evirgen, H. A survey on LoRaWAN architecture, protocol and technologies. *Future Internet* **2019**, *11*, 216. [CrossRef]
59. Dereviankine, G.H.; Guitton, A.; Iova, O.; Ning, B.; Valois, F. Opportunities and Challenges of LoRa 2.4 GHz. *IEEE Commun. Mag.* **2023**, 1–7. [CrossRef]
60. Lyczkowski, E.; Wanjek, A.; Sauer, C.; Kiess, W. Wireless Communication in Industrial Applications. In Proceedings of the 2019 24th IEEE International Conference on Emerging Technologies and Factory Automation (ETFA), Zaragoza, Spain, 10–13 September 2019; pp. 1392–1395. [CrossRef]
61. Lee, J.S.; Su, Y.W.; Shen, C.C. A Comparative Study of Wireless Protocols: Bluetooth, UWB, ZigBee, and Wi-Fi. In Proceedings of the IECON 2007—33rd Annual Conference of the IEEE Industrial Electronics Society, Taipei, Taiwan, 5–8 November 2007; pp. 46–51. [CrossRef]
62. Howard, I. *A Review of Rolling Element Bearing Vibration Detection, Diagnosis and Prognosis*; Defence Science and Technology Organization: Canberra, Australia, 1994.
63. Gangsar, P.; Tiwari, R. Multiclass fault taxonomy in rolling bearings at interpolated and extrapolated speeds based on time domain vibration data by SVM algorithms. *J. Fail. Anal. Prev.* **2014**, *14*, 826–837. [CrossRef]
64. Jain, P.H.; Bhosle, S.P. Analysis of vibration signals caused by ball bearing defects using time-domain statistical indicators. *Int. J. Adv. Technol. Eng. Explor.* **2022**, *9*, 700.
65. Yadav, O.P.; Pahuja, G. Bearing health assessment using time domain analysis of vibration signal. *Int. J. Image Graph. Signal Process.* **2020**, *10*, 27.
66. Lahdelma, S.; Juuso, E. Signal processing in vibration analysis. In Proceedings of the 5th International Conference on Condition Monitoring and Machinery Failure Prevention Technologies, CM 2008 / MFPT 2008, Edinburgh, UK, 15–18 July 2008; pp. 879–889.
67. Vishwakarma, M.; Purohit, R.; Harshlata, V.; Rajput, P. Vibration analysis & condition monitoring for rotating machines: A review. *Mater. Today Proc.* **2017**, *4*, 2659–2664.

68. Williams, T.; Ribadeneira, X.; Billington, S.; Kurfess, T. Rolling element bearing diagnostics in run-to-failure lifetime testing. *Mech. Syst. Signal Process.* **2001**, *15*, 979–993.
69. McFadden, P.; Toozhy, M. Application of synchronous averaging to vibration monitoring of rolling element bearings. *Mech. Syst. Signal Process.* **2000**, *14*, 891–906.
70. Tian, J.; Morillo, C.; Azarian, M.H.; Pecht, M. Motor bearing fault detection using spectral kurtosis-based feature extraction coupled with K-nearest neighbor distance analysis. *IEEE Trans. Ind. Electron.* **2015**, *63*, 1793–1803. [CrossRef]
71. Brigham, E.O.; Morrow, R.E. The fast Fourier transform. *IEEE Spectr.* **1967**, *4*, 63–70. [CrossRef]
72. Patil, S.S.; Gaikwad, J.A. Vibration analysis of electrical rotating machines using FFT: A method of predictive maintenance. In Proceedings of the 2013 Fourth International Conference on Computing, Communications and Networking Technologies (ICCCNT), Tiruchengode, India, 4–6 July 2013; pp. 1–6. [CrossRef]
73. Khadersab, A.; Shivakumar, S. Vibration analysis techniques for rotating machinery and its effect on bearing faults. *Procedia Manuf.* **2018**, *20*, 247–252. [CrossRef]
74. Lin, H.C.; Ye, Y.C.; Huang, B.J.; Su, J.L. Bearing vibration detection and analysis using enhanced fast Fourier transform algorithm. *Adv. Mech. Eng.* **2016**, *8*, 1687814016675080. [CrossRef]
75. Kumar, B.K.; Diwakar, G.; Satynarayana, M. Determination of unbalance in rotating machine using vibration signature analysis. *Int. J. Mod. Eng. Res.* **2012**, *2*, 3415–3421.
76. Saleem, M.A.; Diwakar, G.; Satyanarayana, M. Detection of unbalance in rotating machines using shaft deflection measurement during its operation. *IOSR J. Mech. Civ. Eng* **2012**, *3*, 8–20. [CrossRef]
77. de Sá Só Martins, D.H.C.; Viana, D.P.; de Lima, A.A.; Pinto, M.F.; Tarrataca, L.; Lopes e Silva, F.; Gutiérrez, R.H.R.; de Moura Prego, T.; Monteiro, U.A.B.V.; Haddad, D.B. Diagnostic and severity analysis of combined failures composed by imbalance and misalignment in rotating machines. *Int. J. Adv. Manuf. Technol.* **2021**, *114*, 3077–3092. [CrossRef]
78. Lakis, A. Rotating machinery fault diagnosis using time-frequency methods. In Proceedings of the 7th WSEAS International Conference on Electric Power Systems, High Voltages, Electric Machines, Venice, Italy, 21–23 November 2007.
79. Satyam, M.; Rao, V.S.; Devy, C. Cepstrum analysis: An advanced technique in vibration analysis of defects in rotating machinery. *Def. Sci. J.* **1994**, *44*, 53. [CrossRef]
80. Li, H.; Ai, S. Application of order bi-cepstrum to gearbox fault detection. In Proceedings of the 2008 7th World Congress on Intelligent Control and Automation, Chongqing, China, 25–27 June 2008; pp. 1781–1785. [CrossRef]
81. Randall, R.; Smith, W. New cepstral methods for the diagnosis of gear and bearing faults under variable speed conditions. In Proceedings of the ICSV23 Conference, Athens, Greece, 10–14 July 2016.
82. Liu, Y.; Jiang, Z.; Haizhou, H.; Xiang, J. Asymmetric penalty sparse model based cepstrum analysis for bearing fault detections. *Appl. Acoust.* **2020**, *165*, 107288. [CrossRef]
83. Randall, R.B. A history of cepstrum analysis and its application to mechanical problems. *Mech. Syst. Signal Process.* **2017**, *97*, 3–19. [CrossRef]
84. Randall, R.B.; Antoni, J. Rolling element bearing diagnostics—A tutorial. *Mech. Syst. Signal Process.* **2011**, *25*, 485–520. [CrossRef]
85. Chen, B.; Song, D.; Cheng, Y.; Zhang, W.; Huang, B.; Muhamedsalih, Y. IGIgram: An improved Gini index-based envelope analysis for rolling bearing fault diagnosis. *J. Dyn. Monit. Diagn.* **2022**, 111–124. [CrossRef]
86. Cheng, F.; Qu, L.; Qiao, W.; Wei, C.; Hao, L. Fault diagnosis of wind turbine gearboxes based on DFIG stator current envelope analysis. *IEEE Trans. Sustain. Energy* **2018**, *10*, 1044–1053. [CrossRef]
87. Abboud, D.; Antoni, J.; Sieg-Zieba, S.; Eltabach, M. Envelope analysis of rotating machine vibrations in variable speed conditions: A comprehensive treatment. *Mech. Syst. Signal Process.* **2017**, *84*, 200–226. [CrossRef]
88. Rangel-Magdaleno, J.; Peregrina-Barreto, H.; Ramirez-Cortes, J.; Cruz-Vega, I. Hilbert spectrum analysis of induction motors for the detection of incipient broken rotor bars. *Measurement* **2017**, *109*, 247–255. [CrossRef]
89. Mehrjou, M.R.; Mariun, N.; Marhaban, M.H.; Misron, N. Rotor fault condition monitoring techniques for squirrel-cage induction machine—A review. *Mech. Syst. Signal Process.* **2011**, *25*, 2827–2848. [CrossRef]
90. Radi, A.; Bakalli, G.; Guerrier, S.; El-Sheimy, N.; Sesay, A.B.; Molinari, R. A multisignal wavelet variance-based framework for inertial sensor stochastic error modeling. *IEEE Trans. Instrum. Meas.* **2019**, *68*, 4924–4936. [CrossRef]
91. Huang, Z.; Zhu, J.; Lei, J.; Li, X.; Tian, F. Tool wear monitoring with vibration signals based on short-time fourier transform and deep convolutional neural network in milling. *Math. Probl. Eng.* **2021**, *2021*, 9976939. [CrossRef]
92. Santhoshi, M.S.; Sharath Babu, K.; Kumar, S.; Nandan, D. An investigation on rolling element bearing fault and real-time spectrum analysis by using short-time fourier transform. In Proceedings of the International Conference on Recent Trends in Machine Learning, IoT, Smart Cities and Applications: ICMISC, Hyderabad, India, 29–30 March 2020; Springer: Berlin/Heidelberg, Germany, 2021; pp. 561–567.
93. Ribeiro Junior, R.F.; dos Santos Areias, I.A.; Campos, M.M.; Teixeira, C.E.; da Silva, L.E.B.; Gomes, G.F. Fault detection and diagnosis in electric motors using convolution neural network and short-time fourier transform. *J. Vib. Eng. Technol.* **2022**, *10*, 2531–2542. [CrossRef]
94. Leiber, M.; Marnissi, Y.; Barrau, A.; El Badaoui, M. Differentiable Adaptive Short-Time Fourier Transform with Respect to the Window Length. In Proceedings of the ICASSP 2023—2023 IEEE International Conference on Acoustics, Speech and Signal Processing (ICASSP), Rhodes Island, Greece, 4–10 June 2023; IEEE: Piscataway, NJ, USA, 2023; pp. 1–5.
95. Wang, L.H.; Zhao, X.P.; Wu, J.X.; Xie, Y.Y.; Zhang, Y.H. Motor fault diagnosis based on short-time Fourier transform and convolutional neural network. *Chin. J. Mech. Eng.* **2017**, *30*, 1357–1368. [CrossRef]

96. Al-Badour, F.; Sunar, M.; Cheded, L. Vibration analysis of rotating machinery using time–frequency analysis and wavelet techniques. *Mech. Syst. Signal Process.* **2011**, *25*, 2083–2101. [CrossRef]
97. Teng, W.; Ding, X.; Cheng, H.; Han, C.; Liu, Y.; Mu, H. Compound faults diagnosis and analysis for a wind turbine gearbox via a novel vibration model and empirical wavelet transform. *Renew. Energy* **2019**, *136*, 393–402. [CrossRef]
98. Loparo, K.A.; Adams, M.L.; Lin, W.; Abdel-Magied, M.F.; Afshari, N. Fault detection and diagnosis of rotating machinery. *IEEE Trans. Ind. Electron.* **2000**, *47*, 1005–1014. [CrossRef]
99. Li, Y.; Xu, M.; Wei, Y.; Huang, W. A new rolling bearing fault diagnosis method based on multiscale permutation entropy and improved support vector machine based binary tree. *Measurement* **2016**, *77*, 80–94. [CrossRef]
100. Li, S.; Xin, Y.; Li, X.; Wang, J.; Xu, K. A review on the signal processing methods of rotating machinery fault diagnosis. In Proceedings of the 2019 IEEE 8th Joint International Information Technology and Artificial Intelligence Conference (ITAIC), Chongqing, China, 24–26 May 2019; IEEE: Piscataway, NJ, USA, 2019; pp. 1559–1565.
101. Liao, Y.; Ragai, I.; Huang, Z.; Kerner, S. Manufacturing process monitoring using time-frequency representation and transfer learning of deep neural networks. *J. Manuf. Process.* **2021**, *68*, 231–248. [CrossRef]
102. Sun, K.; Jin, T.; Yang, D. A new reassigned spectrogram method in interference detection for GNSS receivers. *Sensors* **2015**, *15*, 22167–22191. [CrossRef]
103. Ding, C.; Zhao, M.; Lin, J.; Liang, K.; Jiao, J. Kernel ridge regression-based chirplet transform for non-stationary signal analysis and its application in machine fault detection under varying speed conditions. *Measurement* **2022**, *192*, 110871. [CrossRef]
104. Liu, X.; Yan, Y.; Hu, K.; Zhang, S.; Li, H.; Zhang, Z.; Shi, T. Fault diagnosis of rotor broken bar in induction motor based on successive variational mode decomposition. *Energies* **2022**, *15*, 1196. [CrossRef]
105. Pai, P.F.; Huang, L.; Hu, J.; Langewisch, D.R. Time-frequency method for nonlinear system identification and damage detection. *Struct. Health Monit.* **2008**, *7*, 103–127. [CrossRef]
106. Susanto, A.; Liu, C.H.; Yamada, K.; Hwang, Y.R.; Tanaka, R.; Sekiya, K. Application of Hilbert–Huang transform for vibration signal analysis in end-milling. *Precis. Eng.* **2018**, *53*, 263–277. [CrossRef]
107. Bin, G.; Gao, J.; Li, X.; Dhillon, B. Early fault diagnosis of rotating machinery based on wavelet packets—Empirical mode decomposition feature extraction and neural network. *Mech. Syst. Signal Process.* **2012**, *27*, 696–711. [CrossRef]
108. Liu, R.; Yang, B.; Zio, E.; Chen, X. Artificial intelligence for fault diagnosis of rotating machinery: A review. *Mech. Syst. Signal Process.* **2018**, *108*, 33–47. [CrossRef]
109. Zhang, D.; Feng, Z. Enhancement of time-frequency post-processing readability for nonstationary signal analysis of rotating machinery: Principle and validation. *Mech. Syst. Signal Process.* **2022**, *163*, 108145. [CrossRef]

eng

Article

The Learning Curve of People with Complete Spinal Cord Injury Using a NES_S-FES_S Interface in the Sitting Position: Pilot Study

Felipe Augusto Fiorin [1], Larissa Gomes Sartori [1], María Verónica González Méndez [2], Christiane Henriques Ferreira [1], Maria Bernadete de Morais França [3] and Eddy Krueger [1,*]

1 Laboratório de Engenharia Neural e de Reabilitação, Departamento de Anatomia, Universidade Estadual de Londrina, Londrina 86057-970, Brazil; felipe.augusto13@uel.br (F.A.F.); llarissa.sartori@uel.br (L.G.S.); christiane.henriques@uel.br (C.H.F.)
2 Engenharia Elétrica e Informática Industrial, Universidade Tecnológica Federal do Paraná, Curitiba 80230-901, Brazil; mariaveronica@alunos.utfpr.edu.br
3 Laboratório de Automação e Instrumentação Inteligente, Departamento de Engenharia Elétrica, Universidade Estadual de Londrina, Londrina 86057-970, Brazil; mbmoraes@uel.br
* Correspondence: ekrueger@uel.br

Abstract: The use of assistive technologies, such as a non-invasive interface for neuroelectrical signal and functional electrical stimulation (NES_S-FES_S), can mitigate the effects of spinal cord injury (SCI), including impairment of motor, sensory, and autonomic functions. However, it requires an adaptation process to enhance the user's performance by tuning the learning curve to a point of extreme relevance. Therefore, in this pilot study, the learning curves of two people with complete SCI (P_A: paraplegic-T_6, and P_B: quadriplegic-C_4) were analyzed, with results obtained on the accuracy of the classifier ($Ac_{CSP-LDA}$), repetitions of intra-day training, and number of hits and misses in the activation of FES_S for sixteen interventions using the NES_S-FES_S interface. We assumed that the data were non-parametric and performed the Spearman's ρ test (and p-value) for correlations between the data. There was variation between the learning curves resulting from the training of the NES_S-FES_S interface for the two participants, and the variation was influenced by factors both related and unrelated to the individual users. Regardless of these factors, P_A improved significantly in its learning curve, as it presented lower values in all variables in the first interventions compared to the P_B, although only P_A showed statistical correlation (on $Ac_{CSP-LDA}$ values in RLL). It was concluded that despite the variations according to factors intrinsic to the user and the functioning of the equipment used, sixteen interventions were sufficient to achieve a good learning effect to control the NES_S-FES_S interface.

Keywords: paraplegia; electrical stimulation; brain–computer interface

Citation: Fiorin, F.A.; Sartori, L.G.; Méndez, M.V.G.; Ferreira, C.H.; França, M.B.d.M.; Krueger, E. The Learning Curve of People with Complete Spinal Cord Injury Using a NES_S-FES_S Interface in the Sitting Position: Pilot Study. *Eng* **2023**, *4*, 1711–1722. https://doi.org/10.3390/eng4020097

Academic Editor: Antonio Gil Bravo

Received: 29 April 2023
Revised: 15 June 2023
Accepted: 15 June 2023
Published: 17 June 2023

1. Introduction

Spinal cord injury (SCI) is a highly disabling condition that can result in severe impairment of sensory, motor, and autonomic functions. This can lead to several complications that can make it difficult for patients to perform basic activities of daily living, such as sexual and bladder dysfunction, gastrointestinal and respiratory problems, and urinary tract infections [1,2].

In these cases, specific therapies that aim to stimulate or not stimulate neuroplasticity to promote axonal growth have been utilized. While this process is time-consuming, it can increase pre-existing connections and promote the formation of new neuronal circuits [2]. In the presence of a compromised pathway, one form of intervention is through functional neurorecovery, in which training capitalizes on the intrinsic mechanisms of the nervous system (NS) to generate rhythmic movements through sensorimotor pathways [3]. According to Musselman and colleagues [4], this pattern currently used in neurorehabilitation is performed through assistive technologies (AT).

An example of AT in evidence is brain–machine interfaces (BCIs) which link cortical control and electrical stimulation of muscles. Through this mechanism, these interfaces engage part of the spinal cord function, making it possible to restore basic functions such as gripping or muscle stretching [5]. According to Yang et al. [2], such BCIs must contain a mechanism to record the neuroelectric signal (NES_S), such as electroencephalography (EEG), to send signals to a computer, which in turn decodes the expected movement and activates auxiliary devices to perform the expected movement. This stimulus can be performed through functional electrical stimulation devices (FES_S) [6].

In order for the NES from the EEG to be more consistent with the desired task, it is extremely important to use motor imagery (MI) for the desired action, which causes oscillations in sensorimotor rhythms in the motor regions of the brain [7]. In addition, according to Bobrova et al. [8], it has been suggested that BCI systems based on FES_S operate on the Hebbian learning principle, where the simultaneous excitation of the motor zones of the cortex during MI and the spinal cord structure stimulated by FES_S leads to an improvement in the ability to control the movements of the paralyzed limb. With this association between the computer and the effector systems, better secondary functions are observed in intestinal, urinary, and sexual functions in addition to improvements in flexibility and control of fine motor skills using the limbs [2].

However, as BCIs depend on full concentration to perform activities correctly, they can be influenced by situations of anxiety, fatigue, or frustration. Therefore, strategies have emerged to try to mitigate these limitations, which include mindfulness meditation and music training. Of these, there has been greater emphasis on mindfulness for stimulating self-regulation of cognitive and emotional processes [9].

In the use of BCIs, a point of extreme relevance is the learning curve that the user goes through to adapt to the use of the interface and eventually enhance its performance. There is a trend towards gradual improvement in the results as relevant therapies are maintained. However, direct or indirect factors in the execution of neurorehabilitation, alternating between user-specific elements and ways of implementing the methodologies and functioning of the BCI, hinder the development of learning [3,10,11]. This pilot study aims to investigate the learning curves of two people with complete spinal cord injury using the NES_S-FES_S interface in the sitting position. As a hypothesis, it is expected that throughout the interventions the users of the NES_S-FES_S interface will increase the accuracy of the NES_S classifier and reduce the false positive rate during the motor imagery to FES_S activation process.

2. Materials and Methods

This was an uncontrolled longitudinal pilot study with a convenience sample composed of people with motor deficits resulting from spinal cord injury. The research was approved by the Ethics Committee in Research Involving Human Beings of the State University of Londrina (CEP-UEL), with opinion n° 4,060,700. Participants/users were recruited through advertising in physiotherapy clinics in Londrina. The initial contact with the participants took place through a brief interview via a telephone call to collect information about the injury. In addition, the objectives of the project and how the interventions are carried out were briefly explained. Subsequently, the participants were invited to visit the Laboratório de Engenharia Neural e de Reabilitação (LENeR) and sign the form indicating free and informed consent, allowing initiation of the research protocols. The research lasted sixteen weeks for each participant, with one intervention per week (sixteen in total). The transport costs of the participants were paid for by the laboratory's administration section.

2.1. Inclusion Criteria

- Eighteen years of age or older;
- A spinal cord injury between the C_4 and T_{10} levels;

- An injury time greater than twelve months.

2.2. Exclusion Criteria

- Intolerance to discomfort caused by electrical stimulation;
- Presence of neoplastic tissue near or in the lower limbs (electrically stimulated area);
- Metallic implant in regions close to or between electrical stimulation electrodes;
- Any cognitive alteration that might prevent experimental intervention;
- Presence of other neurological or orthopedic disorders that might preclude experimental intervention;
- Presence of infectious disease;
- Presence of a cardiac pacemaker;
- Three consecutive absences without prior warning.

2.3. NES_s

The participant was positioned seated in their wheelchair or an adapted chair in the LENeR facilities. In the interventions, non-woven fabric meshes were used to cover the chair and hide the lower limbs (LL), preventing their visualization by the participant during the BCI calibration. During installation of the equipment care was taken to respect the integrity of the participant in order to ensure their greater comfort.

2.3.1. NES_s Acquisition

EEG_s data were acquired using commercial equipment (Cyton OpenBCI Board®, Brooklyn, NY, USA). OpenVibe® (version 3.0) software was used to acquire the signals; this software is an open-source C^{++} tool that can be customized for different purposes. The acquisition frequency of the NES_s was 250 Hz.

EEG_s Channels

In all sessions, brain activity was recorded using gold electrodes (Maxxi Gold®, Rome, Italy) distributed in a 10–10 system pattern over the scalp with conductive paste (Carbofix®, Herzliya, Israel) and conductive gel (Ultra-gel Eletro®). The channels (unipolar) were positioned in the Cz, C1 (or C2), FCz, and CPz regions and fixed with the aid of a customized EEG_s cap. Reference channels A1 and A2 were fixed bilaterally using Ag/AgCl electrodes on the mastoid processes of the temporal bones.

NES_s Preprocessing

The NES_s was preprocessed using a 5th-order bandpass digital Butterworth filter (8–25 Hz) with an analysis window duration of 500 ms and a passband ripple equal to 1 dB. These settings were necessary to ensure the integrity of events related to desynchronization and synchronization (ERD/ERS) during MI related to the lower limbs [12].

2.3.2. Step I—NES_s Check

In the NES_s checking phase, the feasibility of the signals recorded by the EEG electrodes was investigated. Such signals are composed of a set of local and noise field potentials. In the case of local field potentials, their measurement occurs from the depolarization of a population of superficial neurons. Because it is very weak, this signal is subject to noise, such as the noise caused by the electrical network.

After the signals were obtained, their demonstration was performed graphically under the temporal domain. Afterwards, the data were subjected to Fourier transform in order to decompose the deterministic signals into the frequency components [13] in order to permit graphic demonstration within the spectral domain. This allowed for analysis of noise that might interfere with the signal and the frequency bands that were present according to the brain activity. The absence of wave patterns or cases of recording in rhythmic form indicate errors in capturing the signals in the region by the corresponding channel which need to be corrected before proceeding.

2.3.3. Step II—Calibration of Motor Imagery

The assay methodology was adapted from Yusoff [12], in which participants looked at a reference while simultaneously receiving instruction to perform the MI. An instructor used mirrored and verbal movement to reproduce to the participant the instructions (classes) that were indicated on the monitor, namely, (i) motor imagery and (ii) rest. At these moments, the participant (1) imagined the knee extension movement, or (2) remained inactive. Between 5 and 10 trials were performed, with an average duration of 8 s for the two classes. The sequence of IM classes presented to the individual was random, and the NES_s was recorded by the system. The LL in which the classes started was random in all interventions. MI training was performed at the beginning of every session, and was performed individually for the left leg and right leg. Step II was performed up to three times (lasting approximately 5–7 min in this Step) to achieve a better response before proceeding to Step III.

2.3.4. Step III—Spatial Filtering

Feature extraction was performed using the Common Spatial Pattern (CSP) filter. The equation of this filter is based on the work of Broniera-Junior et al. [14], where the CSP maximizes the variance of the spatially filtered NES_s under one condition (MI) while minimizing it for the other condition (rest). The seventh-order spatial filter was considered ideal for the present work.

2.3.5. Step IV—Motor Imagery Classifier

In this step, the Linear Discriminant Analysis (LDA) classifier was used to classify the results obtained by the CSP filter by reducing the resources to a space of lower dimension and maximizing the separation between the classes (MI and rest) [15]. The result of the ($Ac_{CSP-LDA}$) classifier accuracy is expressed in percentage units (%). The choice of the number of EEG channels and selection of the classifier were based on our previous study [16], which indicated that for LDA four EEG channels presents more satisfactory results compared to multi-layer perceptron and support vector machine approaches.

2.3.6. Step V—Application of the NES_s-FES_s Interface

In this step, the participant was again instructed to imagine the movement indicated by the instructor. When MI activity was requested, the $Ac_{CSP-LDA}$ value was automatically calculated at 1 s. After calculation, if the $Ac_{CSP-LDA}$ value was equal to or above 72% a computational instruction was transferred from the OpenVibe® software to the Virtual-Reality Peripheral Network (VRPN)-based interface developed in C^{++} in the Microsoft Visual Studio® [17] environment. The signal received by the VRPN was then processed and sent in binary form to the electrical stimulator via Bluetooth to activate it.

FES_s

For artificial nerve activation [18], an electrical stimulator was customized exclusively for this work following the criteria proposed in the IEC standard 60601-2-10 [19]. Two self-adhesive electrodes were positioned in the anterior region of the volunteer's thigh according to the methodology of Krueger et al. [20], measuring 5×9 cm. One of the electrodes was positioned with the lower edge 3 cm from the base of the patella and the other over the femoral triangle [21] to stimulate the quadriceps muscle through the femoral nerve [22]. According to Krueger et al. [20], after fixing the electrodes, an interval of 10 min is respected to stabilize the electrode–skin impedance. The biphasic electrical stimulation frequency parameters were: carrier frequency of 1 kHz (positive: 200 μ + negative: 200 μ + off: 600 μ, negative: 200 μ + positive: 200 μ + off: 600 μ). The modulated frequency was 20–40 Hz (active period of 24 ms) to increase and decrease the pulse trains. The amplitude was modulated according to the instant of maximum electrically stimulated extension [20]. During step V, the activation condition of FES_S occurred when the probability of similarity of the classifier output signal is equal to or greater than 72%. The assembly of the equipment on the participant is illustrated in Figure 1.

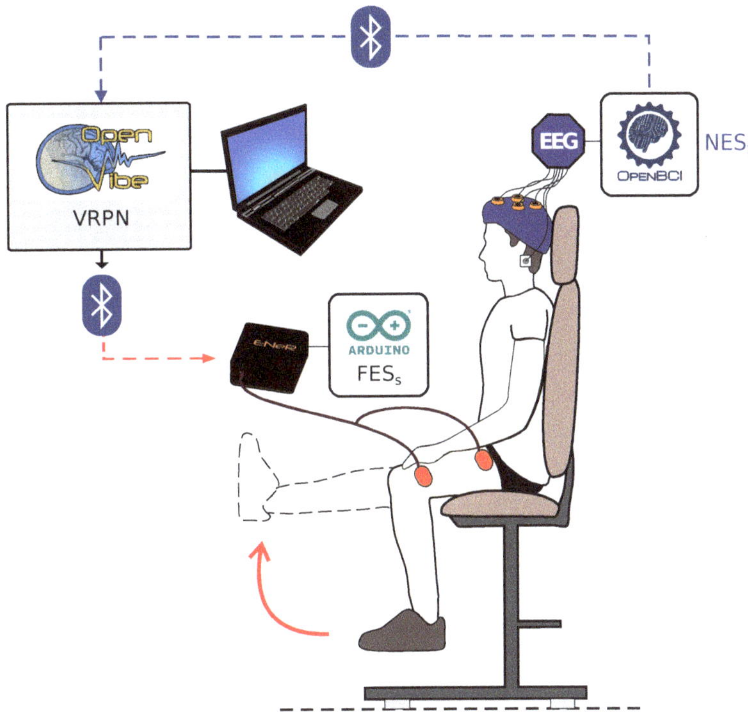

Figure 1. Illustration of participant during intervention. NES_S: Surface neuroelectric signal, FES_S: Surface functional electrical stimulation.

2.3.7. Mindfulness Training

The audio used was developed by instructor Katya Stübing. It was suggested to both participants that they listen to the audio daily at the period of the day that they felt the most comfortable. Each mindfulness training session lasted 149 s. According to Linehan [23], this training has been shown to improve emotional regulation. This daily task aimed to improve the performance of the participant during the sessions when carrying out the training and interventions.

2.4. Learning Curve Assessment

The participants' assessment of the NES_S-FES_S interface learning curve included the following items:

- Repetition of training;
- $Ac_{CSP-LDA}$ values;
- Success percentage of FES_S activation.

2.5. Data Analysis, Presentation, and Statistics

The present study analyzed the learning curves in the application of the NES_S-FES_S interface in two people with complete spinal cord injuries. The participants were P_A, with lesion level T_6 for eight years, and P_B, with lesion level C_4 for fourteen years. The graphics were developed in a customized routine through the Plotly Open Source Graphing

Library for Python®. (version 3.8.2-0ubuntu2) The image's final editing was performed in Inkscape® version 1.2.2. A customized routine of the open-source software Octave® version 5.2.0 was used for statistical tests. Due to the small sample, we assumed that our data were non-parametric and performed the Spearman's ρ correlation (and p-value) test for the sixteen sessions with $Ac_{CSP-LDA}$ and the success percentage of FES_s activation for each participant and each leg.

3. Results and Discussion

3.1. Learning Curve

As the sessions were carried out, an increase in the efficiency of the interventions was observed, with the participants gradually adapting to the interface and the indexes presenting the number of training repetitions ($Ac_{CSP-LDA}$ and success percentage for FES_s activation) improving .

3.1.1. User's Repetition of MI Training

The repetition of user MI training was performed in different situations, such as changes in FES_s or when $Ac_{CSP-LDA}$ values were insufficient for a correct intervention. These repetitions proved to be efficient when the calibration was performed improperly, allowing for better $Ac_{CSP-LDA}$ values for the application of the NES_s-FES_s interface. However, as mentioned by Apicella et al. [11], when the calibrations are performed properly the participants reported mental fatigue despite the benefits of repeating the exercise, which may have reduced their performance.

Therefore, it was essential to look for ways to enhance the performance of the participants in order to minimize repetitions. Removing distractions may have positively affected the learning curve with the interface, as it provided greater comfort for participants during the interventions. Any individuals from the laboratory except the researcher and participant were removed during MI calibration, and any sound sources that could compromise concentration were minimized. These actions are supported by the work of Tianhang Liu [6], who has addressed the importance of mindfulness during such interventions.

However, several factors that interfered with the results persisted, such as the participant's mood and self-reported mental fatigue before and during the calibrations [3,6,11]. Hernandez-Rojas et al. [10] have pointed out that this interference occurs due to the implementation of a two-class classification paradigm in which the classifier aims to discriminate between two highly related and antagonistic conditions, such as "stretch the leg" and "relax the leg". For this reason, interfaces such as the one used in this research require broad user engagement in the completion of the requested tasks, as other cognitive tasks can be potentially interpreted within the pre-established paradigms even if they are not related to movement, ultimately affecting the activation of the NES_s-FES_s interface.

3.1.2. Classifier Accuracy

The Figure 2 shows the $Ac_{CSP-LDA}$ values (0–100%) of both participants during the sixteen-week period, including the left lower limb (LLL) and right lower limb (RLL). A trend towards improvement in $Ac_{CSP-LDA}$ values was observed during the study period; it can be seen in Figure 2 that as better values are obtained, more accurate activation signals are sent to the NES_s-FES_s interface. Nonetheless, high values do not entirely avoid the occurrence of errors during the activation of the FES_s, and may not necessarily represent better performance during the intervention.

P_A showed gradual development in the ease of obtaining $Ac_{CSP-LDA}$ values. At first, the best $Ac_{CSP-LDA}$ results obtained in the first intervention were 61.5% in the LLL and 50% in the RLL. Comparatively, their best results consisted of $Ac_{CSP-LDA}$ values of 91% in LLL and 88% in RLL. However, the $Ac_{CSP-LDA}$ correlation tests for P_A among all interventions were RLL = ρ = 0.59 (p = 0.02) and LLL = ρ = 0.47 (p = 0.07). The significant moderated correlation output to RLL indicates a learning curve along the interventions that does not statistically occur with LLL.

P_B presented high results from the first intervention that remained high and close for the remainder of the study. He showed an initial $Ac_{CSP-LDA}$ of 99% in LLL and 94% in RLL. The best values obtained during the interventions were an $Ac_{CSP-LDA}$ of 99% in both lower limbs. However, the $Ac_{CSP-LDA}$ correlation tests for P_B among all interventions were RLL = [ρ = 0.06 (p = 0.80) and LLL = ρ = −0.21 (p = 0.40). These insignificant correlation outputs indicate that the learning curve for P_B did not increase over the interventions. This may be due to the P_B presenting high $Ac_{CSP-LDA}$ values beginning with the first intervention.

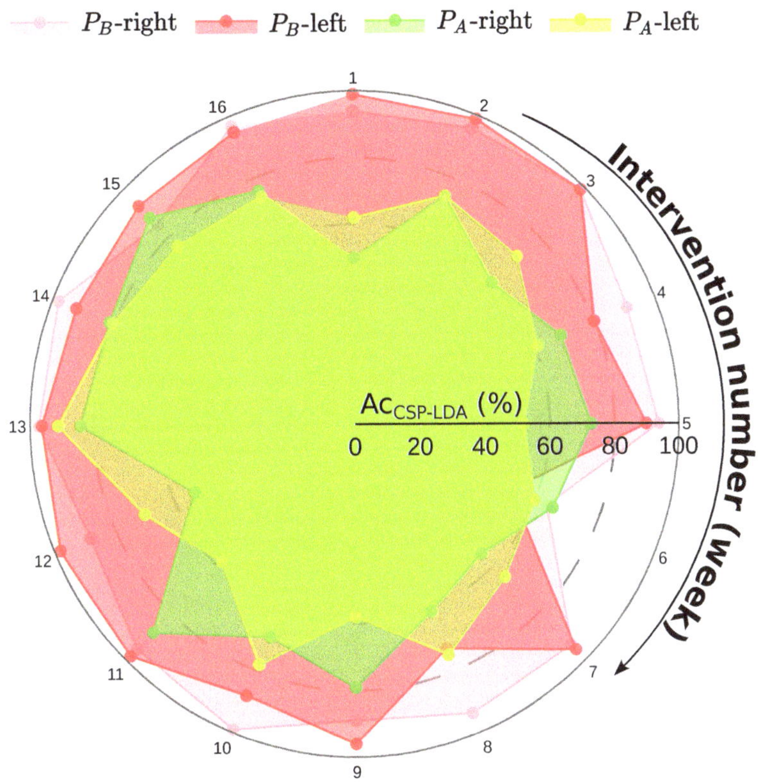

Figure 2. $Ac_{CSP-LDA}$ outputs by participants during interventions.

In the work of Hernandez-Rojas et al. [10], it was established through several studies on BCIs that $Ac_{CSP-LDA}$ in the range of 60% to 70% represents acceptable performance. Based on these parameters, it is possible to analyze the development presented by the users of the NES_s-FES_s interface more efficiently by comparing the performance of P_A and P_B throughout the course of this study.

3.1.3. Success Percentage of FES_s Activation

Figure 3 shows the success percentage of FES_s activation throughout the interventions, with larger circles indicating a greater number of attempts. The image shows the improved adaptation of the participants to the methodology with each new session held. In the first interventions, when the participants were adapting to the routine and the instructions, there was a tendency for lower $Ac_{CSP-LDA}$ values to occur, leading to greater errors in the activation of the FES_s due to the necessity of repeating the training. Performance gradually improved until reaching a more typical result for each of the participants. In Nenadic's study [24] it was found that participants take an average of four to five sessions to achieve their best performance. This suggests that performance may improve over time due to human–computer co-adaptation, and perhaps to other factors such as MI reactivation of dormant cortical areas.

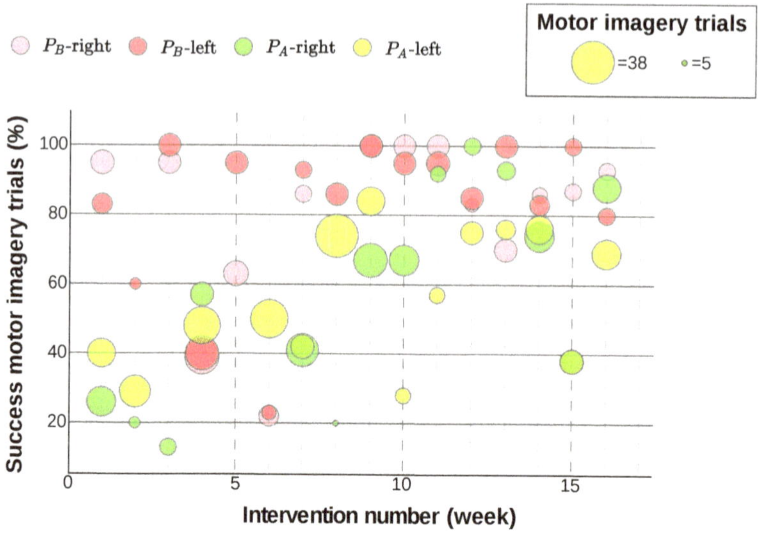

Figure 3. Percentage of correct FES_s activation for P_A and P_B during interventions. Motor imagery trials from 5 (smaller circle) to 38 (greater circle).

Notably, individual performance varied from time to time due to isolated factors on certain days, such as problems with equipment, concentration, or quality in $Ac_{CSP-LDA}$ capture. Therefore, even in instances where several sessions had been performed and there was a full understanding of the researcher's instructions, there were interventions in which the participants' performance was lower than on previous ones. This is corroborated by Liu's work [6], in which it was found that the best performance in NES_s-FES_s interfaces requires long training and a high degree of concentration on the part of participants.

For both participants, it was noticed that a sequence of false positives or omissions in the same period of application of the NES_s-FES_s interface exponentially affected the performance on that day. In these situations, the participants (especially P_A) started to show greater anxiety about seeking to improve their results, even when external factors were interfering. The work of Hernandez-Rojas et al. [10] has a strong relationship with this point, as it highlights the need for the BCI system to be able to minimize the rate of false negatives to avoid frustration on the part of patients when they cannot control the activation of the system. Furthermore, the same study found that the time spent by the BCI system to activate the FES_s device varied between 8.43 s and 13.91 s among participants with SCI. This is compatible with situations in which the participants in the present study presented a lower $Ac_{CSP-LDA}$ than usual, which contrasted with the more common situation where the activation of the interface occurred almost instantly. Similar findings were reported by Nenadic [24] and Liu [6].

Individually, P_A underwent a gradual improvement, improving from hit percentages of 40% (LLL) and 26% (RLL) in the first intervention to a peak of 84% (LLL) and 100% (RLL). The percentage of successful FES_s activation correlation tests for P_A among all the interventions were RLL = ρ = 0.40 (p = 0.11) and LLL = ρ = 0.12 (p = 0.65). These insignificant correlation outputs represent that the learning curve does not increase along the interventions compared to the percentage of successful FES_s activation. P_B, on the other hand, maintained high results, obtaining percentages of 83% (LLL) and 95% (RLL) in the first intervention and reaching a peak of 100% in both lower limbs. The percentage of successful FES_s activation correlation tests for P_B among all interventions was RLL = ρ =

−0.04 (p = 0.88) and LLL = ρ = 0.26 (p = 0.32). These insignificant correlation outputs represent that the learning curve does not increase along the interventions to success percentage of FES_s activation. It should be noted that for certain interventions shown in the images data collection could not be included due to equipment failure.

3.1.4. Inter-Subject Variability

The differences in performance among the participants illustrate the different elements that can affect the learning curve with this interface. P_B presented high performance values much more quickly than P_A, obtaining $Ac_{CSP-LDA}$ greater than 90% in the first interventions, while P_A needed more than half of the sixteen planned interventions to maintain consistent results above 75%. When questioned, P_B claimed to know about and use audio related to mindfulness, including the content indicated when beginning the study. Furthermore, compared to P_A, P_B could be said to have a more restrained and carefree personality; this may be related to greater focus on instructions and less self-pressure related to better performance [6].

Our results are consistent with the findings of Behrman et al. [3] for groups with incomplete SCI, where they found that, in addition to differences in protocol execution, variability in the results may reflect the intrinsic heterogeneity between participants. These individual aspects may be associated with direct consequences of the injury (for example, severity, location, and time of injury, as well as the presence of correlated medications) and personal factors (including personal motivation and family support).

It is possible to notice differences in the methods adopted by the participants in terms of the way they performed the MI. P_B tried to maintain mindfulness with respect to the lower limbs, while P_A typically directed vision to a neutral location. This element is related to the work of Rimbert et al. [25], where it is stated that there is considerable inter-subject and intra-subject variability in the ERD/ERS patterns generated during MI tasks and in interface performances. Considering this, factors such as the nature of the movement or force or the opening and closing of the eyes may influence the modulation of ERD/ERS. Therefore, in the context in which P_A had an MI pattern that does not include the full visualization of the lower limbs in moments of non-execution of movement, a brief observation of the legs could induce the activation of the NES_s-FES_s interface [10].

3.2. *Study Limitations*

Certain situations observed during the study interfered negatively with the achievement of results during the interventions. Occasional problems with the equipment deserve to be highlighted, such as in the capture of EEG by the OpenVibe software and the activation of the FES_s. In this regard, problems in the batteries of the devices involved in the NES_s-FES_s interface were present more frequently. The efficiency of EEG itself has limitations in the implementation of the interface, as discussed by Nenadic [24]; although the possibility of highly accurate control in EEG-based BCIs has been demonstrated, the implementation of these systems outside the experimental environment may require very high levels of performance. This is due to possible reading errors, such as omissions or false positives, which were consistent in certain interventions. It is evident that even with greater adaptation to the equipment by the participants, the capture of signals by the EEG is susceptible to both biological and non-biological artifacts due to the low field potential produced by neuronal depolarization and the limitations of the electrodes [6].

Another important limitation is that this was a pilot study to investigate the use of the NES_s-FES_s interface and the learning curve in individuals with complete SCI; we used a limited sample of two participants, which may not reflect all cases. Furthermore, the results obtained in this study were limited to activating the interface only for the selected specific task (MI), and it was not applied to any other tasks.

One of the possible artifacts that can compromise the performance of an EEG-based BCI is the electrical network, which can cause noise in the capture of signals. Therefore, the use of electronic devices with Bluetooth connection and an internet network close to the equipment involved in the intervention was avoided due to risk of reducing the efficiency of MI calibration. The participant's emotional state at the time of the intervention is another factor that can interfere with the results, either for reasons due to events prior to the intervention or due to frustration at not being able to complete a requested task [3,6,11]. In this aspect, it is important to properly calibrate the system in order to avoid false positive results and prevent frustration on the part of the user [10].

An final point worth mentioning is the uncertainty regarding the reliability of the data, as there is a possibility of the participants contracting the preserved musculature above the injury level during motor imagery training. However, this would lead to a bias in the capture of signals where the activation of FES_s would be triggered by the voluntary contraction of this musculature, resulting in inconsistent data values during integration with the interface, as the command needs to come directly from the brain. A possible correction would be the collection of data involved with this accessory movement, which could be correlated with the results to verify the level of interference.

4. Conclusions

Relevant variability was observed between individuals, with differences in the values of $Ac_{CSP-LDA}$, hit rates, and repetition of training. These differences can be explained by both individual factors and factors directly related to the NES_s-FES_s interface, with the individual factors being more determinative of the variation in the results for each participant. Among the participants, P_A showed gradual development, presenting initial $Ac_{CSP-LDA}$ values of 61.5% (LLL) and 50% (RLL) and reaching high values of 91% (LLL) and 88% (RLL), with statistical correlation only for the RLL. Comparatively, P_B started with good results that remained high, presenting initial $Ac_{CSP-LDA}$ of 99% (LLL) and 94% (RLL) and reaching 99% bilaterally, with no statistical correlation. This may be due P_B presenting high $Ac_{CSP-LDA}$ values from the first intervention. Regarding the percentage of hits, there was variation over the course of the study, although neither participant showed statistical correlation. P_A had initial values of 40% and 26%, respectively, and reached values of 84% and 100%, while P_B initially presented hit rates of 83% and 95%, and eventually reached 100% bilaterally.

Thus, despite the variations between users, sixteen interventions were sufficient to achieve a good learning effect to control the NES_s-FES_s interface for quadriceps femoris muscle activation in the sitting position.

For subsequent studies, our research team is developing a Python® framework in which will make possible:

- Integration with other sensors;
- Use of accelerometer metrics to avoid movement artifacts from head movement;
- Accurate measurement of time spent on FES_s activation;
- Better recognition of false positive FES_s activations;
- Accurate real-time identification of the computational cost of a user performing the most proximal MI task;
- Correlation of obtained data with the neuroanatomical autonomous system through cardiac frequency and electrodermal activity.

Author Contributions: Conceptualization, E.K. and F.A.F.; methodology, F.A.F., E.K., L.G.S., M.B.d.M.F. and M.V.G.M.; writing—original draft preparation, F.A.F. and E.K.; writing—review and editing, L.G.S., M.V.G.M., M.B.d.M.F., C.H.F. and E.K. All authors have read and agreed to the published version of the manuscript.

Funding: This research was funded by Fundação Araucária (research scholarship). The APC was funded by MDPI.

Institutional Review Board Statement: The present study was approved by the Ethics Committee and Research Involving Human Beings of the State University of Londrina, number 4,060,700, on 1 June 2020.

Informed Consent Statement: Informed consent was obtained from all subjects involved in the study. Written informed consent has been obtained from the patient(s) to publish this paper.

Data Availability Statement: The neuroelectrical signals recorded via electroencephalography were not stored; only the accuracy percentage and the percentage of successful FES_s activation were noted. All available data are presented in the images provided in the paper.

Conflicts of Interest: The authors declare no conflict of interest.

Abbreviations

SCI	Spinal Cord Injury
AT	Assistive Technologies
BCI	Brain–Computer Interface
BCIs	Brain–Computer Interfaces
EEG	Electroencephalography
MI	Motor Imagery
LL	Lower Limb
RLL	Right Lower Limb
LLL	Left Lower Limb
CSP	Common Spatial Pattern
LDA	Linear Discriminant Analysis
VRPN	Virtual Reality Peripheral Network

References

1. McDonald, J.W.; Sadowsky, C. Spinal-cord injury. *Lancet* **2002**, *359*, 417–425. [CrossRef] [PubMed]
2. Yang, B.; Zhang, F.; Cheng, F.; Ying, L.; Wang, C.; Shi, K.; Wang, J.; Xia, K.; Gong, Z.; Huang, X.; et al. Strategies and prospects of effective neural circuits reconstruction after spinal cord injury. *Cell Death Dis.* **2020**, *11*, 1–14. [CrossRef] [PubMed]
3. Behrman, A.L.; Bowden, M.G.; Nair, P.M. Neuroplasticity after spinal cord injury and training: An emerging paradigm shift in rehabilitation and walking recovery. *Phys. Ther.* **2006**, *86*, 1406–1425. [CrossRef] [PubMed]
4. Musselman, K.E.; Shah, M.; Zariffa, J. Rehabilitation technologies and interventions for individuals with spinal cord injury: Translational potential of current trends. *J. Neuroeng. Rehabil.* **2018**, *15*, 1–8. [CrossRef] [PubMed]
5. Capogrosso, M.; Milekovic, T.; Borton, D.; Wagner, F.; Moraud, E.M.; Mignardot, J.B.; Buse, N.; Gandar, J.; Barraud, Q.; Xing, D.; et al. A brain–spine interface alleviating gait deficits after spinal cord injury in primates. *Nature* **2016**, *539*, 284–288. [CrossRef] [PubMed]
6. Liu, T. The Application of Brain-computer Interface (BCI) based Functional Electrical Stimulation (FES). In *Proceedings of the Journal of Physics: Conference Series*; IOP Publishing: Bristol, UK, 2021; Volume 2065, p. 012006.
7. Abdulwahab, S.S.; Khleaf, H.K.; Jassim, M.H.; Abdulwahab, S. A Systematic Review of Brain-Computer Interface Based EEG. *Iraq. J. Electr. Electron. Eng.* **2020**, *16*, 1–10. [CrossRef]
8. Bobrova, E.; Reshetnikova, V.; Frolov, A.; Gerasimenko, Y.P. Use of imaginary lower limb movements to control brain–computer interface systems. *Neurosci. Behav. Physiol.* **2020**, *50*, 585–592. [CrossRef]
9. Tan, L.F.; Dienes, Z.; Jansari, A.; Goh, S.Y. Effect of mindfulness meditation on brain–computer interface performance. *Conscious. Cogn.* **2014**, *23*, 12–21. [CrossRef] [PubMed]
10. Hernandez-Rojas, L.G.; Cantillo-Negrete, J.; Mendoza-Montoya, O.; Carino-Escobar, R.I.; Leyva-Martinez, I.; Aguirre-Guemez, A.V.; Barrera-Ortiz, A.; Carrillo-Mora, P.; Antelis, J.M. Brain-Computer Interface Controlled Functional Electrical Stimulation: Evaluation With Healthy Subjects and Spinal Cord Injury Patients. *IEEE Access* **2022**, *10*, 46834–46852. [CrossRef]
11. Apicella, A.; Arpaia, P.; Frosolone, M.; Moccaldi, N. High-wearable EEG-based distraction detection in motor rehabilitation. *Sci. Rep.* **2021**, *11*, 5297. [CrossRef] [PubMed]
12. Abdalsalam M.E.; Yusoff, M.Z.; Kamel, N.; Malik, A.; Meselhy, M. Mental task motor imagery classifications for noninvasive brain computer interface. In Proceedings of the 2014 5th International Conference on Intelligent and Advanced Systems: Technological Convergence for Sustainable Future, ICIAS 2014, Kuala Lumpur, Malaysia, 3–5 June 2014. [CrossRef]
13. Macedo, D.R.; Freitas, F.S.; Nogueira, N.P.; Destro Filho, J.B. Análise espectral de eletroencefalografia para registros patológicos. 2011. Available online: https://seer.ufu.br/index.php/horizontecientifico/article/view/6358/6890 (accessed on 28 April 2023).
14. Broniera-Junior, P.; Nunes, W.R.; Lazzaretti, A.E.; Nohama, P.; Carvalho, A.; Krueger, E.; Teixeira, M.C.M. Classifier for motor imagery during parametric functional electrical stimulation frequencies on the quadriceps muscle. In Proceedings of the IEEE 2019 9th International IEEE/EMBS Conference on Neural Engineering (NER), San Francisco, CA, USA, 20–23 March 2019; pp. 526–529.
15. Cherkassky, V.S.; Mulier, F. *Learning from Data: Concepts, Theory, and Methods*, 1st ed.; John Wiley & Sons, Inc.: New York, NY, USA, 1998.

16. Broniera-Júnior, P.; Campos, D.P.; Lazzaretti, A.E.; Nohama, P.; Carvalho, A.A.; Krueger, E.; Teixeira, M.C.M. Influence of EEG channel reduction on lower limb motor imagery during electrical stimulation in healthy and paraplegic subjects. *Res. Biomed. Eng.* **2022**, *38*, 689–699. [CrossRef]
17. Silva, C.R.; de Araújo, R.S.; Albuquerque, G.; Moioli, R.C.; Brasil, F.L. Interfacing Brains to Robotic Devices—A VRPN Communication Application. In Proceedings of the XXVI Brazilian Congress on Biomedical Engineering, Vitória, Brazil, 26–30 October 2019; Springer: Berlin/Heidelberg, Germany, 2019; pp. 597–603.
18. Nogueira-Neto, G.; Scheeren, E.; Krueger, E.; Nohama, P.; Button, V.L.S. The influence of window length analysis on the time and frequency domain of mechanomyographic and electromyographic signals of submaximal fatiguing contractions. *Open J. Biophys.* **2013**, *3*, 178. [CrossRef]
19. *IEC 60601-2-10:2012*; Medical electrical equipment—Part 2–10: Particular Requirements for the Basic Safety and Essential Performance of Nerve and Muscle Stimulators. International Electrotechnical Commission: Geneva, Switzerland, 2012.
20. Krueger, E.; Scheeren, E.M.; Nogueira-Neto, G.N.; Nohama, P. Mechanomyography energy decreases during muscular fatigue in paraplegics. In Proceedings of the IEEE 2014 36th Annual International Conference of the IEEE Engineering in Medicine and Biology Society, Chicago, IL, USA, 26–30 August 2014; pp. 5824–5827.
21. Rabischong, E. Surface action potentials related to torque output in paraplegics' electrically stimulated quadriceps muscle. *Med. Eng. Phys.* **1996**, *18*, 538–547. [CrossRef] [PubMed]
22. Schiefer, M.A.; Triolo, R.J.; Tyler, D.J. A model of selective activation of the femoral nerve with a flat interface nerve electrode for a lower extremity neuroprosthesis. *IEEE Trans. Neural Syst. Rehabil. Eng.* **2008**, *16*, 195–204. [CrossRef] [PubMed]
23. Linehan, M.M. *Treinamento de Habilidades em DBT: Manual de Terapia Comportamental Dialética para o Terapeuta*; Artmed Editora: Porto Alegre, Brazil, 2018.
24. Nenadic, Z. Brain–computer interfaces for human gait restoration. *Control. Theory Technol.* **2021**, *19*, 516–528. [CrossRef]
25. Rimbert, S.; Trocellier, D.; Lotte, F. Is Event-Related Desynchronization variability correlated with BCI performance? In Proceedings of the 2022 IEEE International Conference on Metrology for Extended Reality, Artificial Intelligence, and Neural Engineering, Rome, Italy, 26–28 October 2022.

eng

MDPI

Article

A Limited-Scope Probabilistic Risk Assessment Study to Risk-Inform the Design of a Fuel Storage System for Spent Pebble-Filled Dry Casks

Joomyung Lee, Havva Tayfur, Mostafa M. Hamza, Yahya A. Alzahrani and Mihai A. Diaconeasa *

Department of Nuclear Engineering, North Carolina State University, Raleigh, NC 27695, USA
* Correspondence: madiacon@ncsu.edu

Abstract: This limited-scope study demonstrates the application of probabilistic risk assessment (PRA) methodologies to a spent fuel storage system for spent pebble-filled dry cask with a focus only on the necessary PRA technical elements sufficient to risk-inform the spent fuel storage system design. A dropping canister scenario in a silo of the spent fuel storage system is analyzed through an initiating event (IE) identification from the Master Logic Diagram (MLD); event sequence analysis (ES) by establishing the event tree; data analysis (DA) for event sequence quantification (ESQ) with uncertainty quantification; mechanistic source term (MST) analysis by using ORIGEN; radiological consequence analysis (RC) by deploying MicroShield, and risk integration (RI) by showing the Frequency-Consequence (F-C) target curve in the emergency area boundary (EAB). Additionally, a sensitivity study is conducted using the ordinary least square (OLS) regression method to assess the impact of variables such as failed pebble numbers, their location in the canister, and building wall thickness. Furthermore, the release categories grouped from the end states in the event tree are verified as safety cases through the F-C curve. This study highlights the implementation of PRA elements in a logical and structured manner, using appropriate methodologies and computational tools, thereby showing how to risk-inform the design of a dry cask system for storing spent pebble-filled fuel.

Keywords: probabilistic risk assessment; spent fuel storage system; spent pebble-filled dry cask; PRA elements; risk quantification

Citation: Lee, J.; Tayfur, H.; Hamza, M.M.; Alzahrani, Y.A.; Diaconeasa, M.A. A Limited-Scope Probabilistic Risk Assessment Study to Risk-Inform the Design of a Fuel Storage System for Spent Pebble-Filled Dry Casks. *Eng* **2023**, *4*, 1655–1683. https://doi.org/10.3390/eng4020094

Academic Editor: Antonio Gil Bravo

Received: 31 March 2023
Revised: 23 May 2023
Accepted: 6 June 2023
Published: 8 June 2023

1. Introduction

1.1. Background of the Dry Cask Storage System and Very High-Temperature Gas Reactor

Since the early 1980s, spent nuclear fuel (SNF) has been managed by the dry cask storage system in the nuclear power plant (NPP) site. As the demand for more spaces and longer storage periods increases, many reactor operators start to utilize dry storage as the existing spent fuel storage pool which not only costs more in the form of high operation and maintenance but also produces radioactive waste [1]. After discharge of the fuel from the reactor, the SNF is stored in an on-site water pool to cool the fuel, which intensively generates radioactivity with heat, for a few years until the released radioactivity decays enough to be moved to dry storage. The main goals of the dry storage system operation are (1) cooling of the fuel to maintain the temperature at a controllable level, (2) prevention of the radioactive release by isolating the fuel with shielding and an enclosed cask, and (3) safety in maintenance from accident scenarios. Compared to wet storage, the dry cask storage method is beneficial because of (1) less corrosion of the stored SNF, (2) good mobility, (3) no concern for cooling water management, and (4) no concern for secondary generated radioactive waste [2]. In virtue of the mentioned advantages, the high requirement for utilizing the dry cask storage system is emphasized at the congressional level. According to the congressional research service report for Congress, Senator Dianne Feinstein kept asking NRC and related institutions to establish regulatory policies for a faster shift process of the SNF to dry cask in March 2011 [3] because the SNF in the on-site water pool might be threatened by an external hazard. In fact, there were no problems with the dry storage in the Fukushima Daiichi accident, while Units 1, 2, and 3 were damaged due to cooling system failures [4]. For these reasons, the importance of the dry cask method as an SNF storage system is enhanced.

There have been several SNF-related Probabilistic Risk Assessment (PRA) research studies conducted to manage the light water reactor (LWR) SNF in dry cask storage safely. In 2007, a pilot PRA study was implemented to provide a guide for assessing and quantifying the risk associated with dry cask storage system operation by examining the feasible events in discharging, transferring, moving, and storing processes [5]. It covered initiating event (IE) identification including internal events and external hazards; potential failure risk due to mechanical, thermal loads, canister, or fuel failure, and radioactive release-related risk with a secondary containment isolation failure. NRC assigned the Idaho National Laboratory (INL) to evaluate the risk significance for the dry cask system operation from the License Amendment Requests (LAR) perspective [6]. Since a high burn-up of the SNF is essential not only for using the fuel economically but also for safe management of the SNF in the dry cask storage, the dry cask was modeled with a thermal load consideration to demonstrate the high burn-up behavior for installation licensing renewals or transportation licensing support [7]. Besides the dry cask storage PRA, the SPF pool PRA research was also performed, and related challenges are listed [8–10]. Previous studies were conducted to evaluate the risks and issue licenses for managing SNF from LWRs, rather than from advanced non-LWR system, such as a very high-temperature gas reactor (VHTR). Despite the anticipated global development of Generation IV nuclear reactor systems, PRA research for SNF management has been confined to the existing LWR domain.

The Generation IV design project aims to develop the next-generation nuclear energy system to be safer in public, more efficient and economical, and a less waste-producing operation [10–12]. The VHTR is a type of proposed next-generation nuclear plant that is a helium-cooled, graphite-moderated, and graphite-reflected reactor with tri-structural isotropic (TRISO)-coated pebble fuel or prismatic block fuel [13]. The VHTR system research plan [14] was initiated to elaborate tasks for fuel and the fuel cycle [15]; materials [16,17]; hydrogen production; computational validation and methodology [18–20], and benchmark project [21]. In China, a high-temperature gas-cooled reactor-pebble bed module 600 (HTR-PM 600) was developed as a commercial version of the HTR-PM, which is a follow-up reactor of the high-temperature gas-cooled reactor-10 (HTR-10) demonstration project [22]. In the United States, as well as the development of VHTR with a small and micro pebble, the advanced reactor demonstration program (ARDP) was launched by INL and Argonne National Laboratory (ANL) with U.S. universities to research and apply advanced computational techniques for system analysis [18], developing the application tool based on thermal-hydraulic simulation codes [19], and coupling codes to validate the experiments [20]. Additionally, from the material management perspective, the Oak Ridge National Laboratory (ORNL) established the material control and counting (MC&A) plan for pebble bed reactors (PBR) to resolve the safeguards and security-related issues [16,17].

Historically, VHTR PRA was performed by similar methods to those used for the LWR, however, in company with the evolution of the coated particle fuel, the VHTR PRA techniques have grown with the evolution of reactor design and licensing issues [23–27]. The licensing modernization project (LMP) established risk-informed and performance-based licensing technical requirements for advanced non-LWRs through the evaluation of a licensing basis event (LBE) and structures, systems, and components (SSC) performance [27–29]. For severe accident progression, source term, and consequence analysis from the LBE and the SSC performance, U.S. NRC plans to improve the capabilities of the existing computational simulation code, including MELCOR, MACCS, and SCALE [30]. Recently, the American Society of Mechanical Engineers/American Nuclear Society (ASME/ANS) joint committee published the PRA standard for non-LWR to announce the PRA technical requirements and application process [31]. Most of the previous projects and plans deal with VHTR design and license-oriented topics that focus on operating a reactor safely and preventing the LBE. On the other hand, the spent fuel-related tasks are stated only in the nonmandatory appendices in the PRA standard report.

1.2. Research Objective

The objective of this paper is to demonstrate the PRA requirements for a spent pebble bed-filled dry cask in order to contribute to the establishment of the PRA of the operation of a dry cask system for the VHTR. While previous research has explored PRA aspects of SNF management and VHTR design and operation, there have been few PRA studies examining the combination of the dry cask method and the spent TRISO fuel from the VHTR. To achieve this goal, several tasks are implemented, including (1) the application of a verified methodology, (2) identification of PRA technical requirements, and (3) risk quantification through the technical requirement process. This study endeavors to employ existing methodologies from non-LWR PRA and available data from dry cask storage operations to identify critical concerns, potential challenges, and limitations associated with the dry cask storage of spent pebble bed-filled canisters. The proposed workflow of this study will serve as a foundation for future investigations aimed at developing an effective PRA framework for non-LWR fuel storage operations. Details of the methodologies are described in the following section.

This paper delimits the scope by considering two aspects: the accident scope and the PRA element scope. For the accident scope, the cask drop scenario is selected because it is a representative mechanical load-related IE during the handling phase and transfer phases for a "moving" dry cask [5]. From the perspective of analyzing the dry cask's performance in response to mechanical loads, the study examines the structural impact resulting from the accidental drop scenario [32]. The interaction between SNF and storage canister due to impact loading [33,34] are examined by employing a finite element model (FEM). For the PRA elements, the ASME/ANS suggested 18 technical requirements [31], and this paper presents initiating event (IE) analysis, event sequence (ESA), data analysis (DA), event sequence quantification (ESQ), mechanistic source term (MST) analysis, radiological consequence analysis (RC), and risk integration (RI).

In this paper, Section 2 introduces general concepts of PRA elements including methodologies. Section 3 illustrates the workflow and dry cask storage system for the spent pebble bed fuel. In Section 4, a case study is implemented, and the conclusion follows in Section 5.

2. Methodology

In this section, the general PRA technical requirements determined in the research scope are introduced, as well as specific demonstrations as to how the spent pebble fuel-filled dry cask system PRA will be performed in the paper.

2.1. Initiating Event Identification

IE is the first disturbance to a normal operation of the NPP, thus, the IE selection is not only the primary step to accomplish the accurate and complete PRA model [35] but also to determine the analysis scope. The advanced non-LWR PRA standard suggests a plant or design-specified systematic approach to ascertain IEs, such as master logic diagram (MLD), failure modes and effort analysis (FMEA), heat balance fault tree (HBFT), or hazard and operability analysis (HAZOP) as a process hazards analysis (PHA) [27,31]. In this paper, the methods for IE identification are briefly introduced and compared, with the most appropriate being employed for the pebble bed-filled dry cask PRA.

- Master Logic Diagram

MLD is a formal logical technique based on a top-down approach to identify the IEs. It decomposes the influential factors from the final consequence, which corresponds to the top event represented by "significant release of radioactive material" until the IEs are identified, and depicts the informative flow about the causes and effects of the risk metrics in a form of a logical block diagram [36,37]. Thanks to the simple formulation of the MLD, the IEs with related event categories are comprehensively disclosed at the system level when collected knowledge and system/component information are provided, despite the fact that the MLD is dependent on the given information status. Therefore, the MLD methodology has been used in a wide range of research, such as the iodine–sulfur (IS) process [38], chemical installations [37], or NPP facilities [36,39].

- Heat Balance Fault Tree

HBFT aims to identify the IE by detecting a deviation from the thermal balance in the system [27]. Since the HBFT is a fault tree combined with technical considerations for heat/energy balance, it is useful to notice where and why the heat/energy imbalance occurs by tracking the cause and effect. Similar to the MLD, the HBFT is built based on a top-down deductive approach with the top event of "occurrence of heat imbalance due to IE" instead of "radioactive release" [40]. On the contrary, compared to the MLD, the HBFT is a methodology to be specifically applied to the NPP system rather than a generally applicable methodology because the NPP retains a steady state in normal operation from the thermal equilibrium perspective through the heat transport over the multiple connected systems including a reactor, a coolant system, and a heat exchange system [40,41].

- Failure Mode and Effects Analysis

FMEA is a risk assessment methodology to detect and reduce or prevent potential error sources in systems. Through the system analysis and failure analysis by focusing on a single equipment or component, the potential root causes with their effects are determined. In other words, the FMEA is a bottom-up inductive process to evaluate the vulnerability of the system. For IE identification, it is employed to assure how specific components can influence the performance of other components, subsystems, and the main systems [31]. The advantage that the impact of the failure is qualitatively analyzed with detailed descriptions enables the FMEA to have been used and improved in techniques in various industries [42]: digital instrumentation and control (I&C) system analysis for nuclear reactor [43,44]; tritium-breeding test blanket module design for fusion reactor [45]; medical radiotherapy research [46,47]; supply chain management [48], and vehicle recall investigation [49]. However, if the object system is complicated or the demonstration of failure mode contains many details, the number of FMEA tasks increases.

- Hazard and Operability Analysis

HAZOP is one of the PHA techniques to identify causes and consequences of potential hazards and operability by examining departure from normal conditions or process variables. Based on the expertise and professional experiences, the HAZOP analysis team divides the system into several sections, which are called HAZOP nodes, according to the inherited features of the nodes or team-defined principles to evaluate the safety-significant incidents [50]. As it is also a bottom-to-top approach methodology, the tasks to define parameters, deviations, cause and effect, and preventive or mitigative action recommendations would be time-consuming and challenging when the system is complicated [27]. From the PRA application perspective, the HAZOP has been used to figure out the IEs or IE groups for the NPP reactor system [27,51–54].

Table 1 summarizes the features of the introduced methodologies. In this paper, as shown in Figure 1, the MLD is used to identify the IEs by the following steps: (1) the final objective determination, (2) source identification, (3) safety barrier or related SSC identification, (4) safety function challenges or the SSC failure identification, (4) failure mode identification, and (5) IE identification. Additionally, steps 3 and 4 are repeated when there are sub-safety systems or sub-safety functions in the system domain. The MLD applied to the spent pebble fuel-filled dry cask PRA is illustrated in the case study section.

Table 1. Comparison of Methodologies for Initiating Event Identification.

	Approach	Features
Master Logic Diagram	Top-down	- Simple and comprehensive - Depicts the cause and effect in general - Challenges in identifying all event types
Heat Balance Fault Tree		- Able to identify the cause and effect due to specific factors - Deductive fault tree analysis procedure - Nuclear power plant applicable method
Failure Mode and Effects Analysis	Bottom-up	- Well organized to understand the impact of the failure - Not friendly for a complicated system - Time-consuming tasks depending on the level of detail
Hazard and Operability Analysis		- Considers various factors - Not applicable to complex systems and time-consuming tasks to identify potential hazards and operability

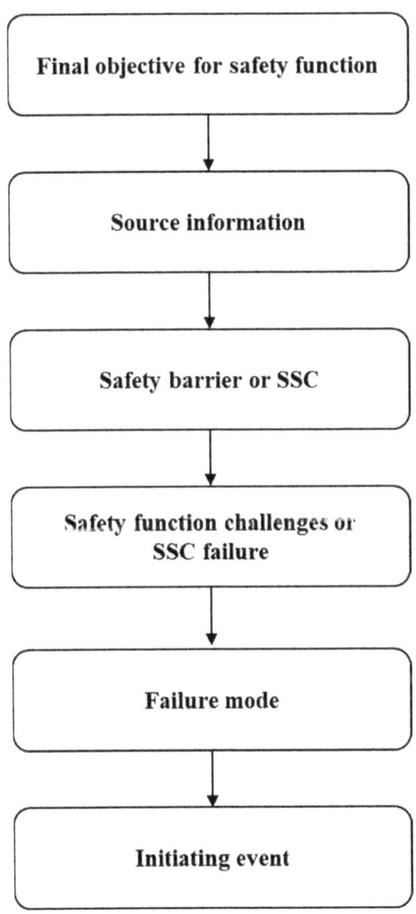

Figure 1. Master Logic Diagram Approach.

2.2. *Event Sequence Analysis*

The purpose of ESA is to delineate the transient or accident scenarios through timely ordered and consecutively enumerated events based on operator actions and system responses. The ESA provides accident progression represented by the discrete intermediate events that depend on the plant-, design-, and site-specific information from the IEs until the established end states, including core damage or radioactive release [31]. It demonstrates how the incidents are expanded or resolved from the IE by safety function challenges, SSC responses, and preventive or mitigative actions. An event tree (ET) is a representative PRA tool used to illustrate the ES analysis. Figure 2 shows the event sequence modeling framework revised from [27].

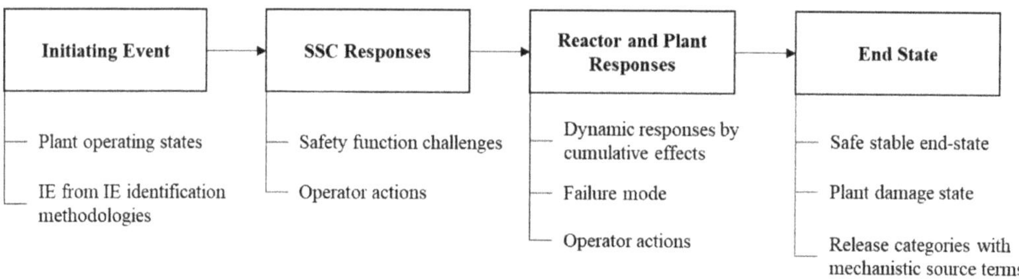

Figure 2. Event Sequence Modeling Framework.

2.3. *Data Analysis*

DA aims to estimate parameter data with consideration of plant- or system-specific evidence. For obtaining intermediate event probability on the ESA, the following are considered (1) analysis of risk-significant failure mode with associated SSCs or grouped components, (2) common cause failure (CCF) parameter, and (3) equipment maintenance, repair, and recovery-related data [31].

2.4. *Event Sequence Quantification*

ESQ indicates calculations of the frequency of the specific accident sequences based on the IE frequency and the failure probabilities of the relevant safety functions from the DA process. The objective of ESQ is achieved with (1) a proper computational model required to integrate individual PRA models, (2) solutions for functional and human dependencies, and (3) uncertainty analysis in the quantification process [31].

2.5. *Mechanistic Source Term Analysis*

MST is defined as the fission product release, led by fuel damage resulting from a particular accident scenario, that the best-estimate model simulates the temporal radionuclide transport in terms of physical and chemical processes [55]. To quantify the source term, the source term characteristics should be determined by consideration of not only released radionuclide transport-related phenomena, such as deposition, condensation, resuspension, and so on, but also system-related information including fuel form, safety engineering system operation states, and component states [56].

2.6. *Radiological Consequence Analysis*

RC analysis quantifies the consequences of an accident that consists of public health impacts and economic impacts. The health impacts of dose populations are fatalities, injuries, and cancer risk, meanwhile, the economic impacts include evacuation, relocation, and recovery for damaged or contaminated assets costs. Therefore, in addition to the MST analysis data, it is required to model the atmospheric transport and dispersion and obtain site-specific data, such as meteorology data, protective action, and emergency plans for the RC analysis [30].

2.7. Risk Integration

The main objective of the RI is to demonstrate risk-significance criteria with the associated uncertainties of the event sequences. To estimate the risk importance, several risk metrics are suggested and measured for expressing risk based on relative and absolute risk significance [57]. The absolute risk metrics are represented by core damage frequency (CDF), large early release frequency (LERF), or Birnbaum Importance, whereas, for ranking the risk order or establishing the regulations, relative risk measures, including Fussell–Vesely, Risk Achievement, or Risk Achievement Worth, can be used [58]. Depending on the ultimate goal of the PRA performance, decision makers or risk takers would decide the proper risk significance or risk metrics [59].

3. Pebble Fuel-Filled Dry Cask PRA

This section demonstrates the workflow and information flow to achieve the objective of each PRA element and introduces the spent fuel storage system (SFSS) for the HTR-PM as a case study target system. Based on the aforementioned PRA elements, the workflow is established for this study. Although the study concentrates on specific scopes of the PRA elements, the workflow can be extended to cover the full scope of each element for future endeavors. Hence, the introduction of the general non-LWR PRA technical requirements in the previous section will provide a foundation for future applications.

3.1. Workflow

Figure 3 demonstrates the workflow for each PRA element to deliver the information to the next PRA element step.

(1) Based on the PRA documents including PRA standard, PRA-related reports, and licensing requirements, various data and information are collected and refined to be used as materials for each PRA element or PRA analysis tool. For example, technical information from the pilot PRA study for a dry cask storage system [5] provides safety function-related information to establish the MLD for IE identification.
(2) From the technical information, the MLD enables to identify the IEs by presenting the causes and effects of the influential failure factors from the final consequences.
(3) Safety function/system information is utilized from the PRA documents to determine the event sequence that maneuvers the event scenario from the IE onto the end states.
(4) Same as in the previous steps, the PRA documents are referred to extract and estimate the failure probability of the determined event sequences.
(5) ESQ is implemented by the Phoenix Architect with the event tree from the ESA and the assumed probability from the DA processes. In this study, CAFTA, PRAQuant, and UNCERT modules are used to develop the fault tree/event tree, quantify the event tree sequences, and perform the uncertainty analysis, respectively
(6) For MST analysis, pebble fuel data and release fraction information from the PRA documents are used to tabulate the fuel inventory data for the consequence analysis. ORGIEN 2.2. is used to calculate the nuclide composition and activity of fuel.
(7) The RCA is performed to simulate the transport of the radioactive nuclides from the source established in the MST analysis according to the release categories by deploying the MicroShield.
(8) Finally, the risk is evaluated with consideration of frequency and consequence in accordance with the release categories.

Details of each step are demonstrated in the case study section.

3.2. System Description

3.2.1. TRISO Particle and Canister

The TRISO-coated particle fuel is composed of an oxide of uranium and mixtures of carbide and oxides and is encompassed by multiple layers to be safely used in the pebble bed reactor. The multi-layered fuel sphere includes ^{235}U-enriched UO_2 or UCO kernel that is encapsulated by (1) porous graphite buffer for excellent integrity and conductivity of the particle, (2) inner pyro-carbon (IPyC) to hold non-metal fission products (FPs), (3) silicon carbide (SiC) for retention of metal FPs, and (4) outer pyro-carbon (OPyC) as the last barrier for FP release and the protector for SiC layer [26,60–62]. Figure 4 illustrates the structure of the TRISO kernel with dimension information [63]. A fuel pebble is typically 60 mm in diameter and consists of randomly distributed 8000 to 18,000 fuel particles embedded in the graphite mix [64].

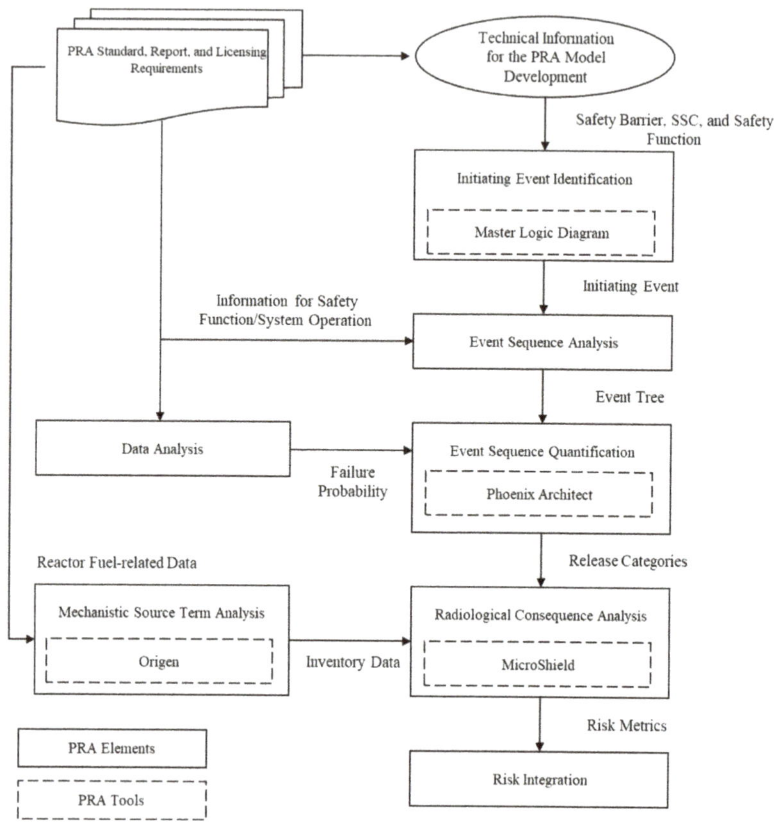

Figure 3. Workflow for Pebble Bed-filled Dry Cask PRA Study.

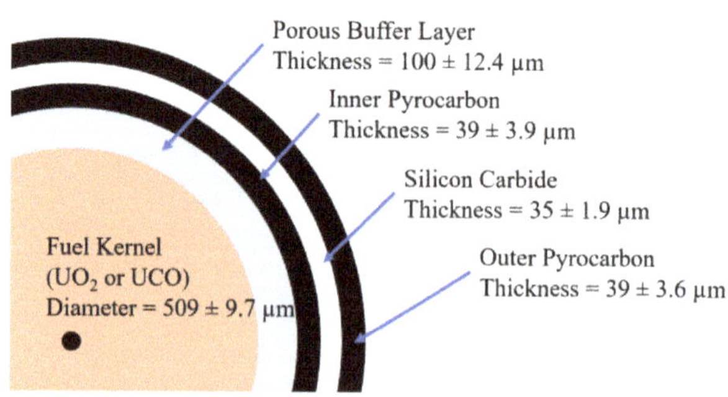

Figure 4. TRISO Kernel Structure.

A canister is a cylindrical stainless-steel structure to accommodate spent fuel elements. In the case of HTR-PM, a canister is 1.74 m in diameter, 4.18 m in height, and 20 mm in thickness to store about 40,000 spent fuel pebbles [65]. Due to the thin thickness, the canister's radiation shielding is vulnerable. Therefore, the spent fuel storage system includes radiation shielding functions when the canisters are transferred, hoisted, and loaded through the conveying and loading system.

3.2.2. Dry Cask Storage System

SFSS for the HTR-PM is a system to transfer the SNF from the reactor to the canister safely through the fuel handling system (FHS) and spent fuel conveying and loading system (SFCLS) and places the spent pebble bed-filled canister in the storage well. The functions of the SFCLS are as follows [66–68]:

- classifying fuel elements into the serviceable fuel element, the spent fuel element, and the graphite element by the direction converter with a burn-up device and retriever device,
- loading the classified elements into the cask or returning them back to the reactor core through the FHS,
- welding the full-filled canister by automatic machine,
- safely stacking the canisters (up to five) into a silo in the storage well by the crane and the hoister,
- and self-cleaning the pipelines by using blowers and iodine/dust filters.

There are several techniques for the operation of the SFSS with SFCLS:

- Safe stacking: A buffer seat at the bottom of the well protects the canister from dropping accident by structure or mechanistic failure. There are guiding rails and rail seats to load the canisters smoothly.
- Residual heat removal: Three cooling modes are operated in the SFSS to remove decay heat from the pebbles: closed loop active mode, open loop active mode, and open loop passive mode. Table 2 and Figure 5 demonstrate the details of the cooling modes. As Figure 5 shows below, the cold inlet air flows between the wall and barrel, then the air flows upward between the barrel and canister to the outlet pipe.
- Radiation shielding: Besides graphite mix within a fuel pebble, a 304 L stainless steel canister and concrete wall ensure the prevention of radioactive release to the environment.

Table 2. Residual Heat Removal Function in the Spent Fuel Storage System.

	Heat Exchanger	Blower	Feature
Closed Loop Active Cooling Mode	Yes	Yes	• Decrease the corrosion speed of the metal material due to dry air flow
Open Loop Active Cooling Mode	No	Yes	• Wet air flow affects corrosion of the metal material • Used when the heat exchanger is failed or under maintenance
Open Loop Passive Cooling Mode	No	No	• Rely on natural convection only • Used when the blowers fail at the same time

The SFSS is carried out for a buffer storage region in which the full-filled canisters are saved for the first three years because of a large amount of decay heat. Therefore, to maintain the storage temperature below a safe temperature, an independent ventilation system for the residual heat removal system mentioned above is operated. Additionally, since the outlet flow used for the open loop cooling mode is connected to the environment [67], the filtered ventilation of the heating, ventilation, and air conditioning (HVAC) system is essential to minimize the impact due to radiation release. To define the MLD for IE identification, the safety functions should be investigated.

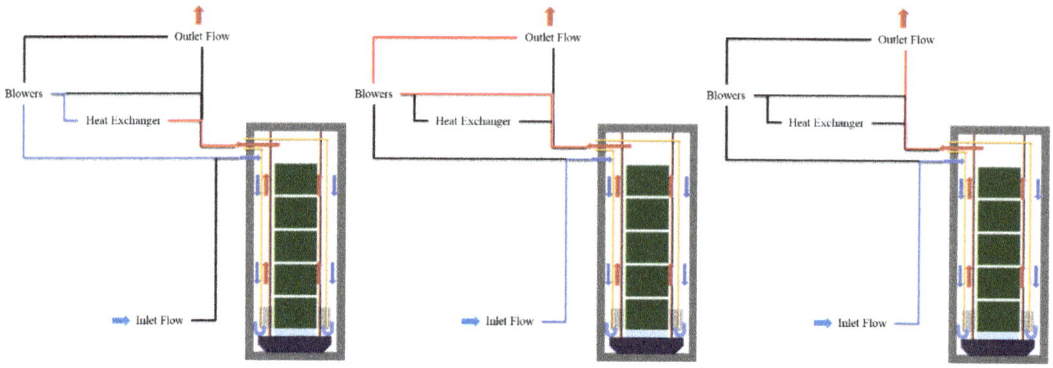

Figure 5. Cooling Modes for the Spent Fuel Storage System ((**Left**): Closed Loop Active Mode, (**Middle**): Open Loop Active Mode, (**Right**): Open Loop Passive Mode).

4. Case Study

A case study is implemented in this section, and each sub-section demonstrates how the PRA techniques are applied to the pebble fuel-filled dry cask PRA thoroughly to achieve its objective with assumptions and utilization of the PRA tools.

4.1. Event Description and Case Study Assumptions

To perform the PRA elements step by step, an event scenario is postulated from the canister drop at the silo of the SFSS. Due to the impact of the canister drop, the canister or spent pebbles in it or both is/are damaged. Therefore, the radioactive material can be released depending on the HVAC system operation status. The following are case study assumptions.

✓ Assumptions for IE identification and ESA

(1) There is no SFCLS operation failure or storage building damage during conveying the spent fuels from the FHS to the canister.

(2) There is no residual heat transfer failure caused by low-quality pebble geometry, canister defect, or SFSS cooling mode failure.

(3) There is no concrete wall (silo well) damage while the canister drops.

(4) The fuel particle coating and the graphite mixture in the pebble are not considered as the separated safety barrier for TRISO fuel failure. As mentioned above, the fuel kernel is protected by pyro-carbon layers with silicon carbide and core graphite in the pebble, however, radionuclide release happens when the TRISO is damaged.

(5) Due to improper crane movement, a canister vertically falls onto the concrete floor in the silo. The drop height is varied: 30 m, 25 m, 20 m, 15 m, and 10 m drop height.

(6) It is assumed that the HVAC system is identical to the HVAC system of the secondary containment isolation system for a dry cask storage system from the NUREG-1864 [5]. Therefore, HVAC failure leads to radioactive release directly into the environment bypassing the containment or building.

✓ Assumptions for DA and ESQ

(1) It is assumed that the HVAC system is identical with the HVAC system of the secondary containment isolation system for a dry cask storage system from the NUREG-1864 [5].

(2) Failure probability due to the impact of the canister drop is assumed by linear interpolation based on given data from [5,69]. Since the failure probability is assumed because of a lack of information, its distribution is induced by the Jeffreys noninformative prior to minimize the influence of the prior input and maximize the influence of the likelihood function [70].

✓ Assumptions for MST

(1) The reactor operates in a steady-state mode, where the neutron flux and power level are constant over time. This assumption simplifies the calculations by allowing the use of averaged parameters and eliminates the need for time-dependent calculations.
(2) The fuel is homogeneous and well-mixed, and the temperature distribution is uniform throughout the fuel. This assumption simplifies the modeling of fuel behavior and allows for a more direct calculation of the isotopic composition of the fuel.
(3) Each pebble does not move during operation, so the geometry of the fuel at the beginning of the cycle remains constant over time. This assumption simplifies the modeling of fuel behavior and allows for a more efficient calculation of the isotopic composition of the fuel.
(4) The fuel resident time in the reactor is assumed to be 3 years at full power.

✓ Assumptions for RCA and RI

(1) The 1 MeV energy level is used as the representative energy level for the modeling and analysis of fission product behavior. The behavior of fission products during undesired release to the environment can be complex and is influenced by several factors such as their physical and chemical properties, release characteristics, atmospheric and meteorological conditions, and energy levels. However, to simplify modeling and analysis of dry cask storage system failure, a single energy level is assumed for all fission products. This assumption allows for a more efficient analysis of fission product transport, retention, and release in the event of dry cask storage system accidents. Therefore, selecting 1 MeV allows for the modeling and analysis of fission product behavior to be simplified, as it provides a suitable approximation for many fission products. However, it should be noted that this assumption may not accurately represent the behavior of all fission products in all scenarios, and more detailed modeling might be necessary to investigate in future work. The use of a 1 MeV energy is discussed in relation to fission product transport and deposition, as well as radiological consequences [71–73].
(2) The release fractions of fission products during the accident are similar to the release fractions used for the LWR spent nuclear fuel. The release fractions used in this study was obtained from NUREG/CR-4982 and NUREG/CR 6451 [74,75].
(3) There are two concrete walls as the external safety barriers: an inside wall and an outside wall. The inside wall indicates the wall of a silo well and the outside wall is the storage building wall.
(4) The failed pebble is located at the bottom-center of the canister. For the sensitivity analysis, the pebble number and the failed pebbles' locations are varied.
(5) The dose point, which is equivalent to the location of the detector, for the absorbed dose rate or exposure rate, is located at 5 km from the source. The 5 km distance is assumed as the exclusion area boundary (EAB) for an advanced reactor [76]. Additionally, for the sensitivity study, another dose point is 10 m from the source which is the vicinity of the storage building.

4.2. Initiating Event Identification by Using MLD

To initiate the PRA analysis, as mentioned above, the MLD is established for identifying IEs of the SFSS operation for the spent pebble bed management. Based on the approach shown in Figure 1, the MLD approach is formulated by considering safety barriers and safety functions which are related to either SFSS operation or spent pebble bed fuel. There are three safety barriers and two main safety challenges: TRISO, canister, and concrete well of the SFSS, and radioactive material release control and heat removal control. Since the heat transfer from the spherical pebbles is dominated by convection, the pebble geometry condition is critical for residual heat removal led by cooling modes [77,78]. Figures 6 and 7 show the MLD approach and MLD of the case study for IE identification, respectively. Among various basic IEs, the canister drop is selected as the IE in accordance with the research scope.

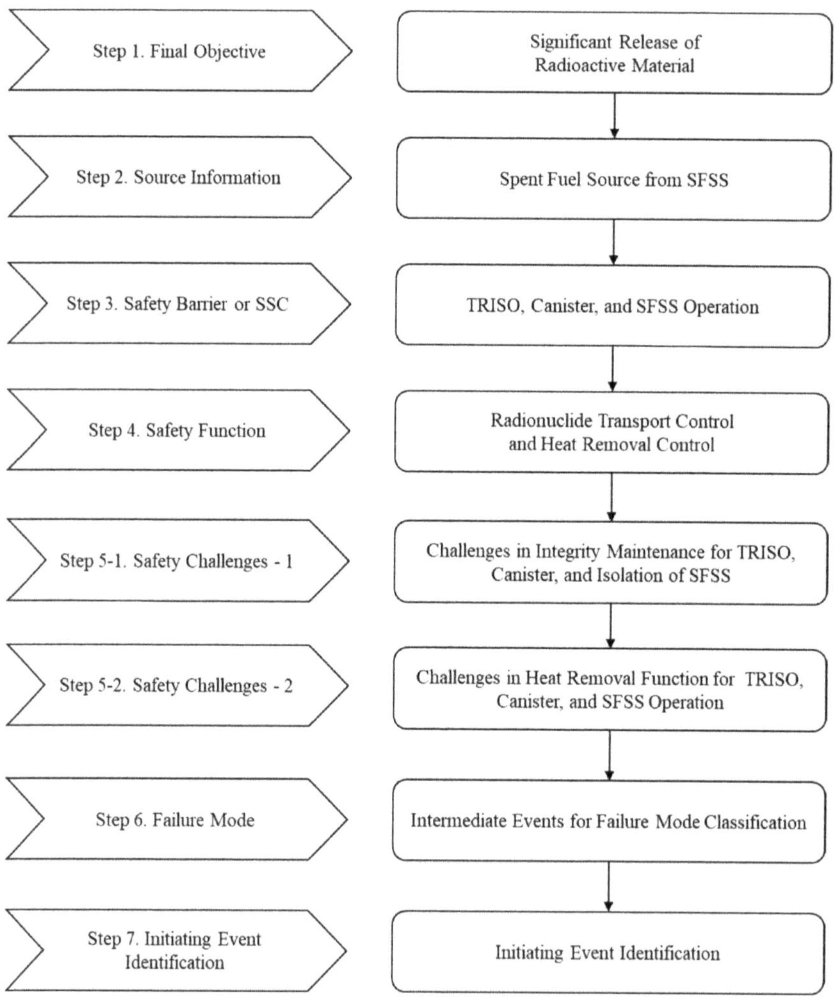

Figure 6. Master Logic Diagram Establishment Approach for the Pebble Bed-Filled Dry Cask PRA.

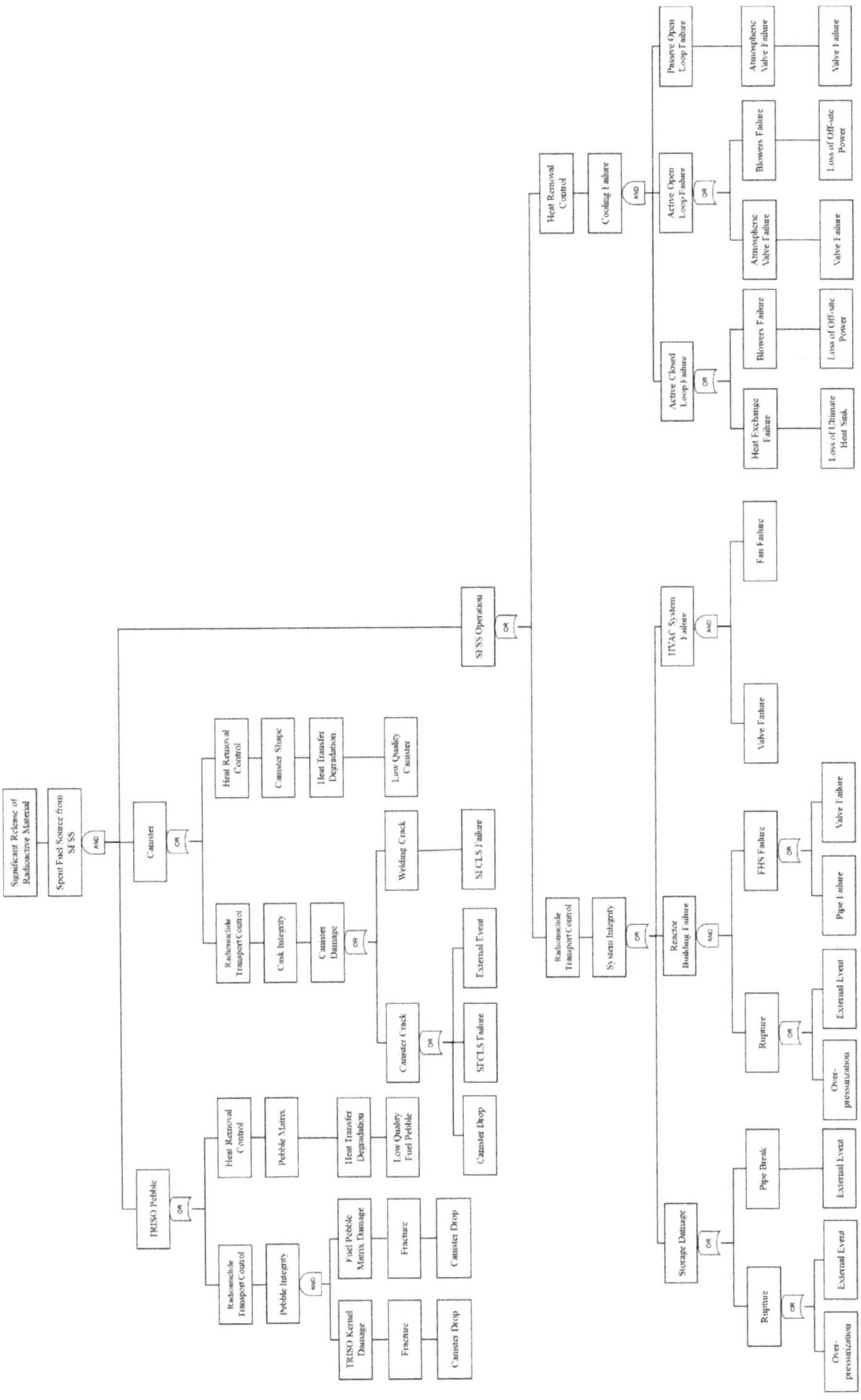

Figure 7. Master Logic Diagram for Spent Fuel Storage System PRA—Event Sequence Analysis for Event Tree Setup.

Event tree is a logical model to describe the event sequences step by step from the IE to the end states for risk analysis. Based on the aforementioned assumptions and safety challenges from the MLD, the event progression is demonstrated by a form of the intermediate events until the determination of the feasible radioactive material release modes. As we can see in Figure 8, only the integrity failure of the pebble bed, canister integrity failure, and HVAC system failure are considered in this case study. The following end states of the event tree are grouped as release categories such as:

- "OK" refers to the no potential risk of release from the SFSS.
- "Direct Exposure" (DE) indicates the event progression that some spent pebbles have failed, but the dry cask is intact. Additionally, the HVAC system operation failure is not considered which means the isolation of the storage is successful.
- "Noble Gas" (NG) is an end state where the release of radionuclide passes through the filtration path of the HVAC system. Since successful HVAC operation enables the filter to retain the radionuclides except the noble gas, only the noble gases, such as Kr and Xe, are released into the environment.
- "Radionuclide Release" (RR) indicates the end state that radioactive material is released to the environment directly without filtration due to HVAC operation failure.

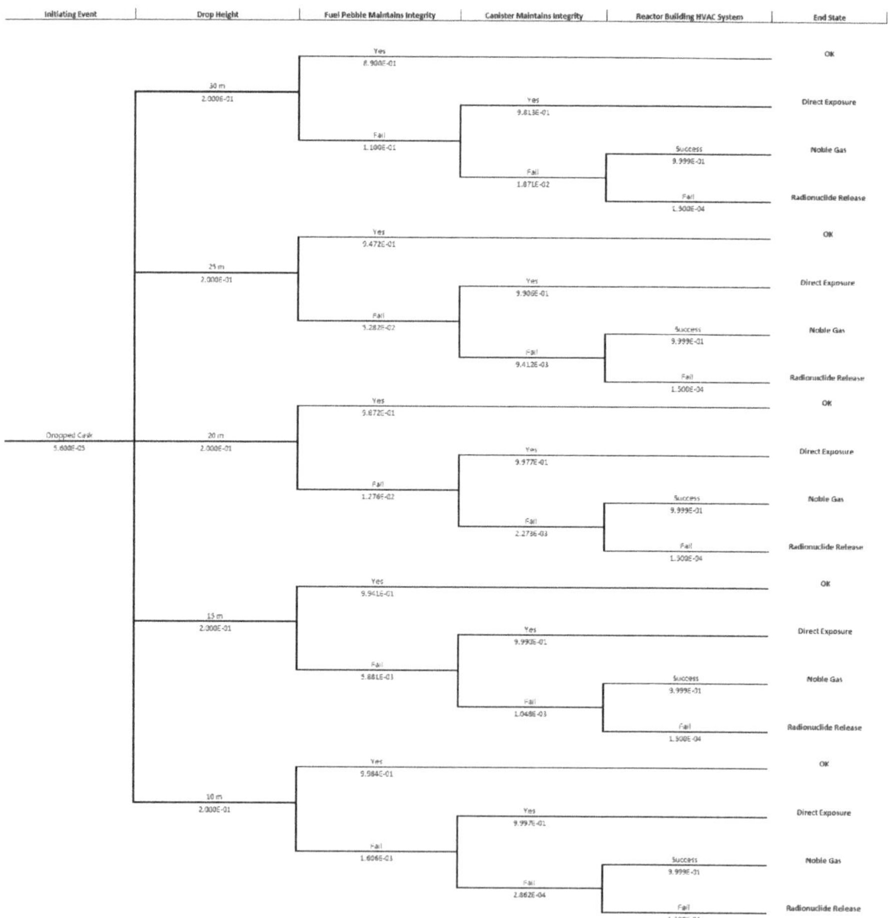

Figure 8. Event Tree for the Case Study (In this work we adopt the "E" notation, which is used to represent "times ten raised to the power of" in the scientific notation).

4.3. *Data Analysis for Failure Probability*

From the given information and data, the failure frequency/probabilities of the IE and intermediate events are estimated.

- IE frequency is a heuristic frequency given from the dropped transfer cask investigation in the United States [5].
- Canister failure probability is given and estimated by linear interpolation from [5].
- Pebble failure probability at 30 m is given from the dynamic analysis and validation experiment [34]. The failure probabilities with dropping height are assumed with the same proportion of canister failure probabilities.
- HVAC failure probability is given from [5].
- The distributions and parameters for the data are given according to the constrained noninformative distribution (CNID) [69]. For the uncertainty quantification, the initiating event and the other failure probabilities are gamma distribution and beta distribution, respectively [70].
- To maintain the uncertainty magnitude of event sequences, the error factors are consistent from the IE to HVAC failure probability. The error factor is defined as the 95th percentile divided by the median (50th percentile).
- Based on the same error factors, alpha and beta are determined by the equations:

$$\lambda_{IE} = \frac{\alpha_\gamma}{\beta_\gamma}$$

where λ_{IE} refers to the frequency of *IE*. α_γ and β_γ indicate alpha and beta parameters for gamma distribution, respectively. For determining alpha (α_β) and beta (β_β) parameters for beta distribution, the linearly interpolated failure probability (λ_{FP}) is used:

$$\lambda_{FP} = \frac{\alpha_\beta}{\alpha_\beta + \beta_\beta}$$

- To quantify the uncertainties along the event sequences, variances for gamma distribution (Var_γ) and beta distribution (Var_β) are calculated:

$$Var_\gamma = \frac{\alpha_\gamma}{{\beta_\gamma}^2}$$

$$Var_\beta = \frac{\alpha_\beta \beta_\beta}{(\alpha_\beta + \beta_\beta)^2(\alpha_\beta + \beta_\beta + 1)}$$

The calculated probability and parameters are shown in Table 3.

4.4. *Event Sequence Quantification by Using Phoenix Architect*

For this paper, the Phoenix Architect modular tools are utilized to quantify the event sequence and corresponding uncertainties. The frequency of the end states is calculated through the PRA Quant module by analyzing the event tree and master fault tree developed by the CAFTA module. For the uncertainty quantification in each sequence, the UNCERT module is used with the database established in the PRA Quant module and a 10,000 Monte Carlo sampling size. Tables 4 and 5 show the quantified frequency and uncertainties for the event sequence/release categories, respectively. In this paper, a sample size of 10,000 is selected as a reasonable balance between computational cost and output from the UNCERT module. The mean frequency converges to a point value as the sample size increases, but beyond certain sample sizes, it is not guaranteed that a larger sample size will be closer to the real value due to the randomness of Monte Carlo methods.

4.5. Mechanistic Source Term Analysis by Using ORIGEN

The design characteristics of a high-temperature gas-cooled reactor (HTGR) allow for high thermal efficiency, resulting in the extraction of more energy from the fuel. In this study, an HTGR with a thermal power of 250 MWth was modeled using ORIGEN 2.2 for depletion calculation. The main input parameters are shown in Table 6. The fuel element is spherical with a diameter of 60 mm and contains approximately 12,000 TRISO particles. The TRISO particles are evenly distributed within a graphite matrix that has a diameter of 50 mm. The spherical fuel zone is surrounded by a fuel-free zone made of pure graphite for mechanical and chemical protection. Each TRISO particle comprises a UO_2 kernel with a diameter of 0.5 mm and three additional layers of PyC and SiC. The heavy metal contained in each spherical fuel element is approximately 7.0 g. The fuel element is designed to have an average burn-up rate of 90 GWd/tU, with a maximum fuel burn-up not exceeding 100 GWd/tU.

Table 3. Data Analysis Results for Event Sequence Quantification.

		Frequency/ Probability	Distribution	Alpha	Beta	Error Factor	Variance
IE Frequency		5.60×10^{-5}	gamma	5.00×10^{-1}	8.93×10^{3}	8.44	6.27×10^{-9}
Canister Failure Probability with Dropping Height (m)	30	1.87×10^{-2}	beta	4.87×10^{-1}	2.56×10^{1}	8.44	6.79×10^{-4}
	25	9.41×10^{-3}	beta	4.94×10^{-1}	5.20×10^{1}	8.44	1.74×10^{-4}
	20	2.27×10^{-3}	beta	4.99×10^{-1}	2.19×10^{2}	8.44	1.03×10^{-5}
	15	1.05×10^{-3}	beta	5.00×10^{-1}	4.76×10^{2}	8.44	2.19×10^{-6}
	10	2.86×10^{-4}	beta	5.00×10^{-1}	1.75×10^{3}	8.44	1.64×10^{-7}
Pebble Failure Probability with Dropping Height (m)	30	1.10×10^{-1}	beta	4.18×10^{-1}	3.38	8.44	2.04×10^{-2}
	25	5.28×10^{-2}	beta	4.63×10^{-1}	8.30	8.44	5.13×10^{-3}
	20	1.28×10^{-2}	beta	4.92×10^{-1}	3.80×10^{1}	8.44	3.18×10^{-4}
	15	5.88×10^{-3}	beta	4.96×10^{-1}	8.38×10^{1}	8.44	6.85×10^{-5}
	10	1.61×10^{-3}	beta	4.99×10^{-1}	3.10×10^{2}	8.44	5.15×10^{-6}
HVAC Failure Probability		1.50×10^{-4}	beta	5.00×10^{-1}	3.33×10^{3}	8.44	4.50×10^{-8}

Table 4. Event Sequence Quantification for Sequence Numbers and Groups.

Sequence Number	Frequency	Sequence Number/Group	Frequency
1	9.968×10^{-6}	13	1.113×10^{-5}
2	1.209×10^{-6}	14	6.580×10^{-8}
3	2.304×10^{-8}	15	6.901×10^{-11}
4	3.457×10^{-12}	16	1.035×10^{-14}
5	1.061×10^{-5}	17	1.118×10^{-5}
6	5.860×10^{-7}	18	1.799×10^{-8}
7	5.567×10^{-9}	19	5.149×10^{-12}
8	8.352×10^{-13}	20	7.725×10^{-16}
9	1.106×10^{-5}	OK	5.395×10^{-5}
10	1.425×10^{-7}	Direct Exposure	2.021×10^{-6}
11	3.246×10^{-10}	Noble Gas	2.901×10^{-8}
12	4.870×10^{-14}	Radionuclide Release	4.352×10^{-12}

Table 5. Uncertainty Quantification Results from the Monte Carlo Sampling.

Sequence Number or Sequence Group	Mean Frequency	Uncertainty (10,000 Monte Carlo Samples)		
		5th Percentile	Median	95th Percentile
1	9.904×10^{-6}	4.230×10^{-8}	4.608×10^{-6}	3.776×10^{-5}
2	1.227×10^{-6}	1.283×10^{-10}	1.475×10^{-7}	6.134×10^{-6}
3	2.269×10^{-8}	2.138×10^{-13}	7.440×10^{-10}	9.862×10^{-8}
4	3.486×10^{-12}	3.697×10^{-18}	3.472×10^{-14}	1.087×10^{-11}
5	1.108×10^{-5}	4.173×10^{-8}	5.074×10^{-6}	4.180×10^{-5}
6	6.071×10^{-7}	9.754×10^{-11}	7.292×10^{-8}	2.838×10^{-6}
7	6.037×10^{-9}	7.098×10^{-14}	2.110×10^{-10}	2.657×10^{-8}
8	8.259×10^{-13}	1.311×10^{-18}	8.864×10^{-15}	2.666×10^{-12}
9	1.119×10^{-5}	4.519×10^{-8}	4.954×10^{-6}	4.249×10^{-5}
10	1.441×10^{-7}	2.748×10^{-11}	1.764×10^{-8}	6.921×10^{-7}
11	3.074×10^{-10}	5.343×10^{-15}	1.222×10^{-11}	1.356×10^{-9}
12	4.458×10^{-14}	9.377×10^{-20}	4.992×10^{-16}	1.424×10^{-13}
13	1.117×10^{-5}	4.078×10^{-8}	5.151×10^{-6}	4.250×10^{-5}
14	6.481×10^{-8}	1.403×10^{-11}	8.575×10^{-9}	3.100×10^{-7}
15	6.506×10^{-11}	1.360×10^{-15}	8.479×10^{-12}	2.865×10^{-10}
16	1.089×10^{-14}	1.958×10^{-20}	1.190×10^{-16}	3.271×10^{-14}
17	1.102×10^{-5}	4.290×10^{-8}	4.959×10^{-6}	4.292×10^{-5}
18	1.849×10^{-8}	3.831×10^{-12}	2.364×10^{-9}	8.844×10^{-8}
19	5.273×10^{-12}	9.506×10^{-17}	1.828×10^{-13}	2.117×10^{-11}
20	7.591×10^{-16}	1.440×10^{-21}	7.601×10^{-18}	2.362×10^{-15}
Ok	3.748×10^{-5}	1.608×10^{-7}	1.721×10^{-5}	1.398×10^{-4}
Direct Exposure	1.971×10^{-6}	3.977×10^{-9}	5.705×10^{-7}	8.682×10^{-6}
Noble Gas	2.928×10^{-8}	1.248×10^{-11}	2.781×10^{-9}	1.350×10^{-7}
Radionuclide Release	4.864×10^{-12}	1.290×10^{-16}	1.229×10^{-13}	1.469×10^{-11}

Table 6. HTGR design parameters.

Parameter	Value	Unit
Thermal power	250	MWth
Number of fuel elements	420,000	-
Number of TRISO per fuel	12,000	-
Fuel type	$U0_2$ TRISO	-
Enrichment	8.9	%
Heavy metal per fuel elements	7	g
Average burn-up	90	GWd/tU
Fuel residence time	1057	Days
Diameter of pebble	60	mm
Fuel zone	50	mm

The end-of-life cycle core inventory activity and the release fractions are given in Table 7. ORIGEN 2.2 is a powerful and versatile tool used in the nuclear industry for the analysis and management of nuclear materials by calculating the nuclide composition and activity of fuel and other irradiated materials. It is used to predict the isotopic evolution of nuclear fuel, radioactive waste, and other materials over time, and to estimate the radioactive source term and decay heat of spent nuclear fuel and other irradiated materials. The ability to generate decay and fission product data for use in other computational codes enables the ORIGEN to be an essential tool for the design and analysis of advanced nuclear fuel cycles, as well as for the management and disposal of nuclear waste. It is available through the Standardized Computer Analysis for Licensing Evaluation package, which is maintained by the Oak Ridge National Laboratory [79–81].

Table 7. End-of-life cycle core inventory activity and the release fraction.

Chemical Group	Element or Isotope	Per Pebble Radioactivity (Ci)	10 Pebble Radioactivity (Ci)	100 Pebble Radioactivity (Ci)	Release Fraction
Noble Gas	Kr85	2.207×10^{-1}	2.207	2.207×10^{1}	1.00
	Kr87	1.939×10^{18}	1.939×10^{19}	1.939×10^{20}	1.00
	Kr88	1.266×10^{18}	1.266×10^{19}	1.266×10^{20}	1.00
	Xe133	4.797×10^{16}	4.797×10^{17}	4.797×10^{18}	1.00
	Xe135	1.335×10^{17}	1.335×10^{18}	1.335×10^{19}	1.00
Halogens	I131	1.553×10^{16}	1.553×10^{17}	1.553×10^{18}	1.00
	I132	7.164×10^{18}	7.164×10^{19}	7.164×10^{20}	2.00×10^{-2}
	I133	3.161×10^{17}	3.161×10^{18}	3.161×10^{19}	1.00
	I134	3.550×10^{19}	3.550×10^{20}	3.550×10^{21}	1.00
	I135	1.099×10^{18}	1.099×10^{19}	1.099×10^{20}	1.00
Alkali Metals	Cs134	1.517×10^{14}	1.517×10^{15}	1.517×10^{16}	1.00
	Cs136	2.650×10^{-1}	2.650	2.650×10^{1}	1.00
	Cs137	2.063	2.063×10^{1}	2.063×10^{2}	1.00
	Rb86	1.014×10^{-2}	1.014×10^{-1}	1.014	1.00
Chalcogens	Te127	3.541×10^{-1}	3.541	3.541×10^{1}	2.00×10^{-2}
	Te129	1.246	1.246×10^{1}	1.246×10^{2}	2.00×10^{-2}
	Te132	1.241×10^{17}	1.241×10^{18}	1.241×10^{19}	2.00×10^{-2}
Alkali Earths	Sr89	4.614	4.614×10^{1}	4.614×10^{2}	2.00×10^{-3}
	Sr90	1.821	1.821×10^{1}	1.821×10^{2}	2.00×10^{-3}
	Sr91	4.531×10^{17}	4.531×10^{18}	4.531×10^{19}	2.00×10^{-3}
	Ba140	8.507×10^{15}	8.507×10^{16}	8.507×10^{17}	2.00×10^{-3}
	Y90	2.010	2.010×10^{1}	2.010×10^{2}	2.00×10^{-3}
	Y91	6.083	6.083×10^{1}	6.083×10^{2}	1.00×10^{-1}
Transition Elements	Zr95	5.271×10^{15}	5.271×10^{16}	5.271×10^{17}	1.00×10^{-2}
	Zr97	2.666×10^{17}	2.666×10^{18}	2.666×10^{19}	1.00×10^{-2}
	Nb95	1.017×10^{16}	1.017×10^{17}	1.017×10^{18}	1.00×10^{-2}
Miscellaneous	Sb127	3.533×10^{-1}	3.533	3.533×10^{1}	1.00
	Sb129	1.260	1.260×10^{1}	1.260×10^{2}	1.00
	Mo99	1.004×10^{16}	1.004×10^{17}	1.004×10^{18}	1.00×10^{-6}
	Ru103	6.429×10^{15}	6.429×10^{16}	6.429×10^{17}	2.00×10^{-5}

Table 7. *Cont.*

Chemical Group	Element or Isotope	Per Pebble Radioactivity (Ci)	10 Pebble Radioactivity (Ci)	100 Pebble Radioactivity (Ci)	Release Fraction
	Ru105	1.093×10^{18}	1.093×10^{19}	1.093×10^{20}	2.00×10^{-5}
	Ru106	2.819	2.819×10^{1}	2.819×10^{2}	2.00×10^{-5}
Lanthanides	La140	4.263×10^{17}	4.263×10^{18}	4.263×10^{19}	1.00×10^{-6}
	Ce141	6.260×10^{15}	6.260×10^{16}	6.260×10^{17}	1.00×10^{-6}
	Ce143	2.527×10^{17}	2.527×10^{18}	2.527×10^{19}	1.00×10^{-6}
	Ce144	3.881×10^{14}	3.881×10^{15}	3.881×10^{16}	1.00×10^{-6}
	Pr143	7.236	7.236×10^{1}	7.236×10^{2}	1.00×10^{-6}
	Nd147	7.265×10^{15}	7.265×10^{16}	7.265×10^{17}	1.00×10^{-6}
	Np239	1.122×10^{18}	1.122×10^{19}	1.122×10^{20}	1.00×10^{-6}
Transuranic	Pu238	7.329×10^{1}	7.329×10^{2}	7.329×10^{3}	1.00×10^{-6}
	Pu239	6.029×10^{-2}	6.029×10^{-1}	6.029	1.00×10^{-6}
	Pu240	4.425	4.425×10^{1}	4.425×10^{2}	1.00×10^{-6}
	Pu241	2.105×10^{-1}	2.105	2.105×10^{1}	1.00×10^{-6}
	Am241	6.758×10^{-1}	6.758	6.758×10^{1}	1.00×10^{-6}

4.6. *Radiological Consequence Analysis by Using MicroShield*

With the tabulated source data from the MST analysis and shielding material information, the radiological consequence can be quantified through the MicroShield. The MicroShield is used to model the source and shielding by determining their geometrical shapes and compositions of them. Additionally, the dose points can be selected to identify the consequences at the specific points. In this paper, the absorbed dose rate and exposure rate at 5 km from the source for three predetermined release categories (RR, NG, and DE) are calculated. Based on multiple independent variables, a sensitivity study is implemented to identify the impact of the input variables in this section.

4.6.1. Case Study 1

The first case study is to calculate the exposure rate and absorbed dose rate of each release category for quantification of the consequences. To simulate the radionuclide transport from the source, the shielding arrangement should be determined. As we can see in Figure 9, the cylinder volume geometry is selected for assuming a radioactive source at the bottom of a pebble fuel-filled canister. As mentioned above, there are two concrete walls as safety barriers: the inside wall between the canister and storage building and the outside wall between the inside wall and the environment. In the case of DE, the safety barriers function as the shielding, however, in the case of RR and NG, the radioactive materials bypass the concrete walls because of the HVAC system's operation status. Table 8 demonstrates the dimensions and material information for the shielding. Table 9 shows the exposure rate and absorbed dose rate as the consequence of each release category when only one pebble has failed at the center of the canister. As mentioned above, it is assumed that the radioactivity is detected at 5 km from the source on the X-axis.

4.6.2. Case Study 2—Sensitivity Study

The second case study is a sensitivity study to analyze the impact of the input variables. As shown in Table 10, 189 case datasets are generated to identify the input variables: the number of failed pebbles, the location of the failed pebbles, the inside wall thickness, and the outside wall thickness. The location of the failed pebble is equivalent to the distance between the canister and the source that depends on the number of failed pebbles because a number of pebbles occupy a large volume in the canister. Therefore, as the number of failed pebbles increases, the shorter the distance between the canister and source in the MicroShield model. The distance is varied from the source to zero which indicates that the location of the failed pebble moves from the center to the canister's surface.

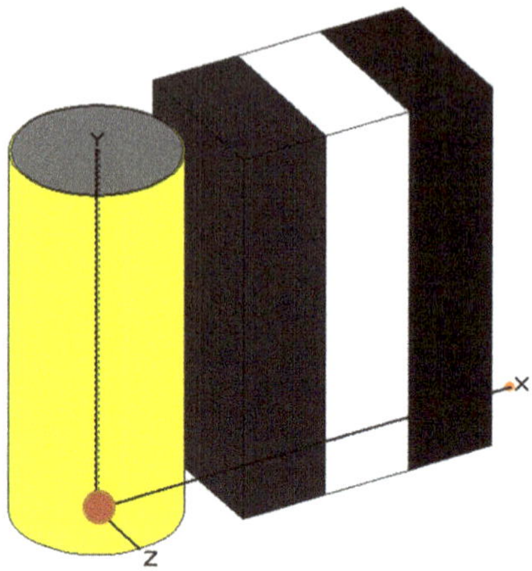

Figure 9. Radiological Consequence Simulation for Spent Pebble Fuel Failure in a Canister in MicroShield.

Table 8. Dimensions and Material Information for the Shielding.

	Material	Height (cm)	Thickness (cm)
Pebbles in the canister	Graphite		89
Canister	304 SL		2.5
Air-gap between the canister and the inside wall	Air		30
Inside wall (silo well)	Barite concrete	418	100
Air-gap between the inside wall and outside wall	Air		100
Outside wall (SFSS building)	NBS concrete		100

Table 9. Exposure Rate and Absorbed Dose Rate for Each Release Category.

Release Categories	Exposure Rate (mR/h)	Absorbed Dose Rate (mrad/h)
Radionuclide Release	1.63×10^{-6}	1.42×10^{-6}
Noble Gas	1.41×10^{-8}	1.23×10^{-8}
Direct Exposure	7.39×10^{-10}	6.45×10^{-10}

Table 10. Dataset Generation from the Input Variables.

Input Variables	Range	Numbers
Number of failed pebbles	1 to 30,000	7
Distance between canister surface and source	From the source to 0	3
Inside wall thickness	[75, 100, 125]	3
Outside wall thickness	[75, 100, 125]	3
Total number of datasets	189	

For the sensitivity analysis method, ordinary least square (OLS) is used to estimate the linearity of the parameters that implies the relationship between independent variables and one dependent variable. The linear regression analysis fitting the coefficients is to directly measure the impact from the sensitivity perspective [82]. The OLS parameter is updated to minimize the sum of the squared residual which is a discrepancy between the observed values and the corresponding fitted values. The equation of the OLS is as follows:

$$y_i = x_i^T \beta + \varepsilon_i$$

where x_i (k by 1 vector) and y_i are regressor (or predictor) and response in i-th observation, respectively. β (k by 1 vector) is a vector of unknown parameters and ε_i is a random variable that refers to the vertical distance between the actual values and fitted values. The residual $(S(b))$ can be calculated by the following equation where b is a candidate value for the β:

$$S(b) = \sum_{i=1}^{n} \left(y_i - x_i^T b \right)^2 = (y - Xb)^T (y - Xb)$$

The parameter $\hat{\beta}$ can be found by minimizing the residual in the equation below:

$$\hat{\beta} = \underset{b \in R}{argmin} S(b) = \left(\frac{1}{n} \sum_{i=1}^{n} x_i x_i^T \right)^{-1} \frac{1}{n} \sum_{i=1}^{n} x_i y_i = \left(X^T X \right)^{-1} X^T y$$

where X is the n by k design matrix. After beta estimation, the predicted value ($\hat{y}$) would be calculated:

$$\hat{y} = X\hat{\beta}$$

To assess the goodness-of-fit of the regression model, R-squared is used by measuring a ratio of 'predicted' variance to the 'actual' of the dependent variable. It represents how closely the observed data points are to the explained regression line. The R-squared is given by:

$$R^2 = \frac{\Sigma \left(\hat{y}_i - \bar{y} \right)^2}{\Sigma \left(y_i - \bar{y} \right)^2}$$

The sensitivity analysis results are summarized in Table 11 below. There are two cases for the sensitivity study: (A) exposure rate at the outside of the storage (10 m from the source) and (B) EAB. Before implementing the OLS method, data points are standardized because the input variables have different units and material characteristics. The results are summarized as:

- R-squared is low in the case of A, whereas, case B has a very high R-squared value (=1).
- The p-values are very low in both cases (p-value < 0.05).
- For case A, both the R-squared and the p-value are low. It indicates that the regression model discloses a significant statistical effect of input variables on response but is not good at predicting the responses from the input variables accurately because of unexplained variance. In other words, the data points are distributed further from the regression line. Whereas, the regression model for case B not only explains the responses well but also is able to predict the output accurately.

Table 11. Ordinary Least Square Method Results.

Exposure Rate	At the Outside of the Storage (A)	At the EAB (B)
Distance from the source	10 m	5 km
R-squared	0.108	1.00
p-value for goodness-of-fit test	2.89×10^{-4}	0

4.6.3. Discussion for Sensitivity Study

The major findings from the sensitivity study results shown in Table 12 are as follows.

Table 12. Sensitivity Analysis Results for Exposure Rate at the Outside of the Storage and EAB.

Impact Variables	At the Outside of the Storage (A)	At the EAB (B)
Number of pebbles	1.438×10^{-1}	1
Distance	-8.78×10^{-2}	-3.557×10^{-5}
Inside wall thickness	-1.889×10^{-1}	1.388×10^{-17}
Outside wall thickness	-1.848×10^{-1}	-1.735×10^{-17}
Sensitivity for case A		Sensitivity for case B

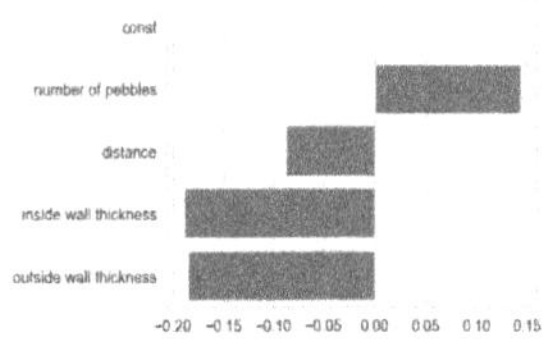

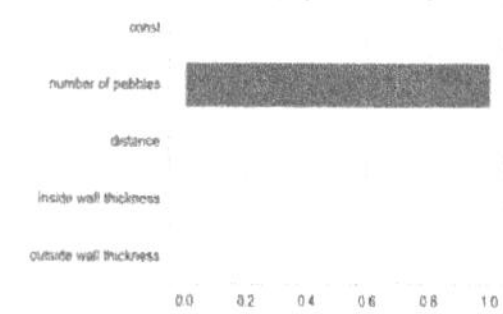

- Findings for case A are:
    - The order of the impact of the input variables is as follows: the inside wall thickness, the outside wall thickness, the number of pebbles, and the distance between the source and the canister surface. However, the difference between the coefficients of the inside wall thickness and the outside wall thickness is small.
    - Therefore, the wall thickness is the most significant variable to determine the exposure rate at the outside of the storage regardless of whether it is the inside wall or the outside wall.
    - Only the number of failed pebbles is positively sensitive to the exposure rate which means more failed pebble numbers and a larger exposure rate. Otherwise, the exposure rate decreases when the variables increase.
- Findings for case B are:
    - The coefficient of the number of pebbles is very high (=1) and the others' coefficients are extremely low. In other words, the number of pebbles is a dominant input variable for case B. The exposure rate at 5 km from the source does not depend on the distance and wall thickness because they are negligible compared to the EAB.

The OLS regression model can be misinterpreted in the impact analysis when there is collinearity between input variables. Thus, it is required to confirm the characteristics of independent variables in the input variable selection process. Moreover, since the OLS model is limited to predicting a single response variable, multi-output regression by using deep learning is recommended when there are two or more output variables.

4.7. Risk Integration for F-C Curve

The measure of risk considering frequency and consequence for an initiating event is given by the following equation [5]:

$$R = f \sum_{n=1}^{m} P_n K_n$$

where R is a risk, f is the frequency of the initiating event, which is a canister drop, and P_n and K_n are conditional probability in a given initiating event and corresponding consequence of n-th event sequence, respectively. Table 13 shows the integrated risk for each release category for one pebble failure case.

Table 13. Risk Integration for Each Release Category.

Release Categories	Consequence (mrad/h)	Consequence (30 Days-REM)	Frequency (/Year)	Risk (REM/Year)
Radionuclide Release	1.42×10^{-6}	1.022×10^{-6}	4.352×10^{-12}	4.449×10^{-18}
Noble Gas	1.23×10^{-8}	8.856×10^{-9}	2.901×10^{-8}	2.569×10^{-16}
Direct Exposure	7.39×10^{-10}	5.321×10^{-10}	2.021×10^{-6}	1.075×10^{-15}

In addition to the integrated risk assessment for the non-LWR PRA approach, Frequency-Consequence (F-C) target is used to evaluate the risk significance by comparing it against evaluation criteria derived from regulatory requirements and NEI 18-04 methodology [27]. The NEI 18-04 F-C target line, a black line shown in Figure 10, is also used to identify the risk margins that could be evidence of defense-in-depth methodology application. The figure below illustrates criterion lines for the F-C target, risk-significant LBE (black dash line under the F-C target line), and LBEs, such as anticipated operational occurrences (AOOs), Design Basis Events (DBEs), and Beyond DBE (BDBEs)—black dash vertical and horizontal grid lines. In this research, release categories' frequency with uncertainties (5th percentile, median, mean, and 95th percentile) and consequences are compared with the target line in the F-C curve when the failed pebbles are 1, 100, and 10,000. Overall, the frequencies and doses, including their uncertainty, for all the scenarios analyzed, meet the NEI 18-04 F-C target and the risk-significant LBE criterion.

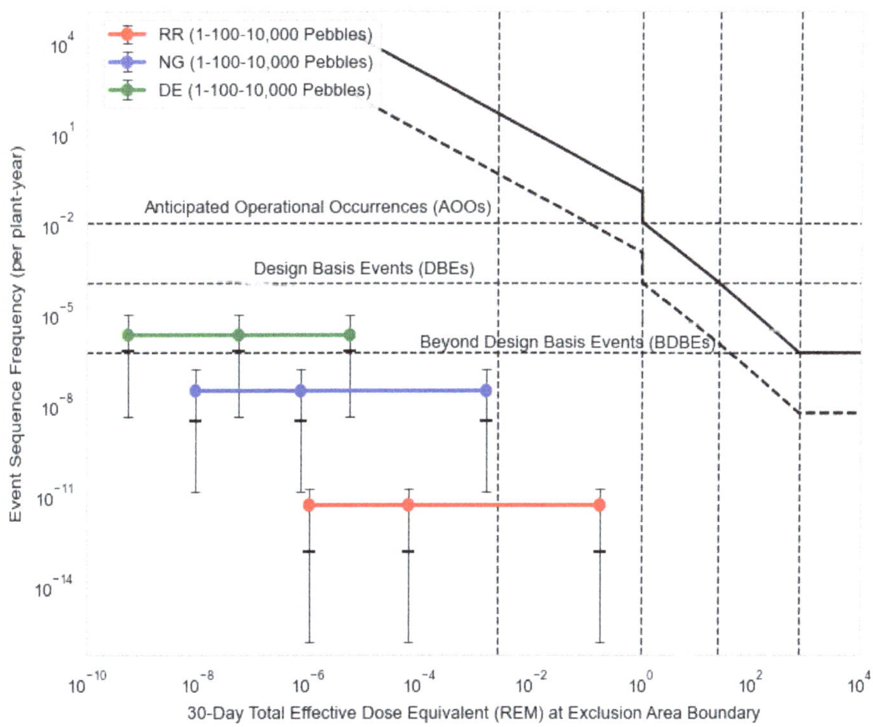

Figure 10. F-C Curve for Spent Pebble fuel-filled Dry Cask PRA.

5. Conclusions

In this research, the spent pebble fuel-filled dry cask PRA is performed in accordance with the PRA standard for the advanced non-LWR power plant. While TRISO-coated pebble fuel is in the spotlight as the fuel for the advanced reactor type, there is a lack of comprehensive study on the dry cask PRA for spent pebble fuel. Thus, this research investigates how the various PRA elements are systematically connected from the IE identification to the risk integration and applied to the pebble-filled dry cask PRA by demonstrating a case study. The ultimate goal of the research is to illustrate how the PRA elements are logically and structurally implemented by using existing methodologies, available data, and tools to contribute to risk-informing the design and operation of a dry cask storage system for spent pebble fuel.

To summarize the workflow in this paper, the MLD identified a canister vertical drop event as the selected IE from the varied height in an SFSS silo. From the given IE, the event tree is established through the pivot events determined by the pebble failure, canister damage, and HVAC system operation system states. The Phoenix Architecture tools are used to quantify the sequence and corresponding uncertainties of the release categories, such as direct exposure, noble gas, and radionuclide release. ORIGEN and MicroShield are used to quantify the radiological consequences of the dropped canister. The OLS methodology is utilized to identify the impact of each independent variable, including the number of failed pebbles, their location, and wall thickness, for exposure rate at the outside of the building and the EAB from the sensitivity study perspective. Finally, the risk is quantified and compared with the NEI 18-04 F-C target to confirm how much the risk would be for the given event scenarios at the EAB. Overall, the frequencies and doses, including their uncertainty, for all the scenarios analyzed, meet the NEI 18-04 F-C target and the risk-significant LBE criterion.

The challenge of this study Is the uncertainty stemming from the data and simulation model due to the lack of information and knowledge. From the uncertainty perspective in this paper, there are two types of uncertainties: epistemic and aleatory uncertainty. Although the CNID method is employed to reduce epistemic uncertainty by quantifying the event sequence frequency, additional studies are needed to determine the failure probabilities of the pebble and canister. Moreover, as the validation study for the number of damaged pebbles resulting from canister drop was implemented by comparing computational simulation and experimental results, it is recommended to conduct further validation studies to reduce uncertainties.

As mentioned earlier, the purpose of this paper is to explicitly demonstrate the PRA workflow through a case study. To simplify the process of examining the PRA elements, the scope of the PRA standard is delimited by selecting one IE and establishing assumptions, such as limiting the canister-dropping scenario and assuming state of sources in the canister. Thus, the PRA workflow can be extended to apply to the full scope of the PRA technical requirements, including the full scope of the iEs, by using not only the MLD but also FMEA or HBFT. Moreover, the CCF parameter estimation utilized the PRA model which could be considered as the extended scope of the PRA elements for SFSS design and operation. By developing a multi-variable and multi-output regression model based on the machine learning algorithm, the sensitivity study would be advanced not only to analyze the impact of the variable but also to evaluate the safety functions or barriers.

Author Contributions: Conceptualization, M.A.D.; Methodology, H.T., M.M.H., Y.A.A. and M.A.D.; Formal analysis, H.T., M.M.H. and Y.A.A.; Investigation, H.T. and M.A.D.; Data curation, J.L.; Writing—original draft, J.L., H.T. and M.A.D.; Writing—review & editing, M.M.H. and Y.A.A.; Visualization, J.L. and H.T.; Supervision, M.A.D.; Funding acquisition, M.A.D. All authors have read and agreed to the published version of the manuscript.

Funding: This research was performed as part of Havva Tayfur's Master of Science degree at North Carolina State University in the Department of Nuclear Engineering supported by the Republic of Turkey. This research was also partially funded through X-energy's subcontracts to North Carolina State University under the Advanced Research Projects Agency-Energy's (ARPA-E) Generating Electricity Managed by Intelligent Nuclear Assets (GEMINA) program and the Advanced Reactor Development Program.

Data Availability Statement: Data available on request.

Conflicts of Interest: The authors declare no conflict of interest.

Abbreviation

ANL	Argonne National Laboratory
AOO	Anticipated Operational Occurrence
BDBA	Beyond Design Basis Accident
CCF	Common Cause Failure
CDF	Core Damage Frequency
CNID	Constrained Noninformative Distribution
DA	Data Analysis
DBA	Design Basis Accident
DE	Direct Exposure
EAB	Exclusion Area Boundary
ESA	Event Sequence Analysis
ESQ	Event Sequence Quantification
ET	Event Tree
F-C	Frequency-Consequence
FEM	Finite Element Model
FHS	Fuel Handling System
FMEA	Failure Modes and Effort Analysis
FP	Fission Product
HAZOP	Hazard and Operability Analysis
HBFT	Heat Balance Fault Tree
HTR-PM	High-Temperature Gas-Cooled Reactor-Pebble Bed Module
HVAC	Heating, Ventilation, and Air Conditioning
IE	Initiating Event
INL	Idaho National Laboratory
IPyC	Inner Pyro-Carbon
LAR	License Amendment Requests
LBE	Licensing Basis Event
LERF	Large Early Release Frequency
LMP	Licensing Modernization Project
LWR	Light Water Reactor
MLD	Master Logic Diagram
MST	Mechanistic Source Term
NG	Noble Gas
NPP	Nuclear Power Plant
NRC	Nuclear Regulatory Commission
OLS	Ordinary Least Square
OPyC	Pyro-Carbon
ORNL	Oak Ridge National Laboratory
PBR	Pebble Bed Reactors
PHA	Process Hazards Analysis
PRA	Probabilistic Risk Assessment
RCA	Radiological Consequence Analysis
RI	Risk Integration
RR	Radionuclide Release
SFCLS	Spent Fuel Conveying and Loading System
SiC	Silicon Carbide

SNF	Spent Nuclear Fuel
SSC	Structures, Systems, and Components
TRISO	Tri-Structural Isotropic
VHTR	Very High-Temperature Gas Reactor

References

1. Bunn, M.; Holdren, J.P.; Macfarlane, A.; Pickett, S.E.; Suzuki, A.; Suzuki, T.; Weeks, J. *Project on Managing the Atom*; Belfer Center for Science and International Affairs: Cambridge, MA, USA, 2012; p. 146.
2. Romanato, L.S. Advantages of dry hardened cask storage over wet storage for spent nuclear fuel. In Proceedings of the 2011 International Nuclear Atlantic Conference—INAC 2011, Belo Horizonte, Brazil, 24–28 October 2011; p. 9.
3. Werner, J.D. *U.S. Spent Nuclear Fuel Storage*; R42513; Congressional Research Service Report for Congress: Washington, DC, USA, 2012; p. 57.
4. Sekiguchi, Y. *Mitigating the Risks of Spent Nuclear Fuel in Japan*; Center for Strategic and International Studies: Washington, DC, USA, 2017; p. 9.
5. Malliakos, A. *A Pilot Probabilistic Risk Assessment of a Dry Cask Storage System at a Nuclear Power Plant (NUREG-1864)*; US NRC: Washington, DC, USA, 2007.
6. JBiersdorf, M.; Eidelpes, E.F. *Development of Dry Cask Risk Tools*; INL/EXT-20-57570-Rev000, 1603757; U.S. Department of Energy: Oak Ridge, TN, USA, 2020. [CrossRef]
7. Fort, J.; Richmond, D.; Cuta, J.; Suffield, S. *Thermal Modeling of the TN-32B Cask for the High Burnup Spent Fuel Data Project*; PNNL-28915, 1566774; U.S. Department of Energy: Oak Ridge, TN, USA, 2019. [CrossRef]
8. Koskenranta, J.; Paavola, I.; Hotakainen, R.; Laato, T. Loviisa Nuclear Power Plant Spent Fuel Storage Risk Analysis. In Proceedings of the Probabilistic Safety Assessment and Management, PSAM 16, Honolulu, HI, USA, 26 June–1 July 2022; p. 12.
9. Olofsson, F.; Olsson, A. Challenges and Lessons Learned from a PSA on a Spent Fuel Pool Facility. In Proceedings of the Probabilistic Safety Assessment and Management, PSAM16, Honolulu, HI, USA, 26 June–1 July 2022; p. 7.
10. Kadak, A.C. A future for nuclear energy: Pebble bed reactors. *IJCIS* **2005**, *1*, 330. [CrossRef]
11. Brey, H.L. *Current Status and Future Development of Modular High Temperature Gas Cooled Reactor Technology*; IAEA-TECDOC-1198; IAEA, Nuclear Power Technology Development Section: Vienna, Austria, 2001.
12. INL. *Modular HTGR Safety Basis and Approach*; INL/EXT-11-22708; Idaho National Laboratory (INL): Idaho Falls, ID, USA, 2011.
13. Del Cul, G.D. *TRISO-Coated Fuel Processing to Support High Temperature Gas-Cooled Reactors*; ORNL/TM-2002/156, 814326; ORNL: Oak Ridge, TN, USA, 2002. [CrossRef]
14. Gen IV International Forum. *GIF Annual Report 2021*; Gen IV International Forum: Brussels, Belgium, 2021.
15. Moses, D.L. *Very High-Temperature Reactor (VHTR) Proliferation Resistance and Physical Protection (PR&PP)*; ORNL/TM-2010/163, 1027406; ORNL: Oak Ridge, TN, USA, 2011. [CrossRef]
16. Kovacic, D.; Gibbs, P.; Worrall, L.; Hunneke, R.; Harp, J.; Hu, J. *Advanced Reactor Safeguards: Nuclear Material Control and Accounting for Pebble Bed Reactors*; Oak Ridge National Laboratory: Oak Ridge, TN, USA, 2021.
17. Kovacic, D.; Gibbs, P.; Scott, L. *Model MC&A Plan for Pebble Bed Reactors*; Oak Ridge National Laboratory: Oak Ridge, TN, USA, 2020.
18. Hu, R.; Hu, G.; Zou, L.; Klingberg, A.; Fei, T.; Nunez, D. *FY20 SAM Code Developments and Validations for Transient Safety Analysis of Advanced Non-LWRs*; ANL/NSE-20/50, 1716517, 162791; U.S. Department of Energy: Oak Ridge, TN, USA, 2020. [CrossRef]
19. Novak, A.J.; Carlsen, R.W.; Schunert, S.; Balestra, P.; Reger, D.; Slaybaugh, R.N.; Martineau, R.C. Pronghorn: A Multidimensional Coarse-Mesh Application for Advanced Reactor Thermal Hydraulics. *Nucl. Technol.* **2021**, *207*, 1015–1046. [CrossRef]
20. Laboure, V.; Ortensi, J.; Hermosillo, A.; Strydom, G.; Balestra, P. *FY21 Status Report on the ART-GCR CMVB and CNWG International Collaborations, ART-M3AT-21IN0603011*; Idaho National Laboratory (INL): Idaho Falls, ID, USA, 2021.
21. IAEA. *Evaluation of High Temperature Gas Cooled Reactor Performance: Benchmark Analysis Related to the PBMR-400, PBMM, GT-MHR, HTR-10 and the ASTRA Critical Facility*; International Atomic Energy Agency: Vienna, Austria, 2013.
22. Zhang, Z.-Y.; Dong, Y.-J.; Shi, Q.; Li, F.; Wang, H.-T. 600-MWe high-temperature gas-cooled reactor nuclear power plant HTR-PM600. *Nucl. Sci. Tech.* **2022**, *33*, 101. [CrossRef]
23. Joksimovic, V.; Houghton, W.J.; Emon, D.E. HTGR Risk Assessment Study. In Proceedings of the International Conference on Nuclear Systems Reliability Engineering and Risk Assessment, Gatlinburg, TN, USA, 20 June 1977; General Atomics: San Diego, CA, USA, 1977. Available online: https://www.osti.gov/biblio/5450842-htgr-risk-assessment-study (accessed on 26 September 2021).
24. Herczeg, J. *Probabilistic Risk Assessment for The Standard Modular High Temperature Gas-Cooled Reactor, DOE-HTGR-86-011*; Civilian Reactor Development, Office of Nuclear Energy: Washington, DC, USA, 1987.
25. Savkina, M.D. *Probabilistic Accident Analysis of the Pebble Bed Modular Reactor for Use with Risk Informed Regulation*; MIT: Cambridge, MA, USA, 2004; Available online: https://dspace.mit.edu/bitstream/handle/1721.1/17748/56504086-MIT.pdf?sequence=2&isAllowed=y (accessed on 29 September 2021).
26. Demkowicz, P.A.; Liu, B.; Hunn, J.D. Coated particle fuel: Historical perspectives and current progress. *J. Nucl. Mater.* **2019**, *515*, 434–450. [CrossRef]

27. Moe, W.L.; Afzali, A. *Modernization of Technical Requirements for Licensing of Advanced Non-Light Water Reactors Selection and Evaluation of Licensing Basis Events*; INL/EXT-19-55513-Rev1; Idaho National Lab. (INL): Idaho Falls, ID, USA, 2020. [CrossRef]
28. Moe, W.L.; Afzali, A. *Modernization of Technical Requirements for Licensing of Advanced Non-Light Water Reactors: Safety Classification and Performance Criteria for Structures, Systems, and Components*; INL/EXT-19-55516-Rev000; SC-29980-102.Rev0; Idaho National Lab. (INL): Idaho Falls, ID, USA; Southern Company Services: Birmingham, AL, USA, 2019. [CrossRef]
29. Fleming, K.; Wallace, E.; Afzali, A. Use of PRA to Select Licensing Basis Events. In *PSAM14*; UCLA: Los Angeles, CA, USA, 2018; p. 12.
30. US NRC. *NRC Non-Light Water Reactor (Non-LWR) Vision and Strategy, Volume 3—Computer Code Development Plans for Severe Accident Progression, Source Term, and Consequence Analysis*; US NRC: Washington, DC, USA, 2020.
31. ASME. *Probabilistic Risk Assessment Standard for Advanced Non-LWR Nuclear Power Plants*; ANS, ASME/ANS RA-S-1.4-2021; ASME: New York, NY, USA, 2021.
32. Braverman, J.I.; Morante, R.J.; Xu, J.; Hofmayer, C.H.; Shaukat, S.K. *Impact Analysis of Spent Fuel Dry Casks under Accidental Drop Scenarios*; U.S. Department of Energy: Oak Ridge, TN, USA, 2003; Volume BNL-NUREG-71196-2003-CP, p. 8.
33. Lin, M.; Li, Y. Analysis of the interactions between spent fuel pebble bed and storage canister under impact loading. *Nucl. Eng. Des.* **2020**, *361*, 110548. [CrossRef]
34. Lin, M.; Wang, J.; Wu, B.; Li, Y. Dynamic analysis of dry storage canister and the spent fuels inside under vertical drop in HTR-PM. *Ann. Nucl. Energy* **2021**, *154*, 108030. [CrossRef]
35. IAEA. *Defining Initiating Events for Purposes of Probabilistic Safety Assessment*; IAEA-TECDOC-719; IAEA: Vienna, Austria, 1993.
36. Garrick, B.J.; Fleming, K.N. *Seabrook Station Probabilistic Safety Assessment*; PLG-0242; Pickard, Lowe and Garrick, Inc.: Newport Beach, CA, USA, 1982.
37. Papazoglou, I.A.; Aneziris, O.N. Master Logic Diagram: Method for hazard and initiating event identification in process plants. *J. Hazard. Mater.* **2003**, *97*, 11–30. [CrossRef]
38. Cho, N.-C.; Jae, M.; Joon-Eon, Y. Initiating Events Identification of the IS Process Using the Master Logic Diagram. In Proceedings of the Korean Nuclear Society Conference, Taejon, Republic of Korea, 11–13 April 2005; pp. 55–56.
39. Purba, J.H. Master Logic Diagram: An Approach to Identify Initiating Events of HTGRs. *J. Phys. Conf. Ser.* **2018**, *962*, 012036. [CrossRef]
40. Lee, H.; Park, J. Analysis of Initiating Events for SMART using Heat Balance Fault Tree Method. In Proceedings of the Korean Nuclear Society, Jeju, Republic of Korea, 14–20 April 2017; p. 4.
41. Han, S.; Park, J.; Jang, S. *Identification of Initiating Events Using the Master Logic Diagram in Low-Power and Shutdown PSA for Nuclear Power Plant*; KAERI/TR-2497/2003; KAERI: Daejeon, Republic of Korea, 2003.
42. Liu, H.-C.; Liu, L.; Liu, N. Risk evaluation approaches in failure mode and effects analysis: A literature review. *Expert Syst. Appl.* **2013**, *40*, 828–838. [CrossRef]
43. Aldemir, T.; Stovsky, M.P.; Kirschenbaum, J.; Mandelli, D.; Bucci, P.; Mangan, L.A.; Miller, D.W.; Sun, X.; Ekici, E.; Guarro, S.; et al. *Dynamic Reliability Modeling of Digital Instrumentation and Control Systems for Nuclear Reactor Probabilistic Risk Assessments*; NUREG/CR-6942; Ohio State University: Columbus, OH, USA, 2007.
44. Gaol, F.L.; Nguyen, Q.V. (Eds.) *Proceedings of the 2011 2nd International Congress on Computer Applications and Computational Science*; Springer: Berlin/Heidelberg, Germany, 2012; Volume 144. [CrossRef]
45. Cadwallader, L.C. *Preliminary Failure Modes and Effects Analysis of the US DCLL Test Blanket Module*; U.S. Department of Energy: Oak Ridge, TN, USA, 2010; p. 158.
46. Bright, M.; Foster, R.; Hampton, C.; Ruiz, J.; Moeller, B. Failure modes and effects analysis for surface-guided DIBH breast radiotherapy. *J. Appl. Clin. Med. Phys.* **2022**, *23*, e13541. [CrossRef] [PubMed]
47. Huq, M.S.; Fraass, B.A.; Dunscombe, P.B.; Gibbons, J.P., Jr.; Ibbott, G.S.; Mundt, A.J.; Mutic, S.; Palta, J.R.; Rath, F.; Thomadsen, B.R.; et al. The report of Task Group 100 of the AAPM: Application of risk analysis methods to radiation therapy quality management: TG 100 report. *Med. Phys.* **2016**, *43*, 4209–4262. [CrossRef] [PubMed]
48. Li, S.; Zeng, W. Risk analysis for the supplier selection problem using failure modes and effects analysis (FMEA). *J. Intell. Manuf.* **2016**, *27*, 1309–1321. [CrossRef]
49. Chi, C.-F.; Sigmund, D.; Astardi, M.O. Classification Scheme for Root Cause and Failure Modes and Effects Analysis (FMEA) of Passenger Vehicle Recalls. *Reliab. Eng. Syst. Saf.* **2020**, *200*, 106929. [CrossRef]
50. Sowder, A. *Program on Technology Innovation: Early Integration of Safety Assessment into Advanced Reactor Design—Preliminary Body of Knowledge and Methodology*; 2018 Technical Report 3002011801; Electric Power Research Institute: Palo Alto, CA, USA, 2018.
51. Marciulescu, C. *Program on Technology Innovation: Early Integration of Safety Assessment into Advanced Reactor Design—Project Capstone Report*; 2019 Technical Report 3002015752; Electric Power Research Institute: Palo Alto, CA, USA, 2019.
52. Rimkevičius, S.; Vaišnoras, M.; Babilas, E.; Ušpuras, E. HAZOP application for the nuclear power plants decommissioning projects. *Ann. Nucl. Energy* **2016**, *94*, 461–471. [CrossRef]
53. Joubert, J.; Kohtz, N.; Coe, I. South African Safety Assessment Framework for the Pebble Bed Modular Reactor. In *Fourth International Topical Meeting on High Temperature Reactor Technology, Volume 2*; ASME: Washington, DC, USA, 2008; pp. 193–203. [CrossRef]

54. Afzali, A. *Molten Salt Reactor Experiment (MSRE) Case Study Using Risk-Informed, Performance—Based Technical Guidance to Inform Future Licensing for Advanced Non-Light Water Reactors*; EPRI AR LR 2019-06; Southern Company: Atlanta, GA, USA, 2019.
55. Taylor, J.M. *Issues Pertaining to the Advanced Reactor (Prism, Mhtgr, And Pius) And Candu 3 Designs and Their Relationship to Current Regulatory Requirements*; SECY-93-092; US NRC: Washington, DC, USA, 1993.
56. Andrews, N.; Nenoff, T.; Luxat, D.; Clark, A.; Leute, J. *Mechanistic Source Term Considerations for Advanced Non-LWRs*; SAND--2020-6730, 1638572, 687170; U.S. Department of Energy: Oak Ridge, TN, USA, 2020. [CrossRef]
57. Moe, W.L. *Risk-Informed Performance-Based Technology Inclusive Guidance for Advanced Reactor Licensing Basis Development*; INL/EXT-19-55375-Rev000; Idaho National Lab. (INL): Idaho Falls, ID, USA, 2019. [CrossRef]
58. US NRC. *White Paper on Options for Risk Metrics for New Reactors*; US NRC: Washington, DC, USA, 2009.
59. Boyer, R.L. *Probabilistic Risk Assessment (PRA): Analytical Process for Recognizing Design and Operational Risks*; Safety & Mission Assurance (S&MA), NASA Johnson Space Center: Houston, TX, USA, 2017.
60. Phillips, J.A.; Nagley, S.G.; Shaber, E.L. Fabrication of uranium oxycarbide kernels and compacts for HTR fuel. *Nucl. Eng. Des.* **2012**, *251*, 261–281. [CrossRef]
61. Seibert, R.L.; Jolly, B.C.; Balooch, M.; Schappel, D.P.; Terrani, K.A. Production and characterization of TRISO fuel particles with multilayered SiC. *J. Nucl. Mater.* **2019**, *515*, 215–226. [CrossRef]
62. Bruna, G.B.; Bourgois, T.; Ivanov, E.; Monhardt, D. *Overview of Generation IV (Gen IV) Reactor Designs*; IRSN Report 2012/158; IRSN: Paris, France, 2012.
63. Morris, R.N.; Petti, D.A.; Powers, D.A.; Boyack, B.E. *TRISO-Coated Particle Fuel Phenomenon Identification and Ranking Tables (PIRTs) for Fission Product Transport Due to Manufacturing, Operations and Accidents*; NUREG/CR-6844; US NRC: Washington, DC, USA, 2004; Volume 1.
64. Kwapis, E.H.; Liu, H.; Hartig, K.C. Tracking of individual TRISO-fueled pebbles through the application of X-ray imaging with deep metric learning. *Prog. Nucl. Energy* **2021**, *140*, 103913. [CrossRef]
65. Moormann, R.; Kemp, R.S.; Li, J. Caution Is Needed in Operating and Managing the Waste of New Pebble-Bed Nuclear Reactors. *Joule* **2018**, *2*, 1911–1914. [CrossRef]
66. Wu, B.; Wang, J.; Li, Y.; Wang, H.; Ma, T. Design, Experiment, and Commissioning of the Spent Fuel Conveying and Loading System of HTR-PM. *Sci. Technol. Nucl. Install.* **2022**, *2022*, e1817191. [CrossRef]
67. Wang, J.; Zhang, Z.; Wu, B.; Li, Y. Design of the HTR-PM Spent Fuel Storage Facility. In Proceedings of the 7th International Topical Meeting on High Temperature Reactor Technology: The Modular HTR Is Advancing towards Reality Papers and Presentations, Weihai, China, 27–31 October 2014; p. v.
68. Wang, J.; Wang, B.; Wu, B.; Li, Y. Design of the Spent Fuel Storage Well of HTR-PM. In Proceedings of the 2016 24th International Conference on Nuclear Engineering, Charlotte, NC, USA, 26–30 June 2016. [CrossRef]
69. Ma, Z.; Kvarfordt, K.; Wierman, T. *Industry-Average Performance for Components and Initiating Events at U.S. Commercial Nuclear Power Plants: 2020 Update*; INL/EXT-21-65055-Rev000, 1847110; Idaho National Lab. (INL): Idaho Falls, ID, USA, 2022. [CrossRef]
70. Atwood, C.L.; LaChance, J.; Martz, H.F.; Anderson, D.J.; Englehardt, M.; Whitehead, D.; Wheeler, T. *Handbook of Parameter Estimation for Probabilistic Risk Assessment*; NUREG/CR-6823; Sandia National Lab.: Albuquerque, NM, USA, 2003.
71. IAEA. Performance of Engineered Barrier Materials in Near Surface Disposal Facilities for Radioactive Waste. International Atomic Energy Agency, Text. 2001. Available online: https://www.iaea.org/publications/6297/performance-of-engineered-barrier-materials-in-near-surface-disposal-facilities-for-radioactive-waste (accessed on 10 March 2023).
72. Severe Accident Risks: An Assessment for Five U.S. Nuclear Power Plants—Final Summary Report (NUR. NRC Web). Available online: https://www.nrc.gov/reading-rm/doc-collections/nuregs/staff/sr1150/v1/index.html (accessed on 6 January 2023).
73. IAEA. Handbook of Parameter Values for the Prediction of Radionuclide Transfer in Terrestrial and Freshwater Environments. International Atomic Energy Agency, Text. 2010. Available online: https://www.iaea.org/publications/8201/handbook-of-parameter-values-for-the-prediction-of-radionuclide-transfer-in-terrestrial-and-freshwater-environments (accessed on 10 March 2023).
74. Sailor, V.L.; Perkins, K.R.; Weeks, J.R.; Connell, H.R. *Severe Accidents in Spent Fuel Pools in Support of Generic Safety, Issue 82*; Div. of Reactor and Plant Systems, NUREG/CR-4982; BNL-NUREG-52093; Brookhaven National Lab. (BNL): Upton, NY, USA; Nuclear Regulatory Commission: Washington, DC, USA, 1987. [CrossRef]
75. Travis, R.J.; Davis, R.E.; Grove, E.J.; Azarm, M.A. *A Safety and Regulatory Assessment of Generic BWR and PWR Permanently Shutdown Nuclear Power Plants*; Div. of Regulatory Applications; NUREG/CR-6451; BNL-NUREG-52498; US Nuclear Regulatory Commission (NRC): Washington, DC, USA; Brookhaven National Lab. (BNL): Upton, NY, USA, 1997. [CrossRef]
76. U.S. Nuclear Regulatory Commission. *Population-Related Siting Considerations for Advanced Reactors*; NRC Staff Prepared White Paper; U.S. Nuclear Regulatory Commission: Rockville, MD, USA, 2019.
77. Moore, R.L.; Oh, C.H.; Merrill, B.J.; Petti, D.A. *Studies on Air Ingress for Pebble Bed Reactors*; IAEA: Vienna, Austria, 2002.
78. Hadad, Y.; Jafarpur, K. Modeling of Laminar Forced Convection Heat Transfer in Packed Beds with Pebbles of Arbitrary Geometry. *J. Por. Media* **2013**, *16*, 1049–1061. [CrossRef]
79. Zheng, M.; Tian, W.; Wei, H.; Zhang, D.; Wu, Y.; Qiu, S.; Su, G. Development of a MCNP–ORIGEN burn-up calculation code system and its accuracy assessment. *Ann. Nucl. Energy* **2014**, *63*, 491–498. [CrossRef]
80. Bowman, S.M. SCALE 6: Comprehensive Nuclear Safety Analysis Code System. *Nucl. Technol.* **2011**, *174*, 126–148. [CrossRef]

81. Ismail, A.S.; Takip, K.M.; Mustafa, M.K.A.; Anwar, A. RTP: Radionuclides Inventories Calculation Using Origen Code. In Proceedings of the R and D Seminar 2012: Research and Development Seminar 2012, Bangi, Malaysia, 26–28 September 2012.
82. Mohantyl, S. Sensitivity analysis methods for identifying influential parameters in a problem with a large number of random variables. In *Risk Analysis III*; WIT Press: Southampton, UK, 2002.

eng

MDPI

Article

A Deep Learning-Based Visual Map Generation for Mobile Robot Navigation

Carlos A. García-Pintos [1], Noé G. Aldana-Murillo [1], Emmanuel Ovalle-Magallanes [2,*] and Edgar Martínez [3,*]

[1] Departamento de Ingeniería, Universidad Iberoamericana León, Blvd. Jorge Vértiz Campero 1640, Leon 37238, Mexico; 184430-3@iberoleon.edu.mx (C.A.G.-P.); noe.aldana@iberoleon.mx (N.G.A.-M.)
[2] Telematics (CA), Engineering Division of the Campus Irapuato-Salamanca (DICIS), University of Guanajuato, Carretera Salamanca-Valle de Santiago km 3.5 + 1.8 km, Comunidad de Palo Blanco, Salamanca 36885, Mexico
[3] Tecnológico Nacional de Mexico/ITS de Guanajuato, Guanajuato 36262, Mexico
* Correspondence: e.ovallemagallanes@ugto.mx (E.O.-M.); emartinez@itesg.edu.mx (E.M.)

Abstract: Visual map-based robot navigation is a strategy that only uses the robot vision system, involving four fundamental stages: learning or mapping, localization, planning, and navigation. Therefore, it is paramount to model the environment optimally to perform the aforementioned stages. In this paper, we propose a novel framework to generate a visual map for environments both indoors and outdoors. The visual map comprises key images sharing visual information between consecutive key images. This learning stage employs a pre-trained local feature transformer (LoFTR) constrained with a 3D projective transformation (a fundamental matrix) between two consecutive key images. Outliers are efficiently detected using marginalizing sample consensus (MAGSAC) while estimating the fundamental matrix. We conducted extensive experiments to validate our approach in six different datasets and compare its performance against hand-crafted methods.

Keywords: deep learning; local feature matching; mobile robot; monocular vision; visual map

Citation: García-Pintos, C.A.; Aldana-Murillo, N.G.; Ovalle-Magallanes, E.; Martínez, E. A Deep Learning-Based Visual Map Generation for Mobile Robot Navigation. *Eng* **2023**, *4*, 1616–1634. https://doi.org/10.3390/eng4020092

Academic Editor: Antonio Gil Bravo

Received: 11 April 2023
Revised: 31 May 2023
Accepted: 2 June 2023
Published: 6 June 2023

1. Introduction

Humans can observe their environment and extract, describe, and store high-consistency areas of what they see, which can be understood as compositions of objects forming structures. This process, which has a spatial and relational component, is known as the visual memory scheme. It refers to an image's mental representations stored in the human brain's memory cells. This ability to remember reference images of previous locations allows humans to locate themselves and move in the environment.

In mobile robotics, autonomous navigation is crucial, as it allows robots to navigate and adapt to complex and dynamic environments. There are three main types of navigation [1]. The first is reactive navigation, where the robot has no information about the environment and is limited to avoiding obstacles. The second requires an a priori representation of the environment, such as a map. The third scenario involves the robot building a map as it moves through the environment. Typically, map-building algorithms use landmarks or features extracted by a sensor on the robot, such as measurements obtained by a laser sensor or visual features extracted from images captured by a camera.

A robot navigation strategy that uses a vision system as the only sensor is based on a map of images (visual map) scheme similar to human visual memory [2]. The representation used in this strategy can be seen as a topological map, where each node of the environment is represented through the use of images called key images. Recent works have used visual maps for localization, route planning, and navigation [3–7].

Visual map-based navigation systems involve four fundamental stages [2,8]:

1. Learning stage: The visual map is constructed using key images from an unknown environment. The images are taken during human-guided navigation or autonomous exploration.

2. Localization stage: Recognition algorithms attempt to find correspondence between the current image of the robot and the most similar image in the visual map.
3. Visual route planning stage: The route allows the robot to reach a desired position through landmarks and image comparison within the visual map.
4. Autonomous navigation stage: During this phase, the robot should theoretically reach the desired position by following the visual path defined in the previous stages.

In this work, we will focus on the learning stage to obtain an optimal visual map from the environment to allow visual localization and autonomous navigation. Figure 1 illustrates this visual map representation.

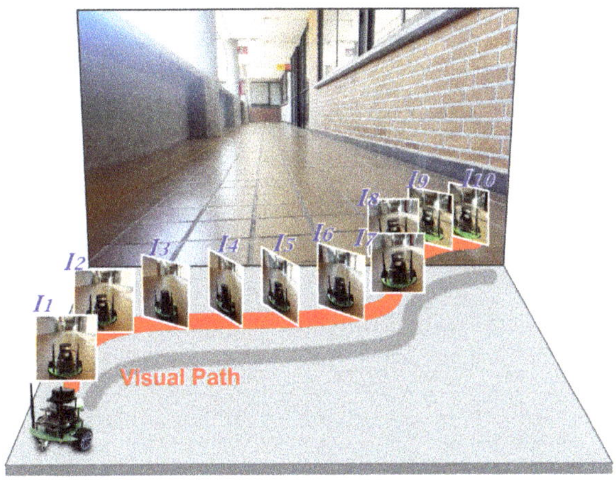

Figure 1. The visual map $\mathcal{M}$ comprises a set of key images that model an environment where the robot can be localized and navigated.

We propose a novel framework based on a detector-free deep neural network that enables the generation of feature-reach visual maps for indoor and outdoor environments. Our deep learning approach relies on pre-trained LoFTR (local feature transformer), including a geometry constraint to ensure that two consecutive key images share features to generate a visual control policy between them. We performed exhaustive experiments and made numerical and visual comparisons between the local hand-crafted feature extraction and matching of Oriented FAST and Rotated BRIEF (ORB).

The paper is organized as follows. First, the related work is presented in Section 2. Subsequently, the theoretical background and the proposed method are described in Section 3. The numerical results are shown and interpreted in Section 4, and finally, the conclusions are given in Section 5.

2. Related Work

Our work is related to visual map schemes for robotics navigation and deep learning-based methods to build a compact representation of the environment using only images. Our work proposes a framework based on a detector-free deep neural network to generate feature-reach visual maps for multiple environments.

Reference [9] presents the construction of a visual map suited for humanoid robot navigation. It proposes a genetic algorithm that estimates the epipolar geometry to dive into the image-matching problem in the construction process of a visual map.

The work of [10] proposes VLMaps. This spatial map representation fuses pre-trained visual-language features with a 3D reconstruction of the physical world to enable natural language indexing of the map without additional effort.

In [11], a memory construction module that builds a topological graph is proposed, based on supervised learning and a graph attention module that extracts guided attention features. The deep reinforcement learning-based control module makes decisions based on visual observations and guided attention features.

In [12], the authors present a survey of the use of semantic knowledge in robot navigation. The article discusses the use of behavioral mechanisms based on human psychology and the processes associated with thinking. The authors explore how semantic knowledge has opened new paths in robot navigation, allowing a higher level of abstraction in the representation of information.

The scheme presented in [6] integrates humanoid localization in the visual map for autonomous navigation. A pure vision-based localization algorithm is used to find the key image that best fits the current image. The visual map is updated when a new obstacle is detected.

In the framework of [13], a graph-based unsupervised semantic clustering method and cluster matching create a multilayer semantic visual memory map robust to illumination changes. They use a community detection algorithm (CDA) to test the visual data obtained from an unmanned aerial robot and public datasets. Then, metric information is obtained by a hierarchical agglomerative clustering algorithm to refine the extracted communities.

A wide diversity of methods have been proposed for image-matching tasks, which are based on deep learning techniques to be implemented. In [14], a full review and analysis are developed to compare the classical and the latest techniques, taking into count the feature detection, description, and matching techniques from hand-crafted methods, and introducing the reader to typical image matching-base applications. Recent works use LoFTR, a detector-free deep learning matching model, and a matching process with MAGSAC++ estimator to propose a transformer-based feature matching approach to find the same key points from two images with different perspectives of the shot [15].

3. Methodology

3.1. Fundamental Matrix

Visual constraints relate two views from monocular images that share some part of their field of view. They can be estimated from the images when there are correspondences between visual features identified in both images. Visual constraints can be used for the autonomous navigation of mobile robots through a model-based predictive feedback control (MPC) scheme [3]. Two visual constraints exist between two images: the homography matrix and the fundamental matrix [16]. The homography matrix can be estimated if points of the image are detected in a plane of a scene. To estimate the fundamental matrix, it is necessary to find points in different depths of a scene. The fundamental matrix is proposed since it can be estimated from points detected at different depths, which is the case in real environments. The visual map must be chosen to estimate the fundamental matrix between each pair of images in the map.

The fundamental matrix $\mathbf{F}$ encodes the epipolar geometry (see Figure 2) and relates corresponding image points $\mathbf{x}_1, \mathbf{x}_2$ between views through the epipolar constraint:

$$\mathbf{x}_1^T \mathbf{F} \mathbf{x}_2 = 0. \tag{1}$$

The fundamental matrix can be computed from the current and target images (images captured on the final position) with at least 7-point correspondences [16].

Methods for feature matching can give bad matches (outliers), and a wrong fundamental matrix can be found. One strategy to remove outliers when estimating a fundamental matrix is to use the random sampling consensus (RANSAC) [17] technique. The idea behind RANSAC is that the probability of selecting only inliers in a random sample increases as the sample size increases. By repeating the process multiple times, RANSAC can find a fundamental matrix that is robust to outliers.

Another robust and efficient algorithm for estimating a fundamental matrix from corresponding points in two images, even in the presence of outliers, is the marginalizing sample consensus (MAGSAC) [18]. The MAGSAC algorithm works by first randomly selecting a minimal set of points and estimating the fundamental matrix using the normalized eight-point algorithm [16]. Next, the algorithm evaluates the quality of the estimated fundamental matrix by counting the number of points that satisfy the epipolar constraint defined by the fundamental matrix (inliers). The MAGSAC algorithm repeats this process multiple times, each time selecting a new random set of points and estimating a new fundamental matrix. The algorithm then selects the fundamental matrix with the largest number of inliers as the final estimate.

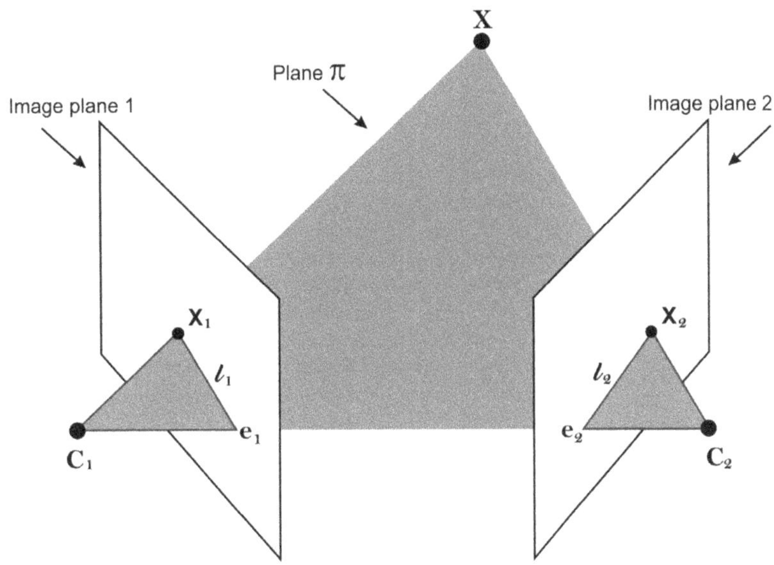

Figure 2. The fundamental matrix: two-view geometry in the case of non-planar scenes. An **X** point in 3D is shown in two images, $\mathbf{x}_1$ in the first and $\mathbf{x}_2$ in the second. The camera centers ($\mathbf{C}_1$ and $\mathbf{C}_2$), the point **X**, and its projections $\mathbf{x}_1$ and $\mathbf{x}_2$ lie in a common plane π. $\mathbf{e}_1$ and $\mathbf{e}_2$ are the epipolar points. The epipole is the point of intersection of the line joining the camera centers with the image plane. The fundamental matrix is the algebraic representation of epipolar geometry.

3.2. Local Descriptors: ORB and LOFTR

Deep learning methods are now the de facto starting point for wide computer vision tasks. However, many renowned works with traditional hand-crafted local features have achieved good performance with low computational costs for feature extraction and matching. Moreover, it is desired that local features be robust against variations in scale, orientation, lighting, occlusions, and viewpoint changes regarding the locations of the features and terms of the local descriptors. In such a way, from the existing local detectors/descriptors, we found two outstanding methods: Oriented FAST and Rotated BRIEF (ORB) [19], and LOFTR (local feature transformer) [20].

ORB is a feature detection and description algorithm that improves FAST (features from accelerated segment test) [21] and BRIEF (binary robust independent elementary features) [22] algorithms. The ORB algorithm has two main steps: feature detection and feature description. In the feature detection step, it involves detecting distinctive features (key points) in the image. As a key point detector, ORB uses FAST points by comparing the gray levels along a circle of radius three to the gray level of the circle center. In the feature description, a descriptor is generated for each key point once the key points are detected. The ORB descriptor is a binary vector ($\mathbf{d}_i \in \mathbb{R}^{256}$), where each bit results from an intensity comparison between some pairs of pixels within a patch around the detected key points that compute a dominant orientation between the center of the key point and the intensity centroid of its patch. Notice that the ORB descriptors represent image features by binary strings. Thus, the extracted information is very compact, occupies less memory, and can be compared faster.

The brute force matcher is a simple feature-matching algorithm in computer vision and image processing. It takes the descriptor of one feature in the first set and matches it with all other features in the second set using some distance calculation. The closest match is returned as a result. This process is repeated for all the features in the first set. The algorithm is called "brute force" because it compares all possible matches between the two groups of features. The brute force matcher is used to match the features of one image with another.

On the other hand, LoFTR (local feature transformer) [20] is a detector-free deep neural network architecture that performs local feature matching end-to-end. It comprises four main components: patch embedding, multi-scale transformer network, local feature descriptor, and geometric verification. Firstly, the patch embedding component takes an image patch and computes a multi-level feature vector representation of it. Secondly, the multi-scale transformer network processes the feature vector of image patches to capture local feature information at different scales. Then, the local feature descriptor takes the transformer network output as input and maps it to a fixed-length feature descriptor ($d_i \in \mathbb{R}^{1024}$) that captures the local geometry and appearance information of the i-th image patch. Finally, geometric verification estimates a homography matrix that aligns two images based on the matched features after extracting feature descriptors from both images. The pre-trained model has 11.56 million parameters and the inference runs at 116 milliseconds to process a 640 $\times$ 480 image pair on an NVIDIA RTX2080Ti with 11 GB of GPU memory.

Regardless of the framework used (hand crafted or deep learning based), the process for matching two images remains the same: feature detection, feature description, feature matching, and geometry verification. This is illustrated in Figure 3.

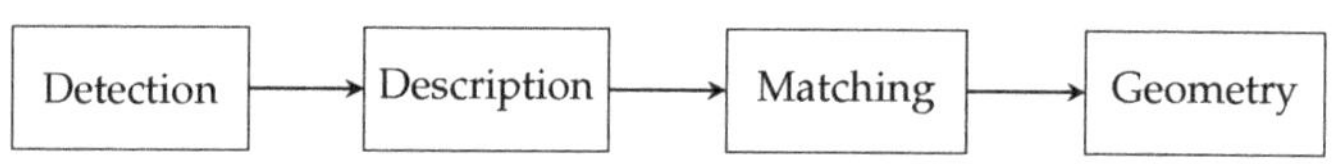

Figure 3. Image matching general pipeline. Both hand-crafted and deep learning-based approaches use feature detection, description, matching, and geometry verification.

3.3. *Visual Map Generation*

A visual map $\mathcal{M}$ encodes a set of $\mathcal{I}_i$ key images, representing an unknown environment taken by the robot during previous human-guided navigation. During this learning stage, the robot stores a video $\mathcal{V}$ of n frames taken with a pre-defined video rate and image size, such as

$$\mathcal{V} = \{I_j \mid j = 1, 2, \cdots, n\}. \tag{2}$$

Formally, the visual map is defined as

$$\mathcal{M} = \{\mathcal{I}_i \mid i = 1, 2, \cdots, m\}. \tag{3}$$

There is a double objective in the visual map generation process. It is wanted to minimize the number of key images selected (separating them as much as possible). However, at the same time, consecutive key images continue to share enough visual information that connects them i.e., a fundamental matrix can be computed. Figure 4 shows the visual constraint between consecutive key images.

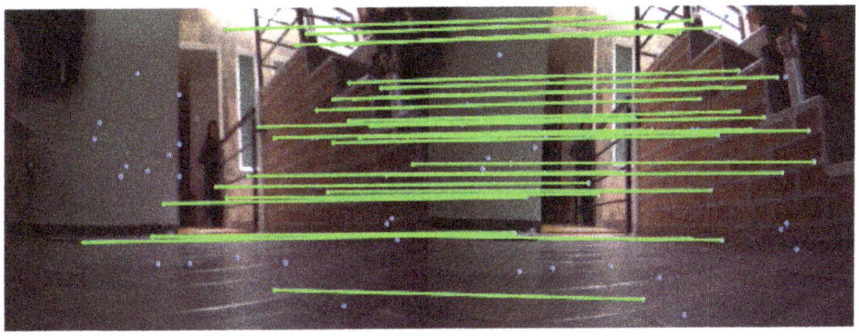

Figure 4. Matching process between two consecutive key images. The matches that meet the fundamental matrix constraint are shown in green lines and in blue points otherwise.

We need to determine whether a pair of images share visual information to select each key image. Local descriptors are used to encode an image I_i with a set of $\mathbf{D}_i$ descriptors, represented as

$$\mathbf{D}_i = \{\mathbf{d}_k | k = 1, \cdots, K\}, \tag{4}$$

where $\mathbf{d}_k \in \mathbb{R}^d$ is the k-th descriptor of the image. Therefore, we consider two images I_i and I_j to be similar if at least $M \leq K$ descriptors match between $\mathbf{D}_i$ and $\mathbf{D}_j$. In this way, we define a similarity measure ratio given by

$$S(I_i, I_j) = \frac{f_{\text{match}(\mathbf{D}_i,\mathbf{D}_j)}}{\max(\mathbf{D}_i, \mathbf{D}_j)}, \tag{5}$$

where $f_{\text{match}}(\cdot,\cdot)$ is a match selection function, i.e., mutual nearest neighbor (MNN) to establishing coarse-level matches in LoFTR or brute force in ORB, and $\max(\cdot)$ retrieves the maximum number of descriptors between $\mathbf{D}_i$ and $\mathbf{D}_j$. It is noteworthy that the match selection function passes through a geometry verification process; thus, the matching is refined by the fundamental matrix.

By construction, we select the first frame of the input video sequence (I_1) as the first key image $\mathcal{I}_i, i = 1$; therefore, we choose a new key image $\mathcal{I}_{i+1}$ if the similarity ratio between $\mathcal{I}_0$ and the current frame $I_j, j = 2, \cdots, n$ is below a threshold τ. This is defined as

$$\mathcal{I}_{i+1} = I_{j-1} \textbf{ if } S(\mathcal{I}_i, I_j) < \tau. \tag{6}$$

Notice that feature matching will fail if there are not enough features to match and satisfy the fundamental matrix constraint. In this scenario, we also select as a new key image $\mathcal{I}_{i+1}$ the previous frame I_{j-1}. Algorithm 1 summarizes the visual map framework.

Algorithm 1: Visual map generation.

Input:
A video sequence $\mathcal{V}$ of n frames
Similarity threshold τ
Output:
A visual map $\mathcal{M}$ of m key images
$\mathcal{M} \leftarrow I_1$;
$i \leftarrow 1$;
for $j \leftarrow 2$ **to** n **do**
  if $S(\mathcal{I}_i, I_j)] < \tau$ **then**
    $\mathcal{M} \leftarrow I_j$;
    $i \leftarrow i + 1$;
  end
end
return $\mathcal{M}$;

We propose a quality metric that calculates the harmonic mean of two components: the key images ratio, which measures the proportion of images selected as key images out of the total number of frames, and the mean similarity of the key images. Mathematically, this can be expressed as follows:

$$Q(\mathcal{M}) = \frac{1}{2}\left((1 - \frac{m}{n}) + (\frac{1}{m}\sum_{i=0}^{m-1} S(\mathcal{I}_i, \mathcal{I}_{i+1}))\right) \in [0,1], \quad (7)$$

where m is the total of key images and n is the video sequence length (number of frames). The quality metric combines these two factors to provide a comprehensive evaluation of the visual map generation approach. We want to maximize the quality of the visual map; thus, in the expression, the first term penalizes keeping more key images, encouraging a more selective approach. This helps to ensure that only the most relevant images are selected as key images. On the other hand, the second term penalizes low similarity ratios between consecutive key images, emphasizing the importance of maintaining high similarity (matches) within the visual map.

4. Experimental Results and Discussion

We evaluated the visual map generation framework mentioned in Section 3.3 on six distinct datasets. Three of these datasets correspond to indoor environments (D1, D3, and D4), and three to outdoor environments (D2, D5, and D6). We compared our deep learning approach with ORB, using RANSAC and MAGSAC as outlier detectors for the fundamental matrix. Moreover, we employed different similarity thresholds for the key image selection decision.

4.1. Experimental Setup

The datasets were acquired in human-guided navigation using an open-source JetBot AI (artificial intelligence) robot powered by the Jetson Nano Developer Kit. This robot, along with relevant components for experimentation, can be seen in Figure 5.

Figure 5. Relevant JetBot components.

After pondering on all the available pre-designed development kits, we settled on the Professional Version ROS (robotics operating system) AI Kit B, offered by WaveShare [23]. Relevant specifications for this version of JetBot can be seen in Table 1:

Table 1. JetBot professional ROS AI Kit characteristics.

Software DevKit	Linux Distribution	ROS Distribution	Microcontroller
Jetpack 4.5	Ubuntu 18.04	Melodic Morenia [24]	Raspberry Pi RP2040

While multiple models can oversee complex operations, such as the visual processing methods employed throughout our research, the chosen kit provides multiple advantages, such as better mechanical components, preinstalled libraries, wheel odometry data availability right after assembly, and a Jetson Nano with 4 GB of GPU. Employed datasets are captured using an eight-megapixel, 160° field-of-view IMX219-160 CSI camera.

4.2. Description of the Datasets

Images used in the datasets for the visual map were obtained using the aforementioned camera. Test images were obtained and processed with a Python script and OpenCV [25] at a 640 × 480 resolution at thirty frames per second (FPS); a live video feed to verify that the developed algorithm that generates the visual map worked as intended. A sample image of the datasets can be seen in Figure 6, and Table 2 summarizes the main characteristics of each dataset.

(**a**) Dataset D1 (**b**) Dataset D2 (**c**) Dataset D3

(**d**) Dataset D4 (**e**) Dataset D5 (**f**) Dataset D6

Figure 6. Samples of images taken from the robot camera for each dataset.

Table 2. Datasets used for the visual map generation.

Dataset	Environment	Number of Frames	Image Size (px × px)	Description
D1	Indoor	3703	640 × 480	Room
D2	Outdoor	3525	640 × 480	Backyard
D3	Indoor	3477	640 × 480	Corridor
D4	Indoor	3004	640 × 480	Traveling between two rooms
D5	Outdoor	3238	640 × 480	Roof
D6	Outdoor	3330	640 × 480	Courtyard

Dataset D1 was acquired indoors within a room using JetBot. This dataset contains good-quality images, blurred images, and others affected by lighting changes. Dataset D2 was captured in an outdoor setting in a backyard. Images within this dataset are of good quality, some with different lighting conditions and a few blurred images. Dataset D3 was captured within the corridors of the Engineering Department of Universidad Iberoamericana León. This indoors dataset has good-quality images and some blurry images. Dataset D4 was also obtained from an indoor environment, where the robot traveled between two rooms. In general, it contains images with constant illumination, but some images have blur effects. Dataset D5, another outdoor environment, was acquired from a house rooftop on its second floor. It presents blur effects due to the wrinkled concrete affecting the movement of the robot. Finally, Dataset D6 presents a courtyard in an outdoor environment. This dataset in particular shows the same blurring as dataset D5, derived from similar wrinkles in the concrete where the robot navigates.

4.3. Visual Map Evaluation

After completing the human-guided navigation to capture images of each environment, we proceeded to the learning stage for building the visual map. In this stage, we compare the visual map generation using hand-crafted feature extraction and matching using ORB and the deep learning approach within LoFTR. Since this first exploratory navigation and the learning stage were not performed simultaneously, the experiments were implemented in Python using Kornia [26], an Open Source Computer Vision Library for PyTorch, and executed in Google Colaboratory [27]. Geometric validation utilized the fundamental matrix with RANSAC and MAGSAC as feature-matching refinement (outlier detection). Moreover, we used different similarity threshold values from 0.1 to 0.9 to study the effect of the feature matching ratio between the last key image and a current frame. Thus, the lower the similarity, the fewer key images are selected; on the other hand, the higher the thresholds, the greater the number of key images.

We conducted extensive experiments using six datasets, combining ORB and LoFRT algorithms with varying similarity thresholds, for both ORB and LoFTR detectors. Table 3 presents the results for ORB with MAGSAC, specifically for dataset D1. It can be seen that for thresholds 0.1 to 0.5, the algorithm yields 1303 key images, that is, 33% of the total images, with similar average match values. For the threshold of 0.9, there are more key images, maintaining 66% of the images; hence, there are more matched points (an average of 244), as expected. The maximum quality was obtained with a threshold of 0.8; thus, 1434 key images were selected, which is 39% of the images. A mean of 43 matches was computed between consecutive key images.

Table 3. Visual map generation with ORB+MAGSAC for Dataset D1.

Threshold	Number of Key Images	Min. Matches	Max. Matches	Mean Matches	Quality
0.1	1303	14	56	29.8	0.71
0.2	1303	14	56	29.8	0.71
0.3	1303	14	56	29.8	0.71
0.4	1303	14	56	29.8	0.71
0.5	1303	14	56	29.8	0.71
0.6	1305	14	79	29.9	0.71
0.7	1322	14	314	19.8	0.70
0.8	**1434**	**14**	**576**	**43.0**	**0.74**
0.9	2505	17	501	244.8	0.62

We evaluated the ORB and LoFTR descriptors. Figure 7 illustrates examples of only 50 matches between consecutive key images for the best quality visual map. The LoFTR+MAGSAC has more and better-distributed points than ORB+MAGSAC.

Table 4 exhibits the experiment conducted using LoFTR and MAGSAC for dataset D1. This table shows a better distribution of the key images selected by the algorithm. Specifically, at a threshold of 0.7, the algorithm identifies the best-quality visual map, with 269 key images, that is, 7% of the total images. This threshold gives a mean of matched points of 2180 instead of 19 achieved with ORB. Thus, a reliable fundamental matrix can be estimated.

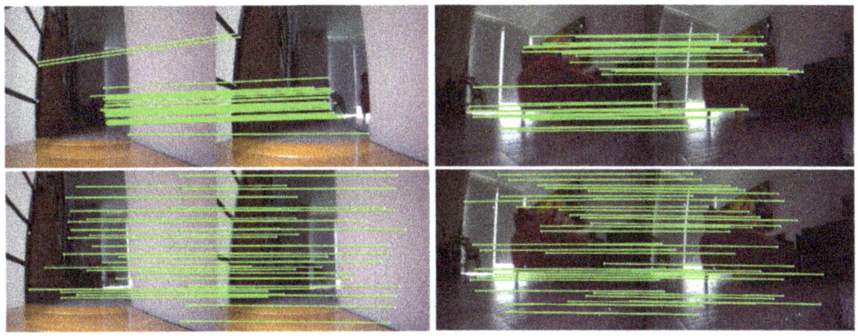

Figure 7. Examples of matched points between key images for Dataset D1. (**Top row**): ORB+MAGSAC. (**Bottom row**): LoFTR+MAGSAC.

Table 4. Visual map generation with LoFTR+MAGSAC for Dataset D1.

Threshold	Number of Key Images	Min. Matches	Max. Matches	Mean Matches	Quality
0.1	15	11	34	17.7	0.77
0.2	15	12	122	40.0	0.71
0.3	22	11	450	178.1	0.69
0.4	33	90	886	446.4	0.71
0.5	60	247	1413	892.8	0.76
0.6	122	460	2491	1551.7	0.81
0.7	**269**	**979**	**2942**	**2180.9**	**0.84**
0.8	2177	15	3587	1493.1	0.63
0.9	3566	2970	4041	3416.3	0.48

Figure 8 shows three consecutive key images obtained with LoFTR+MAGSAC.

(**a**) key image 49 (**b**) key image 50 (**c**) key image 51

Figure 8. Examples of key images obtained with LoFTR+MAGSAC for Dataset D1.

Now, we evaluate our framework using the RANSAC algorithm as an outlier detector. Table 5 presents the results of ORB with RANSAC for Dataset D1. This combination yielded the best quality at the threshold of 0.7, where 1533 key images were obtained. The execution time of the algorithms was faster with MAGSAC.

Executing the algorithms using Jetbot's GPU and RAM yields similar results. Tests were conducted within a virtual environment using Python 3.10, along with Jupyter Notebook. Dataset D1 was employed to obtain the average execution time per video frame. To keep the Google Colaboraty time factors consistent, we kept the original Notebook in the web server and established a local runtime WebSocket connection to use the available resources of the robot. An average time of 1.62 min per frame was obtained for MAGSAC, with a slightly slower time of 1.72 min per frame for RANSAC.On the other hand, executing our algorithm in the Cloud GPU resulted in an average time of 1.4 s per frame for the LoFTR and 0.05 s for ORB. Thus, all experiments were conducted in the offline mode.

Table 6 shows the results of LoFTR with RANSAC, which select more key images than LoFTR with MAGSAC. For instance, the optimal threshold is 0.5, at which we retrieved

152 key images; thus, 4% of the images were selected against the 7% selected with MAGSAC. Additionally, the mean matches were reduced with RANSAC, with 1356 vs. 2180 with the MAGSAC configuration.

It is worth noting that MAGSAC is more restrictive than RANSAC with higher thresholds, making it more convenient to create a visual map, in which fewer key images are selected but with higher similarity ratios. Based on these results, MAGSAC was chosen for subsequent experiments.

Table 5. Visual map generation with ORB+RANSAC for Dataset D1.

Threshold	Number of Key Images	Min. Matches	Max. Matches	Mean Matches	Quality
0.1	1303	12	49	29.7	0.71
0.2	1303	12	49	29.7	0.71
0.3	1303	12	49	29.7	0.71
0.4	1303	12	49	29.7	0.71
0.5	1303	12	49	29.7	0.71
0.6	1330	12	214	18.0	0.70
0.7	**1533**	**14**	**383**	**38.2**	**0.73**
0.8	2781	14	507	204.6	0.53
0.9	3598	39	607	359.6	0.47

Table 6. Visual map generation with LoFTR+RANSAC for Dataset D1.

Threshold	Number of Key Images	Min. Matches	Max. Matches	Mean Matches	Quality
0.1	15	10	27	16.0	0.74
0.2	21	11	344	118.4	0.65
0.3	43	153	662	428.7	0.65
0.4	77	303	1274	863.3	0.70
0.5	**152**	**803**	**1843**	**1356.2**	**0.75**
0.6	995	12	2988	2030.6	0.69
0.7	2192	12	3552	1597.7	0.57
0.8	3597	24	3327	2447.8	0.46
0.9	3625	3377	3823	3822.9	0.48

In the case of Dataset D2, Tables 7 and 8 show the comparative study between ORB and LoFTR, respectively. With the ORB+MAGSAC combination, the number of key images remains constant at 454 for similarity thresholds ranging from 0.1 to 0.7. The average number of matches is around 20. The best memory quality, with a score of 0.86, was achieved using a threshold of 0.9. In this case, the memory map comprises 796 key images with a mean of 441 matches. When using LoFTR+MAGSAC, the number of key images increases as the threshold value grows. For instance, for the threshold of 0.6 that maximizes the quality, 45 key images were selected with 1122 mean matches, three times more matches than the best mean matches obtained with ORB. This corresponds to approximately only 1.2% of the total images comprising the visual map.

Table 7. Visual map generation with ORB+MAGSAC for Dataset D2.

Threshold	Number of Key Images	Min. Matches	Max. Matches	Mean Matches	Quality
0.1	454	12	42	21.7	0.85
0.2	454	12	42	21.7	0.85
0.3	454	12	42	21.7	0.85
0.4	454	12	42	21.7	0.85
0.5	454	12	42	21.7	0.85
0.6	454	12	42	21.7	0.85
0.7	468	13	65	22.7	0.84
0.8	624	16	407	75.6	0.84
0.9	**796**	**17**	**950**	**441.6**	**0.86**

Table 8. Visual map generation with LoFTR+MAGSAC for Dataset D2.

Threshold	Number of Key Images	Min. Matches	Max. Matches	Mean Matches	Quality
0.1	10	26	44	37.2	0.62
0.2	10	26	44	37.2	0.62
0.3	13	36	103	66.2	0.67
0.4	15	55	458	157.6	0.72
0.5	21	164	784	396.9	0.77
0.6	**45**	**199**	**1894**	**1122.3**	**0.82**
0.7	604	728	2583	1771.0	0.77
0.8	723	1319	3414	2716.7	0.82
0.9	3236	2382	3628	3240.0	0.52

Figure 9 displays four examples of the first 50 matched points in consecutive key images employing the best visual map obtained with LoFTR+MAGSAC, and Figure 10 presents three examples of three consecutive key images from Dataset D2.

Figure 9. Examples of matched points between key images for Dataset D2. (**Top row**): ORB+MAGSAC. (**Bottom row**): LoFTR+MAGSAC.

(**a**) key image 01 (**b**) key image 02 (**c**) key image 03

Figure 10. Examples of key images obtained with LoFTR+MAGSAC for Dataset D2.

Evaluating our approach for Dataset D3, we observe a similar behavior to the previous datasets, where the ORB+MAGSAC was only available to match around 30 features. Meanwhile, the LoFTR+MAGSAC matches more than 500 features for threshold values higher than 0.5. This can be seen in Tables 9 and 10, respectively. Additionally, the number of key images in the LoFTR+MAGSAC is lower than in the ORB+MAGSAC. Particularly, the best-quality memory of 0.53 for ORB+MAGSAC was obtained with a threshold of 0.9, obtaining 3000 key images. In total, 86% of the images were selected. On the other hand, LoFTR+MAGSAC with a similarity threshold of 0.2 obtained a quality of 0.77, around 49 mean matches, and only 32 key images (0.9% of the total images). Figure 11 shows four examples of matched points in key images from the best-quality visual maps, and Figure 12 presents three examples of consecutive key images from the D3 dataset.

Table 9. Visual map generation with ORB+MAGSAC for Dataset D3.

Threshold	Number of Key Images	Min. Matches	Max. Matches	Mean Matches	Quality
0.1	2866	13	126	28.1	0.48
0.2	2866	13	126	28.1	0.48
0.3	2866	13	126	28.1	0.48
0.4	2866	13	126	28.1	0.48
0.5	2866	13	126	28.1	0.48
0.6	2866	13	126	28.1	0.48
0.7	2865	15	275	102.7	0.49
0.8	2894	15	310	36.7	0.50
0.9	**3000**	**23**	**469**	**200.4**	**0.53**

Table 10. Visual map generation with LoFTR+MAGSAC for Dataset D3.

Threshold	Number of Key Images	Min. Matches	Max. Matches	Mean Matches	Quality
0.1	31	10	231	49.1	0.76
0.2	**32**	**10**	**231**	**49.4**	**0.77**
0.3	50	9	360	122.9	0.71
0.4	108	12	894	345.0	0.72
0.5	386	12	1953	919.6	0.74
0.6	955	12	2491	971.9	0.69
0.7	3006	12	2771	1266.5	0.43
0.8	3423	14	3857	1450.7	0.43
0.9	3465	2840	2840	2840.0	0.47

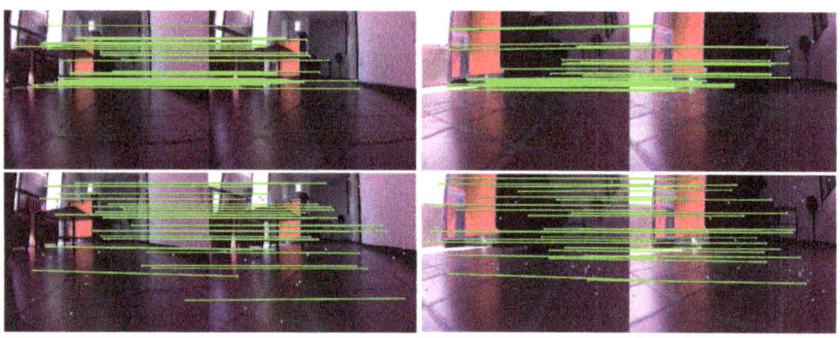

Figure 11. Examples of matched points between key images for Dataset D1. (**Top row**): ORB+MAGSAC. (**Bottom row**): LoFTR+MAGSAC.

(**a**) key image 03 (**b**) key image 04 (**c**) key image 05

Figure 12. Examples of key images obtained with LoFTR+MAGSAC for Dataset D3.

Table 11 shows the experiment results with LoFTR and MAGSAC with Dataset D4. This table shows a better distribution of the key images selected by the algorithm. At the threshold of 0.5, the best quality was obtained; it returned 63 key images. This threshold gives a mean of matched points of 1030 compared to ORB, whose mean of matched points is 28 with a threshold range of 0.1 to 0.5. At these thresholds, the number of key images was 2372, that is, 37 times more key images than LoFTR (see Table 12).

Table 11. Visual map generation with LoFTR+MAGSAC for Dataset D4.

Threshold	Number of Key Images	Min. Matches	Max. Matches	Mean Matches	Quality
0.1	11	12	65	27.0	0.73
0.2	12	11	57	30.5	0.70
0.3	17	13	545	144.5	0.70
0.4	24	16	3053	502.0	0.74
0.5	**63**	**59**	**1963**	**1030.0**	**0.78**
0.6	396	14	3053	1443.0	0.76
0.7	1983	12	3729	2860.1	0.60
0.8	2834	15	3909	2655.3	0.45
0.9	2984	3376	4175	3399.3	0.46

Table 12. Visual map generation with ORB+MAGSAC for Dataset D4.

Threshold	Number of Key Images	Min. Matches	Max. Matches	Mean Matches	Quality
0.1	**2372**	**13**	**228**	**28.2**	**0.56**
0.2	2372	13	228	28.2	0.56
0.3	2372	13	228	28.2	0.56
0.4	2372	13	228	28.2	0.56
0.5	2372	13	228	28.2	0.56
0.6	2373	13	228	28.2	0.55
0.7	2378	15	228	24.6	0.49
0.8	2395	16	257	26.4	0.54
0.9	2466	30	600	136.4	0.54

Figure 13 shows examples of matches between consecutive key images for the best threshold for both detectors. Additionally, Figure 14 shows three consecutive key images obtained with LoFTR+MAGSAC.

Table 13 shows the experiment results with LoFTR and MAGSAC with Dataset D5. A top quality of 0.77 was obtained with the 0.5 threshold value, selecting 54 key images with a mean matches of 640. The number of key images represents the 1.6% of the total images. On the scheme that used ORB (see Table 14, the best achieved quality was 0.96 but with only 20 matches in the 39 key images selected.

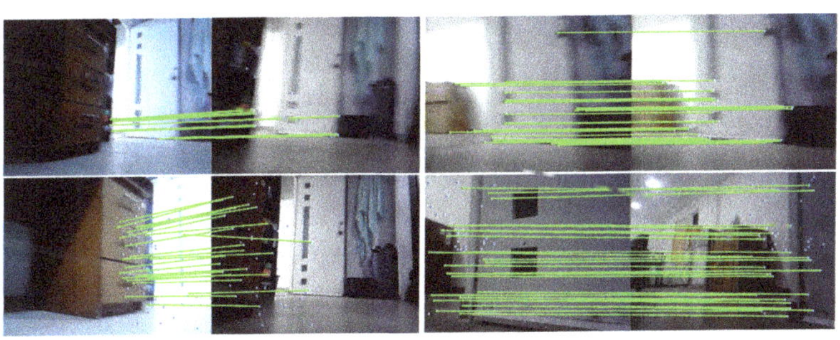

Figure 13. Examples of matched points between key images for Dataset D4. (**Top row**): ORB+MAGSAC. (**Bottom row**): LoFTR+MAGSAC.

(**a**) key image 1028

(**b**) key image 1029

(**c**) key image 1030

Figure 14. Examples of key images obtained with LoFTR+MAGSAC for Dataset D4 .

Table 13. Visual map generation with LoFTR+MAGSAC for Dataset D5.

Threshold	Number of Key Images	Min. Matches	Max. Matches	Mean Matches	Quality
0.1	7	24	34	28.7	0.58
0.2	9	37	120	68.4	0.63
0.3	12	59	440	151.7	0.70
0.4	20	121	556	286.1	0.74
0.5	**54**	**176**	**2002**	**640.2**	**0.77**
0.6	1083	275	3020	913.4	0.65
0.7	2987	1182	3204	2095.4	0.40
0.8	3034	2307	3012	2572.5	0.44
0.9	3039	2901	3568	3567.6	0.49

Table 14. Visual map generation with ORB+MAGSAC for Dataset D5.

Threshold	Number of Key Images	Min. Matches	Max. Matches	Mean Matches	Quality
0.1	39	14	39	20.9	0.96
0.2	39	14	39	20.9	0.96
0.3	39	14	39	20.9	0.96
0.4	39	14	39	20.9	0.96
0.5	39	14	39	20.9	0.96
0.6	39	14	39	20.9	0.96
0.7	**39**	**14**	**39**	**20.9**	**0.96**
0.8	53	15	192	53.6	0.94
0.9	246	16	530	185.8	0.93

Figure 15 shows some images with the matches between consecutive key images for the optimal threshold of 0.7 and 0.5 for ORB and LoFTR, respectively. Additionally, Figure 16 depicts three successive key images obtained with LoFTR+MAGSAC.

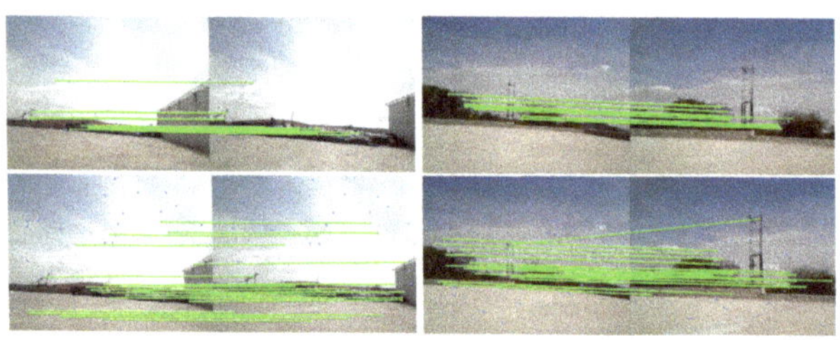

Figure 15. Examples of matched points between key images for Dataset D5. (**Top row**): ORB+MAGSAC. (**Bottom row**): LoFTR+MAGSAC.

Table 15 shows the experiment results with LoFTR and MAGSAC with Dataset D6. In particular, using a threshold of 0.4, it achieved the best memory quality (0.75) and returned 32 key images, a mean of matched points of 319. For ORB, in comparison, whose results can be seen in Table 16, the optimal threshold of 0.7 obtained a mean of matched points of 18 and 3023 key images from 3330. Only seven images were discarded.

(**a**) key image 1535 (**b**) key image 1536 (**c**) key image 1537

Figure 16. Examples of key images obtained with LoFTR+MAGSAC for Dataset D5.

Table 15. Visual map generation with LoFTR+MAGSAC for Dataset D6.

Threshold	Number of Key Images	Min. Matches	Max. Matches	Mean Matches	Quality
0.1	4	18	31	23.3	0.57
0.2	19	39	182	95.9	0.64
0.3	27	65	488	174.7	0.69
0.4	**32**	**135**	**1672**	**319.9**	**0.75**
0.5	3023	521	1442	1170.0	0.42
0.6	2367	224	2987	1628.3	0.52
0.7	3104	1042	2274	2199.0	0.43
0.8	3145	2432	2432	2432.0	0.48
0.9	3147	2725	3188	2725.1	0.50

Table 16. Visual map generation with ORB+MAGSAC for Dataset D6.

Threshold	Number of Key Images	Min. Matches	Max. Matches	Mean Matches	Quality
0.1	3023	17	18	18.0	0.55
0.2	3023	17	18	18.0	0.55
0.3	3023	17	18	18.0	0.55
0.4	3023	17	18	18.0	0.55
0.5	3023	17	18	18.0	0.55
0.6	3023	17	18	18.0	0.55
0.7	**3023**	**17**	**18**	**18.0**	**0.55**
0.8	3023	16	19	16.0	0.49
0.9	3037	19	201	29.4	0.50

Figure 17 shows the images with the matches between consecutive key images for the best ORB and LoFTR detectors. Figure 18 presents an example of three key images obtained with LoFTR+MAGSAC.

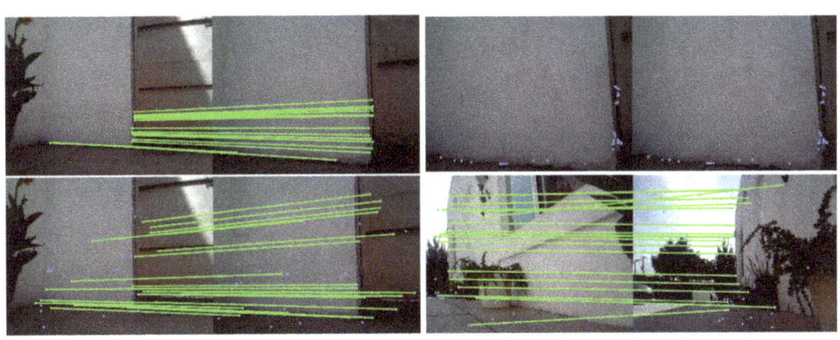

Figure 17. Examples of matched points between key images for Dataset D6. (**Top row**): ORB+MAGSAC. (**Bottom row**): LoFTR+MAGSAC.

(**a**) key image 545 (**b**) key image 546 (**c**) key image 547

Figure 18. Examples of key images obtained with LoFTR+MAGSAC for Dataset D6.

4.4. Discussion

First, an ablation study was presented in the indoor Dataset D1 to select the best outlier algorithm between RANSAC and MAGSAC. We observed that MAGSAC was more restrictive with higher similarity thresholds, allowing us to create the visual map with fewer key images and higher similarity ratios. In general, with the datasets presented for indoor environments (D1, D3, and D4) and outdoor environments (D2, D5, and D6), we present extensive visual memory comparisons for ORB+MAGSAC and LoFTR+MAGSAC. The LoFTR+MAGSAC scheme presents the best results for the visual map quality. Additionally, LoFTR+MAGSAC reached the visual memory aim to obtain fewer key images with more matches between consecutive key images.

The LoFTR descriptors perform better under various conditions, such as lighting changes, viewpoint changes, and weather conditions. ORB uses variants of FAST as a detector, i.e., they detect key points by comparing gray levels along a circle of radius 3 to the gray level of the circle center. ORB uses oriented FAST key points, an improved version of FAST, including an adequate orientation component. The ORB descriptor is a binary vector of user-choice length. Each bit results from an intensity comparison between some pixels within a patch around the detected key points. The patches are previously smoothed with a Gaussian kernel to reduce noise. ORB computes a dominant orientation between the center of the key point and the intensity centroid of its patch to be robust to orientation changes. The key points are not robust to lighting changes present in the outdoor scenarios. On the other hand, LoFTR was trained with a large dataset, including images under different lighting conditions, view changes, stopovers, and hours of the day. For such a reason, LoFTR is more robust than ORB in these scenarios.

It is possible to extend the framework to any type of robot, for this is necessary to adapt the design of movement controllers to move the robot from an initial configuration to the final configuration keeping the key points on the images of the visual map. According to the kinematics model of the robot and the controller design, it is necessary to change the camera position on different parts of the robot to maintain specific parts of the environment in the camera field of view (fov) or avoid singularities with the controller. We planned to use an MPC; with this scheme, the camera location is not relevant. Using other sensors can help to deal with obstacles that lie out of the fov or the visual map generated and problems related to circumstances of the environment, such as the lighting or reflective or transparent surfaces.

5. Conclusions

We proposed a novel framework for generating a visual map to both indoor and outdoor environments. Our approach involves selecting key images sharing visual information between consecutive key images. Moreover, we introduced a quality measure to identify the optimal similarity threshold. This will allow visual localization and model predictive control for the localization and autonomous navigation stages. We employed a pre-trained LoFTR, constrained with a fundamental matrix between two consecutive key images; also, the MAGSAC algorithm effectively detects outliers during fundamental matrix estimation, improving the key image selection. Our proposed approach outperforms traditional hand-crafted methods, demonstrating the effectiveness of our visual map generation process, minimizing the number of key images selected, maximizing the memory quality, and ensuring that key images share enough visual information between them. For instance, from the ORB+MAGSAC approach, more than 30% of the images were selected as key images, with qualities around 0.5 to 0.7 with fewer than 100 mean matching features. For the LoFTR+MAGSAC, we obtained more than 1000 mean matching features and qualities greater than 0.7. Additionally, the key images only represent, at most, 8% of the total images. This improvement was also noticeable not only in indoor but also in outdoor environments, where ORB+MAGSAC obtained more key images and fewer matching features than LoFTR+MAGSAC.

Future work will include the evaluation of other deep learning-based approaches and the implementation of more efficient models in order to perform simultaneous localization and mapping in the onboard Jetson Nano of the Jetbot.

Author Contributions: Conceptualization, E.O.-M. and N.G.A.-M.; methodology, E.O.-M.; software, E.O.-M., C.A.G.-P., E.M. and N.G.A.-M.; validation, E.M., C.A.G.-P. and N.G.A.-M.; investigation, E.O.-M., C.A.G.-P., E.M. and N.G.A.-M.; resources, E.O.-M., C.A.G.-P., E.M. and N.G.A.-M.; writing—original draft preparation, E.O.-M., C.A.G.-P., E.M. and N.G.A.-M.; visualization, C.A.G.-P., E.M. and N.G.A.-M.; supervision, N.G.A.-M.; funding acquisition, E.O.-M., E.M. and N.G.A.-M. All authors have read and agreed to the published version of the manuscript.

Funding: This work was supported in part by Universidad Iberoamericana León, Tecnológico Nacional de México/ITS de Guanajuato (Instituto Tecnológico Superior de Guanajuato), and by the University of Guanajuato Grant NUA 147347. Partially by the Mexican Council of Science and Technology CONACyT Grant no. 626154/755609.

Institutional Review Board Statement: Not applicable.

Informed Consent Statement: Not applicable.

Data Availability Statement: Data available under formal demand.

Conflicts of Interest: The authors declare no conflict of interest.

Abbreviations

The following abbreviations are used in this manuscript:

AI	Artificial Intelligence
BRIEF	Binary Robust Independent Elementary Features
FAST	Features from Accelerated Segment Test
FPS	Frames per Second
LOFTR	Detector-Free Local Feature Matching with Transformers
MAGSAC	Marginalizing Sample Consensus
MNN	Mutual Nearest Neighbor
MPC	Model Predictive Control
ORB	Oriented FAST and rotated BRIEF
RANSAC	Random Sample Consensus
ROS	Robot Operating System

References

1. Siegwart, R.; Nourbakhsh, I.R.; Scaramuzza, D. *Introduction to Autonomous Mobile Robots*; MIT Press: Cambridge, MA, USA, 2011.
2. Courbon, J.; Mezouar, Y.; Martinet, P. Autonomous Navigation of Vehicles from a Visual Memory Using a Generic Camera Model. *IEEE Trans. Intell. Transp. Syst.* **2009**, *10*, 392–402. [CrossRef]
3. Aldana-Murillo, N.G.; Sandoval, L.; Hayet, J.B.; Esteves, C.; Becerra, H.M. Coupling humanoid walking pattern generation and visual constraint feedback for pose-regulation and visual path-following. *Robot. Auton. Syst.* **2020**, *128*, 103497. [CrossRef]
4. Aldana-Murillo, N.G.; Hayet, J.B.; Becerra, H.M. Evaluation of Local Descriptors for Vision-Based Localization of Humanoid Robots. In Proceedings of the Pattern Recognition, Mexico City, Mexico, 24–27 June 2015; Springer International Publishing: Berlin/Heidelberg, Germany, 2015; pp. 179–189.
5. Aldana-Murillo, N.G.; Hayet, J.B.; Becerra, H.M. Comparison of Local Descriptors for Humanoid Robots Localization Using a Visual Bag of Words Approach. *Intell. Autom. Soft Comput.* **2018**, *24*, 471–481. [CrossRef]
6. Delfin, J.; Becerra, H.M.; Arechavaleta, G. Humanoid navigation using a visual memory with obstacle avoidance. *Robot. Auton. Syst.* **2018**, *109*, 109–124. [CrossRef]
7. Ovalle-Magallanes, E.; Aldana-Murillo, N.G.; Avina-Cervantes, J.G.; Ruiz-Pinales, J.; Cepeda-Negrete, J.; Ledesma, S. Transfer Learning for Humanoid Robot Appearance-Based Localization in a Visual Map. *IEEE Access* **2021**, *9*, 6868–6877. [CrossRef]
8. Diosi, A.; Segvic, S.; Remazeilles, A.; Chaumette, F. Experimental Evaluation of Autonomous Driving Based on Visual Memory and Image-Based Visual Servoing. *IEEE Trans. Intell. Transp. Syst.* **2011**, *12*, 870–883. [CrossRef]
9. López-Martínez, A.; Cuevas, F.; Sosa-Balderas, J. Visual Memory Construction for Autonomous Humanoid Robot Navigation. In *Progress in Optomechatronic Technologies: Proceedings of International Symposium on Optomechatronic (2018)*; Springer: Berlin/Heidelberg, Germany, 2019; pp. 103–109.
10. Huang, C.; Mees, O.; Zeng, A.; Burgard, W. Visual Language Maps for Robot Navigation. *arXiv* **2022**, arXiv:2210.05714.

11. Li, D.; Zhang, Q.; Zhao, D. Graph attention memory for visual navigation. In Proceedings of the 2022 4th International Conference on Data-driven Optimization of Complex Systems (DOCS), Chengdu, China, 28–30 October 2022; pp. 1–7.
12. Crespo, J.; Castillo, J.C.; Mozos, O.M.; Barber, R. Semantic information for robot navigation: A survey. *Appl. Sci.* **2020**, *10*, 497. [CrossRef]
13. Papapetros, I.T.; Balaska, V.; Gasteratos, A. Multi-layer map: Augmenting semantic visual memory. In Proceedings of the 2020 International Conference on Unmanned Aircraft Systems (ICUAS), Athens, Greece, 1–4 September 2020; pp. 1206–1212.
14. Ma, J.; Jiang, X.; Fan, A.; Jiang, J.; Yan, J. Image matching from handcrafted to deep features: A survey. *Int. J. Comput. Vis.* **2021**, *129*, 23–79. [CrossRef]
15. Tian, L. Matching Images from Different Viewpoints with Deep Learning Based on LoFTR and MAGSAC++. In Proceedings of the 2023 5th International Conference on Image Processing and Machine Vision, New York, NY, USA, 13–15 January 2023; pp. 18–23.
16. Hartley, R.; Zisserman, A. *Multiple View Geometry in Computer Vision*, 2nd ed.; Cambridge University Press: Cambridge, UK, 2006.
17. Fischler, M.A.; Bolles, R.C. Random Sample Consensus: A Paradigm for Model Fitting with Applications to Image Analysis and Automated Cartography. *Commun. ACM* **1981**, *24*, 381–395. [CrossRef]
18. Barath, D.; Matas, J.; Noskova, J. MAGSAC: Marginalizing Sample Consensus. In Proceedings of the IEEE/CVF Conference on Computer Vision and Pattern Recognition (CVPR), Long Beach, CA, USA, 15–20 June 2019.
19. Rublee, E.; Rabaud, V.; Konolige, K.; Bradski, G. ORB: An efficient alternative to SIFT or SURF. In Proceedings of the IEEE International Conference on Computer Vision, Barcelona, Spain, 6–13 November 2011; pp. 2564–2571.
20. Sun, J.; Shen, Z.; Wang, Y.; Bao, H.; Zhou, X. LoFTR: Detector-free local feature matching with transformers. In Proceedings of the IEEE/CVF Conference on Computer Vision and Pattern Recognition, Nashville, TN, USA, 15–20 June 2021; pp. 8922–8931.
21. Rosten, E.; Drummond, T. Machine learning for high-speed corner detection. In Proceedings of the European Conference on Computer Vision, Graz, Austria, 7–13 May 2006; pp. 430–443.
22. Calonder, M.; Lepetit, V.; Strecha, C.; Fua, P. Brief: Binary robust independent elementary features. *Eur. Conf. Comput. Vis.* **2010**, 778–792. [CrossRef]
23. JetBot Ros Ai Kit. Available online: https://www.waveshare.com/wiki/JetBot_ROS_AI_Kit (accessed on 10 April 2023).
24. Quigley, M.; Conley, K.; Gerkey, B.P.; Faust, J.; Foote, T.; Leibs, J.; Wheeler, R.; Ng, A.Y. Robotic Operating System . Available online: https://www.ros.org/ (accessed on 10 April 2023).
25. Bradski, G. The OpenCV Library. *Dr. Dobb's J. Softw. Tools* **2000**. Available online: http://opencv.org/ (accessed on 10 April 2023).
26. Riba, E.; Mishkin, D.; PONSA, D.; Rublee, E.; Bradski, G. Kornia: An Open Source Differentiable Computer Vision Library for PyTorch. In Proceedings of the Winter Conference on Applications of Computer Vision, Snowmass Village, CO, USA, 1–5 March 2020.
27. Bisong, E. Google Colaboratory. In *Building Machine Learning and Deep Learning Models on Google Cloud Platform: A Comprehensive Guide for Beginners*; Apress: Berkeley, CA, USA, 2019; pp. 59–64.

eng MDPI

Review

Skid Resistance of Asphalt Pavements

Szabolcs Rosta [1,2] and László Gáspár [3,4,*]

1 Multidisciplinary Technical Scientific Doctoral School, Széchenyi István University, 9026 Győr, Hungary; rosta.szabolcs@dunaaszfalt.hu
2 Duna Group, 1138 Budapest, Hungary
3 KTI Institute for Hungarian Transport Sciences and Logistics Non-Profit Ltd., 1119 Budapest, Hungary
4 Department of Transport Construction and Water Management, Széchenyi István University, 9026 Győr, Hungary
* Correspondence: gaspar.laszlo@kti.hu; Tel.: +36-30-9401-288

Abstract: Skid resistance of a road pavement surface is the force developed when a tyre is prevented from rotating and slides along the pavement surface. This property comes from the combination of the macro- and micro-texture of pavement. The skid resistance of an asphalt pavement is an important parameter influencing driving safety on a road since there is a proven relationship between skid resistance and accident parameters. This paper deals with the measurement principle of pavement skid resistance (surface friction) including longitudinal and transverse friction. A high number of measuring devices of skid resistance are also introduced, highlighting their advantages and limitations. Moreover, the measurement policies in the European Union and in Hungary are outlined. Pavement surface texture is investigated, dealing with the levels of surface texture, the most common measuring techniques, the macro-texture features of asphalt types, as well as the Hungarian regulation in the field. As a related topic, the aggregate properties and their implication in the relevant Hungarian specification are introduced briefly as well. Some outcomes of the EU's COST Action 354 on the development of unified European macro-roughness and skid resistance performance indicators and indices are also presented.

Keywords: asphalt pavements; skid resistance; surface texture; COST Action 354; macro-texture

Citation: Rosta, S.; Gáspár, L. Skid Resistance of Asphalt Pavements. *Eng* **2023**, *4*, 1597–1615. https://doi.org/10.3390/eng4020091

Academic Editor: Antonio Gil Bravo

Received: 25 April 2023
Revised: 24 May 2023
Accepted: 1 June 2023
Published: 6 June 2023

1. Introduction

Skid resistance of a road pavement surface is the force developed when a tyre is prevented from rotating and slides along the pavement surface. This is a property that characterises the macro- and micro-texture of pavement, and shows the effect on frictional forces when the surface receives wheel loading. One of the main reasons for the spread of asphalt is that it provides favourable resistance to sliding and driving conditions for each road vehicle type. However, safety need to be considered under different environmental and construction conditions. The skid resistance of an asphalt pavement is an important parameter influencing driving safety on a road.

The fact that the slip resistance of pavement surfaces and the frequency of crashes are closely related is worldwide and accepted. However, several studies concluded that when the friction of the surfaces falls below a site-specific threshold, the number and severity of collision accidents on wet pavement surfaces increase significantly [1,2]. An example of such a relationship developed for single carriageways by English researchers shows that crash (road accident) risk approximately halves as pavement friction doubles over normal ranges (Figure 1) [3].

Asphalt pavements usually provide good driving conditions for all road vehicles, and are therefore widespread worldwide. Nevertheless, there are safety problems that need to be investigated and possibly solved. There is a good relationship between the skid resistance of pavement surfaces and accidents (crashes). The exact nature of the relationship between pavement friction and wet crashes is site-specific, as it is influenced by not only pavement friction but other factors as well. Therefore, the relationship between pavement friction and wet crashes should be determined for a given pavement network.

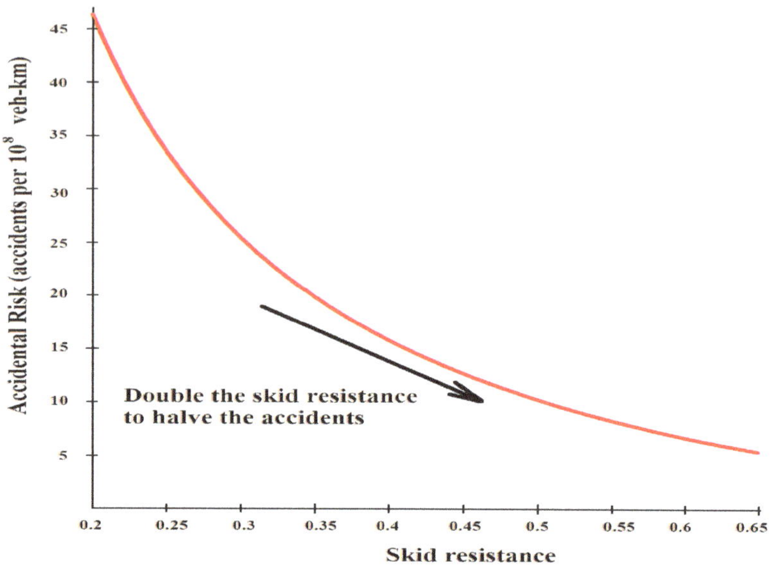

Figure 1. Relationship between pavement friction and crash risk, after [3].

Skid resistance/friction have two components, namely, adhesion and hysteresis. These components largely depend on two properties of the pavement surface: macro-texture and micro-texture. The macro-texture and micro-texture of a pavement surface largely depends on the type, size, shape, and quality of coarse aggregates used in the asphalt mixture. The widely applied hot mix asphalt (HMA) types generally have similar (or at least comparable) micro-texture values. Their frictional performance follows the same order as their macro-textures. The skid resistance of an asphalt pavement changes (typically worsens) over time, and is affected by various environmental factors. At the same time, differences in the pavement surface temperatures can significantly influence the reliability of skid resistance measurements [4].

2. Pavement Friction

2.1. Pavement Friction Generalities

Pavement friction is the force that resists the relative motion between a road vehicle tyre and a pavement surface. The resistive force, characterized by the non-dimensional friction coefficient, μ, is the ratio of the tangential friction force (F) between the tyre tread rubber and the horizontal travelled pavement surface to the perpendicular force or vertical load (F_W) and is calculated as:

$$\mu = F/F_W, \tag{1}$$

Pavement friction plays a vital role in keeping vehicles on the road, as it provides drivers the ability to control/manoeuvre their vehicles in a safe manner, in both the longitudinal and the lateral directions. It is a key input for highway geometric design, as it is used in determining the adequacy of the minimum stopping sight distance, minimum horizontal radius, minimum radius of crest vertical curves, and maximum super-elevation in horizontal curves. So, it can be concluded that the higher the friction available at the pavement–tyre interface, the more control the driver has over the road vehicle [5].

2.2. Longitudinal Frictional Forces

Longitudinal frictional forces occur between a rolling pneumatic tyre of a road vehicle (in the longitudinal direction) and the road surface when operating in the free rolling or constant-braked mode. In the free rolling mode (no braking), the relative speed between the tyre circumference and the pavement, referred to as the slip speed, is zero. In the constant-braked mode, the slip speed increases from zero to the potential maximum of the speed of the vehicle. The following mathematical relationship explains slip speed [6]:

$$S = V - Vp = V - 0.68\ \omega r, \tag{2}$$

where

S is the slip speed, mi/h;
V is the vehicle speed, mi/h;
V_p is the average peripheral speed of the tyre, mi/h;
ω is the angular velocity of the tyre, rad/s;
r is the average radius of the tyre, ft.

The coefficient of friction between a tyre and the pavement surface changes with varying slip, as shown in Figure 2 [7]. The coefficient of friction increases rapidly with increasing slip to a peak value that usually occurs between 10 and 20% slip (critical slip). The friction then decreases to a value (coefficient of sliding friction), which occurs at 100% slip. The difference between the peak and sliding coefficients of friction may equal up to 50% of the sliding value, and is much greater on wet pavement surfaces than on dry ones.

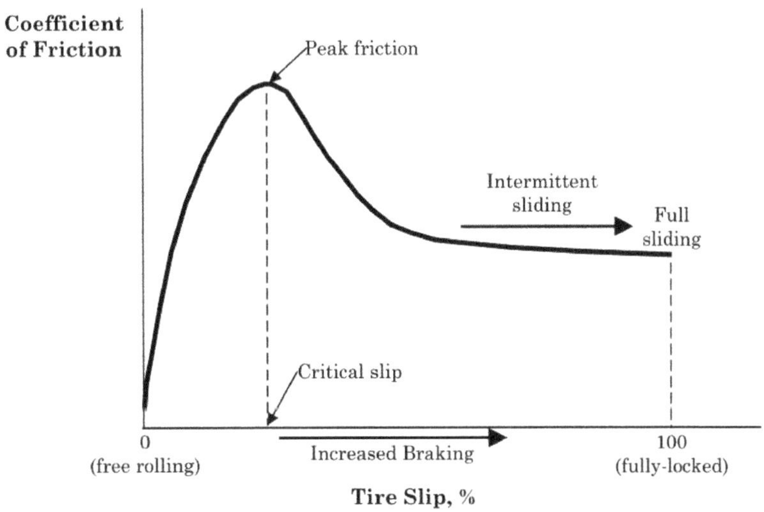

Figure 2. Pavement friction versus tyre slip, after [7].

The relationship shown in Figure 2 is the basis for the anti-locking brake system (ABS), which takes advantage of the front side of peak friction, and minimizes the loss of side/steering friction due to sliding action. Road vehicles with ABS are designed to apply the brakes on and off (i.e., pump the brakes) repeatedly, such that the slip is held near the peak. The braking is turned off before the peak is reached, and turned on at a set time or slip % below the peak. The actual timing is a proprietary design of the manufacturer [6].

2.3. Lateral Frictional Forces

Another important aspect of friction is the lateral (side-force) friction, which occurs as a vehicle changes direction or compensates for pavement cross-slope and/or cross-wind effects. The relationship between the forces acting on the vehicle tyre and the pavement surface as the vehicle steers around a curve, changes lanes, or compensates for lateral forces is as follows:

$$F_s = V^2/15R - e, \quad (3)$$

where

F_s is the slide friction;
V is the vehicle speed, mi/h;
R is the radius of the path of the vehicle's centre of gravity (also, the radius of curvature in curve), ft;
e is the pavement super-elevation, ft/ft.

2.4. Combined Braking and Cornering

In the case of combined braking and cornering, the driver of the vehicle either risks not stopping as rapidly or losing control due to reduced lateral/side forces. When operating at the limits of tyre grip, the interaction of the longitudinal and lateral forces is such that as one force increases, the other must decrease by a proportional amount. The application of longitudinal braking decreases the lateral force significantly. Similarly, the application of high lateral force leads to less effective longitudinal braking [8].

Many researchers call it friction circle or friction ellipse (Figure 3) [9], the vector sum of the two combined forces remains constant (circle) or near constant (ellipse). When operating within the limits of tyre grip, the amount of braking and turning friction components can change independently as long as the vector sum of these components does not exceed the limits of tyre grip (coming from friction circle or friction ellipse). The degree of ellipse depends on several features of tyre and road pavement.

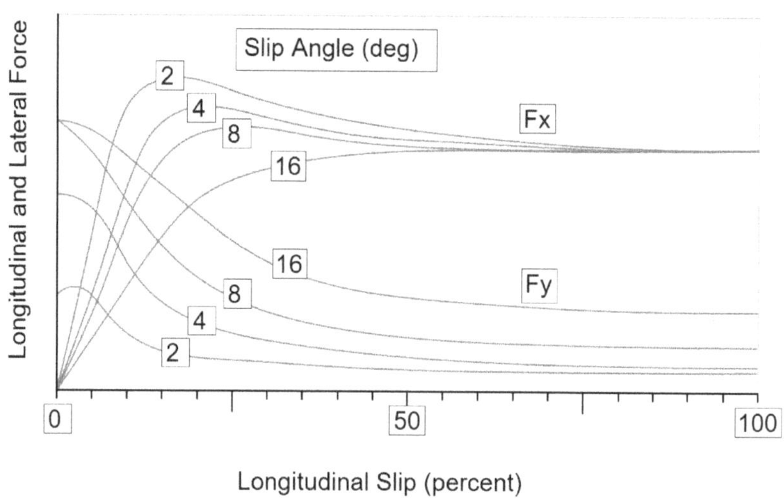

Figure 3. Brake (Fx) and lateral (Fy) forces as a function of longitudinal slip, after [9]. Numbers in figure mean the Slip Angles in degrees.

3. Measuring Surface Friction (Skid Resistance)

3.1. Measurement Principles

The three major operating principles of frictional measuring equipment are: longitudinal friction coefficient (LFC), sideway force coefficient (SFC), and sliders that can be stationary or slow-moving (slow speed) devices.

The longitudinal friction coefficient (LFC) measurement principle is applied when the vehicle is travelling in a straight line, and the brake pedal is pressed; braking forces with the help of the braking system are transmitted to the wheels of road vehicle; the angular speed of the wheel decreases, kinetic vehicle rolling energy converts to thermal energy, and the vehicle slows down. However, in the case of too high braking forces, the vehicle wheels block and the vehicle starts to slide on the pavement surface, friction forces are generated in the tyre, and road contact surface slowing down the vehicle. Longitudinal friction coefficient measurement devices create this wheel blocking process by producing a controlled sliding process. Slip ratio is also applied to assess and compare wheel and vehicle speeds. Slip ratio varies from 0 to 1:

- 0: wheel speed is the same as vehicle speed, wheel rolls freely;
- 1: wheel is fully blocked, and slides on the pavement surface.

The optimal slip ratio value varies from 0.15 to 0.20. In this interval is ensured the highest LFC and shortest vehicle stopping distance. When the slip ratio is higher, the vehicle wheel starts to block and slide. The majority of longitudinal friction measurement devices apply fixed slip ratio, which is determined automatically. However, some of the devices utilize variable slip ratio, increasing friction force until the wheel starts to block. Devices with fixed slip ratio are more appropriate for monitoring purposes because they measure continuously, while variable slip ratio devices can measure short lengths of road, which is why they are more suitable for various research purposes [10,11].

Transverse friction measurement principle is applied when the vehicle is travelling in a horizontal curve, and the wheels of the road vehicle are turned. The angle between vehicle and turned wheel direction is called slip angle (δ). Slip angle induces friction between tyre and road, which in turn generates a centripetal force opposing the centrifugal force exerted on the vehicle in the bend, allowing the vehicle to follow the curve. When the braking force grows, the wheel starts to slip over the surface of the road pavement. The sideways friction coefficient (SFC) varies with the slip angle of the vehicle tyre. Normally, the maximum SFC value occurs at a slip angle between 4° and 7° for cars, and between 6° and 10° for trucks. Skid resistance measurement devices commonly use fixed slip angle; they are more appropriate for pavement monitoring purposes since they measure continuously while variable slip ratio devices can measure just short road sections, and consequently, are more appropriate for research tasks [10,11].

The stationary or slow-moving equipment measurement principle is used by devices, which are light, simple, portable, and mostly used for laboratory or stationary testing. Static devices utilise rubber sliders to make contact with the road surface with a mechanism that initiates relative motion between the slider and the road. Two devices are used: the pendulum arm, which swings and with a rubber slider contacts the surface until friction forces slow down the slider; and the rotating head with sliders, which is lowered on to the road so that friction between the sliders and the road causes the head to slow down. Slow-moving devices are used to measure friction at a very low speed or when stationary, but these devices are good for measuring friction for special purposes (e.g., road marking) [12].

3.2. Measurement Devices

The measurement of pavement friction is rather hard, because the frictional forces being measured are sensitive to a number of factors, which are difficult to control. Such factors include texture, temperature, chemistry of materials, etc., of the road pavement. Others are connected with the tyre (its tread design, rubber composition, sliding velocity temperature, etc.) and fluid/contaminant (viscosity, density, thickness of film, etc.). However, during friction measurement, all other factors except the road surface are kept constant. The pavement surface is then wetted with a specified amount of water and a standardized measuring tyre is used on friction measuring equipment to characterise the pavement friction. Therefore, a high number of skid measurement devices was developed and currently are used in various countries [13]. This equipment utilises different measurement speeds, tyres, wheel loads, and water film thickness. Some measurement devices are similar due to the measurement principle, but the measurement method is not the same.

Table 1 introduces the skid resistance measuring devices that are being used in in the EU. It can be seen that more devices use the LFC method than the SFC method.

Table 1. List of skid resistance measurement devices [10–13].

Title	Measuring Principle	Main Parameter	Tyre and Wheel Load
ADHERA	Longitudinal friction coefficient (LFC)	Slip ratio: 1.0; Water film thickness: 1.0 mm; Measures macro-texture; Measurement speed: 40, 60, 90, 120 km/h; Measurement interval: 20 m.	PIARC smooth profile tyre 165R15 (180 kPa); Wheel load: 2500 N.
ASFT	Longitudinal friction coefficient (LFC)	Slip ratio: 0.12; Water film thickness:0–1.5 mm (ideally 0.9 mm) Measurement speed: 50, 70, 90 km/h Measurement interval: 10 m	Trelleborg type 4.00-8″ tyre (700 ± 5 kPa); Wheel load: 2500 N.
BV-11	Longitudinal friction coefficient (LFC)	Slip ratio: 0.17; Water film thickness: 0.5–1.0 mm; Measurement speed: 70 km/h; Measurement interval: 20 m.	Trelleborg type T49 tyre (140 kPa); Wheel load: 1000 N.
GripTester	Longitudinal friction coefficient (LFC)	Slip ratio: 0.15; Water film thickness: 0.5 mm; Measurement speed: 5–100 km/h; Measurement interval: 10–20 m or other.	254 mm diameter smooth profile ASTM-tyre (140 kPa); Wheel load: 250 N.
RoadSTAR	Longitudinal friction coefficient (LFC)	Slip ratio: 0.18; Water film thickness: 0.5 mm; Measures macro-texture; Measurement speed: 30, 60 km/h; Measurement interval: 50 m.	PIARC tyre with tread; Wheel load: 3500 N.
ROAR DK	Longitudinal friction coefficient (LFC)	Slip ratio: 0.2; Water film thickness: 0.5 mm; Measures macro-texture; Measurement speed: 60, 80 km/h; Measurement interval: >5 m.	ASTM 1551 tyre (207 kPa); Wheel load: 1200 N.
ROAR NL	Longitudinal friction coefficient (LFC)	Slip ratio: 0.86; Water film thickness: 0.5 mm; Measures macro-texture; Measurement speed: 50, 70 km/h; Measurement interval: 5–100 m.	ASTM 1551 tyre (200 kPa); Wheel load: 1200 N.
RWS NL Skid Resistance Trailer	Longitudinal friction coefficient (LFC)	Slip ratio: 0.86; Water film thickness: 0.5 mm; Measurement speed: 50, 70 km/h; Measurement interval: 5–100 m.	PIARC smooth profile tyre 165R15 (200 kPa); Wheel load: 1962 N.
Skiddometer BV-8	Longitudinal friction coefficient (LFC)	Slip ratio: 1.0 or 0.14; Water film thickness: 0.5 mm; Measurement speed: 40, 60, 80 km/h; Measurement interval: 30–50 m.	AIPCR tyre with longitudinal tread 165R15; Wheel load: 3500 N.
SRM	Longitudinal friction coefficient (LFC)	Slip ratio: 0.15; Water film thickness: 0.5 mm; Measurement speed: 40, 60, 80 km/h; Measurement interval: 20 m or other.	AIPCR tyre with longitudinal tread 165R15; Wheel load: 3500 N.
TRT	Longitudinal friction coefficient (LFC)	Slip ratio: 0.15; Water film thickness: 0.5 mm; Measurement speed: 40–140 km/h; Measurement interval: 20 m or other.	Smooth profile ASTM-tyre; Wheel load: 1000 N.
SRT-3	Longitudinal friction coefficient (LFC)	Slip ratio: 1.0; Water film thickness: 0.5 mm; Measurement speed: 60 km/h.	Tyre with tread (200 kPa).
IMAG	Longitudinal friction coefficient (LFC)	Slip ratio: 1.0; Water film thickness: 1.0 mm; Measurement speed: 65 km/h;	PIARC smooth profile tyre; Wheel load: 1500 N.
SCRIM	Sideway friction coefficient (SFC)	Slip angle: 20°; Water film thickness: 0.5 mm; Measures macro-texture; Measurement speed: 50 km/h; Measurement interval: >10 m.	Avon SCRIM smooth profile tyre 76/508 (350 kPa); Wheel load: 1960 N.
SKM	Sideway friction coefficient (SFC)	Slip angle: 20°; Water film thickness: 0.5 mm; Measurement speed: 50 km/h; Measurement interval: 100 m or other.	Smooth profile tyre; Wheel load: 1960 N.

Table 1. *Cont.*

Title	Measuring Principle	Main Parameter	Tyre and Wheel Load
DFT Dynamic Friction Tester	Rotating friction	For stationary measurements	
SRT Pendulum	Pendulum test	For stationary measurements	
T2GO	Slow-moving measurement; Longitudinal friction coefficient (LFC)	Slip ratio: 0.2; Used for pedestrian/bicycle paths, road marking	Two 75 mm width tyres.
VTI Portable Friction Tester (PFT)	Slow-moving measurement; Longitudinal friction coefficient (LFC)	Used for pedestrian/bicycle paths	

Devices based on the LFC principle can measure LFC at different speeds (typically between 40 and 140 km/h), are relatively small, and have better mobility. They are commonly completed as trailers. However, it is always required to have an additional water metering system, which is mounted in an accompanying vehicle. At the same time, transverse friction measurement devices perform measurement at a medium speed, the devices are bigger, and their operation costs are significantly higher. However, these devices are supplemented with large water tanks allowing them relatively long-distance measurements [12]. Measurement devices with installed pavement macro-texture measurement equipment allow also measuring mean profile depth (MPD) for further comparisons and analysis of pavement surface condition and identifying the interrelationship between skid resistance and the MPD of the road surface to be characterized. Stationary and slow-moving devices are slow and unproductive since these measurements are carried out mechanically. However, these devices are useful when performing measurements in specific locations. However, stationary and slow-moving devices are cheap, and can be easily transported and operated.

3.3. Measurement Policy in EU

In the European Union, most countries have national skid resistance measurement policies, such as methodologies, purposes, devices, thresholds, and frequency measurements for road categories during the monitoring system such as a condition index. In many EU countries, measurements for routine monitoring are made between spring and autumn.

An important factor in skid resistance measurement policy is the threshold levels, which are the values that are compared with the skid resistance characteristic values and can initiate some kind of action when the skid resistance falls to the threshold level. The threshold can be a fixed or "warning" type. A fixed threshold is the value of skid resistance, and when it falls below this value, action to improve the surface must be taken. Warning thresholds are a good option because if skid resistance falls below this level it provides a warning that accident risk may be increasing, and therefore an investigation should be made to assess whether treatment is necessary or appropriate. In some EU practice, threshold levels are mostly based on statistics, accident analysis, and theoretical calculations (accident risk calculations). For countries that are introducing skid resistance policy, a good option is to set thresholds utilizing another country's regulation, which has networks with similar characteristics and comparable traffic "culture".

Many EU countries mostly apply measurements for secondary roads, primary roads, and motorways. Measurement frequency also varies due to the use of different devices, different size of road networks, and different allocated funds. Some countries measure twice a year while other countries measure annually, every 2nd, 3rd, or 5th year. Regarding measurement frequency due to the significance of roads, it is observed that in most of EU countries, skid resistance measurements in motorways are performed annually, in primary roads annually or every 2nd year, and in secondary roads every 2nd year [12].

3.4. Measurement Policy in Hungary

In Hungary, tyre and pavement surface friction measurements need to be measured on newly built motorways and main roads after building new roads or reconstruction works during the quality control (acceptance test). However, skid resistance is currently not a qualification but a data collection feature. In the case of the construction of asphalt pavements, this kind of measurement of the surface layers (wearing courses) must be performed 3 months after road opening. The skid resistance must be measured on the pavement surface using the Sideway-force Coefficient Routine Investigation Machine (SCRIM) road survey vehicle in accordance with e-UT 09.02.23:1999 [14] or using ASFT equipment based on e-UT 09.02.27:2009 Hungarian Road Regulation (UME) [15]. Measurements shall be made at a speed of 50 km/h per road traffic lane, at a distance of 0.8 to 1.2 m from the right lane edge, in the same direction as traffic. ASFT value should be converted to SCRIM value using the following empirical formula:

$$\alpha_{SFC} = 0.65x^2 - 0.35x + 0.42, \quad (4)$$

where

x is the measured ASFT value;
α_{SFC} is the coefficient of skid resistance determined by SCRIM surface friction tester.

The skid resistance of a surface layer can also be determined by the SRT (pendulum) test in accordance with EN 13036-4:2012 [16] measured every 500 m per traffic lane.

The e-UT 09.02.23:1999 UME [14] specifies surface course skid resistance (SFC) thresholds measured by the SCRIM apparatus for various asphalt concrete types (Table 2) based on Hungarian measurement results.

Table 2. SCRIM- sideways force coefficient (SFC) limits for the surface of pavement type course.

Pavement Category	Hot Rolled Asphalt	Asphalt Concrete	Cement Concrete	Surface Dressing
I.	0.80–0.50	0.75–0.50	0.75–0.50	0.90–0.50
II.	0.80–0.45	0.70–0.45	0.65–0.45	0.80–0.45
III.	0.70–0.40	0.64–0.40	0.64–0.40	0.80–0.40
IV.		0.64–0.33	0.64–0.33	0.64–0.33

I. Motorways and main roads with a traffic above 3000 unit vehicle/day, as well as on urban main roads at the approach of signs, pedestrian crossings and similar dangerous points, in curves with a radius of less than 100 m. II. Excluding motorway, approach to major junctions on roads with a traffic above 3000 unit vehicle/day, on slopes, slopes above 5%, curves with a radius below 150 m, transition curves, before railway light barriers, bridges. III. Main roads with a traffic above 3000 unit vehicle/day on straight sections, urban main roads in curves with a radius above 100 m, slopes below 5%. IV. Local roads with a traffic below 3000 unit vehicle/day.

The upper limit values of Table 2 (e.g., 0.80) characterize the skid resistance at the new pavements (in 3 months after road opening), while the lower ones (e.g., 0.50) are related to heavily worn road surfaces. The e-UT 09.02.27:2009 UME [15] recommended skid resistance limits for SFT values measured by ASFT method based on Hungarian and international experience, and measurement results (Table 3).

Table 3. Informative limits ASFT-SFT skid resistance values for all pavement types.

Pavement Category	New Pavement Threshold	Warning Threshold	Intervention Level
I.	0.84	0.75	0.70
II.	0.80	0.65	0.60
III.	0.70	0.55	0.45
IV.	0.65	0.45	0.35

I. Motorways and main roads with a traffic above 3000 unit vehicle/day, as well as on urban main roads at the approach of signs, pedestrian crossings and similar dangerous points, in curves with a radius of less than 100 m. II. Excluding motorway, approach to major junctions on roads with a traffic above 3000 unit vehicle/day, on slopes, slopes above 5%, curves with a radius below 150 m, transition curves, before railway light barriers, bridges. III. Main roads with a traffic above 3000 unit vehicle/day on straight sections, urban main roads in curves with a radius above 100 m, slopes below 5%. IV. Local roads with a traffic below 3000 unit vehicle/day.

There are differences in principle between the two regulations currently in Hungarian practice [14,15]. The specification for SCRIM measurement defines ranges for different types of pavement. The Road Technical Directives for ASFT apparatus differentiates only the road category regardless of the actual type of pavement. Here, however, it is possible to determine the thresholds that can be defined in accordance with the actual European practice. When choosing limit values, there is always more favourable and less favourable specified values due to empirical formulas. It is also obvious that measurements based on two different principles are rather difficult to apply at the same time because the comparison of measuring results of different measuring techniques creates an almost unsolvable problem. Using the empirical formula, it can be concluded that the interventional threshold values are in fairly close agreement in the proposed categories of both procedures.

4. Pavement Surface Texture

4.1. Measurement Principles

As described in the previous chapter, there are several factors influencing pavement friction forces. An NCHRP report [17] summarised these factors, dividing them into four categories (pavement surface characteristics, vehicle operational parameters, tyre properties, and environmental factors) (Table 4). Friction should be viewed as a process instead of an inherent property of the pavement. It is only when all these factors are fully specified that friction takes on a definite value.

Table 4. Factors affecting actual pavement friction [5].

Pavement Surface Characteristics	Vehicle Operating Parameters	Tire Properties	Environment
Micro-texture **Macro-texture** Mega-texture/unevenness **Material properties** Temperature	**Slip speed** (vehicle speed, braking action) Driving manoeuvre (turning, overtaking)	Footprint **Tread design and condition** Rubber composition and hardness **Inflation pressure** Load Temperature	Climate (wind **temperature; rainfall/condensation; snow and ice)** **Contaminants** (Anti-skid material; dirt, mud, debris)

Note: Critical (most important) factors are shown in bold.

Among these factors, micro-texture and macro-texture, as well as pavement materials properties (and slip speed), can be managed in a pavement management system, and can be controlled by regulations.

4.2. Measurement Principles

Pavement friction is occasioned by the combination of two mechanisms: adhesion and hysteresis (Figure 4). Adhesion comes from molecular bonds in areas of high local pressure resulting from pavement unevenness. Hysteresis is the consequence of energy loss resulting from the deformation of the tyre rubber around the protuberances and depressions in the road surface. However, there are also two other less important components of tyre friction: frictional contributions from rubber wear, and surface (or micro) hysteresis as opposed to bulk hysteresis. The cohesion loss component of friction from rubber wear was verified through testing recently [18].

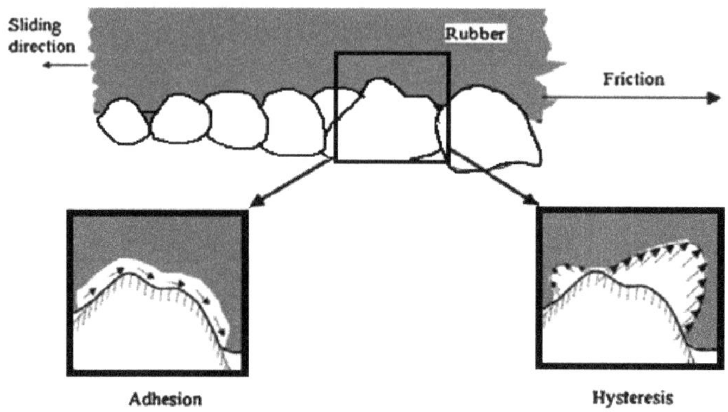

Figure 4. Key mechanisms of tyre–pavement friction, after [5].

4.3. Measurement Principles

Pavement surface texture is defined as the deviations of the road surface from a true flat one. These deviations occur at three distinct levels of scale, each defined by the wavelength (λ) and peak-to-peak amplitude (A) of its components. The three levels of texture are as follows [5,19]:

- Micro-texture (λ < 0.5 mm, A = 1 to 500 μm): Surface roughness quality at the sub-visible or microscopic level. It depends on the surface properties of the aggregate grains in the asphalt or concrete paving material;
- Macro-texture (λ = 0.5 to 50 mm, A = 0.1 to 20 mm): Surface roughness quality de-fined by the mixture properties (shape, size, and gradation of aggregate) of asphalt paving mixtures and the method of finishing/texturing used on a concrete surface;
- Mega-texture (λ = 50 to 500 mm, A = 0.1 to 50 mm): Texture with wavelengths in the same order of size as the pavement–tyre interface. It is mainly affected by the distress, defects, or "waviness" on the surface of road pavement;
- Wavelengths above 500 mm are defined as roughness (USA) or unevenness (UK).

Figure 5 illustrates the three texture ranges, as well as a fourth level, rough-ness/unevenness, representing wavelengths with more than 500 mm of mega-texture.

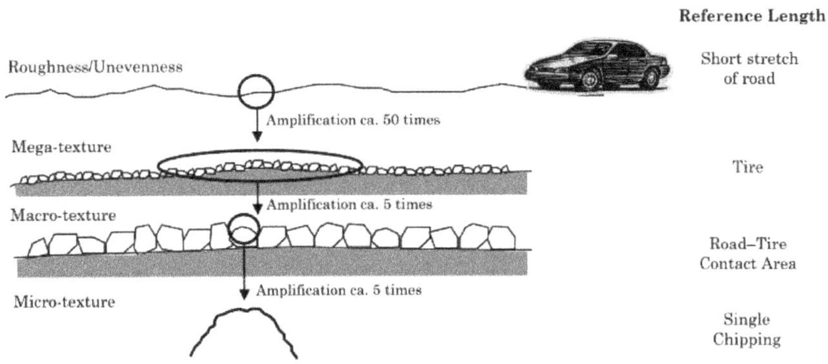

Figure 5. Various texture ranges for a pavement surface [19].

Figure 6 shows the relative effects of micro-texture, macro-texture, and measuring speed on pavement friction. It is obvious, that the micro-texture of pavement surface affects the magnitude of tyre friction, while macro-texture has an impact on the friction–speed gradient. At low speeds, micro-texture dominates the wet and dry friction level. At higher speeds, the presence of high macro-texture facilitates the drainage of water so that the adhesive component of friction afforded by micro-texture is re-established by being above the water. Hysteresis grows with speed exponentially, and at speeds above 65 mi/h (105 km/h) accounts for more than 95% of the pavement friction [17].

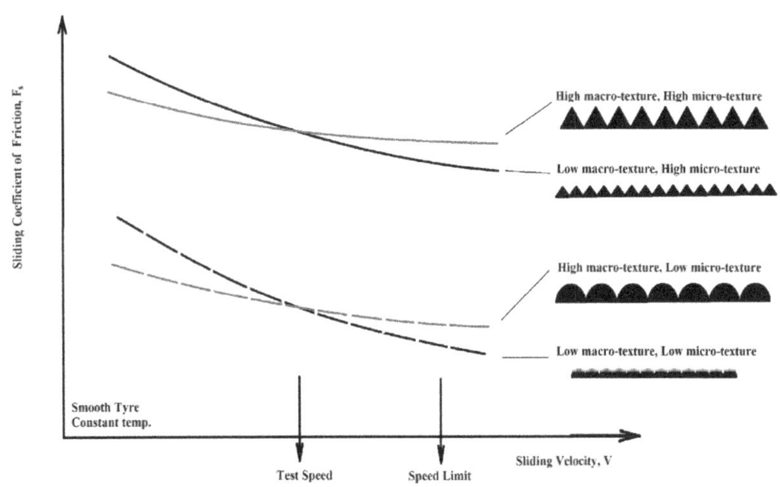

Figure 6. Various texture ranges for a pavement surface, after [19].

4.4. Pavement Texture Measurement

The majority of the generally used methods for pavement texture measurement are related to macro-texture. These are mostly rubber-on-road techniques. More recent developments in non-contact measurements are geared towards micro-texture measurements. The most common macro-texture measuring equipment are volumetric sand patch method, circular texture meter and the outflow meter.

The sand patch method (SPM) [20,21] is a volumetric-based spot test method that assesses pavement surface macro-texture through the spreading of a known volume of glass beads or sand of controlled grain size in a circle onto a cleaned pavement surface area, and through the measurement of the average diameter of the resulting circle. The volume of sand or glass bead spread divided by the area of the circle is reported as the mean texture depth (MTD):

$$MTD = 4V/(\pi D^2) \quad (5)$$

The outflow meter method (OFM) [22] is a volumetric test method measuring the water drainage rate through surface texture and interior voids. It provides information about the hydroplaning potential of a pavement surface by relating to the escape time of water under a moving vehicle tyre. The equipment consists of a cylinder with a rubber ring on the bottom and an open top. Sensors measure the time required for a known volume of water to pass under the seal or into the pavement. The measurement parameter, outflow time (OFT), defines the macro-texture; high time indicates smooth macro-texture, while low time rough one.

The circular texture meter (CTM) [23] is a non-contact laser device measuring the profile of pavement surface along a circular path of 11.25 in (286 mm) diameter at intervals of 0.034 in (0.868 mm). The texture meter device rotates at 20 ft/min (6 m/min) and generates profile traces of the pavement surface. These data are transmitted, and stored on a portable computer. Two kinds of macro-texture indices can be computed from these profiles: mean profile depth (MPD) and root mean square (RMS). MPD is actually a two-dimensional estimate of the three-dimensional MTD [17].

High-speed methods for characterizing pavement surface texture are typically based on non-contact surface profiling techniques, such as the Road Surface Analyzer (ROSAN V) developed by the FHWA [24]. It is a portable, vehicle-mounted, automated system for the measurement of pavement texture at highway speeds along a linear path. The device incorporates a laser sensor mounted on the vehicle's front bumper and the device can be operated at speeds of up to 70 mi/h (113 km/h). The system calculates both MPD and estimated mean texture depth (EMTD) using a transformation equation [5].

The portable Model 9300 laser texture scanner (LTS) [25] is a laser-based equipment. It is capable of determining a range of texture parameters also. It precisely characterises surface elevations at intervals <0.25 mm. Its more recent version is capable of measuring texture profiles with wavelengths down to 0.05 mm. The system can compute MPD, estimated texture depth (ETD), texture profile index (TPI), RMS, and band-passed filtered elevation.

4.5. Macro-Texture Properties of Various Asphalt Mixture Types

Based on the asphalt binder manufacturing process and temperatures used, there are two types of asphalt mixes: hot mix asphalt (HMA) or warm mix asphalt (WMA). HMA is heated and laid at temperatures between 150 °C and 190 °C, while WMA is manufactured and laid under the HMA's temperature by 15–40 °C. HMA is the conventionally used mix and predates the WMA. Production of WMA is cheaper, and can remain workable for longer period than HMA. No significant difference can be detected between the early slip resistance of the two mixture types. HMA can be classified further into three categories, based mainly on different aggregate grading, which has an important effect on surface macro-texture [26].

Dense graded HMA: well- or continuously graded mixture of coarse and fine aggregates, mineral filler, and 5–6% bituminous binder are categorized by nominal maximum aggregate sizes (ranging from 9.5 mm to 19.0 mm) into fine-graded and coarse-graded. The former has a higher % of sand and small stones than the latter one. Proper design and placement leads to relatively impermeable mixes. Mixes are suitable for all pavement layers and traffic conditions. Gap- and open-graded mixes have similar micro-texture values showing lower texture depths (typically 0.4–0.6 mm for fine-graded and 0.6–1.2 mm for coarse-graded) than gap- and open-graded mixture types.

Gap-graded HMA or Stone mastix asphalt (SMA): aimed at creating stone-on-stone contact within mixture to improve tyre grip and rutting (deformation) resistance. It contains more durable aggregates, higher (polymer-modified) bitumen content (6–9%), fillers, and fibres. Hence, it is more expensive than the other options. Their wet weather friction and lower tyre noise due to its coarser surface texture are among favourable features. It also shows less severe cracking. Their macro-texture depths typically exceed 0.8 mm, usually higher than those of dense graded mixes.

Open-graded HMA or porous asphalt (PA): designed to be water permeable. Hence, it uses mostly coarse aggregates, small percentage of sand/mineral filler, and 3–6% bituminous binder. The layer must contain >15% air voids. It shows similar micro-texture values to other mixture types but larger macro-texture (typically of 1.0–3.0 mm depth). Importantly, this version shows higher friction values than gap- and dense-graded mixes [5,19,27].

4.6. Hungarian Regulation for Measuring the Macro-Texture

In accordance with the relevant Hungarian road technical directives [28], asphalt surface course layers must meet certain macro-texture requirements during quality assurance. On the surface of the wearing courses, the macro-texture shall be measured in all traffic lanes, in the outer and inner wheel tracks, in every 6000 m^2. The measured values of the macro-texture of the wearing course in accordance with relevant EN standard [20] shall meet the requirements presented in Table 5.

Table 5. Macro-texture compliance limit values for all pavement types.

Asphalt Mixture Types	AC 8 top (F), AC 8 top (mF), AC 11 top (F), AC 11 top (mF), BBTM 4 A (mF), BBTM 8 A (mF), BBTM 8 B (mF), SMA 8 (mF), SMA 8 (mI), MA 11 (F), MA 11 (mF)	AC 16 top (F), AC 16 top (mF), BBTM 11 A (mF), BBTM 11 B (mF), SMA 11 (mF), SMA 11 (mI)
Macro-texture compliance limit, mm, min.	0.40	0.50

In the case of cement concrete pavement surfaces, according to the Hungarian directives [28], the requirements for macro-textures for different texture levels are provided in Table 6. The primary role of macro-texture, characterized by measuring sand depth, is to prevent the formation of a coherent water film, in the case of intense rain, its achievable value depends on the method of surface formation [29]. It must be measured during the quality assurance every 500 m in the outer and inner wheel tracks per lane.

Table 6. Requirements for the macro-texture levels of concrete pavement surfaces.

Road Type	Texture Depth, mm
Motorways	>0.5
Main road	>0.4
Local roads	>0.3
Urban roads, parking areas	none

In Hungary, Road Survey Tester (RST) [30] is used for quality assurance measurement, and for systematic road condition monitoring by determining several pavement condition parameters. Thanks to its Laser Scanning technique and software background, this method also allows the vehicle to determine MTD value, derived from MPD using a transformation equation. The surface texture shall be measured in the following wave length ranges: 1 to 10 mm (fine macro-texture), 10–100 mm (coarse macro-texture), and 100–500 mm (mega-texture). The measured values are the root mean square of the pavement surface wave heights.

5. Aggregates Characteristics

5.1. Aggregates Characteristics Generalities

The coefficient of friction is significantly affected by the grading of aggregates used for the manufacturing of bituminous mixtures. The extent to which the pavement surface provides adequate skid resistance depends largely on the aggregate, especially as the micro-texture plays a key role in the development of tyre–pavement friction, and it is mainly governed by aggregate properties [13]. Asphalt binder may have some measure of influence on micro-texture soon after the laying of asphalt. Nevertheless, aggregates make up the bulk of asphalt mixtures and serve as the primary contact medium with vehicle tyres. Hence, for adequate frictional performance of the pavement surface, the coarse aggregates for asphalt mixtures should be carefully chosen [5].

5.2. Mechanical and Physical Properties of Aggregates (Hungarian Specifications)

Of the numerous worldwide spread aggregate tests, several investigate the physical properties of the coarse aggregate in terms of slip resistance, which can mainly affect the micro-texture. In Hungary, the coarse aggregate can be used in wearing coarse asphalt mixture if it can meet certain physical limit values, among others, those influencing long-term skid resistance.

In accordance with the relevant specification used in Hungary [31], depending on the expected traffic load of the asphalt layer in question, there is an increasingly strict regulation on the PSV values of coarse aggregates (>4.0 mm) used for surface layers (wearing courses).

The categorization is based on the Design Traffic (DT) and Equivalent Standard Axle repetition number (F100) of the road section. Below 10^5 F100 during the expected life, there is no PSV value requirement. Between 0.1×10^6 and 0.3×10^6 F100 expected heavy traffic load, min. PSV44 polished stone value category is needed for the coarse aggregate used. While above 0.3×10^6 F100, the coarse aggregate selected must comply with the requirements of the PSV50 category. In general, the PSV 50 category can usually be fulfilled by rocks of volcanic origin. In Hungary, basalt and andesite type crushed rocks can primarily by taken into account.

The other two physical properties are the fragmentation resistance in accordance with the standard Los Angeles measurement [32] and the wear resistance in accordance with the Micro Deval measurement [33]. These parameters (LA and MDE) indicate the resistance of aggregate to mechanical degradation. In the Hungarian specification, in the case of normal traffic load (below 0.3×10^6 F100), the coarse aggregate of asphalt mixtures must meet LA25 criterion of Los Angeles fragmentation, while in the increased and intensive traffic category (above 0.3×10^6 F100) LA 20 is the compliance limit. In terms of Micro Deval wear resistance, there is a requirement for coarse aggregate in BBTM asphalt mixtures (which are typically used for increased category 0.3×10^6–3.0×10^6 F100) and for SMA (usually used in intensive traffic category above 3.0×10^6 F100). In these cases, the prescribed limit amounts to 18 M_{DE}.

5.3. Geometric Properties in Hungarian Specification

Another rather important aggregate property that can affect skid resistance of the pavement surface is the grain structure. Sharp and angular coarse aggregate particles interlock and provide higher texture depth. While flat and elongated particles tend to be horizontally oriented, resulting in lower texture depth.

According to Hungarian practice, the most commonly used aggregate material for asphalt mixtures of surface layers is crushed stone. Crushed gravel is only permitted in the case of normal traffic load (below 0.3×10^6 F100). In these cases, the share of crushed and broken aggregate grain surfaces should be at least 90% in accordance with the relevant standard [34].

Determination of the particle shape-flakiness index determines the incidence of flat and elongated aggregates relative to cubic aggregates. This can be provided by a reference number called the flakiness index (FI) in accordance with the relevant standard [35]. As for the Hungarian national specification on flakiness index [36], the specified value is FI20 for the coarse aggregates (KZ 4/8 and KZ 8/11) of SMA-type asphalt mixtures that can be used in the intensive traffic category (above 3.0×10^6 F100). In the case of lower travel category, the specified value for crushed rock aggregates amounts to be FI35. In the case of crushed gravel, the requirement of the category of FI25 must be met.

6. COST Action 354

6.1. COST Action 354 Generalities

The main objective of COST Action 354 is the development of unified European performance indicators and indices for road users and managers [37–39]. The numerical evaluation of performance indicators is of major importance since it provides information for national and European road pavement design and maintenance. The actual intervention limit values of indicators supply the minimally achievable quality requirements for planned and existing road pavement surfaces. In the case of various pavement structure and road types, different performance indicator values prove as appropriate. These performance indicators can also be inputs of pavement management system (PMSs) inputs [40] when the maintenance needs of relevant road sections or networks are calculated (forecasted).

Based on the replies of the comprehensive COST 354 questionnaire [38], the development of the following performance indices at a unitless scale was aimed at:

- Longitudinal unevenness;
- Transverse unevenness
- Macro-texture (macro-roughness in the USA);
- Skid resistance;
- Rolling noise;
- Air pollution;
- Pavement structure;
- Load bearing capacity.

6.2. Macro-Texture Performance Indicators

Pavement macro-texture is directly related to traffic safety and road user comfort. It influences indirectly pavement skid resistance by the actual connection between (wet) pavement surface and vehicle tyre. Based on the questionnaire survey of COST 354, it was revealed that two texture measurement principles are used in Europe: the Volumetric and the Laser method.

- The volumetric method (sand patch test) is actually the spreading of a given amount of sand or glass bead on the pavement surface with help of a standard flat disk, then measuring the diameter of sand by a steel scale, taking five readings. The mean diameter can be used for the calculation of the mean texture depth;
- The laser method is where the result of the texture measurement performed with a laser is independent of the measuring speed. Although numerous apparatus types (ARAN, Rav, Roadstar, Roar, RST, Rugolaser, and SCRIM) are available, the measurement is performed almost exactly the same way in each. The devices apply one or more laser beams aimed at various points of pavement surface, often in both wheel paths. The measurement is carried out always on a given line longitudinally, and each profile contains numerous, defined periodically performed level equalizations.

Depending on the device and software used, the measurement results can be provided in the form of the following performance indices: "mean profile depth (MPD)" [39], "mean texture depth (MTD)" [20,40] or "sensor measured texture depth (SMTD)". This latter one does not have any international specification. Eight European countries use the MPD, there are three that prefer MTD, while two countries prefer SMTD. None of these indicators can be measured by a single device type.

The laser-based method is reliable and fully automated, the differences among its results in the case of calibrated lasers are negligible. The quite simple sand patch test is a proven method, which makes it sufficiently reliable.

The traffic speed measuring MPD and SMTD is safe at a network level, while the sand patch method is slow in a closed area of the pavement surface causing some traffic hazard.

Transformation functions were determined in the Czech Republic in order to convert MPD into indices (PIT) at a unitless scale (Table 7).

Table 7. Macro-texture limit values in the Czech Republic [41].

Index (Note)	Verbal Classification	MPD (mm) in the Road Class of	
		Motorway, Primary Road	Secondary Road
1	very good	min. 0.89	min. 0.79
2	good	0.74–0.88	0.64–0.78
3	appropriate	0.64–0.73	0.54–0.63
4	poor	0.54–0.63	0.44–0.53
5	very poor	max. 0.53	max. 0.43

The conversion algorithm developed between the texture indicator MPD and the index PIT [21]:

- Motorways and main roads:

$$PI_T = 5.3MPD - 1.6; \tag{6}$$

- Local roads:

$$PI_T = 6.9MPD - 2.0. \tag{7}$$

6.3. *Macro-Texture Performance Indicators*

The questionnaire answers of COST 354 prove that the measurement of pavement skid resistance is usually (in 66%) carried out based on national specifications. The reason is that the measuring results are basically affected by the measuring principle and other factors (measuring speed, load of measuring wheel, water film thickness, etc.). The harmonization efforts at the European level have so far been unsuccessful [42,43]; thus, no valid international standard is available.

The distribution of the skid resistance measuring principles among the 26 questionnaire responses was as follows: 10 longitudinal force on measuring wheel, 10 lateral force on measuring wheel, 3 laser-based, 1 SRT-pendulum, 2 no answer.

Thirteen of 26 questionnaires returned have some information on skid resistance indicators, and just 4 of them contain conversion algorithms (Table 8).

Table 8. Conversion algorithms for pavement skid resistance indicators.

Country	Index	Conversion Function	Remarks
Austria	Skid resistance index	9.9286–14.236 TP *	network level
Belgium	Skid resistance index	4(SFC **-0.1)/3	
Poland	Skid parameter	100–180 TP	

Legend: * TP: technical parameter; ** SFC: sideways force coefficient.

For the sideways force coefficient, the following limit values were recommended by COST 354 WP2 [38]:

$$TV = -0.31OS + 59, \tag{8}$$

$$WV = -0.31OS + 64 \quad (9)$$

$$TAV = -0.31OS + 75, \quad (10)$$

where

TV is the threshold value;
WV is the warning value;
TAV is the target value;
OS is the operating speed (km/h).

In case of longitudinal frictional force, the defined limit values can be found in Table 9.

Table 9. Limit values for longitudinal frictional force [38].

Limits	Main Roads	Local Roads
Threshold values	TV = −0.23OS + 48	TV = −0.23OS + 45
Warning values	WV = −0.23OS + 56	WV = −0.23OS + 54

7. Summary

Safe and reliable pavements are of paramount importance for both human losses and economic growth, especially with the intensive growth of traffic systems. Traffic safety is basically influenced by the skidding parameters of a road pavement surface. Key parameters influencing measurements of skid resistance of asphalt pavements were identified and reviewed. A brief overview of some of the most common approaches to modelling the major aspects of tyre–pavement friction was also provided. Skid resistance is one of the substantial factors which ensure safe travelling on automobile roads, there is a need to improve the safety situation. Road pavements with high and long-lasting skid resistance can be a solution. It is important to develop a Hungarian skid resistance policy covering skid resistance measurements not only for acceptance tests but also for skid resistance routine monitoring of the whole Hungarian road network, and include skid resistance measurement data into pavement management systems. Designing hot mix asphalt (AC) pavements meeting frictional "demand" requires selecting mix designs and aggregate types, as well as properties that can adequately provide long-term friction. In Hungary, slightly simplified, the regulations for aggregates include these elements for micro-texture. Aggregate types are differentiated and selected by Polish Value (PSV), Los Angeles Fragmentation Value (LA), Micro Deval Wear Resistance Value (MDE), Crushed Particle Ratio (C), and Flakiness Index [31–35]. Some results of a COST Action are also presented, highlighting the typical European practice on macro-texture and skid resistance measurement and pavement surface evaluation as an important input of Pavement Management Systems.

Funding: The research was carried out as part of the project numbered 2020-1.1.2-PIACI-KFI-2020-00060, with the support of the National Research, Development and Innovation Office of the Ministry of Innovation and Technology, and was funded by the 2020-1.1.2-PIACI KFI tender program.

Institutional Review Board Statement: Informed consent was obtained from all subjects involved in the study. Ethical review and approval were waived for this study due to the fact that its topic was concentrating on road pavements.

Informed Consent Statement: No informed consent was needed since no subjects had been involved in the study.

Data Availability Statement: Being a review paper, no new data were created by the authors.

Conflicts of Interest: Authors have no conflict of interest to declare.

References

1. Wallman, C.-G.; Åström, H. *Friction Measurement Methods and the Correlation Between Road Friction and Traffic Safety—A Literature Review*; Swedish National Road and Traffic Institute: Linköping, Sweden, 2001.
2. Kuttesch, J.S. Quantifying the Relationship between Skid Resistance and Wet Weather Accidents for Virginia Data. Master's Thesis, Department of Civil Engineering, Virginia Polytechnic Institute and State University, Blacksburg, VA, USA, 2004; 147p.
3. Viner, H.; Sinhal, R.; Parry, T. Review of UK Skid Resistance Policy. In Proceedings of the 5th International Symposium on Pavement Surface Characteristics—Road and Airports, Toronto, ON, Canada, 6–10 June 2004. 12p.
4. Anupam, K.; Srirangam, S.; Scarpas, A.; Kasbergen, C. Influence of Temperature on Tire-Pavement Friction: Analyses, Transportation Research Record. *J. Transp. Res. Board* **2013**, *2369*, 114–124. [CrossRef]
5. Hall, J.W.; Smith, K.L.; Titus-Glover, L.; Evans, L.D.; Wambold, J.C.; Yager, T.J.; Rado, Z. *Guide for Pavement Friction. Contractor's Final Report for National Cooperative Highway Research Program (NCHRP) Project 01-43*; Transportation Research Board of the National Academies: Washington, DC, USA, 2009; 257p, Available online: http://onlinepubs.trb.org/onlinepubs/nchrp/nchrp_w108.pdf (accessed on 21 June 2015).
6. Meyer, W.E. *Synthesis of Frictional Requirements Research*; Report, No. FHWA/RD-81/159; Federal Highway Administration (FHWA): Washington, DC, USA, 1982.
7. Henry, J.J. *Evaluation of Pavement Friction Characteristics*; NCHRP Synthesis 291; National Cooperative Highway Research Program (NCHRP): Washington, DC, USA, 2000; 72p.
8. Gillespie, T.D. *Fundamentals of Vehicle Dynamics*; Society of Automotive Engineers (SAE): Warrendale, PA, USA, 1992; 519p.
9. Radt, H.S.; Milliken, W.F. *Motions of Skidding Automobiles*; Paper No. 600133 (205A); Society of Automotive Engineers (SAE): Warrendale, PA, USA, 1960.
10. Do, M.-T.; Roe, P. Report on State-of-the-Art of Test Methods. In *TYROSAFE Project Deliverable D04*; FEHRL: Brussels, Belgium, 2008; 89p.
11. Descornet, G.; Schmidt, B.; Boulet, M.; Gothie, M.; Do, M.-T.; Fafie, J.; Alonso, M.; Roe, P.; Forest, R.; Viner, H. Harmonization of European Routine and Research Measuring Equipment for Skid Resistance. In *HERMES Final Report*; Institut Francais des Sciences et Technologies des Transports, de l'Aménagement et des Réseaux (IFSTTAR): Champs-sur-Marne, France, 2006; 161p.
12. Andriejauskas, T.; Vorobjovas, V.; Mielonas, V. Evaluation of Skid Resistance Characteristics and Measurement Methods. In Proceedings of the 9th International Conference Environmental Engineering, Vilnius, Lithuania, 22–23 May 2014; Cygas, D., Tollazzi, T., Eds.; VGTU Press: Vilnius, Lithuania, 2014. 9p. [CrossRef]
13. Kogbara, R.B.; Masad, E.A.; Kassem, E.; Scarpas, A.T.; Anupam, K. A State-of-the-Art Review of Parameters Influencing Measurement and Modeling of Skid Resistance of Asphalt Pavements. *Constr. Build. Mater.* **2016**, *114*, 602–617. [CrossRef]
14. *e-UT 09.02.23:1999*; Az Útburkolat-Felület Csúszásellenállásának Vizsgálata. Mérés SCRIM-Mérőkocsival (Skid Resistance Testing of Road Surfaces. Measurement with SCRIM Testing Vehicles). Útügyi Műszaki Előírás (Road Technical Directives). Ministry for Economy and Transport: Budapest, Hungary, 1999; 15p.
15. *e-UT 09.02.27:2009*; Az Útburkolat-Felület Csúszásellenállásának Vizsgálata. Mérés ASF-Berendezéssel (Skid Resistance Testing of Road Surfaces. Measurement by ASFT Equipment). Útügyi Műszaki Előírás (Road Technical Directives). Ministry for Economy and Transport: Budapest, Hungary, 2009; 15p.
16. *EN 13036-4: 2011*; Road and Airfield Surface Characteristics—Test Methods—Part 4: Method for Measurement of Slip/Skid Re-sistance of a Surface: The Pendulum Test. SIST—Slovenian Institute for Standardization: Ljubljana, Slovenia, 2011; 32p.
17. Hall, J.W.; Smith, K.L.; Titus-Glover, L.; Wambold, J.C.; Yager, T.J.; Rado, Z. *NCHRP Web-Only Document 108: Guide for Pavement Friction*; Final Report for NCHRP Project 01-43; National Cooperative Highway Research Program: Washington, DC, USA, 2009.
18. Flintsch, G.W.; De Leon, E.; McGhee, K.K.; Al-Qadi, I.L. Pavement Surface Macrotexture Measurement and Application. *Transp. Res. Rec. J. Transp. Res. Board* **2003**, *1860*, 168–177. [CrossRef]
19. Permanent International Association of Road Congresses (PIARC). Report of the Committee on Surface Characteristics. In Proceedings of the 18th World Road Congress, Brussels, Belgium, 13–19 September 1987. 108p.
20. *EN 13036-1*; Road and Airfield Surface Characteristics—Test Methods—Part 1: Measurement of Pavement Surface Macrotexture Depth Using a Volumetric Patch Technique. SIST—Slovenian Institute for Standardization: Ljubljana, Slovenia, 2010; 11p.
21. *ASTM E 965*; Standard Test Method for Measuring Pavement Macrotexture Depth Using a Volumetric Technique. ASTM International: West Conshohocken, PA, USA, 2015; 4p.
22. *ASTM E 2380*; Standard Test Method for Measuring Pavement Texture Drainage Using an Outflow Meter. ASTM International: West Conshohocken, PA, USA, 2010; 4p.
23. *ASTM E 2157-5*; Standard Test Method for Measuring Pavement Macrotexture Properties Using the Circular Track Meter. ASTM International: West Conshohocken, PA, USA, 2019; 5p.
24. U.S. Department of Transportation. *ROSAN V—Technical Notes*; Federal Highway Administration: Washington, DC, USA, 1997; 6p.
25. Bittelli, G.; Simone, A.; Girardi, F.; Lantieri, C. Laser Scanning on Road Pavements: A New Approach for Characterizing Surface Texture. *Sensors* **2012**, *12*, 9110–9128. [CrossRef] [PubMed]
26. Flintsch, G.W.; Al-Qadi, I.L.; Davis, R.; McGhee, K.K. Effect of HMA Properties on Pavement Surface Characteristics. In Proceedings of the Pavement Evaluation Conference, Roanoke, VA, USA, 21–25 October 2002. 16p.

27. Sandburg, U. *Influence of Road Surface Texture on Traffic Characteristics Related to Environment, Economy, and Safety: A State-of-the-Art Study Regarding Measures and Measuring Methods*; VTI Report 53A-1997; Swedish National Road Administration: Borlange, Sweden, 1998; 94p. Available online: https://trid.trb.org/view/1149891 (accessed on 28 February 2022).
28. *e-UT 06.03.21:2018*; Útpályaszerkezetek Aszfaltburkolati Rétegeinek Követelményei (Requirements of the Asphalt Layers of Road Pavement Structures). Ministry for Economy and Transport: Budapest, Hungary, 2018; 57p. (In Hungarian)
29. Ongel, A.; Lu, Q.; Harvey, J. Frictional Properties of Asphalt Concrete Mixes. In Proceedings of the Institution of Civil Engineers-Transport; Thomas Telford Ltd: London, UK, 2009; Volume 162, pp. 19–26. [CrossRef]
30. Ramboll RST. *RST—Road Survey Technology*; Ramboll RST: Malmö, Sweden, 2019; 5p.
31. *EN 1097-8:2020*; Tests for Mechanical and Physical Properties of Aggregates—Part 8: Determination of the Polished Stone Value. SIST—Slovenian Institute for Standardization: Ljubljana, Slovenia, 2020; 35p.
32. *EN 1097-2:2020*; Tests for Mechanical and Physical Properties of Aggregates—Part 2: Methods for the Determination of Resistance to Fragmentation. SIST—Slovenian Institute for Standardization: Ljubljana, Slovenia, 2020; 43p.
33. *EN 1097-1:2011*; Tests for Mechanical and Physical Properties of Aggregates—Part 1: Determination of the Resistance to Wear (Micro-Deval). SIST—Slovenian Institute for Standardization: Ljubljana, Slovenia, 2011; 14p.
34. *EN 933-5:2022*; Tests for Geometrical Properties of Aggregates—Part 5: Determination of Percentage of Crushed Particles in Coarse and All-In Natural Aggregates. SIST—Slovenian Institute for Standardization: Ljubljana, Slovenia, 2022; 20p.
35. *EN 933-3:2012*; Tests for Geometrical Properties of Aggregates—Part 3: Determination of Particle Shape—Flakiness Index. SIST—Slovenian Institute for Standardization: Ljubljana, Slovenia, 2012; 11p.
36. *e-UT 05.01.15:2018*; Útépítési Kőanyaghalmazok (Road Building Stone Masses). Ministry for Economy and Transport: Budapest, Hungary, 2018; 74p. (In Hungarian)
37. COST 354. *Performance Indicators for Road Pavements*; Memorandum of Understanding; COST 354; European Union: Brussels, Belgium, 2004.
38. COST 354. *Performance Indicators for Road Pavements. WP2 Selection and Assessment of Individual Performance Indicators*; Final Report; COST 354; European Union: Brussels, Belgium, 2007; 170p.
39. Gáspár, L.; Károly, R. Útpályaszerkezetek Makroérdességi és Csúszásellenállási Mutatói (Macro Roughness and Skid Resistance Indicators of Road Pavements). *Közlekedéstudományi Szle.* **2007**, *12*, 442–449. (In Hungarian)
40. Gáspár, L. *Útgazdálkodás (Road Management)*; Akadémiai Kiadó: Budapest, Hungary, 2003; 361p. (In Hungarian)
41. *EN ISO 13473-1:2019*; Characterization of Pavement Texture by Use of Surface Profiles—Part 1: Determination of Mean Profile Depth. ISO: Geneva, Switzerland, 2019; 43p.
42. Wambold, J.C.; Antle, C.E.; Henry, J.; Rado, Z.; Descornet, G.; Sandburg, U.; Gothié, M.; Huschek, S. *International Experiment to Compare and Harmonize Skid Resistance and Texture Measurements*; PIARC Technical Committee 1 Surface Characteristics; PIARC: Paris, France, 1995; 430p, ISBN 84-87825-96-6.
43. Descornet, G.; Schmidt, B.; Boulet, M.; Gothie, M.; Do, M.T.; Fafie, J.; Alonso, M.; Roe, P.; Viner, H. *HERMES Project (Harmonization of European Routine and Research Measuring Equipment for Skid Resistance)*; FEHRL Report 2006/01; FEHRL: Brussels, Belgium, 2006; 24p.

eng

MDPI

Article

An Overview of Smart Materials and Technologies for Concrete Construction in Cold Weather

Jonny Nilimaa * and Vasiola Zhaka

Department of Civil, Environmental and Natural Resources Engineering, Luleå University of Technology, SE-97187 Luleå, Sweden
* Correspondence: jonny.nilimaa@ltu.se

Abstract: Cold weather conditions pose significant challenges to the performance and durability of concrete materials, construction processes, and structures. This paper aims to provide a comprehensive overview of the material-related challenges in cold weather concrete construction, including slow setting, reduced curing rate, and slower strength development, as well as frost damage, early freezing, and freeze–thaw actions. Various innovative materials and technologies may be implemented to address these challenges, such as optimizing the concrete mix proportions, chemical admixtures, supplementary cementitious materials, and advanced construction techniques. The paper also examines the impact of weather-related challenges for personnel, equipment, and machinery in cold environments and highlights the importance of effective planning, communication, and management strategies. Results indicate that the successful implementation of appropriate strategies can mitigate the challenges, reduce construction time, and enhance the performance, durability, and sustainability of concrete structures in cold and freezing temperatures. The paper emphasizes the importance of staying updated about the latest advancements and best practices in the field. Future trends include the development of smart and functional concrete materials, advanced manufacturing and construction techniques, integrated design, and optimization of tools, all with a strong focus on sustainability and resilience.

Keywords: concrete; cold construction; engineering challenges; freezing; concrete setting; construction materials

Citation: Nilimaa, J.; Zhaka, V. An Overview of Smart Materials and Technologies for Concrete Construction in Cold Weather. *Eng* **2023**, *4*, 1550–1580. https://doi.org/10.3390/eng4020089

Academic Editor: Antonio Gil Bravo

Received: 17 May 2023
Revised: 29 May 2023
Accepted: 30 May 2023
Published: 31 May 2023

1. Introduction

Concrete is a the most commonly used construction material [1,2]. It plays an important role in the development and construction of sustainable infrastructure, buildings, and various other structures. It is well known for its versatility, durability, and cost-effectiveness [3–7]. However, concrete construction in cold weather conditions presents numerous challenges that can significantly impact the performance, durability, and sustainability of the structures [8–12]. As cold weather construction becomes more prevalent due to expanding urbanization, expedited by rapid population growth [13,14] and the increased requirements for infrastructure development in colder regions, understanding the material-related challenges and identifying effective strategies and technologies to address these challenges is essential. Cold weather concrete construction has been an active area of research for several years, with numerous studies focusing on the challenges related to setting [15], curing [16], strength development [17], and drying [18], as well as frost damage [19], early freezing [20], and freeze–thaw cycles [21]. Despite these research efforts, there is still a need for a better understanding of the material-related challenges and technologies that can be employed to address them. This study contributes by filling the knowledge gap in the literature and providing a holistic understanding of the challenges, strategies, and performance enhancement opportunities for concrete construction. A comprehensive understanding of the challenges and strategies associated with concrete materials, construction processes, and structures, can be used to support construction professionals in making informed decisions and implementing effective solutions that improve the performance and sustainability of concrete structures in harsh environments.

The aim of this paper is to provide an in-depth analysis of the material-related challenges in cold weather concrete construction as well as explore innovative materials and the technologies to implement them, as illustrated in Figure 1. The study encompasses a comprehensive review of the existing literature, including research studies and case examples related to cold weather concrete construction. This review focuses on understanding the material-related challenges associated with slow setting, reduced curing rate, and slower strength development, as well as frost damage due to low temperatures, early freezing, and freeze–thaw actions. Furthermore, the paper explores innovative strategies and technologies that can be employed to address the various challenges, such as optimizing the concrete mix proportions [22], the use of chemical admixtures [23], supplementary cementitious materials [24], and advanced construction techniques [25]. The study emphasizes the emerging materials and technologies in the field, including the development of smart [26] and functional concrete materials [27], advanced manufacturing [28] and construction techniques [29], integrated design [30], and optimization of tools [31], all with a focus on sustainability and resilience. The study also highlights the potential impact of weather-related challenges on personnel, equipment, and machinery during construction. Furthermore, it emphasizes the importance of effective planning, communication, and management strategies to ensure successful completion of construction projects in low temperatures. As climate change continues to result in more unpredictable and extreme weather, understanding the challenges associated with cold weather construction and developing effective strategies and technologies to address them becomes increasingly important. This research not only contributes to the existing body of knowledge on this topic but also serves as a foundation for future research and development efforts in the field of concrete construction. The findings of this study are not limited to the specific challenges and strategies discussed but can also be extended to other construction materials and processes that may be affected by low temperatures. By fostering a culture of continuous learning and innovation, the construction industry can better adapt to the changing climate [32] and ensure the resilience and sustainability of infrastructure and buildings worldwide [33].

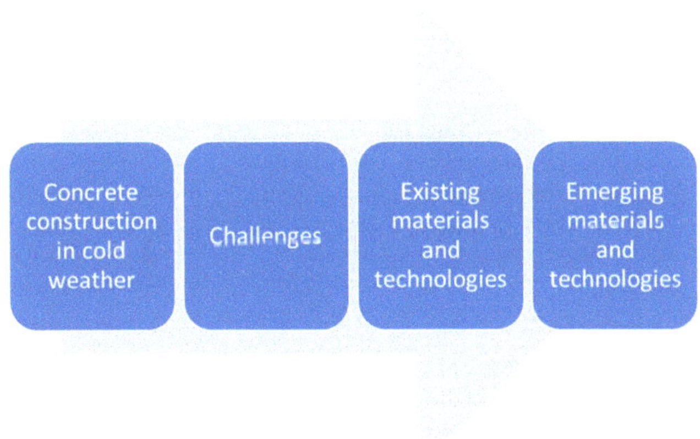

Figure 1. Parts included in this review.

2. Challenges in Cold Weather Concrete Construction

2.1. Setting and Curing

Setting and curing are two critical processes that ultimately determine the strength growth, durability, and long-term performance of concrete materials. Low temperatures can have a large negative effect on the development of concrete properties, as these processes can be significantly slowed down, leading to various material, structural, and management issues. The challenges of longer setting times and slower curing are discussed in the following sections, and counteracting engineering strategies are proposed in Section 2.1.3.

2.1.1. Setting

Setting refers to a specific time when concrete changes from a fluid, workable state to a solid, rigid state [34]. It involves the initial chemical reactions between cement [35] or other binders [36] and water, a process called hydration [37]. In cold conditions, the rate of hydration decreases due to the lower temperature, causing the concrete to set more slowly [38]. Slow setting can lead to problems affecting the construction procedure, the material properties, the durability of the structure, and the long-term sustainability. A longer setting time for concrete in cold and freezing temperatures can lead to various challenges and consequences, including increased construction time and costs [39]. As the concrete sets more slowly, the formwork and temporary supports must remain in place for a longer period to ensure the concrete can support its own weight and any applied loads [40]. This extended support time can cause delays in the construction schedule [41], as it may prevent other activities from taking place in parallel. The cost of formwork typically accounts for a major part of the total construction costs [42], and the prolonged need for supporting formwork can lead to high additional costs. Longer setting times may also necessitate more on-site personnel for a longer duration to monitor the concrete [43], maintain temperature control measures [44], and perform finishing tasks [45]. This increased labor requirement can lead to higher labor costs for the project. As the concrete sets more slowly in cold weather, it can be challenging to achieve a smooth, even surface finish [46]. This may also require additional effort and resources to rectify. Slower setting times may require the use of additional equipment, such as heated enclosures [11] or insulated formwork [47], to maintain optimal temperature conditions and accelerate the setting process. Any damage to the concrete due to the extended setting time may necessitate repairs or even replacement, which can further increase the complexity and add to the material and labor costs.

Construction projects often have strict deadlines, and any delays in the schedule can result in different types of financial penalties for the builder [48]. The longer setting time and its impact on the construction timeline may in the end lead to the activation of such penalties, increasing the overall cost of the project. With the extended setting time, the concrete also becomes more susceptible to damage from adverse weather conditions [49], such as rain, snow, or freezing temperatures, which can negatively affect its properties and structural performance. This may require additional measures to protect the concrete, adding to the overall cost and potentially causing further delays. If the slower setting time results in reduced strength or other undesirable properties in the concrete, it may be necessary to modify the structural design or add extra reinforcement, which will increase costs and extend the construction timeline. It is therefore important for the engineers to always consider the setting time and choose a concrete that fulfills the structural requirements in the ambient environment of the project.

2.1.2. Curing

Curing of concrete is the process of maintaining proper moisture and temperature conditions within the concrete after it has been placed and finished [50]. This process allows the concrete to develop its desired strength, durability, and long-term performance through the continued hydration of cement or other types of alternative binders [51]. Hydration is the chemical reaction between binders and water; it forms new compounds and crystalline structures, binding the aggregate particles together and giving concrete its strength. Adequate curing is essential to ensure that the hydration process continues for an extended period, allowing the concrete to reach its full potential in terms of strength and durability. By preventing the evaporation of water from the concrete, the curing process ensures that there is enough water available for the hydration of cement, as illustrated in Figure 2 [52]. This can be achieved using various methods, such as applying curing compounds [53], covering the concrete with plastic sheets [54], or using wet coverings like burlap [55] or cotton mats [56]. Curing also involves maintaining a suitable temperature range within the concrete to facilitate the hydration process [57]. This is especially important in extreme weather conditions, such as cold or hot temperatures, where external measures may be required to regulate the concrete's temperature. These measures may include using insulated blankets or enclosures [58], heated water [59], heating cables [60], or steam curing [61]. Curing helps reduce the potential for shrinkage and cracking in concrete by controlling the rate at which it dries and contracts [62]. Proper curing can minimize the development of microcracks, leading to improved durability and performance. Curing typically begins immediately after the concrete has been placed and finished, and it continues for a certain period, depending on the type of concrete, the intended use of the structure, and the environmental conditions. The curing duration can range from a few days to several weeks, with the most critical period being the first few days after placement [63]. By ensuring proper curing, contractors can optimize the performance, durability, and service life of concrete structures, making it a critical aspect of the overall construction process.

Figure 2. Hydration process of cement [52]: (**a**) Unhydrated cement particles (2000× magnification); (**b**) Partially hydrated cement particles (4000× magnification); (**c**) Hydrated cement particle (11,000× magnification).

A reduced curing rate in cold weather can lead to various challenges and consequences, including increased construction time, safety concerns, compromised durability, and increased costs [64]. Slower curing rates mean that it takes longer for the concrete to achieve its desired strength and be ready for subsequent construction stages, such as removing formwork, loading, or applying finishes. This can delay the overall construction schedule and affect the project timeline. If the concrete has not fully cured, it may not be able to support its self-weight or the loads applied during construction. This can lead to structural instability and increase the risk of accidents or failures on the construction site. Inadequate curing can result in incomplete hydration, which can cause the concrete to be more porous and susceptible to freeze–thaw damage, chemical attack, and other forms of deterioration [65]. This can compromise the long-term durability of the structure and lead to a reduced service life. Slower curing rates can result in higher labor and equipment costs, as additional resources are needed to maintain optimal curing conditions and monitor the concrete's progress. Additionally, any damage or structural failures due to insufficient curing may require repairs or even replacement, which can further increase material and labor costs. To prevent curing-related issues, it is crucial to implement appropriate cold-weather concreting practices, such as maintaining optimal temperature conditions, using admixtures to accelerate the curing rate [66], and closely monitoring the concrete's curing progress. By taking these measures, contractors can minimize the impact of reduced curing rates on construction time, safety, durability, and costs, while ensuring the quality and performance of concrete structures in cold and freezing temperatures.

2.1.3. Strategies to Prevent Setting and Curing Challenges

To prevent issues related to longer setting and slower curing of concrete in cold weather, various measures can be implemented to ensure the quality, performance, and durability. Chemical admixtures, such as accelerators, can be added to the concrete mix to accelerate the hydration process and reduce the setting time. These admixtures can help the concrete to achieve the desired strength faster, minimizing the impact of cold weather [67]. Maintaining the concrete temperature within an appropriate range is crucial to ensure proper setting and curing. This can be achieved through various methods, such as pre-heating [59], insulation, and heated enclosures [58]. Heating the water and/or aggregates before mixing can help maintain a suitable temperature for the concrete mix, ensuring a timely setting. Insulated formwork or blankets can be used to maintain the concrete's temperature during setting and curing, protecting it from cold weather and reducing heat loss and rapid cooling. Temporary, heated enclosures can be built around the construction site to maintain a controlled environment with adequate temperature and humidity levels.

Employing appropriate curing techniques in cold weather is essential to maintain the moisture and temperature conditions necessary for the hydration process [50]. Insulated curing blankets can be placed over the concrete surface to minimize heat loss and maintain the desired curing temperature. Circulating heated water [59] through pipes embedded in the concrete can provide consistent and controlled heat to maintain the curing temperature. Steam can be introduced into an enclosed space to maintain the temperature and humidity required for curing. Modifying the concrete mix design to include more cement or incorporating supplementary cementitious materials (SCMs), such as fly ash or slag [51], can help increase the rate of strength development and counteract the effects of cold temperatures on setting time. Regularly or continuously monitoring the temperature and strength development of the concrete is crucial during cold weather construction. Field testing, such as the use of maturity meters or penetration resistance tests [68], can help ensure the concrete achieves the desired performance characteristics. Effective planning and scheduling of construction activities can help minimize the impact of cold weather on setting and curing times. Scheduling concrete placement during warmer periods of the day or coordinating with weather forecasts can help optimize the construction process. By implementing these preventing or proactive measures, contractors can mitigate the challenges associated with longer setting times in cold weather, ensuring the quality, durability, and performance of concrete structures.

2.2. *Strength Development*

Cold weather conditions can significantly affect the strength development of concrete [69], which is a critical aspect of its overall performance and durability. The setting, curing, and strength development processes of concrete are all temperature-dependent, and lower air temperatures can slow down these processes [70,71], leading to challenges in construction and potential long-term issues in the final structure. The rate of hydration, and consequently the rate of strength development, is heavily influenced by the concrete temperature [72]. As temperature decreases, the rate of hydration reduces, resulting in a slower strength gain. The rate of hydration is roughly reduced by half in 10 °C temperatures compared to 20 °C. Consequently, concrete placed and cured in cold conditions will exhibit a slower rate of strength development compared to concrete placed in warmer temperatures.

2.2.1. Factors Affecting the Strength Development in Cold Weather

The hydration process is highly dependent on temperature [69]. Low air temperatures can reduce the concrete temperature and hydration rate [70], ultimately affecting the material properties. The reduced hydration rate directly impacts the strength development, as it takes more time for the concrete to achieve its full potential strength in cold climates [73]. By understanding the factors affecting the material properties, contractors can take appropriate measures to prevent problems associated with slower strength development.

Cold weather prolongs the setting, which delays the strength development process. An extended setting time increases the vulnerability of concrete to external factors, such as harsh weather or mechanical damage, before it reaches an adequate level of strength. The curing process plays an important role in the strength development of concrete by maintaining appropriate moisture and temperature conditions [50]. The curing process takes more time in low temperatures, which can result in incomplete hydration and slower strength gain. If the curing process is not carefully managed in cold conditions, the concrete may be exposed to freezing temperatures, which can lead to freeze–thaw damage and significantly compromise the structure's durability and performance.

The use of Supplementary Cementitious Materials, SCMs, such as fly ash or slag, is becoming more and more common in concrete mix designs. These materials can be used for several reasons, such as enhancing the concrete's properties or reducing its environmental impact [74]. SCMs can also impact the hydration process and, consequently, the strength development. In cold weather, the pozzolanic reactions between SCMs and cement hydration products can be slower [75], further contributing to the reduced strength gain. Chemical admixtures, such as accelerators and water reducers, are often used to modify the properties of the concrete mix [76]. In cold weather, the effectiveness of some admixtures may be reduced due to the lower temperatures. During mixing and transportation, the concrete's temperature can decrease [77], particularly in cold weather, and the effects can be significant when long distances separate concrete factories and construction sites, which is not uncommon in regions with cold climate. Delays in placing the concrete due to logistical challenges can exacerbate this issue, as the concrete may experience further temperature drops and reduced workability.

2.2.2. Consequences of Slow Strength Development

Understanding the potential consequences of slower strength development is essential to managing the challenges associated with cold weather concreting effectively. By implementing appropriate measures to mitigate these consequences, contractors can ensure the successful completion of their projects and maintain high quality, durability, and performance of the constructed concrete structures. Slow strength development can ultimately result in long construction times, as it takes more time for the concrete to achieve the required strength [77] for subsequent construction activities. This can lead to delays in the overall construction schedule, potentially causing a domino effect on the project timeline and increasing costs related to labor and equipment. A slow concrete is more susceptible to damage during construction, such as premature loading, weather exposure, or other construction activities [78]. Early-age damage can result in defects that compromise the structural integrity, require costly repairs, or even necessitate a complete replacement of the affected concrete elements [79]. If the concrete does not achieve its required strength within the expected timeframe, it may not be able to support the intended loads or perform as designed [80]. This can lead to potential safety concerns, a reduced service life, and increased likelihood of structural failures. This may, for example, be a major concern for deciding a safe time for form stripping, which typically requires a minimum strength of 5 MPa for vertical structural members and 70% of the final strength for horizontal members [81].

Slower strength development can lead to increased permeability and microcracking within the concrete [82], making it more susceptible to various forms of deterioration, such as freeze–thaw damage [83], alkali–silica reactions [84], and chloride-induced corrosion of reinforcing steel [85]. These forms of deterioration can compromise the long-term durability of the structure and result in increased maintenance and repair costs over the structure's life. A reduced hydration rate can complicate the quality control process during construction. As the concrete takes longer to achieve its target strength, it becomes more difficult to accurately assess its performance characteristics and ensure it meets the project's specifications. In some cases, this can lead to aesthetic issues [86], such as an uneven or poor-quality finish on the concrete surface. The longer the construction process takes, the more resources are typically consumed, such as energy, water, and raw materials [87]. This can ultimately contribute to a larger environmental footprint for the construction project.

2.2.3. Measures to Mitigate Slow Strength Development

Understanding the impact of harsh weather on concrete strength development is essential for successful construction projects in such conditions. By implementing appropriate measures to mitigate the challenges associated with slow strength gain, contractors can ensure the quality, durability, and performance of concrete structures, even in cold and freezing temperatures. Proper planning, mix design adjustments [88], temperature control [89], and curing methods [50] can all contribute to a more efficient construction process and a more resilient final structure, ultimately leading to long-lasting and high-performing concrete structures.

Chemical admixtures [90], such as accelerators, can be added to the concrete mix to increase the rate of cement hydration and reduce the setting time, helping the concrete to achieve its desired strength more quickly. Modifying the concrete mix design to include more cement or incorporating supplementary cementitious materials [51], such as fly ash or slag, can help increase the rate of strength development and counteract the effects of low temperatures on the hydration process [57]. Some methods for temperature control include preheated materials [59], insulation, or heated enclosures [58]. Heating the water and/or aggregates before mixing can help maintain a suitable temperature for the concrete mix. Insulated formwork or blankets can be used to maintain the concrete's temperature during setting and curing, protecting it from cold weather and reducing heat loss. Temporary enclosures can be built around the construction site to maintain a controlled environment with adequate temperature and humidity levels.

Employing appropriate curing techniques in cold weather is essential to maintain the moisture and temperature conditions necessary for the hydration process [62]. Some common cold weather curing methods include insulated curing blankets, heated water curing, and steam curing, as discussed in Section 2.1.3. Insulated curing blankets can be placed over the concrete surface to minimize the heat loss and maintain the desired curing temperature. Heated water curing implies circulating heated water through pipes embedded in the concrete, which can provide consistent and controlled heat to maintain the curing temperature. Steam curing can be introduced into an enclosed space to maintain the temperature and humidity required for curing.

Regularly or continuously monitoring the temperature and strength development of the concrete is crucial during cold weather construction [72]. By following the temperature history of the concrete, good estimations of the strength can be calculated using maturity equations [91]; see Equation (1). The equation shows that the maturity (M) depends on the temperature (T) and the time (t). The reference temperature (T_0) is typically 20 °C.

$$M(t, T) = \sum(T - T_0)\Delta t, \quad (1)$$

Field testing, including the use of maturity meters or penetration resistance tests, can also help to ensure that the concrete achieves the desired performance characteristics [92]. Effective planning of construction activities can help minimize the impact of cold weather on strength development.

2.3. *Freezing of Concrete*

Freezing temperatures pose significant challenges to the performance and durability of concrete structures. When concrete is subjected to cold conditions, it can experience different types of frost damage, which can lead to a reduction in the concrete's strength, integrity, and service life. Early freezing and freeze–thaw actions can enhance and accelerate the negative effects of concrete freezing. The following sections will discuss these phenomena and their impacts on concrete and will detail some preventative measures that can be employed to mitigate the issues of concrete freezing. A wider analysis of preventive measures is presented in Sections 3 and 4.

2.3.1. Frost Damage

Frost-related deterioration and damage is a significant concern for concrete structures in cold climates, as it can lead to a decrease in strength, durability, and overall performance [93]. Frost damage occurs when water present in the concrete's porous structure freezes and expands [94]. This expansion generates internal pressures that can exceed the tensile strength of the concrete, resulting in various forms of deterioration [95]. The primary factors that contribute to frost damage in concrete include aspects such as the water–cement ratio [96], air entrainment [97], permeability [98] and exposure to de-icing chemicals [99]. The water–cement ratio plays a critical role in determining the porosity and permeability of the concrete [100]. A higher water–cement ratio leads to increased porosity, which allows more water to infiltrate the concrete, raising the risk of frost damage. Air entrainment is the deliberate incorporation of microscopic air voids within the concrete mix. These voids provide space for the expansion of freezing water, helping to alleviate the internal pressures caused by ice formation. Insufficient air entrainment can increase the susceptibility of concrete to frost damage [101]. The permeability of concrete refers to its ability to allow water to penetrate its structure. Concrete with higher permeability is more prone to frost damage, as it permits a greater amount of water to enter and become trapped within the material [102]. The use of de-icing chemicals, such as salts, can exacerbate frost damage by increasing the saturation of water within the concrete and facilitating freeze–thaw cycles.

The primary consequences of frost damage in concrete include cracking, scaling, spalling, and a reduced overall durability [103]. The internal pressures generated by the expansion of freezing water can cause cracks to form and propagate in the concrete. These cracks can weaken the structure and provide pathways for further water ingress, leading to additional frost damage and other forms of deterioration. Scaling [104] is the flaking or peeling of the concrete surface; it occurs when the surface layer is subjected to frost damage. Scaling can result in an unsightly appearance and increased surface roughness, which can be particularly problematic in architectural concrete applications. Spalling [105] refers to the breaking away of large fragments of concrete, typically caused by the expansion of freezing water within the material. Spalling can compromise the structural integrity and aesthetics of the concrete. Frost can lead to a reduction in the durability of concrete structures, as the associated cracking, scaling, and spalling can expose the reinforcing steel to corrosion and facilitate other forms of deterioration, as seen in Figure 3. Understanding the causes and consequences of these damage processes is essential for designing, constructing, and maintaining resilient and long-lasting concrete structures [106]. By implementing appropriate mix designs, curing practices, and preventative measures, engineers and contractors can effectively manage the challenges associated with frost damage and ensure the successful completion of their projects. Regular inspection and maintenance of concrete structures exposed to freezing temperatures are also crucial in mitigating the risks of frost damage and extending the service life of the structures. Ultimately, a comprehensive understanding of frost damage and the application of appropriate mitigation strategies can help to maintain the structural integrity, performance, and aesthetics of concrete structures in cold weather conditions.

Figure 3. Spalling of concrete due to cracking and corrosion in a cold climate [107].

2.3.2. Early Freezing

Early freezing is a significant concern for concrete structures in cold climates, as it can lead to a reduction in strength, durability, and overall performance [108]. Early freezing occurs when the concrete is exposed to freezing temperatures before it has achieved sufficient strength, typically during the initial setting and curing stages [109]. Several concrete guidelines and research articles have defined threshold limits for the concrete compressive strength of 5–10 MPa [110] in terms of the necessity to avoid early freezing. When the ambient temperature falls below the freezing point, the water within the concrete mix can freeze, interrupting the cement hydration process and affecting the concrete's strength development [111]. Insufficient protection of the concrete during the setting and curing stages, such as the use of inadequate insulating materials or heated enclosures, can expose the concrete to freezing temperatures and result in early freezing. A concrete mix design that does not consider the specific requirements for cold weather concreting, such as the use of admixtures designed to accelerate setting and hardening, can increase the risk of early freezing [112].

The implications of early freezing on concrete can be detrimental, with several potential consequences. When the cement hydration process is hindered due to early freezing, the concrete may not achieve its full potential strength [111]. This can result in a weaker structure that is unable to support the intended loads or perform as designed. The formation of ice lenses within the concrete during early freezing can lead to increased porosity and permeability [113]. This makes the concrete more susceptible to further freeze–thaw damage as well as other forms of deterioration [83]. Figures 4–6 show how early freezing can affect the material properties of concrete [114]. The figures show that the time when the concrete is first exposed to the freezing temperatures (frost onset time) is an important parameter, and that freezing within the first day has the biggest negative impact on the strength. The notation "Ref" in Figures 5 and 6 represents the unfrozen reference specimen.

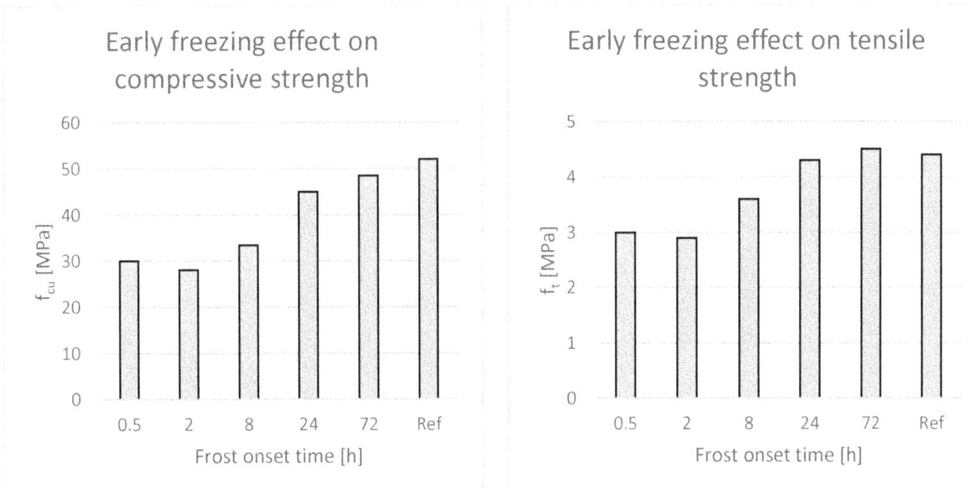

(**a**) (**b**)

Figure 4. Effect of early freezing with different frost onset times on concrete strength properties. The frost duration was 8 h, with a temperature of −5 °C. Reproduced with data from [114]. (**a**) Compressive strength of early frost affected concrete with onset times between 0.5 and 72 h, compared to unfrozen concrete (Ref). (**b**) Tensile strength of early frost affected concrete with onset times between 0.5 and 72 h, compared to unfrozen concrete (Ref).

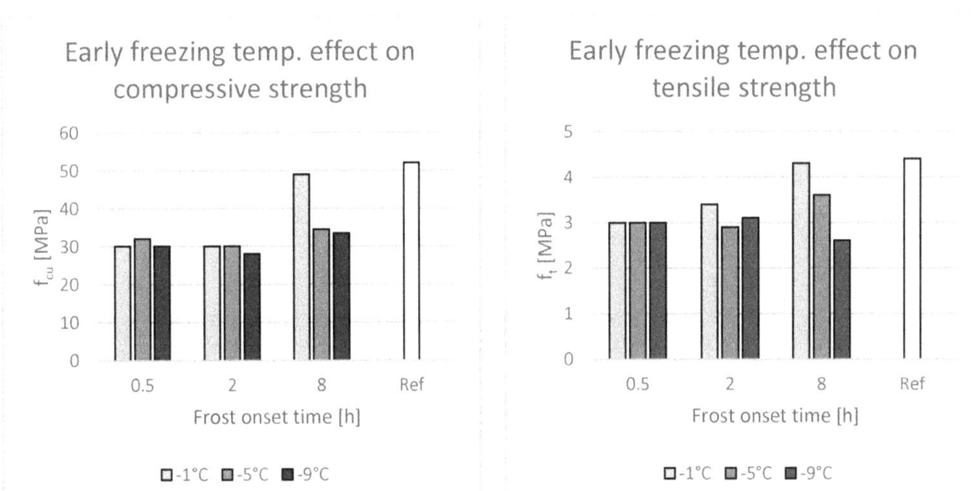

(**a**) (**b**)

Figure 5. Effect of early freezing with different frost onset times and temperatures on concrete strength properties. The frost duration was 8 h, and the temperatures were between −1 and −9 °C. Reproduced with data from [114]. (**a**) Compressive strength of early frost affected concrete with onset times between 0.5 and 8 h, compared to unfrozen concrete (Ref). (**b**) Tensile strength of early frost affected concrete with onset times between 0.5 and 8 h, compared to unfrozen concrete (Ref).

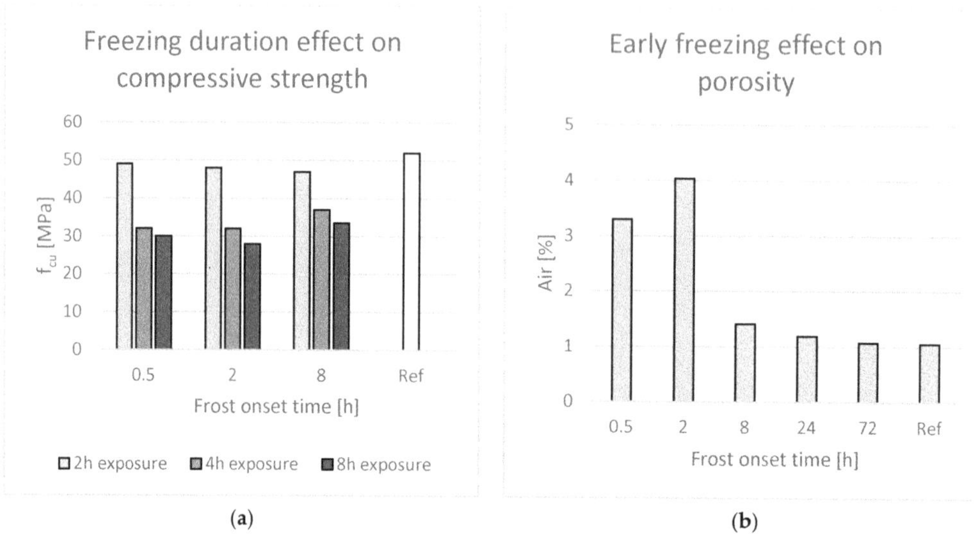

Figure 6. Effect of early freezing with different frost onset times and durations on concrete strength properties. The frost duration was 8 h and the temperatures were between −1 and −9 °C. Reproduced with data from [114]. (**a**) Compressive strength of early frost affected concrete with onset times between 0.5 and 8 h, compared to unfrozen concrete (Ref). (**b**) Porosity of concrete affected by early freezing with onset times between 0.5 and 72 h, compared to unfrozen concrete (Ref).

The combination of reduced strength, increased permeability, and surface damage [115] caused by early freezing can compromise the long-term durability of the concrete structure, potentially leading to a shorter service life and increased maintenance and repair costs [116]. Early freezing poses a significant challenge to the performance and durability of concrete structures in cold climates [117]. Ongoing research and development in concrete technology, such as the introduction of new admixtures, materials, and construction techniques, can help improve the industry's ability to manage early freezing and other related challenges [118]. This will ultimately contribute to the construction of more durable, resilient, and sustainable concrete structures, even in harsh and freezing environments.

2.3.3. Freezing and Thawing

Repeated freezing and thawing cycles can have significant adverse effects on the performance and durability of concrete structures in cold climates [119]. Freezing and thawing cycles occur when the concrete is exposed to fluctuating temperatures that repeatedly cause the water within its porous structure to freeze and thaw. The primary factors contributing to freezing and thawing in concrete include temperature fluctuations, water absorption, and inadequate air entrainment [120]. Frequent changes in temperature above and below the freezing point can lead to multiple freeze–thaw cycles, increasing the risk of damage to the concrete. The presence of water in the concrete is necessary for freeze–thaw damage to occur [121]. Concrete with high porosity and permeability is more likely to absorb water and be susceptible to freeze–thaw cycles [122]. Insufficient air entrainment in the concrete mix can lead to a lack of air voids, which serve as a relief mechanism for the pressures generated by freezing water [123].

The expansion of freezing water within the concrete can cause internal pressures that exceed the tensile strength of the material, leading to various forms of deterioration such as cracking, scaling, and spalling, as seen in Figure 7 [124]. The damage caused by freeze–thaw cycles can lead to a reduction in the compressive and tensile strength of the concrete, as well as reduced E-modulus, as shown in Figure 8, compromising its structural integrity and ability to support the intended loads. The cumulative effects of freezing and thawing can result in a decrease in the concrete's durability, leading to a shorter service life and increased maintenance and repair costs [125]. Freeze–thaw cycles can also cause a loss of bond between the concrete and reinforcing steel, which may compromise the integrity and performance of the structure [126].

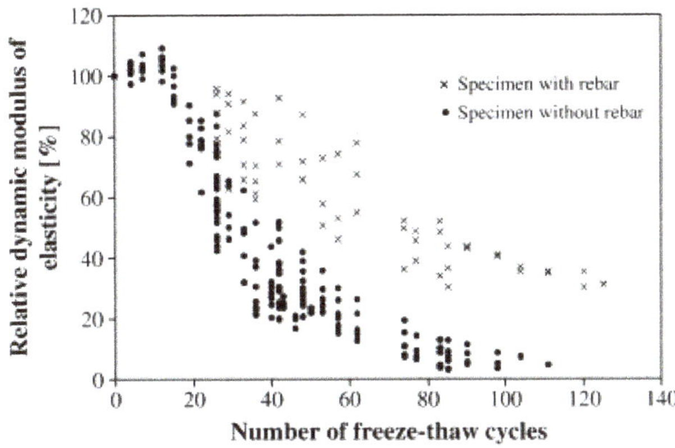

Figure 7. Reduced E-modulus due to repeated freeze–thaw cycles [124].

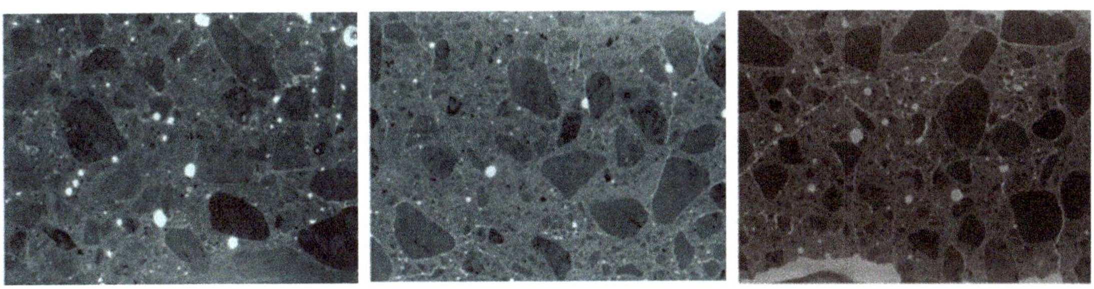

(**a**) (**b**) (**c**)

Figure 8. Cracking caused by repeated freeze–thaw cycles [124]. (**a**) Uncracked reference concrete. (**b**) Cracking of concrete with damage level 1, i.e., 25% reduction of compressive strength. (**c**) Cracking of concrete with damage level 2, i.e., 50% reduction of compressive strength.

2.3.4. Preventive Measures

Implementing appropriate preventive measures and strategies can help mitigate the risks of frost damage, ensuring the durability, performance, and longevity of the concrete structures [93]. Regular inspection and maintenance of concrete structures are essential in identifying potential issues related to frost damage, early freezing, and freeze–thaw cycles [127]. Timely intervention can address these problems before they escalate and compromise the structure's integrity [128]. Maintenance activities, such as repairing cracks, sealing joints, and applying protective coatings, can help maintain the durability and performance of the concrete structure. Educating engineers, contractors, and other stakeholders about the potential risks and consequences of frost-related problems is critical in ensuring the successful implementation of preventive measures and strategies. Preventing frost damage in concrete requires a comprehensive approach, including mix design optimization, cold weather concreting practices, protection and insulation, drainage and waterproofing, and regular inspection and maintenance [129]. By implementing these preventive measures and strategies, engineers and contractors can effectively manage the challenges associated with cold weather concreting and ensure the successful completion of their projects. A detailed discussion on different preventive approaches is given in Section 3.

2.4. Weather-Related Challenges in Construction

Cold weather construction presents unique challenges to personnel, equipment, and site management, including low temperatures, snow, and wind [130]. These challenges can lead to increased costs, reduced productivity [131], and potential safety risks. By understanding and addressing these challenges, construction professionals can minimize their impact on project schedules, budgets, and outcomes.

2.4.1. Challenges for Personnel

Low temperatures can result in cold stress, frostbite, and hypothermia, posing significant risks to worker health and safety [132]. Cold weather can affect workers' dexterity, coordination, and overall productivity [130]. Providing appropriate personal protective equipment (PPE), such as insulated clothing, gloves, and headgear, can help protect workers from cold-related illnesses and injuries. Snow and wind can reduce visibility on the construction site, making it difficult for workers to communicate and coordinate their efforts effectively [133]. Ensuring proper communication tools and establishing clear communication protocols can help mitigate this challenge. Proper training in cold weather construction practices is crucial in ensuring worker safety and productivity. Workers should be educated on the risks associated with cold weather construction and be provided with the necessary tools and resources to manage these risks effectively. A list of potential health challenges in cold weather is presented in Table 1.

2.4.2. Challenges for Equipment and Machinery

Low temperatures can affect the performance and efficiency of construction equipment and machinery. Engines may be more challenging to start, lubricants may become less effective, and hydraulic systems may become less responsive [134]. Regular maintenance and the use of cold-weather-specific lubricants and fluids can help minimize these performance issues. Cold temperatures can significantly reduce battery life for construction equipment and tools, affecting their performance and efficiency [135]. Ensuring proper battery storage and charging practices as well as using cold-weather-specific batteries can help mitigate this issue. Cold weather can also affect the properties of construction materials, making them more brittle and susceptible to damage [136]. Proper handling and storage techniques, such as protecting materials from moisture and temperature fluctuations, can help maintain their integrity and performance. Regular maintenance and inspection of construction equipment and machinery are crucial in cold weather conditions to prevent breakdowns and ensure efficient operation. Proper winterization applications, including engine block heaters, antifreeze, and cold-weather lubricants, can help protect equipment and machinery from the effects of low temperatures [137].

2.4.3. Challenges for Site Management

Accumulated snow and ice can pose significant safety risks and impede construction operations. Implementing a comprehensive snow and ice removal plan, including the use of snowplows, snow blowers, and de-icing agents, can help ensure a safe and efficient construction site [138]. Windy conditions on the construction site can create safety hazards, reduce worker productivity, and cause damage to structures and materials. Erecting wind barriers, such as windbreaks or temporary enclosures, can help protect the construction site and its occupants from the effects of strong winds [139]. Snow and ice can make access to and transportation within the construction site more difficult, leading to delays and increased costs. Ensuring proper site access and transportation planning, including snow removal and the use of appropriate vehicles, can help minimize these challenges. Cold weather conditions can result in construction delays and increased costs. Developing a comprehensive scheduling and contingency plan that accounts for potential weather-related challenges can help ensure the project remains on track and within budget. Cold weather construction can have an impact on the surrounding environment, including the potential for erosion, sedimentation, and harm to wildlife habitats. Implementing appropriate environmental protection measures, such as sediment control and erosion prevention, can help minimize the environmental impact of construction activities in cold weather conditions [140]. A list of management strategies to ensure worker safety and prevent production interference is presented in Table 2.

Table 1. Physical health challenges for personnel on cold weather conditions [130].

Physical Health Challenge	Cause	Symptoms
Numbness in exposed body parts	Exposed extremities in cold	Increased nerve pain
Increase in number of injuries	Continuous exertion in cold weather	Tiredness
Hobbling effect and uneasiness	Tight thermal clothing	Reduced ability to move
Physical fatigue	Performing for longer duration in cold weather	Tiredness
Hypothermia	Excessive loss of heat	Weak pulse, lack of consciousness
Vasoconstriction of blood vessels	Prolonged exposure to cold	Increase in blood pressure
Frostbite	Freezing of tissues	Long-term numbness in affected region
Necrosis	Lack of blood supply to tissue	Malfunctioning of cells
Upper and lower respiratory issues	Inhaling cold, dry air	Shortness of breath
Musculoskeletal disorders such as wrist, neck, back and overall body pain and inflammation	Increased muscular load	Carpal tunnel syndrome
Increase in number of accidents caused by slips and falls	Icy and slippery surfaces in workplace	Major and minor injuries in body parts
Trench foot	Performing activities in cold water	Blisters, blotchy skin
Increase in onset of fatigue due to personal protective clothing	Increase in metabolic energy	Tiredness, weakness in muscles
Reduced dexterity	Impaired response from hand receptors	Inability to handle tools and equipment

2.4.4. Additional Weather-Related Challenges

Freezing and thawing cycles can cause frost heave, leading to soil instability and potential damage to foundations and other structural elements. Ensuring proper site preparation, including soil stabilization techniques and the use of frost-protected shallow foundations, can help mitigate the risk of frost heave and soil instability [141]. Cold weather can also lead to increased condensation and moisture accumulation within buildings and structures, potentially resulting in mold growth and structural damage [142]. Implementing proper moisture control measures, such as vapor barriers and adequate ventilation, can help prevent condensation-related issues. Low temperatures often lead to increased energy consumption for heating, equipment operation, and other site activities. Implementing energy-efficient practices, such as using energy-efficient equipment and optimizing construction processes, can help reduce energy consumption and associated costs [143]. Cold weather construction may also involve additional permitting and regulatory requirements related to environmental protection, safety, and other factors [144]. Ensuring compliance with all relevant regulations and obtaining the necessary permits can help prevent potential delays, fines, and other complications.

Table 2. Strategies for mitigating challenges of working in cold weather [130].

Type of Control	Strategies for Cold Weather Conditions
Engineering control	Encourage the use of smart clothing inserted with infrared, humidity, and temperature sensors Wear a Peltier-embedded cooling jacket Heat exchange masks Protective coverings and insulation Anti-slip shoes Provide workers with infrared heaters Provide warming facilities/local shelters with heating mechanisms
Administrative control	Cold protection plan and cold management Place warning signs on slippery surfaces
Personal protective equipment (PPE) control	Ensure personal protective clothing (PPC) fits properly

3. Materials, Technologies, and Strategies for Cold Weather

Cold weather conditions can, as discussed in previous sections, have a significant impact on concrete construction and the performance of concrete structures, with challenges such as reduced setting time, slower strength development, and increased vulnerability to frost damage. However, technological advances and innovative construction strategies can help mitigate the challenges and improve the performance of concrete structures in cold weather [145]. By employing advanced concrete mix designs [146], appropriate concreting techniques [147], protective measures and insulation [107], accelerating admixtures and chemical additives [148], advanced monitoring and quality control systems [149], and prefabrication and modular construction [150], construction professionals can overcome the challenges associated with cold weather construction. Water is one of the biggest risks related to concrete construction in hazardous environments experiencing low and freezing temperatures; ensuring proper drainage and waterproofing is therefore extremely important to avoid water-related issues and risks, as discussed in Table 3.

Table 3. Drainage and waterproofing as damage-preventive strategies for concrete construction in cold weather.

Drainage and Waterproofing	
Drainage	Ensuring proper drainage around the concrete structure can help prevent the accumulation of water and reduce the risk of freeze–thaw cycles [151]. Effective drainage systems, such as well-designed slopes, drains, and gutters, can help minimize water ingress and mitigate the risk of frost damage and early freezing [152].
Waterproofing and protective coatings	Applying waterproofing membranes [153] or protective coatings [154] to the concrete surface can help prevent water absorption and reduce the risk of frost damage, early freezing, and freeze–thaw cycles. These treatments can improve the concrete's resistance to moisture ingress [155] and thereby enhance its overall durability.

The following sections will discuss various technologies and construction strategies that can enhance the performance of concrete construction and concrete structures in cold weather conditions.

3.1. Advanced Concrete Mix Designs

Innovative concrete mix designs offer diverse opportunities to improve the performance of concrete in cold weather by addressing challenges such as slow setting, reduced strength development, and frost susceptibility [156]. Material measures that can be taken to mitigate frost-related problems are, for example, adjustments of the water–cement ratio, increasing the air content, and addition of supplementary cementitious materials, as discussed in Table 4. Some other important or recent advancements in concrete mix design include the development of new concrete materials such as ultra-high-performance concrete [157], self-consolidating concrete [158], fiber-reinforced concrete [159], and nanotechnologically enhanced concrete [160].

Table 4. Approaches for mix-design adjustments as damage-preventive strategies for concrete construction in cold weather.

Adjusting the Mix Design	
Water–cement ratio	Controlling the water–cement ratio is crucial to producing a dense and durable concrete mix with low permeability, reducing the risk of frost damage and freeze–thaw cycles [161]. A lower water–cement ratio reduces the porosity of the concrete, making it more resistant to water ingress and freezing [162].
Air entrainment	Incorporating air-entraining admixtures into the concrete mix creates small, evenly distributed air voids within the concrete [163]. These air voids provide space for the expansion of freezing water, reducing internal pressure and preventing frost damage, early freezing, and freeze–thaw deterioration [164].
Supplementary Cementitious Materials (SCMs)	The use of SCMs, such as fly ash, slag, or silica fume, can improve the concrete's resistance to freezing and thawing cycles by increasing the water–binder ratio [165]. SCMs can thereby reduce the permeability of concrete and enhance its durability, making it less susceptible to frost damage and early freezing [166].

High-performance concrete (HPC) is a type of concrete with enhanced strength, durability, and resistance to environmental factors, including cold weather conditions [167]. The use of HPC or ultra-high-performance concrete (UHPC) can improve the performance of concrete structures by reducing permeability, increasing resistance to freeze–thaw cycles, and enhancing overall durability [157]. It may also offer enhanced cold weather opportunities such as faster setting and strength development, shorter times requirements for supportive formwork, and ultimately higher construction rates due to its rapid development of material and mechanical properties [168]. Self-consolidating concrete (SCC) is a type of concrete that flows and consolidates under its own weight, eliminating the need for mechanical consolidation [169]. SCC can be advantageous in cold weather construction, as it can be placed more quickly and with less labor, reducing the risk of early freezing and the need for additional heating and protection measures [170]. Additionally, it improves the working environment for construction workers as it eliminates the need for harsh work tasks such as vibration and enables the construction of complicated shapes and geometries that would not be possible to construct by using traditional vibrated concrete [171].

Fiber-reinforced concrete (FRC) is a type of concrete that incorporates fibers, such as steel or synthetic fibers, into the concrete mix [172]. The incorporation of fibers in concrete can improve the material's tensile strength, ductility, and resistance to cracking, making it more resilient in cold weather conditions [173]. The increased cracking resistance is especially important in harsh environments where cracks must be avoided, for example, power plants, tunnels, marine structures, and dams [174]. Fiber-reinforced concrete can also help reduce the risk of frost damage and freeze–thaw deterioration due to the reduced crack risk. Nanotechnologically enhanced concrete (NEC) refers to the use of nanotechnology and nanoparticles in the mix design of concrete [175]. This innovative material can help improve the performance of concrete in cold weather by enhancing its strength, durability, and resistance to environmental factors [176]. Nanoparticles, such as nano-silica or nano-titanium dioxide, can help reduce porosity, increase strength development, and improve resistance to freeze–thaw cycles [177].

3.2. Cold Weather Concreting Techniques

Innovative concreting techniques can help ensure the successful placement, setting, and curing of concrete in cold weather conditions. Some of these techniques include the use of precooled or preheated materials, accelerating admixtures, or real-time temperature monitoring for accurate strength estimations, as discussed in Table 5. Proper thermal management during concrete placement and curing is always crucial in low temperatures. Innovative methods for managing concrete temperature, such as the use of electric heating cables [18,60], hydronic heating systems [178], or insulated formwork [179], can help maintain the required temperature for optimal curing and strength development.

Table 5. Innovative concreting practices as damage-preventive strategies for concrete construction in cold weather.

Cold Weather Concreting Practices	
Preheated ingredients	Preheating the concrete ingredients, such as aggregates and water, can help maintain the concrete's temperature during placement and reduce the risk of early freezing [180]. This practice ensures the proper setting and curing of the concrete in cold weather conditions [181]. The preheating technique can be useful in extreme cold conditions or when using mass concrete.
Accelerating admixtures	Chemical additives and accelerating admixtures can help improve the performance of concrete in cold weather by reducing setting time and shrinkage [182], promoting faster strength development, and enhancing the durability [183,184]. Non-chloride accelerators, such as calcium nitrate or calcium formate, can help speed up the setting and strength development of concrete in cold weather without the risk of corrosion associated with chloride-based accelerators [185].
Temperature monitoring and control	Monitoring and controlling the concrete temperature during placement and curing is critical in preventing early freezing and frost damage. Maintaining the concrete temperature within certain limits, typically between 5 and 35 °C, is recommended for proper curing and strength development [186]. Advanced temperature monitoring systems, such as wireless sensors or thermocouples, can provide real-time information on concrete temperature during placement and curing, helping the concrete to maintain the necessary temperature for optimal curing and strength development [187].

3.3. Protective Measures and Insulation

Innovative protective measures and insulation techniques can be used to protect concrete from the effects of low ambient temperatures and ensure proper curing and strength development, as explained in Table 6. Advances in insulation materials have led to the development of lightweight, reusable insulating boards, blankets, or covers that can provide better thermal performance and durability than traditional insulation methods [188]. These blankets can help the concrete maintain the required temperature and moisture levels for proper curing and strength development and can prevent early freezing [189].

Table 6. Insulation and protection techniques as damage-preventive strategies for concrete construction in cold weather.

Protection and Insulation	
Insulating blankets or covers	Providing adequate insulation for the concrete during setting and curing can help maintain the necessary temperature and moisture levels for optimal curing [189]. Insulating blankets or covers can protect the concrete from freezing temperatures, preventing early freezing and thereby preventing frost damage [188].
Heated enclosures	In very low, freezing temperatures, enclosures can be used to provide a controlled environment for concrete placement and curing [190]. The enclosures are typically equipped with heating systems, such as propane or electric heaters [191]. These enclosures can thereby maintain the temperature and humidity required for proper curing, minimizing the risk of frost-related problems.
Insulated concrete forms	Insulated formwork systems, such as insulated concrete forms (ICFs) or insulated sandwich panels, can provide a protective thermal barrier for concrete during placement and curing [192]. These systems can help maintain the necessary temperature for proper curing and strength development while also improving the energy efficiency of the finished structure.

3.4. *Advanced Monitoring and Quality Control*

Innovative monitoring and quality control systems can help ensure the successful completion of concrete construction projects in cold weather by providing real-time information on critical factors, such as temperature, humidity, and strength development [193]. Advanced wireless sensor systems can provide real-time data on concrete temperature [194], humidity [195], and strength development [196], helping to ensure that the concrete maintains the necessary conditions for optimal curing and performance. These systems can also help identify potential issues early, allowing for timely corrective action [197]. Digital image correlation (DIC) techniques involve the use of high-resolution cameras and advanced image processing algorithms to measure the deformation and strain of concrete structures during curing and service life [198]. This information can be used to assess the performance of concrete in cold weather and identify potential issues related to cracking, shrinkage, or other forms of damage [199].

3.5. *Prefabrication and Modular Construction*

Prefabrication and modular construction techniques can help improve the efficiency and performance of concrete construction in cold weather by reducing the time and labor required for on-site placement and curing [200]. Some of the advantages of prefabrication and modular construction include the possibility of a controlled environment, faster construction, and improved quality control [201]. Prefabricated concrete elements can be produced in a controlled environment, ensuring optimal curing conditions, and reducing the risk of early freezing, frost damage, or other cold-weather-related issues. Prefabricated and modular elements can be assembled on site more quickly than traditional cast-in-place construction, reducing the time and labor required for concrete placement and curing in low temperatures [202]. The use of prefabrication and modular construction techniques can help improve quality control by allowing for more precise and consistent production of concrete elements.

4. Emerging Materials, Technologies, and Strategies

As the construction industry continues to evolve in response to changing environmental conditions, new technologies, and increasing demand for energy-efficient and resilient infrastructure, the future of concrete materials, construction, and structures in cold weather environments is also expected to undergo significant changes. This section will discuss some of the key future trends and opportunities in the field of concrete materials, construction, and structures in cold weather, with a focus on enhancing their durability, performance, and sustainability.

4.1. Smart Concrete Materials and Their Production

The development and implementation of smart and functional concrete materials are expected to play a significant role in the future of cold weather construction. These advanced materials can provide enhanced performance, durability, and resilience in cold weather conditions, as well as offering new functionalities and capabilities [203]. Some potential smart and functional concrete materials include self-healing concrete [204] and phase change materials [205]. The sustainability can also be promoted in the production phase by adapting carbon capture, utilization, and storage technologies [206]. These technologies are discussed in Table 7.

Table 7. Smart concrete materials and production for improved sustainability.

Smart Concrete Materials and Production Technologies	
Self-healing concrete	Self-healing concrete is an innovative type of building material that can autonomously repair cracks and damage, thereby improving the durability and longevity of concrete structures [207]. This technology typically relies on the use of bacteria or microcapsules containing healing agents, which are activated when cracks form, releasing the healing agent and promoting the formation of new concrete material [208].
Phase change materials	Phase change materials (PCMs) involve the incorporation of phase transitioning materials into concrete mixes, which can help improve the thermal performance of concrete structures in cold weather [209]. PCMs can store and release thermal energy as they undergo phase transitions, effectively acting as thermal batteries that help regulate the temperature of concrete structures and reduce the risk of frost damage or freeze–thaw deterioration [64].
Carbon capture, utilization, and storage technologies	Carbon capture, utilization, and storage (CCUS) technologies offer the potential to reduce the environmental impact of concrete production and use by capturing carbon dioxide emissions and incorporating them into concrete materials [210]. These technologies can help create more sustainable concrete materials and construction practices [211].

4.2. Advanced Manufacturing and Construction Technologies

The adoption of advanced manufacturing and construction techniques, such as additive manufacturing (3D printing) and robotics, is expected to transform the way concrete materials and structures are produced and constructed [212]. These techniques can improve the overall efficiency, reduce waste, and enhance the quality and performance of the structures. Some current trends and future opportunities in this area include 3D-printed concrete [213], robotic construction [214], and prefabrication [215], discussed in Table 8. These technologies offer possibilities for improved sustainability for all types of concrete construction, including work in harsh environments.

Table 8. Advanced manufacturing and construction technologies for sustainable concrete.

Advanced Manufacturing and Construction Technologies for Concrete	
3D printing	3D printing technology offers the potential to revolutionize the production of concrete elements and structures [216]. By enabling precise and automated fabrication of complex or custom-designed components, 3D printing provides opportunities to reduce labor costs, minimize material waste, and improve the overall quality and performance of concrete structures [217].
Robotic construction	The use of robotic systems in the construction industry can help improve efficiency, reduce labor costs, and enhance quality and performance [218]. Robotic systems can be used for a range of construction tasks, such as concrete placement [219], reinforcement installation [220], and formwork assembly [221], helping to streamline construction processes and ensure consistent quality and performance.
Modular and prefabricated construction	Modular and prefabricated construction techniques involve the off-site production and assembly of concrete components [222]. These production techniques can help improve the efficiency and performance of concrete construction in low temperature environments [223]. By applying modular or prefabrication technologies, the construction industry can reduce on-site labor requirements, minimize weather-related delays, and ensure consistent quality and performance [224].

4.3. *Integrated Design and Optimization Technologies*

The development and adoption of integrated design and optimization technologies, such as building information modeling (BIM) and artificial intelligence (AI), are expected to play a significant role in the future of concrete construction and structures [225]. These tools can help streamline design and construction processes [226], improve collaboration and communication between project stakeholders [227], and enhance the performance and durability of concrete structures, both in normal weather conditions and in cold environments [228]. Some integrated design and optimization technologies and their applications are shown in Table 9.

Table 9. Integrated design and optimization technologies for improved sustainability.

Integrated Design and Optimization Technologies	
Building information modeling	Building information modeling (BIM) is a digital representation of the physical and functional characteristics of a building or infrastructure, enabling the integration of design, construction, and management processes [229]. BIM can help improve the efficiency, performance, and sustainability of concrete construction in cold weather by facilitating better coordination and communication among project stakeholders [227], optimizing material selection and construction techniques [229], and predicting potential issues related to frost damage, freeze–thaw cycles, or other cold weather-related challenges [230].
Artificial intelligence and machine learning	Artificial intelligence (AI) and machine learning technologies offer significant potential for improving the efficiency, performance, and durability of concrete construction and structures [231]. These technologies can help to optimize concrete mix designs [232], predict the performance of concrete materials and structures under various environmental conditions [233], and develop more efficient construction processes and techniques [234].
Digital twins	The digital twin technology involves the creation of a virtual replica of a physical asset or system [235], allowing for real-time monitoring, analysis, and optimization of its performance [236]. Digital twins can be used to model and predict the behavior of concrete structures in cold weather environments, enabling the use and development of more resilient and efficient construction techniques and materials [237].

4.4. Sustainability and Resilience in Cold Weather Concrete Construction

As climate change and environmental concerns continue to drive the need for more sustainable and resilient infrastructure [238], the future of cold weather concrete construction is expected to focus increasingly on enhancing the sustainability and resilience of concrete materials, structures, and construction practices. The development and adoption of green concrete materials [239], such as those incorporating alternative binders like fly ash, slag, or geopolymers, can help reduce the environmental impact of concrete construction [240] while also improving the performance and durability of structures in cold weather environments [241]. These materials can offer enhanced material properties such as better resistance to freeze–thaw cycles [242], reduced permeability [243], and improved thermal performance [244].

Carbon capture, utilization, and storage technologies offer the potential to significantly reduce the carbon footprint of concrete production and use by capturing, storing, and utilizing carbon dioxide emissions [245]. These technologies can help create more sustainable and resilient concrete materials and structures while also addressing the global challenge of climate change [246]. As climate change leads to more extreme and variable weather conditions [247], the need for climate-adaptive design and construction practices is becoming increasingly important. In the context of cold weather concrete construction, this may involve designing and constructing structures that can withstand more frequent and severe freeze–thaw cycles [248], incorporating advanced materials and technologies to enhance resilience, and implementing construction practices that minimize the environmental impact. The future of concrete construction in cold weather environments is expected to be shaped by several key trends and opportunities, including the development of smart and functional concrete materials [249], the adoption of advanced manufacturing and construction techniques [250], the use of integrated design and optimization tools [251], and an increased focus on sustainability and resilience [252]. By staying informed about these trends and seeking to implement innovative solutions and best practices, construction professionals can continue to improve the overall performance, durability, and long-term sustainability of concrete structures in cold weather, ultimately benefiting the construction industry and society.

5. Conclusions

Various challenges, strategies, and performance enhancement techniques related to cold weather concrete construction were discussed in this paper. Cold weather conditions can significantly impact the setting, curing, and strength development of concrete, as well as increase the risk of frost damage, early freezing, and freeze–thaw deterioration. These challenges can lead to increased construction time, higher costs, and potential safety and durability concerns for concrete structures in cold weather environments. Several innovative materials and technologies can be employed to improve the performance and durability of concrete in cold and freezing temperatures. These strategies include adjusting and optimizing the concrete mix proportions, using chemical admixtures and supplementary cementitious materials, modifying the construction practices, and employing innovative materials and construction techniques. Weather-related problems and challenges in construction include the impact of low temperatures, snow, and wind, which affects personnel, equipment, and machinery. The implementation of effective planning, communication, and management strategies can help mitigate the weather-related challenges and ensure the successful completion of construction projects in cold weather environments. The future of cold weather concrete construction is expected to be shaped by several key trends and opportunities, including the development of smart and functional materials, the adoption of advanced manufacturing and construction techniques, the use of integrated design and optimization tools, and an increased focus on sustainability and resilience. These trends and opportunities offer a potential to further enhance the performance, durability, and sustainability of concrete structures in cold weather conditions.

The successful construction of durable, high-performing, and sustainable concrete structures in harsh environments requires a comprehensive understanding of the material-related challenges and the appropriate strategies and technologies to address them. By staying informed about the latest advancements and best practices in the field, construction professionals can continue to develop and implement more effective solutions to the unique challenges posed by cold weather concrete construction, ultimately benefiting both the construction industry and the end-users of these structures.

Author Contributions: J.N.: conceptualization, methodology, writing—original draft; V.Z.: conceptualization, writing—reviewing and editing. All authors have read and agreed to the published version of the manuscript.

Funding: This research was supported by Formas, SBUF and LTU.

Institutional Review Board Statement: The study was conducted in accordance with the Declaration of Helsinki, and approved by the Institutional Review Board Luleå University of Technology.

Informed Consent Statement: Not applicable.

Data Availability Statement: All data are available upon request.

Conflicts of Interest: The authors declare no conflict of interest.

References

1. Statista. Global Cement Production from 1995 to 2022. Available online: https://www.statista.com/statistics/219343/cement-production-worldwide/ (accessed on 27 April 2023).
2. Monteiro, P.J.; Miller, S.A.; Horvath, A. Towards sustainable concrete. *Nat. Mater.* **2017**, *16*, 698–699. [CrossRef]
3. Demissew, A. Comparative analysis of selected concrete mix design methods based on cost-effectiveness. *Adv. Civ. Eng.* **2022**, *2022*, 4240774. [CrossRef]
4. Tang, S.W.; Yao, Y.; Andrade, C.; Li, Z.J. Recent durability studies on concrete structure. *Cem. Concr. Res.* **2015**, *78*, 143–154. [CrossRef]
5. Mather, B. Concrete durability. *Cem. Concr. Compos.* **2004**, *26*, 3–4. [CrossRef]
6. Mathews, M.E.; Nandhagopal, M.; Anand, N.; Arulraj, G.P. Versatility at its best: An integrated review on development of self-compacting concrete. *Int. J. Sci. Tech. Res.* **2019**, *8*, 513–519.
7. Hegger, J.; Curbach, M.; Stark, A.; Wilhelm, S.; Farwig, K. Innovative design concepts: Application of textile reinforced concrete to shell structures. *Struct. Concr.* **2018**, *19*, 637–646. [CrossRef]
8. Kozikowski, R.L.; McCall, W.C.; Suprenant, B.A. Cold weather concreting strategies. *Concr. Internat.* **2014**, *36*, 45–49.
9. Korhonen, C. New developments in cold-weather concreting. In Proceedings of the 11th International Conference on Cold Regions Engineering, Anchorage, AK, USA, 20–22 May 2002.
10. Nmai, C.K. Cold weather concreting admixtures. *Cem. Concr. Compos.* **1998**, *20*, 121–128. [CrossRef]
11. Barna, L.A.; Seman, P.M.; Korhonen, C.J. Energy-efficient approach to cold-weather concreting. *J. Mater. Civ. Eng.* **2011**, *23*, 1544–1551. [CrossRef]
12. Chen, S.; Lu, W.; Olofsson, T.; Dehghanimohammadabadi, M.; Emborg, M.; Nilimaa, J.; Wang, Y.; Feng, K. Concrete Construction: How to Explore Environmental and Economic Sustainability in Cold Climates. *Sustainability* **2020**, *12*, 3809. [CrossRef]
13. Ritchie, H.; Roser, M. Urbanization. Available online: https://ourworldindata.org/urbanization (accessed on 29 April 2023).
14. Roser, M.; Ritchie, H.; Ortiz-Ospina, E.; Rodés-Guirao, L. World Population Growth. Available online: https://ourworldindata.org/world-population-growth (accessed on 29 April 2023).
15. Ryou, J.S.; Lee, Y.S. Properties of early-stage concrete with setting-accelerating tablet in cold weather. *Mater. Sci. Eng. A* **2012**, *532*, 84–90. [CrossRef]
16. Liu, Y.; Tian, W.; Ma, G. Electric activation curing behaviour of reinforced concrete beam under severely-cold environment: Breakthrough of rapid concrete manufacturing at cold region. *Constr. Build. Mater.* **2023**, *384*, 131443. [CrossRef]
17. Ortiz, J.; Aguado, A.; Agulló, L.; Garcia, T. Influence of environmental temperatures on the concrete compressive strength: Simulation of hot and cold weather conditions. *Cem. Concr. Res.* **2005**, *35*, 1970–1979. [CrossRef]
18. Nilimaa, J.; Zhaka, V. Material and Environmental Aspects of Concrete Flooring in Cold Climate. *Constr. Mater.* **2023**, *3*, 180–201. [CrossRef]
19. Şahin, Y.; Akkaya, Y.; Taşdemir, M.A. Effects of freezing conditions on the frost resistance and microstructure of concrete. *Constr. Build. Mater.* **2021**, *270*, 121458. [CrossRef]
20. Sang, Y.; Yang, Y. Assessing the freezing process of early age concrete by resistivity method. *Constr. Build. Mater.* **2020**, *238*, 117689. [CrossRef]
21. Wang, R.; Hu, Z.; Li, Y.; Wang, K.; Zhang, H. Review on the deterioration and approaches to enhance the durability of concrete in the freeze–thaw environment. *Constr. Build. Mater.* **2022**, *321*, 126371. [CrossRef]

22. Chen, H.; Cao, Y.; Liu, Y.; Qin, Y.; Xia, L. Enhancing the durability of concrete in severely cold regions: Mix proportion optimization based on machine learning. *Constr. Build. Mater.* **2023**, *371*, 130644. [CrossRef]
23. Demirboğa, R.; Karagöl, F.; Polat, R.; Kaygusuz, M.A. The effects of urea on strength gaining of fresh concrete under the cold weather conditions. *Constr. Build. Mater.* **2014**, *64*, 114–120. [CrossRef]
24. Kothari, A.; Habermehl-Cwirzen, K.; Hedlund, H.; Cwirzen, A. A Review of the Mechanical Properties and Durability of Ecological Concretes in a Cold Climate in Comparison to Standard Ordinary Portland Cement-Based Concrete. *Materials* **2020**, *13*, 3467. [CrossRef]
25. Qu, Z.; Yu, Q.; Ong, G.P.; Cardinaels, R.; Ke, L.; Long, Y.; Geng, G. 3D printing concrete containing thermal responsive gelatin: Towards cold environment applications. *Cem. Concr. Compos.* **2023**, *140*, 105029. [CrossRef]
26. Han, B.; Wang, Y.; Dong, S.; Zhang, L.; Ding, S.; Yu, X.; Ou, J. Smart concretes and structures: A review. *J. Intell. Mater. Syst. Struct.* **2015**, *26*, 1303–1345. [CrossRef]
27. Golroo, A.; Tighe, S.L. Pervious concrete pavement performance modeling: An empirical approach in cold climates. *Can. J. Civ. Eng.* **2012**, *39*, 1100–1112. [CrossRef]
28. Provost-Smith, D.J.; Elsayed, M.; Nehdi, M.L. Effect of early-age subfreezing temperature on grouted dowel precast concrete wall connections. *Constr. Build. Mater.* **2017**, *140*, 385–394. [CrossRef]
29. Shlyakhtina, T.F.; Ivankova, E.A.; Popov, N.A. Winter concrete: Problems and solutions. *IOP Conf. Ser. Earth Environ. Sci.* **2021**, *75*, 012081. [CrossRef]
30. Sakai, K.; Banthia, N. Integrated Design of Concrete Structures and Technology Development. In *Concrete Technology for a Sustainable Development in the 21st Century*; E&FN Spon: London, UK, 2000; pp. 14–26.
31. Shelehov, I.J.; Dorofeeva, N.L.; Shelehov, M.I.; Dorofeev, I.A. Optimization of concrete works technological process in the winter period. *IOP Conf. Ser. Earth Environ. Sci.* **2021**, *751*, 012080. [CrossRef]
32. Sanz-Valle, R.; Naranjo-Valencia, J.C.; Jiménez-Jiménez, D.; Perez-Caballero, L. Linking organizational learning with technical innovation and organizational culture. *J. Knowl. Manage.* **2011**, *15*, 997–1015. [CrossRef]
33. Santiago, P. Achieving Sustainability in Construction Through Digitally Informed Decisions. In *Industry 4.0 for the Built Environment: Methodologies, Technologies and Skills*; Springer: Berlin/Heidelberg, Germany, 2021; pp. 247–269. [CrossRef]
34. Schindler, A.K. Prediction of concrete setting. In Proceedings of the RILEM International Symposium on Advances in Concrete through Science and Engineering, Evanston, IL, USA, 25 March 2004.
35. Hu, J.; Ge, Z.; Wang, K. Influence of cement fineness and water-to-cement ratio on mortar early-age heat of hydration and set times. *Constr. Build. Mater.* **2014**, *50*, 657–663. [CrossRef]
36. Yang, K.H.; Moon, G.D.; Jeon, Y.S. Implementing ternary supplementary cementing binder for reduction of the heat of hydration of concrete. *J. Clean. Product.* **2016**, *112*, 845–852. [CrossRef]
37. Schindler, A.K. Concrete Hydration, Temperature Development, and Setting at Early-Ages. Ph.D. Thesis, The University of Texas at Austin, Austin, TX, USA, May 2002.
38. Wade, S.A.; Nixon, J.M.; Schindler, A.K.; Barnes, R.W. Effect of temperature on the setting behavior of concrete. *J. Mater. Civ. Eng.* **2010**, *22*, 214–222. [CrossRef]
39. Chen, Y.; Okudan, G.E.; Riley, D.R. Sustainable performance criteria for construction method selection in concrete buildings. *Autom. Constr.* **2010**, *19*, 235–244. [CrossRef]
40. Lee, T.; Lee, J.; Kim, Y. Effects of admixtures and accelerators on the development of concrete strength for horizontal form removal upon curing at 10 C. *Constr. Build. Mater.* **2020**, *237*, 117652. [CrossRef]
41. Carette, J.; Staquet, S. Monitoring the setting process of eco-binders by ultrasonic P-wave and S-wave transmission velocity measurement: Mortar vs concrete. *Constr. Build. Mater.* **2016**, *110*, 32–41. [CrossRef]
42. Li, W.; Lin, X.; Bao, D.W.; Xie, Y.M. A review of formwork systems for modern concrete construction. *Structures* **2022**, *38*, 52–63. [CrossRef]
43. Krawczyńska-Piechna, A. Comprehensive approach to efficient planning of formwork utilization on the construction site. *Proc. Eng.* **2017**, *182*, 366–372. [CrossRef]
44. Schindler, A.K.; Frank McCullough, B. Importance of concrete temperature control during concrete pavement construction in hot weather conditions. *Transp. Res. Rec.* **2002**, *1813*, 3–10. [CrossRef]
45. Liu, J.; Lee, Y.C.; Bard, J. Material characterization of workability and process imaging for robotic concrete finishing. *Constr. Robotics* **2021**, *5*, 73–85. [CrossRef]
46. Karagöl, F.; Demirboğa, R.; Kaygusuz, M.A.; Yadollahi, M.M.; Polat, R. The influence of calcium nitrate as antifreeze admixture on the compressive strength of concrete exposed to low temperatures. *Cold Reg. Sci. Tech.* **2013**, *89*, 30–35. [CrossRef]
47. Aslani, A.; Hachem-Vermette, C. Energy and environmental assessment of high-performance building envelope in cold climate. *Energy Build.* **2022**, *260*, 111924. [CrossRef]
48. Lin, M.C.; Tserng, H.P.; Ho, S.P.; Young, D.L. Developing a construction-duration model based on a historical dataset for building project. *J. Civ. Eng. Managem.* **2011**, *17*, 529–539. [CrossRef]
49. Saravanan, T.J.; Balamonica, K.; Priya, C.B.; Reddy, A.L.; Gopalakrishnan, N. Comparative performance of various smart aggregates during strength gain and damage states of concrete. *Smart Mater. Struct.* **2015**, *24*, 085016. [CrossRef]
50. Taylor, P.C. *Curing Concrete*; CRC Press: Boca Raton, FL, USA, 2013.

51. Hassan, A.; Arif, M.; Shariq, M. Effect of curing condition on the mechanical properties of fly ash-based geopolymer concrete. *SN Appl. Sci.* **2019**, *1*, 1694. [CrossRef]
52. Hover, K.C. The influence of water on the performance of concrete. *Constr. Build. Mater.* **2011**, *25*, 3003–3013. [CrossRef]
53. Chyliński, F.; Michalik, A.; Kozicki, M. Effectiveness of Curing Compounds for Concrete. *Materials* **2022**, *15*, 2699. [CrossRef]
54. Al-Gahtani, A.S. Effect of curing methods on the properties of plain and blended cement concretes. *Constr. Build. Mater.* **2010**, *24*, 308–314. [CrossRef]
55. Raheem, A.A.; Soyingbe, A.A.; Emenike, A.J. Effect of curing methods on density and compressive strength of concrete. *Int. J. Appl. Sci. Technol.* **2013**, *3*, 55–64.
56. Huo, X.S.; Wong, L.U. Experimental study of early-age behavior of high performance concrete deck slabs under different curing methods. *Constr. Build. Mater.* **2006**, *20*, 1049–1056. [CrossRef]
57. Tang, Y.; Su, H.; Huang, S.; Qu, C.; Yang, J. Effect of curing temperature on the durability of concrete under highly geothermal environment. *Adv. Mater. Sci. Eng.* **2017**, *2017*, 7587853. [CrossRef]
58. Kodur, V.K.R.; Bhatt, P.P.; Soroushian, P.; Arablouei, A. Temperature and stress development in ultra-high performance concrete during curing. *Constr. Build. Mater.* **2016**, *122*, 63–71. [CrossRef]
59. Li, W.; Huang, Z.; Hu, G.; Duan, W.H.; Shah, S.P. Early-age shrinkage development of ultra-high-performance concrete under heat curing treatment. *Constr. Build. Mater.* **2017**, *131*, 767–774. [CrossRef]
60. Nilimaa, J.; Hösthagen, A.; Emborg, M. Thermal Crack Risk of Concrete Structures: Evaluation of Theoretical Models for Tunnels and Bridges. *Nordic Concr. Res.* **2017**, *56*, 55–69. Available online: http://urn.kb.se/resolve?urn=urn:nbn:se:ltu:diva-64967 (accessed on 16 May 2023).
61. Zeyad, A.M.; Tayeh, B.A.; Adesina, A.; de Azevedo, A.R.; Amin, M.; Hadzima-Nyarko, M.; Agwa, I. SReview on effect of steam curing on behavior of concrete. *Clean. Mater.* **2022**, *3*, 100042. [CrossRef]
62. Duan, Y.; Wang, Q.; Yang, Z.; Cui, X.; Liu, F.; Chen, H. Research on the effect of steam curing temperature and duration on the strength of manufactured sand concrete and strength estimation model considering thermal damage. *Constr. Build. Mater.* **2022**, *315*, 125531. [CrossRef]
63. Otieno, M.; Ikotun, J.; Ballim, Y. Experimental investigations on the effect of concrete quality, exposure conditions and duration of initial moist curing on carbonation rate in concretes exposed to urban, inland environment. *Constr. Build. Mater.* **2017**, *246*, 118443. [CrossRef]
64. Yu, K.; Jia, M.; Yang, Y.; Liu, Y. A clean strategy of concrete curing in cold climate: Solar thermal energy storage based on phase change material. *Appl. Ener.* **2021**, *331*, 120375. [CrossRef]
65. Jana, D. Concrete scaling—A critical review. In Proceedings of the 29th Conference on Cement Microscopy, Quebec City, QC, Canada, 20–24 May 2007.
66. Wang, W.C.; Duong, H.T.H.; Zhang, C.H. Influence of accelerating admixtures on high early strength cement performance using heat curing method. *Case Stud. Constr. Mater.* **2023**, *18*, e01746. [CrossRef]
67. Ren, G.; Tian, Z.; Wu, J.; Gao, X. Effects of combined accelerating admixtures on mechanical strength and microstructure of cement mortar. *Constr. Build. Mater.* **2021**, *304*, 124642. [CrossRef]
68. Tan, K.; Gjorv, O.E. Performance of concrete under different curing conditions. *Cem. Concr. Res.* **1996**, *26*, 355–361. [CrossRef]
69. Bagheri-Zadeh, S.H.; Kim, H.; Hounsell, S.; Wood, C.R.; Soleymani, H.; King, M. Field study of concrete maturity methodology in cold weather. *J. Constr. Eng. Manage.* **2007**, *133*, 827–835. [CrossRef]
70. Pushpalal, D.; Danzandorj, S.; Bayarjavkhlan, N.; Nishiwaki, T.; Yamamoto, K. Compressive strength development and durability properties of high-calcium fly ash incorporated concrete in extremely cold weather. *Constr. Build. Mater.* **2022**, *316*, 125801. [CrossRef]
71. Wang, D.; Shi, C.; Wu, Z.; Xiao, J.; Huang, Z.; Fang, Z. A review on ultra high performance concrete: Part II. Hydration, microstructure and properties. *Constr. Build. Mater.* **2015**, *96*, 368–377. [CrossRef]
72. Huang, H.; Huang, T.; Yuan, Q.; Zhou, D.; Deng, D.; Zhang, L. Temperature dependence of structural build-up and its relation with hydration kinetics of cement paste. *Constr. Build. Mater.* **2019**, *201*, 553–562. [CrossRef]
73. El-Mir, A.; Assaad, J.J.; Nehme, S.G.; El-Hassan, H. Correlating strength and durability to time-temperature profiles of high-performance mass concrete. *Case Stud. Constr. Mater.* **2022**, *16*, e01055. [CrossRef]
74. Alqarni, A.S. A comprehensive review on properties of sustainable concrete using volcanic pumice powder ash as a supplementary cementitious material. *Constr. Build. Mater.* **2022**, *323*, 126533. [CrossRef]
75. Wang, T.; Ishida, T.; Gu, R.; Luan, Y. Experimental investigation of pozzolanic reaction and curing temperature-dependence of low-calcium fly ash in cement system and Ca-Si-Al element distribution of fly ash-blended cement paste. *Constr. Build. Mater.* **2021**, *267*, 121012. [CrossRef]
76. Shobeiri, V.; Bennett, B.; Xie, T.; Visintin, P. A comprehensive assessment of the global warming potential of geopolymer concrete. *J. Cleaner Prod.* **2021**, *297*, 126669. [CrossRef]
77. Hooton, R.D. Future directions for design, specification, testing, and construction of durable concrete structures. *Cem. Concr. Res.* **2019**, *124*, 105827. [CrossRef]
78. Asamoto, S.; Kurashige, I.; Chun, P.J. Effect of high temperature at early ages on drying shrinkage and creep of concrete focusing on microcracking damage. *Mater. Struct.* **2023**, *56*, 67. [CrossRef]

79. Alexander, M.; Beushausen, H. Durability, service life prediction, and modelling for reinforced concrete structures–review and critique. *Cem. Concr. Res.* **2019**, *122*, 17–29. [CrossRef]
80. Zheng, Y.; Yang, J.; Liao, B.; Ma, S.; Zhong, H.; Lei, J. Investigation on the concrete strength performance of underlying tunnel structure subjected to train-induced dynamic loads at an early age. *Constr. Build. Mater.* **2022**, *337*, 127622. [CrossRef]
81. Lee, T.; Lee, J. Setting time and compressive strength prediction model of concrete by nondestructive ultrasonic pulse velocity testing at early age. *Constr. Build. Mater.* **2020**, *252*, 119027. [CrossRef]
82. Bui, D.C.; Nakarai, K.; Nishikawa, H. Effects of early-age thermal microcracking on material properties and structural performance of limestone aggregate concrete. *Cem. Concr. Res.* **2021**, *124*, 104267. [CrossRef]
83. Zheng, X.; Wang, Y.; Zhang, S.; Xu, F.; Zhu, X.; Jiang, X.; Zhou, L.; Shen, Y.; Chen, Q.; Yan, Z.; et al. Research progress of the thermophysical and mechanical properties of concrete subjected to freeze-thaw cycles. *Constr. Build. Mater.* **2022**, *330*, 127254. [CrossRef]
84. Figueira, R.B.; Sousa, R.; Coelho, L.; Azenha, M.; de Almeida, J.M.; Jorge, P.A.S.; Silva, C.J.R. Alkali-silica reaction in concrete: Mechanisms, mitigation and test methods. *Constr. Build. Mater.* **2019**, *222*, 903–931. [CrossRef]
85. Zhang, F.; Hu, Z.; Dai, L.; Wen, X.; Wang, R.; Zhang, D.; Song, X. Study on Corrosion Mechanism of Different Concentrations of Na_2SO_4 Solution on Early-Age Cast-In-Situ Concrete. *Materials* **2021**, *14*, 2018. [CrossRef]
86. Chen, Y.; Matalkah, F.; Rankothge, W.; Balachandra, A.; Soroushian, P. Improvement of the surface quality and aesthetics of ultra-high-performance concrete. *Proc. Inst. Civ. Eng. Constr. Mater.* **2019**, *172*, 246–255. [CrossRef]
87. Zhang, Y.; Yan, D.; Hu, S.; Guo, S. Modelling of energy consumption and carbon emission from the building construction sector in China, a process-based LCA approach. *Energy Policy* **2019**, *134*, 110949. [CrossRef]
88. Sharifi, E.; Sadjadi, S.J.; Aliha, M.R.M.; Moniri, A. Optimization of high-strength self-consolidating concrete mix design using an improved Taguchi optimization method. *Constr. Build. Mater.* **2020**, *236*, 117547. [CrossRef]
89. Nguyen, D.H.; Nguyen, V.T.; Lura, P.; Dao, V.T. Temperature-stress testing machine–A state-of-the-art design and its unique applications in concrete research. *Cem. Concr. Res.* **2019**, *102*, 28–38. [CrossRef]
90. Choi, H.; Inoue, M.; Choi, H.; Kim, J.; Sudoh, Y.; Kwon, S.; Lee, B.; Yoneyama, A. Physicochemical Study on the Strength Development Characteristics of Cold Weather Concrete Using a Nitrite–Nitrate Based Accelerator. *Materials* **2019**, *12*, 2706. [CrossRef]
91. Nandhini, K.; Karthikeyan, J. The early-age prediction of concrete strength using maturity models: A review. *J. Build. Rehabil.* **2021**, *6*, 1–10. [CrossRef]
92. Nepomuceno, M.C.S.; Bernardo, L.F.A. Evaluation of Self-Compacting Concrete Strength with Non-Destructive Tests for Concrete Structures. *Appl. Sci.* **2019**, *9*, 5109. [CrossRef]
93. Kurihashi, Y.; Konno, H.; Hama, Y. Effects of frost-damaged reinforced concrete beams on their impact resistance behavior. *Constr. Build. Mater.* **2020**, *274*, 122089. [CrossRef]
94. Wang, Z.; Gong, F.; Ueda, T. Modeling and simulation on static and fatigue behaviors of intact and frost damaged concrete with ice-strengthening effects. *J. Adv. Concr. Technol.* **2021**, *19*, 346–358. [CrossRef]
95. Mao, J.; Jia, H.; Wu, K.; Wang, Q.; Li, S.; Qian, W.; Xiong, F. A novel method to prevent frost cracking of perforated concrete components in cold regions. *Cold Reg. Sci. Technol.* **2023**, *211*, 103848. [CrossRef]
96. Wang, Z.; Gong, F.; Zhang, D.; Wang, Y.; Ueda, T. RBSM based analysis on mechanical degradation of non-air entrained concrete under frost action–A general prediction with various water cement ratio, lowest temperatures and FTC numbers. *Constr. Build. Mater.* **2019**, *211*, 744–755. [CrossRef]
97. Deng, X.; Gao, X.; Wang, R.; Gao, M.; Yan, X.; Cao, W.; Liu, J. Investigation of microstructural damage in air-entrained recycled concrete under a freeze–thaw environment. *Constr. Build. Mater.* **2021**, *268*, 121219. [CrossRef]
98. Wang, Y.; Liu, Z.; Fu, K.; Li, Q.; Wang, Y. Experimental studies on the chloride ion permeability of concrete considering the effect of freeze–thaw damage. *Constr. Build. Mater.* **2020**, *236*, 117556. [CrossRef]
99. Yuan, J.; Du, Z.; Wu, Y.; Xiao, F. Freezing-thawing resistance evaluations of concrete pavements with deicing salts based on various surfaces and air void parameters. *Constr. Build. Mater.* **2019**, *204*, 317–326. [CrossRef]
100. Liu, R.; Xiao, H.; Liu, J.; Guo, S.; Pei, Y. Improving the microstructure of ITZ and reducing the permeability of concrete with various water/cement ratios using nano-silica. *J. Mater. Sci.* **2019**, *54*, 444–456. [CrossRef]
101. Zhi, D.; Gong, F.; Wang, Z.; Zhao, Y.; Ueda, T. RBSM-based mesoscale study of frost deterioration for recycled concrete considering air-entrainment in old and new mortar. *J. Build. Eng.* **2023**, *68*, 106210. [CrossRef]
102. Yan, W.; Wu, Z.; Niu, F.; Wan, T.; Zheng, H. Study on the service life prediction of freeze–thaw damaged concrete with high permeability and inorganic crystal waterproof agent additions based on ultrasonic velocity. *Constr. Build. Mater.* **2020**, *259*, 120405. [CrossRef]
103. Zhang, G.; Yu, H.; Li, H.; Yang, Y. Experimental study of deformation of early age concrete suffering from frost damage. *Constr. Build. Mater.* **2019**, *215*, 410–421. [CrossRef]
104. Konzilia, J.; Egger, M.; Feix, J. Experimental investigation on salt frost scaling of textile-reinforced concrete. *Struct. Concr.* **2022**, *23*, 954–969. [CrossRef]
105. Zeiml, M.; Leithner, D.; Lackner, R.; Mang, H.A. How do polypropylene fibers improve the spalling behavior of in-situ concrete? *Cem. Concr. Res.* **2006**, *36*, 929–942. [CrossRef]

106. Rosenqvist, M. Frost-Induced Deterioration of Concrete in Hydraulic Structures: Interactions between Water Absorption, Leaching and Frost Action. Ph.D. Thesis, Lund University, Lund, Sweden, 14 September 2016. Available online: https://lup.lub.lu.se/record/dfac7b31-229a-4d1d-a02e-43258ce060f4 (accessed on 27 April 2023).
107. Adachi, Y.; Hirakawa, H.; Fukushima, A.; Uematsu, T.; Kikuta, K.; Taniguchi, M. Investigation of the Deterioration of Medium-Rise-Wall Type Reinforced Concrete Buildings with External Insulation in Snowy Cold Districts. *Buildings* **2022**, *12*, 2048. [CrossRef]
108. Wang, R.; Zhang, Q.; Li, Y. Deterioration of concrete under the coupling effects of freeze–thaw cycles and other actions: A review. *Constr. Build. Mater.* **2022**, *319*, 126045. [CrossRef]
109. Zhang, L.; Ma, R.; Lai, J.; Ruan, S.; Qian, X.; Yan, D.; Qian, K.; Wang, S. Performance buildup of concrete cured under low-temperatures: Use of a new nanocomposite accelerator and its application. *Constr. Build. Mater.* **2022**, *335*, 127529. [CrossRef]
110. Yuan, J.; Huang, X.; Chen, X.; Ge, Q.; Zhang, Z. Early-age mechanical properties and hydration degrees of magnesium phosphate cement paste in freezing winter of cold regions. *Constr. Build. Mater.* **2022**, *345*, 128337. [CrossRef]
111. Jamolovich, T.U. Modern methods for controlling the strength of concrete in the structures of monolithic buildings. *Acad. Globe Indersci. Res.* **2022**, *3*, 588–595. [CrossRef]
112. Wang, Y.; Lei, L.; Liu, J.; Ma, Y.; Liu, Y.; Xiao, Z.; Shi, C. Accelerators for normal concrete: A critical review on hydration, microstructure and properties of cement-based materials. *Cem. Concr. Compos.* **2022**, *134*, 104762. [CrossRef]
113. Zhang, C.; Liu, T.; Jiang, C.; Chen, Z.; Zhou, K.; Chen, L. The Freeze-Thaw Strength Evolution of Fiber-Reinforced Cement Mortar Based on NMR and Fractal Theory: Considering Porosity and Pore Distribution. *Materials* **2022**, *15*, 7316. [CrossRef] [PubMed]
114. Hu, X.; Peng, G.; Niu, D.; Zhao, N. Damage study on service performance of early-age frozen concrete. *Constr. Build. Mater.* **2019**, *210*, 22–31. [CrossRef]
115. Regan, E.E.; O'Brien, S.M. Exposing Exposed Concrete—Design, Construction, and Maintenance Considerations for Success. In Proceedings of the Ninth Congress on Forensic Engineering, Denver, CO, USA, 4–7 November 2022; pp. 111–120. [CrossRef]
116. Guler, S.; Akbulut, Z.F. Effect of freeze-thaw cycles on strength and toughness properties of new generation 3D/4D/5D steel fiber-reinforced concrete. *J. Build.Eng.* **2022**, *51*, 104239. [CrossRef]
117. Xu, Y.; Ye, H.; Yuan, Q.; Shi, C.; Gao, Y.; Fu, Q. The durability of concrete subject to mechanical load coupled with freeze–thaw cycles: A review. *Arch. Civ. Mech. Eng.* **2022**, *22*, 47. [CrossRef]
118. Al-Rashed, R.; Al-Jabari, M. Managing Thermal Effects in Waterproofed Concrete with Multi-Crystallization Enhancer. *Cement* **2022**, *10*, 100050. [CrossRef]
119. Qiu, W.L.; Teng, F.; Pan, S.S. Damage constitutive model of concrete under repeated load after seawater freeze-thaw cycles. *Constr. Build. Mater.* **2020**, *236*, 117560. [CrossRef]
120. Lu, Z.; Feng, Z.G.; Yao, D.; Li, X.; Ji, H. Freeze-thaw resistance of Ultra-High performance concrete: Dependence on concrete composition. *Constr. Build. Mater.* **2021**, *293*, 123523. [CrossRef]
121. Al-Kheetan, M.J.; Al-Tarawneh, M.A.; Ghaffar, S.H.; Chougan, M.; Jweihan, Y.S.; Rahman, M.M. Resistance of hydrophobic concrete with different moisture contents to advanced freeze–thaw cycles. *Struct. Concr.* **2021**, *22*, E1050–E1061. [CrossRef]
122. Santana Rangel, C.; Amario, M.; Pepe, M.; Martinelli, E.; Toledo Filho, R.D. Durability of Structural Recycled Aggregate Concrete Subjected to Freeze-Thaw Cycles. *Sustainability* **2020**, *12*, 6475. [CrossRef]
123. Ma, H.; Yu, H.; Da, B.; Tan, Y. Study on failure mechanism of concrete subjected to freeze-thaw condition in airport deicers. *Constr. Build. Mater.* **2021**, *313*, 125202. [CrossRef]
124. Hanjari, K.Z.; Utgenannt, P.; Lundgren, K. Experimental study of the material and bond properties of frost-damaged concrete. *Cem. Concr. Res.* **2011**, *41*, 244–254. [CrossRef]
125. Song, H.; Yao, J.; Xiang, J. The role of aggregate and cement paste in the deterioration of the transitional interface zone of pervious concrete during freeze-thaw cycles. *Case Stud. Constr. Mater.* **2022**, *16*, e01086. [CrossRef]
126. Jiang, J.; Yang, H.; Deng, Z.; Li, Z. Bond performance of deformed rebar embedded in recycled aggregate concrete subjected to repeated loading after freeze–thaw cycles. *Constr. Build. Mater.* **2022**, *318*, 125954. [CrossRef]
127. Suzuki, T.; Nishimura, S.; Shimamoto, Y.; Shiotani, T.; Ohtsu, M. Damage estimation of concrete canal due to freeze and thawed effects by acoustic emission and X-ray CT methods. *Constr. Build. Mater.* **2020**, *245*, 118343. [CrossRef]
128. Faheem, A.; Hasholt, M.T. The effect of temperature distribution in mortar on frost scaling. *Cem. Concr. Res.* **2022**, *157*, 106824. [CrossRef]
129. Liu, Z.; Li, L.; Gao, J.; Li, Y.; Lou, B.; Maria Barbieri, D.; Ye, T.; Sha, A. Frost-Related Damage of Portland Cement Pastes at Early Age. *J. Mater. Civ. Eng.* **2023**, *35*, 04023020. [CrossRef]
130. Karthick, S.; Kermanshachi, S.; Pamidimukkala, A.; Namian, M. A review of construction workforce health challenges and strategies in extreme weather conditions. *Int. J. Occup. Safety Ergo.* **2023**, *29*, 773–784. [CrossRef]
131. Ballesteros-Pérez, P.; Rojas-Céspedes, Y.A.; Hughes, W.; Kabiri, S.; Pellicer, E.; Mora-Melià, D.; del Campo-Hitschfeld, M.L. Weather-wise: A weather-aware planning tool for improving construction productivity and dealing with claims. *Autom. Constr.* **2017**, *84*, 81–95. [CrossRef]
132. Karthick, S.; Kermanshachi, S.; Ramaji, I. Health and safety of construction field workforce active in extreme weather conditions. In Proceedings of the Construction Research Congress, Arlington, TX, USA, 9–12 March 2022; pp. 737–747. [CrossRef]
133. Apipattanavis, S.; Sabol, K.; Molenaar, K.R.; Rajagopalan, B.; Xi, Y.; Blackard, B.; Patil, S. Integrated framework for quantifying and predicting weather-related highway construction delays. *J. Constr. Eng. Mana.* **2010**, *136*, 1160–1168. [CrossRef]

134. Raymond, D.L. A Synopsis of Cold Weather Operation Problems of Engine Powered Equipment Experienced in the Northern States and Border Areas of Canada. *SAE Tech. Paper* **1965**, 650935. [CrossRef]
135. Jaguemont, J.; Boulon, L.; Dubé, Y. A comprehensive review of lithium-ion batteries used in hybrid and electric vehicles at cold temperatures. *Appl. Energy* **2016**, *164*, 99–114. [CrossRef]
136. Edwards, K.L. Selecting materials for optimum use in engineering components. *Mater. Design* **2005**, *26*, 469–473. [CrossRef]
137. Huang, D.; Chen, Z.; Zhou, S. Self-powered heating strategy for lithium-ion battery pack applied in extremely cold climates. *Energy* **2022**, *239*, 122095. [CrossRef]
138. Rahman, M.L.; Malakooti, A.; Ceylan, H.; Kim, S.; Taylor, P.C. A review of electrically conductive concrete heated pavement system technology: From the laboratory to the full-scale implementation. *Constr. Build. Mater.* **2022**, *329*, 127139. [CrossRef]
139. Shin, Y.S.; Kim, G.H. A comparison of the economic feasibility of the protecting tent method and SCSFM method for heat curing of cold-weather concrete. *Appl. Mech. Mater.* **2014**, *651*, 1311–1314. [CrossRef]
140. Harbor, J. Engineering geomorphology at the cutting edge of land disturbance: Erosion and sediment control on construction sites. *Geomorphology* **1999**, *31*, 247–263. [CrossRef]
141. Nan, J.; Liu, J.; Chang, D.; Li, X. Mechanical characteristics and microstructure study of saline soil stabilized by quicklime after curing and freeze-thaw cycle. *Cold Reg. Sci. Tech.* **2022**, *201*, 103625. [CrossRef]
142. Brambilla, A.; Sangiorgio, A. Mould growth in energy efficient buildings: Causes, health implications and strategies to mitigate the risk. *Renew. Sust. Energy Rev.* **2020**, *132*, 110093. [CrossRef]
143. Pokhrel, S.; Amiri, L.; Poncet, S.; Sasmito, A.P.; Ghoreishi-Madiseh, S.A. Renewable heating solutions for buildings; a techno-economic comparative study of sewage heat recovery and solar borehole thermal energy storage system. *Energy Build.* **2022**, *259*, 111892. [CrossRef]
144. Tsagarakis, K.P.; Efthymiou, L.; Michopoulos, A.; Mavragani, A.; Anđelković, A.S.; Antolini, F.; Bacic, M.; Bajare, D.; Baralis, M.; Bogusz, W.; et al. A review of the legal framework in shallow geothermal energy in selected European countries: Need for guidelines. *Renew. Energy* **2020**, *147*, 2556–2571. [CrossRef]
145. Patil, M.; Boraste, S.; Minde, P. A comprehensive review on emerging trends in smart green building technologies and sustainable materials. *Mater. Today Proc.* **2022**, *65*, 1813–1822. [CrossRef]
146. Akeed, M.H.; Qaidi, S.; Ahmed, U.; Faraj, R.H.; Majeed, S.S.; Mohammed, A.S.; Emad, W.; Tayeh, B.A.; Azevedo, A.R. Ultra-high-performance fiber-reinforced concrete. Part V: Mixture design, preparation, mixing, casting, and curing. *Case Stud. Constr. Mater.* **2022**, *17*, e01363. [CrossRef]
147. Ikumi, T.; Salvador, R.P.; Aguado, A. Mix proportioning of sprayed concrete: A systematic literature review. *Tunnel. Undergr. Space Tech.* **2022**, *124*, 104456. [CrossRef]
148. Zhu, Z.G.; Hou, K.P.; Sun, H.F.; Cheng, Y.; Yang, B.J.; Sun, W. Research on the Effect of Curing Temperature, Steel Fiber, and Admixture Content on Concrete Performance Based on Orthogonal Test in Cold Region. *Adv. Mater. Sci. Eng.* **2022**, *2022*, 3671823. [CrossRef]
149. Fu, X.; Zhao, G.; Wang, M.; Wang, J.; Xu, Y.; Gu, C. Comprehensive evaluation method for structural behavior of concrete dams in cold regions. *Eng. Struct.* **2023**, *278*, 115435. [CrossRef]
150. Pan, W.; Zhang, Z. Benchmarking the Sustainability of Concrete and Steel Modular Construction for Buildings in Urban Development. *Sus. Cities Soc.* **2023**, *90*, 104400. [CrossRef]
151. Kia, A.; Wong, H.S.; Cheeseman, C.R. Freeze–thaw durability of conventional and novel permeable pavement replacement. *J. Transp. Eng. B Pavem.* **2022**, *148*, 04022051. [CrossRef]
152. Huang, W.; Wang, H. Multi-aspect engineering properties and sustainability impacts of geopolymer pervious concrete. *Compos. B Eng.* **2022**, *242*, 110035. [CrossRef]
153. Kavitha, R.; Vivekananthan, M.R.; Dhanagopal, K.; Divyabharathi, P.; Prabhakaran, M.; Ramprathap, M. An overview of water proofing system in concrete structures. *Mater. Today Proc.* 2023; *in press*. [CrossRef]
154. Yin, B.; Xu, H.; Fan, F.; Qi, D.; Hua, X.; Xu, T.; Liu, C.; Hou, D. Superhydrophobic coatings based on bionic mineralization for improving the durability of marine concrete. *Constr. Build. Mater.* **2023**, *362*, 129705. [CrossRef]
155. Ayyoob, S.; Zaki, M.; Khan, M.S.; Khan, S.A.; Kamal, M.A. The Efficiency of Waterproofing Agents in Concrete: Assessment of Compressive Strength and Permeability. *Amer. J. Civ. Eng. Arch.* **2022**, *10*, 137–146. [CrossRef]
156. Tang, C.W.; Cheng, C.K.; Ean, L.W. Mix Design and Engineering Properties of Fiber-Reinforced Pervious Concrete Using Lightweight Aggregates. *Appl. Sci.* **2022**, *12*, 524. [CrossRef]
157. Lu, J.X.; Shen, P.; Ali, H.A.; Poon, C.S. Mix design and performance of lightweight ultra high-performance concrete. *Mater. Design* **2022**, *216*, 110553. [CrossRef]
158. Sharbaf, M.; Najimi, M.; Ghafoori, N. A comparative study of natural pozzolan and fly ash: Investigation on abrasion resistance and transport properties of self-consolidating concrete. *Constr. Build. Mater.* **2022**, *346*, 128330. [CrossRef]
159. Ganapathy, G.P.; Keshav, L.; Ravindiran, G.; Razack, N.A. Strength prediction of self-consolidating concrete containing steel fibre with different fibre aspect ratio. *J. Nanomater.* **2022**, *2022*, 7604383. [CrossRef]
160. Liu, X.; Liu, L.; Lyu, K.; Li, T.; Zhao, P.; Liu, R.; Zuo, J.; Fu, F.; Shah, S.P. Enhanced early hydration and mechanical properties of cement-based materials with recycled concrete powder modified by nano-silica. *J. Build. Eng.* **2022**, *50*, 104175. [CrossRef]
161. Liu, Y.; Yang, S.; Li, J.; Wang, F.; Hu, S. Effect of w/c ratio and antifreeze admixture on the frost damage of sulfoaluminate cement concrete at −20 °C. *Const. Build. Mater.* **2022**, *347*, 128457. [CrossRef]

162. Wang, Z.; Wang, X.; Zhu, P.; Liu, H.; Yan, X.; Wei, D. Mix-Proportion Design Methods and Sustainable Use Evaluation of Recycled Aggregate Concrete Used in Freeze–Thaw Environment. *J. Mater. Civ. Eng.* **2023**, *35*, 04022401. [CrossRef]
163. Mardani, A.; Emin, A. Utilization of high-range water reducing admixture having air-entraining agents in cementitious systems. *J. Build. Eng.* **2022**, *64*, 105565. [CrossRef]
164. Zeng, H.; Lai, Y.; Qu, S.; Yu, F. Exploring the effect of graphene oxide on freeze–thaw durability of air-entrained mortars. *Constr. Build. Mater.* **2022**, *324*, 126708. [CrossRef]
165. Zhang, D.; Zhang, T.; Yang, Q. Mechanical Properties and Damage Layer Thickness of Green Concrete under a Low-Temperature Environment. *Materials* **2022**, *15*, 7409. [CrossRef] [PubMed]
166. Zhang, S.; Chen, B.; Tian, B.; Lu, X.; Xiong, B. Effect of Fly Ash Content on the Microstructure and Strength of Concrete under Freeze–Thaw Condition. *Buildings* **2022**, *12*, 2113. [CrossRef]
167. Wu, X.; Zheng, S.; Feng, Z.; Chen, B.; Qin, Y.; Xu, W.; Liu, Y. Prediction of the frost resistance of high-performance concrete based on RF-REF: A hybrid prediction approach. *Constr. Build. Mater.* **2022**, *333*, 127132. [CrossRef]
168. Ullah, R.; Qiang, Y.; Ahmad, J.; Vatin, N.I.; El-Shorbagy, M.A. Ultra-High-Performance Concrete (UHPC): A State-of-the-Art Review. *Materials* **2022**, *15*, 4131. [CrossRef] [PubMed]
169. Kanaan, D.; Safhi, A.e.M.; Suleiman, A.R.; Soliman, A.M. Fresh, Hardened, and Microstructural Properties of Ambient Cured One-Part Alkali-Activated Self-Consolidating Concrete. *Sustainability* **2023**, *15*, 2451. [CrossRef]
170. Prithiviraj, C.; Saravanan, J.; Ramesh Kumar, D.; Murali, G.; Vatin, N.I.; Swaminathan, P. Assessment of Strength and Durability Properties of Self-Compacting Concrete Comprising Alccofine. *Sustainability* **2022**, *14*, 5895. [CrossRef]
171. Bhuva, P.; Bhogayata, A. A review on the application of artificial intelligence in the mix design optimization and development of self-compacting concrete. *Mater. Today Proc.* **2022**, *65*, 603–608. [CrossRef]
172. Zhang, P.; Wang, C.; Gao, Z.; Wang, F. A review on fracture properties of steel fiber reinforced concrete. *J. Build. Eng.* **2023**, *67*, 105975. [CrossRef]
173. Behnia, B.; Askarinejad, P.; LaRussa-Trott, N. Investigating Low-Temperature Cracking Behavior of Fiber-Reinforced Asphalt Concrete Materials. *Int. J. Pavement Res. Tech.* **2023**, 1–12. [CrossRef]
174. Cao, Y.; Bao, J.; Zhang, P.; Sun, Y.; Cui, Y. A state-of-the-art review on the durability of seawater coral aggregate concrete exposed to marine environment. *J. Build. Eng.* **2022**, *60*, 105199. [CrossRef]
175. Saleh, A.N.; Khalil, O.A.; Attar, A.A. An overview of mechanical and thermal properties of nano concrete. *AIP Conf. Proc.* **2022**, *2398*, 020025. [CrossRef]
176. Eissa, A.; Yasien, A.M.; Bassuoni, M.T.; Alfaro, M. Performance of treated soft clay with nano-modified cementitious binders at reference and cold temperatures. *Int. J. Geosynth. Ground Eng.* **2022**, *8*, 52. [CrossRef]
177. Vijayan, D.S.; Devarajan, P.; Sivasuriyan, A. A review on eminent application and performance of nano based silica and silica fume in the cement concrete. *Sust. Energy Tech. Ass.* **2023**, *56*, 103105. [CrossRef]
178. Hurley, M.; Habibzadeh-Bigdarvish, O.; Lei, G.; Yu, X. Laboratory study of a hydronic concrete deck heated externally in a controlled sub-freezing environment. *Energy Built Environ.* 2022, *in press*. [CrossRef]
179. Kildashti, K.; Nash, S.; Samali, B. In-plane lateral behaviour of PVC modular concrete form squat walls: Experimental and numerical study. *J. Build. Eng.* **2022**, *52*, 104494. [CrossRef]
180. Mazumder, E.A.; Prasad, M.L.V. Performance Enhancement of Fly Ash-Based Self Compacting Geopolymer Concrete Using Pre-heating Technique. *Iran. J. Sci. Tech. Trans. Civ. Eng.* **2023**, 1–13. [CrossRef]
181. Eren, F.; Keskinateş, M.; Felekoğlu, B.; Felekoğlu, K.T. The role of Pre-Heating and mineral additive modification on Long-Term strength development of calcium aluminate cement mortars. *Constr. Build. Mater.* **2022**, *340*, 127720. [CrossRef]
182. Zhan, P.M.; He, Z.H. Application of shrinkage reducing admixture in concrete: A review. *Constr. Build. Mater.* **2019**, *201*, 676–690. [CrossRef]
183. Abubakri, S.; Lomboy, G.R.; Kennedy, D.; Watts, B. Development of Electrically Conductive Cold Weather Cement Concrete. *J. Cold Reg. Eng.* **2023**, *37*, 04022016. [CrossRef]
184. Su, Y.; Luo, B.; Luo, Z.; Huang, H.; Li, J.; Wang, D. Effect of Accelerators on the Workability, Strength, and Microstructure of Ultra-High-Performance Concrete. *Materials* **2022**, *15*, 159. [CrossRef]
185. Reddy, P.N.; Naqash, J.A. Development of high early strength in concrete incorporating alccofine and non-chloride accelerator. *SN Appl. Sci.* **2019**, *1*, 1–11. [CrossRef]
186. Hamada, H.; Alattar, A.; Tayeh, B.; Yahaya, F.; Almeshal, I. Influence of different curing methods on the compressive strength of ultra-high-performance concrete: A comprehensive review. *Case Stud. Constr. Mater.* **2022**, *17*, e01390. [CrossRef]
187. Taheri, S. A review on five key sensors for monitoring of concrete structures. *Constr. Build. Mater.* **2019**, *204*, 492–509. [CrossRef]
188. Abu-Jdayil, B.; Mourad, A.H.; Hittini, W.; Hassan, M.; Hameedi, S. Traditional, state-of-the-art and renewable thermal building insulation materials: An overview. *Constr. Build. Mater.* **2019**, *214*, 709–735. [CrossRef]
189. Surahyo, A.; Surahyo, A. Hot and cold weather concreting. *Concr. Constr. Pract. Prob. Sol.* **2019**, 257–272. [CrossRef]
190. Jun, H.; Ke, G. Analysis on Energy Saving Potential of Residential Building Enclosure Based on Orthogonal Experimental Analysis. *J. Civ. Eng. Urb. Plan.* **2022**, *4*, 27–33.
191. Viegas, G.M.; Jodra, J.I.; San Juan, G.A.; Díscoli, C.A. Heat storage wall made of concrete and encapsulated water applied to mass construction social housing in temperate climates. *Energy Build.* **2018**, *159*, 346–356. [CrossRef]

192. Ekrami, N.; Garat, A.; Fung, A.S. Thermal analysis of insulated concrete form (ICF) walls. *Energy Proc.* **2015**, *75*, 2150–2156. [CrossRef]
193. Zuo, Z.; Huang, Y.; Pan, X.; Zhan, Y.; Zhang, L.; Li, X.; Zhu, M.; Zhang, L.; De Corte, W. Experimental research on remote real-time monitoring of concrete strength for highrise building machine during construction. *Measurement* **2021**, *178*, 109430. [CrossRef]
194. Afzal, J.; Yihong, Z.; Afzal, U.; Aslam, M. A complex wireless sensors model (CWSM) for real time monitoring of dam temperature. *Heliyon* **2023**, *9*, e13371. [CrossRef]
195. Zhang, H.; Li, J.; Kang, F. Real-time monitoring of humidity inside concrete structures utilizing embedded smart aggregates. *Constr. Build. Mater.* **2022**, *331*, 127317. [CrossRef]
196. John, S.T.; Roy, B.K.; Sarkar, P.; Davis, R. IoT enabled real-time monitoring system for early-age compressive strength of concrete. *J. Constr. Eng. Mana.* **2020**, *146*, 05019020. [CrossRef]
197. Gamil, Y.; Nilimaa, J.; Najeh, T.; Cwirzen, A. Formwork pressure prediction in cast-in-place self-compacting concrete using deep learning. *Auto. Constr.* **2023**, *151*, 104869. [CrossRef]
198. Gehri, N.; Mata-Falcón, J.; Kaufmann, W. Refined extraction of crack characteristics in Large-scale concrete experiments based on digital image correlation. *Eng. Struct.* **2022**, *251*, 113486. [CrossRef]
199. Kinda, J.; Bourdot, A.; Charpin, L.; Michel-Ponnelle, S.; Benboudjema, F. Investigation of drying shrinkage of cement-based materials assisted by digital image correlation. *J. Mater. Civ. Eng.* **2022**, *34*, 04021477. [CrossRef]
200. Wong, R.W.; Loo, B.P. Sustainability implications of using precast concrete in construction: An in-depth project-level analysis spanning two decades. *J. Clean. Prod.* **2022**, *378*, 134486. [CrossRef]
201. Yang, S. The Development and Application of Prefabricated Buildings. *High. Sci. Eng. Tech.* **2022**, *18*, 162–167. [CrossRef]
202. Bischof, P.; Mata-Falcon, J.; Kaufmann, W. Fostering innovative and sustainable mass-market construction using digital fabrication with concrete. *Cem. Concr. Res.* **2022**, *161*, 106948. [CrossRef]
203. Li, Z.; Zhou, X.; Ma, H.; Hou, D. *Advanced Concrete Technology*, 2nd ed.; John Wiley & Sons: Hoboken, NJ, USA, 2022.
204. Onyelowe, K.C.; Ebid, A.M.; Riofrio, A.; Baykara, H.; Soleymani, A.; Mahdi, H.A.; Jahangir, H.; Ibe, K. Multi-Objective Prediction of the Mechanical Properties and Environmental Impact Appraisals of Self-Healing Concrete for Sustainable Structures. *Sustainability* **2022**, *14*, 9573. [CrossRef]
205. Sawadogo, M.; Benmahiddine, F.; Hamami, A.E.A.; Belarbi, R.; Godin, A.; Duquesne, M. Investigation of a novel bio-based phase change material hemp concrete for passive energy storage in buildings. *Appl. Therm. Eng.* **2022**, *212*, 118620. [CrossRef]
206. Monteiro, J.; Roussanaly, S. CCUS scenarios for the cement industry: Is CO_2 utilization feasible? *J. CO2 Util.* **2022**, *61*, 102015. [CrossRef]
207. Hossain, M.R.; Sultana, R.; Patwary, M.M.; Khunga, N.; Sharma, P.; Shaker, S.J. Self-healing concrete for sustainable buildings. A review. *Environ. Chem. Lett.* **2022**, *20*, 1265–1273. [CrossRef]
208. Amran, M.; Onaizi, A.M.; Fediuk, R.; Vatin, N.I.; Muhammad Rashid, R.S.; Abdelgader, H.; Ozbakkaloglu, T. Self-Healing Concrete as a Prospective Construction Material: A Review. *Materials* **2022**, *15*, 3214. [CrossRef] [PubMed]
209. Wu, D.; Rahim, M.; El Ganaoui, M.; Bennacer, R.; Liu, B. Multilayer assembly of phase change material and bio-based concrete: A passive envelope to improve the energy and hygrothermal performance of buildings. *Energy Conv. Mana.* **2022**, *257*, 115454. [CrossRef]
210. Hargis, C.W.; Chen, I.A.; Devenney, M.; Fernandez, M.J.; Gilliam, R.J.; Thatcher, R.P. Calcium Carbonate Cement: A Carbon Capture, Utilization, and Storage (CCUS) Technique. *Materials* **2021**, *14*, 2709. [CrossRef]
211. Kong, Y.K.; Ruan, S.; Kurumisawa, K. Recycling of calcined carbonated cement pastes as cementitious materials: Proposed CCUS technology for calcium looping. *J. Environ. Chem. Eng.* **2022**, *10*, 108247. [CrossRef]
212. Guimarães, A.S.; Delgado, J.M.P.Q.; Lucas, S.S. Advanced Manufacturing in Civil Engineering. *Energies* **2021**, *14*, 4474. [CrossRef]
213. Classen, M.; Ungermann, J.; Sharma, R. Additive Manufacturing of Reinforced Concrete—Development of a 3D Printing Technology for Cementitious Composites with Metallic Reinforcement. *Appl. Sci.* **2020**, *10*, 3791. [CrossRef]
214. Xiao, H.; Muthu, B.; Kadry, S.N. Artificial intelligence with robotics for advanced manufacturing industry using robot-assisted mixed-integer programming model. *Intell. Serv. Robot.* **2020**, 1–10. [CrossRef]
215. Gunawardena, T.; Mendis, P. Prefabricated Building Systems—Design and Construction. *Encyclopedia* **2022**, *2*, 70–95. [CrossRef]
216. El Inaty, F.; Baz, B.; Aouad, G. Long-term durability assessment of 3D printed concrete. *J. Adh. Sci. Tech.* **2022**, 1–16. [CrossRef]
217. Bazli, M.; Ashrafi, H.; Rajabipour, A.; Kutay, C. 3D printing for remote housing: Benefits and challenges. *Aut. Constr.* **2023**, *148*, 104772. [CrossRef]
218. Dörfler, K.; Dielemans, G.; Lachmayer, L.; Recker, T.; Raatz, A.; Lowke, D.; Gerke, M. Additive Manufacturing using mobile robots: Opportunities and challenges for building construction. *Cem. Concr. Res.* **2022**, *158*, 106772. [CrossRef]
219. Meisel, N.A.; Watson, N.; Bilén, S.G.; Duarte, J.P.; Nazarian, S. Design and system considerations for construction-scale concrete additive manufacturing in remote environments via robotic arm deposition. *3D Print. Add. Manuf.* **2022**, *9*, 35–45. [CrossRef]
220. Momeni, M.; Relefors, J.; Khatry, A.; Pettersson, L.; Papadopoulos, A.V.; Nolte, T. Automated fabrication of reinforcement cages using a robotized production cell. *Aut. Constr.* **2022**, *133*, 103990. [CrossRef]
221. Ou, Y.; Bao, D.W.; Zhu, G.Q.; Luo, D. Additive Fabrication of Large-Scale Customizable Formwork Using Robotic Fiber-Reinforced Polymer Winding. *3D Print. Add. Manuf.* **2022**, *9*, 109–121. [CrossRef]
222. Han, Y.; Yan, X.; Piroozfar, P. An overall review of research on prefabricated construction supply chain management. Engineering. *Constr. Arch. Mana.* 2022; *in press*. [CrossRef]

223. Aurier, L.; Hassan, M.; Jaworski, J.; Guizani, L. Review of Accelerated Bridge Construction Systems for Bridge Superstructures and Their Adaptability for Cold Weather. *CivilEng* **2023**, *4*, 83–104. [CrossRef]
224. Zhao, X.; Cao, L. Application Strategy of BIM Technology in Virtual Construction of Prefabricated Buildings. *J. Arch. Res. Dev.* **2022**, *6*, 40–46. [CrossRef]
225. Jiang, J.; Xiong, Y.; Zhang, Z.; Rosen, D.W. Machine learning integrated design for additive manufacturing. *J. Intel. Manuf.* **2022**, *33*, 1073–1086. [CrossRef]
226. Anane, W.; Iordanova, I.; Ouellet-Plamondon, C. BIM-driven computational design for robotic manufacturing in off-site construction: An integrated Design-to-Manufacturing (DtM) approach. *Auto. Constr.* **2023**, *150*, 104782. [CrossRef]
227. Ng, M.S.; Graser, K.; Hall, D.M. Digital fabrication, BIM and early contractor involvement in design in construction projects: A comparative case study. *Arch. Eng. Design Mana.* **2023**, *19*, 39–55. [CrossRef]
228. Liu, Y.; Cao, Y.; Wang, L.; Chen, Z.S.; Qin, Y. Prediction of the durability of high-performance concrete using an integrated RF-LSSVM model. *Constr. Build. Mater.* **2022**, *356*, 129232. [CrossRef]
229. Sherif, M.; Nassar, K.; Hosny, O.; Safar, S.; Abotaleb, I. Automated BIM-based structural design and cost optimization model for reinforced concrete buildings. *Sci. Rep.* **2022**, *12*, 21616. [CrossRef]
230. Mousavi, M.; TohidiFar, A.; Alvanchi, A. BIM and machine learning in seismic damage prediction for non-structural exterior infill walls. *Auto. Constr.* **2022**, *139*, 104288. [CrossRef]
231. Baduge, S.K.; Thilakarathna, S.; Perera, J.S.; Arashpour, M.; Sharafi, P.; Teodosio, B.; Shringi, A.; Mendis, P. Artificial intelligence and smart vision for building and construction 4.0: Machine and deep learning methods and applications. *Auto. Constr.* **2022**, *141*, 104440. [CrossRef]
232. Sadrossadat, E.; Basarir, H.; Karrech, A.; Elchalakani, M. Multi-objective mixture design and optimisation of steel fiber reinforced UHPC using machine learning algorithms and metaheuristics. *Eng. Comput.* **2022**, *38*, 2569–2582. [CrossRef]
233. Moein, M.M.; Saradar, A.; Rahmati, K.; Mousavinejad, S.H.G.; Bristow, J.; Aramali, V.; Karakouzian, M. Predictive models for concrete properties using machine learning and deep learning approaches: A review. *J. Build. Eng.* **2022**, *63*, 105444. [CrossRef]
234. Zabin, A.; González, V.A.; Zou, Y.; Amor, R. Applications of machine learning to BIM: A systematic literature review. *Adv. Eng. Inf.* **2022**, *51*, 101474. [CrossRef]
235. Opoku, D.G.J.; Perera, S.; Osei-Kyei, R.; Rashidi, M.; Famakinwa, T.; Bamdad, K. Drivers for Digital Twin Adoption in the Construction Industry: A Systematic Literature Review. *Buildings* **2022**, *12*, 113. [CrossRef]
236. Madubuike, O.C.; Anumba, C.J.; Khallaf, R. A review of digital twin applications in construction. *J. Inf. Tech. Constr.* **2022**, *27*, 145–172. [CrossRef]
237. Hosamo, H.H.; Nielsen, H.K.; Kraniotis, D.; Svennevig, P.R.; Svidt, K. Digital Twin framework for automated fault source detection and prediction for comfort performance evaluation of existing non-residential Norwegian buildings. *Energy Build.* **2023**, *281*, 112732. [CrossRef]
238. Ying, J.; Zhang, X.; Zhang, Y.; Bilan, S. Green infrastructure: Systematic literature review. *Econ. Res.* **2022**, *35*, 343–366. [CrossRef]
239. Coffetti, D.; Crotti, E.; Gazzaniga, G.; Carrara, M.; Pastore, T.; Coppola, L. Pathways towards sustainable concrete. *Cem. Concr. Res.* **2022**, *154*, 106718. [CrossRef]
240. Das, S.; Saha, P.; Jena, S.P.; Panda, P. Geopolymer concrete: Sustainable green concrete for reduced greenhouse gas emission—A review. *Mater. Today Proc.* **2022**, *60*, 62–71. [CrossRef]
241. Coffetti, D.; Cabrini, M.; Crotti, E.; Gazzaniga, G.; Lorenzi, S.; Pastore, T.; Coppola, L. Durability of Mortars Manufactured with Low-Carbon Binders Exposed to Calcium Chloride-Based De-Icing Salts. *Key Eng. Mater.* **2022**, *919*, 151–160. [CrossRef]
242. Alhassan, M.; Alkhawaldeh, A.; Betoush, N.; Alkhawaldeh, M.; Huseien, G.F.; Amaireh, L.; Elrefae, A. Life Cycle Assessment of the Sustainability of Alkali-Activated Binders. *Biomimetics* **2023**, *8*, 58. [CrossRef] [PubMed]
243. Manjunatha, M.; Seth, D.; Balaji, K.V.G.D.; Bharath, A. Engineering properties and environmental impact assessment of green concrete prepared with PVC waste powder: A step towards sustainable approach. *Case Stud. Constr. Mater.* **2022**, *17*, e01404. [CrossRef]
244. Forero, J.A.; Bravo, M.; Pacheco, J.; de Brito, J.; Evangelista, L. Thermal Performance of Concrete with Reactive Magnesium Oxide as an Alternative Binder. *Sustainability* **2022**, *14*, 5885. [CrossRef]
245. Roy, S.; Roy, S. Carbon Dioxide Capture—A Challenge for Green Economy. *Amer. J. Appl. Bio-Tech. Res.* **2022**, *3*, 30–49. [CrossRef]
246. Guo, Y.; Luo, L.; Liu, T.; Hao, L.; Li, Y.; Liu, P.; Zhu, T. A review of low-carbon technologies and projects for the global cement industry. *J. Environ. Sci.* **2023**, *136*, 682–697. [CrossRef]
247. Clarke, B.; Otto, F.; Stuart-Smith, R.; Harrington, L. Extreme weather impacts of climate change: An attribution perspective. *Environ. Res. Clima.* **2022**, *1*, 012001. [CrossRef]
248. Zaqout, T.; Andradóttir, H.Ó.; Arnalds, Ó. Infiltration capacity in urban areas undergoing frequent snow and freeze–thaw cycles: Implications on sustainable urban drainage systems. *J. Hydr.* **2022**, *607*, 127495. [CrossRef]
249. Zhang, H.; Li, J.; Kang, F.; Zhang, J. Monitoring and evaluation of the repair quality of concrete cracks using piezoelectric smart aggregates. *Constr. Build. Mater.* **2022**, *317*, 125775. [CrossRef]
250. Katman, H.Y.B.; Khai, W.J.; Bheel, N.; Kırgız, M.S.; Kumar, A.; Benjeddou, O. Fabrication and Characterization of Cement-Based Hybrid Concrete Containing Coir Fiber for Advancing Concrete Construction. *Buildings* **2022**, *12*, 1450. [CrossRef]

251. Azam, R.; Riaz, M.R.; Farooq, M.U.; Ali, F.; Mohsan, M.; Deifalla, A.F.; Mohamed, A.M. Optimization-Based Economical Flexural Design of Singly Reinforced Concrete Beams: A Parametric Study. *Materials* **2022**, *15*, 3223. [CrossRef] [PubMed]
252. Junaid, M.F.; Rehman, Z.U.; Kuruc, M.; Medved', I.; Bačinskas, D.; Čurpek, J.; Čekon, M.; Ijaz, N.; Ansari, W.S. Lightweight concrete from a perspective of sustainable reuse of waste byproducts. *Constr. Build. Mater.* **2022**, *319*, 126061. [CrossRef]

eng

MDPI

Review

Integrating Multi-Criteria Decision-Making Methods with Sustainable Engineering: A Comprehensive Review of Current Practices

Anđelka Štilić [1,*] and Adis Puška [2]

[1] The College of Tourism, Academy of Applied Studies Belgrade, Bulevar Zorana Đinđića 152a, 11070 Belgrade, Serbia
[2] Department of Public Safety, Government of Brčko District, 76100 Brčko, Bosnia and Herzegovina; adispuska@yahoo.com
* Correspondence: andelka.stilic@assb.edu.rs

Abstract: Multi-criteria decision-making (MCDM) methods have gained increased attention in sustainable engineering, where complex decision-making problems require consideration of multiple criteria and stakeholder perspectives. This review paper provides a comprehensive overview of the different MCDM methods, their applications in sustainable engineering, and their strengths and weaknesses. The paper discusses the concept of sustainable engineering, its principles, and the different areas where MCDM methods have been applied, including energy, manufacturing, transportation, and environmental engineering. Case studies of real-world applications are presented and analyzed, highlighting the main findings and implications for engineering practice. Finally, the challenges and limitations of MCDM methods in sustainable engineering are discussed, and future research directions are proposed. This review contributes to the understanding of the role of MCDM methods in sustainable engineering and provides guidance for researchers and practitioners.

Keywords: sustainable engineering; MCDM; AHP; TOPSIS; fuzzy sets; ANP; DEMATEL; BWM; VIKTOR; GRA; entropy

Citation: Štilić, A.; Puška, A. Integrating Multi-Criteria Decision-Making Methods with Sustainable Engineering: A Comprehensive Review of Current Practices. *Eng* **2023**, *4*, 1536–1549. https://doi.org/10.3390/eng4020088

Academic Editor: Antonio Gil Bravo

Received: 3 May 2023
Revised: 24 May 2023
Accepted: 30 May 2023
Published: 31 May 2023

1. Introduction

Multi-criteria decision-making (MCDM) methods have become a necessary tool throughout contemporary engineering practice [1]. They enable decision-makers to assess complex problems involving multiple criteria, trade-offs, and uncertainties. The methods used in MCDM are especially beneficial when applied to sustainable engineering, where decision-making requires balancing economic, environmental, and social considerations [2,3].

Sustainable engineering aims to design and implement engineering solutions that are environmentally friendly, socially acceptable, and economically viable [4]. Achieving sustainability requires the consideration of multiple criteria, such as resource conservation, pollution prevention, energy efficiency, economic feasibility, social equity, and stakeholder participation [5–7]. Sustainable engineering challenges decision-makers to balance these criteria and make trade-offs among them to identify the best solutions [8].

MCDM methods can assist in the decision-making process by offering an organized and transparent framework for assessing alternative solutions based on a variety of criteria. These methods enable decision-makers to identify optimal solutions by conducting quantitative and qualitative assessments, taking into account the preferences and priorities of different stakeholders [9].

The purpose of this review paper is to provide a thorough overview of the different MCDM methods and their applications in sustainable engineering. The review explores how MCDM methods could be utilized in various areas of sustainable engineering, including energy, manufacturing, transportation, and environmental engineering. The review also analyzes case studies of real-world applications of MCDM methods and highlights the strengths and weaknesses of each approach.

The motivation for this review is to address the need for a comprehensive and up-to-date overview related to MCDM methods regarding sustainable engineering. The need arises from the increasing importance of sustainability in contemporary engineering practice [1] and the growing complexity of decision-making problems [10]. This review paper seeks to contribute to the existing literature by highlighting the most effective MCDM methods for sustainable engineering problems and by proposing future research directions.

Its primary objectives are:

- To offer a comprehensive overview of different MCDM approaches and how they are used in sustainable engineering;
- To analyze case studies of real-world applications of MCDM methods in sustainable engineering and to highlight their outcomes;
- To identify the strengths and weaknesses of each MCDM technique in sustainable engineering and to compare and contrast them;
- To propose future research directions and discuss how MCDM methods can be further developed to enhance their effectiveness and applicability in sustainable engineering.

This review paper aims to contribute to the ongoing efforts to develop sustainable engineering solutions by providing decision-makers with a framework for selecting the most effective MCDM methods for their specific problems. By doing so, the review paper aims to enhance the effectiveness and applicability of MCDM methods in sustainable engineering and grow what is currently the state-of-the-art in the field.

To achieve the objectives, the review paper is organized in the following manner. After the Introduction, Section 2 is presented. The Primary Results are presented, followed by a presentation of the Detailed Review Results. A summary of the principles and applications of MCDM methods is presented, highlighting their importance and relevance to sustainable engineering. Next, the specific areas of sustainable engineering where MCDM methods have been applied are discussed, and the outcomes achieved are reviewed. Case studies of real-world applications of MCDM methods in sustainable engineering are then presented, and the main findings and implications for engineering practice are analyzed. Following this, the challenges and limitations of MCDM methods in sustainable engineering are then discussed, and prospective research recommendations are proposed. Finally, the main contributions of this review paper are summarized, and the implications for decision-making in sustainable engineering are suggested.

2. Materials and Methods

The knowledge used to conduct this research was obtained from various sources, including published research papers, technical reports, and case studies in academic journals. A thorough search was conducted using the Web of Science (WoS) Core Collection Database, as well as the online EBSCO Discovery Service engine. The search terms used included MCDM, sustainability, and sustainable engineering, with various combinations of these keywords also utilized. The search was limited to papers published between 2018 and 2023, written in English, and focused on MCDM methodologies employed for sustainable engineering.

After an initial screening of search results based on the titles and abstracts, a total of 36,490 articles associated with MCDM methods across different disciplines was identified, with 12,879 of these articles specifically addressing MCDM methods in sustainable engineering.

The case studies presented in this paper were selected based on their relevance and representativeness to the applications of MCDM methodologies in sustainable engineering. The cases were analyzed using a systematic approach to identify the decision-making problems, the criteria used, and the outcomes achieved. The case studies were also used to illustrate the strengths and limitations of various MCDM methods, as well as to identify the challenges and opportunities for future research in this area.

3. Primary Results

The publishers of articles pertaining to the use of MCDM methods across various fields of sustainable engineering are diverse and include well-known names such as Springer Nature, Elsevier, Wiley-Blackwell, and Taylor & Francis Ltd. (Figure 1). However, the publisher with the most articles published in this field is MDPI, with 4264 articles. Other publishers with a significant number of articles include Hindawi Limited, Emerald Publishing Limited, and IOS Press. The list also includes smaller publishers such as the Rural Outreach Program and Dr. M.N. Khan, indicating a wide range of contributors to this field. The diversity of publishers reflects the multidisciplinary nature of sustainable engineering, where different fields intersect and collaborate to achieve sustainable solutions.

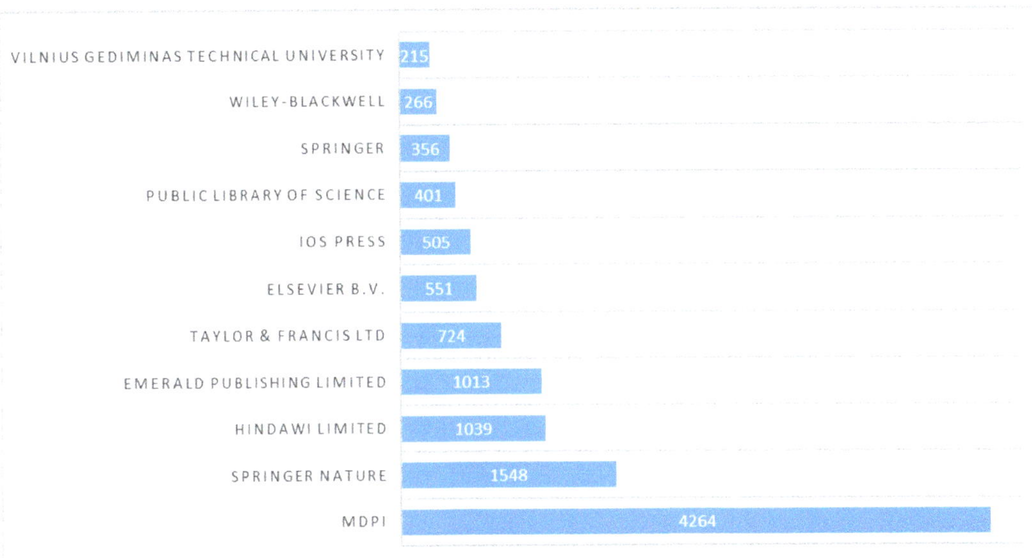

Figure 1. Publishers of articles pertaining to the use of MCDM methods in sustainable engineering.

The journal Sustainability has the most publications on the topic, with 2290 articles, followed by the Journal of Intelligent & Fuzzy Systems and Mathematical Problems in Engineering, with 440 articles each (Figure 2). PLoS ONE and Energies also have a significant number of publications, with 401 and 220 articles, respectively. The topics covered by the publications include environmental management, energy production and consumption, transportation, water management, and quality and reliability management, among others. The use of MCDM methods allows for the consideration of multiple criteria in decision-making, which is considered essential in achieving sustainability in engineering practices.

The assortment of literature pertaining to the application of MCDM methods in sustainable engineering covers a wide range of subjects, as shown in the list of the most frequent keywords (Figure 3). Decision-making and MCDM are the most common subjects, with a total of 1433 and 1318 articles, respectively, followed closely by the analytic hierarchy process (AHP), with 1293 articles. Sustainability, sustainable development, fuzzy sets, and supply chains are also important subjects, with over 500 articles each. Other notable subjects include risk assessment, renewable energy sources, geographic information systems, fuzzy logic, and multi-criteria decision-making. The literature also covers specific applications such as company business management, construction projects, logistics, waste management, water supply, and power resources. The research regarding the use of MCDM methods in sustainable engineering is diverse and covers a wide range of subjects, reflecting the broad scope of sustainable engineering as a field of study.

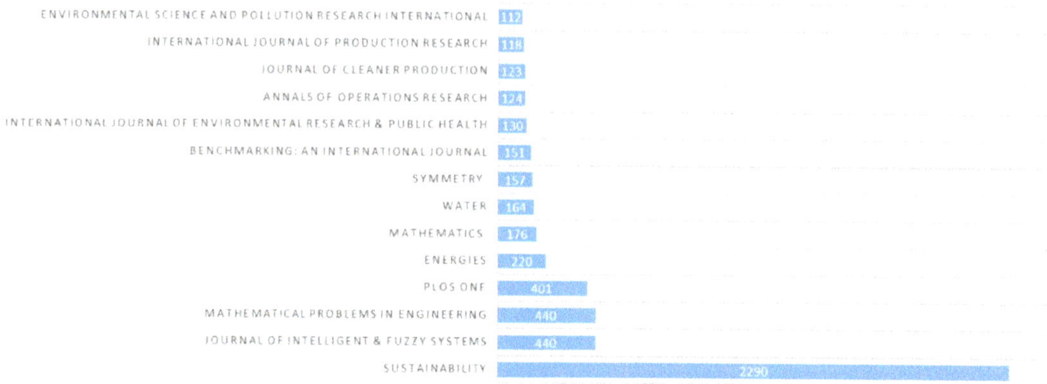

Figure 2. Publications of articles pertaining to the the use of MCDM methods in sustainable engineering.

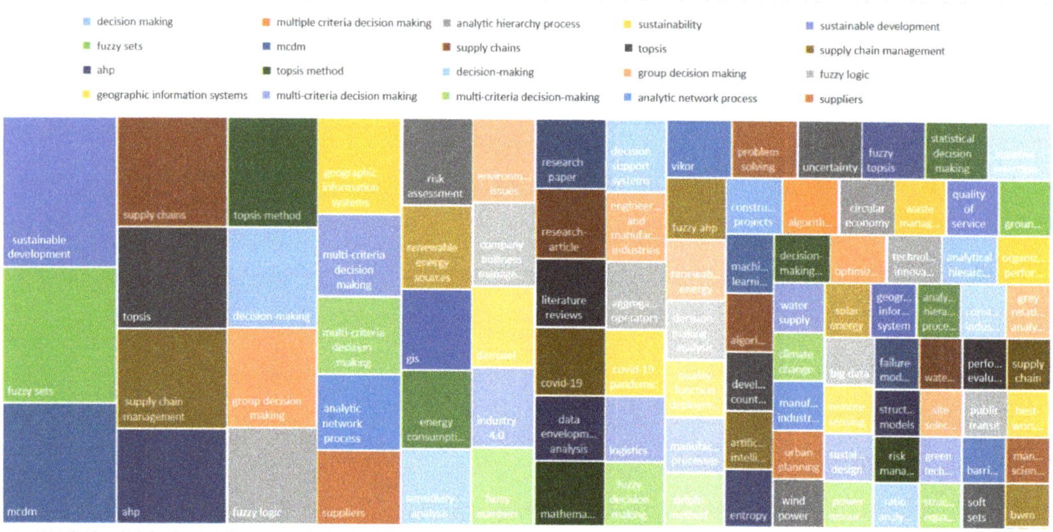

Figure 3. Subjects of articles pertaining to the use of MCDM methods in sustainable engineering.

In sustainable engineering, the researched literature demonstrates that the researchers have frequently employed MCDM methods to address complex decision-making challenges. AHP has been the most frequently used method, with 1986 articles published, followed by TOPSIS (939), ANP (281), and DEMATEL (227). DEMATEL, BWM, and VIKTOR have been used in sustainable engineering, with 227, 174, and 168 articles published, respectively. Finally, Fuzzy sets have been widely used in various fields, with 1471 articles published, and Fuzzy AHP and Fuzzy TOPSIS are also popular. These methods (Figure 4) have been used to address various decision-making problems in sustainable engineering, ranging from environmental management to energy management.

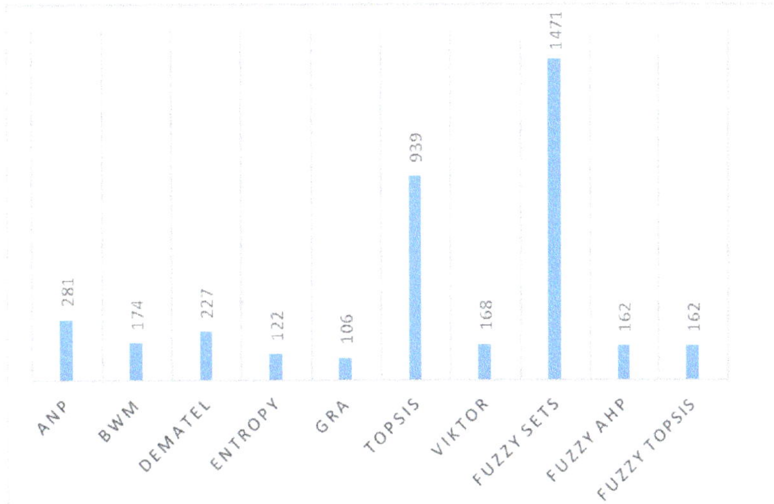

Figure 4. The most commonly used methods in the sustainable engineering articles.

4. Detailed Review Results

4.1. Sustainable Engineering

Sustainable engineering is a multidisciplinary approach [7] to designing and managing engineering systems that meets the demands of present-day society without jeopardizing future generations' ability to meet their specific requirements. The concept of sustainability has its roots in environmentalism and conservationism [11], but it has evolved to encompass social and economic aspects as well [12]. Sustainable engineering considers the environmental, social, and economic impacts of engineering systems throughout their entire life cycle, from design and construction to operation and decommissioning [13,14]. The goal of sustainable engineering is to create systems that are resilient [15–17], adaptive [18], and regenerative [19], and that contribute to the well-being of humans and the planet.

The principles of sustainable engineering include minimizing resource use and waste generation, reducing carbon emissions and other environmental impacts, enhancing social equity and inclusion, promoting economic prosperity and resilience, and embracing systems thinking and innovation [20–22]. Sustainable engineering is essential in contemporary engineering practice as it addresses the challenges of climate change, resource depletion, population growth, and urbanization, and contributes to the fulfillment of the Sustainable Development Goals established by the United Nations [23,24].

The use of MCDM methods in sustainable engineering is motivated by the need to make informed decisions that balance environmental, social, and economic considerations, and that account for the interdependencies of different criteria and stakeholders. MCDM methods provide a systematic and transparent approach to evaluating alternatives and trade-offs, considering multiple criteria and preferences, and identifying the most preferred options [25]. The use of MCDM methods in sustainable engineering has increased in recent years due to advances in computing power, data availability, and stakeholder engagement, as well as the growing recognition of the significance of sustainability in engineering practice [26].

MCDM methodologies have been adopted in various areas of sustainable engineering, including energy systems [27–31], transportation systems [32–36], water and wastewater systems [37–41], building design and construction [42–46], and industrial processes [47–51]. These applications aim to identify the most sustainable options among a set of alternatives, considering various criteria and stakeholders' preferences. For example, different MCDM methods were put to use to identify the most adequate renewable energy technology for a given location, considering technical, economic, environmental, and social criteria [31]. MCDM methods have also been used to identify the sustainability performance of buildings and infrastructure projects, considering criteria such as energy efficiency, carbon emissions, water use, and social and economic impacts [44].

The outcomes of these applications have shown that MCDM methods can provide valuable insights into the sustainability trade-offs and synergies among different criteria and alternatives, and can support informed decision-making that balances environmental, social, and economic considerations. However, the success of MCDM methods in sustainable engineering depends on the quality and availability of data, the validity and reliability of the criteria and indicators used [52–54], and the participation and engagement of stakeholders in the decision-making process [55–57]. To address these challenges, ongoing research is focused on developing more sophisticated MCDM methods that can handle complex and uncertain data, incorporate dynamic and feedback processes, and integrate qualitative and quantitative information [58–62].

4.2. MCDM Methods

MCDM methods are a set of tools used to evaluate alternatives that satisfy multiple criteria or objectives. In sustainable engineering, MCDM methods are widely used to support decision-making processes that involve complex and conflicting criteria such as environmental impact, economic viability, social equity, and technological feasibility. MCDM methods aim to provide an organized and transparent framework for assessing alternatives by considering a variety of criteria and identifying the most preferred alternative.

There are many different MCDM methods, each with its own set of theoretical foundations and applications. Some of the most commonly used MCDM methods in sustainable engineering are listed hereinafter.

The Analytic Hierarchy Process (AHP), developed by Thomas L. Saaty in the 1970s, is a widely used method for decomposing a complex decision problem into a hierarchy of simpler sub-problems and evaluating the relative importance of each criterion and alternative [63]. AHP is particularly useful when the decision problem is complex and involves a large number of criteria and alternatives [64]. AHP has been used in 1986 articles pertaining to the application of MCDM in sustainable engineering, indicating its popularity.

The Technique for Order Preference by Similarity to Ideal Solution (TOPSIS), developed by Hwang and Yoon [65], is a method for ranking alternatives based on their distance from the ideal solution to the worst solution. The TOPSIS has been widely used in various fields, including sustainable engineering, with 939 articles on its application in MCDM methods.

Fuzzy sets are used to represent imprecise and uncertain information in decision-making processes [66]. Fuzzy logic, which is based on fuzzy sets theory, is useful for handling uncertainty and imprecision in decision-making [67]. Fuzzy sets and fuzzy logic have been used in 1471 articles pertaining to the application of MCDM in sustainable engineering. Fuzzy AHP and Fuzzy TOPSIS are two commonly used extensions of AHP and TOPSIS that incorporate fuzzy sets.

The Analytic Network Process (ANP), developed by Saaty [68], is a generalization of the AHP that can model feedback and dependence among criteria and alternatives [69]. The ANP has been used in 281 articles pertaining to the application of MCDM in sustainable engineering.

The Decision-Making Trial and Evaluation Laboratory (DEMATEL), developed by Gabus and Fontela [70], is a method for modeling and analyzing the causal relationships between criteria and alternatives [71]. The DEMATEL has been used in 227 articles pertaining to the application of MCDM in sustainable engineering.

The Best Worst Method (BWM), developed by Rezaei [72], is a method for evaluating and ranking alternatives based on their best and worst performance with respect to a set of criteria [73]. The BWM has been used in 174 articles pertaining to the application of MCDM in sustainable engineering.

The VlseKriterijumska Optimizacija I Kompromisno Resenje (VIKOR) method, published by Duckstein and Opricovic [74], is a method for ranking and selecting alternatives based on their proximity to the ideal and anti-ideal solutions. VIKOR has been used in 168 articles pertaining to the application of MCDM in sustainable engineering.

Grey Relational Analysis (GRA), developed by Deng [75], is a method for analyzing and ranking alternatives based on their similarities and differences with respect to a reference alternative. GRA has been used in 106 articles pertaining to the application of MCDM in sustainable engineering.

The entropy method is a method for weighting criteria based on their relative importance and uncertainty [76]. It has been used in 122 articles pertaining to the application of MCDM in sustainable engineering.

Each MCDM technique has its own strengths and weaknesses, depending on the specific problem and application. The choice of MCDM technique depends on the specific problem and application, as well as the availability of data, and stakeholder preferences [77]. Therefore, it is essential to carefully evaluate the strengths and weaknesses of each technique and select the most appropriate one for the given problem and context.

In addition to the specific problem and application, the availability of data and expertise can also influence the choice of MCDM technique. For example, some MCDM methods, such as AHP and TOPSIS, require pairwise comparison matrices [78] that may be difficult to obtain or may involve subjective judgments. Other methods, such as fuzzy logic and entropy, can handle uncertain and imprecise information, but may require significant expertise in fuzzy set theory or information theory [79]. Therefore, it is important to consider the availability and quality of data and expertise when selecting an MCDM technique.

Stakeholder preferences can also influence the choice of MCDM methods. Different methods may be more suitable for different types of stakeholders or decision contexts. For example, PROMETHEE and ELECTRE are particularly useful for handling conflicting preferences and priorities among stakeholders [80], while fuzzy logic can be useful for representing vague or ambiguous preferences [81]. Therefore, it is important to include stakeholders in the decision-making process and consider their preferences and perspectives when selecting an MCDM technique. By carefully evaluating the strengths and weaknesses of each technique and selecting the most appropriate one for the given problem and context, decision-makers can make more informed and effective decisions that balance multiple criteria and objectives.

4.3. Case Studies: Applications of MCDM Methods in Sustainable Engineering

In this section, exemplary case studies of real-world applications of MCDM methods in sustainable engineering are presented. The presented case studies aim to provide a deeper understanding of how MCDM methods could be utilized to address complex decision-making problems in different areas of sustainable engineering.

MCDM methods have been frequently employed in various areas of sustainable engineering to support decision-making that considers multiple criteria and stakeholders' preferences. In this section, we will discuss some specific examples of MCDM applications in energy, manufacturing, transportation, and environmental engineering, highlighting the criteria considered and the outcomes achieved.

For example, energy engineering is an area where MCDM methods have been extensively utilized to determine the most sustainable options among different renewable and non-renewable energy sources [82–84], considering technical, economic, environmental, and social criteria. For instance, MCDM methods have been used to select the most appropriate renewable energy technology for a given location, taking into account factors such as resource availability, technical feasibility, economic viability, and social acceptance. These methods have been applied to various renewable energy sources, including solar, wind, geothermal, and hydroelectric power. The outcomes of these applications have shown that MCDM methods can provide valuable insights into the trade-offs among different criteria and help identify the most sustainable options. Additionally, a study by

Alhakami [85] addresses the need for a comprehensive security evaluation approach and proposes an MCDM methodology to assess security risks in power control technology and communication networks of energy management and control systems.

Manufacturing and production engineering is another area in which MCDM methods have been used to support sustainable decision-making. For example, MCDM methods have been used to evaluate the sustainability performance of manufacturing processes, considering criteria such as energy efficiency, waste generation, water use, and social and economic impacts [86–88]. These methods have also been applied to support product design and development [89–91], considering criteria such as material selection, energy consumption, and end-of-life disposal. The outcomes of these applications have shown that MCDM methods can help to identify the most sustainable manufacturing processes and products and support the transition towards a circular economy. Furthermore, applying these methods in the specific context [92] can enhance sustainable decision-making practices.

Transportation engineering is a critical area for sustainable engineering as transportation systems are responsible for a significant portion of greenhouse gas emissions and other environmental impacts [93]. MCDM methods have been used to support decision-making in transportation engineering [94–96], considering criteria such as energy efficiency, emissions reduction, safety, and social and economic impacts. For example, MCDM methods have been used to evaluate the sustainability performance of different modes of transportation, such as cars [97], buses [98], trains [99], and airplanes [100], and to identify the most sustainable options for a given transportation problem. These methods can also be applied to support the design and planning of transportation infrastructure, such as roads [101], bridges [102], and airports [103], considering criteria such as energy consumption, environmental impacts, and social and economic benefits, and even assess potential suppliers based on their ability to address specific challenges such as the COVID-19 epidemic [104]. Furthermore, MCDM methodologies are also proposed to monitor customer satisfaction in the airline service industry, aiming to enhance service quality and meet consumer expectations [105].

Environmental engineering is a broad area that encompasses various disciplines, such as water and wastewater treatment, air pollution control, and solid waste management. MCDM methods have been used to support decision-making in environmental engineering, considering criteria such as environmental impacts, economic costs, and social benefits. For example, MCDM methods have been used to identify the most sustainable options for water and wastewater treatment [106–110], considering criteria such as treatment efficiency, energy consumption, and social acceptance. These methods can also be applied to support the management of solid waste [111], considering criteria such as waste reduction, recycling, and disposal options. In a similar manner, the evaluation of environmental quality in specific contexts [112] utilizes fuzzy MCDM methods incorporating multiple factors to guide decision-making in environmental protection research and future renovation planning, and the compatibility between MCDM methods in assessing erosion risk highlights the fuzzy methods as an effective tool for evaluating erosion risk in semi-arid areas and guiding erosion prevention actions [113].

The applications of MCDM methods in sustainable engineering have shown that these methods can provide valuable insights into the sustainability trade-offs and synergies among different criteria and alternatives, and can support informed decision-making that balances environmental, social, and economic considerations. However, the success of these applications depends on the quality and availability of data, the validity and reliability of the criteria and indicators used, and the participation and engagement of stakeholders in the decision-making process.

5. Challenges and Future Directions

MCDM methods face several challenges when applied to sustainable engineering problems. One of the key challenges of MCDM methods is the availability and quality of data. Sustainable engineering problems often involve multiple criteria and sources of information [114], and it can be difficult to obtain reliable data that represent the complexity of the problem. In addition, the data may be incomplete, inconsistent, or subjective, which can affect the decision-making process' reliability and accuracy [115].

Another challenge of MCDM methods is model uncertainty. Many of these methods are based on mathematical models that may not accurately reflect the complexity and dynamics of sustainable engineering problems [116]. This can lead to errors in the estimation of criteria weights, rankings, and overall scores, which can affect the credibility and acceptability of the decision-making process.

MCDM methods also face challenges related to stakeholder engagement. Sustainable engineering problems often involve multiple stakeholders with different perspectives, values, and interests [117]. It can be difficult to engage stakeholders effectively in the decision-making process and to ensure that their voices are heard and their concerns are addressed. In addition, stakeholders may have different levels of expertise and understanding of the decision-making process and the MCDM approaches employed, which can affect the quality and acceptability of the decision.

Despite these challenges, MCDM methods have an opportunity to be vital in sustainable engineering practice. One promising research direction is the blend of MCDM methods, artificial intelligence (AI), and machine learning (ML) methods. AI and ML may enhance the accuracy and efficiency of the decision-making process by enabling the automated processing and analysis of large and complex data sets [118]. The combination of MCDM methods with AI and ML can also facilitate the incorporation of expert knowledge, uncertainty, and risk into the decision-making process.

Another future research direction is the incorporation of dynamic and complex systems into the decision-making process. Many sustainable engineering problems involve complex systems that are characterized by non-linear relationships, feedback loops, and emergent properties. MCDM methods can be further developed to account for these complexities by incorporating methods such as system dynamics, agent-based modeling, and network analysis.

A third future research direction is the enhancement of multi-stakeholder decision-making. This involves developing MCDM methods that can facilitate effective stakeholder engagement by incorporating methods such as participatory decision-making, collaborative modeling, and multi-criteria deliberation. The development of user-friendly and transparent decision support tools can also help to enhance stakeholder engagement and improve the acceptability the process of making decisions.

Finally, upcoming research can concentrate on the development of user-friendly and transparent decision support tools. MCDM methods can be complex and difficult to understand for non-experts, which can limit their use in practice. User-friendly and transparent decision support tools can help to bridge this gap by providing intuitive and accessible interfaces, visualizations, and explanations.

MCDM methods have an important function in encouraging sustainable engineering practices. However, to realize their full potential, it is essential to address the challenges and limitations they face and explore new research directions that can enhance their effectiveness and applicability in real-world decision-making contexts.

6. Conclusions

This review has provided a comprehensive overview of the applications of MCDM methods in sustainable engineering. The review discussed the theoretical foundations and applications of various MCDM methods, including their strengths, weaknesses, and comparisons. It also highlighted the importance of sustainable engineering and discussed the different areas in which MCDM methods have been applied, such as energy, manufacturing, transportation, and environmental engineering. Furthermore, this review presented case studies of real-world applications of MCDM methods in sustainable engineering and analyzed the main findings and implications for engineering practice. Finally, the review discussed the challenges and limitations of MCDM methods in sustainable engineering and proposed future research directions to enhance their effectiveness and applicability.

The review has demonstrated that MCDM methods have the potential to address complex decision-making problems in sustainable engineering by considering multiple criteria and stakeholder perspectives. However, the effective implementation of these methods requires issues related to data availability, model uncertainty, and stakeholder engagement to be addressed. Future research directions include the development of more robust and transparent MCDM models, the integration of new data sources, and the incorporation of emerging technologies such as artificial intelligence and machine learning. The findings of this review have important implications for engineering practice and research and can inform the development of more sustainable and efficient engineering solutions in the future.

Author Contributions: Conceptualization, A.Š. and A.P.; methodology, A.Š.; writing—original draft preparation, A.Š. and A.P.; writing—review and editing, A.Š. and A.P.; visualization, A.Š.; supervision, A.P.; project administration, A.Š. and A.P.; funding acquisition, A.Š. All authors have read and agreed to the published version of the manuscript.

Funding: This research received no external funding.

Institutional Review Board Statement: Not applicable.

Informed Consent Statement: Not applicable.

Data Availability Statement: Data are contained within the article.

Conflicts of Interest: The authors declare no conflict of interest.

References

1. Stojcic, M.; Zavadskas, E.K.; Pamučar, D.; Stević, Ž.; Mardani, A. Application of MCDM Methods in Sustainability Engineering: A Literature Review 2008–2018. *Symmetry* **2019**, *11*, 350. [CrossRef]
2. Banasik, A.; Bloemhof-Ruwaard, J.M.; Kanellopoulos, A.; Claassen, G.; Van Der Vorst, J.G. Multi-criteria decision making approaches for green supply chains: A review. *Flex. Serv. Manuf. J.* **2018**, *30*, 366–396. [CrossRef]
3. Guarnieri, P.; Trojan, F. Decision making on supplier selection based on social, ethical, and environmental criteria: A study in the textile industry. *Resour. Conserv. Recycl.* **2019**, *141*, 347–361. [CrossRef]
4. Kharat, M.G.; Murthy, S.; Kamble, S.J.; Raut, R.D.; Kamble, S.S.; Kharat, M.G. Fuzzy multi-criteria decision analysis for environmentally conscious solid waste treatment and disposal technology selection. *Technol. Soc.* **2019**, *57*, 20–29. [CrossRef]
5. Alhama, C.C.; Igual-Antón, D. Corporate Social Responsibility Strategies in Spanish Electric Cooperatives. Analysis of Stakeholder Engagement. *Sustainability* **2021**, *13*, 6810. [CrossRef]
6. Geng, Y.; Fujita, T.; Bleischwitz, R.; Chiu, A.S.; Sarkis, J. Accelerating the transition to equitable, sustainable, and livable cities: Toward post-fossil carbon societies. *J. Clean. Prod.* **2019**, *239*, 118020. [CrossRef]
7. Glavič, P. Updated Principles of Sustainable Engineering. *Processes* **2022**, *10*, 870. [CrossRef]
8. Durmić, E.; Herzegovina Stević, Ž.; Chatterjee, P.; Vasiljević, M.; Tomašević, M. Sustainable supplier selection using combined FUCOM—Rough SAW model. *Rep. Mech. Eng.* **2020**, *1*, 34–43. [CrossRef]
9. Raut, R.D.; Kharat, M.G.; Kamble, S.J.; Kumar, C.M. Sustainable evaluation and selection of potential third-party logistics (3PL) providers. *Benchmarking Int. J.* **2018**, *25*, 76–97. [CrossRef]
10. Tang, M.; Liao, H. From conventional group decision making to large-scale group decision making: What are the challenges and how to meet them in big data era? A state-of-the-art survey. *Omega* **2021**, *100*, 102141. [CrossRef]
11. Restrepo, J.M.; Morales-Pinzón, T. Urban metabolism and sustainability: Precedents, genesis and research perspectives. *Resour. Conserv. Recycl.* **2018**, *131*, 216–224. [CrossRef]
12. Toli, A.M.; Murtagh, N. The Concept of Sustainability in Smart City Definitions. *Front. Built Environ.* **2020**, *6*, 77. [CrossRef]
13. Iskandar, M.; Nelson, D.; Tehrani, F.M. Managing Sustainability and Resilience of the Built Environment in Developing Communities. *CivilEng* **2022**, *3*, 427–440. [CrossRef]
14. Jamieson, M.V.; Lefsrud, L.; Sattari, F.; Donald, J.A. Sustainable leadership and management of complex engineering systems: A team based structured case study approach. *Educ. Chem. Eng.* **2021**, *35*, 37–46. [CrossRef]
15. Marchese, D.; Reynolds, E.; Bates, M.; Morgan, H.; Clark, S.J.; Linkov, I. Resilience and sustainability: Similarities and differences in environmental management applications. *Sci. Total Environ.* **2018**, *613–614*, 1275–1283. [CrossRef]
16. Mihelcic, J.R.; Zimmerman, J.B. *Environmental Engineering: Fundamentals, Sustainability, Design*; John Wiley & Sons: Hoboken, NJ, USA, 2021.

17. Sharma, N.; Tabandeh, A.; Gardoni, P. Resilience analysis: A mathematical formulation to model resilience of engineering systems. *Sustain. Resilient Infrastruct.* **2018**, *3*, 49–67. [CrossRef]
18. Thacker, S.; Adshead, D.; Fay, M.; Hallegatte, S.; Harvey, M.S.; Meller, H.; O'Regan, N.; Rozenberg, J.; Watkins, G.; Hall, J.W. Infrastructure for sustainable development. *Nat. Sustain.* **2019**, *2*, 324–331. [CrossRef]
19. Suárez-Eiroa, B.; Fernández, E.M.; Méndez-Martínez, G.; Soto-Oñate, D. Operational principles of circular economy for sustainable development: Linking theory and practice. *J. Clean. Prod.* **2019**, *214*, 952–961. [CrossRef]
20. Dogaru, L. Green Economy and Green Growth—Opportunities for Sustainable Development. *Proceedings* **2021**, *63*, 70. [CrossRef]
21. Lehmann, S. Implementing the Urban Nexus approach for improved resource-efficiency of developing cities in Southeast-Asia. *City Cult. Soc.* **2017**, *13*, 46–56. [CrossRef]
22. Sachs, J.D.; Schmidt-Traub, G.; Mazzucato, M.; Messner, D.; Nakicenovic, N.; Rockström, J. Six Transformations to achieve the Sustainable Development Goals. *Nat. Sustain.* **2019**, *2*, 805–814. [CrossRef]
23. Štilić, A.; Puška, A.; Đurić, A.; Božanić, D.K. Electric Vehicles Selection Based on Brčko District Taxi Service Demands, a Multi-Criteria Approach. *Urban Sci.* **2022**, *6*, 73. [CrossRef]
24. United Nations [UN]. THE 17 GOALS | Sustainable Development. United Nations, Department of Economic and Social Affairs, Sustainable Development. Available online: https://sdgs.un.org/goals (accessed on 2 May 2023).
25. Bhardwaj, A.; Joshi, M.; Khosla, R.; Dubash, N.K. More priorities, more problems? Decision-making with multiple energy, development and climate objectives. *Energy Res. Soc. Sci.* **2019**, *49*, 143–157. [CrossRef]
26. Tseng, M.; Tran, T.H.; Ha, H.M.; Bui, T.; Lim, M.K. Sustainable industrial and operation engineering trends and challenges Toward Industry 4.0: A data driven analysis. *J. Ind. Prod. Eng.* **2021**, *38*, 581–598. [CrossRef]
27. Bohra, S.S.; Shafie-Khah, M. A comprehensive review on applications of multicriteria decision-making methods in power and energy systems. *Int. J. Energy Res.* **2021**, *46*, 4088–4118. [CrossRef]
28. Cao, Q.; Esangbedo, M.O.; Bai, S.; Esangbedo, C.O. Grey SWARA-FUCOM Weighting Method for Contractor Selection MCDM Problem: A Case Study of Floating Solar Panel Energy System Installation. *Energies* **2019**, *12*, 2481. [CrossRef]
29. Jahangiri, M.; Shamsabadi, A.A.; Mostafaeipour, A.; Rezaei, M.; Yousefi, Y.; Pomares, L.A. Using fuzzy MCDM technique to find the best location in Qatar for exploiting wind and solar energy to generate hydrogen and electricity. *Int. J. Hydrogen Energy* **2020**, *45*, 13862–13875. [CrossRef]
30. Lee, H.; Chang, C. Comparative analysis of MCDM methods for ranking renewable energy sources in Taiwan. *Renew. Sustain. Energy Rev.* **2018**, *92*, 883–896. [CrossRef]
31. Siksnelyte-Butkiene, I.; Zavadskas, E.K.; Streimikiene, D. Multi-Criteria Decision-Making (MCDM) for the Assessment of Renewable Energy Technologies in a Household: A Review. *Energies* **2020**, *13*, 1164. [CrossRef]
32. Ferreira, J.J.; Ilander, G.O.P.; Ferreira, J.J. MCDM/A in practice: Methodological developments and real-world applications. *Manag. Decis.* **2019**, *57*, 295–299. [CrossRef]
33. Görçün, Ö.F. Evaluation of the selection of proper metro and tram vehicle for urban transportation by using a novel integrated MCDM approach. *Sci. Prog.* **2021**, *104*, 003685042095012. [CrossRef] [PubMed]
34. Kiciński, M.; Solecka, K. Application of MCDA/MCDM methods for an integrated urban public transportation system—Case study, city of Cracow. *Arch. Transp.* **2018**, *46*, 71–84. [CrossRef]
35. Moradi, S.; Sierpiński, G.; Masoumi, H.E. System Dynamics Modeling and Fuzzy MCDM Approach as Support for Assessment of Sustainability Management on the Example of Transport Sector Company. *Energies* **2022**, *15*, 4917. [CrossRef]
36. Wang, C.; Le, T.; Chang, K.; Dang, T. Measuring Road Transport Sustainability Using MCDM-Based Entropy Objective Weighting Method. *Symmetry* **2022**, *14*, 1033. [CrossRef]
37. Ali, Y.; Pervez, H.; Khan, J. Selection of the Most Feasible Wastewater Treatment Technology in Pakistan Using Multi-Criteria Decision-Making (MCDM). *Water Conserv. Sci. Eng.* **2020**, *5*, 199–213. [CrossRef]
38. Gichamo, T.; Gökçekuş, H.; Ozsahin, D.U.; Gelete, G.; Uzun, B. Ranking of Natural Wastewater Treatment Techniques by Multi-criteria Decision Making (MCDM) Methods. In *Professional Practice in Earth Sciences*; Springer International Publishing: Berlin/Heidelberg, Germany, 2021; pp. 87–100. [CrossRef]
39. Narayanamoorthy, S.; Brainy, J.V.; Sulaiman, R.; Ferrara, M.; Ahmadian, A.; Kang, D. An integrated decision making approach for selecting a sustainable waste water treatment technology. *Chemosphere* **2022**, *301*, 134568. [CrossRef]
40. Radmehr, A.; Bozorg-Haddad, O.; Loáiciga, H.A. Developing Strategies for Agricultural Water Management of Large Irrigation and Drainage Networks with Fuzzy MCDM. *Water Resour. Manag.* **2022**, *36*, 4885–4912. [CrossRef]
41. Zolfaghary, P.; Zakerinia, M.; Kazemi, H. A model for the use of urban treated wastewater in agriculture using multiple criteria decision making (MCDM) and geographic information system (GIS). *Agric. Water Manag.* **2021**, *243*, 106490. [CrossRef]
42. Chalekaee, A.; Turskis, Z.; Khanzadi, M.; Amiri, G.G.; Keršulienė, V. A New Hybrid MCDM Model with Grey Numbers for the Construction Delay Change Response Problem. *Sustainability* **2019**, *11*, 776. [CrossRef]
43. Haruna, A.; Shafiq, N.; Montasir, O. Building information modelling application for developing sustainable building (Multi criteria decision making approach). *Ain Shams Eng. J.* **2021**, *12*, 293–302. [CrossRef]
44. Mathiyazhagan, K.; Gnanavelbabu, A.; Prabhuraj, B.L. A sustainable assessment model for material selection in construction industries perspective using hybrid MCDM approaches. *J. Adv. Manag. Res.* **2019**, *16*, 234–259. [CrossRef]
45. Matić, B.; Jovanovic, S.; Das, D.K.; Zavadskas, E.K.; Stević, Ž.; Sremac, S.; Marinković, M. A New Hybrid MCDM Model: Sustainable Supplier Selection in a Construction Company. *Symmetry* **2019**, *11*, 353. [CrossRef]

46. Zolfani, S.H.; Pourhossein, M.; Yazdani, M.; Zavadskas, E.K. Evaluating construction projects of hotels based on environmental sustainability with MCDM framework. *Alex. Eng. J.* **2017**, *57*, 357–365. [CrossRef]
47. Chandra, M.; Shahab, F.; Kek, V.; Rajak, S. Selection for additive manufacturing using hybrid MCDM technique considering sustainable concepts. *Rapid Prototyp. J.* **2022**, *28*, 1297–1311. [CrossRef]
48. Marhavilas, P.K.; Filippidis, M.; Koulinas, G.K.; Koulouriotis, D.E. A HAZOP with MCDM Based Risk-Assessment Approach: Focusing on the Deviations with Economic/Health/Environmental Impacts in a Process Industry. *Sustainability* **2020**, *12*, 993. [CrossRef]
49. Nguyen, T.H.O.; Nguyen, P.H.; Pham, H.T.; Nguyen, T.; Nguyen, D.K.; Tran, T.; Le, H.; Phung, H. A Novel Integrating Data Envelopment Analysis and Spherical Fuzzy MCDM Approach for Sustainable Supplier Selection in Steel Industry. *Mathematics* **2022**, *10*, 1897. [CrossRef]
50. Nguyen, V.T. Sustainable Energy Source Selection for Industrial Complex in Vietnam: A Fuzzy MCMD Approach. *IEEE Access* **2022**, *10*, 50692–50701. [CrossRef]
51. Van Thanh, N.; Lan, N.P.H. A New Hybrid Triple Bottom Line Metrics and Fuzzy MCDM Model: Sustainable Supplier Selection in the Food-Processing Industry. *Axioms* **2022**, *11*, 57. [CrossRef]
52. Balaei, B.; Wilkinson, S.; Potangaroa, R.; Hassani, N.; Alavi-Shoshtari, M. Developing a Framework for Measuring Water Supply Resilience. *Nat. Hazards Rev.* **2018**, *19*, 04018013. [CrossRef]
53. Bhat, S.; Antony, J.; Gijo, E.; Cudney, E.A. Lean Six Sigma for the healthcare sector: A multiple case study analysis from the Indian context. *Int. J. Qual. Reliab. Manag.* **2019**, *37*, 90–111. [CrossRef]
54. Yontar, E. Assessment of the logistics activities with a structural model on the basis of improvement of sustainability performance. *Environ. Sci. Pollut. Res.* **2022**, *29*, 68904–68922. [CrossRef]
55. Fritz, M.M.; Rauter, R.; Baumgartner, R.J.; Dentchev, N. A supply chain perspective of stakeholder identification as a tool for responsible policy and decision-making. *Environ. Sci. Policy* **2018**, *81*, 63–76. [CrossRef]
56. Reed, M.; Vella, S.; Challies, E.; De Vente, J.; Frewer, L.; Hohenwallner-Ries, D.; Huber, T.B.; Neumann, R.K.; Oughton, E.; Del Ceno, J.S.; et al. A theory of participation: What makes stakeholder and public engagement in environmental management work? *Restor. Ecol.* **2018**, *26*, S7–S17. [CrossRef]
57. Sharpe, L.M.; Harwell, M.C.; Jackson, C.A. Integrated stakeholder prioritization criteria for environmental management. *J. Environ. Manag.* **2021**, *282*, 111719. [CrossRef]
58. Chai, N.; Zhou, W. A novel hybrid MCDM approach for selecting sustainable alternative aviation fuels in supply chain management. *Fuel* **2022**, *327*, 125180. [CrossRef]
59. Chung, H.; Chang, K. A Novel General Data Envelopment Analysis Based Approach for MCDM Issues of Hydrogen Energy under a Fuzzy Environment. *Systems* **2022**, *10*, 176. [CrossRef]
60. Le, M.D.; Nhieu, N. An Offshore Wind–Wave Energy Station Location Analysis by a Novel Behavioral Dual-Side Spherical Fuzzy Approach: The Case Study of Vietnam. *Appl. Sci.* **2022**, *12*, 5201. [CrossRef]
61. Wang, C.; Dang, T.; Nguyen, N.; Chou, C.; Hsu, H.; Dang, L. Evaluating Global Container Shipping Companies: A Novel Approach to Investigating Both Qualitative and Quantitative Criteria for Sustainable Development. *Axioms* **2022**, *11*, 610. [CrossRef]
62. Wang, X.; Zhang, C.; Deng, J.; Su, C.; Gao, Z. Analysis of Factors Influencing Miners' Unsafe Behaviors in Intelligent Mines using a Novel Hybrid MCDM Model. *Int. J. Environ. Res. Public Health* **2022**, *19*, 7368. [CrossRef]
63. Štilić, A.; Njeguš, A. Primena metoda višekriterijumske analize u odabiru kandidata za rad u turističkoj privredi. In *Sinteza 2019—International Scientific Conference on Information Technology and Data Related Research*; Singidunum University: Belgrad, Serbia, 2019. [CrossRef]
64. Tuljak-Suban, D.; Bajec, P. Integration of AHP and GTMA to Make a Reliable Decision in Complex Decision-Making Problems: Application of the Logistics Provider Selection Problem as a Case Study. *Symmetry* **2020**, *12*, 766. [CrossRef]
65. Hwang, C.; Yoon, K. *Multiple Attribute Decision Making: Methods and Applications a State-of-the-Art Survey*; Springer Science & Business Media: Berlin/Heidelberg, Germany, 2012.
66. Garg, H. Linguistic Pythagorean fuzzy sets and its applications in multiattribute decision-making process. *Int. J. Intell. Syst.* **2018**, *33*, 1234–1263. [CrossRef]
67. Tavana, M.; Hajipour, V. A practical review and taxonomy of fuzzy expert systems: Methods and applications. *Benchmarking Int. J.* **2019**, *27*, 81–136. [CrossRef]
68. Saaty, T.L. *Decision Making with Dependence and Feedback: The Analytic Network Process: The Organization and Prioritization of Complexity*; Rws Publications: Pittsburgh, PA, USA, 1996.
69. Jayawardena, T.S.S.; Jayawardena, C.S. Analytical Network Process in Decision Making. In *Advances in Knowledge Acquisition, Transfer and Management Book Series*; Springer Nature: Berlin/Heidelberg, Germany, 2023; pp. 180–196. [CrossRef]
70. Gabus, A.; Fontela, E. *World Problems an Invitation to Further Thought within the Framework of DEMATEL*; Battelle Geneva Research Centre: Geneva, Switzerland, 1972.
71. Sang, X.; Yu, X.; Chang, C.; Liu, X. Electric bus charging station site selection based on the combined DEMATEL and PROMETHEE-PT framework. *Comput. Ind. Eng.* **2022**, *168*, 108116. [CrossRef]
72. Rezaei, J. Best-worst multi-criteria decision-making method. *Omega* **2015**, *53*, 49–57. [CrossRef]

73. Mohammadi, M.; Rezaei, J. Bayesian best-worst method: A probabilistic group decision making model. *Omega* **2020**, *96*, 102075. [CrossRef]
74. Duckstein, L.; Opricovic, S. Multiobjective optimization in river basin development. *Water Resour. Res.* **1980**, *16*, 14–20. [CrossRef]
75. Deng, J. Introduction to Grey system theory. *J. Grey Syst.* **1989**, *1*, 1–24.
76. Brito-Parada, P.R. A multiple criteria decision making method to weight the sustainability criteria of renewable energy technologies under uncertainty. *Renew. Sustain. Energy Rev.* **2020**, *127*, 109891. [CrossRef]
77. Ghaleb, A.M.; Kaid, H.; Al-Samhan, A.M.; Mian, S.H.; Hidri, L. Assessment and Comparison of Various MCDM Approaches in the Selection of Manufacturing Process. *Adv. Mater. Sci. Eng.* **2020**, *2020*, 4039253. [CrossRef]
78. Shaikh, S.A.; Memon, M.A.; Prokop, M.; Kim, K. An AHP/TOPSIS-Based Approach for an Optimal Site Selection of a Commercial Opening Utilizing GeoSpatial Data. In Proceedings of the International Conference on Big Data and Smart Computing, Busan, Republic of Korea, 19–22 February 2020. [CrossRef]
79. Gül, S.; Aydoğdu, A. Novel distance and entropy definitions for linear Diophantine fuzzy sets and an extension of TOPSIS (LDF-TOPSIS). *Expert Syst.* **2022**, *40*, e13104. [CrossRef]
80. Kokaraki, N.; Hopfe, C.J.; Robinson, E.P.; Nikolaidou, E. Testing the reliability of deterministic multi-criteria decision-making methods using building performance simulation. *Renew. Sustain. Energy Rev.* **2019**, *112*, 991–1007. [CrossRef]
81. Salim, F.S.; Bakar, Z.A.; Noor, N.M.M.; Mohemad, R.; Sabri, I.A.A. Aesthetic user interfaces ranking using fuzzy analytic hierarchy process (FAHP) approach. In *AIP Conference Proceedings*; American Institute of Physics: College Park, MD, USA, 2023. [CrossRef]
82. Asakereh, A.; Soleymani, M.; Ardebili, S.M.S. Multi-criteria evaluation of renewable energy technologies for electricity generation: A case study in Khuzestan province, Iran. *Sustain. Energy Technol. Assess.* **2022**, *52*, 102220. [CrossRef]
83. Bilgili, F.; Zarali, F.; Ilgün, M.F.; Dumrul, C.; Dumrul, Y. The evaluation of renewable energy alternatives for sustainable development in Turkey using intuitionistic fuzzy-TOPSIS method. *Renew. Energy* **2022**, *189*, 1443–1458. [CrossRef]
84. Yazdani, H.; Baneshi, M.; Yaghoubi, M. Techno-economic and environmental design of hybrid energy systems using multi-objective optimization and multi-criteria decision making methods. *Energy Convers. Manag.* **2023**, *282*, 116873. [CrossRef]
85. Alhakami, W. Computational Study of Security Risk Evaluation in Energy Management and Control Systems Based on a Fuzzy MCDM Method. *Processes* **2023**, *11*, 1366. [CrossRef]
86. Favi, C.; Marconi, M.; Mandolini, M.; Germani, M. Sustainable life cycle and energy management of discrete manufacturing plants in the industry 4.0 framework. *Appl. Energy* **2022**, *312*, 118671. [CrossRef]
87. Ghosh, S.; Mandal, M.C.; Ray, A. Strategic sourcing model for green supply chain management: An insight into automobile manufacturing units in India. *Benchmarking Int. J.* **2021**, *29*, 3097–3132. [CrossRef]
88. Saeidi, P.; Mardani, A.; Mishra, A.R.; Cajas, V.E.C.; Carvajal, M.G. Evaluate sustainable human resource management in the manufacturing companies using an extended Pythagorean fuzzy SWARA-TOPSIS method. *J. Clean. Prod.* **2022**, *370*, 133380. [CrossRef]
89. Batwara, A.; Sharma, V.; Makkar, M.; Giallanza, A. An Empirical Investigation of Green Product Design and Development Strategies for Eco Industries Using Kano Model and Fuzzy AHP. *Sustainability* **2022**, *14*, 8735. [CrossRef]
90. Feng, C.; Huang, Y.; Chen, X. Sustainable Design for Transforming Sustainability Requirements to Design Parameters Based on Multi-criteria Decision-Making Methodology. In *Mechanisms and Machine Science*; Springer Nature: Dordrecht, The Netherlands, 2022; pp. 933–959. [CrossRef]
91. Hameed, A.; Sultan, M.T.H.; Raj, S.A.; Baghdadi, M.A.; Shahzad, M. Sustainable Product Development Using FMEA ECQFD TRIZ and Fuzzy TOPSIS. *Sustainability* **2022**, *14*, 14345. [CrossRef]
92. Keshavarz-Ghorabaee, M. Sustainable Supplier Selection and Order Allocation Using an Integrated ROG-Based Type-2 Fuzzy Decision-Making Approach. *Mathematics* **2023**, *11*, 2014. [CrossRef]
93. Oladunni, O.J.; Mpofu, K.; Olanrewaju, O.A. Greenhouse gas emissions and its driving forces in the transport sector of South Africa. *Energy Rep.* **2022**, *8*, 2052–2061. [CrossRef]
94. Kokkinos, K.; Nathanail, E.; Gerogiannis, V.C.; Moustakas, K.; Karayannis, V. Hydrogen storage station location selection in sustainable freight transportation via intuitionistic hesitant decision support system. *Energy* **2022**, *260*, 125008. [CrossRef]
95. Saraji, M.K.; Streimikiene, D.; Čiegis, R. A novel Pythagorean fuzzy-SWARA-TOPSIS framework for evaluating the EU progress towards sustainable energy development. *Environ. Monit. Assess.* **2021**, *194*, 42. [CrossRef]
96. Wei, Q.; Zhou, C. A multi-criteria decision-making framework for electric vehicle supplier selection of government agencies and public bodies in China. *Environ. Sci. Pollut. Res.* **2022**, *30*, 10540–10559. [CrossRef]
97. Peng, X.; Huang, H.; Luo, Z. Fuzzy dynamic MCDM method based on PRSRV for financial risk evaluation of new energy vehicle industry. *Appl. Soft Comput.* **2023**, *136*, 110115. [CrossRef]
98. Goyal, S.; Agarwal, S.; Singh, N.; Mathur, T.; Mathur, N. Analysis of Hybrid MCDM Methods for the Performance Assessment and Ranking Public Transport Sector: A Case Study. *Sustainability* **2022**, *14*, 15110. [CrossRef]
99. Liu, Z.; Zhang, Y. Comprehensive Sustainable Assessment and Prioritization of Different Railway Projects Based on a Hybrid MCDM Model. *Sustainability* **2022**, *14*, 12065. [CrossRef]
100. Markatos, D.N.; Pantelakis, S.G. Implementation of a Holistic MCDM-Based Approach to Assess and Compare Aircraft, under the Prism of Sustainable Aviation. *Aerospace* **2023**, *10*, 240. [CrossRef]

101. Chen, F.; Li, Y.; Feng, Q.; Dong, Z.; Qian, Y.; Yan, Y.; Ho, M.S.; Ma, Q.; Zhang, D.; Jin, Y. Road safety performance rating through PSI-PRIDIT: A planning tool for designing policies and identifying best practices for EAS countries. *Socio-Econ. Plan. Sci.* **2022**, *85*, 101438. [CrossRef]
102. Khan, S.; Kabir, G.; Billah, M.; Dutta, S. An integrated framework for bridge infrastructure resilience analysis against seismic hazard. *Sustain. Resilient Infrastruct.* **2022**, *8* (Suppl. 1), 5–25. [CrossRef]
103. Badi, I.; Alosta, A.; Elmansouri, O.; Abdulshahed, A.M.; Elsharief, S. An application of a novel grey-CODAS method to the selection of hub airport in North Africa. *Decis. Mak.* **2023**, *6*, 18–33. [CrossRef]
104. Dang, T.; Nguyen, N.; Nguyen, V.T.; Dang, L. A Two-Stage Multi-Criteria Supplier Selection Model for Sustainable Automotive Supply Chain under Uncertainty. *Axioms* **2022**, *11*, 228. [CrossRef]
105. Awadh, M.A. Assessing the Quality of Sustainable Airline Services Utilizing the Multicriteria Decision-Making Approach. *Sustainability* **2023**, *15*, 7044. [CrossRef]
106. Chaisar, M.; Garg, S.K. Selection of Sewage Treatment Technology using Analytic Hierarchy Process. *Mater. Today Proc.* **2021**, *56*, 3433–3440. [CrossRef]
107. Demircan, B.G.; Yetilmezsoy, K. A Hybrid Fuzzy AHP-TOPSIS Approach for Implementation of Smart Sustainable Waste Management Strategies. *Sustainability* **2023**, *15*, 6526. [CrossRef]
108. Dewalkar, S.V.; Shastri, S.S. Integrated Life Cycle Assessment and Life Cycle Cost Assessment based fuzzy multi-criteria decision-making approach for selection of appropriate wastewater treatment system. *J. Water Process Eng.* **2022**, *45*, 102476. [CrossRef]
109. Garcia-Garcia, G. Using Multi-Criteria Decision-Making to optimise solid waste management. *Curr. Opin. Green Sustain. Chem.* **2022**, *37*, 100650. [CrossRef]
110. Kabirifar, K.; Ashour, M.; Yazdani, M.; Mahdiyar, A.; Malekjafarian, M. Cybernetic-parsimonious MCDM modeling with application to the adoption of Circular Economy in waste management. *Appl. Soft Comput.* **2023**, *139*, 110186. [CrossRef]
111. Van Thanh, N. Optimal Waste-to-Energy Strategy Assisted by Fuzzy MCDM Model for Sustainable Solid Waste Management. *Sustainability* **2022**, *14*, 6565. [CrossRef]
112. Yang, J.; Qiao, L.; Li, C. Fuzzy Comprehensive Evaluation Method for Geological Environment Quality of Typical Heavy Metal Mines. *Pol. J. Environ. Stud.* **2023**, *32*, 1877–1886. [CrossRef]
113. Kum, G.; Sönmez, M.; Kargın, A. An Alternative Process for Determining Erosion Risk: The Fuzzy Method. *Coğrafya Derg.* **2022**, *44*, 219–229. [CrossRef]
114. Baumann, M.; Weil, M.; Peters, J.F.; Chibeles-Martins, N.; Moniz, A.B. A review of multi-criteria decision making approaches for evaluating energy storage systems for grid applications. *Renew. Sustain. Energy Rev.* **2019**, *107*, 516–534. [CrossRef]
115. Hariri, R.H.; Fredericks, E.M.; Bowers, K.M. Uncertainty in big data analytics: Survey, opportunities, and challenges. *J. Big Data* **2019**, *6*, 44. [CrossRef]
116. Willard, J.; Jia, X.; Xu, S.; Steinbach, M.; Kumar, V. Integrating scientific knowledge with machine learning for engineering and environmental systems. *ACM Comput. Surv.* **2022**, *55*, 1–37. [CrossRef]
117. Freudenreich, B.; Lüdeke-Freund, F.; Schaltegger, S. A stakeholder theory perspective on business models: Value creation for sustainability. *J. Bus. Ethics* **2020**, *166*, 3–18. [CrossRef]
118. Bag, S.; Gupta, S.; Kumar, A.; Sivarajah, U. An integrated artificial intelligence framework for knowledge creation and B2B marketing rational decision making for improving firm performance. *Ind. Mark. Manag.* **2021**, *92*, 178–189. [CrossRef]

eng

Review

Applications of Machine Learning in Mechanised Tunnel Construction: A Systematic Review

Feng Shan, Xuzhen He *, Haoding Xu, Danial Jahed Armaghani and Daichao Sheng

School of Civil and Environmental Engineering, University of Technology Sydney, Sydney, NSW 2007, Australia; feng.shan@student.uts.edu.au (F.S.); haoding.xu@student.uts.edu.au (H.X.); danial.jahedarmaghani@uts.edu.au (D.J.A.); daichao.sheng@uts.edu.au (D.S.)
* Correspondence: xuzhen.he@uts.edu.au

Abstract: Tunnel Boring Machines (TBMs) have become prevalent in tunnel construction due to their high efficiency and reliability. The proliferation of data obtained from site investigations and data acquisition systems provides an opportunity for the application of machine learning (ML) techniques. ML algorithms have been successfully applied in TBM tunnelling because they are particularly effective in capturing complex, non-linear relationships. This study focuses on commonly used ML techniques for TBM tunnelling, with a particular emphasis on data processing, algorithms, optimisation techniques, and evaluation metrics. The primary concerns in TBM applications are discussed, including predicting TBM performance, predicting surface settlement, and time series forecasting. This study reviews the current progress, identifies the challenges, and suggests future developments in the field of intelligent TBM tunnelling construction. This aims to contribute to the ongoing efforts in research and industry toward improving the safety, sustainability, and cost-effectiveness of underground excavation projects.

Keywords: tunnel boring machine; machine learning; TBM performance; surface settlement; time series forecasting

Citation: Shan, F.; He, X.; Xu, H.; Armaghani, D.J.; Sheng, D. Applications of Machine Learning in Mechanised Tunnel Construction: A Systematic Review. *Eng* **2023**, *4*, 1516–1535. https://doi.org/10.3390/eng4020087

Academic Editor: António Gil Bravo

Received: 24 April 2023
Revised: 15 May 2023
Accepted: 29 May 2023
Published: 30 May 2023

1. Introduction

The Tunnel of Eupalinos, the oldest known tunnel, was constructed in the 6th century BC in Greece for transporting water. The Industrial Revolution brought about a significant increase in tunnel construction used for various purposes including mining, defensive fortification, and transportation. The technology continued to evolve in modern times, and tunnel boring machines (TBMs) became widespread for tunnel excavation projects, including transportation tunnels, water and sewage tunnels, and mining operations. TBMs typically consist of a rotating cutterhead that breaks up the rock or soil and a conveyor system that removes the excavated material. TBMs are preferred over traditional drill and blast techniques due to their higher efficiency, safer working conditions, minimal environmental disturbance, and reduced project costs [1–3]. The continuous cutting, mucking, and lining installation process enables TBMs to excavate tunnels efficiently. However, the high cost of building and operating TBMs, as well as the need for regular maintenance, remains a significant concern. Most importantly, tunnel collapse, rock bursting, water inrush, squeezing, or machine jamming can pose major challenges in complex geotechnical conditions. Therefore, optimising tunnelling operations is critical for project time management, cost control, and risk mitigation.

Traditionally, TBM operators rely primarily on empiricism based on site geology, operational parameters, and tunnel geometry. While theoretical models enhance a fundamental understanding of TBM cutting mechanics, they fail to reasonably predict field behaviour [4,5]. Empirical models study regressive correlations between TBM performance and related parameters in the field but are limited to similar geological conditions [6–8].

The accuracy of theoretical or empirical models is acceptable, but not sufficiently high to meet the demands for safe and efficient construction.

The abundance of data collected by the data acquisition system provides an opportunity for the application of machine learning (ML) in TBM tunnelling. ML techniques are known for their high effectiveness and versatility in capturing complex, non-linear relationships, and have been successfully applied in this field. We conducted a comprehensive analysis of research on ML techniques and TBM tunnelling using the Web of Science search engine. Figure 1 shows 254 published papers, indicating little interest before 2018 but growing popularity between 2018 and 2022. The increasing trend in published papers signifies the growing interest and recognition of the benefits of ML techniques in TBM tunnelling.

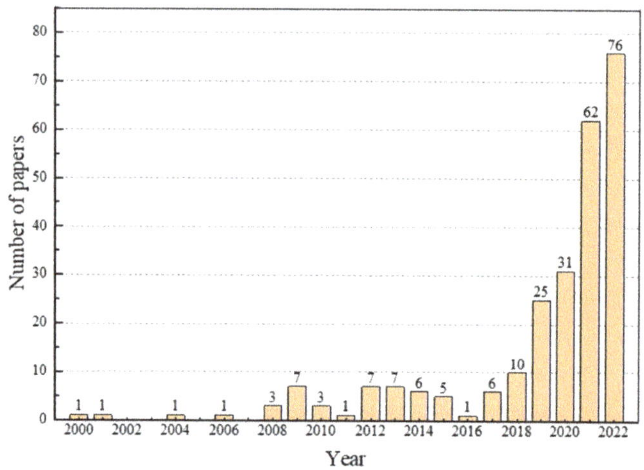

Figure 1. Number of papers using ML models in TBM tunnelling.

Regarding this, literature reviews on soft computing techniques for TBM tunnelling were conducted. Shreyas and Dey [9] mainly introduced ML techniques and investigated their characteristic and limitations. Shahrour and Zhang [10] discussed predictive issues related to surface settlement, tunnel convergence, and TBM performance. They highlighted the importance of feature selection, model architecture, and data repartition to choose an optimal algorithm. Sheil et al. [11] investigated four main applications–TBM performance prediction, surface settlement prediction, geological forecasting, and cutterhead design optimisation. It is found that the sharing of a complete and high-quality database remains a major challenge in the development of ML techniques in TBM tunnelling [12]. In addition, no paper clearly identified the difference between prediction and time series forecasting, and the latter is much more complex due to known inputs being current and historical information.

In this study, we present a typical framework for ML modelling and review the methodology for data processing, ML algorithm, hyperparameter tuning, and evaluation metrics in Section 2. We then focus on three research topics in TBM tunnelling in Section 3—prediction of TBM performance, prediction of surface settlement, and time series forecasting. Section 4 summarises the application of ML in tunnelling including the current progress, challenges, and future development. The goal is to provide guidance for future research and industry on intelligent TBM tunnelling construction.

2. Machine Learning Modelling

Figure 2 depicts a basic flowchart for building a near-optimal model using machine learning. Prior to modelling, a dataset is processed to select relevant data by outlier detection, interpolation, data smoothing, and feature selection. The processed data are randomly split into training, validation, and test sets. The training and validation set is used to train the model, while the test set is used to evaluate its performance. The choice of algorithm is crucial as it can significantly impact the model's accuracy and reliability. Hyperparameter tuning aims to find the best combination of hyperparameters, which helps to fine-tune the model's performance. As a result, the near-optimal model is built and evaluated in the test set. In the ML modelling process, data processing, ML algorithm selection, hyperparameter tuning, and evaluation metrics, four main components, will be briefly described.

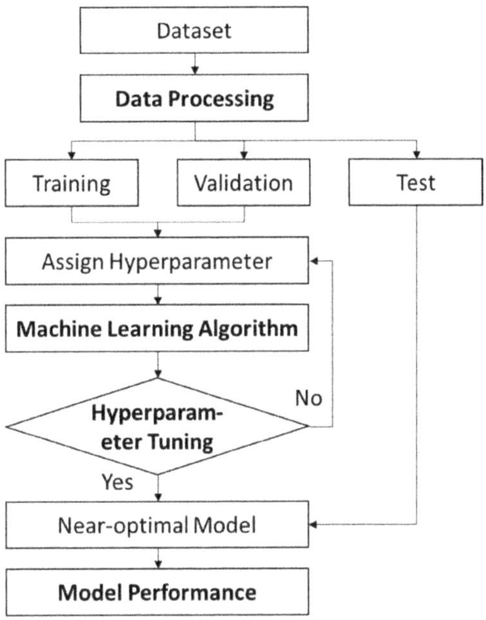

Figure 2. Typical flowchart of the modelling process in ML studies.

2.1. Data Processing

The data generated during tunnel construction is extensive and diverse, encompassing geological and geotechnical survey data, operational parameters, and monitoring data of surface settlement and structure deformation. For example, the data acquisition system recorded 199 operational parameters per second in the Yingsong water diversion project, accounting for 86,400 data points per day. Since the quality and quantity of data heavily influence the performance of ML models, data processing is essential to delete outliers, interpolate missing values, remove noise, and select features for better applicability.

Outliers are data points that significantly differ from other observations in a dataset which are considered errors and should be removed. There are several methods available to detect and remove outliers in a dataset. Assuming the dataset follows a normal distribution, data points that fall outside the range of the mean plus or minus three standard deviations can be removed according to the three-sigma rule [13,14]. Another method is the interquartile range (IQR) method, which sets up a minimum and maximum fence based on the first quartile (Q1) and the third quartile (Q3), respectively [15]. Any observations that exceed 1.5 times the IQR below Q1 or above Q3 are considered outliers and should be removed. In addition to statistical methods, isolation forest is an unsupervised decision-tree-based algorithm used for outlier detection [16]. It generates partitions recursively by randomly selecting an attribute and a split value between the minimum and maximum values to isolate the data point.

Interpolation is a common technique in data analysis used to estimate unknown values between two known data points. It involves constructing a function that approximates the behaviour of the data within the range of the known values, such as linear, polynomial, and spline interpolation [13]. Another interpolation method, called kriging interpolation, takes into account the spatial correlation between the locations to estimate the value of a variable at an unsampled location using the values at sampled locations [17,18]. Kriging interpolation is particularly useful in geology, hydrology, and environmental science fields for modelling and predicting spatial data.

Data smoothing is a data analysis technique commonly used to eliminate noise and fine-grained variation in time series data to reveal the underlying information. The simple moving average method creates a smoothed version by averaging observations within a specific period, assigning equal weight to each observation [14,19]. In contrast, the exponential moving average method assigns greater weight and significance to the recent data points while gradually reducing the weight of older data points. Wavelet transform is a technique used to decompose signals into basic functions by contracting, expanding, and translating a wavelet function. Wavelet denoising, which applies a threshold to the wavelet coefficients, reduces the contribution of the noisy components in time series data [17,20–23]. The denoised signal is then reconstructed from the remaining wavelet coefficients, resulting in a signal with reduced noise and preserved features.

Feature selection is a crucial process for handling high-dimensional data, where the primary objective is to identify the most relevant features that can offer valuable insights into the underlying patterns and relationships within the data. However, many selected features are based on prior experience of laboratory tests and field studies, resulting in ignoring the effects of uncertain factors. The variance threshold method removes features that do not meet a specified threshold [2], including zero-variance features that have the same value across all samples. Pearson correlation coefficient (PCC) measures the linear relationship between two or more variables on a scale of -1 to 1 [24,25], where features with a higher absolute value indicate a stronger relationship with the target variable. Alternatively, Principal component analysis (PCA) is a dimensionality reduction method that can transform a large set of variables into a smaller set that contains most of the information in the original set [26–29].

2.2. Machine Learning Algorithms

'Artificial intelligence', 'Machine Learning (ML)', and 'Deep Learning' are commonly used interchangeably to describe software that can demonstrate intelligent behaviour. Artificial intelligence involves creating algorithms and computational models that enable machines to imitate human cognitive abilities such as decision-making, learning from experience, and adapting to new situations. ML is a subfield of artificial intelligence to develop relationships between inputs and outputs, performing specific tasks without explicit programming. Deep learning is a further subfield of ML that utilises 'deep' neural networks with multiple hidden layers to learn from large amounts of data.

Artificial neural networks (ANNs) are deep learning algorithms inspired by the structure and function of the human brain [30–32]. ANNs comprise many interconnected nodes, or neurons, that work together to perform a specific task. Each neuron receives input from one or more other neurons and applies a mathematical function to that input to generate an output. The output of one neuron becomes the input to other neurons, and this process continues until the final output is produced. Various ANN variants are developed to improve model accuracy, including the wavelet neural network [33], radial basis function network [24], general regression neural network [34], and extreme learning machine [35].

Convolutional neural networks (CNNs) are powerful deep learning algorithms that are particularly well-suited for image recognition/classification tasks [14,36]. The input to the network is usually an image or a set of images. Convolutional layers apply a set of filters to extract relevant features, with the filters typically being small squares of pixels that slide over the image. Pooling layers reduce the dimensionality of the data to make the network more robust. The output of the final layer is then passed through one or more fully connected layers, which perform a final classification or regression task.

Recurrent neural networks (RNNs) are deep learning algorithms for sequential problems such as speech recognition, natural language processing, or time series forecasting [19,37]. RNNs are characterised by recurrent connections to maintain an internal state or memory, which enables them to capture temporal dependencies. However, RNNs can encounter the vanishing gradient problem when input sequences are too long. Long short-term memory (LSTM) is a type of RNN that includes additional memory cells and gating mechanisms, which selectively store and retrieve information over long periods [2,17,21,23,25,38,39]. In an LSTM, the input gate controls how much new information is stored in the cell, the forget gate controls how much old information is discarded from the cell, and the output gate controls how much information is passed to the next time step.

Fuzzy logic (FL) is a branch of mathematics that deals with reasoning with imprecision and uncertainty [40,41]. FL allows for degrees of truth or falsity to be represented as values between 0 and 1, in contrast to traditional logic that operates on the binary true/false principle. FL is ideal for artificial intelligence, control systems, and decision-making applications. An adaptive neuro-fuzzy inference system (ANFIS) is a hybrid system that combines fuzzy logic and neural network techniques to represent the input–output relationship with a set of fuzzy if-then rules [26,27,42]. The input data is first fuzzified, then the fuzzy rules are applied to generate an output. The neural network component of ANFIS is used to adjust the parameters of the fuzzy rules.

Support vector machine (SVM) is a type of ML algorithm with the ability to handle high-dimensional data and produce accurate results with relatively small datasets [3,17,43,44]. SVM uses a kernel function to map input data into a high-dimensional space where a hyperplane is used to separate the data into different classes. The main goal of SVM is to find the best-fitting hyperplane that maximises the margin between the predicted and actual values. However, SVM can be computationally expensive for large datasets and sensitive to the choice of kernel function and hyperparameters.

Decision tree (DT) is a widely used ML algorithm for classification and regression analysis, which is very effective because of its simplicity, interpretability, and accuracy in handling complex datasets [15,45]. It is based on a hierarchical structure where each node represents a decision or test of a specific feature. The tree is built by recursively splitting the data into smaller subsets based on the feature that provides the most information gain or reduction in entropy. Once the tree is built, it can be used to classify new data by following the path from the root node to the appropriate leaf node.

Random forest (RF) is an extension of the decision tree that uses multiple trees to produce more robust results and reduce the risk of overfitting [18,46,47]. RF combines the results of many decision trees to obtain a more accurate prediction. It has been successfully applied to a wide range of problems including remote sensing, object recognition, and cancer diagnosis. A classification and regression tree (CART) is a decision tree that simultaneously handles categorical and continuous variables [48,49]. A CART output is a decision tree where each fork is split into a predictor variable, and each node at the end has a prediction for the target variable. Another popular extension of the decision tree is extreme gradient boosting (XGBoost), which uses a gradient boosting framework to improve the accuracy and speed of prediction [36,50]. XGBoost uses a regularisation term to prevent overfitting and can handle missing values and sparse data. It also includes several advanced features such as cross-validation, early stopping, and parallel processing, which make it a popular choice for large-scale datasets and competitions on platforms.

2.3. Hyperparameter Tuning

Hyperparameter optimisation techniques aim to find the optimal combination of hyperparameters that are not learned from data but are instead specified by the user before training the model, such as the learning rate, batch size, and the number of hidden sizes.

Grid search is a simple and effective way to find good hyperparameter values for a model, but it can be computationally expensive. With grid search, a range of values is specified for each hyperparameter, and the model is trained and evaluated for every possible combination of hyperparameter values. The optimal combination of hyperparameters is chosen based on the highest performance in a validation set.

Particle swarm optimisation (PSO) is a population-based optimisation algorithm that is inspired by the behaviour of swarms of birds or insects [22,25,35,51,52]. In PSO, a population of candidate solutions, represented by particles, are evaluated according to a fitness function. Each particle moves towards its own best position and the best position found by the swarm, with the speed and direction of each particle being determined by its current position and velocity, as well as the position and velocity of the best particle in the swarm. This process is repeated until a satisfactory solution is found, or a stopping criterion is met.

Bayesian optimisation (BO) is a method used to optimise expensive, black box functions that lack an explicit mathematical form, particularly useful when the evaluation of the function is time-consuming or expensive [44,49,53,54]. It combines the previous knowledge of the target function with the results of previously evaluated points to determine the next point to be evaluated. It works by constructing a probabilistic model of the target function and updating the model as new observations. It allows the algorithm to balance exploration and exploitation to converge to the optimum of the function quickly.

Imperialist competitive algorithm (ICA) is a metaheuristic optimisation algorithm inspired by the concept of empires and colonies in history [30,55,56]. In ICA, each solution in the population represents a colony and the best solution is designated as imperialist. The imperialist expands its territory by attracting other colonies towards it, while the weaker colonies are forced to merge with the stronger ones. ICA balances exploration and exploitation, as weaker colonies explore new regions of the search space while stronger colonies exploit promising regions.

2.4. Evaluation Metrics

Model performance in testing is an indicator of the quality of the trained model. Equations (1)–(6) show various evaluation metrics that quantitatively evaluate prediction errors. In expressions, y_i, $\hat{y}_i$ are measured and predicted values, and $\overline{y}$ is the mean of measured values. Mean absolute error (MAE), mean squared error (MSE), and root mean squared error (RMSE) are dimensional and assess the errors between measured and predicted values, while mean absolute percentage error (MAPE) is non-dimensional and expressed as a percentage. The coefficient of determination (R^2) and variance account for (VAF) represent the proportion of the variance in the dependent values between 0 and 1, where a larger value indicates a higher accuracy between predicted and measured values, and vice versa.

$$\text{MAE} = \frac{1}{n}\sum_{i=1}^{n}|y_i - \hat{y}_i|, \quad (1)$$

$$\text{MSE} = \frac{1}{n}\sum_{i=1}^{n}(y_i - \hat{y}_i)^2, \quad (2)$$

$$\text{RMSE} = \sqrt{\frac{1}{n}\sum_{i=1}^{n}(y_i - \hat{y}_i)^2}, \quad (3)$$

$$\text{MAPE} = \frac{1}{n}\sum_{i=1}^{n}\left|\frac{y_i - \hat{y}_i}{y_i}\right|, \quad (4)$$

$$\text{R}^2 = 1 - \frac{\sum_{i=1}^{n}(y_i - \hat{y}_i)^2}{\sum_{i=1}^{n}(y_i - \overline{y})^2}, \quad (5)$$

$$\text{VAF} = 1 - \frac{\text{Var}(y_i - \hat{y}_i)}{\text{Var}(y_i)}. \quad (6)$$

3. Application in TBM Tunnelling

Figure 3 summarises the number of studies that have utilised different ML algorithms to address the challenges of predicting TBM performance, predicting surface settlement, and time series forecasting. Specifically, ANN is the most widely used algorithm used in 19 studies to predict TBM performance and surface settlement, followed by SVM in 11 studies and RF in 10 studies. Given the time-dependent nature of the TBM tunnelling process, RNN, LSTM, and gated recurrent unit (GRU) are widely utilised in time series forecasting with studies of 8, 12, and 3, respectively. RNN, LSTM, and GRU models in time series forecasting are highly effective because of a loop structure to capture temporal dependencies, enabling them to outperform SVM and RF models.

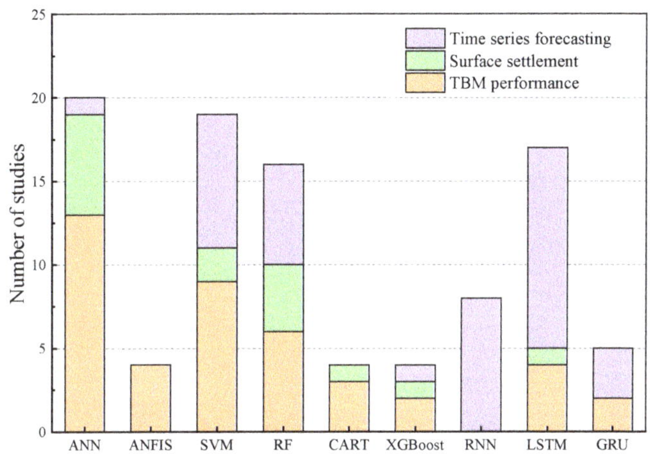

Figure 3. Summary of ML algorithms in TBM performance, surface settlement and time series forecasting.

Predicting TBM performance or surface settlement is a function of input parameters in Equation (7), while time series forecasting is expressed in Equation (8).

$$Y = \sigma(WX, b) \tag{7}$$

$$Y = X_{n+1} = \sigma[W(X_1, X_2, \ldots, X_n), b], \tag{8}$$

where X is the input vector and Y is the output vector. The weight matrix W and bias b are the arguments to be trained by the activation function σ using ML algorithms. For time series forecasting, input vector comprises historical sequential data $X_1, X_2, \ldots, X_n$ and output vector is the target value in the next step X_{n+1}.

Typically, penetration rate, revolutions per minute, thrust force, and cutterhead torque are considered as feature vectors in ML models [13,57,58]. In addition to these four operational parameters, Lin et al. [25] used PCC to identify mutually independent parameters such as face pressure, screw conveyor speed, foam volume, and grouting pressure. Zhang et al. [29] applied PCA to reduce dimensionality and found the first eight principal components can capture the main information of 33 input parameters.

3.1. TBM Performance

Extensive research has been conducted on employing ML algorithms to investigate TBM performance in Table 1. TBM performance refers to the effectiveness and efficiency of the machine in excavating a tunnel and involves various indicators such as penetration rate, advance rate, field penetration index, thrust force, and cutterhead torque. Understanding and optimising TBM performance is crucial for project time management, cost control, and risk mitigation.

Table 1. Summary of literature on ML algorithms and predicting TBM performance.

Literature	Data Processing [a]	Algorithms [b]	Hyperparameter Tuning [c]	Targets [d]	Data Size and Data Set
Grima et al. [26]	PCA	MR, ANN, ANFIS	-	PR, AR	640 tunnel project
Benardos and Kaliampakos [31]	-	ANN	-	AR	11-Athens metro
Tiryaki [28]	PCA	MR, ANN	-	specific energy	44-Three tunnel projects
Mikaeil et al. [41]	-	FL	-	Penetrability	151-Queens water tunnel
Yagiz [59]	PCC	MR, ANN	-	PR	151-Queens water tunnel
Javad and Narges [60]	-	ANN	-	PR	185-Three tunnel projects
Mahdevari et al. [43]	-	MR, SVM	-	PR	151-Queens water tunnel
Salimi et al. [27]	PCA	MR, SVM, ANFIS	-	FPI	75-Zagros lot 1B and 2
Armaghani et al. [30]	-	ANN	PSO, ICA	PR	1286-Pahang-Selangor raw water transfer
Armaghani et al. [61]	-	MR, GEP	-	PR	1286-Pahang-Selangor raw water transfer
Sun et al. [18]	Kriging interpolation, rate of change	RF	-	TH, TO, PR	88-Shenzhen metro
Armaghani et al. [55]	-	ANN	PSO, ICA	AR	1286-Pahang-Selangor raw water transfer
Koopialipoor et al. [62]	-	ANN, DNN	-	PR	1286-Pahang-Selangor raw water transfer
Salimi et al. [48]	PCA	MR, CART, GP	-	FPI	580-Seven tunnel projects
Zhang et al. [47]	PCC	RF	PSO	TO, TH, PR, FP	294-Changsha metro line 4
Koopialipoor et al. [63]	-	ANN	firefly algorithm	PR	1200-Pahang-Selangor raw water transfer
Mokhtari and Mooney [44]	PCC, Relief	SVM	BO	PR	Northgate Link tunnel
Wang et al. [64]	-	ANN, LSTM, RF, SVM	-	AR	806-Nanning metro line 1
Zhang et al. [49]	-	SVM, CART, RF, bagging, Ada boosting	BO	PR	151-Queens water tunnel
Zhang et al. [22]	WT, MD, GRG	LSTM, RF	PSO	TH, TO, PR, RPM, CP	3549-Changsha metro line 4 and Zhengzhou metro line 2
Zhou et al. [65]	-	ANN, GP	-	AR	1286-Pahang-Selangor raw water transfer
Bai et al. [45]	PCC, Seasonal-trend decomposition	MR, SVM, DT, GBoost	-	TO, TH, FP	450-Xi'an metro
Bardhan et al. [66]	-	hybrid ensemble model	-	PR	185-Three tunnel project
Harandizadeh et al. [56]	-	ANFIS-PNN	ICA	PR	209-Pahang-Selangor raw water transfer
Lin et al. [67]	-	MR, ANN, SVM, LSTM, GRU, EML		PR	1000-Shenzhen railway
Parsajoo et al. [42]	-	ANFIS	artificial bee colony	FPI	150-Queens water tunnel
Zeng et al. [35]	-	EML	PSO	AR	1286-Pahang-Selangor raw water transfer
Zhou et al. [54]	-	XGBoost	BO	AR	1286-Pahang-Selangor raw water transfer

Table 1. *Cont.*

Literature	Data Processing [a]	Algorithms [b]	Hyperparameter Tuning [c]	Targets [d]	Data Size and Data Set
Zhou et al. [50]	-	ANN, RF, XGBoost, SVM	GWO, PSO, SCA, SSO, MVO, MFO	PR	1286-Pahang-Selangor raw water transfer
Lin et al. [25]	-	LSTM	PSO	TH	1500-Shenzhen railway
Lin et al. [68]	-	GRU	PSO	TO	1500-Shenzhen railway
Salimi et al. [69]	-	MR, CART	-	FPI	666-Eight tunnel projects
Yang et al. [3]	-	SVM	GWO, biogeography-based optimisation	PR	503-Shenzhen metro line

[a] WT, wavelet transform; MD, Mahalanobis distance; GRG, grey rational grade. [b] MR, multiple regression (linear/non-linear); GP, genetic programming; GBoost, gradient boosting; GEP, gene expression programming; EML, extreme machine learning; PNN, polynomial neural network; DNN, deep neural network. [c] GWO, grey wolf optimiser; SCA, sine cosine algorithm; SSO, social spider optimisation; MVO, multi-verse optimisation; MFO, moth flame optimization. [d] FP, face pressure; RPM, revolutions per minute; CP, chamber pressure.

Since ML models are data-driven, the quality of datasets (e.g., availability to the public, number of samples, input parameters used, etc.) is crucial. Table 2 displays three types of models corresponding to three typical datasets and their respective limitations. It is worth noting that models are categorised according to their input parameters: Model A includes geological conditions, operational parameters, and TBM type and size, Model B only includes geological conditions, and Model C includes geological conditions and operational parameters.

Table 2. Three types of models based on input parameters and their limitations.

Model Type	Dataset	Data Size	Parameters	Open Access	Limitations
Model A	640 tunnel projects	-	Geological conditions, Operational parameters, TBM type and size	No	hard to access
Model B	Queen water tunnel	151	Geological conditions	Yes	overfitting or lack of generalisability
Model C	Pahang-Selangor raw water transfer	1286	Geological conditions, Operational parameters	Yes	hard to apply in practice

The penetration rate (PR) measures the speed of boring distance divided by the working time, typically quantified in m/h or mm/min. PR plays a crucial role in tunnelling operations as it directly affects overall productivity. A higher penetration rate results in faster tunnel excavation, ultimately reducing project time and costs. For predicting PR, ANIFS, ANN, and SVM models have shown promising results in various studies. For instance, the ANIFS model [26] demonstrated better performance than multiple regression and empirical methods based on a database of 640 TBM projects in rock. The ANIFS model (Model A) is adaptable as it takes into account geological conditions, operational parameters, and even TBM type and size, but most TBM datasets are not available for public access.

The ANN and SVM models [43,59] outperformed linear and non-linear regression when applied to the publicly available Queen water tunnel dataset with 151 samples. In the sensitivity analysis, interestingly, the brittleness index was found to be the least effective parameter in the SVM model [43] but the most sensitive parameter in the RF model [70]. These contrasting results can be attributed to a limited number of samples for training, which leads to overfitting or lack of generalisability of Model B.

In the project of Pahang–Selangor raw water transfer with 1286 samples, ML models for predicting PR were robust and reliable because of more data and adding operational parameters [30,50,63]. However, TBM performance is a real-time operational parameter that cannot be obtained before the start of a project, making it infeasible to apply Model C in practice. For example, although the average thrust force is an effective parameter for predicting PR [63], it is an operational input in Model C and is unavailable as it is collected in real-time as well as PR itself.

Given the expression for predicting PR using statistical analysis, optimisation techniques can be applied to optimise the correlations of weighting in multiple regression [52]. On the other hand, optimisation techniques can be used to fine-tune the hyperparameters of ML models, such as the XGBoost model by Zhou et al. [50]. Figure 4 compares the model performance using different optimisation techniques, with Figure 4a showing the MR model and Figure 4b showing the XGBoost model. The accuracy improves by utilising optimisation techniques, but the difference between different optimisation techniques is small.

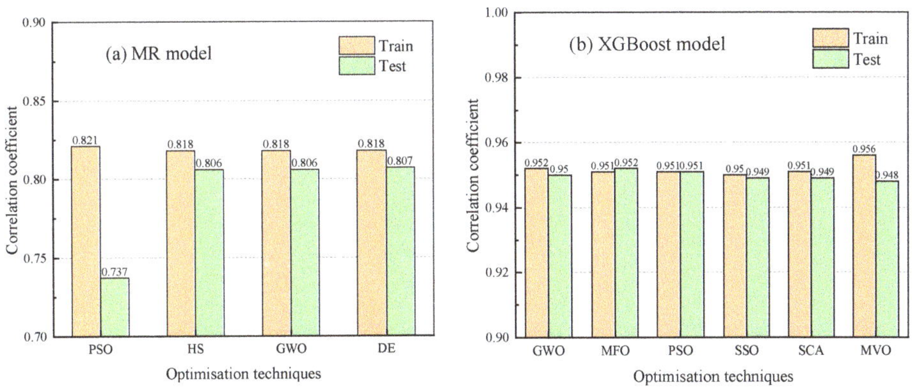

Figure 4. Comparison optimisation techniques (**a**) MR model based on dataset from Queen water tunnel; (**b**) XGBoost model based on dataset from Pahang–Selangor raw water transfer.

Advance rate (AR) is a crucial indicator in tunnelling operations, calculated as the boring distance divided by the working time and stoppages. Compared with PR, AR additionally considers stoppages due to TBM maintenance, cutters change, breakdowns, or tunnel collapses. Comparing AR prediction models, the ANN model by Benardos and Kaliampakos [31] was limited by the small size of the Athens metro dataset. In contrast, the Pahang–Selangor raw water transfer dataset allowed for the development of more robust and reliable ML models for AR prediction [35,54,55,65].

Field penetration index (FPI) evaluates TBM efficiency in the field calculated as the average cutter force divided by penetration per revolution. For predicting FPI, ANIFS and RF models performed well when applied to the Queen water tunnel dataset [3,42]. Furthermore, Salimi et al. [27,48,69] successfully developed ML models to predict FPI in different rock types and conducted a sensitivity analysis to better understand the relationship between FPI and input parameters.

Thrust force (TH) refers to the force that TBM exerts on the excavation face, whereas cutterhead torque (TO) refers to the twisting force applied to the cutterhead. The amount of TH or TO depends on the hardness and strength of the material being excavated and the size and type of TBM being used. Regarding the prediction of TH and TO, Sun et al. [18] built RF models for heterogeneous strata, while Lin et al. [25,68] utilised PSO-LSTM and PSO-GRU models based on the dataset from Shenzhen intercity railway. Bai et al. [45] utilised an SVM classifier to identify the location of interbedded clay or stratum interface and subsequently developed ML models to predict TH, TO, and FP.

Although these ML models offer high accuracy in predicting TBM performance, their applicability is limited due to their project-specific nature (Model B and Model C) and lack of generalisability across different TBM types and geological conditions [71]. Despite these limitations, ML models remain highly flexible in adding or filtering related parameters and implicitly capturing the impact of uncertain parameters, providing valuable insights into TBM performance optimisation.

3.2. *Surface Settlement*

The surface settlement, the subsidence of the ground surface above a tunnel due to excavation, poses risks to surrounding structures and utilities. Accurate prediction of surface settlement is essential for mitigating potential damages during tunnel construction. Engineers can minimise ground movement and reduce the risk of damage by adjusting excavation parameters and support structures. Table 3 reviews papers on settlement induced by TBM tunnelling and excludes construction methods such as drilling, blasting, and the new Austrian Tunnelling Method [72–74].

Table 3. Summary of literature on ML algorithms and predicting surface settlement.

Literature	Data Processing	Algorithms [a]	Hyperparameter Tuning	Targets	Data Size and Data Set
Suwansawat and Einstein [32]	-	ANN	-	S_{max}	49-Bangkok subway project
Boubou et al. [75]	-	ANN	-	S(X)	432-Toulouse subway line B
Pourtaghi and Lotfollahi-Yaghin [33]	-	Wavelet-ANN	-	S_{max}	49-Bangkok subway project
Dindarloo and Siami-Irdemoosa [76]	PCC	CART	-	S_{max}	34-Various tunnel projects
Goh et al. [77]	-	MARS	-	S_{max}	148-Three Singapore MRT projects
Chen et al. [24]	PCC	ANN, RBF, GRNN	-	S_{max}	200-Changsha metro line 4
Zhang et al. [47]	PCC	RF	PSO	S_{max}	294-Changsha metro line 4
Zhang et al. [34]	PCC	ANN, SVM, RF, EML, GRNN	PSO	S_{max}	294-Changsha metro line 4
Zhang et al. [22]	WT, MD, GRG	LSTM, RF	PSO	S_{max}	423-Changsha metro line 4
Zhang et al. [78]	PCC	XGBoost, ANN, SVM, MARS	-	S_{max}	148-Three Singapore MRT projects
Kannangara et al. [46]	PCC, sequential feature selection, Boruta algorithm	RF	-	S_{max}	264-Hangzhou metro line 2 and line 6

[a] MARS, multivariate adaptive regression spline; RBF, radial basis function; GRNN, general regression neural network.

Suwansawat and Einstein [32] were among the first to use ANN to predict the maximum settlement (S_{max}) for the Bangkok subway project, considering tunnel geometry, geological conditions, and operational parameters. Pourtaghi and Lotfollahi-Yaghin [33] improved the ANN model by adopting wavelets as activation functions, resulting in higher accuracy than traditional ANN models. In contrast, Goh et al. [77] utilised MARS and Zhang et al. [78] utilised XGBoost to predict S_{max} for Singapore mass rapid transport lines with 148 samples. Interestingly, the mean standard penetration test showed opposite sensitivities in these two models. It further highlights the unreliability and unrobustness of ML models with limited samples, which may lead to overfitting or lack of generalisability. A comprehensive dataset from Changsha metro line 4, including geometry, geological conditions, and real-time operational parameters, has been used to compare the performance of various ML models such as ANN, SVM, RF, and LSTM [22,24,34,47].

Since the observed settlement showed a Gaussian shape in the transverse profile, Boubou et al. [75] incorporated the distance from the tunnel axis as an input parameter in their ANN model. They identified advance rate, hydraulic pressure, and vertical guidance parameter as the most influential factors in predicting surface settlement.

Various ML models have been employed to predict surface settlement induced by TBM tunnelling. The choice of ML algorithms and feature selection can significantly impact prediction accuracy, and researchers should carefully consider these factors when applying ML to surface settlement prediction in TBM tunnelling.

3.3. Time Series Forecasting

Time series forecasting is a real-time prediction using current and historical data to forecast future unknown values, which means input parameters are available and it does not have the practical problem of Model C. It is crucial in TBM tunnelling for predicting TBM performance, surface settlement, and moving trajectory in real time because operators can make necessary adjustments when potential issues are detected. Several studies using ML techniques for time series forecasting are shown in Table 4. Since the quality and quantity of data heavily influence model performance, moving average or wavelet transform are employed to eliminate noise and fine-grained variation to reveal the underlying information in time series data [14,17,19,21].

High-frequency data is collected directly from the data acquisition system every few seconds or minutes. High-frequency prediction of next-step TBM performance can be achieved with high accuracy using RNN, LSTM, and GRU. These ML algorithms have been found to outperform others by incorporating both current and historical parameters [21,36,37,53,82]. However, it is less meaningful to predict TBM performance just a few seconds or millimetres in advance, as shown in Table 5. Therefore, multi-step forecasts were explored, and it was found that errors increase significantly with an increasing forecast horizon [39,81,84].

Table 4. Summary of literature on ML algorithms and time series forecasting.

Literature	Data Processing [a]	Algorithms [b]	Hyperparameter Tuning	Targets [c]	Data Size and Data Set
Guo et al. [20]	WT	Elman RNN	PSO	longitudinal settlement	Jiangji subway tunnel
Zhang et al. [79]	WT	ANN, SVM	-	daily settlement	60-Wuhan metro line 2
Gao et al. [37]	-	RNN, LSTM, GRU, SVM. RF, Lasso	-	TO, TH, AR, CP	3000-Shenzhen metro
Zhou et al. [23]	WT	ARIMA, LSTM, CNN-LSTM	-	HDSH, HDST, VDSH, VDST, roll, pitch	5005-Sanyang Road Tunnel
Gao et al. [80]	3-sigma rule, MA, GRG	GRU	genetic algorithm	earth pressure	1538-Luoyang metro line 2
Erharter and Marcher [81]	PCC	LSTM, RF, SVM	-	TO	200,000-Brenner base tunnel
Feng et al. [13]	3-sigma rule, WT	DBN	-	FPI	8915-Yingsong water diversion project
Gao et al. [82]	-	ARIMA, RNN, LSTM	-	PR	Hangzhou second water source project
Li et al. [2]	PCC	LSTM	-	TO, TH	4650-Yingsong water diversion project
Qin et al. [36]	cosine similarity	CNN-LSTM, XGBoost, RF, SVM, LSTM, RNN, CNN	-	TO	150,000-Singapore metro T225 project
Shi et al. [39]	WT, variational mode decomposition	LSTM, CNN, RNN, SVM, RF	-	TO	60,000-Singapore metro T225 project
Wang et al. [21]	WT, light gradient boosting machine	LSTM	-	PR, TO	25,543-Sutong gas transmission line
Xu et al. [14]	3-sigma rule, MA, PCC	SVM, RF, CNN, LSTM, GBoost, KNN, Bayesian ridge regression	-	PR, TO, TH, RPM	7000-Yingsong water diversion project
Zhang et al. [83]	-	RF	-	S_{max}	386-Changsha Metro Line 4
Huang et al. [53]	SelectKBest	LSTM	BO	TO	Yingsong water diversion project
Shan et al. [19]	MA	RNN, LSTM	-	PR	463-Changsha metro line 4 and Zhengzhou metro line 2
Shen et al. [17]	WT, Kriging interpolation	LSTM, SVM, RNN	-	HDSH, HDST, VDSH, VDST, roll, pitch	1200-Shenzhen intercity railway
Zhang et al. [29]	PCA, PCC	GRU, RNN, SVM	-	HDSH, HDST, VDSH, VDST,	22,010-Guang-Fo intercity railway

[a] MA, moving average. [b] DBN, deep belief network; KNN, k-nearest neighbours. [c] HDSH, horizontal deviation of shield head; HDST, horizontal deviation of shield tail; VDSH, vertical deviation of shield head; VDST, vertical deviation of shield tail.

Table 5. Comparison of time series forecasting on historical data and forecast horizon.

Literature	Category	Historical Data		Forecast Horizon	
		Step behind	Distance behind	Step ahead	Distance ahead
Gao et al. [37]	high-frequency	5 steps	1.25 mm [a]	1 step	0.25 mm [a]
Qin et al. [36]		10 steps	-	1 step	-
Huang et al. [53]		6 steps	22.4 mm [a]	1 step	3.73 mm [a]
Erharter and Marcher [81]		50 steps	2.75 m	1 or 100 steps	0.055 or 5.5 m
Shi et al. [39]		10 steps	-	1–5 steps	-
Gao et al. [80]	low-frequency	5 steps	7.5 m	1 step	1.5 m
Feng et al. [13]		7 steps	7 m	1 step	1 m
Shan et al. [19]		5 steps	7.5 m	1–5 steps	1.5–7.5 m

[a] The distance is estimated based on the time step, sampling period, and average penetration rate.

High-frequency data can be preprocessed into low-frequency data, where each data point represents a fixed segment or working cycle spanning 1–2 m. Low-frequency data, such as that from the Yingsong water diversion project, have been used to forecast average operational parameters [2,14] and predict next-step TBM performance in different geological conditions [13]. In contrast, Shan et al. [19] employed RNN and LSTM to predict near-future TBM performance (1.5–7.5 m ahead), focusing on the difference in geological conditions between training data and test data. While one-step forecasts are highly accurate, predictions decrease in accuracy as the forecast horizon increases.

Regarding the number of steps back required to predict future TBM performance, Table 5 demonstrates that the number of steps used for training ranges from 5 to 10, except for those who used data from the last 50 steps. High-frequency prediction normally uses data just a few millimetres beforehand for training, while low-frequency prediction uses data up to seven metres beforehand. Nevertheless, these data are collected a few millimetres to a few meters away from the current cutterhead location and essentially reflect the current operation of the TBM [85].

To account for the surface settlement developing over time in a single point, Guo et al. [20] used an Elman RNN to predict the longitudinal settlement profile, while Zhang et al. [79] integrated wavelet transform and SVM to forecast daily surface settlement. Zhang et al. [83] used historical geometric and geological parameters to build an RF model to predict operational parameters in the next step. They then combined predicted operational parameters with geometric and geological parameters to estimate S_{max} in the next step based on another RF model.

To improve moving trajectory, current, and historical parameters have been used to predict real-time TBM movements such as horizontal deviation of shield head, horizontal deviation of shield tail, vertical deviation of shield head, vertical deviation of shield tail, roll, and pitch [17,23,29]. When deviations reach the alarm value, the TBM route can be regulated by fine-tuning the thrust force and strokes in the corresponding positions.

Time series forecasting techniques vary in effectiveness depending on the frequency of data collection, the forecast horizon, and the specific application in TBM tunnelling. Understanding these differences and selecting the appropriate ML algorithm is essential for optimising tunnelling operations.

4. Summary and Perspectives

Many studies have reported successful applications of ML techniques in TBM tunnelling, with an increasing trend in Figure 1. This trend is likely to persist as the volume of data continues to grow and the use of ML becomes more common. This paper presents a systematic literature review on using ML techniques in TBM tunnelling. A framework of ML modelling is presented, highlighting the importance of data processing before modelling, ML algorithms, and optimisation techniques used to build near-optimal models, and evaluation metrics for model performance. Furthermore, it identifies three main issues in TBM tunnelling: predicting TBM performance, predicting surface settlement, and time series forecasting.

ANN, SVM, and RF are the most popular algorithms adopted in the prediction of TBM performance and surface settlement. The model performance heavily depends on the selection of ML algorithms and hyperparameter tuning. Availability to the public, number of samples, and input parameters for training are also crucial in ML modelling when applied to tunnel projects. Optimisation techniques can effectively enhance the performance of both multiple regression and ML models.

Given the time-dependent nature of the TBM tunnelling process, RNN, LSTM, and GRU are widely utilised to deal with time series problems. However, high-frequency prediction is less meaningful as it only provides a few seconds or millimetres advance warning, while low-frequency prediction is limited by the number of samples after data preprocessing. One-step forecasts have proven to be highly accurate and play a practical role in warning of possible accidents. However, the accuracy of multi-step forecasts decreases significantly with an increasing forecast horizon, mainly due to the decreased impact of parameters farther away from the TBM cutterhead.

The black box problem is a significant limitation of ML models as they lack interpretability. While ML models are able to make predictions based on complex patterns and relationships within data, it can be difficult to interpret how the model arrived at its results. To address the limitation, researchers are developing more interpretable ML models. Specifically, decision tree-based algorithms can provide insights into the model's decision-making process through probabilistic sensitivity analysis. Theory-guided machine learning and physics-informed neural networks can incorporate theoretical knowledge or physical laws into the learning process, facilitating the capture of optimal solutions and effective generalisation, even with limited training samples.

Another important challenge is that ML models are developed and validated using only one dataset or similar datasets, which limits their applicability to different projects. Validation and generalisation of ML models across various datasets are necessary for the industry to gain confidence in their effectiveness. As tunnelling data become more accessible, it may be possible to interrogate larger data for training data. This would allow the reliability and robustness of ML models on future projects to improve feedback in the industry.

Author Contributions: Conceptualisation, F.S.; methodology, F.S.; writing—original draft preparation, F.S. and H.X.; writing—review and editing, X.H., D.J.A. and D.S.; supervision, X.H. and D.S.; funding acquisition, D.S. All authors have read and agreed to the published version of the manuscript.

Funding: This research received no external funding.

Institutional Review Board Statement: Not applicable.

Informed Consent Statement: Not applicable.

Data Availability Statement: No new data were created or analysed in this study. Data sharing is not applicable to this article.

Conflicts of Interest: The authors declare that they have no known competing financial interest or personal relationships that could have appeared to influence the work reported in this paper.

Abbreviations

TBM	Tunnel boring machine
ML	Machine learning
PR	Penetration rate
AR	Advance rate
FPI	Field penetration index
TH	Thrust force
TO	Cutterhead torque

S_{max}	Maximum surface settlement
PCA	Principal component analysis
PCC	Pearson correlation coefficient
ANN	Artificial neural network
CNN	Convolutional neural network
RNN	Recurrent neural network
LSTM	Long short-term memory
GRU	Gated recurrent unit
FL	Fuzzy logic
ANIFS	Adaptive neuro-fuzzy inference system
SVM	Support vector machine
DT	Decision tree
RF	Random forest
CART	Classification and regression tree
XGBoost	Extreme gradient boosting
PSO	Particle swarm optimisation
BO	Bayesian optimisation
ICA	Imperialism competitive algorithm

References

1. Adoko, A.C.; Gokceoglu, C.; Yagiz, S. Bayesian prediction of TBM penetration rate in rock mass. *Eng. Geol.* **2017**, *226*, 245–256. [CrossRef]
2. Li, J.; Li, P.; Guo, D.; Li, X.; Chen, Z. Advanced prediction of tunnel boring machine performance based on big data. *Geosci. Front.* **2021**, *12*, 331–338. [CrossRef]
3. Yang, H.; Wang, Z.; Song, K. A new hybrid grey wolf optimizer-feature weighted-multiple kernel-support vector regression technique to predict TBM performance. *Eng. Comput.* **2022**, *38*, 2469–2485. [CrossRef]
4. Ozdemir, L. *Development of Theoretical Equations for Predicting Tunnel Boreability*; Colorado School of Mines: Golden, CO, USA, 1977.
5. Rostami, J. *Development of a Force Estimation Model for Rock Fragmentation with Disc Cutters through Theoretical Modeling and Physical Measurement of Crushed Zone Pressure*; Colorado School of Mines: Golden, CO, USA, 1997.
6. Barton, N.R. *TBM Tunnelling in Jointed and Faulted Rock*; Crc Press: Boca Raton, FL, USA, 2000.
7. Bruland, A. Hard Rock Tunnel Boring. Ph.D. Thesis, Norwegian University of Science and Technology, Trondheim, Norway, 1998.
8. Yagiz, S. Utilizing rock mass properties for predicting TBM performance in hard rock condition. *Tunn. Undergr. Space Technol.* **2008**, *23*, 326–339. [CrossRef]
9. Shreyas, S.; Dey, A. Application of soft computing techniques in tunnelling and underground excavations: State of the art and future prospects. *Innov. Infrastruct. Solut.* **2019**, *4*, 46. [CrossRef]
10. Shahrour, I.; Zhang, W. Use of soft computing techniques for tunneling optimization of tunnel boring machines. *Undergr. Space* **2021**, *6*, 233–239. [CrossRef]
11. Sheil, B.B.; Suryasentana, S.K.; Mooney, M.A.; Zhu, H. Machine learning to inform tunnelling operations: Recent advances and future trends. *Proc. Inst. Civ. Eng. -Smart Infrastruct. Constr.* **2020**, *173*, 74–95.
12. Li, J.-B.; Chen, Z.-Y.; Li, X.; Jing, L.-J.; Zhangf, Y.-P.; Xiao, H.-H.; Wang, S.-J.; Yang, W.-K.; Wu, L.-J.; Li, P.-Y. Feedback on a shared big dataset for intelligent TBM Part I: Feature extraction and machine learning methods. *Undergr. Space* **2023**, *11*, 1–25. [CrossRef]
13. Feng, S.; Chen, Z.; Luo, H.; Wang, S.; Zhao, Y.; Liu, L.; Ling, D.; Jing, L. Tunnel boring machines (TBM) performance prediction: A case study using big data and deep learning. *Tunn. Undergr. Space Technol.* **2021**, *110*, 103636. [CrossRef]
14. Xu, C.; Liu, X.; Wang, E.; Wang, S. Prediction of tunnel boring machine operating parameters using various machine learning algorithms. *Tunn. Undergr. Space Technol.* **2021**, *109*, 103699. [CrossRef]
15. Hou, S.; Liu, Y.; Yang, Q. Real-time prediction of rock mass classification based on TBM operation big data and stacking technique of ensemble learning. *J. Rock Mech. Geotech. Eng.* **2022**, *14*, 123–143. [CrossRef]
16. Wu, Z.; Wei, R.; Chu, Z.; Liu, Q. Real-time rock mass condition prediction with TBM tunneling big data using a novel rock–machine mutual feedback perception method. *J. Rock Mech. Geotech. Eng.* **2021**, *13*, 1311–1325. [CrossRef]
17. Shen, S.-L.; Elbaz, K.; Shaban, W.M.; Zhou, A. Real-time prediction of shield moving trajectory during tunnelling. *Acta Geotech.* **2022**, *17*, 1533–1549. [CrossRef]
18. Sun, W.; Shi, M.; Zhang, C.; Zhao, J.; Song, X. Dynamic load prediction of tunnel boring machine (TBM) based on heterogeneous in-situ data. *Autom. Constr.* **2018**, *92*, 23–34. [CrossRef]
19. Shan, F.; He, X.; Armaghani, D.J.; Zhang, P.; Sheng, D. Success and challenges in predicting TBM penetration rate using recurrent neural networks. *Tunn. Undergr. Space Technol.* **2022**, *130*, 104728. [CrossRef]
20. Guo, J.; Ding, L.; Luo, H.; Zhou, C.; Ma, L. Wavelet prediction method for ground deformation induced by tunneling. *Tunn. Undergr. Space Technol.* **2014**, *41*, 137–151. [CrossRef]
21. Wang, R.; Li, D.; Chen, E.J.; Liu, Y. Dynamic prediction of mechanized shield tunneling performance. *Autom. Constr.* **2021**, *132*, 103958. [CrossRef]

22. Zhang, P.; Wu, H.; Chen, R.; Dai, T.; Meng, F.; Wang, H. A critical evaluation of machine learning and deep learning in shield-ground interaction prediction. *Tunn. Undergr. Space Technol.* **2020**, *106*, 103593. [CrossRef]
23. Zhou, C.; Xu, H.; Ding, L.; Wei, L.; Zhou, Y. Dynamic prediction for attitude and position in shield tunneling: A deep learning method. *Autom. Constr.* **2019**, *105*, 102840. [CrossRef]
24. Chen, R.-P.; Zhang, P.; Kang, X.; Zhong, Z.-Q.; Liu, Y.; Wu, H.-N. Prediction of maximum surface settlement caused by earth pressure balance (EPB) shield tunneling with ANN methods. *Soils Found.* **2019**, *59*, 284–295. [CrossRef]
25. Lin, S.; Zhang, N.; Zhou, A.; Shen, S. Time-series prediction of shield movement performance during tunneling based on hybrid model. *Tunn. Undergr. Space Technol.* **2022**, *119*, 104245. [CrossRef]
26. Grima, M.A.; Bruines, P.; Verhoef, P. Modeling tunnel boring machine performance by neuro-fuzzy methods. *Tunn. Undergr. Space Technol.* **2000**, *15*, 259–269. [CrossRef]
27. Salimi, A.; Rostami, J.; Moormann, C.; Delisio, A. Application of non-linear regression analysis and artificial intelligence algorithms for performance prediction of hard rock TBMs. *Tunn. Undergr. Space Technol.* **2016**, *58*, 236–246. [CrossRef]
28. Tiryaki, B. Application of artificial neural networks for predicting the cuttability of rocks by drag tools. *Tunn. Undergr. Space Technol.* **2008**, *23*, 273–280. [CrossRef]
29. Zhang, N.; Zhang, N.; Zheng, Q.; Xu, Y.-S. Real-time prediction of shield moving trajectory during tunnelling using GRU deep neural network. *Acta Geotech.* **2022**, *17*, 1167–1182. [CrossRef]
30. Armaghani, D.J.; Mohamad, E.T.; Narayanasamy, M.S.; Narita, N.; Yagiz, S. Development of hybrid intelligent models for predicting TBM penetration rate in hard rock condition. *Tunn. Undergr. Space Technol.* **2017**, *63*, 29–43. [CrossRef]
31. Benardos, A.; Kaliampakos, D. Modelling TBM performance with artificial neural networks. *Tunn. Undergr. Space Technol.* **2004**, *19*, 597–605. [CrossRef]
32. Suwansawat, S.; Einstein, H.H. Artificial neural networks for predicting the maximum surface settlement caused by EPB shield tunneling. *Tunn. Undergr. Space Technol.* **2006**, *21*, 133–150. [CrossRef]
33. Pourtaghi, A.; Lotfollahi-Yaghin, M. Wavenet ability assessment in comparison to ANN for predicting the maximum surface settlement caused by tunneling. *Tunn. Undergr. Space Technol.* **2012**, *28*, 257–271. [CrossRef]
34. Zhang, P.; Wu, H.-N.; Chen, R.-P.; Chan, T.H. Hybrid meta-heuristic and machine learning algorithms for tunneling-induced settlement prediction: A comparative study. *Tunn. Undergr. Space Technol.* **2020**, *99*, 103383. [CrossRef]
35. Zeng, J.; Roy, B.; Kumar, D.; Mohammed, A.S.; Armaghani, D.J.; Zhou, J.; Mohamad, E.T. Proposing several hybrid PSO-extreme learning machine techniques to predict TBM performance. *Eng. Comput.* **2021**, *38*, 3811–3827. [CrossRef]
36. Qin, C.; Shi, G.; Tao, J.; Yu, H.; Jin, Y.; Lei, J.; Liu, C. Precise cutterhead torque prediction for shield tunneling machines using a novel hybrid deep neural network. *Mech. Syst. Signal Process.* **2021**, *151*, 107386. [CrossRef]
37. Gao, X.; Shi, M.; Song, X.; Zhang, C.; Zhang, H. Recurrent neural networks for real-time prediction of TBM operating parameters. *Autom. Constr.* **2019**, *98*, 225–235. [CrossRef]
38. Jin, Y.; Qin, C.; Tao, J.; Liu, C. An accurate and adaptative cutterhead torque prediction method for shield tunneling machines via adaptative residual long-short term memory network. *Mech. Syst. Signal Process.* **2022**, *165*, 108312. [CrossRef]
39. Shi, G.; Qin, C.; Tao, J.; Liu, C. A VMD-EWT-LSTM-based multi-step prediction approach for shield tunneling machine cutterhead torque. *Knowl.-Based Syst.* **2021**, *228*, 107213. [CrossRef]
40. Acaroglu, O.; Ozdemir, L.; Asbury, B. A fuzzy logic model to predict specific energy requirement for TBM performance prediction. *Tunn. Undergr. Space Technol.* **2008**, *23*, 600–608. [CrossRef]
41. Mikaeil, R.; Naghadehi, M.Z.; Sereshki, F. Multifactorial fuzzy approach to the penetrability classification of TBM in hard rock conditions. *Tunn. Undergr. Space Technol.* **2009**, *24*, 500–505. [CrossRef]
42. Parsajoo, M.; Mohammed, A.S.; Yagiz, S.; Armaghani, D.J.; Khandelwal, M. An evolutionary adaptive neuro-fuzzy inference system for estimating field penetration index of tunnel boring machine in rock mass. *J. Rock Mech. Geotech. Eng.* **2021**, *13*, 1290–1299. [CrossRef]
43. Mahdevari, S.; Shahriar, K.; Yagiz, S.; Shirazi, M.A. A support vector regression model for predicting tunnel boring machine penetration rates. *Int. J. Rock Mech. Min. Sci.* **2014**, *72*, 214–229. [CrossRef]
44. Mokhtari, S.; Mooney, M.A. Predicting EPBM advance rate performance using support vector regression modeling. *Tunn. Undergr. Space Technol.* **2020**, *104*, 103520. [CrossRef]
45. Bai, X.-D.; Cheng, W.-C.; Li, G. A comparative study of different machine learning algorithms in predicting EPB shield behaviour: A case study at the Xi'an metro, China. *Acta Geotech.* **2021**, *16*, 4061–4080. [CrossRef]
46. Kannangara, K.P.M.; Zhou, W.; Ding, Z.; Hong, Z. Investigation of feature contribution to shield tunneling-induced settlement using Shapley additive explanations method. *J. Rock Mech. Geotech. Eng.* **2022**, *14*, 1052–1063. [CrossRef]
47. Zhang, P.; Chen, R.; Wu, H. Real-time analysis and regulation of EPB shield steering using Random Forest. *Autom. Constr.* **2019**, *106*, 102860. [CrossRef]
48. Salimi, A.; Rostami, J.; Moormann, C. Application of rock mass classification systems for performance estimation of rock TBMs using regression tree and artificial intelligence algorithms. *Tunn. Undergr. Space Technol.* **2019**, *92*, 103046. [CrossRef]
49. Zhang, Q.; Hu, W.; Liu, Z.; Tan, J. TBM performance prediction with Bayesian optimization and automated machine learning. *Tunn. Undergr. Space Technol.* **2020**, *103*, 103493. [CrossRef]
50. Zhou, J.; Qiu, Y.; Armaghani, D.J.; Zhang, W.; Li, C.; Zhu, S.; Tarinejad, R. Predicting TBM penetration rate in hard rock condition: A comparative study among six XGB-based metaheuristic techniques. *Geosci. Front.* **2021**, *12*, 101091. [CrossRef]

51. Yagiz, S.; Karahan, H. Prediction of hard rock TBM penetration rate using particle swarm optimization. *Int. J. Rock Mech. Min. Sci.* **2011**, *48*, 427–433. [CrossRef]
52. Yagiz, S.; Karahan, H. Application of various optimization techniques and comparison of their performances for predicting TBM penetration rate in rock mass. *Int. J. Rock Mech. Min. Sci.* **2015**, *80*, 308–315. [CrossRef]
53. Huang, X.; Zhang, Q.; Liu, Q.; Liu, X.; Liu, B.; Wang, J.; Yin, X. A real-time prediction method for tunnel boring machine cutter-head torque using bidirectional long short-term memory networks optimized by multi-algorithm. *J. Rock Mech. Geotech. Eng.* **2022**, *14*, 798–812. [CrossRef]
54. Zhou, J.; Qiu, Y.; Zhu, S.; Armaghani, D.J.; Khandelwal, M.; Mohamad, E.T. Estimation of the TBM advance rate under hard rock conditions using XGBoost and Bayesian optimization. *Undergr. Space* **2021**, *6*, 506–515. [CrossRef]
55. Armaghani, D.J.; Koopialipoor, M.; Marto, A.; Yagiz, S. Application of several optimization techniques for estimating TBM advance rate in granitic rocks. *J. Rock Mech. Geotech. Eng.* **2019**, *11*, 779–789. [CrossRef]
56. Harandizadeh, H.; Armaghani, D.J.; Asteris, P.G.; Gandomi, A.H. TBM performance prediction developing a hybrid ANFIS-PNN predictive model optimized by imperialism competitive algorithm. *Neural Comput. Appl.* **2021**, *33*, 16149–16179. [CrossRef]
57. Chen, Z.; Zhang, Y.; Li, J.; Li, X.; Jing, L. Diagnosing tunnel collapse sections based on TBM tunneling big data and deep learning: A case study on the Yinsong Project, China. *Tunn. Undergr. Space Technol.* **2021**, *108*, 103700. [CrossRef]
58. Liu, B.; Wang, R.; Zhao, G.; Guo, X.; Wang, Y.; Li, J.; Wang, S. Prediction of rock mass parameters in the TBM tunnel based on BP neural network integrated simulated annealing algorithm. *Tunn. Undergr. Space Technol.* **2020**, *95*, 103103. [CrossRef]
59. Yagiz, S. Assessment of brittleness using rock strength and density with punch penetration test. *Tunn. Undergr. Space Technol.* **2009**, *24*, 66–74. [CrossRef]
60. Javad, G.; Narges, T. Application of artificial neural networks to the prediction of tunnel boring machine penetration rate. *Min. Sci. Technol.* **2010**, *20*, 727–733. [CrossRef]
61. Armaghani, D.J.; Faradonbeh, R.S.; Momeni, E.; Fahimifar, A.; Tahir, M. Performance prediction of tunnel boring machine through developing a gene expression programming equation. *Eng. Comput.* **2018**, *34*, 129–141. [CrossRef]
62. Koopialipoor, M.; Tootoonchi, H.; Jahed Armaghani, D.; Tonnizam Mohamad, E.; Hedayat, A. Application of deep neural networks in predicting the penetration rate of tunnel boring machines. *Bull. Eng. Geol. Environ.* **2019**, *78*, 6347–6360. [CrossRef]
63. Koopialipoor, M.; Fahimifar, A.; Ghaleini, E.N.; Momenzadeh, M.; Armaghani, D.J. Development of a new hybrid ANN for solving a geotechnical problem related to tunnel boring machine performance. *Eng. Comput.* **2020**, *36*, 345–357. [CrossRef]
64. Wang, Q.; Xie, X.; Shahrour, I. Deep learning model for shield tunneling advance rate prediction in mixed ground condition considering past operations. *IEEE Access* **2020**, *8*, 215310–215326. [CrossRef]
65. Zhou, J.; Yazdani Bejarbaneh, B.; Jahed Armaghani, D.; Tahir, M. Forecasting of TBM advance rate in hard rock condition based on artificial neural network and genetic programming techniques. *Bull. Eng. Geol. Environ.* **2020**, *79*, 2069–2084. [CrossRef]
66. Bardhan, A.; Kardani, N.; GuhaRay, A.; Burman, A.; Samui, P.; Zhang, Y. Hybrid ensemble soft computing approach for predicting penetration rate of tunnel boring machine in a rock environment. *J. Rock Mech. Geotech. Eng.* **2021**, *13*, 1398–1412. [CrossRef]
67. Lin, S.; Shen, S.; Zhang, N.; Zhou, A. Modelling the performance of EPB shield tunnelling using machine and deep learning algorithms. *Geosci. Front.* **2021**, *12*, 101177. [CrossRef]
68. Lin, S.-S.; Shen, S.-L.; Zhou, A. Real-time analysis and prediction of shield cutterhead torque using optimized gated recurrent unit neural network. *J. Rock Mech. Geotech. Eng.* **2022**, *14*, 1232–1240. [CrossRef]
69. Salimi, A.; Rostami, J.; Moormann, C.; Hassanpour, J. Introducing Tree-Based-Regression Models for Prediction of Hard Rock TBM Performance with Consideration of Rock Type. *Rock Mech. Rock Eng.* **2022**, *55*, 4869–4891. [CrossRef]
70. Yang, J.; Yagiz, S.; Liu, Y.-J.; Laouafa, F. Comprehensive evaluation of machine learning algorithms applied to TBM performance prediction. *Undergr. Space* **2022**, *7*, 37–49. [CrossRef]
71. Zhang, W.; Li, H.; Li, Y.; Liu, H.; Chen, Y.; Ding, X. Application of deep learning algorithms in geotechnical engineering: A short critical review. *Artif. Intell. Rev.* **2021**, *54*, 5633–5673. [CrossRef]
72. Ahangari, K.; Moeinossadat, S.R.; Behnia, D. Estimation of tunnelling-induced settlement by modern intelligent methods. *Soils Found.* **2015**, *55*, 737–748. [CrossRef]
73. Neaupane, K.M.; Adhikari, N. Prediction of tunneling-induced ground movement with the multi-layer perceptron. *Tunn. Undergr. Space Technol.* **2006**, *21*, 151–159. [CrossRef]
74. Santos, O.J., Jr.; Celestino, T.B. Artificial neural networks analysis of Sao Paulo subway tunnel settlement data. *Tunn. Undergr. Space Technol.* **2008**, *23*, 481–491. [CrossRef]
75. Boubou, R.; Emeriault, F.; Kastner, R. Artificial neural network application for the prediction of ground surface movements induced by shield tunnelling. *Can. Geotech. J.* **2010**, *47*, 1214–1233. [CrossRef]
76. Dindarloo, S.R.; Siami-Irdemoosa, E. Maximum surface settlement based classification of shallow tunnels in soft ground. *Tunn. Undergr. Space Technol.* **2015**, *49*, 320–327. [CrossRef]
77. Goh, A.T.C.; Zhang, W.; Zhang, Y.; Xiao, Y.; Xiang, Y. Determination of earth pressure balance tunnel-related maximum surface settlement: A multivariate adaptive regression splines approach. *Bull. Eng. Geol. Environ.* **2018**, *77*, 489–500. [CrossRef]
78. Zhang, W.; Li, H.; Wu, C.; Li, Y.; Liu, Z.; Liu, H. Soft computing approach for prediction of surface settlement induced by earth pressure balance shield tunneling. *Undergr. Space* **2021**, *6*, 353–363. [CrossRef]
79. Zhang, L.; Wu, X.; Ji, W.; AbouRizk, S.M. Intelligent approach to estimation of tunnel-induced ground settlement using wavelet packet and support vector machines. *J. Comput. Civ. Eng.* **2017**, *31*, 04016053. [CrossRef]

80. Gao, M.-Y.; Zhang, N.; Shen, S.-L.; Zhou, A. Real-time dynamic earth-pressure regulation model for shield tunneling by integrating GRU deep learning method with GA optimization. *IEEE Access* **2020**, *8*, 64310–64323. [CrossRef]
81. Erharter, G.H.; Marcher, T. On the pointlessness of machine learning based time delayed prediction of TBM operational data. *Autom. Constr.* **2021**, *121*, 103443. [CrossRef]
82. Gao, B.; Wang, R.; Lin, C.; Guo, X.; Liu, B.; Zhang, W. TBM penetration rate prediction based on the long short-term memory neural network. *Undergr. Space* **2021**, *6*, 718–731. [CrossRef]
83. Zhang, P.; Chen, R.; Dai, T.; Wang, Z.; Wu, K. An AIoT-based system for real-time monitoring of tunnel construction. *Tunn. Undergr. Space Technol.* **2021**, *109*, 103766. [CrossRef]
84. Qin, C.; Shi, G.; Tao, J.; Yu, H.; Jin, Y.; Xiao, D.; Zhang, Z.; Liu, C. An adaptive hierarchical decomposition-based method for multi-step cutterhead torque forecast of shield machine. *Mech. Syst. Signal Process.* **2022**, *175*, 109148. [CrossRef]
85. Shan, F.; He, X.; Armaghani, D.J.; Zhang, P.; Sheng, D. Response to Discussion on "Success and challenges in predicting TBM penetration rate using recurrent neural networks" by Georg H. Erharter, Thomas Marcher. *Tunn. Undergr. Space Technol.* **2023**, 105064. [CrossRef]

eng

MDPI

Article

A Comparison of Personalized and Generalized LSTM Neural Networks for Deriving VCG from 12-Lead ECG

Prashanth Shyam Kumar [1,*], Mouli Ramasamy [1] and Vijay K. Varadan [1,2]

[1] The Department of Engineering Science and Mechanics, The Pennsylvania State University, 212 Earth-Engineering Sciences Bldg., State College, PA 16802, USA
[2] The Department of Neurosurgery, Milton S. Hershey Medical Center, 500 University Dr., Hershey, PA 17033, USA
* Correspondence: pshyamku@alumni.psu.edu

Abstract: Vectorcardiography (VCG) is a valuable diagnostic tool that complements the standard 12-lead ECG by offering additional spatiotemporal information to clinicians. However, due to the need for additional measurement hardware and too many electrodes in a clinical scenario if performed along with a standard 12-lead, there is a need to find methods to derive the VCG from the ECG. We have evaluated the use of Long Short-term Memory (LSTM) neural networks to learn the transformation from 12-lead ECG to VCG that is applicable across subjects and for each subject. We refer to these networks as generalized and personalized, respectively. We calculated the Root Mean Square Error (RMSE), R^2, and Pearson correlation coefficient to compare waveforms of derived and actual VCG. We also extracted and compared diagnostic parameters from VCG, namely the QRS-loop magnitude, T-loop magnitude, and QRS-T spatial angle, from actual and derived VCGs using the Pearson correlation coefficient and Bland Altman limits of agreement. The personalized models performed better than generalized models in waveform comparisons and in the error of extracted diagnostic parameters from VCG waveforms. The use of personalized transformations for the derivation of VCG from standard 12-lead has the potential to improve and augment the diagnostic yield and accuracy of a standard 12-lead interpretation.

Keywords: ECG; vectorcardiography; LSTM networks; personalized medicine; Bayesian optimization

Citation: Shyam Kumar, P.; Ramasamy, M.; Varadan, V.K. A Comparison of Personalized and Generalized LSTM Neural Networks for Deriving VCG from 12-Lead ECG. *Eng* **2023**, *4*, 1337–1355. https://doi.org/10.3390/eng4020078

Academic Editor: Antonio Gil Bravo

Received: 2 February 2023
Revised: 27 April 2023
Accepted: 6 May 2023
Published: 10 May 2023

1. Background

Clinical ECG consists of 12 leads (S12)—namely limb leads I, II, and III, augmented leads aVR, aVL, aVF, and precordial leads V1 through V6. Vectorcardiography (VCG) [1] is complementary to the S12. It is essentially the spatiotemporal representation of the cardiac vector in 3 orthogonal planes—namely vertical, transverse, and sagittal planes. S12 is the standard whereas VCG is rarely acquired. However, several conditions have more prominent VCG changes than S12, so it is a useful complement to S12. Furthermore, dynamic spatial and temporal information that can be derived from VCGs is unavailable from an ECG, which may enhance the automatic assessments of cardiovascular diseases [2].

Ernest Frank introduced the XYZ lead system known as vectorcardiography to provide a 3-dimensional representation of the cardiac vector. Figure 1 shows the electrode placement for vectorcardiogram and an example of a vectorcardiogram tracing for a healthy male subject.

Following the illustration in Figure 1, a vector tracing the boundary of this 3D object circumscribed by the vectorcardiography is the cardiac vector. Ideally, these ECG leads—X, Y, and Z—would be orthogonal to each other and form a basis for the cartesian space spanned by the cardiac vector. Figure 2 illustrates how the temporal X, Y, and Z lead waveforms translate to a spatiotemporal VCG.

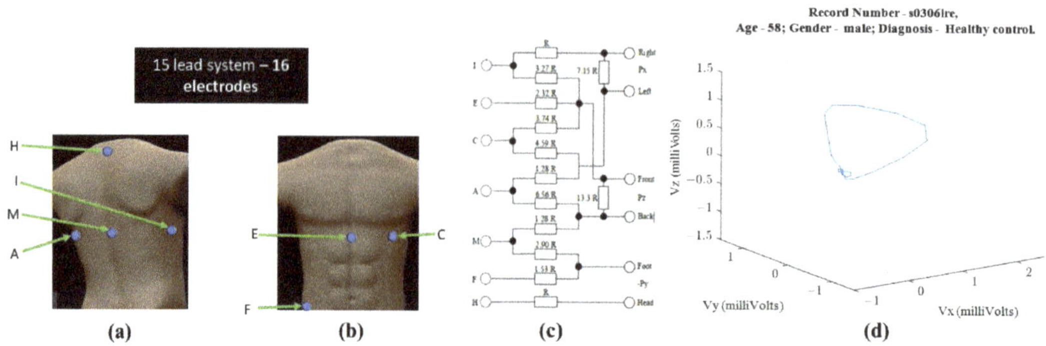

Figure 1. For obtaining VCG—(**a**) ventral electrode positions, (**b**) dorsal electrode positions, (**c**) resistor network needed to compensate for non-homogenous human tissue—additional instrumentation beyond standard 12-lead ECG equipment [1] (**d**) 3-D illustration of a single heartbeat from a 58-year-old healthy male [3,4].

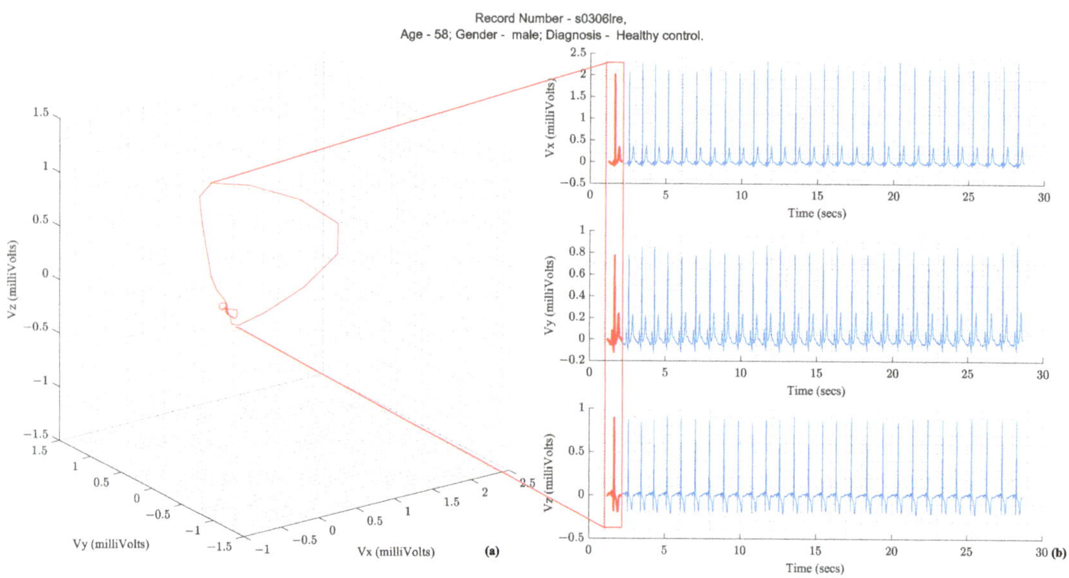

Figure 2. (**a**) Vectorcardiography of one heartbeat (**b**) Tracing of Vx, Vy, and Vz, which are the three leads of a Vectorcardiogram [3,4].

The S12 requires ten electrodes on the skin while the Frank XYZ requires only 7 electrodes. There is only one electrode position in common (i.e., the left leg). Suppose all 15 leads are to be recorded with an ECG acquisition system, then sixteen electrodes should be placed on the patient's skin. Another practical issue with the location of the Frank XYZ electrodes is the rear electrodes. Patients can sleep on their backs, but having cables on their backs can be uncomfortable.

1.1. Diagnostic Importance of VCG and Its Complementarity to ECG

Over several decades of research, three parameters extracted from the VCG waveform are considered diagnostically important. They are the QRS amplitude, T-loop magnitudes, and Spatial QRS-T loop angles. Figure 3 illustrates these parameters on a vectorcardiogram.

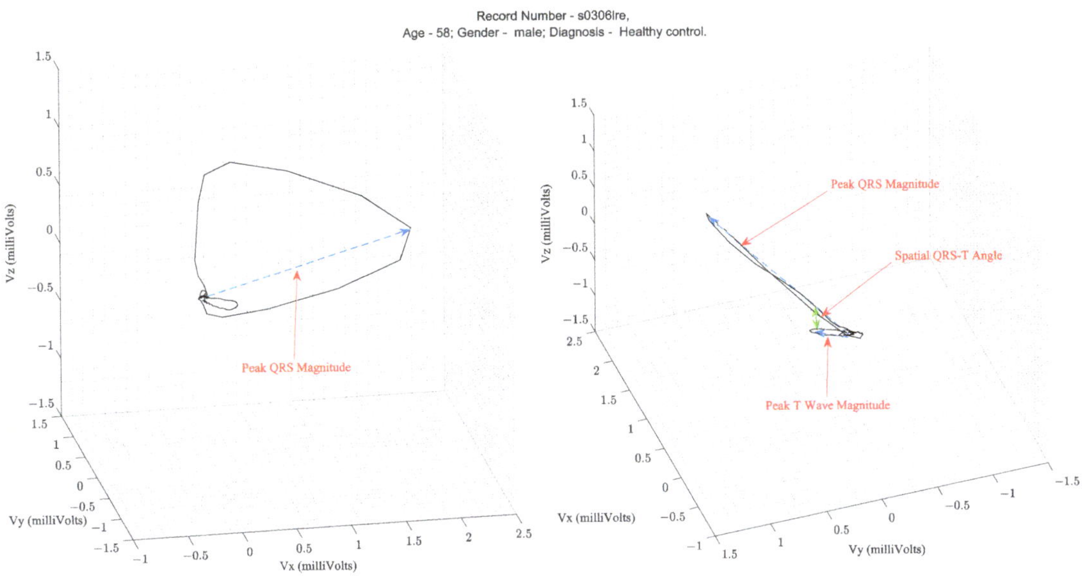

Figure 3. Illustration of the parameters that are extracted from a VCG peak QRS magnitude, peak T wave magnitude, and spatial QRS to T angle [3,4].

Table 1 lists the clinical applications of these parameters that have been validated in the literature. The spatial QRS-T angle parameter has been shown to be useful for risk stratification for cardiac events, evaluation of incident coronary disease and heart failure, and efficacy of therapy for adult hypertension and diabetes mellitus [5]. For example, in the PTB diagnostic database [3,4] ECG used in this study, the mean and standard deviation of the Spatial QRS-T angles from patients with MI and healthy controls were $87.9^\circ \pm 46.84^\circ$ and $52.95^\circ \pm 35.76^\circ$, respectively, as computed using the VCG parameter extraction algorithms described in this study.

Table 1. Diagnostic parameters of interest computed from Vectorcardiograms.

VCG Signal Feature	Clinical Application
QRS magnitude	Predicts ventricular arrhythmia in selected cohorts
T magnitude	Lower values are associated with an increased risk of cardiac events
Spatial QRS-T angle	In addition to assisting with risk stratification for cardiac events, the angle is also useful for the evaluation of • Incident coronary heart disease • Heart failure • Efficacy of therapy for • Adult hypertension • Diabetes mellitus

Additionally, there are specific conditions where the VCG is considered superior to the ECG. VCG is more sensitive and specific than ECG in detecting atrial and ventricular enlargements. Due to the greater spatially localized information in a VCG, the suspicion of electrically inactive areas in the septal or anteroseptal walls of the left ventricle can be assessed with a VCG. The left ventricular mass, which is currently assessed with an echocardiogram, can be assessed with a VCG. VCG findings are better correlated to echocardiography findings than ECG findings. VCG has a greater diagnostic sensitivity than ECG for AMI when associated with a left anterior fascicular block [6]. Lastly, the myocardial damage caused by Chagas disease can be assessed with VCG findings complementing ECG findings [7]. Figure 4 illustrates the conditions and the location of the affected heart anatomy.

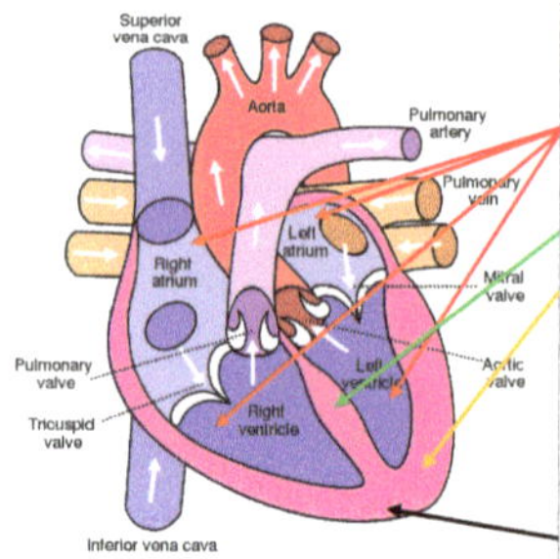

Clinical condition	Attribute of VCG that can complement ECG
Detection of atrial enlargement and diagnosis of left ventricular enlargement	Greater sensitivity and specificity for VCG
Suspicion of electrically inactive area in the septal or anteroseptal wall of the left ventricle	Greater spatial information in VCG
Determining left ventricular mass	VCG is better correlated to echocardiography findings than ECG
Acute myocardial infarction (AMI), when associated with left anterior fascicular block (LAFB)	VCG has a greater diagnostic sensitivity in comparison to the ECG
Degree of myocardial damage in patients with chagas disease	VCG derived parameters can be used as a complementary diagnostic technique

Figure 4. List of conditions where VCG is more effective than ECG for diagnosis and the corresponding anatomic location of the affected region of the heart. Authored by Wapcaplet and shared under Creative Commons (CC BY-NC-SA 3.0). The white arrows indicate the direction of blood flow.

Since VCGs are not acquired during regular clinical settings, but standard 12-lead waveforms are acquired, there is a need to derive VCG from the 12-lead ECG. Specifically, an arbitrarily complex transformation mapping the 12-lead ECG to the VCG is needed. Several research efforts are focused on arriving at the linear transformation of ECGs from standard 12-lead to Frank XYZ. However, the transformation is likely to be arbitrarily complex due to multiple underlying variabilities from person to person in terms of the distribution of fat, muscle, and organs in the torso where the ECG leads are measured. These complex variations suggest that we need methods capable of approximating arbitrarily complex transformations, such as neural networks [8,9]. Therefore, we used a class of neural networks, namely Long Short-term Memory (LSTM) [10,11], that might be best suited for time-series data regression tasks, such as transforming leads. Moreover, most recently, Sohn et al. [12] reported the successful use of LSTM networks to achieve accurate lead transformations. The following are the original contributions of this research:

- We apply LSTM networks to the task of deriving VCG from 12-lead ECGs. Since LSTM networks require the pre-specification of several hyperparameters, we apply Bayesian global optimizations to find the combination of these parameters that is optimal to obtain the least error between derived VCG and actual VCG;
- We apply transfer learning to obtain personalized transformations for each subject as part of the data set;
- We compare the accuracy of extraction of VCG diagnostic parameters from derived VCG and actual VCG.

1.2. Related Work

Linear regression has been explored in the literature for lead transformation. Some studies have used open, publicly available data sets, whereas others have used closed data sets or data sets acquired with custom built hardware devices. Between 1986 and 2009, the lead transformation of interest was from S12 to Frank XYZ. Closed data sets were used for some studies [13–16] and open for others [17,18]. A neural network-based transformation was first proposed in 2010 [19]. Table 2 summarizes the works in the literature that focused on obtaining Frank XYZ from S12. Since then, several efforts have been made in reducing leads required to be monitored while retaining the diagnostic power of S12. Most of the studies have tried to derive S12 from a three-lead ECG [12,19–24].

Table 2. List of related works that evaluated lead transformations from S12 to Frank XYZ. (N is the number of samples per ECG channel, y is the actual acquired ECG, and $\hat{y}$ is the output ECG from the transformation).

Publication	Data Availability/Transformation Method	Reported Performance Metrics
Bjerle P et al., 1986 [11]	closed/Linear regression	Amplitudes of ECG waves QRS, ST and T
Edenbrandt L et al., 1988 [12]		Amplitude of R wave
Kors J. A et al., 1990 [13]		Distance Measure $D = \frac{1}{K}\sum_{k=1}^{K} \frac{\lvert V_k - V_k^* \rvert}{\lvert V_k \rvert}$
Hyttinen J et al., 1995 [14]		$Pearson\ Correlation\ Coefficient = \left\{ \frac{\sum_{i=1}^{N} \hat{y}[i] * y[i]}{\left(\sum_{i=1}^{N} y[i]^2 * \sum_{i=1}^{N} \hat{y}[i]^2\right)^{\frac{1}{2}}} \right\}$
Guillem MS et al., 2006 [15]	open/Linear Regression	$RMSE = \sqrt{\frac{\sum_{i=1}^{N} (y[i] - \hat{y}[i])^2}{N}}$; Pearson correlation coefficient
Dawson D et al., 2009 [16]		$R^2 = \left\{1 - \frac{\sum_{i=1}^{N} [\hat{y}[i] - y[i]]^2}{\sum_{i=1}^{N} [y[i]]^2}\right\} * 100.$
This work	PTB diagnostic ECG [3,4]. Open/LSTM	RMS error; Correlation coefficient, R2, QRS magnitude, T magnitude, and Spatial QRS-T angle

Several studies have used closed data sets that are unavailable to other researchers. We used the PTB diagnostic ECG repository for this study [3,4].

The Root Mean Square (RMS) and correlation coefficient are the most commonly reported metrics used to evaluate the error between generated or derived ECG compared to the ground truth waveform. In the literature, R-squared is used. Table 2 includes the definitions and equations of these metrics. Some clinically relevant VCG-derived parameters can also be compared between the derived ECG leads and the ground truth waveform. Therefore, the RMS error, correlation coefficient, R^2, QRS amplitude or magnitude, T wave amplitude or magnitude, and spatial QRS-T angle form a complete assessment.

The coefficients of the transformations reported in the literature are presented in Table S1 in the Supplementary Materials.

2. Materials and Methods

2.1. Experimental Setup

We had previously presented the methodology of training a generalized model and then applying transfer learning for a different problem, which was for the S12 lead ECG derivation from a subset of leads, namely Lead II, V2, and V6 [25]. However, in this paper, we evaluate the performance of the transformations from S12 lead to Frank XYZ lead. We present the methodology here for the convenience of the reader. All data analysis programs and applications were implemented using MATLAB 2021a Update 5 version 9.10.0.1739362 (MathWorks Inc., Natick, MA, USA) on a system with an Intel processor (i7-7820X), RTX 3090 graphics processing unit (NVIDIA Corp., Santa Clara, CA, USA), and 32 GB of RAM.

2.1.1. Source of Data and Data Preparation

The PTB ECG database available on Physionet [3,4] contains fifteen lead ECGs sampled at 1 kHz from 249 patients. Some patients have multiple records, bringing the total number of ECGs to 549. Figure 5 plots the histograms that summarize the data set's characteristics.

This data set contains only one diagnosis per patient. As shown in Figure 5, a large proportion of the data set is MI patients and healthy controls.

The ECG signals were band pass filtered using a second order Butterworth filter with a passband from 0.05 Hz to 45 Hz, which is the bandwidth used for long-term rhythm monitoring according to AAMI standards [26]. Following filtering and suppression of frequencies beyond 40 Hz, the signal was down sampled using decimation to avoid aliasing effects. First, ECG signal content in adults was below 100 Hz [26], so 200 Hz satisfied the

Nyquist rate requirements to avoid aliasing. Second, lower sampling rates reduced the amount of data so iterations could be faster. Figure 6 shows an example of a recording from the data set before and after applying the above-stated data preparation steps. The data processing steps cause no visible distortion.

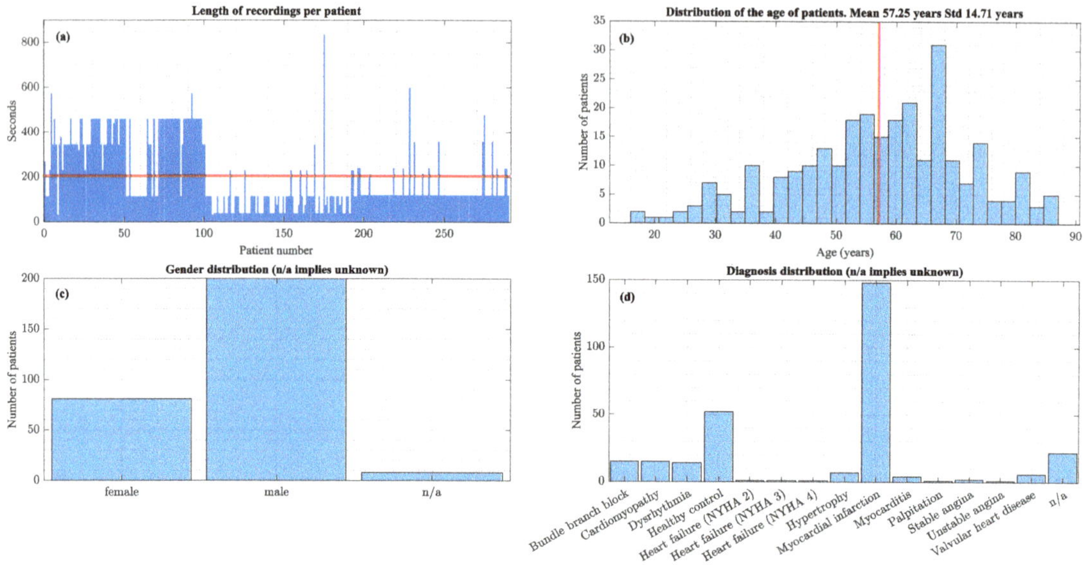

Figure 5. (**a**) Distribution of recording lengths (**b**) Age distribution, (**c**) Gender Distribution, (**d**) Diagnosis distribution. The red lines indicate the mean values.

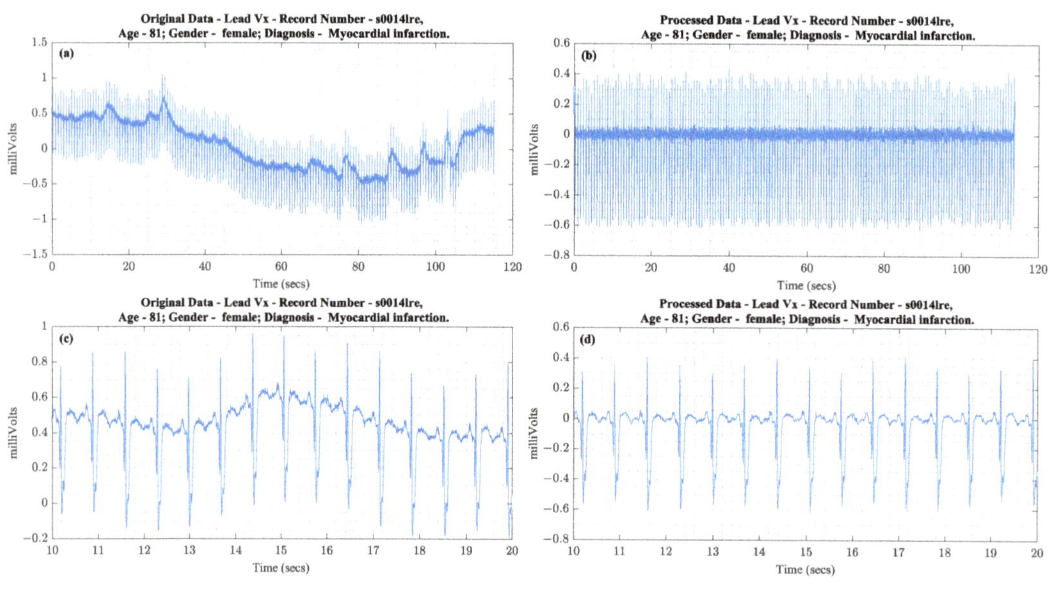

Figure 6. (**a**) Complete raw Vx lead of ECG waveform, (**b**) Complete processed Vx waveform, (**c**) 10 s of the raw waveform, (**d**) 10 s of the processed waveform.

Three recordings were removed from the data set because of missing data or complete data corruption by noise. Table 3 lists them and the reason for exclusion.

Table 3. List of recordings that were excluded due to low signal quality or no signal.

Rejected Recording	Reason for Exclusion
Record 291 from patient 095	V1 lead missing
Record 537 from patient 285	No ECG data
Record 453 from patient 220	Lead III data missing

2.1.2. Personalized Training Data Preparation

Data augmentation was performed using the sliding window method as used in [12]. Each sliding window was 17 s, and the overlap size was 16 s. We chose a window size of 17 s solely for formatting and initial review purposes for the S12 leads. We required 12 s of data to chart S12 in a standard clinical ECG format. We also had symmetrically cropped 2.5 s of data on both ends of each 17-s-long segment so that we could be consistent across all segments. The beginning and end of several records had settling noise, such as baseline wander or powerline noise, so we removed these segments during data preparation. The 16 s overall was chosen to maximize overlap and number of training samples available following similar approaches in the literature that showed good performance using LSTM.

2.2. Neural Network Architecture

As mentioned earlier in Section 1.2, we used LSTM neural networks to learn a transfer function from S12 to Frank XYZ. The principal constituents of the LSTM network were the input gate (i), forget gate (f), and output gate (o). In addition, each LSTM network consisted of a cell state that was updated upon each timestep of input presented to the network.

The LSTM cell is a type of recurrent neural network. The output at time $t-1$ influences the output at time t. Figure 7 presents a depiction of a single LSTM cell.

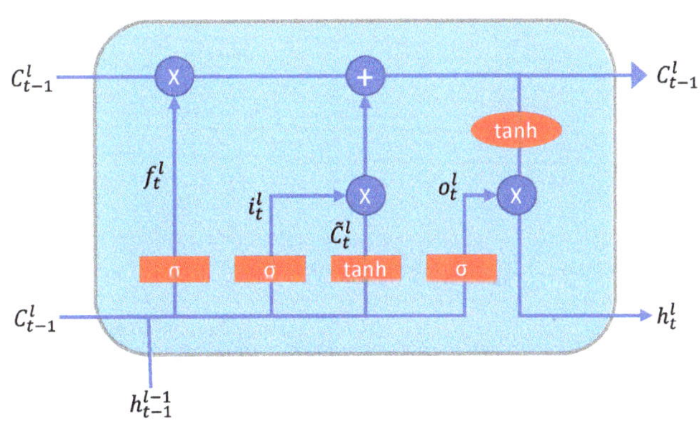

Figure 7. Depiction of the computation occurring in a unit LSTM cell.

The training process for the LSTM neural network involved a standard four-step sequence: forward propagation, cost computation, backward propagation, and weight update. This process was repeated for each item in the training set multiple times. The loss function was the mean squared error without normalization of the number of output dimensions or the number of ECG channels. Weight updates were performed using Adam optimizer [27].

$$loss = \frac{1}{2S}\sum_{i=1}^{S}\sum_{j=1}^{R}\left(\hat{y}_{ij} - y_{ij}\right)^2 \quad (1)$$

where S was sequence length of each ECG channel, R equaled the number of ECG channels in the output, $\hat{y}$ was the instantaneous estimated output, and y was the instantaneous actual sample of ECG.

The LSTM network required selection of the following list of hyperparameters and architecture specifications prior to training—number of layers, number of hidden units per layer, learning rate, minibatch size, learning rate schedule (i.e., periodic changes as training progresses or fixed with no changes), and finally, the weight optimizer parameters—momentum coefficient (β_1) and root mean square (RMS) propagation coefficient (β_2).

The LSTM network architecture requires the specification of a list of hyperparameters. Bayesian optimization (BO) is a global optimization approach that is preferred in the literature for computationally intensive functions like the training of neural network [28].

2.3. *Network Training Options*

The 546 available records were split 80/20 between training and testing. The training set had 437 records, and the test had 109. Network training was performed over 100 epochs for all networks, including the personalized networks.

BO did not include number of layers as an optimizable variable, so we were able to infer the meaning of the number of layers in a controlled way rather than as part of a probabilistic search like BO. Therefore, independent hyperparameter tuning was performed for 1- through 5-layer networks and the results were compared across the number of layers to understand the impact of multiple LSTM layers on performance.

Hyperparameter Optimization Using BO

BO was used to find the optimal combination of values for the hyperparameters needed for the LSTM networks. The application method of BO included three key elements:

1. A Gaussian Process Model ($Q(f|x,\ y)$)—final validation RMSE was the objective function $f(x)$. The kernel function for the model was ARD Matérn 5/2;
2. A procedure for updating ($Q(f|x,\ y)$) after each iteration;
3. An acquisition function $a(x)$ that was 'expected improvement' [29].

$$(Expected\ Improvement)EI(x,\ Q) = E_Q\left[\max\left(0,\ \mu_Q\left(x_{Optimal}\right) - f(x)\right)\right] \quad (2)$$

where $\mu_Q\left(x_{Optimal}\right)$ was the minimum of the posterior mean and $x_{Optimal}$ was the location of this minimum in hyperparameter space. To boost the inclination for sampling x and prevent over-sampling of a region within the hyperparameter space around a local minimum of x, another criterion was added in addition to the one used to select $a(x)$. This condition was implemented as an additional restriction when choosing the subsequent x for evaluation. A candidate x had to satisfy the criteria in (3) to be selected as the subsequent point to be evaluated.

$$\sigma_{f(x)} \geq 0.8 * \sigma \quad (3)$$

where $\sigma_{f(x)}$ represented the standard deviation of the posterior objective function at x and σ the additive noise's posterior standard deviation.

The optimizable variables or the hyperparameters had to be defined in terms of bounds and the type of transformation to be applied prior to sampling. Table S2 in the Supplementary Materials lists the hyperparameters optimized for networks ranging from 1-layer to 5-layer. For each objective function evaluation, networks were trained for 100 epochs to allow adequate iterations to reach the lowest final RMSE.

2.4. *Training Personalized Networks*

Transfer learning is the process of further training a pre-trained neural network using a different data set or subset of data [30]. We trained a personalized neural network for each patient using transfer learning with the optimal network architecture and hyperparameter combinations found by BO. The data set had 549 ECG recordings from 290 patients, averaging 200 s per recording, with a few patients having only 100 s of data.

As described in Section 2.1.2, network training was performed over 100 epochs with personalized data.

2.5. Evaluation of Extracted VCG Parameters

As described previously in Section 1.1, three VCG extracted parameters—peak QRS magnitude, peak T wave magnitude, and spatial QRS-T angle—were of diagnostic importance. These parameters were computed from the actual Frank XYZ leads and the derived Frank XYZ leads from all the transformations. The algorithm for calculating these parameters began with the detection of the R wave of the ECG in the Vx lead. The QRS duration and the R wave durations were defined relative to the corresponding RR interval, as depicted in Figure 8.

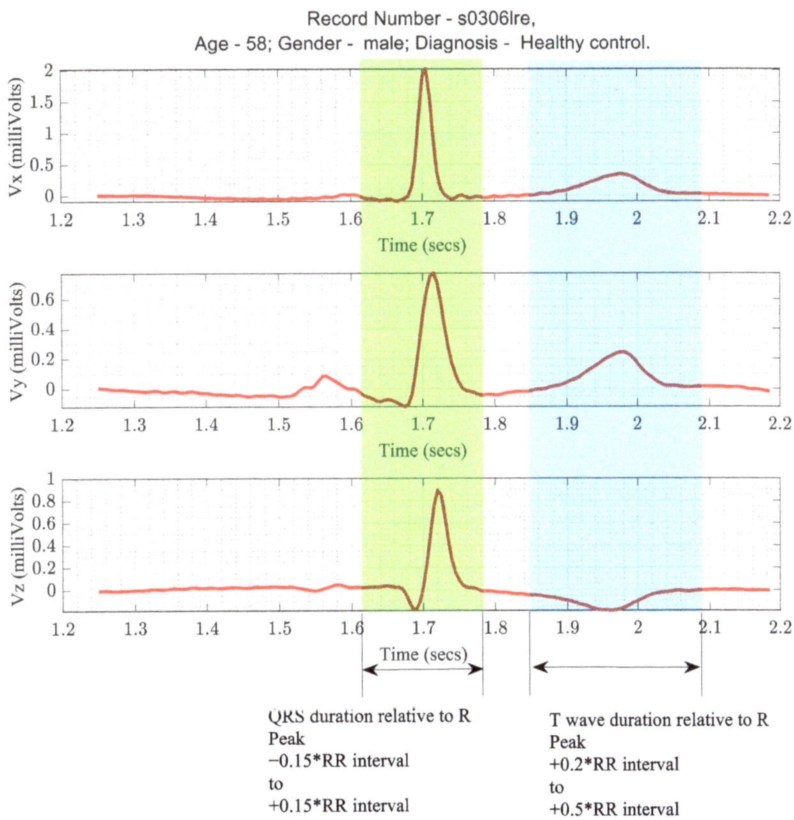

Figure 8. Timing criteria for the segmentation of QRS and T wave patterns in ECG.

The peak magnitudes of QRS were calculated as the maximum of the L2 norm of (V_x, V_y, V_z) within the QRS duration time window. Similarly, the maximum within the T wave duration time window was the peak T wave magnitude. The QRS-T angle was computed using Equations (4) through (6).

$$|QRS_{area}| = \sqrt{(QRS_x)^2 + (QRS_y)^2 + (QRS_z)^2} \tag{4}$$

$$|T_{area}| = \sqrt{(T_x)^2 + (T_y)^2 + (T_z)^2} \quad (5)$$

$$Mean\ Spatial\ QRS - T\ angle = cos\left(\frac{QRS_xT_x + QRS_yT_y + QRS_zT_z}{|QRS_{area}||T_{area}|}\right)^{-1} \quad (6)$$

where QRS_x, QRS_y, and QRS_z were the area under the curve of the QRS complex in the X, Y, and Z leads, respectively, and T_x, T_y, and T_z were the area under the curve of the T wave in the X, Y, and Z leads, respectively. Several possible integration methods (for example, the Trapezoidal rule, the Simpson's rule, or the Simpson's 3/8) could be used to calculate the area [31]. In this implementation, we used the trapezoidal rule. We compared the extracted parameters using Pearson's correlation coefficient and Bland Altman limits of agreement.

3. Results

As part of the BO experiment, we trained 250 neural networks: 50 networks each for 1- through 5-layer networks.

Figure 9 shows the overall results of the BO for 1-to-5-layer neural networks to transform ECG leads from S12 to Frank XYZ leads. The following were the set of hyperparameters that resulted in the optimal Final RMSE: Number of Hidden Units = 47; Mini Batch Size = 27; Learning rate schedule = Piecewise; β_1 = 0.90025; β_2 = 0.90035; and Learning Rate = 0.062561.

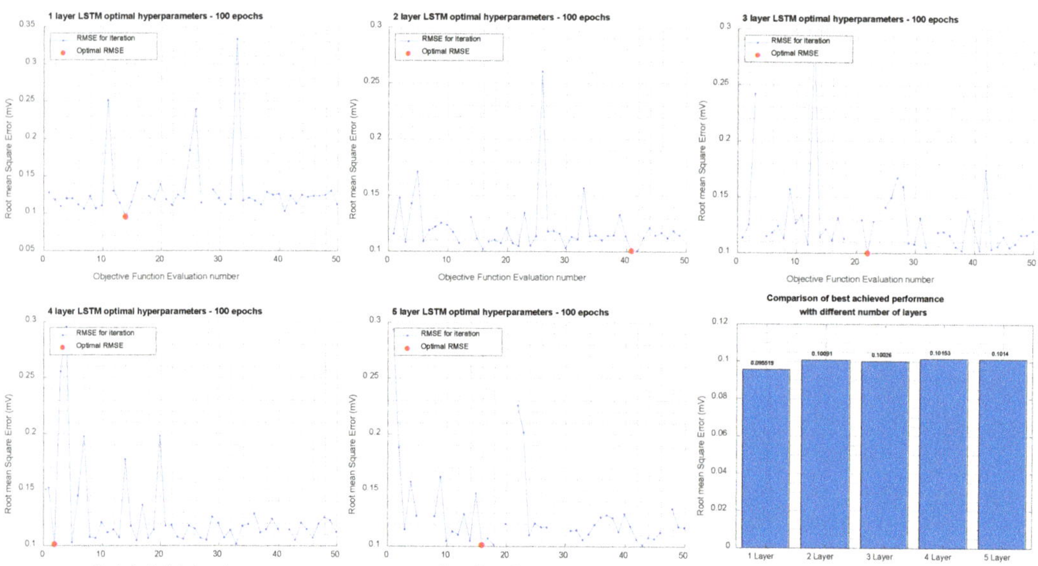

Figure 9. Results of Bayesian Optimization for 1-, 2-, 3-, 4-, and 5-Layer networks to transform Standard 12-lead to Frank XYZ.

The 1-layer network was found to have the lowest validation RMSE (0.0955 mV). Overall, there is an insignificant difference between the RMSE across the number of layers. The best RMSE and the worst RMSE differ by only 5 µV.

3.1. Comparison of Performance Metrics

The metrics for quantitative comparison of waveforms in this study were RMSE, R^2, and Pearson correlation coefficient. Table 4 provides the results for all the methods of transformation implemented in this study.

Table 4. Performance and error metrics for comparison of actual VCG and derived VCG.

Method	Lead	RMSE Mean ± std (μV)	R^2 Mean ± std (μV)	Correlation Coefficient Mean ± std (μV)
Bjerle quasi-orthogonal	Vx	67.21 ± 59.68	76.91 ± 34.72	0.91 ± 0.14
	Vy	145.51 ± 114.00	−6.12 ± 60.38	0.90 ± 0.15
	Vz	108.90 ± 100.54	41.30 ± 60.08	0.81 ± 0.20
Dawson control	Vx	34.65 ± 41.77	96.00 ± 9.33	0.98 ± 0.06
	Vy	46.67 ± 43.95	83.53 ± 34.44	0.94 ± 0.11
	Vz	46.08 ± 16.34	85.19 ± 14.48	0.94 ± 0.06
Dawson post-MI	Vx	32.50 ± 43.01	88.87 ± 26.42	0.96 ± 0.08
	Vy	37.93 ± 56.19	84.54 ± 27.17	0.92 ± 0.21
	Vz	49.24 ± 42.21	79.79 ± 25.53	0.92 ± 0.14
General LSTM Model S12 → XYZ	Vx	36.76 ± 56.04	92.82 ± 16.87	0.97 ± 0.09
	Vy	38.08 ± 51.57	86.60 ± 27.40	0.93 ± 0.14
	Vz	53.98 ± 81.02	83.24 ± 22.75	0.94 ± 0.11
Guillem	Vx	53.14 ± 70.52	88.87 ± 15.38	0.96 ± 0.07
	Vy	66.87 ± 74.99	73.06 ± 34.71	0.87 ± 0.23
	Vz	69.24 ± 62.15	68.98 ± 38.38	0.92 ± 0.12
Hyttinen	Vx	468.15 ± 411.65	−214.80 ± 137.96	−0.26 ± 0.48
	Vy	378.53 ± 292.08	−176.77 ± 140.56	0.33 ± 0.47
	Vz	283.42 ± 283.15	−164.97 ± 140.74	−0.18 ± 0.47
Inverse Dower	Vx	918.51 ± 594.24	−384.11 ± 93.65	0.81 ± 0.26
	Vy	493.56 ± 388.43	−351.66 ± 96.28	0.84 ± 0.20
	Vz	603.08 ± 418.76	−334.54 ± 97.95	0.92 ± 0.12
Kors quasi-orthogonal	Vx	64.13 ± 57.08	76.55 ± 40.03	0.91 ± 0.14
	Vy	66.69 ± 70.20	68.60 ± 43.75	0.89 ± 0.20
	Vz	103.74 ± 99.84	43.57 ± 61.47	0.82 ± 0.20
Kors regression-related	Vx	42.87 ± 64.66	90.85 ± 19.77	0.97 ± 0.07
	Vy	49.39 ± 57.39	78.55 ± 39.98	0.92 ± 0.17
	Vz	73.94 ± 58.37	62.79 ± 47.35	0.91 ± 0.14
Personalized LSTM Model S12 → XYZ	Vx	24.10 ± 53.10	96.13 ± 16.87	0.98 ± 0.06
	Vy	26.28 ± 53.42	92.97 ± 15.78	0.96 ± 0.09
	Vz	30.79 ± 89.51	95.32 ± 11.13	0.98 ± 0.07
Personalized Linear Regression S12 → XYZ	Vx	28.71 ± 79.29	94.46 ± 17.63	0.97 ± 0.12
	Vy	27.65 ± 33.81	89.61 ± 19.60	0.95 ± 0.10
	Vz	34.07 ± 32.33	87.87 ± 24.26	0.94 ± 0.14

3.2. Comparison of Extracted Diagnostic Parameters

As described in Section 2.5, three diagnostic features of the VCG waveform were computed from the actual and derived XYZ waveform data. The features computed from the actual data are treated as actual measurements, and those computed from the derived data are treated as measurements from a test device or methodology. In this case, the methodology is the transformation of ECG leads through a general and patient-specific personalized model. The metrics used for comparison are Pearson's correlation coefficient and the Bland-Altman limits of agreement [32]. We further present the effect size for comparison of VCG parameters from each transformation method and actual VCG. The effect size metrics include Cohen's U1, U3 [33], and common language effect sizes [34]. We also present *t*-test results with t-statistic and the associated *p*-value. The *t*-test *p*-values in this case should be greater than 0.05 if we are to accept the null hypothesis that the difference in means is not significant (i.e., the transformation method yielded results for VCG parameters that were comparable or similar to those obtained from the actual VCG).

3.2.1. Peak QRS-Loop Magnitude

The personalized models showed the highest correlation coefficient values and the smallest limits of agreement, indicating that the derived peak QRS magnitudes were closest to the values computed from the actual data. Table 5 lists the correlation coefficients for the different transformation methods in descending order along with the statistical measures of comparison and effect sizes. Table 6 presents the Bland–Altman limits of agreement between QRS-loop magnitudes extracted from actual and derived VCG waveforms. Figure 10 presents the Bland-Altman plots for QRS-loop magnitude comparison.

Table 5. Lists transformation methods and the correlation between QRS-loop magnitudes extracted from actual and derived VCG waveforms.

Method	Correlation	*t*-Statistic	*t*-Test *p* Value	Cohens U1 [95% CI]	Cohens U3 [95% CI]	Common Language Effect Sizes [95% CI]
Personalized LSTM S12 to XYZ	0.9985	0.22	0.8251	0.0027 [0.0018, 0.0073]	0.4945 [0.4890, 0.5073]	0.5038 [0.5024, 0.5052]
LSTM S12 to XYZ	0.9616	1.12	0.2649	0.0055 [0.0027, 0.0110]	0.4670 [0.4451, 0.4872]	0.5190 [0.5125, 0.5263]
Personalized Linear Regression S12 to XYZ	0.9484	−0.30	0.7643	0.0027 [0.0018, 0.0073]	0.5000 [0.4817, 0.5266]	0.4949 [0.4872, 0.5018]
Kors regression-related	0.9283	−2.97	0.0030	0.0018 [0.0018, 0.0165]	0.5897 [0.5568, 0.6145]	0.4494 [0.4390, 0.4592]
Guillem	0.8934	2.46	0.0142	0.0082 [0.0037, 0.0137]	0.4579 [0.4286, 0.4908]	0.5419 [0.5319, 0.5533]
Kors Quasi-orthogonal	0.8701	−2.42	0.0158	0.0046 [0.0018, 0.0101]	0.5440 [0.5128, 0.5879]	0.4588 [0.4474, 0.4713]
Bjerle Quasi-orthogonal	0.8467	−8.11	<0.001	0.0064 [0.0037, 0.0211]	0.6703 [0.6374, 0.7051]	0.3643 [0.3492, 0.3776]
Hyttinen	0.8408	−20.36	<0.001	0.1200 [0.1067, 0.2184]	0.9542 [0.9359, 0.9707]	0.1918 [0.1407, 0.2345]
Inverse Dower	0.7934	−46.67	<0.001	0.5668 [0.5504, 0.8755]	1.0000 [1.0000, 1.0000]	0.0229 [0.0151, 0.0311]

3.2.2. Peak T-loop Magnitude

The personalized models show the highest correlations and the smallest limits of agreement. The generalized models perform comparably to the better performing transforms from the literature. Table 7 shows the methods' respective correlation coefficients sorted in descending order along with the statistical measures of comparison and effect sizes. Table 8 presents Bland–Altman Limits of Agreement between the peak T-loop magnitudes computed from the actual VCG waveforms and the derived waveforms across different methods of derivation. Figure 11 presents the Bland -Altman plots for the comparison of Peak T-loop magnitude.

Table 6. Lists transformation methods and the Bland–Altman limits of agreement between QRS-loop magnitudes extracted from actual and derived VCG waveforms.

Method	Mean Differences	Limits of Agreement
Personalized LSTM S12 to XYZ	0.0068	−0.0483 to 0.0618
LSTM S12 to XYZ	0.0334	−0.2397 to 0.3065
Personalized Linear Regression S12 to XYZ	−0.0094	−0.3392 to 0.3204
Kors Regression-Related	−0.0907	−0.4650 to 0.2836
Guillem	0.0691	−0.3832 to 0.5215
Kors Quasi-Orthogonal	−0.0786	−0.6264 to 0.4691
Bjerle Quasi-orthogonal	−0.3125	−1.1248 to 0.4997
Hyttinen	−1.3295	−3.3853 to 0.7264
Inverse Dower	−4.2346	−7.5364 to −0.9327

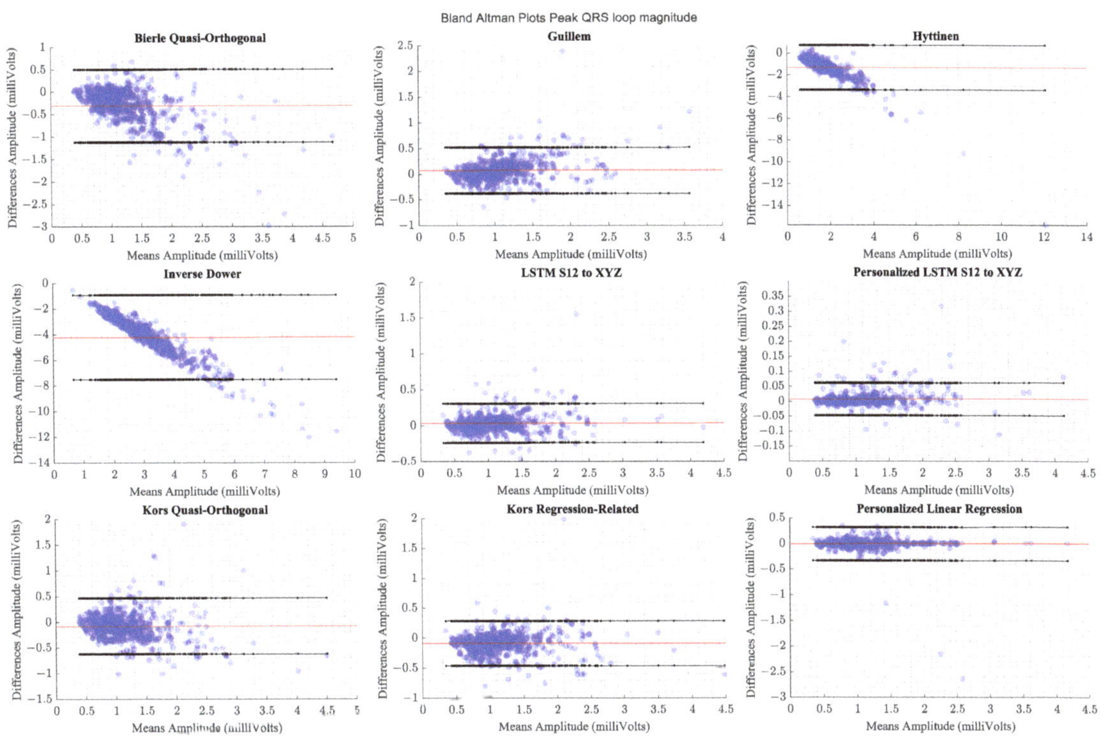

Figure 10. Comparison of Bland–Altman limits of agreement for peak QRS-loop magnitudes across transformation methods. The red horizontal line indicates the mean of differences.

Table 7. Correlation coefficients between the peak T-loop magnitudes computed from the actual VCG waveforms and the derived waveforms across different methods of derivation.

Method	Correlation	*t*-Statistic	*t* Test *p* Value	Cohens U1 [95% CI]	Cohens U3 [95% CI]	Common Language Effect Sizes [95% CI]
Personalized LSTM S12 to XYZ	0.9988	0.43	0.665	0.0018 [0.0018, 0.0064]	0.4927 [0.4817, 0.5000]	0.5074 [0.5061, 0.5088]
Personalized Linear Regression S12 to XYZ	0.9861	−0.09	0.924	0.0018 [0.0018, 0.0064]	0.5092 [0.4963, 0.5311]	0.4984 [0.4944, 0.5024]
LSTM S12 to XYZ	0.9706	1.53	0.126	0.0018 [0.0018, 0.0064]	0.4744 [0.4524, 0.4945]	0.5261 [0.5196, 0.5332]
Kors regression-related	0.9483	−2.90	0.004	0.0055 [0.0027, 0.0119]	0.5842 [0.5440, 0.6117]	0.4506 [0.4408, 0.4591]
Guillem	0.9192	0.33	0.738	0.0046 [0.0027, 0.0133]	0.4982 [0.4634, 0.5330]	0.5057 [0.4956, 0.5152]
Inverse Dower	0.9012	−31.39	<0.001	0.6026 [0.5870, 0.7326]	0.9982 [0.9945, 1.0000]	0.0896 [0.0749, 0.1039]
Kors Quasi-orthogonal	0.8887	−2.37	0.018	0.0037 [0.0018, 0.0110]	0.5916 [0.5476, 0.6172]	0.4595 [0.4467, 0.4720]
Bierle Quasi-orthogonal	0.8330	−8.43	<0.001	0.0082 [0.0046, 0.0247]	0.6905 [0.6630, 0.7271]	0.3592 [0.3410, 0.3770]
Hyttinen	0.8271	−15.94	<0.001	0.0330 [0.0256, 0.1081]	0.8901 [0.8645, 0.9185]	0.2476 [0.2059, 0.2800]

Table 8. Bland–Altman Limits of Agreement between the peak T-loop magnitudes computed from the actual VCG waveforms and the derived waveforms across different methods of derivation.

Method	Mean Differences	Limits of Agreement
Personalized LSTM S12 to XYZ	0.0049	−0.0137 to 0.0235
Personalized Linear Regression S12 to XYZ	−0.0011	−0.0621 to 0.0600
LSTM S12 to XYZ	0.0169	−0.0719 to 0.1056
Kors Regression-Related	−0.0338	−0.1565 to 0.0889
Guillem	0.0037	−0.1413 to 0.1487
Kors Quasi-Orthogonal	−0.0265	−0.1974 to 0.1444
Bierle Quasi-orthogonal	−0.1070	−0.3567 to 0.1427
Hyttinen	−0.3398	−0.9754 to 0.2958
Inverse Dower	−1.5289	−3.4052 to 0.3473

3.2.3. Mean Spatial QRS-T Angle

The personalized models show the highest correlations and the smallest limits of agreement. The generalized models perform comparably to the better performing transforms from the literature. Table 9 shows the methods' respective correlation coefficients sorted in descending order along with the statistical measures of comparison and effect sizes. Table 10 presents Bland–Altman limits of agreement between the mean QRS and T-loop spatial angle magnitudes computed from the actual VCG waveforms and the derived waveforms across different methods of derivation. Figure 12 presents Bland-Altman plots for comparison of spatial QRS-T angle.

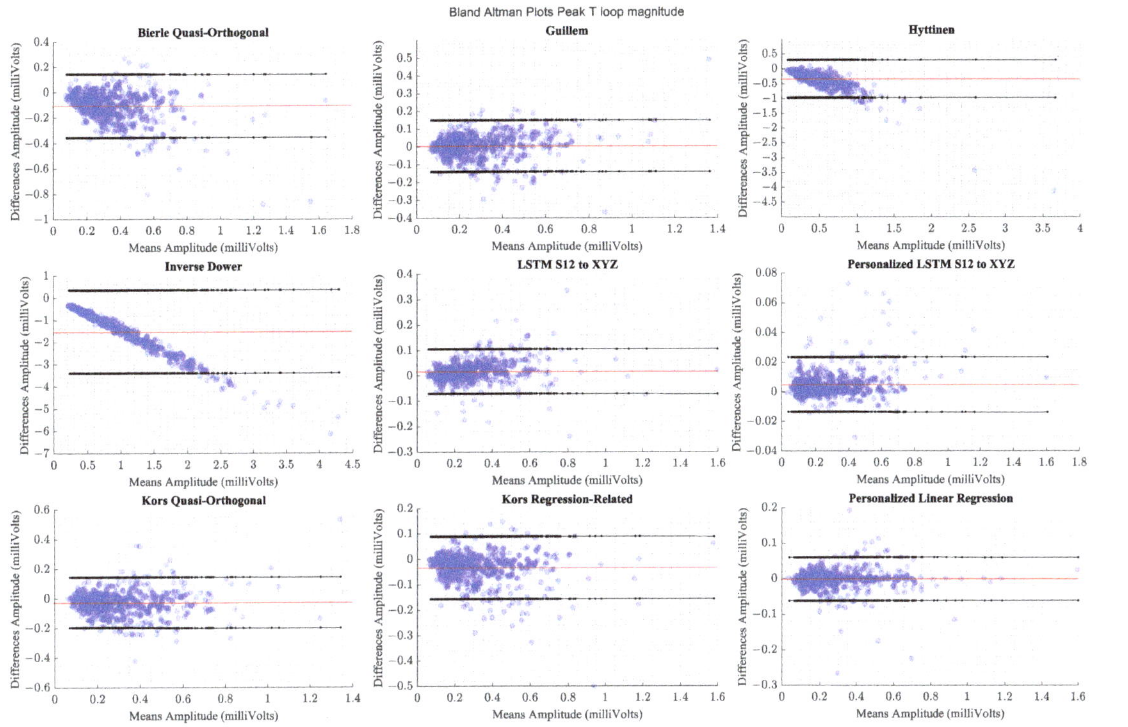

Figure 11. Comparison of Bland–Altman limits of agreement for peak T-loop magnitudes across transformation methods. The red horizontal line indicates the mean of differences.

Table 9. Correlation coefficients between the mean QRS and T-loop spatial angle magnitudes computed from the actual VCG waveforms and the derived waveforms across different methods of derivation.

Method	Correlation	*t*-Statistic	*t* Test P Value	Cohens U1 [95% CI]	Cohens U3 [95% CI]	Common Language Effect Sizes [95% CI]
Personalized LSTM S12 to XYZ	0.9956	−0.07	0.944	0.0027 [0.0018, 0.0064]	0.5055 [0.4890, 0.5156]	0.4988 [0.4966, 0.5010]
Personalized Linear Regression	0.9746	0.21	0.836	0.0037 [0.0018, 0.0082]	0.4908 [0.4689, 0.5178]	0.5035 [0.4986, 0.5088]
LSTM S12 to XYZ	0.9461	−0.23	0.821	0.0027 [0.0018, 0.0101]	0.4908 [0.4615, 0.5165]	0.4961 [0.4878, 0.5044]
Kors regression-related	0.9182	−1.79	0.074	0.0037 [0.0018, 0.0110]	0.5238 [0.4963, 0.5604]	0.4695 [0.4597, 0.4792]
Guillem	0.8672	−3.18	0.002	0.0018 [0.0018, 0.0092]	0.5623 [0.5366, 0.6044]	0.4458 [0.4329, 0.4586]
Inverse Dower	0.8584	−0.11	0.911	0.0027 [0.0018, 0.0092]	0.5092 [0.4799, 0.5522]	0.4981 [0.4847, 0.5107]
Hyttinen	0.851	1.2	0.23	0.0046 [0.0027, 0.0137]	0.4762 [0.4377, 0.5037]	0.5205 [0.5084, 0.5328]
Bjerle Quasi-orthogonal	0.812	−2.78	0.006	0.0027 [0.0018, 0.0110]	0.5641 [0.5348, 0.6117]	0.4527 [0.4378, 0.4677]
Kors Quasi-orthogonal	0.8013	−2.37	0.018	0.0027 [0.0018, 0.0096]	0.5623 [0.5348, 0.6035]	0.4596 [0.4434, 0.4740]

Table 10. Bland–Altman limits of agreement between the mean QRS and T-loop spatial angle magnitudes computed from the actual VCG waveforms and the derived waveforms across different methods of derivation.

Method	Mean Differences	Limits of Agreement
Personalized LSTM S12 to XYZ	−0.1760	−7.7905 to 7.4386
Personalized Linear Regression S12 to XYZ	0.5202	−17.8583 to 18.8987
LSTM S12 to XYZ	−0.5558	−26.7171 to 25.6055
Kors Regression-Related	−4.4134	−36.7731 to 27.9462
Guillem	−7.9564	−49.6889 to 33.7760
Inverse Dower	−0.2839	−44.0710 to 43.5032
Hyttinen	3.0254	−41.5211 to 47.5720
Bjerle Quasi-orthogonal	−6.8745	−56.0440 to 42.2950
Kors Quasi-Orthogonal	−6.0598	−58.3992 to 46.2796

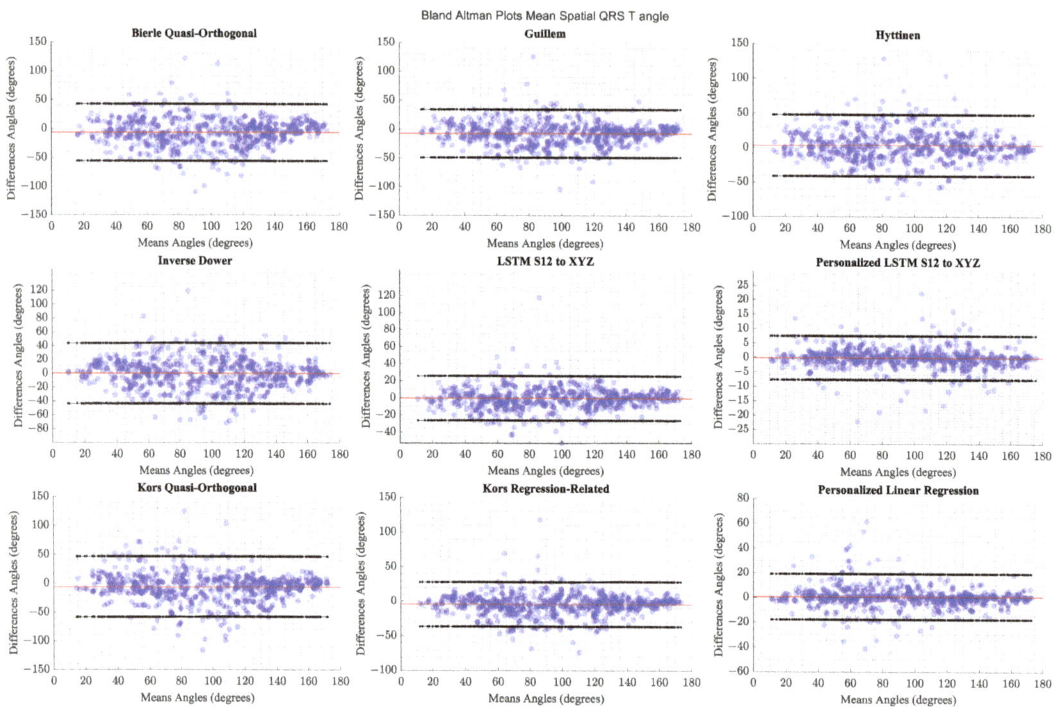

Figure 12. Comparison of Bland-Altman limits of agreement for QRS-T loop angle magnitudes across transformation methods. The red horizontal line indicates the mean of differences.

4. Discussion

The findings in this study indicate that personalized transformation models are preferable, but there are limitations to interpreting the results and practical considerations. The data set for this research is widely available, supporting further research and reproducibility of these results. However, the amount of data available is restricted to a small population that is not geographically or ethnically diverse. There is potential for overly optimistic results obtained in this study due to this aspect. Future studies should evaluate additional data sources from other geographic regions to confirm that these inferences are valid. Furthermore, there is only one diagnosis available per patient as the reason for hospitalization. ECG and VCG interpretations are unavailable. Comparisons on diagnostic yield and outcomes will require specific waveform interpretations, as well as longitudinal follow-up with patients so that we can evaluate how the clinical management of patients were impacted by the availability of VCG in addition to S12. In the absence of this information within the current data set, we could only evaluate the performance in terms of quantitative measures. We have presented effect sizes as statistical measures that could help with the evaluation of various transformation methods. However, there is no reasonable or equivalent comparison available in the literature thus far regarding an absolute interpretation of these results. The effect sizes can be compared across transformations in this study and reveal that generalized LSTM, personalized LSTM, and personalized linear regression perform better than other methods in that they have the least effect sizes when compared to the actual VCG in terms of values obtained for the VCG diagnostic parameters.

The findings indicate that personalized LSTM and personalized linear regression methods lead to nearly identical results with marginally better performance for personalized LSTM. Since the S12 leads cover the ventral plane of the body, it is plausible that the association between S12 and at least X and Y leads of VCG are nearly linear so that they can be derived using linear models. The comparisons between Z leads of VCG derived from the methods reveal a larger difference than X and Y leads. A future avenue of research may be to specifically explore Z-lead comparisons to understand if there is further scope for an improvement of performance with other lead transformation methods.

It is possible that neural network architectures other than LSTM may lead to better lead transformation performance. This study only explores LSTM and not the variants of LSTM. The choice of LSTM for this work was based on recent findings in the literature that demonstrated the use of this architecture to obtain acceptable results for the problem of lead transformations [12]. Alternatively, this work explores personalization and its impact on lead transformation performance and LSTM architecture was evaluated. Several architectures can be explored in this manner for future research.

We had chosen to down sample and filter the ECG waveforms as part of the preprocessing step. There could be different findings if the ECGs were retained at the 1 kHz sampling rate and without filtering. Since the entire data set was preprocessed in the same manner, and all transformation methods were evaluated on the same data, there is no expectation that there would be bias in the results presented herein. However, an empirical evaluation of the impact of pre-processing may be beneficial to explore in a future study with the evaluation of sampling rate and signal conditioning approaches as the goal.

From a practical perspective, implementing the personalized models would require acquiring 15-lead ECGs for each patient, which is not currently part of standard care and would result in added costs and work for healthcare professionals. Furthermore, the data available in this data set is not longitudinal because no recordings span a time frame before and after a significant cardiovascular event. Longitudinal data of this kind must be used to validate the hypothesis LSTM networks as trained have adequately inferred the lead transformations following the subject's anatomy.

Regarding the evaluation involving the extraction of VCG parameters, there is an underlying assumption that the algorithms were accurate. Therefore, we did not evaluate the performance of the algorithms alone. The use of the same algorithm for all data eliminates potential biases in comparisons, but further testing against a labeled VCG data set is necessary to assess the performance of the algorithms.

5. Conclusions

The personalized transformations performed better than generalized transformations in waveform comparisons and error performance of extracted diagnostic parameters from VCG waveforms. The use of personalized transformations for the derivation of VCG from S12 has the potential to improve and augment the diagnostic yield and accuracy of a S12 interpretation. The differences between personalized LSTM and linear regression transformation were marginally in favor of personalized LSTM. There were no statistically significant differences in the performance between them. A study focused on outcomes for patients and diagnostic yield is needed to evaluate the clinical impact of using such an approach for the derivation of VCG from S12 and using it as part of the patient management plan for a broader population. On balance, the results in this study suggest that personalization should be the preferred approach for ECG lead derivations.

Supplementary Materials: The following supporting information can be downloaded at: https://www.mdpi.com/article/10.3390/eng4020078/s1, Table S1. Lists the coefficients for the linear transformation between Standard 12-lead ECG and Frank XYZ VCG; Table S2. Hyperparameter optimization variables, bounds, and sampling transformations.

Author Contributions: Conceptualization, P.S.K., M.R. and V.K.V.; methodology, data analysis, and visualizations P.S.K. and M.R.; writing and original draft—P.S.K.; writing review, editing and supervision—M.R. and V.K.V.; The final version of this manuscript was read and approved by all authors. All authors have read and agreed to the published version of the manuscript.

Funding: This research received no external funding.

Institutional Review Board Statement: Not required. Study uses an open and free database.

Informed Consent Statement: Not applicable. No subjects were recruited.

Data Availability Statement: Free and open public repository.

Conflicts of Interest: The authors declare no conflict of interest.

References

1. Frank, E. An Accurate, Clinically Practical System for Spatial Vectorcardiography. *Circulation* **1956**, *13*, 737–749. [CrossRef] [PubMed]
2. Yang, H.; Bukkapatnam, S.T.S.; Komanduri, R. Spatiotemporal representation of cardiac vectorcardiogram (VCG) signals. *BioMed. Eng. OnLine* **2012**, *11*, 16. [CrossRef] [PubMed]
3. Bousseljot, R.; Kreiseler, D.; Schnabel, A. Nutzung der EKG-Signaldatenbank CARDIODAT der PTB über das Internet. *Biomed. Tech. Biomed. Eng.* **1995**, *40*, 317–318. [CrossRef]
4. Goldberger, A.L.; Amaral, L.A.; Glass, L.; Hausdorff, J.M.; Ivanov, P.C.; Mark, R.G.; Mietus, J.E.; Moody, G.B.; Peng, C.K.; Stanley, H.E. PhysioBank, PhysioToolkit, and PhysioNet: Components of a new research resource for complex physiologic signals. *Circulation* **2000**, *101*, E215–E220. [CrossRef]
5. Chaudhry, U.; Cortez, D.; Platonov, P.G.; Carlson, J.; Borgquist, R. Vectorcardiography Findings Are Associated with Recurrent Ventricular Arrhythmias and Mortality in Patients with Heart Failure Treated with Implantable Cardioverter-Defibrillator Device. *Cardiology* **2020**, *145*, 784–794. [CrossRef]
6. Pérez Riera, A.R.; Uchida, A.H.; Filho, C.F.; Meneghini, A.; Ferreira, C.; Schapacknik, E.; Dubner, S.; Moffa, P. Significance of Vectorcardiogram in the Cardiological Diagnosis of the 21st Century. *Clin. Cardiol.* **2007**, *30*, 319–323. [CrossRef]
7. Correa, R.; Laciar, E.; Arini, P.; Jané, R. Analysis of QRS loop in the Vectorcardiogram of patients with Chagas' disease. In Proceedings of the 2010 Annual International Conference of the IEEE Engineering in Medicine and Biology, Buenos Aires, Argentina, 31 August–4 September 2010; pp. 2561–2564.
8. Andoni, A.; Panigrahy, R.; Valiant, G.; Zhang, L. Learning Polynomials with Neural Networks. In Proceedings of the 31st International Conference on Machine Learning, Proceedings of Machine Learning Research, Beijing, China, 21–26 June 2014; pp. 1908–1916.
9. Tianping, C.; Hong, C. Universal approximation to nonlinear operators by neural networks with arbitrary activation functions and its application to dynamical systems. *IEEE Trans. Neural Netw.* **1995**, *6*, 911–917. [CrossRef]
10. Greff, K.; Srivastava, R.K.; Koutník, J.; Steunebrink, B.R.; Schmidhuber, J. LSTM: A Search Space Odyssey. *IEEE Trans. Neural Netw. Learn. Syst.* **2017**, *28*, 2222–2232. [CrossRef]
11. Hochreiter, S.; Schmidhuber, J. Long Short-Term Memory. *Neural Comput.* **1997**, *9*, 1735–1780. [CrossRef]
12. Sohn, J.; Yang, S.; Lee, J.; Ku, Y.; Kim, H.C. Reconstruction of 12-Lead Electrocardiogram from a Three-Lead Patch-Type Device Using a LSTM Network. *Sensors* **2020**, *20*, 3278. [CrossRef]
13. Bjerle, P.; Arvedson, O. Comparison of Frank vectorcardiogram with two different vectorcardiograms derived from conventional ECG-leads. *Proc. Eng. Found* **1986**, *11*, 13–26.
14. Edenbrandt, L.; Pahlm, O. Vectorcardiogram synthesized from a 12-lead ECG: Superiority of the inverse Dower matrix. *J. Electrocardiol.* **1988**, *21*, 361–367. [CrossRef]

15. Kors, J.A.; van Herpen, G.; Sittig, A.C.; van Bemmel, J.H. Reconstruction of the Frank vectorcardiogram from standard electrocardiographic leads: Diagnostic comparison of different methods. *Eur. Heart J.* **1990**, *11*, 1083–1092. [CrossRef] [PubMed]
16. Hyttinen, J.A.; Viik, J.J.; Eskola, H.; Malmivuo, J.A. Optimization and comparison of derived Frank VECG lead systems employing an accurate thorax model. In Proceedings of the Computers in Cardiology 1995, Vienna, Austria, 10–13 September 1995; pp. 385–388.
17. Guillem, M.S.; Sahakian, A.V.; Swiryn, S. Derivation of orthogonal leads from the 12-lead ECG. accuracy of a single transform for the derivation of atrial and ventricular waves. In Proceedings of 2006 Computers in Cardiology, Valencia, Spain, 17–20 September 2006; pp. 249–252.
18. Dawson, D.; Yang, H.; Malshe, M.; Bukkapatnam, S.T.S.; Benjamin, B.; Komanduri, R. Linear affine transformations between 3-lead (Frank XYZ leads) vectorcardiogram and 12-lead electrocardiogram signals. *J. Electrocardiol.* **2009**, *42*, 622–630. [CrossRef] [PubMed]
19. Atoui, H.; Fayn, J.; Rubel, P. A Novel Neural-Network Model for Deriving Standard 12-Lead ECGs From Serial Three-Lead ECGs: Application to Self-Care. *IEEE Trans. Inf. Technol. Biomed.* **2010**, *14*, 883–890. [CrossRef]
20. Tomasic, I.; Trobec, R.; Lindén, M. Can the Regression Trees Be Used to Model Relation Between ECG Leads? In Proceedings of the Internet of Things, Rome, Italy, 27–29 October 2015; IoT Infrastructures: Rome, Italy, 2015; pp. 467–472.
21. Trobec, R.; Tomašić, I. Synthesis of the 12-Lead Electrocardiogram From Differential Leads. *IEEE Trans. Inf. Technol. Biomed.* **2011**, *15*, 615–621. [CrossRef]
22. Zhu, H.; Pan, Y.; Cheng, K.-T.; Huan, R. A lightweight piecewise linear synthesis method for standard 12-lead ECG signals based on adaptive region segmentation. *PLoS ONE* **2018**, *13*, e0206170. [CrossRef] [PubMed]
23. Zhang, Q.; Frick, K. All-ECG: A Least-number of Leads ECG Monitor for Standard 12-lead ECG Tracking during Motion*. In Proceedings of the 2019 IEEE Healthcare Innovations and Point of Care Technologies, (HI-POCT), Bethesda, MD, USA, 20–22 November 2019; pp. 103–106.
24. Lee, D.; Kwon, H.; Lee, H.; Seo, C.; Park, K. Optimal Lead Position in Patch-Type Monitoring Sensors for Reconstructing 12-Lead ECG Signals with Universal Transformation Coefficient. *Sensors* **2020**, *20*, 963. [CrossRef] [PubMed]
25. Shyam Kumar, P.; Ramasamy, M.; Kallur, K.R.; Rai, P.; Varadan, V.K. Personalized LSTM Models for ECG Lead Transformations Led to Fewer Diagnostic Errors Than Generalized Models: Deriving 12-Lead ECG from Lead II, V2, and V6. *Sensors* **2023**, *23*, 1389. [CrossRef]
26. Kligfield, P.; Gettes, L.S.; Bailey, J.J.; Childers, R.; Deal, B.J.; Hancock, E.W.; Herpen, G.v.; Kors, J.A.; Macfarlane, P.; Mirvis, D.M.; et al. Recommendations for the Standardization and Interpretation of the Electrocardiogram. *Circulation* **2007**, *115*, 1306–1324. [CrossRef]
27. Kingma, D.P.; Ba, J. Adam: A Method for Stochastic Optimization. *Int. Conf. Learn. Represent.* **2014**. [CrossRef]
28. Snoek, J.; Larochelle, H.; Adams, R.P. Practical Bayesian optimization of machine learning algorithms. *Proc. Adv. Neural Inf. Process. Syst* **2012**, *2*, 2951–2959.
29. Mockus, J. On the Bayes Methods for Seeking the Extremal Point. *IFAC Proc. Vol.* **1975**, *8*, 428–431. [CrossRef]
30. Pan, S.J.; Yang, Q. A Survey on Transfer Learning. *IEEE Trans. Knowl. Data Eng.* **2010**, *22*, 1345–1359. [CrossRef]
31. Cortez, D.; Sharma, N.; Devers, C.; Devers, E.; Schlegel, T.T. Visual transform applications for estimating the spatial QRS–T angle from the conventional 12-lead ECG: Kors is still most Frank. *J. Electrocardiol.* **2014**, *47*, 12–19. [CrossRef] [PubMed]
32. King, A.P.; Eckersley, R.J. Chapter 2—Descriptive Statistics II: Bivariate and Multivariate Statistics. In *Statistics for Biomedical Engineers and Scientists*; King, A.P., Eckersley, R.J., Eds.; Academic Press: Cambridge, MA, USA, 2019; pp. 23–56. [CrossRef]
33. Cohen, J. *Statistical Power Analysis for the Behavioral Sciences*; L. Erlbaum Associates: Mahwah, NJ, USA, 1988.
34. Lakens, D. Calculating and reporting effect sizes to facilitate cumulative science: A practical primer for t-tests and ANOVAs. *Front. Psychol.* **2013**, *4*, 863. [CrossRef]

eng

MDPI

Article

Pressure Anomalies Beneath Solitary Waves with Constant Vorticity

Marcelo V. Flamarion [1,*,†], Eduardo M. Castro [2,†] and Roberto Ribeiro-Jr [2,†]

1 Unidade Acadêmica do Cabo de Santo Agostinho, UFRPE/Rural Federal University of Pernambuco, BR 101 Sul, Cabo de Santo Agostinho 54503-900, PE, Brazil
2 Departament of Mathematics, UFPR/Federal University of Paraná, Centro Politécnico, Jardim das Américas, Curitiba 81531-980, PR, Brazil
* Correspondence: marcelo.flamarion@ufrpe.br
† These authors contributed equally to this work.

Abstract: While some studies have investigated the particle trajectories and stagnation points beneath solitary waves with constant vorticity, little is known about the pressure beneath such waves. To address this gap, we investigate numerically the pressure beneath solitary waves in flows with constant vorticity. Through a conformal mapping that flats the physical domain, we develop a numerical approach that allows us to compute the pressure and the velocity field in the fluid domain. Our experiments indicate that there exists a threshold vorticity such that pressure anomalies and stagnation points occur when the intensity of the vorticity is greater than this threshold. Above this threshold, the pressure on the bottom boundary has two points of local maxima and there are three stagnation points in the flow, and below it the pressure has one local maximum and there is no stagnation point.

Keywords: constant vorticity; solitary water waves; Euler equations; pressure anomalies; stagnation points

Citation: Flamarion, M.V.; Castro, E.M.; Ribeiro-Jr, R. Pressure Anomalies Beneath Solitary Waves with Constant Vorticity. *Eng* **2023**, *4*, 1306–1319. https://doi.org/10.3390/eng4020076

Academic Editor: Antonio Gil Bravo

Received: 6 February 2023
Revised: 21 April 2023
Accepted: 24 April 2023
Published: 27 April 2023

1. Introduction

The study of water waves and their interactions with underline currents is a topic of research that has piqued the curiosity of engineers, mathematicians, physicists and oceanographers over the centuries. Although many advances have already been achieved, there are a number of basic questions that are still open.

Currents are caused mainly by density differences in the water, tidal forces and by wind [1]. Mathematically, wave–current interaction has been widely investigated under the assumption that the current is linearly sheared, i.e., it flows with constant vorticity. Physically, this can be representative of a realistic flow when waves are long compared with the depth or when waves are short compared with the length scale of the vorticity distribution [2].

Flows with constant vorticity are mainly characterized by the existence of overhanging waves, the appearance of stagnation points and the emergence of pressure anomalies.

Overhanging waves are free surface waves that are not a graph of a function. Among the numerical studies in this direction, the works of Vanden-Broeck [3,4] stand out, in which the author finds periodic and solitary overhanging waves, and more recently, the works of Dyachenko and Hur [5,6]. The existence of overhanging waves is proved rigorously by Constantin et al. [7] for periodic waves with constant vorticity, and more recently by Hur and Wheeler [8] for large or infinite depths. Although some theoretical works have already allowed overhanging solitary wave profiles in their approach [9], the rigorous proof of such wave types is still an open problem.

Stagnation points can be understood as fluid particles that are stationary in the wave moving frame. For irrotational flows, they occur at a sharp crest [10], and in flows with constant vorticity they can emerge within the bulk of the fluid, forming a recirculation zone whose profile resembles Kelvin's cat's eye flow. The literature on stagnation points is extensive, starting with the work of Teles da Silva and Peregrine [2], the reader is referred to Ribeiro-Jr et al. for a detailed study on the appearance of stagnation points beneath periodic waves with constant vorticity. An overview of the works on stagnation point is given by Flamarion and Ribeiro-Jr [11]. More recently, Ige and Kalisch [12] investigated the particle trajectories associated with the propagation of periodic waves with constant vorticity in the framework of a new Benjamin–Bona–Mahony equation.

In irrotational flows, the pressure exerted in the bulk of the fluid beneath a Stokes wave (a periodic travelling wave with a monotone profile from the crest to the trough) attains its maximum on the bottom of the channel and below the crest. Moreover, the pressure strictly increases with the depth and strictly decreases horizontally away from a crest to a trough [13]. Notable exceptions to these features arise in rotational flows with constant vorticity: (i) the maxima and minima of the pressure may occur within the bulk of the fluid and (ii) the pressure on the bottom can be out of phase with the surface elevation [2,14–17]. The characteristics (i) and (ii) of the pressure are defined as pressure anomalies.

Although many advances have been accomplished in understanding the flow structure beneath waves with constant vorticity, it is unknown whether the pressure anomalies known for periodic waves with constant vorticity also occur for solitary waves. Strauss and Wheeler [16] proved that overhanging periodic or solitary waves must have a pressure sink, i.e., the pressure achieves its minimum within the bulk of the fluid and not on the free surface. However, this is still an open question for free surface waves that are graph of a function. This issue was raised recently by Kozlov et al. [18]. In their words, the following question was raised: "Is the pressure beneath a solitary wave in a flow with constant vorticity different from the one in the irrotational case?"

In this work, we address the question above. The novelty is twofold: (i) we find numerically that, when the vorticity crosses a threshold, the pressure on the bottom boundary caused by a solitary wave on the free surface can have two points of local maxima; (ii) we analyse in detail the appearance of stagnation points beneath solitary waves. Thus, the paper at hand responds to the question raised by Kozlov et al. [18]. Moreover, it complements the studies carried out by Vasan and Oliveras [17] and Ribeiro-Jr et al. [15], who have showed numerically the occurrence of pressure anomalies beneath periodic waves with constant vorticity and analysed the appearance of stagnation points beneath such waves. The approach used to compute the pressure and the stagnation points consists of determining a conformal mapping under which the physical domain is the image of a strip (canonical domain), then all calculations are performed through pseudo-spectral methods.

In summary, the results presented in this work are of interest to theorists and experimentalists. For a theorist, it can provide physical insights on a rigorous proof of the pressure anomalies. Likewise, it may inspire more experimental studies on this topic, since the phenomenon can be observed by manipulating the intensity of the underlying current.

For reference, this article is organized as follows: The governing equations of water waves in flows with constant vorticity are presented in Section 2. In Section 3, we describe the conformal mapping and the numerical method. Then, we present the results in Section 4 and proceed to our final considerations.

2. Governing Equations

We consider an incompressible flow of an inviscid fluid with constant density (ρ) in a two-dimensional channel with finite depth (d) under the force of gravity (g). Moreover, we assume that the flow is in the presence of a linearly sheared current (constant vorticity). Denoting the velocity field in the bulk of the fluid by $\vec{U}(x,y,t) = (u(x,y,t), v(x,y,t))$, and the free surface by $\zeta(x,t)$, this free boundary problem can be described by the Euler equations

$$\vec{U}_t + (\vec{U} \cdot \nabla)\vec{U} = -\frac{\nabla p}{\rho} - g\vec{j} \quad \text{in } -d < y < \zeta(x,t), \tag{1}$$

$$\nabla \cdot \vec{U} = 0 \quad \text{in } -d < y < \zeta(x,t), \tag{2}$$

$$p = P_{atm} \quad \text{at } y = \zeta(x,t), \tag{3}$$

$$v = \zeta_t + u\zeta_x \quad \text{at } y = \zeta(x,t), \tag{4}$$

$$v = 0 \quad \text{at } y = -d, \tag{5}$$

where $\vec{j}$ is the unitary vector $(0,1)$ and P_{atm} is the atmospheric pressure.

The assumption of constant vorticity enables us to write the velocity field as

$$\vec{U} = \vec{U_0} + \nabla\overline{\phi}, \tag{6}$$

where

$$\vec{U_0} = (ay + f, 0), \quad f \in \mathbb{R},$$

is a linear shear flow solution of (1)–(5) characterized by the flat surface $\zeta \equiv 0$ and constant vorticity $-a$. Here, $\overline{\phi}$ is the velocity potential of an irrotational perturbation of the shear flow.

Equations (1)–(5) are written in terms of $\overline{\phi}$, then non-dimensionalised via transformation (7)

$$\begin{array}{lll} x = dx', & \zeta = d\zeta', & \Omega = \dfrac{ad}{\sqrt{dg}}, \\ y = dy', & \overline{\phi} = d\sqrt{dg}\,\overline{\phi}', & p = P_0 + \rho g d p', \\ t = \sqrt{\frac{d}{g}}\,t', & \overline{\psi} = d\sqrt{dg}\,\overline{\psi}', & F = \dfrac{f}{\sqrt{dg}}. \end{array} \tag{7}$$

Dropping the prime notation, this gives us the dimensionless version of the governing equations

$$\Delta\overline{\phi} = 0 \text{ in } -1 < y < \zeta(x,t), \tag{8}$$

$$\zeta_t + (\Omega\zeta + F + \overline{\phi}_x)\zeta_x = \overline{\phi}_y \text{ at } y = \zeta(x,t), \tag{9}$$

$$\overline{\phi}_t + \frac{1}{2}(\overline{\phi}_x^2 + \overline{\phi}_y^2) + (\Omega\zeta + F)\overline{\phi}_x + \zeta - \Omega\overline{\psi} = B(t) \text{ at } y = \zeta(x,t), \tag{10}$$

$$\overline{\phi}_y = 0 \text{ at } y = -1, \tag{11}$$

where $-\Omega$ is the dimensionless vorticity, F is the Froude number and the pressure in the fluid body is given by

$$p = -\left(\overline{\phi}_t + \frac{1}{2}(\overline{\phi}_x^2 + \overline{\phi}_y^2) + (\Omega\zeta + F)\overline{\phi}_x + \zeta - a\overline{\psi} - B(t)\right). \tag{12}$$

For the study of traveling wave solutions, it is convenient to eliminate time from the problem by passing to a moving frame

$$X = x - ct \quad \text{and} \quad Y = y,$$

where c is the wave speed, to be determined a posteriori. In this new moving reference frame, the wave is stationary and the flow is steady. Taking this new frame of reference into account, Equations (8)–(12) are written as

$$\Delta\overline{\phi} = 0 \quad \text{in } -1 < Y < \zeta(X), \tag{13}$$
$$-c\zeta_X + (F + \Omega\zeta + \overline{\phi}_X)\zeta_X = \overline{\phi}_Y \quad \text{at } Y = \zeta(X), \tag{14}$$
$$-c\overline{\phi}_X + \frac{1}{2}(\overline{\phi}_X^2 + \overline{\phi}_Y^2) + (\Omega\zeta + F)\overline{\phi}_X + \zeta - \Omega\overline{\psi} = B \quad \text{at } Y = \zeta(X), \tag{15}$$
$$\overline{\phi}_Y = 0 \quad \text{at } Y = -1, \tag{16}$$

and

$$p = -\left(-c\overline{\phi}_X + \frac{1}{2}(\overline{\phi}_x^2 + \overline{\phi}_y^2) + (\Omega\zeta + F)\overline{\phi}_x + \zeta - \Omega\overline{\psi} - B\right). \tag{17}$$

We assume that $\zeta(X)$ is a solitary wave whose crest is located at $X = 0$ and satisfies

$$\zeta(X) \to 0 \quad \text{as} \quad |X| \to \infty. \tag{18}$$

In the following, we present a numerical scheme to compute the solutions of the system (13)–(16) and to calculate the pressure in the fluid body via Formula (17).

3. Conformal Mapping and the Numerical Method

Since $\zeta(X)$ decays to zero as $|X| \to \infty$, we can truncate its infinite domain to a finite one $[-\lambda/2, \lambda/2]$ with $\lambda > 0$, and approximate the boundary conditions by periodic conditions. Then, we can solve Equations (13)–(16) through the conformal mapping technique introduced by Dyachenko et al. [19], which has been widely applied in free boundary problems [15,20,21]. This strategy consists of using conformal mapping from a strip of length L and width D (canonical domain) onto the flow domain of the solitary wave $\{(X,Y) \in \mathbb{R}^2, -\lambda/2 \leq X \leq \lambda/2 \text{ and } -1 \leq Y \leq \zeta(X)\}$. This map is such that in the canonical domain, the free boundary problem (13)–(16) can be solved numerically by the use of a spectral collocation method and Newton's method.

3.1. Conformal Mapping

Consider the conformal mapping

$$Z(\xi, \eta) = X(\xi, \eta) + iY(\xi, \eta), \tag{19}$$

under which the strip $\{(\xi, \eta) \in \mathbb{R}^2;\ -L/2 \leq \xi \leq L/2 \text{ and } -D \leq \eta \leq 0\}$ is mapped onto the flow domain, as in Figure 1. The constant D will be determined so that the canonical and the physical domain have the same length. Since Z is taken to be conformal, thus analytical, X and Y are actually conjugate harmonic functions, whereas the mapping's Jacobian is given by

$$J = X_\xi^2 + Y_\xi^2. \tag{20}$$

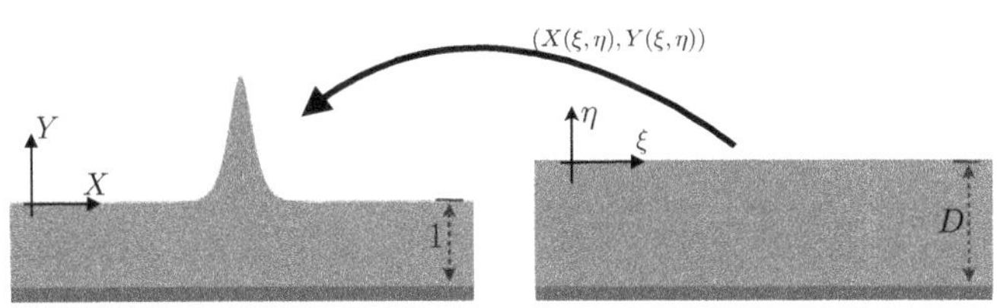

Figure 1. Illustrative depiction of the conformal mapping. The free surface is flattened out in the canonical domain.

A central characteristic of this mapping is given by the way that the boundary curves from each domain are related

$$\begin{cases} Y(\xi,0) = \zeta(X(\xi,0)), \\ Y(\xi,-D) = -1, \end{cases} \tag{21}$$

which serves as Dirichlet data for the Laplace equation for $Y(\xi,\eta)$. By denoting $\mathbf{Y}(\xi) = Y(\xi,0)$ and $\mathbf{X}(\xi) = X(\xi,0)$, the traces of the respective harmonic functions along $\eta = 0$, we have

$$Y(\xi,\eta) = \mathcal{F}^{-1}\left[\frac{\sinh(k(\eta+D))}{\sinh(kD)}\mathcal{F}(\mathbf{Y})\right] + \frac{(\eta+D)\langle\mathbf{Y}\rangle+\eta}{D}, \quad k \neq 0, \tag{22}$$

where $k = k(j) = (\pi/L)j$, for $j \in \mathbb{Z}$, $\mathcal{F}$ is the Fourier transform in ξ-variable given by

$$\mathcal{F}(f(\xi)) = \hat{f}(k) = \frac{1}{L}\int_{-L/2}^{L/2} f(\xi)e^{-ik\xi}d\xi,$$

$$\mathcal{F}^{-1}(\hat{f}(k)) = f(\xi) = \sum_{j\in\mathbb{Z}} \hat{f}(k)e^{ik\xi},$$

and $\langle\cdot\rangle$ denotes the average defined by

$$\langle\mathbf{Y}\rangle = \frac{1}{L}\int_{-L/2}^{L/2} \mathbf{Y}(\xi)d\xi.$$

By differentiating Equation (22) with respect to η and integrating the Cauchy–Riemann equation $X_\xi = Y_\eta$, we obtain

$$X(\xi,\eta) = \left(\frac{1+\langle\mathbf{Y}\rangle}{D}\right)\xi - \mathcal{F}^{-1}\left[\frac{i\cosh(k(\eta+D))}{\sinh(kD)}\mathcal{F}(\mathbf{Y})\right], \quad k \neq 0. \tag{23}$$

The canonical depth D can be fixed if we require that both canonical and physical domains have the same length. Let L and λ be the respective lengths, thus

$$\mathbf{X}(\xi = L/2) - \mathbf{X}(\xi = -L/2) = \lambda.$$

It follows from (23) that this restriction leads to the relation

$$D = 1 + \langle\mathbf{Y}\rangle. \tag{24}$$

A reader interested in further details on the conformal mapping presented here should consult Flamarion and Ribeiro-Jr [22] for conformal mapping in the context of uneven topographies and its accuracy.

By denoting $\phi(\xi,\eta) = \overline{\phi}(X(\xi,\eta),Y(\xi,\eta))$ and $\psi(\xi,\eta) = \overline{\psi}(X(\xi,\eta),Y(\xi,\eta))$, the governing Equations (13)–(16) can be written in the canonical coordinate system. Subsequently, by combining Equations (14) and (15), a single equation for the free surface wave can be obtained, namely,

$$\begin{aligned} -\frac{c^2}{2} + \frac{c^2}{2J} + \mathbf{Y} + \frac{(\mathcal{C}[(\Omega\mathbf{Y}+F)\mathbf{Y}_\xi])^2}{2J} - \frac{\mathcal{C}[(\Omega\mathbf{Y}+F)\mathbf{Y}_\xi]}{J}(c - (\Omega\mathbf{Y}+F)\mathbf{X}_\xi) \\ -\frac{(\Omega\mathbf{Y}+F)^2\mathbf{Y}_\xi^2}{2J} - \frac{c(\Omega\mathbf{Y}+F)\mathbf{X}_\xi}{J} + Fc + \Omega\left(\frac{\Omega\mathbf{Y}}{2}+F\right)\mathbf{Y} + \Omega M = B, \end{aligned} \tag{25}$$

where $\mathcal{C}$ is the periodic Hilbert transform on a strip whose Fourier symbol is $icoth(kD)$ and

$$M = \left\langle c\mathbf{Y} - \left(\frac{\Omega\mathbf{Y}^2}{2} + F\mathbf{Y}\right)\right\rangle.$$

Details of the derivation of this equation are given in the Appendix B.

Observe that $\mathbf{X}_\xi = 1 - \mathcal{C}_0[\mathbf{Y}_\xi]$ and $J = \mathbf{X}_\xi^2 + \mathbf{Y}_\xi^2$ are given in terms of $\mathbf{Y}(\xi)$. Consequently, Equation (25) has as unknowns $\mathbf{Y}(\xi)$, c, D and B. It is the aim of the next section to describe a numerical approach for computing solitary waves.

3.2. Numerical Method

Up to this point, we have transformed the free boundary problem (13)–(16) into a nonlinear system of two equations ((24) and (25)) and four unknowns $\mathbf{Y}(\xi)$, c, D and B. In order to obtain a system that can be handled by Newton's method, we add two extra equations.

We fix the amplitude A of the wave through

$$Y(0) - Y(-L/2) = A, \tag{26}$$

and based on the limit (18), we impose that

$$Y(-L/2) = 0. \tag{27}$$

Consider a discrete version of Equations (24)–(27) as follows. Let us take an evenly spaced grid in the ξ-axis in the canonical domain, say

$$\xi_j = -L/2 + (j-1)\Delta\xi, \quad j = 1, ..., N, \quad \text{where } \Delta\xi = L/N, \tag{28}$$

with N being even. We impose symmetry about $\xi = 0$ so that $Y_j = Y_{N-j+2}$, where $Y_j = \mathbf{Y}(\xi_j)$. Fixing Ω and F, we have $N/2+4$ unknowns: $Y_1, \cdots, Y_{N/2+1}$, c, D and B. We satisfy Equation (25) at the grid points (28). The Fourier modes are computed by the Fast Fourier Transform (FFT) and derivatives in the ξ-variable are performed spectrally [23]. This yields a system with $N/2+1$ equations

$$\mathcal{G}_j(Y_1, \cdots, Y_{N/2+1}, c, D, B) = 0 \quad j = 1, \cdots, N/2+1.$$

Equation (24) is discretized using the trapezoidal rule, which leads to the equation

$$\mathcal{G}_{N/2+2}(Y_1, \cdots, Y_{N/2+1}, c, D, B) = \frac{Y_1 + Y_{N/2+1}}{2} + \sum_{j=2}^{N/2} Y_j + 1 - D = 0.$$

Finally, we satisfy Equations (26) and (27), resulting in a system of $N/2+4$ equations and $N/2+4$ unknowns,

$$\mathcal{G}_{N/2+3}(Y_1, \cdots, Y_{N/2+1}, c, D, B) = Y_{N/2+1} - Y_1 - A = 0,$$

$$\mathcal{G}_{N/2+4}(Y_1, \cdots, Y_{N/2+1}, c, D, B) = Y_1 = 0.$$

The system is solved by Newton's method, where our initial guess is taken to be the well-known solitary wave solution for the classical (irrotational) Korteweg–de Vries equation, that is

$$\mathbf{Y}(\xi) = A_0 \operatorname{sech}^2\left(\sqrt{3A_0/4}\xi\right), \; c = 1 + \frac{A_0}{2},$$

where A_0 is chosen to be small. From there, the idea is to make use of the continuation technique in A and Ω, where the prior converged solution is fed as an initial guess to a new solution. The Jacobian matrix of the system is computed by finite difference and the stopping criterion for the Newton's method is

$$\frac{\sum_{j=1}^{N/2+4} |\mathcal{G}_j|}{N/2+4} < 10^{-10}.$$

In all experiments performed we used $L = 1500$, which is important to make sure that the method indeed converges to a solitary wave solution.

4. Results

In Section 4.1, we present some solitary waves computed through our numerical method. A comparison between such waves with a weakly nonlinear KdV equation is performed in order to provide a validation of our numerical procedure. Then, the main results of the paper are discussed in Section 4.2.

4.1. Steady Waves

Several numerical computations are available and provide a detailed characterization of the shape of the free surface wave in flows with constant vorticity. More specifically, it is known that the crests of the waves become rounder as Ω decreases. This has been shown for periodic travelling waves [2,3,6,15,24] and for solitary waves [4].

Figure 2 displays various wave profiles for different vorticity values. As can be seen, the numerical method captures the well-known characteristics of waves with vorticity: more rounded or cuspidate profiles depending on the Ω sign. Although the computational domain used was $L = 1500$, for visualization purposes the plot window was chosen to be 50 units long. Moreover, for each choice of Ω, the Froude number was fixed as $F = \Omega/2$. This implies cancelling the average mass flow of the stream $\vec{U}_0 = (\Omega Y + F, 0)$. The choice of F has no impact on the shape of wave nor on the location of the stagnation points and the appearance of pressure anomalies.

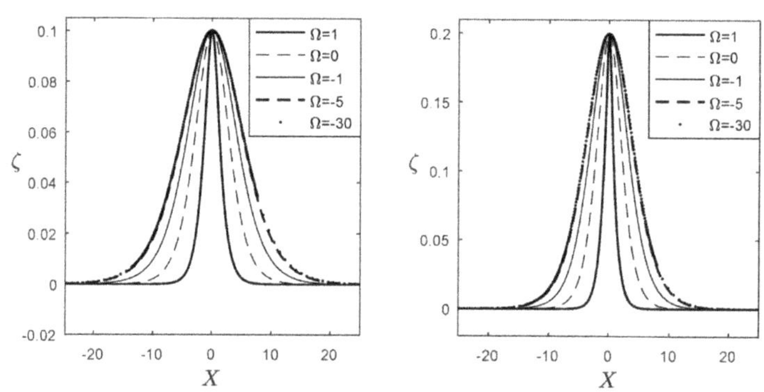

Figure 2. Wave profiles with amplitudes $A = 0.1$ (**left**) and $A = 0.2$ (**right**).

Furthermore, vorticity also has a straightforward and expected effect on the velocity of the waves: greater vorticity implies greater velocity across the amplitude spectrum, a trend that matches with the well-known dispersion relation for linear long waves, as depicted in Figure 3. From that same figure, it is also notable that even though the method captures waves with negative Ω, which are considerably large in modulus, convergence stops earlier in the positive direction. This phenomenon is in a large part explained by the loss of solution regularity in a neighbourhood of $X = 0$ when Ω becomes more positive, something hinted at by Figure 2. The closed formula for the velocity shown in dashed lines in Figure 3 is given by

$$c_{lin} = F - \frac{\Omega}{2} + \sqrt{\frac{\Omega^2}{4} + 1}.$$

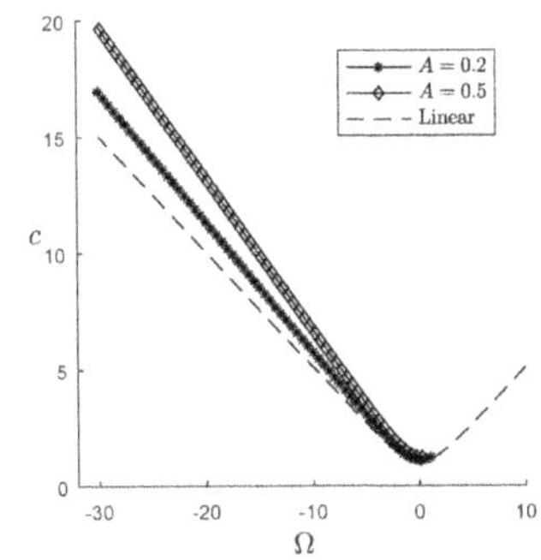

Figure 3. The wave speed as function of Ω for different values of A.

Beyond the linear theory, another model that can be used for comparison purposes is the weakly nonlinear KdV equation. In what follows, we are interested in investigating how the velocities are influenced by the increase in amplitude for a fixed vorticity. For small amplitudes, it is expected that the waves computed should be similar to the sech^2-type solution of the KdV equation.

Regarding the analysis of the KdV model in the presence of vorticity, we refer to the work of Guan [25]. The formulation presented by this author is used as a benchmark for our numerical solutions.

For a given choice of parameters Ω and A, Figure 4 indicates the distance between our solutions and the analytical solution determined by the KdV equation. The dashed line displays the wave speed from the KdV solution after scaling to the Euler regime. As expected, we see a very close wave speed whenever A is small, but the overall pattern of speed/amplitude relation in the case of the Euler solutions present a clear deviation from the linear distribution found in KdV. In particular, around $A = 0.15$ and $A = 0.2$ we see a slight takeoff from the Euler regime in comparison to the KdV, while it is interesting to observe that the general aspect of this "takeoff curve" remains unchanged when we vary Ω. For the interested reader, a study of the resolution of the numerical method is presented in Appendix A.

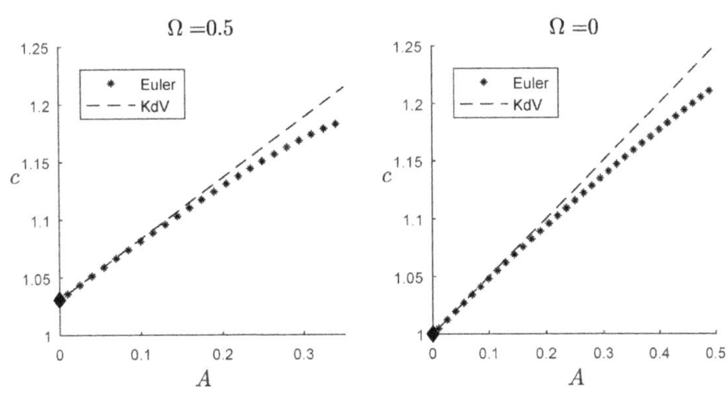

Figure 4. *Cont.*

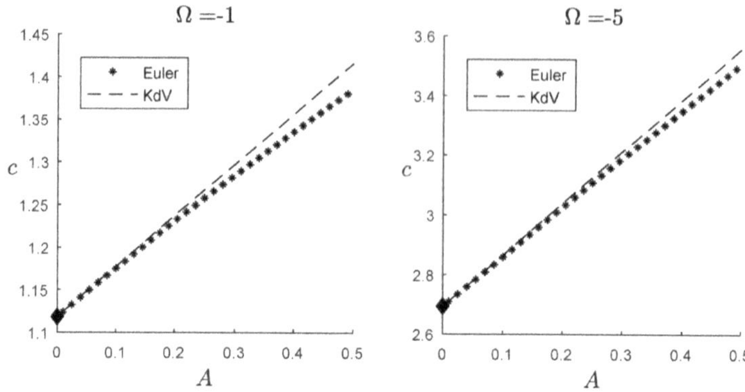

Figure 4. The wave speed as function of the wave amplitude for different choices of vorticity. Diamond denotes the linear speed c_{lin}.

4.2. Pressure in the Bulk of the Fluid

It is well known in the literature that pressure anomalies beneath nonlinear periodic waves are connected to the emergence of stagnation points [2,15,17]. Starting from this point, we first investigated the appearance of stagnation points in terms of the intensity of the vorticity parameter (Ω), then analysed the pressure within the bulk of the fluid. For this purpose, we fixed solitary waves with amplitude $A = 0.2$, $F = \Omega$ and let the vorticity vary. This choice of F leads to a background flow $(\Omega Y + F, 0)$ with zero velocity at the bottom.

Our first numerical essay consists of computing the phase portrait for different values of the vorticity parameter—this is depicted in Figure 5. The markers represent the position of the stagnation points. We find that the stagnation points first appear on the bottom and below the crest for a critical value $\Omega^{\star} \approx -2.4967$. For $\Omega > \Omega^{\star}$, there are no stagnation points in the bulk of the fluid. Nonetheless, for $\Omega < \Omega^{\star}$, we obtain a flow with three stagnation points: two saddles located at the bottom and one centre located within the bulk of the fluid and below the crest, forming a region with closed streamlines which is described as a single Kelvin cat's eye structure. As the vorticity becomes stronger, this structure becomes wider, i.e, the saddles remain on the bottom moving away from each other and the centre moves upwards.

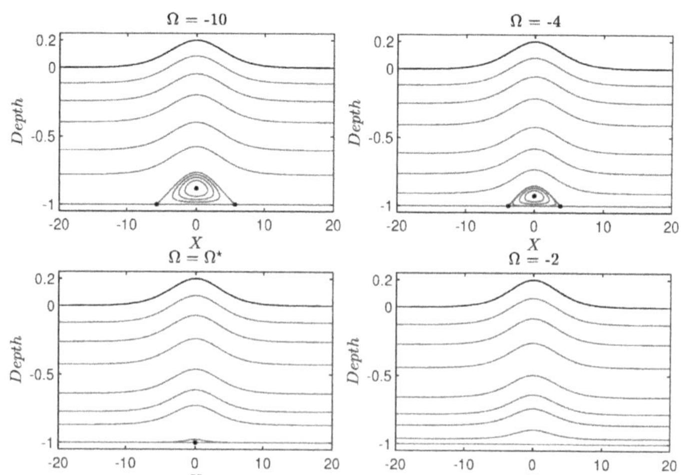

Figure 5. Phase portraits for different values of the vorticity parameter. Circles correspond to the location of the stagnation points. $\Omega^{\star} \approx -2.4967$.

Figure 6 shows the pressure contours and the pressure on the bottom boundary for the same waves depicted in Figure 5. We notice that, according to the value of Ω, the following anomalies occur: (i) the maximum pressure value may not be attained at the bottom and below the crest; (ii) the minimum pressure value may be attained within the bulk of the fluid; and (iii) the pressure on the bottom boundary may have two local maxima. These anomalies have been observed for nonlinear periodic waves [2,15,17] and for overhanging solitary waves [16], but to the best of our knowledge this the first time that such results have been reported for solitary waves that are graph of a function. These results indicate that the anomalies in the pressure and the stagnation points are somehow related. Moreover, pressure anomalies and stagnation points occur when a threshold vorticity is achieved. However, a detailed theoretical study is necessary to explain such a phenomenon.

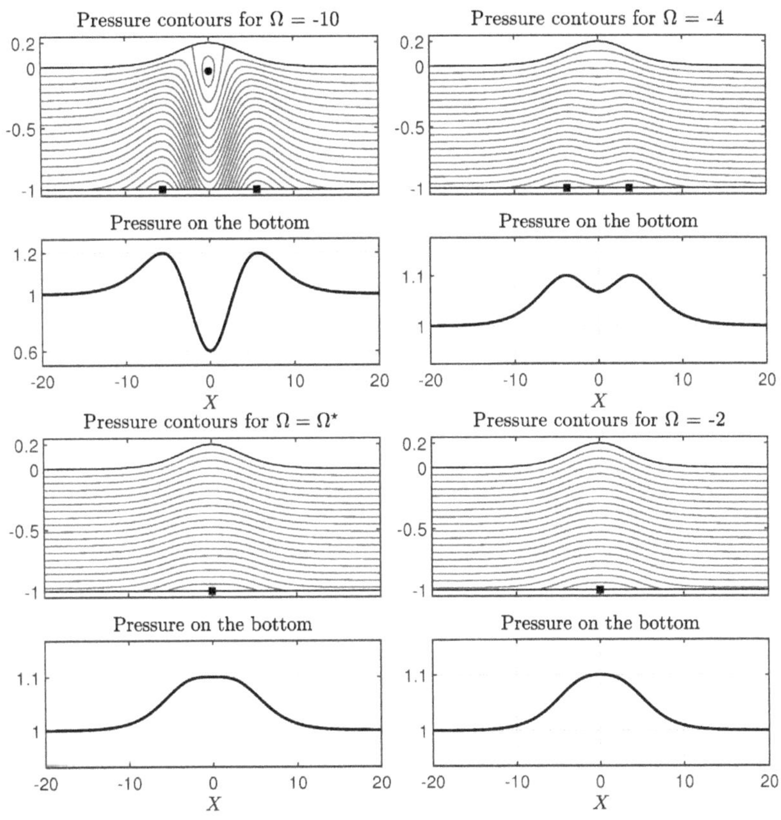

Figure 6. Pressure beneath the solitary wave with amplitude $A = 0.2$ and its corresponding pressure on the bottom boundary. Circle and square markers indicate the location of global minima and maxima of the pressure, respectively.

5. Conclusions

In the present work, we have studied the pressure beneath solitary waves in flows with constant vorticity. Our results indicate that there exists a threshold vorticity such that pressure anomalies and stagnation points occur when the intensity of the vorticity is greater than the threshold. More specifically, when the vorticity is below this threshold, the pressure on the bottom boundary has one local maximum and there is no stagnation point in the flow. Once the vorticity crosses this threshold, the pressure on the bottom boundary has two local maxima and the flow has three stagnation points (one centre and two saddles).

Author Contributions: Conceptualization, M.V.F., E.M.C. and R.R.-J.; Methodology, M.V.F., E.M.C. and R.R.-J.; Writing—original draft, M.V.F., E.M.C. and R.R.-J.; Writing—review & editing, M.V.F., E.M.C. and R.R.-J. All authors have read and agreed to the published version of the manuscript.

Funding: The author E.M.C. is grateful for the financial support provided by CAPES Foundation (Coordination for the Improvement of Higher Education Personnel) during part of the development of this work.

Institutional Review Board Statement: Not applicable.

Informed Consent Statement: Informed consent was obtained from all subjects involved in the study.

Data Availability Statement: Not applicable.

Acknowledgments: M.V.F. and R.R.J. are grateful to IMPA for hosting them as visitors during the 2023 Post-Doctoral Summer Program.

Conflicts of Interest: The authors declare no conflict of interest.

Abbreviations

In this section, we provide a list of the main symbols that have been utilized throughout the text.

Symbol	Meaning
B	Bernoulli constant
c	Wave speed
$\mathcal{C}[\cdot]$	The periodic Hilbert transform on a strip of width D
D	Width of strip that corresponds to the canonical domain
F	Froude number
L	Length of the canonical domain
p	pressure in fluid body
(ξ,η)	Coordinate system in the canonical domain
$(X(\xi,\eta),Y(\xi,\eta))$	Conformal mapping that applies a strip of width D in the physical domain.
$(\mathbf{X}(\xi),\mathbf{Y}(\xi)) = (X(\xi,0),Y(\xi,0))$	Free surface wave profile written in terms of the conformal mapping
$-\Omega$	Dimensionless vorticity
(x,y)	Laboratory frame of reference
$\zeta(X)$	Free surface wave profile in the moving frame $X = x - ct$ and $Y = y$
$\overline{\phi}$	Velocity potential for the irrotational part of the velocity field
$\overline{\psi}$	Harmonic conjugate function of $\overline{\phi}$
$\phi(\xi,\eta) = \overline{\phi}(X(\xi,\eta),Y(\xi,\eta))$	Potential $\overline{\phi}$ written in the coordinate system (ξ,η)
$\psi(\xi,\eta) = \overline{\psi}(X(\xi,\eta),Y(\xi,\eta))$	Function $\overline{\psi}$ written in the coordinate system (ξ,η)
$\mathbf{\Phi}(\xi) = \psi(\xi,0)$	Potential ϕ evaluated at $\eta = 0$
$\mathbf{\Psi}(\xi) = \psi(\xi,0)$	Function ψ evaluated at $\eta = 0$

Appendix A. Resolution Study

In what follows, we show that the method is independent of the grid size by calculating the distance between outputs for different choices of $\Delta\xi$. These experiments were performed for waves with amplitude $A = 0.2$. We take the reference grid as $\Delta\xi^* = 0.0458$, the finest resolution computed.

In Table A1, we denote by $\zeta_{\Delta\xi}$ the wave profile and by $c_{\Delta\xi}$ the wave speed obtained from Newton's method using a grid with size $\Delta\xi$. In addition, we consider ζ^* and c^* as the wave profile and its speed computed in the finest grid. These experiments were performed for waves with amplitude $A = 0.2$. Note that for $\Omega = 1$, the numerical scheme requires more resolution to approximate the solution with more accuracy. This can be explained by a combination of two factors: (i) the emergence of cusps and (ii) the issue of crowding phenomenon present in conformal mappings. For this reason, finer grids are necessary to accurately compute waves in the presence of currents where Ω is positive.

Table A1. Resolution study for waves of amplitude $A = 0.2$.

Ω	$\Delta\xi$	$\frac{\|\zeta_{\Delta\xi} - \zeta^*\|_2}{\|\zeta^*\|_2}$	$\frac{\lvert c_{\Delta\xi} - c^*\rvert}{\lvert c^*\rvert}$
0	0.0916	1.6×10^{-10}	2.5×10^{-12}
	0.1831	4.6×10^{-10}	7.4×10^{-12}
	0.3662	6.7×10^{-8}	5.6×10^{-10}
	0.7324	1.2×10^{-4}	5.2×10^{-6}
−1	0.0916	5.4×10^{-13}	2.1×10^{-14}
	0.1831	5.7×10^{-13}	1.0×10^{-14}
	0.3662	1.5×10^{-12}	1.3×10^{-14}
	0.7324	1.2×10^{-7}	1.8×10^{-9}
1	0.0916	6.8×10^{-5}	1.0×10^{-6}
	0.1831	0.0027	1.7×10^{-4}
	0.3662	0.0477	0.0015
	0.7324	0.1164	0.0102

Finally, to provide an indication of the accuracy of our numerical method, we conducted a comparison between our numerical results and those obtained by Teles da Silva and Peregrine [2] for the relationship between the wave speed and wave amplitude. Specifically, we set the Froude number F = 0 and compared the graphs of the wave speed as a function of wave amplitude for flows with $\Omega = -1, 0, 0.5$ with Figures 9 and 10 presented in Teles da Silva and Peregrine's work. The comparison is illustrated in Figure A1, which demonstrates a strong agreement between the two methods.

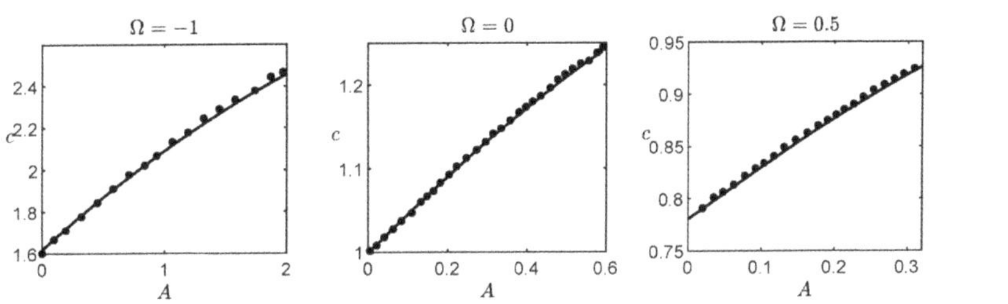

Figure A1. Comparison between our numerical results and those obtained by Teles da Silva and Peregrine [2] for the wave speed as a function of wave amplitude. The solid line represents the results from our numerical method, while the dots correspond to the results computed by Teles da Silva and Peregrine.

Appendix B. The Free Surface Wave in the Canonical Coordinate System

In this section, we give more details on the derivation of Equation (25). We start by noticing that the Laplace equation is conformally invariant. So, denoting by $\phi(\xi,\eta) = \overline{\phi}(X(\xi,\eta), Y(\xi,\eta))$ and $\psi(\xi,\eta) = \overline{\psi}(X(\xi,\eta), Y(\xi,\eta))$ the potential and its harmonic conjugate in the canonical coordinates, one can easily obtain the following:

$$\begin{aligned} &\phi_{\xi\xi} + \phi_{\eta\eta} = 0 && \text{in } -D < \eta < 0, \\ &\phi = \mathbf{\Phi}(\xi) && \text{at } \eta = 0, \\ &\phi_\eta = 0 && \text{at } \eta = -D, \end{aligned}$$

and

$$\begin{aligned} &\psi_{\xi\xi} + \psi_{\eta\eta} = 0 && \text{in } -D < \eta < 0, \\ &\psi = \mathbf{\Psi}(\xi) && \text{at } \eta = 0, \\ &\psi = Q && \text{at } \eta = -D, \end{aligned}$$

where Q is an arbitrary constant. The formulas for $\phi(\xi,\eta)$ and $\psi(\xi,\eta)$ can be found in a similar fashion to that worked out for $X(\xi,\eta)$ and $Y(\xi,\eta)$, which yields

$$\phi(\xi,\eta) = \mathcal{F}^{-1}\left[\frac{\cosh(k(\eta+D))}{\cosh(kD)}\mathcal{F}(\mathbf{\Phi})\right],$$

$$\psi(\xi,\eta) = \mathcal{F}^{-1}\left[\frac{\sinh(k(\eta+D))}{\sinh(kD)}\mathcal{F}(\mathbf{\Psi})\right] - Q\frac{\eta}{D}.$$

Using the Cauchy–Riemman equation $\phi_\xi = \psi_\eta$ and evaluating along $\eta = 0$, we find that

$$\mathbf{\Phi}_\xi(\xi) = \mathcal{F}^{-1}\left[-i\coth(kD)\mathcal{F}_k(\mathbf{\Psi}_\xi)\right]. \tag{A1}$$

For simplicity, we make use of the Fourier operator $\mathcal{C}[\cdot]$ defined as follows: given a function $h(\xi)$,

$$\mathcal{C}[h(\xi)] = \mathcal{C}_0[h(\xi)] + \lim_{k\to 0} i\coth(kD)\hat{h}(k), \tag{A2}$$

where $\mathcal{C}_0[\,\cdot\,] = \mathcal{F}^{-1}\mathcal{H}\mathcal{F}[\,\cdot\,]$, with $\mathcal{H}$ given by

$$\mathcal{H}(k) = \begin{cases} i\coth(kD),\ k\neq 0 \\ 0,\ k=0. \end{cases} \tag{A3}$$

For the particular case of $\mathcal{C}[\cdot]$ evaluated at $h_\xi(\xi)$, we have

$$\mathcal{C}[f_\xi(\xi)] = \mathcal{C}_0[h_\xi(\xi)] - \frac{\hat{h}(0)}{D}, \tag{A4}$$

With this notation, we obtain from relations (23), (24) and (A1) that

$$\mathbf{X}_\xi = 1 - \mathcal{C}_0[\mathbf{Y}_\xi] \tag{A5}$$

$$\mathbf{\Phi}_\xi = -\mathcal{C}_0[\mathbf{\Psi}_\xi] + \frac{\hat{\mathbf{\Psi}}(0)}{D}. \tag{A6}$$

Performing straightforward calculations, we obtain that the kinematic condition (14) and Bernoulli law (15) in canonical coordinates are given by

$$\mathbf{\Psi}_\xi = c\mathbf{Y}_\xi - (\Omega\mathbf{Y} + F)\mathbf{Y}_\xi, \tag{A7}$$

$$-c\frac{\mathbf{\Phi}_\xi\mathbf{X}_\xi + \mathbf{\Psi}_\xi\mathbf{Y}_\xi}{J} + \frac{1}{2J}(\mathbf{\Phi}_\xi^2 + \mathbf{\Psi}_\xi^2) + \mathbf{Y} + (\Omega\mathbf{Y} + F)\frac{\mathbf{\Phi}_\xi\mathbf{X}_\xi + \mathbf{\Psi}_\xi\mathbf{Y}_\xi}{J} - \Omega\mathbf{\Psi} = 0. \tag{A8}$$

Then, integrating (A7), we obtain

$$\mathbf{\Psi} = c\mathbf{Y} - \left(\frac{\Omega\mathbf{Y}^2}{2} + F\mathbf{Y}\right) + M, \tag{A9}$$

where M is an arbitrary constant. In order to simplify the use of the formula (A6), we choose $\mathbf{\Psi}$ so that $\hat{\mathbf{\Psi}}(0) = 0$. This leads naturally to

$$M = \left\langle c\mathbf{Y} - \left(\frac{\Omega\mathbf{Y}^2}{2} + F\mathbf{Y}\right)\right\rangle.$$

Hence,

$$\mathbf{\Phi}_\xi = -\mathcal{C}_0[\mathbf{\Psi}_\xi]. \tag{A10}$$

By substituting Equation (A9) and (A10) into (A8), then Equation (A7) into the resulting equation, we obtain a single equation for the free surface

$$-\frac{c^2}{2}+\frac{c^2}{2J}+\mathbf{Y}+\frac{(\mathcal{C}[(\Omega\mathbf{Y}+F)\mathbf{Y}_{\xi}])^2}{2J}-\frac{\mathcal{C}[(\Omega\mathbf{Y}+F)\mathbf{Y}_{\xi}]}{J}(c-(\Omega\mathbf{Y}+F)\mathbf{X}_{\xi})$$
$$-\frac{(\Omega\mathbf{Y}+F)^2\mathbf{Y}_{\xi}^2}{2J}-\frac{c(\Omega\mathbf{Y}+F)\mathbf{X}_{\xi}}{J}+Fc+\Omega\left(\frac{\Omega\mathbf{Y}}{2}+F\right)\mathbf{Y}+\Omega M=B,$$

References

1. Dalrymple, R.A. *Water Wave Models and Wave Forces with Shear Currents*; Technical Report No. 20; Coastal and Oceanographic Engineering Laboratory, University of Florida: Gainesville, FL, USA, 1973.
2. Teles Da Silva, A.F.; Peregrine, D.H. Steep, steady surface waves on water of finite depth with constant vorticity. *J. Fluid Mech.* **1988**, *195*, 281–302. [CrossRef]
3. Vanden-Broeck, J.-M. Periodic waves with constant vorticity in water of infinite depth. *IMA J. Appl. Maths* **1996**, *56*, 207–217. [CrossRef]
4. Vanden-Broeck, J.-M. Steep solitary waves in water of finite depth with constant vorticity. *J. Fluid Mech.* **1994**, *274*, 339–348. [CrossRef]
5. Dyachenko, S.A.; Hur, V.M. Stokes waves with constant vorticity: Folds, gaps and fluid bubbles. *J. Fluid Mech.* **2019**, *878*, 502–521. [CrossRef]
6. Dyachenko, S.A.; Hur, V.M. Stokes waves with constant vorticity: I. Numerical computation. *Stud. Appl. Maths* **2019**, *142*, 162–189. [CrossRef]
7. Constantin, A.; Strauss, W.; Vărvărucă, E. Global bifurcation of steady gravity water waves with critical layers. *Acta Math.* **2016**, *217*, 195–262. [CrossRef]
8. Hur, V.M.; Wheeler, M. Overhanging and touching waves in constant vorticity flows. *J. Differ. Equations* **2022**, *338*, 572–590. [CrossRef]
9. Haziot, S.V.; Wheeler, M. Large-amplitude steady solitary water waves with constant vorticity. *arXiv* **2021**, arXiv:2110.04901.
10. Varvaruca, E. Singularities of Bernoulli free boundaries. Commun. Partial Differ. *Equations* **2006**, *31*, 1451–1477.
11. Flamarion, M.V.; Ribeiro, R., Jr. Solitary Waves on Flows with an Exponentially Sheared Current and Stagnation Points. *Q. J. Mech. Appl. Math.* **2023**, *76*, 79–91. [CrossRef]
12. Ige, O.E.; Kalisch, H. Particle trajectories in a weakly nonlinear long-wave model on a shear flow. *Appl. Numer. Math.* **2023**, *in press*. [CrossRef]
13. Constantin, A.; Strauss, W. Pressure beneath a Stokes wave. *Comm. Pure. Appl. Math.* **2010**, *63*, 533–557. [CrossRef]
14. Ali, A.; Kalisch, H. Reconstruction of the pressure in long-wave models with constant vorticity. *Eur. J. Mech. B Fluids* **2013**, *37*, 187–194. [CrossRef]
15. Ribeiro, R., Jr.; Milewski, P.A.; Nachbin, A. Flow structure beneath rotational water waves with stagnation points. *J. Fluid. Mech.* **2017**, *812*, 792–814. [CrossRef]
16. Strauss, W.A.; Wheeler, M. Bound on the slope of steady water waves with favorable vorticity. *Arch. Rat. Mech. Anal.* **2016**, *222*, 1555–1580. [CrossRef]
17. Vasan, V.; Oliveras, K. Pressure beneath a travelling wave with constant vorticity. *DSDC-A* **2014**, *34*, 3219–3239.
18. Kozlov, V.; Kuznetsov, N.; Lokharu, E. Solitary waves on constant vorticity flows with an interior stagnation point. *J. Fluid. Mech.* **2020**, *904*, A4. [CrossRef]
19. Dyachenko, A.; Zakharov, V.; Kuznetsov, E. Nonlinear dynamics of the free surface of an ideal fluid. *Plasma Phys.* **1996**, *22*, 916–928.
20. Choi, W. Strongly nonlinear long gravity waves in uniform shear flows. *Phys. Rev. E* **2003**, *68*, 026305. [CrossRef]
21. Milewski, P.; Vanden-Broeck, J.; Wang, Z. Dynamics of steep two-dimensional gravity-capillary solitary waves. *J. Fluid Mech.* **2010**, *664*, 466–477. [CrossRef]
22. Flamarion, M.V.; Ribeiro, R., Jr. An iterative method to compute conformal mappings and their inverses in the context of water waves over topographies. *Int. J. Numer. Methods Fluids* **2021**, *93*, 3304–3311. [CrossRef]
23. Trefethen, L.N. *Spectral Methods in MATLAB*; SIAM: Philadelphia, PA, USA, 2001.
24. Ko, J.; Strauss, W. Large-amplitude steady rotational water waves. *Eur. J. Mech.* **2008**, *27*, 96–109. [CrossRef]
25. Guan X. Particle trajectories under interactions between solitary waves and a linear shear current. *Theor. Appl. Mech. Lett.* **2020**, *10*, 125–131. [CrossRef]

eng

MDPI

Article

A Numerical Study on the Response of a Very Large Floating Airport to Airplane Movement

Taro Kakinuma [1,*] and Masaki Hisada [2]

[1] Graduate School of Science and Engineering, Kagoshima University, Kagoshima 890-0065, Japan
[2] Faculty of Engineering, Kagoshima University, Kagoshima 890-0065, Japan
* Correspondence: taro@oce.kagoshima-u.ac.jp; Tel.: +81-99-285-8467

Abstract: Numerical simulations were generated to investigate the response of a floating airport to airplane movement using the nonlinear shallow water equations of velocity potential for water waves interacting with a floating thin plate. First, in the 1D calculations, the airplanes were B747 and B737. At touch-and-go, when the airplane speed is closer to the water wave speed, even B737 produced large waves based on the resonance. The impacts due to both the touchdown and leaving of the airplanes generated other forward and backward waves. At landing, when the airplane speed approached the water wave speed, a forced wave was generated and amplified, with many free waves ahead. At takeoff, a wave clump, generated shortly after starting to run, propagated in front of the airplanes. Although the wave height increased from superposition with the reflected waves, the wave reflectance was reduced by lowering the flexural rigidity near the airport edge. Second, in the 2D calculations, B787 performed landing and takeoff. When the still water depth is shallower, a grid-like pattern was formed at the floating airport and appeared more remarkably in landing than in takeoff. The effective amplification occurred from a sufficient load applied when the airplane speed approached the water wave speed. Furthermore, the maximum upslope gradient beneath the airplane increased as the still water depth decreased, and it was larger in takeoff than in landing.

Keywords: very large floating structure; VLFS; offshore airport; landing; takeoff; touch-and-go; hydroelasticity; resonance; wave reflection

Citation: Kakinuma, T.; Hisada, M. A Numerical Study on the Response of a Very Large Floating Airport to Airplane Movement. *Eng* **2023**, *4*, 1236–1264. https://doi.org/10.3390/eng4020073

Academic Editor: Antonio Gil Bravo

Received: 12 February 2023
Revised: 11 April 2023
Accepted: 14 April 2023
Published: 21 April 2023

1. Introduction

A very large floating structure, a VLFS, which is designed to be an offshore airport, storage facility, wind/solar power plant, evacuation area in disaster events, or others, has advantages such as towability and environmental friendliness due to the ability of seawater to flow under the structure. In recent years, techniques to obtain sustainable energy by converting water wave energy using floating flexible structures have also been devised, e.g., [1,2]. Such floating structures exist at sea, i.e., on a fluid, and because of their large size, they deform and vibrate based on hydroelasticity. Therefore, to design a VLFS, it is necessary to understand the interaction between the oscillation of the structure and the motion of the fluid. For the design of a flexible VLFS interacting with wind waves including typhoon-driven waves, various models have been developed, e.g., [3–9].

Even in nature, the interaction between flexible platforms and a fluid can be observed when ice plates float on the sea surface [10]. The effects of floating ice in different forms on water waves have been studied [11–14], and the response of an ice plate to a moving load on it has also been investigated [15–17]. These results can be referred to when designing a VLFS.

Regarding the response of a VLFS to long waves including tsunamis, the Boussinesq-type equations for surface waves were solved numerically using a finite difference method (FDM) to examine the relationship between the bending moment and flexural rigidity of a floating thin plate on a progressing solitary wave [18]. The interaction of a thin plate with an incident solitary wave was also investigated by coupling a finite element method (FEM) and boundary element method (BEM) in two vertical dimensions [19]. This interaction was also reproduced by hydraulic experiments, and the solitary waves were disintegrated by the floating thin plate, as their nonlinearity was strong [19]. The results—the wave height of the incident waves decreased because of the generation of floating-body waves—suggest that the wave height of a huge tsunami decreases after propagating through an offshore VLFS. The tsunami-height reduction using a VLFS was discussed based on the numerical simulation of water waves interacting with a floating thin plate using an FDM [20].

When density stratification is developed under a floating structure, its oscillation may generate internal waves, leading to a change in seawater salinity and temperature, especially in coastal environments, through the propagation, shoaling, and breaking of the internal waves. To study the response of a floating thin plate in a coexisting field of surface and internal waves, a vertical two-dimensional problem was formulated with the framework of a linear potential theory [21]. The surface/internal waves due to a moving load on a VLFS in two vertical dimensions were examined using an FDM, considering both the nonlinearity and dispersion of the water waves [22].

One of the artificial loads on a VLFS is the movement of airplanes on a floating airport, and cases of an airplane moving on a floating airport have also been studied. For example, the drag against an airplane taking off from a floating airport of infinite length was evaluated numerically, using the Fourier transform theory, for different flexural rigidities of the airport [23]. Conversely, the transient response of a floating airport under the load of a landing airplane was studied using an FEM [24]. In the coexistence field of linear waves and a current, a BEM was applied to simulate the thin plate response to a moving weight [25]. Under the combined loads of water waves and an airplane landing or taking off, the time variation of an airport profile and the drag induced on an airplane by the deformed runway were obtained using both an FEM-scheme-based method and Wilson's θ method [26]. The horizontally two-dimensional linear response of a floating airport to the landing and takeoff of an airplane was investigated using the time-domain mode expansion method [27]. Moreover, a combination of a BEM and moving element method (MEM) was utilized to study the hydroelastic response of floating composite plates subjected to moving loads [28].

In the present paper, an offshore airport is assumed as a floating thin plate, and its oscillation due to airplane movement—touch-and-go, landing, and takeoff—is discussed based on numerical simulations with an FDM. The governing equations were the nonlinear shallow water equations on velocity potential, which were obtained by reducing the equations obtained based on a variational principle for water waves, considering the flexibility of floating thin plates [29]. In the numerical calculations, the flexural rigidity is given at the location of a thin plate to express the thin plate covering part of the water area. With this method, it is possible to consider the reflection and transmission of waves at the edges of the thin plate, so we also discuss the reflection and transmission of floating-body waves, which were not discussed for infinite length airports. First, we investigate the response of a very large floating airport to the movement of two sizes of jetliners in one-dimensional problems. Thereafter, we discuss horizontally two-dimensional problems of the behavior of the floating airport due to the movement of a medium-sized jetliner.

2. Numerical Calculation Method

2.1. Governing Equations

The illustration in Figure 1 is our schematic for a system consisting of multilayer fluids and thin plates, where the fluid layers and thin plates are represented as the i-layers and i-plates ($i = 1, 2, \ldots, I$) from top to bottom, respectively.

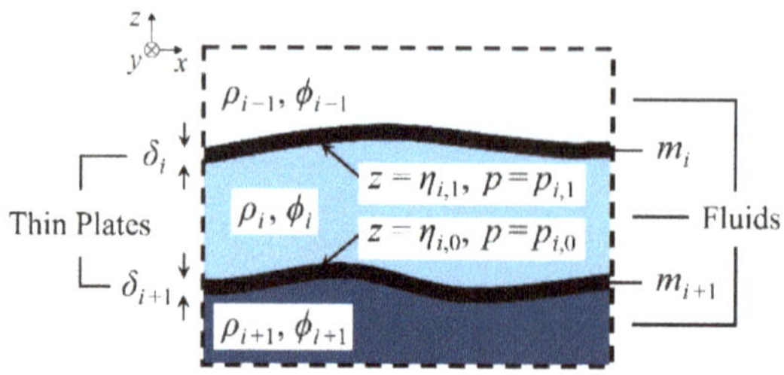

Figure 1. Schematic for a system consisting of multilayer fluids and thin plates.

We assume that none of the fluids mix even in motion without plates, and the density ρ_i ($\rho_1 < \rho_2 < \ldots < \rho_I$) of the i-layer is spatially uniform and temporally constant in each layer. The thickness of the i-layer is $h_i(\boldsymbol{x})$ in still water, where $\boldsymbol{x}$ is the coordinate in the horizontal plane, namely (x, y). The origin of the vertical axis z is located at the top surface of the system in the stationary state, and the positive direction of z is vertically upward. The elevations of the lower and upper interfaces of the i-layer are expressed by $z = \eta_{i,0}\ (\boldsymbol{x}, t)$ and $z = \eta_{i,1}\ (\boldsymbol{x}, t)$, respectively, and the pressures at the lower and upper interfaces of the i-layer are defined as $p_{i,0}\ (\boldsymbol{x}, t)$ and $p_{i,1}\ (\boldsymbol{x}, t)$, respectively.

The thin plate touching the upper interface of the i-layer is called the i-plate. The density and vertical width of the i-plate are m_i and δ_i, respectively. When m_i, δ_i, and the flexural rigidity of the i-plate are zero, the plate yields no resistance to fluid motion, where two immiscible fluids touch each other directly without any plate. Both surface tension and capillary action are ignored, and friction is also ignored for simplicity. Moreover, the energy attenuation inside the thin plates is not considered.

We assume that the fluids are inviscid and incompressible, and fluid motion is irrotational, resulting in the existence of velocity potential ϕ_i in the i-layer, and ϕ_i is expanded into a power series of z with weightings $f_{i,\alpha}$ as

$$\phi_i(\boldsymbol{x}, z, t) = \sum_{\alpha=0}^{N_i-1} \left[f_{i,\alpha}(\boldsymbol{x}, t) \cdot z^{\alpha} \right], \tag{1}$$

where N_i is the number of terms for an expanded velocity potential in the i-layer.

In the i-layer, when both the displacement of one interface, $z = \eta_{i,1-j}(\boldsymbol{x}, t)$ ($j = 0$ or 1), and the pressure on the other interface, $p_{i,j}(\boldsymbol{x}, t)$, are known, the unknown variables are the velocity potential $\phi_i(\boldsymbol{x}, z, t)$ and interface displacement $\eta_{i,j}(\boldsymbol{x}, t)$. Then, the definition of the functional for the variational problem in the i-layer, F_i, is as follows [29]:

$$F_i\left[\phi_i, \eta_{i,j}\right] = \int_{t_0}^{t_1} \iint_A \int_{\eta_{i,0}}^{\eta_{i,1}} \left[\frac{\partial \phi_i}{\partial t} + \frac{1}{2}(\nabla \phi_i)^2 + \frac{1}{2}\left(\frac{\partial \phi_i}{\partial z}\right)^2 + gz + \frac{p_{i,j} + P_i + W_i}{\rho_i} \right] dz\, dA\, dt, \tag{2}$$

where $\nabla = (\partial/\partial x, \partial/\partial y)$ is a horizontal partial differential operator, and the gravitational acceleration g is 9.8 m/s^2. The plane A, which is the orthogonal projection of the object domain on to the x-y plane, is assumed to be independent of time.

In comparison with the functional referred to in [30] for the rotational motion of a fluid, Equation (2) introduces an additional term of the integral of $(p_{i,j} + P_i + W_i)/\rho_i$ as an interfacial pressure term, without the terms related to vorticity. Using the functional of [30], after omitting the vorticity terms, the set of nonlinear equations for one-layer problems without thin plates was derived by [31].

P_i and W_i in Equation (2) are expressed by

$$P_i = \sum\nolimits_{k=1}^{i-1} \left[(\rho_i - \rho_k) g h_k \right], \tag{3}$$

$$W_i = \sum\nolimits_{k=1}^{i} (-m_k g \delta_k), \tag{4}$$

respectively, in the case of no buoyancy of the structures.

After substituting the velocity potential ϕ_i expanded in Equation (1) into Equation (2), the Euler–Lagrange equations on $\eta_{i,j}$ and $f_{i,\alpha}$ are derived as

$$\eta_{i,1}^{\alpha}\frac{\partial\eta_{i,1}}{\partial t}-\eta_{i,0}^{\alpha}\frac{\partial\eta_{i,0}}{\partial t}+\nabla\left[\left(\eta_{i,1}^{\alpha+\beta+1}-\eta_{i,0}^{\alpha+\beta+1}\right)\nabla f_{i,\beta}\right]-\frac{\alpha\beta}{\alpha+\beta-1}\left(\eta_{i,1}^{\alpha+\beta-1}-\eta_{i,0}^{\alpha+\beta-1}\right)f_{i,\beta}=0, \quad (5)$$

$$\eta_{i,j}^{\beta}\frac{\partial f_{i,\beta}}{\partial t}+\frac{1}{2}\eta_{i,j}^{\beta+\gamma}\nabla f_{i,\beta}\nabla f_{i,\gamma}+\frac{1}{2}\beta\gamma\eta_{i,j}^{\beta+\gamma-2}f_{i,\beta}f_{i,\gamma}+g\eta_{i,j}+\frac{p_{i,j}+P_i+W_i}{\rho_i}=0\ (j=0\text{ or }1), \quad (6)$$

where the sum rule of product is adopted for subscripts β and γ. For example, $f_{2,3}$ is the weighting of z^3 in the 2-layer.

For long surface waves in one-layer problems without thin plates, the accuracy of the above equations was investigated by [32]: when the maximum order of equations is $2n$, the order of error in the set of Equations (5) and (6) is σ^{4n+2}, where σ is the representative ratio of water depth to wavelength. Conversely, the order of error in the extended Green–Naghdi equation [33] is σ^{2n+2}. Therefore, especially when $O(\sigma) \ll 1$, the accuracy of the former is significantly higher than that of the latter for $n \geq 1$.

Regarding the i-plate, the horizontal length scale is assumed to be much larger than the thickness of the thin plate, so the differences in curvature between the upper surface, neutral plane, and lower surface of the thin plate are ignored. Therefore, the governing equation of motion for the i-plate is the following classical equation to describe the oscillation of an elastic thin plate as

$$m_i\delta_i\frac{\partial^2\eta_{i,1}}{\partial t^2}+B_i\nabla^2\nabla^2\eta_{i,1}+m_ig\delta_i+p_{i-1,0}-p_{i,1}=0, \quad (7)$$

where B_i is the flexural rigidity of the i-plate between the $(i-1)$- and i-layers. Although both the plate density m_i and vertical width δ_i are assumed to be constant throughout the i-plate for simplicity, the flexural rigidity B_i can be distributed along the thin plate.

When the representative values of wave height, wavelength, fluid depth, and density are H, l, d, and ρ, respectively, the dimensionless quantities are

$$\left.\begin{array}{l} x^*=\frac{x}{l},\quad y^*=\frac{y}{l},\quad t^*=\frac{\sqrt{gd}}{l}t,\quad \nabla^*=l\nabla,\quad \frac{\partial}{\partial t^*}=\left(\frac{\partial}{\partial t}\right)^*=\frac{l}{\sqrt{gd}}\frac{\partial}{\partial t},\\ \eta_{i,e}^*=\frac{\eta_{i,e}}{H},\quad \delta_i^*=\frac{\delta_i}{H},\quad m_i^*=\frac{m_i}{\rho},\quad B_i^*=\frac{B_i}{\rho g l^4},\quad p_{i,e}^*=\frac{p_{i,e}}{\rho g d}\end{array}\right\}, \quad (8)$$

where $e = 0$ and 1.

We substitute Equation (8) into Equation (7) and obtain

$$\varepsilon^2\sigma^2m_i^*\delta_i^*\frac{\partial^2\eta_{i,1}^*}{\partial t^{*2}}+\varepsilon B_i^*\nabla^{*2}\nabla^{*2}\eta_{i,1}^*+\varepsilon m_i^*\delta_i^*+p_{i-1,0}^*-p_{i,1}^*=0, \quad (9)$$

where $\varepsilon = H/d$ and $\sigma = d/l$ are the representative ratio of wave height to water depth and that of water depth to wavelength, respectively. In a manner similar to that of [18], each layer is assumed to be relatively shallow, so the orders of the parameters are $O(\varepsilon) = O(\sigma^2) \ll 1$. Thus, the first term on the left-hand side of Equation (9) can be ignored. Without this term, we obtain the i-plate equation for the dimensional quantities as

$$B_i\nabla^2\nabla^2\eta_{i,1}+m_ig\delta_i+p_{i-1,0}-p_{i,1}=0. \quad (10)$$

In the present paper, the interaction between surface water waves and a flexible platform floating at the sea surface is discussed, so the velocity potential for the one layer is described as $\phi(\boldsymbol{x}, z, t) = f_\alpha z^\alpha$. Thus, the unknown values are the weighting factors f_α and the surface displacement $\eta_{1,1}(\boldsymbol{x}, t)$, which is simply described as $\zeta(\boldsymbol{x}, t)$ for the horizontally two-dimensional cases and $\eta(x, t)$ for the one-dimensional cases.

2.2. Numerical Method

The governing equations—Equations (5), (6), and (10)—were transformed to finite difference equations and solved to study the interaction of surface water waves with a floating thin plate. An implicit scheme to solve the present equations without thin plates was developed by [34], in which the two-layer problems between two fixed horizontal plates were solved to simulate one-dimensional propagations of internal waves. We applied this scheme to the equations with a thin plate floating at the sea surface to simulate the propagations of surface waves. This model has been applied for several numerical simulations [20,22,35].

In the initial state at t = 0 s, the weighting coefficients $f_{i,\alpha}(x, 0 \text{ s})$ of the expanded velocity potential in Equation (1) were all zero, so the initial velocity was zero everywhere. In this paper, the values are written without considering significant digits, although the calculations were conducted using 64-bit floating-point numbers. In the present study, the number of terms for the velocity potential expanded as in Equation (1), i.e., $N_1 = N$ was one, so the governing equations were reduced to nonlinear shallow water equations for velocity potential considering the flexural rigidity of a floating thin plate. The numerical calculation method described above is also applicable in this case.

In order to verify the accuracy of this numerical model, the reproducibility of the response of a floating thin plate is examined by comparing the results of numerical computation and those of existing hydraulic experiments. A floating thin plate with a flexural rigidity of 450 N·m^2 was installed in a wave channel [19], as sketched in Figure 2.

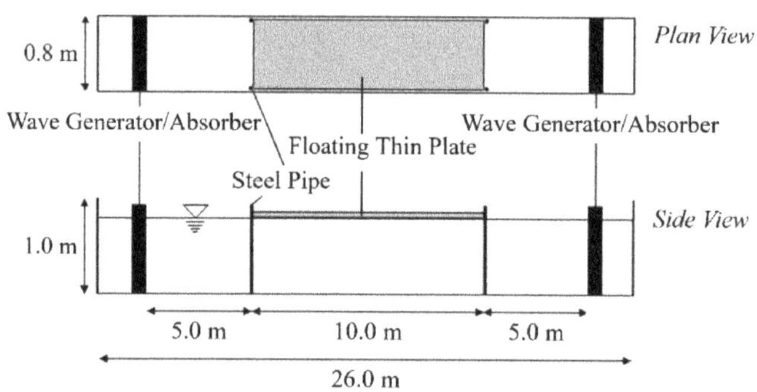

Figure 2. Wave flume used in the hydraulic experiments [19].

In the numerical computation, the computational domain is illustrated in Figure 3. It should be noted that discretization in the z-axis direction is not performed because the governing equations are vertically integrated equations.

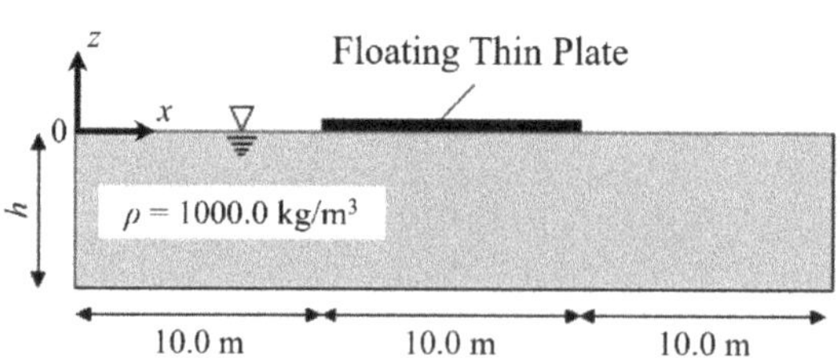

Figure 3. Side view of the computational domain with a floating thin plate.

As depicted in Figure 4, the distribution of the flexural rigidity B was given at the location of the structure floating at the water surface to express the thin plate covering part of the water area.

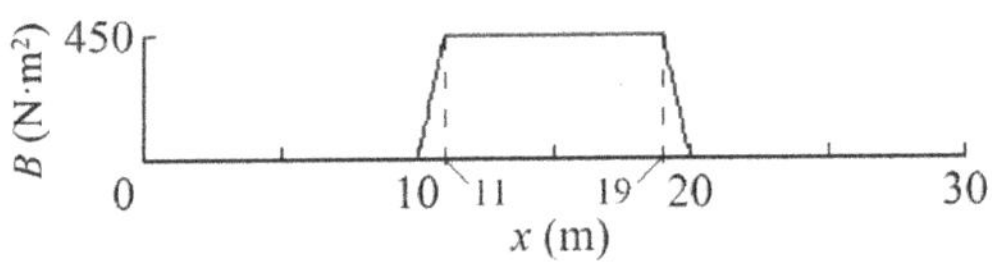

Figure 4. Distribution of the flexural rigidity at the surface of the computational domain depicted in Figure 3.

The grid sizes and time interval were $\Delta x = \Delta y = 5.0 \times 10^{-2}$ m and $\Delta t = 2.5 \times 10^{-5}$ s, respectively. When a solitary wave is incident, Figure 5 presents the experimental and numerical displacements ζ of the floating thin plate or water surface at $x = 7.0$ m and 14.5 m. Based on the results, it is confirmed that the surface displacements in the almost one-dimensional wave propagation were simulated successfully.

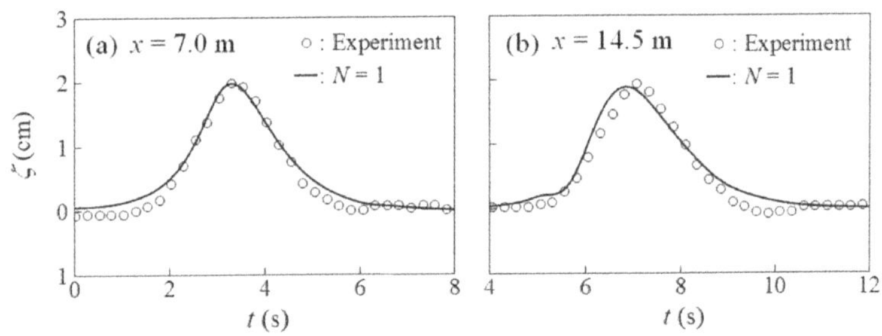

Figure 5. Surface displacements ζ at $x = 7.0$ m (**a**) and 14.5 m (**b**). The still water depth was 0.4 m and the incident wave height was 0.02 m.

In the present study, we generated numerical calculations for one- and two-dimensional propagations of surface waves. In the 1D calculations, we obtained surface waves generated by two sizes of airplanes with different weights at touch-and-go, takeoff, and landing. When the hydroelastic runway is not so wide compared to the spacing of the aircraft's left and right landing gears, 1D wave propagation can be dominant. Conversely, in the 2D calculations, we obtained surface wave patterns for different water depths when a medium-sized airplane lands at or takes off from a floating airport.

3. Calculation Conditions

3.1. 1D Calculations

3.1.1. Common Conditions

In the 1D calculations, an airport with a length of 5 km was floating at the sea surface within 0.5 km $\leq x \leq$ 5.5 km, and the still water depth h was uniform. An airplane ran on the floating airport in the positive direction of the x-axis. The grid size Δx was 20 m and the time interval Δt was 0.01 s.

We considered the movement of airplanes of two sizes—B747-400 and B737-800. The B747 series airplanes, nicknamed "jumbo jet", continued growth in sales from the 1990s to the early 2000s, playing a leading role in long-haul international flights. In recent years, however, for reasons such as energy saving, more compact airplanes have become mainstream, including A320 and B737 jetliners, especially in the routes connecting domestic regional cities. Figure 6 shows the photographs of airplane models for size comparison.

For simplicity, we call the B747-400 and B737-800 airplanes B747 and B737, respectively, in this paper. The masses of B747 and B737 were set to 397,000 kgs and 79,000 kgs, respectively, referring to the maximum takeoff weights [36]. The unit "kgs" is often used for the mass of airplanes in aviation industry and is the same as "kg" in physics. In the 1D calculations, the tire contact distances of both B747 and B737 were assumed to be 9.8 m, considering the unit width of 1 m. In all 1D cases of touch-and-go, landing, and takeoff,

B747 ran from $x = 1$ km to 4 km, whereas B737 ran from $x = 1$ km to 3 km, on the floating airport. To simplify the conditions, we assumed that the point load due to the airplanes was constant while the airplanes were running on the airport, unlike in the following 2D calculations.

Figure 6. Photographs of the airplane models for size comparison. The upper model is B747-400 and the lower model is B737-800, which were produced by SkyMarks Models and Solaseed Air Inc., respectively. The length scale is 1/200.

3.1.2. Conditions for Touch-and-Go

In touch-and-go, an airplane lands on an airport, keeps running while preparing for takeoff, and takes off from the runway. This is part of basic training called "circuits and bumps". In touch-and-go, both B747 and B737 were assumed to run at their constant speeds. The calculation conditions for touch-and-go are listed in Table 1. The values of flexural rigidity B were determined with reference to those obtained during the prototype test at the Mega-Float airport [37]. It should be noted that the airport length of 15 km is an unrealistic value set to ignore wave reflections at an edge of the airport.

Table 1. Calculation conditions for touch-and-go in the 1D calculations.

Case *	Airplane				Airport		Water Depth
	Type (Mass)	Running Speed	Running Distance	Run Time	Length L	Flexural Rigidity B	h
GA-L1	B747-400	83 m/s	3 km	36.1 s	15 km	1×10^{11} N·m	10 m
GA-L2	(397,000 kgs)						10 m to 50 m
GA-S1					5 km		10 m, 20 m
GA-S2						1×10^{10} N·m	10 m
GB-L1	B737-800	78 m/s	2 km	25.6 s	15 km		50 m
GB-L2	(79,000 kgs)						10 m to 50 m
GB-S1					5 km		50 m

* In the case names, GA and GB indicate the touch-and-go of B747 and B737, respectively, and L and S indicate that the airport is long and short, respectively.

3.1.3. Conditions for Landing

When an airplane lands on a floating airport, we assume that the airplane runs at a constant deceleration on the airport. In 1D problems, we consider the cases in which an airplane quickly shifts to taxiing when the running speed becomes sufficiently slow and leaves the runway, heading for a terminal. Therefore, after the run time, both the running speed and load of the airplane are assumed to be zero. The calculation conditions are listed in Table 2.

Table 2. Calculation conditions for landing in the 1D calculations.

Case *	Airplane					Airport		Water Depth
	Type (Mass)	Landing Speed **	Running Deceleration	Running Distance	Run Time	Length L	Flexural Rigidity B	h
LA-L	B747-400	72 m/s	0.86 m/s^2	3 km	83.7 s	15 km	1×10^{11} N·m	10 m
LA-S	(397,000 kgs)					5 km		
LB-L	B737-800		1.3 m/s^2	2 km	55.4 s	15 km	1×10^{10} N·m	50 m
LB-S	(79,000 kgs)					5 km		

* In the case names, LA and LB indicate the landing of B747 and B737, respectively, and L and S after hyphen indicate that the airport is long and short, respectively. ** The landing speed is the airplane speed when the airplane touches down at the airport.

3.1.4. Conditions for Takeoff

When an airplane takes off from a floating airport, we assume that the airplane runs at a constant acceleration on the airport. In 1D problems, we consider the cases in which an airplane plays a rolling start—the airplane starts running to take off immediately after arriving at the starting point. Therefore, at the starting time, both the running speed and load of the airplane are assumed to be zero. The calculation conditions are listed in Table 3.

Table 3. Calculation conditions for takeoff in the 1D calculations.

Case *	Airplane					Airport		Water Depth
	Type (Mass)	Takeoff Speed **	Running Acceleration	Running Distance	Run Time	Length L	Flexural Rigidity B	h
TA-L	B747-400	83 m/s	1.15 m/s^2	3 km	72.2 s	15 km	1×10^{11} N·m	10 m
TA-S	(397,000 kgs)					5 km		
TA-S-B							1×10^{9} N·m to 1×10^{11} N·m	50 m
TB-L	B737-800	78 m/s	1.52 m/s^2	2 km	51.3 s	15 km	1×10^{10} N·m	
TB-S	(79,000 kgs)					5 km		

* In the case names, TA and TB indicate the takeoff of B747 and B737, respectively, and L and S indicate that the airport is long and short, respectively. ** The takeoff speed is the airplane speed when the airplane leaves the airport.

3.2. 2D Calculations

3.2.1. Common Conditions

In 2D problems, we move a point load on a floating airport with a finite width. As sketched in Figure 7, the computational domain covers the area of 0 km $\leq x \leq$ 6 km and 0 km $\leq y \leq$ 1 km, and an airport is floating at the sea surface within 0.5 km $\leq x \leq$ 5.5 km and 0 km $\leq y \leq$ 0.5 km.

The perfect reflection condition was adopted along the x-axis, considering that the phenomena are axisymmetric with respect to the x-axis when an airplane runs along the x-axis. Therefore, the length and width of the airport were assumed to be 5 km and 1 km, respectively.

The still water depth h was uniformly 10 m, 20 m, or 100 m in the computational domain. The flexural rigidity of the airport, B, which was uniformly 1.0×10^{11} N·m^2, was given in the area covered by the airport, and it was possible to consider both the reflection and transmission of waves at the edges of the floating airport, including the side edges. The Sommerfeld open boundary condition was adopted at the lateral boundaries other than the x-axis. The grid sizes and time interval were $\Delta x = \Delta y$ = 20 m and $\Delta t = 2.0 \times 10^{-4}$ s, respectively.

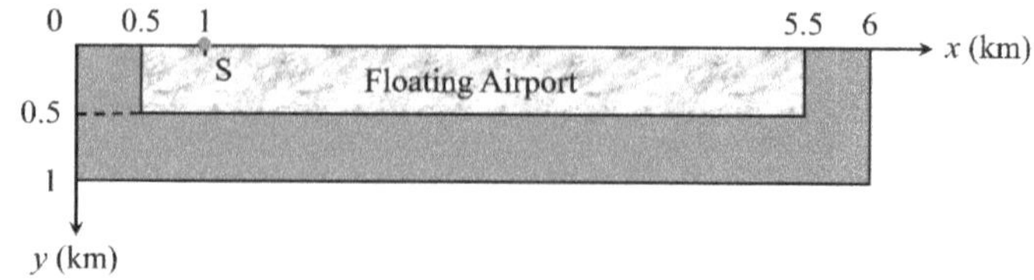

Figure 7. Computational domain for the 2D calculations. The floating airport is located in the marble-colored area. The perfect reflection condition is adopted along the x-axis, so the x-axis is the neutral axis of the domain, assuming that the length and width of the airport are 5 km and 1 km, respectively. At Point S, located at $x = 1$ km and $y = 0$ km, an airplane touches down for landing and starts running for takeoff.

In the 2D problems, we selected a medium-sized passenger airplane, namely the B787-8 Dreamliner, which we simply call B787. B787 can be operated with approximately 20% less fuel than its predecessors, bringing the economy of large jetliners to the medium-sized airplane market. The mass of B787, M, was set to 228,400 kgs, referring to the maximum takeoff weight [36]. B787 touches down for landing and starts running for takeoff at Point S, located at $x = 1$ km and $y = 0$ km in Figure 7, and runs on the floating airport in the positive direction of the x-axis from Point S.

When normal stress ω is applied to square grids in the horizontal plane, we assume that the distribution of the normal stress is a cone, the bottom of which is a circle of radius Δx centered at a grid point. Therefore, the normal stress ω at the central grid point is applied, satisfying

$$M = \pi \Delta x^2 \omega / (3g). \tag{11}$$

3.2.2. Conditions for Landing

B787 lands on a floating airport in Case LC, in which the landing speed is 75 m/s. The deceleration is assumed to be constant during landing. To understand the phenomena more clearly, we take an example with a larger value of deceleration, namely 3 m/s^2, so that the running distance and run time are 0.938 km and 25 s, respectively. The conditions of Case LC are listed in Table 4.

Table 4. Calculation conditions in the 2D calculations.

Case *	Airplane					Airport		Water Depth
	Type (Mass)	Landing/Takeoff Speed **	Running Acceleration	Running Distance	Run Time	Length L	Flexural Rigidity B	h
LC TC	B787 (228,400 kgs)	75 m/s	-3 m/s^2 3 m/s^2	0.938 km	25 s	5 km	1×10^{11} N·m^2	10 m, 20 m, or 100 m

* In the case names, LC and TC indicate the landing and takeoff of B787, respectively. ** The landing and takeoff speeds are the airplane speeds when the airplane touches down at the airport and leaves the airport, respectively.

In the 2D calculations for landing, the load on the airport due to the airplane was zero at the touchdown time, i.e., $t = 0$ s, whereafter it increased linearly with time, and the full weight of the airplane was applied on the airport when the airplane stopped at $t = 25$ s.

3.2.3. Conditions for Takeoff

B787 takes off from a floating airport in Case TC, in which the takeoff speed is 75 m/s. The acceleration is assumed to be constant during takeoff. For takeoff, we also consider a large value of acceleration, namely 3 m/s^2, so the running distance and run time are 0.938 km and 25 s, respectively, which are relatively short values. The conditions of Case TC are listed in Table 4.

In the 2D cases for takeoff, we assume that B787 has stopped and the field is in a steady state at $t = 0$ s. Therefore, the full weight of the airplane is applied on the airport in a steady state at $t = 0$ s, and we numerically solve Equation (10) to obtain the initial displacement of the floating airport. After $t = 0$ s, the airplane gradually accelerates and the load due to the airplane is assumed to decrease linearly with time, to become zero at the leaving time, i.e., $t = 25$ s, at which the airplane is completely away from the airport. After $t = 25$ s, no loading onto the floating airport is carried out.

4. 1D Response of a Floating Airport to Airplane Movement

4.1. Touch-and-Go

We numerically simulated the motion of a floating airport when an airplane performs touch-and-go, by assuming that the airplane ran at a constant speed and the point load due to the airplane was constant while running. The calculation conditions are described in Table 1.

In Case GA-L1, B747 performed touch-and-go on a long-enough floating airport. The time variation of the floating airport and water surface profiles is depicted in Figure 8, in which the black dotted line indicates the location of the airplane running on the floating airport.

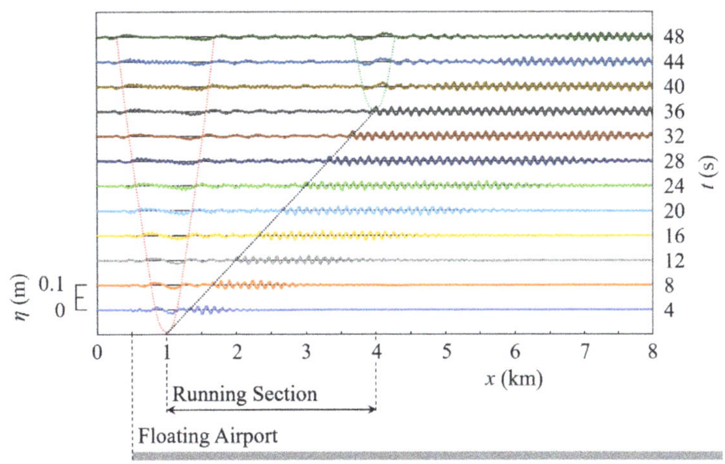

Figure 8. Profiles of the floating airport and water surface at every 4 s when B747 performed touch-and-go in Case GA-L1, the conditions of which are described in Table 1. The still water depth h was 10 m, the flexural rigidity of the airport, B, was 1×10^{11} N·m, and the airport length L was 15 km. The black dotted line indicates the location of the airplane running on the floating airport. The red and green dotted lines indicate the waves generated by the touchdown and leaving of the airplane, respectively.

When the airplane landed on the airport at $t = 0$ s, the touchdown impact due to the airplane generated forward and backward waves, the propagation of which is indicated by the red dotted curves in Figure 8. While the airplane was running, a floating-body wave train with larger wave heights appeared in front of the airplane. However, the maximum wavelength of the floating-body waves produced by the running airplane was shorter than that generated by the touchdown impact. The generated floating-body waves traveled at velocities greater than the running velocity of the airplane. When the airplane took off at $t = 36.1$ s, this leaving impact due to the airplane generated other waves, which are indicated by the green dotted curves in the figure. The maximum wavelength of these waves generated by the leaving impact was also longer than that of the waves due to the running airplane.

In Case GA-S1, the time variation of the floating airport and water surface profiles is depicted in Figure 9, in which the black dotted line indicates the location of the airplane running on the floating airport.

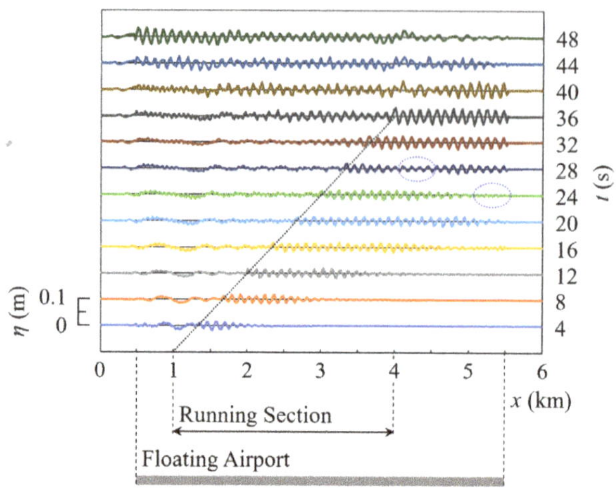

Figure 9. Profiles of the floating airport and water surface at every 4 s when B747 performed touch-and-go in Case GA-S1, the conditions of which are described in Table 1. The still water depth h was 10 m, the flexural rigidity of the airport, B, was 1×10^{11} N·m, and the airport length L was 5 km. The black dotted line indicates the location of the airplane running on the floating airport. The two blue dotted ellipses indicate examples of wave height reduction.

A floating-body wave generated by an airplane and reaching an end of a floating airport is divided into a reflected wave and a transmitted wave. The former is a floating-body wave that travels in the negative direction of the x-axis after being reflected at the airport edge, whereas the latter is a water wave that propagates in the positive direction of the x-axis. The wave height ratio between the former and the wave before reaching the airport edge is called reflectance in the present paper. Especially at t = 36 s, large wave heights appeared based on the superposition of the waves reflected at the airport edge and the waves newly generated by the running airplane. There are two reasons why the reflectance of floating-body waves increases: (1) the wave energy attenuation is not effective, and (2) the difference between the traveling velocity of the floating-body waves and that of the water surface waves is large. Regarding the former, the wave energy attenuation was ignored in the present computation. A case study will be required because wave attenuation in a floating body depends on the type of structure and members. Conversely, to discuss the latter, we consider the dispersion relation of floating-body waves. When the draft of a floating thin plate is assumed to be zero, the linear dispersion relation of floating-body waves is expressed by

$$\theta^2 = \left(\frac{Bk^4}{m} + g \right) k \tanh(kh), \tag{12}$$

where θ and k are the angular frequency and wavenumber of floating-body waves, respectively, and m is the density of the floating body [38]. As an example, Figure 10 displays the relationship between the traveling speed C and wavelength λ of floating-body waves evaluated by Equation (12) when $B = 1.0 \times 10^{11}$ N·m, m = 1000 kg/m^3, and h = 20 m. As depicted in Figure 10, we also obtained the corresponding numerical result of the linearized present model when the number of expansion terms for the velocity potential, N, is one, by considering the wave dispersion only due to the flexural rigidity of the floating body. For comparison, the phase velocity of linear water waves is depicted in the figure, which indicates that the difference in traveling velocity C between floating-body waves and water surface waves increases as the wavelength λ decreases. In Case GA-S1, the wavelength of the floating-body waves generated by the running airplane was less than 120 m based on

Figures 8 and 9. Moreover, the phase velocity of water waves decreases as the still water depth decreases in shallow water. Therefore, in Case GA-S1, the traveling velocity ratio between the floating-body waves and water waves was large, and the wave reflectance at the airport edges was large.

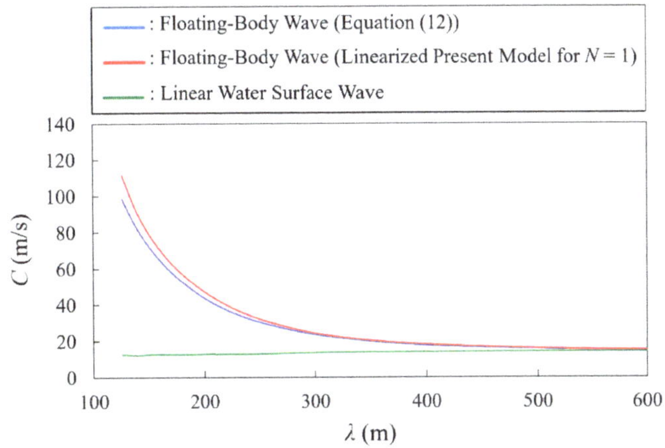

Figure 10. Traveling velocities of floating-body waves from Equation (12) and the linearized present model when the number of expansion terms for the velocity potential, N, is one. The flexural rigidity B and density m are 1.0×10^{11} N·m and 1000 kg/m^3, respectively. The still water depth h is 20 m. The phase velocity of linear water waves is also depicted for comparison.

The relationships between the traveling velocity of floating-body waves, C, and the flexural rigidity of the floating body, B, is presented in Figure 11, using Equation (12), when the density of the floating body, m, is 1000 kg/m^3, the wavelength of the floating-body waves is 100 m, and the still water depth h is 10 m. Based on the figure, as B is increased, C increases remarkably when $B > 10^9$ N·m. When the flexural rigidity B decreases, the difference between the decreased traveling velocity of floating-body waves and the phase velocity of water waves decreases, so the wave reflectance is reduced. Thus, the reflectance of floating-body waves depends on both the flexural rigidity of an airport and the wavelength-to-water-depth ratio.

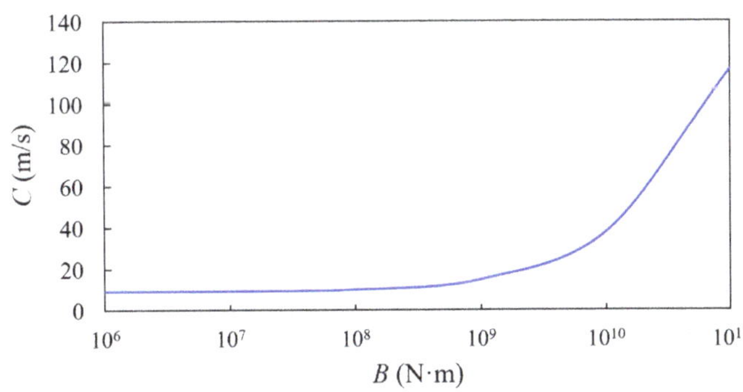

Figure 11. Relationship between the traveling velocity of floating-body waves, C, and the flexural rigidity of the floating body, B, using Equation (12). The density of the floating body, m, is 1000 kg/m^3, the wavelength of the floating-body waves is 100 m, and the still water depth h is 10 m.

Moreover, the relationship between the traveling velocity of floating-body waves, C, and the still water depth h is depicted in Figure 12, using Equation (12), when $B = 1.0 \times 10^{11}$ N·m,

m = 1000 kg/m^3, and the wavelength of the floating-body waves is 100 m. Figure 12 indicates that C decreases as h is increased for the same wavelength.

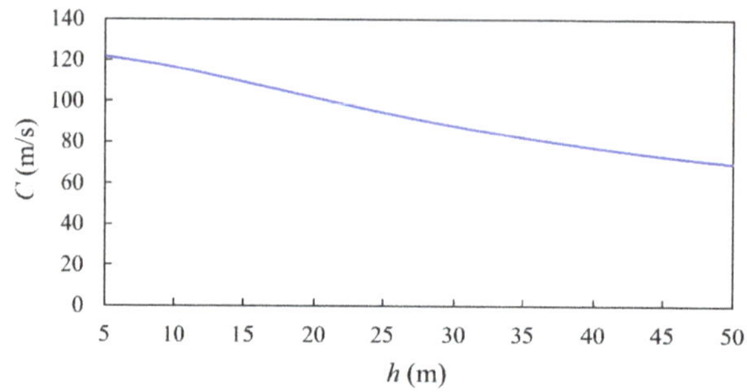

Figure 12. Relationship between the traveling velocity of floating-body waves, C, and the still water depth h, using Equation (12). The flexural rigidity B and density m of the floating body are 1.0×10^{11} N·m and 1000 kg/m^3, respectively. The wavelength of the floating-body waves is 100 m.

In Figure 9, at t = 48 s, the reflected waves mentioned above, the re-reflected waves at another edge of the airport, and the waves produced by the touchdown impact due to the airplane are superposed. Conversely, within the two blue dotted ellipses depicted in the figure, the wave height is reduced because the phase of the waves newly produced by the running airplane and that of the reflected waves happen to be close to opposite. If the reflectance of floating-body waves at airport edges is large, it is necessary to pay attention to the long-lasting vibrations of the floating airport for flight landing and takeoff. A method of reducing the reflectance will be described in Section 4.3.

Figure 13 depicts the time variation of the floating airport and water surface profiles in Case GA-S2, which was obtained by reducing the flexural rigidity B from Case GA-S1. Comparing Figure 13 with Figure 9 for Case GA-S1, both the wavelength and traveling velocity of the floating-body waves generated by the running airplane decreased, so the total length of the floating-body wave train decreased.

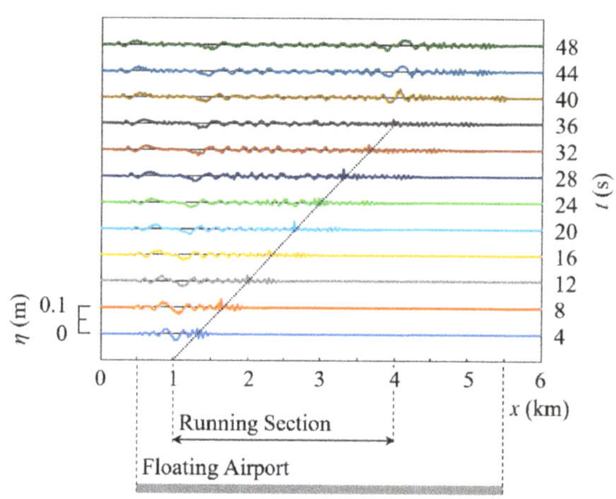

Figure 13. Profiles of the floating airport and water surface at every 4 s when B747 performed touch-and-go in Case GA-S2, the conditions of which are described in Table 1. The still water depth h was 10 m, the flexural rigidity of the airport, B, was 1×10^{10} N·m, and the airport length L was 5 km. The black dotted line indicates the location of the airplane running on the floating airport.

Conversely, when B737 performed touch-and-go in Cases GB-L1 and GB-S1, the time variations of the floating airport and water surface profiles are depicted in Figures 14 and 15, respectively. Comparing these results, in Case GB-S1 with an airport of limited length, the wave heights of the floating-body waves were also larger near the edge of the airport because of wave superposition based on the high reflectance of the floating-body waves at the airport edge.

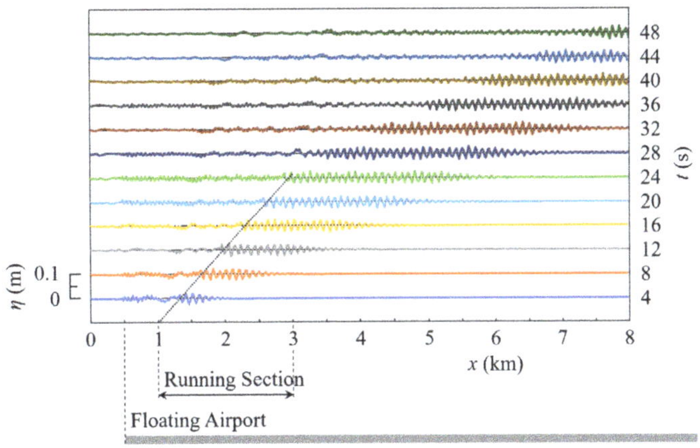

Figure 14. Profiles of the floating airport and water surface at every 4 s when B737 performed touch-and-go in Case GB-L1, the conditions of which are described in Table 1. The still water depth h was 50 m, the flexural rigidity of the airport, B, was 1×10^{10} N·m, and the airport length L was 15 km. The black dotted line indicates the location of the airplane running on the floating airport.

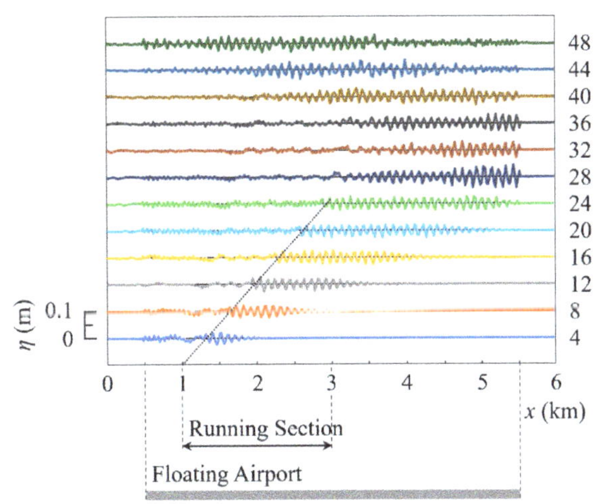

Figure 15. Profiles of the floating airport and water surface at every 4 s when B737 performed touch-and-go in Case GB-S1, the conditions of which are described in Table 1. The still water depth h was 50 m, the flexural rigidity of the airport, B, was 1×10^{10} N·m, and the airport length L was 5 km. The black dotted line indicates the location of the airplane running on the floating airport.

Comparing Figures 14 and 15 with Figures 8 and 9, B737 produced floating-body waves with wave heights comparable to those of B747 because the water depth was larger for the former than for the latter. When the still water depth is larger, i.e., h = 20 m, the time variation of the floating airport and water surface profiles in Case GA-S1 is depicted in

Figure 16. When the still water depth was increased, the wave heights of both the floating-body waves generated by the running airplane and those produced by the touchdown and leaving shocks due to the airplane increased in comparison with the results depicted in Figure 9 for h = 10 m.

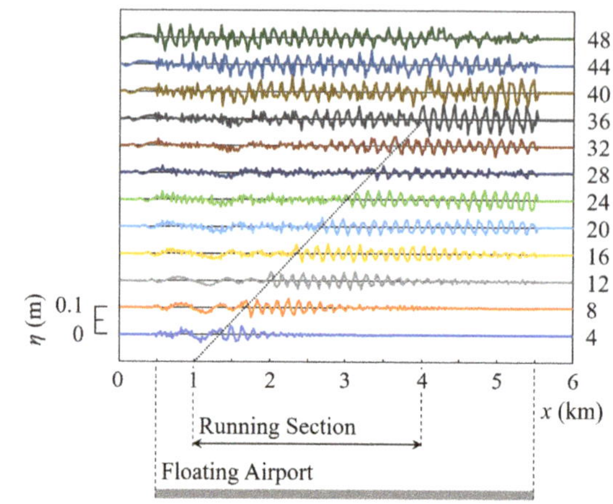

Figure 16. Profiles of the floating airport and water surface at every 4 s when B747 performed touch-and-go in Case GA-S1, the conditions of which are described in Table 1. The still water depth h was 20 m, the flexural rigidity of the airport, B, was 1×10^{11} N·m, and the airport length L was 5 km. The black dotted line indicates the location of the airplane running on the floating airport.

Figure 17 presents two examples of the relationships between the maximum displacement of the floating airport, η_{max}, and the still water depth h in Cases GA-L2 and GB-L2, when B747 and B737 left the long-enough airport in touch-and-go, respectively. These relationships are linearly approximated by $\eta_{max} = 0.0025h + 0.013$ and $\eta_{max} = 0.0009h + 0.0087$ (unit length in meter), respectively. When installing a floating airport in deeper water, it should be noted that the wave height of floating-body waves may increase because of airplane movement. The reason why larger water depths produce larger floating-body waves due to a running airplane will be discussed in Section 4.2.

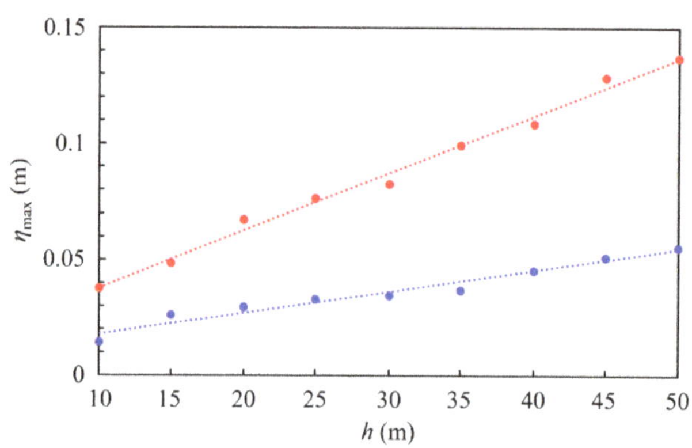

Figure 17. Relationships between the maximum displacement of the floating airport, η_{max}, and the still water depth h in Cases GA-L2 (red) and GB-L2 (blue), when B747 and B737 left the airport in touch-and-go, respectively. The conditions of these cases are described in Table 1. The flexural rigidity of the airport, B, was 1×10^{11} N·m in Case GA-L2, whereas B was 1×10^{10} N·m in Case GB-L2. The airport length L was 15 km in both cases.

4.2. Landing

We numerically simulated the motion of a floating airport when an airplane lands, by assuming that the airplane ran at a constant deceleration and the point load due to the airplane was constant while running. The calculation conditions are described in Table 2.

When B747 landed on a long-enough floating airport in Case LA-L, the time variation of the floating airport and water surface profiles is depicted in Figure 18. As the running speed of the airplane slowed down, both the wave height and wavelength of the floating-body waves generated by the running airplane increased. Based on the results of the one-dimensional propagation calculations [22], floating-body waves are significantly amplified when the moving speed of a point load on a floating thin plate is close to the phase velocity of the linear shallow-water waves, i.e., $\sqrt{gh}$, in shallow-water conditions. This is due to resonance similar to that occurring in tsunami generation due to atmospheric-pressure waves, e.g., [39], based on the Proudman resonance [40]. Such resonance phenomena, often with a tail including waves of short wavelengths, are also known in other transient waves, e.g., [41–46]. In Case LA-L, $\sqrt{gh}$ is approximately 9.9 m/s, which corresponds to the running speed of the airplane at $t \simeq 72$ s, near its stop time. Therefore, the airplane produced larger floating-body waves as it approached a stop. As depicted in Figure 17 for touch-and-go, the wave height of the floating-body waves increases as the still water depth increases, the reason for which is that the traveling speeds of the airplane and water waves become closer.

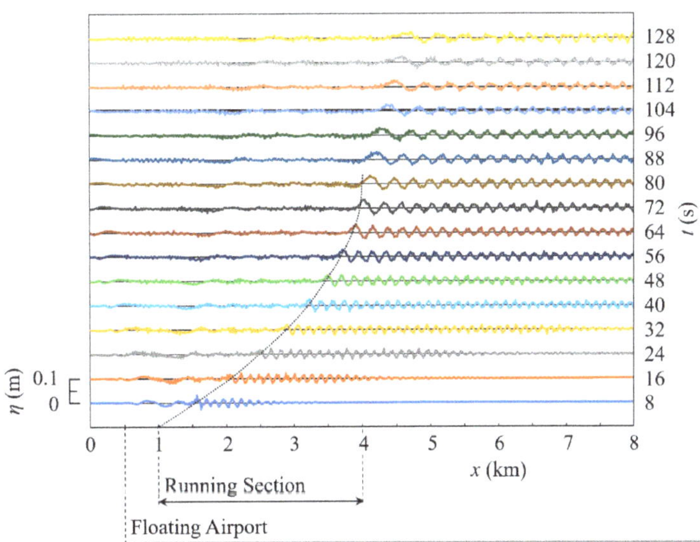

Figure 18. Profiles of the floating airport and water surface at every 8 s when B747 landed in Case LA-L, the conditions of which are described in Table 2. The still water depth h was 10 m, the flexural rigidity of the airport, B, was 1×10^{11} N·m, and the airport length L was 15 km. The black dotted line indicates the location of the airplane running on the floating airport.

In more detail, at landing, when the running speed of an airplane approaches the phase velocity of water waves, a "forced wave" is generated and amplified. In the present paper, a wave that follows the airplane is called a forced wave even if it contains a free-wave component in the resonance process. The generated forced waves do not satisfy the dispersion relation of floating-body waves, so the waves are disintegrated to produce "free waves" that satisfy the dispersion relation with traveling velocities larger than the airplane speed. The restraint by the airplane is released at the stop time of the airplane, whereafter the floating-body wave with the maximum wave height also propagates as a free wave.

Conversely, in Case GA-L1, in which B747 performed touch-and-go as described above, effective amplification did not occur because the airplane speed was too fast, and it was not close to the water wave speed at the water depth. While energy was being supplied by the running airplane, modest forced waves were generated and free waves continued to occur so as to satisfy the dispersion relation. Thus, large floating-body waves were not generated, not even by the jumbo jet.

Furthermore, although the landing impact due to the airplane at $t = 0$ s generated floating-body waves, as in the above cases of touch-and-go, no shock wave appeared when the airplane came to a stop because the airplane slowed down gradually.

When B747 landed on a floating airport of limited length in Case LA-S, the time variation of the floating airport and water surface profiles is depicted in Figure 19. The figure indicates that the floating-body waves reflected at the airport edge were superposed with the waves newly created by the running airplane. However, after $t = 72$ s, transmitted waves clearly appeared from the airport edge. Immediately after landing, the airplane speed was larger and the wavelength of the generated floating-body waves was shorter, so the wave reflectance was larger. Thereafter, as the airplane speed slowed down, the wavelength of the floating-body waves increased, and the wave reflectance decreased, as described above for Figure 10. Thus, some time after landing, transmitted waves may begin to appear.

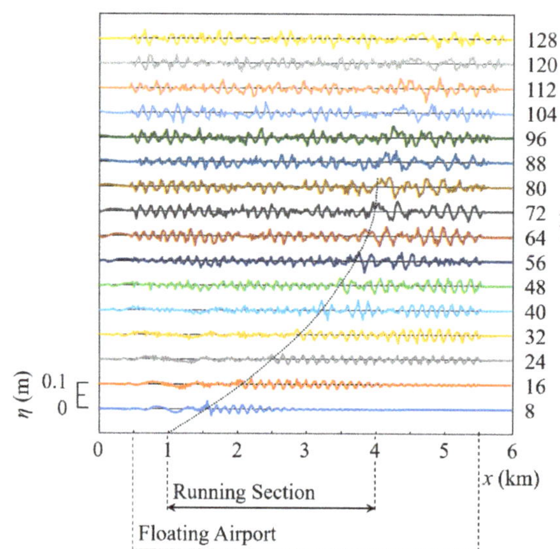

Figure 19. Profiles of the floating airport and water surface at every 8 s when B747 landed in Case LA-S, the conditions of which are described in Table 2. The still water depth h was 10 m, the flexural rigidity of the airport, B, was 1×10^{11} N·m, and the airport length L was 5 km. The black dotted line indicates the location of the airplane running on the floating airport.

When B737 landed on a long-enough floating airport in Case LB-L, the time variation of the floating airport and water surface profiles is depicted in Figure 20. Based on the figure, as the airplane speed decreased, the wave height of the generated floating-body waves increased, as in the cases of B747. It should be noted that the amplification factor of the wave height due to B737 in Figure 20 is larger than that due to B747 in Figure 18. As far as the flexural rigidity of the floating airport, B, is concerned, when B decreases, the traveling velocity of floating-body waves, C, decreases, as indicated in Figure 11. In the case of Figure 20, B decreased and C decreased, so the wavelength decreased, also depending on the relationship between C and the airplane speed. Moreover, the deceleration of B737 was larger than that of B747; hence, the floating-body waves were generated more effectively by B737 running at a slower speed. Consequently, combined with the deeper water, the maximum wave height generated by B737 became larger than that due to B747.

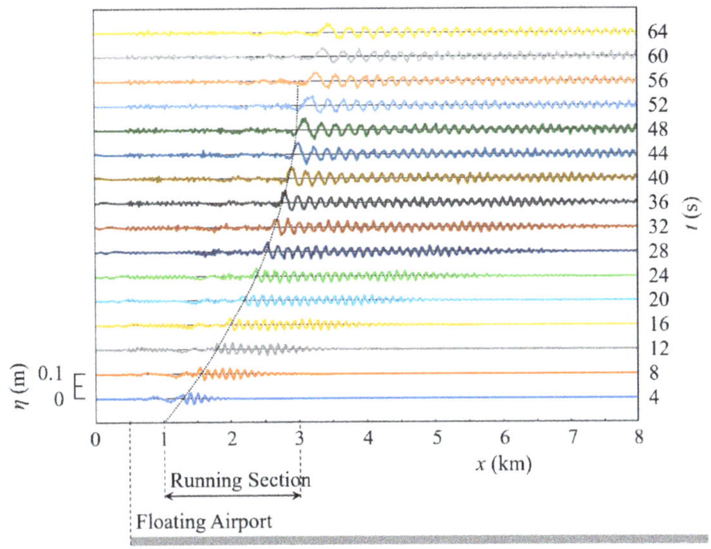

Figure 20. Profiles of the floating airport and water surface at every 4 s when B737 landed in Case LB-L, the conditions of which are described in Table 2. The still water depth h was 50 m, the flexural rigidity of the airport, B, was 1×10^{10} N·m, and the airport length L was 15 km. The black dotted line indicates the location of the airplane running on the floating airport.

Furthermore, when B737 landed on a floating airport of limited length in Case LB-S, the time variation of the floating airport and water surface profiles is depicted in Figure 21. Comparing this figure with Figure 19, because of the reduced flexural rigidity, the traveling velocity of the floating-body waves generated by the running airplane decreased, as described above, whereas the water wave speed increased in the deeper water. Thus, the transmitted waves from the airport edge started appearing at t = 52 s, earlier than in the case for Figure 19.

In the present paper, the acceleration of the light airplane was increased using the values close to the current actual conditions. However, in less congested airports, reverse thrust can be reduced to save fuel, allowing airplanes to land over longer distances. Future studies will also consider the cases of the same traveling speed but different airplane types.

4.3. Takeoff

We numerically simulated the motion of a floating airport when an airplane takes off, by assuming that the airplane ran at a constant acceleration and the point load due to the airplane was constant while running. The calculation conditions are described in Table 3.

When B747 took off from a long-enough floating airport in Case TA-L, the time variation of the floating airport and water surface profiles is depicted in Figure 22. Moreover, when B747 took off from a floating airport of limited length in Case TA-S, the result is depicted in Figure 23.

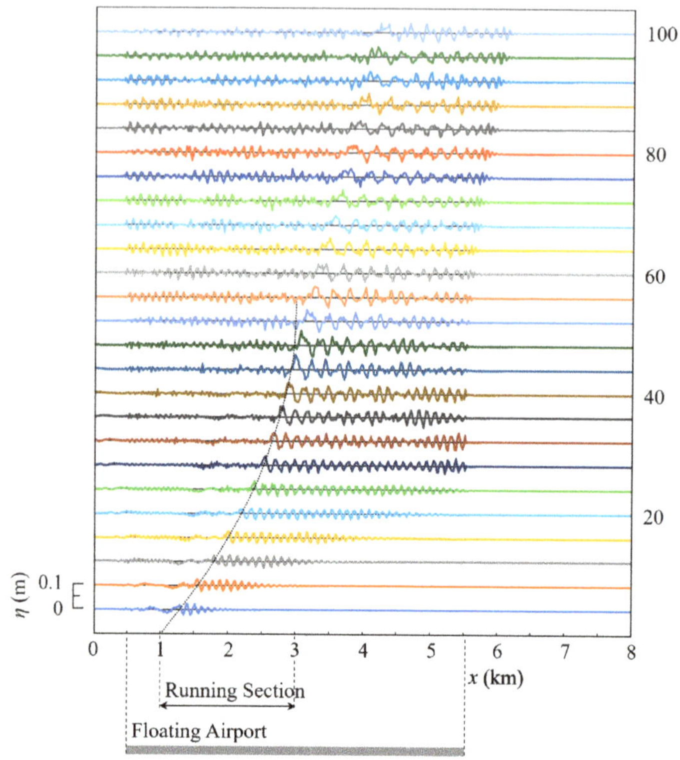

Figure 21. Profiles of the floating airport and water surface at every 4 s when B737 landed in Case LB-S, the conditions of which are described in Table 2. The still water depth h was 50 m, the flexural rigidity of the airport, B, was 1×10^{10} N·m, and the airport length L was 5 km. The black dotted line indicates the location of the airplane running on the floating airport.

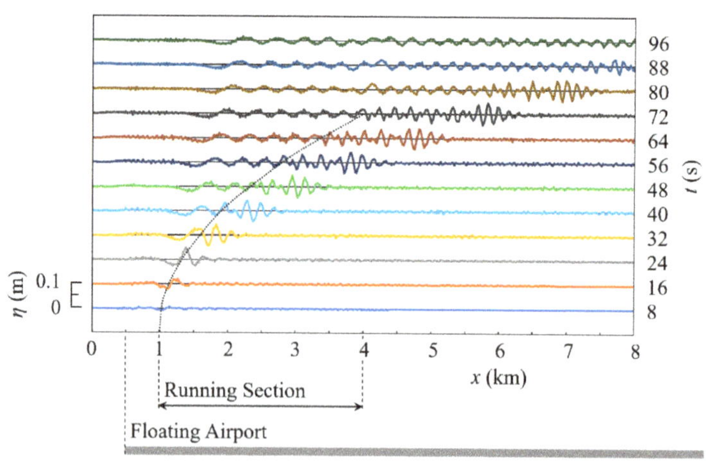

Figure 22. Profiles of the floating airport and water surface at every 8 s when B747 took off in Case TA-L, the conditions of which are described in Table 3. The still water depth h was 10 m, the flexural rigidity of the airport, B, was 1×10^{11} N·m, and the airport length L was 15 km. The black dotted line indicates the location of the airplane running on the floating airport.

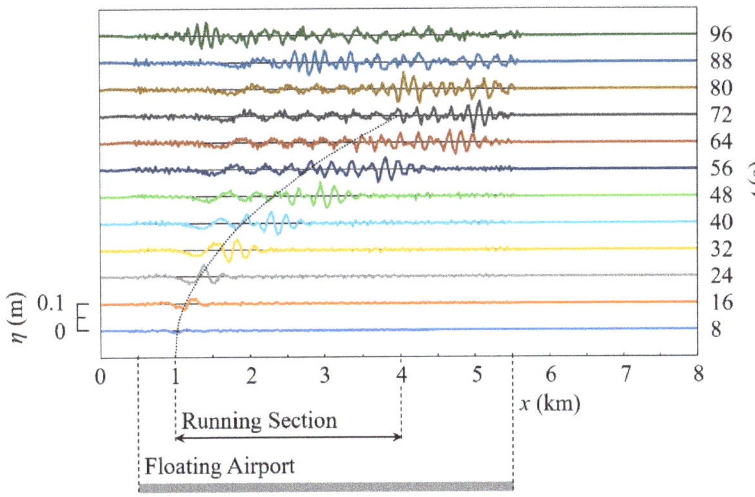

Figure 23. Profiles of the floating airport and water surface at every 8 s when B747 took off in Case TA-S, the conditions of which are described in Table 3. The still water depth h was 10 m, the flexural rigidity of the airport, B, was 1×10^{11} N·m, and the airport length L was 5 km. The black dotted line indicates the location of the airplane running on the floating airport.

Based on Figures 22 and 23, during the takeoff, much of the floating-body wave energy generated by the running airplane was localized in a short interval, resulting in a "wave clump" or wave group with a waveform like a discrete breather. The wavelength of the waves in the clump was approximately 170 m, and the traveling speed of the wave clump was approximately 43 m/s from Equation (12). The airplane reached this speed at $t \simeq 37$ s after departure. Although the traveling velocity of the wave clump did not reach 43 m/s for a while after its generation, the wave clump, which was produced by the slow-moving airplane shortly after its departure, traveled faster than the airplane. Thus, the wave clump traveling ahead of the airplane was followed by the waves newly generated by the running airplane.

To explain the above based on resonance, at takeoff, while airplane speed is close to water wave speed, the airplane generates and amplifies a forced wave based on the resonance. Thereafter, as the airplane speed increases and the difference in traveling speed between the airplane and the water waves increases, the restraint by the airplane is gradually released, and the amplified wave propagates as a free wave at a traveling velocity greater than the airplane speed. Therefore, the wave clump, which has been formed by changing the wave profile of the forced wave generated and amplified in a limited time, propagates in front of the airplane.

Conversely, when B737 took off from a long-enough floating airport and one of limited length in Cases TB-L and TB-S, respectively, the time variations of the floating airport and water surface profiles are depicted in Figures 24 and 25, respectively. Comparing these figures with Figures 22 and 23, when B737 takes off in deeper water, it may produce floating-body waves with a wave height similar to that of B747 in shallower water, depending on the flexural rigidity of the airport.

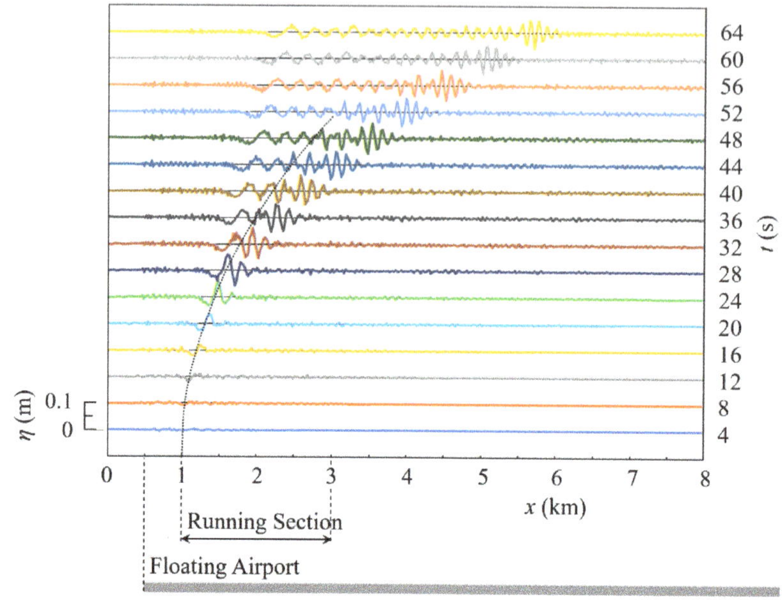

Figure 24. Profiles of the floating airport and water surface at every 4 s when B737 took off in Case TB-L, the conditions of which are described in Table 3. The still water depth h was 50 m, the flexural rigidity of the airport, B, was 1×10^{10} N·m, and the airport length L was 15 km. The black dotted line indicates the location of the airplane running on the floating airport.

To mitigate the hydroelastic vibration of a VLFS under wave action, several methods have been proposed: for example, the introduction of floating breakwaters [47], aircushions [48], and member connectors [49]. The reduction of the resonance phenomena due to the presence of a breakwater near a VLFS was also investigated [50]. If airplanes land or take off while floating airport vibration remains, unexpected large floating-body waves may occur because of wave superposition. As one of the methods of lowering the wave reflectance at the edges of a floating airport, we consider reducing the flexural rigidity of the airport, B, near the airport edges. In Case TA-S-B, the airplane to take off is B747 and the still water depth is 50 m. We numerically evaluated the wave reflectance R from the maximum wave heights before and after the first reflection at the airport edge. When B was 1×10^{11} N·m throughout the airport for 0.5 km $\leq x \leq$ 5.5 km, R was approximately 0.67. Conversely, when B was 1×10^{11} N·m for 0.5 km $\leq x <$ 5 km, and 1×10^{10} N·m for 5 km $\leq x \leq$ 5.5 km, R was approximately 0.55. This reduction in wave reflectance was due to two-stage reflection, which suppressed the wave height of the reflected waves. Furthermore, when B was 1×10^{11} N·m for 0.5 km $\leq x <$ 5 km, and B decreased linearly from 1×10^{11} N·m at x = 5 km to 1×10^{9} N·m at x = 5.5 km, R was approximately 0.27 because of successive reflection near the airport edge. Consequently, the wave reflectance was reduced by lowering B near the airport edge. Even if the flexural rigidity is structurally or economically fixed for most of a floating airport, the wave reflectance can be reduced by modifying the structure or installing accessories to lower the flexural rigidity only near the airport edges, leading to an increase in the calmness of the floating airport.

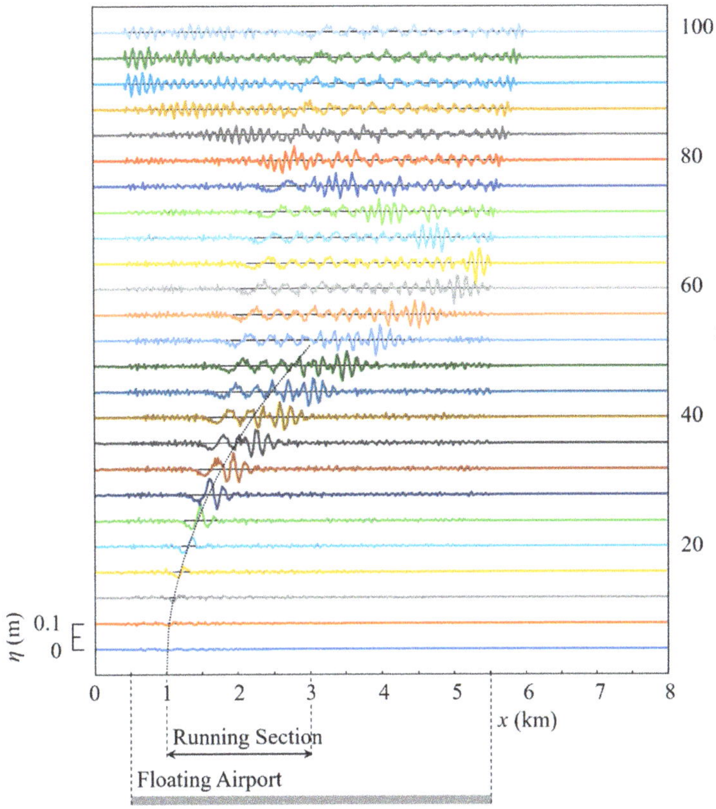

Figure 25. Profiles of the floating airport and water surface at every 4 s when B737 took off in Case TB-S, the conditions of which are described in Table 3. The still water depth h was 50 m, the flexural rigidity of the airport, B, was 1×10^{10} N·m, and the airport length L was 5 km. The black dotted line indicates the location of the airplane running on the floating airport.

5. 2D Response of a Floating Airport to Airplane Movement

5.1. Landing

We numerically simulated the two-dimensional motion of a floating airport when B787 lands in Case LC, by assuming that the airplane ran at a constant deceleration and the point load due to the airplane increased linearly with time while running. The calculation conditions are described in Table 4.

Figure 26 presents the surface level distributions at t = 25 s, at which B787 stopped after landing, for different still water depths h. As indicated in the figure, numerous surface waves exhibiting directional dispersion propagated and reflected at the airport edges. As a result, the superposition of the floating-body waves formed grid-like surface profiles at the airport. As the still water depth increases, the difference in propagation speed between the floating-body waves and water waves is reduced, so the wave reflectance at the airport edges decreases. Therefore, when h = 100 m in Case LC, the grid-like pattern was suppressed at the floating airport compared to when the still water depth is shallower.

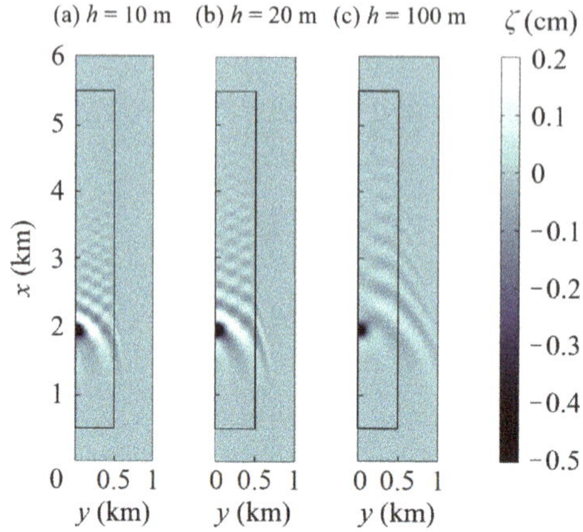

Figure 26. Surface-level distributions at t = 25 s, at which B787 stopped after landing on the floating airport in Case LC. The still water depth h values were 10 m, 20 m, and 100 m for the figures on the left (**a**), middle (**b**), and right (**c**), respectively. The flexural rigidity of the airport, B, was 1×10^{11} N·m^2.

Figure 27 depicts the surface profiles along the x-axis, which includes the longitudinal centerline of the floating airport, at the stop time t = 25 s, for different still water depths in Case LC. The order of the results was the same as that of the existing horizontally two-dimensional results for a jumbo jet [27], although the maximum values of the present results were less because of the different airplane weights, decelerations, etc. When the still water depth h is 20 m, the phase velocity of linear shallow-water waves, i.e., $\sqrt{gh}$, is 14 m/s, which corresponds to the airplane speed at $t \simeq 20$ s. At around this time, the point load due to the airplane was approximately 4/5ths of the maximum value and large enough to generate floating-body waves, resulting in a relatively large amplitude. Conversely, when h is 100 m, $\sqrt{gh}$ is approximately 31.3 m/s, which corresponds to the airplane speed at $t \simeq 15$ s. At around this time, the point load was still only approximately 3/5ths of the maximum value, so the floating-body waves were not generated effectively.

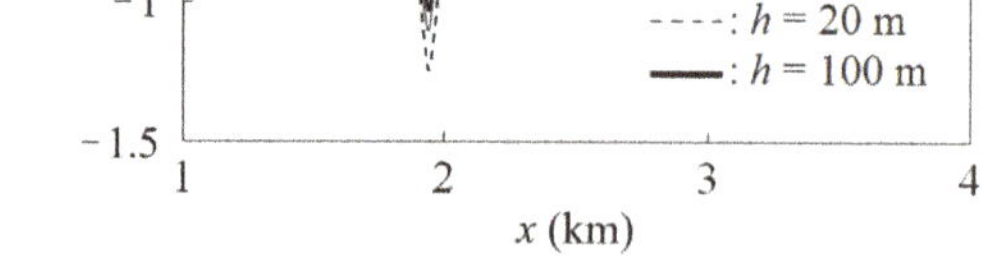

Figure 27. Surface profiles along the x-axis at t = 25 s, at which B787 stopped after landing in Case LC. The still water depth h values were 10 m, 20 m, and 100 m, and the flexural rigidity of the airport, B, was 1×10^{11} N·m^2.

Figure 28a,b depict the time variations of the vertical positions of B787 and the surface gradients at the location of B787, respectively, for different still water depths in Case LC. As indicated in Figure 28a, the airplane position was gradually lowered. The surface-gradient graphs are zigzag in Figure 28b because the values of the surface displacements are discretized at the grid points.

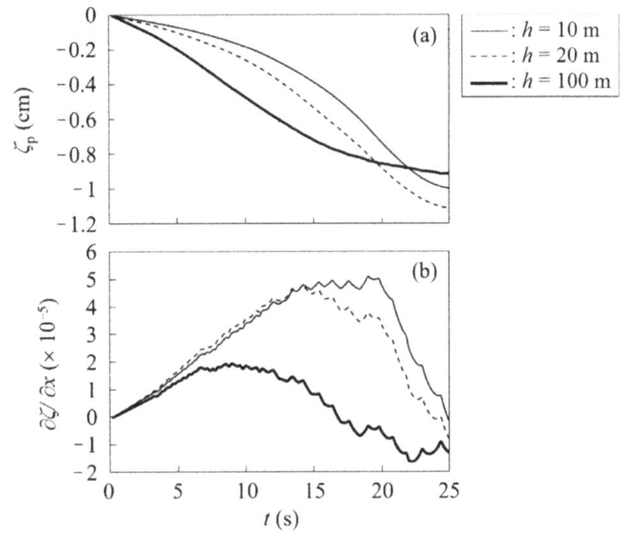

Figure 28. Time variations of the vertical positions of B787, ζ_p, (**a**), and the surface gradient $\partial\zeta/\partial x$ at the location of B787 (**b**), in Case LC, in which B787 stopped at t = 25 s after landing. The still water depth h values were 10 m, 20 m, and 100 m, and the flexural rigidity of the airport, B, was 1×10^{11} N·m^2.

During landing, the floating-body waves were generated by the airplane, as described above for the one-dimensional calculations, and Figure 28b indicates that the airport deflection caused the airplane to run uphill while the running speed of the airplane was large. The maximum upslope gradient of the floating airport beneath the airplane increased as the still water depth h decreased. When h = 10 m, the airplane was on an upslope most of the time during landing. When h = 20 m, the airplane ran downhill just before coming to a stop. Conversely, when h = 100 m, the airplane was on a downslope for approximately 1/3rd of the landing time. It is necessary to pay attention to the gradient change of the runway in rolling an airplane.

5.2. *Takeoff*

We numerically simulated the two-dimensional motion of a floating airport when B787 takes off in Case TC, by assuming that the airplane ran at a constant acceleration and the point load due to the airplane decreased linearly with time while running. The calculation conditions are described in Table 4.

Figure 29 depicts the surface-level distributions at t = 25 s, at which B787 left the airport, for different still water depths h. Figure 29a,b indicate that grid-like patterns were also produced at the floating airport in takeoff when h = 10 m and 20 m, respectively, although the grid-like patterns were not so remarkable as those in Figure 26a,b for landing. Conversely, when h = 100 m, such a grid-like vibration was not formed at the floating airport. The grid-like oscillation did not appear in the existing results for relatively deep water, e.g., [27], and this phenomenon, which occurs based on the high wave reflectance at airport edges, is peculiar to floating airports with large flexural rigidity installed in shallower water.

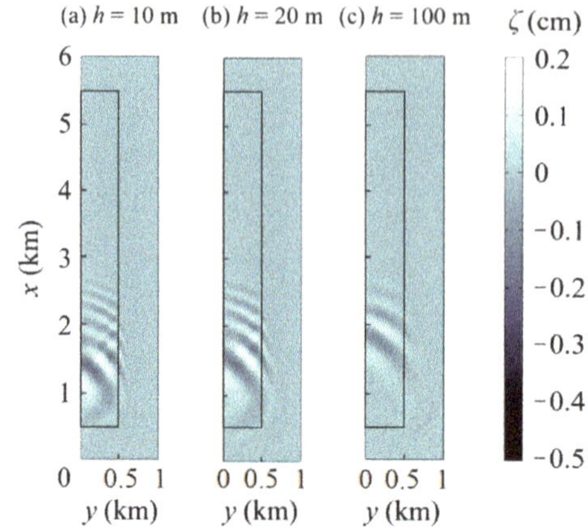

Figure 29. Surface-level distributions at t = 25 s, at which B787 left the floating airport in Case TC. The still water depth h values were 10 m, 20 m, and 100 m for the figures on the left (**a**), middle (**b**), and right (**c**), respectively. The flexural rigidity of the airport, B, was 1×10^{11} N·m^2.

Figure 30 depicts the surface profiles along the x-axis at t = 25 s, at which B787 left the airport, for different still water depths in Case TC. When the still water depth h is 100 m, the phase velocity of linear shallow-water waves, i.e., $\sqrt{gh}$, is approximately 31.3 m/s, which corresponds to the airplane speed at $t \simeq 10$ s. Before this time, the point load due to the airplane was larger than 3/5ths of the maximum value. In addition, the wave clumps were produced when the running speed of the airplane was low, as indicated in Figures 22–25. Therefore, when h = 100 m, larger floating-body waves were generated in takeoff than in landing.

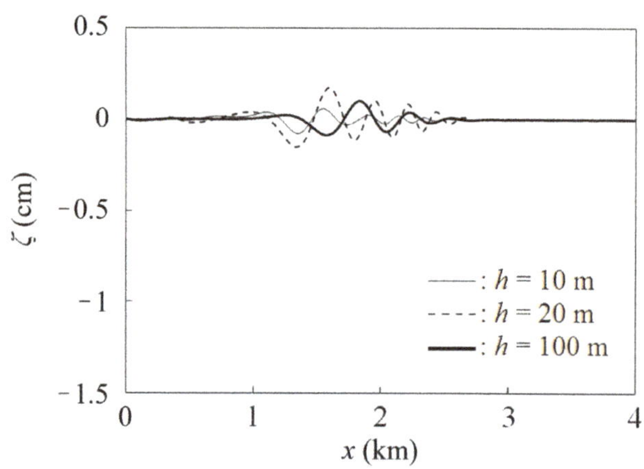

Figure 30. Surface profiles along the x-axis at t = 25 s, at which B787 left the airport in Case TC. The still water depth h values were 10 m, 20 m, and 100 m, and the flexural rigidity of the airport, B, was 1×10^{11} N·m^2.

Moreover, Figure 31a,b present the time variations of the vertical positions of B787 and the surface gradients at the location of B787, respectively, for different still water depths in Case TC. As indicated in Figure 31a, although the airplane passed the peak of height when h = 10 m and 20 m, the airplane position was increased most of the time during takeoff, which was different than during landing as depicted in Figure 28a.

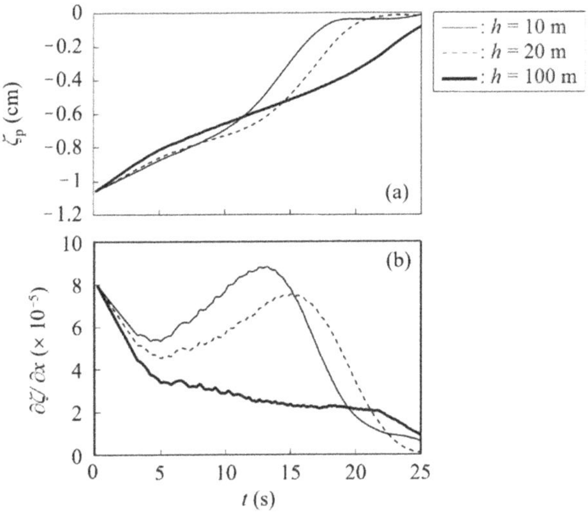

Figure 31. Time variations of the vertical positions of B787, ζ_p, (**a**), and the surface gradient $\partial\zeta/\partial x$ at the location of B787 (**b**), in Case TC, in which B787 left the airport at $t = 25$ s. The still water depth h values were 10 m, 20 m, and 100 m, and the flexural rigidity of the airport, B, was 1×10^{11} N·m^2.

Figure 31b indicates that when taking off, the airplane ran on an upslope in a trough of the floating airport because we assumed that the floating airport was bending under the weight of the stationary airplane at $t = 0$ s in Case TC. During takeoff, when $h = 100$ m, the upslope gradient beneath the airplane decreased with time, whereas it dropped, then increased, and then dropped again when $h = 10$ m and 20 m. For all these still water depths, the peak values of the upslope gradient were larger than those in landing depicted in Figure 28b. Therefore, during takeoff, the airplane ran on steeper slopes than during landing. For example, in Japan, the maximum longitudinal slope of a long airport runway is allowed up to 0.8% [51], and the calculated values of the slopes indicated in Figure 31b, as well as Figure 28b, are within the allowable range. However, if the floating-body waves from the previous landing/takeoff remain, or if the wave height of the reflected waves is large, the runway gradient may increase, not only in the traveling direction of the airplanes but also in the transverse direction. When the wave reflectance at the edges of a floating airport is large, it is necessary to take a sufficient time interval between landings and takeoffs.

6. Conclusions

Numerical simulations were generated to investigate the response of a very large floating airport to airplane movement using the set of nonlinear shallow water equations of velocity potential for water waves interacting with a floating thin plate.

First, the one-dimensional motion of a floating airport was simulated numerically when B747 and B737 performed touch-and-go, landing, and takeoff. During touch-and-go, when the running speed of the airplanes is much faster than the phase velocity of the water waves, effective amplification did not occur, and modest forced waves were generated while the airplanes were running, with many free waves ahead that satisfied the dispersion relation. However, when the airplane speed is closer to the water wave speed, even B737 produced large floating-body waves based on the resonance. Moreover, the impacts due to both the touchdown and leaving of the airplanes generated other forward and backward waves.

Conversely, during landing, when the airplane speed approached the water wave speed, a forced wave was generated and amplified based on the resonance. When the airplanes stopped, the restraint from the airplanes was released, whereafter the amplified floating-body wave with the maximum wave height also propagated as a free wave.

During takeoff, while the airplane speed was close to the water wave speed, the airplanes also generated and amplified a forced wave. Thereafter, as the running speed of the airplanes increased and the difference in traveling speed between the airplanes and the water waves increased, the restraint from the airplanes was gradually released and the amplified wave propagated as a free wave at a traveling velocity greater than those of the airplanes. Therefore, the wave clump, which was formed by changing the wave profile of the forced wave generated and amplified shortly after starting to run, propagated in front of the airplanes.

If the reflectance of floating-body waves at airport edges is high, prolonged vibrations may interfere with the operation of airplanes and cause structural fatigue. When we tried lowering the flexural rigidity of the airport near its edge, the wave reflectance was reduced. In the present study, we ignored the attenuation of wave energy in the floating airport, so future work is required to consider wave energy attenuation by modifying the present model.

Second, the horizontally two-dimensional motion of a floating airport was simulated numerically when B787 performed landing and takeoff. When the still water depth is shallower, a grid-like pattern was formed on the floating airport. This pattern was due to the reflection of the floating-body waves with directional dispersion at the edges of the floating airport and appeared more remarkably in landing than in takeoff. In the 2D calculations, we assumed that the load of the airplane changed while running, so the effective amplification of waves occurred from the sufficient load applied when the airplane speed approached the water wave speed. Furthermore, the maximum upslope gradient beneath the airplane increased as the still water depth decreased, and it was larger in takeoff than in landing.

Although we used the numerical scheme for the water wave equations considering both wave nonlinearity and dispersion, we ignored the dispersion of water waves in the present study. In the future, we should simulate floating-body waves with more expansion terms of velocity potential when the frequency dispersion is significant, especially for an airport installed in deep water. It is also necessary to consider whether the plate equation can be linear since there may be cases where the nonlinearity of floating-body motion that does not interfere with airplane operation cannot be ignored.

Author Contributions: Conceptualization, T.K.; Methodology, T.K.; Validation, M.H.; Investigation, T.K.; Writing—original draft, T.K. All authors have read and agreed to the published version of the manuscript.

Funding: This work was supported by JSPS KAKENHI Grant Number JP22H01136, the Collaborative Research Program of Research Institute for Applied Mechanics, Kyushu University, and the Research Institute for Mathematical Sciences, an International Joint Usage/Research Center located in Kyoto University.

Acknowledgments: Sincere gratitude is extended to Nakayama, K., Kobe University, who built the original program code for internal wave calculations. The authors thank Yamashita, K., the Nuclear Regulation Authority, for his contribution to the 2D calculations when he was a member of our laboratory. The authors would like to express our thanks to the anonymous reviewers for their valuable comments that improved the article.

Conflicts of Interest: The authors declare no conflict of interest.

References

1. Renzi, E.; Michele, S.; Zheng, S.; Jin, S.; Greaves, D. Niche applications and flexible devices for wave energy conversion: A review. *Energies* **2021**, *14*, 6537. [CrossRef]
2. Michele, S.; Zheng, S.; Greaves, D. Wave energy extraction from a floating flexible circular plate. *Ocean Eng.* **2022**, *245*, 110275. [CrossRef]

3. Kashiwagi, M. Research on hydroelastic responses of VLFS: Recent progress and future work. *Int. J. Offshore Polar Eng.* **2000**, *10*, 81–90.
4. Watanabe, E.; Utsunomiya, T.; Wang, C.M. Hydroelastic analysis of pontoon-type VLFS: A literature survey. *Eng. Struct.* **2004**, *26*, 245–256. [CrossRef]
5. Montiel, F.; Bonnefoy, F.; Ferrant, P.; Bennetts, L.; Squire, V.; Marsault, P. Hydroelastic response of floating elastic discs to regular waves. Part 1. Wave basin experiments. *J. Fluid Mech.* **2013**, *723*, 604–628. [CrossRef]
6. Zhang, X.; Lu, D.; Gao, Y.; Chen, L. A time domain discrete-module-beam-bending-based hydroelasticity method for the transient response of very large floating structures under unsteady external loads. *Ocean Eng.* **2018**, *164*, 332–349. [CrossRef]
7. Ding, J.; Wu, Y.; Xie, Z.; Yang, W.; Wang, S.; Yu, J.; Yu, T. Overview: Research on hydroelastic responses of VLFS in complex environments. *Mar. Struct.* **2021**, *78*, 102978. [CrossRef]
8. Michele, S.; Zheng, S.; Buriani, F.; Borthwick, A.G.L.; Greaves, D.M. Floating hydroelastic circular plate in regular and irregular waves. *Eur. J. Mech. B Fluids* **2023**, *99*, 148–162. [CrossRef]
9. Li, W.; Ke, S.; Chen, J.; Zhu, T.; Ren, H. Hydrodynamic response and energy analysis in a very large floating structure supporting a marine airport under typhoon-driven waves. *Ocean Eng.* **2022**, *266*, 112987. [CrossRef]
10. Squire, V.A.; Dugan, J.P.; Wadhams, P.; Rottier, P.J.; Liu, A.K. Of ocean waves and sea ice. *Annu. Rev. Fluid Mech.* **1995**, *27*, 115–168. [CrossRef]
11. Meylan, M.H.; Squire, V.A. Response of a circular ice floe to ocean waves. *J. Geophys. Res. Oceans* **1996**, *101*, 8869–8884. [CrossRef]
12. Sakai, S.; Hanai, K. Empirical formula of dispersion relation of waves in sea ice. In Proceedings of the Ice in the Environment: Proceedings of the 16th IAHR International Symposium on Ice, Dunedin, New Zealand, 2–6 December 2002; pp. 327–335. Available online: https://www.iahr.org/library/infor?pid=19937 (accessed on 10 February 2023).
13. Meylan, M.H.; Bennetts, L.; Cavaliere, C.; Alberello, A.; Toffoli, A. Experimental and theoretical models of wave-induced flexure of a sea ice floe. *Phys. Fluids* **2015**, *27*, 041704. [CrossRef]
14. Waseda, T.; Alberello, A.; Nose, T.; Toyota, T.; Kodaira, T.; Fujiwara, Y. Observation of anomalous spectral downshifting of waves in the Okhotsk Sea Marginal Ice Zone. *Philos. Trans. R. Soc. A Math. Phys. Eng. Sci.* **2022**, *380*, 256. [CrossRef] [PubMed]
15. Părău, E.; Dias, F. Nonlinear effects in the response of a floating ice plate to a moving load. *J. Fluid Mech.* **2002**, *460*, 281–305. [CrossRef]
16. Dinvay, E.; Kalisch, H.; Părău, E.I. Fully dispersive models for moving loads on ice sheets. *J. Fluid Mech.* **2019**, *876*, 122–149. [CrossRef]
17. Ţugulan, C.; Trichtchenko, O.T.; Părău, E. Three-dimensional waves under ice computed with novel preconditioning methods. *J. Comput. Phys.* **2022**, *459*, 111129. [CrossRef]
18. Takagi, K. Interaction between solitary wave and floating elastic plate. *J. Waterw. Port Coastal Ocean Eng.* **1997**, *123*, 57–62. [CrossRef]
19. Sakai, S.; Liu, X.; Sasamoto, M.; Kagesa, T. Experimental and numerical study on the hydroelastic behavior of VLFS under tsunami. In *Hydroelasticity in Marine Technology*; Kashiwagi, M., Koterayama, W., Ohkusu, M., Eds.; Yomei Printing Cooperative Society: Fukuoka, Japan, 1998; pp. 385–392. ISBN 4-87780-001-8.
20. Kakinuma, T.; Ochi, N. Tsunami-height reduction using a very large floating structure. In *Mathematical Analysis of Continuum Mechanics and Industrial Applications III, CoMFoS 2018, Mathematics for Industry 34*; Itou, H., Hirano, S., Kimura, M., Kovtunenko, V.A., Khludnev, A.M., Eds.; Springer: Singapore, 2020; pp. 193–202. [CrossRef]
21. Xu, F.; Lu, D.Q. Wave scattering by a thin elastic plate floating on a two-layer fluid. *Int. J. Eng. Sci.* **2010**, *48*, 809–819. [CrossRef]
22. Kakinuma, T.; Yamashita, K.; Nakayama, K. Surface and internal waves due to a moving load on a very large floating structure. *J. Appl. Math.* **2012**, *2012*, 830530. [CrossRef]
23. Kim, J.W.; Webster, W.C. The drag on an airplane taking off from a floating runway. *J. Mar. Sci. Technol.* **1998**, *3*, 76–81. [CrossRef]
24. Watanabe, E.; Utsunomiya, T.; Tanigaki, S. A transient response analysis of a very large floating structure by finite element method. *Struct. Eng./Earthq. Eng. JSCE* **1998**, *15*, 155s–163s. [CrossRef]
25. Hermans, A.J. A boundary element method for the interaction of free-surface waves with a very large floating flexible platform. *J. Fluids Struct.* **2000**, *14*, 943–956. [CrossRef]
26. Endo, H. The behavior of a VLFS and an airplane during takeoff/landing run in wave condition. *Mar. Struct.* **2000**, *13*, 477–491. [CrossRef]
27. Kashiwagi, M. Transient responses of a VLFS during landing and take-off of an airplane. *J. Mar. Sci. Technol.* **2004**, *9*, 14–23. [CrossRef]
28. Nguyen, X.V.; Luong, V.H.; Cao, T.N.T.; Lieu, X.Q.; Nguyen, T.B. Hydroelastic responses of floating composite plates under moving loads using a hybrid moving element-boundary element method. *Adv. Struct. Eng.* **2020**, *23*, 2759–2775. [CrossRef]
29. Kakinuma, T. Nonlinear interaction of surface and internal waves with very large floating or submerged structures. In *Fluid Structure Interaction II*; Chakrabarti, S.K., Brebbia, C.A., Almorza, D., Gonzalez-Palma, R., Eds.; WIT Press: Billerica MA, USA, 2003; pp. 117–126.
30. Luke, J.C. A variational principle for a fluid with a free surface. *J. Fluid Mech.* **1967**, *27*, 395–397. [CrossRef]
31. Isobe, M. Time-dependent mild-slope equations for random waves. In *Coastal Engineering 1994*; Edge, B.L., Ed.; ASCE: Reston, VA, USA, 1995; pp. 285–299. [CrossRef]

32. Iguchi, T. A Mathematical justification of the Isobe–Kakinuma model for water waves with and without bottom topography. *J. Math. Fluid Mech.* **2018**, *20*, 1985–2018. [CrossRef]
33. Matsuno, Y. Hamiltonian formulation of the extended Green–Naghdi equations. *Phys. D Nonlinear Phenom.* **2015**, *301–302*, 1–7. [CrossRef]
34. Nakayama, K.; Kakinuma, T. Internal waves in a two-layer system using fully nonlinear internal-wave equations. *Int. J. Numer. Methods Fluids* **2010**, *62*, 574–590. [CrossRef]
35. Kakinuma, T. A numerical study on distant tsunami propagation considering the strong nonlinearity and strong dispersion of waves, or the plate elasticity and mantle fluidity of Earth. *Fluids* **2022**, *7*, 150. [CrossRef]
36. Boeing Japan Homepage. Available online: https://www.boeing.jp (accessed on 10 February 2023).
37. Andrianov, A.O.I. Hydroelastic Analysis of Very Large Floating Structures. Ph.D. Thesis, Electrical Engineering, Mathematics and Computer Science, TU Delft, Delft, The Netherlands, 2005. Available online: http://resolver.tudelft.nl/uuid:85cef785-c17c-41d7-9a58-6a183c468523 (accessed on 11 April 2023).
38. Tsubogo, T. A basic investigation on deflection wave propagation and strength of very large floating structures (the 2nd report). *J. Soc. Nav. Arch. Jpn.* **1997**, *1997*, 381–390. [CrossRef]
39. Kakinuma, T. Tsunamis generated and amplified by atmospheric pressure waves due to an eruption over seabed topography. *Geosciences* **2022**, *12*, 232. [CrossRef]
40. Proudman, J. The effects on the sea of changes in atmospheric pressure. *Geophys. J. Int.* **1929**, *2*, 197–209. [CrossRef]
41. Whitham, G.B. *Linear and Nonlinear Waves*; John Wiley & Sons, Inc.: New York, NY, USA, 1974; pp. 511–532. [CrossRef]
42. Wu, T. Generation of upstream advancing solitons by moving disturbances. *J. Fluid Mech.* **1987**, *184*, 75–99. [CrossRef]
43. Lee, S.; Yates, G.; Wu, T. Experiments and analyses of upstream-advancing solitary waves generated by moving disturbances. *J. Fluid Mech.* **1989**, *199*, 569–593. [CrossRef]
44. Kakinuma, T.; Akiyama, M. Numerical analysis of tsunami generation due to seabed deformation. In *Coastal Engineering 2006*; Smith, J.M., Ed.; World Scientific Publishing Co., Pte. Ltd.: Singapore, 2007; pp. 1490–1502. [CrossRef]
45. Dalphin, J.; Barros, R. Optimal shape of an underwater moving bottom generating surface waves ruled by a forced Korteweg-de Vries equation. *J. Optim. Theory Appl.* **2018**, *180*, 574–607. [CrossRef]
46. Michele, S.; Renzi, E.; Borthwick, A.; Whittaker, C.; Raby, A. Weakly nonlinear theory for dispersive waves generated by moving seabed deformation. *J. Fluid Mech.* **2022**, *937*, 94. [CrossRef]
47. Hong, D.C.; Hong, S.Y.; Hong, S.W. Reduction of hydroelastic responses of a very-long floating structure by a floating oscillating-water-column breakwater system. *Ocean Eng.* **2006**, *33*, 610–634. [CrossRef]
48. van Kessel, J.L.F.; Pinkster, J.A. The effect of aircushion division on the motions of large floating structures. In Proceedings of the ASME 2007 26th International Conference on Offshore Mechanics and Arctic Engineering, San Diego, CA, USA, 10–15 June 2007; pp. 677–686. [CrossRef]
49. Gao, R.P.; Wang, C.M.; Koh, C.G. Reducing hydroelastic response of pontoon-type very large floating structures using flexible connector and gill cells. *Eng. Struct.* **2013**, *52*, 372–383. [CrossRef]
50. Ohmatsu, S. Numerical calculation method for the hydroelastic response of a pontoon-type very large floating structure close to a breakwater. *J. Mar. Sci. Technol.* **2000**, *5*, 147–160. [CrossRef]
51. Regulation for Enforcement of the Civil Aeronautics Act. *Order Minist. Land Infrastruct. Transp. Tour.* **2008**, *73*, 79. Available online: https://www.japaneselawtranslation.go.jp/en/laws/view/4052 (accessed on 10 February 2023).

eng MDPI

Article

Rapid Prediction of Leaf Water Content in Eucalypt Leaves Using a Handheld NIRS Instrument

Joel B. Johnson

School of Health, Medical and Applied Sciences, Central Queensland University, North Rockhampton, QLD 4701, Australia; joel.johnson@cqumail.com

Abstract: Leaf water content (LWC) is a crucial physiological parameter that plays a limiting role in the efficiency of photosynthesis and biomass production in many plants. This study investigated the use of diffuse reflectance near-infrared spectroscopy (NIRS) for the rapid prediction of the gravimetric LWC in eucalypt leaves from *Eucalyptus* and *Corymbia* genera. The best-performing model for LWC gave a R^2_{pred} of 0.85 and RMSEP of 2.32% for an independent test set, indicating that the handheld NIR instrument could predict the LWC with a high level of accuracy. The use of support vector regression gave slightly more accurate results compared with partial least squares regression. Prediction models were also developed for leaf thickness, although these were somewhat less accurate (R^2_{pred} of 0.58; RMSEP of 2.7 μm). Nevertheless, the results suggest that handheld NIR instruments may be useful for in-field screening of LWC and leaf thickness in Australian eucalypt species. As an example of its use, the NIR method was applied for rapid analysis of the LWC and leaf thickness of every leaf found on an *E. populnea* sapling.

Keywords: dry matter content; in-field analysis; near infrared spectroscopy (NIRS)

Citation: Johnson, J.B. Rapid Prediction of Leaf Water Content in Eucalypt Leaves Using a Handheld NIRS Instrument. *Eng* **2023**, *4*, 1198–1209. https://doi.org/10.3390/eng4020070

Academic Editor: Antonio Gil Bravo

Received: 23 March 2023
Revised: 18 April 2023
Accepted: 18 April 2023
Published: 19 April 2023

1. Introduction

The leaf water content (LWC) is an important physiological parameter in the discipline of plant science [1]. The water content of leaves and their transpiration rates control the extent of stomata opening [2], thus influencing the rate of photosynthesis [3]. In turn, this limits the growth rates and biomass productivity of the plant [4]. Furthermore, LWC also plays a role in determining salinity tolerance [5] and even affects oviposition preferences of herbivorous insects [6,7].

Consequently, measuring LWC may provide information on various physiological aspects of plants such as their current levels of water stress [8–10], drought tolerance [11,12], salinity tolerance [5], flammability [13,14], and photosynthetic rates [15]. Survey studies have indicated that genetics (particularly classification at a family or order level) have the largest impact on LWC, while precipitation and climate have smaller effects [16]. Nevertheless, there is significant spatial variation in LWC between different habitat types. Researchers have also highlighted that LWC will be an important variable to measure in future studies of large-scale trait variations [16].

The most common method of measuring LWC is by oven-drying the leaves to a constant mass and determining the loss of mass. However, this method is destructive and can be time consuming, particularly when large numbers of leaves must be analysed. Consequently, in order to better understand the distribution of LWC within plants and ecosystems, rapid methods of measuring this parameter are required. This is particularly significant for the eucalypts (tribe Eucalypteae; principally comprising the *Eucalyptus*, *Corymbia*, and *Angophora* genera), as these species are a crucial part of many Australian ecosystems [17].

Rapid, non-invasive analytical techniques have been reported for measuring LWC [1,10,18], principally using near-infrared spectroscopy (NIRS) [19,20]. This analytical technique uses light with a longer wavelength than visible light to investigate the presence of key functional groups (e.g., OH, CH, NH) in a sample matrix. The benefits of NIRS include its speed (no sample preparation), low cost (no ongoing expenses), portability, and broad applicability to a range of sample types [8,21]. Recent technological advances in reducing the size and cost of NIR instruments have made them highly suited to use in field surveys investigating various analytes [20,22–24].

However, there are limited studies applying NIRS for the prediction of LWC in eucalypt species. This tribe of plants is somewhat unique in possessing a thick waxy cuticle layer on their leaves [25], which acts to reduce water loss [26]. This physiological feature may potentially complicate the development of NIRS models for the prediction of LWC. Yang, et al. [27] used NIRS to predict the leaf water potential (Ψ_{leaf}) in *E. camaldulensis*, but not the absolute LWC. Similarly, Datt [28] was able to predict the equivalent water thickness in several eucalypt species using NIRS, but not the gravimetric water content (i.e., the absolute LWC). In contrast, Kumar [29] was able to predict LWC in six *Eucalyptus* species by using a laboratory NIR instrument. This appears to be the only study to date reporting NIRS calibration for LWC that can be utilised across different eucalypt species. Furthermore, there are no comparable studies incorporating the *Corymbia* genus. Consequently, the aim of this study was to apply portable NIRS instrumentation for the rapid prediction of gravimetric LWC in eucalypt leaves from several species across two genera (*Eucalyptus* and *Corymbia*).

As a secondary aim, NIRS was also trialled for the prediction of leaf thickness in these species. Although this parameter is not usually measured with NIRS, it was thought that the leaf thickness could be indirectly predicted from the magnitude of absorbance produced from all of the compounds (e.g., structural carbohydrates, proteins, and water) present in the leaves. Leaves with a smaller thickness would consequently be expected to show less absorbance overall, and vice versa for thicker leaves. This is supported by previous work demonstrating that leaf thickness was the best predictor of NIR reflectance and internal light scattering [30]. Therefore, NIRS might be able to rapidly screen for leaf thickness at the same time as measuring leaf water content.

2. Materials and Methods

2.1. *Eucalyptus Leaf Samples*

This study was conducted in May 2022 on a grazing property in Central Queensland, comprising an open woodland of mixed eucalypt species. Leaves of varying maturity stages were collected from six different *Eucalyptus* and *Corymbia* species (*E. populnea*, *C. citriodora*, *E. platyphylla*, *E. tereticornis*, *E. melanophloia*, and *C. tessellaris*). Between 2 and 4 trees were sampled for each species; 20 leaves were collected for each species (only 19 for *E. tereticornis*). Effort was made to collect leaves from multiple canopy levels in each species. The leaves sampled were from the 'intermediate' and 'adult' maturity stages [31].

The NIR spectra were collected from the fresh leaf samples as soon as practicable after collection (approx. 10–20 min).

2.2. *Collection of NIR Spectra*

Spectra between 908–1676 nm were collected directly from the eucalypt leaves using a MicroNIR OnSite handheld spectrometer (Viavi, Santa Rosa, CA, USA). This instrument has a diffuse reflectance geometry. The instrument was calibrated using dark and light reference measurements every 10 min. The following parameters were used: 6 nm resolution; 100 ms integration time; 1 scan per spectra. Spectra were collected in triplicate, from different locations on each leaf. No tile was placed behind the leaves during measurement, to simulate the situation that would take place during in-field measurements. The adaxial and abaxial sides of each leaf were both randomly sampled, again to simulate a simplified method for potential in-field use. Previous work indicated some variation in the spectral absorbance from the adaxial and abaxial sides, albeit relatively minor [32].

In addition to collecting spectra from all of the fresh leaves, NIR spectra were also collected from 20 *E. populnea* leaves half-way through the drying process, and another 5 *E. populnea* leaves when they were completely oven-dried (0% moisture). Furthermore, spectra were collected from 30 *E. populnea* leaves that had naturally dried (to varying extents) after falling off the trees. Inclusion of these samples provided a wider range of moisture contents for the creation of a robust prediction model.

In total, 174 leaf samples were scanned, each in triplicate (n = 522 spectra). One leaf sample (n = 3 spectra) was excluded due to outlier values. The spectra were not averaged prior to data analysis.

2.3. Measurement of LWC and Leaf Thickness

The leaf water content was measured by oven drying the leaves at 65 °C until they reached a constant mass. The loss in mass was recorded for each leaf and the moisture content was calculated and expressed as a percent of the fresh weight as per the following formula:

$$\text{LWC}\ (\%) = (\text{mass}_{\text{fresh}} - \text{mass}_{\text{dried}})/\text{mass}_{\text{fresh}} \times 100$$

Leaf thickness (recorded to ±0.1 μm) was measured near the centre of each leaf, using an engineer's micrometer (RS PRO External Micrometer; item code 705–1213; range 0–25 mm; accuracy ±0.004 mm). Care was taken to ensure that the thickness was not measured through the central vein, which could have influenced the results.

2.4. Model Development

The NIR spectra were exported in ASCII (*.csv) format and subsequently imported into Unscrambler X software (version 10; Camo Analytics; Oslo, Norway) for chemometric analysis.

The spectra were split into 2 sets: the calibration set (comprising all 5 eucalypt species, but excluding all *E. platyphylla* spectra), and the independent test set (comprising all of the *E. platyphylla* spectra). Each model was built and cross-validated using the entire calibration set, then applied to the independent test set to assess its performance. Full cross-validation was performed on each calibration model using the leave-one-out (LOO) approach. All models were limited to a maximum of 7 factors to avoid the potential of over-fitting.

Partial least squares regression (PLS-R) was performed in Unscrambler X, using leave-one-out (LOO) cross-validation. Various spectral pre-processing treatments were trialled, including standard normal variate (SNV) normalisation, multiplicative scatter correction (MSC), and first and second derivative treatments using differing numbers of smoothing points for the Savitzky–Golay algorithm. Abbreviations for these pre-processing methods indicate the derivative and number of Savitzky–Golay smoothing points used. For example, 1d5 indicates 1st derivative with 5 smoothing points. Each pre-processing method was trialled on the entire calibration set.

Support vector regression (SVR) was also trialled as an alternative algorithm method to PLS-R. An Epsilon SVR model was used, with a radial basis function, gamma value of 0.008, and 2 classes.

Graphs were drawn in R Studio, running R version 4.0.5 [33].

2.5. In-Field Application

Finally, the NIR method was applied in-field to measure the LWC and leaf thickness of every leaf found on a small *E. populnea* sapling (~2.5 m height; 67 leaves) in central Queensland. The process of NIR spectra collection was quite rapid, with up to 7 leaves scanned per minute.

In addition to collecting the NIR spectra, the height of each leaf above the ground was measured using a laser distance measurer (Ozito LMR-025), to determine whether there were any correlations between leaf height and LWC/thickness.

3. Results and Discussion

3.1. Descriptive Statistics

As shown in Table 1, the size (as measured by fresh mass), thickness, and water content of the leaves varied significantly between the six eucalypt species. The leaf thickness did not vary significantly between the four *Eucalyptus* species; however, it was significantly lower for the two *Corymbia* species (Figure 1). There was no clear difference in LWC between *Corymbia* and *Eucalyptus* (Figure 2), with *E. melanophloia* and *C. citriodora* showing the lowest LWC, and *E. platyphylla* and *C. tessellaris* the highest. This agreed with observations by Wang et al. [16] that genetic factors had the largest impact on leaf moisture content, rather than environmental or climatic factors.

Table 1. Physical parameters and leaf water contents for the fresh eucalypt leaf samples. The one-way ANOVA row shows the results of ANOVA testing between the six species. Results followed by the same superscript letter were not significantly different according to post-hoc Tukey testing at $\alpha = 0.05$.

Species	Leaf Fresh Mass (g)	Leaf Thickness (μm)	Leaf Water Content (%)
C. citriodora (n = 20)	0.72 ± 0.24^{c}	28.4 ± 2.0^{b}	46.1 ± 1.3^{d}
C. tessellaris (n = 20)	0.39 ± 0.15^{d}	24.6 ± 3.7^{c}	51.2 ± 1.5^{ab}
***Corymbia* average (n = 40)**	**0.55 ± 0.26**	**26.5 ± 3.5**	**48.6 ± 3.0**
E. melanophloia (n = 20)	1.29 ± 0.43^{a}	34.8 ± 2.7^{a}	45.3 ± 2.5^{d}
E. platyphylla (n = 20)	1.30 ± 0.48^{a}	34.0 ± 3.7^{a}	53.1 ± 4.6^{a}
E. populnea (n = 20)	0.96 ± 0.34^{bc}	34.8 ± 3.6^{a}	49.7 ± 5.2^{bc}
E. tereticornis (n = 19)	1.04 ± 0.29^{ab}	32.2 ± 3.2^{a}	47.8 ± 3.6^{cd}
***Eucalyptus* average (n = 79)**	**1.15 ± 0.42**	**34.0 ± 3.4**	**49.0 ± 5.0**
One-way ANOVA	***	***	***

*** $p < 0.001$.

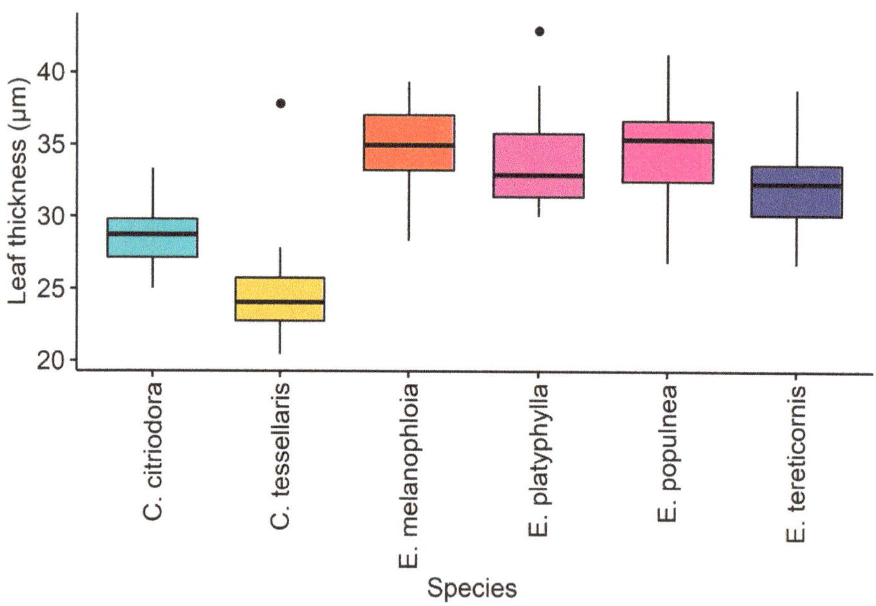

Figure 1. Boxplot of leaf thickness for the six eucalypt species. The points are outlier samples.

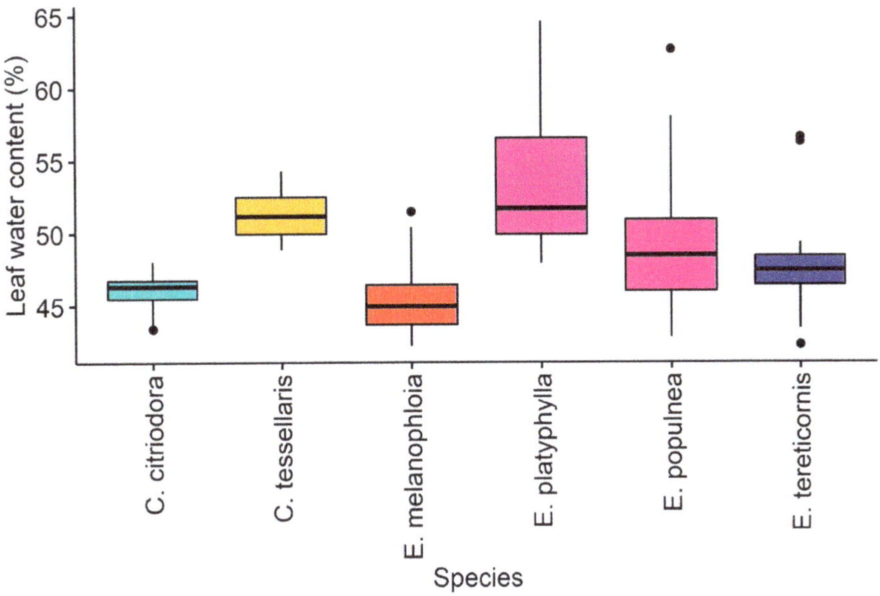

Figure 2. Boxplot of leaf water content for the fresh leaves from the six eucalypt species. The points are outlier samples.

There was a positive correlation between the fresh leaf mass and the leaf thickness ($r_{117} = 0.58$, $p < 0.001$), but not with the LWC (Table 2). There was also a moderate negative relationship between the leaf thickness and the LWC ($r_{117} = -0.25$, $p < 0.01$).

Table 2. Correlation between the fresh leaf mass, thickness, and water content of the eucalypt leaves from five different species. Values are given as Pearson R correlation coefficients (n = 119 leaves).

Parameter	Mass (g)	Thickness (µm)	LWC (%)
Mass (g)	-	-	-
Thickness (µm)	0.58 ***	-	-
LWC (%)	−0.13 NS	−0.25 **	-

NS = $p > 0.05$, ** $p < 0.01$, *** $p < 0.001$.

3.2. NIR Spectra

The NIR spectra of the leaf samples showed a large peak at 1454 nm, a shoulder at approximately 1350 nm, and a smaller peak centred at 1196 nm (Figure 3). These may be attributed to the OH first overtone, CH first overtone combination, and CH second overtone of the stretching mode, respectively [34,35]. Only a very small peak was observed in the OH second overtone region (around 970 nm). Overall, the spectra were very similar to those previously reported for *E. pellita* leaves [36] and mulberry (*Morus alba*) leaves [37].

Additionally, there were clear visual differences in the leaf samples with different moisture contents, with lower absorption in the ca. 1450 nm region (first overtone of the stretching mode (v_s) of -OH) in samples with less water content.

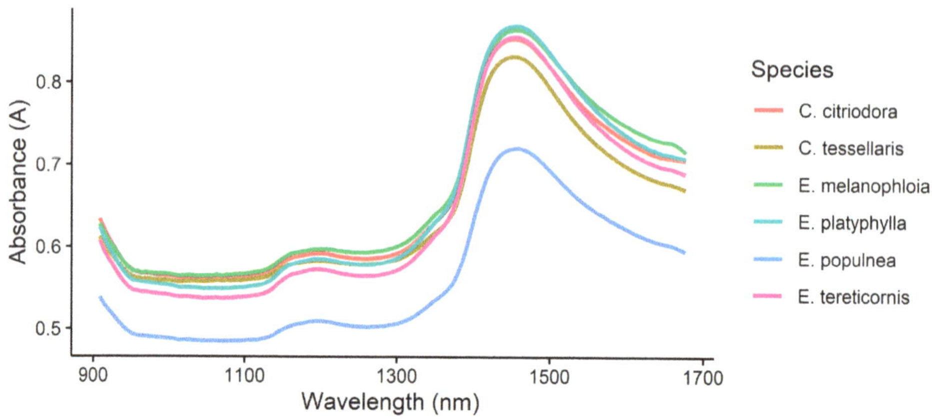

Figure 3. NIR spectra of the fresh and dried leaf samples, averaged by species.

3.3. Prediction of Leaf Water Content

As shown in Table 3, most of the analyte values for the independent test set fell within the range of values included in the calibration set. Moderately good results were found for the prediction of LWC from the NIR spectra. The best pre-processing method was SNV smoothing (Table 4), which gave an excellent R^2_{CV} of 0.97 and RMSECV of 2.21% (Figure 4). Furthermore, the performance on the independent test set was acceptable, with an R^2_{pred} of 0.80 and RMSEP of 2.46%.

Table 3. Descriptive statistics for the calibration and test sets. Values are given as mean ± SD, with the range provided in brackets.

Sample Set	Leaf Water Content (%)	Leaf Thickness (µm)
Calibration—five different species (n = 153)	41.1 ± 13.8 (0–62.6)	31.1 ± 4.9 (20.4–41.3)
Test set—*E. platyphylla* (n = 20)	53.1 ± 4.6 (47.9–64.6)	34.0 ± 3.7 (30.0–42.9)

Note that all sample numbers refer to the number of samples, not spectra; i.e., 20 samples = 60 spectra.

Table 4. Statistical results for the prediction of leaf water content (LWC) in eucalypt leaves using different pre-processing methods combined with PLS regression.

Pre-Processing	Factors	R^2_{CV}	RMSECV (% *w*/*w*)	RPD	R^2_{pred}	RMSEP (% *w*/*w*)	Bias (% *w*/*w*)	Slope	Intercept
None	3	0.939	3.41	4.05	0.567	4.38	−3.12	0.43	26.9
MSC	5	0.962	2.68	5.15	0.738	2.76	−0.90	0.51	25.1
SNV	6	0.974	2.21	6.24	0.803	2.46	−0.40	0.53	24.4
1d5	3	0.943	3.30	4.18	0.546	3.49	−1.68	0.50	25.1
1d11	4	0.955	2.93	4.71	0.788	2.85	−1.77	0.63	17.7
1d15	3	0.946	3.21	4.30	0.651	3.54	−2.32	0.64	16.8
2d5	5	0.953	3.00	4.60	0.478	3.99	−2.27	0.43	27.9
2d11	6	0.961	2.73	5.05	0.582	2.95	−1.34	0.58	21.1
2d15	4	0.947	3.17	4.35	0.576	2.99	−2.42	0.51	23.7

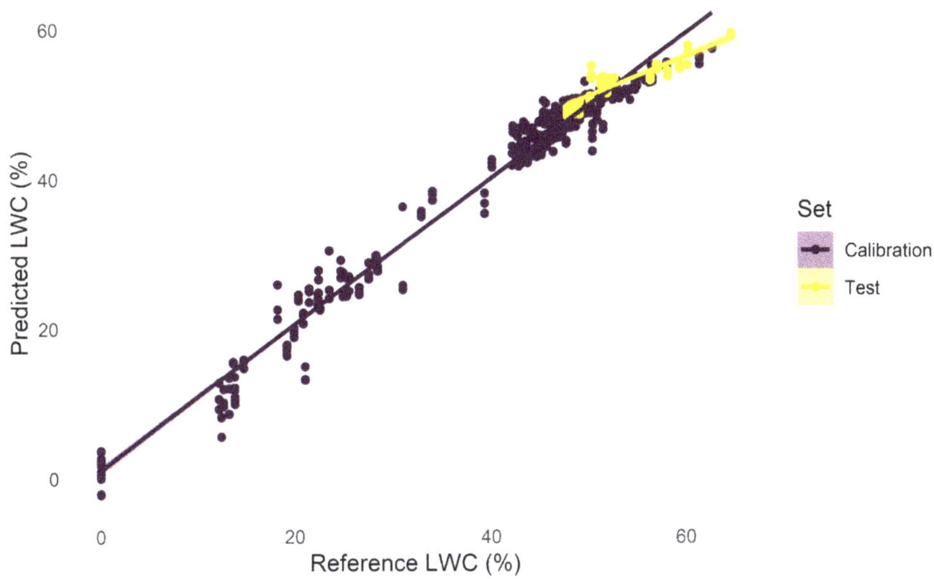

Figure 4. Predicted vs. reference leaf water content in the calibration set (purple) and test set (yellow), using PLSR with SNV pre-processing.

The loadings plot for the LWC model showed a strong positive loading at around 1404 nm, with a smaller shoulder at 1348 nm (Figure 5). This corresponded to the first overtone of the OH bond in water, with potentially a more minor contribution from the CH combination overtone.

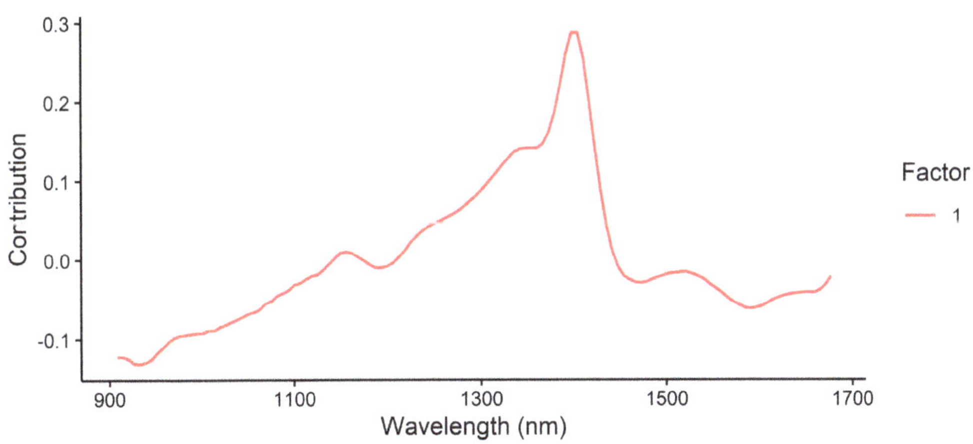

Figure 5. Loadings plot (first factor) for the prediction of LWC using NIRS, with SNV pre-processing.

The use of support vector regression (SVR) was also trialled for the prediction of LWC in the leaf spectra. Again, the pre-processing method had a strong impact on the model accuracies, with the lowest RMSEP found using 2d11 pre-processing. A similar RMSEP and higher linearity (R^2_{pred}) were found using 2d15 pre-processing. As shown in Table 5, the SVR model with 2d15 pre-processing showed a moderate improvement in performance over the PLSR method, with an R^2_{pred} of 0.85 and RMSEP of 2.32%.

Table 5. Statistical results for the prediction of leaf water content (LWC) in eucalypt leaves, obtained using different pre-processing methods combined with support vector regression.

Pre-Processing	R^2_{CV}	RMSECV (% *w*/*w*)	RPD	R^2_{pred}	RMSEP (% *w*/*w*)	Bias (% *w*/*w*)	Slope	Intercept
None	0.897	4.48	3.08	0.264	5.93	−4.26	0.10	43.3
MSC	0.961	2.76	4.99	0.72	3.67	−2.26	0.41	28.8
SNV	0.965	2.62	5.26	0.735	3.47	−2.01	0.43	28.2
1d5	0.973	2.31	5.96	0.837	2.69	−1.18	0.52	24.5
1d11	0.970	2.42	5.69	0.807	3.01	−1.48	0.47	26.8
1d15	0.968	2.49	5.53	0.784	3.16	−1.74	0.47	26.4
2d5	0.979	2.09	6.59	0.778	2.53	−0.72	0.55	23.1
2d11	0.979	2.09	6.59	0.800	2.22	−0.45	0.64	18.7
2d15	0.973	2.31	5.96	0.845	2.32	−0.71	0.58	21.7

3.4. Prediction of Leaf Thickness

Compared to the predictions of LWC, somewhat poorer prediction results were found for the predictions of leaf thickness (Table 6). This is because the leaf thickness was indirectly sensed by the magnitude of absorbance resulting from all IR-active compounds present in the leaves. In other words, thinner leaves showed less absorbance from compounds such as structural carbohydrates, proteins, and water, and therefore had weaker NIR signals.

Table 6. Prediction results for leaf thickness in eucalypt leaves using different pre-processing methods combined with PLS regression.

Pre-Processing	Factors	R^2_{CV}	RMSECV (μm)	RPD	R^2_{pred}	RMSEP (μm)	Bias (μm)	Slope	Intercept
None	5	0.543	3.29	1.48	0.546	2.98	−1.46	0.73	7.8
MSC	7	0.54	3.31	1.47	0.488	2.89	−1.10	0.61	12.1
SNV	6	0.524	3.36	1.45	0.571	2.83	−0.88	0.56	14.2
1d5	7	0.569	3.19	1.52	0.483	2.85	−0.25	0.70	10.1
1d11	7	0.557	3.24	1.50	0.500	2.83	−0.46	0.71	9.3
1d15	5	0.529	3.38	1.44	0.479	2.97	−0.40	0.73	8.7
2d5	7	0.602	3.07	1.58	0.487	2.90	−0.88	0.66	10.6
2d11	7	0.626	2.97	1.64	0.583	2.73	−1.11	0.76	7.0
2d15	7	0.608	3.05	1.59	0.480	3.14	−0.37	0.80	6.4

The best-performing PLSR model for leaf thickness was found using 2d11 pre-processing, with an R^2_{CV} of 0.63, RMSECV of 3.0 μm, R^2_{pred} of 0.58, and RMSEP of 2.7 μm. This appears to be the best pre-processing combination for smoothing out the background noise while retaining the informative wavelengths for the prediction of LWC. While a greater number of smoothing points does necessarily remove a proportionate number of wavelengths from both ends of the spectrum, the wavelengths found in these regions were not considered to contain informative peaks arising from OH bonds, so should not significantly influence the results. The loadings plot for leaf thickness indicated the strongest contribution was from approximately 1390 nm (Figure 6), again corresponding to the combination of the OH and CH first overtones.

In contrast to the results observed for the prediction of LWC, the use of support vector regression gave slightly poorer results for the prediction of leaf thickness, with an R^2_{pred} of 0.53 and RMSEP of 2.8 μm (Table 7).

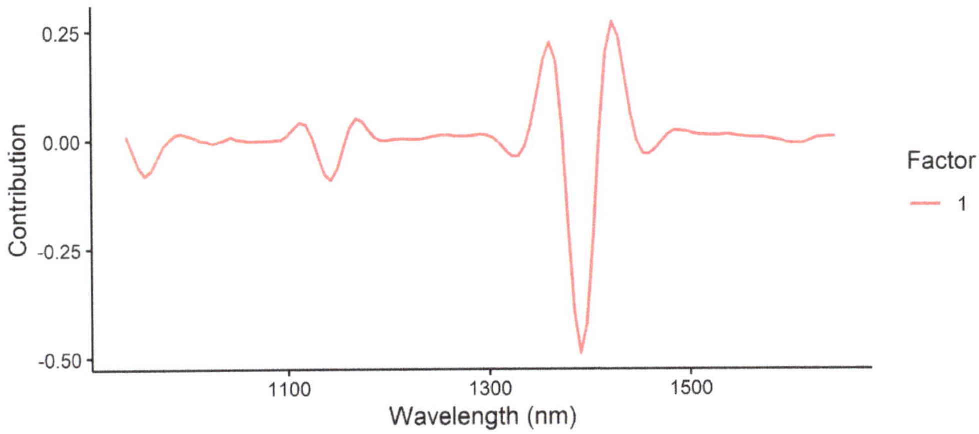

Figure 6. Loadings plot (first factor) for the prediction of leaf thickness using NIRS, with 2d11 pre-processing.

Table 7. Prediction results for leaf thickness in eucalypt leaves using different pre-processing methods combined with support vector regression.

Pre-Processing	R^2_{CV}	RMSECV (µm)	RPD	R^2_{pred}	RMSEP (µm)	Bias (µm)	Slope	Intercept
None	0.257	4.22	1.15	0.312	3.18	−0.87	0.23	25.3
MSC	0.448	3.61	1.35	0.438	2.84	−0.67	0.52	15.6
SNV	0.461	3.57	1.36	0.443	2.85	−0.68	0.54	15.0
1d5	0.589	3.12	1.56	0.469	2.88	−0.20	0.68	10.7
1d11	0.539	3.30	1.47	0.441	2.97	−0.22	0.66	11.5
1d15	0.525	3.35	1.45	0.454	2.90	−0.33	0.65	11.7
2d5	0.673	2.79	1.74	0.522	2.70	−0.38	0.70	9.7
2d11	0.684	2.74	1.77	0.527	2.75	−0.62	0.72	8.9
2d15	0.595	3.09	1.57	0.461	2.87	−0.10	0.66	11.6

Despite the lower accuracy compared with the prediction of LWC, the results may be useful for general screening of leaf thickness in some settings. For example, leaf thickness could be another piece of information extracted from leaf NIR spectra collected during field surveys.

3.5. Correlations between Leaf Height and LWC/Thickness

Finally, to demonstrate the potential in-field use of this technique, NIR spectra were collected from all 67 leaves on a 2.5-metre-tall *E. populnea* sapling. The predicted LWCs ranged from 44.0–53.9% (mean 47.8 ± 2.3%), while the leaf thickness was between 7.1–89.9 µm (mean 42.4 ± 19.3 µm).

There was no significant correlation between leaf height and LWC or thickness (Figure 7; $p > 0.05$ for both). However, LWC and thickness were positively correlated with one another in this sapling ($r_{65} = 0.65$, $p < 0.001$; Figure 7). This concurred with previous research by Búrquez [38], who found a positive correlation between leaf thickness and water potential in four non-woody species.

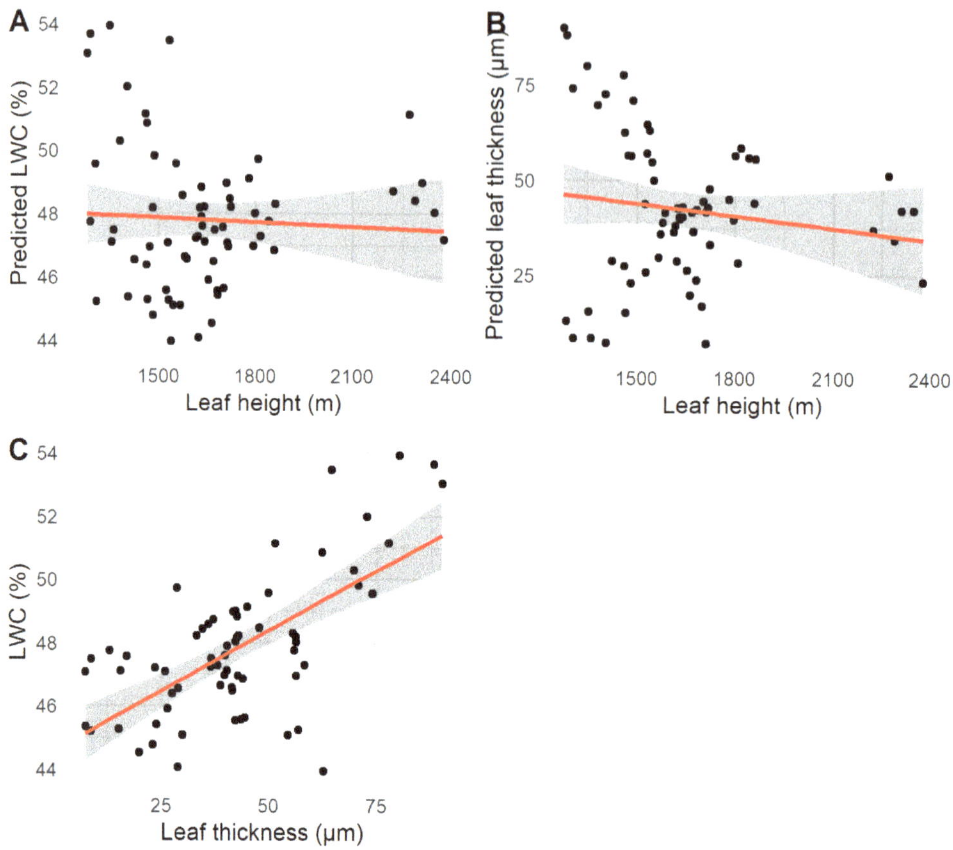

Figure 7. (**A**) Correlation between sapling leaf height and NIR-predicted leaf water content. (**B**) Correlation between sapling leaf height and NIR-predicted leaf thickness. (**C**) Correlation between leaf thickness and leaf water content.

4. Conclusions

This study developed for the first time a rapid NIRS model for the prediction of leaf water content across *Corymbia* and *Eucalyptus* genera. High accuracy was obtained using NIRS for the prediction of leaf water content, with an R^2_{pred} of 0.85 and RMSEP of 2.32. Somewhat poorer results were found for the prediction of leaf thickness; however, the results could still be useful for screening purposes where high accuracy is not required. Handheld NIR instrumentation with a diffuse reflectance geometry—such as the instrument used in this study—may be particularly useful for the rapid, in-field measurement of LWC during survey studies.

Funding: This research received no external funding.

Institutional Review Board Statement: Not applicable.

Informed Consent Statement: Not applicable.

Data Availability Statement: The datasets supporting the findings of this paper are available from the corresponding author upon request.

Acknowledgments: Thanks to Kerry Walsh for providing access to the MicroNIR instrument.

Conflicts of Interest: The author declares no conflict of interest.

References

1. Li, B.; Zhao, X.; Zhang, Y.; Zhang, S.; Luo, B. Prediction and monitoring of leaf water content in soybean plants using terahertz time-domain spectroscopy. *Comput. Electron. Agric.* **2020**, *170*, 105239. [CrossRef]
2. Buckley, T.N. How do stomata respond to water status? *New Phytol.* **2019**, *224*, 21–36. [CrossRef] [PubMed]
3. Lawlor, D.W.; Tezara, W. Causes of decreased photosynthetic rate and metabolic capacity in water-deficient leaf cells: A critical evaluation of mechanisms and integration of processes. *Ann. Bot.* **2009**, *103*, 561–579. [CrossRef] [PubMed]
4. Nautiyal, P.C.; Rachaputi, N.R.; Joshi, Y.C. Moisture-deficit-induced changes in leaf-water content, leaf carbon exchange rate and biomass production in groundnut cultivars differing in specific leaf area. *Field Crops Res.* **2002**, *74*, 67–79. [CrossRef]
5. Suriya-arunroj, D.; Supapoj, N.; Toojinda, T.; Vanavichit, A. Relative leaf water content as an efficient method for evaluating rice cultivars for tolerance to salt stress. *Sci. Asia* **2004**, *30*, 411–415. [CrossRef]
6. Ishino, M.N.; De Sibio, P.R.; Rossi, M.N. Leaf trait variation on *Erythroxylum tortuosum* (Erythroxylaceae) and its relationship with oviposition preference and stress by a host-specific leaf miner. *Austral Ecol.* **2011**, *36*, 203–211. [CrossRef]
7. Steinbauer, M.J. Oviposition preference and neonate performance of *Mnesampela privata* in relation to heterophylly in *Eucalyptus dunnii* and *E. globulus*. *Agric. For. Entomol.* **2002**, *4*, 245–253. [CrossRef]
8. Torres, I.; Sánchez, M.-T.; Benlloch-González, M.; Pérez-Marín, D. Irrigation decision support based on leaf relative water content determination in olive grove using near infrared spectroscopy. *Biosyst. Eng.* **2019**, *180*, 50–58. [CrossRef]
9. Zhou, H.; Zhou, G.; He, Q.; Zhou, L.; Ji, Y.; Lv, X. Capability of leaf water content and its threshold values in reflection of soil–plant water status in maize during prolonged drought. *Ecol. Indic.* **2021**, *124*, 107395. [CrossRef]
10. Zygielbaum, A.I.; Gitelson, A.A.; Arkebauer, T.J.; Rundquist, D.C. Non-destructive detection of water stress and estimation of relative water content in maize. *Geophys. Res. Lett.* **2009**, *36*, L12403. [CrossRef]
11. Omae, H.; Kumar, A.; Egawa, Y.; Kashiwaba, K.; Shono, M. Midday Drop of Leaf Water Content Related to Drought Tolerance in Snap Bean (*Phaseolus vulgaris* L.). *Plant Prod. Sci.* **2005**, *8*, 465–467. [CrossRef]
12. Ashok, K.; Sharma, K.D. Leaf water content—A simple indicator of drought tolerance in crop plants. *Indian J. Agric. Sci.* **2010**, *80*, 1095–1097.
13. De Lillis, M.; Bianco, P.M.; Loreto, F. The influence of leaf water content and isoprenoids on flammability of some Mediterranean woody species. *Int. J. Wildland Fire* **2009**, *18*, 203–212. [CrossRef]
14. Yebra, M.; Chuvieco, E.; Riaño, D. Estimation of live fuel moisture content from MODIS images for fire risk assessment. *Agric. For. Meteorol.* **2008**, *148*, 523–536. [CrossRef]
15. Song, X.; Zhou, G.; He, Q. Critical Leaf Water Content for Maize Photosynthesis under Drought Stress and Its Response to Rewatering. *Sustainability* **2021**, *13*, 7218. [CrossRef]
16. Wang, R.; He, N.; Li, S.; Xu, L.; Li, M. Spatial variation and mechanisms of leaf water content in grassland plants at the biome scale: Evidence from three comparative transects. *Sci. Rep.* **2021**, *11*, 9281. [CrossRef] [PubMed]
17. Bennett, A.F. Eucalypts, wildlife and nature conservation: From individual trees to landscape patterns. *Proc. R. Soc. Vic.* **2016**, *128*, 71–86. [CrossRef]
18. Zheng, L.; Wang, Z.; Sun, H.; Zhang, M.; Li, M. Real-time evaluation of corn leaf water content based on the electrical property of leaf. *Comput. Electron. Agric.* **2015**, *112*, 102–109. [CrossRef]
19. Gillon, D.; Dauriac, F.; Deshayes, M.; Valette, J.C.; Moro, C. Estimation of foliage moisture content using near infrared reflectance spectroscopy. *Agric. For. Meteorol.* **2004**, *124*, 51–62. [CrossRef]
20. Santos, A.O.; Kaye, O. Grapevine leaf water potential based upon near infrared spectroscopy. *Sci. Agric.* **2009**, *66*, 287–292. [CrossRef]
21. Johnson, J.B.; Naiker, M. Seeing red. A review of the use of near-infrared spectroscopy (NIRS) in entomology. *Appl. Spectrosc. Rev.* **2019**, *55*, 810–839. [CrossRef]
22. Maraphum, K.; Saengprachatanarug, K.; Wongpichet, S.; Phuphaphud, A.; Posom, J. In-field measurement of starch content of cassava tubers using handheld vis-near infrared spectroscopy implemented for breeding programmes. *Comput. Electron. Agric.* **2020**, *175*, 105607. [CrossRef]
23. Wang, J.; Shen, C.; Liu, N.; Jin, X.; Fan, X.; Dong, C.; Xu, Y. Non-Destructive Evaluation of the Leaf Nitrogen Concentration by In-Field Visible/Near-Infrared Spectroscopy in Pear Orchards. *Sensors* **2017**, *17*, 538. [CrossRef] [PubMed]
24. Cozzolino, D. Use of Infrared Spectroscopy for In-Field Measurement and Phenotyping of Plant Properties: Instrumentation, Data Analysis, and Examples. *Appl. Spectrosc. Rev.* **2014**, *49*, 564–584. [CrossRef]
25. Steinbauer, M.J.; Davies, N.W.; Gaertner, C.; Derridj, S. Epicuticular waxes and plant primary metabolites on the surfaces of juvenile *Eucalyptus globulus* and *E. nitens* (Myrtaceae) leaves. *Aust. J. Bot.* **2009**, *57*, 474–485. [CrossRef]
26. Kane, C.N.; Jordan, G.J.; Jansen, S.; McAdam, S.A.M. A Permeable Cuticle, Not Open Stomata, Is the Primary Source of Water Loss From Expanding Leaves. *Front. Plant Sci.* **2020**, *11*, 774. [CrossRef]
27. Yang, G.; Lu, W.; Lin, Y.; Luo, J.; Wang, C.; Meder, R.; Warburton, P.; Arnold, R. Monitoring water potential and relative water content in *Eucalyptus camaldulensis* using near infrared spectroscopy. *J. Trop. For. Sci.* **2017**, *29*, 121–128.
28. Datt, B. Remote Sensing of Water Content in Eucalyptus Leaves. *Aust. J. Bot.* **1999**, *47*, 909–923. [CrossRef]

29. Kumar, L. High-spectral resolution data for determining leaf water content in Eucalyptus species: Leaf level experiments. *Geocarto Int.* **2007**, *22*, 3–16. [CrossRef]
30. Knapp, A.K.; Carter, G.A. Variability in leaf optical properties among 26 Species From A Broad Range Of Habitats. *Am. J. Bot.* **1998**, *85*, 940–946. [CrossRef] [PubMed]
31. Boland, D.J.; Brooker, M.I.H.; Chippendale, G.M.; Hall, N.; Hyland, B.P.M.; Johnston, R.D.; Kleinig, D.A.; Turner, J.D. *Forest Trees of Australia*; Nelson–CSIRO: Melbourne, Australia, 1984.
32. De Bei, R.; Cozzolino, D.; Sullivan, W.; Cynkar, W.; Fuentes, S.; Dambergs, R.; Pech, J.; Tyerman, S. Non-destructive measurement of grapevine water potential using near infrared spectroscopy. *Aust. J. Grape Wine Res.* **2011**, *17*, 62–71. [CrossRef]
33. R Core Team. *R: A Language and Environment for Statistical Computing*, Version 4.0.2; R Foundation for Statistical Computing: Vienna, Austria, 2020.
34. Workman, J.; Weyer, L. *Practical Guide to Interpretive Near-Infrared Spectroscopy*; CRC Press: Boca Raton, FL, USA, 2007.
35. Lapcharoensuk, R.; Danupattanin, K.; Kanjanapornprapa, C.; Inkawee, T. Combination of NIR spectroscopy and machine learning for monitoring chili sauce adulterated with ripened papaya. In Proceedings of the 13th Thai Society of Agricultural Engineering International Conference (TSAE 2020), Nakhon Ratchasima, Thailand, 30–31 July 2020; Volume 187, p. 04001. [CrossRef]
36. Alwi, A.; Meder, R.; Japarudin, Y.; Hamid, H.A.; Sanusi, R.; Yusoff, K.H.M. Near infrared spectroscopy of *Eucalyptus pellita* for foliar nutrients and the potential for real-time monitoring of trees in fertiliser trial plots. *J. Near Infrared Spectrosc.* **2021**, *29*, 158–167. [CrossRef]
37. Ma, Y.; Zhang, G.-Z.; Rita-Cindy, S.A.-A. Quantification of Water, Protein and Soluble Sugar in Mulberry Leaves Using a Handheld Near-Infrared Spectrometer and Multivariate Analysis. *Molecules* **2019**, *24*, 4439. [CrossRef] [PubMed]
38. Búrquez, A. Leaf Thickness and Water Deficit in Plants: A Tool for Field Studies. *J. Exp. Bot.* **1987**, *38*, 109–114. [CrossRef]

eng MDPI

Article

Microservice-Oriented Architecture for Industry 4.0

Ricardo Pasquati Pontarolli [1,*], Jeferson André Bigheti [2], Lucas Borges Rodrigues de Sá [3] and Eduardo Paciencia Godoy [3,*]

1 Federal Institute of Education, Science and Technology of São Paulo (IFSP), Boituva 18552-252, Brazil
2 National Service of Industrial Training (SENAI), Lencois Paulista 18685-730, Brazil
3 São Paulo State University (UNESP), Sorocaba 18087-180, Brazil
* Correspondence: pasquati@ifsp.edu.br (R.P.P.); eduardo.godoy@unesp.br (E.P.G.)

Abstract: Industry 4.0 (I4.0) is characterized by the integration of digital technologies into manufacturing processes and highlights new requirements for industrial systems such as greater interoperability, decentralization, modularization, and independence. The traditional hierarchical architecture of Industrial Automation Systems (IAS) does not fulfill these requirements and is evolving to incorporate information technologies in order to support I4.0 applications. The integration among these technologies, equipment, and systems at different industry levels requires a migration from the legacy vertical architecture to a flat architecture based on services. Service-oriented architecture (SOA) and, more recently, microservices play a critical role in I4.0 by providing a framework for integrating complex systems and meeting those requirements. This paper presents the development of a Microservice-Oriented Architecture for Industry 4.0 (MOAI), initially focused on evolving IAS to the I4.0. The objective is to describe the development, deployment, and testing of an IAS architecture based on microservices prepared for I4.0 applications. On the contrary to developing the whole software for the industrial SOA, the MOAI was developed on top of the Moleculer framework, which allowed focusing on creating services and applications for the automation and process control industry context. The development of several microservices and security mechanisms for the MOAI is presented, as is the deployment of IAS applications as services such as process control, SCADA, discrete automation, among others. The MOAI was implemented in a process control pilot plant for experimentation. Experimental results of the MOAI for IAS applications are investigated, the microservice communication performance is evaluated, and the pros and cons of microservices for I4.0 are discussed.

Keywords: service-oriented architecture; industrial automation system; industrial internet of things; Moleculer framework

Citation: Pontarolli, R.P.; Bigheti, J.A.; de Sá, L.B.R.; Godoy, E.P. Microservice-Oriented Architecture for Industry 4.0. *Eng* **2023**, *4*, 1179–1197. https://doi.org/10.3390/eng4020069

Academic Editors: Antonio Gil Bravo and Lech Bolesław Dobrzański

Received: 27 January 2023
Revised: 28 March 2023
Accepted: 13 April 2023
Published: 19 April 2023

1. Introduction

The goal of Industry 4.0 (I4.0) is to create a more connected, digitalized, intelligent, and autonomous manufacturing system that can improve efficiency, reduce costs, and increase quality. Most of the technologies needed for the implementation of I4.0 already exist, such as augmented reality, IPV6, wireless networks, artificial intelligence, robotics, radio frequency identification, the Industrial Internet of Things (IIoT), cloud computing, cyber-physical systems, and big data. Therefore, the biggest obstacle relies on the integration and joint use of these technologies to obtain a new manufacturing reality, where all the participants involved will be connected so that the best production decisions, safety, and values are made on demand [1–3].

The requirements for industrial systems in the context of IIoT and I4.0 must go through the recent demand for greater decentralization, modularity, and autonomy of the systems. The traditional Industrial Automation Systems (IAS) do not fulfill these requirements and need to advance to promote cooperation among different technologies, equipment, and systems at different industry levels [4,5]. The paradigm of collaborative automation has emerged as a solution to provide it through a networked architecture based on the use and sharing of services [6,7].

Operational Technology and Information Technologies (OT-IT) convergence in IAS was not possible decades ago, starting with the mainframes and direct digital control in Figure 1. The development of industrial networks and programable controllers enabled OT-IT convergence through hierarchical architectures such as the ISA-95 pyramid model. The need for data interoperability among vertical layers was met through a parallel link with the NAMUR Open Architecture (NOA). Nowadays, greater scalability and total integration can be achieved using cloud-based services or Microservice-Oriented Architectures (MOA), as shown in Figure 1 [8].

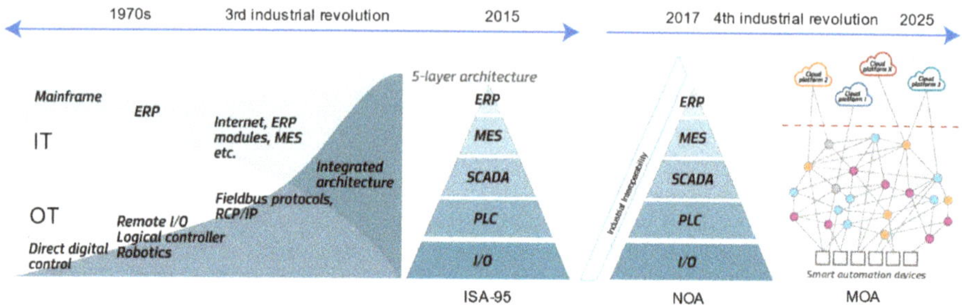

Figure 1. Convergence of OT and IT architectures adapted from [8].

References [9,10] present the NAMUR Open Architecture (NOA) to enable innovative solutions for existing process industries and adapt them to I4.0. The process control core remains largely unchanged. The basic idea is an open interface, for example OPC UA, between the existing layers of the main process control domain and the optimization monitoring domain [11]. Alternatively, a second communication channel, shown in Figure 1 (industrial interoperability), can provide direct and standard access to the existing solutions.

With I4.0, industries are using technology to create highly interconnected systems, digitalized processes, and data-driven decision-making. However, these systems often involve multiple applications and devices that need to communicate with each other seamlessly. This is where Service Oriented Architecture (SOA) comes in, as it allows for the development of modular, loosely coupled services that can be combined to create larger systems. This approach reduces complexity and increases flexibility, making it easier to adapt to changing business needs. SOA enables cross-layer integration and interoperability for decentralized systems by providing cloud services to various devices, gateways, or systems and standardizing the communication among them [7,12].

Considering the SOA application in I4.0, there is a new trend toward using microservices [13]. Microservice Oriented Architecture (MOA) represents an evolution of SOA and involves breaking up traditional applications into smaller, independent services that work together to provide the original functionality. Microservices are more independent, simple, and lightweight services with improved network communication and greater granularity (functionality) than in SOA, increasing the modularity and scalability of the application [13,14]. MOA also facilitates the reuse, redundancy, and composition of services, which can reduce development time and cost. Overall, it is a critical enabler of the principles at the core of I4.0.

The main principles of I4.0 are interoperability, virtualization, decentralization, service orientation, and modularity [15,16]. This paper presents the development of a Microservice Oriented Architecture for Industry 4.0 (MOAI), initially focused on evolving the IAS to properly support I4.0 applications. The objective is to describe the development, deployment, and testing of an IAS architecture based on microservices and discuss how it covers those principles and is prepared to support I4.0 applications.

This paper is organized as follows: This introduction presented an overview of the evolution of industrial architectures to support I4.0 and IIoT applications, with a focus on the adoption of service orientation. Section 2 presents a literature review about industrial SOA and MOA architectures and their differences from the MOAI proposal. A detailed explanation of service-oriented concepts and the Moleculer framework is given in Section 3. Section 4 describes the MOAI approach, microservices, networking, and security mechanisms. An application example of the MOAI is presented in Section 5 along with results discussions. Finally, Section 7 summarizes the conclusions of the paper.

2. Related Works

The IAS architecture has evolved to enable the integration of OT and IT technologies and support I4.0. Industrial-Ethernet networking allows integration of IAS with standard IT services. IAS covers some industry levels such as field level (devices, data acquisition), control level (programmable equipment, process continuous control, and discrete automation), supervisory level (monitoring, historian, and process management), management level (production and execution systems), and enterprise level (resources and planning systems). With the SOA's adoption, all elements (technologies, equipment, and systems) in these levels are able to communicate with each other without restrictions or hierarchical concerns [17]. These elements are services with standardized communication among them and are able to be combined, or used together, in order to create an industrial application [6].

Even with all the benefits provided by SOA for industry applications, the biggest bottleneck relies on how to enable and standardize communication and the discovery of services to integrate heterogeneous equipment and systems. Some technologies have already been applied for SOA in I4.0, such as Open Platform Communication Unified Architecture (OPC UA), Device Profile for Web Services (DPWS), and Representational State Transfer (REST) [5,7].

The OPC UA protocol is service-oriented and provides in its part four (Services) a group of essential services for networking [11]. Even though OPC UA has been widely used in I4.0 applications, the complete development of application-oriented services is required to complement the basic ones provided by the standard [13]. Reference [18] proposes an OPC UA-based SOA with three elements: a group of services for device integration, a group of application-oriented services with the necessary functionalities for the application, and a composition service in charge of controlling the sequence of service execution necessary to create an application.

The SOA development in IAS and I4.0 applications has been investigated in several projects sponsored by the European Union [12,19–21], such as the IMC-AESOP, PerFORM, and Arrowhead. In these projects, the whole software framework and network protocol to support SOA have been developed. The service development covered IAS applications ranging from the first level of devices to the level of control and supervision. DPWS communication and integration of services were used, and case studies applied to factory IAS were used to validate the projects [19,20]. The Simple Object Access Protocol (SOAP) in DPWS had undesirable performance due to protocol verbosity, even though it provided standard and secure communication and discovery capabilities for SOA. In order to overcome that, the REST standard was used in communication and service integration, resulting in better network performance (response time and jitter) for SOA [22].

The Arrowhead project [21] was based on the results of the IMC-AESOP project. It developed the concept of an on-premises automation cloud, in which IAS components and devices were made available as services, providing full interoperability for I4.0 [23]. In this project, the entire integration and communication infrastructure used SOA, structured in the format of a framework. Usually, industrial SOA architectures are based on service orchestration for the composition of applications and ease of deployment. In [24], the communication between devices using the Arrowhead framework through service choreography was presented and investigated.

The MOA has also been recently developed and tested for industry applications. Two European projects are focusing on the development of MOA for Cyber-Physical Systems (CPS), IIoT, and I4.0 applications. Reference [25] presents a MOA proposal to foster the implementation of the digital factory concept. This architecture is part of the MAYA (Multi-Disciplinary Integrated Simulation and Forecasting Tools) project, whose focus is the deployment and testing of CPS and I4.0 applications with support for the conception of digital twins [26]. In the architecture, five main groups of services are proposed that communicate via the REST standard with HTTP and WebSocket via TCP/IP. A highlight can be given to the orchestrator and scheduler microservices, which coordinate and organize the other services to enable the composition and creation of high-level services and process applications. A cloud-based MOA proposal to build a collaborative platform for I4.0 and IoT that enables real-time monitoring and optimization in manufacturing was presented in [27]. This architecture is part of the NIMBLE project [28].

Considering the literature review, the majority of SOA projects initially focused on developing the SOA software and communication protocol for industrial applications. In addition, the services and applications focused on adapting or migrating legacy IAS to the I4.0 requirements. The MOA projects are more recent and are also focusing on supporting I4.0 applications in addition to providing the required IAS functionalities. However, most of the development results of these projects are complex and even proprietary, hindering their usage. This is the main difference between them and the MOAI proposal. The MOAI was built on top of the Moleculer framework for microservices, which is open source, leveraging its development and simplifying its usage. Therefore, the first contribution of the paper is investigating the use of a non-industrial framework for microservices in industrial applications. The second difference is related to the MOAI proposal. Even though the MOAI initially covers the same required IAS functionalities as other projects in the literature, the idea was not to compare the architectures. Nevertheless, the second contribution of the paper is describing how those IAS functionalities and applications were developed using microservices and what the pros and cons of that proposal are. Considering this, this research collaborates with a MOAI, presenting the development and results of microservices initially applied to IAS [29–33].

3. Service Oriented Paradigm

In the computational context, services refer to software components that perform specific functions and are designed to be accessed by other software applications over a network through standard protocols. Examples of computational services include web services, which allow different software applications to communicate with each other over the Internet, and cloud services, which provide on-demand access to computing resources such as storage, processing power, and databases. Services are often designed to be modular and composable, allowing them to be combined with other services to build more complex applications and systems [34].

Microservices represent the evolution of the service-oriented paradigm. It refers to a style of software architecture where applications are built as a collection of small, independent, and loosely coupled services. Each microservice is designed to perform specific functionality and can be developed, deployed, and scaled independently. Microservices typically communicate with each other through well-defined, standardized, and lightweight protocols. They can be developed using different programming languages, databases, and technology stacks, as long as they conform to a common set of interfaces and standards [14].

3.1. Composition of Services

The composition of services refers to the process of integrating multiple individual services to create a more complex application or system that provides higher-level functionality. Service composition involves defining the interactions among the services and the sequence of executing them to achieve a specific goal or business process [34].

The composition of services can be achieved using different approaches, including orchestration, choreography, and an API gateway. The specific approach chosen depends on the requirements of the application or system being developed. The orchestration approach involves using a central service or middleware component to coordinate the interactions between different services. In orchestration, there is a master (composer), such as a maestro in an orchestra, that can be a service or an application that coordinates the requests of other services to make up a more complex function. The services that are requested are unaware of the complete composition, and the only one that holds this information is the master.

The choreography approach involves allowing the services to communicate with each other directly, without the need for a central orchestrator. Each service is responsible for managing its own interactions with other services based on events and messages received and predefined rules and workflows. In this type of approach, intelligence is distributed among the services involved, with each member having a part of the application's knowledge. Because the output of a service is used in the subsequent service, it is possible to use more than one means of communication.

3.2. Moleculer Framework

The Moleculer framework is an open-source and fast microservices framework for Node.js, designed to simplify the development of scalable and fault-tolerant microservices-based applications. It provides a set of features and tools to help developers build distributed systems. Microservices are developed in individual nodes with no hierarchy or priority and are designed to be agnostic to the underlying transport layer, as Moleculer supports various network protocols with the communication being automatic and transparent (no coding is needed). Moleculer also provides built-in support for service discovery, load balancing, and circuit breakers, which helps to ensure the availability and reliability of the system through the reuse and replication of microservices [35]. The microservices can provide one or multiple actions, which are specific functionalities developed that can be requested by other microservices or applications.

4. Microservice-Oriented Architecture for Industry 4.0 (MOAI)

This section presents the development of the MOAI, presented in Figure 2, describing the microservices created, security mechanisms, and microservices deployment and management. MOAI is built on the of the Moleculer framework and supports IAS functionalities in I4.0. The microservices are separated at abstract industry levels for organization and isolation of functionalities because the architecture itself is flat (without level hierarchy). Until now, microservices have been developed up to the third level (I/O, Control, and SCADA). However, future work may develop some for the top industry levels (MES, ERP).

The basic architecture infrastructure is composed of two types of components: microservices and applications. Infrastructure microservices provide functionalities for high-level composed services, and the following are available: Transporter (networking), Machine to Machine (M2M), and Data Acquisition (DAQ). In addition, microservices responsible for business or processes provide high-level composite resources such as supervision and control, and the following are available: Database, Process Control (PID4.0), and Programmable Logic Controller (PLC). IAS functionalities and applications can be created using and composing the Infrastructure and Process microservices. Both orchestration and choreography may be used for the microservice composition.

In the MOAI (Figure 2), the applications are composed of other microservices and can be Internal (blue square) or External (purple square). The Internal is based on platforms supported by Moleculer (Python, JavaScript, etc.), where the composition of microservices occurs only through the Transporter. The external is compatible with any platforms that support REST communication, which is interesting for the integration of industrial software and systems. The composition of microservices occurs initially through the API gateway and next through the Transporter for microservice communication.

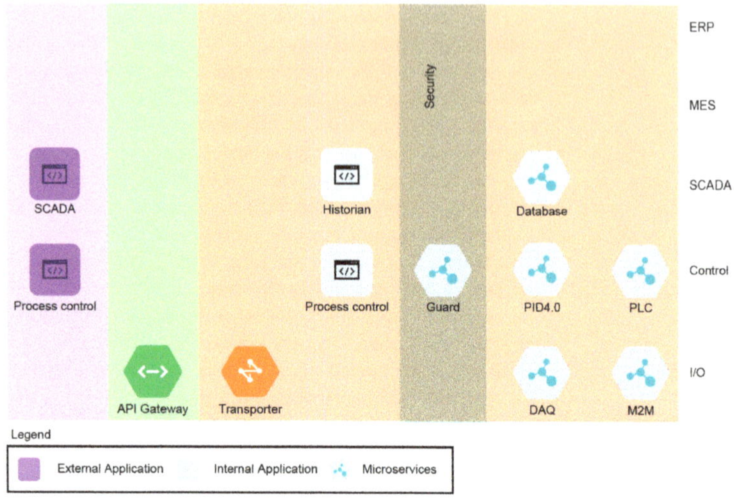

Figure 2. Microservice-Oriented Architecture for Industry 4.0.

4.1. Microservices Description

This subsection describes the microservices of the MOAI. The DAQ is in charge of accessing data from process variables through I/O hardware and has two main actions: reading the inputs (sensors) and updating the outputs (actuators). Because these actions are made transparently available by Moleculer discovery, registration, and networking, there is no need to conform to the hardware and software (coding) required to obtain the input and output data. This information is simply requested over the network.

The M2M performs the same functionalities as the DAQ but allows network communication with legacy devices, being middleware that allows obtaining information from other protocols (Modbus TCP/IP, CoAP, etc.) and making it accessible to the MOAI microservices and applications.

The PLC enables the functionalities of a programmable logic controller (PLC) at the MOAI. Actually, it comprises in a microservice the functionalities of softPLC on an open hardware and software platform. This microservice allows the execution of PLC programs (IEC 61131-3) and monitoring of input and output data and is built on top of the OpenPLC project [36].

The PID4.0 is in charge of controlling processes and is able to implement redundant controllers (running replicas of the same microservice). As a high-level microservice, it needs to be composed with other low-level microservices such as DAQ or M2M in order to obtain the input and output data. This microservice is based on a modified version of the PIDPlus control algorithm developed for network control [37]. However, it was changed to run as a microservice with redundancy for the MOAI.

The database is a microservice adapter for the InfluxDB database, which is an open-source time series database used to record historical industrial process data. The data can be requested for creating any necessary applications. IAS-based applications (internal or external) were developed through the composition of microservices such as Process Control, SCADA (supervision), and Process Historian.

The microservices in MOAI are able to be hosted on different platforms as necessary, such as embedded systems, computers, local servers, or Platform as a Service (PaaS), with different Windows, Linux, or MAC-based operating systems. All the microservices have the ServiceBroker, which is the core of the microservices and is in charge of various parameterizations as well as networking configuration (Transporter and API gateway) for communication between microservices.

4.2. Networking

The digitalization of processes and cross-layer integration demanded by I4.0 can be provided by the powerful networking mechanisms available in SOA and MOA. In the case of the MOAI, it is carried out by the networking microservice that is called Transporter (orange hexagon) and by the API Gateway (green hexagon) in Figure 2. Internal calls for microservices are handled by the Transporter and external calls are handled by the API Gateway.

The API Gateway is responsible for interfacing the MOAI microservices with external applications through HTTP(s). All actions (functionalities) from all microservices are automatically mapped to a RESTful API, which standardizes the external communication with the MOAI. It is an important characteristic for enabling integration with other IT-OT solutions. The Transporter is responsible for several networking functions, such as message communication (events, requests, and responses) among services, microservice discovery and registry, microservice load balancing, and status checking. In the MOAI, the Transporter can be seen as a broker element in a producer-consumer communication architecture.

The Transporter provides different protocols for message communication between microservices (TCP/IP, NATS, MQTT, AMQP, etc.), which are mapped to the Moleculer protocol. There is also the possibility of creating customized protocols or adding other existing protocols to them. The Transporter is defined and configured in the microservices. After configuring it, the communication between microservices and the communication between microservices and the API Gateway occurs transparently. It is also possible to change between transporters. Therefore, no matter what communication protocol is used, no additional configuration or coding is required for all message communication in the MOAI. It is an important characteristic for standardizing communication and enabling flexibility. As communication is transparent, only changes in microservices and actions are required in order to provide other functionalities that are automatically made available on the network.

Service discovery is the process of automatically detecting and locating available services in a distributed system and is an important subject in SOA and MOA. The MOAI service registry and discovery are also carried out by the Transporter. Microservice discovery is dynamic, meaning that a microservice or application does not need to know anything about other ones. When it starts, it will announce its presence to the Transporter and consequently to all the other microservices for service registry. In the event of a microservice failure (or stop), it is detected and removed from the service registry. As a result, Transporter can route calls or requests to live microservices.

Another important characteristic of the Transporter is the load balancing mechanism, as it provides availability and microservice redundancy. In the MOAI, microservices can be reused or even replicated (copies). If multiple instances of a microservice are running on different nodes (location/hardware), the Transporter balances the calls/requests between them according to the defined strategy. Available strategies are Round-Robin, Random, CPU (it is requested the microservice with the lowest CPU usage), and Latency (it is requested the microservice with the lowest network latency).

4.3. Security Mechanisms

In this section, the security mechanisms developed for the MOAI are described [31]. Figure 3 summarizes the security layers developed in the MOAI. An explanation of the security mechanism is given for each of the three layers presented: API Gateway HTTPs and Authentication, Transporter Authentication, and Microservice Authorization. In layer 1 (green), the API Gateway security is guaranteed by using authentication through the URL and encryption with HTTPs. In layer 2 (orange), the Transporter also uses authentication for microservices connections. In layer 3 (blue), the access control of microservices is guaranteed using the Guard microservice.

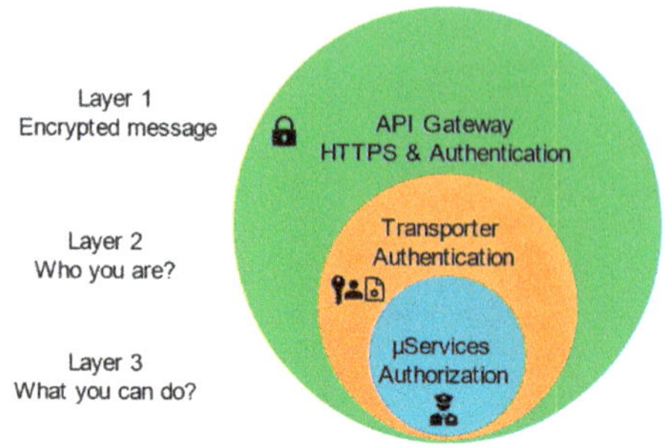

Figure 3. Security layers.

1. API Gateway (Layer 1)

Hyper Text Transfer Protocol Secure (HTTPS) is a security-enhanced version of the HTTP protocol that adds an extra layer of security using the Secure Socket Layer (SSL)/Transport Layer Security (TLS) protocols. This allows for the secure transmission of data over an encrypted connection and ensures that the identities of both the server and client are verified through the use of digital certificates.

This layer in MOAI ensures security in the access of external applications, preventing attacks (man in the middle) and the capture of the information contained in the packages. An authentication procedure has also been developed, where it is required that the client enter a token together with the URL address of the requested microservice.

2. Transporter (Layer 2)

Authentication is a security procedure in which a client's connection to a server is allowed by checking an id (ID) and password. Without this procedure in the MOAI, any microservice could use the Transporter to communicate with or access other microservices without any security. The Transporter authentication mechanism was implemented in order to enable the microservice id/password verification against the Transporter registered list.

As a result, only registered microservices will be allowed to connect to the Transporter and be part of the MOAI. Connection requests not allowed by the Transporter will be triggered, indicating the attempted security breach, and registered in a log. This security layer enables controlling the connections to the Transporter and the access to the MOAI microservices.

3. Microservice Guard (Layer 3)

Authorization is a security mechanism that determines whether a user has permission to access a particular resource, perform an action, or execute a specific operation within a system or application. It is a process that verifies if a user has the necessary privileges or permissions to perform a specific action or access a particular resource. In the MOAI, the Guard microservice is responsible for authorization.

The Guard microservice acts as a surveillance system for messages exchanged between microservices. It intercepts messages being sent before they reach their intended microservice. The authorization procedure uses a JSON Web Token (JWT) to safeguard the actions of the microservices. This involves generating an encoded token using a secret or private key that is internally defined in the system. Once the key is validated and accepted, the encoded token is returned. This process is repeated for each microservice, with each one having its own unique token.

If a microservice or application requests an action from another microservice, the Guard microservice verifies whether the client token is valid or not, controlling the client's access to this action. If it is valid, the action will be accessed normally through the Transporter, if not, the desired action will be blocked by returning an error.

4.4. Microservices Management

Managing microservice-based architectures may be a challenging task, mainly when considering the conservative industrial area. Considering that, a management structure for the MOAI was proposed. The management systems are based on container images and network-based deployment and monitoring tools. The container is an application-level abstraction that wraps code and dependencies together, illustrated as a generic microservice (blue hexagon in server 3) in Figure 4, showing the layers that this service encapsulation contains (Container layers). Containers can be allocated on the same server, sharing resources with other containers running autonomous processes, occupying reduced storage spaces relative to virtual machines (VMs), and dealing with a larger number of applications without requiring much from the operating system [38–40].

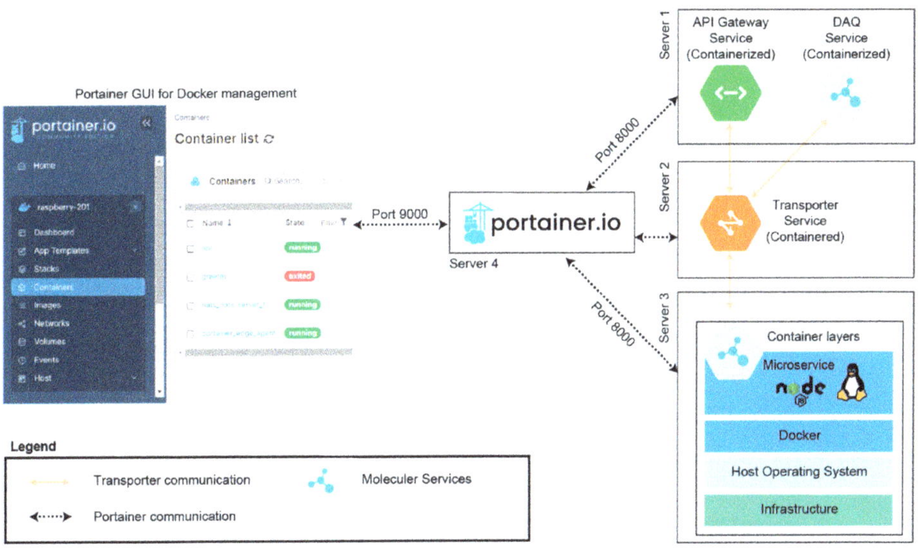

Figure 4. Services management through Portainer GUI in the MOAI.

A container can contain multiple applications (App A, App B, and App C) functioning in a monolithic manner, or just one application (App A) that contains a simple function of a more granularized (microservice) functioning as a service, as represented in Server 3 in Figure 4. Dockerfile is a simple text file with a set of commands or instructions to support the creation of a container in Docker [38]. For example, it informs that a Moleculer framework microservice developed in JavaScript will run in Node.js and all project dependencies must be installed. These commands are then executed for the construction (build) of an image (a Docker Image) for multiple CPU architectures. This static image can be stored in a local or remote registry (Docker Hub) that functions as a library of images available simply by downloading it and running it (run) on the desired host [39].

Multiple microservices (Transporter, API Gateway, and DAQ) in Figure 4 were created and distributed in different environments (Server 1, Server 2, and Server 3) on the same network, making the management task arduous. Portainer's graphical interface, together with Portainer Edge Agent, facilitates this task by abstracting and removing the need to use the Command-Line Interface (CLI) [40]. The remote environments (Server 1, Server 2, and Server 3) in Figure 4 need to be able to access the Portainer Server (Server 4). This communication is performed by an encrypted TLS tunnel (Port 8000) with keys between the Portainer Edge Agent of each remote environment and the Portainer Server.

To manage these environments, the Portainer Server GUI is accessed through a browser (Port 9000). This GUI enables controlling the operation of containers (microservices), in addition to loading environment variables and new services directly from the Docker HUB registry according to the application. As a result, it simplifies the management into just one dashboard (Portainer GUI for Docker management) in Figure 4, which contains all microservices and applications.

5. MOAI Application Examples

The MOAI was developed and implemented in a real pilot plant focused on process control. The difference of this pilot plant is that the MOAI is used for all activities. Different IAS applications were developed to validate the MOAI and demonstrate its flexibility and advantages. To simplify the presentation of the developments, it was decided to describe in detail one of these applications, which is closed-loop process control (number 1). This explanation and comprehension provide the necessary information to understand how the MOAI works and can be further applied to understand any application (including the other examples from numbers 2 to 5 briefly presented).

An industrial pilot plant for process control in the laboratory was used to test the MOAI (Figure 5). The P&ID of the plant is shown in Figure 6, comprising a tank (TQ02) at the bottom responsible for storing the fluid, which can be pumped to the upper tank (TQ01) and to the reservoir (R01) through the pumps (P2 and P1). The frequency inverters (PZ and UZ) and the solenoid valve (LV 122) represent the manipulated variables of the control loop. The process or controlled variables are the control (LIT125), the pipe flow (FIT116), the pipe pressure (PIT118), and the reservoir pressure (PIT129).

1. Closed-loop Process Controller.

Closed-loop control is the basis of any IAS. The necessary microservices to develop this external application are API Gateway, Transporter, DAQ, PID4.0, and Guard (security) in Figure 2. The orchestration of each of the plant's microservices can be seen in the sequence diagram of Figure 7.

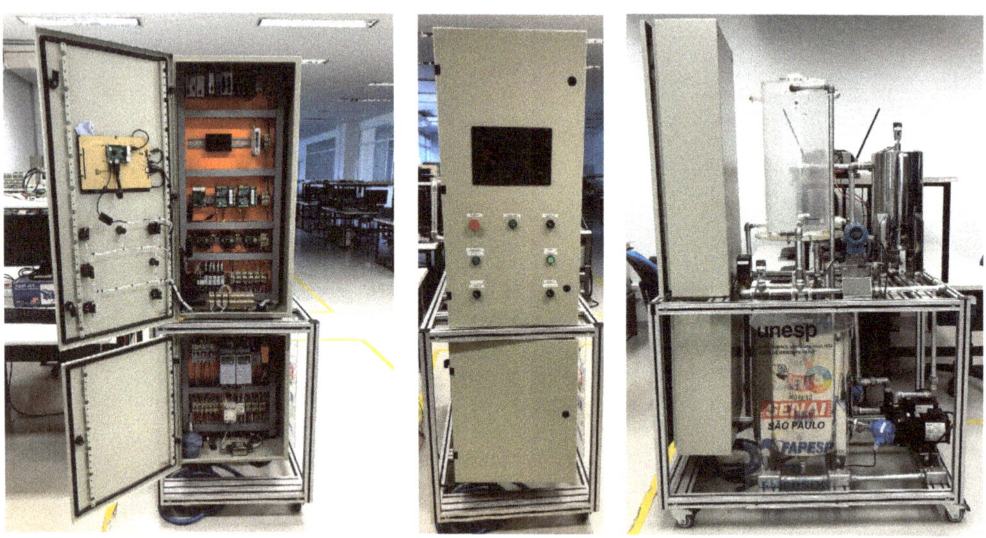

(a) (b) (c)

Figure 5. Industrial pilot plant: (**a**) Front part with open panel; (**b**) front part with the panel closed; (**c**) right side.

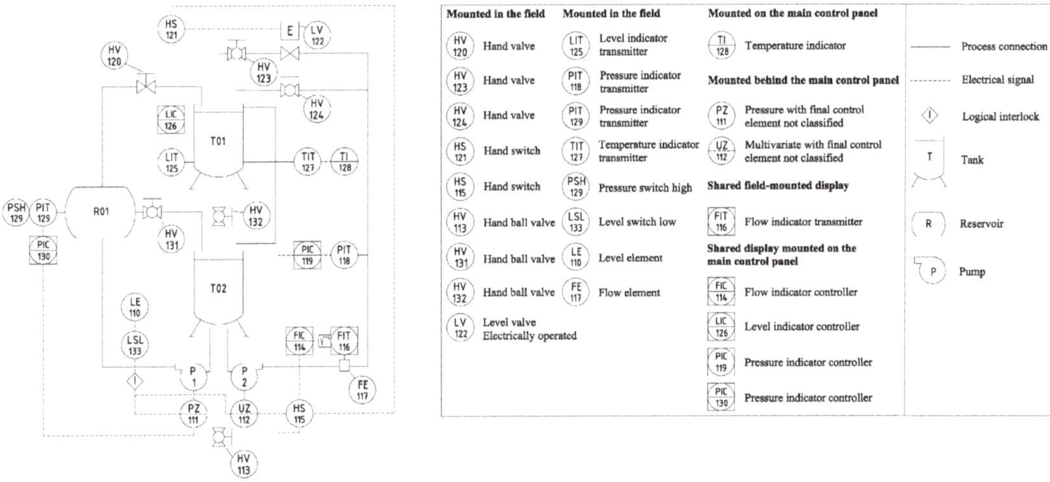

(a) (b)

Figure 6. Industrial pilot plant: (**a**) P&ID; (**b**) legend.

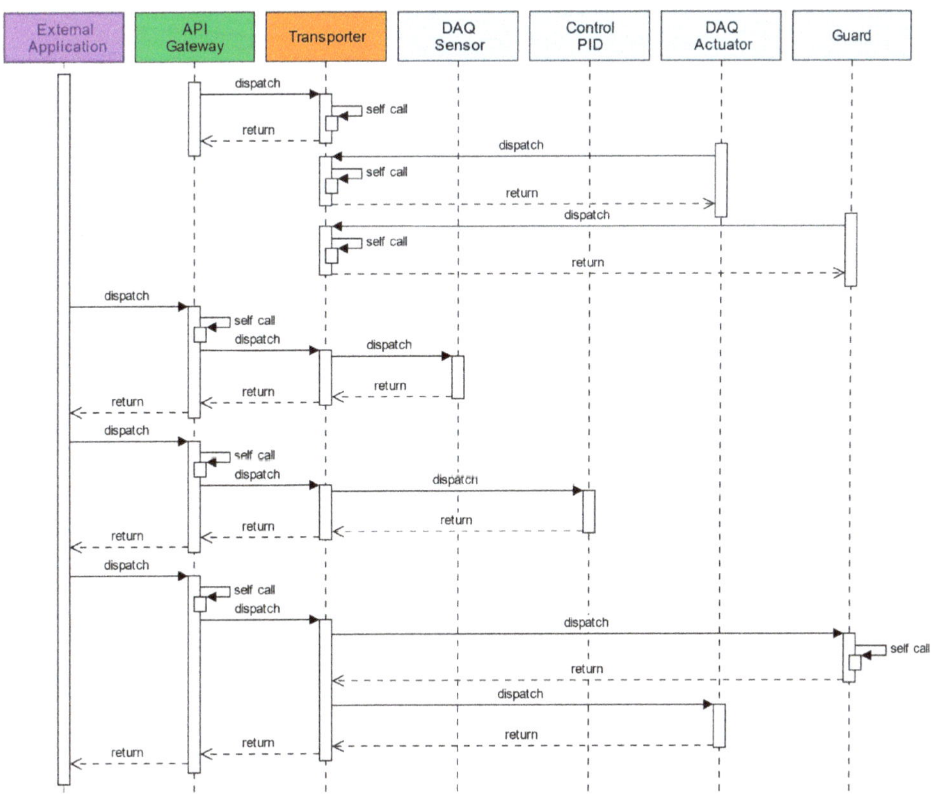

Figure 7. Diagram of the external orchestration sequence with all security mechanisms enabled.

The industrial software (LabVIEW) is represented by the external application (purple). This application orchestrates microservices through the API Gateway (green) and the communication service Transporter (orange). The microservices used are blue. This example also includes the usage of all security mechanisms implemented in the MOAI. This sequence diagram is important as its compression is the basis for understanding how any other application operates in the MOAI.

The orchestration of microservices initiates with an external application: The process control application (LabVIEW), Figure 7, cyclically requests sensor data to the DAQ service via API. If this request is successful, it receives a header with the code "200 OK", and the value of all sensors in the plant in the response. The API gateway service makes an internal call to check the token passed in the request for each request made. The HTTPs encryption steps occur but have been omitted to simplify the figure. All services must authenticate to the Transporter only once; this process was also omitted for better understanding of Figure 7.

The sensor data are sent to the PID4.0 control microservice, which is responsible for calculating the control signal to be applied to the plant. Finally, the control signal is sent to the DAQ microservice, which is responsible for updating the plant's actuator. However, before that happens, this request goes through the guard service. The Guard microservice was used to ensure secure access to the actuators, such as the actuator (output) of the DAQ service. This way, only authorized services will be able to actuate on the plant, thus assuring safety in the three layers of the IAS application.

A compilation of the security mechanisms used in each test can be seen in Table 1, where in the first test no security mechanism was used. The other tests enabled one security mechanism at a time to see its impact individually. In test 2, it only enabled authentication (Authen) in the API gateway. In test 3, it used only HTTPS encryption. In test 4, only the Transporter requires authentication. In test 5, only the Guard check was enabled (Authorization), and in test 6, all security mechanisms were enabled together.

Table 1. Security mechanisms used in each test with statistics of all communication tests performed.

Expt. Number	API Authen.	API HTTPS	Transporter Authn.	Guard Authz.	Mean (ms)	Median (ms)	Mode (ms)	SD	Variance
1	No	No	No	No	68.14	63.97	61.72	14.32	205.19
2	Yes	No	No	No	75.70	71.99	67.57	16.96	287.55
3	No	Yes	No	No	197.27	201.06	202.47	18.23	332.18
4	No	No	Yes	No	74.33	69.01	66.60	17.10	292.31
5	No	No	No	Yes	93.36	91.01	91.44	15.12	228.56
6	Yes	Yes	Yes	Yes	263.31	267.51	274.11	16.86	284.35

In Table 1, in all tests, the standard deviation (SD) ranged from 14 to 18, having a symmetrical distribution profile. The variance of the values was between 200 and 300 and cannot be disregarded in processes where latency is significant. Regarding the control loop orchestration times, the best result obtained was without the use of any security mechanism (Expt. 1), with an average time of 68.14 ms for a closed loop control cycle. The security mechanisms that have the least impact, respectively, are the authorization in the API gateway (Expt. 2) with 75.70 ms, the authentication in the Transporter (Expt. 4) with 74.33 ms, and finally the guard service (Expt. 5) with 93.36 ms.

The communication performance of all experiments (Expt.) was analyzed. The control experiment interval duration was 300 s (600 control cycles). The objective was to measure the communication time of the control loop orchestration (execution sequence of microservices in Figure 7). All times were compiled and presented through statistical data using a boxplot, as shown in Figure 8 and its respective summary table, Table 1.

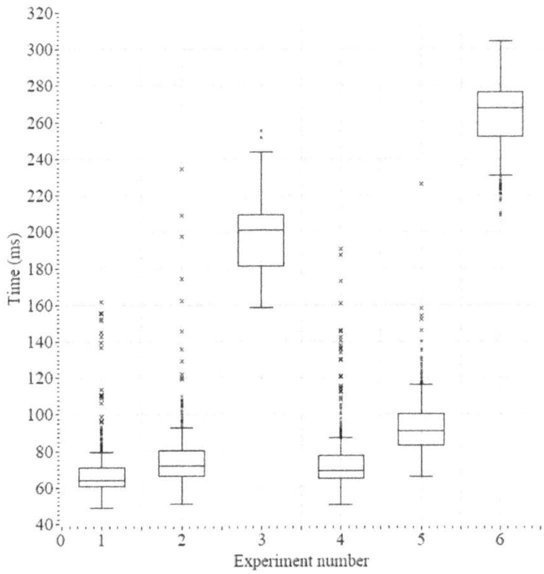

Figure 8. Boxplot statistics of all experiments performed.

The security mechanism that most impacted the communication was enabling HTTPs (Expt. 3), where the total time rose to 197.27 ms. Considering the worst-case scenario with all mechanisms enabled, the control loop orchestration time was 263.31 ms, which means an increase of 4x times to the test without any security. However, even the worst-case result is within the acceptable limits for controlling processes, for which a control cycle of 500 ms (2 Hz) is an acceptable standard in industries today. The communication performance of the MOAI was better (for example, faster than in [22]) or at least similar to the other reviewed SOA/MOA in industrial applications.

During the experiments, the following process variables of the pilot plant were controlled: pipe pressure, reservoir pressure, and pipe flow. The objective was not to evaluate or optimize the control performance, nor to compare the response curves of the processes, because each control loop has its own characteristic dynamics. The objective was to investigate the composition (orchestration) of microservices and their capacity to create a composite application to control different processes. Figure 9 presents the control responses, in which red is the desired value (setpoint) and the process variables are in other colors. The output curves can track the setpoint; in addition, there is no control saturation and the variation is smooth. However, the results show in general that the process control is stable and feasible to perform using the MOAI.

2. Distributed I/O

Data acquisition is a necessary step in any industry application. IAS has traditionally been used for SCADA and remote I/O. Remote I/O is responsible for acquiring data from sensors and actuators in industrial processes. Distributed I/O is an evolution of remote I/O used for networked data acquisition. IIoT and I4.0 applications require interoperability and vertical interaction among devices and systems. To fulfill that, a distributed I/O has been developed as a microservice, and some applications have shown its effectiveness [32].

The necessary microservices to develop the distributed I/O were Transporter and DAQ in Figure 2. The distributed I/O operation is equal to the traditional solution, but the network communication is automatic through the Transporter. Another difference is that it is programmable, so customer-specific functions can be added. A great advantage over the traditional solution is the ease of deployment of distributed I/O redundancy as a result of the replication of microservices. Redundancy in IAS means keeping redundant (duplicate or triplicate) systems in order to ensure the availability of critical processes or devices. The redundancy of distributed I/O is achieved when two or more replicas of the same distributed I/O communicate through the Transporter.

Several use cases were presented demonstrating that the distributed I/O using microservices enabled the I/O data acquisition in IAS to function in a networked, standardized, and scalable manner [32].

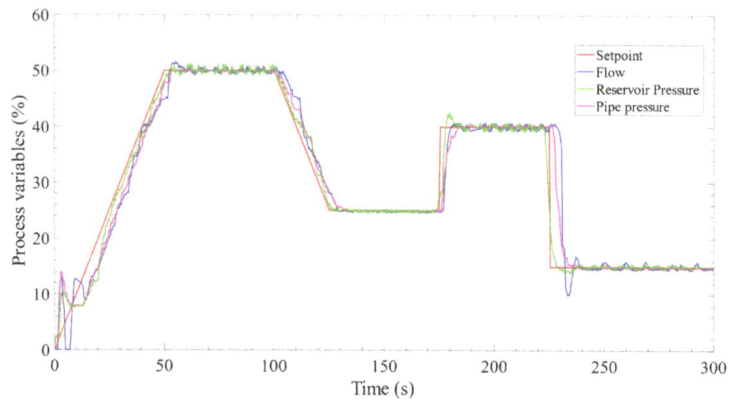

Figure 9. Control response from all pilot plant control loops.

3. Control as a Microservice and Controllers' Redundancy

The I4.0 is also pushing forward process control in IAS. Beyond interoperability and vertical integration, process control needs to be flexible, modular, and easy to deploy. In order to fulfill that, a process controller has been developed as a microservice [34]. As a result, the control algorithm can be developed according to the application's needs and made scalable through the reuse of the microservice. Considering the redundancy scenario cited before, this approach also eases and enables the redundancy of process controllers in the industry.

The necessary microservices to develop a process controller (redundant or not) were the Transporter, PID4.0, and the DAQ or M2M for I/O data acquisition in Figure 2. PID4.0 implements a network-based PID control algorithm (PIDPlus) as a microservice. The difference is that when the controller is made available as a microservice, it can be easily reused and replicated on multiple platforms, generating great hardware savings and ease of deployment. That is interesting because different versions of hardware, OS, and programming languages can be used, which is very different from a traditional PLC with dedicated hardware controllers. As a result, it reduces the overall complexity of implementing redundancy in the IAS.

Another advantage is that the Transporter communication uses load balancing strategies available in the Moleculer framework, reducing CPU overload or latency according to the selected strategy. Several use cases were presented with service compositions demonstrating that the controller as a microservice enabled using and sharing microservices in order to obtain a flexible and distributed control architecture [30,34].

4. PLC with IEC 61131-3 standard

IAS are usually based on programmable logic controllers (PLC) and development software based on IEC 61131-3 PLC programming languages. In order to make the MOAI fully compatible with IAS, a PLC microservice has been developed, and function blocks have been added to the programming IDE to make PLC programs integrated with the Transporter communication [32]. The PLC microservice integrates the OpenPLC project [38] into the Moleculer framework. OpenPLC brings together the functionalities of a PLC with the advantages of an open software and hardware architecture.

The necessary microservices to develop PLC-based automation and process control are Transporter, PLC, and the DAQ or M2M for I/O data acquisition shown in Figure 2. The PLC microservice can perform logic and sequencing of tasks that are performed on conventional PLCs, as well as the acquisition and monitoring of data from a local controller or distributed inputs and outputs (DAQ and M2M microservices). However, it is available as a microservice, which provides the same flexibility advantages cited before, such as reusing and deployment on multiple platforms.

The working logic of the controller process is developed in the OpenPLC Editor, a software that allows to write PLC programs according to IEC 61131-3. Considering the MOAI, it is also possible to create applications using the OpenPLC Editor that orchestrate the microservices available. Several use cases were presented with the PLC microservice usage in discrete automation and process control and with the microservice composition of applications using IEC 61131-3 (Ladder and Function Block Diagram) [33].

5. Process Historian and SCADA

Monitoring and understanding the behavior of industrial processes is fundamental to increasing efficiency in the production line and enabling management decision-making. Process historians or operational historians are being applied to store increasing amounts of data obtained from a broader variety of sources, including control and monitoring, enterprise resource planning (ERP), and asset management systems. With that in mind, a SCADA application was developed with a historian service for the acquisition, storage, and analysis of structured data from industrial processes [32].

The necessary microservices to develop the SCADA and historian were the Transporter, Database, and DAQ or M2M for I/O data acquisition in Figure 2. The solution is based on using the Transporter for automatically acquiring data through the DAQ or M2M, storing it in a time-series database (InfluxDB), and feeding the process historian (Grafana). Several IAS use cases were presented for SCADA and historian usage [32].

6. Discussions and Final Remarks

The MOAI enabled the use and sharing of microservices for the design of a scalable, flexible, interoperable, and distributed industrial architecture for IAS. The implementation of IAS as microservices contributes to a new model of interactions between different industrial systems, equipment, users, and applications that meets the changes and fulfills the requirements of I4.0 applications.

On the contrary of the SOA/MOA for industrial applications available in the literature, the MOAI was built on top of the Moleculer framework for microservices. It was a very positive development choice for the MOAI. Many mandatory development steps have been avoided, such as service-oriented software, service communication, service registry, and discovery, because the framework already provides them. The functionalities of the MOAI and other architectures in the literature are very similar, as they were developed for the same proposal of evolving IAS to support I4.0 applications. As a result, a simple and direct comparison among them is not fair, but rather some aspects and functionalities can be confronted. Most of the services and applications cover the same tasks. Service composition by orchestration is most commonly used in industrial applications and is supported by all architectures in the literature as it provides better execution control and management of the application. The MOAI, as also described in [22], additionally supports choreography, which can improve decentralization and aims to keep services loosely coupled and more autonomous.

The great difference between the MOAI and the other architectures is the service communication mechanism (Transporter). In addition to providing the basic microservice networking functionalities, such as communication, discovery and registry, and load balancing and status checking, it also enables the unique functionality of microservice redundancy. Microservice redundancy (through replication of the same service) is an important feature for IAS, as described in the distributed I/O and controller redundancy examples. In this case, the Transporter verifies if there are two or more (redundant) instances of the same microservice, and when there is a request for that service, it automatically chooses the best instance of them for execution at that time, enabling the redundancy. As the Transporter itself is also a microservice in the MOAI, it is also possible to make it redundant. In this case, the microservice networking will change from one Transporter (active) to another (redundant) in case of failure.

Considering the pros and cons of the MOAI, although there are disadvantages, the advantages are greater and more relevant. The MOAI is scalable because the microservices are modular and independently deployable and manageable, allowing them to be easily changed on demand. In addition, microservices can be reused for scaling applications (ex: a set of DAQ-sensor, PID4.0, and DAQ-actuator microservices reused for each control loop of the plant) and replicated in order to provide redundancy or availability (ex: two or more redundant PID4.0 microservices for the same control loop or redundant DAQ microservices for data acquisition as described in the distributed I/O example).

The MOAI is flexible as it is able to adapt to changing requirements and can be modified or extended to meet new needs without requiring significant changes to the architecture. New microservices (functionalities) can be easily created, as well as new applications that can be composed using the microservices.

The MOAI is interoperable because the networking functionalities of automatic, standardized, and transparent communication between microservices and applications overcome data integration problems between different hierarchical industry levels. The MOAI is fundamentally distributed as microservices are deployed in individual nodes that communicate through the network (Transporter). It enables decentralization of control tasks or decision-making, as each node or microservice has its own functionalities and responsibilities. In addition, it provides resilience and fault tolerance as the MOAI continues to operate in the case of individual microservice failures and failures of replicated microservices.

Microservices are a different paradigm from traditional IAS monolithic architectures in terms of development, maintenance, and commissioning/deployment. The disadvantages of the MOAI or difficulties are more related to learning new concepts, tools, and development environments and their development in the industry than to the complexity of using microservices. The development of microservices, such as those in the MOAI, must be based on standardized structures (containers) and management tools. Maintaining the applications and commissioning microservices require a repository structure for better version management and operational control, which is not common in IAS. Another point to be highlighted is the complexity of management in large applications containing many microservices. Due to the networked characteristics of MOAI, fault maintenance or error checking is more complex. On the other hand, these activities can be carried out entirely online and remotely over the network on the MOAI.

A limitation of the MOAI is related to the network protocols available for communication. As the Moleculer is a computing framework, the supported protocols are not industrial standards. As a result, direct communication with legacy industrial devices is not possible, and an intermediary service is necessary for bridging them to the Transporter. However, it is a known issue of SOA/MOA in industrial applications that was handled in this research with the M2M microservice, using the same idea as other approaches in the literature as the mediator service in [20] and the translation and gateway services in [22].

In terms of security, microservices introduce additional concerns, as each service's access needs to be secured and there needs to be secure communication between services. The common security mechanisms available in other SOAs/MOAs, such as encryption, authentication, and authorization, have also been developed in the MOAI. As an additional security layer in the MOAI, the Guard microservice mechanism checks all microservice requests before they reach their recipient, verifying permissions and blocking unallowed requests.

7. Conclusions

Even though traditional IAS architecture is still present in the industry, great efforts are being made to evolve it, especially using service-based architectures. SOA and MOA overcome the main problems of interoperability and vertical integration of heterogeneous systems in the IAS, fulfilling the requirements for I4.0 applications.

The paper described the development, deployment, and testing of a Microservice Oriented Architecture for Industry 4.0 (MOAI) based on the Moleculer framework. The first contribution was investigating and demonstrating that a non-industrial framework can be applied to IAS and I4.0 applications. In addition to simplifying MOAI development, it provided a flexible, interoperable, and distributed architecture.

The functionalities, pros, and cons of the MOAI were discussed. The highlight and differential point of the MOAI is the networking microservice (Transporter). In addition to offering fundamental microservice networking capabilities, it enabled the distinctive feature of microservice redundancy. Operational details of the architecture were explained, as well as the communication and service composition through orchestration and choreography. The development of microservices and applications as well as security mechanisms were also discussed.

Considering the MOAI microservices, some can be highlighted. The PLC microservice can be applied in applications where determinism is not required and is advantageous because it is not necessary to deploy dedicated physical controllers in the plant. The DAQ microservice resembles the operation of a networked remote I/O, with the advantage of programing capability and incorporating customized functionality. The PID4.0 microservice evolved traditional PID control into a network-based, redundant-ready controller for IAS.

Application examples of the MOAI were discussed, and the results demonstrated the feasibility of supporting IAS in the context of the I4.0. The MOAI proved to be valuable both at the device and application levels by providing a high level of loose coupling between the various components of the system.

Future work will focus on expanding the experiments to evaluate the benefits and drawbacks of the MOAI developed. It is expected to automate the composition of services by taking advantage of microservice discovery and using the concept of Plug and Play.

Author Contributions: Conceptualization, J.A.B. and E.P.G.; Methodology, R.P.P. and L.B.R.d.S.; Software, R.P.P., J.A.B. and L.B.R.d.S.; Validation, R.P.P.; Investigation, R.P.P.; Writing–original draft, R.P.P.; Writing–review & editing, E.P.G.; Supervision, E.P.G.; Funding acquisition, E.P.G. All authors have read and agreed to the published version of the manuscript.

Funding: Research supported by grant 2018/19984-4, São Paulo Research Foundation (FAPESP).

Data Availability Statement: There is no research data available in this article.

Acknowledgments: We are thankful for the participation of Sergio L. Risso, Douglas P. Fernandez, João P. da Silva Cavalcante, João V. Rocha Santana, Vitor I. Guermandi, Felipe O. Domingues, and Michel M. Fernandes in the development of this project.

Conflicts of Interest: The authors declare no conflict of interest.

References

1. Lu, Y. Industry 4.0: A Survey on Technologies, Applications and Open Research Issues. *J. Ind. Inf. Integr.* **2017**, *6*, 1–10. [CrossRef]
2. Borangiu, T.; Trentesaux, D.; Thomas, A.; Leitão, P.; Barata, J. Digital Transformation of Manufacturing through Cloud Services and Resource Virtualization. *Comput. Ind.* **2019**, *108*, 150–162. [CrossRef]
3. Sisinni, E.; Saifullah, A.; Han, S.; Jennehag, U.; Gidlund, M. Industrial Internet of Things: Challenges, Opportunities, and Directions. *IEEE Trans. Ind. Inf.* **2018**, *14*, 4724–4734. [CrossRef]
4. Colombo, A.W.; Karnouskos, S.; Bangemann, T. Towards the next Generation of Industrial Cyber-Physical Systems. In *Industrial Cloud-Based Cyber-Physical Systems: The IMC-AESOP Approach*; Springer International Publishing: Berlin/Heidelberg, Germany, 2014; Volume 9783319056241, pp. 1–22. ISBN 9783319056241.
5. Wollschlaeger, M.; Sauter, T.; Jasperneite, J. The Future of Industrial Communication: Automation Networks in the Era of the Internet of Things and Industry 4.0. *IEEE Ind. Electron. Mag.* **2017**, *11*, 17–27. [CrossRef]

6. Givehchi, O.; Landsdorf, K.; Simoens, P.; Colombo, A.W. Interoperability for Industrial Cyber-Physical Systems: An Approach for Legacy Systems. *IEEE Trans. Ind. Inf.* **2017**, *13*, 3370–3378. [CrossRef]
7. Jammes, F.; Karnouskos, S.; Bony, B.; Nappey, P.; Colombo, A.W.; Delsing, J.; Eliasson, J.; Kyusakov, R.; Stluka, P.; Tilly, M.; et al. Promising Technologies for SOA-Based Industrial Automation Systems. In *Industrial Cloud-Based Cyber-Physical Systems: The IMC-AESOP Approach*; Springer: Cham, Switzerland, 2014; pp. 89–109. ISBN 978-3-319-05623-4. [CrossRef]
8. ISA beyond the Pyramid: Using ISA95 for Industry 4.0 and Smart Manufacturing. Available online: https://www.isa.org/intech-home/2021/october-2021/features/beyond-the-pyramid-using-isa95-for-industry-4-0-an (accessed on 10 January 2023).
9. Klettner, C.; Tauchnitz, T.; Epple, U.; Nothdurft, L.; Diedrich, C.; Schröder, T.; Großmann, D.; Banerjee, S.; Krauß, M.; Iatrou, C.; et al. Namur Open Architecture: Die Namur-Pyramide wird geöffnet für Industrie 4.0. *Atp Mag.* **2017**, *59*, 20–37. [CrossRef]
10. de Caigny, J.; Tauchnitz, T.; Becker, R.; Diedrich, C.; Schröder, T.; Großmann, D.; Banerjee, S.; Graube, M.; Urbas, L. NOA–Von Demonstratoren Zu Pilotanwendungen. *Atp Mag.* **2019**, *61*, 44–55. [CrossRef]
11. OPC Foundation Unified Architecture—OPC Foundation. Available online: https://opcfoundation.org/developer-tools/specifications-unified-architecture/part-4-services/ (accessed on 21 December 2022).
12. Leitão, P.; Colombo, A.W.; Karnouskos, S. Industrial Automation Based on Cyber-Physical Systems Technologies: Prototype Implementations and Challenges. *Comput. Ind.* **2016**, *81*, 11–25. [CrossRef]
13. di Francesco, P.; Malavolta, I.; Lago, P. Research on Architecting Microservices: Trends, Focus, and Potential for Industrial Adoption. In Proceedings of the 2017 IEEE International Conference on Software Architecture, ICSA 2017, Gothenburg, Sweden, 5–7 April 2017; pp. 21–30. [CrossRef]
14. Xiao, Z.; Wijegunaratne, I.; Qiang, X. Reflections on SOA and Microservices. In Proceedings of the 4th International Conference on Enterprise Systems: Advances in Enterprise Systems, ES 2016, Online, 2–3 November 2016; IEEE: Piscataway, NJ, USA, 2017; pp. 60–67.
15. da Xu, L.; Xu, E.L.; Li, L. Industry 4.0: State of the Art and Future Trends. *Int. J. Prod. Res.* **2018**, *56*, 2941–2962. [CrossRef]
16. Mrugalska, B.; Wyrwicka, M.K. Towards Lean Production in Industry 4.0. *Procedia. Eng.* **2017**, *182*, 466–473. [CrossRef]
17. Delsing, J.; Rosenqvist, F.; Carlsson, O.; Colombo, A.W.; Bangemann, T. Migration of Industrial Process Control Systems into Service Oriented Architecture. In Proceedings of the IECON 2012—38th Annual Conference on IEEE Industrial Electronics Society 2012, Montreal, QC, Canada, 25–28 October 2012; pp. 5786–5792. [CrossRef]
18. Girbea, A.; Suciu, C.; Nechifor, S.; Sisak, F. Design and Implementation of a Service-Oriented Architecture for the Optimization of Industrial Applications. *IEEE Trans. Ind. Inf.* **2014**, *10*, 185–196. [CrossRef]
19. Cannata, A.; Gerosa, M.; Taisch, M. SOCRADES: A Framework for Developing Intelligent Systems in Manufacturing. In Proceedings of the 2008 IEEE International Conference on Industrial Engineering and Engineering Management, IEEM 2008, Singapore, 8–11 December 2008; pp. 1904–1908. [CrossRef]
20. Karnouskos, S.; Colombo, A.W.; Bangemann, T.; Manninen, K.; Camp, R.; Tilly, M.; Sikora, M.; Jammes, F.; Delsing, J.; Eliasson, J.; et al. The IMC-AESOP Architecture for Cloud-Based Industrial Cyber-Physical Systems. In *Industrial Cloud-Based Cyber-Physical Systems: The IMC-AESOP Approach*; Springer International Publishing: Berlin/Heidelberg, Germany, 2014; Volume 9783319056241, pp. 49–88. ISBN 9783319056241.
21. ARROWHEAD Eclipse Arrowhead—Eclipse Arrowhead Framework and Implementation Platform. Available online: https://arrowhead.eu/ (accessed on 10 January 2023).
22. Espí-Beltrán, J.V.; Gilart-Iglesias, V.; Ruiz-Fernández, D. Enabling Distributed Manufacturing Resources through SOA: The REST Approach. *Robot. Comput. Integr. Manuf.* **2017**, *46*, 156–165. [CrossRef]
23. Delsing, J. Local Cloud Internet of Things Automation: Technology and Business Model Features of Distributed Internet of Things Automation Solutions. *IEEE Ind. Electron. Mag.* **2017**, *11*, 8–21. [CrossRef]
24. Paniagua, C.; Eliasson, J.; Delsing, J. Efficient Device-to-Device Service Invocation Using Arrowhead Orchestration. *IEEE Internet Things J.* **2020**, *7*, 429–439. [CrossRef]
25. Ciavotta, M.; Alge, M.; Menato, S.; Rovere, D.; Pedrazzoli, P. A Microservice-Based Middleware for the Digital Factory. *Procedia. Manuf.* **2017**, *11*, 931–938. [CrossRef]
26. MAYA Multi-DisciplinArY Integrated SimulAtion and Forecasting Tools, Empowered by Digital Continuity and Continuous Real-World Synchronization, towards Reduced Time to Production and Optimization | MAYA Project | Fact Sheet | H2020 | CORDIS | European Commission. Available online: https://cordis.europa.eu/project/id/678556 (accessed on 19 January 2023).
27. Innerbichler, J.; Gonul, S.; Damjanovic-Behrendt, V.; Mandler, B.; Strohmeier, F. NIMBLE Collaborative Platform: Microservice Architectural Approach to Federated IoT. In Proceedings of the GIoTS 2017—Global Internet of Things Summit, Geneva, Switzerland, 6–9 June 2017; Institute of Electrical and Electronics Engineers Inc.: Piscataway, NJ, USA, 2017.
28. NIMBLE The Novel, Federated Approach for Industry B2B Platforms—Nimble Project—The Novel, Federated Approach for Industry B2B Platforms. Available online: https://www.nimble-project.org/ (accessed on 10 January 2023).
29. Pontarolli, R.P.; Bigheti, J.A.; Fernandes, M.M.; Domingues, F.O.; Risso, S.L.; Godoy, E.P. Microservice Orchestration for Process Control in Industry 4.0. In Proceedings of the 2020 IEEE International Workshop on Metrology for Industry 4.0 and IoT, MetroInd 4.0 and IoT 2020, Roma, Italy, 3–5 June 2020; pp. 245–249. [CrossRef]
30. Pontarolli, R.P.; Bigheti, J.A.; de Sá, L.B.R.; Godoy, E.P. Towards Security Mechanisms for an Industrial Microservice-Oriented Architecture. In Proceedings of the 2021 14th IEEE International Conference on Industry Applications, INDUSCON 2021, São Paulo, Brazil, 15–18 August 2021; pp. 679–685. [CrossRef]

31. Pontarolli, R.P.; Bigheti, J.A.; Domingues, F.O.; de Sá, L.B.R.; Godoy, E.P. Distributed I/O as a Service: A Data Acquisition Solution to Industry 4.0. *HardwareX* **2022**, *12*, e00355. [CrossRef] [PubMed]
32. Fernandes, M.D.M.; Bigheti, J.A.; Pontarolli, R.P.; Godoy, E.P. Industrial Automation as a Service: A New Application to Industry 4.0. *IEEE Lat. Am. Trans.* **2021**, *19*, 2046–2053. [CrossRef]
33. Bigheti, J.A.; Fernandes, M.M.; Godoy, E. Paciencia. Control as a Service: A Microservice Approach to Industry 4.0. In Proceedings of the 2019 IEEE International Workshop on Metrology for Industry 4.0 and IoT, MetroInd 4.0 and IoT 2019, Naples, Italy, 4–6 June 2019; Institute of Electrical and Electronics Engineers Inc.: Piscataway, NJ, USA, 2019; pp. 438–443.
34. Dustdar, S.; Papazoglou, M.P. Services and Service Composition-An Introduction. *IT Inf. Technol.* **2008**, *50*, 86–92.
35. Moleculer Services | Moleculer—Progressive Microservices Framework for Node.Js. Available online: https://moleculer.services/docs/0.14/services.html#Internal-Services (accessed on 17 October 2022).
36. OpenPLC OpenPLC—Open-Source PLC Software. Available online: https://openplcproject.com/ (accessed on 30 December 2022).
37. Song, J.; Mok, A.K.; Chen, D.; Nixon, M.; Blevins, T.; Wojsznis, W. Improving PID Control with Unreliable Communications. In Proceedings of the ISA Expo Technical Conference, Houston, TX, USA, 17–19 October 2006; pp. 105–116.
38. Docker. Docker What Is a Container? Available online: https://www.docker.com/resources/what-container/ (accessed on 28 December 2022).
39. Moleculer Deploying. Moleculer—Progressive Microservices Framework for Node.Js. Available online: https://moleculer.services/docs/0.14/deploying.html (accessed on 28 December 2022).
40. Portainer Portainer Architecture—Portainer Documentation. Available online: https://docs.portainer.io/start/architecture (accessed on 28 December 2022).

eng

MDPI

Article

Are Implant Threads Important for Implant Stability? An In Vitro Study Using Low-Density Polyurethane Sheets

Luca Comuzzi [1,†], Margherita Tumedei [2,†], Natalia Di Pietro [3,4,*], Tea Romasco [3,4], Lorenzo Montesani [5], Adriano Piattelli [6] and Ugo Covani [7]

1 Independent Researcher, 31015 Conegliano, Italy; luca.comuzzi@gmail.com
2 Department of Medical, Surgical and Dental Sciences, University of Milan, 20122 Milan, Italy; margherita.tumedei@unimi.it
3 Department of Medical, Oral and Biotechnological Sciences, "G. d'Annunzio" University of Chieti-Pescara, 66013 Chieti, Italy; tea.romasco@unich.it
4 Center for Advanced Studies and Technology—CAST, "G. d'Annunzio" University of Chieti-Pescara, 66013 Chieti, Italy
5 Independent Researcher, 00187 Rome, Italy; lomonte@bu.edu
6 School of Dentistry, Saint Camillus International University of Health and Medical Sciences, 00131 Rome, Italy; apiattelli51@gmail.com
7 Tuscan Dental Institute, Versilia General Hospital, 55041 Lido di Camaiore, Italy; covani@covani.it
* Correspondence: natalia.dipietro@unich.it
† These authors equally contributed to this work.

Abstract: Different strategies are known to improve implant primary stability (PS) and the bone-to-implant contact in post-extractive conditions, such as the macro-geometry of screws and apical threads, which can enhance the mechanical characteristics. In any case, the role of the apical area design in maintaining or improving the PS, especially in low-quality bone, still remains unclear. Thus, the present study aimed at evaluating in vitro the Insertion Torque (IT), Removal Torque (RT), and Resonance Frequency Analysis (RFA) of different implant apical threads in a cylindrical (EE) and in three conical implants (T3, TAC, Intra-lock) inserted in simulated post-extraction conditions on low-density polyurethane foams of 10 and 20 pounds per cubic foot (PCF), with and without the addition of a cortical sheet of 30 PCF in density. The IT, RT, and RFA values of all the implants tested were directly proportional to the polyurethane density and to the presence of the cortical sheet, but TAC and Intra-lock implants, which had the latest-generation thread design, always showed significantly higher values (e.g., IT: 18.6 and 18.6 Ncm, RT: 10.8 and 13.7 Ncm, RFA mean: 46 and 43 ISQ, in the 20 PCF density with the cortical sheet for TAC and Intra-lock, respectively). In particular, TAC implants also reached the highest RFA values in the lowest-density foam (19 ISQ). In conclusion, the present in vitro study demonstrated that TAC and Intra-lock apical designs and macro-geometries could play a key role in determining the PS and the polyurethane-implant contact in simulated post-extraction conditions in low-density artificial bone.

Keywords: conical implants; cylindrical implants; implant apex; implant stability quotient; implant threads; insertion torque; polyurethane; post-extraction; primary stability; removal torque

Citation: Comuzzi, L.; Tumedei, M.; Di Pietro, N.; Romasco, T.; Montesani, L.; Piattelli, A.; Covani, U. Are Implant Threads Important for Implant Stability? An In Vitro Study Using Low-Density Polyurethane Sheets. *Eng* **2023**, *4*, 1167–1178. https://doi.org/10.3390/eng4020068

Academic Editor: Antonio Gil Bravo

Received: 31 March 2023
Revised: 15 April 2023
Accepted: 17 April 2023
Published: 19 April 2023

1. Introduction

Achieving an adequate primary stability (PS) is crucial for the long-term survival and success of dental implants, as well to obtain a fast implant osseointegration [1]. As a matter of fact, before proceeding with the functional loading of the dental implant, an intimate contact between the implant surface and the surrounding bone has to be established [2]. Currently, different clinical strategies have been adopted to increase PS and the bone-to-implant contact in a post-extractive condition, such as site under-preparation by means of surgical rotating drills, piezo electric or electromagnetic devices, osseodensifying burs, bone expanders, osteotomes, and cutters [3–5]. Besides these, the implant micro- and macro-geometry, the surface characteristics, the thread profile and pitch together with their cutting ability have all been reported to affect the biomechanics of the bone-to-implant contact and the bone anchorage, improving implant stability [6–8].

However, the effects of the shape of implant apical threads on osseointegration and PS are not still completely elucidated [9,10] and the relationship between the implant apical design and clinical results does not appear to be clear [11]. For example, limited data are present in literature about the influence of cutting flute implant designs on PS and on mineralized bone formation at the interface with implants can be found [12]. In their study, Hsieh et al. [11] demonstrated that self-tapping implants with self-cutting flutes may determine an improvement of implant stability, especially in immediate post-extraction sites, when the implant stability is generated by the apex anchoring of the bone only in a few mm (3–5 mm). Moreover, new implant macro-designs, also with innovative apical portions, have been proposed to improve PS and long-term implant survival and success [13].

Bone density is also recognized to represent another key factor in determining implants PS, being correlated with the bone-to-implant contact. In this regard, Misch classification [14] categorized natural bone, according to the density and the cortical and trabecular microstructure:

- D1: dense cortical bone and poor/absent trabecular bone (mandible symphysis);
- D2: cortical and dense trabecular bone tissue (mandible and anterior maxilla);
- D3: thin cortical and trabecular bone (mandible and anterior/posterior maxilla);
- D4: poor/absent cortical and thin trabecular tissues (posterior maxilla).

In recent years, the polyurethane study model has been widely used as an alternative material to perform biomechanical tests in different medical fields, and also to test dental implants [15–17]. Indeed, this artificial bone has been considered to be similar to natural bone tissue by the American Society for Testing and Materials (ASTM F-1839-08 [18]), as it presents consistent mechanical and physical characteristics that closely resemble those of different bone tissue densities, with high reliability and stability, without any special handling or human variables. In this way, easy and clinically applicable non-invasive tests could be performed in order to evaluate implant stability and the relationship between the fixture and different qualities of bone [5,6,17,19].

Therefore, assessing the influence of the shape and structure of different implant apexes and threads on the biomechanical properties of implants inserted in different polyurethane foam densities could be useful to corroborate other preclinical and clinical studies. For this reason, the aim of the present in vitro study was an evaluation of the Insertion Torque (IT), the Removal Torque (RT), and the Resonance Frequency Analysis (RFA) of different latest generation implant apical threads in cylindrical and conical implants inserted in low-density polyurethan foam sheets.

2. Materials and Methods

2.1. Dental Implants

The characteristics of the different dental implants used for this in vitro study are listed below:

- Cylindrical implants with V-shaped threads, flat apex, and an external hexagon connection (EE: 3i Osseotite, Biomax SpA, Vicenza, Italy) (Figure 1);
- Conical tapered implants with V-shaped threads, round apex, and an external hexagon connection (T3: 3i T3 Implant, Biomax SpA, Vicenza, Italy) (Figure 2);
- Conical implants with Smooth Design threads, a cutting apex with a reduced diameter, and a Cone Morse self-locking connection (TAC: Total Approach Concept, Aon Implants, Grisignano di Zocco, Italy) (Figure 3);
- Conical cutting tapered implants with the CT cutting design of the threads, an atraumatic apex with a reduced diameter, and a Cone Morse connection (Intra-lock: Intralock System Europa SpA, Salerno, Italy) (Figure 4).

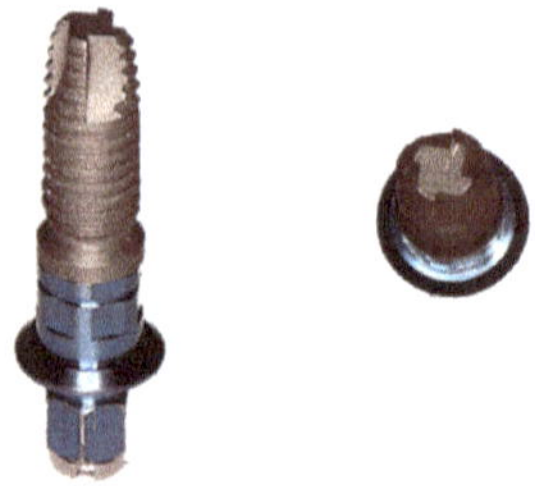

Figure 1. Details of EE implants. From the left: lateral and top (apex) views.

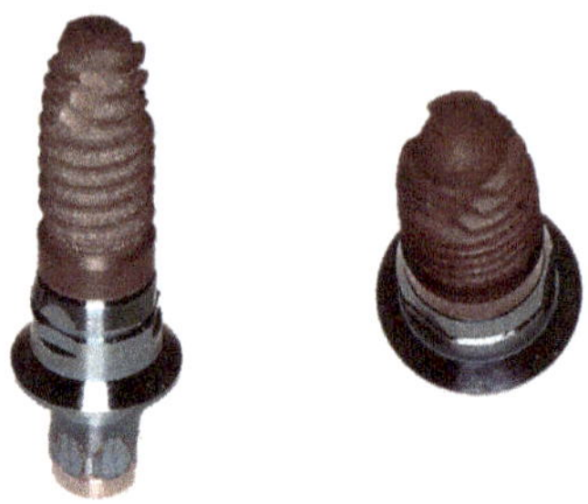

Figure 2. Details of T3 implants. From the left: lateral and top (apex) views.

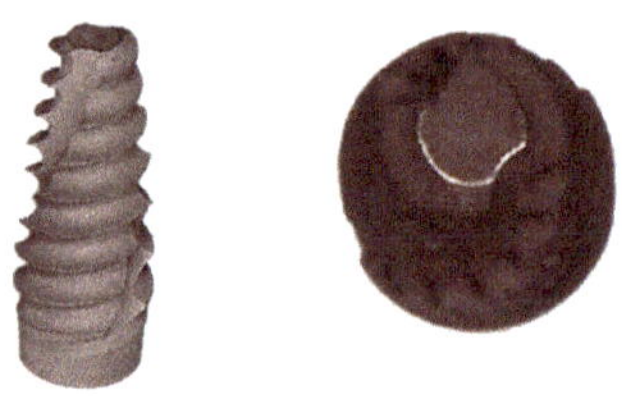

Figure 3. Details of TAC implants. From the left: lateral and top (apex) views.

Figure 4. Details of Intra-lock implants. From the left: lateral and top (apex) views.

These two latter implants had new apical threads designs. The Smooth Design of TAC threads and the reduced apex were designed to perfectly adapt to the mini-invasive preparation of the implant site and to respect as much as possible the hard tissue. On the other hand, the conical shape, reduced apex, and the Blossom threadform of Intra-lock implants have been developed to have a high cutting action, distributing the native bone over the entire implant surface, and with an ideal angulation for type III and IV bones.

All the implants used had the same dimensions: a diameter of 4 mm and a length of 10 mm. A representation of all the implants tested is reported in Figure 5.

Figure 5. Different designs (lateral and top views) of the four implants used in the present in vitro study. From the left: EE, T3, TAC, and Intra-lock implants.

2.2. Study Design

A total of 40 dental implants (10 EE, 10 T3, 10 TAC, and 10 Intra-lock) were used in this investigation to perform a total of 160 drilling sites on 10 and 20 PCF polyurethane foam sheets, with or without the presence of a cortical foam sheet of 30 PCF in density (10 osteotomies per implant for each experimental condition) (Figure 6).

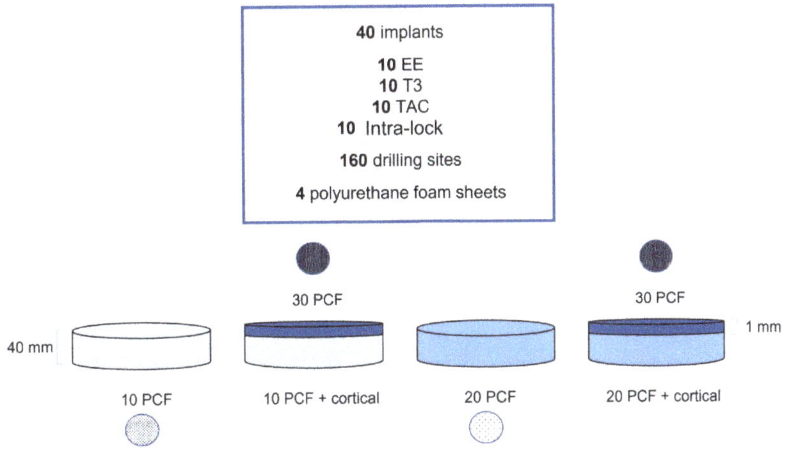

Figure 6. Study design of the in vitro experimental plan.

Different 40 mm thickness polyurethane foam sheets with densities of 10 and 20 PCF, with and without the addition of a 1 mm thickness cortical sheet with a density of 30 PCF (13 cm $\times$ 18 cm, Sawbones Europe AB, Malmö, Sweden), were used to perform this in vitro test. The 10 PCF density sheet corresponds to a density of 0.16 g/cm^3 and has been used to mime the D3 bone quality, whereas the 20 PCF density sheet has a density of 0.32 g/cm^3, similar to a D2 bone type. Moreover, a 1 mm thickness sheet with a density of 0.48 g/cm^3 (30 PCF) was added in order to simulate a layer of D1 cortical bone [15,16] (Figure 7).

Figure 7. An example of sheets of 20 PCF in density with (on the **left**) and without (on the **right**) the cortical sheet of 30 PCF in density.

The polyurethane foams are available with different proportions of cortical and trabecular components, which reproduce the bone microstructure, obtaining a realistic simulation of the clinical condition.

2.3. Drilling Protocol

The drilling and the insertion of the implants into all the experimental polyurethane sheets were conducted only for the first 4–5 mm in length, in order to properly evaluate the apical portion of all the implants (Figures 8 and 9).

Figure 8. The four implants inserted in the sheets of 10 PCF in density without (on the **left**) and with (on the **right**) the cortical sheet.

Implants were positioned in the polyurethane sheets of any densities, following the corresponding manufacturer's instructions and kit, but standardizing the use of the initial lanceolate drill at 300 rpm. After this, a 2 mm bur and a 2.8 mm bur were used for TAC and Intra-lock implants, respectively, while for EE and T3 implants a further 3 mm bur was used, always using a surgical implant motor at 300 rpm (Chiropro, Bien Air Dental

SA, Bienne, Switzerland), following the manufacturer's protocol established for critical conditions such as the present test.

Figure 9. The four implants inserted in the sheets of 20 PCF in density without (on the **left**) and with (on the **right**) the cortical sheet.

The investigation was conducted by a single operator (Dr. Luca Comuzzi), measuring the IT, RT, and RFA values of all implants tested. Particularly, the final implant insertion was performed at 30 rpm with a torque set at 50 N and then, the final 1mm IT and RT values were recorded by dynamometric analysis using a calibrated torque meter during screw positioning and removal. At the end, the RFA was measured recording the ISQ values in two different orientations at 90 degrees (bucco-lingual: BL, and mesio-distal: MD) (Figure 10), by using the SmartPeg n.78 (Ostell Inc., Göteborg, Schweden) for TAC and Intra-lock implants, and the SmartPeg n.1 (Ostell Inc., Göteborg, Schweden) for EE and T3 implants.

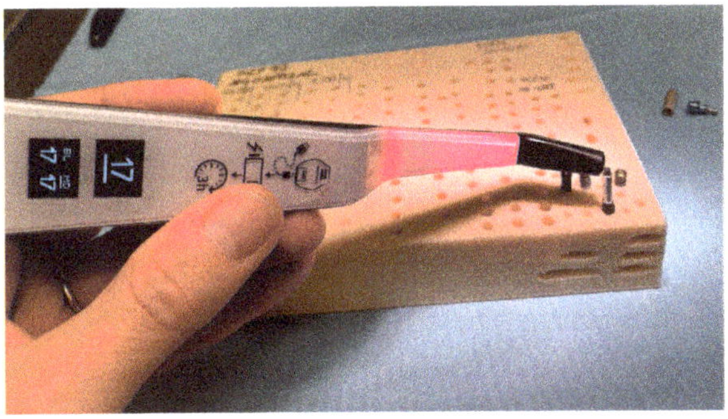

Figure 10. An example of RFA measurements (BL and MD) evaluated on a TAC implant after the insertion in the sheet of 10 PCF in density without the cortical sheet.

2.4. Statistical Analysis

Power analysis and sample size calculation were performed using the ANCOVA statistical test (effect size: 0.4, α err: 0.05; power $(1-\beta)$: 0.95; numerator df: 9; number of groups: 4; number of covariates: 4), using the program G*Power 3.1.9.7. The minimum total sample size necessary to achieve a statistically significant output was 157 implant sites. The Kolmogorov-Smirnov test was used to assess the normality distribution of data. Then, Kruskal-Wallis nonparametric test was performed to evaluate the IT, RT, and RFA values, considering a p-value < 0.05 statistically significant. These data were analyzed by using the statistical software package GraphPad 9.0 (Prism, San Diego, CA, USA).

3. Results

3.1. Insertion Torque Evaluation

The IT values of all the implants tested proportionally increased with the increase of the polyurethane density and with the presence of the cortical sheet, especially for TAC and Intra-lock implants that showed statistically significant higher values in both the thickest sheets when compared to a 10 PCF density sheet without the cortical. Indeed, the lowest values were reported by EE implants in the 10 PCF density sheet without the cortical sheet (2 Ncm), while the highest values for TAC and Intra-lock implants were in the 20 PCF density sheet with the cortical sheet (18.6 Ncm).

There was not a significant difference among IT values of the four implants tested in the same polyurethane condition, except for TAC and Intra-lock implants in comparison with EE implants when inserted in the 20 PCF density sheet with the cortical sheet, although in all the experimental conditions the first two implants have always showed higher results. Moreover, also T3 implants reported statistically significant higher values in the thickest sheet in comparison with the 10 PCF density sheet without the cortical sheet, even if the highest IT value was slightly sufficient (11.8 Ncm).

Overall, the worst results were found in EE implants when tested in all the polyurethane conditions. The highest value reached by this implant was 6.9 Ncm in the 20 PCF density sheet with the addition of the cortical sheet (Figure 11).

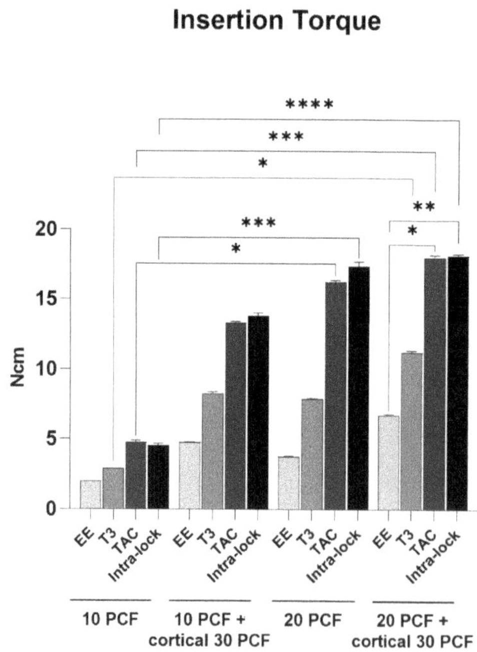

Figure 11. Evaluation of the IT values expressed by the implants tested in the different artificial bone conditions. Data are expressed as means ± Standard Error of Mean (SEM). * $p < 0.05$; ** $p < 0.01$; *** $p < 0.001$; **** $p < 0.0001$.

3.2. Removal Torque

The RT values were proportional to the sheet densities and to the presence of the cortical sheet as well. In fact, all the implants, except for EE, have shown significantly higher results in presence of the cortical sheet on the 10 and 20 PCF density sheets.

The highest values were always expressed by TAC and Intra-lock implants, with a RT of 10.8 Ncm for TAC implants and 13.7 Ncm for Intra-lock ones in the 20 PCF density sheet with the cortical sheet. On the other hand, the lowest results were always reported by EE and T3 implants, with 2 Ncm for both in the 10 PCF density sheet without the cortical sheet. In particular, Intra-lock implants showed statistically significant higher values in respect to EE implants in almost all the conditions, except for the 10 PCF density sheet without the cortical sheet, as well as TAC implants in the 20 PCF density cortical sheet without the cortical sheet.

The RT values were always lower than the corresponding IT values for all the implants tested, with the highest difference registered for TAC implants (Figure 12).

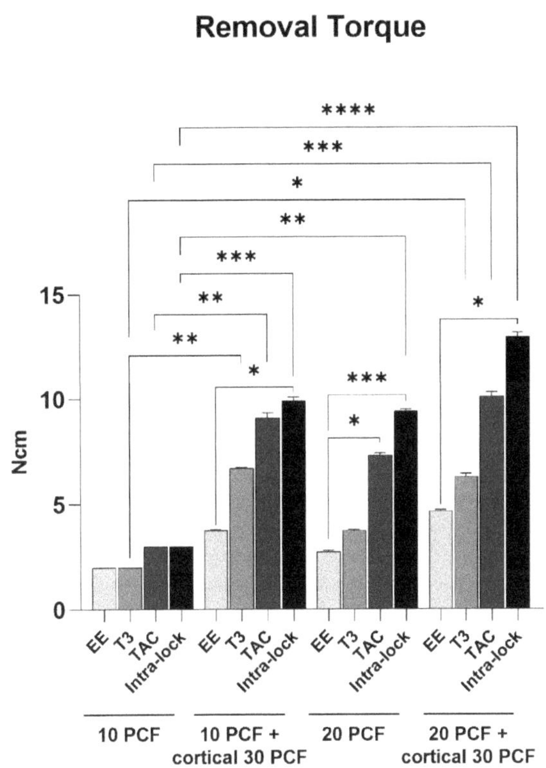

Figure 12. Evaluation of the RT values expressed by the implants tested in the different artificial bone conditions. Data are expressed as means + SEM. * $p < 0.05$; ** $p < 0.01$; *** $p < 0.001$; **** $p < 0.0001$.

3.3. Resonance Frequency Analysis

The BL and MD RFA values were expressed by the Implant Stability Quotient (ISQ). TAC and Intra-lock values never showed significant differences, but they were always higher in respect to EE and T3 implants, reaching statistical significances in the 10 PCF density with the cortical sheet (TAC implants) and the 20 PCF density sheet without the cortical sheet (Intra-lock and TAC implants) in respect to EE implants.

The highest values were reported by TAC implant in the 20 PCF density sheet with the cortical sheet (46 ISQ), while in the lowest density sheet (10 PCF without the cortical sheet), the EE implants could be not even measurable by the device.

All the ISQ values registered by the four implants in the thickest sheet were significantly higher in respect to the corresponding values found in the lowest density sheet (Figure 13).

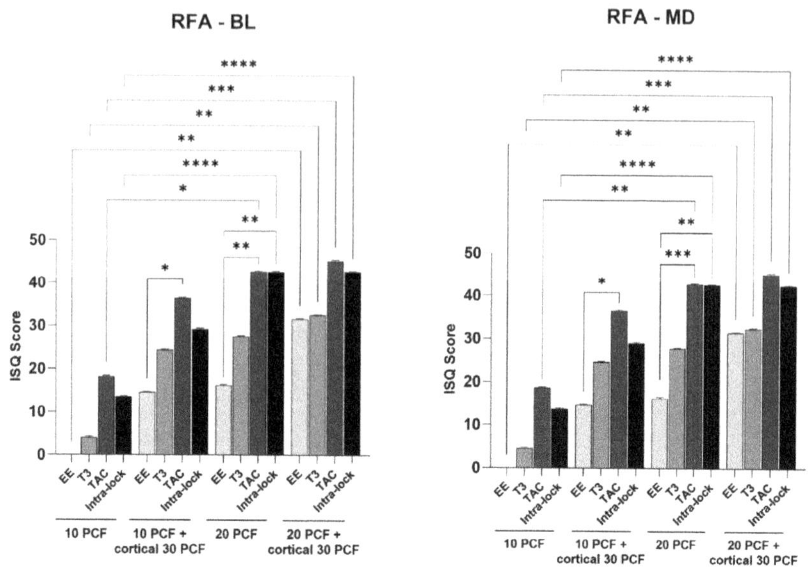

Figure 13. Evaluation of the RFA-BL (on the **left**) and RFA-MD (on the **right**) values expressed by the implants tested in the different artificial bone conditions. Data are expressed as means ± SEM. * $p < 0.05$; ** $p < 0.01$; *** $p < 0.001$; **** $p < 0.0001$.

4. Discussion

The implant design and the surgical technique are important parameters that can be improved, possibly influencing implant stability [4]. Nowadays several novel implant designs that may facilitate the achievement of initial implant PS have been proposed [13]. Nevertheless, the effect of the apical portion on the initial implant stability has not been completely elucidated yet [9,10,20].

In relation to this, homogeneous synthetic solid rigid polyurethane foam sheets have been produced as an alternative test medium for human cancellous/cortical bone. Even though polyurethane foam cannot perfectly replicate the same human bone structure, the ASTM F-1839-08 standard stated that it exhibits mechanical properties in the range of human cancellous bone. Artificial bones are regularly used to perform pre-clinical tests on dental implants and several studies have already shown findings on the influence of the threads design on PS [15,16,21].

For example, Romanos et al. [20] in a similar study on polyurethane foam sheets reported that the apical implant stability contributed to nearly 30–43% of the entire implant stability and thus it was clear that the apical portion of an implant may play a fundamental role. In particular, their study evaluated the role of the apical threads design on implant stability after being placed in type II dense artificial bone (D2). The implants were divided into two groups: a test group (with the aim to evaluate the apical stability of implants inserted with only 3 mm of the apical portion into the artificial bone and without lateral contact between the residual implant length and the osteotomy walls) and a control group (for the assessment of full implant stability with implants inserted with full contact between the osteotomy walls and the implant surface). Another polyurethane in vitro study was conducted by Sciasci et al. [22] to evaluate the mechanical behavior of geometric changes in the apex region of dental implants. Implants without apical cut (Wc), apical bi-split cut, apical tri-split cut, apical quadri-split cut (Qs), and without any modifications, have been evaluated. At the end it was demonstrated that implants without any modifications and Qs had shown a significant increase of the IT, but those without any modifications and Wc had the greatest ISQ values. In this way, the authors concluded that the proposed geometries at the apical third of implants have greatly influenced their IT and PS. In animal experimental studies as well [8], it was reported that the apical portion design of implants had an effect on PS and on the amount of mineralized bone present at the implant-bone interface. In particular, immediately loaded fully tapered implants with deep apical threads had similar high survival rates and bone-to-implant contact when inserted in healed sites and fresh extraction sockets [13].

In the present study, T3 conical tapered implants seemed to have an insufficient IT and PS in all the experimental conditions, reporting higher results only in the 20 PCF density sheet with the cortical sheet (11.8 Ncm and 33 ISQ for the IT and RFA, respectively) and, together with EE cylindrical implants, registering slightly measurable results in the lowest density sheet. Indeed, the EE macro-geometry appeared to not properly suit the post-extraction condition and thus to neither allow implant stabilization nor the correct implant positioning into the osteotomy. All the tested measurements confirmed this data with very low IT, RT, and RFA values in all the artificial bone densities. The cylindrical design and the apical shape of these implants may be more suitable for higher mandibular symphysis densities (D1–D2) and less for low-quality and quantity of bone, like in post-extraction sites, where implant orientation is also important to consider. The T3 implant shape helped these implants to improve some parameters showed by the EE design, but the round apex did not suit with the post-extraction condition, because a spinning effect that stripped the osteotomy bottom was noticed during the insertion in the 20 PCF density sheets, when the operator had to force the implant insertion, thereby jeopardizing the IT and RFA values. Thus, a high cutting design was needed to increase implant penetration without stripping and spinning, and without the loss of all the mechanical characteristics. Indeed, in a finite elements study on the possible relationship between implant apex and maxillary sinus floor, Yan et al. [23] showed that the different positions of the apex (in contact with, piercing or breaking through bone) had an important effect on the implant PS.

Moreover, other clinical studies investigated the influence of a cutting flute shape on implant stability, reporting a higher PS for implants without cutting flutes and an important effect to withstand lateral loads with the straight flute design [11], whereas other ex vivo works compared the PS of implants with apically located screw threads or a flat tip, demonstrating a significantly higher stability for implants with apically located screw threads [9]. These and other authors [8,10] concluded that factors, such as implant geometry (also at the implant tip), should be taken into account for biomechanical evaluations. In this regard, the present in vitro study observed a clear different behavior between old generation apical shapes and the latest generation ones tested in post-extraction conditions. In particular, the threads design and the conical macro-geometries of TAC and Intra-lock implants showed a better stability in anchoring the polyurethane material (up to 46 ISQ), even during the first inserting rotations, and the particular tip designs resulted to be favorable in the management of the direction, stability, and depth of the implants.

Jimbo et al. [12] evaluated implants with two different apical geometries, one with a cutting flute design and the other with a self-tapping design. These authors found that implants with the modified cutting flute design had a significantly reduced IT with an improved organization of the peri-implant bone in an animal study. In the present work, the RFA values were proportional to the IT values, thus TAC and Intra-lock implants always presented a higher PS, even if in the lowest density sheet the apical portion was able to direct the insertion of the implant, but not to guarantee a good initial stability and an intimate contact of the fixture to the artificial bone (19 and 14 ISQ for TAC and Intra-lock, respectively). In any case, in the other experimental conditions their apical and threads designs allowed a fast and efficient implant insertion and a good contact with the material, progressively increasing the IT with the thickness of the sheet (up to 18.6 Ncm in the 20 PCF density sheet with the cortical sheet). The RT values were always lower than the corresponding IT with higher results for Intra-lock implants (up to 13.7 Ncm in the 20 PCF density sheet with the cortical sheet). Overall, although the only apical portion of TAC and Intra-lock implants did not guarantee an adequate ISQ value to allow the immediate loading of these implants, it showed an ISQ score that adequately favored the implant insertion until its final placement and reduced micromovements, which may promote osseointegration. These findings could be clinically relevant, because they proved that the apical morphology allows a good bone-to-implant contact from the first mm of insertion, without compromising the artificial bone structure, stressing the implant connection or the insertion at high IT values.

Nevertheless, the in vitro nature of this study evidently constitutes a strong limitation, although it provided substantial in vitro mechanical knowledge on the influence of different apical and thread implant shapes on PS. There is need of further experimental animal and clinical studies to corroborate these results on implant stability with histomorphometric and biomechanical measurements that also reflect the condition of the peri-implant bone [10,11,13]. In addition, the IT, RT, and RFA values expressed by different implants may be influenced in vivo by the biological conditions and by multiple variabilities that could affect the density and nature of the bone, such as the presence of graft materials or the concurrent medical conditions of the patient affecting bone density.

In future, it would be interesting to verify if high ISQ values could be detected and maintained also in clinical situations up to the first month, a crucial moment for the biological stability, when ISQ values could lower and sometimes lead to implant stability failures. This data could allow the immediate loading of implants only by means of the RFA measurement, as an alternative to the current prosthetic loading protocols.

5. Conclusions

In conclusion, the higher ISQ values detected for TAC and Intra-lock implants placed in simulated post-extraction conditions showed a higher and adequate stability with the only latest generation apical cutting properties, even if they may not allow such implants to be loaded immediately after placing in clinical situations. On the other hand, the low ISQ values registered by T3 and EE groups did not demonstrate the best mechanical stability, to be even not detectable in case of EE implants.

The present in vitro polyurethane experimental study shed light on the importance of the apical implant thread design in determining the entire implant stability in simulated post-extraction sites and in guaranteeing the implant anchorage at the apical area.

Author Contributions: Conceptualization, L.C. and U.C.; methodology, L.C. and M.T.; validation, L.M., N.D.P. and A.P.; formal analysis, T.R. and L.M.; investigation, L.C.; resources, L.C. and U.C.; data curation, T.R. and M.T.; writing—original draft preparation, A.P., M.T., T.R. and N.D.P.; writing—review and editing, N.D.P. and T.R.; visualization, U.C. and A.P.; supervision, L.C.; project administration, L.C. and U.C. All authors have read and agreed to the published version of the manuscript.

Funding: This research received no external funding.

Institutional Review Board Statement: Not applicable.

Informed Consent Statement: Not applicable.

Data Availability Statement: Data is contained within the article and available on request from the corresponding author.

Conflicts of Interest: The authors declare no conflict of interest.

References

1. Calvo-Guirado, J.L.; Delgado-Peña, J.; Maté-Sánchez, J.E.; Mareque Bueno, J.; Delgado-Ruiz, R.A.; Romanos, G.E. Novel Hybrid Drilling Protocol: Evaluation for the Implant Healing-Thermal Changes, Crestal Bone Loss, and Bone-to-Implant Contact. *Clin. Oral Implants Res.* **2015**, *26*, 753–760. [CrossRef] [PubMed]
2. Colnot, C.; Romero, D.M.; Huang, S.; Rahman, J.; Currey, J.A.; Nanci, A.; Brunski, J.B.; Helms, J.A. Molecular analysis of healing at a bone-implant interface. *J. Dent. Res.* **2007**, *9*, 862–867. [CrossRef] [PubMed]
3. Maglione, M.; Bevilacqua, L.; Dotto, F.; Costantinides, F.; Lorusso, F.; Scarano, A. Observational Study on the Preparation of the Implant Site with Piezosurgery vs. Drill: Comparison between the Two Methods in Terms of Postoperative Pain, Surgical Times, and Operational Advantages. *Biomed. Res. Int.* **2019**, *2019*, 8483658. [CrossRef]

4. Gehrke, S.A. Evaluation of the Cortical Bone Reaction Around of Implants Using a Single-Use Final Drill: A Histologic Study. *J. Craniofac. Surg.* **2015**, *26*, 1482–1486. [CrossRef]
5. Gehrke, S.A.; Guirado, J.L.C.; Bettach, R.; Del Fabbro, M.; Martínez, C.P.A.; Shibli, J.A. Evaluation of the Insertion Torque, Implant Stability Quotient and Drilled Hole Quality for Different Drill Design: An in Vitro Investigation. *Clin. Oral Implants Res.* **2018**, *29*, 656–662. [CrossRef] [PubMed]
6. Falco, A.; Berardini, M.; Trisi, P. Correlation Between Implant Geometry, Implant Surface, Insertion Torque, and Primary Stability: In Vitro Biomechanical Analysis. *Int. J. Oral Maxillofac. Implants* **2018**, *33*, 824–830. [CrossRef]
7. Eliasa, C.N.; Rocha, F.A.; Nascimento, A.L.; Coelho, P.G. Influence of Implant Shape, Surface Morphology, Surgical Technique and Bone Quality on the Primary Stability of Dental Implants. *J. Mech. Behav. Biomed. Mater.* **2012**, *16*, 169–180. [CrossRef] [PubMed]
8. Gehrke, S.A.; Pérez-Albacete Martínez, C.; Piattelli, A.; Shibli, J.A.; Markovic, A.; Calvo Guirado, J.L. The Influence of Three Different Apical Implant Designs at Stability and Osseointegration Process: Experimental Study in Rabbits. *Clin. Oral Implants Res.* **2017**, *28*, 355–361. [CrossRef]
9. Staedt, H.; Heimes, D.; Lehmann, K.M.; Ottl, P.; Bjelopavlovic, M.; Wagner, W.; Al-Nawas, B.; Kämmerer, P.W. Does the Modification of the Apical Geometry of a Dental Implant Affect Its Primary Stability? A Comparative Ex Vivo Study. *Materials* **2021**, *14*, 1728. [CrossRef]
10. Gehrke, S.A.; Eliers Treichel, T.L.; Pérez-Díaz, L.; Calvo-Guirado, J.L.; Aramburú Júnior, J.; Mazón, P.; De Aza, P.N. Impact of Different Titanium Implant Thread Designs on Bone Healing: A Biomechanical and Histometric Study with an Animal Model. *J. Clin. Med.* **2019**, *8*, 777. [CrossRef]
11. Hsieh, M.C.; Huang, C.H.; Hsu, M.L. Effect of Cutting Flute Design Features on Primary Stability of Immediate Implant Placement and Restoration: A Dynamic Experimental Analysis. *Med. Biol. Eng. Comput.* **2023**, *61*, 475–484. [CrossRef] [PubMed]
12. Jimbo, R.; Tovar, N.; Marin, C.; Teixeira, H.S.; Anchieta, R.B.; Silveira, L.M.; Janal, M.N.; Shibli, J.A.; Coelho, P.G. The Impact of a Modified Cutting Flute Implant Design on Osseointegration. *Int. J. Oral Maxillofac. Surg.* **2014**, *43*, 883–888. [CrossRef] [PubMed]
13. Hadaya, D.; Pi-Anfruns, J.; Bellon, B.; Pippenger, B.E.; Aghaloo, T.L. Immediate Loading of a Fully Tapered Implant with Deep Apical Threads Placed in Healed Alveolar Ridges vs. Immediate Extraction Sockets. *Clin. Oral Implants Res.* **2022**, *33*, 501–510. [CrossRef] [PubMed]
14. Misch, C.E. Bone Density: A Key Determinant for Clinical Success. *Contemp. Implant Dent.* **1999**, *8*, 109–118.
15. Di Stefano, D.A.; Arosio, P.; Gastaldi, G.; Gherlone, E. The Insertion Torque-Depth Curve Integral as a Measure of Implant Primary Stability: An in Vitro Study on Polyurethane Foam Blocks. *J. Prosthet. Dent.* **2018**, *120*, 706–714. [CrossRef]
16. Tsolaki, I.N.; Tonsekar, P.P.; Najafi, B.; Drew, H.J.; Sullivan, A.J.; Petrov, S.D. Comparison of Osteotome and Conventional Drilling Techniques for Primary Implant Stability: An In Vitro Study. *J. Oral Implantol.* **2016**, *42*, 321–325. [CrossRef]
17. Comuzzi, L.; Tumedei, M.; Romasco, T.; Petrini, M.; Afrashtehfar, K.I.; Inchingolo, F.; Piattelli, A.; Di Pietro, N. Insertion Torque, Removal Torque, and Resonance Frequency Analysis Values of Ultrashort, Short, and Standard Dental Implants: An In Vitro Study on Polyurethane Foam Sheets. *J. Funct. Biomater.* **2023**, *14*, 10. [CrossRef]
18. *ASTM F-1839-08*; Standard Specification for Rigid Polyurethane Foam for Use as a Standard Material for Testing Orthopedic Devices and Instruments. ASTM International: West Conshohocken, PA, USA, 2021.
19. Arosio, P.; Arosio, F.; Di Stefano, D.A. Implant Diameter, Length, and the Insertion Torque/Depth Integral: A Study Using Polyurethane Foam Blocks. *Dent. J.* **2020**, *2*, 56. [CrossRef]
20. Romanos, G.E.; Delgado-Ruiz, R.; Kuyunov, O.; Sacks, D.; Calvo-Guirado, J.L. Apical Stability of Implants with Progressive Thread Design in Vitro, Based on Clinicians with Different Levels of Experience. *J. Periodontol.* **2019**, *90*, 1320–1324. [CrossRef]
21. Comuzzi, L.; Tumedei, M.; De Angelis, F.; Lorusso, F.; Piattelli, A.; Iezzi, G. Influence of the Dental Implant Macrogeometry and Threads Design on Primary Stability: An in Vitro Simulation on Artificial Bone Blocks. *Comput. Methods Biomech. Biomed. Eng.* **2021**, *24*, 1242–1250. [CrossRef]
22. Sciasci, P.; Casalle, N.; Vaz, L.G. Evaluation of Primary Stability in Modified Implants: Analysis by Resonance Frequency and Insertion Torque. *Clin. Implant Dent. Relat. Res.* **2018**, *20*, 274–279. [CrossRef] [PubMed]
23. Yan, X.; Zhang, X.; Chi, W.; Ai, H.; Wu, L. Association between Implant Apex and Sinus Floor in Posterior Maxilla Dental Implantation: A Three-Dimensional Finite Element Analysis. *Exp. Ther. Med.* **2015**, *9*, 868–876. [CrossRef] [PubMed]

eng

MDPI

Article

Transit Behaviour and Sociodemographic Interrelation: Enhancing Urban Public-Transport Solutions

Aisha Hasan [1], Umair Hasan [1,*], Hamad AlJassmi [1,2] and Andrew Whyte [3]

1 Emirates Centre for Mobility Research, United Arab Emirates University, Al Ain P.O. Box 15551, United Arab Emirates; aysha.u.hasan@gmail.com
2 Department of Civil and Environmental Engineering, United Arab Emirates University, Al Ain P.O. Box 15551, United Arab Emirates
3 School of Civil and Mechanical Engineering, Curtin University, Perth, WA 6845, Australia
* Correspondence: umair.hasan@uaeu.ac.ae

Abstract: Recent years have seen a considerable shift in the focus of public investment agencies from extensive roadway networks to a more planned approach that meets environmental, cost, and social dimensions more aptly. Past research has mainly explored the engineering aspect and cost parameters, while the human or social component is often neglected. This study aims to identify the trip-making behaviour of residents in an urban area towards bus transport network enhancement. Abu Dhabi, the location of study, is heavily dependent upon car travel, creating much congestion, which the local government seeks to address by enhanced public transport. This work examined eight public-transport routes in two zones, with data collected on both weekdays (n = 751) and weekends (n = 769). Multinomial logistic regression models showed that respondents highlighted overcrowded buses and traffic congestion as two of the main hurdles pertinent to urban routes in the bus network influencing their mode choice. Proposals pertinent to the local authority for further consideration need to factor in current low satisfaction with bus transit network coverage, low satisfaction with the quality of bus rides, inhibiting a mode shift from cars/taxis towards buses, cumulative income profiles of public-transport users, with findings that the low-income bracket is already at saturation, and that reducing congestion needs innovative (sociodynamic rather than technical road network) public-transport solutions.

Keywords: public-transport solutions; sociodemographic parameters; travel satisfaction; transport quality; low-income communities

Citation: Hasan, A.; Hasan, U.; AlJassmi, H.; Whyte, A. Transit Behaviour and Sociodemographic Interrelation: Enhancing Urban Public-Transport Solutions. *Eng* **2023**, *4*, 1144–1155. https://doi.org/10.3390/eng4020066

Academic Editor: Antonio Gil Bravo

Received: 12 March 2023
Revised: 12 April 2023
Accepted: 12 April 2023
Published: 14 April 2023

1. Introduction

The situation of urban roadway asset management is complicated by the social, environmental, political, and budgetary constraints of transportation agencies, making sustainability the primary concern [1]. Current research on life-cycle assessment of roadways has focused on either material type or pavement overlay (e.g., Hasan et al. [2], Santero et al. [3] and AzariJafari et al. [4]). The policies and decisions involved with investments on transportation infrastructures interact with exogenous variables of urban density, traffic congestion due to specific roadway design (e.g., single vs. multiple carriageways, intersections and length, and public-transport lanes), and number of lanes [5]; however, complimentary facilities of on-street parking and adjacent parking zones [6] must also be considered. Shoup [6] and Hawas et al. [7] noted that the decision of commuters to choose between private vehicles and public transport is affected by these factors. The choice of travellers to lean towards any mode choice, regardless of its private or shared nature, is affected by the trip purpose, perceived service quality, and service attributes, which are sensitive to the individuals and are largely impacted by the transit options available in a region, social and cultural norms, and the trade-off between perceived service and quality [8]. Mass-transit systems also open a window of opportunity for any urban area to reduce its transportation-related cost and environmental burdens [9]. Yet, due to the high investment cost requirement and involvement of the public, administrative agencies demand that the perception of transportation system functionality and attributes is an integral part of planning.

A number of studies [10,11] have suggested that the attitudes of people towards transportation systems are increasingly becoming complex as their understanding of daily commute, safety, travel duration, and ride quality change. This is especially significant since transport authorities have to optimise the public system against multiple competing private, short-term rental, shared, and on-demand options. In a changing transit mode choice environment, micro-mobility integration can affect public-transport attraction compared to other modes, particularly in the wake of a return to normalcy after the lifting of COVID-19-related mobility restrictions, where increased mobility is expected [12]. The importance of an accessibility-based approach in this context was recently explored in a study by Ali et al. [13], which highlighted that focusing on accessibility to plan transport solutions is especially significant for resilient transport planning, and that transit solutions should be gauged through travel time.

Nonetheless, the association between public-transit accessibility and usage frequency is not a recent topic in transport policy research and has been addressed in multiple studies soliciting usage patterns and underlying contributory variables from questionnaire surveys. As one of the earliest explorations, an empirical study on the association between travel behaviour of urban travellers and scale of the urban neighbourhood was conducted by Krizek [14]. They found that the reported vehicle miles travelled were reduced if the accessibility of the neighbourhood increased. Another study focusing on the effect of urban form on the variation of travel behaviour was conducted by Pan et al. [15] in four selected Shanghai neighbourhoods. They proposed that, if neighbourhoods are designed with denser street networks, the increased reliance on private vehicle travel induced by higher incomes may be replaced by bicycle/pedestrian trips.

The reasons for a traveller to choose to drive to work in a small urban English area were investigated by Gardner and Abraham [16] through 19 semi-structured interviews with private vehicle users. They found that the decision may be primarily driven by monetary costs, effort minimisation, origin destination, and time spent in journey as the public transport was perceived to be comparatively slower. Delays in the public-transport system and lack of reliance on its schedule, strikes, and perceived safety were noted as critical factors. Private vehicle commuters also highlighted the importance of the public-transport system in dealing with the problem of parking, which has been noted by researchers [17,18] as responsible for traffic congestion, as well as the cost and environmental burdens of a roadway system.

Another early qualitative study to identify the mode choice attitude of car and public-transport users was conducted by Beirão and Cabral [8]. They found that the mode choice is affected by situational variables, perceived performance, journey type, and user lifestyle and characteristics. They proposed that the policy-making process should accommodate customer expectations so that the usage of public-transport systems can be increased. In the context of transitioning economies, Grdzelishvili and Sathre [19] investigated the travel behaviour of Tbilisi residents. They identified perceived safety, comfort, frequency, and time as the most important factors that tended to skew the survey respondents towards private vehicle ownership and usage.

The service quality attributes of public transport and the social dynamics associated with car use and ownership factors were also found to influence the travellers' mode choice in a study by Javid et al. [20], where it was also observed that, in order for a public-transport authority to motivate more users towards public bus transport, survey responses and service quality attribute optimisation should be conducted. In a follow-up study [21], the authors argued that, in a mixed-mode use environment where multiple competitive transport choices including private cars, public bus transport, shared bus transport and shared taxis, and car-rental services are present, it is absolutely imperative to analyse the user satisfaction according to service quality attributes. Data on bus transport service riders using flexible on-demand service options of two app-based bus services were collected using a questionnaire survey. The authors found that waiting time at the bus stop, income profile and profession of the travellers, vehicle ownership, and trip purpose were significant predictors of mode choices by applying factor analysis and structural Equation modelling techniques. The study also noted that a positive perception of the users affects their usage tendency for bus transport. Students and privately employed individuals were more inclined to use bus services, whereas increases in bus-stop waiting time and travel time, as well as low coverage/accessibility, negatively affected usage.

The City of Abu Dhabi has witnessed an increase in population, accompanied by an increasing dependence of commuters on private vehicle use, resulting in traffic congestion in urban Abu Dhabi localities [22]. Most studies in the country have focused on road transport from the infrastructural [23,24], environmental [25,26], cost [27,28], or traffic safety [29] perspectives, while some [30] extended the research to operational and facility management issues for providing connected pathways from urban communities to city centres and central business hubs and districts. However, the above-referred studies hinted that the urban transportation network cannot be solely evaluated by conventional cost and environmental aspects; a social aspect should also be considered. However, very few studies have addressed the travel behaviour and perception of urban public-transport network in Abu Dhabi city.

The literature review conducted above highlights that, in order to promote public transport among urban travellers in a mixed-mode choice environment, the critical variables of service quality, accessibility, and travel time need to be optimised. However, the definition of the former two attributes is scattered in literature and regional-dependent, including aspects such as safety, onboard facilities, connection (waiting stops/stations) facilities, and cost for service quality and network coverage, service frequency, seating, and community inclusion for accessibility. Additionally, most studies were conducted in the European region, where the public-transport system is well-developed, and car-free precincts exist that regulatorily and culturally promote sustainable transit options (walking and public transport) over car use. Studies conducted in the developing world either focused on on-demand bus mobility options or did not include income- and employment-related variables, which affect the affordability and lack of choice parameters, potentially rendering travellers incapable of choosing costlier options over public-transport services. Abu Dhabi provides a unique opportunity to investigate the travel mode choice patterns of multicultural residents belonging to different income groups and sociodemographic categories in a highly developed infrastructural yet car-dependent urban setting, where enhanced public-transport planning informed by public preferences can trigger a positive shift. This study attempts to address this by soliciting public responses on a questionnaire survey to establish a clear definition of service quality, accessibility, and travel time attributes for Abu Dhabi (and, by extension, the multicultural countries in the Gulf Cooperation Council), which can result in public-transport uptake, as well as produce insights into the various sociodemographic classes that utilise this mode for their transit needs.

2. Method

This work was conducted to form the basis of a pilot study exploring the application of innovative mass transit over the lifecycle of a transportation infrastructure asset. The strategy of this study was intended to primarily focus on capturing the use of public transport, specifically bus transport, in the urban area of Abu Dhabi. Travel behaviour, user demographics, attitude towards travel, and trip distribution are emphasised.

2.1. Questionnaire Design, Bus Routes, and Sample Size Selection

The questionnaire used for data collection was designed for soliciting the travel information of bus users, perception of existing bus network, demographic profile of the service benefiters, and their respective attitude towards travelling attributes: network coverage, quality and satisfaction, perception of congestion, and potential improvement strategies. The questionnaire was limited to 11 multiple-choice questions designed to take less than 5 min. The detailed questionnaire including the sub-questions and options is provided in the Supplementary Materials. The survey was administered using the CAPI-based surveying methodology for on-site data collection. Teams of multilingual surveyors administrated questionnaires in both Arabic and English, along with the capabilities to assist passengers from various ethnic and lingual backgrounds.

The purpose of study was to target the steady growth in urban Abu Dhabi; as such, according to the DoT observations, the area between Corniche and Hazaa Bin Zayed the First Street was selected. To increase the range of collected samples, interviews were conducted on both weekdays and weekends during two 8 h shifts. The collected data were collected and tabulated using MS Excel. A total of 769 interviews on the weekend and 751 interviews on weekdays were completed, and all responses with missing or incomplete responses were disregarded according to the exclusion criteria set by the local transport authority responsible for data collection, curation, and management.

2.2. Data Analysis

An analysis was performed of the survey data collected for the Public-Transport Usage Study of the Abu Dhabi Department of Transport, as part of the Abu Dhabi government's initiatives to reduce travel dependency on cars and reduce the increasing traffic congestion problems currently being observed in the city.

Logical checking of data consistency was performed of the raw data MS Excel files to address data sparseness, outliers, and missing data. Interlinking of passenger demographics against travel attributes resulted in minor data revisions. The revised data were broken down into three different sections: distribution of generated trips for each mode (i.e., bus and car travel) and current level of network coverage. The literature review section highlighted that accessibility, service quality, and travel time affect the mode choice of travellers; however, the exact distribution and inclusion of variables within these attributes differed across studies. In order to investigate this further and estimate the effect of including service frequency, perceived congestion, onboard situation, trip purpose, and bus-stop and coverage facilities on public-transport uptake over competing modes (particularly private cars), four different regression models were tested with two primary objectives: if inclusion of the variables in TSC, TA, and SDV blocks improved model fit, and which parameter is a strong predictor.

Three different variable blocks were created for statistical analysis, with the variable description presented in Table 1 and comparison method explained in the last three rows. The analysis was conducted in SPSS v22. Models were controlled for the reference category in ordinal regression analysis (i.e., very satisfied for NetCovSat, five or more times a week for FBT, and first time for FCT). Reference categories were selected to identify the comparative influence of independent variable blocks on decreasing the satisfaction level of the public-transport system and increasing reliance on travel by taxis and private cars.

Table 1. Summary of the variables selected and their respective nomenclature.

Variables	Coding
	Dependent variables (DVs)
Frequency of bus travel	FBT
Frequency of car travel	FCT
Network coverage satisfaction	NetCovSat
	Independent variables (IDVs)
Transport system characteristics (TSC)	

Table 1. *Cont.*

Variables	Coding
Time to the bus stop	TimeBusS
Quality of ride	QoR
Journey time	JourT
Frequency of buses	FreqofBuses
Travel attributes (TA)	
I like to live close to where I work	C_WRK
I like to live close to my family and friends	C_FF
Travel by bus is the easiest way for me	B_ESY
I cannot afford to travel by car or taxi	NOCAR
The buses are too crowded	B_CRWD
I would pay more to travel by bus if I always had a seat	B_ST
My journey is often delayed by traffic congestion	TR_CONG
Sociodemographic variables (SDV)	
Nationality	Nat
Income	Inc
Age	Age
	Model hypothesis improvement test
	$\lvert -2$ log likelihood of TSC and TA $\rvert - \lvert -2$ log likelihood of TSC $\rvert$
	$\lvert -2$ log likelihood of TSC, TA, and SDV $\rvert - \lvert -2$ log likelihood of TSC and TA $\rvert$

3. Results and Discussion

The results of the analysis were tabulated in separate sheets according to respective occurrence on the weekdays and weekends. Statistical analysis suggests that the majority (57%) of the survey respondents were South Asians, regardless of weekday or weekend. Moreover, the younger (i.e., 25–34 years old; 48% for both weekdays and weekends) male population (weekdays: 86% and weekends: 89%) largely working in the full-time workforce formed the largest (83%) proportion of the respondents. According to the previously recorded statistical distribution of the Abu Dhabi city residents, these results are representative of the local population, which is predominantly (62%) male in the under 34 years old (66%) age group, with over 50% being of South Asian descent [31,32]. The income profile captured in the survey showed that the majority earned a gross monthly salary of AED 1000–5000, which is also in line with the findings of these previous statistical studies which found the majority to be full-time workers earning an average monthly salary of AED 3500.

Regarding the statistical response distribution of the qualitative data variables, the majority perceived bus travel as an uneasy transit mode, yet found them to not be very crowded; however, respondents were unsatisfied with the current distribution of the bus stops on the surveyed networks as they reported spending over 15 min to reach the nearest stop. Additionally, the majority had a neutral perception of current travel time while using public bus services and either had a good or neutral perception of the current conditions of bus stops.

To address the research question of travel behaviour patterns and what variables define service quality, accessibility, and the eventual mode choice, three multinomial dependent variables (MDVs) were identified: frequency of bus travel (FBT), frequency of car travel (FCT), and network coverage satisfaction (NetCovSat). NetCovSat was originally recorded on a Likert-type scale in the order of decreasing likeability of the DV, whereas FCT and FBT were arranged with "1" representing more frequent travel (i.e., five or more times a week) and "6" representing the least travel first time. The percentage distribution of respondents on each scale was used to reverse-recode FBT and NetCovSat so as to represent a higher occurrence with increasing numeral order.

The probability of mode choice for a traveller was affected by several parameters and factors of transportation system characteristics, travel attributes, and sociodemographic variables, as shown in the multistage multinomial logistic regression models summarised in Table 2. Results from the weekday analysis are presented first, followed by the weekend analytical analysis results. Similarly, variables from each block were carried forwards to subsequent analysis except for SDV-independent variable block, which was separately performed, and the variables were then added in the logistic regression equations to perform the final analysis.

Odds ratios, i.e., the probability that a certain variable may influence the outcome of the model when all other variables are controlled, as well as model fit and significance level, of the regression models for recorded polychotomous variables are provided in Table 2.

3.1. Factors in Traveller Satisfaction from Public-Transport Network Coverage

Analysis results showed strong correlation between the transportation system characteristics and satisfaction of public-transport system users, as also reported in the literature. Results were generally similar across weekdays and weekends. Distance to nearby bus stop was only negligibly identified as an obstacle across all four variable blocks, with the OR remaining in the range of 0.98–1.004, signifying a relatively unimportant association. Strong correlation of traveller satisfaction with frequency of buses and network coverage was also noticed, with odds ratios < 1 (Table 2) for all variable blocks with high significance, implying that, as users perceived buses to be more frequent, the probability of respondents being satisfied with the network coverage also increased. These results are partially supported by findings from similar cultural contexts in literature, where service frequency [33] and network coverage [34] were found to define the accessibility parameter, yet the direct correlation between parameters was not estimated.

The second main concern of public-transport users was the journey time, where increasing satisfaction with time spent on a trip was associated with a higher rating of network coverage (OR $\approx$ 0.586–0.642 > 1). Figure 1 also shows this strong association, whereby 37% of weekday users and 48% of weekend users were satisfied with the coverage of the public-transport network. Most respondents (40% weekday, 53% weekend) were satisfied with the frequency of buses and journey time (47% weekday, 53% weekend). In general, public-transport users were more satisfied with network as these two factors became more satisfactory. These results somewhat comply with the findings of Gibson et al. [35] which compared rapid bus lanes against mixed traffic, finding that savings in the user time represented one of the most important benefits, and that its relation with network coverage was similar to service frequency following response curves that displayed an exponential or power model style trend.

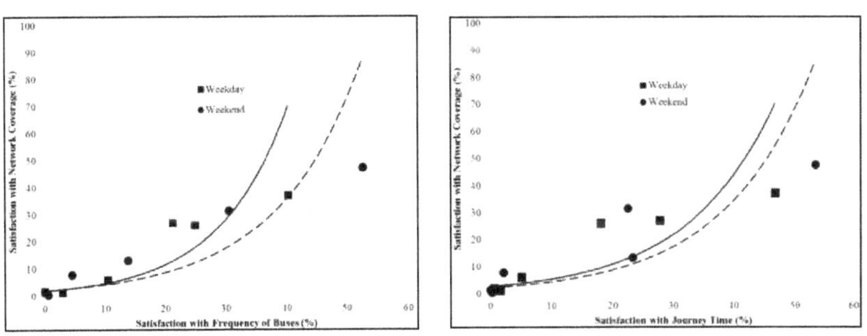

Figure 1. Correlation between percentile distribution of respondent satisfaction with journey time and public-transport network coverage (*solid line = weekends trendline, dotted line = weekdays trendline*).

Table 2. Multistage multinomial logistic regression equations for each dependent variable with different independent variable blocks.

Statistical Summary						Weekdays				Weekends			
Number of independent variable blocks						5				5			
Number of cases (n)						751				769			
Dependent variables		Network coverage satisfaction				Frequency of car travel				Frequency of bus travel			
Multinomial regression model blocks		TSC OR (CI)	TSC and TA OR (CI)	SDV OR (CI)	TSC, TA, and SDV OR (CI)	TSC OR (CI)	TSC and TA OR (CI)	SDV OR (CI)	TSC, TA, and SDV OR (CI)	TSC OR (CI)	TSC and TA OR (CI)	SDV OR (CI)	TSC, TA, and SDV OR (CI)
TimeBusS	Weekday	1.003	1.002	-	1.004	1.02 *	1.02 *	-	1.02 *	1.03 *	1.02 **	-	1.024 **
	Weekend	0.985 ***	0.985 ***	-	0.984 ***	1.007	1.004	-	1.007	1.027 **	1.024 **	-	1.023 **
JourT	Weekday	0.586 *	0.589 *	-	0.587 *	1.079	1.045	-	1.059	1.091	1.102	-	1.155
	Weekend	0.609 *	0.637 *	-	0.642 *	1.096	1.048	-	1.003	1.020	0.888	-	0.873
FreqofBuses	Weekday	0.394 *	0.398 *	-	0.408 *	-	-	-	-	-	-	-	-
	Weekend	0.265 *	0.286 *	-	0.298 *	-	-	-	-	-	-	-	-
QoR	Weekday	-	-	-	-	0.991	0.991	-	0.970	0.979	0.916	-	0.885
	Weekend	-	-	-	-	1.182 **	1.26 **	-	1.325 *	0.984	0.910	-	0.974
C_WRK	Weekday	-	0.943	-	0.960	-	1.376 **	-	1.251	-	1.688 **	-	1.636 **
	Weekend	-	0.717 ***	-	0.616 **	-	1.594 **	-	1.80 *	-	1.519 **	-	1.392 ***
C_FF	Weekday	-	1.77 *	-	1.613 **	-	0.779 ***	-	0.842	-	0.479 *	-	0.536 *
	Weekend	-	20.02 *	-	1.96 *	-	0.58 **	-	0.628 **	-	0.416 *	-	0.396 *
B_ESY	Weekday	-	1.495 **	-	1.454 **	-	1.195	-	1.246	-	1.349 ***	-	1.374 ***
	Weekend	-	1.205	-	1.102	-	1.594	-	1.574 **	-	1.989 *	-	1.845 *
NOCAR	Weekday	-	1.342 ***	-	1.31 ***	-	0.944	-	0.962	-	1.253	-	1.262
	Weekend	-	1.268	-	1.192	-	1.86 *	-	1.726 **	-	0.719 ***	-	0.543 *
B_CRWD	Weekday	-	0.777	-	0.819	-	1.295 ***	-	1.307 ***	-	0.733 ***	-	0.798 ***
	Weekend	-	0.844	-	0.835	-	0.519 *	-	0.435 *	-	0.586 **	-	0.637 **
B_ST	Weekday	-	1.578 **	-	1.621 **	-	0.965	-	0.955	-	0.570 *	-	0.534 *
	Weekend	-	1.442 **	-	1.618 **	-	0.447 *	-	0.557 **	-	0.545 *	-	0.664 **
TR_CONG	Weekday	-	0.415 *	-	0.381 *	-	0.955	-	0.946	-	0.929 ***	-	0.865 ***
	Weekend	-	0.721 **	-	0.713 **	-	1.363 ***	-	1.174	-	1.99 *	-	20.158 *
Nat	Weekday	-	-	0.959	1.009	-	-	1.088	1.084	-	-	1.31 *	1.224 **
	Weekend	-	-	1.076	1.010	-	-	1.100	1.137	-	-	1.162 **	1.205 **
Inc	Weekday	-	-	0.910	0.966	-	-	0.714 *	0.714 *	-	-	0.829 ***	0.822 ***
	Weekend	-	-	1.057	1.038	-	-	0.717 *	0.721 *	-	-	0.946	0.901
Age	Weekday	-	-	0.905	0.889	-	-	1.184 **	1.152 **	-	-	1.186 ***	1.111
	Weekend	-	-	0.867	0.994	-	-	1.314 **	1.285 **	-	-	1.146	1.028
−2 log likelihood of parametrised model	Weekday	844.924 *	1520.076 *	485.823	1462.945 *	1148.718 **	2333.436 **	955.376 *	2418.929 *	798.527 **	1493.358 *	617.505 *	1387.309 *
	Weekend	660.633 *	1320.876 *	594.173	1324.868 *	898.288 **	1849.275 *	676.541 *	1791.023 *	763.622 **	1591.866 *	616.994 ***	1570.052 *
Chi-square of parametrised model	Weekday	253.740 *	333.801 *	4.947	305.214 *	14.158 **	26.732 **	31.536 *	56.381 *	15.292 **	63.958 *	22.685 *	76.140 *
	Weekend	259.796 *	288.997 *	3.567	240.281 *	9.577 **	77.723 *	28.693 *	94.504 *	8.255 **	97.114 *	6.161 ***	89.509 *
Goodness of fit; pseudo R-squared (McFadden)	Weekday	0.132	0.173	0.003	0.172	0.006	0.011	0.013	0.023	0.009	0.039	0.015	0.052
	Weekend	0.142	0.158	0.002	0.151	0.004	0.035	0.015	0.049	0.004	0.050	0.004	0.053

* Significance $p < 0.001$, ** significance $p < 0.05$, *** significance $p < 0.1$. The coding nomenclature is provided in Table 1.

Sociodemographic variables of nationality and income showed little effect on the probability of a user to respond favourably with regard to network coverage (OR $\approx$ 1), and only a slight influence of age (OR $\approx$ 0.8, for SDV and third-stage models). Users were also asked if their mode choice was influenced by travel attributes and closeness to work, and family was reported by all users as most important. The main reasons stated for dissatisfaction with PT network coverage were crowded buses (67% and 70%) and traffic congestion (~50%). This suggests capacity distribution in public buses and traffic congestion on roads as critical issues, as also noted by Tyrinopoulos and Antoniou [36]. Further illustrating this, Figure 2 shows that, for the travellers that were largely dissatisfied with the current network coverage of the public-transport service in the studied region, onboard crowding and traffic congestion were noted as significant variables influencing their perception of the public transport. On the other hand, Figure 2 also shows that the satisfied traveller groups largely considered bus travel as the easier transit mode for their work- and family-related trips.

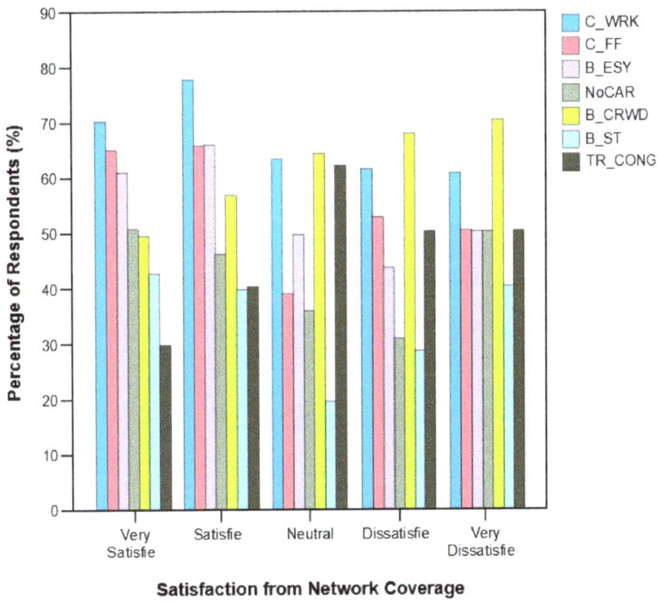

Figure 2. Perceived obstacles in public-transport user satisfaction level. The variable description for the legend is described in Table 1.

3.2. Crosslink between Travel Mode Choice and IDV Blocks

Anticipated yet contrasting results were obtained for the transportation system characteristics across all variable blocks of travel mode choice models. A user's choice of mode was relatively unaffected by the distance from the bus station (OR $\approx$ 1), while journey time adversely influenced travel by both car and bus. Users reported that, in bus travel, the likelihood of trip frequency tended to decrease with increasing traffic (OR generally >1). On the other hand, increasing quality of ride positively affected the frequency of bus travels (OR generally <1) as also validated by the trendline shown in Figure 3, despite the scattered nature of traveller percentage.

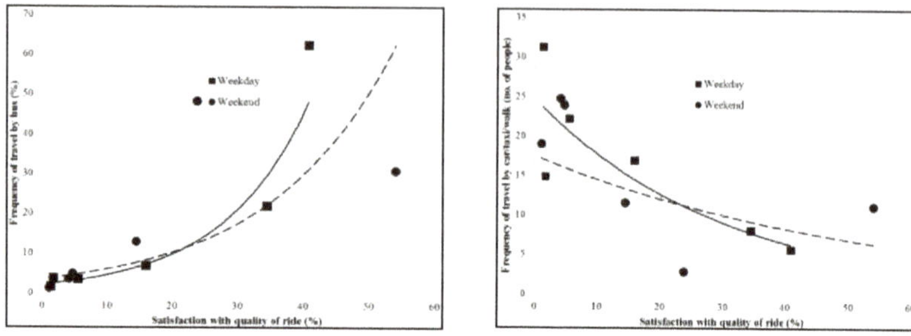

Figure 3. Scatter plot of passengers' response (percentage) to quality of ride against travel mode choice (*solid line = weekends trendline, dotted line = weekdays trendline*).

Urban populations tended to be unevenly distributed towards transport usage patterns as public-transport use tended to skew towards lower-income brackets. The Abu Dhabi population exhibited similar trends when analysed for sociodemographic variables, as shown in Figure 4, with the majority of users from the lower–middle-income bracket (1000–3000 AED/month) for both weekdays (~28.3%) and weekends (~21.9%). The results were also characterised by the observance that most users (~42%) for both also reported that they either did not own a car or could not afford to travel by taxis. Results displayed in Figure 4 also show that, regardless of bus travel frequency, respondents highlighted traffic congestion as the main obstacle. This finding may further extend the range of critical public-transport service attributes to include not only the quantitative travel time attribute, as also noted heavily in the literature [34,37], but also the qualitative perceived traffic congestion variable, which is comparatively less explored.

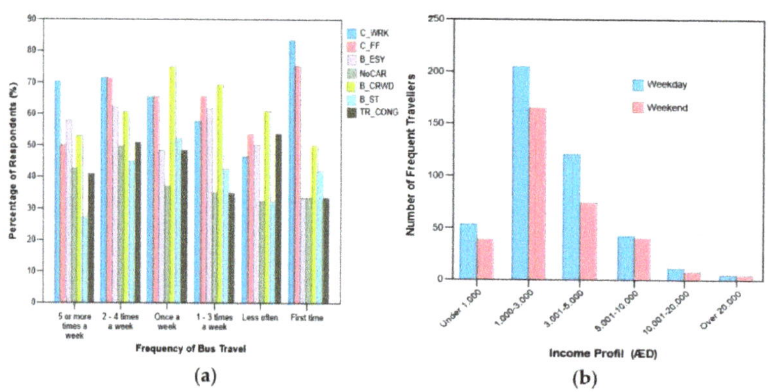

Figure 4. (**a**) Correlation between travel attributes and willingness to travel by bus; (**b**) bus travel distribution by income. The variable description for the legend is described in Table 1.

3.3. Hypothesis Model Improvement Tests across TSC, TA, and SDV Blocks

The previous sections exhibited that the inclusion of income profile, service frequency, and perceived traffic congestion were significant variables influencing public-transport uptake over competing modes (particularly private cars), in addition to the conventional accessibility, quality, and travel time attributes. This was further investigated in the regression modelling stage as four different models were tested with two primary objectives: if inclusion of variables improved model fit, and which parameter was a strong predictor. In the case of network coverage satisfaction, compared with the null hypothesis, adding transport service characteristics (TSC) variables (see Table 1) improved the model, as the -2 log likelihood decreased (weekday: $\chi^2 = 253.74$, $p < 0.0001$; weekend: $\chi^2 = 259.79$, $p < 0.0001$), showing relatively good fit (weekday: $\rho^2 = 0.132$; weekend: $\rho^2 = 0.142$).

Further addition of travel attribute variables improved model fit as the −2LL further decreased (weekday: $\chi^2 = 333.801$, $p < 0.0001$; weekend: $\chi^2 = 288.979$, $p < 0.0001$), also improving the goodness of fit (weekday: $\rho^2 = 0.173$; weekend: $\rho^2 = 0.158$). When both variable blocks were removed from the regression model and only the effect of SDV block was tested, the parameterised model showed a small improvement (weekday: $\chi^2 = 4.94$, $p < 0.0001$; weekend: $\chi^2 = 3.56$, $p < 0.0001$) while the McFadden ρ^2 also decreased. As can be anticipated, adding all three variable blocks simultaneously in the regression equation produced adverse effects on model fit (Table 2). The results show that, while both TSC and TA variable blocks were significant predictors of a respondent's satisfaction with transport network coverage, and even though some SDVs may have also been successful in prediction, their effect may have been nullified once TSC and TA variables were present in the logistic regression equations, showing that postulating the perceived congestion and frequency for estimating accessibility and network coverage variables improved the prediction abilities of the model.

Mode choice models exhibited slightly different behaviour to the network coverage models, where similar effects of the SDV model and expansion of the "TSC and TA model" to include SDV block were found for the weekend data. On the other hand, models based on weekday data tended to display optimum fitness for the final models that included all three variable blocks. For example, models investigating the frequency of bus travel found that the −2LL of the parametrised model containing all three variables was lower than the null hypothesis (weekday: $\chi^2 = 76.14$, $p < 0.0001$; weekend: $\chi^2 = 89.509$, $p < 0.0001$), supported by a higher goodness of fit (weekday: $\rho^2 = 0.052$; weekend: $\rho^2 = 0.053$). Although the McFadden ρ^2 was comparatively lower for mode choice models, the values were higher for parameterised model with three variable blocks.

4. Conclusions

The analysis results of the collected urban travel survey exhibited that travel attributes, especially service frequency, closeness to trip origin/destination, and traffic congestion, as well as characteristics of the transportation system, are predictors of the accessibility, network coverage, service quality, and by extension, the mode choice. This shows that, while optimising public-transport services, particularly in the multicultural, developed infrastructure yet car-centric context of the rapidly developing countries in the Gulf Cooperation Council, it may not be sufficient to limit the definition of accessibility to extending network coverage or service quality to onboard seating or bus-stop quality, as the perception of a more comprehensive network itself may be affected by underlying variables of trip purpose, sociodemographic characteristics, traffic congestion while travelling on public transport, and service frequency, in addition to the more convention ride quality, onboard crowding, and travel time variables.

The regression results for the CAPI-based questionnaire survey data responses of urban Abu Dhabi residents showed that, within the TSC block, distance of a traveller from the bus station was comparatively unimportant, although past research covered in Section 1 noted it as a significant factor. Comparisons of different variable blocks in regression models supported by objective responses of travellers showed that, across all datasets, network coverage satisfaction was reported to be only influenced by the TA and TSC blocks, where increasing congestion and frequency of buses correlated with traveller satisfaction. When mode choice behaviour was evaluated, expanded models containing all three variable blocks were more suited to explain the survey responses.

The findings of this study are important for gaining useful information about the perceived importance of several factors in the functionality of a public-transport system as postulated by the system users. This Abu Dhabi-based study suggests that there may be a hypothesised relationship between the ultimate decision of a user to travel via urban public-transport network instead of private vehicles and its attributes. Further research conducted in the field may be more supportive of this association between variables. At this stage, it should be noted that, although this study is one of the few studies analysing the sociodemographic trends and public-transport usage situation in the car-centric traveller mode choice situation of the Gulf Cooperation Countries, the United Arab Emirates, and particularly the City of Abu Dhabi, there are several shortcomings and limitations of the study that can be addressed in future work. Firstly, this study only collected responses about public bus transport usage compared to car use and did not consider a mixed-mode transit option where multiple competing public-transport modes can be compared to cars as the preferred mode choice. Secondly, it did not consider first- and last-mile choices, and responses captured relative to the provision of micro-mobility options supporting a large-scale public-transport network were not considered. This might greatly affect the tendency of respondents to lean towards private or public transport regardless of frequency or network coverage, as micro-mobility integration might bridge gaps in the current system.

It may also be noteworthy that some interaction between different variables and curvilinearity may also exist, which this study did not address. These shortcomings are acknowledged by the authors, and we aim to address them in future research. Nonetheless, this study showed that a future public-transport system needs to target the adverse effects of traffic congestion and crowded buses, as well as improve the quality of ride and increase the frequency of buses on the investigated travel routes. As such, investment decisions taken by stakeholders in public-transport agencies should consider the attributes of the trip, as well as the characteristics of the transportation system itself.

Supplementary Materials: The following supporting information can be downloaded at: https://www.mdpi.com/article/10.3390/eng4020066/s1.

Author Contributions: Conceptualization, U.H., H.A. and A.W.; Methodology, A.H., U.H. and H.A.; Software, A.H.; Validation, A.H.; Formal analysis, A.H., U.H. and H.A.; Investigation, A.W.; Resources, H.A.; Data curation, H.A.; Writing—original draft, A.H., U.H., H.A. and A.W.; Writing—review & editing, A.H.; Supervision, U.H. and H.A.; Project administration, U.H. and H.A.; Funding acquisition, A.W. All authors have read and agreed to the published version of the manuscript.

Funding: This research project is supported by an Australian Government Research Training Program (RTP) scholarship.

Institutional Review Board Statement: Not applicable.

Informed Consent Statement: Not applicable.

Data Availability Statement: Data will be provided upon request.

Conflicts of Interest: The authors declare no conflict of interest.

References

1. Hasan, A.; Hasan, U.; Whyte, A.; Al Jassmi, H. Lifecycle analysis of recycled asphalt pavements: Case study scenario analyses of an urban highway section. *CivilEng* **2022**, *3*, 242–262. [CrossRef]
2. Hasan, U.; Whyte, A.; Al Jassmi, H. Life cycle assessment of roadworks in United Arab Emirates: Recycled construction waste, reclaimed asphalt pavement, warm-mix asphalt and blast furnace slag use against traditional approach. *J. Clean. Prod.* **2020**, *257*, 120531. [CrossRef]
3. Santero, N.J.; Masanet, E.; Horvath, A. Life-cycle assessment of pavements Part II: Filling the research gaps. *Resour. Conserv. Recycl.* **2011**, *55*, 810–818. [CrossRef]
4. AzariJafari, H.; Yahia, A.; Ben Amor, M. Life cycle assessment of pavements: Reviewing research challenges and opportunities. *J. Clean. Prod.* **2016**, *112*, 2187–2197. [CrossRef]
5. Hasan, U.; Whyte, A.; Al Jassmi, H. Framework for delivering an AV-based mass mobility solution: Integrating government-consumer actors and life-cycle analysis of transportation systems. In Proceedings of the 46th European Transport Conference, Dublin, Ireland, 10–12 October 2018.
6. Shoup, D.C. *The High Cost of Free Parking*; Planners Press, American Planning Association: Chicago, IL, USA, 2011; p. 808.
7. Hawas, Y.E.; Hassan, M.N.; Abulibdeh, A. A multi-criteria approach of assessing public transport accessibility at a strategic level. *J. Transp. Geogr.* **2016**, *57*, 19–34. [CrossRef]
8. Beirão, G.; Cabral, J.S. Understanding attitudes towards public transport and private car: A qualitative study. *Transp. Policy* **2007**, *14*, 478–489. [CrossRef]
9. Hasan, U.; Whyte, A.; Al Jassmi, H. A microsimulation modelling approach to quantify environmental footprint of autonomous buses. *Sustainability* **2021**, *13*, 15657. [CrossRef]

10. Hasan, U.; Whyte, A.; Al Jassmi, H. Public bus transport service satisfaction: Understanding its value to urban passengers towards improved uptake. *Trans. Transp. Sci.* **2021**, *12*, 25–37. [CrossRef]
11. Ponte, C.; Melo, H.P.M.; Caminha, C.; Andrade, J.S.; Furtado, V. Traveling heterogeneity in public transportation. *EPJ Data Sci.* **2018**, *7*, 42. [CrossRef]
12. Hasan, U.; Al Jassmi, H.; Tridane, A.; Stanciole, A.; Al-Hosani, F.; Aden, B. Modelling the effect of non-pharmaceutical interventions on COVID-19 transmission from mobility maps. *Infect. Dis. Model.* **2022**, *7*, 400–418. [CrossRef]
13. Ali, N.; Abdullah, M.; Javid, M.A. Accessibility-based approach: Shaping travel needs in pandemic situation for planners' perspectives. *Eng. J.* **2021**, *25*, 15–22. [CrossRef]
14. Krizek, K.J. Residential Relocation and Changes in Urban Travel: Does Neighborhood-Scale Urban Form Matter? *J. Am. Plan. Assoc.* **2003**, *69*, 265–281. [CrossRef]
15. Pan, H.; Shen, Q.; Zhang, M. Influence of urban form on travel behaviour in four neighbourhoods of Shanghai. *Urban Stud.* **2009**, *46*, 275–294.
16. Gardner, B.; Abraham, C. What drives car use? A grounded theory analysis of commuters' reasons for driving. *Transp. Res. Part F Traffic Psychol. Behav.* **2007**, *10*, 187–200. [CrossRef]
17. Shoup, D.C. Cruising for parking. *Transp. Policy* **2006**, *13*, 479–486. [CrossRef]
18. Chester, M.; Horvath, A.; Madanat, S. Parking infrastructure: Energy, emissions, and automobile life-cycle environmental accounting. *Environ. Res. Lett.* **2010**, *5*, 034001. [CrossRef]
19. Grdzelishvili, I.; Sathre, R. Understanding the urban travel attitudes and behavior of Tbilisi residents. *Transp. Policy* **2011**, *18*, 38–45. [CrossRef]
20. Javid, M.A.; Okamura, T.; Nakamura, F.; Tanaka, S.; Wang, R. People's behavioral intentions towards public transport in Lahore: Role of situational constraints, mobility restrictions and incentives. *KSCE J. Civ. Eng.* **2016**, *20*, 401–410. [CrossRef]
21. Javid, M.A.; Ali, N.; Hussain Shah, S.A.; Abdullah, M. Travelers' attitudes toward mobile application–based public transport services in Lahore. *Sage Open* **2021**, *11*, 2158244020988709. [CrossRef]
22. Hasan, U.; Whyte, A.; Al-Jassmi, H. A life-cycle decision-making framework to assess the need for autonomous mobility. *Transp. Res. Procedia* **2019**, *42*, 32–43. [CrossRef]
23. Hasan, U. *Experimental Study on Bentonite Stabilisation Using Construction Waste and Slag*; Curtin University: Perth, WA, Australia, 2015.
24. Al-Saadi, R.; Abdou, A. Factors critical for the success of public-private partnerships in UAE infrastructure projects: Experts' perception. *Int. J. Constr. Manag.* **2016**, *16*, 234–248. [CrossRef]
25. Hasan, U. *Development of a Multi-Criteria Decision-Making Framework for Sustainable Road Transport Systems: Integrating Stakeholder-Cost-Environment-Energy Lifecycle Impacts*; Curtin University: Perth, WA, Australia, 2019.
26. Maraqa, M.; Albuquerque, F.; Alzard, M.; Chowdhury, R.; Kamareddine, L.; El Zarif, J. GHG Emission Reduction Opportunities for Road Projects in the Emirate of Abu Dhabi: A Scenario Approach. *Sustainability* **2021**, *13*, 7367. [CrossRef]
27. Hasan, U.; Whyte, A.; Al Jassmi, H.; Hasan, A. Lifecycle cost analysis of recycled asphalt pavements: Determining cost of recycled materials for an urban highway section. *CivilEng* **2022**, *3*, 316–331. [CrossRef]
28. Al Hosani, I.I.A.; Dweiri, F.T.; Ojiako, U. A study of cost overruns in complex multi-stakeholder road projects in the United Arab Emirates. *Int. J. Syst. Assur. Eng. Manag.* **2020**, *11*, 1250–1259. [CrossRef]
29. AlKetbi, L.M.B.; Grivna, M.; Al Dhaheri, S. Risky driving behaviour in Abu Dhabi, United Arab Emirates: A cross-sectional, survey-based study. *BMC Public Health* **2020**, *20*, 1541. [CrossRef]
30. Hasan, U.; Whyte, A.; Al Jassmi, H. Life-cycle asset management in residential developments building on transport system critical attributes via a data-mining algorithm. *Buildings* **2018**, *9*, 1. [CrossRef]
31. SCAD. *Statistical Yearbook of Abu Dhabi*; Statistics Centre—Abu Dhabi (SCAD): Abu Dhabi, United Arab Emirates, 2020. Available online: https://www.scad.gov.ae/Release%20Documents/Statistical%20Yearbook%20of%20Abu%20Dhabi_2020_Annual_Yearly_en.pdf (accessed on 5 March 2023).
32. Daleure, G.; Daleure, G. Wages, Salaries, and Expatriate Labor. In *Emiratization in the UAE Labor Market: Opportunities and Challenges*; Springer: Berlin/Heidelberg, Germany, 2017; pp. 57–71.
33. Adhvaryu, B.; Chopde, A.; Dashora, L. Mapping public transport accessibility levels (PTAL) in India and its applications: A case study of Surat. *Case Stud. Transp. Policy* **2019**, *7*, 293–300. [CrossRef]
34. Stępniak, M.; Pritchard, J.P.; Geurs, K.T.; Goliszek, S. The impact of temporal resolution on public transport accessibility measurement: Review and case study in Poland. *J. Transp. Geogr.* **2019**, *75*, 8–24. [CrossRef]
35. Gibson, J.; Munizaga, M.A.; Schneider, C.; Tirachini, A. Estimating the bus user time benefits of implementing a median busway: Methodology and case study. *Transp. Res. Part A Policy Pract.* **2016**, *84*, 72–82. [CrossRef]
36. Tyrinopoulos, Y.; Antoniou, C. Public transit user satisfaction: Variability and policy implications. *Transp. Policy* **2008**, *15*, 260–272. [CrossRef]
37. Belgiawan, P.F.; Ilahi, A.; Axhausen, K.W. Influence of pricing on mode choice decision in Jakarta: A random regret minimization model. *Case Stud. Transp. Policy* **2017**, *7*, 87–95. [CrossRef]

eng

MDPI

Article

Frequency Analysis of Equivalent Property-Damage-Only (EPDO) Crashes at Intersections

Mostafa Sharafeldin [1,*], Khaled Ksaibati [1] and Ken Gerow [2]

1 Wyoming Technology Transfer Center (WYT2/LTAP), Department of Civil and Architectural Engineering, University of Wyoming, Laramie, WY 82071, USA
2 Department of Mathematics and Statistics, University of Wyoming, Laramie, WY 82071, USA
* Correspondence: msharafe@uwyo.edu

Abstract: Traffic crashes are burdening societies with severe costs in terms of public health and economic loss. Intersection-related crashes are responsible for a large proportion of these losses due to their higher frequency and severity. Understanding the factors contributing to intersection crash frequency and severity is essential to mitigate their grave consequences. This study covered the analysis of roadway risk factors, influencing the frequency of equivalent property-damage-only (EPDO) crashes at intersections. The study included developing a negative binomial modeling framework to examine nine years of intersection crash records in the state of Wyoming. The modeling results revealed the key role of pavement friction in intersection safety and EPDO frequency. The findings also demonstrated that intersection location, grade, road functional classification, road surface type, the presence of guardrails, right shoulder type, and horizontal curvature all influence the EPDO crash frequency at intersections.

Keywords: EPDO crashes; pavement friction; intersection safety; intersection crash analysis; roadway geometric characteristics; roadway functional classification

Citation: Sharafeldin, M.; Ksaibati, K.; Gerow, K. Frequency Analysis of Equivalent Property-Damage-Only (EPDO) Crashes at Intersections. *Eng* **2023**, *4*, 1116–1126. https://doi.org/10.3390/eng4020064

Academic Editor: Antonio Gil Bravo

Received: 7 March 2023
Revised: 3 April 2023
Accepted: 6 April 2023
Published: 10 April 2023

1. Introduction

The estimated annual economic cost for motor vehicle crashes in the state of Wyoming is USD 788 million [1]. This figure reflects the financial burden of traffic crashes on the state of Wyoming and the well-being of its residents. Intersections are recognized among the most high-risk locations in roadway networks. Intersections are considered crash-prone locations due to the interaction between different transportation models and complex traffic movements. Crashes occurring at intersections are responsible for over 20% and 40% of traffic-related fatalities and injuries, respectively. The study of crash economic burden requires combining the effects of crash frequency and crash severity. Therefore, transportation professionals commonly use equivalent property-damage-only (EPDO) crashes, which are crash counts magnified by crash costs according to their severity [2–9]. EPDO crash analysis is also used for network screening and crash blackspot identification [10,11].

Traffic crashes are affected by the driver, vehicle, environmental, and roadway attributes. Intersection characteristics such as traffic control, type, location, roadway classification, median and shoulder attributes, grade, and curvature influence the crash frequency and severity. The impact of roadway geometric attributes on crash count and severity has been widely studied. However, other roadway characteristics related to the pavement are yet to be fully examined. Pavement friction is rarely considered in traffic safety studies and its influence on roadway safety is yet to be thoroughly explained. The studies which examined pavement friction demonstrated that friction can be a significant risk factor affecting the severity and frequency of certain crash types [12–18]. Pavement friction, also commonly known as skid resistance, is the resisting force to relative motion between traveling vehicle tires and the pavement surface. This force is essential for vehicle maneuvering and stopping, which directly affects the crash frequency and severity [19,20].

The pavement friction supply tends to decline over time due to the aggregate polishing in the surface layer under moving traffic. Therefore, consistent monitoring of friction levels is needed to ensure the sufficiency of pavement friction supply. Friction monitoring can be even more significant at special roadway locations with higher friction demand and rapid friction deterioration rates, such as curves, ramps, and intersections. A limited number of agencies would collect friction data at such locations with the need to investigate safety concerns, usually involving the increase of wet crashes [21–23]. Maintaining adequate friction levels can be achieved by applying pavement surface treatments, particularly at special locations in the network. These treatments include chip seals, hot-mix asphalt (HMA) overlays, open-graded friction courses (OGFC), micro-milling, and high-friction surface treatments (HFST), which are ideal for spot applications [19,24,25].

As Federal and state departments are working to mitigate the severe impacts of traffic crashes on society, further investigation is needed to examine the factors contributing to traffic crashes. The objective of this study was to further investigate the geometric and roadway characteristics, including pavement friction, that contribute to crash frequency and severity at intersections. The study included developing a negative binomial model to explore the influence of intersection attributes on the EPDO crash frequency at intersections in the state of Wyoming.

2. Literature Review

This section reviews multiple studies relevant to EPDO crash analysis and the pavement friction impact on intersection safety. The limitations of the reviewed studies are identified and the research contribution to the literature is discussed.

Son et al. [26] proposed new performance measures to identify high-risk urban intersections using data from South Korea. The study utilized random forest and extreme gradient boosting (XGB) approaches to estimate the severity factor weights. The authors used the estimated factor weights to develop safety performance functions (SPFs) following Poisson and negative binomial techniques. The study included crash, vehicle, driver, and environmental characteristics as potential risk factors. Even though the study considered the road surface condition (dry, wet, etc.), the authors did not include roadway geometric attributes or pavement friction in the model.

Roshandeh et al. [27] utilized a random-parameter Poisson model to investigate the factors contributing to crash frequency at signalized intersections in Chicago, Illinois. The study examined the impact of traffic, geometric, and environmental characteristics. The study included the pavement condition as a potential influencing factor. The results identified the main factors influencing crash frequency as traffic volume at evening peak, intersection lighting, and pavement condition. The findings highlighted the strong correlation between deteriorated pavement conditions and increased crash frequency at intersections. However, the authors did not incorporate the crash severity and did not consider pavement friction in the analysis.

Poch and Mannering [28] developed a negative binomial model to analyze the factors contributing to intersection crash frequency in Bellevue, Washington. The results identified several traffic-related and geometric attributes influencing the frequency of different types of intersection crashes. However, the study did not consider crash injury severity and did not include any of the pavement characteristics in the model.

Sharafeldin et al. [29] utilized a Bayesian ordered probit model to examine the factors influencing the injury severity of intersection crashes. The analysis incorporated crash, driver, environmental, and roadway risk factors, including pavement friction. The findings showed that pavement friction is a significant contributing factor to crash severity and that lower friction numbers are associated with increasing severity levels. Even though the study incorporated pavement friction in the analysis, the study examined a limited dataset, did not include other roadway characteristics, and did not consider crash frequency.

Claros et al. [30] investigated the challenges of traffic safety management in small jurisdictions and small agencies. The study developed local crash prediction models for intersections in the Madison metropolitan area, Wisconsin. The authors utilized performance measures including EPDO crash frequency and level of service of safety (LOSS) to screen the network and rank the examined sites. The benefit–cost ratio was estimated for intersections using potential treatments. Sites with the highest safety benefit were selected for implementation. The study considered traffic volume, intersection type, and traffic control as risk factors for crash severity and frequency. However, the study did not include other intersection attributes or pavement friction in the analysis.

Afghari et al. [31] employed a joint model of crash count and crash severity to identify the high-risk crash spots on the roadway network in Queensland, Australia. The study examined several traffic characteristics and roadway geometric attributes. The analysis included pavement seal condition and pavement roughness as independent variables. The results showed that unsealed pavements and higher pavement roughness increased crash counts and crash severity. The results also showed that sharper horizontal curves and urban roads were associated with less severe crashes. In addition, rolling terrain was linked to a decreased crash count. However, the study did not include pavement friction in the analysis and did not examine intersection crashes.

Sharafeldin et al. [32] developed an ordinal probit model to examine the roadway attributes contributing to the injury severity of intersection crashes. The study investigated the influence of environmental conditions and a set of roadway characteristics, including pavement friction. The results demonstrated that pavement friction significantly influences intersection crash severity. The study also identified other significant roadway attributes, including location, functional classification, and right shoulder width. However, the study did not consider crash frequency nor the financial burden of crashes.

Roy et al. [33] developed safety performance functions (SPFs) for different classes of roads in Wyoming. The authors examined the effect of traffic, roadway, environmental, and vehicle characteristics on total crashes and EPDO crash frequencies. Pavement friction was investigated in this study. The findings demonstrated that pavement friction is a significant factor, as lower friction numbers increased the frequency of total and EPDO crashes at most of the examined roadway functional classes. The study also identified several significant factors, including traffic volume, grade, horizontal curvature, and road surface condition. This study did not examine crashes at intersections and only examined crashes at road segments.

While there is a growing interest in exploring the relationship between pavement surface friction and roadway safety, the knowledge of pavement friction's impact on intersection safety is insufficient. To the best of the authors' knowledge, this is the first study to explore the influence of pavement friction on crash frequency and severity considering the financial burden of crashes. This work included developing a negative binomial modeling framework to investigate the influence of roadway geometric elements and intersection attributes on the EPDO crash frequency at intersections.

3. Research Methodology

This study developed a negative binomial (NB) model to explain the influence of intersection attributes, as independent variables, and EPDO crash count as the response, which is a non-negative integer. The NB modeling framework was utilized instead of Poisson as the NB model assumptions allow for overdispersion, unlike the Poisson model. The response variable, EPDO, was over-dispersed. Overdispersion is when the variance of the count variable is greater than its mean.

Following the NB model, the risk, P_i, of observing y_i EPDO crashes on a specific intersection, i, is computed as follows [34]:

$$P_i(y_i) = \frac{\Gamma\left(y_i + \frac{1}{k_i}\right)}{y_i! * \Gamma\left(\frac{1}{k_i}\right)} \times \left(\frac{\frac{1}{k_i}}{\frac{1}{k_i} + \lambda_i}\right)^{\frac{1}{k_i}} \times \left(\frac{\lambda_i}{\frac{1}{k_i} + \lambda_i}\right)^{y_i} \quad (1)$$

where λ_i and k_i represent the mean EPDO frequency and dispersion parameter, while $\Gamma(.)$ is the gamma function. Dispersion parameters measure the sample fluctuation around the mean value. The Gamma function is typically used as an extension of the factorial function to complex numbers [33]. The variance function of the NB model is expressed as follows [35]:

$$\text{Var}(y_i) = \lambda_i(1 + \lambda_i k_i) \quad (2)$$

The mean EPDO frequency is a function of the intersection characteristics, x_i, and is expressed as follows [35]:

$$\lambda_i = \exp(\beta_0 + \beta_1 X_1 + \beta_2 X_2 + \ldots + \beta_p X_p + \theta_i) \quad (3)$$

The regression coefficients, β's, are obtained by using the maximum likelihood estimation and are known as MLE estimates. The parameter, θ_i, is expressed as $\exp(\theta_i) \sim \Gamma(1, k_i)$ [35]. The crash injury severity was included in the analysis to properly reflect the impact of the intersection attributes, which could not be captured by considering only the crash counts. The study implemented the EPDO approach to combine the crash severity with the crash count.

The EPDO crash frequencies are estimated by taking the ratio of crash cost to property-damage-only (PDO) crash cost, according to WYDOT crash cost estimates. The crash costs include medical costs, legal expenses, property damage costs, etc. The EPDO crash frequencies were calculated as follows for each crash severity level:

- 277 for fatal crashes
- 13 for suspected serious injury crashes
- 4 for suspected minor injury, possible injury, or unknown severity crashes
- 1 for PDO crashes

4. Data Preparation

The crash data analyzed in this study were obtained from the Critical Analysis Reporting Environment (CARE) package. The CARE package is supported by the Wyoming Department of Transportation (WYDOT). The dataset is developed by collecting the crash records details from the police crash reports.

The data included EPDO crash frequency at 359 intersections in Wyoming from January 2007 through December 2017, except for the years 2010 and 2011 due to the unavailability of friction data. The examined intersections were mapped, and the crashes that occurred within a 250 feet buffer (76.2 m) from the center of the intersection were considered intersection crashes. This approach follows the guidelines of the American Association of State Highway and Transportation Officials [36]. The data were prepared such that each observation represented the EPDO crash frequency at an intersection in one year. The dataset comprised 1594 observations. EPDO crash frequencies were estimated as follows [37]:

$$\begin{array}{c}\text{EPDO count} = 277 \times \text{fatal crashes} + 13 \times \text{suspected serious injury} + \\ 4 \times \text{suspected minor injury} + 4 \times \text{possible injury} + 4 \times \text{unknown severity} + \text{PDO}\end{array} \quad (4)$$

Table 1 presents the number of crashes, crash costs, and EPDO counts per crash severity level for the examined dataset.

Each observation included EPDO crash count as the response and the intersection characteristics as the independent variables, including pavement friction measured in that year at the intersection location. The examined intersection attributes were pavement friction, intersection type, and location, traffic control, grade, horizontal curvature, road functional classification, number of lanes, type of road surface, median type and width, right shoulder type and width, and guardrail presence. No multicollinearity was detected between the independent variables. Table 2 presents a summary of the data.

Table 1. Summary of crash counts, costs, and EPDO frequency.

Crash Severity	Number of Crashes	Crash Costs (USD)	EPDO Counts
Property-damage-only (PDO)	6807	$306,315,000	6807
Suspected minor injury/possible injury/unknown	2140	$385,200,000	8560
Suspected serious injury	144	$84,240,000	1872
Fatal injury	17	$211,905,000	4709

Table 2. Descriptive data statistics.

Continuous Variables				
	Mean	Standard Deviation	Minimum	Maximum
EPDO crash count (response)	13.77	30.271	1	314
Pavement friction	41.48	9.186	18.65	71
Number of lanes	3.481	0.877	2	4
Median width	13.87	21.078	0	120
Right shoulder width	5.296	2.908	0	10
Binary Variable			Count	Percentage
Type: Four legs or more (1 if yes or 0 otherwise)			1270	79.7
Location: Urban (1 if yes or 0 otherwise)			1319	82.7
Traffic control: Signalized (1 if yes or 0 otherwise)			1298	81.4
Grade: Uphill or downhill (1 if yes or 0 otherwise)			170	10.7
Functional classification: Interstate (1 if yes or 0 otherwise)			22	1.4
Functional classification: Principal arterial (1 if yes or 0 otherwise)			1204	75.5
Functional classification: Minor arterial (1 if yes or 0 otherwise)			180	11.3
Functional classification: Collector (1 if yes or 0 otherwise)			31	1.9
Road surface: Concrete (1 if yes or 0 otherwise)			654	41.0
Guardrail (1 if yes or 0 otherwise)			30	1.9
Median: Depressed (1 if yes or 0 otherwise)			169	10.6
Median: Raised (1 if yes or 0 otherwise)			973	61.0
Paved right shoulder (1 if yes or 0 otherwise)			1450	91.0
Slight horizontal curve (>1500 ft radius) (1 if yes or 0 otherwise)			245	15.4
Heavy horizontal curve (<1500 ft radius) (1 if yes or 0 otherwise)			195	12.2

Table 2 demonstrates the overdispersion of the EPDO crash count across the observations. The pavement friction values ranged from 19 to 71, with a mean of 41. The location of each intersection was identified as urban or rural following the US Census Bureau maps [38]. The examined attributes included the functional classification of the major roadway at the intersection. The category "Interstate" referred to the intersections at the interstate interchange, with on/off ramps.

WYDOT collected pavement friction data using the locked-wheel testing device. The locked-wheel tester is a two-wheel trailer fitted with two standard tires, which can be either smooth or ribbed. To measure the longitudinal pavement friction, one or both wheels are locked to achieve the fully locked state with 100% slip and report the average sliding force. Accordingly, the friction can be measured using the locked-wheel tester only at regular intervals to meet the full-lock requirement [39]. It is recommended by the Federal Highway

Administration (FHWA) to use continuous pavement friction measurement (CPFM) for continuous friction data collection for road segments and special locations, such as curves and intersections [40].

The collected friction data were then calibrated in the WYDOT regional calibration center. For this study, the friction data were allocated to the intersections based on the milepost of the intersections and the recorded measurements. When not directly available, the two nearest measurements (before and after the intersection milepost) were averaged to estimate pavement friction at the intersection. For years with missing friction data, friction numbers were estimated by averaging the measurements from previous and subsequent years for the same location. This approach was followed only for locations with decreasing friction numbers, which indicates the absence of maintenance work for this period. This approach assumed a consistent deterioration rate for pavement friction over three years for the same location. The friction number (FN) is recorded as FN40R, which represents measuring friction using a standard ribbed tire on a locked wheel device, at a testing speed of 40 miles per hour.

5. Empirical Analysis

Negative binomial modeling results of intersection crashes are presented in Table 3. All explanatory variables from Table 2 were included in the model. The statistically insignificant predictors at the 95th percentile confidence level were excluded using backward elimination.

Table 3. Negative binomial model results.

Coefficients	Estimate	Standard Error	*p*-Value
Constant	2.803	0.194	<0.001
Pavement friction	−0.015	0.003	<0.001
Location: Urban	−0.231	0.081	<0.001
Grade: Uphill or downhill	−0.465	0.091	<0.001
Functional classification: Interstate	1.031	0.239	<0.001
Functional classification: Principal arterial	0.397	0.068	<0.001
Functional classification: Collector	−0.936	0.229	<0.001
Road surface: Concrete	0.287	0.057	<0.001
Guardrail	−0.483	0.215	0.025
Paved right shoulder	0.234	0.096	0.015
Slight horizontal curve (>1500 ft radius)	0.179	0.076	0.019
Heavy horizontal curve (<1500 ft radius)	−0.241	0.085	0.004
Log-likelihood		−11,438.892	
Log-likelihood of constant-only model		−17,558.649	
ρ^2		0.349	

The modeling results demonstrated that several risk factors, including pavement friction, significantly influenced the EPDO crash frequency at intersections. The marginal effects of the significant influencing factors are presented in Table 4.

Table 4. Marginal effects of EPDO crash frequency factors.

Marginal Effects	Estimate	Standard Error	*p*-Value
Pavement friction	−0.199	0.042	<0.001
Location: Urban	−3.286	1.214	0.007
Grade: Uphill or downhill	−5.096	0.821	<0.001
Functional classification: Interstate	23.137	8.513	0.007
Functional classification: Principal arterial	4.611	0.727	<0.001
Functional classification: Collector	−7.881	1.226	<0.001
Road surface: Concrete	3.737	0.775	<0.001
Guardrail	−5.044	1.715	0.003
Paved right shoulder	2.629	1.024	0.010
Slight horizontal curve (>1500 ft radius)	5.417	1.819	0.003
Heavy horizontal curve (<1500 ft radius)	−2.228	1.056	0.035

6. Discussion

As per the modeling results, pavement friction was a significant factor in reducing the EPDO crash frequency at intersections. Increasing friction numbers (FN40R) by one unit is predicted to reduce EPDO crashes in a given year by 0.199, while controlling for all other parameters. This finding emphasized the key role of friction in intersection safety and the importance of maintaining adequate friction supply on roads, particularly in locations with a higher friction demand and a higher crash risk. This finding can be related to the key role of pavement friction in safe vehicle maneuvering and stopping. Insufficient friction levels increase the crash frequency and the chances of crashing at higher speeds, which results in more severe injuries and higher EPDO frequencies. Several studies have reported relevant findings [29,32,33]. This finding also highlighted the safety benefit of improving pavement friction from an economic perspective, as the cost of one PDO crash in Wyoming is estimated as USD 45,000 [33], while the cost of applying a skid-resistant surface for the intersection is between USD 20,000 and 50,000 [30].

The following is an example of a four-legged, urban intersection on a principal arterial road and a level grade. The intersection is on a slight horizontal curve, the road surface is concrete, and it has a paved right shoulder with no installed guardrails. In a given year, this intersection experienced 73 EPDO, while pavement friction at the intersection was measured at 32.

Assuming application of a Chip Seal treatment at the intersection influence area (250 ft from the center of the intersection) with a service life of 5 years, the estimated annual cost of the treatment will be approximately USD 10,000 [41]. The friction provided by the treatment is expected to decline over its service life and the friction number is expected to be 60 (FN40R) for the newly applied treatment [42]. As per the modeling results, increasing the pavement friction at this intersection from 32 to 60 would reduce the EPDO crashes by 5.57 for the first year. This EPDO cost reduction is estimated to be over USD 250,000, which will result in a benefit–cost ratio for friction improvement of approximately 25.

Urban intersections were found to experience fewer EPDO crashes. An intersection in an urban area would have 3.286 fewer EPDO crashes per year, compared to a rural intersection, while controlling for all other characteristics. This finding can be attributed to the higher traveling speeds on rural roads, driver's susceptibility to distraction, and higher chances of driver fatigue due to longer travel distances. This result emphasized the additional attention needed for rural intersections as they experience higher-severity crashes and a higher EPDO frequency [43,44]. This is particularly important to the state of Wyoming since it has a high number of intersections in rural and semi-rural areas. Intersections on non-level grades were found to witness less frequent EPDO crashes. An uphill or downhill intersection would experience 5.096 fewer EPDO crashes per year, compared to intersections on a level grade, while controlling for all other variables. This finding can be related to the fact that drivers are more cautious while traveling on rolling terrains [31].

Two functional classes of the roadway were found to be associated with a higher EPDO frequency. Intersections linked to interstates and principal arterial roads would experience 23.137 and 4.611 more EPDO crashes per year, compared to local roads, assuming all other factors are controlled. These findings can be attributed to the larger traffic volumes, higher speeds, higher percentage of trucks, and more complex traffic movements on these higher-class roads compared to local roads. On the contrary, intersections on collector roads would have 7.881 fewer EPDO crashes per year, compared to local roads, while controlling for all other parameters. This interesting finding can be related to several reasons. Non-compliance with safety measures and traffic control rules is more common on local roads compared to collectors. Speeding and driving under the influence can be more dangerous on narrower local roads. Driver distraction and overconfidence can be observed more on local roads with lower traffic volumes and lower speed limits.

Intersections on concrete roads would have 3.737 more EPDO crashes per year, compared to asphalt roads, while controlling for all other characteristics. This result can be related to the fact that most concrete road segments in Wyoming are higher-class roads with higher posted speeds and traffic volumes. Installing guardrails at intersections was found to significantly affect the EPDO crash frequency. Installing guardrails would reduce EPDO crashes at an intersection by 5.044 per year, assuming all other factors are controlled. This finding would emphasize the role of guardrails in mitigating crash severity at intersections. Intersections with a paved right shoulder would experience 2.629 more EPDO crashes per year, compared to intersections with no right shoulder, while controlling for all other variables. This finding can be attributed to the improper use of the paved and wide shoulder for overtaking or making a right turn, which increases the risk of sideswiping and rear-end crashes [45].

Intersections on slight horizontal curves (>1500 ft radius) would have 5.417 more EPDO crashes per year, compared to an intersection at non-curved segments, while controlling for all other variables. On the contrary, intersections on heavy horizontal curves (<1500 ft radius) would have 2.228 fewer EPDO crashes per year, compared to an intersection at non-curved segments, assuming all other factors are controlled. These findings are related to cautious driving behavior under complex road geometry and the fact that sharper horizontal curves will require the drivers to travel at lower speeds. Afghari reported similar conclusions [31]. Several studies suggested the use of perceptual measures, and delineation treatments to warn drivers of the presence of curves, which will impact their behavior and improve traffic safety at such locations [46–48].

7. Conclusions and Recommendations

This study presented a negative binomial model of intersection crashes to investigate the factors contributing to the EPDO crash frequency at intersections. The study explored the role of pavement friction, intersection attributes, and roadway characteristics. The analysis results revealed that several factors significantly influence the EPDO frequency. Pavement friction was found to be a significant parameter as increasing friction numbers reduced the EPDO crash frequency at intersections. Intersections in urban areas, or on uphill/downhill grades, were found to experience fewer EPDO crashes. The number of EPDO crashes at intersections linked to interstates or principal arterial roads was higher than those local road intersections, while the EPDO crash frequency was lower for intersections on collector roads. The presence of a paved right shoulder or concrete roadway surface was associated with higher EPDO counts. Guardrails were found to reduce EPDO counts at intersections. Horizontal curvature was found to influence EPDO differently according to its radius. Intersections on slight horizontal curves (>1500 ft radius) would experience higher EPDO counts, while the EPDO frequency would be lower on intersections on heavy horizontal curves (<1500 ft radius).

This study intended to examine the role of pavement friction on intersection crashes, incorporating both crash count and severity by analyzing the EPDO crash frequency. The findings of this work would help to define desirable friction thresholds for roadway intersections, considering other intersection attributes and geometric elements. Additionally, these conclusions can be used during the planning stage for intersection construction or rebuilding, and they can also be applied while reviewing high-risk intersections. The findings of this study can be used in support of road safety assessments and Safety Index (SI) validation, as the findings identified roadway risk factors impacting the EPDO crash frequency at intersections [49].

8. Study Limitations and Future Research

Underreporting of PDO crashes may affect the estimation accuracy. This study examined data only from the state of Wyoming, which may limit the transferability of the findings to other states with different related conditions.

Future research may focus on developing economical solutions and spot applications for improving pavement friction.

Author Contributions: Conceptualization, M.S., K.K. and K.G.; data curation, M.S.; formal analysis, M.S.; methodology, M.S., K.K. and K.G.; resources, K.K.; supervision, K.K. and K.G.; writing—original draft, M.S.; writing—review and editing, M.S., K.K. and K.G. All authors have read and agreed to the published version of the manuscript.

Funding: This research was funded by the Wyoming Department of Transportation (WYDOT), grant number: RS05221.

Institutional Review Board Statement: Not applicable.

Informed Consent Statement: Not applicable.

Data Availability Statement: The data were collected from the Critical Analysis Reporting Environment (CARE) package, supported by the Wyoming Department of Transportation (WYDOT).

Acknowledgments: The authors gratefully acknowledge the effective financial support of the Wyoming Department of Transportation (WYDOT) through grant number: RS05221. All opinions are solely of the authors.

Conflicts of Interest: The authors declare that they have no conflict of interest.

References

1. U.S. Department of Transportation. State by State Crash Data and Economic Cost Index. 2022. Available online: https://www.transportation.gov/research-and-technology/state-state-crash-data-and-economic-cost-index (accessed on 8 November 2022).
2. PDO; TPDO. *The National Intersection Safety Problem*; U.S. Department of Transportation, Federal Highway Administration, Office of Safety: Washington, DC, USA, 2004.
3. Arafat, M.; Hadi, M.; Raihan, A.; Iqbal, S.; Tariq, M.T. Benefits of connected vehicle signalized left-turn assist: Simulation-based study. *Transp. Eng.* **2021**, *4*, 100065. [CrossRef]
4. Reza, I.; Ratrout, N.T.; Rahman, S.M. Calibration protocol for PARAMICS microscopic traffic simulation model: Application of neuro-fuzzy approach. *Can. J. Civ. Eng.* **2016**, *43*, 361–368. [CrossRef]
5. Dobrota, N.; Stevanovic, A.; Mitrovic, N. Development of Assessment Tool and Overview of Adaptive Traffic Control Deployments in the U.S. In *Transportation Research Record*; SAGE Publications Ltd.: New York, NY, USA, 2020; pp. 464–480.
6. Cvijovic, Z.; Zlatkovic, M.; Stevanovic, A.; Song, Y. Conditional Transit Signal Priority for Connected Transit Vehicles. *Transp. Res. Rec. J. Transp. Res. Board* **2021**, *2676*, 490–503. [CrossRef]
7. Stevanovic, A.; Dobrota, N.; Baek, C.; Mason, J.M.; Thai, J.H. *NCHRP 20-07, Task 414 Benefits of Adaptive Traffic Control Deployments—A Review of Evaluation Studies*; AASHTO Standing Committee on Highways: Washington, DC, USA, 2019.
8. Nazneen, S.; Rezapour, M.; Ksaibati, K. Application of Geographical Information System Techniques to Determine High Crash-Prone Areas in the Fort Peck Indian Reservation. *Open Transp. J.* **2020**, *14*, 174–185. [CrossRef]
9. Assi, K. Prediction of Traffic Crash Severity Using Deep Neural Networks: A Comparative Study. In Proceedings of the 2020 International Conference on Innovation and Intelligence for Informatics, Computing and Technologies (3ICT), Sakheer, Bahrain, 20–21 December 2020; pp. 1–6. [CrossRef]

10. Alrejjal, A.; Ksaibati, K. Impact of Combined Alignments and Different Weather Conditions on Vehicle Rollovers. *KSCE J. Civ. Eng.* **2021**, *26*, 893–906. [CrossRef]
11. Alrejjal, A.; Ksaibati, K. Impact of mountainous interstate alignments and truck configurations on rollover propensity. *J. Saf. Res.* **2021**, *80*, 160–174. [CrossRef]
12. Cvijovic, Z.; Zlatkovic, M.; Stevanovic, A.; Song, Y. Multi-Level Conditional Transit Signal Priority in Connected Vehicle Environments. *J. Road Traffic Eng.* **2021**, *67*, 1–12. [CrossRef]
13. Sharafeldin, M.; Farid, A.; Ksaibati, K. Injury Severity Analysis of Rear-End Crashes at Signalized Intersections. *Sustainability* **2022**, *14*, 13858. [CrossRef]
14. Sharafeldin, M.; Farid, A.; Ksaibati, K. A Random Parameters Approach to Investigate Injury Severity of Two-Vehicle Crashes at Intersections. *Sustainability* **2022**, *14*, 13821. [CrossRef]
15. Nazneen, S.; Farid, A.; Ksaibati, K. Examining Seat Belt Use Habits in Wyoming: A Bayesian Hierarchical Logit Approach. *Transp. Res. Rec. J. Transp. Res. Board* **2022**, *2677*, 927–945. [CrossRef]
16. Sharafeldin, M.; Farid, A.; Ksaibati, K. Examining the Risk Factors of Rear-End Crashes at Signalized Intersections. *J. Transp. Technol.* **2022**, *12*, 635–650. [CrossRef]
17. Assi, K.; Rahman, S.; Mansoor, U.; Ratrout, N. Predicting Crash Injury Severity with Machine Learning Algorithm Synergized with Clustering Technique: A Promising Protocol. *Int. J. Environ. Res. Public Health* **2020**, *17*, 5497. [CrossRef]
18. Reza, I.; Ratrout, N.T.; Rahman, S.M. Artificial Intelligence-Based Protocol for Macroscopic Traffic Simulation Model Development. *Arab. J. Sci. Eng.* **2021**, *46*, 4941–4949. [CrossRef]
19. Hafez, M.; Farid, A.; Ksaibati, K.; Rickgauer, S.; Carlson, M. *Managing Pavement Friction of Wyoming's Roads Considering Safety*; Wyoming Department of Transportation: Cheyenne, WY, USA, 2020.
20. Hall, J.W.; Smith, K.L.; Titus-Glover, L.; Wambold, J.C.; Yager, T.J.; Rado, Z. *Guide for Pavement Friction National Cooperative Highway Research Program NCHRP*; NCHRP: Washington, DC, USA, 2009.
21. Elkhazindar, A.; Hafez, M.; Ksaibati, K. Incorporating Pavement Friction Management into Pavement Asset Management Systems: State Department of Transportation Experience. *Civileng* **2022**, *3*, 541–561. [CrossRef]
22. Abdalla, A.; Faheem, A.F.; Walters, E. Life cycle assessment of eco-friendly asphalt pavement involving multi-recycled materials: A comparative study. *J. Clean. Prod.* **2022**, *362*, 132471. [CrossRef]
23. FHWA. Evaluation of Pavement Safety Performance. Available online: https://www.fhwa.dot.gov/publications/research/safety/14065/003.cfm (accessed on 31 July 2022).
24. Abaza, O.A.; Chowdhury, T.D.; Arafat, M. Comparative Analysis of Skid Resistance for Different Roadway Surface Treatments. *Am. J. Eng. Appl. Sci.* **2017**, *10*, 890–899. [CrossRef]
25. Abaza, O.A.; Arafat, M.; Uddin, M.S. Physical and economic impacts of studded tyre use on pavement structures in cold climates. *Transp. Saf. Environ.* **2021**, *3*, tdab022. [CrossRef]
26. Son, S.-O.; Park, J.; Lee, G.; Abdel-Aty, M. Development of New Performance Measures Based on Data Mining Weights for Hotspot Identification. In *Transportation Research Record*; SAGE Publications Ltd.: New York, NY, USA, 2022; pp. 633–647. [CrossRef]
27. Roshandeh, A.M.; Agbelie, B.R.; Lee, Y. Statistical modeling of total crash frequency at highway intersections. *J. Traffic Transp. Eng. (Engl. Ed.)* **2016**, *3*, 166–171. [CrossRef]
28. Poch, M.; Mannering, F. Negative Binomial Analysis of Intersection-Accident Frequencies. *J. Transp. Eng.* **1996**, *122*, 105–113. [CrossRef]
29. Sharafeldin, M.; Albatayneh, O.; Farid, A.; Ksaibati, K. A Bayesian Approach to Examine the Impact of Pavement Friction on Intersection Safety. *Sustainability* **2022**, *14*, 12495. [CrossRef]
30. Claros, B.; Chitturi, M.; Bill, A.; Noyce, D.A. Roadway Safety Management in Small Municipalities. *Transp. Res. Rec. J. Transp. Res. Board* **2020**, *2674*, 341–354. [CrossRef]
31. Afghari, A.P.; Haque, M.; Washington, S. Applying a joint model of crash count and crash severity to identify road segments with high risk of fatal and serious injury crashes. *Accid. Anal. Prev.* **2020**, *144*, 105615. [CrossRef] [PubMed]
32. Sharafeldin, M.; Farid, A.; Ksaibati, K. Investigating The Impact of Roadway Characteristics on Intersection Crash Severity. *Eng* **2022**, *3*, 412–423. [CrossRef]
33. Roy, U.; Farid, A.; Ksaibati, K. Effects of Pavement Friction and Geometry on Traffic Crash Frequencies: A Case Study in Wyoming. *Int. J. Pavement Res. Technol.* **2022**. [CrossRef]
34. Farid, A.; Abdel-Aty, M.; Lee, J. A new approach for calibrating safety performance functions. *Accid. Anal. Prev.* **2018**, *119*, 188–194. [CrossRef]
35. Xu, P.; Huang, H. Modeling crash spatial heterogeneity: Random parameter versus geographically weighting. *Accid. Anal. Prev.* **2015**, *75*, 16–25. [CrossRef]
36. AASHTO. *Highway Safety Manual*, 1st ed.; AASHTO: Washington, DC, USA, 2010.
37. Rezapour, M.; Ksaibati, K. Comprehensive Evaluation of a Sparse Dataset, Assessment and Selection of Competing Models. *Signals* **2020**, *1*, 157–169. [CrossRef]
38. U.S. Census Bureau. 2010 Census Urban Area Reference Maps. Available online: https://www.census.gov/geographies/reference-maps/2010/geo/2010-census-urban-areas.html (accessed on 19 July 2022).

39. Federal Highway Administration. *Federal Highway Administration Pavement Friction Management (PFM) Support Program Locked-Wheel and Sideway-Force Continuous Friction Measurement Equipment Comparison and Evaluation Report*; Federal Highway Administration: Washington, DC, USA, 2019.
40. FHWA. Pavement Friction Management. Federal Highway Administration (FHWA). Available online: https://safety.fhwa.dot.gov/roadway_dept/pavement_friction/cpfm/ (accessed on 31 July 2022).
41. Merritt, D.K.; Lyon, C.; Persaud, B. *Evaluation of Pavement Safety Performance*; (No. FHWA-HRT-14-065); Federal Highway Administration: Washington, DC, USA, 2015.
42. Li, S.; Noureldin, S.; Jiang, Y.; Sun, Y. *Evaluation of Pavement Surface Friction Treatments*; (No. FHWA/IN/JTRP-2012/04); Purdue University, Joint Transportation Research Program: West Lafayette, IN, USA, 2012. [CrossRef]
43. Abbasi, E.; Li, Y. A Comprehensive Review of Driver's Attention and the Evaluation Methods. *Comput. Res. Prog. Appl. Sci. Eng.* **2021**, *7*, 1–4. [CrossRef]
44. Abbasi, E.; Li, Y.; Wu, X.; Craig, B. Using Classification and Regression Trees (CART) to Identify Factors Contributing to Vehicle Crash Severity in a Port City. *Int. J. Transp. Syst.* **2021**, *6*, 29–38.
45. Sharafeldin, M. Pavement Safety Management of Wyoming's Intersections. Ph.D. Dissertation, University of Wyoming, Laramie, WY, USA, 2022.
46. Galante, F.; Mauriello, F.; Pernetti, M.; Riccardi, M.R.; Montella, A. Effects of Traffic Control Devices on Rural Curve Lateral Position. *Transp. Res. Rec. J. Transp. Res. Board* **2021**, *2676*, 162–180. [CrossRef]
47. Awan, H.H.; Pirdavani, A.; Houben, A.; Westhof, S.; Adnan, M.; Brijs, T. Impact of perceptual countermeasures on driving behavior at curves using driving simulator. *Traffic Inj. Prev.* **2019**, *20*, 93–99. [CrossRef]
48. Wang, B.; Hallmark, S.; Savolainen, P.; Dong, J. Crashes and near-crashes on horizontal curves along rural two-lane highways: Analysis of naturalistic driving data. *J. Saf. Res.* **2017**, *63*, 163–169. [CrossRef]
49. Riccardi, M.R.; Augeri, M.G.; Galante, F.; Mauriello, F.; Nicolosi, V.; Montella, A. Safety Index for evaluation of urban roundabouts. *Accid. Anal. Prev.* **2022**, *178*, 106858. [CrossRef]

eng MDPI

Review

A Critical Review and Bibliometric Analysis on Applications of Ground Penetrating Radar in Science Based on Web of Science Database

Nehal Elshaboury [1,2], Eslam Mohammed Abdelkader [1,3,*], Abobakr Al-Sakkaf [4,5] and Tarek Zayed [1]

1 Department of Building and Real Estate, Faculty of Construction and Environment, The Hong Kong Polytechnic University, Kowloon 999077, Hong Kong
2 Construction and Project Management Research Institute, Housing and Building National Research Centre, Giza 12311, Egypt
3 Structural Engineering Department, Faculty of Engineering, Cairo University, Giza 12613, Egypt
4 Department of Buildings, Civil and Environmental Engineering, Concordia University, Montreal, QC H3G 1M8, Canada
5 Department of Architecture and Environmental Planning, College of Engineering and Petroleum, Hadhramout University, Mukalla 50512, Yemen
* Correspondence: eslam.abdelkader@polyu.edu.hk

Abstract: Ground-penetrating radar (GPR) is an established technology with a wide range of applications for civil engineering, geological research, archaeological studies, and hydrological practices. In this regard, this study applies bibliometric and scientometric assessment to provide a systematic review of the literature on GPR-related research. This study reports the publication trends, sources of publications and subject categories, cooperation of countries, productivity of authors, citations of publications, and clusters of keywords in GPR-related research. The Science Citation Index Expanded (SCI-EXPANDED) and the Social Sciences Citation Index (SSCI), which can be accessed through the Web of Science Core Collection, are used as references. The findings report that the number of publications is 6880 between 2001 and 2021. The number of annual publications has increased significantly, from 139 in 2001 to 576 in 2021. The studies are published in 894 journals, and the annual number of active journals increased from 68 in 2001 to 215 in 2021. Throughout the study, the number of subject categories involved in GPR-related research fluctuated, ranging from 38 in 2001 to 68 in 2021. The research studies originated from 118 countries on 6 continents, where the United States and the People's Republic of China led the research articles. The top five most common keywords are ground-penetrating radar, non-destructive testing, geophysics, electrical resistivity tomography, and radar. After investigating the clusters of keywords, it is determined that civil engineering, geological research, archaeological studies, and hydrological practices are the four main research fields incorporating GPR utilization. This study offers academics and practitioners an in-depth review of the latest research in GPR research as well as a multidisciplinary reference for future studies.

Keywords: ground penetrating radar; bibliometric and scientometric assessment; bibliographic coupling; co-citation analysis

Citation: Elshaboury, N.; Mohammed Abdelkader, E.; Al-Sakkaf, A.; Zayed, T. A Critical Review and Bibliometric Analysis on Applications of Ground Penetrating Radar in Science Based on Web of Science Database. *Eng* **2023**, *4*, 984–1008. https://doi.org/10.3390/eng4010059

Academic Editor: Antonio Gil Bravo

Received: 14 February 2023
Revised: 20 March 2023
Accepted: 21 March 2023
Published: 22 March 2023

1. Introduction

In the twenty-first century, mapping existing structures was a big issue for civil engineers. A crucial part of this procedure was extracting relevant information about the position, shape, and type of materials for embedded parts (e.g., sewers and reinforcing rebar). There are varieties of non-destructive testing (NDT) methods available, with ground-penetrating radar (GPR) being the most widely used in the civil engineering field [1,2]. GPR is a relatively new geophysical technology that has made significant progress in the recent decade. It can identify embedded things in structures without being destroyed since it uses electromagnetic waves and radargram processing technologies. Furthermore, it is portable equipment that can manage the full scanning procedure with just one operator [3]. The two key criteria of GPR are depth range and resolution. The signal may propagate farther with a lower center frequency of the GPR antenna, but the resolution in the shallower layers decreases. Higher frequencies do not penetrate deeply but provide higher resolution in shallower layers. Moreover, depth affects the size of the observable item. At shallow depths, little things can be observed, but as the depth range increases, an object's physical size must be big in order to be detected.

The antenna, storage unit, display unit, control unit, and various auxiliary devices (e.g., battery, car, and global positioning system (GPS)) are all part of the GPR system [4,5]. Figure 1 depicts the structure of a typical GPR system. The antenna is made up of a transmitter that sends electromagnetic waves into structures and monitors for echoes caused by changes in the material characteristics of the structure. The GPR signal has a wide variety of frequency components and commonly operates in the 10–5000 MHz range. The GPR receiver detects these reflected signals, which serve as the foundation for imaging inside the invisible structure [6]. A control unit delivers commands for sample time, repetition time, frequency, and other parameters. A graphical user interface (GUI) is included in the display unit, allowing numerous parameters to be visualized and adjusted. There can be a storage unit that can deliver data onto a PC or other processing units for additional analysis. Finally, depending on the type and technical requirements of the system, accessories such as GPS and wheels may be provided.

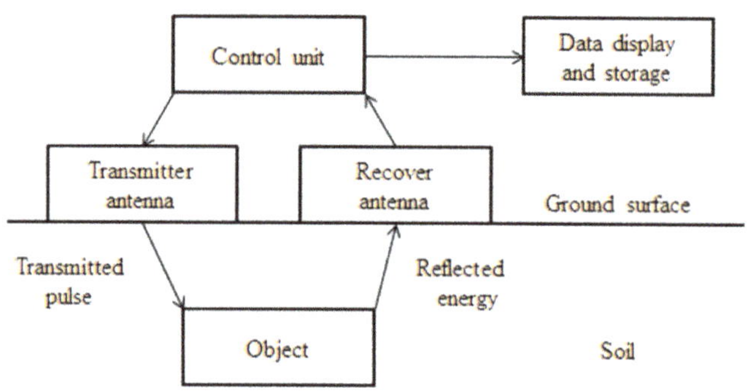

Figure 1. Components of a typical GPR system.

One of the most important applications of GPR in civil engineering is scanning buildings and common structural components [7]. GPR is also used to assess roads and pavements since it is one of the most used non-destructive approaches to obtain subsurface information on the structure of roads and pavements [8,9]. It may also be used for underground utility (e.g., pipes, tunnels, and sewers) detection [10,11]. Mineral exploration has major difficulty with complex geological conditions that could be investigated using this emerging technology [12]. In geology, GPR is used to detect permafrost, locate fractures or water-bearing zones, map shallow formations, and so on [13,14]. In archaeology, GPR is commonly employed for mapping buildings in historical sites [15,16].

Notwithstanding the significance of this geophysical technology, there is a dearth of comprehensive literature reviews to analyze the global research topics and future trends of GPR from a statistical standpoint. Bibliometric studies rely on the analysis of journals, authors, publications, author keywords, and collaboration between countries or institutions. The method is gaining popularity as a research tool for examining the knowledge domain or visualizing networks to offer a more comprehensive view of the subject [17]. Furthermore, such investigations aim to evaluate how research has evolved and provide some insights into future research trends [18–20].

The primary goal of this study is to perform a comprehensive literature analysis to study current and future global research trends in GPR research. The following sub-objectives are carried out to attain the primary goal: (1) establishing a framework for the reviewed literature; (2) conducting a science mapping analysis to identify time and geographical distribution, authorship, sources, keywords, and citations of publications in the field of GPR; (3) summarizing emerging research themes and determining current research gaps; and (4) proposing future directions in GPR research. The novelty of this study is as follows: (1) conducting objective review-based research in the GPR domain by applying bibliometric and scientometric assessment; (2) providing a more comprehensive analysis of research papers published across 20 years, from 2001 to 2021; (3) performing qualitative analysis that identifies current research status and emerging research trends; and (4) expanding the previous research study related to GPR research [21].

2. Research Methodology

In this review-based study, a holistic analytical approach that integrates quantitative and qualitative evaluations is applied to gain a better understanding of the study area and remove biased findings [22]. The flowchart of this review-based research study is divided into three primary steps, as illustrated in Figure 2. In this study, bibliometric and scientometric assessment has been applied as quantitative tools for examining GPR research from a variety of perspectives, including publication time and citations, author collaboration and productivity, subject categories and journals, relevant countries and institutions, and author keywords [23]. The final element of the study framework is qualitative analysis, which offers a thorough knowledge of the major topics in the GPR research study.

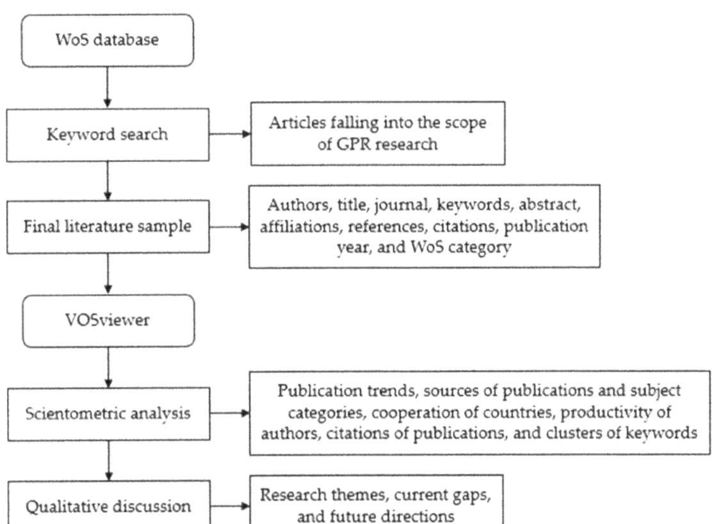

Figure 2. Review process for relevant papers.

Web of Science (WoS) is the most frequently used scientific literature database platform, with over 12,000 high-impact publications. Furthermore, this database is frequently used by scholars to gather accurate data for bibliometric studies [24,25]. As a result, the literature for this study is retrieved from the WoS database. Various phrases are examined to search for the targeted publications from the database. The utilized search phrases are GPR radar*, ground probing radar*, ground penetration radar*, ground penetrating radar*, GPR micro-wave*, GPR microwave*, geo-radar*, and georadar*. The asterisk indication guarantees that the search includes all relevant keywords. The targeted research and review publications are those published between 2001 and 2021. The conference papers are omitted because they lack the comprehensiveness of scientific content available in journal publications [26]. Science Citation Index Expanded (SCI-EXPANDED) and the Social Sciences Citation Index (SSCI) are chosen as citation indexes, and English is selected as the publishing language. Records related to the authors, article title, source title, author keywords, abstract, cited references, citations count, publisher, publication year, and WoS category of 6880 papers are downloaded as plain text (on 23 December 2021) from the database.

Bibliometric and scientometric assessment can be accomplished using many software applications such as Bibexcel, CiteSpace, VOSviewer, and VantagePoint. In this research, MS Excel is used to do the standard data analysis (such as publication trends, subject categories, journals, authors, nations, institutions, and keywords). The co-occurrence, bibliographic coupling, and co-authorship networks are developed using VOSviewer software (version 1.6.17) because of its aptitude for knowledge mining and visualization of vast networks (www.vosviewer.com; accessed on 23 December 2021 [27]). Kessler [28] was the first to use the term bibliographic coupling to characterize the thematic closeness between two research studies. Although bibliographic coupling was designed to locate articles with comparable research viewpoints, it may also be extended to other sources, such as authors and journals [29,30].

VOSviewer software visualizes three different formats of maps: network, overlay, and density visualization. The elements in the network visualization are represented by nodes where the size of each node reflects its weight and degree of importance. In addition, a node's color is determined by the cluster to which it belongs. The correlation between nodes is inversely proportional to their distances. The overlay visualization is similar to the network visualization, except that the elements are colored using a scale bar that displays the scores of clusters/items with respect to specific aspects (e.g., publication year). The density visualization shows how dense an object is at a specific point [31]. The density can be presented individually for each cluster to which the items belong (cluster density view) or without taking this distinction into account (item density view). However, network visualization is used in this study for brevity and visualization.

3. Results and Discussion

3.1. Publication Trends

The search yields 6880 publications during the twenty years from 2001 to 2021. The number of annual publications has increased significantly, from 139 in 2001 to 576 in 2021. Between 2001 and 2005, the annual number of articles was fewer than 200, with an average of around 160 each year. The annual number of publications climbed in 2006, reaching 219 compared to 173 in the previous year. There are around 258, 388, and 525 articles each year on average in the years 2007–2011, 2012–2016, and 2017–2021, respectively (Figure 3). In terms of the yearly total number of citations, Figure 3 shows an uneven trend: the three greatest values were 7659, 7548, and 7284 citations in 2007, 2013, and 2014, respectively. However, after 2014, the annual number of citations has steadily fallen, achieving 494 in 2021. This can be attributed to the fact that the more recent papers have had less time to be referenced.

3.2. Sources of Publications and Subject Categories

GPR papers were published in 894 journals between 2001 and 2021, illustrating the many disciplines and areas involved. Annually, the number of active journals increased from 68 in 2001 to 215 in 2021. As depicted in Table 1, the influence of these journals is quantified using four indicators; the number of publications, the average publication year, the average citations, and the average normal citations. The top five active journals in GPR-related research are the *Journal of Applied Geophysics*, *IEEE Transactions on Geoscience and Remote Sensing*, *Near Surface Geophysics*, *Remote Sensing*, and *Geophysics*, with 401, 316, 231, 199, and 198 total publications, respectively. These journals publish a total of 19.5% of the total articles, demonstrating the importance of these publications for GPR-related research. Figure 4 illustrates the annual number of publications for the five most productive journals. It demonstrates that each journal experiences yearly fluctuations in the number of published articles, with multiple peaks that tend to converge in the same periods, such as those in 2003, 2005, and 2010.

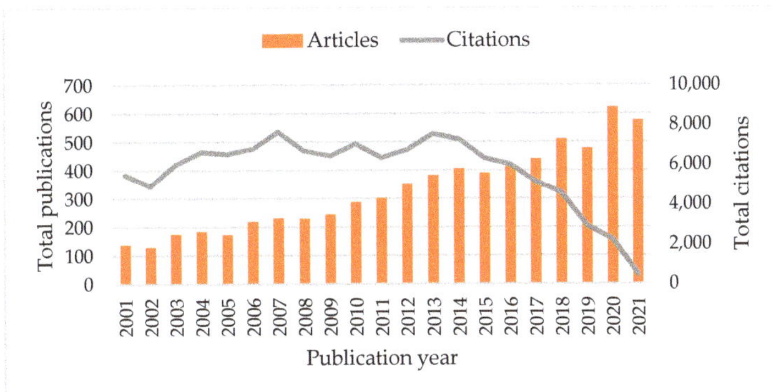

Figure 3. Trends of GPR-related publication outputs from 2001 to 2021.

Table 1. Top 5 most productive journals in GPR research between 2001 and 2021.

Journal	Publisher	Impact Factor (2021)	Number of Publications	Average Publication Year	Total Citations	Average Citations	Average Normal Citations
Journal of Applied Geophysics	Elsevier	2.121	401	2012.66	8052	20.08	1.04
IEEE Transactions on Geoscience and Remote Sensing	IEEE	5.600	316	2011.79	9365	29.64	1.42
Near Surface Geophysics	Wiley	2.033	231	2011.80	2525	10.93	0.54
Remote Sensing	MDPI	4.848	199	2019.11	1276	6.41	0.98
Geophysics	SEG Library	2.928	198	2011.19	4421	22.33	0.97

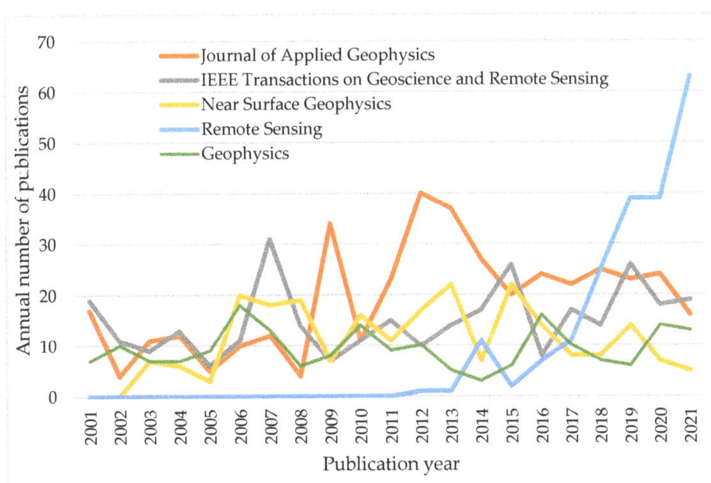

Figure 4. Number of annual publications by the five most active journals.

The top two most productive journals also have the highest number of total citations. For instance, the *IEEE Transactions on Geoscience and Remote Sensing* journal achieves the highest total citations of 9365, followed by the *Journal of Applied Geophysics* (8052 citations). However, journals with the highest average citations per article are not among the top three most productive or cited documents: *Computer Physics Communication* has the most (109.67), followed by *Tree Physiology* (94.75) and *Earth-Science Reviews* (80.67). On the other hand, articles published in *IEEE Transactions on Geoscience and Remote Sensing* have garnered an average of 29.64 citations per article, showing that GPR research papers published in these journals have had a significant impact on this subject.

Based on the average normal citations indicator, *IEEE Transactions on Industrial Informatics* (10.33) is shown to be the most significant journal in this study subject. This indicator is computed by dividing the total number of citations by the average number of citations published each year. It is used to deduce that older texts are not always cited more frequently than newer releases [32]. Furthermore, while having the largest research output, the *Journal of Applied Geophysics* is not regarded as the most fruitful journal since it does not have the highest average normal citations. The most recent studies are published in *IEEE Access*, *Water*, and *Remote Sensing* journals. The *IEEE Aerospace and Electronic Systems Magazine* and *IEE Proceedings—Radar, Sonar* and, *Navigation* journals, on the other hand, are no longer active in this study arena.

Figure 5 presents the prominent journals on the topic of GPR. The minimal threshold values are established at three articles and thirty citations. It was found that 251 out of 894 journals matched these criteria. Each journal is distinguished by a circle whose size is proportionate to the number of published articles. The distance between circles indicates the strength of the link with the other journal, such that the shorter distance indicates a stronger connection. Furthermore, the colors of clusters refer to the research topics such that journals belonging to the same cluster cover similar themes. For instance, 77 journals that primarily publish articles on earth sciences, geology, landslide, and sedimentology make up the largest red cluster. A total of 67 Journals that publish articles on applied geophysics, structure, and infrastructure are represented by the green cluster. The blue cluster, comprising 51 journals, includes research on glaciology, atmosphere, hydrology, and planetary and space science in the GPR-related domain.

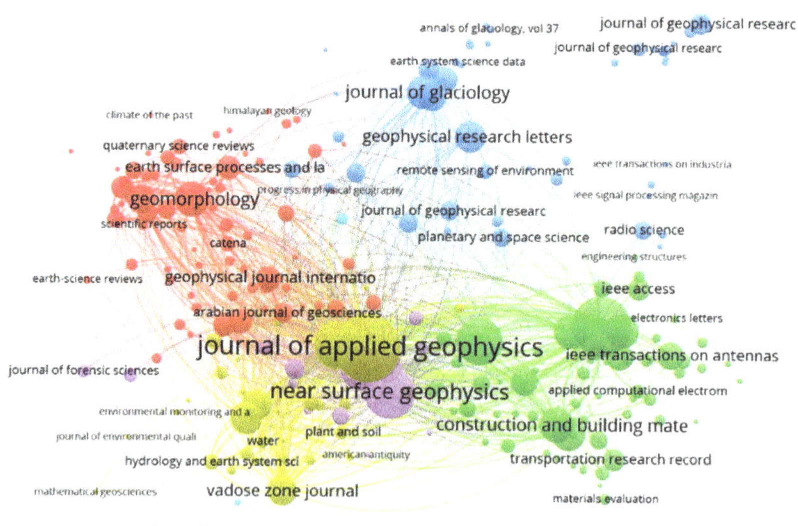

Figure 5. Bibliographic coupling analysis for the active journals.

Throughout the study, the number of WoS categories involved in GPR-related research fluctuated, ranging from 38 in 2001 to 68 in 2021. A large number of involved categories is dependent on the fact that publishing journals might cover numerous WoS categories. As shown in Figure 6, Geosciences, Multidisciplinary is the most important (2491 articles, or 17.51% of the total), followed by Geochemistry and Geophysics (1393 articles, or 9.79%), Engineering, Electrical, and Electronic (1152 articles, or 8.10%), Remote Sensing (857 articles, or 6.02%), and Imaging Science and Photographic Technology (803 articles, or 5.64%). These five categories account for 45.66% of all GPR-related research articles.

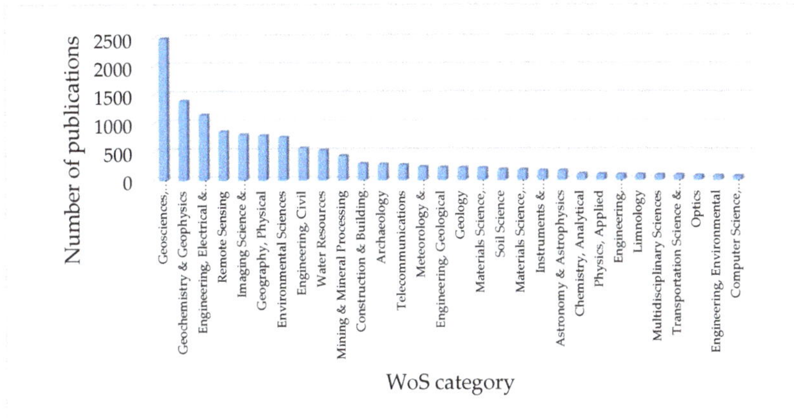

Figure 6. Distribution of top thirty WoS categories involved in GPR-related research.

After examining the growth trends of the five most important categories, it is clear that, while the trends show a general increase in the number of articles published, the scientific interest in each category has shifted over time (Figure 7). The Geosciences, Multidisciplinary and Geochemistry, and Geophysics categories have switched places multiple times over the years. The same applies to Geochemistry and Geophysics and Engineering, Electrical, and Electronic categories. In addition, there have been some significant changes in the number of articles published, such as the peaks in the Geochemistry and Geophysics category in 2007 and the sudden increase in the number of articles published in the Geosciences, Multidisciplinary category from 2011 onwards.

3.3. *Country and Institution of Publications*

Due to the lack of author address information in 24 (0.34%) of the 6880 papers, these data have been removed from the analysis of the publishing country and institution of research outputs. In the period 2001–2021, the GPR research encompasses 118 countries on six continents from all over the world. The six continents are arranged in the following order: Europe (43 nations), Asia (34 nations), Africa (18 nations), South America (10 nations), North America (9 nations), and Oceania (4 nations). It is determined that 59 countries (50.0%), 14 countries (11.86%), and 45 countries (38.14%) produced fewer than 10 articles, between 10 and 20 articles, and more than 20 articles, respectively. With 2002 research outputs (20.16%), the United States is the top productive country (see Table 2). With 959 articles (9.66%), the People's Republic of China comes in second but is still a long way behind the first. Italy (732, 7.37%), Germany (583, 5.87%), England (540, 5.44%), France (463, 4.66%), and Canada (408, 4.11%) are among the nations with at least 400 publications. Because both developing and developed countries are rated among the top ten countries, academic contributions are not solely dependent on economic progress. Furthermore, the articles have a global reach because they are scattered across three continents: North America, Asia, and Europe.

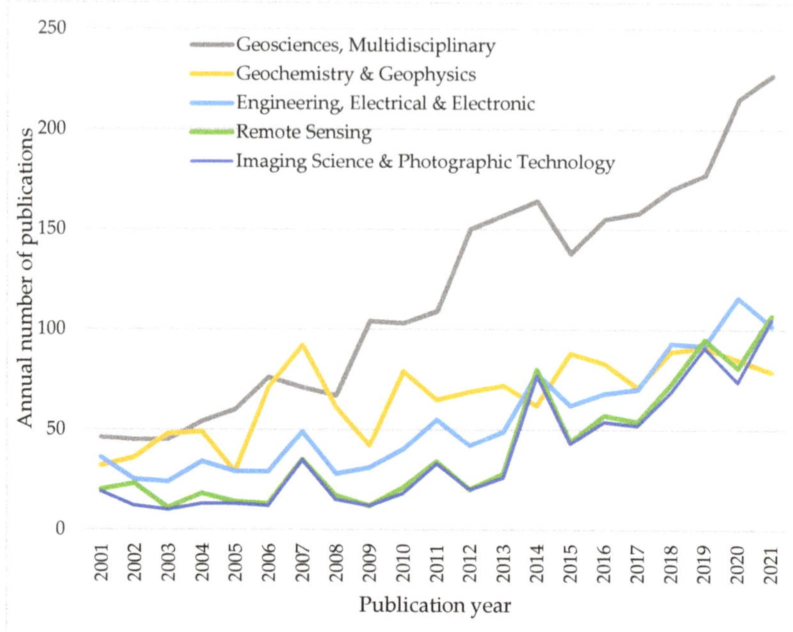

Figure 7. Number of annual publications in the top five WoS subject categories.

Table 2. Top 5 most active countries in GPR research between 2001 and 2021.

Country	Number of Publications	Average Publication Year	Total Citations	Average Citations	Average Normal Citations
United States	2002	2012.58	41,744	20.85	1.11
People's Republic of China	959	2016.75	9372	9.77	1.08
Italy	732	2013.69	12,193	16.66	1.02
Germany	583	2013.47	12,825	22.00	1.24
England	540	2013.49	12,054	22.32	1.50

Regarding the average publication year, the most recent studies are published in Qatar (2018.75), Cyprus (2018.71), and Vietnam (2018.64). The United States (41744), Germany (12825), Italy (12193), England (12054), and the People's Republic of China (9372) attain the highest total citations among other countries. Despite that, Syria, the country with the highest average citations (42.00), is not among the top-five most productive or cited countries. Concerning the average normal citations indicator, Kyrgyzstan (2.06) is the most significant country in this study subject.

The bibliographic coupling analysis for the active countries involved in GPR research is illustrated in Figure 8. Countries with at least three articles and thirty citations are included in the study. The network comprises 81 countries from a total of 118. It is worth noting that the countries are represented by nodes, where the size of each node indicates the total number of articles produced by each country. Meanwhile, the line thickness represents the strength of the cooperation link between the two countries. Malaysia (1.88), Oman (1.80), and Scotland (1.70) have the greatest influence on the average normal citation indicator, followed by Portugal (1.50), England (1.50), and Northern Ireland (1.45). The United States partners with the majority of nations, including the People's Republic of China, France, and Switzerland. Italy, which co-authored papers with the top productive countries, worked more closely with the People's Republic of China, Greece, and Algeria. Similarly, Switzerland, which coordinated with most of the other nations in the network, has Norway, Sweden, and Austria as its primary collaborators. All of these findings suggest that GPR research greatly encourages cross-national collaboration.

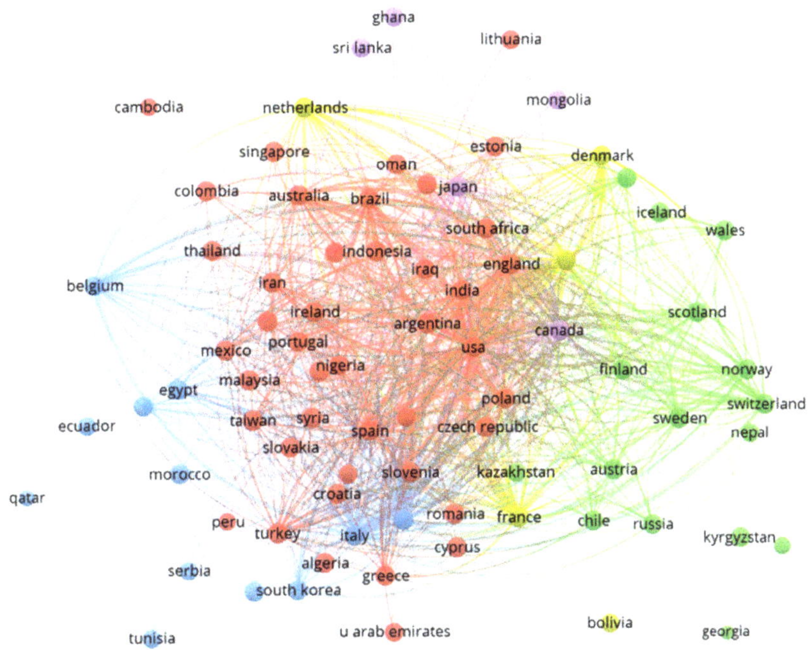

Figure 8. Bibliographic coupling analysis for the active countries.

The contribution of the institution has been assessed based on the affiliation of the article's authors. The institutions are determined by counting the total articles ascribed to a certain institution. GPR-related research is supported by 4714 institutions in total. The analysis includes organizations with at least three papers and thirty citations. Out of the total institutions, 889 organizations satisfy these criteria and are incorporated into the network (see Figure 9). It should be noted that the organizations are represented by nodes, with the size of each node indicating the total number of articles published by each organization. Meanwhile, the thickness of the line symbolizes the strength of the two organizations' cooperation relationship. The Chinese Academy of Sciences works with the vast majority of organizations, including Penn State University, the University of Waterloo, and Ohio State University. Meanwhile, the University of Wisconsin collaborates more closely with Boston University and the University of Copenhagen. All of these data indicate that GPR research promotes cross-national collaboration.

Table 3 depicts the top-five institutions that have been active in the last 20 years. The Chinese Academy of Sciences is the most productive among the main organizations, with 169 papers. The other four organizations came from four different countries (i.e., Italy, Holland, United States, and Germany); National Research Council is the most productive, with 116 papers, followed by Delft University of Technology (91), University of Illinois (88), and Forschungszentrum Jülich (85). With regard to the average publication year, Sun Yat-sen University (2019.92), Chongqing Jiaotong University (2019.80), and Central South University (2019.70) published the most recent studies. Delft University of Technology (2784), University of Leeds (2583),

Forschungszentrum Jülich (2427), Chinese Academy of Sciences (2357), and Université Catholique de Louvain (2343) obtain the highest total citations among other organizations. However, organizations with the highest average citations are not among the top five most productive institutions: the University of North Dakota (50.50), the University of California, Berkeley (47.69), and the National Center for Atmospheric Research (46.29). Based on the average normal citations indicator, the Nanjing University of Aeronautics and Astronautics (25.72) is the most significant organization in this study subject. Furthermore, while having the largest research output, the Chinese Academy of Sciences (1.26) is not regarded as the most fruitful organization since it does not have the highest average normal citations.

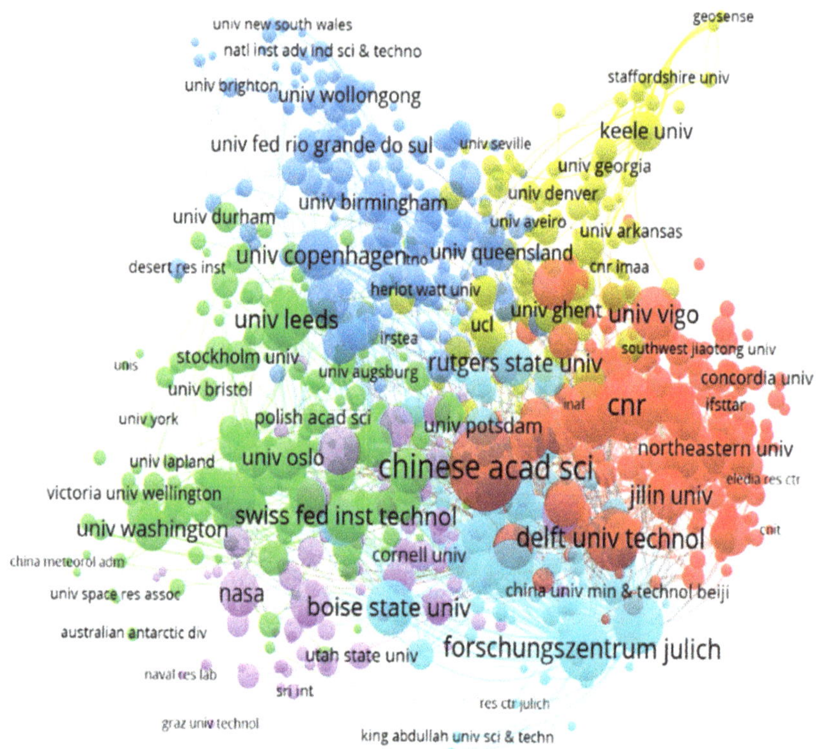

Figure 9. Bibliographic coupling analysis for the active organizations.

Table 3. Top 5 most active institutions in GPR research between 2001 and 2021.

Institution	Country	Number of Publications	Average Publication Year	Total Citations	Average Citations	Average Normal Citations
Chinese Academy of Sciences	China	169	2016.35	2357	13.95	1.26
National Research Council	Italy	116	2012.61	2339	20.16	1.05
Delft University of Technology	Holland	91	2010.84	2784	30.59	1.29
University of Illinois Urbana-Champaign	United States	88	2012.77	2207	25.08	1.43
Forschungszentrum Jülich	Germany	85	2013.94	2427	28.55	1.57

3.4. Productivity of Authors

Nineteen thousand thirty-seven individual authors are participating in GPR-related research, and the average cooperation index (i.e., number of authors per article) is 4.4. Table 4 depicts the quantitative measures of the most significant scholars. S. Lambot from the Université Catholique de Louvain and F. Soldovieri from the Italian National Research Council are the most productive authors, with 63 articles, followed by H. Vereecken from the Forschungszentrum Jülich, with 47 articles. According to the average publication year, X. Liu and D. Kumlu are among the most recently active researchers. Based on the total citations, S. Lambot, F. Soldovieri, H. Vereecken, J. Van Der Kruk, and Y. Rubin receive a total of 1547, 1271, 1118, 1027, and 1022 citations, respectively. Several scholars in the same cluster receive the same average normal citations. For example, C. Le Bastard and Y. Wang have an average normal citation of 0.74. This shows that these scholars made equal contributions to the scientific field. Furthermore, different groups of researchers differ in terms of the average normal citations. This suggests that these researchers collaborated with other teams or worked alone to develop new collisions.

Table 4. Top 5 most productive authors in GPR research between 2001 and 2021.

Scholar	Affiliation	Number of Publications	Average Publication Year	Total Citations	Average Citations	Average Normal Citations
S. Lambot	Université catholique de Louvain	63	2013.97	1547	24.56	1.37
F. Soldovieri	Italian National Research Council	63	2013.06	1271	20.17	1.18
H. Vereecken	Forschungszentrum Jülich	47	2014.64	1118	23.79	1.33
R. Persico	Institute for Archaeological and Monumental Heritage	43	2012.55	747	17.37	0.85
J. Van Der Kruk	Forschungszentrum Jülich	42	2014.29	1027	24.25	1.39

The bibliographic coupling for the most productive authors is illustrated in Figure 10. 1477 out of 19,037 authors meet the minimal criterion of three publications and thirty citations. The node size and color represent an author's number of articles and the membership cluster, respectively. Meanwhile, the distance between circles indicates the strength of the relationship between authors. In general, the shorter the distance, the stronger the connection based on bibliographic coupling. In other words, authors who are close to each other tend to cite the same publications and vice versa. The lines connecting between nodes are represented such that the thicker line indicates a greater bibliographic coupling between the two authors. The researchers are classified into seven clusters, and the size of each cluster ranges from 15 to 324. These clusters reflect the research network of academics in GPR research, such as the research group of G. Leucci, X. Comas, and E. Forte.

The co-authorship map for the influential researchers is illustrated in Figure 11. Because not all of these researchers collaborated, the cooperation network has only 721 authors. There exist 38 clusters that range in size from 3 to 77 scholars. There is a clear distinction between co-authorship and bibliographic coupling networks. The researchers in a particular co-authorship cluster may belong to a larger cluster of bibliographic coupling that comprises authors from other co-authorship clusters. For instance, F. Soldovieri is a member of the green cluster in the co-authorship map, with 41 authors. The same author, on the other hand, is a member of a larger green cluster in the author bibliographic coupling, with 266 researchers.

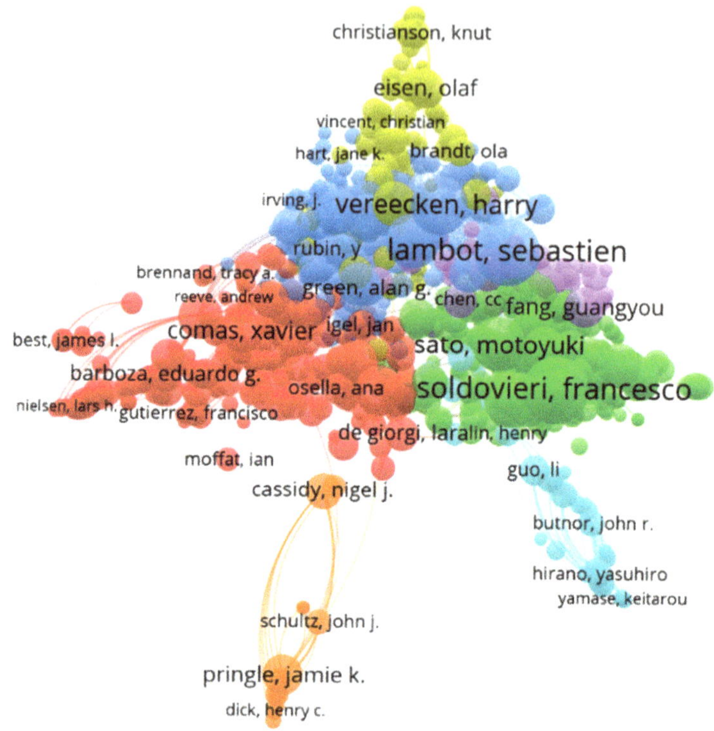

Figure 10. Bibliographic coupling analysis for the active authors.

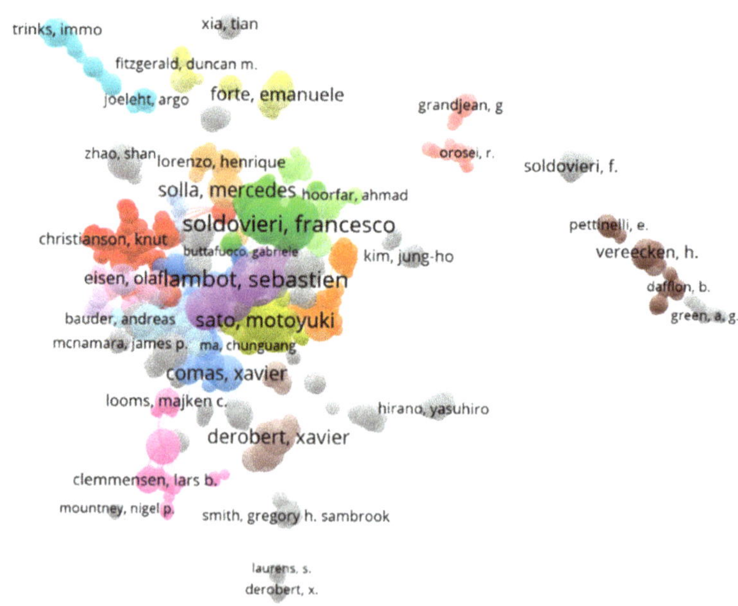

Figure 11. Co-authorship analysis for the most productive authors.

3.5. Citations of Publications

The total citation count is gathered from the Web of Sciences Core Collection (23 December 2021). A total of 470 out of 6880 papers receive at least 50 citations. The top five most-cited publications are listed in Table 5. Giannopoulos [33] is the first most referenced article (409 citations). The paper discussed the foundations of GPR operation, in addition to presenting a software tool for modeling GPR responses from complex targets (GprMax). Yoshikawa and Hinzman [34] is the second most referenced article (334 citations). Lambot et al. [35], which was referenced 299 times, is the third most cited article. Warren et al.'s [36] publication entitled "gprMax: Open source software to simulate electromagnetic wave propagation for ground penetrating radar" came in fourth place with 293 citations. Finally, Gurbuz et al.'s [37] article, which was referenced 234 times, is the fifth most cited article.

Table 5. Top 5 most influential publications in GPR research between 2001 and 2021.

Scholars	Title	Journal	Impact Factor 2021	Publication Year	Total Citations	Normal Citations
Giannopoulos [33]	"Modelling ground penetrating radar by GprMax"	*Construction and Building Materials*	6.141	2005	409	11.32
Yoshikawa and Hinzman [34]	"Shrinking thermokarst ponds and groundwater dynamics in discontinuous permafrost near Council, Alaska"	*Permafrost and Periglacial Processes*	4.368	2003	334	10.29
Lambot et al. [35]	"Modeling of ground-penetrating radar for accurate characterization of subsurface electric properties"	*IEEE Transactions on Geoscience and Remote Sensing*	5.600	2004	299	8.74
Warren et al. [36]	"gprMax: Open source software to simulate electromagnetic wave propagation for ground penetrating radar"	*Computer Physics Communications*	4.390	2016	293	20.97
Gurbuz et al. [37]	"Compressive sensing for subsurface imaging using ground penetrating radar"	*Remote Sensing*	4.662	2009	234	3.93

3.6. Cited References

When two studies cite one or more documents in common, this is known as bibliographic coupling. The bibliographic coupling strength is higher when there are more common citations in the referring works, detecting the subject similarity between the two studies. On the other side, co-citation analysis overcomes the shortcomings of bibliographic coupling by considering document citations that change over time to assess the similarity between articles. Two documents are said to be co-cited when they acquire a citation from the same third document [38].

Out of 149,276, 1004 cited references are extracted and grouped into seven clusters based on a criterion of at least 20 co-citations. Table 6 presents the top five most co-cited references. Davis and Annan [39] is the most cited reference (741 citations). The article demonstrated the ability to apply radar to map the stratigraphy of soil and rock. Neal [40] is the second most cited reference (491 citations). This research was concerned with studying the principles and problems of applying GPR in sedimentology. Daniels [41], which was referenced 453 times, is the third most cited reference. This article discussed the general system considerations, modeling aspects, applicability in different soil types, modulation techniques, and various technology applications. Jol's [42] article, which was referenced 423 times, is the fourth most cited reference. This research discussed the fundamental theory and current developments of GPR for different applications. Finally, Huisman et al.'s [43] publication came in fifth place with 334 citations. This article provided a detailed review of GPR technologies for measuring soil water content.

Table 6. Top 5 most cited references in GPR research between 2001 and 2021.

Scholars	Title	Publication Year	Total Citations
Davis and Annan [39]	"Ground-penetrating radar for high-resolution mapping of soil and rock stratigraphy"	1989	741
Neal [40]	"Ground-penetrating radar and its use in sedimentology: Principles, problems and progress"	2004	491
Daniels [41]	"Ground penetrating radar"	2004	453
Jol [42]	"Ground penetrating radar: Theory and applications"	2009	423
Huisman et al. [43]	"Measuring soil water content with ground penetrating radar: A review"	2003	334

3.7. Author Keywords in Publications

Author keywords can aid in grasping the patterns in a certain subject [44]. As a result, the author's keywords are investigated to examine the most important themes in the articles. A portion as large as 5117 (74.4%) of the 6880 total articles features one or more keywords, whereas the remaining 1763 (25.6%) articles do not incorporate any keywords. The majority of articles (1574; 22.9%) have five keywords (Figure 12).

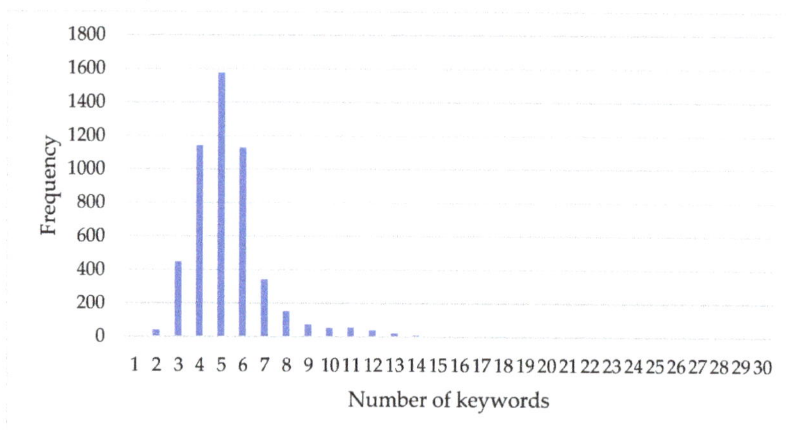

Figure 12. Statistical distribution of the keywords in GPR research publications.

There are a total of 13,626 occurrences of 27,499 unique author keywords. The 50 most common terms (0.37%) are thoroughly examined to obtain a more detailed look. These terms refer to some major "hot" issues, such as geophysics (e.g., electrical resistivity tomography), archaeological investigations (e.g., archaeology and cultural heritage), civil engineering (e.g., landmine detection and concrete), geology (e.g., stratigraphy and geomorphology), and hydrological research (e.g., hydrogeophysics, soil water content, soil, water content, and permafrost).

Keyword co-occurrence networks are one of the most prominent linguistic networks examined in the past. These networks are used to discover the semantic similarity between phrases [45]. In addition, they are used to reveal information about knowledge structures and their temporal evolution in a changing research topic [46]. The author keywords' and 'full counting' approaches are used to display the co-occurrence network. Some parameters in the VOSviewer, such as the minimum number of term occurrences, are adjusted to ensure building the map with several terms that represent the article content. When the frequency of keywords in the network map is adjusted to 10,230, 13,626 of the keywords are chosen. In addition, a thesaurus text file is inserted for combining multiple spellings of the same phrase (e.g., antennas and antenna, electrical resistivity tomography (ERT), and (FDTD) finite-difference time-domain). Furthermore, it could be useful for deleting irrelevant words (e.g., Egypt and Italy) or merging different terms referring to the same concept (e.g., numerical models and numerical modeling).

A co-occurrence map that includes 230 keywords for the whole period (2001–2021) is developed in Figure 13. This map provides an overview of GPR research and subfields with their interdependencies. The diameter of a circle and the size of labels show the occurrences of phrases. Meanwhile, the colors of the nodes reflect the clusters such that the phrases that often co-occur are clustered together on the map. Finally, the distances between nodes show the connections between the keywords. Furthermore, the keywords are organized into twelve clusters. Leading words identify each of these clusters. The keywords in the largest five clusters are described in this sub-section. The largest red category (36 items) includes keywords such as landmine detection, buried object detection, and permittivity. The second green cluster (31 items) includes phrases such as archeological prospection, geoarchaeology, monitoring, and geographic information system (GIS). The third blue cluster (27 items) comprises many keywords such as finite difference time domain, image processing, and tomography. The fourth yellow cluster (24 items) is responsible for topics related to mountain permafrost, rock glacier, soil, and seismic reflection. The fifth purple cluster (24 items) includes terms such as glacier, hydrology, ice, and snow. Within the same cluster, geophysics and electrical resistivity tomography are all significantly connected. On the contrary, there is no relationship between other terms in the same cluster, such as ice and radar signal processing. Furthermore, significant relationships between terms from different clusters, such as non-destructive testing, concrete, and condition assessment, may exist. This demonstrates the capacity of co-occurrence networks to determine the extent of a given area.

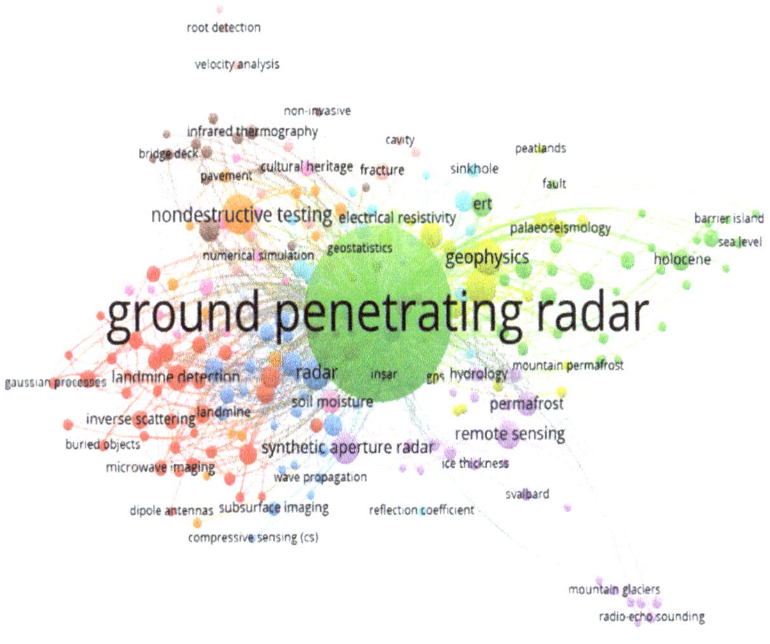

Figure 13. Co-occurrence of keywords for GPR-related studies.

With 2960 occurrences, ground-penetrating radar is the most frequent keyword (see Table 7). With 160 occurrences, the non-destructive testing keyword comes in second but is still a long way behind the first. Geophysics, electrical resistivity tomography, and radar are among the top five most frequent keywords. Regarding the average publication year, the most recent keywords include machine learning (2019.59), predictive models (2020.10), data models (2020.25), and deep learning (2020.36). On the other hand, traditional keywords include buried objects (2007.56), borehole radar (2008.80), rough surfaces (2009.25), and subsurface (2009.80). Inverse modeling (52.83), frequency domain (50.60), full-waveform inversion (43.08), seismic refraction (40.88), hydrology and buried objects (40.31) attain the highest average citations, among other keywords. Concerning the average normal citations indicator, the predictive model's keyword (13.69) is the most significant keyword in this study subject. This indicates the different applications of prediction modeling from GPR in various fields.

Table 7. Top 5 most cited keywords in GPR research between 2001 and 2021.

Keywords	Total Occurrences	Average Publication Year	Average Citations	Average Normal Citations
Ground penetrating radar	2960	2014.25	15.53	1.00
Non-destructive testing	160	2014.80	18.69	1.32
Geophysics	128	2013.50	18.34	1.02
Electrical resistivity tomography	122	2014.31	11.81	0.63
Radar	110	2010.65	21.04	0.87

4. Qualitative Analysis

The scientometric analysis provides the readers with keyword clusters without identifying the current gaps and future directions in the studied research area [47]. As a result, a qualitative discussion of the papers used in the scientometric review is conducted to provide a comprehensive classification and summarization of GPR research. It also aims at presenting the current gaps and future trends in this research field. In this research, GPR-related studies are classified based on the dedicated application. It is determined that GPR systems could be utilized in many diverse applications, including civil engineering, geological research, archaeological studies, and hydrological applications, as follows:

4.1. Civil Engineering

GPR could be widely utilized in civil engineering applications [48], including buildings [49], foundations [2], roads [50], bridges [51], railways [52], tunnels [53], landmine detection [54], pavements [8,9,55–61], and underground utilities [62–64], as follows:

4.1.1. Buildings

Buildings can be classified into different types: cultural heritage structures and modern structures. GPR is used to assess buildings and common structural components for (1) heritage preservation and building code compliance, (2) deterioration mapping, which can be used as a decision-making tool for preventive maintenance, and (3) determining the extent of structural damage (i.e., detect fractures, voids, moisture, and rebar) caused by natural disasters such as earthquakes, floods, and landslides [6]. The GPR application in building deterioration is extremely useful, especially for occupied structures. This can be attributed to the fact that GPR does not disrupt residents' and tenants' everyday activities (i.e., less intrusive), unlike other assessment methods. In addition, because building maintenance and repair are likewise costly, GPR is regarded as a useful approach to identify early problems before damage or failure is visible [65]. Despite that, the GPR application to detect the probable damage causes and support the rehabilitation of buildings after natural disasters is still very limited. Instead, the buildings are demolished or rehabilitated without using non-destructive techniques [6].

4.1.2. Foundations

There are a few studies of GPR application in substructures that examine the interaction between building foundations and the ground. Examples of these applications include: (1) detecting foundations and assessing their structural safety and integrity [66] and

(2) identifying water tables and wet ground that may induce settlement [67]. However, because of the difficulties of accessing such structures through an antenna, the number of these applications is currently limited [68].

4.1.3. Road Pavements and Bridges

Another unique application of GPR is providing subsurface information for the transport infrastructure, including roads, pavements, and bridges [69]. For road pavements, GPR surveys are conducted on flexible, rigid, semi-rigid, and composite pavements. Nowadays, the scope of these surveys is not only focused on assessing steel bars or the thickness of layers, but it is extended to conducting a structural assessment, detecting water infiltration, subsidence, voids, cracks, and anomalies [70,71]. On the other side, GPR bridge surveys are conducted either from the bridge deck or from specific bridge parts such as girders and columns [72–75]. The surveys are mostly used to diagnose bridges for detecting problems related to embedded reinforcement (e.g., bars and post-tensioned or pre-stressed tendons), corrosion and cracks [76,77], and poor compaction [78]. For the shortcomings, it is worth mentioning that GPR is still being used on an ad-hoc basis rather than regularly. Furthermore, the integration between building information modeling (BIM) and pavement management systems (PMS) is still being researched for developing integrated management and decision-making system [79,80].

4.1.4. Underground Utilities

GPR may also be used to detect the invisible and sophisticated network of underground utilities such as water supply pipes, stormwater drainage, sewers, gas pipes, power cables, communications cables, and traffic lights cables [81,82]. The mapping and scanning of underground utilities are one of the most difficult GPR activities of all civil engineering applications. This is due to the following facts: (1) radargram patterns of utility depths, orientations, and material types are frequently non-typical when compared to other infrastructures [6], and (2) the location and status of these underground utilities remain mysterious in most cities in contrast to the evident and visible damage to above-ground assets (e.g., roads and bridges). The condition deterioration of these utilities will be apparent by the occurrence of road collapse or traffic safety hazards from water leaks and seepage from water utilities, soil wash-out, and gas explosions [83,84]. Future research shall focus on GPR interpretation, particularly in extracting the hyperbolas pattern, to forecast unanticipated disturbances [82]. Another research direction incorporates examining combinations of different underground hazards under the utility networks.

4.2. Geological Studies

Identifying geological layers is necessary for locating drinking water supplies and other natural resources, as well as identifying risk zones. Furthermore, various stratigraphic and geological studies have reported significant findings concerning the Earth's surface [40,85]. Excavations are time-consuming, expensive, and frequently impossible to be conducted because some study sites are protected. As an alternative, GPR, a non-invasive geophysical technology, could be utilized to provide high-resolution subsurface imaging and map shallow formations, detect permafrost, and locate fracture or water-bearing zones [13,14,86]. It is typically used in conjunction with electrical resistivity tomography to offer comprehensive geological information. However, these geophysical technologies have not been generally evaluated and deployed because of the following reasons [6]: (1) the lack of knowledge and understanding of the capabilities of GPR in geophysics by the geotechnical engineers, and (2) the preference of the geological community to believe in the soil and rock that they can see (borehole log) over what they cannot see (radar signal). Therefore, the potential of applying GPR in geotechnical studies needs to be further examined in the future.

4.3. Archeological Applications

In archeology, GPR is capable of mapping historical buildings as well as detecting cracks, fractures, and cavities in historical sites [15,16,87]. Unlike conventional excavation methods that might destroy important archaeological structures, this subsurface imaging approach is very effective [88]. The capacity of GPR to detect the targeted buried objects is influenced by the geometry of the object of interest, subsurface geometric features, and the existence of complicated stratigraphy [89]. It has been demonstrated that soil type and density, sediment mineralogy, and moisture and clay content have a significant impact on data processing parameters and detection accuracy [90]. Other variables include topography, burial depth, and vegetation cover [89]. Therefore, it is critical to investigate and examine the burial conditions when detecting buried archaeological remains.

4.4. Hydrological Research

GPR is a potential technology for the characterization and monitoring of hydrological systems at high resolution and on a broad scale [91,92]. Object detection in the ground is affected by geological elements such as mineralogical clay, saline water, hot water, and soils [93]. The application of GPR is viable for assessing soil water content with an accurate vertical resolution and an increased spatial resolution [94]. It could also be applied in the future to examine different soil types of different sizes and gradations [95]. Furthermore, radar data may be utilized to detect the presence of liquid organic pollutants in contaminant hydrology applications [96].

5. New Avenues of GPR Applications

GPR, ERT, shallow seismic refraction (SSR), and very low frequency electromagnetic (VLFEM) are examples of the several available geophysical methods that could be utilized in the fields of civil engineering, geological research, archaeological studies, and hydrological applications. Geotechnical risks are the primary determinants of building decisions in structurally challenging zones. Different techniques, such as geotechnical techniques, geophysical tools, and remote sensing, are used and integrated to study the subsurface structures and locate any geological formations that might obstruct the development of new communities [97–99]. In the archaeological field, geophysical methods are commonly used to highlight differences in the physical behavior of the subsurface caused by the presence of buried remains [16,100,101]. Understanding the complexities of the interaction between archaeological features and their geophysical reaction will consume considerable effort. The studies revealed that integrating geophysical approaches can offset these limitations and improve the reported findings. For hydrological applications, geophysical investigations are carried out to demonstrate the capacity of technologies for detecting cavities, sinkholes, and water infiltration pathways [102].

It is demonstrated that satellite radar remote sensing systems can be applied to complement non-destructive ground-based techniques (e.g., GPR), paving the way for the smart monitoring of infrastructure assets. The combination of these approaches enables the high resolution, flexibility, and capacity of GPR to detect the sources of shallow defects to be paired with the ability of satellite remote sensing to simulate the evolution trend of distresses on a broader scale. Indeed, the increased precision can help to increase a facility's resistance to both external catastrophes and internal degradation, thus leading to infrastructure resilience [103–105]. The groundwater potential of complex areas characterized by moderate to steep slopes of topography, strong heterogeneity, multiple intrusions, and repetitive deformations could be assessed using a joint venture of satellite remote sensing, geoelectrical resistivity, and GPR techniques [106]. The integrated approach of remote sensing, sedimentological, and geophysical approaches has been proven to be accurate and successful in the mapping of paleochannels and accomplishing sustainable groundwater development goals [107]. Another important application has been reported by the combination of field observations, geophysical tools, and satellite remote sensing for landslide characterization [108]. The integration of various remote sensing techniques could be utilized for detecting and inspecting buried archaeological remains as well as assessing their preservation degrees [109,110].

The classic GPR system suffers from low detection efficiency and high labor costs when the detection field is vast. Furthermore, the GPR application could be dangerous for field investigations in harsh weather and terrain conditions. In an attempt to overcome these shortcomings, an integrated system of the unmanned aerial vehicle (UAV)-mounted GPR is developed to examine regions without being in direct contact with the Earth. This system is beneficial for spotting and detecting destructive objects such as landmines or archaeological surveys [111–114].

Future studies shall focus on automating GPR scanning operations. At the current time, several parts of the scanning process are performed manually, being labor-intensive and consuming considerable time. Furthermore, the output formats of different scanners are not uniform, and the data cannot be simply imported into a general-purpose 3D modeling software such as Autodesk Revit. This calls for the necessity of exploring other options for solving this problem [3]. In civil applications, future studies might explore using GPR in pavement design and maintenance procedures. The focus shall be given to examining the drawbacks, including the method's accuracy and the difficulty in interpreting recorded signals [115]. In the archaeological field, it is suggested to integrate geophysical data to acquire volumetric and planimetric structures in the subsurface, necessitating the use of advanced algorithms (i.e., machine learning algorithms and image data fusion) [100].

In the recent few years, satellite-based interferometry has been leveraged for analyzing and monitoring structural deformation in bridges. Wang et al. [116] used persistent scatters interferometric synthetic aperture radar (PS-InSAR) technology to scrutinize the collected time series data and detect differential deformation between piers. In this regard, they built a three-dimensional deformation model using green's function-based interpolation method. In another study, Schlögl et al. [117] experimented with the use of airborne laser scanning (ALS), vehicle-mounted mobile laser scanning (MLS), and satellite radar interferometry (InSAR) for identifying structural deformation trends. They elucidated that three non-invasive technologies were able to monitor deformation, with ALS offering a more flexible and cost-effective approach than MLS. In addition, InSAR was found to stand out as a more efficacious technology for long-term deformation assessment of bridge structures. Impact echo was lately exploited by Hu et al. [118] for defect detection in ballastless tracks. They utilized the finite-difference time-domain (FDTD) technique to emulate the propagation of elastic waves in ballastless tracks, and an improved synthetic aperture focusing technique (SAFT) was presented for the visualization of defects. In a study by Stüwe et al. [119], impact echo and ultrasonic contact testing were implemented to investigate scaling growth in geothermal pipelines. They evinced that both tests are applicable, while impact echo offered a rapid and more cost-efficient scaling monitoring alternative.

Electrical resistivity tomography (ERT) is another non-destructive technique that was newly deployed by Abudeif et al. [120] to find groundwater pathways and observe their level rise. In this context, ERT was able to create 3D profiles and 3D voxel interpretations for subsurface geoelectrical zones. As well, Guo et al. [121] jointly employed self-potential and electrical resistivity tomography for the sake of seepage detection in earth-filled dams. They managed to reestablish pseudo-3D seepage pathways by combining the measurements of electrical resistivity alongside inversion outcomes of self-potential data. An infrared thermography camera is a contactless non-destructive technique that was mounted by unmanned aerial vehicles (UAV) and used by Zhou et al. [122] for the sake of automated detection of earth embankment leakage. In their study, an AlexNet-based transfer learning framework was created for the classification of infrared images into either cold slope leakage, warm slope leakage, normal slope, normal ponding, cold piping, and warm piping. Moreover, Loiotine et al. [123] utilized airborne infrared thermography for characterization of rock mass in complex conditions. By analyzing thermograms, they succeeded in mapping the correlation exhibited between rock mass properties and temperature profiles.

6. Conclusions

The research study applied a holistic approach of bibliometric and scientometric assessment to present a global overview of ground penetrating radar (GPR) research from 2001 to 2021. The Web of Science database produced 6880 publications that were examined with respect to the publication trends, sources of publications and subject categories, cooperation of countries, the productivity of authors, citations of publications, and clusters of keywords. According to the findings, there has been a shift in the development and promotion of the field of GPR. The number of annual publications had climbed from 139 in 2001 to 576 in 2021. Specifically, the publishing output has risen rapidly since 2006, with a multidisciplinary and multi-regional approach distinguishing it. The research studies were published in 894 journals, with the number of active journals rising from 68 in 2001 to 215 in 2021. The number of subject categories included in GPR-related research fluctuated during the study, ranging from 38 in 2001 to 68 in 2021. The GPR research studies involved 118 countries from all around the world. The United States and the People's Republic of China made the most significant contributions to the research community. The Chinese Academy of Sciences, China, was the most prolific institution, followed by the National Research Council, Italy. "Modelling ground penetrating radar by GprMax" article that was authored by Giannopoulos [33] in the *Construction and Building Materials* journal and was the top most cited article (409 citations). Ground-penetrating radar, non-destructive testing, geophysics, electrical resistivity tomography, and radar ranked first through fifth in terms of emerging keywords. GPR was widely applied in four different fields; civil engineering (landmine detection, bridge deck, and asphalt pavement), geological research (sedimentology, stratigraphy, and Holocene), archaeological studies (archeology, cultural heritage, and geoarchaeology), and hydrological practices (soil moisture, soil, and moisture content). All of these findings and conclusions have been interpreted in light of the Web of Science database. This review article could assist academics in identifying the most prestigious journals and researchers with whom to collaborate or publish in the future. It also aided in recognizing current hotspots in order to gain a comprehensive understanding of the subject at hand.

Author Contributions: Conceptualization, N.E., A.A.-S. and E.M.A.; methodology, N.E. and E.M.A.; formal analysis, N.E., A.A.-S. and E.M.A.; data curation, N.E., A.A.-S. and E.M.A.; investigation, N.E., A.A.-S., E.M.A. and T.Z.; resources, N.E., A.A.-S., E.M.A. and T.Z.; writing—original draft preparation, N.E., A.A.-S. and E.M.A.; writing—review and editing, N.E., A.A.-S., E.M.A. and T.Z. All authors have read and agreed to the published version of the manuscript.

Funding: This research received no external funding.

Institutional Review Board Statement: Not applicable.

Informed Consent Statement: Informed consent was obtained from all subjects involved in the study.

Data Availability Statement: Some or all data, models, or code that support the findings of this study are available from the corresponding author upon reasonable request.

Conflicts of Interest: The authors declare no conflict of interest.

References

1. Rhee, J.Y.; Park, K.T.; Cho, J.W.; Lee, S.Y. A study of the application and the limitations of GPR investigation on underground survey of the Korean expressways. *Remote Sens.* **2021**, *13*, 1805. [CrossRef]
2. Zhou, D.; Zhu, H. Application of ground penetrating radar in detecting deeply embedded reinforcing bars in pile foundation. *Adv. Civ. Eng.* **2021**, *2021*, 4813415. [CrossRef]
3. Xiang, Z.; Rashidi, A.; Ou, G.G. States of practice and research on applying GPR technology for labeling and scanning constructed facilities. *J. Perform. Constr. Facil.* **2019**, *33*, 03119001. [CrossRef]
4. Dabous, S.A.; Yaghi, S.; Alkass, S.; Moselhi, O. Concrete bridge deck condition assessment using IR thermography and ground penetrating radar technologies. *Autom. Constr.* **2017**, *81*, 340–354. [CrossRef]
5. Teoh, Y.J.; Bruka, M.A.; Idris, N.M.; Ismail, N.A.; Muztaza, N.M. Introduction of a ground penetrating radar system for subsurface investigation in Balik Pulau, Penang Island. *J. Phys. Conf. Ser.* **2018**, *995*, 012098. [CrossRef]

6. Lai, W.W.L.; Derobert, X.; Annan, P. A review of ground penetrating radar application in civil engineering: A 30-year journey from locating and testing to imaging and diagnosis. *NDT E Int.* **2018**, *96*, 58–78.
7. Rathod, H.; Debeck, S.; Gupta, R.; Chow, B. Applicability of GPR and a rebar detector to obtain rebar information of existing concrete structures. *Case Stud. Constr. Mater.* **2019**, *11*, e00240. [CrossRef]
8. Wang, S.; Zhao, S.; Al-Qadi, I.L. Real-time monitoring of asphalt concrete pavement density during construction using ground penetrating radar: Theory to practice. *Transp. Res. Rec.* **2019**, *2673*, 329–338. [CrossRef]
9. Joshaghani, A.; Shokrabadi, M. Ground penetrating radar (GPR) applications in concrete pavements. *Int. J. Pavement Eng.* **2022**, *23*, 4504–4531. [CrossRef]
10. Tabarro, P.G.; Pouliot, J.; Losier, L.M.; Fortier, R. Detection and location of buried infrastructures using ground penetrating radar: A new approach based on GIS and data integration. *Int. J. 3D Inf. Model. IJ3DIM* **2018**, *7*, 57–77. [CrossRef]
11. Li, S.; Zhang, Y.; Han, S. Safety inspection system and comprehensive evaluation method for concrete structure of gas pipeline tunnel based on fuzzy mathematics. *Adv. Mech. Eng.* **2021**, *13*, 16878140211046098. [CrossRef]
12. Kgarume, T.; Van Schoor, M.; Nontso, Z. The use of 3D ground penetrating radar to mitigate the risk associated with falls of ground in Bushveld Complex platinum mines. *J. S. Afr. Inst. Min. Metall.* **2019**, *119*, 973–982. [CrossRef]
13. Fedorova, L.; Lejzerowicz, A.; Kulyandin, G.; Savvin, D.; Fedorov, M. Ground penetrating radar investigations of the geological structure of loose sediments at solid mineral deposits. *E3S Web Conf.* **2020**, *192*, 04005. [CrossRef]
14. Zheng, L.; Li, X.; Liu, Z.; Huang, D.; Tang, Z. Accuracy evaluation of advanced geological prediction based on improved AHP and GPR. *Math. Probl. Eng.* **2020**, *2020*, 8617165. [CrossRef]
15. Ebraheem, M.O.; Ibrahim, H.A. Contributions of ground-penetrating radar in research of some predynastic and dynastic archaeological sites at the eastern and western banks of the River Nile, Assiut, Egypt. *Archaeol. Prospect.* **2022**, *29*, 177–189.
16. El Emam, A.E.; Lethy, A.; Radwan, A.M.; Awad, A. Archaeological investigation and hazard assessment using magnetic, ground-penetrating radar, and GPS tools at Dahshour area, Giza, Egypt. *Front. Earth Sci.* **2021**, *9*, 437. [CrossRef]
17. Huang, Y.; Ding, X.H.; Liu, R.; He, Y.; Wu, S. Reviewing the domain of technology and innovation management: A visualizing bibliometric analysis. *SAGE Open* **2019**, *9*, 2158244019854644. [CrossRef]
18. de Oliveira, O.J.; da Silva, F.F.; Juliani, F.; Barbosa, L.C.F.M.; Nunhes, T.V. Bibliometric Method for Mapping the State-of-the-Art and Identifying Research Gaps and Trends in Literature: An Essential Instrument to Support the Development of Scientific Projects. In *Scientometrics Recent Advances*; IntechOpen: London, UK, 2019.
19. Donthu, N.; Kumar, S.; Mukherjee, D.; Pandey, N.; Lim, W.M. How to conduct a bibliometric analysis: An overview and guidelines. *J. Bus. Res.* **2021**, *133*, 285–296. [CrossRef]
20. Mejia, C.; Wu, M.; Zhang, Y.; Kajikawa, Y. Exploring topics in bibliometric research through citation networks and semantic analysis. *Front. Res. Metr.* **2021**, *6*, 742311. [CrossRef] [PubMed]
21. Gizzi, F.T.; Leucci, G. Global research patterns on ground penetrating radar (GPR). *Surv. Geophys.* **2018**, *39*, 1039–1068. [CrossRef]
22. Snyder, H. Literature review as a research methodology: An overview and guidelines. *J. Bus. Res.* **2019**, *104*, 333–339. [CrossRef]
23. Moral Muñoz, J.A.; Herrera Viedma, E.; Santisteban Espejo, A.; Cobo, M.J. Software tools for conducting bibliometric analysis in science: An up-to-date review. *Prof. Inf.* **2020**, *29*, e290103. [CrossRef]
24. Birkle, C.; Pendlebury, D.A.; Schnell, J.; Adams, J. Web of Science as a data source for research on scientific and scholarly activity. *Quant. Sci. Stud.* **2020**, *1*, 363–376. [CrossRef]
25. Butt, N.S.; Malik, A.A.; Shahbaz, M.Q. Bibliometric analysis of statistics journals indexed in web of science under emerging source citation index. *SAGE Open* **2021**, *11*, 2158244020988870. [CrossRef]
26. Butler, L.; Visser, M.S. Extending citation analysis to non-source items. *Scientometrics* **2006**, *66*, 327–343. [CrossRef]
27. Waltman, L.; Van Eck, N.J.; Noyons, E.C. A unified approach to mapping and clustering of bibliometric networks. *J. Informetr.* **2010**, *4*, 629–635. [CrossRef]
28. Kessler, M.M. Bibliographic coupling between scientific papers. *Am. Doc.* **1963**, *14*, 10–25. [CrossRef]
29. Rehman, S.U.; Al-Almaie, S.M.; Haq, I.U.; Ahmad, S.; Ahmad, S.; Al-Shammari, M.A.; Darwish, M.; Mustafa, T. Journal of family and community medicine: A scientometric analysis 1994–2020. *J. Fam. Community Med.* **2021**, *28*, 164.
30. Yanhui, S.; Lijuan, W.; Shiji, C. An exploratory study of the all-author bibliographic coupling analysis: Taking scientometrics for example. *J. Inf. Sci.* **2022**, *48*, 767–782. [CrossRef]
31. Biresselioglu, M.E.; Demir, M.H.; Solak, B.; Kayacan, A.; Altinci, S. Investigating the trends in arctic research: The increasing role of social sciences and humanities. *Sci. Total Environ.* **2020**, *729*, 139027. [CrossRef]
32. Van Eck, N.J.; Waltman, L. Visualizing bibliometric networks. In *Measuring Scholarly Impact*; Springer: Cham, Switzerland, 2014; pp. 285–320.
33. Giannopoulos, A. Modelling ground penetrating radar by GprMax. *Constr. Build. Mater.* **2005**, *19*, 755–762. [CrossRef]
34. Yoshikawa, K.; Hinzman, L.D. Shrinking thermokarst ponds and groundwater dynamics in discontinuous permafrost near Council, Alaska. *Permafr. Periglac. Process.* **2003**, *14*, 151–160. [CrossRef]
35. Lambot, S.; Slob, E.C.; van den Bosch, I.; Stockbroeckx, B.; Vanclooster, M. Modeling of ground-penetrating radar for accurate characterization of subsurface electric properties. *IEEE Trans. Geosci. Remote Sens.* **2004**, *42*, 2555–2568. [CrossRef]
36. Warren, C.; Giannopoulos, A.; Giannakis, I. gprMax: Open source software to simulate electromagnetic wave propagation for ground penetrating radar. *Comput. Phys. Commun.* **2016**, *209*, 163–170. [CrossRef]

37. Gurbuz, A.C.; McClellan, J.H.; Scott, W.R., Jr. Compressive sensing for subsurface imaging using ground penetrating radar. *Signal Process.* **2009**, *89*, 1959–1972. [CrossRef]
38. Mas-Tur, A.; Roig-Tierno, N.; Sarin, S.; Haon, C.; Sego, T.; Belkhouja, M.; Porter, A.; Merigó, J.M. Co-citation, bibliographic coupling and leading authors, institutions and countries in the 50 years of technological forecasting and social change. *Technol. Forecast Soc. Chang.* **2021**, *165*, 120487. [CrossRef]
39. Davis, J.L.; Annan, A.P. Ground-penetrating radar for high-resolution mapping of soil and rock stratigraphy. *Geophys. Prospect.* **1989**, *37*, 531–551. [CrossRef]
40. Neal, A. Ground-penetrating radar and its use in sedimentology: Principles, problems and progress. *Earth Sci. Rev.* **2004**, *66*, 261–330. [CrossRef]
41. Daniels, D.J. *Ground Penetrating Radar*, 2nd ed.; IEE Radar, Sonar and Navigation Series; The Institution of Electrical Engineers: London, UK, 2004.
42. Jol, H.M. *Ground Penetrating Radar: Theory and Applications*; Elsevier: Amsterdam, The Netherlands, 2009; Volume 509.
43. Huisman, J.A.; Hubbard, S.S.; Redman, J.D.; Annan, A.P. Measuring soil water content with ground penetrating radar: A review. *Vadose Zone J.* **2003**, *2*, 476–491. [CrossRef]
44. Tripathi, M.; Kumar, S.; Sonker, S.K.; Babbar, P. Occurrence of author keywords and keywords plus in social sciences and humanities research: A preliminary study. *CJSIM* **2018**, *12*, 215–232. [CrossRef]
45. Bordag, S. A comparison of co-occurrence and similarity measures as simulations of context. In *International Conference on Intelligent Text Processing and Computational Linguistics*; Springer: Berlin/Heidelberg, Germany, 2008; pp. 52–63.
46. Radhakrishnan, S.; Erbis, S.; Isaacs, J.A.; Kamarthi, S. Novel keyword co-occurrence network-based methods to foster systematic reviews of scientific literature. *PLoS ONE* **2017**, *12*, e0172778.
47. Hussein, M.; Zayed, T. Crane operations and planning in modular integrated construction: Mixed review of literature. *Autom. Constr.* **2021**, *122*, 103466. [CrossRef]
48. Benedetto, A.; Pajewski, L. *Civil Engineering Applications of Ground Penetrating Radar*; Springer: Cham, Switzerland, 2015; p. 371.
49. Pérez-Gracia, V.; Caselles, O.; Clapés, J.; Santos-Assunçao, S. GPR building inspection: Examples of building structures assessed with ground penetrating radar. In Proceedings of the 2017 9th International Workshop on Advanced Ground Penetrating Radar (IWAGPR), Edinburgh, UK, 28–30 June 2017.
50. Benedetto, A.; Tosti, F.; Ciampoli, L.B.; D'amico, F. An overview of ground-penetrating radar signal processing techniques for road inspections. *Signal Process.* **2017**, *132*, 201–209. [CrossRef]
51. Alani, A.M.; Aboutalebi, M.; Kilic, G. Applications of ground penetrating radar (GPR) in bridge deck monitoring and assessment. *J. Appl. Geophy.* **2013**, *97*, 45–54. [CrossRef]
52. Benedetto, A.; Tosti, F.; Ciampoli, L.B.; Calvi, A.; Brancadoro, M.G.; Alani, A.M. Railway ballast condition assessment using ground-penetrating radar–An experimental, numerical simulation and modelling development. *Constr. Build. Mater.* **2017**, *140*, 508–520. [CrossRef]
53. Alani, A.M.; Tosti, F. GPR applications in structural detailing of a major tunnel using different frequency antenna systems. *Constr. Build. Mater.* **2018**, *158*, 1111–1122. [CrossRef]
54. Lopera, O.; Slob, E.C.; Milisavljevic, N.; Lambot, S. Filtering soil surface and antenna effects from GPR data to enhance landmine detection. *IEEE Trans. Geosci. Remote Sens.* **2007**, *45*, 707–717. [CrossRef]
55. Al-Qadi, I.L.; Lahouar, S. Measuring layer thicknesses with GPR–Theory to practice. *Constr. Build. Mater.* **2005**, *19*, 763–772. [CrossRef]
56. Benedetto, A.; Pensa, S. Indirect diagnosis of pavement structural damages using surface GPR reflection techniques. *J. Appl. Geophy.* **2007**, *62*, 107–123. [CrossRef]
57. Loizos, A.; Plati, C. Accuracy of pavement thicknesses estimation using different ground penetrating radar analysis approaches. *NDT E Int.* **2007**, *40*, 147–157. [CrossRef]
58. Loizos, A.; Plati, C. Accuracy of ground penetrating radar horn-antenna technique for sensing pavement subsurface. *IEEE Sens. J.* **2007**, *7*, 842–850. [CrossRef]
59. Lahouar, S.; Al-Qadi, I.L. Automatic detection of multiple pavement layers from GPR data. *NDT E Int.* **2008**, *41*, 69–81. [CrossRef]
60. Al-Qadi, I.L.; Leng, Z.; Lahouar, S.; Baek, J. In-place hot-mix asphalt density estimation using ground-penetrating radar. *Transp. Res. Rec.* **2010**, *2152*, 19–27. [CrossRef]
61. Plati, C.; Loizos, A. Estimation of in-situ density and moisture content in HMA pavements based on GPR trace reflection amplitude using different frequencies. *J. Appl. Geophy.* **2013**, *97*, 3–10. [CrossRef]
62. Yan, J.; Jaw, S.W.; Soon, K.H.; Wieser, A.; Schrotter, G. Towards an underground utilities 3D data model for land administration. *Remote Sens.* **2019**, *11*, 1957. [CrossRef]
63. Iftimie, N.; Savin, A.; Steigmann, R.; Dobrescu, G.S. Underground pipeline identification into a non-destructive case study based on ground-penetrating radar imaging. *Remote Sens.* **2021**, *13*, 3494. [CrossRef]
64. Bernatek-Jakiel, A.; Kondracka, M. Detection of soil pipes using ground penetrating radar. *Remote Sens.* **2019**, *11*, 1864. [CrossRef]
65. Pérez-Gracia, V.; García García, F.; Rodriguez Abad, I. GPR evaluation of the damage found in the reinforced concrete base of a block of flats: A case study. *NDT E Int.* **2008**, *41*, 341–353. [CrossRef]

66. Ortega-Ramírez, J.; Bano, M.; Cordero-Arce, M.T.; Villa-Alvarado, L.A.; Fraga, C.C. Application of non-invasive geophysical methods (GPR and ERT) to locate the ancient foundations of the first cathedral of Puebla, Mexico. A case study. *J. Appl. Geophys.* **2020**, *174*, 103958. [CrossRef]
67. Salako, A.O.; Osotuyi, A.G.; Adepelumi, A.A. Seepage investigations of heterogeneous soils beneath some buildings using geophysical approaches: Example from southwestern Nigeria. *Int. J. Geo-Eng.* **2019**, *10*, 11. [CrossRef]
68. Pérez-Gracia, V.; Solla, M. Inspection procedures for effective GPR surveying of buildings. In *Civil Engineering Applications of Ground Penetrating Radar*; Benedetto, A., Pajewski, L., Eds.; Springer: Berlin/Heidelberg, Germany, 2015; pp. 97–123.
69. Solla, M.; Pérez-Gracia, V.; Fontul, S. A review of GPR application on transport infrastructures: Troubleshooting and best practices. *Remote Sens.* **2021**, *13*, 672. [CrossRef]
70. Pedret Rodés, J.; Martínez Reguero, A.; Pérez-Gracia, V. GPR spectra for monitoring asphalt pavements. *Remote Sens.* **2020**, *12*, 1749. [CrossRef]
71. Liu, X.; Hao, P.; Wang, A.; Zhang, L.; Gu, B.; Lu, X. Non-destructive detection of highway hidden layer defects using a ground-penetrating radar and adaptive particle swarm support vector machine. *PeerJ Comput. Sci.* **2021**, *7*, e417. [CrossRef] [PubMed]
72. Alsharqawi, M.; Zayed, T.; Dabous, S.A. Integrated condition rating and forecasting method for bridge decks using visual inspection and ground penetrating radar. *Autom. Constr.* **2018**, *89*, 135–145. [CrossRef]
73. Dinh, K.; Gucunski, N.; Duong, T.H. An algorithm for automatic localization and detection of rebars from GPR data of concrete bridge decks. *Autom. Constr.* **2018**, *89*, 292–298. [CrossRef]
74. Abdelkader, E.M.; Marzouk, M.; Zayed, T. An optimization-based methodology for the definition of amplitude thresholds of the ground penetrating radar. *Soft Comput.* **2019**, *23*, 12063–12086. [CrossRef]
75. Dinh, K.; Gucunski, N.; Zayed, T. Automated visualization of concrete bridge deck condition from GPR data. *NDT E Int.* **2019**, *102*, 120–128. [CrossRef]
76. Bachiri, T.; Khamlichi, A.; Bezzazi, M. Bridge deck condition assessment by using GPR: A review. *MATEC Web Conf.* **2018**, *191*, 00004. [CrossRef]
77. Zaki, A.; Megat Johari, M.A.; Wan Hussin, W.M.A.; Jusman, Y. Experimental assessment of rebar corrosion in concrete slab using ground penetrating radar (GPR). *Int. J. Corros.* **2018**, *2018*, 5389829. [CrossRef]
78. Plati, C.; Loizos, A.; Gkyrtis, K. Integration of non-destructive testing methods to assess asphalt pavement thickness. *NDT E Int.* **2020**, *115*, 102292. [CrossRef]
79. Oreto, C.; Massotti, L.; Biancardo, S.A.; Veropalumbo, R.; Viscione, N.; Russo, F. BIM-based pavement management tool for scheduling urban road maintenance. *Infrastructures* **2021**, *6*, 148. [CrossRef]
80. Oreto, C.; Biancardo, S.A.; Viscione, N.; Veropalumbo, R.; Russo, F. Road pavement information modeling through maintenance scenario evaluation. *J. Adv. Transp.* **2021**, *2021*, 8823117. [CrossRef]
81. Gabryś, M.; Ortyl, Ł. Georeferencing of multi-channel GPR—Accuracy and efficiency of mapping of underground utility networks. *Remote Sens.* **2020**, *12*, 2945. [CrossRef]
82. Ali, H.; Ideris, N.S.M.; Zaidi, A.A.; Azalan, M.Z.; Amran, T.T.; Ahmad, M.R.; Rahim, N.A.; Shukor, S.A. Ground penetrating radar for buried utilities detection and mapping: A review. *J. Phys. Conf. Ser.* **2021**, *2107*, 012056. [CrossRef]
83. Kuliczkowska, E. An analysis of road pavement collapses and traffic safety hazards resulting from leaky sewers. *Balt. J. Road Bridge Eng.* **2016**, *11*, 251–258. [CrossRef]
84. Chen, S.Y.; Lin, M.S.; Hsiao, G.L.K.; Wang, T.C.; Kao, C.S. Underground pipeline leakage risk assessment in an urban city. *Int. J. Environ. Res. Public Health* **2020**, *17*, 3929. [CrossRef]
85. Slob, E.; Sato, M.; Olhoeft, G. Surface and borehole ground-penetrating-radar developments. *Geophysics* **2010**, *75*, A103–A175. [CrossRef]
86. Proulx-McInnis, S.; St-Hilaire, A.; Rousseau, A.N.; Jutras, S. A review of ground-penetrating radar studies related to peatland stratigraphy with a case study on the determination of peat thickness in a northern boreal fen in Quebec, Canada. *Prog. Phys. Geogr.* **2013**, *37*, 767–786. [CrossRef]
87. Corradini, E.; Wilken, D.; Zanon, M.; Groß, D.; Lübke, H.; Panning, D.; Dörfler, W.; Rusch, K.; Mecking, R.; Erkul, E.; et al. Reconstructing the palaeoenvironment at the early Mesolithic site of Lake Duvensee: Ground-penetrating radar and geoarchaeology for 3D facies mapping. *Holocene* **2020**, *30*, 820–833. [CrossRef]
88. Conyers, L.B. *Ground-Penetrating Radar for Geoarchaeology*; John Wiley & Sons: Hoboken, NJ, USA, 2016.
89. Conyers, L.B.; Cameron, C.M. Ground-penetrating radar techniques and three-dimensional computer mapping in the American Southwest. *J. Field Archaeol.* **1998**, *25*, 417–430.
90. Annan, A.P.; Cosway, S.W.; Redman, J.D. Water table detection with ground-penetrating radar. In *SEG Technical Program Expanded Abstracts*; Society of Exploration Geophysicists: Tulsa, OK, USA, 1991; pp. 494–496.
91. Mahmoudzadeh, M.R.; Francés, A.P.; Lubczynski, M.; Lambot, S. Using ground penetrating radar to investigate the water table depth in weathered granites—Sardon case study, Spain. *J. Appl. Geophys.* **2012**, *79*, 17–26. [CrossRef]
92. Lambot, S.; Weihermüller, L.; Huisman, J.A.; Vereecken, H.; Vanclooster, M.; Slob, E.C. Analysis of air-launched ground-penetrating radar techniques to measure the soil surface water content. *Water Resour. Res.* **2006**, *42*, 1–12. [CrossRef]
93. Maryanto, S.; Suciningtyas, I.K.L.N.; Dewi, C.N.; Rachmansyah, A. Integrated resistivity and ground penetrating radar observations of underground seepage of hot water at Blawan-Ijen geothermal field. *Int. J. Geophys.* **2016**, *2016*, 6034740. [CrossRef]

94. Liu, X.; Chen, J.; Cui, X.; Liu, Q.; Cao, X.; Chen, X. Measurement of soil water content using ground-penetrating radar: A review of current methods. *Int. J. Digit. Earth* **2019**, *12*, 95–118. [CrossRef]
95. Zhang, S.; Zhang, L.; Ling, T.; Fu, G.; Guo, Y. Experimental research on evaluation of soil water content using ground penetrating radar and wavelet packet-based energy analysis. *Remote Sens.* **2021**, *13*, 5047. [CrossRef]
96. Knight, R. Ground penetrating radar for environmental applications. *Annu. Rev. Earth Planet Sci.* **2001**, *29*, 229–255. [CrossRef]
97. Mesbah, H.; Shokry, M.; Soliman, M.; Atya, M. Integrated geophysical investigations to detect the shallow subsurface settings at new Sohag city, Egypt. *Int. J. Geosci.* **2017**, *8*, 364–377. [CrossRef]
98. Sulaiman, N.; Nordiana, M.M.; Azwin, I.N.; Taqiuddin, Z.M.; Maslinda, U.; Hisham, H.; Amalina, M.N.; Saharudin, M.A.; Nordiana, A.N. Integration of ground penetrating radar (GPR) and 2-D resistivity imaging methods for soil investigation. *IOP Conf. Ser. Earth Environ. Sci.* **2017**, *62*, 012007. [CrossRef]
99. Araffa, S.A.S.; Gobashy, M.M.; Khalil, M.H.; Abdelaal, A. Integration of geophysical techniques to detect geotechnical hazards: A case study in Mokattam, Cairo, Egypt. *Bull. Eng. Geol. Environ.* **2021**, *80*, 8021–8041. [CrossRef]
100. Capozzoli, L.; Giampaolo, V.; De Martino, G.; Perciante, F.; Lapenna, V.; Rizzo, E. ERT and GPR prospecting applied to unsaturated and subwater analogue archaeological site in a full scale laboratory. *Appl. Sci.* **2022**, *12*, 1126. [CrossRef]
101. Murín, I.; Neumann, M.; Brady, C.; Bátora, J.; Čapo, M.; Drozd, D. Application of magnetometry, georadar (GPR) and geoelectrical methods in archaeo-geophysical investigation of a Napoleonic battlefield with fortification at Pressburg (Bratislava, Slovakia). *J. Appl. Geophys.* **2022**, *196*, 104493. [CrossRef]
102. Hussain, Y.; Uagoda, R.; Borges, W.; Nunes, J.; Hamza, O.; Condori, C.; Aslam, K.; Dou, J.; Cárdenas-Soto, M. The potential use of geophysical methods to identify cavities, sinkholes and pathways for water infiltration. *Water* **2020**, *12*, 2289. [CrossRef]
103. Clementini, C.; Latini, D.; Gagliardi, V.; Ciampoli, L.B.; D'Amico, F.; Del Frate, F. Synergistic Monitoring of Transport Infrastructures by Multi-Temporal InSAR and GPR Technologies: A Case Study in Salerno, Italy. In Proceedings of the Earth Resources and Environmental Remote Sensing/GIS Applications XII, Online, 13–18 September 2021.
104. D'Amico, F.; Gagliardi, V.; Clementini, C.; Latini, D.; Del Frate, F.; Bianchini Ciampoli, L.; Di Benedetto, A.; Fiani, M.; Benedetto, A. Integrated Health Monitoring of Transport Assets by Ground-Based Non-Destructive Technologies (NDTs) and Satellite Remote Sensing Analysis. In Proceedings of the EGU General Assembly Conference Abstracts, Online, 19–30 April 2021.
105. Tosti, F.; Gagliardi, V.; Ciampoli, L.B.; Benedetto, A.; Threader, S.; Alani, A.M. Integration of Remote Sensing and Ground-Based Non-Destructive Methods in Transport Infrastructure Monitoring: Advances, Challenges and Perspectives. In Proceedings of the 2021 IEEE Asia-Pacific Conference on Geoscience, Electronics and Remote Sensing Technology (AGERS), Jakarta Pusat, Indonesia, 29–30 September 2021.
106. Maury, S.; Tiwari, R.K.; Balaji, S. Joint application of satellite remote sensing, ground penetrating radar (GPR) and resistivity techniques for targeting ground water in fractured Ophiolites of South Andaman Island, India. *Environ. Earth Sci.* **2016**, *75*, 237. [CrossRef]
107. Upadhyay, R.K.; Kishore, N.; Sharma, M. Delineation and mapping of palaeochannels using remote sensing, geophysical, and sedimentological techniques: A comprehensive approach. *Water Sci.* **2021**, *35*, 100–108. [CrossRef]
108. Kannaujiya, S.; Chattoraj, S.L.; Jayalath, D.; Bajaj, K.; Podali, S.; Bisht, M.P.S. Integration of satellite remote sensing and geophysical techniques (electrical resistivity tomography and ground penetrating radar) for landslide characterization at Kunjethi (Kalimath), Garhwal Himalaya, India. *Nat. Hazards* **2019**, *97*, 1191–1208. [CrossRef]
109. Agapiou, A.; Lysandrou, V.; Sarris, A.; Papadopoulos, N.; Hadjimitsis, D.G. Fusion of satellite multispectral images based on ground-penetrating radar (GPR) data for the investigation of buried concealed archaeological remains. *Geosciences* **2017**, *7*, 40. [CrossRef]
110. Zong, X.; Wang, X.Y.; Luo, L. The integration of VHR satellite imagery, GPR survey and boring for archaeological prospection at the Longcheng Site in Anhui Province, China. *Archaeometry* **2018**, *60*, 1088–1105. [CrossRef]
111. Fernández, M.G.; López, Y.Á.; Arboleya, A.A.; Valdés, B.G.; Vaqueiro, Y.R.; Andrés, F.L.H.; García, A.P. Synthetic aperture radar imaging system for landmine detection using a ground penetrating radar on board a unmanned aerial vehicle. *IEEE Access* **2018**, *6*, 45100–45112. [CrossRef]
112. Garcia-Fernandez, M.; Alvarez-Lopez, Y.; Las Heras, F.; Gonzalez-Valdes, B.; Rodriguez-Vaqueiro, Y.; Pino, A.; Arboleya-Arboleya, A. GPR system onboard a UAV for non-invasive detection of buried objects. In Proceedings of the 2018 IEEE International Symposium on Antennas and Propagation & USNC/URSI National Radio Science Meeting, Boston, MA, USA, 8–13 July 2018.
113. Šipoš, D.; Gleich, D. A lightweight and low-power UAV-borne ground penetrating radar design for landmine detection. *Sensors* **2020**, *20*, 2234. [CrossRef] [PubMed]
114. Hou, J.; Yan, Y.; Cong, P. Application of technology of UAV-mounted ground penetrating radar in the study of the thickness of soil plow layer. *IOP Conf. Ser. Earth Environ. Sci.* **2021**, *719*, 042074. [CrossRef]
115. Li, Y.; Liu, C.; Yue, G.; Gao, Q.; Du, Y. Deep learning-based pavement subsurface distress detection via ground penetrating radar data. *Autom. Constr.* **2022**, *142*, 1–11. [CrossRef]
116. Wang, R.; Zhang, J.; Liu, X. A most-unfavorable-condition method for bridge-damage detection and analysis using PSP-inSAR. *Remote Sens.* **2022**, *14*, 137. [CrossRef]
117. Schlögl, M.; Dorninger, P.; Kwapisz, M.; Ralbovsky, M.; Spielhofer, R. Remote sensing techniques for bridge deformation monitoring at millimetric scale: Investigating the potential of satellite radar interferometry, airborne laser scanning and ground-based mobile laser scanning. *PFG-J. Photogramm. Remote Sens. Geoinf. Sci.* **2022**, *90*, 391–411. [CrossRef]

118. Hu, M.; Xu, Y.; Li, S.; Lu, H. Detection of defect in ballastless track based on impact echo method combined with improved SAFT algorithm. *Eng. Struct.* **2022**, *269*, 114779. [CrossRef]
119. Stüwe, I.; Zacherl, L.; Grosse, C.U. Ultrasonic and impact-echo testing for the detection of scaling in geothermal pipelines. *J. Nondestruct. Eval.* **2023**, *42*, 18. [CrossRef]
120. Abudeif, A.M.; Abdel Aal, G.Z.; Masoud, A.M.; Mohammed, M.A. Detection of groundwater pathways to monitor their level rise in Osirion at Abydos archaeological site for reducing deterioration hazards, Sohag, Egypt using electrical resistivity tomography technique. *Appl. Sci.* **2022**, *12*, 10417. [CrossRef]
121. Guo, Y.; Cui, Y.; Xie, J.; Luo, Y.; Zhang, P.; Liu, H.; Liu, J. Seepage detection in earth-filled dam from self-potential and electrical resistivity tomography. *Eng. Geol.* **2022**, *306*, 106750. [CrossRef]
122. Zhou, R.; Wen, Z.; Su, H. Automatic recognition of earth rock embankment leakage based on UAV passive infrared thermography and deep learning. *ISPRS J. Photogramm. Remote Sens.* **2022**, *191*, 85–104. [CrossRef]
123. Loiotine, L.; Andriani, G.F.; Derron, M.H.; Parise, M.; Jaboyedoff, M. Evaluation of infraRed thermography supported by UAV and field surveys for rock mass characterization in complex settings. *Geosciences* **2022**, *12*, 116. [CrossRef]

eng

MDPI

Review

Development and Applications of 3D Printing-Processed Auxetic Structures for High-Velocity Impact Protection: A Review

Raí Felipe Pereira Junio, Pedro Henrique Poubel Mendonça da Silveira *, Lucas de Mendonça Neuba, Sergio Neves Monteiro and Lucio Fabio Cassiano Nascimento

Department of Materials Science, Military Institute of Engineering-IME, Praça General Tibúrcio, 80, Urca, Rio de Janeiro 22290-270, Brazil
* Correspondence: pedroo.poubel@gmail.com

Abstract: Auxetic structures (AXSs) are a novel class of materials with unique mechanical deformation behavior associated with negative Poisson ratio. The combination of AXS configurations with various types of materials has unveiled a wide field of applications, including military high-velocity protection against explosions and ballistic projectiles. However, the characteristic geometric re-entrant model of AXSs imposes limitations and difficulties when using conventional manufacturing methods to assemble the structure lattice. Additive manufacturing (AM) has recently been explored as a more efficient and cost-effective method to fabricate AXSs, regardless of the type of material. This review paper focuses on the development and applications of AM processed AXSs. The review highlights the significance and great potential for this class of materials that can be produced relatively fast and at a low cost. The advantages of AXS/AM are expected to extend to important industrial sectors, particularly for military ballistic armor, where the feasibility for products with improved properties is critical. The use of AM offers a viable solution to overcome the difficulties associated with the conventional manufacturing methods, and thus offers greater design flexibility, cost efficiency, and reduced material waste. This review paper aims to contribute to the understanding of the current state-of-the-art and future research prospects for the production and applications of AXS/AM.

Keywords: honeycomb; sandwich panel; auxetic structures; auxetic materials; additive manufacturing; high-velocity impact; ballistic armor; anti-blast application

Citation: Junio, R.F.P.; da Silveira, P.H.P.M.; Neuba, L.d.M.; Monteiro, S.N.; Nascimento, L.F.C. Development and Applications of 3D Printing-Processed Auxetic Structures for High-Velocity Impact Protection: A Review. *Eng* **2023**, *4*, 903–940. https://doi.org/10.3390/eng4010054

Academic Editor: Antonio Gil Bravo

Received: 25 January 2023
Revised: 24 February 2023
Accepted: 3 March 2023
Published: 8 March 2023

1. Introduction

Owing to the continuous and pressing need for lighter and stronger structures for space systems, aircraft and military equipment, the development of a novel easy to fabricate and cost-effective sandwich panel composite stands as the most appropriated structural material. Indeed, the sandwich panel is one type belonging to the main class division of "Structural Composites" designed to be lightweight with high stiffness and strength [1], as well as remarkable energy absorption characteristics under the impact from blast or projectiles [2].

A sandwich panel is composed of two outer plates' face sheets bonded by adhesive to an inner core as schematically illustrated in Figure 1. The outer plate in this figure must be made of materials strongly sufficient to resist the expected applied load in order to protect the inner core. According to Callister and Rethwisch [1], the core presents several functions: (i) provide support for the outer plates; (ii) hold them together with the adhesive contribution; (iii) have sufficient strength to withstand shear stresses; and (iv) be thick enough to resist buckling imposed by bending forces to the panel.

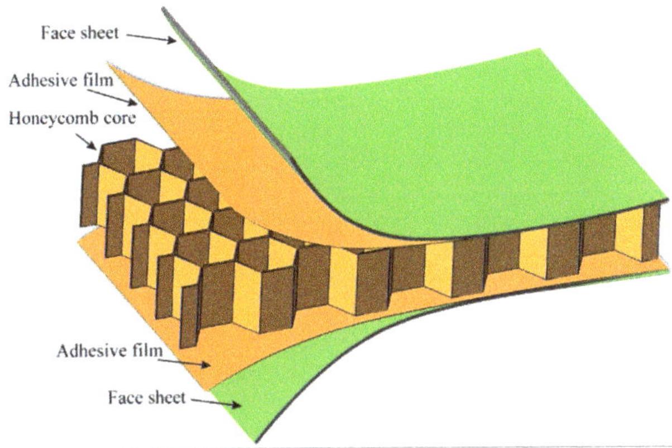

Figure 1. Schematic of basic sandwich panel configuration. Reprinted with permission from [3]. Copyright 2012, MDPI AG.

The typical honeycomb structure for sandwich panels is fabricated with relatively thin foils arranged into interlocking cells exhibiting a proper configuration, which can display different geometrical patterns [4,5]. A characteristic hexagonal configuration exhibited in Figure 1 was the first proposed [6] and assisted as a basic model for other bio-inspired patterns [7,8], as depicted in Figure 2.

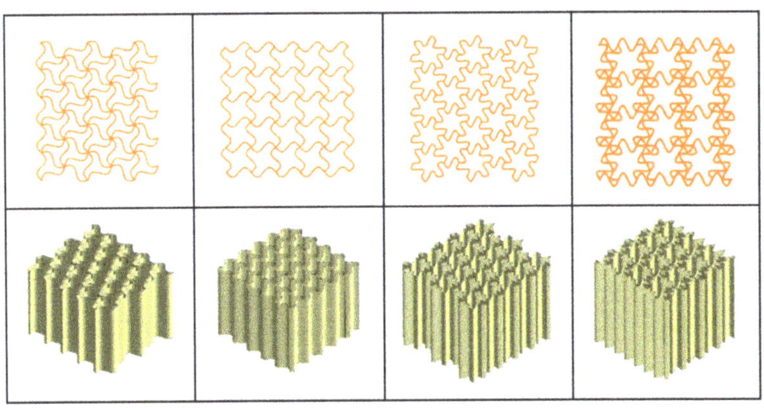

Figure 2. Horse-shaped honeycombs with different cross sections. Adapted with permission from Ref. [7]. Copyright 2018, Elsevier.

It is important to emphasize that these honeycomb structures were also investigated for their energy absorption applications [9,10], which might be a fundamental aspect for the presently reviewed ballistic application. In particular, auxetic structures (AXSs) possessing honeycomb patterns are being investigated as sandwich panels' cores for intrinsically higher energy absorption performance. As discussed in a prominent article, Yan et al. [11] indicated that this higher energy absorption of AXSs might be associated with both low-velocity [12–15] and high-velocity impact caused either by the blast shockwave from bomb explosions [16–18] or by ballistic projectiles [19,20]. This latter will be included in the present review regarding complementary aspects to the work of Yan et al. [11].

In spite of comparative improvement in the physical properties of sandwich panels with an AXS core, owing to their negative Poisson's ratio (NPR), inherent fabrication difficulties and consequently higher costs were, until recently, a practical challenge for their common-scale engineering applications. The very recent remarkable review on anti-blast and impact performance of AXSs by Bohara et al. [21] discussed different manufacturing techniques and, in principle, recommended additive manufacturing (AM), also known as 3D printing, as it is currently the most used for small-scale pieces by enabling the construction of any kind of complex cellular structure such as those demonstrated in Figure 2. Furthermore, these authors indicate that future improvement in AM technology is necessary to produce large-scale AXSs for armor sandwich panels protecting personnel and vehicles from blast and impact projectiles.

Based on the aforementioned preliminary information, the objective of this article is to review the actual development and expected applications of 3D printing-processed AXSs intended for high-velocity impact protection as core sandwich panels. First, a deep discussion of relevant papers on AXSs, especially those not yet reviewed, is presented. Second, the current status of using the diversity of AM techniques for the fabrication of AXSs is critically discussed. Third, novel papers investigating the ballistic performance of 3D-printed AXSs are summarized. The following section discloses the existing and proposed ballistic and anti-blast applications for AXS. Final remarks on possible directions for large-scale production of AXSs, concerning specific interest as a material for ballistic armor, will provide the reader with an ongoing trajectory of a subject with indisputable future interest.

2. Properties of Auxetic Materiais

A novel class of materials, commonly known as "auxetics", possess peculiar structures that give them distinctive deformation characteristics [22]. In fact, materials deform in different ways when forces act upon them. The vast majority behaves in a traditional manner, in which an object expands axially under tension stress while contracting in the transverse direction, as illustrated in Figure 3a. The inverse ratio of transverse strain to axial strain, indicating a measure of how a material deforms under load, is known as the Poisson's ratio [23]. A material that stretches axially but contracts transversely in response to a tensile force, or enlarges transversely while axially contracting under compressive force has a positive Poisson's coefficient. The material owning auxetic behavior expands in the axial direction in one or more orthogonal directions under tension, exhibiting a negative Poisson's ratio (NPR), Figure 3b [24]. Under compressive forces, AXSs contract in both the axial and transverse directions, Figure 3d.

In the late 1800s, some discoveries about AXSs were made; however, they were not significant at that time. In 1882, the existence of AXSs cellular structures in iron mineral pyrite was reported, verified by torsion and flexural tests [25,26]. In 1893, Voigt [27] documented the auxetic effect and negative Poisson's coefficient in some materials. Until then, already known AXSs were limited to natural materials only, such as wood, and did not draw much attention to synthetic materials research. In 1987, almost a century later, Lakes [22] described for the first time an auxetic foam structure made of polyurethane (PU) that could be easily manufactured by compression and heating processes and that had an NPR equal to -0.6. Other works reported the existence of isolated materials possessing an NPR, such as bismuth, antimony and arsenic [28], cadmium [29], α-cristobalite [30], as well as metals that crystallize in a cubic structure [31]. Some research has also demonstrated the existence of skin-like biological materials that exhibit auxetic behavior, such as cat skin [32], ox skin [33], and salamander skin [34], as well as human bones [35]. However, Lakes' study [22] assisted as a basis for future research on AXSs acquired synthetically.

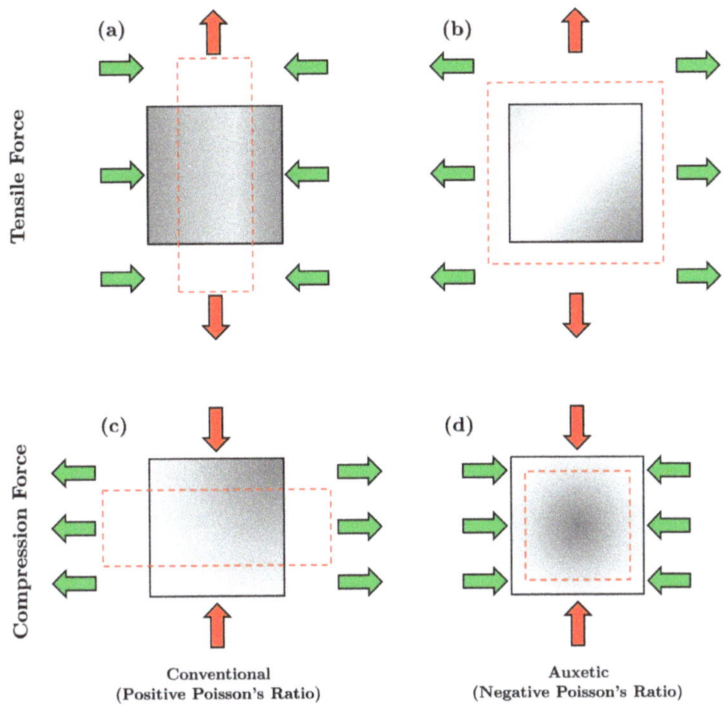

Figure 3. Comparison of the mechanical behavior of conventional and auxetic materials under tensile and compressive forces: (**a**) Tensile forces acting on a conventional material; (**b**) Tensile forces acting on an auxetic material; (**c**) Compressive forces acting on a conventional material; (**d**) Compressive forces acting on an auxetic material.

After this discovery, research on AXSs has been developed for several sectors. The range of applications for these materials is quite wide, for example: sports artifacts [36,37], auxetic foams [38–40] and military use [41–43], biomedicine [44–47], textiles [48–50], etc. Auxetic materials are found in basically all known material classes: metals [51–53], crystals [54–56], polymers [57–59], ceramics [60–62], and composites [63–65]. Table 1 presents a brief list based on the most recent known AXSs and their respective NPRs.

Table 1. AXSs previously investigated and their Poisson ratios.

Material	Poisson's Ratio	Year	Reference
AA1060 aluminum	−0.82~−1.01	2023	[66]
Cellulose	−0.2	2023	[67]
Carbon Fiber/Aluminum	−0.13	2023	[68]
LDPE	−1.5~−3.5	2022	[69]
ABS	−0.65~−1.56	2022	[70]
316L stainless steel	−0.4~−0.8	2022	[71]
Graphene/Cu	−0.19~−0.25	2022	[72]
$Ba_{0.85}Ca_{0.15}Zr_{0.1}Ti_{0.9}O_3$	−1~−2	2022	[73]
Cement/PU	−0.21~−0.47	2022	[74]
PLA	−0.50~−0.78	2022	[75]
$MgHfN_2$ and $MgZrN_2$	−0.2	2022	[76]
PU/PDA/Graphene	−2.36	2022	[77]

AXSs diverges from conventional materials (CMs), presenting a peculiar behavior associated with unique mechanical properties. When experimentally investigated, AXS materials display superior properties reported as uncommon, when compared to CMs. An unusual deformation pattern under compression and tension is found for AXSs bringing forth many desirable properties, such as superior shear resistance [78], indentation resistance [79,80], fracture toughness [81], synclastic behavior [22], variable permeability [82] and improved energy absorption performance [83–85]. These properties are listed in the following subsections.

2.1. Resistance to Indentation

The indentation resistance property occurs when a non-AXS is subjected to indentation, e.g., hardness testing. The load applied at the indentation location naturally compresses the material. To compensate for this localized pressure, it spreads perpendicular to the applied load, as shown in Figure 4a [86]. On the other hand, the AXSs behavior is different; when an indentation occurs on it, a local shrinkage is observed. Instead of spreading it perpendicularly, a flow occurs and accumulates under the indented region. Then, a denser area with higher indentation resistance is created below the indented zone, as depicted by Figure 4b [87]. As such, they have a higher resistance to indentation, when compared to CMs [79,88].

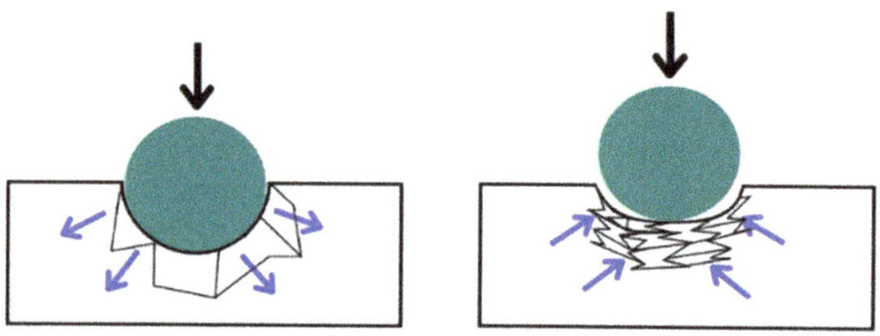

Figure 4. Deformation behavior and indentation resistance of: (**a**) conventional material and (**b**) auxetic material. Reprinted with permission from Ref. [89]. Copyright 2021, IOP.

The increased indentation resistance can be explained by the theory of elasticity. The indentation resistance is associated with the hardness of the material ($H\alpha$) [90]. This property is correlated with the Poisson's ratio (υ) given by:

$$H\alpha = \left[\frac{E}{(1-\upsilon^2)}\right]^{\gamma} \tag{1}$$

where E is the Young's modulus, υ is the Poisson's ratio of the base materials, and γ is the constant that assumes the value 1 or 2/3 in the case of uniform pressure distribution or Hertzian indentation, respectively.

Based on Equation (4), it can be seen that, for isotropic three-dimensional (3D) materials, when the Poisson's ratio decreases to extreme values, for instance near -1, their hardness tends to infinity. Since the upper limit of Poisson's ratio for 3D isotropic solids is 0.5, the observed values are substantially lower. By contrast, the upper limit of the Poisson's ratio for two-dimensional (2D) isotropic systems is 1. Materials with such positive Poisson's values can also have infinite hardness values [24,91,92].

2.2. Shear Resistance

As demonstrated for indentation resistance, AXSs are more resistant to shear forces than CMs [90]. The classical theory of elasticity for 3D isotropic solids shows that the elastic behavior of a body can be described by two out of four constants: the Young's modulus (E), the shear modulus (G), the bulk modulus (K) and the Poisson's ratio (v) [93]. In 3D solids, the relationship between these constants is given by the following equations [88]:

$$G = \frac{3K(1-2v)}{2\cdot(1+v)} \tag{2}$$

$$G = \frac{E}{2\cdot(1+v)} \tag{3}$$

From these equations, it is inferred that the value of shear modulus increases as a function of the reduction in the Poisson's ratio, resulting in a shear strength increase. The range of elastic modulus corresponding to instability and stability under different conditions is shown in the map of Figure 5. It indicates that the Poisson's ratio of the isotropic solid should be ranging from -1 to 0.5. When it approaches -1, the shear modulus tends to infinity.

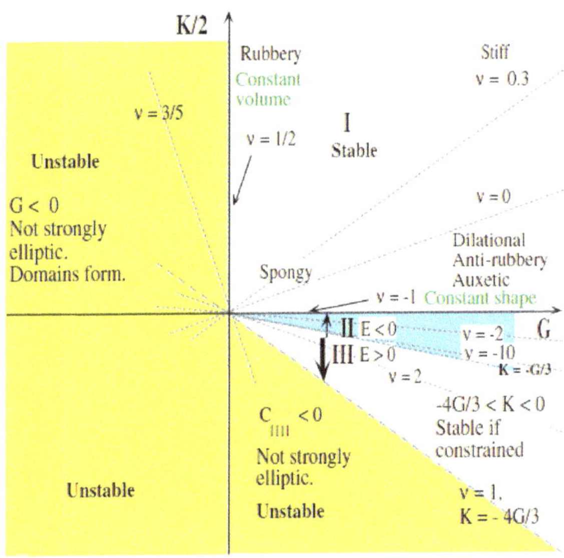

Figure 5. Map showing the elastic material properties corresponding to different values of K-modulus and shear modulus G. Reprinted with permission from Ref. [94]. Copyright 2007, Wiley.

2.3. Fracture Resistance

AXSs display better fracture resistance than CMs [81]. AXSs also exhibit low crack propagation, requiring a greater amount of energy to expand than CMs [95]. Thus, auxetic materials present brittle fracture. Some works concerning the fracture toughness behavior of AXSs are worth reviewing. Maiti et al. [96] demonstrated that the stress intensity factor (K^*_{IC}) for conventional foams is proportional to the normalized density and can be described by:

$$\frac{K^*_{\rm IC}}{\sigma_{\rm f}\sqrt{\pi l}} = 0.19\cdot\left(\frac{\rho*}{\rho_{\rm s}}\right) \tag{4}$$

where $\sigma_{\rm f}$ is the fracture stress of the cell rib, l is the rib length, ρ^* is the foam density, and $\rho_{\rm s}$ is the density of the foam based material.

Some works reported increased fracture toughness of AXSs, while CMs did not disclose the same behavior. Yang et al. [97] have produced epoxy/Kevlar auxetic laminate composites and correlated the mechanical properties with plain Kevlar and polyurethane (PU) treated Kevlar fabrics. From the fracture initiation toughness (G_{IC}) test, the authors observed that the auxetic fabrics disclose superior performance as compared to plain and PU treated kevlar fabrics, as illustrated in Figure 6. Some conditions of the auxetic epoxy/Kevlar composite disclosed a G_{IC} value up to 225% higher than untreated woven Kevlar and up to 125% higher than PU-treated Kevlar.

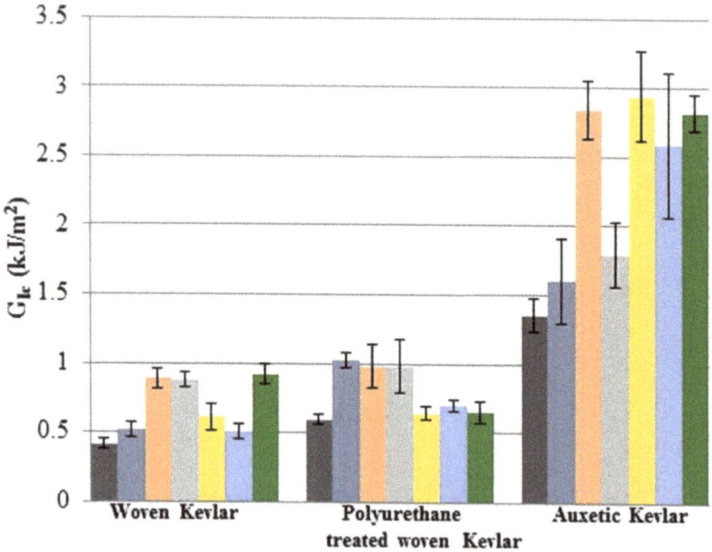

Figure 6. Fracture initiation toughness (G_{IC}) for all types of composites. Reprinted with permission from Ref. [97]. Copyright 2017, Elsevier.

2.4. Energy Absorption

AXSs manifest superior properties regarding energy absorption. Ultrasonic, acoustic and damping energy absorption are characteristics in which AXSs have an advantage over CMs. It is reported that auxetic foams can display better sound absorption ability than conventional foams at low frequencies [83]. Dynamic crushing properties were remarkable in the auxetic foam, while the conventional foam did not show significant resilience at a high constant stress rate [98].

Oh et al. [99] developed an auxetic foam made of PU and filled with wrinkled 2D graphene oxide (GO). The authors investigated the properties of acoustic energy absorption and damping absorption. Auxetic foam reinforced with GO was set side by side with another AXS without filling. Graphene-polyurethane oxide auxetic foam exhibited high sound absorption performance, with a range of 98.4% at 2 kHz due to the hierarchically porous structure. The improved mechanical compression stability is attributed to the superior energy absorption properties resulting from superb compressibility and internal auxeticity. This greatly increases the shock energy absorption time by 189% in the low-velocity impact test in contrast to the value of the pure PU foam. The blending of a 2D corrugated GO sheet and a flexible 3D AXS promoted greater acoustic and vibration absorption as a consequence of multiple energy dissipation mechanisms.

Najafi et al. [100] analyzed the mechanical performance and energy absorption in order to show differences and likenesses of three different AXSs produced in ABS by the 3D printing method: re-entrant, arrowhead and anti-tetra chiral, as depicted in Figure 7a–c. From the compression test results, the authors observed that the anti-tetra chiral structure presented a greater deformation (Figure 7d–f), when set against the arrowhead and re-entrant structures. Nevertheless, this structure also brought up a significantly higher specific energy absorption (SEA), illustrated in Figure 7g.

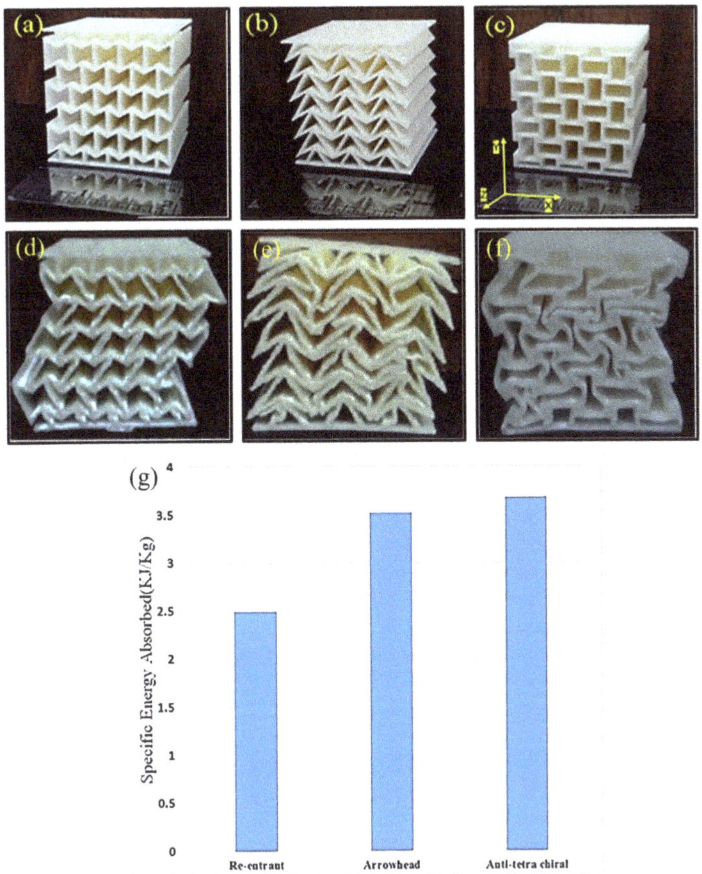

Figure 7. Auxetic 3D printed ABS structures: (**a**) re-entrant; (**b**) arrowhead; (**c**) anti-tetrachiral; (**d**) re-entrant after compression test; (**e**) arrowhead after compression test; (**f**) anti-tetra chiral after compression test; and (**g**) specific energy absorption of the tested structures. Reprinted with permission from Ref. [100]. Copyright 2021, Elsevier.

2.5. Synclastic Behavior

Synclastic behavior is the ability of an AXS to suffer deformation into a dome shaped curvature when subjected to bending stresses [101]. Contrastingly, CMs uncover a stretching saddle-shaped on their surface when subjected to bending stresses, originating a shrinkage in their perpendicular direction. Therefore, it presents an anticlastic curvature, as shown in Figure 8a. On the other hand, the AXs, when bent, will present a dome shape because, throughout the application of stress, there is an expansion and contraction of the material both on the outside and inside, respectively [102]. Thus, the result of the expansion of the tensioned part and the shrinkage of the compressed part is the dome shape, shown in Figure 8b. This property is desired in materials for application in the biomedical and sports sectors, such as stents, prostheses, sports accessories, etc. [23,36,37,103–106].

Figure 8. Curvature in bending: (**a**) anticlastic curvature of conventional, non-auxetic materials and (**b**) synclastic curvature of auxetic materials. Reprinted with permission from Ref. [37]. Copyright 2018, MDPI AG.

2.6. Variable Permeability

Currently, it is already known that AXSs own a porous microstructure and that its dimensions can vary under tensile or compressive stresses. Given this characteristic, a potential use that emerges from AXS porosity is the application in filters. Alderson et al. [107] detailed in their work that AXSs may offer superior performance in filter application at both the macro and nano-scale in view of their unique pore opening properties and characteristics. Variable permeability is paramount for medical applications such as smart stents, which are used to deliver pharmaceutical drugs into the human body [108,109]. In Figure 9, a schematic view of how variable permeability works is illustrated, in which a stress applied to the AXS develops the expansion of the pore structure, allowing a particle to go across it.

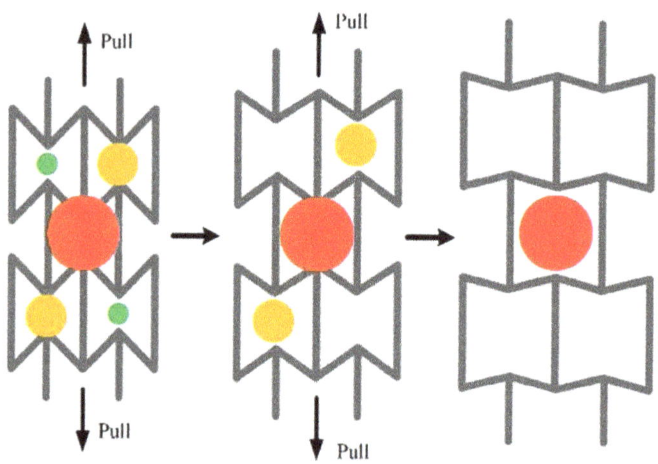

Figure 9. Schematic of variable permeability in auxetic structures. Reprinted with permission from Ref. [110]. Copyright 2020, Wiley.

2.7. Shape Memory Auxetics

Another important property related to AXSs is shape memory. It occurs when a material subjected to deformation, whether plastic or elastic, has the ability to return to its original state after undergoing thermal excitation [111,112]. Currently, there are several studies reporting the shape memory effects in materials of different classes—for instance, NiTi alloy [113–115], PLA/TPU/Fe_3O_4 [116], SMEP/GNP [117], ZrO_2/Y_2O_3 shape memory ceramic [118], among several others performing this singular characteristic. Studies demonstrate that it is possible to obtain auxetic foams that can be reverted to conventional foams several times without a loss of mechanical properties [119]. Therefore, shape memory is useful in situations that require variable auxetic and non-auxetic mechanical properties, being changed as a function for the temperature.

In Figure 10, a schematic of this effect in an AXS developed by Kamyab et al. [120] is illustrated for an acrylate shape memory polymer (SMP). In the schematic, it is possible to observe that, with the heating of the material, tensile stresses start to act, as a consequence. After this process, the SMP is cooled, and then its load is removed, inducing it to return to its first state. Finally, it is reheated in order to completely return to its original state.

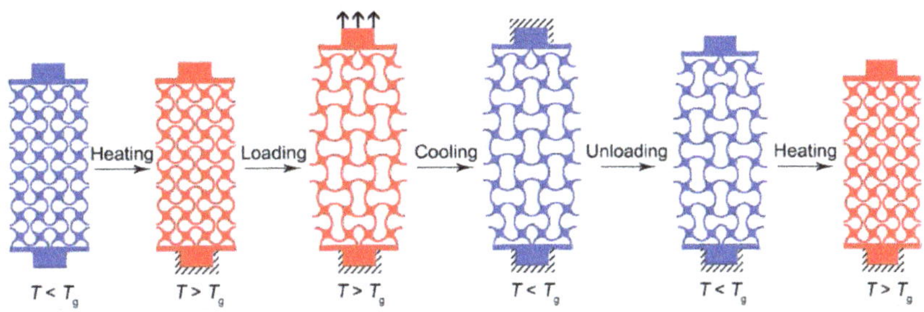

Figure 10. Schematic illustrating the thermo-mechanical loading cycle under tensile mode for shape memory polymer (SMP). Reprinted with permission from Ref. [120]. Copyright 2022, ProQuest.

2.8. AXSs for Ballistic Protection

In Section 2, the properties of auxetic panels were described, which are made possible by the complex structures that compose them. In Section 3, these structures will be presented in more detail. The development of auxetic panels with these structures allows for the absorption of impact energy, which can be applied in various areas, such as protection of military vehicles against explosions and ballistic panels. According to Bohara et al. [21], energy absorption occurs in AXSs due to the deformation of the cell wall and the subsequent formation of plastic hinges in the wall. The inherent porosity of AXSs also contributes to greater energy absorption. The authors cite studies that show the use of AXSs in blast armor, vehicle bumpers, impact shields, and protective helmets.

An interesting example is the study by Yan et al. [11], who evaluated AXS honeycomb sandwich panels impacted by projectiles, made with carbon fiber-reinforced polymer. Figure 11 schematically illustrates the penetration of the bullet into the AXS. The authors observed that, as the bullet perforates the front steel plate, the honeycomb cells in its path are directly cut, resulting in brittle failure and delamination. The external steel plate had little effect on reducing the bullet's velocity but contributed to reducing the crushing area. On the other hand, the internal core of the honeycomb played an important role in reducing the penetration velocity and, consequently, the impact energy.

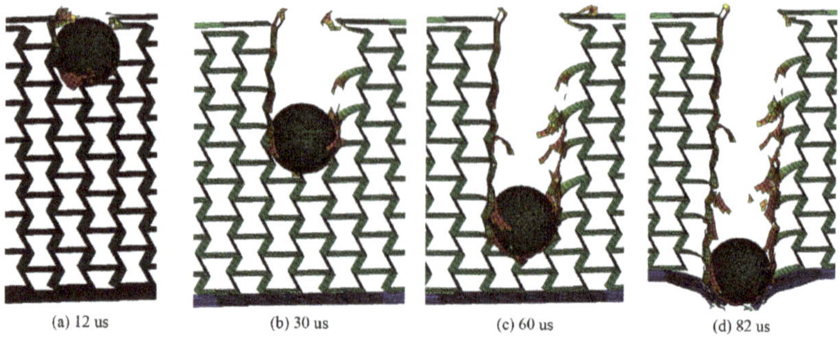

Figure 11. The perforation process of bullet penetration into AXS honeycomb sandwich plate. Reprinted with permission from Ref. [11]. Copyright 2022, Elsevier.

3. Auxetic Structures

The term auxetic is used for both structures and materials. The AXS plays a key role in this material expansion and mechanical behavior. Therefore, several structural models have been investigated, and their auxetic behaviors have been discussed. For this topic, the main models of auxetic-related structures will be described.

3.1. Re-Entrant Structures

The re-entrant pattern is one of the most well-known structures used for developing AXSs. It was originally proposed in 1982 by Gibson et al. [121], where, from the deformation of a hexagonal honeycomb structure, it was possible to obtain 2D re-entrant structures, as schematically shown in Figure 12. When a load is applied, the diagonal ribs are stretched, moving the structure and giving rise to the auxetic effect.

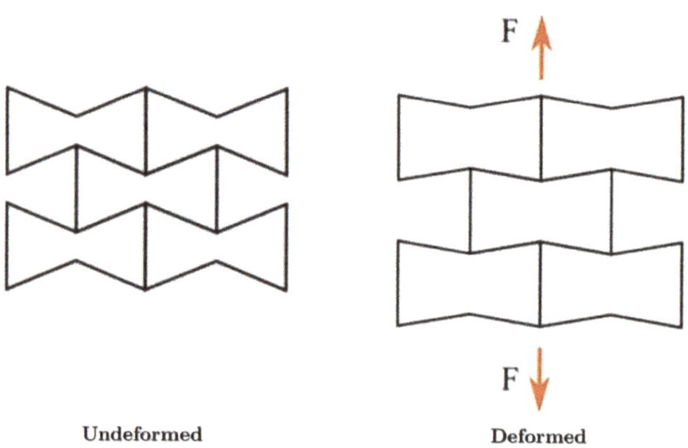

Figure 12. Auxetic behavior of the re-entrant structures. Reprinted with permission from Ref. [122]. Copyright 2022, Wiley.

As a result of the development of this structure, advances in the development of AXSs have been significant. From the original re-entrant structure, others have been proposed, as illustrated in Figure 13. Similar to the re-entrant structure, the arrowhead and star models (Figure 13b,c) have the auxetic behavior coming from the re-entrant parts, as highlighted by the blue color in the figure. Using rhombus-shaped lines, the hexagonal re-entrant and grid-like models were established (Figure 13d,e). In Figure 13f, a sinusoidal structure is illustrated, in which one can observe the elongation occurring and, for this reason, the auxetic effect arises. In addition, links on the sinusoidal curves can be modified following the articulated linear ligaments.

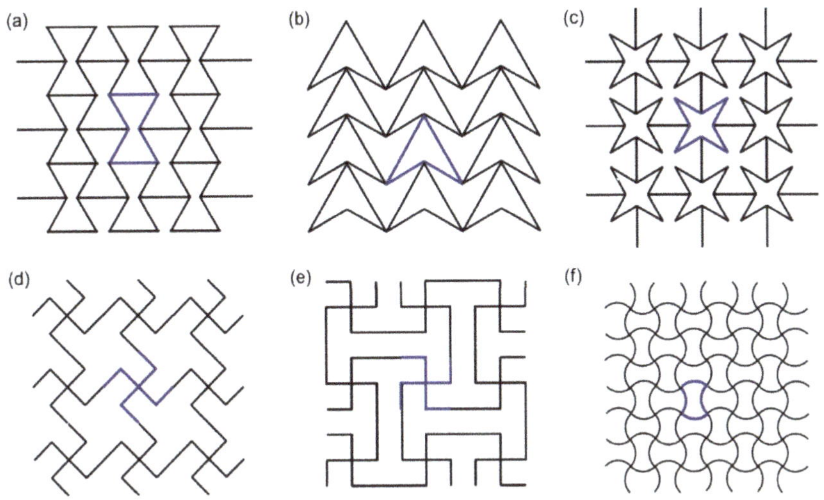

Figure 13. Re-entrant structures. (**a**) re-entrant honeycomb; (**b**) double arrowhead; (**c**) star honeycomb; (**d**) structurally hexagonal re-entrant honeycomb; (**e**) lozenge grids; (**f**) sinusoidal ligaments. Reprinted with permission from Ref. [104]. Copyright 2021, MDPI.

3.2. Rotating Unit Structures

Another type of AXSs is related to rotating unit structures. In this model, different uniform geometric shapes such as rigid/semi-rigid squares, triangles, rectangles and tetrahedra are configured at their vertices [123–125]. The auxetic behavior of this structure occurs through the rotation of these shapes when stretched in a certain direction. This model was firstly proposed by Grima et al. [126], when they verified the occurrence of AXS in an inorganic material. The concept of rotating squares with rigid units was shown to be independent of the initial geometry and the direction of loading applied to the material, keeping the NPR near -1 [127]. Several studies have been conducted in order to better understand the deformational characteristics of this AXS.

The square rotating unit, Figure 14a, is the simplest 2D structure among a variety of 2D unit structures available to be manufactured. It is possible to acquire rectangle or parallelogram units by scaling and tilting square structures, as presented in Figure 14b [128,129]. It is also possible to acquire a rotating unit with two or more different types. Typical examples of these heterogeneous rotating units are the trans-rectangular, Figure 14c, and the two-square structures, Figure 14d. In addition, for rotating frames possessing a square section, there are 2D rotating frames with triangular units, owning three hinge points. They were manufactured similarly to the square type, but their structure is simpler, and might be developed based on simple geometrical units such as triangles, as shown in Figure 14e–g. The rotating structure that has a hexagonal unit, Figure 14h, can be used in two ways: (i) the hexagon attached to a square or triangular unit or (ii) triangular unit. However, it cannot be used individually to manufacture a rotating structure, without square or rectangular units. A hexagonal is employed with other types of units to create a heterogeneous rotating 2D structure.

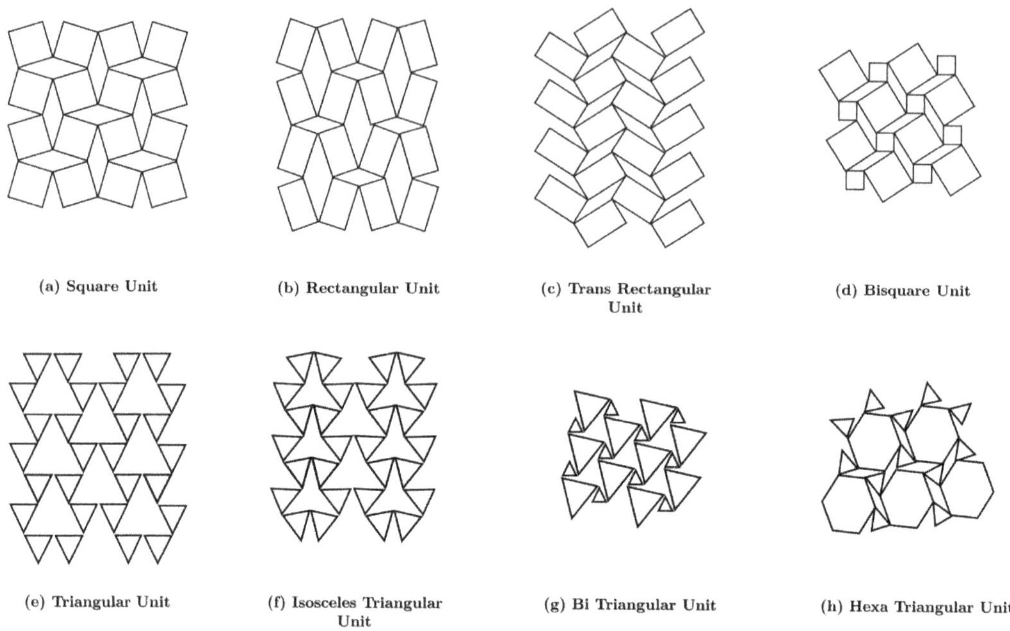

Figure 14. AXSs Rotating units: (**a**) square; (**b**) rectangular; (**c**) trans-rectangular; (**d**) bisquare; (**e**) triangular; (**f**) isosceles triangular; (**g**) bi-triangular; and (**h**) hexa-triangular.

3.3. Chiral Structures

Chiral structures are manufactured by straight ligament connections (ribs) joined to central nodes that can be circles or other geometric shapes. The auxetic behavior regarding this structure is achieved by wrapping or unwrapping the ligaments around the nodes when a load is applied.

The unit circles present in chiral structures are uniformly aligned in tri-, tetra-, or hexahedral arrangements, where the ribs wrap around the circles and connect them to others. When a load is applied to a chiral structure, it is transferred to the circular units via chiral ribs, and the transmitted load is deflected from the center of the circle, producing a rotational torque. Thus, it rotates in a particular direction pulling and pushing adjacent units amongst ribs connected in a different direction of the load [130]. Two-dimensional chiral structures are composed of circular units and ribs, and are limited to five different basic types: tetrachiral, anti-tetrachiral, trichiral, anti-trichiral, and hexachiral, as illustrated in Figure 15 [131].

Chiral structures are produced from narrow designs, which have similar auxetic characteristics to re-entrant structures. Even though the high porosity of chiral structures allows them to be lighter than re-entrant models, this reduction in weight ultimately obstructs the durability and stability. Moreover, even if the chiral structure exhibits NPR, it becomes vulnerable to local buckling if compressive loads between circular units are transmitted through thin ribs. Furthermore, due to the geometric arrangement, they have limited structural variation and narrow design space, unlike the re-entrant and rotating structures.

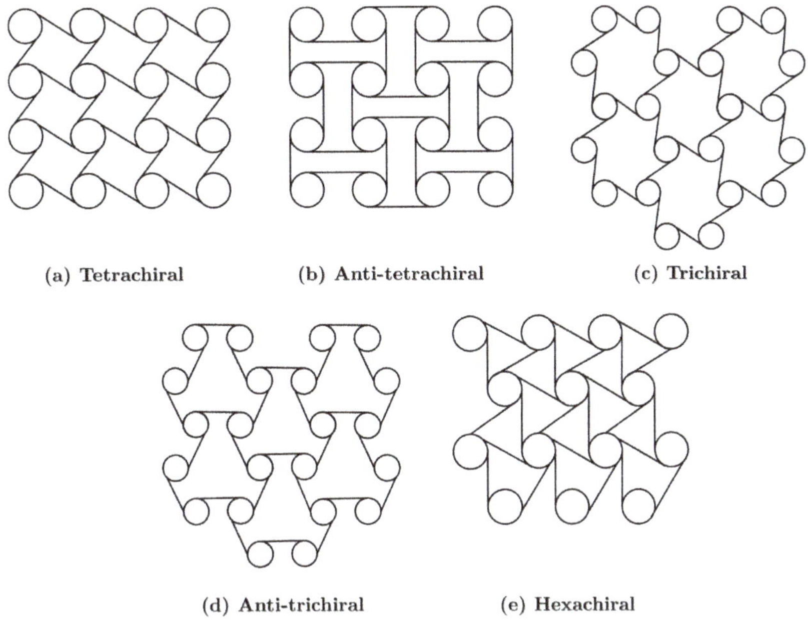

Figure 15. Chiral AXSs: (**a**) tetrachiral; (**b**) anti-tetrachiral; (**c**) trichiral; (**d**) anti-trichiral; and (**e**) hexachiral structures.

3.4. Nodule and Fibril Structures

Nodule and fibril structures consist of a rigid unit connected among fibrils. These fibrils are responsible for the load transfer to the rigid structures. The working scheme is schematically illustrated in Figure 16; without a load acting upon them, they are twisted around the rigid unit. When a tensile force is applied to the fibril structure, they tend to align and push it in a perpendicular direction, inducing an auxetic behavior. Initially, Alderson and Evans [132] developed a model for a polymer featuring auxetic micropores. These AXSs can be subdivided into classes based on how the rigid units and fibrils are connected: (i) network type; with (ii) circular or (iii) square nodules; and (iv) bundle type [133].

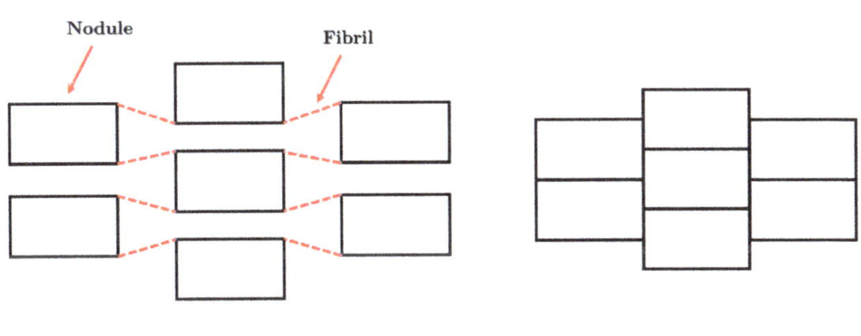

Figure 16. Typical shape of fibril–nodule structures: (**a**) before compression; (**b**) after compression.

The fibril–nodule model was developed early in studies concerning AXSs, to be employed in polymeric materials. An example of an auxetic polymer is the liquid crystalline, where its behavior can be defined by the fibril–nodule mechanism. This material has a crystalline fibril–nodule structure when it has no load acting upon the material. However, when a stress is applied, it presents an amorphous AXS due to the decrystallization with its stretching [134]. Given this concept, the auxeticity of the fibril–nodule structures only occurs for a tensile condition, disappearing under the action of compressive loads.

3.5. Interlocking Structure

The auxetic interlocking structure was first proposed by Ravirala et al. [135], and its associated micromechanical model is schematically illustrated in Figure 17. The structure without a stress state is all connected and interlocked, Figure 17a. At the moment, a load is applied, it expands, exhibiting auxetic behavior, Figure 17b. The conventional hexagonal geometry shown in this model ($0 < \alpha < 90$) displays an auxetic behavior regarding its structure, while the reentrant hexagons ($90 < \alpha < 180$) disclose a positive Poisson's coefficient.

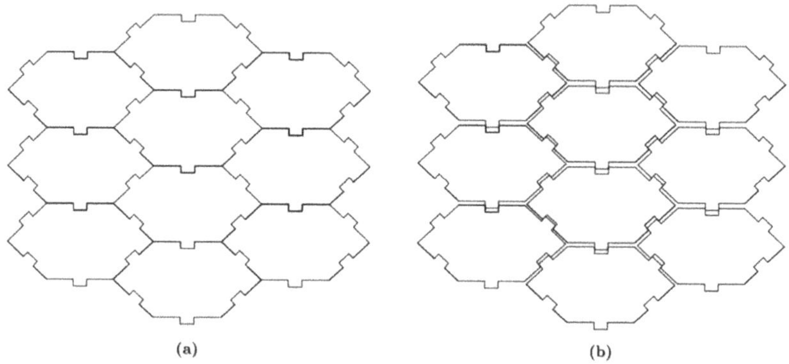

Figure 17. Interlocking hexagon structure: (**a**) fully densified; (**b**) partially expanded.

3.6. Keyed Brick Structure

The keyed brick structure is designed to withstand the horizontal components of the forces generated throughout an earthquake, while allowing a motion that assists with differential thermal movements between the graphite core and the steel supporting the edges [86]. The free movement of the core was also designed to sustain the expansion and shrinkage of the graphite bricks when radiation exposure occurs. Furthermore, the structure needs to have high shear deformation resistance in the horizontal plane and low resistance to volume changes. As such, it expands in all radial directions when subjected to a tensile load in the horizontal plane and keeps the square mesh over deformation. In this way, this type of model becomes auxetic, obtaining $\upsilon = \pm 1$ in the horizontal plane [86]. They are arranged in columns of 10 ± 12 bricks, making the entire core possible to be approximated to a continuous anisotropic solid material. The scheme of the keyed brick structure is illustrated in Figure 18.

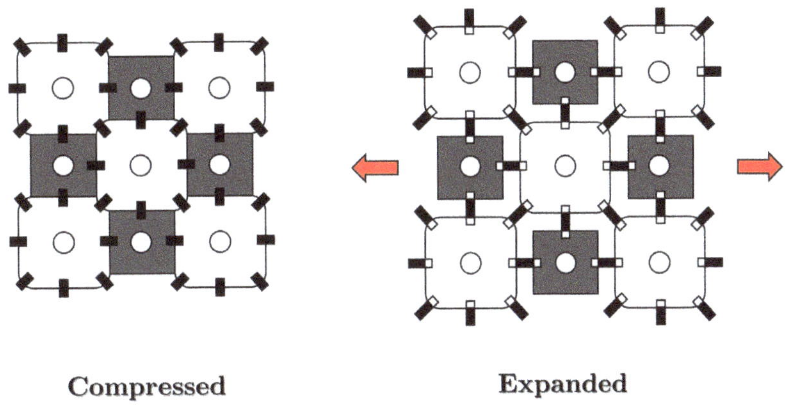

Figure 18. Schematic of a keyed brick structure.

3.7. Other Geometries

In addition to the most well-known AXSs that were presented in the previous items, there are several other cellular structures that present auxetic characteristics. These structures end up forming new classifications, and it is often not possible to classify them according to the models already known.

Bohara et al. [85] developed three new AXSs for application in military vehicles protection by absorbing blast energy. They were named: (i) hourglass structure (HGS); (ii) braced cross-petal structure (BCPS); and (iii) cross-petal structure (CPS). The finite element method (FEM) was employed to analyze these novel ASXs using a variety of parameters such as stress–strain behavior, Poisson's ratio, energy absorption capacity and others. From the simulations, the authors found that, for demands requiring high energy absorption, these three AXSs have shown superior performance contrasting the traditional ones. Figure 19 depicts a comparison of the specific energy absorbed (SEA) for all AXSs analyzed.

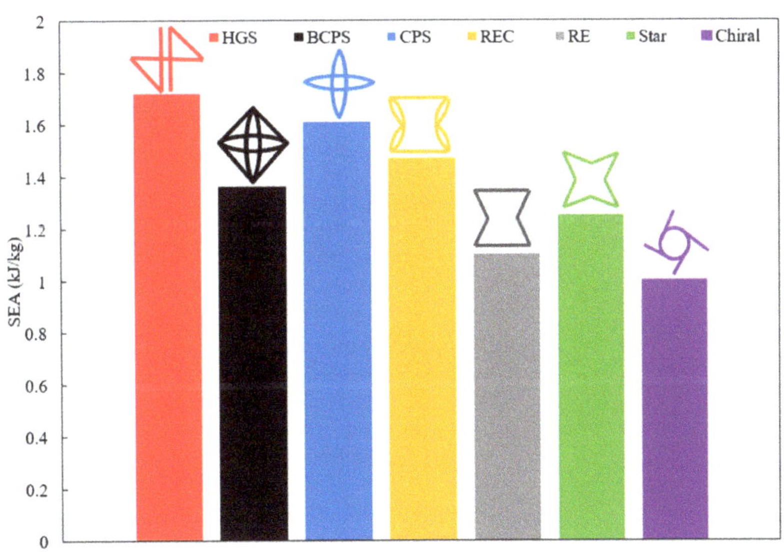

Figure 19. SEA values for different AXSs. Reprinted with permission from Ref. [85]. Copyright 2021, Elsevier.

Moreover, another approach for acquiring novel structures is the union of two AXSs. Etemadi et al. [136] developed hybrid structures attaching re-entrants and arrowhead structures (RAH), Figure 20a, as well as star and arrowhead structures (SAH), Figure 20b. The authors produced the samples by the 3D fused deposition method using a standard polylatic acid (PLA) resin for the manufacture. FEM analyses were performed, as well as tensile and compression tests, in order to evaluate their mechanical properties. From the analyses, it was found that the hybrid models have shown higher E and v than re-entrant and arrowhead ones. RAH and SAH achieved E and v values of 66.21 and 86.24 GPa, and -1.34 and -0.98, respectively. On the other hand, the re-entrant and arrowhead conventional structures presented the following values: E = 22.86 and 34.67 GPa and $v = -1.24$ and -1.00, respectively.

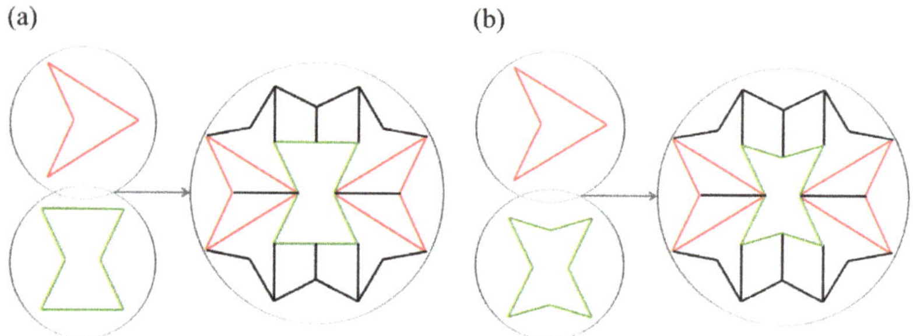

Figure 20. Hybrid structures: (**a**) re-entrant arrowhead (RAH); (**b**) star arrowhead (SAH). The re-entrant, star, and arrowhead structures are illustrated in colors for better understanding. Reprinted with permission from Ref. [136]. Copyright 2023, Elsevier.

3.8. Brief Summary

In this section, it is worth noticing the impressive number of distinct models of auxetic structures that have been investigated so far, from reentrant [121,136] to cross-petals [85] for absorbing blast energy. It predicts a vast field of opportunity for the development of novel auxetic structures.

4. Manufacturing Techniques of Auxetic Structures

The special properties obtained through the development of AXSs are associated with their origin in the manufacturing of a complex 3D structure. Additionally, small details such as the reentrance dimensions of a unit cell are crucial to this auxetic behavior. However, these details also make it difficult to fabricate these materials using conventional techniques. To overcome this challenge, additive manufacturing (AM) techniques are employed in the fabrication of these structures. AM provides a quick and precise solution in the manufacturing of AXSs sandwich panels that present complex geometries at their core. However, conventional techniques are also used, albeit less commonly, in the fabrication of AXSs [137]. Currently, there are several manufacturing techniques available for the production of AXSs, including both conventional and AM/3D printing techniques such as:

- Self-Propagating Photopolymer Waveguide (SPPW);
- Microstereolitography (μSLA);
- Direct Laser Writing (DLW);
- Self-Assembly
- Selective Laser Melting (SLM).

4.1. Self-Propagating Photopolymer Waveguides (SPPW)

The self-propagating photopolymer waveguides (SPPW) technique assembles a 3D polymeric structure in a cellular microscale arrangement, interconnected periodically [138]. It consists of an ultraviolet (UV) light beam enlightening a 2D mask, which has an aperture inside a photomonomer container, as demonstrated in Figure 21 [139]. The SPPW is based on the incident light self-trapping effect upon the polymer, being induced by a change in the refraction index among the liquid monomers and the rigid polymeric material. Owing to this event, the SPPW is able to create high aspect ratio beams having a continuous cross-section, being perfect for the 3D topologies construction based on beams. Furthermore, it is possible to control the topologies by an aperture pattern on the mask and orientation of the incident UV light. While the unit cell dimension and the smaller size details, regarding the topologies, rely on the aperture diameter and mask space. The material thickness is controlled mostly by the maximum waveguide length propagation. Moreover, a multilayer strategy could be applied in order to acquire structures displaying a higher thickness [140].

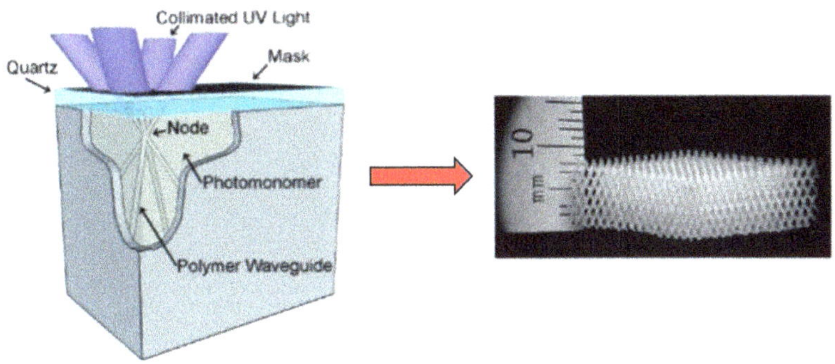

Figure 21. Operational scheme of the self-propagating polymer waveguide (SPPW) technique. Reprinted with permission from Ref. [138]. Copyright 2007, Wiley.

However, in contrast to other AM techniques, the SPPW is limited regarding the topologies randomness, since it is just possible to manufacture variations of beam-based topologies. The extended UV light exposure requested to achieve the maximum waveguide length often enlarges the original diameter, developing thicker materials [138]. On the other hand, the great advantage of it is the high manufacture speed, being able to produce microscale AXS in a few minutes in addition to a great scalability capacity, yielding manufacture rates superior to 1 m^2/min [141]. The aforementioned characteristics make it an attractive method for large-scale fabrication.

4.2. Microstereolithography (μSLA)

Microstereolithography (μSLA) is an AM technique classified according to the ISO/ASTM 52900:2021 standard [142] as belonging to the Vat Photopolymerization (VPP) category. μSLA uses a UV light beam to cure the resin, resulting in the rapid assembly of structures. The operational mechanism of the SLA technique is based on the local polymerization of a photosensitive resin, as shown in Figure 22. Each layer is obtained by moving the UV light beam over a surface, which is guided on the x- and y-axes by galvanometric mirrors. The solidified object is then immersed in a resin reservoir, and a new layer is reflected onto the already polymerized layer, allowing the manufacture of the next layer [137,143,144].

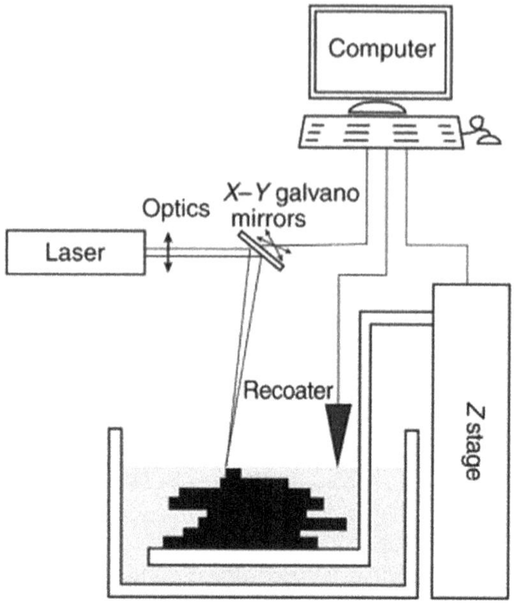

Figure 22. Schematic of microstereolithography technique. Reprinted with permission from Ref. [145]. Copyright 2020, Elsevier.

Furthermore, the AM techniques yielded a strong influence on the production of industrial parts because they significantly reduced the processing time, entailing a manufacturing economy regarding objects with complex geometry. The μSLA was one of the first to be developed, and due to its velocity, is, hitherto, the most applied in industry. The manufacturing costs for the μSLA pieces are affordable, and the surface finishing is appropriate, without a requirement for further operations. The typical resolution range for μSLA is 150 mm for the three space directions. However, some parameters, such as: resolution, precision, and surface roughness may change since they are based on over layering. In particular, the vertical resolution (over the construction axis) is associated with the overlapped layer thickness, whereas the lateral resolution (on the plane) is determined by the light beam dimension used to design each layer shape on the resin surface.

In comparison to others' AM, the paramount μSLA advantage is the production of 3D microscale objects holding complex structures and supporting a high velocity fabrication. On account of these compelling characteristics, μSLA is assigned to several fields such as: biomedicine, tissue engineering, micro-optic devices, bioinspired materials, microelectromechanical, among other systems [146–149].

The μSLA is relevant regarding the development of AXSs since it can produce with ease auxetic cores for sandwich panels, and therefore apply them on a large scale. Alomarah et al. [150] manufactured re-entrant AXSs and re-entrant chiral auxetic (RCA) composed of a photopolymer composed by polypropylene (PP) by μSLA. In a more recent work, Varas et al. [151] also used PP to produce different types of AXSs and evaluate the properties of each structure.

4.3. Direct Laser Writing (DLW)

Direct laser writing (DLW) is a well-known AM method for the fabrication of complex structures on a nanoscale up to 100 nm [152]. In this method, a beam laser is focused through an objective lens in order to cure photopolymers by means of a single or multi-photon absorption [153,154]. The DLW by single-photon absorption is limited to the manufacture of 2D structures because its absorption occurs inside a whole area of photopolymers exposed to light. The multi-photon absorption, also known as two-photon polymerization, takes place in a small voxel on the focal point of the laser beam, where the light intensity is remarkably high [155]; thus, arbitrary 3D structures possessing details excessively smaller could be fabricated by the voxel [156,157]. As a consequence, this process is an attractive tool for applications in several fields, such as micro-optics, supercapacitors, microfluidics, biomedical implants and tissue engineering [158–164]. In Figure 23, the schematic diagram of how the DLW works is illustrated.

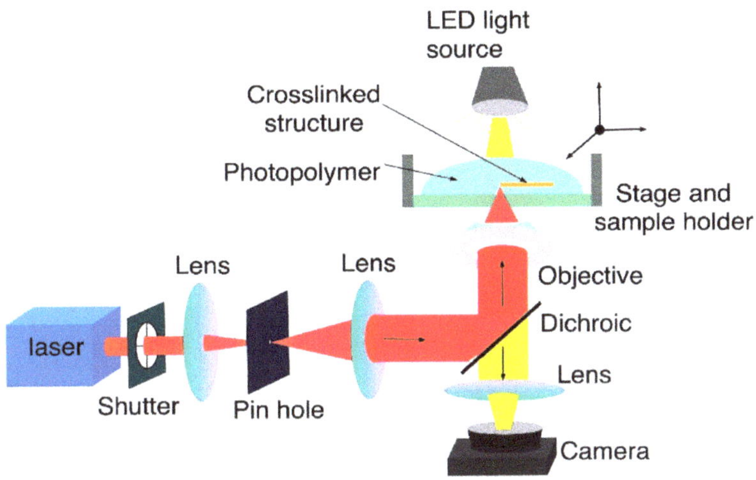

Figure 23. Setup of the direct laser writing system. Reprinted with permission from Ref. [153]. Copyright 2020, ACS.

The principle of the DLW is based on a laser beam that is pulsed at a wavelength close to infrared (λ = 780 nm). This laser is focused on a photoresist material, transparent to the wavelength, through a high aperture objective lens. The photoresist contains photoinitiators for the absorption of two-photons. Due to the focus through the objective lens, the photons are absorbed within the focal volume. Thus, initiators cause a chemical reaction when excited, which enables the formation of chemical bonds, for example, by polymerization or bond breaking. During sample manufacturing, moving the focal spot in different directions enables the fabrication of complex 3D structures at high fabrication speeds [165].

The DLW is able to manufacture materials without the necessity of supporting matrices or a layer by layer process, as the SPPW, owing to the capacity to precisely induce the polymerization within a particular spatial position of a photoresist thick-film [166]. Despite the aforementioned benefits, the DLW is not widely applied for mass production in all industries because it is much less scalable than other methods. Additionally, the costs of acquiring and conserving DLW systems are very high because accessories such as optical systems and femtosecond lasers are expensive and difficult to obtain. Thus, the process becomes unfeasible for large-scale fabrication of AXSs. Another factor that makes this technique inaccessible in the production of AXSs is the choice of photoresist materials. Because the transparency to the infrared beam is necessary for the processing to occur, and the selection of materials is limited to transparent polymers, other materials such as metallic and ceramic particles are not attainable in the process because they inhibit the penetration of the laser.

4.4. Self-Assembly

The Self-Assembly technique is characterized by the mechanism of polymeric phase separation, which can be in the form of emulsion or colloidal suspension. Although previously not considered an AM technique, due to its technological unfeasibility for such classification, since 2018, research involving this manufacturing method has been growing exponentially, allowing the technique to gain technological maturity to be classified as an AM technique [167]. This is due to its ability to rapidly produce complex structures in a distributed manner, making it a method of great importance during the COVID-19 pandemic between 2020 and 2022 [168–170]. The interactions between the components in the fabrication present van der Waals bonds, hydrogen bonding, electrostatic attraction, as well as hydrophobic and hydrophilic interactions. These interactions can create a self-assembly system owing to a thermodynamic condition of non-equilibrium, establishing a stable state within well arranged hydrophobic structures [171–173]. It is possible to develop materials with organized nanostructures from the separation of the microphases present in the material. Self-assembly is very competitive when compared to other AM techniques owing to the ability to produce parts with customizable morphology and functionalities [137].

In order to manufacture materials on a large scale, a block copolymer (BCP) phase separation technique was developed. BCPs are divided by the management of chemical properties and molecular weight of each polymer. The self-assembly by BCP separation occurs by the thermodynamic incompatibility of the copolymer phases, where a separation of microphases with a variety of nanoscale morphologies is generated, such as: lamellas, spheres, cylinders and others [174–176]. In Figure 24, a schematic of the separation and arrangement of the BCPs microphases is illustrated.

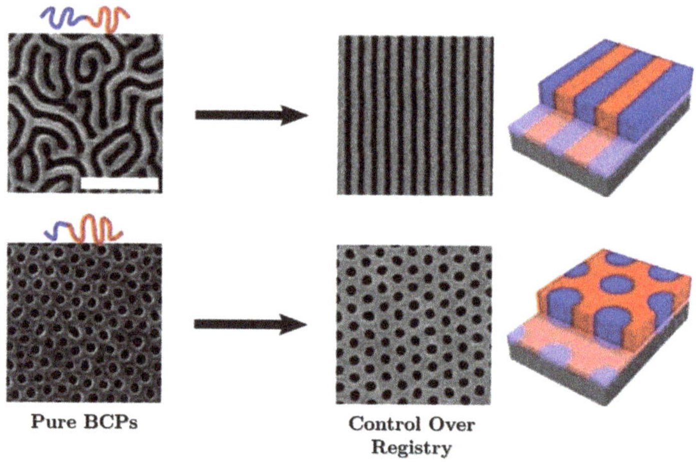

Figure 24. Schematic of self-assembly. The AM method uses a substrate template to give long-range order to lamellar and cylindrical self-assembled block copolymer films. Reprinted with permission from Ref. [177]. Copyright 2016, Nature.

Manufacturing nanostructured materials by self-assembly is not limited only to polymers. Metallic and ceramic structures can also be processed by this technique; for instance, it has already been developed nanostructures of gold [178], CeO_2/MnO_x@C [179] CoPC/CNTs [180], Fe_3O_4/rGO [181] among others for several applications.

In comparison to other AM, self-assembly yields a unique low-cost opportunity, highly scalable and a fast manufacture for 3D structures possessing micro and nanoscale. However, this process's ability for the fabrication of several morphologies, such as octet or cubic plate lattices, is still very limited. In addition, up to now, it is difficult to manage a self-assembled final topology [173]. As a consequence, defects are acquired in large-scale fabrications, and further investigations are required to solve these problems.

4.5. Selective Laser Melting (SLM)

The selective laser (SLM) is another method for the construction of 3D metallic structures. It uses computational support and a high energy laser in order to bond the metallic particles [182]. SLM makes viable the manufacture of metallic components layer by layer following a 3D-computer aided design (CAD) model. Thus, this enables an almost unlimited fabrication of complex geometries without the necessity for pre-production costs or specific tools [183]. The SLM mechanism is demonstrated in Figure 25, and occurs according to the following steps: (i) the 3D-CAD model is decomposed in layers and sent to a selective fusion laser equipment; (ii) a powder material is laid upon the substrate, creating a thin layer; (iii) each piece of geometry information is transferred by the laser beam to the powder bed where the regions having the solid materials are scanned under an inert atmosphere; thus, another layer can be manufactured; (iv) after a layer is constructed upon the substrate, the equipment bases connected to a plunger go down, so a new layer is inserted above the first one. Therefore, the manufacturing steps are repeated layer by layer until the sample is constructed [184,185]. The printed samples can reach a density value close to 100% because metallic powders are used, disclosing suitable mechanical properties combined with a rigorous management of the material composition [186].

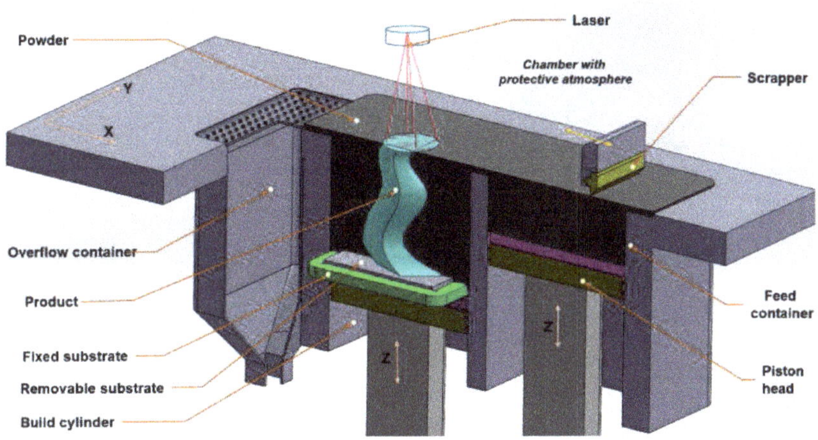

Figure 25. Schematic of selective laser melting 3D printing method. Reprinted with permission from Ref. [187]. Copyright 2017, Elsevier.

The AXSs production by SLM allows the acquisition of materials with superior mechanical properties compared to the auxetic ones composed of polymers. A great majority of AM techniques process polymer structures. Although they present acceptable properties, they still exhibit limited characteristics, in comparison to metals in terms of a greater thermal and mechanical resistance. Singh et al. [188] processed re-entrant AXSs by SLM. They adjusted parameters and mechanical tests and were able to attain a Poisson's Ratio of −3.1. Gao et al. [189] brought on auxetic contacted cube (ACC) structures by SLM using an AlSi10Mg alloy. The authors observed a superior density to 99% for the specimens; in addition, they displayed NPR and a good deformation distribution for all the samples.

4.6. Other Techniques

In addition to the previously mentioned techniques, there are several others for the production of AXSs. Each method has its own process particularities, making it viable to construct complex structures fast and with ease. As mentioned in the text, AXSs can be produced by both conventional techniques and more advanced techniques, such as AM techniques. Among all the available techniques, it is worth mentioning: extrusion [190–192], fused deposition modeling (FDM) [193–195], inkjet printing [196,197], selective electron beam melting (SEBM) [198–201], selective laser sintering (SLS) [202–205], aerosol jet printing [206–208], chemical vapor deposition (CVD) [209–211], and others. Classifying these techniques is a complex task because the research areas for the AM and AXSs present an accelerated expansion. However, certain methods may show similar features but manufacture distinct materials classes. Therefore, it is not possible to classify them in order to expand a range of applications and materials.

4.7. Brief Summary

In this section, the relevance of combining AXS and AM was emphasized. Indeed, the numerous proposed AXS, despite their complex shapes, certainly have a proper technique for cost-effective fabrication of armor protection against high-velocity projectile and blast protection.

5. Applications

The sandwich panels having a core assembled by AXSs present remarkable properties, allowing them to be applied in a variety of distinctive applications, such as biomedical, textiles, sports and military. The AM techniques enable fast processing for materials holding complex structures, while they display great precision and replicability. For this reason, these materials become attractive for large-scale demands. Table 2 presents the main sector and applications for auxetic materials manufactured by AM methods.

Table 2. Summary of the main applications for auxetic materials [87,212].

Field	Application
Aerospace	Vanes for engine, thermal protection, aircraft nose-cones, wing panel, vibration absorber
Automotive	Bumper, cushion, thermal protection, sounds and vibration absorber parts, fastener
Biomedical	Bandage, wound pressure pad, dental floss, artificial blood vessel, artificial skin, drug-release unit, ligament anchors, surgical implants
Composite	Fiber reinforcement
Military	Helmet, bulletproof vest, knee pad, glove, protective gear
Sensors/Actuators	Hydrophone, piezoelectric devices
Textile	Fibers, functional fabric, color-change straps or fabrics, threads

Among the specified applications in Table 2, the area showing promising potential is in the military field. The AXSs are outstanding because they have the capacity to absorb high-velocity impacts, i.e., shock waves and blasts, commonly applied to bullet vests and ballistic helmets [213–215]. Moreover, auxetic materials enable the production of more lightweight and greater uniform shock wave energy absorption; thus, soldiers acquire higher mobility on the battlefield [36,216]. An example is a shoe based on an auxetic material, capable of reducing the body weight impact upon the ground, preventing injuries and providing more comfort [217].

High-speed military applications encompass ballistic protection, in which armor must absorb the energy of projectiles traveling at speeds above 300 m/s and prevent the dispensed projectile and/or shock wave from causing injury to the combatant. These speeds are classified in different ways according to the NIJ 0101.06 standard [218]. From these speeds, it is possible to classify light armor and heavy armor. Soft armor is classified in the lower and medium levels in the NIJ standard, IIA, II, and IIIA, in which protection requirements are against low-energy projectiles, such as the bullets shot from handguns [219,220]. The higher levels must include hard armor inserts in what is called "in conjunction design". Those plates are destined for high-risk military applications, since they are subjected to higher energy projectiles, such as the bullets from AK-47s. Table 3 indicates the relationship of impact velocities of different projectile types to armor levels according to NIJ 0101.06 [218].

Table 3. Test requirements from NIJ 0101.06 standard [218].

Armor Type	Test Bullets	Bullet Mass (g)	Armor Test Velocity (m/s)
IIA	9 mm FMJ RN 0.40 S & W FMJ RN	8.0 11.7	373 352
II	9 mm FMJ RN 0.357 Magnum JSP	8.0 10.0	398 436
IIIA	0.357 SIG FMJ FN 0.44 Magnum SJHP	8.1 15.6	448 436
III	7.62 mm NATO FMJ	9.6	847
IV	0.30 Calibre M2 AP	10.8	878

Considering the application of AXSs in anti-blast and high-velocity (impact and ballistic), Table 4 shows a summary of the most recent research investigating the behavior of auxetic materials for military demands.

Table 4. Summary of recent papers discussing the auxetic sandwich panels for anti-blast and ballistic applications.

Author	Year	Material of AXS	Structure	Application
Liu et al. [221]	2023	Carbon Fiber/PU	Double-Arrow	Ballistic Protection
Jiang et al. [222]	2023	Aluminum AA6061/Grade 800 HSS	Re-entrant Circular (REC)	Blast Protection
Lan et al. [223]	2023	AlSi10Mg aluminum alloy/5052 aluminum alloy	Hybrid Chiral	Ballistic Protection
Haq et al. [224]	2023	Aluminum AA2024/Armor Steel	Re-Entrant/Hexagonal Star/Re-Entrant	Ballistic Protection
Yan et al. [225]	2022	AlSi10Mg aluminum alloy/Q345 steel	Hexagonal and Re-Entrant	Blast Protection
Critchley et al. [226]	2022	TPU	Re-entrant	Blast Protection
Usta et al. [227]	2022	CFRP/PLA	Re-entrant	Ballistic Protection
Wu et al. [64]	2022	Kevlar	Hexagonal and Re-entrant	Impact Resistance

A relevant observation regarding the works previously mentioned in Table 4 is that most of them present a re-entrant structure. From the economic perspective, it is simpler to be designed, consumes less material for its manufacture, and less processing time is required. However, other models are being investigated for ballistic and anti-blast applications, such as the double-arrow [221], re-entrant circular [222] and hybrid structures [223,224].

Liu et al. [221] have developed sandwich panels consisting of double-arrow AXS, aiming to ballistic demands. The researchers fabricated the sandwich structure by the prepreg method, using carbon fiber reinforced polymer (CFRP) as the material for the panels. The auxetic core was also made of CFRP and manufactured by the computer numerical control (CNC) method, attached to the panels employing an epoxy resin and hot pressing process. Subsequently, the auxetic cores were filled with PU foam. The residual velocity tests were performed with a projectile's incident velocity ranging from 75 to 178 m/s. The experiments were carried out by correlating the auxetic sandwich panel filled with PU and another specimen without the foam. From the ballistic test results, the authors discovered that, when the incident velocity was up to 105 m/s, the residual velocity for both panels conditions was the same. However, it was observed that the auxetic panel filled with PU foam presented a decrease in the residual velocity when there was an increase in the incident velocity. Thus, it shows a higher energy absorption throughout the ballistic impact.

Jiang et al. [222] published a study regarding the development of an auxetic sandwich panel for anti-blast protection that may be used in automobiles, ships, buildings and others. The novel material was based on a re-entrant structure. The authors modified it and produced the re-entrant circular (REC) model. It consists of a circular reinforcement attached to each side of the cellular wall. The panels were manufactured using 800 HSS steel, and the auxetic core was fabricated with AA6061 aluminum alloy. Explosion tests were performed, allowing the investigation concerning the absorption power of this AXS. From the acquired numerical results, the researchers verified a higher energy absorption capacity for the REC structure in comparison to the conventional ones because a reduction in the maximum displacement of 28% was observed for the FEM simulations and the experimental tests. In addition, the absorbed impact energy increased by approximately 2%. Therefore, the REC model presents a relevant potential for applications requiring structures resistant against explosions.

The work by Lan et al. [223] proposes the development of AXSs for ballistic applications based on a sandwich panel having its core built from a chiral hybrid structure. The sandwich panel core was fabricated by the selective laser melting (SLM) technique. and AlSi10Mg aluminum alloy powder was used. The panels were built using a 5052 aluminum alloy, while the auxetic re-entrant cores were filled with a PU foam. The ballistic performance was based on the comparison of a model without the foam and the other one filled with it. The ballistic test was performed with a projectile created from aluminum foam, with an incident velocity between 206 and 296 m/s, and a mass ranging from 302 to 364 g. The impact moment was recorded by a high-velocity camera, illustrated in Figure 26. The ballistic tests results have shown that the sandwich panel filled with PU presented severe delamination on the foam after the impact. However, it prevented greater delamination for the panels. On the other hand, the panel without a filling has shown severe delamination on its surface. The indentation region for both specimens exhibited bending, although the specimen filled with PU disclosed less bending due to the energy absorption.

Haq et al. [224] proposed a comparison of the auxetic sandwich panels' dynamic behavior against an explosion. The researchers performed FEM simulations, adopting hexagonal cores based on conventional honeycomb and auxetic star structures composed of AA2024 aluminum alloy, while the panels were made of a material known as "Armor Steel". The explosion simulations were performed using trinitrotoluene (TNT), in concentrations of 1.5, 3 and 5 kg, and dropped over the sandwich panels on heights of 100, 300 and 500 mm. The FEM explosion simulations' results have shown that the auxetic cellular structures absorbed 28% more energy than the conventional honeycomb model. The auxetic cores applied on sandwich panels have demonstrated effectiveness in absorbing a huge amount of energy displaying a relatively smaller deformation for the back panel. The cellular wall structure provides an increase in the energy absorption capacity by the uniform dissipation of impact energy. For the load of 1.5 kg, the findings revealed that delamination did not occur on the structure; on the other hand, for loads of 3 and 5 kg, the simulation identified a delamination.

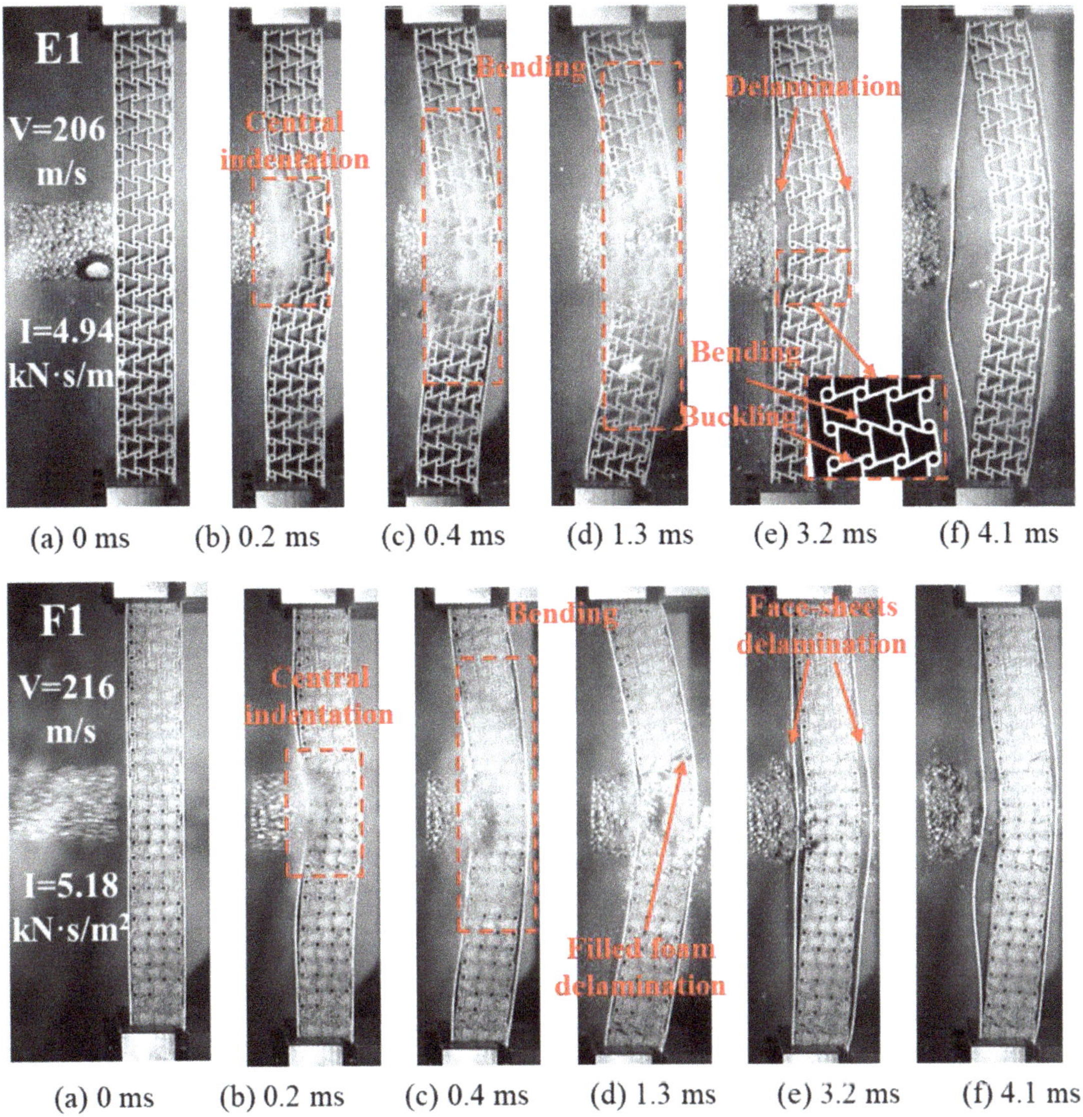

Figure 26. Dynamic response recorded by high-speed camera of auxetic sandwich panels under impact: (E1) Panel without PU foam filling; (F1) Panel filled with PU foam. Reprinted with permission from Ref. [223]. Copyright 2023, Elsevier.

Yan et al. [225] manufactured sandwich panels with an auxetic core focusing on anti-blast systems applications. The sandwich panel was fabricated from Q345 steel sheets, and the core based on a re-entrant structure was made of AlSi10Mg aluminum and constructed by the SLM method. An epoxy resin was applied to attach the core and the panels. The blasting test was performed using 500 g of TNT, placing it 10, 15 and 25 cm above the panel, as demonstrated in Figure 27a. A correlation was carried out with a structure manufactured by the same method. However, the core was a honeycomb non-auxetic model. The blasting test revealed that the auxetic core panel presented less deformation, as depicted in Figure 27b, attaining a maximum value of 19.58 mm, while the hexagonal core panel disclosed 26.29 mm after the explosion. This difference is associated with the amount of absorbed energy, where the AXS material displayed a higher absorption without fracturing. Figure 27c illustrates the panel after the explosion. It is possible to notice the damaged core, in which a re-entrant structure was completely disfigured near the explosion location, but the panel was not fractured. Moreover, the hexagonal core panel exhibited in Figure 27d presented a higher deformation since it absorbed less energy. In addition, the hexagonal core was fractured and bent out of shape while the explosion occurred, and delamination on the front panel has also been detected. These results unveil that the AXS is capable of absorbing greater amounts of energy, and, for this reason, it is suitable for anti-blast system applications.

Critchley et al. [226] investigated the cellular wall angulation effect upon the explosion resistance for re-entrants' AXSs. The samples were manufactured by the fused deposition modeling (FDM) method using a thermoplastic polyurethane (TPU). Five specimens were processed, with a range of cellular wall angulation and, as a result, the following structures were developed: 60° and 75° re-entrant, 90° square, and 105° and 120° hexagonal. Blasting tests were performed in a shocking tube, which simulates a controlled explosion on a lower scale and records the internal explosion reactions. The test findings have shown that the AXS cellular wall angulation is of utmost importance for the blast mitigation, where the 75° (re-entrant) angulation disclosed a mitigation of 23%, whereas the 60° model displayed a value of 12%. The hexagonal structures yielded a blast amplification effect, where the 105° and 120° angulation have shown values of 60% and 277%, respectively.

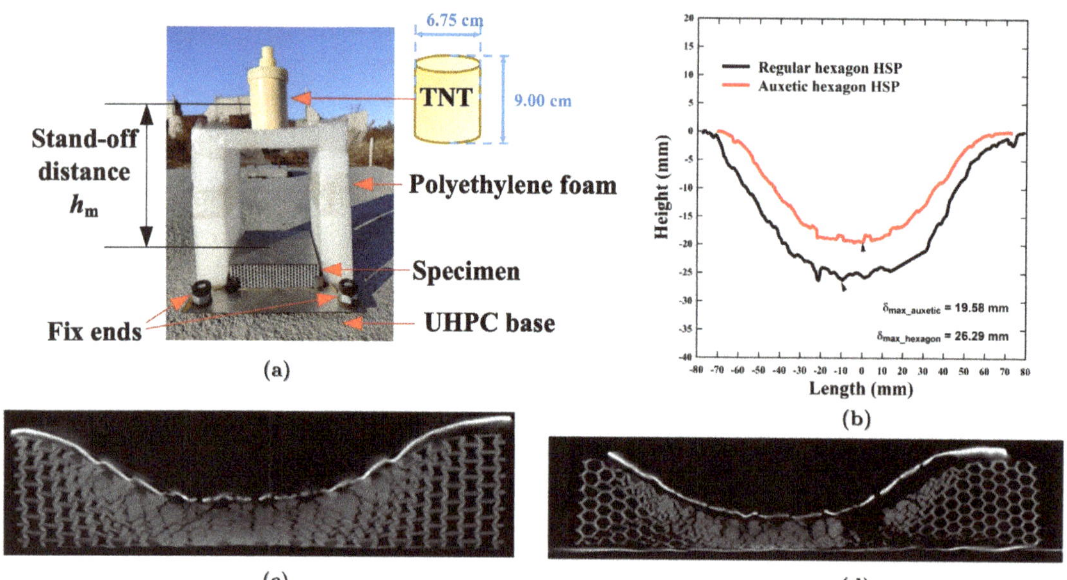

Figure 27. (a) Blasting test setup; (b) comparison of the core deformation; (c) damage mode of auxetic sandwich; (d) damage mode of honeycomb sandwich. Reprinted with permission from Ref. [225]. Copyright 2023, Elsevier.

Usta et al. [15] portrayed the high velocity impact behavior for curved sandwich panels. They were fabricated from CFRP sheets, and the distinct cores were based on no-auxetic PU foams manufactured by CNC, with auxetic PLA presenting a re-entrant structure produced by FDM. High velocity impact tests were performed on a single stage air gas machine that simulates a gun. A spherical steel projectile was used with a radius and velocity of 5 mm and 100 m/s, respectively. The impact test results exposed an increase in the energy absorption as the curvate diminished for the panels and re-entrant cores. The sandwich panels having PU foams disclosed a superior impact resistance as a consequence of their synergistic interaction with the CFRP sheets. The failure mechanisms for the AXS and the PU foam are characterized by a fracture on the superior sheets, where the fibers rupture and matrix cracks have an influence. The foam cores are exposed to shear and crush deformation, while the auxetic core presents bends and cellular wall ruptures. Nevertheless, AXSs portrayed superior SEA values in comparison to the PU foam structures, suggesting that AXSs may have a relevant role in ballistic resistance. Meanwhile, other variables must be investigated in order to improve these metamaterials' resistance.

Wu et al. [64] investigated the AXSs impact resistance developed with Kevlar foils based on the Kirigami sheets, whilst the specimens were cut and rearranged on a 3D hexagonal and re-entrant design. Among the hollow spaces of the structure, shear stiffening gel (SSG) was introduced. The tests revealed that the AXS disclosed higher flexibility and resistance than the hexagonal model. The re-entrant model joined with the SSG filing developed and increased up to 76% in comparison to the hexagonal model also filled with the same material. This result is a consequence of the mechanical behavior provided by an NPR, but also the thickening behavior supplied by the SSG. This research does not make a ballistic application approach. However, the employment of SSG as a filling for AXS might provide great findings on absorption energy and ballistic resistance investigations [61,228]. Furthermore, the SSG is able to enhance its hardness when a shear load is increased because of its thickening that raises the mechanical resistance [229,230]. The same phenomenon is found for the shear thickening fluid (STF), which is extensively used on fabrics for ballistic demands [231–233].

Brief Summary

A brief summary of this section is intended to reveal that, among the several sectors in Table 2, from aerospace to textile, AXS for military, high-velocity impact protection is included. However, only a few investigations [225,226] have used 3D-printed AXS, which is another open opportunity for future research works.

6. Conclusions

An impressive range of types and models of auxetic structures (AXSs), from the initially proposed re-entrant pattern to the novel cross-petal AXS structure for military applications requiring absorbing blast energy, are presently being investigated and industrially developed.

The numerous proposed AXS structures, independent of their complex shapes, can now be more cost-effectively fabricated by several techniques, either by traditional techniques or by AM, also known as 3D printing. In addition to being cheaper, 3D-printed AXS structures have the advantage of fast and easy fabrication.

Looking ahead, there is significant potential for the continued growth and development of 3D-printed AXS structures for use in high-velocity impact applications. As such, we can expect to see continued research and product development in this area in the coming years.

7. Final Remarks

Sandwich panels, particularly those with honeycomb structure inner core, have received considerable attention in the past few decades with an exponential growth in research works. Applications of these honeycomb panels extend to diversified industrial sectors, including military owing to their higher energy absorption. In particular, AXSs honeycomb patterns fabricated by AM have recently been the subject of relevant investigation on their ballistic characteristics, as in the remarkable work of Yan et al. [11] and a prominent review paper on anti-blast performance by Bohara et al. [21].

In the present review, the main points of published AXS/3D-printed articles discussing high-velocity impact were critically discussed. All of these reviewed articles emphatically indicated the superior resistance of AXS/3D-printed sandwich panels to shockwave blast and ballistic impact. One condition, however, is still uncertain in the opinion of Bohara et al. [21]. Under very high impact loads, the blast shockwave propagates much faster than the stress wave. As a result, the AXSs might not have time to deform with an NPR. The authors also recommended further research on improvements in 3D printing technology to produce large-scale AXSs to be applied in armors for high-velocity impact protection. The combination of AM and discrete assembly methods is suggested as a solution for today's size limitation of 3D printers, in order to improve, optimize and fabricate AXS-based ballistic armor. As for the specific case of honeycomb AXS, constructed by SLM/AM, Yan et al. [11] concluded that carbon fiber reinforced polymer, as an outer sheet, is a potential reinforcement for associated sandwich panels with excellent ballistic performance.

Although the commercial use of AXS/AM-based high-velocity impact protective armor is still in its infancy [21], all presently reviewed works strongly support indisputable future military interest.

Author Contributions: Conceptualization, R.F.P.J. and L.F.C.N.; methodology, R.F.P.J., S.N.M. and L.F.C.N.; validation, S.N.M. and L.F.C.N.; formal analysis, R.F.P.J., P.H.P.M.d.S. and L.d.M.N.; investigation, R.F.P.J., P.H.P.M.d.S. and L.d.M.N.; resources, S.N.M.; data curation, P.H.P.M.d.S. and S.N.M.; writing—original draft preparation, R.F.P.J., P.H.P.M.d.S., S.N.M. and L.d.M.N.; writing—review and editing, P.H.P.M.d.S., R.F.P.J. and L.d.M.N.; visualization, L.F.C.N. and S.N.M.; supervision, L.F.C.N. and S.N.M.; project administration, S.N.M.; funding acquisition, S.N.M. All authors have read and agreed to the published version of the manuscript.

Funding: This study was financed in part by the Brazilian Agency CAPES (Coordenação de Aperfeiçoamento de Pessoal de Nível Superior), Protocol #001.

Institutional Review Board Statement: Not applicable.

Informed Consent Statement: Not applicable

Data Availability Statement: Not applicable.

Conflicts of Interest: The authors declare no conflict of interest.

Abbreviations

The following abbreviations are used in this manuscript:

υ	Poisson's Ratio
λ	Wavelength
ABS	Acrylonitrile Butadiene Styrene
ACC	Auxetic Contacted Cube
Ag	Silver
AM	Additive Manufacturing
Au	Gold
AXS	Auxetic Structures
BCP	Block Copolymers
BCPS	Braced Cross-Petal Structure
CFRP	Carbon Fiber Reinforced Polymer
CM	Conventional Material
CNC	Computer Numeric Control
Cu	Copper
CPS	Cross-Petal Structure
CVD	Chemical Vapor Deposition
DLW	Direct Laser Writing
E	Young's Modulus
FDM	Fused Deposition Modeling

FEM	Finite Elements Method
Fe_3O_4	Magnetite
G	Shear Modulus
G_{IC}	Fracture Initiation Toughness
GNP	Graphene Nanoplatelets
GPa	Gigapascal
H	Hardness
HGS	Hourglass Structure
K	Bulk Modulus
K_{IC}^*	Fracture Toughness
LDPE	Low Density Polyethylene
LRP	Laser Rapid Prototyping
MPP	Multi-Photon Polymerization
NIJ	National Institute of Justice
NiTi	Nickel Titanium
NPR	Negative Poisson's Ratio
PDA	Polydiacetylene
PLA	Polylactic Acid
PP	Polypropylene
PU	Polyurethane
RAH	Re-entrant Arrowhead Structure
RCA	Re-entrant Chiral Auxetic
REC	Re-entrant Circular
SAH	Star Arrowhead Structure
SEA	Specific Energy Absorbed
SEBM	Selective ELectron Beam Melting
SLA	Microstereolithography
SLM	Selective Melting Laser
SLS	Selective Laser Sintering
SMEP	Shape Memory Epoxy Polymer
SMP	Shape Memory Polymers
SPPW	Self-propagationg Photopolymer Waveguides
SSG	Shear Stiffening Gel
STF	Shear Thickening Fluid
TNT	Trinitrotoluene
TPU	Thermoplastic Polyurethane
UV	Ultraviolet
VPP	Vat Photopolymerization
ZrO_2	Zirconium Oxide

References

1. Callister, W.D., Jr.; Rethwisch, D.G. *Fundamentals of Materials Science and Engineering: An Integrated Approach*; John Wiley & Sons: Hoboken, NJ, USA, 2020.
2. Li, X.; Zhang, P.; Wang, Z.; Wu, G.; Zhao, L. Dynamic behavior of aluminum honeycomb sandwich panels under air blast: Experiment and numerical analysis. *Compos. Struct.* **2014**, *108*, 1001–1008. [CrossRef]
3. Han, X.; Cai, H.; Sun, J.; Wei, Z.; Huang, Y.; Wang, A. Numerical Studies on Failure Mechanisms of All-Composite Sandwich Structure with Honeycomb Core under Compression and Impact Loading Conditions. *Polymers* **2022**, *14*, 4047. [CrossRef] [PubMed]
4. Zhang, Q.; Yang, X.; Li, P.; Huang, G.; Feng, S.; Shen, C.; Han, B.; Zhang, X.; Jin, F.; Xu, F.; et al. Bioinspired engineering of honeycomb structure—Using nature to inspire human innovation. *Prog. Mater. Sci.* **2015**, *74*, 332–400. [CrossRef]
5. Wahl, L.; Maas, S.; Waldmann, D.; Zürbes, A.; Frères, P. Shear stresses in honeycomb sandwich plates: Analytical solution, finite element method and experimental verification. *J. Sandw. Struct. Mater.* **2012**, *14*, 449–468. [CrossRef]
6. Bitzer, T. *Honeycomb Technology: Materials, Design, Manufacturing, Applications and Testing*; Springer Science & Business Media: Berlin/Heidelberg, Germany, 1997.
7. Yang, X.; Sun, Y.; Yang, J.; Pan, Q. Out-of-plane crashworthiness analysis of bio-inspired aluminum honeycomb patterned with horseshoe mesostructure. *Thin-Walled Struct.* **2018**, *125*, 1–11. [CrossRef]
8. He, Q.; Feng, J.; Chen, Y.; Zhou, H. Mechanical properties of spider-web hierarchical honeycombs subjected to out-of-plane impact loading. *J. Sandw. Struct. Mater.* **2020**, *22*, 771–796. [CrossRef]

9. Ha, N.S.; Lu, G. A review of recent research on bio-inspired structures and materials for energy absorption applications. *Compos. Part B Eng.* **2020**, *181*, 107496. . [CrossRef]
10. Haq, A.U.; Reddy, N.S.K. A brief review on various high energy absorbing materials. *Mater. Today Proc.* **2021**, *38*, 3198–3204. [CrossRef]
11. Yan, J.; Liu, Y.; Yan, Z.; Bai, F.; Shi, Z.; Si, P.; Huang, F. Ballistic characteristics of 3D-printed auxetic honeycomb sandwich panel using CFRP face sheet. *Int. J. Impact Eng.* **2022**, *164*, 104186. [CrossRef]
12. Yang, C.; Vora, H.D.; Chang, Y. Behavior of auxetic structures under compression and impact forces. *Smart Mater. Struct.* **2018**, *27*, 025012. [CrossRef]
13. Hou, S.; Li, T.; Jia, Z.; Wang, L. Mechanical properties of sandwich composites with 3d-printed auxetic and non-auxetic lattice cores under low velocity impact. *Mater. Des.* **2018**, *160*, 1305–1321. [CrossRef]
14. Li, Y.; Chen, Z.; Xiao, D.; Wu, W.; Fang, D. The Dynamic response of shallow sandwich arch with auxetic metallic honeycomb core under localized impulsive loading. *Int. J. Impact Eng.* **2020**, *137*, 103442. [CrossRef]
15. Usta, F.; Türkmen, H.S.; Scarpa, F. Low-velocity impact resistance of composite sandwich panels with various types of auxetic and non-auxetic core structures. *Thin-Walled Struct.* **2021**, *163*, 107738. [CrossRef]
16. Jin, X.; Wang, Z.; Ning, J.; Xiao, G.; Liu, E.; Shu, X. Dynamic response of sandwich structures with graded auxetic honeycomb cores under blast loading. *Compos. Part B Eng.* **2016**, *106*, 206–217. [CrossRef]
17. Qi, C.; Remennikov, A.; Pei, L.Z.; Yang, S.; Yu, Z.H.; Ngo, T.D. Impact and close-in blast response of auxetic honeycomb-cored sandwich panels: Experimental tests and numerical simulations. *Compos. Struct.* **2017**, *180*, 161–178. [CrossRef]
18. Imbalzano, G.; Linforth, S.; Ngo, T.D.; Lee, P.V.S.; Tran, P. Blast resistance of auxetic and honeycomb sandwich panels: Comparisons and parametric designs. *Compos. Struct.* **2018**, *183*, 242–261. [CrossRef]
19. Qi, C.; Yang, S.; Wang, D.; Yang, L.J. Ballistic resistance of honeycomb sandwich panels under in-plane high-velocity impact. *Sci. World J.* **2013**, *2013*, 892781. [CrossRef]
20. Wang, Y.; Yu, Y.; Wang, C.; Zhou, G.; Karamoozian, A.; Zhao, W. On the out-of-plane ballistic performances of hexagonal, reentrant, square, triangular and circular honeycomb panels. *Int. J. Mech. Sci.* **2020**, *173*, 105402. [CrossRef]
21. Bohara, R.P.; Linforth, S.; Nguyen, T.; Ghazlan, A.; Ngo, T. Anti-blast and-impact performances of auxetic structures: A review of structures, materials, methods, and fabrications. *Eng. Struct.* **2023**, *276*, 115377. [CrossRef]
22. Lakes, R. Foam structures with a negative Poisson's ratio. *Science* **1987**, *235*, 1038–1040. [CrossRef]
23. Mardling, P.; Alderson, A.; Jordan-Mahy, N.; Le Maitre, C.L. The use of auxetic materials in tissue engineering. *Biomater. Sci.* **2020**, *8*, 2074–2083. [CrossRef] [PubMed]
24. Critchley, R.; Corni, I.; Wharton, J.A.; Walsh, F.C.; Wood, R.J.; Stokes, K.R. A review of the manufacture, mechanical properties and potential applications of auxetic foams. *Phys. Status Solidi* **2013**, *250*, 1963–1982. [CrossRef]
25. Love, A. *A Treatise on the Mathematical Theory of Elasticity*; Cambridge University Press: Cambridge, UK, 1892.
26. Greenhill, A. A treatise on the mathematical theory of elasticity. *Nature* **1893**, *47*, 529–530. [CrossRef]
27. Voigt, W. Bestimmung der Elasticitätsconstanten für das chlorsaure Natron. *Ann. Phys.* **1893**, *285*, 719–723. [CrossRef]
28. Gunton, D.; Saunders, G. The Young's modulus and Poisson's ratio of arsenic, antimony and bismuth. *J. Mater. Sci.* **1972**, *7*, 1061–1068. [CrossRef]
29. Li, Y. The anisotropic behavior of Poisson's ratio, Young's modulus, and shear modulus in hexagonal materials. *Phys. Status Solidi* **1976**, *38*, 171–175. [CrossRef]
30. Yeganeh-Haeri, A.; Weidner, D.J.; Parise, J.B. Elasticity of α-cristobalite: A silicon dioxide with a negative Poisson's ratio. *Science* **1992**, *257*, 650–652. [CrossRef]
31. Baughman, R.H.; Shacklette, J.M.; Zakhidov, A.A.; Stafström, S. Negative Poisson's ratios as a common feature of cubic metals. *Nature* **1998**, *392*, 362–365. [CrossRef]
32. Veronda, D.; Westmann, R. Mechanical characterization of skin—Finite deformations. *J. Biomech.* **1970**, *3*, 111–124. [CrossRef]
33. Lees, C.; Vincent, J.F.; Hillerton, J.E. Poisson's ratio in skin. *Bio-Med. Mater. Eng.* **1991**, *1*, 19–23. [CrossRef]
34. Frolich, L.; LaBarbera, M.; Stevens, W. Poisson's ratio of a crossed fibre sheath: The skin of aquatic salamanders. *J. Zool.* **1994**, *232*, 231–252. [CrossRef]
35. Williams, J.; Lewis, J. Properties and an anisotropic model of cancellous bone from the proximal tibial epiphysis. *J. Biomech. Eng.* **1982**, *104*, 50–56. [CrossRef]
36. Sanami, M.; Ravirala, N.; Alderson, K.; Alderson, A. Auxetic materials for sports applications. *Procedia Eng.* **2014**, *72*, 453–458. [CrossRef]
37. Duncan, O.; Shepherd, T.; Moroney, C.; Foster, L.; Venkatraman, P.D.; Winwood, K.; Allen, T.; Alderson, A. Review of auxetic materials for sports applications: Expanding options in comfort and protection. *Appl. Sci.* **2018**, *8*, 941. [CrossRef]
38. Du, L.; Luo, S.; Xu, Y. Understanding nonlinear behaviors of auxetic foams using X-ray tomography and pore structure analysis. *Mech. Mater.* **2022**, *165*, 104196. [CrossRef]
39. Zhang, Q.; Scarpa, F.; Barton, D.; Zhu, Y.; Lang, Z.Q.; Zhang, D.; Peng, H.X. Impact properties of uniaxially thermoformed auxetic foams. *Int. J. Impact Eng.* **2022**, *163*, 104176. [CrossRef]
40. Nazari, Z.; Ahmadi, H.; Liaghat, G.; Vahid, S. Investigation on the compressive properties of auxetic foams under different loading rates. *Polym. Eng. Sci.* **2022**, *62*, 1720–1730. [CrossRef]

41. Shah, I.A.; Khan, R.; Koloor, S.S.R.; Petru, M.; Badshah, S.; Ahmad, S.; Amjad, M. Finite Element Analysis of the Ballistic Impact on Auxetic Sandwich Composite Human Body Armor. *Materials* **2022**, *15*, 2064. [CrossRef]
42. Hassanin, H.; Abena, A.; Elsayed, M.A.; Essa, K. 4D printing of NiTi auxetic structure with improved ballistic performance. *Micromachines* **2020**, *11*, 745. [CrossRef]
43. Imbalzano, G.; Tran, P.; Ngo, T.D.; Lee, P.V. Three-dimensional modelling of auxetic sandwich panels for localised impact resistance. *J. Sandw. Struct. Mater.* **2017**, *19*, 291–316. [CrossRef]
44. Scarpa, F. Auxetic materials for bioprostheses [In the Spotlight]. *IEEE Signal Process. Mag.* **2008**, *25*. [CrossRef]
45. Kolken, H.M.; Janbaz, S.; Leeflang, S.M.; Lietaert, K.; Weinans, H.H.; Zadpoor, A.A. Rationally designed meta-implants: A combination of auxetic and conventional meta-biomaterials. *Mater. Horizons* **2018**, *5*, 28–35. [CrossRef]
46. Zadpoor, A.A. Mechanical performance of additively manufactured meta-biomaterials. *Acta Biomater.* **2019**, *85*, 41–59. [CrossRef] [PubMed]
47. Dusfour, G.; LeFloc'h, S.; Cañadas, P.; Ambard, D. Heterogeneous mechanical hyperelastic behavior in the porcine annulus fibrosus explained by fiber orientation: An experimental and numerical approach. *J. Mech. Behav. Biomed. Mater.* **2020**, *104*, 103672. [CrossRef] [PubMed]
48. Li, Y.; Yu, W.R. Development of a textile structure for multi-directional auxetic deformation. *Mater. Des.* **2022**, *223*, 111237. [CrossRef]
49. Shah, A.A.; Shahid, M.; Siddiqui, N.A.; Nawab, Y.; Iqbal, M. Effect of Geometric Arrangement on Mechanical Properties of 2D Woven Auxetic Fabrics. *Textiles* **2022**, *2*, 606–623. [CrossRef]
50. Yuping, C.; Liu, Y.; Hong, H. Deformation behavior of auxetic laminated fabrics with rotating square geometry. *Text. Res. J.* **2022**, *92*, 4652–4665. [CrossRef]
51. Shen, H.S.; Xiang, Y. Assessment of negative Poisson's ratio effect on thermal post-buckling of FG-GRMMC laminated cylindrical panels. *Adv. Nano Res.* **2021**, *10*, 423–435.
52. Kolken, H.; De Jonge, C.; Van der Sloten, T.; Garcia, A.F.; Pouran, B.; Willemsen, K.; Weinans, H.; Zadpoor, A. Additively manufactured space-filling meta-implants. *Acta Biomater.* **2021**, *125*, 345–357. [CrossRef]
53. Muslov, S. Characteristics of elastic anisotropy of Ti-Ni crystals with shape memory and their relation to extreme values of the Poisson's ratio. *Lett. Mater.* **2021**, *11*, 28–32. [CrossRef]
54. Guo, L.; Zhao, S.; Guo, Y.; Yang, J.; Kitipornchai, S. Bandgaps in functionally graded phononic crystals containing graphene origami-enabled metamaterials. *Int. J. Mech. Sci.* **2023**, *240*, 107956. [CrossRef]
55. Liao, M.E.; Li, C.; Shah, N.; Hsiao, Y.H.; Bauchy, M.; Sant, G.; Goorsky, M.S. Experimental evidence of auxeticity in ion implanted single crystal calcite. *Sci. Rep.* **2022**, *12*, 6071. [CrossRef]
56. Jiang, Z.; Xu, N. Emergent Physics on the Route from Crystals to Amorphous Solids. *Chin. J. Chem.* **2022**, *40*, 1091–1098. [CrossRef]
57. Amaya-Amaya, V.; de Icaza-Herrera, M.; Martínez-Hernández, A.L.; Martínez-Barrera, G.; Velasco-Santos, C. Experimental approximation of the sound absorption coefficient (∝) for 3D printed reentrant auxetic structures of poly lactic acid reinforced with chicken keratin materials. *Mater. Lett.* **2021**, *283*, 128757. [CrossRef]
58. Magesh, M.; Jawahar, P. Examination of shape memory polymer-auxetic landing gears on landing approach for quadcopter. *Mater. Today Proc.* **2021**, *47*, 471–479. [CrossRef]
59. Verma, P.; He, C.; Griffin, A.C. Implications for Auxetic Response in Liquid Crystalline Polymers: X-Ray Scattering and Space-Filling Molecular Modeling. *Phys. Status Solidi* **2020**, *257*, 2000261. [CrossRef]
60. Zhang, J.; Dong, B.; Zhang, W. Dynamic Crushing of Gradient Auxetic Honeycombs. *J. Vib. Eng. Technol.* **2021**, *9*, 421–431. [CrossRef]
61. Tang, H.; Jiang, X.; Li, L.; Ling, L.; Hu, Y. Electromechanical properties of ultra-low porous auxetic piezocomposite: From the perspective of Poisson's ratio. *J. Am. Ceram. Soc.* **2021**, *104*, 2628–2645. [CrossRef]
62. Lantada, A.D.; de Blas Romero, A.; Schwentenwein, M.; Jellinek, C.; Homa, J. Lithography-based ceramic manufacture (LCM) of auxetic structures: Present capabilities and challenges. *Smart Mater. Struct.* **2016**, *25*, 054015. [CrossRef]
63. Zhang, X.G.; Jiang, W.; Zhang, Y.; Luo, C.; Zhang, X.Y.; Han, D.; Hao, J.; Teng, X.C.; Xie, Y.M.; Ren, X. Energy absorption properties of composite tubes with hexagonal and re-entrant honeycomb fillers. *Constr. Build. Mater.* **2022**, *356*, 129298. [CrossRef]
64. Wu, L.; Zhao, F.; Lu, Z.; Lin, J.H.; Jiang, Q. Impact energy absorption composites with shear stiffening gel-filled negative poisson's ratio skeleton by Kirigami method. *Compos. Struct.* **2022**, *298*, 116009. [CrossRef]
65. Li, Z.Y.; Wang, X.T.; Ma, L.; Wu, L.Z. Study on the mechanical properties of CFRP composite auxetic structures consist of corrugated sheets and tubes. *Compos. Struct.* **2022**, *292*, 115655. [CrossRef]
66. Jiang, F.; Yang, S.; Zhang, Y.; Qi, C.; Chen, S. Fabrication and crushing response of graded re-entrant circular auxetic honeycomb. *Int. J. Mech. Sci.* **2023**, *242*, 107999. [CrossRef]
67. Reyes, G.; Ajdary, R.; Kankuri, E.; Kaschuk, J.J.; Kosonen, H.; Rojas, O.J. Cellulose gelation in NaOH (aq) by CO_2 absorption: Effects of holding time and concentration on biomaterial development. *Carbohydr. Polym.* **2023**, *302*, 120355. [CrossRef] [PubMed]
68. Zhao, C.; Goh, K.L.; Lee, H.P.; Yin, C.; Zhang, K.; Zhong, J. Experimental study and finite element analysis on energy absorption of carbon fiber reinforced composite auxetic structures filled with aluminum foam. *Compos. Struct.* **2023**, *303*, 116319. [CrossRef]
69. Chen, X.Y.; Hamdi, O.; Rodrigue, D. Conversion of low density polyethylene foams into auxetic metamaterials. *Polym. Adv. Technol.* **2022**, *34*, 228–237. [CrossRef]

70. Xue, B.Z.; Li, J.W.; Huang, R.; Yang, Y.Z.; Gong, H.S.; Zhang, Q.M.; Kong, L.H.; Li, D. A Study of Negative Poisson's Ratio of 3D Printed Auxetic Structures. *Mech. Solids* **2022**, *57*, 1–10. [CrossRef]
71. Zhou, Y.; Li, Y.; Jiang, D.; Chen, Y.; Xie, Y.M.; Jia, L.J. In-plane impact behavior of 3D-printed auxetic stainless honeycombs. *Eng. Struct.* **2022**, *266*, 114656. [CrossRef]
72. Fan, Y.; Shen, H.S. Non-symmetric stiffness of origami-graphene metamaterial plates. *Compos. Struct.* **2022**, *297*, 115974. [CrossRef]
73. Köllner, D.; Simon, S.; Niedermeyer, S.; Spath, I.; Wolf, E.; Kakimoto, K.i.; Fey, T. Relation between Structure, Mechanical and Piezoelectric Properties in Cellular Ceramic Auxetic and Honeycomb Structures. *Adv. Eng. Mater.* **2022**, *25*, 2201387. [CrossRef]
74. Fan, G.Z.; Ren, X.; Wang, S.L.; Luo, C.; Xie, Y.M. A novel cement-based auxetic foam composite: Experimental study. *Case Stud. Constr. Mater.* **2022**, *17*, e01159. [CrossRef]
75. Yolcu, D.A.; Baba, B.O. Measurement of Poisson's ratio of the auxetic structure. *Measurement* **2022**, *204*, 112040. [CrossRef]
76. Li, P.; Xu, Y.; Liang, C.; Zeng, X.C. MgXN2 (X = Hf/Zr) Monolayers: Auxetic Semiconductor with Highly Anisotropic Optical/Mechanical Properties and Carrier Mobility. *J. Phys. Chem. Lett.* **2022**, *13*, 10534–10542. [CrossRef]
77. Fan, D.; Li, N.; Li, M.; Wang, S.; Li, S.; Tang, T. Polyurethane/polydopamine/graphene auxetic composite foam with high-efficient and tunable electromagnetic interference shielding performance. *Chem. Eng. J.* **2022**, *427*, 131635. [CrossRef]
78. Choi, J.; Lakes, R. Nonlinear properties of metallic cellular materials with a negative Poisson's ratio. *J. Mater. Sci.* **1992**, *27*, 5375–5381. [CrossRef]
79. Argatov, I.I.; Guinovart-Díaz, R.; Sabina, F.J. On local indentation and impact compliance of isotropic auxetic materials from the continuum mechanics viewpoint. *Int. J. Eng. Sci.* **2012**, *54*, 42–57. [CrossRef]
80. Coenen, V.; Alderson, K. Mechanisms of failure in the static indentation resistance of auxetic carbon fibre laminates. *Phys. Status Solidi* **2011**, *248*, 66–72. [CrossRef]
81. Choi, J.; Lakes, R. Fracture toughness of re-entrant foam materials with a negative Poisson's ratio: Experiment and analysis. *Int. J. Fract.* **1996**, *80*, 73–83. [CrossRef]
82. Alderson, A.; Rasburn, J.; Ameer-Beg, S.; Mullarkey, P.G.; Perrie, W.; Evans, K.E. An auxetic filter: A tuneable filter displaying enhanced size selectivity or defouling properties. *Ind. Eng. Chem. Res.* **2000**, *39*, 654–665. [CrossRef]
83. Francisco, M.B.; Pereira, J.L.J.; Oliver, G.A.; Roque da Silva, L.R.; Cunha, S.S., Jr.; Gomes, G.F. A review on the energy absorption response and structural applications of auxetic structures. *Mech. Adv. Mater. Struct.* **2022**, *29*, 5823–5842. [CrossRef]
84. Gao, Q.; Liao, W.H. Energy absorption of thin walled tube filled with gradient auxetic structures-theory and simulation. *Int. J. Mech. Sci.* **2021**, *201*, 106475. [CrossRef]
85. Bohara, R.P.; Linforth, S.; Nguyen, T.; Ghazlan, A.; Ngo, T. Novel lightweight high-energy absorbing auxetic structures guided by topology optimisation. *Int. J. Mech. Sci.* **2021**, *211*, 106793. [CrossRef]
86. Evans, K.E.; Alderson, A. Auxetic materials: Functional materials and structures from lateral thinking! *Adv. Mater.* **2000**, *12*, 617–628. [CrossRef]
87. Prawoto, Y. Seeing auxetic materials from the mechanics point of view: A structural review on the negative Poisson's ratio. *Comput. Mater. Sci.* **2012**, *58*, 140–153. [CrossRef]
88. Carneiro, V.H.; Meireles, J.; Puga, H. Auxetic materials—A review. *Mater. Sci. Poland* **2013**, *31*, 561–571. [CrossRef]
89. Wallbanks, M.; Khan, M.F.; Bodaghi, M.; Triantaphyllou, A.; Serjouei, A. On the design workflow of auxetic metamaterials for structural applications. *Smart Mater. Struct.* **2021**, *31*, 023002. [CrossRef]
90. Liu, Q. *Literature Review: Materials with Negative Poisson's Ratios and Potential Applications to Aerospace and Defence*; Defence Science and Technology Organisation: Victoria, Australia, 2006.
91. Tretiakov, K.V.; Wojciechowski, K.W. Elasticity of two-dimensional crystals of polydisperse hard disks near close packing: Surprising behavior of the Poisson's ratio. *J. Chem. Phys.* **2012**, *136*, 204506. [CrossRef]
92. Narojczyk, J.; Alderson, A.; Imre, A.; Scarpa, F.; Wojciechowski, K. Negative Poisson's ratio behavior in the planar model of asymmetric trimers at zero temperature. *J. Non-Cryst. Solids* **2008**, *354*, 4242–4248. [CrossRef]
93. Yang, W.; Li, Z.M.; Shi, W.; Xie, B.H.; Yang, M.B. Review on auxetic materials. *J. Mater. Sci.* **2004**, *39*, 3269–3279. [CrossRef]
94. Xinchun, S.; Lakes, R.S. Stability of elastic material with negative stiffness and negative Poisson's ratio. *Phys. Status Solidi* **2007**, *244*, 1008–1026. [CrossRef]
95. Donoghue, J.; Alderson, K.; Evans, K. The fracture toughness of composite laminates with a negative Poisson's ratio. *Phys. Status Solidi* **2009**, *246*, 2011–2017. [CrossRef]
96. Maiti, S.; Ashby, M.; Gibson, L. Fracture toughness of brittle cellular solids. *Scr. Metall.* **1984**, *18*, 213–217. [CrossRef]
97. Yang, S.; Chalivendra, V.B.; Kim, Y.K. Fracture and impact characterization of novel auxetic Kevlar®/Epoxy laminated composites. *Compos. Struct.* **2017**, *168*, 120–129. [CrossRef]
98. Scarpa, F.; Yates, J.; Ciffo, L.; Patsias, S. Dynamic crushing of auxetic open-cell polyurethane foam. *Proc. Inst. Mech. Eng. Part C J. Mech. Eng. Sci.* **2002**, *216*, 1153–1156. [CrossRef]
99. Oh, J.H.; Kim, J.S.; Oh, I.K. Auxetic graphene oxide-porous foam for acoustic wave and shock energy dissipation. *Compos. Part B Eng.* **2020**, *186*, 107817. [CrossRef]
100. Najafi, M.; Ahmadi, H.; Liaghat, G. Experimental investigation on energy absorption of auxetic structures. *Mater. Today Proc.* **2021**, *34*, 350–355. [CrossRef]

101. Grima, J. Modelling Auxetic Materials. In Proceedings of the 2009 Computational Nanotechnology Intensive Programme (CoNan 2009, EU Life-Long Learning Programme), Gdansk, Poland, 2–14 August 2009.
102. Chen, Y.; Scarpa, F.; Farrow, I.R.; Liu, Y.; Leng, J. Composite flexible skin with large negative Poisson's ratio range: Numerical and experimental analysis. *Smart Mater. Struct.* **2013**, *22*, 045005. [CrossRef]
103. Veerabagu, U.; Palza, H.; Quero, F. Auxetic polymer-based mechanical metamaterials for biomedical applications. *ACS Biomater. Sci. Eng.* **2022**, *8*, 2798–2824. [CrossRef]
104. Kim, Y.; Son, K.H.; Lee, J.W. Auxetic structures for tissue engineering scaffolds and biomedical devices. *Materials* **2021**, *14*, 6821. [CrossRef]
105. Xue, H.; Luo, Z.; Brown, T.; Beier, S. Design of self-expanding auxetic stents using topology optimization. *Front. Bioeng. Biotechnol.* **2020**, *8*, 736. [CrossRef]
106. Asadi, A.; Hedayat, D.; Ghofrani, S.; Abouei Mehrizi, A.; Shadalooyi, G.; Kadkhodapour, J.; Pourkamali Anaraki, A. Modification of hexachiral unit cell to enhance auxetic stent performance. *Mech. Adv. Mater. Struct.* **2022**, 1–15. [CrossRef]
107. Alderson, A.; Rasburn, J.; Evans, K.; Grima, J. Auxetic polymeric filters display enhanced de-fouling and pressure compensation properties. *Membr. Technol.* **2001**, *2001*, 6–8. [CrossRef]
108. Dolla, W.J.S.; Fricke, B.A.; Becker, B.R. Structural and drug diffusion models of conventional and auxetic drug-eluting stents. *J. Med. Devices.* **2007**, *1*, 47–55. [CrossRef]
109. Shirdel, M.; Ghofrani, S.; Mehrizi, A.A. Effect of increasing porosity as drug reservoirs for local drug delivery in an auxetic stent. In Proceedings of the 2021 28th National and 6th International Iranian Conference on Biomedical Engineering (ICBME), Tehran, Iran, 25–26 November 2021; pp. 65–69.
110. Wang, Z.; Luan, C.; Liao, G.; Liu, J.; Yao, X.; Fu, J. Progress in auxetic mechanical metamaterials: Structures, characteristics, manufacturing methods, and applications. *Adv. Eng. Mater.* **2020**, *22*, 2000312.
111. Otsuka, K.; Wayman, C.M. *Shape Memory Materials*; Cambridge University Press: Cambridge, UK, 1999.
112. Huang, W.; Ding, Z.; Wang, C.; Wei, J.; Zhao, Y.; Purnawali, H. Shape memory materials. *Mater. Today* **2010**, *13*, 54–61. [CrossRef]
113. Le Sénéchal, N.V.; Teixeira, R.; Rodrigues, P.F.; Ribeiro, S.B.; dos Santos Paula, A. The influence of aging treatment on the microstructural and mechanical behavior by ultra-micro hardness tester in Ni-rich NiTi alloy. *Cadernos UniFOA* **2021**, *16*, 3810. [CrossRef]
114. Freitas Rodrigues, P.; Teixeira, R.S.; Le Sénéchal, N.V.; Braz Fernandes, F.M.; Paula, A.S. The Influence of the Soaking Temperature Rotary Forging and Solution Heat Treatment on the Structural and Mechanical Behavior in Ni-Rich NiTi Alloy. *Materials* **2021**, *15*, 63. [CrossRef]
115. Song, D.; Yu, C.; Zhang, C.; Kang, G. Superelasticity degradation of NiTi shape memory alloy in wide ranges of temperature and loading level: Experimental observation and micromechanical constitutive model. *Int. J. Plast.* **2023**, *161*, 103487. [CrossRef]
116. Liu, H.; Wang, F.; Wu, W.; Dong, X.; Sang, L. 4D printing of mechanically robust PLA/TPU/Fe3O4 magneto-responsive shape memory polymers for smart structures. *Compos. Part B Eng.* **2023**, *248*, 110382. [CrossRef]
117. Idowu, A.; Thomas, T.; Boesl, B.; Agarwal, A. Cryo-Assisted Extrusion Three-Dimensional Printing of Shape Memory Polymer–Graphene Composites. *J. Manuf. Sci. Eng.* **2023**, *145*, 041003.
118. Lotfolahpour, A.; Huber, W.; Zaeem, M.A. A phase-field model for interactive evolution of phase transformation and cracking in superelastic shape memory ceramics. *Comput. Mater. Sci.* **2023**, *216*, 111844. [CrossRef]
119. Bianchi, M.; Scarpa, F.; Smith, C.; Whittell, G.R. Physical and thermal effects on the shape memory behaviour of auxetic open cell foams. *J. Mater. Sci.* **2010**, *45*, 341–347. [CrossRef]
120. Kamyab, A.; Ghasemi-Ghalebahman, A.; Fereidoon, A.; Khonakdar, H.A. Investigation into the shape memory behavior of peanut-pattern auxetic structures. *Express Polym. Lett.* **2022**, *16*, 679–693. [CrossRef]
121. Gibson, L.J.; Ashby, M.F.; Schajer, G.; Robertson, C. The mechanics of two-dimensional cellular materials. *Proc. R. Soc. Lond. A Math. Phys. Sci.* **1982**, *382*, 25–42.
122. Bronder, S.; Herter, F.; Röhrig, A.; Bähre, D.; Jung, A. Design Study for Multifunctional 3D Re-entrant Auxetics. *Adv. Eng. Mater.* **2022**, *24*, 2100816. [CrossRef]
123. Grima, J.N.; Evans, K.E. *Auxetic Behavior from Rotating Squares*; Kluwer Academic Publishers: Amsterdam, The Netherlands, 2000.
124. Grima, J.N.; Evans, K.E. Auxetic behavior from rotating triangles. *J. Mater. Sci.* **2006**, *41*, 3193–3196. [CrossRef]
125. Alderson, A.; Evans, K.E. Molecular origin of auxetic behavior in tetrahedral framework silicates. *Phys. Rev. Lett.* **2002**, *89*, 225503. [CrossRef]
126. Grima, J.N.; Alderson, A.; Evans, K. Auxetic behaviour from rotating rigid units. *Phys. Status Solidi* **2005**, *242*, 561–575. [CrossRef]
127. Wang, Z.; Hu, H. Auxetic materials and their potential applications in textiles. *Text. Res. J.* **2014**, *84*, 1600–1611. [CrossRef]
128. Grima, J.N.; Gatt, R.; Alderson, A.; Evans, K.E. On the auxetic properties of 'rotating rectangles' with different connectivity. *J. Phys. Soc. Jpn.* **2005**, *74*, 2866–2867. [CrossRef]
129. Grima, J.N.; Farrugia, P.S.; Gatt, R.; Attard, D. On the auxetic properties of rotating rhombi and parallelograms: A preliminary investigation. *Phys. Status Solidi* **2008**, *245*, 521–529. [CrossRef]
130. Prall, D.; Lakes, R. Properties of a chiral honeycomb with a Poisson's ratio of—1. *Int. J. Mech. Sci.* **1997**, *39*, 305–314. [CrossRef]
131. Grima, J.N.; Gatt, R.; Farrugia, P.S. On the properties of auxetic meta-tetrachiral structures. *Phys. Status Solidi* **2008**, *245*, 511–520. [CrossRef]
132. Alderson, A.; Evans, K. Microstructural modelling of auxetic microporous polymers. *J. Mater. Sci.* **1995**, *30*, 3319–3332. [CrossRef]

133. Cho, H.; Seo, D.; Kim, D.N. Mechanics of Auxetic Materials. In *Handbook of Mechanics of Materials*; Schmauder, S., Chen, C.S., Chawla, K.K., Chawla, N., Chen, W., Kagawa, Y., Eds.; Springer: Singapore, 2019; pp. 733–757. [CrossRef]
134. He, C.; Liu, P.; Griffin, A.C. Toward negative Poisson ratio polymers through molecular design. *Macromolecules* **1998**, *31*, 3145–3147. [CrossRef]
135. Ravirala, N.; Alderson, A.; Alderson, K. Interlocking hexagons model for auxetic behaviour. *J. Mater. Sci.* **2007**, *42*, 7433–7445. [CrossRef]
136. Etemadi, E.; Gholikord, M.; Zeeshan, M.; Hu, H. Improved mechanical characteristics of new auxetic structures based on stretch-dominated-mechanism deformation under compressive and tensile loadings. *Thin-Walled Struct.* **2023**, *184*, 110491. [CrossRef]
137. Lu, C.; Hsieh, M.; Huang, Z.; Zhang, C.; Lin, Y.; Shen, Q.; Chen, F.; Zhang, L. Architectural Design and Additive Manufacturing of Mechanical Metamaterials: A Review. *Engineering* **2022**, *17*, 44–63. [CrossRef]
138. Jacobsen, A.J.; Barvosa-Carter, W.; Nutt, S. Micro-scale truss structures formed from self-propagating photopolymer waveguides. *Adv. Mater.* **2007**, *19*, 3892–3896. [CrossRef]
139. Erdeniz, D.; Schaedler, T.A.; Dunand, D.C. Deposition-based synthesis of nickel-based superalloy microlattices. *Scr. Mater.* **2017**, *138*, 28–31. [CrossRef]
140. Bauer, J.; Meza, L.R.; Schaedler, T.A.; Schwaiger, R.; Zheng, X.; Valdevit, L. Nanolattices: An emerging class of mechanical metamaterials. *Adv. Mater.* **2017**, *29*, 1701850. [CrossRef]
141. Jacobsen, A.; Kolodziejska, J.; Doty, R.; Fink, K.; Zhou, C.; Roper, C.; Carter, W. Interconnected self-propagating photopolymer waveguides: An alternative to stereolithography for rapid formation of lattice-based open-cellular materials. In Proceedings of the 2010 International Solid Freeform Fabrication Symposium, Austin, TX, USA, 9–11 August 2010.
142. *ISO/ASTM 52900*; Additive Manufacturing—General Principles—Fundamentals and Vocabulary. International Organization for Standardization: Geneva, Switzeland, 2021.
143. Gibson, I.; Rosen, D.W.; Stucker, B.; Khorasani, M.; Rosen, D.; Stucker, B.; Khorasani, M. *Additive Manufacturing Technologies*; Springer: Berlin/Heidelberg, Germany , 2021; Volume 17.
144. Choi, J.W.; MacDonald, E.; Wicker, R. Multi-material microstereolithography. *Int. J. Adv. Manuf. Technol.* **2010**, *49*, 543–551. [CrossRef]
145. Bertsch, A.; Renaud, P. Microstereolithography. In *Three-Dimensional Microfabrication Using Two-Photon Polymerization*; Elsevier: Amsterdam, The Netherlands, 2020; pp. 25–56.
146. Sun, Q.; Fang, F.; Wang, W.; Yin, J.; Liu, Q.; Hao, L.; Peng, Y. Stereolithography 3D printing of transparent resin lens for high-power phosphor-coated WLEDs packaging. *J. Manuf. Process.* **2023**, *85*, 756–763. [CrossRef]
147. Kim, K.; Yeatts, A.; Dean, D.; Fisher, J.P. Stereolithographic bone scaffold design parameters: Osteogenic differentiation and signal expression. *Tissue Eng. Part B Rev.* **2010**, *16*, 523–539. [CrossRef]
148. Mondschein, R.J.; Kanitkar, A.; Williams, C.B.; Verbridge, S.S.; Long, T.E. Polymer structure-property requirements for stereolithographic 3D printing of soft tissue engineering scaffolds. *Biomaterials* **2017**, *140*, 170–188. [CrossRef]
149. Deng, X.; Zhang, G.; Yu, Z.; Shao, G.; Li, L. Manufacturing of mesoscale non-assembly mechanism with water-soluble support in projection stereolithography process. *J. Manuf. Process.* **2023**, *85*, 658–665. [CrossRef]
150. Alomarah, A.; Ruan, D.; Masood, S.; Sbarski, I.; Faisal, B. An investigation of in-plane tensile properties of re-entrant chiral auxetic structure. *Int. J. Adv. Manuf. Technol.* **2018**, *96*, 2013–2029. [CrossRef]
151. Varas, D.; Pernas-Sánchez, J.; Fjeldberg, N.; Martín-Montal, J. Experimental analysis at different loading rates of 3D printed polymeric auxetic structure based on cylindrical elements. *Polym. Test.* **2023**, *119*, 107930. [CrossRef]
152. Von Freymann, G.; Ledermann, A.; Thiel, M.; Staude, I.; Essig, S.; Busch, K.; Wegener, M. Three-dimensional nanostructures for photonics. *Adv. Funct. Mater.* **2010**, *20*, 1038–1052. [CrossRef]
153. Kunwar, P.; Soman, P. Direct laser writing of fluorescent silver nanoclusters: A review of methods and applications. *ACS Appl. Nano Mater.* **2020**, *3*, 7325–7342. [CrossRef]
154. Wang, R.; Wei, J.; Fan, Y. Chalcogenide phase-change thin films used as grayscale photolithography materials. *Opt. Express* **2014**, *22*, 4973–4984. [CrossRef]
155. Fischer, J.; Wegener, M. Three-dimensional direct laser writing inspired by stimulated-emission-depletion microscopy. *Opt. Mater. Express* **2011**, *1*, 614–624. [CrossRef]
156. Fanyaeu, I.; Mizeikis, V. Realisation of 3D metamaterial perfect absorber structures by direct laser writing. In Proceedings of the Advanced Fabrication Technologies for Micro/Nano Optics and Photonics X, San Francisco, CA, USA, 29 January–1 February 2017; Volume 10115, pp. 177–185.
157. Debnath, S.; Zhang, X.; Guney, D.; Soukoulis, C. Two-dimensionally isotropic optical metamaterial feasible for stimulated emission depletion microscopy inspired direct laser writing. In Proceedings of the 2013 7th International Congress on Advanced Electromagnetic Materials in Microwaves and Optics, Bordeaux, France, 16–21 September 2013; pp. 427–429.
158. Melissinaki, V.; Gill, A.; Ortega, I.; Vamvakaki, M.; Ranella, A.; Haycock, J.; Fotakis, C.; Farsari, M.; Claeyssens, F. Direct laser writing of 3D scaffolds for neural tissue engineering applications. *Biofabrication* **2011**, *3*, 045005. [CrossRef] [PubMed]
159. Selimis, A.; Mironov, V.; Farsari, M. Direct laser writing: Principles and materials for scaffold 3D printing. *Microelectron. Eng.* **2015**, *132*, 83–89. [CrossRef]

160. Srikanth, S.; Dudala, S.; Raut, S.; Dubey, S.K.; Ishii, I.; Javed, A.; Goel, S. Optimization and characterization of direct UV laser writing system for microscale applications. *J. Micromech. Microeng.* **2020**, *30*, 095003. [CrossRef]
161. Somers, N.; Jean, F.; Lasgorceix, M.; Preux, N.; Delmotte, C.; Boilet, L.; Petit, F.; Leriche, A. Fabrication of doped β-tricalcium phosphate bioceramics by Direct Ink Writing for bone repair applications. *J. Eur. Ceram. Soc.* **2023**, *43*, 629–638. [CrossRef]
162. Samartzis, N.; Bhorkar, K.; Athanasiou, M.; Sygellou, L.; Dracopoulos, V.; Ioannides, T.; Yannopoulos, S.N. Direct laser-assisted fabrication of turbostratic graphene electrodes: Comparing symmetric and zinc-ion hybrid supercapacitors. *Carbon* **2023**, *201*, 941–951. [CrossRef]
163. Zhan, T.; Gao, S.; Xu, B. Microfluidic cloth-based analytical devices through simple correction pen direct-writing. *Sensors Actuators B Chem.* **2023**, *375*, 132947. [CrossRef]
164. Schmid, M.; Thiele, S.; Herkommer, A.; Giessen, H. Adjustment-free two-sided 3D direct laser writing for aligned micro-optics on both substrate sides. *Opt. Lett.* **2023**, *48*, 131–134. [CrossRef]
165. Waller, E.H.; von Freymann, G. From photoinduced electron transfer to 3D metal microstructures via direct laser writing. *Nanophotonics* **2018**, *7*, 1259–1277. [CrossRef]
166. Mao, M.; He, J.; Li, X.; Zhang, B.; Lei, Q.; Liu, Y.; Li, D. The emerging frontiers and applications of high-resolution 3D printing. *Micromachines* **2017**, *8*, 113. [CrossRef]
167. Sola, A.; Trinchi, A.; Hill, A.J. Self-assembly meets additive manufacturing: Bridging the gap between nanoscale arrangement of matter and macroscale fabrication. *Smart Mater. Manuf.* **2023**, *1*, 100013. [CrossRef]
168. Tareq, M.S.; Rahman, T.; Hossain, M.; Dorrington, P. Additive manufacturing and the COVID-19 challenges: An in-depth study. *J. Manuf. Syst.* **2021**, *60*, 787–798. [CrossRef]
169. Choong, Y.Y.C.; Tan, H.W.; Patel, D.C.; Choong, W.T.N.; Chen, C.H.; Low, H.Y.; Tan, M.J.; Patel, C.D.; Chua, C.K. The global rise of 3D printing during the COVID-19 pandemic. *Nat. Rev. Mater.* **2020**, *5*, 637–639. [CrossRef]
170. Colorado, H.A.; Mendoza, D.E.; Lin, H.T.; Gutierrez-Velasquez, E. Additive manufacturing against the Covid-19 pandemic: A technological model for the adaptability and networking. *J. Mater. Res. Technol.* **2022**, *16*, 1150–1164. [CrossRef]
171. Soukoulis, C.M.; Wegener, M. Optical metamaterials—More bulky and less lossy. *Science* **2010**, *330*, 1633–1634. [CrossRef]
172. Whitesides, G.M.; Grzybowski, B. Self-assembly at all scales. *Science* **2002**, *295*, 2418–2421. [CrossRef]
173. Li, L.; Sun, R.; Zheng, R. Tunable morphology and functionality of multicomponent self-assembly: A review. *Mater. Des.* **2021**, *197*, 109209. [CrossRef]
174. Kriksin, Y.A.; Neratova, I.V.; Khalatur, P.G.; Khokhlov, A.R. Pattern multiplication by template-guided self-assembly of cylinder-forming copolymers: Field-theoretic and particle-based simulations. *Chem. Phys. Lett.* **2010**, *492*, 103–108. [CrossRef]
175. Li, L.; Wang, Q. Spontaneous self-assembly of silver nanoparticles into lamellar structured silver nanoleaves. *Acs Nano* **2013**, *7*, 3053–3060. [CrossRef] [PubMed]
176. Lin, S.H.; Wu, S.J.; Ho, C.C.; Su, W.F. Rational design of versatile self-assembly morphology of rod–coil block copolymer. *Macromolecules* **2013**, *46*, 2725–2732. [CrossRef]
177. Stein, A.; Wright, G.; Yager, K.G.; Doerk, G.S.; Black, C.T. Selective directed self-assembly of coexisting morphologies using block copolymer blends. *Nat. Commun.* **2016**, *7*, 1–7. [CrossRef] [PubMed]
178. Suzuki, Y.; Cardone, G.; Restrepo, D.; Zavattieri, P.D.; Baker, T.S.; Tezcan, F.A. Self-assembly of coherently dynamic, auxetic, two-dimensional protein crystals. *Nature* **2016**, *533*, 369–373. [CrossRef] [PubMed]
179. Wang, K.; Qin, M.; Wang, C.; Yan, T.; Zhen, Y.; Sun, X.; Wang, J.; Fu, F. CeO_2/MnOx@ C hollow cathode derived from self-assembly of Ce-Mn-MOFs for high-performance aqueous zinc-ion batteries. *J. Colloid Interface Sci.* **2023**, *629*, 733–743. [CrossRef]
180. Wang, H.; Zhao, J.; Yu, J.; Wang, Z. Metal organic framework-derived hierarchical 0D/1D CoPC/CNTs architecture interlaminated in 2D MXene layers for superior absorption of electromagnetic waves. *Synth. Met.* **2023**, *292*, 117215. [CrossRef]
181. Liu, C.; Liu, S.; Feng, X.; Zhu, K.; Lin, G.; Bai, Z.; Wang, L.; Liu, X. Phthalocyanine-mediated interfacial self-assembly of magnetic graphene nanocomposites toward low-frequency electromagnetic wave absorption. *Chem. Eng. J.* **2023**, *452*, 139483. [CrossRef]
182. Waqar, S.; Guo, K.; Sun, J. FEM analysis of thermal and residual stress profile in selective laser melting of 316L stainless steel *J. Manuf. Process.* **2021**, *66*, 81–100. [CrossRef]
183. Bremen, S.; Meiners, W.; Diatlov, A. Selective laser melting: A manufacturing technology for the future? *Laser Tech. J.* **2012**, *9*, 33–38. [CrossRef]
184. Korkmaz, M.E.; Gupta, M.K.; Robak, G.; Moj, K.; Krolczyk, G.M.; Kuntoğlu, M. Development of lattice structure with selective laser melting process: A state of the art on properties, future trends and challenges. *J. Manuf. Process.* **2022**, *81*, 1040–1063. [CrossRef]
185. Tan, J.H.; Wong, W.L.E.; Dalgarno, K.W. An overview of powder granulometry on feedstock and part performance in the selective laser melting process. *Addit. Manuf.* **2017**, *18*, 228–255. [CrossRef]
186. Vieira de Oliveira, R.; Pereira de Lima, Y.; Hoisler Sallet, E.; Abílio Corrêa Gonçalves, D.; Vieira Le Sénèchal, N.; Alves Oliveira Melo, E.; Teixeira, R.; Freitas Rodrigues, P.; Inforçatti Neto, P.; Vicente Lopes da Silva, J.; et al. Production of Cylindrical Specimens Based on the Ni-Ti System by Selective Laser Melting from Elementary Powders. *J. Mater. Eng. Perform.* **2021**, *30*, 5477–5490. [CrossRef]
187. Park, H.S.; Nguyen, D.S. Study on flaking behavior in selective laser melting process. *Procedia CIRP* **2017**, *63*, 569–572. [CrossRef]
188. Singh, D.; Tobin, D.; Dowling, L.; Trimble, D. Optimization of Cobalt Chrome 3D re-entrant Auxetics fabricated using Selective Laser Melting. *Eng. Struct.* **2023**, *278*, 115542. [CrossRef]

189. Gao, M.; He, D.; Guo, X.; Wu, H.; Tan, Z.; Wu, X.; Shao, W.; Wang, G. Design, preparation and characterization of a 3D hierarchical auxetic lattice structure produced by selective laser melting. In *Proceedings of the Structures*; Elsevier: Amsterdam, The Netherlands, 2022; Volume 44, pp. 1219–1231.
190. Alderson, K.; Nazaré, S.; Alderson, A. Large-scale extrusion of auxetic polypropylene fibre. *Phys. Status Solidi* **2016**, *253*, 1279–1287. [CrossRef]
191. Vyavahare, S.; Teraiya, S.; Kumar, S. Auxetic structures fabricated by material extrusion: An experimental investigation of gradient parameters. *Rapid Prototyp. J.* **2021**, *27*, 1041–1058. [CrossRef]
192. Chikkanna, N.; Logakannan, K.P.; Krishnapillai, S.; Ramachandran, V. Quasi-static compression performance of material extrusion enabled re-entrant diamond auxetic metamaterial: Fabrication, tuning the geometrical parameters and fibre reinforcements. *Thin-Walled Struct.* **2022**, *179*, 109550. [CrossRef]
193. Khadem-Reza, L.; Etemadi, E.; Abbaslou, M.; Hu, H. Design of novel 3D auxetic structures based on S-shaped unit-cells. *Smart Mater. Struct.* **2022**, *31*, 075024. [CrossRef]
194. Li, B.; Liang, W.; Zhang, L.; Ren, F.; Xuan, F. TPU/CNTs flexible strain sensor with auxetic structure via a novel hybrid manufacturing process of fused deposition modeling 3D printing and ultrasonic cavitation-enabled treatment. *Sens. Actuators A Phys.* **2022**, *340*, 113526. [CrossRef]
195. Shin, E.J.; Park, Y.; Jung, Y.S.; Choi, H.Y.; Lee, S. Fabrication and characteristics of flexible thermoplastic polyurethane filament for fused deposition modeling three-dimensional printing. *Polym. Eng. Sci.* **2022**, *62*, 2947–2957. [CrossRef]
196. Walther, M.; Ortner, A.; Meier, H.; Löffelmann, U.; Smith, P.J.; Korvink, J.G. Terahertz metamaterials fabricated by inkjet printing. *Appl. Phys. Lett.* **2009**, *95*, 251107. [CrossRef]
197. Kashiwagi, K.; Xie, L.; Li, X.; Kageyama, T.; Miura, M.; Miyashita, H.; Kono, J.; Lee, S.S. Inkjet-printed silver-nanoparticle THz metamaterial. In Proceedings of the 2016 41st International Conference on Infrared, Millimeter, and Terahertz waves (IRMMW-THz), Copenhagen, Denmark, 25–30 September 2016; pp. 1–2.
198. Schwerdtfeger, J.; Schury, F.; Stingl, M.; Wein, F.; Singer, R.; Körner, C. Mechanical characterisation of a periodic auxetic structure produced by SEBM. *Phys. Status Solidi* **2012**, *249*, 1347–1352. [CrossRef]
199. Schwerdtfeger, J.; Heinl, P.; Singer, R.; Körner, C. Auxetic cellular structures through selective electron-beam melting. *Phys. Status Solidi* **2010**, *247*, 269–272. [CrossRef]
200. Schwerdtfeger, J.; Wein, F.; Leugering, G.; Singer, R.; Körner, C.; Stingl, M.; Schury, F. Design of auxetic structures via mathematical optimization. *Adv. Mater.* **2011**, *23*, 2650–2654. [CrossRef] [PubMed]
201. Mitschke, H.; Schwerdtfeger, J.; Schury, F.; Stingl, M.; Körner, C.; Singer, R.F.; Robins, V.; Mecke, K.; Schröder-Turk, G.E. Finding auxetic frameworks in periodic tessellations. *Adv. Mater.* **2011**, *23*, 2669–2674. [CrossRef]
202. Pan, Y.; Zhang, X.G.; Han, D.; Li, W.; Xu, L.F.; Zhang, Y.; Jiang, W.; Bao, S.; Teng, X.C.; Lai, T.; et al. The out-of-plane compressive behavior of auxetic chiral lattice with circular nodes. *Thin-Walled Struct.* **2023**, *182*, 110152. [CrossRef]
203. Jiang, Y.; Shi, K.; Zhou, L.; He, M.; Zhu, C.; Wang, J.; Li, J.; Li, Y.; Liu, L.; Sun, D.; et al. 3D-printed auxetic-structured intervertebral disc implant for potential treatment of lumbar herniated disc. *Bioact. Mater.* **2023**, *20*, 528–538. [CrossRef]
204. Li, Q.; Wu, L.; Hu, L.; Miao, X.; Liu, X.; Zou, T. A sinusoidal beam lattice structure with negative Poisson's ratio property. *Aerosp. Sci. Technol.* **2023**, *133*, 108103. [CrossRef]
205. Zhang, M.; Zhao, C.; Li, G.; Luo, K. Mechanical properties of the composite lattice structure with variable density and multi-configuration. *Compos. Struct.* **2023**, *304*, 116405. [CrossRef]
206. Gamba, L.; Johnson, Z.T.; Atterberg, J.; Diaz-Arauzo, S.; Downing, J.R.; Claussen, J.C.; Hersam, M.C.; Secor, E.B. Systematic Design of a Graphene Ink Formulation for Aerosol Jet Printing. *ACS Appl. Mater. Interfaces* **2023**, *15*, 3325–3335 [CrossRef]
207. Konstantinou, X.; Lan, X.; Nguyen, K.; Escorcia, A.; Sandhu, R.; Tice, J.; Spain, W.; Albrecht, J.D.; Papapolymerou, J. Fully Aerosol Jet-Printed Passive Components for 140–220-GHz Operation. *IEEE Microw. Wirel. Technol. Lett.* **2023**, 1–4. [CrossRef]
208. Ratnayake, D.; Curry, A.; Qu, C.; Wei, D.; Gerber, E.; Walsh, K. Optimizing the Conductivity of a New Nano-particle Silver Ink for Aerosol Jet Printing and Demonstrating its use as a Strain Gauge. *IEEE J. Flex. Electron.* **2023**, 1. [CrossRef]
209. Chen, Z.; Li, Z.; Wang, H. Two-dimensional auxetic GeSe2 material with ferroelasticity and flexoelectricity. *J. Phys. Chem. C* **2021**, *125*, 19666–19672. [CrossRef]
210. Coluci, V.R.; Hall, L.J.; Kozlov, M.E.; Zhang, M.; Dantas, S.O.; Galvao, D.S.; Baughman, R.H. Modeling the auxetic transition for carbon nanotube sheets. *Phys. Rev. B* **2008**, *78*, 115408. [CrossRef]
211. Yuan, J.H.; Xue, K.H.; Wang, J.F.; Miao, X.S. Gallium thiophosphate: An emerging bidirectional auxetic two-dimensional crystal with wide direct band gap. *J. Phys. Chem. Lett.* **2019**, *10*, 4455–4462. [CrossRef]
212. Ren, X.; Das, R.; Tran, P.; Ngo, T.D.; Xie, Y.M. Auxetic metamaterials and structures: A review. *Smart Mater. Struct.* **2018**, *27*, 023001. [CrossRef]
213. Ahmed, H.I.; Umair, M.; Nawab, Y.; Hamdani, S.T.A. Development of 3D auxetic structures using para-aramid and ultra-high molecular weight polyethylene yarns. *J. Text. Inst.* **2021**, *112*, 1417–1427. [CrossRef]
214. Steffens, F.; Rana, S.; Fangueiro, R. Development of novel auxetic textile structures using high performance fibres. *Mater. Des.* **2016**, *106*, 81–89. [CrossRef]
215. Khoshgoftar, M.; Abbaszadeh, H. Experimental and finite element analysis of the effect of geometrical parameters on the mechanical behavior of auxetic cellular structure under static load. *J. Strain Anal. Eng. Des.* **2021**, *56*, 131–138. [CrossRef]

216. Janus-Michalska, M.; Jasińska, D.; Smardzewski, J. Comparison of contact stress distribution for foam seat and seat of auxetic spring skeleton. *Int. J. Appl. Mech. Eng.* **2013**, *18*, 55–72. [CrossRef]
217. Dehaghani, M.R.; Nourani, A.; Arjmand, N. Effects of auxetic shoe on lumbar spine kinematics and kinetics during gait and drop vertical jump by a combined in vivo and modeling investigation. *Sci. Rep.* **2022**, *12*, 18326. [CrossRef]
218. *NIJ 0101.06*; Ballistic Resistance of Body Armor. National Institute of Justice: Washington, DC, USA, 2008.
219. Nayak, R.; Crouch, I.; Kanesalingam, S.; Ding, J.; Tan, P.; Lee, B.; Miao, M.; Ganga, D.; Wang, L. Body armor for stab and spike protection, Part 1: Scientific literature review. *Text. Res. J.* **2018**, *88*, 812–832. [CrossRef]
220. Mawkhlieng, U.; Majumdar, A.; Laha, A. A review of fibrous materials for soft body armour applications. *RSC Adv.* **2020**, *10*, 1066–1086. [CrossRef]
221. Liu, J.; Yang, W.; Liu, J.; Liu, J.; Huang, W. Ballistic impact analyses of foam-filled double-arrow auxetic structure. *Thin-Walled Struct.* **2023**, *182*, 110173. [CrossRef]
222. Jiang, F.; Yang, S.; Qi, C.; Liu, H.T.; Remennikov, A.; Pei, L.Z. Blast response and multi-objective optimization of graded re-entrant circular auxetic cored sandwich panels. *Compos. Struct.* **2023**, *305*, 116494. [CrossRef]
223. Lan, X.; Huang, G.; Bian, X.; Liu, Y. Impact resistance of foam-filled hybrid-chiral honeycomb beam under localized impulse loading. *Int. J. Impact Eng.* **2023**, *173*, 104477. [CrossRef]
224. Ul Haq, A.; Gunashekar, G.; Narala, S.K.R. The Dynamic Response of AuxHex and Star-Reentrant Honeycomb Cored Sandwich Panels Subject to Blast Loading. *Arab. J. Sci. Eng.* **2023**, 1–17. [CrossRef]
225. Yan, Z.; Liu, Y.; Yan, J.; Wang, B.; Bai, F.; Shi, Z.; Huang, F. Anti-blast performance of 3D-printed sandwich panels with auxetic hexagonal and regular hexagonal honeycomb cores. *Eng. Struct.* **2022**, *272*, 114996. [CrossRef]
226. Critchley, R.; Hazael, R.; Bhatti, K.; Wood, D.; Peare, A.; Johnson, S.; Temple, T. Blast mitigation using polymeric 3D printed auxetic re-entrant honeycomb structures: A preliminary study. *Int. J. Prot. Struct.* **2021**, *13*, 469–486. [CrossRef]
227. Usta, F.; Türkmen, H.S.; Scarpa, F. High-velocity impact resistance of doubly curved sandwich panels with re-entrant honeycomb and foam core. *Int. J. Impact Eng.* **2022**, *165*, 104230. [CrossRef]
228. Zhao, C.; Wang, Y.; Cao, S.; Xuan, S.; Jiang, W.; Gong, X. Conductive shear thickening gel/Kevlar wearable fabrics: A flexible body armor with mechano-electric coupling ballistic performance. *Compos. Sci. Technol.* **2019**, *182*, 107782. [CrossRef]
229. Zhang, Q.; Qin, Z.; Yan, R.; Wei, S.; Zhang, W.; Lu, S.; Jia, L. Processing technology and ballistic-resistant mechanism of shear thickening fluid/high-performance fiber-reinforced composites: A review. *Compos. Struct.* **2021**, *266*, 113806. [CrossRef]
230. Zhao, C.; Gong, X.; Wang, S.; Jiang, W.; Xuan, S. Shear stiffening gels for intelligent anti-impact applications. *Cell Rep. Phys. Sci.* **2020**, *1*, 100266. [CrossRef]
231. Ribeiro, M.P.; da Silveira, P.H.P.M.; de Oliveira Braga, F.; Monteiro, S.N. Fabric Impregnation with Shear Thickening Fluid for Ballistic Armor Polymer Composites: An Updated Overview. *Polymers* **2022**, *14*, 4357. [CrossRef]
232. Zhang, J.; Wang, Y.; Zhou, J.; Wu, J.; Liu, S.; Sang, M.; Liu, B.; Pan, Y.; Gong, X. Multi-functional STF-based yarn for human protection and wearable systems. *Chem. Eng. J.* **2023**, *453*, 139869. [CrossRef]
233. Sheikhi, M.R.; Gürgen, S. Deceleration behavior of multi-layer cork composites intercalated with a non-Newtonian material. *Arch. Civ. Mech. Eng.* **2023**, *23*, 1–11. [CrossRef]

eng

MDPI

Article

The Present and Future of a Digital Montenegro: Analysis of C-ITS, Agriculture, and Healthcare

Peter A. Kara [1,2,*], Ivana Ognjanovic [3], Ingo Maindorfer [4], John Mantas [5], Andras Wippelhauser [1], Ramo Šendelj [6], Luka Laković [3], Milovan Roganović [7], Christoph Reich [4], Aniko Simon [8] and Laszlo Bokor [1]

1 Department of Networked Systems and Services, Faculty of Electrical Engineering and Informatics, Budapest University of Technology and Economics, Műegyetem rkp. 3, 1111 Budapest, Hungary
2 Wireless Multimedia and Networking Research Group, Department of Computer Science, School of Computer Science and Mathematics, Faculty of Science, Engineering and Computing, Kingston University, Penrhyn Road Campus, Kingston upon Thames, London KT1 2EE, UK
3 Faculty for Information Systems and Technologies, University of Donja Gorica, Oktoih 1, 81000 Podgorica, Montenegro
4 Institute of Data Science, Cloud Computing and IT Security, Furtwangen University, 78120 Furtwangen, Germany
5 Faculty of Nursing, School of Health Sciences, National and Kapodistrian University of Athens, Papadiamantopoulou 123, 11527 Goudi, Greece
6 Humanistic Studies, University of Donja Gorica, Oktoih 1, 81000 Podgorica, Montenegro
7 Clinic for Neurology, Clinical Centre of Montenegro, Ljubljanska bb, 81000 Podgorica, Montenegro
8 Sigma Technology, 1093 Budapest, Hungary
* Correspondence: kara@hit.bme.hu

Abstract: The digitization and general industrial development of Montenegro is a great challenge for engineering and science due to its special characteristics. As the accession of Montenegro to the European Union has been an ongoing agenda for over a decade now, and the accession of the country is expected by 2025, adapting the interconnectivity and smart automation of Industry 4.0 plays an essential role in reducing the current gap between Montenegro and EU member states. In this paper, we investigate the present and potential future digitization efforts in the fields of Cooperative Intelligent Transport Systems (C-ITS), agriculture, and healthcare in Montenegro. Our work takes into consideration the characteristics of the country and analyzes the considerations and implications regarding the deployment of state-of-the-art technologies in the investigated fields.

Keywords: cooperative intelligent transportation systems (C-ITS); digital agriculture; smart farming; digital health; eHealth/mHealth

Citation: Kara, P.A.; Ognjanovic, I.; Maindorfer, I.; Mantas, J.; Wippelhauser, A.; Šendelj, R.; Laković, L.; Roganović, M.; Reich, C.; Simon, A.; et al. The Present and Future of a Digital Montenegro: Analysis of C-ITS, Agriculture, and Healthcare. *Eng* **2023**, *4*, 341–366. https://doi.org/10.3390/eng4010021

Academic Editor: Antonio Gil Bravo

Received: 31 December 2022
Revised: 21 January 2023
Accepted: 25 January 2023
Published: 29 January 2023

1. Introduction

Humanity has come a long way since the dawn of the Industrial Revolution in the 1760s. Today, we embrace the Fourth Industrial Revolution—commonly known as Industry 4.0—which fundamentally builds on the extensive interconnectivity of smart systems. Together, they create complete smart ecosystems. However, these ecosystems are not limited to the modern urban areas of smart cities but also extend to rural areas in the form of smart agriculture.

The concept of Industry 4.0 encompasses numerous technologies and fields of practical utilization, ultimately affecting a vast portion of our everyday lives. Its basic ideology [1–3] is built around interconnectivity and smart automation, and its terminology was made widespread and popular by Klaus Martin Schwab [4], the founder of the World Economic Forum (WEF). Some argue that Industry 4.0 is the new German ideology [5], while others already point toward Industry 5.0 [6–9], or deny its existence in the first place [10] ("there is no such phenomenon as a Fourth

Industrial Revolution"). It can be approached from the associated changes in industrial capitalism [11]—or rather, "post-capitalism" [12–14], considering work and education in a post-work era, or transhumanism [15–17], focusing on the significant potential increases in life expectancy and cognitive capabilities; however, in the scope of this article, we focus on technology.

Among others, these technological fields include transportation, agriculture, and healthcare. Advanced transportation solutions not only improve the safety and efficiency of personal vehicles [18,19] but also significantly affect public transportation [20], emergency services (e.g., ambulance [21]), the transportation of goods [22], and many more. Additionally, such transformation of the transportation ecosystem may have an immensely positive impact on the environment (e.g., reduction of harmful emissions and more efficient usage of available resources) [23]. Regarding agriculture, the demand for consumable substances is expected to rise parallel with the rapid growth of Earth's population [24], which, at the time of writing this paper, is already above 8 billion; yet, at the same time, various factors (e.g., climate change) limit the adequate areas for producing food [25]. Therefore, efficiency in agriculture is more vital than ever [26]. Finally, the importance of healthcare must never be underestimated, as it is one of the keys to the quality and length of human life [27,28].

The digital transformation of these fields happens all over the face of the Earth. Of course, the pace at which such forms of transformation occur may vary greatly between the different regions of the world [29–31]. It is not only influenced by technological readiness and economic support but by the challenges imposed by regional characteristics as well.

In this paper, we analyze the present presence and the potential future of digitization efforts in the fields of transportation, agriculture, and healthcare in the country of Montenegro. Montenegro is a particularly interesting target of analysis in the context of Industry 4.0, both due to its rather unique natural characteristics (e.g., intensely rugged terrains) and its rapidly developing modern history. After all, the Montenegro we have on the world's map today was formed via the independence referendum of 2006, and since 2008, the accession of Montenegro to the European Union has been in progress. This latter fact comes with strict requirements and the need for compliance with regulations, but also means development aid in various forms, such as funding mechanisms (e.g., Instrument for Pre-Accession Assistance). Additionally, assistive programs and projects have been created to support the digitization of the country. One such program is the Digital Entrepreneurial Nest and Industry 4.0 in Montenegro, also known as DigNest [32]. A great number of the authors of this paper are contributing members of the DigNest project.

The rationale of this paper is to contribute to the driving forces of Montenegrin digitization. The comprehensive analyses and thorough assessments of this work aim to provide expert insights and know-how support regarding the investigated fields, and to assist paving the next steps toward short-term and long-term goals. In the scientific literature, there are already published works about the different aspects of the current state and the advancement of Montenegro. Most of these case studies focus on specialized topics of development. Digital transformation as a whole is addressed by Golubović et al. [33]. Beyond the challenges and the opportunities, the authors also draw conclusions regarding the new mindset in the Montenegrin organizational culture, education, customer satisfaction, and other social impacts. Transformation is viewed through the necessities mandated by the pandemic of the recent years. The work of Melović et al. [34] narrows down the focus of digital transformation to digital marketing and electronic business in Montenegro. The large-scale research encompassed a survey, in which data were collected from 172 companies. The authors conclude that "the more a company relies on the use of digital marketing in its business, the more significant its impact on promotion and brand positioning". In the areas of transportation, the Montenegrin railway is studied by Bošković et al. [35], Jakovljevic et al. [36], and Bramo and Llaci [37] on the topic of sustainability, bridge rehabilitation, and its connection to

Albanian agribusiness. Maritime transportation is a highly relevant field of research in Montenegro due to the country's geography. The works of Nikčević and Škurić [38], Kapidani and Kočan [39], and Hasaj et al. [40] also address sustainability, and similarly focus on cross-border cooperation. Agriculture is investigated by the works of Rajović and Bulatović [41,42], Ćorić and Popović [43], and Radonjić and Hrnčić [44]. Apart from sustainability and e-agriculture, the impact of non-native pests, which can be a major issue in Montenegro for various plants and crops (e.g., olives [45] and raspberries [46]), is also studied. Regarding healthcare-related works, Ivanović and Raković [47] performed a Montenegrin case study for the country's e-health card information system, while Safet [48] dealt with questions of management, and Mantas et al. [49] focused on modern healthcare education. The technological trends of Montenegrin manufacturing are reviewed by Vujanović [50], and Boban et al. [51] studied the competitiveness of homemade consumables. Labović et al. [52] studied food safety in Montenegrin restaurants and other food service establishments, Rajović and Bulatović [53] investigated plant- and animal-based food production, and Martinovic and Mirecki [54] analyzed their connection to health in the country. Tourism is addressed by the works of Jaksic-Stojanovic et al. [55], Moric et al. [56,57], and Bigović [58], highlighting sports, cultural and rural tourism, and its seasonality.

The methodology of this work was based on various pillars. First of all, the authors consulted with numerous Montenegrin authorities, organizations, and institutions, including the Chamber of Economy of Montenegro (https://www.privrednakomora.me/en (accessed on 21 January 2023)); the Ministry of Agriculture, Forestry and Water Management (https://www.gov.me/mpsv (accessed on 21 January 2023)); Montenegro Ministry of Health (https://www.gov.me/mzd (accessed on 21 January 2023)); Clinical Centre of Montenegro (https://kccg.me/ (accessed on 21 January 2023)); Mljekara Lazine, Montenegro (https://mljekaralazine.me/ (accessed on 21 January 2023)); the University of Montenegro (https://www.ucg.ac.me/ (accessed on 21 January 2023)); and the University of Donja Gorica, Montenegro (https://www.udg.edu.me/ (accessed on 21 January 2023)). The authors of this work also investigated the Montenegrin state of the respective fields in the scope of the DigNest project. Additionally, the recent regional advances were studied from published scientific works, as well as from technological, industrial, and economic reports.

The remainder of this paper is structured as follows: Transportation (i.e., cooperative intelligent transportation systems), agriculture (i.e., digital agriculture and smart farming), and healthcare (digital healthcare and eHealth) are discussed in Sections 2–4, respectively. Each section begins with an introduction to the field and the related concepts. This is followed by a brief overview of the current state in Montenegro and the rest of the world. Then, the involved technologies and the potential use cases are elaborated. After the relevant features and attributes of Montenegro are revised, the considerations and implications of these properties are analyzed from a technological perspective, followed by a concise summary. The analysis of transportation is detailed from all these sections in order to reflect its importance regarding the long-lasting financial and geopolitical implications of the ongoing investments, and the associated recent events. The paper is concluded in Section 5, highlighting the most important discussions of the investigated fields and the future steps toward a digital Montenegro.

2. Cooperative Intelligent Transportation Systems

2.1. Introduction and Concepts

Present transportation systems face various challenges. In the past decades, the number of traffic participants has grown significantly. The spread of just-in-time supply chains requires highly available and reliable transportation systems. Such systems' safe and economical operation is an essential building block of modern economies.

Intelligent Transportation Systems (ITS) try to solve some of the above-mentioned problems. ITS technology covers various use cases and methods in order to address current and upcoming issues. These use cases include data collection from the road network, data provisioning to the traffic participants, law enforcement with smart devices, and communication among the various traffic participants. From a road operator's perspective, ITS can use induction loop detectors, intelligent cameras, radars, or other devices to count the traffic and measure the speed or other attributes of the passing vehicles. The road operator can use this information to create long-term statistics, optimize traffic flows, oversee and control the traffic network in real-time, analyze traffic disturbances, and increase drivers' awareness via smart traffic signs or other communication technologies. ITS also covers efficient ways of toll collection. Vehicle drivers also benefit from ITS via the more-efficient road networks or the valid and up-to-date information on the roads. ITS can also reduce the load of law enforcement. For example, an intelligent camera deployed at an intersection can detect and charge violations of traffic rules. The speed cameras and radars can see violations of speed limits, whereas the weighing types of equipment built into the road surface can help to enforce the special limitations applicable to the particular road section. As we can see, ITS uses and facilitates digitalization in order to provide more efficient, safe, and sustainable transportation.

Cooperative Intelligent Transportation Systems (C-ITS) provide the next generation of ITS. In C-ITS, the vehicles and the infrastructure are equipped with communication devices. These devices offer two-way communication methods to further enhance existing, or introduce new, ITS use cases. We call this kind of communication Vehicle-to-Everything (V2X). Based on the direction of the communication, we can categorize the use cases. In Vehicle-to-Vehicle (V2V) use cases, the vehicles communicate with each other directly. Such use cases are traffic jams ahead or V2V safety services such as collision and "do not pass" warnings. In Infrastructure-to-Vehicle (I2V) use cases, the infrastructure can provide information to the vehicles; for example, it can warn the drivers about the applicable speed limit or ongoing road works. These use cases help the road operators to influence and actively increase traffic participants' awareness. In Vehicle-to-Infrastructure (V2I) use cases, vehicles can provide statistical data to the roadside infrastructure or ask for traffic sign prioritization. Vehicle-to-Network (V2N) use cases implement services such as teleoperated driving and automated valet parking.

The evolution of C-ITS splits into three phases that rely on different standards. Day 1 use cases mainly focus on awareness services where the infrastructure notifies the vehicles about speed limits, road works, traffic light status, etc. Further, vehicles share their presence and status/attribute information. Such Day 1 systems are currently deployed in various locations, e.g., under the coordination of the European C-Roads platform [59]. From the road-user point of view, the Golf VIII was the first mass-produced vehicle in Europe with V2X as a standard feature. Since then, various other models have been equipped with this technology. The currently available V2X solutions mainly support Day 1 use cases.

Day 2 focuses on perception sharing and advanced use cases. On Day 1, the C-ITS stations share their status/attribute information only. Contrary to this, perception sharing enables the exchange of perceived information. This overcomes a massive challenge in the C-ITS world—the penetration issue—meaning that the stakeholders are unwilling to invest in the technology due to the low number of initially deployed devices. This is mainly a chicken-and-egg problem, but several strategies exist to overcome this issue. One of them is providing governmental funding for deployment projects (such as the C-Roads project). Another C-ITS technology enabler can reward Original Equipment Manufacturers (OEMs) in various benchmarks such as Euro NCAP, which will include V2X technology in its evaluation methodology. The third possible solution for this problem is an appropriate strategy to have non-V2X-equipped users in the system. Perception-sharing technologies are based on advanced sensor infrastructures and are expected to significantly mitigate the concerns about penetration by putting unequipped vehicles on the C-ITS horizon. The standardization of Day 2 enabler V2X protocols and related message formats is currently under development.

Day 3 systems further enhance the quality and volume of the shared information in the C-ITS domain. The main goal of Day 3 is to enable the so-called cooperative driving use cases. They are expected to further enhance the safety and efficiency of traffic. Cooperative driving enables sharing the future state or the intention of the equipped vehicles. These use cases also assume some level of automation for the vehicles. Currently, the use cases are under development for Day 3 systems.

2.2. Current State in Montenegro

In the European Union, the Platform for the Deployment of Cooperative Intelligent Transportation Systems (C-ITS Platform) was created by the European Commission services (DG MOVE) in 2014 with the intention to facilitate the emergence of a shared vision across all actors involved in the C-ITS value chain. Based on the results, the EC adopted the European Strategy on Cooperative Intelligent Transport Systems in 2016 with the primary goal of helping the convergence of investments and regulatory frameworks across the EU, including the adoption of the appropriate legal structures, initiating various programs for EU-funded projects, continuing the C-ITS Platform and increasing the volume of international cooperation with other regions of the world on the aspects of C-ITS. It also involves the C-Roads platform [59], the flagship initiative of European Member States, road operators, and associated partners for testing and implementing C-ITS services with cross-border harmonization and interoperability in the main scope. Within this learning-by-doing approach, the EC gathers real-life deployment experiences through Member State pilot activities and continuous coordination with C-Roads.

Since its official launch in 2016, C-Roads proved that C-ITS has arrived into the everyday life of road users and that there are vast benefits to exploit in the near future. With the help of Europe-wide harmonization activities and relying on strategic partners, such as the CAR 2 CAR Communication Consortium (C2C-CC), ASECAP, CEDR, EU EIP, and the European Commission, C-ITS deployments reached a total of 20,000 km of road sections equipped with mature ad hoc short-range communication units (ETSI ITS-G5). Moreover, 100,000 km overall are covered with wide-area cellular communication technologies—such as 4G and 5G—and 35 cities are already actively working on their C-ITS service implementation, with more than 50 to be expected by the year 2023 [59].

The closest C-Roads Core Member states to Montenegro are Slovenia, Hungary, and Greece, while Croatia is an Associated Member of the platform [59]. In comparison, at the time of writing this paper, ITS technology in Montenegro is very much in its infancy. ITS is a part of the long-term goals of the country, according to the Transport Development Strategy until 2035 [60]. However, it is important to highlight that C-ITS is not a part of the current strategy (i.e., it is not being considered for the following 13 years). Regarding ITS, the primary considerations for enhanced traffic safety are intersections and railway crossings. In accordance with Directive 2010/40/EU [61], the installation of ITS devices for traffic signalization, highway control, and management is planned for the first-ever Montenegrin motorway and the Sozina tunnel—the longest tunnel in the country. Yet, the analysis provided by the strategy also states that "Montenegro lacks experience and appropriate structures for coordination and management of the introduction of ITS, especially in road and rail transport sectors" and that the "structure of state administration's transport sector lacks dedicated bodies for intermodality and co-modality, intelligent transport systems and interurban public transportation". Additionally, the recent case study of Prelevic [62] highlights that the lack of information-technology-related education and greater confidence in more conventional approaches contribute to the current state.

2.3. Technologies and Use Cases

C-ITS use cases and V2X communication pose several technical challenges [63,64]. The most critical ones are listed below.

- The communication parties are highly mobile. The relative speeds can easily exceed 250 km/h.
- The network topology changes frequently and rapidly.
- Message distribution has to be handled in a spatially aware way while the nodes are moving.
- The messages typically have broadcast destination addresses.
- The node density can be high. Channel contention needs to be avoided by appropriate countermeasures.
- The communication scheme has to support nomadic devices.
- ITS stations shall trust each other.
- Privacy protection methods need to be applied to avoid sensitive data collection.
- Distributed operation is favored over centralized due to the mission-critical use cases.
- Accurate absolute positioning is required.

Various standardization bodies participated in the harmonization and the conceptual and technical work in order to meet the multiple requirements and successfully address the challenges.

ISO created the CALM standard [65], which defines the architectural aspects of ITS stations. The standard introduced four horizontal and two vertical layers, in contrast with the classical 7-layer ISO OSI standard [66]. The four horizontal layers provide a simplified yet optimized architecture for C-ITS use cases, whereas the two vertical layers provide cross-layer functionalities.

The first layer among the horizontal layers is the Access layer. This layer covers physical access and media access control functionalities. The IEEE defined the 802.11p standard [67] in this layer in 2010, which was the enabler of V2X use cases. This standard was created as a modification/customization of the popular Wi-Fi standard; thus, it uses the same access control approach (CSMA/CA). The current mass production deployments (e.g., VW Golf VIII) are typically based on this standard. The successor of the 802.11p standard (802.11bd) is currently under development, aiming to feature backward compatibility, increased range and bandwidth, and various other advanced features [68]. An alternative to the Wi-Fi-based Access layer evolution path is provided by the standardization body of 3GPP, which developed cellular V2X (C-V2X) communication technologies [69], incorporating a slightly different channel access approach. Instead of CSMA/CA, 3GPP applied a so-called semi-persistent scheduling [70] to ensure the proper and efficient usage of the channel. This standard was introduced in Release 14 and was initially based on the LTE proximity services. 3GPP also works on the successor of the standard for the fifth generation of mobile cellular systems called 5G NR V2X [71]. ETSI also has standards that adopt these access layer solutions in the ETSI ITS standard family.

The second layer in the architecture is the Network and Transport layer. Here, IEEE and ETSI were the key contributors. IEEE developed the WAVE standard [72], which offers a lightweight transport protocol with single-hop broadcast support. On the other hand, ETSI standardized GeoNet [73], which enables single-hop broadcast, geographically scoped broadcast, anycast, and unicast transmission schemes. GeoNet defines algorithms to forward and distribute messages in the rapidly changing ad hoc network environment, by using geographic addressing and routing. On top of the GeoNet, ETSI defined the Basic Transport Protocol (BTP) [74] for simple multiplexing/demultiplexing of higher-layer protocols.

The third layer is the Facilities layer. This layer's standardization was handled mainly by ETSI, ISO, SAE, and CSAE. Besides others, ETSI defined CAM [75] for ITS station status/attribute information and DENM [76] for event notification exchange. ETSI adopted the IVI standard [77] from ISO, which tries to provide a digital representation of traffic signs precisely. ETSI also adopted SAE standards for signalized intersection digitalization [78]. This mutual influence between SAE and ETSI has a long history, resulting in similar message types and protocols. SAE facility messages are typically deployed in the USA, whereas ETSI focuses on Europe. CSAE typically covers Chinese standards. All the standardization bodies mentioned above consider the ITS domain very seriously and continuously work on future extensions of their outputs.

The Application layer defines and controls how the facility services should behave; thus, they are typically covered by the same standardization body as the related facility layer service. For example, ETSI has service specifications for Cooperative Awareness and other facility services. The SAE developed the J2945 standard family, which contains system requirements for the V2X services. This document defines the requirements for BSM transmission [79]. These standards are designed in a way to respect the addressed use case requirements and the communication channel peculiarities.

Besides the horizontal layers, there are two vertical layers in the ITS architecture—more precisely, the Security and the Management layers. These layers are cross-layer entities because they have interfaces to multiple horizontal layers. The Security layer needs to implement various requirements, which are different from the regular client–server-based systems. C-ITS is a distributed system; thus, the trust between two nomadic ITS stations must be ensured. The implementation [80] is based on conventional PKI systems. Each ITS station signs the transmitted messages with their private keys using asymmetric cryptography. The public key of the key-pair is signed by an upper-level certification authority, which ensures trustworthiness. It is important to note that the messages are not encrypted; they are only signed. They are typically broadcast messages and are relevant to all receivers in the dissemination area; thus, the encryption of the messages is typically not justified. However, the traceability of the clients raises privacy issues. In order to avoid such issues, pseudonymity techniques are implemented. This means that regular vehicle ITS stations need to change their unique identifiers in all protocol layers after a certain time and traveled distance. The pseudonyms include the MAC address, the GeoNet address, the StationID, and the path history. The certificate used to sign the messages needs to be exchanged as well because the public key is also a unique identifier; thus, the vehicular ITS stations need certificate provisioning. Technically, the certificates are implemented in the Networking layer and typically use the IEEE 1609.2 security certificates—in both EU and US regions [72].

Naturally, the distributed system also expects some kind of common requirement set. The applied techniques are described below. If an ITS station fulfills the set of requirements, it can go through the so-called enrollment process. In this process, a long-term identification token is registered with the related authority (Enrollment Authority) [81]. Then, the ITS station is able to request short-term certificates via the proper authority, which ensures the valid registration of the requester ITS station in the Enrollment Authority during the certificate authorization. The PKI usually applies the butterfly key expansion technique to ensure scalability. The Enrollment and Authorization Authorities are validated by a Root Certificate Authority. The public keys of the trustworthy Root Certificate Authorities—and, via those keys, the individual ITS stations—are distributed via the European Certificate Trust List (ECTL) [82]. A Root Certificate Authority can be added to the ECTL after various audits. The revocation is handled via the European Certificate Revocation List (ECRL). The names and the number of levels of the certificate authorities in the PKI architecture can be different, but the base roles are the same in all regions.

For successful C-ITS deployments, the understanding of the security architecture is crucial. From a business perspective, Montenegro belongs to the European vehicle market. This means that besides the bitwise compatibility, which can be ensured by the adoption of the proper ITS standards, the harmonization of the PKI systems is also essential to effectively support C-ITS use cases.

The Management layer is responsible for some cross-layer operations. It enables distributed congestion control and multi-channel operation use cases by facilitating communication across the architecture layers.

In order to further specify and concretize the operation of the protocols, agree on triggering conditions, and provide commonly agreed requirements, profiling activities were performed. The C2C-CC [83] and the C-Roads project [59] created triggering conditions and standard profiles for V2V, I2V, and V2I use cases. These profiles specify performance and functional requirements for the ITS stations. They also define use cases and triggering conditions for the use cases. Some use cases extended with V2V use cases are listed in Table 1. These profiles typically define transmission-side behavior because the reception side has to be defined by the OEMs. The Common Criteria document [84] also specifies security requirements.

Table 1. Sample V2X use cases.

Use Case	Type	Most Relevant Deployment Location	Expected Impact (1–5)
Do not pass warning	V2V	Rural main roads	5
Slow moving vehicle	V2V	Rural main roads	3
Roadworks warning	I2V	Rural main roads, Urban	3
Hazardous location notification	I2V	Rural main roads, Urban	4
In-vehicle signage	I2V	Rural main roads	3
Signalized intersections	I2V	Urban	3
Signalized intersections with preemption	I2V, V2I	Urban	4
Adverse weather conditions	I2V	Rural main roads	4
Probe vehicle data	V2I	Urban	1

2.4. Relevant Features and Attributes of Montenegro

Montenegro is a small (13,812 km^2 total area), yet greatly diverse country, particularly in terms of geography. While it has a 293.5 km-long coastline at the Adriatic Sea—the altitude of which is by definition close to sea level—it is generally a mountainous country. In fact, the native name of Montenegro is "Crna Gora", which translates as "black mountain", referring to Mount Lovćen, a mountain in the southwestern region of the country [85]. However, it is not a single mountain that defines the terrain configuration of Montenegro. There are 50 peaks within the modest area of Montenegro that reach over 2000 m. The highest peak of the country is Zla Kolata at 2534 m, located on the border of Montenegro and Albania; although, historically, it was thought to be Bobotov Kuk at 2522 m, located in the Durmitor mountain range. Furthermore, even with such a long coastline and 17 islands, the average elevation is 1086 m.

The mountainous nature of Montenegro affects its transportation, as the roads connecting the major cities tend to span across mountains. One may think that the roads near the coastline form an exception to this statement; however, even those roads may be situated on mountains, as it is common that the elevation steeply rises from the coast towards the inland. A typical example of this is Budva, where the main road circumventing the town is actually up in the mountains and not on the coastline. Additionally, there are numerous tunnels in Montenegro. The longest tunnel in the country is Sozina, as mentioned earlier, which has a length of 4189 m.

The overall length of paved roads in Montenegro is 1729 km. This includes the recently (July 2022) inaugurated 41 km-long section of the Bar-Boljare motorway, which contributes to the motorway connection between Podgorica and Belgrade. At the time of writing this paper, this is the only motorway section of the country. Among the paved roads, there are 12 main roads (i.e., highways) and 31 regional roads. Furthermore, Montenegro has 3548 km of unpaved roads [86].

The paved main roads are typically composed of one lane for each direction, built as a single carriageway (i.e., no physical separation between the lanes). This structure is sometimes extended with a third lane for overtaking. In the mountains, overtaking is often prohibited, as the terrain may occlude the line of sight of the driver. The main roads are often steep, winding, and narrow. There are no guardrails or barriers to prevent vehicles from falling off the road at several places.

A well-known dangerous area for road transport in Montenegro can be found north of Podgorica. The road through the Morača Canyon is a twisting, two-lane highway, usually overcrowded during summer, and is subject to frequent rockslides. During winter, this road segment and northern parts of Montenegro, in general, are covered with snow, seriously slowing down traffic and creating hazardous road situations. The roads leading to the country's coastal regions are in better condition but are overcrowded in summer. Extreme caution is required because local drivers can be reckless and often attempt to pass on winding roads and hills.

The total number of towns in Montenegro is 20, having populations between 136,473 (the capital, Podgorica) and 1073 (the smallest city, Andrijevica) [87]. Podgorica traffic lights are coordinated, and in some areas, the city has implemented smart traffic lights that can adapt to traffic conditions in real-time. With only a few exceptions, the cities are relatively tiny, often comprising only one or two traffic lights and roundabouts. The most commonly used vehicles in the country are cars and motorcycles; together, they make up nearly 88% of the vehicles on the roads. The remaining fractures consist of other vehicles used but less frequently than the two main types, such as buses, vans/minivans, and trucks.

There is a set of international standards increasingly accepted as basic minimum specifications for vehicle manufacture/assembly. The WHO works for the global implementation of those, and several countries' regulations now consider them as basic requirements for all vehicles. Unfortunately, Montenegrin regulations implement zero out of the eight key vehicle standards [88]. Surrounding countries—Bosnia and Herzegovina, Serbia, North Macedonia, and Albania—have the same situation, but countries in the broader region—Greece, Bulgaria, Romania, Hungary, Croatia, Slovenia, and Italy—implement eight out of eight.

The existing railway network in Montenegro consists of single-track rails of standard width in three main lines: (1) Vrbnica–Bar (part of the Belgrade–Bar railway through Montenegro); (2) Podgorica–Tuzi state border; and (3) Podgorica–Nikšić. The total length of the three lines is 248.6 km—together with station tracks, it is 327.6 km—of which 167.4 km are electrified. The number of level crossings (i.e., intersections where a railway line crosses a road at the same level) is 23, where 19 are equipped with signal-safety solutions (automatic barriers, light/sound signals) and 4 with road horizontal/vertical signalization (without barriers) [89]. The railway network density is 1.8 km/100 km^2, a condition that is unsatisfactory regarding its density and quality, and with a permanent danger of system-level vulnerability due to the concentration of the road and rail traffic in the same corridor passing through challenging terrains. In Montenegro, 204 accidents/incidents were reported in the five years between 2014–2018, with 21 accidents in level crossings (around 10%) [89].

Road traffic in Montenegro is considered relatively safe. According to current WHO data, the total number of deaths on the roads per year reported by Montenegro is 65, meaning 10.7 deaths in a year per 100,000 inhabitants [88]. Detailed statistics are depicted in Figure 1 [90–92].

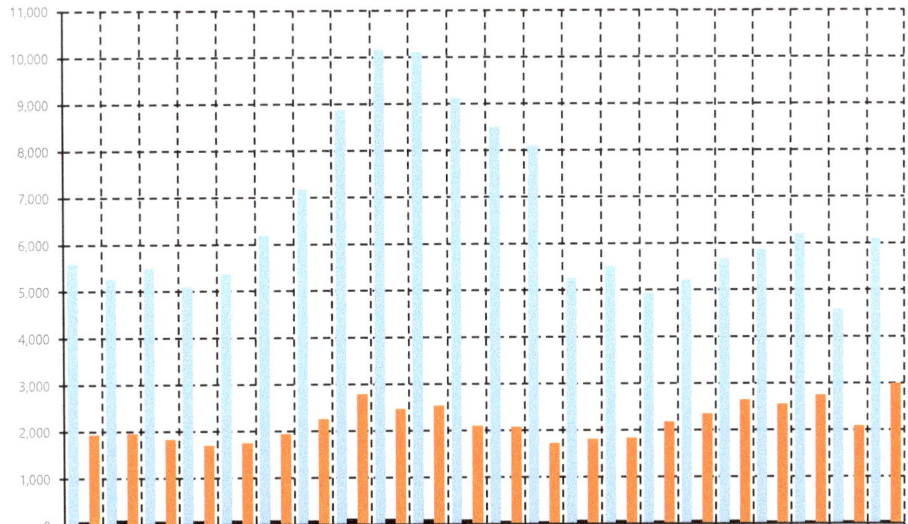

	2000	2001	2002	2003	2004	2005	2006	2007	2008	2009	2010	2011	2012	2013	2014	2015	2016	2017	2018	2019	2020	2021
Total number of road accidents	5597	5275	5503	5094	5377	6192	7185	8882	10170	10112	9138	8519	8103	5264	5531	4944	5229	5678	5872	6210	4595	6109
Number of killed	81	105	81	84	91	95	85	122	111	100	95	58	46	74	65	51	65	63	48	47	48	55
Number of injured	1933	1957	1834	1702	1750	1942	2257	2796	2473	2542	2099	2075	1722	1812	1835	2173	2358	2648	2563	2754	2087	3003

Figure 1. Annual statistics of road traffic accidents and casualties in Montenegro [90–92].

2.5. Considerations and Implications

The peculiarities of the Montenegrin topography—and thus, the road network—imply the need for special V2X use cases.

Due to the mountainous terrain in Montenegro, the road network contains several tunnels. A tunnel means an additional traffic-related risk (e.g., due to the rapidly changing environment, namely, the brightness, temperature, or wind speed), which can be mitigated by using V2X techniques. Brož et al. [93] presented the challenges of V2X use cases in tunnels, especially the coverage and positioning issues. V2X systems increase the awareness of the drivers so that they can prepare for changing conditions, such as standing vehicles, road works, strong wind, black ice, etc.

The mountainous terrain also results in tortuous roads and serpentines. On these roads, the visibility is restricted, and the roads are sometimes possibly narrow. Thus, road users' awareness is especially important. For example, road works, road closures, or slow-moving vehicles (e.g., snow plows) after a curve can surprise drivers, which can result in dangerous situations. The execution of an overtaking maneuver is also cumbersome due to restricted visibility. In order to avoid accidents caused by unforeseeable traffic, "do not pass" use cases can be implemented. In these use cases, the overtaking vehicle gets notified about the upcoming traffic, so it can avoid dangerous maneuvers.

A further potential feature use case could be the Cooperative Intersection Collision Warning (CICW) [94–96]. This C-ITS application would improve safety at intersections, which are known to be high-risk areas for collisions, particularly in Podgorica, as the city is known to have heavy traffic and a high volume of vehicles passing through, especially during peak hours. The CICW system would use V2X communication to transmit real-time information about the location and speed of vehicles approaching an intersection, allowing other vehicles and traffic signals to anticipate and respond to potential conflicts. This could reduce the number of accidents caused by human error, such as running red lights or failing to yield.

Another C-ITS feature that could be implemented in Montenegro is Real-Time Traffic Information (RTTI) [97–99]. This system would use V2X-based data exchange to gather and disseminate information about the current traffic conditions on the roads. This would allow drivers to make more informed decisions about their routes and could help to reduce congestion and travel times. RTTI could also be used by traffic management authorities to better understand and respond to traffic patterns and incidents, improving the efficiency of the transportation system overall.

The hilly terrain also affects the deployment questions. In European C-ITS standards, message dissemination is technically handled by the GeoNet layer. Practically, the facility layer service defines a dissemination area and the GeoNet protocol tries to distribute the messages in the particular area. The GeoNet, however, does not support altitude information. This has to be considered in service deployments at serpentines or roads with high elevation differences. For example, in the case of in-level roads where one Roadside Unit (RSU) could cover a particular area, multiple RSUs could be needed. This is particularly true because most RSUs do not have omnidirectional antennas—instead, directional antennas are more widespread with a focused, narrowed radio wave beam. Therefore, the signal strength below the RSUs' might not be sufficient. Hence, the antenna placement and direction/gain parameters and settings are also critical to achieving the proper service quality. The terrain can also hinder line-of-sight communication, which reduces the effective range of the RSUs. This means that more RSUs with more sophisticated placement design considerations might be needed to cover a certain road section.

2.6. Summary

V2X and C-ITS have great perspectives to make road traffic safer and more efficient. The road network of Montenegro can also benefit from these developments. However, in order to successfully launch V2X-based deployments, the proper developments and harmonization activities have to be performed.

In Montenegro, the immaturity of the ITS, in general, can severely limit C-ITS V2I/I2V use case deployments. In urban scenarios, the deployment of traffic light controller-based V2X scenarios could be limited by immature traffic light systems. Most V2I/I2V use cases assume the proper amount and quality of data availability on the road operator side. However, the parallel design of legacy ITS with the advanced C-ITS schemes could facilitate advanced solutions. This is also true for the V2X use cases aiming to support safer level crossings; future road-rail infrastructure optimization in the Montenegrin network can be designed with the potential benefits of near-future C-ITS capabilities in mind.

According to our assumptions, the nomadic use cases could be the easiest to be deployed in the first phases. Such use cases are "do not pass" warnings, slow-moving vehicles (road operator), road works with mobile RSUs, and classic V2V applications.

It is important to note that harmonizing Montenegrin C-ITS deployment with European projects, especially with the C-Roads project, is essential. If the harmonization is not performed, then the vehicles produced for the Montenegrin market will likely not accept information from the road infrastructure. Therefore, as an initial step towards C-ITS applications, it is essential for Montenegro to join C-Roads and start the contribution to the platform first, at least as an associated partner.

In the short term, the deployment of C-ITS in Montenegro—considering the current state of transportation in the Western Balkans—may face a number of challenges. One of the main issues is the heterogeneous quality of the transportation infrastructure in the region, meaning the lack of well-developed road networks and the immaturity of supporting ITS system components in several areas of the country, which could seriously impede the deployment of C-ITS. Another challenge that may arise is the lack of standardization and interoperability among countries in the Western Balkans. This could lead to difficulties in integrating C-ITS systems across borders and could hinder the full potential of the technology to improve transportation in the region. On the other hand, the deployment of C-ITS in Montenegro could also bring some benefits in the short term. Besides the expected improvements in the safety of road traffic and enhancements in the efficiency of transportation in the country, it could also pave the way for future infrastructure improvements in the Western Balkans.

The introduction of C-ITS in Montenegro has the potential to not only improve transportation efficiency and safety, reduce congestion, and lower emissions within the country, but also to take a leading role in the harmonization and deployment of C-ITS in the Western Balkans region. With the proper investment and support, Montenegro's C-ITS deployment could serve as a model for other countries in the region, promoting the development of a more integrated and efficient transportation system across the Western Balkans. However, to achieve this potential leading role, Montenegro may need to overcome some challenges, such as the lack of infrastructure and funding, limited technical expertise, and the lack of standardization and interoperability across the region. Nevertheless, with the successful deployment of C-ITS, Montenegro could pave the way for a more connected and sustainable transportation system throughout the Western Balkans.

3. Digital Agriculture and Smart Farming

3.1. Introduction and Concepts

Especially in the north of Montenegro—the mountainous part of the country—agriculture is one of the most important sectors and represents a significant source of employment and income. The share of agriculture in Montenegro's GDP [100] should not be underestimated at around 7.6%.

For the past 25 years, European beekeepers have been reporting decreasing bee numbers and colony losses, and the situation is worsening. According to the EU Reference Laboratory (EURL) for honey-bee health, some countries in the EU are losing up to a third of their colonies every year. The most important contribution bees make to agriculture is the pollination service they provide. Pollinators' decline brings along a significant loss of pollination services, which has negative ecological and economic impacts, affecting the maintenance of wild plant diversity and large-scale ecosystem stability with potentially harmful effects on crop production, food security, and human welfare.

Smart farming and digital transformation in agriculture describe the idea of making all digital technologies and innovations usable for agriculture to make it more efficient, more resilient, more resource-efficient, and even more sustainable. "GNSS (GPS) and automatic steering systems ensure that agricultural machinery can operate in the field with an accuracy of 2–3 cm. Site-specific farming with section control and variable rate control enables the emergence of strong and healthy crops on each individual subplot of a field" [101].

Livestock has been used in many ways and served humanity for the production of food and consumer goods such as wool, eggs, and meat. While animals' needs—such as general welfare, the urge to move, and social relationships—have taken a back seat, there is a rethinking taking place in today's society.

Now, one of the biggest challenges our society has to master is the ability to feed a growing population while minimizing environmental impacts, ensuring human health, as well as addressing animal health and welfare. Global meat production is expected to double by 2050. This increase in production might be achieved by a combination of expansion in animal numbers and increased productivity. This increase in animal numbers makes their management more challenging, especially if the number of farmers continues to decrease [102]. To meet these challenges, the development of new technologies has gained importance. The largest potential lies in individual animal monitoring and analysis, which is referred to as precision livestock farming (PLF). PLF technologies are designed to support farmers in livestock management by monitoring and controlling animal productivity, environmental impacts, as well as health and welfare parameters in a continuous, real-time, and automated manner.

3.2. Current State in Montenegro

Montenegro has a land area of 255,564 ha/13,812 km^2 (2021). About 38% of the country's land is used for agriculture, 88% of which is used for meadows and extensive pastures. Arable land, fruit growing, and viticulture are practiced on only 58,200 ha. Over 60% of agricultural production comes from livestock production. Agriculture—including hunting, forestry, as well as food production—plays an important role in the economy of Montenegro and contributes to about 10% of the gross domestic product [103]. In Montenegro, most olive trees are grown in a conventional way, without normal pruning and with replacement yields. Due to the uneven landscape, agricultural production is limited to the valley systems and the narrow beach belt. Crop production is carried out only in some valleys and is restricted by scarce water resources. A lot of households still have small family plots for the production of fruit and vegetables. The dominant agricultural system is extensive grazing of cattle, sheep, and goats on semi-natural pastures [104].

Since the production methods and techniques in the Montenegrin viticulture sector mainly employ traditional approaches, there is plenty of room for improvement. The current irrigation practice, for example, is based on experience and visual inspection of the vineyard. Based on this, the plants are irrigated every fifth day, providing 50 liters of water. However, this does not take into account different soil compositions, and hence, more often than not results in over- or under-irrigation [105]. A long tradition has been beekeeping in Montenegro [106], which has been changing more and more to organic beekeeping in recent years. A joint effort from parties around the world emphasizes the development and integration of different technologies to monitor pollinators and their environment, as detailed in the next subsection.

3.3. Technologies and Use Cases

In the recent years, the digitization of agriculture has increased throughout the world. Figure 2 shows the number of articles counted by Google Scholar in 2022 for the various topics of digital agriculture. Although digital beekeeping and vineyards are less investigated in comparison, these are two highly relevant topics to Montenegro.

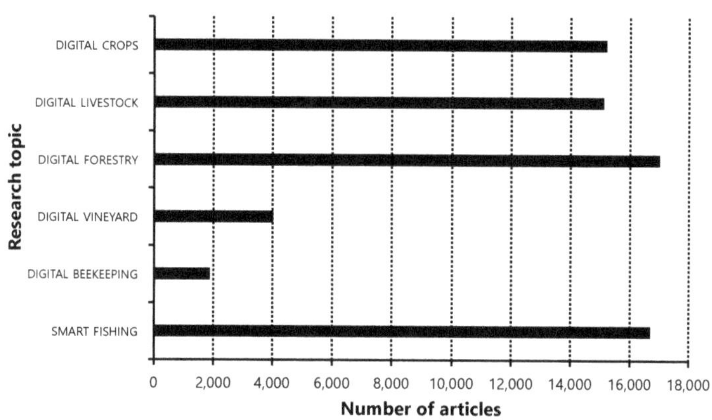

Figure 2. Number of articles on Google Scholar in 2022 for the topics of digital agriculture.

In order to improve the well-being of the bees and make the beekeepers' work easier, the Montenegro-based startup BeeAndme [107] equips beehives with intelligent sensors measuring all significant parameters such as local humidity, temperature, collected amount of honey, and bees' health through sound. Artificial intelligence is also used. The company counts how many bees fly in and out with the help of TensorFlow and Open CV [108]. Furthermore, data mining and machine learning algorithms are collected in order to help to increase the number of bees. The IoBee project [109], for example, addresses in-hive and in-field monitoring, as well as the implementation of satellite imagery and Spatial Decision Support Systems (SDSS). BeeLife [110] is a solution-oriented non-profit organization working to improve conditions for bees and pollinators.

In PLF, tools and sensors are used to continuously and automatically monitor key performance indicators (KPIs) of livestock in the areas of animal health, productivity, and environmental load. A variety of systems using technologies such as sensors, cameras, or microphones can directly alert farmers via connected devices (e.g., phones, computers, or tablets) about detected anomalies, allowing farmers to intervene at an early stage. Research is pointing toward the great potential for these smart technologies to help livestock farmers in monitoring the welfare of their animals, and several countries are already investing in their development, reflecting their potential to be part of strategies to move toward sustainable agriculture. A metastudy classifying the different technologies was undertaken by Schillings et al. [111].

Like all technologies, PLF technologies are constantly evolving with every passing day. Many are adopted and highly successful across numerous farms everywhere, while some are in their early developmental stages. A few of the PLF technologies in application today are the following: automated weighing systems, water meters, and different types of feed intake sensors. Automated-weighing systems include "step-on scales" and cameras that allow calculating the weight of individuals through machine-learning analysis of images and videos. However, weight is just one of the many data that can be obtained from imaging solutions. Due to the coverage of a wide range of vital data and their affordability, imaging solutions are the most common form of precision livestock monitoring and one of the initial steps of a stepwise PLF adoption approach on a farm. An extensive literature review listing the main systems available on the market, consisting of combined sensors to monitor and manage livestock, was provided by Aquilani et al. [112].

Vineview [113] uses high-tech drones in precision viticulture to help winegrowers monitor vineyards better and more efficiently, and to plan the grape harvest in a more differentiated and precise way. The drones are equipped with infrared sensors and reconnoitre the vineyard in regular reconnaissance flights. From the air, water stress and ripeness can be observed and the vines counted. The unmanned aerial vehicles (UAVs) from VisioVitis are also used to optimize fertilizer, irrigation, and pest control.

3.4. Considerations and Implications

While their potential is promising, the use of these new PLF technologies also raises ethical concerns, such as their potential impact on the human–animal relationship, the objectification of animals, the notion of care, and farmers' identity as animal keepers. The human–animal relationship is an important aspect that can influence both animal welfare and productivity. Some of the challenges and risks associated with PLF are listed below:

- Affordability is still a major challenge associated with the integration of expensive technologies on farms. Although studies show that PLF technologies make a farm more profitable, the diverse nature of each farm makes it a concern worth considering thoroughly before deciding to adopt PLF.
- The major risk of PLF is that since it is often integrated and automatic, a system failure can cause devastating impacts, especially if the system is fully automatic.
- Another associated risk is when the unit of animals is not individuals but a group of individuals such as poultry where flocks are measured. In such cases, special individual needs can be overlooked.
- The use of intrusive tags is a risk to animal welfare, which is still used in many PLF practices and technologies.

3.5. Summary

Many aspects have to be considered when it comes to digitalization. Farmers can readjust systems when technology detects irregularities but, still, they have to be mindful that not every problem can be detected. To be able to intervene, they need not only a sound understanding of their profession, but also good knowledge of the technologies that are in use on their farms. Yet, these tools can be successful in the long term only if they are seamlessly integrated into everyday work [101].

Short-term efforts for Montenegro are considered in the area of precision viticulture, since there are bigger farms that can invest into digitization. The precision farming applied to optimize vineyard performance—in particular, maximizing grape yield and quality while minimizing environmental effects—has impact in the long run. Precision agriculture emphasizes "doing the right thing, in the right place, at the right time", and it is practical for viticulture because of the high local variability of conditions within vineyards. The long-term efforts for Montenegro are shown in the research conducted on the Ćemovsko field [114]. Concerning beekeeping, the Montenegro-based startup BeeAndme [107] is focusing on the improvement of the well-being of the bees. A long-term need to a thriving startup scene is to support farms in digitalization. The first step into building up a startup ecosystem is in the early stage of forming.

The greatest benefits of digitalization in agriculture are seen in the potential increase in sustainability and productivity, as well as in labor time savings and facilitation. As a result, this leads to reduced use of fertilizers, pesticides, and energy, as well as improvements in animal welfare [115].

4. Digital Healthcare and eHealth

4.1. Introduction and Concepts

Digital health technologies may aid in the provision of health and social care to an aging population with growing healthcare demands. However, numerous challenges must be overcome, such as ensuring accessibility and inclusivity; digital literacy and skills; sustained patient and clinician engagement; scalability and successful integration into healthcare systems; the development of the necessary evidence base; and effective regulation [116]. The current digital landscape in healthcare delivery and clinical research conduct includes the following [117]:

- Use of digital health technologies as a diagnostic tool—detection of heart rhythm disorders (e.g., atrial fibrillation), detection of retinopathy, metastases, metabolic disorders in tumor cells, etc.—today is possible with the use of digital technologies.
- Digital health as a disease management and decision support tool—several applications related to a specific disease or condition have been established so far. Some of them are more related to diagnostic approaches, but there are a lot of them that aid in treatment.
- Digital health to improve research recruitment—MyHeart Counts and Health eHeart [118] are examples of randomized clinical trials in which digital technologies were used as a recruitment tool.

Implementing digital healthcare requires not only the development of electronic clinical guidelines but also the education of all parties involved, from physicians and healthcare workers to patients and their carers [119]. Moreover, digitalization in the field of healthcare enables not only diagnostics and treatment but also enables focusing on preventive medicine. "Prevention is better than cure" is a main principle, the fulfillment of which is made possible by the application of digital technologies in healthcare. A systematic review [120] evaluated the potential benefits of digital health interventions on cardiovascular disease outcomes and risk factors and evaluated the potential benefits of telemedicine, web-based strategies, email, mobile phones, mobile applications, text messaging, and sensor-based monitoring. These interventions reduced cardiovascular events, hospitalizations, and mortality interventions compared to usual cardiological healthcare.

The implementation of new and improvement of existing digital solutions in the healthcare system of Montenegro must be based on the following EU-based pillars [121]:

- Political commitment;
- Normative and regulatory frameworks;
- Technical infrastructure;
- Economic investments;
- Training and education;
- Research;
- Monitoring and evaluation.

4.2. Current State in Montenegro

The public health system of Montenegro is organized hierarchically: the primary level of healthcare consists of health centers (i.e., elected doctors and emergency services); the secondary level of healthcare is realized through general hospitals; and at the tertiary level of healthcare, the following institutions function—Clinical Center of Montenegro, Institute of Public Health of Montenegro, Blood Transfusion Institute of Montenegro, and others. The Institute for Medicines and Medical Devices of Montenegro, the Montefarm Pharmacy Institution, and a number of other special hospitals are also part of the health system in Montenegro. In the private sector, there are several clinics, polyclinics, and hospitals, and some of them are integrated into the public health system. The Ministry of Health creates and the Health Insurance Fund finances healthcare in Montenegro.

Chronic non-communicable diseases are the leading causes of illness, disability, and premature (before the age of 65) death of the inhabitants of Montenegro. The number of doctors is lower than the European average.

In the previous period, a significant step forward was made in the digital technologies domain. First of all, an electronic service called eHealth is available to all health-insured individuals. Within the same framework, several services are available: eScheduling (a service for online scheduling of visits with doctors in primary health centers), eRecipe (an electronic service that enables patients to view prescribed and implemented prescriptions), eResults (an electronic service that enables patients to view the results of biochemical laboratory analyses), ePharmacy (an electronic service intended for patients, developed with the aim of providing information on the availability of medicines in all pharmacies on the territory of Montenegro), eInsurance (an electronic service that enables citizens to view the status of their health insurance [47]), and more.

The emergence of the COVID-19 pandemic significantly accelerated the process of digital transformation in the health system of Montenegro; new digital services were developed (e.g., the COVID-19 vaccination certificate) and, in parallel, during the COVID-19 pandemic, a number of scientific and research projects were implemented, which aimed at improvement in this area.

4.3. Technologies and Use Cases

COVID-19 influenced the development of digital solutions worldwide. Society, more than ever before, understood the necessity of digitalization, especially in the health sector. The Montenegrin health system was not an exception [122]. Digitalization of critical processes in the health system of Montenegro started before the outbreak of COVID-19, but the most significant results and popularization of the services were achieved during the crisis [122]. Digitalization of procedures of healthcare—to which digitalization was applicable—resulted in a severe reduction in human-to-human interaction [123]. This was beneficial not only by preventing the spread of the virus but also by ensuring efficient time management for both patients and doctors.

One of the main eHealth features beneficial for both the public Health Insurance Fund and the people it insures is ePharmacy, as it ensures visibility of the status of the supplies of medicines in pharmacies and their availability. This service is accessible through both eHealth and eGovernment portals, allowing all relevant stakeholders to optimize procedures of medicine procurement. On the other hand, authorities can supervise the logistics of the supply chain and optimize all inputs related to it. Such a technology created a potential for stable distribution of medicine even in times of the highest demand and market disruption.

An addition to ePharmacy, paperless prescription issuing was adopted by both citizens and medical staff through ePrescription. This service resulted in less contact between parties involved in procedures, which is especially important in times of viral diseases. Additionally, it allows supervision of patients' medication history and realization of the previous prescriptions, preventing any unwanted abuse or misuse.

In an attempt to prevent waiting for the results of biochemical laboratory analysis, the eResult service was developed for all institutions within the public healthcare system. By providing digitized reports, patients' time spent within the institution is minimized, and the system stores and thoroughly explores previous results of biochemical analysis of the patients for optimal healthcare. Such systems minimize human-to-human interaction in institutions, providing analysis and maintaining patients' safety through persevering reporting.

Healthcare system digitalization consists of improving the administrative procedures in order to optimize the processes within [124]. Providing citizens with relevant information about their Health Insurance is an administrative task, and its digitalization was conducted through the eInsurance service, which allows interested citizens to obtain relevant information without interacting with administration offices in a time-efficient manner. Additionally, patients were offered the eOrdering service, through which most of the ordering services were made available online. Through eOrdering, ordering of common reports (i.e., sick leave, reports for calculations of salary compensation during temporary incapacity for work, etc.) became a seamless effort, displaced from healthcare-providing institutions.

Making appointments to selected doctors also migrated to a digital environment, allowing patients to use the service regardless of their current location [125]. Visits to institutions providing healthcare just to schedule an appointment were completely replaced by the service, allowing safer environments and faster procedures. This service was recognized as crucial within the health infrastructure by both patients and healthcare providers.

Described technologies optimized the procedures in terms of money, time efficiency, safety, and security. These technologies are here to stay, as COVID-19 exposed the impracticality and instability of previously established and conducted practices [126].

4.4. Relevant Features and Attributes of Montenegro

Healthcare service (HCS) providers in Montenegro are structured within three main groups: public HCS providers; private HCS providers with an agreement signed with the public Health Insurance Fund; and other private HCS providers. Private HCS providers with agreement are predominantly stomatology- (175) and pharmacy-related (189) institutions. Besides those, there are general and special hospitals (2), and institutions providing radiology (2), histology (2), gynecology (3), and ophthalmology (8). There are 20 health centers, 114 special ambulances, and 32 medical–technical providers classified as other private HCS providers with no agreement. Public HCS providers are classified as follows: health centers (18); general hospitals (7); special hospitals (3); and one of each clinical center, public health, medical rehabilitation, blood transfusion, emergency medical assistance, pharmacy, and medical–technical provider.

The medical information systems described in the previous chapter support processes in all public HCS providers and private HCS providers with an agreement. The importance of such technologies and features was especially emphasized during the COVID-19 outbreak, as the greatest extent of service use peaked with the first wave of the COVID-19 pandemic. Figure 3 (provided by the Montenegrin territorial health services) illustrates the frequency of the usage of the different eHealth services in Montenegro in 2020.

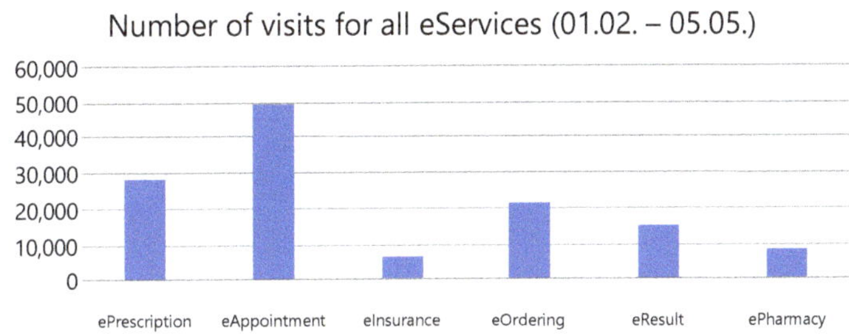

Figure 3. Frequency of usage of the eHealth services in Montenegro in 2020.

The most used service by far is eAppointments, with 49,183 registered applications. Patients showed the least interest in eInsurance and ePharmacy, while ePrescription, eOrdering, and eResult are gaining traction daily. Figure 4 (provided by the Montenegrin territorial health services) shows the number of visits for all eServices in Montenegro in 2020.

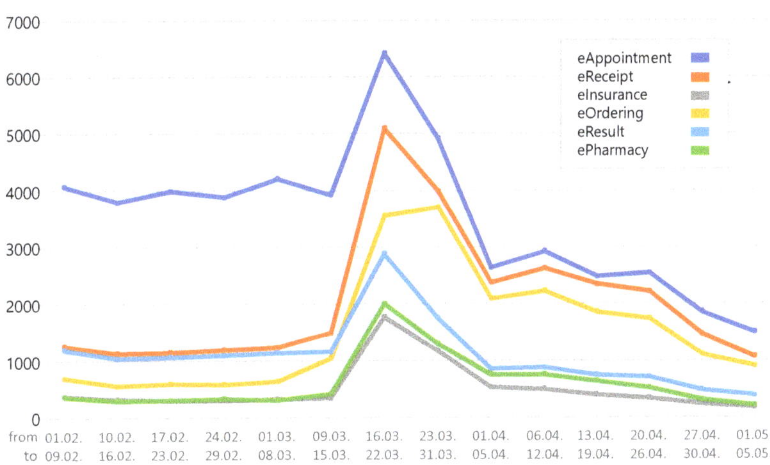

Figure 4. The number of visits for all eServices in Montenegro in 2020.

Services and features such as these prevented countless unnecessary direct contacts. Additionally, time management on both sides, patients and HSC providers, were more efficient than ever. Established digital procedures will grow in the number of users as most of the HSC providers migrate completely to certain solutions, such as prohibiting the old-fashioned means of scheduling appointments, ordering reports, or reporting the results.

4.5. Considerations and Implications

Healthcare systems are recognized as carriers of critical information infrastructure, and in the light of current cyber-attacks in Montenegro, it is of particular importance to be specially devoted to ensuring the security of the system [127]. The challenges are numerous in that regard: existing legal frameworks define boundaries that should be met with security mechanisms, as well as the future legal frameworks, while the application of IoT in healthcare—including mobile devices, smart watches, and different sensors—is rapidly growing, therefore increasing the cyber vulnerabilities of the system. The least attention is paid to checking the safety of the users and how they can be endangered.

On the other hand, the COVID-19 pandemic has imposed that the future guidelines for the development and improvement of health—not only at national but also at global level—should be focused on meeting the lessons learned during the pandemic and preparation of systems for possible future large-scale health issues [128].

The key challenges can be identified as follows: support of ecosystem coordination [129], the development of services based on the integration of a large number of devices and data, and the application of artificial intelligence—especially with regard to disease prediction (not only early diagnosis and treatment as before)—all with a high degree of user-oriented experience.

4.6. Summary

Montenegro, as a small country, recognized the importance of developing an integral health information system and started with the development of modern eHealth services. However, modern trends in development of health technologies impose a transition to a paradigm focused on health instead of healthcare, which is a task that will be necessary to achieve with the application of modern technologies, while ensuring a high level of safety and cybersecurity.

The future efforts at national level are guided by the Health Information System Development Strategy 2018–2023 and the corresponding Action Plan, and their new adoptions for the period of 2024–2029, as well as the operational program for the implementation of the Smart Specialization Strategy 2021–2024 [130]. All these documents identified the necessity of new developments in the Electronic Health Record (EHR), as an umbrella project that should be connected to all information systems in public health and in the future to all private health institutions and systems. This is a system that is designed as a "silo" of all important medical data for patients/citizens, who are the sole owners of their data and have the ability to manage the data (e.g., granting access, control and monitoring of access). With an equally important impact, continual education of staff members as well as general improvement of economic and social conditions for healthcare workers are essential for the proper implementation of ICT services in healthcare. In short-term efforts, creation of cohesion between different sectors—especially academia and business [131]—towards supporting innovation in the field of health are identified as key mechanisms for the creation of a digitally enhanced ecosystem in healthcare.

5. Conclusions

In this paper, we provided a comprehensive analysis of the digitization of Montenegrin transportation, agriculture, and healthcare. We investigated the relevant technologies and use cases based on the properties of the country and formulated the appropriate considerations and implications for each field separately.

C-ITS is still in its infancy stage in Montenegro, and there are as yet no focused strategies, regulations, or plans for preparing any pilot deployments. We believe that the first step on the long road towards initial pilot C-ITS services in Montenegro should be becoming an associated C-Roads partner and starting the collection of expertise within the platform at a national level. The necessary know-how can be built-up in the appropriate decision-making bodies relatively fast by relying on the experiences gathered in C-Roads, and the tested deployment approaches can be adapted to the Montenegrin peculiarities. With proper planning and considering the milestones on the path foreseen in the intelligent transport systems evolution already discovered and made available by Core Member pilots, Montenegro could tremendously benefit from the transportation digitalization possibilities the continent-wide harmonized C-ITS ecosystem offers.

The need to feed a growing population not only leads to the development of smart technologies for agriculture but also for livestock management. Honey bees, along with pigs and cattle, are some of the most important farm animals. Modern technologies offer beekeepers new solutions to protect these valuable creatures. Precision livestock farming technologies in Montenegro may improve both production efficiency and preservation.

Our work presented the current state of development of eServices in the healthcare of Montenegro, with a special focus on services developed during the COVID-19 pandemic, as well as records of their use and exploitation. As the imperatives in the development of modern health systems are focused on prevention and the improvement of the services themselves, it is necessary in the near future to integrate scientific results, available data, and technologies in order to ensure the preservation of the well-being of the inhabitants, and the early detection and proactive intervention of various diseases that will certainly not disappear as long as science and society progress.

Summa summarum, Montenegro still has a long way to go on the road towards digitization, but these efforts contribute to a significantly more advanced country and society. By adapting to the changes that Industry 4.0 brings, Montenegro may embrace the vast benefits that engineering and science in these fields can provide.

It is important to note the most pertinent limitations of this work. First of all, while transportation in Montenegro is composed of many vital means and modes—particularly the usage of waterborne vessels enabled by the long shoreline—our research effort narrows its transportation-related focus to land-vehicle-based C-ITS. One reason is that in the EU—to which Montenegro aims to ascend within the upcoming years—C-ITS is one of the hottest research areas in transportation science, and regulation/harmonization has already begun on the levels of both the EC and standardization organizations. Regarding agriculture, autonomous tractors or cow milking robots—typical digitization applications for huge farms—have not been addressed, since most Montenegrin farms are typically small, operate on much thinner margins, and are thus usually less willing to spend money on ventures that might not work out for such small farms. As for healthcare, while the presented work is focused on providing a comprehensive picture related to its digitization in Montenegro—with a particular emphasis on the COVID-19 period—there are, in fact, certain limitations of the analysis, lacking themes such as specialized topics of surgery (e.g., efforts of neurologists to pilot AI use for stroke patients [132]) and deeper investigation of Montenegrin telemedicine [133].

Due to the well-defined scope and the above-listed limitations of this work, there is still much to be covered by such research efforts. Beyond the unaddressed means of transportation, future work should study education and training, manufacturing, production, all forms of logistics, cyber-physical systems, business models, human–machine interfaces, the various applications of IoT, privacy and trust, and many more, all of which are fundamentally affected by the transformative power of Industry 4.0 and digitization in general.

Author Contributions: Conceptualization, P.A.K., I.O., C.R. and L.B.; methodology, P.A.K., I.O., C.R. and L.B.; investigation, P.A.K., I.O., I.M., J.M., A.W., R.Š., L.L., M.R., C.R., A.S. and L.B.; resources, P.A.K., I.O., I.M., J.M., A.W., R.Š., L.L., M.R., C.R., A.S. and L.B.; writing—original draft preparation, P.A.K., I.O., I.M., J.M., A.W., R.Š., L.L., M.R., C.R., A.S. and L.B.; writing—review and editing, P.A.K., I.O., C.R., A.S. and L.B.; visualization, P.A.K., I.O., C.R. and L.B.; supervision, P.A.K., I.O., C.R. and L.B.; project administration, P.A.K., I.O., C.R. and L.B.; funding acquisition, P.A.K., I.O., R.Š., C.R. and L.B. All authors have read and agreed to the published version of the manuscript.

Funding: The work was supported by the European Commission Erasmus+ Project Digital Entrepreneurial Nest and Industry 4.0 in Montenegro (DigNest) 619099-EPP-1-2020-1-ME-EPPKA2-CBHE-JP, and by the Cooperative Doctoral Programme–National Defence Subprogramme of the National Research, Development, and Innovation Fund of Hungary (NVKDP-2021).

Institutional Review Board Statement: Not applicable.

Informed Consent Statement: Not applicable.

Data Availability Statement: Not applicable.

Conflicts of Interest: The authors declare no conflict of interest.

Abbreviations

The following abbreviations are used in this manuscript:

3GPP	3rd Generation Partnership Project
BSM	Basic Safety Message
BTP	Basic Transport Protocol
C2C-CC	CAR 2 CAR Communication Consortium
CALM	Communications Access for Land Mobiles
CAM	Cooperative Awareness Message
CEDR	Conference of European Directors of Roads
CICW	Cooperative Intersection Collision Warning
C-ITS	Cooperative Intelligent Transportation System
C-V2X	Cellular Vehicle-to-Everything
CSMA/CA	Carrier-Sense Multiple Access with Collision Avoidance
DENM	Decentralized Environmental Notification Message
DG MOVE	Directorate-General for Mobility and Transport
ECRL	European Certificate Revocation List
ECTL	European Certificate Trust List
EHR	Electronic Health Record
ETSI	European Telecommunications Standards Institute
Euro NCAP	European New Car Assessment Programme
GNSS	Global Navigation Satellite System
GPS	Global Positioning System
HCS	Healthcare service
I2V	Infrastructure-to-Vehicle
IEEE	Institute of Electrical and Electronics Engineers
IoT	Internet-of-Things
ISO	International Organization for Standardization
ITS	Intelligent Transportation System
IVI	In-Vehicle Information
LTE	Long-Term Evolution
MAC	Media Access Control
OEM	Original Equipment Manufacturer
OSI	Open Systems Interconnection
PKI	Public Key Infrastructure
PLF	Precision Livestock Farming
RSU	Roadside Unit
RTTI	Real-time Traffic Information
SAE	System Architecture Evolution
SDSS	Spatial Decision Support Systems
UAV	Unmanned Aerial Vehicle
V2I	Vehicle-to-Infrastructure
V2N	Vehicle-to-Network
V2V	Vehicle-to-Vehicle
V2X	Vehicle-to-Everything
WAVE	Wireless Access in Vehicular Environments
WEF	World Economic Forum

References

1. El Hamdi, S.; Abouabdellah, A.; Oudani, M. Industry 4.0: Fundamentals and main challenges. In Proceedings of the 2019 International Colloquium on Logistics and Supply Chain Management (LOGISTIQUA), Paris, France, 12–14 June 2019; pp. 1–5.
2. Rainnie, A.; Dean, M. Industry 4.0 and the future of quality work in the global digital economy. *Labour Ind. J. Soc. Econ. Relat. Work* **2020**, *30*, 16–33. [CrossRef]
3. Rudrapati, R. Industry 4.0: Prospects and challenges leading to smart manufacturing. *Int. J. Ind. Syst. Eng.* **2022**, *42*, 230–244. [CrossRef]
4. Schwab, K. *The Fourth Industrial Revolution*; Crown Business: New York, NY, USA, 2017.
5. Fuchs, C. Industry 4.0: The digital German ideology. *Triplec Commun. Capital. Crit.* **2018**, *16*, 280–289. [CrossRef]

6. Skobelev, P.; Borovik, S.Y. On the way from Industry 4.0 to Industry 5.0: From digital manufacturing to digital society. *Ind. 4.0* **2017**, *2*, 307–311.
7. Callaghan, C.W. Transcending the threshold limitation: A fifth industrial revolution? *Manag. Res. Rev.* **2019**, *43*, 447–461. [CrossRef]
8. Sarfraz, Z.; Sarfraz, A.; Iftikar, H.M.; Akhund, R. Is COVID-19 pushing us to the fifth industrial revolution (society 5.0)? *Pak. J. Med. Sci.* **2021**, *37*, 591. [CrossRef] [PubMed]
9. Zizic, M.C.; Mladineo, M.; Gjeldum, N.; Celent, L. From industry 4.0 towards industry 5.0: A review and analysis of paradigm shift for the people, organization and technology. *Energies* **2022**, *15*, 5221. [CrossRef]
10. Moll, I. The myth of the fourth industrial revolution. *Theoria* **2021**, *68*, 1–38. [CrossRef]
11. Philbeck, T.; Davis, N. The fourth industrial revolution. *J. Int. Aff.* **2018**, *72*, 17–22.
12. Morgan, J. Will we work in twenty-first century capitalism? A critique of the fourth industrial revolution literature. *Econ. Soc.* **2019**, *48*, 371–398. [CrossRef]
13. Avis, J. Post-Work, Post-Capitalism and the Fourth Industrial Revolution. In *Vocational Education in the Fourth Industrial Revolution*; Springer: Berlin/Heidelberg, Germany, 2020; pp. 73–102.
14. Avis, J. Education and Employment in a Post-Work Age. In *Vocational Education in the Fourth Industrial Revolution*; Springer: Berlin/Heidelberg, Germany, 2021.
15. Giesen, K.G. The Transhumanist Ideology and the International Political Economy of the Fourth Industrial Revolution. In *Ideologies in World Politics*; Springer: Berlin/Heidelberg, Germany, 2020; pp. 143–156.
16. Bloom, P. Shared consciousness: Toward a world of transhuman relations. In *Identity, Institutions and Governance in an AI World*; Springer: Berlin/Heidelberg, Germany, 2020; pp. 247–264.
17. Degtyareva, V.; Lyapina, S.; Tarasova, V.; Yeremyan, L. Development of the National Qualifications System: The Impact of Transhumanism. *Wisdom* **2022**, *2*, 30–42. [CrossRef]
18. Dimitrakopoulos, G.; Demestichas, P. Intelligent transportation systems. *IEEE Veh. Technol. Mag.* **2010**, *5*, 77–84. [CrossRef]
19. Lamssaggad, A.; Benamar, N.; Hafid, A.S.; Msahli, M. A survey on the current security landscape of intelligent transportation systems. *IEEE Access* **2021**, *9*, 9180–9208. [CrossRef]
20. Rajkumar, S.; Deborah, L.J. An improved public transportation system for effective usage of vehicles in intelligent transportation system. *Int. J. Commun. Syst.* **2021**, *34*, e4910. [CrossRef]
21. Liu, Q.; Kang, B.; Yu, K.; Qi, X.; Li, J.; Wang, S.; Li, H.A. Contour-maintaining-based image adaption for an efficient ambulance service in intelligent transportation systems. *IEEE Access* **2020**, *8*, 12644–12654. [CrossRef]
22. Gechter, F.; Aglzim, E.H.; Senouci, S.M.; Rodet-Kroichvili, N.; Cappelle, C.; Fass, D. Transportation of goods in inner-city centers: can autonomous vehicles in platoon be a suitable solution? In Proceedings of the 2017 IEEE Vehicle Power and Propulsion Conference (VPPC), Belfort, France, 11–14 December 2017; pp. 1–5.
23. Dontu, A.; Gaiginschi, L.; Barsanescu, P. Reducing the urban pollution by integrating weigh-in-motion sensors into intelligent transportation systems. State of the art and future trends. *IOP Conf. Ser. Mater. Sci. Eng.* **2019**, *591*, 012087. [CrossRef]
24. Schneider, U.A.; Havlík, P.; Schmid, E.; Valin, H.; Mosnier, A.; Obersteiner, M.; Böttcher, H.; Skalský, R.; Balkovič, J.; Sauer, T.; et al. Impacts of population growth, economic development, and technical change on global food production and consumption. *Agric. Syst.* **2011**, *104*, 204–215. [CrossRef]
25. Corwin, D.L. Climate change impacts on soil salinity in agricultural areas. *Eur. J. Soil Sci.* **2021**, *72*, 842–862. [CrossRef]
26. Kumar, S.; Meena, R.S.; Jhariya, M.K. *Resources Use Efficiency in Agriculture*; Springer: Berlin/Heidelberg, Germany, 2020.
27. Deshpande, N.; Kumar, A.; Ramaswami, R. The Effect of National Healthcare Expenditure on Life Expectancy. 2014. Available online: https://smartech.gatech.edu/handle/1853/51648 (accessed on 21 January 2023).
28. van den Heuvel, W.J.; Olaroiu, M. How important are health care expenditures for life expectancy? A comparative, European analysis. *J. Am. Med. Dir. Assoc.* **2017**, *18*, 276.e9–276.e12. [CrossRef]
29. Liao, Y.; Loures, E.R.; Deschamps, F.; Brezinski, G.; Venâncio, A. The impact of the fourth industrial revolution: A cross-country/region comparison. *Production* **2018**, *28*, e20180061. [CrossRef]
30. Sutherland, E. The fourth industrial revolution–The case of South Africa. *Politikon* **2020**, *47*, 233–252. [CrossRef]
31. Nankervis, A.; Connell, J.; Cameron, R.; Montague, A.; Prikshat, V. 'Are we there yet?'Australian HR professionals and the Fourth Industrial Revolution. *Asia Pac. J. Hum. Resour.* **2021**, *59*, 3–19. [CrossRef]
32. DigNest: Digital Entrepreneurial Nest and Industry 4.0 in Montenegro. Available online: https://dignest.me/ (accessed on 21 January 2022).
33. Golubović, V.; Mirković, M.; Mićunović, N.; Srića, V. Digital Transformation in Montenegro–Current Status, Issues and Proposals for Improvement. *J. Comput. Sci.* **2021**, *9*, 1–12. [CrossRef]
34. Melović, B.; Jocović, M.; Dabić, M.; Vulić, T.B.; Dudic, B. The impact of digital transformation and digital marketing on the brand promotion, positioning and electronic business in Montenegro. *Technol. Soc.* **2020**, *63*, 101425. [CrossRef]
35. Bošković, B.; Nuhodžić, R.; Bugarinović, M. The Sustainability of Small Countries' Railway Sector Institutions in Liberalized Market—Case Study Montenegro. In Proceedings of the World Congress on Railway Research (WCRR), Milan, Italy, 29 May–2 June 2016.
36. Jakovljevic, N.; Jonic, I.; Nesic, K. Design for rehabilitation of 7 steel railway bridges in Montenegro. *Procedia Eng.* **2016**, *156*, 164–171. [CrossRef]

37. Bramo, F.; Llaci, S. The impact of railway (corridor VIII and the line Albania-Montenegro) on the agribusiness development in Albania. *Albanian J. Agric. Sci.* **2012**, *11*, 271–274.
38. Nikčević, J.; Škurić, M. A contribution to the sustainable development of maritime transport in the context of blue economy: The Case of Montenegro. *Sustainability* **2021**, *13*, 3079. [CrossRef]
39. Kapidani, N.; Kočan, E. Implementation of national maritime single window in Montenegro. In Proceedings of the 2015 23rd Telecommunications Forum Telfor (TELFOR), Belgrade, Serbia, 24–26 November 2015; pp. 17–20.
40. Hasaj, A.; Krymbi, E.; Kruja, D. Cross-border cooperation on water transportation, between region of Shkodra and Montenegro. In Proceedings of the 7th International Conference on Maritime Transport, Barcelona, Spain, 27–29 June 2016; pp. 535–544.
41. Rajović, G.; Bulatović, J. Zoning as a condition of sustainable agriculture northeastern Montenegro: A case study. *Rocz. Ochr. Środowiska* **2016**, *18*, 65–88.
42. Rajović, G. Important social factors for development of agriculture in North-Eastern Montenegro. *Ekon. J. Econ. Theory Pract. Soc. Issues* **2011**, *57*, 62–72.
43. Ćorić, D.; Popović, M. E-Agriculture: Montenegro Case. *Entren. Enterp. Res. Innov.* **2015**, *1*, 507–513.
44. Radonjić, S.; Hrnčić, S. A review of new alien arthropod pests and their impact on agriculture crops in Montenegro. *Acta Zool. Bulg.* **2017**, *9*, 203–210.
45. Hrncic, S. A survey of olive pests in Montenegro. In *Acta Horticulturae*; ISHS: Leuven, Belgium, 2002; Volume 2, pp. 819–821.
46. Hrnčić, S.; Radonjić, S. A survey of raspberry pests in Montenegro. In Proceedings of the X International Rubus and Ribes Symposium 946, Zlatibor, Serbia, 22–26 June 2011; pp. 243–246.
47. Ivanović, A.; Raković, P. E-health Card Information System: Case Study Health Insurance Fund of Montenegro. In Proceedings of the 2019 8th Mediterranean Conference on Embedded Computing (MECO), Budva, Montenegro, 10–14 June 2019; pp. 1–5.
48. Lješnjanin, S. Management in healthcare institutions of Montenegro. *Zdr. Zaštita* **2019**, *48*, 33–40. [CrossRef]
49. Mantas, J.; Sendelj, R.; Ammenwerth, E. Multidisciplinary Approach for Education in Healthcare Management: Case Study from Montenegro. *Importance Health Inform. Public Health Dur. Pandemic* **2020**, *272*, 330.
50. Vujanović, N. Technological Trends in the Manufacturing and Service Sectors. The Case of Montenegro. *South East Eur. J. Econ. Bus.* **2021**, *16*, 120–133. [CrossRef]
51. Boban, M.; Slavica, M.; Stevan, M.; Radovan, P.; Ore, E. Research of consumption and competitiveness of homemade products for manufacturing improvement: A case study from Montenegro. *Afr. J. Agric. Res.* **2012**, *7*, 3757–3764.
52. Labović, S.B.; Joksimović, I.; Galić, I.; Knežević, M.; Mimović, M. Food Safety Behaviours among Food Handlers in Different Food Service Establishments in Montenegro. *Int. J. Environ. Res. Public Health* **2023**, *20*, 997. [CrossRef]
53. Rajović, G.; Bulatović, J. Plant and Animal production in Montenegro with Overview of the Food Industry. *Int. Lett. Soc. Humanist. Sci.* **2015**, *63*, 7–16. [CrossRef]
54. Martinovic, A.; Mirecki, S. Food, nutrition, and health in Montenegro. In *Nutritional and Health Aspects of Food in the Balkans*; Elsevier: Amsterdam, The Netherlands, 2021; pp. 165–186.
55. Jaksic-Stojanovic, A.; Jankovic, M.; Seric, N. Montenegro as high-quality sports tourism destination-trends and perspectives. *Sport Mont.* **2019**, *17*, 93–95. [CrossRef]
56. Moric, I. Clusters as a factor of rural tourism competitiveness: Montenegro experiences. *Bus. Syst. Res. Int. J. Soc. Adv. Innov. Res. Econ.* **2013**, *4*, 94–107. [CrossRef]
57. Moric, I.; Pekovic, S.; Janinovic, J.; Perovic, Đ.; Griesbeck, M. Cultural tourism and community engagement: Insight from Montenegro. *Bus. Syst. Res. Int. J. Soc. Adv. Innov. Res. Econ.* **2021**, *12*, 164–178. [CrossRef]
58. Bigović, M. Quantifying seasonality in tourism: A case study of Montenegro. *Acad. Tur. Tour. Innov. J.* **2011**, *4*, 15–32.
59. C-ROADS. The Platform of harmonised C-ITS deployment in Europe. 2022. Available online: https://www.c-roads.eu/platform.html (accessed on 21 January 2023).
60. Ministry of Capital Investments: Transport Development Strategy—Montenegro 2019–2035. 2019. Available online: https://www.gov.me/dokumenta/a080d54d-0b87-4d8c-bfbf-bdc8ae5dc8bb (accessed on 21 January 2022).
61. Directive 2010/40/EU of the European Parliament and of the Council of 7 July 2010 on the Framework for the Deployment of Intelligent Transport Systems in the Field of Road Transport and for Interfaces with Other Modes of Transport. 2010. Available online: https://eur-lex.europa.eu/legal-content/EN/ALL/?uri=CELEX%3A32010L0040 (accessed on 21 January 2022).
62. Prelevic, M. Management of Transport Processes with the use of Intelligent Transport Systems—A Case Study of the Company Vertigo Montenegro Ltd.-Podgorica. *Glob. J. Res. Eng.* **2020**, *20*, 11–15. [CrossRef]
63. Strom, E.G. On Medium Access and Physical Layer Standards for Cooperative Intelligent Transport Systems in Europe. *Proc. IEEE* **2011**, *99*, 1183–1188. [CrossRef]
64. Boban, M.; Manolakis, K.; Ibrahim, M.; Bazzi, S.; Xu, W. Design aspects for 5G V2X physical layer. In Proceedings of the 2016 IEEE Conference on Standards for Communications and Networking (CSCN), Berlin, Germany, 31 October–2 November 2016; pp. 1–7. [CrossRef]
65. *ISO 21217:2020*; Intelligent Transport Systems—Station and Communication Architecture. International Organization for Standardization: Geneva, Switzerland, 2020.
66. *ISO/IEC 7498-1:1994*; Information Technology–Open Systems Interconnection–Basic Reference Model: The Basic Model. International Organization for Standardization: Geneva, Switzerland, 1994.

67. *IEEE Std 802.11-2016 (Revision of IEEE Std 802.11-2012)*; IEEE Standard for Information Technology—Telecommunications and Information Exchange between Systems Local and Metropolitan Area Networks—Specific Requirements—Part 11: Wireless LAN Medium Access Control (MAC) and Physical Layer (PHY) Specifications. Institute of Electrical and Electronics Engineers: New York, NY, USA, 2016; pp. 1–3534. [CrossRef]
68. Jacob, R.; Anwar, W.; Schwarzenberg, N.; Franchi, N.; Fettweis, G. System-level Performance Comparison of IEEE 802.11p and 802.11bd Draft in Highway Scenarios. In Proceedings of the 2020 27th International Conference on Telecommunications (ICT), Bali, Indonesia, 5–7 October 2020; pp. 1–6. [CrossRef]
69. 3GPP; Digital Cellular Telecommunications System (Phase 2+) (GSM); Universal Mobile Telecommunications System (UMTS); LTE; 5G. Release 14 Description; Summary of Rel-14 Work Items. Technical Specification (TS) 21.914, 3rd Generation Partnership Project (3GPP). 2018. Version 14.0.0. Available online: https://portal.3gpp.org/desktopmodules/Specifications/SpecificationDetails.aspx?specificationId=3179 (accessed on 21 January 2023).
70. Dayal, A.; Shah, V.K.; Choudhury, B.; Marojevic, V.; Dietrich, C.; Reed, J.H. Adaptive Semi-Persistent Scheduling for Enhanced On-road Safety in Decentralized V2X Networks. In Proceedings of the 2021 IFIP Networking Conference (IFIP Networking), Espoo and Helsinki, Finland, 21–24 June 2021; pp. 1–9. [CrossRef]
71. Garcia, M.H.C.; Molina-Galan, A.; Boban, M.; Gozalvez, J.; Coll-Perales, B.; Şahin, T.; Kousaridas, A. A Tutorial on 5G NR V2X Communications. *IEEE Commun. Surv. Tutor.* **2021**, *23*, 1972–2026. [CrossRef]
72. *IEEE Std 1609.3-2020 (Revision of IEEE Std 1609.3-2016)—Redline*; IEEE Standard for Wireless Access in Vehicular Environments (WAVE)–Networking Services–Redline. Institute of Electrical and Electronics Engineers: New York, NY, USA, 2021; pp. 1–344. [CrossRef]
73. *ETSI EN 302 636-4-1 V1.4.1*; Intelligent Transport Systems (ITS); Vehicular Communications; GeoNetworking; Part 4: Geographical Addressing and Forwarding for Point-to-Point and Point-to-Multipoint Communications; Sub-Part 1: Media-Independent Functionality. European Telecommunications Standards Institute: Sophia Antipolis, France, 2020. Available online: https://www.etsi.org/deliver/etsi_en/302600_302699/3026360401/01.04.01_60/en_3026360401v010401p.pdf (accessed on 21 January 2023).
74. *ETSI EN 302 636-5-1 V2.2.1 (2019-05)*; Intelligent Transport Systems (ITS); Vehicular Communications; GeoNetworking; Part 5: Transport Protocols; Sub-Part 1: Basic Transport Protocol. European Telecommunications Standards Institute: Sophia Antipolis, France, 2019. Available online: https://www.etsi.org/deliver/etsi_en/302600_302699/3026360501/02.02.01_60/en_3026360501v020201p.pdf (accessed on 21 January 2023).
75. *ETSI EN 302 637-2 V1.4.1(2019-04)*; European Standard, Intelligent Transport Systems (ITS); Vehicular Communications; Basic Set of Applications; Part 2: Specification of Cooperative Awareness Basic Service. 2019. Available online: https://www.etsi.org/deliver/etsi_en/302600_302699/30263702/01.04.01_60/en_30263702v010401p.pdf (accessed on 21 January 2023).
76. *ETSI EN 302 637-3 V1.3.1 (2019-04)*; Intelligent Transport Systems (ITS); Vehicular Communications; Basic Set of Applications; Part 3: Specifications of Decentralized Environmental Notification Basic Service. 2019. Available online: https://www.etsi.org/deliver/etsi_en/302600_302699/30263703/01.03.01_60/en_30263703v010301p.pdf (accessed on 21 January 2023).
77. *ISO/TS 19321:2020*; Intelligent Transport Systems—Cooperative ITS—Dictionary of in-Vehicle Information (IVI) Data Structures; Standard, International Organization for Standardization: Geneva, Switzerland, 2020. Available online: https://www.iso.org/standard/76974.html (accessed on 21 January 2023).
78. SAE International. *J2735_202211-V2X Communications Message Set Dictionary*; SAE International: Warrendale, PA, USA, 2022. Available online: https://www.sae.org/standards/content/j2735_202211/ (accessed on 21 January 2023).
79. SAE International. *On-Board System Requirements for V2V Safety Communications*; SAE International: Warrendale, PA, USA, 2022. Available online: https://www.sae.org/standards/content/j2945/1_202004/ (accessed on 21 January 2023).
80. *ETSI TS 102 941 V2.2.1 (2022-11)*; Intelligent Transport Systems (ITS); Security; Trust and Privacy Management. European Telecommunications Standards Institute: Sophia Antipolis, France, 2021. Available online: https://www.etsi.org/deliver/etsi_ts/102900_102999/102941/02.02.01_60/ts_102941v020201p.pdf (accessed on 21 January 2023).
81. *ETSI TS 102 940 V2.1.1*; Intelligent Transport Systems (ITS); Security; ITS Communications Security Architecture and Security Management; Release 2. European Telecommunications Standards Institute: Sophia Antipolis, France, 2021. Available online: https://www.etsi.org/deliver/etsi_ts/102900_102999/102940/02.01.01_60/ts_102940v020101p.pdf (accessed on 21 January 2023).
82. The European Commission. ANNEX III and IV to the Commission Delegated Regulation supplementing Directive 2010/40/EU of the European Parliament and of the Council with regard to the Deployment and Operational Use of Cooperative Intelligent Transport Systems. 2019. Available online: https://eur-lex.europa.eu/legal-content/EN/TXT/?uri=PI_COM%3AC%282019%291789 (accessed on 21 January 2023).
83. Car2Car Communication Consortium. Basic System Profile. 2022. Available online: https://www.car-2-car.org/documents/basic-system-profile (accessed on 21 January 2023).
84. *ISO/IEC 15408-1:2022*; Information Security, Cybersecurity and Privacy Protection—Evaluation Criteria for IT Security—Part 1: Introduction and General Model. International Organization for Standardization: Geneva, Switzerland, 2022.
85. Ministry of Economic Development. Spatial Plan of Montenegro Until 2020. 2008. Available online: https://wapi.gov.me/download-preview/b57f5f50-6f9c-4c27-ba86-122f5166b126?version=1.0 (accessed on 21 January 2023).
86. Preparation of the Transport Development Strategy–Montenegro (Contract No. 829-4147). Transport Development Strategy Report. 2017. Available online: https://va.mite.gov.it/File/Documento/228159 (accessed on 21 January 2023).

87. List of Cities in Montenegro by Population. 2022. Available online: https://all-populations.com/en/me/list-of-cities-in-montenegro-by-population.html (accessed on 21 January 2022).
88. WHO. Death on the Roads–Based on the WHO Global Status Report on Road Safety 2018. Available online: https://extranet.who.int/roadsafety/death-on-the-roads/#country_or_area/MNE (accessed on 21 January 2022).
89. Level crossing–Current Situation in the Western Balkan Region, Statistics, Classification and Comparison on LCs within Western Balkan. Available online: https://www.transport-community.org/wp-content/uploads/2019/12/Railroad-Level-Crossings-Current-State_TCS.pdf (accessed on 21 January 2022).
90. Pajkovic, A.P.V.; Grdinic, M. Road traffic safety performance in Montenegro. *Mach. Technol. Mater.* **2014**, *3*, 32–34.
91. Lučić, M. Characteristics and state of road traffic safety in Montenegro. *Trans Motauto World* **2021**, *6*, 51–53.
92. MONSTAT. Annual Statistic of Transport Storage and Communications, 2021. 2022. Available online: https://www.monstat.org/eng/novosti.php?id=3487 (accessed on 21 January 2023).
93. Brož, J.; Tichý, T.; Angelakis, V.; Bělinová, Z. Usage of V2X Applications in Road Tunnels. *Appl. Sci.* **2022**, *12*, 4624. [CrossRef]
94. Chen, L.; Englund, C. Cooperative Intersection Management: A Survey. *IEEE Trans. Intell. Transp. Syst.* **2016**, *17*, 570–586. [CrossRef]
95. Buinevich, M.; Spirkina, A.; Elagin, V.; Tarakanov, S.; Vladyko, A. V2X-based Intersection Priority Management. In Proceedings of the 2021 Systems of Signals Generating and Processing in the Field of on Board Communications, Moscow, Russia, 16–18 March 2021; pp. 1–7. [CrossRef]
96. Lee, J.W.; Lee, S.K.; Lee, H.M.; Jang, J.A.; Oh, H.S. Effectiveness Analysis of Warning Service using V2X Communication Technology at Intersection. In Proceedings of the 2018 International Conference on Information and Communication Technology Convergence (ICTC), Jeju, Republic of Korea, 17–19 October 2018; pp. 1506–1509. [CrossRef]
97. Ito, K.; Hirakawa, G.; Shibata, Y. Experimentation of V2X Communication in Real Environment for Road Alert Information Sharing System. In Proceedings of the 2016 IEEE 30th International Conference on Advanced Information Networking and Applications (AINA), Crans-Montana, Switzerland, 23–25 March 2016; pp. 711–716. [CrossRef]
98. Aissaoui, R.; Menouar, H.; Dhraief, A.; Filali, F.; Belghith, A.; Abu-Dayya, A. Advanced real-time traffic monitoring system based on V2X communications. In Proceedings of the 2014 IEEE International Conference on Communications (ICC), Sydney, NSW, Australia, 10–14 June 2014; pp. 2713–2718. [CrossRef]
99. Billot, R.; El Faouzi, N.E.; Guériau, M.; Monteil, J. Can C-ITS lead to the emergence of Traffic Management 2.0? In Proceedings of the 17th International IEEE Conference on Intelligent Transportation Systems (ITSC), Qingdao, China, 8–11 October 2014; pp. 483–488. [CrossRef]
100. Susic-Radovanovic, B.; Vojinovic, N.; Remikovic, S. Montenegro in Figures 2021. Statistical Office of Montenegro-Monstat. 2021. Available online: http://monstat.org/uploads/files/publikacije/Monstat%20-%20CG%20u%20Brojkama%20ENG_WEB.pdf (accessed on 21 January 2023).
101. Smart Farming and the Future of Agriculture. Available online: https://www.lll.tum.de/interview-smart-farming-and-the-future-of-agriculture/ (accessed on 21 January 2023).
102. Publications Office of the European Union: Agriculture, Forestry, and Fishery Statistics—2020 Edition. Available online: https://ec.europa.eu/eurostat/web/products-statistical-books/-/ks-fk-20-001 (accessed on 21 January 2023).
103. Statistical Office of Montenegro–Monstat. Statistical Yearbook 2022. Available online: http://monstat.org/eng/novosti.php?id=3646 (accessed on 21 January 2023).
104. Stjepana Filipovića. Sustainable Agriculture for Sustainable Balkans. Available online: http://www.sasb-eu.org/en/nature/the-western-balkans/montenegro (accessed on 21 January 2023).
105. Maraš, V.; Popovic, T.; Gajinov, S.; Mugosa, M.; Popovic, V.; Savović, A.; Pavicevic, K.; Mirovic, V. Precision Viticulture Using Wireless Sensor Network. In Proceedings of the 2020 9th Mediterranean Conference on Embedded Computing (MECO), Budva, Montenegro, 8–11 June 2020; pp. 1–6. [CrossRef]
106. Meda, K. Association of Beekeeping Organizations of Montenegro. Available online: https://pcelarstvo.me/ (accessed on 21 January 2023).
107. Bee and Me. Available online: https://beeandme.com/ (accessed on 21 January 2023).
108. Smart Beehive: Species Protection with the Cloud and IoT. Available online: https://open-telekom-cloud.com/en/blog/references/magenta-bieneninitiative (accessed on 21 January 2023).
109. IoBee. Available online: https://io-bee.eu/ (accessed on 21 January 2023).
110. BeeLife. Available online: https://www.bee-life.eu/about (accessed on 21 January 2023).
111. Schillings, J.; Bennett, R.; Rose, D.C. Exploring the Potential of Precision Livestock Farming Technologies to Help Address Farm Animal Welfare. *Front. Anim. Sci* **2021**, *2*, 639678. [CrossRef]
112. Aquilani, C.; Confessore, A.; Bozzi, R.; Sirtori, F.; Pugliese, C. Review: Precision Livestock Farming technologies in pasture-based livestock systems. *Animal* **2022**, *16*, 100429. [CrossRef]
113. Vineview. Vineview: Monitoring the Wineyard by Drones. Available online: https://vineview.com/ (accessed on 21 January 2023).
114. Maraš, V.; Popović, T.; Gajinov, S.; Mugoša, M.; Popović, V.; Savović, A.; Pavićević, K.; Mirović, V. Optimal Irrigation as a tool of Precision Agriculture. In Proceedings of the 2019 8th Mediterranean Conference on Embedded Computing (MECO), Budva, Montenegro, 10–14 June 2019; pp. 1–4. [CrossRef]

115. DigNest. WP1. Analysis and Road-Mapping: Supporting Digitization of Industry in Montenegro; DELIVERABLE 1.2 Cooperation between Higher Education Institutions (HEI) and Economics and Social Environment in EU Countries. 2021. Available online: https://dignest.me/#/page/wp1-analysis-and-road-mapping-supporting-digitization-of-industry-in-montenegro (accessed on 21 January 2023).
116. Watson, A.; Wilkinson, T.M. Digital healthcare in COPD management: A narrative review on the advantages, pitfalls, and need for further research. *Ther. Adv. Respir. Dis.* **2022**, *16*, 17534666221075493. [CrossRef]
117. Sharma, A.; Harrington, R.A.; McClellan, M.B.; Turakhia, M.P.; Eapen, Z.J.; Steinhubl, S.; Mault, J.R.; Majmudar, M.D.; Roessig, L.; Chandross, K.J.; et al. Using digital health technology to better generate evidence and deliver evidence-based care. *J. Am. Coll. Cardiol.* **2018**, *71*, 2680–2690. [CrossRef] [PubMed]
118. McConnell, M.V.; Shcherbina, A.; Pavlovic, A.; Homburger, J.R.; Goldfeder, R.L.; Waggot, D.; Cho, M.K.; Rosenberger, M.E.; Haskell, W.L.; Myers, J.; et al. Feasibility of obtaining measures of lifestyle from a smartphone app: The MyHeart Counts Cardiovascular Health Study. *JAMA Cardiol.* **2017**, *2*, 67–76. [CrossRef] [PubMed]
119. Rabello, G.M.; Pêgo-Fernandes, P.M.; Jatene, F.B. Are We Preparing for the Digital Healthcare Era? *Sao Paulo Med. J.* **2022**, *140*, 161–162. [CrossRef] [PubMed]
120. Widmer, R.J.; Collins, N.M.; Collins, C.S.; West, C.P.; Lerman, L.O.; Lerman, A. Digital health interventions for the prevention of cardiovascular disease: A systematic review and meta-analysis. In *The Mayo Clinic Proceedings*; Elsevier: Amsterdam, The Netherlands, 2015; Volume 90, pp. 469–480.
121. Odone, A.; Buttigieg, S.; Ricciardi, W.; Azzopardi-Muscat, N.; Staines, A. Public health digitalization in Europe: EUPHA vision, action and role in digital public health. *Eur. J. Public Health* **2019**, *29*, 28–35. [CrossRef] [PubMed]
122. Šćepanović, L.; Terzić, N. Changes in primary health care in response to the COVID-19 pandemic in Montenegro. *Srp. Med. časopis Lek. Komore* **2022**, *3*, 149–158. [CrossRef]
123. Milenkovic, A.; Jankovic, D.; Rajkovic, P. Extensions and adaptations of existing medical information system in order to reduce social contacts during COVID-19 pandemic. *Int. J. Med. Inform.* **2020**, *141*, 104224. [CrossRef]
124. Erakovic, J.; Milikic, D.; Radulovic, L.; Perunicic, S.; Idrizovic, Z.; Roganovic, M. Reorganization of multiple sclerosis health care system in Clinical Centre of Montenegro during the COVID-19 pandemic. *Eneurologicalsci* **2020**, *21*, 100263. [CrossRef] [PubMed]
125. Rudrapati, R. Using industrial 4.0 technologies to combat the COVID-19 pandemic. *Ann. Med. Surg.* **2022**, *78*, 103811. [CrossRef]
126. Dobrović, D.; Peličić, D. Telemedicine in the COVID-19 pandemic. *Zdr. Zaštita* **2020**, *49*, 67–74. [CrossRef]
127. Sendelj, R.; Ognjanovic, I. Cybersecurity Challenges in Healthcare. In *Achievements, Milestones and Challenges in Biomedical and Health Informatics*; IOS Press: Amsterdam, The Netherlands, 2022; pp. 190–202.
128. Kumpunen, S.; Webb, E.; Permanand, G.; Zheleznyakov, E.; Edwards, N.; van Ginneken, E.; Jakab, M. Transformations in the landscape of primary health care during COVID-19: Themes from the European region. *Health Policy* **2022**, *126*, 391–397. [CrossRef] [PubMed]
129. Djurovic, G.; Djurovic, V.; Bojaj, M.M. The macroeconomic effects of COVID-19 in Montenegro: A Bayesian VARX approach. *Financ. Innov.* **2020**, *6*, 1–16. [CrossRef] [PubMed]
130. Smart Specialization Strategy 2021-2024. Available online: https://s3.me/pametna-specijalizacija-u-crnoj-gori/ (accessed on 21 January 2023).
131. Roganović, M.; Ognjanović, I.; Šendelj, R.; Reich, C.; Bokor, L.; Mantas, J.; Golob, M.; Šimšić, N.; Orović, I.; Radusinović, T.; et al. Digital Entrepreneurial Nest: Supporting Digitization of Healthcare at National Level in Montenegro. *Stud. Health Technol. Inform.* **2022**, *289*, 491–493. [PubMed]
132. Stojanovic, S.; Pavlicic, M.; Stojanovic, R.; Krivokapic, S. "TeleCG"–Pilot Telemedicine Network of Montenegro. In Proceedings of the INFOFEST2004, Budva, Montenegro, 26 September–2 October 2004.
133. Ognjanović, I.; Šendelj, R.; Mantas, J.; Roganović, M. Development of ICT Enhanced Person-Centred Care Services for Stroke Outpatient Rehabilitation. In Proceedings of the 2021 10th Mediterranean Conference on Embedded Computing (MECO), Budva, Montenegro, 7–10 June 2021; pp. 1–4.

eng MDPI

Review

Amazon Natural Fibers for Application in Engineering Composites and Sustainable Actions: A Review

Pedro Henrique Poubel Mendonça da Silveira [1,2,*], Bruno Figueira de Abreu Ferreira Cardoso [2], Belayne Zanini Marchi [2] and Sergio Neves Monteiro [2]

1 West Zone Campus, Rio de Janeiro State University—UERJ, Avenida Manuel Caldeira de Alvarenga, 1203, Campo Grande, Rio de Janeiro 23070-200, RJ, Brazil
2 Department of Materials Science, Military Institute of Engineering—IME, Praça General Tibúrcio 80, Urca, Rio de Janeiro 22290-270, RJ, Brazil; brunofigueira1991@gmail.com (B.F.d.A.F.C.); belayne@ime.eb.br (B.Z.M.); snevesmonteiro@gmail.com (S.N.M.)
* Correspondence: pedroo.poubel@gmail.com

Abstract: The Amazon rainforest, spanning multiple countries in South America, is the world's largest equatorial expanse, housing a vast array of relatively unknown plant and animal species. Encompassing the planet's greatest flora, the Amazon offers a tremendous variety of plants from which natural lignocellulosic fibers (NLFs) can be extracted. In this century, NLFs, which have long been utilized by indigenous populations of the Amazon, have garnered interest as potential reinforcements for composites, whether polymer- or cement-based, in various technical applications such as packaging, construction, automotive products, and ballistic armor. A comparison with synthetic materials like glass, carbon, and aramid fibers, as well as other established NLFs, highlights the cost and specific property advantages of Amazon natural fibers (ANFs). Notably, the sustainable cultivation and extraction of ANFs, as alternatives to deforestation and livestock pasture, contribute to the preservation of the Amazon rainforest. This review article provides a comprehensive examination of recent studies directly related to ANF-reinforced polymer matrix composites. The specific advantages, proposed applications, and reported challenges are highlighted, shedding light on the potential of these unique natural fibers.

Keywords: natural lignocellulosic fibers; NLFs; Amazon rainforest; composite materials; engineering applications; sustainability; properties

Citation: da Silveira, P.H.P.M.; Cardoso, B.F.d.A.F.; Marchi, B.Z.; Monteiro, S.N. Amazon Natural Fibers for Application in Engineering Composites and Sustainable Actions: A Review. *Eng* **2024**, *5*, 133–179. https://doi.org/10.3390/eng5010009

Academic Editor: Antonio Gil Bravo

Received: 24 November 2023
Revised: 29 December 2023
Accepted: 2 January 2024
Published: 12 January 2024

1. Introduction

Nature has provided a comprehensive diversity of natural fibers throughout the ages. With a remarkably extensive history, these fibers have roots that date back to prehistoric times, unquestionably meeting human needs throughout history [1–3]. Vegetable fibers have been used for various purposes in different eras and regions of the world. Primitive humans used them to keep warm, protect themselves, and as materials for household utensils such as ropes and fishing nets [1,4–6].

The use of vegetable fibers for textiles dates back to the end of the Paleolithic period (12,000–8000 BC). Discoveries, like palm leaf fragments in a 10,000 BC Mexican cave and net bags in a Utah cave, reveal Native Americans' skills in processing vegetable fibers [1]. Table 1 outlines the history of natural fiber use in prehistoric eras.

Vegetable fibers played a crucial role in human history, influencing agricultural, fishing, artisanal, and commercial activities [7,8]. Civilizations aimed to enhance their quality of life, driving improvements in living standards, technology, economics, and societal strength [6,8]. Technological progress in vegetable fiber production surged after the 18th-century Industrial Revolution, fostering economic growth. However, it brought challenges like increased resource consumption and pollution, impacting global sustainability efforts [9,10].

Table 1. Timeline of natural fiber use in antiquity [1].

Period (Year)	Fact
20,000 BC	Humans making ropes and cords during the Paleolithic Age
12,000 BC	Evidence of the existence of cotton in Egypt
10,000 BC	First reports of the cultivation of wild plants and the manufacture of fabrics from natural fibers
9000 BC	The discovery of a net bag in Utah indicated that the American Indians had advanced skills in processing plant fibers in approximately 9000 BC.
8000 BC	The Swiss Lake Dwellers of the Stone Age cultivated flax and wove it into linen fabric.
6000 BC	Hemp, believed by some to be the oldest cultivated fiber plant, originated in Southeast Asia and spread to China
5700 BC	Evidence of cotton fabrics produced in Mexico during this period
5000 BC	The first evidence of weaving, through the manufacture of baskets using plant fibers, dates from this period. Cotton was cultivated and used in the Tehuacan Valley of Mexico.
3500 BC	Heavy, strong ropes were used to drag heavy objects in Egypt. The ropes were made by twisting strips cut from hides or fibers from papyrus reeds
3400 BC	The art of spinning and weaving linen was well developed in Egypt by 3400 BC, suggesting flax had been cultivated prior to that date
3000 BC	Spinning and weaving of cotton is practiced in Pakistan, evidenced by the discovery of cotton fabrics and string from excavations at Mohenjo-Daro
2900 BC	The Emperor Shen Nung encouraged the growth of hemp in China
2500 BC	Cotton and cotton textiles existed in Peru
2300 BC	Cotton was cultivated in the Indus Valley
1400 BC	A Hindu hymn describes the manufacture of cotton yarns and the weaving of cotton cloth
445 BC	Herodotus wrote of trees which grow wild in India, the fruit of which is a wool exceeding in beauty and goodness that of sheep and from which the natives make clothing
63 BC	Lentullus Spinther introduced cotton awnings in the theater at the Appolinarian games

The pursuit of a better quality of life is an important demand, as is the technological progress of a nation. This means that integrating technology and sustainability represents a low-environmental-impact option for fully viable engineering applications. Thus, in the face of growing environmental concerns and the possibility of depleting oil reserves, natural fibers can be an environmentally friendly alternative, in line with the principles of sustainable development. These materials are known to be renewable and biodegradable, and they have been widely used in various industrial and engineering applications [11–16]. According to the annual report of the Discover Natural Fibers Initiative (DNFI), shown in Table 2, in the year 2023, global production of natural fibers was estimated at 31.9 million tons, surpassing the production of the year 2020. The utilization of fibers in the industry may result in the global fiber production in 2023 being higher than that of the years 2021 and 2022 [17].

Given the exponential growth in the use of natural fibers, a promising class of engineering materials has emerged from the need to improve the properties of conventional materials. An example of this is the composite materials made from raw natural fibers and polymeric matrices, which, when combined, result in a new material with unique properties and an enhanced combination of characteristics compared to the individual materials that compose them. These materials have become one of the most investigated research topics in recent times [12,13,18]. This trend is not only due to environmental concerns but also to the favorable mechanical properties offered by the fibers at a lower cost, which has attracted the interest of the industry, seeking to introduce alternative materials into the market to replace synthetic components [19–21].

One of the industries that extensively uses natural fibers in the production of parts is the automotive sector. Manufacturers such as Audi, Volkswagen, Toyota, Mercedes-Benz, Volvo, and Ford utilize these composites with natural fibers. These materials are primarily employed in non-structural components of car bodies, such as door panels, package trays, hat trays, instrument panels, internal engine covers, sun visors, luggage liners, oil and air filters, and are even progressing towards more demanding structural parts like seat backs and external floor panels [22]. In addition to the automotive sector, these composites are also being applied in other areas such as maritime structures, military vest production, sports, and general engineering, with potential for use in aerospace, wind energy, and even space applications like satellites [11,19,23–27]. The applications of these composites are diverse, and there is a variety of new industrial applications that can be fully realized using the concept of sustainability, including recycling [19,25].

Table 2. Global production of natural fibers estimated by DNFI—Discover Natural Fibers Initiative in the period 2020–2023. Updated on 9 November 2023 [17].

Fiber	Global Production (Ton)			
	2020	2021	2022	2023
Abaca	75,889	83,501	72,000	66,000
Agave Fibers	40,625	40,743	41,000	41,000
Coir	1,101,498	1,115,349	1,145,000	1,175,000
Cotton	23,989,000	25,176,000	25,314,609	24,515,567
Other fiber crops	739,145	755,326	733,000	742,000
Flax, processed but not spun	974,806	896,636	851,805	851,805
True hemp, raw or retted	251,062	302,318	272,000	272,000
Jute, Kenaf and Allied fibers	2,874,000	3,175,600	3,095,000	2,700,000
Kapok	78,674	82,150	80,000	80,000
Ramie, raw or retted	62,228	10,138	10,000	10,000
Sisal, Henequen and similar hard fibers	280,800	281,400	273,000	278,000
Silk, raw	91,765	86,311	91,221	90,000
Total Natural Fibers	**31,606,868**	**33,069,866**	**33,100,000**	**31,900,000**

Since 2015, the United Nations has established a technical document signed by more than 190 countries, based on the sustainable development of their societies: the 2030 Agenda. One of the Sustainable Development Goals (SDGs) is dedicated to recyclable plastic (goal 12) and the elimination of the use of plastic bags (goal 14). Therefore, one of the greatest solutions to minimize the global problem of plastic pollution is to use sustainable, biodegradable, and recyclable plastics and composites [28]. Recycling plays a crucial role within a circular economy, an economic model that aims to minimize waste and optimize the use of resources [29]. Unlike the traditional linear model (production, use, and disposal), the circular economy seeks to keep materials and products in use for as long as possible, promoting reuse, repair, and recycling at the end of their life cycle [30,31]. Reuse and recycling contribute to environmental sustainability because the reduction in the extraction of new resources decreases pressure on the environment and, at the same time, reduces the amount of waste that can cause negative impacts associated with production [25]. Thus, the incorporation of recyclable and/or bio-based polymer matrices in composites can allow for recyclability, non-toxicity, and lower environmental impacts [19]. In addition to the use of recyclable matrices, the use of natural fibers allows for greater recyclability of the composite, as well as easier degradation in nature. Therefore, new sustainable industrial applications can be fully realized using the concept of sustainability design for Natural Fiber-Reinforced Polymer Composites (NFRPCs) [32], as shown in Figure 1.

Additionally, unused fiber residues from composite production can also be utilized. These lignocellulosic residues emerge as a promising option to serve as an alternative source of energy, being repurposed as raw materials in a differentiated manner. For instance, they can be used for heat or electricity generation, or for the production of materials and chemicals, following the principles of green chemistry that advocate for the use of renewable raw materials [33]. Another alternative to address the excessive accumulation of plant residues is the utilization of these residues as fillers in polymeric composites and other conventional engineering materials. Within this context, waste management brings benefits to both organizations and society, promoting the commercialization of these materials and generating income through recycling and sustainability practices [16]. This is highly advantageous, considering that Brazil is currently one of the main producers of various natural fibers, thanks to the great diversity of these materials in the country, especially those originating from the Amazon [34–36].

Figure 1. Diagram presenting the sustainable process of natural fiber-reinforced plastic composites. Reprinted with permission from ref. [32]. Copyright 2022, MDPI AG. Licensed under CC BY 4.0.

Considering the context of Brazil in the production of natural lignocellulosic fibers, the country is home to the Amazon region. The Amazon rainforest boasts the highest biodiversity on the planet, housing a wide variety of plant species. This region's biome stands out as the largest preserved forest in the world and the largest biome in Brazil, covering approximately 49.29% of the national territory and about 40% of the South American continent. With over 13,000 tree species, including 2956 endemic species, it is an incredibly rich ecosystem that contributes to a unique and unparalleled diversity in this region [37,38].

Encompassing all this diversity, Brazil demonstrates considerable potential to position itself as a global leader in the production of lignocellulosic fibers for use in composite materials [39]. This is already evident in the production of certain fibers, such as sisal, in which the country is the largest global producer, accounting for 44.7% of the world's total production of this fiber [35]. In addition to sisal, the country also stands out as one of the main global producers of cotton fiber.

Based on the information presented in this introduction, it can be inferred that lignocellulosic natural fibers are an attractive option for use in composite materials as substitutes for synthetic fibers. The objective of this article is to conduct a review on natural fibers from the Amazon region, highlighting some of the various fibers of Brazilian origin, describing their main characteristics and areas of occurrence, as well as providing an overview of the current use of these fibers in engineering applications. In this way, the aim is to promote knowledge and utilization of these Amazon fibers, with the intention of encouraging further studies in this field.

2. Characteristics and Properties of Natural Lignocellulosic Fibers

In this section, the concepts related to lignocellulosic natural fibers will be addressed in order to provide a better understanding to the reader. Definitions, classifications, characteristics, and properties of these fibers will be presented.

Fibers can be classified into natural or artificial fibers. In recent decades, there has been an increase in the use of natural fibers as a replacement for artificial fibers due to advantages such as low cost, low density, and reduced tool wear. Additionally, natural fibers exhibit similar or even superior properties in various applications [40,41]. The extensive utilization of natural fibers as reinforcement in composites has driven research in a wide range of fibers. There are three main types of natural fibers based on their origins: plant fibers, mineral fibers, and animal fibers. However, animal fibers such as hair and silk [42–44] and mineral fibers like asbestos and basalt [45–49] are not as widely used as reinforcement compared to plant fibers [50–52]. On the other hand, several plant fibers have been extensively employed in biocomposites for automotive, maritime, aerospace, and construction applications [53–61]. Plant fibers can also be classified based on the region of the plant from which they are extracted, categorized as bast, fruit, grass, seed, leaf, stalk, and wood fibers [62,63]. Figure 2 illustrates the classification of plant fibers according to their extraction source.

Figure 2. Classification of natural fibers based on the part of the plant of origin.

Natural fibers have satisfactory mechanical performance when used as reinforcement agents in composite materials [64]. Although they have lower tensile strength compared to synthetic fibers, they offer several significant advantages. Additionally, natural fibers are typically rigid and do not fracture during processing, exhibiting specific strength and stiffness comparable to glass fibers []. They also have lower density and competitive Young's modulus or elasticity [65]. The performance of polymer composites reinforced with natural fibers depends on various factors, including chemical composition, cell dimensions, microfibril angle, defects, structure, and the physical and mechanical properties of the fiber, as well as the interaction between the fiber and the polymer [66].

Natural fibers can be considered as natural composites, primarily composed of crystalline cellulose fibrils incorporated in an amorphous lignin matrix. These cellulose fibrils are aligned along the length of the fiber, and the effectiveness of natural fiber as reinforcement is related to the nature of cellulose and its crystallinity [67]. Figure 3, illustrated below, depicts the structure of a natural fiber.

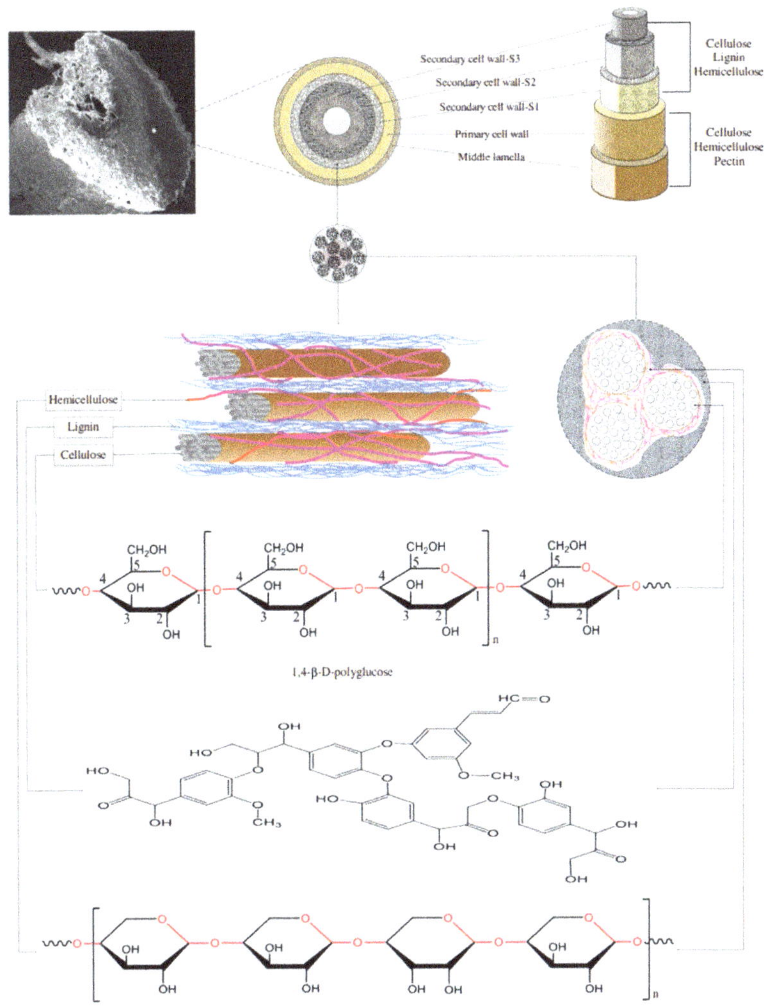

Figure 3. Schematic diagram of the natural fiber cell wall and its molecular structure. Reprinted with permission of ref. [68]. Copyright 2022, Elsevier.

Natural fibers, as shown in Figure 3, are composed of hollow cellulose fibrils interconnected by a matrix of lignin and hemicellulose [69]. The cell wall of a fiber is not uniform and consists of a complex layered structure. Each fibril has a thin primary wall, which is the first layer deposited during cell growth, surrounding a secondary wall. The secondary wall is composed of three layers, with the thick middle layer determining the mechanical properties of the fiber. This middle layer is formed by a series of helically twisted cellulose microfibrils, composed of long cellulose molecules [67].

These fibrils have a diameter ranging from 10 to 30 nm and are composed of 30 to 100 cellulose molecules in an extended chain conformation, providing mechanical strength to the fiber. The amorphous phase in a cell wall consists of hemicellulose, lignin, and in some cases, pectin. Hemicellulose molecules are linked by hydrogen bonds to cellulose and act as a cement matrix between cellulose microfibrils, forming the cellulose–hemicellulose network, which is considered the main structural component of the fiber cell. The hydrophobic network of lignin affects the properties of other networks, acting as a coupling agent and increasing the stiffness of the cellulose/hemicellulose composite [70].

The structure, microfibril angle, cell dimensions, defects, and chemical composition of the fibers are the most important variables determining their properties [71]. Typically, the tensile strength and Young's modulus of the fibers increase with increasing cellulose content. The microfibril angle determines the stiffness of the fibers. Vegetable fibers are more ductile when the microfibrils have a spiral orientation relative to the fiber axis. If the microfibrils are oriented parallel to the fiber axis, the fibers will be rigid, inflexible, and have high tensile strength [67]. Cellulose has a positive relationship with the tensile strength and Young's modulus of natural fibers, while lignin impairs these properties. In addition to cellulose, hemicellulose and wax contents have shown a positive correlation with the Young's modulus, while lignin and pectin reduce the values of the Young's modulus. Moisture absorption is affected by hemicellulose and lignin, where higher contents of these two components result in increased absorption [72,73]. Table 3 presents the relationship of each component in the natural fiber composition to their main characteristics and properties.

Table 3. The influence of chemical composition on the mechanical and physical properties of natural fibers. The (+) symbol represents a positive correlation and (-) symbol represents a negative correlation [73].

Chemical Component of Natural Fibers	Parameters of Mechanical Properties				Parameter of Physical Properties		
	Tensile Strength	Specific Young's Modulus	Failure Strain	Microfibril Angle (MFA)	Diameter	Density	Moisture Gain
Cellulose	+++	++	-	-	+	+++	-
Hemicellulose	-	+++	++	-	+	-	++
Lignin	-	-	+++	+++	-	-	++
Pectin	-	-	++	+++	-	+++	-
Wax	-	++	-	-	-	-	+

The properties of fibers are influenced by the microfibril angle and the arrangement within the cell wall [74]. Although plants are composed of lignocellulosic material, there are significant differences between different types of plants that affect how different plant materials can be used in the manufacturing process [75]. Figure 3 illustrates the position of the chemical components, while Table 4 presents the chemical properties of some natural fibers. The amount of cellulose increases from the primary layer (S1) to the secondary (S2) and tertiary (S3) layers, while the hemicellulose content remains constant in each layer and the lignin content decreases proportionally to cellulose. Hemicellulose binds to cellulose and together they form a network with lignin and pectin, providing adhesive quality. The S2 layer is responsible for the physical and mechanical strength of the fibers. Additionally, better strength properties are achieved with high cellulose content and lower microfibril angle [76,77].

Table 4. Composition of some fibers reported in literature.

Fiber	Cellulose (wt.%)	Lignin (wt.%)	Hemicellulose (wt.%)	Pectin (wt.%)	Wax (wt.%)	Reference
Flax	60–81	2	14–21	2–5	1–2	[78,79]
Hemp	57–78	3–13	11–22	1	0–3	[80–82]
Ramie	68–75	0.8–1.5	13–16	4–5	1–2	[83,84]
Kenaf	45–66	14–20	12–20	0.4–2.7	0.3–3	[40,85]
Guaruman	39–40	10–12	40–41	-	-	[86]
Jute	61–72	12–13	13–20	0.2	-	[87]
Sisal	67–78	8–12	10–14	10	2	[88,89]
Cabuya	48–84	8.3–17	0.5–11	-	2	[90,91]
Abacca	56–66	7–13	21–30	1–3	-	[91,92]
Betelnut	53	7	33	-	0.6	[93,94]
Banana	64–82	5–8.5	19	-	-	[95,96]
Coir	32–50	30–45	0.15–15	1.8–4	-	[97,98]
Bamboo	26–75	10–31	12–16	0.37	-	[99,100]
Bagasse	32–55	19–25	27–32	-	-	[101,102]
Sponge gourd	62	11.2	20	-	-	[102,103]
Rice husk	35–57	21	12–33	-	-	[104]
Wheat Straw	47–63	5.5–18.5	12–32	-	-	[105]
Oat	31–48	16–19	-	-	-	[91]
Napier Grass	45–59	20–24	20–33	-	-	[106]
Curaua	73.6	9.9	5.5	-	-	[107,108]
Henequen	60	8	25	-	2	[109]
Cotton	77–96	2–5	3	0.8–2.5	0.6	[91,110,111]
Nettle	72–84	2.2–7.5	6–12	-	-	[112,113]
Pineapple	49–82	5–31	6–13	-	-	[88,114,115]
Hard Wood	70–74	2.6–5.2	0.5–0.7	-	-	[116]
Soft Wood	40–45	25–34	20–30	-	-	[117,118]
Piassava	28–32	45–48	25–26	-	-	[119–121]
Açai	45–47	31–34	10–15	-	-	[122]
Phormium tenax	67	11	30	-	-	[123]
Sansevieria ehrenberg	80	3.8	10	-	0.1	[124]
Sea Grass	40–77	5–11	14–38	10	-	[91,125]
Isora	71–75	21–23	3.1	-	-	[126,127]
Oil Palm	60	11	-	-	-	[128]
Rachis	43–45	26	28–31	-	-	[129]
Rachilla	42	16	-	-	-	[91,130]
Coconut	26–50	49–53	6–43	-	-	[131,132]
Barley	31–45	14–15	-	-	-	[91]
Pigeon Pea	55	18	-	2.4	-	[133]
Arundo donax L.	75.3	4.3	-	-	-	[134]
Rye	33–50	16–31	16	-	-	[91,135]
Esparto	42–44.5	12–17	25.6–27.5	-	-	[136]
Sabai	43–67	14–18	13–21		-	[137]
Phragmites communis	43–48	10–11	33–36	-	-	[138]
Coniferous	40–45	26–34	-	-	-	[91]
Deciduous	38–49	23–30	-	-	-	[91]
Cytostachys renda	42–49	17–22	19–23	-	-	[139]
Phychosperma macarthurii	39	18.2	19.1	-	-	[91]
Petiole bark	29–48	23–42	-	-	-	[140]
Kudzu	43–78	18–42	1–18	-	-	[141]

Based on the data presented in Table 4, a variation in the composition of different species of natural fibers can be observed, as reported in previous studies in the literature. Although there are sometimes significant differences in fiber composition, one factor that remains constant is the high cellulose content, always higher than the content of other fiber components. However, it is important to note that the cellulose level alone is not the sole determinant of the properties of natural fibers, but rather a series of interconnected factors.

These factors include environmental growth conditions, extraction methods, harvesting timing, and harvesting methods [142]. Plant age, harvesting method, and fiber extraction method are crucial in determining fiber quality and, consequently, the quality of composites produced with these fibers.

Natural fibers with a relatively high cellulose content include jute, pineapple, flax, and ramie. Basically, a high cellulose content and a lower amount of lignin provide high tensile strength, as can be observed in the example of these fibers in Table 5. However, the correlation is not always linear due to various factors that affect tensile strength. Cellulose, especially in its crystalline parts, exerts a significant influence on the tensile strength value, as higher cellulose crystallinity leads to greater fiber strength. The position of lignin in the biomass also affects tensile strength, making it lower, as lignin is located between cellulose and hemicellulose and exhibits lower resistance.

As noted in Table 5, most natural fibers have a maximum density close to 1.60 g/cm^3, making them heavier than water. Although some natural fibers, such as vakka, bamboo, and rachila abaca, among others, are hollow and have low densities in their natural state, they are often densified during processing. Thus, the density of natural fibers is considerably lower than that of synthetic fibers such as glass and carbon fibers.

The low density of natural fibers makes them attractive as reinforcements in engineering applications, as weight is a crucial factor in certain applications. Natural fibers are added to matrices, usually plastics, to improve mechanical performance, such as stiffness and strength, without significantly increasing density [143]. The generally lower impact performance of natural fiber composites compared to synthetic fiber composites tends to limit their use, and addressing this issue is an active area of research. The low density of natural fibers offers more flexibility in composite structure design [,[144]]. Thus, low density plays a significant role in reducing the weight of biocomposites, making them competitive in terms of mechanical properties [145]. In general, the use of natural fibers allows for a reduction in composite weight between 40 and 50%, while maintaining good tensile strength and modulus values [146].

To enhance the properties of natural fibers, interfacial adhesion with composites, and consequently, the properties of the composites, natural fibers undergo treatments, either physical or chemical, that improve their properties [147]. Natural fibers typically exhibit poor hydrophilic properties, resulting in low chemical resistance, inferior mechanical properties, and a porous structure, limiting their engineering applications [148,149]. The hydrophilic nature also reduces the applicability of textile products, especially in transportation and packaging [150]. Therefore, various treatments are applied to the fibers, generally leading to improvements in their properties. Among the physical treatments used to improve the properties of fibers, Corona Discharge, Plasma Treatment, Ultraviolet (UV) Treatment, Fiber Beating, and Heat Treatment are notable. Corona Discharge is particularly effective in surface oxidation of fibers, resulting in a change in the surface energy of cellulose fibers and improving compatibility with hydrophobic matrices [151]. Plasma treatment, on the other hand, has proven to be an efficient option for removing impurities and dust particles from fibers, leading to an enhanced fiber surface. Precise control of gas type, pressure, and concentration is crucial for effective processing [152]. Regarding UV treatment, this relatively new approach stands out for its ability to remove dust particles from the surface of plant fibers. However, it is important to note that certain factors, such as flow and gas type, are not controlled during this treatment [148,153].

Table 5. Comparison of the tensile properties of various natural fibers with synthetic fibers [154].

Fiber	Density (g/cm^3)	Diameter (μm)	Tensile Strength (MPa)	Tensile Modulus (GPa)	Elongation (%)
Jute	1.46	-	393–800	10–30	1.5–1.8
Sisal	1.45	30–300	227–400	9–20	2–14
Pineapple	1.44	20–80	413–1627	345–825	0.8–1
Kenaf	1.40	81	250	4.3	-
Red Banana	-	-	482–567	-	30.6
Nendranbanana	-	-	407–505	-	28.3
Rasthalybanana	-	-	304–388	-	27.8
Morrisbanana	-	-	222–282	-	24.2
Poovanbanana	-	-	144–206	-	21.8
Vakka	0.81	175–230	549	1.5–8.5	3.46
Abaca	0.83	114–130	418–486	12–13.8	-
Alfa	0.89	-	35	22	5.8
Softwood kraft pulp	1.5	-	1000	40	4.4
Viscose	-	-	593	11	11.4
Wool	-	-	120–174	2.3–3.4	25–35
Nettle	-	-	650	38	1.7
Flax	1.50	-	345–1500	27.6–80	1.2–3.2
Hemp	1.48	-	550–900	70	1.6
Banana	1.35	80–250	529–759	8–20	1–3.5
Coir	1.15	100–460	108–252	4–6	15–40
Root	1.15	100–650	157	6.2	3
Palymyrah	1.09	70–1300	180–215	7.4–604	7–15
Date	0.99	-	309	11.3	2.7
Bamboo	0.91	-	503	35–91	1.4
Talipot	0.89	200–700	143–294	9.3–13	3.2–5
Snake Grass	0.88	45–250	279	9.7	2.9
Elephant Grass	0.81	70–400	185	7.4	2.5
Petiole Bark	0.69	250–650	185	15	2.1
Spatha	0.69	150–400	75.6	3.1	6
Rachila	0.65	200–400	61	2.8	8.1
Rachis	0.61	350–408	73	2.5	13.5
Coconut tree leafs heath	-	-	119.8	18	5.5
Sansevieria ehrenbergii	0.88	20–250	50–585	1.5–7.7	2.8–21.7
Sanseveria rifasciata	0.89	83–93	526–598	13.5–15.3	-
Sanseveria cylindrica	0.91	230–280	585–676	0.2–11.2	11–14
Palm	1.03	400–490	377	2.75	13.7
Agave	1.20	126–344	-	-	-
Henequen	1.20	-	430–470	11.1–16.3	3.7–5.9
Bagasse	1.25	200–400	290	11	-
Curaua	1.40	170	158–729	-	5
Sea Grass	1.50	5	453–692	3.1–3.7	13–26.6
Oil Palm	0.70–1.55	150–500	80–248	0.5–3.2	17–25
Piassava	1.4	-	134–143	1.07–4.59	7.8–21.9
PALF	0.80–1.60	20–80	180–1627	1.44–82.5	1.6–14.5
Ramie	1.00–1.55	20–80	400–1000	24.5–128	1.2–4.0
Isora	1.20–1.30	-	500–600	-	5–6
Hivernal	-	12.9 ± 3.3	1111 ± 544	71.7 ± 23.3	1.7 ± 0.6
Alaska	-	15.8 ± 4.1	733 ± 271	49.5 ± 3.2	1.7 ± 0.6
Niagara	-	15.6 ± 2.3	741 ± 400	45.6 ± 16.7	1.7 ± 0.6
Oliver	-	13.7 ± 3.7	899 ± 461	55.5 ± 20.9	1.7 ± 0.8
Cotton	1.60	-	287–597	5.5–12.6	3–10
E—glass	2.55	17	3400	73	3.4
S—glass	2.50	-	4580	85	4.6
Aramid	1.4	11.9	300	124	2.5
HS Carbon	1.82	8.2	2550	200	1.3
Carbon (Std. PAN-based)	1.4	-	4000	230–240	1.4–1.8

During the UV treatment process, the fibers are placed in a chamber for surface oxidation. Additionally, UV treatment increases the polarity of the fiber surface, improving fiber wettability and resulting in higher strength of the NFRPCs [151,152].

Chemical treatments have a significant impact on the mechanical properties of NFRPCs [155–157]. This is due to the presence of hydroxyl groups in cellulose and lignin [158]. Chemical treatment strategies involve the use of reagents or active groups capable of interacting with the structures of natural fibers, removing non-cellulosic materials [159]. Additionally, the hydroxyl groups resulting from chemical treatments can form hydrogen bonds within the cellulose fibers, limiting their movement towards the matrix [78,155]. As a result, chemical modifications activate these groups or introduce new structures that effectively bond with the matrix, promoting good adhesion [160]. Table 6 provides a summary of the main chemical treatments and their primary effects on the natural fibers.

Table 6. Different chemical treatments and their effect on natural fibers [147].

Chemical Treatment	Improvement in Natural Fibers
Alkaline treatment	Adhesion
Silane treatment	Control Fiber Swelling
Acetylation treatment	Moisture absorption
Benzoylation treatment	Thermal stability
Peroxide treatment	Adhesion
Maleated coupling agents	Bonding between fibers and matrix
Sodium chlorite treatment	Moisture absorption
Acrylation and acrylonitrile grafting	Coupling
Isocyanate treatment	Bonding
Oleoyl chloride treatment	Wettability
Stearic acid treatment	Water resistance
Permanganate treatment	Adhesion
Fungal treatment	Remove lignin
Triazine treatment	Adhesion

In addition to characteristics such as recyclability, low density, mechanical properties, and treatments to improve fiber adhesion and properties, another important factor in choosing natural fibers is cost. Natural fibers have significantly lower purchasing costs compared to synthetic fibers [161,162]. However, a more accurate comparison of composite production costs should be performed specifically during the processing of each individual fiber and its use in a particular matrix. Factors such as the lifespan of the component can interfere with large-scale production costs, where logically, synthetic fibers would have an advantage due to their greater durability. However, each application will determine the requirements that need to be evaluated to determine the feasibility of raw material and production costs for NFRPCs. Huda et al. [163] investigated the costs of natural fibers and synthetic fibers for application in the automotive industry. According to the results obtained by the authors, natural fibers require much less energy to produce, which leads to advantages in terms of cost and energy compared to traditional reinforcement fibers such as fiberglass and carbon fiber. This comparison is shown in Table 7.

Table 7. Comparison of cost and energy expenditure for the production of natural and synthetic fibers [163].

Fibers	Cost (USD/Ton)	Energy (GJ/Ton)
Natural fibers	200–1000	4
Glass fiber	1200–1800	30
Carbon fiber	12,500	130

3. Amazon Natural Fibers

The Amazon region is globally recognized for its immense natural and cultural diversity. Located in South America, the Amazon spans eight countries: Brazil, Bolivia, Colombia, Ecuador, Guyana, Peru, Venezuela, and Suriname. However, the majority of its expanse is located in Brazil. The Amazon stands out for harboring the greatest fauna and flora on the planet, representing approximately 20% of the world's biodiversity [164]. Figure 4 illustrates a map of South America, highlighting the Brazilian Amazon region.

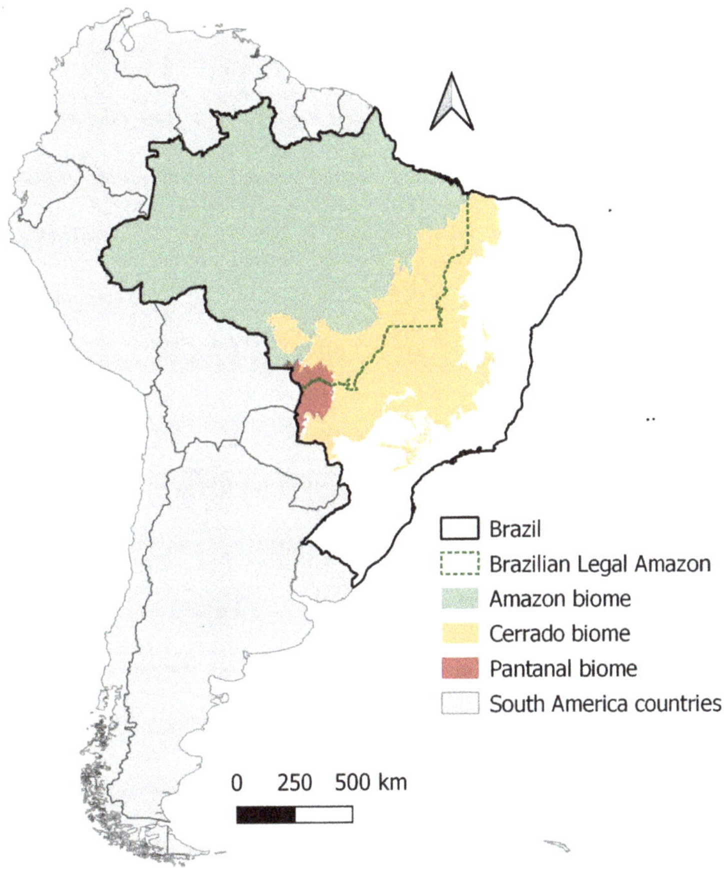

Figure 4. Map of South America, highlighting Brazil and its main biomes: Amazon, colored in green; Cerrado, colored in orange; Pantanal, colored in red; and the Legal Amazon Region, highlighted with green dashed lines. Reprinted with permission from ref. [165]. Copyright 2023, Nature. Licensed under CC BY 4.0.

The Amazon rainforest is rich in plant species that produce high-quality fibers used for a variety of purposes. Among the most well-known fibers are ubim, jute, buriti, piassava, and tucum. Each of these fibers has unique characteristics and interesting properties that make them valuable for different applications.

These natural fibers from the Amazon are widely used by local communities, both for traditional crafts such as baskets, mats, and nets, and for the construction of rural dwellings. Additionally, these fibers generate interest in the global market, being used in the textile industry, paper production, furniture manufacturing, and the creation of sustainable products.

In this section, we will discuss some plant fibers from the Amazon region that have applications in engineering composites. Among the countless plants from which fiber can be extracted, whether from leaves, stems, fruits, or roots, 10 fibers have been selected for this study. Throughout the review, we will address topics such as plant aspects, occurrence regions, fiber extraction characteristics, examples of fiber property characterization, and their application in composite materials, aiming for sustainability in the production of engineering materials with excellent properties.

3.1. Açaí

The açaí palm, scientifically known as *Euterpe oleraceae Mart.*, is a palm tree belonging to the Arecaceae family and is widely cultivated in the Brazilian Amazon region (Figure 5a). This plant is prominently featured due to its economic significance in regional fruit cultivation, particularly in the state of Pará, where the production and commercialization of açaí pulp generate a significant market [166]. The fruit holds considerable nutritional value and is a fundamental part of the diet in the states of Pará and Amapá. Its composition is characterized by high levels of lipids, proteins, fibers, and anthocyanins [167]. The primary cultivation areas for this species are located in the estuary region of the Amazon River, considered its center of origin. In this area, dense and diverse populations inhabit periodically flooded lands due to tides [168].

The mature fruit exhibits a color ranging from purple to almost black, as illustrated in Figure 5b. The pulp can be obtained through the pulping process, which can be performed manually or mechanically. This pulp is consumed fresh or used in the production of various products such as cream, liqueur, jelly, porridge, ice cream, and sweets [169–171].

Figure 5. Açaí (*Euterpe oleraceae Mart.*): (**a**) palm tree; (**b**) açaí fruit; (**c**) fruit after drying; (**d**) fiber extracted from the fruit. Figures (**a**,**b**) reprinted with permission from ref. [172]. Licensed under CC BY-NC 4.0. Figures (**c**,**d**) reprinted from ref. [173].

The açaí fruit has a rounded shape, and approximately 70% of the fruit consists of residues, with only 3% of these residues being composed of lignocellulosic fibers. Although these residues can be used in bioenergy production, it is advisable to separate the fibers from the seeds since burning these components together can result in charcoal with varied chemical composition and thermal behavior, potentially altering the physical and mechanical properties of a specific material in which açaí fiber has been used [174].

The fibers from the açaí mesocarp are by-products of pulp extraction and adhere to the fruit's seed, as shown in Figure 5c,d [174]. These fibers are lignocellulosic in nature and have an elliptical shape with an average thickness of 130 µm and a length of about 18 mm. They have a slightly higher density than water, approximately 1.11 g/cm^3. Generally, açaí fibers are underutilized due to their toxic residues, leading to various environmental issues, and the extracted fiber yield is low [175–177].

Despite açaí being well-known, primarily for consumption, the properties of its fiber are relatively unexplored. Nevertheless, there are studies in the literature that examine the fiber's properties and its application in composite materials. Castro et al. [178] conducted a study on the production of composites using two distinct polymeric matrices, namely, polypropylene (PP) and high-impact polystyrene (HIPS), both derived from recycling processes. In this study, pressed açaí fibers were employed as reinforcement agents in the composites. The manufacturing of the composites took place through the hot compression method, and their properties were subsequently evaluated through tensile, compression, and impact tests. The tensile test results revealed significantly superior performance for the PP/açaí composite compared to the HIPS/açaí composite. Furthermore, the PP/açaí composite demonstrated higher impact resistance when contrasted with the HIPS/açaí composite. Notably, the HIPS/açaí composite exhibited superior properties only in terms of compression resistance, indicating an overall inferior mechanical behavior. This phenomenon is attributed to the low interfacial adhesion present in the HIPS/açaí composite. Thus, this study emphasizes the importance of the choice of polymeric matrix and the quality of the interface between components in determining the mechanical properties of composites.

Bastos et al. [179] conducted a study in which they developed panels made solely of pressed açaí fibers for sustainable application in the acoustic insulation of a classroom. The authors extracted the açaí fibers from the fruit and used a binder to keep the panels fixed during the cold compression process. Acoustic parameter measurements were taken, simulating the effects in a classroom. After simulating and testing the panels, the authors demonstrated that açaí fiber panels are a highly attractive solution for schools in the Northern Region of Brazil, as they combine low cost and good acoustic performance.

Martins et al. [180] analyzed the morphological characteristics and thermal stability of manually extracted açaí fiber from the fruit. Through thermogravimetric analysis conducted in a nitrogen (N_2) atmosphere and an oxidative atmosphere, the authors reported that the fiber exhibited good thermal stability when the test was conducted in a nitrogen atmosphere. However, thermal stability was compromised when heating was carried out in an oxidative atmosphere. The morphological analysis illustrated in Figure 6 showed that the fibers completely cover the açaí seed, and the fiber surface has numerous pores that may facilitate interfacial adhesion in a polymeric matrix.

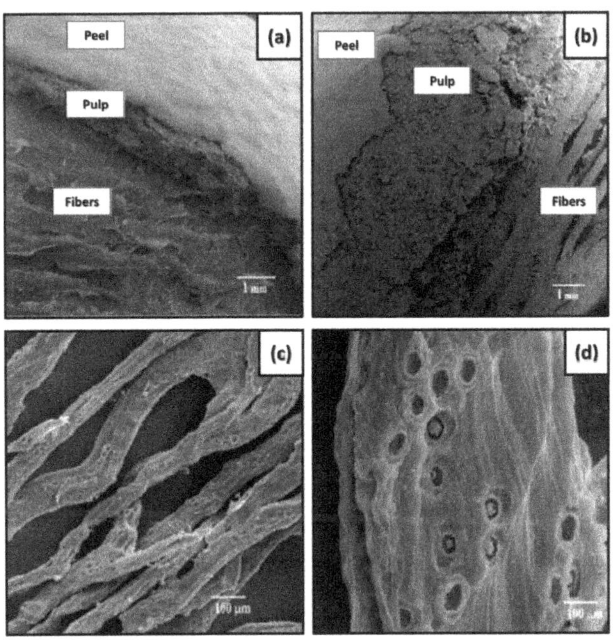

Figure 6. SEM micrographs of different regions of açaí: (**a**) fruit micrograph indicating the regions of peel, pulp, and fibers; (**b**) another region indicating the position of peel, pulp, and fibers; (**c**) micrograph of fine açaí fibers; (**d**) micrograph of a wider açaí fiber, showing the presence of pores and imperfections on the surface. Adapted with permission from ref. [180]. Licensed under CC BY 4.0.

3.2. *Babassu*

The babassu, belonging to the Arecaceae family and the Attalea genus, has the Brazilian species *Attalea speciosa*, a palm tree that can reach up to 20 m in height, as illustrated in Figure 7a. Its fruit contains oleaginous and edible seeds, with a high number of coconuts per cluster (between 150 and 250) and an average of four clusters per palm tree [181]. The fruits, as shown in Figure 7b, are ellipsoidal, measuring 8 to 15 cm in length and 5 to 7 cm in diameter, weighing between 90 and 280 g [182,183]. In Brazil, there are numerous babassu groves distributed from the southern Amazon region to the northeast, with occurrences also in Bolivia for the *Attalea speciosa* species [184].

In the states of Maranhão, Piauí, and Tocantins, the largest expanses of babassu forests in Brazil are found, forming homogeneous, dense, and naturally dark clusters due to the proximity of the large babassu palm trees [185]. This region is recognized as the world's largest concentration of oil-producing plants and the primary source of extractive plant production, known as the "Mata dos Cocais" [186].

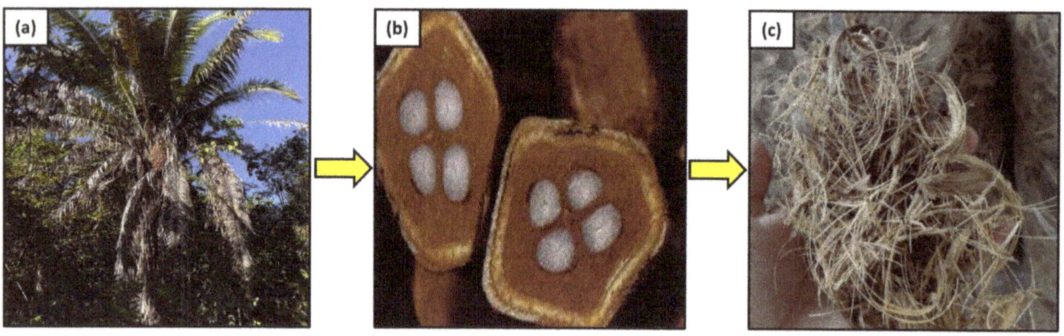

Figure 7. Babassu (*Attalea speciosa Mart ex Spreng.*): (**a**) palm tree; (**b**) babassu fruit; (**c**) fiber extracted from the fruit. Figures reprinted with permission from ref. [187]. Licensed under CC BY-NC 4.0.

The production of babassu nuts is of significant importance for generating income for thousands of families that rely on babassu nut harvesting, with estimates suggesting that over 300,000 women depend on this activity. After maturation, the babassu nut falls to the ground, where it is harvested by workers. It is also collected by climbing the palm tree. When collected, it is transported in straw baskets, typically on the backs of animals. When not possible, the nut cracking is carried out at the base of the palm tree. The fruits are broken in a rudimentary manner, usually by women, using a machete as a cutting tool and a wooden bar for mechanical action. The babassu nut is a fruit that can be fully utilized [188].

Unlike other plants that yield natural fibers, babassu has a distinctive characteristic: the practicality of utilizing almost all parts of the plant. Its trunk is used for structural support in the construction of houses in these regions, and the leaves are used for roofing houses, fences, and in the fabrication of small utensils such as baskets and fans. From the babassu nut, almonds are extracted and used in the production of oil known as "azeite". The mesocarp is used to prepare flour with medicinal properties, and the husk is employed in charcoal production. These products are used in the daily lives of families [189]. It is possible to obtain more than 60 products from babassu, many manufactured from the nut, such as oil, "azeite", milk, for both fresh consumption and industries like food, cleaning materials, personal hygiene, cosmetics, as well as charcoal, fertilizers, and other by-products [190–192].

Babaçu fibers are manually obtained, extracted from both the coconut and the palm tree trunk. However, the most common extraction method is from the coconut. Chaves et al. [193] conducted the extraction of babaçu fibers with the aim of characterizing their properties for potential application in composites. Throughout the study, the authors followed a sequence of steps to obtain the fibers. Initially, the babaçu coconut is left to dry for 48 h for dehydration. Subsequently, the coconut shell is placed in a container for washing, where it remains for a period of 7 days. This phase is crucial to facilitate the extraction of fibers in the subsequent defibration process. Figure 8 illustrates the steps of fiber extraction carried out by the authors.

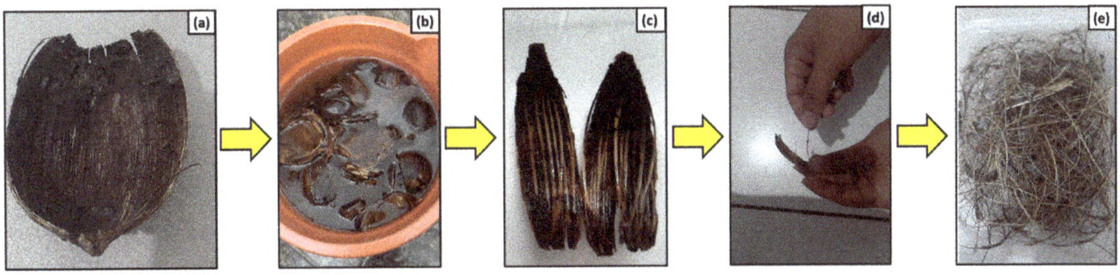

Figure 8. Steps of babassu fiber extraction: (**a**) babassu coconut in drying process for 48 h; (**b**) washing babassu; (**c**) babassu after washing, ready for defibrillation; (**d**) manual defibrillation process; (**e**) extracted babassu fibers. Reprinted with permission from ref. [193]. Licensed under CC BY-NC-ND 4.0.

The babassu palm already has a significant history of applications, as mentioned earlier, including the use of its fibers in composites for engineering applications. In addition to the analysis of the properties of the fiber in isolation [193–195], the investigation of composite materials using this fiber as a reinforcing agent enables its application in various engineering fields while maintaining a sustainable approach to plastic materials.

An example is highlighted in the work of Furtado et al. [196], in which the authors produced low-density polyethylene (LDPE) composites reinforced with babassu fibers at concentrations ranging from 5 to 20 wt.%. They conducted a comprehensive analysis of the mechanical and photodegradative properties, simulating solar and ultraviolet (UV) radiation conditions, aiming for potential use in biodegradable plastic bags. The results obtained by the authors were promising for the composites, revealing good interfacial adhesion between the fiber and the polymeric matrix. Additionally, photodegradation tests indicated degradation under exposure to sunlight and UV, without the release of CO_2 groups into the atmosphere. These results highlight babassu as a viable and highly potential option for the production of biodegradable and cost-effective plastic bags.

Dourado et al. [197] used babassu fibers as reinforcement in cementitious mortars, applying an alkaline treatment to the fibers to enhance the interfacial adhesion between the reinforcement phase and the matrix. Scanning Electron Microscopy images revealed the removal of a superficial layer, reducing the fiber's adhesion to the matrix. Consequently, through tensile tests, the authors observed that the treated fibers exhibited higher tensile strength compared to untreated fibers. Additionally, liquid absorption tests indicated that the treated fibers showed lower total absorption. The incorporation of fibers into the mortar significantly increased the compressive strength of the composites, coupled with a reduction in porosity and water absorption. Overall, the addition of babassu fibers resulted in enhanced strength of the cementitious composites, making this material attractive for the construction industry.

Marinho et al. [198] investigated the biodegradation properties of polyhydroxybutyrate (PHB) composites reinforced with 20 wt.% of babassu fibers and different types of stabilizers. The authors conducted soil biodegradation tests and observed that composites with various stabilizers exhibited a more significant mass variation, indicating a better biodegradation capacity, as illustrated in Figure 9. Figure 9b–d present the results of the modulus of elasticity, tensile strength, and elongation at break properties obtained through tensile tests for both biodegraded and non-biodegraded samples. The results indicated a drastic reduction in the properties of the composites after biodegradation, emphasizing the ease of degradation of these materials in the environment and their utility as eco-friendly materials.

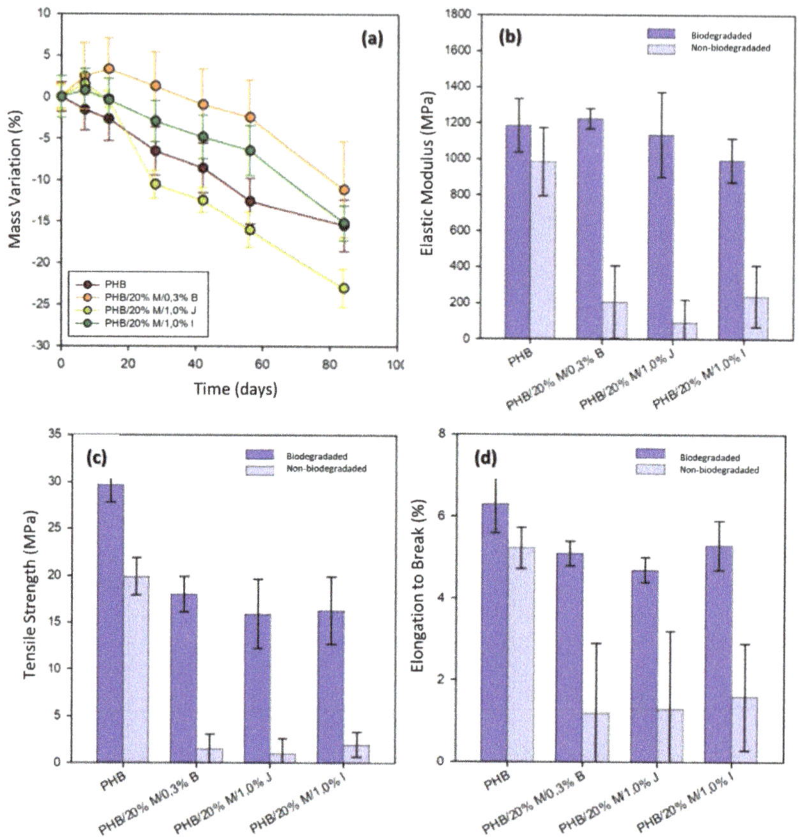

Figure 9. Results of the properties of PHB/babassu composites: (**a**) biodegradation results; (**b**) effect of biodegradation on the elastic modulus of the composites; (**c**) effect of biodegradation on the tensile strength of the composites; (**d**) effect of biodegradation on the elongation at break of the composites. Adapted with permission from ref. [198]. Licensed under CC BY-NC-ND 4.0.

3.3. Buriti

The fiber from the Buriti palm (*Mauritia Flexuosa*) originates from a plant that is widely found in different regions of Brazil, with the main occurrence in the Amazon. Although buriti is also found in the Brazilian central region, as well as in the states of Bahia, Ceará, Maranhão, Minas Gerais, and Piauí, it is found predominantly in regions with a tropical climate, with an annual average temperature of between 26 °C and 30 °C and a rainfall of between 200 mm and 400 mm [199,200].

Buriti palm trees (Figure 10a) display some notable characteristics, reaching significant heights of up to 40 m, with a stem diameter between 50 and 60 cm. The leaves, which are over 15 cm long, remain attached to the stem after death, before eventually falling off. These leaves are widely used to make handicrafts and as roofing material in community dwellings. The fruit of the Buriti palm tree has horny scales with a reddish-brown hue, while the inner pulp displays an orange color. This pulp proves to be versatile, serving as human food, bait for hunting, a source of oil, and with potential medicinal applications. This diversity of uses highlights the ecological and socioeconomic significance of this species in local communities [201].

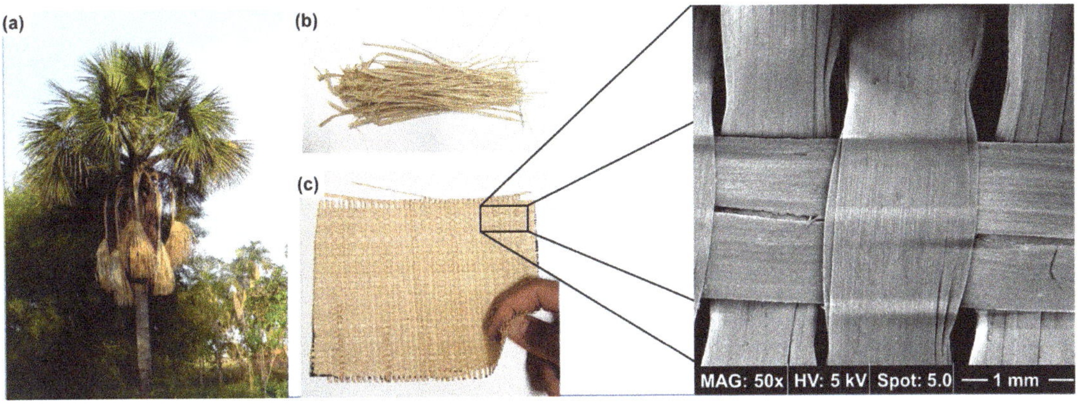

Figure 10. *Buriti (Mauritia flexuosa.)*: (**a**) palm tree; (**b**) bundle of fibers extracted from the leaf; (**c**) fabric produced from the extracted fibers with an inset for viewing the weave from an SEM micrograph. Reprinted with permission from ref. [202]. Copyright 2020, Elsevier. Licensed under CC BY-NC-ND 4.0.

Products derived from buriti have gained high market value, and the practice of destructive harvesting of palm trees is a growing concern. The felling of palm trees to collect fruit in the Peruvian Amazon has been documented as a threat since the late 1980s [203]. Other products, such as young leaves and oil extracted from the buriti palm, are rapidly gaining economic value, presenting potential challenges of overexploitation [204]. Considering that a buriti palm produces, on average, one leaf per month [205], the intensive collection of young leaves, as opposed to the collection of fallen fruit and extraction of mature leaves for subsistence, can result in significant negative impacts on the sustainability of the buriti palm. In addition to its commercial value, buriti plays a vital role in indigenous communities, being one of the most relevant plant species for their subsistence needs, such as food, shelter, building material, and ornaments [206].

The extraction of buriti fiber (Figure 10b,c) is carried out in an artisanal manner, where residents of rural areas climb the trees, remove the green leaves, and cut the fibers, which can be obtained from both the leaves (known as linen) and the petiole [207]. The petiole, or stem of the leaf, can reach up to 3m long and its fibers have a high cellulose content (77.8%) and a low lignin content (24.0%). These fibers, extracted from the epidermis of the petiole, are useful for making mats and curtains [208]. However, a deep understanding of the inherent physical and chemical properties and characteristics of buriti fiber is essential to anticipate the behavior of this material when used as reinforcement in polymer matrix composites. Based on the properties and characteristics exhibited by buriti fibers, several researchers suggest their application as reinforcement in polymer matrix composites [209–212].

The buriti fibers exhibit a comparatively low density ranging from 0.63 to 1.12 g/cm^3, coupled with a moderate tensile strength within the range of 129 to 254 MPa. This characteristic renders them suitable as reinforcement for polymer composites characterized by lower density yet relatively weaker strength [213].

The epoxy composite reinforced with buriti fabric at 10 vol.% showed promising results as a second layer of MAS against high-speed level III NIJ ammunition [214,215]. The work evaluated the ballistic performance of composite fabrics reinforced with synthetic and natural fabrics as a second layer of MAS with the same thickness (10 mm). Figure 11 shows a comparison in terms of indentation depth caused by the impact of a 7.62 mm projectile against a clay witness simulating a human body. However, this buriti fabric composite was unable to maintain its integrity after the impact of the 7.62 mm projectile. In practice, this constitutes a ballistic failure in a multiple-impact test, which requires the second MAS layer to be a whole number after six shots, according to the [215]. As a second MAS layer, the smaller the indentation depth, the better the ballistic performance. A depth greater than 44 mm is considered lethal trauma for a human being.

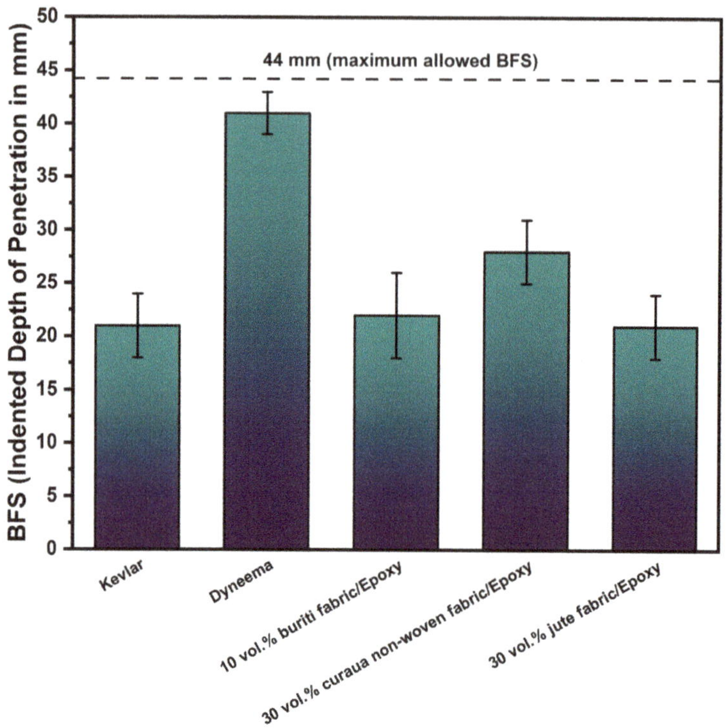

Figure 11. Comparison between the ballistic performance of both synthetic fabrics and natural fiber-reinforced polymer composites as the MAS second layer with the same thickness. Adapted from ref. [208].

According to Demonsthenes et al. [208], the buriti composites reinforced with 10 vol.% epoxy failed to maintain their integrity and are not recommended according to the NIJ Standard [215]. Those reinforced with 20 vol.% buriti showed cracks, indicating that they also do not comply with the NIJ Standard. However, the composites reinforced with 30 vol.% showed partial damage caused by the cloud of fragments in the center of the plate, but did not manifest any open failure, unlike the 10 vol% and 20 vol% composites. This suggests that the 30 vol% proportion can be used as a material for ballistic vests, resisting level III 7.62 mm ammunition.

Cattani and Ramos [216] explored differents treatments for fibers extracted from a buriti palm tree, separated into five different groups: the group of fibers in natura; the fibers boiled by the original community; the fibers boiled in bleach (2–2.5 % sodium hypochlorite, boiled in 500 mL of water and 30 mL of bleach for 15 min); the fibers boiled in fabric softener for 15 min (boiled in 500 mL of water and 30 mL of fabric softener for 15 min); and the fibers boiled in lemon juice (500 mL of water with half a Tahiti lemon for 15 min). The results of the tensile properties of these fibers are shown in Table 8. According to the authors [216], there is no significant difference in the tensile properties associated with the treatment.

Table 8. Results of tensile tests. The values are expressed as average of 20 determinations, including standard deviation and variation coefficient [216].

Set of Samples	Count Number (tex)	Rupture Strength (N)	Toughness (cN/tex)	Elongation (%)	Young Modulus (N/tex)
In natura fibers	223.4 ± 77.7 (34.8%)	64.1 ± 28.0 (43.6%)	28.4 ± 5.5 (19.6%)	8.3 ± 0.5 (6.8%)	6.1 ± 0.8 (13.1%)
Fibers boiled by the origin community	196.9 ± 71.7 (36.4%)	60.4 ± 25.7 (42.6%)	31.1 ± 7.6 (24.7%)	8.3 ± 0.7 (9.2%)	7.2 ± 0.6 (9.5%)
Fibers boiled in bleach	199 ± 81.8 (41.1%)	55.3 ± 28.6 (51.7%)	27.6 ± 7.1 (25.9%)	7.8 ± 0.5 (7.4%)	5.9 ± 1.1 (19.9%)
Fibers boiled in softener	208.7 ± 83.8 (40.1%)	49.9 ± 31.8 (63.7%)	22.0 ± 8.7 (39.7%)	8.6 ± 1.8 (21.5%)	4.9 ± 0.9 (19.7%)
Fibers boiled in lemom juice	194.6 ± 67.2 (34.5%)	58.4 ± 25.9 (44.3%)	29.7 ± 6.4 (21.7%)	8.5 ± 0.5 (6.4%)	5.8 ± 0.7 (13.4%)

Buriti fiber is also used in civil engineering. Castro et al. [217] studied the use of buriti fibers treated with NaOH in mortar in civil construction. The author used a 1:2:0.5 mortar mix, adding 2% fiber treated with 2% NaOH. The mortar underwent a wet curing process for 28 days. The composites reinforced with buruti fibers obtained an 87–96% increase in flexural strength compared to pure mortar. Owing to its versatility and good properties, buriti stands out as a high-potential fiber for use in composite materials, which can be applied to both polymer and cementitious matrices.

3.4. Carnauba

The Carnauba tree, illustated in Figure 12a, is classified as a palm of the Arecaceae family and has xerophytic characteristics. Its scientific name is *Copernícia prunífera* and it originated in Brazil. The term "carnauba" comes from the indigenous language and means "the tree that scratches", an allusion to the 44 thorns distributed along the stem. In addition to its primary name, the plant is also known by variations such as carnaúva, carnaba, carandaúba, and carnaíba. The genus Copernicia comprises approximately 28 species, distributed in regions of India and South America. On the South American continent, species such as Copernícia tectorum (found in Venezuela and Colombia), Copernícia alba (found in Bolivia, Argentina, and Paraguay), and Copernícia prunífera are predominant in Brazil [218].

Figure 12. Carnauba (*Copernícia prunífera*): (**a**) carnauba tree; (**b**) leaf stalks of carnauba tree; (**c**) leaf stalks of carnauba tree; (**d**) carnauba fibers extracted from leaf stalks. Adapted with permission from ref. [219]. Copyright 2020, MDPI AG. Licensed under CC BY 4.0.

It is estimated that the carnauba tree can reach a height between 10 and 15 m, with a productive life expectancy of around 200 years. Demonstrating remarkable resistance, this species adapts effectively to adverse climatic phenomena, such as severe droughts and floods. Its ideal habitat includes floodplains and riverbanks, and the plant thrives at altitudes ranging from 45 m above sea level to around 500 m. Natural propagation takes place predominantly in its native environment, especially in sandy, moist soils. The palm has opaque green leaves, arranged in a spiral around the stem, concentrated in the crown of the plant [220].

Every part of the carnauba palm finds utility; its roots are employed for medicinal purposes, the fruits serve as a significant component in animal nutrition, and the trunk constitutes a valuable source of timber for civil construction. Carnauba wax, derived from the palm, is extensively utilized in the manufacturing of lubricants and anti-corrosive agents. Additionally, the leaves of the carnauba palm find application in various domains, including house roofing, handicraft production, and the extraction of fibers [218,220].

Carnauba fiber exhibits highly promising outcomes for its application as reinforcement in epoxy matrix composites or within a biodegradable polyhydroxybutyrate (PHB) matrix. Noteworthy mechanical properties of carnauba fibers include significant elongation (1.7–2.6%), impressive tensile strength (205–264 MPa), and a substantial Young's modulus (8.2–9.2 GPa) [219].

The quantification of chemical constituents within carnauba fibers constitutes a crucial aspect for comprehending the thermal and mechanical properties of these materials. As elucidated by Monteiro et al. [221], several factors exert direct influence on the properties of natural fibers, including cultivation and storage location, plant age, porosity, and internal defects. Ref. [218] conducted a comprehensive analysis encompassing chemical composition, thermal behavior, and morphological characteristics. The investigation revealed a moisture content of 7.2%, a notably lower value compared to other natural fibers (NLFs). Specifically, the examined carnauba fibers comprised 4.8% wax, 36.9% lignin, 40.9% hemicellulose, and 20.2% cellulose. Notably, when compared to other NLFs, carnauba fibers exhibited a lower percentage of cellulose in their composition. Through SEM micrographs, it is possible to identify surface defects present on the longitudinal surfaces of the fibers. Defects such as cracks, roughness, and surface porosity are identified. Furthermore, XRD analysis of carnauba fibers enabled the calculation of the crystallinity index (86.9%) and microfibril angle (7.48%). These parameters serve as indicative measures of favorable mechanical properties, emphasizing the comprehensive insights derived from the integrated analysis of various characterization techniques [218].

According to Junio et al. [222], the negative factors that influence the quality and properties of carnauba fibers can be minimized by selecting the average fiber diameter. NLFs in general show a tendency to reduce their properties as the average diameter increases, a fact related to the reduction in voids present in fibers with smaller diametrical intervals, giving the smaller fibers more resistance [223].

Junio et al. [222] studied the influence of the average diameter of carnauba fiber. When conducting studies with fibers with an average of 0.765 ± 0.22 mm, the author observed that the average density of the fibers was 1.13 ± 0.22 g/cm^3, and there was a 54.5% reduction in density with increasing fiber diameter compared to the beginning and end of the fiber, as well as an 87.3% reduction in Young's modulus and 83.4% in tensile strength.

Melo et al. [220] investigated the effect of chemical modification on the development of biodegradable composites of polyhydroxybutyrate (PHB) reinforced with carnauba fibers. Enhanced interfacial bonding was evident with alkali, peroxide, potassium permanganate, and acetylation treatments. Composites treated with hydrogen peroxide exhibited superior tensile strength compared to those utilizing untreated fibers or other fiber treatments. Scanning electron microscopy (SEM) observations further disclosed improved fiber–matrix adhesion after this treatment, contributing to heightened mechanical properties. Dynamic mechanical thermal analysis indicated an augmentation in storage modulus at elevated temperatures.

Eduard et al. [224] investigated the effects of incorporating carnauba fibers, at concentrations of 3% and 5%, with lengths of 20, 40, and 60 mm, into mortar. The study scrutinized the behavior of the mortar in both fresh and hardened states. In terms of the fresh state, the addition of fibers resulted in decreased consistency, indicating reduced fluidity and workability. This effect was attributed to the porous surface of carnauba fibers, leading to heightened water absorption and subsequently diminishing the flow capacity of the mortars [225]. Concerning the hardened state, compression tests demonstrated a reduction in compressive strength due to the incorporation of fibers, which increased the void content in the mixture, rendering the mortar less compact. Interestingly, the introduction of 60 mm long fibers contributed to enhanced ductility, acting as a retarder in crack initiation [226]. Additionally, the inclusion of 3% fibers (20, 40 mm) and 5% fibers (20 mm, 40 mm, and 60 mm) resulted in a reduction in flexural strength by 12–20%. However, the incorporation of 3% fiber (60 mm) led to a notable 10% improvement in flexural strength. Pellegrin et al. [227] posit that longer fibers impart superior mechanical performance to mortars, particularly in terms of flexural strength.

3.5. Curaua

Curauá (*Ananas Erectifolius*) is a hydrophilic species native to the Amazon region, from which lignocellulosic fibers are extracted, known for their excellent mechanical properties [228]. In the Amazon, curauá fibers are widely recognized in the Amazon River basin region, particularly in the western part of the state of Pará, where the first commercial plantations of this plant were pioneeringly established [229,230].

Distinctive characteristics of curauá include hard, flat, and erect leaves, with an average length of 1–1.5 m, a width of approximately 40 mm, and a thickness of 5 mm. Each curauá plant exhibits a remarkable leaf production, averaging 50–60 per year, weighing about 150 g each [231]. This yield results in an annual production of 3–9 tons of dry fibers per hectare, notably relying on natural irrigation from rainfall throughout the year [230,232].

Beyond its economic significance, curauá fibers play a crucial role in the traditional practices of indigenous peoples. Indigenous communities use these fibers to craft ropes, hammocks, and fishing lines, requiring materials with high strength and deformability [233]. This application underscores the versatility of curauá fibers, combining remarkable mechanical properties with a sustainable origin. Figure 13 illustrates the curauá plant and the fiber resulting from the extraction of the plant's leaves.

Figure 13. Curauá (*Ananas Erectifolius*): (**a**) curauá plant; (**b**) bundle of manually extracted fibers; (**c**) SEM micrograph of a cross-section of a curauá fiber. Reprinted with permission from ref. [234]. Copyright 2020, Elsevier. Licensed under CC BY-NC-ND 4.0.

Curauá fibers are obtained from the leaves of the plant, which are manually cut. These leaves undergo a process called decortication, in which rudimentary machines equipped with rotating cutting blades remove the mucilage, extracting the fibers. Subsequently, the extracted fibers undergo a mercerization process in tanks, lasting 36 h, followed by washing to remove mucilage residues. Finally, the fibers are dried in an oven at 50 °C for 5 h or in the open air for 2 days before being baled. Each curauá leaf produces between 3 and 8% of dry fibers. The majority of curauá fiber production is still carried out by small farmers in the city of Santarém and the state of Amazonas, characterizing a traditional process without the use of advanced technology and appropriate safety measures [229,231].

Since the early 2000s, the production of curauá fibers has experienced exponential growth, driven by their use in automotive components. Renowned companies such as Volkswagen and Mercedes Benz have played a crucial role in this advancement, adopting curauá fibers as reinforcement in polypropylene matrix composites [235]. This substitution of glass fibers has become particularly notable in the manufacturing of automotive parts, including bumpers, interior panels, trunk lids, and various other components. The success of this transition is evident in the successful integration of curauá fibers in the construction of vehicle parts, notably in Volkswagen's VW Fox and VW Polo models [236,237]. This significant milestone has further propelled research into the application of curauá fibers not only in the automotive sector but also in various areas such as construction [238–240], ballistic armor [241–244], and biodegradable packaging [245–247], among other applications.

An example illustrating the application of curauá fiber is found in the work of Barbalho et al. [248]. In this study, the authors produced and characterized composites of biodegradable polyethylene derived from sugarcane alcohol (B-HDPE), reinforced with curauá in fractions of 0.1, 3, and 5 wt.%, along with 10 wt.% of maleic anhydride (PP-g-MA). This composition results in a fully biodegradable and sustainable composite. From the results obtained through thermal, chemical, and mechanical characterization, the authors revealed that the addition of small amounts of curauá fibers to this specific matrix contributed to a synergistic effect. This synergistic effect led to an increase in the mechanical properties and improvement in the thermal stability of the composite compared to pure B-HDPE. These findings highlight the significant potential of this material for applications in the automotive sector, emphasizing that it is biodegradable and sustainable.

Significant research related to curauá fiber has explored enhancements in its properties and those of the composite through functionalization with carbonaceous materials such as graphene, graphene oxide, and carbon nanotubes. The study conducted by Neto et al. [249] investigated the impact of functionalization using multi-walled carbon nanotubes (MWCNTs) on the thermal and morphological properties of epoxy matrix composites reinforced with curauá fibers. Both the epoxy matrix and the curauá fiber were independently functionalized. The influence of carbon nanotubes in each composite configuration was assessed and compared with the control group, which did not contain MWCNTs. Functionalizing the curauá fiber resulted in a significant increase in mechanical strength compared to the "in natura" fiber, as illustrated in Figure 14a. The addition of MWCNTs promoted a considerable increase in the flexural strength of the composites, as shown in Figure 14b.

When evaluating the composite configurations, it was observed that resin functionalization contributed to an increase in mechanical strength. However, functionalizing the curauá fiber resulted in higher flexural strength values. This observation was supported by scanning electron microscopy (SEM) analyses of the curauá fiber, as presented in Figure 14c,d. In Figure 14c, the fiber without the presence of MWCNTs exhibits a smooth and less rough surface, which may hinder interfacial adhesion with the matrix. In contrast, functionalization allowed the fiber, as depicted in Figure 14d, to obtain a rougher surface due to the MWCNT adherence in the fiber spaces, favoring interlocking between curaua fibers and epoxy matrix.

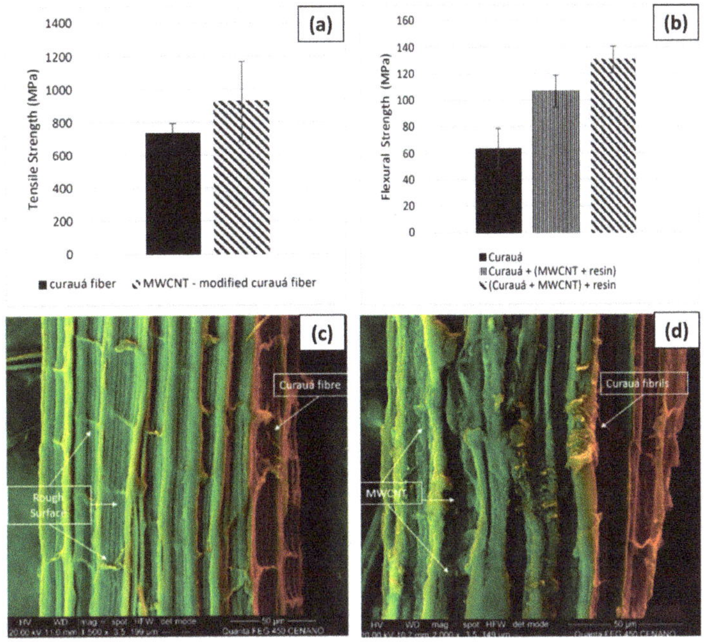

Figure 14. Results of epoxy matrix composites reinforced with curauá fibers and functionalized with MWCNTs: (**a**) tensile test results of the fibers; (**b**) flexural test results of the composites; (**c**) SEM micrograph of the "in natura" fiber; (**d**) SEM micrograph of the fiber functionalized with MWCNTs. Adapted with permission from ref. [249]. Copyright 2023, MDPI AG. Licensed under CC BY 4.0.

3.6. Guaruman

The guaruman plant (*Ischnoshiphon Koern*) is frequently found along the riverbanks in the Amazon region, especially in the Salgado Paeaense area in the state of Pará. Extracted from this region, these plants play a crucial role as raw material for handicrafts [250]. The Amazon is renowned for its vast diversity of native plant species, which play fundamental roles in food, medicine, construction, and fiber production. In the specific context of guaruman, this plant holds significant importance in the culture of riverside caboclos and various indigenous tribes. It is widely used in crafting, particularly in the creation of the famous straw weaving, a highly popular practice in the Para region [250,251].

Guaruman, also known as arumã, belongs to the Marantaceae family and is typically found in flooded várzea areas along riverbanks [252]. Barcarena, in the state of Pará, and more specifically, the Utinga-Açu community, are the main hubs for artisanal production of products made from guaruman fibers. The extraction process involves processing the guaruman stem, resulting in flexible, durable fibers with a distinctive golden hue, as illustrated in Figure 15 [250].

The crafting of handicrafts from these fibers often constitutes the main source of economic sustenance for the surrounding riverside communities. Similar to other non-timber materials used in artisanal production, the goal is always to transform these resources into higher value-added goods [250,252]. Research focused on the development of new products from guaruman not only contributes to the technological innovation of the country but also addresses the specific needs of the regional population that relies on the trade of these fibers.

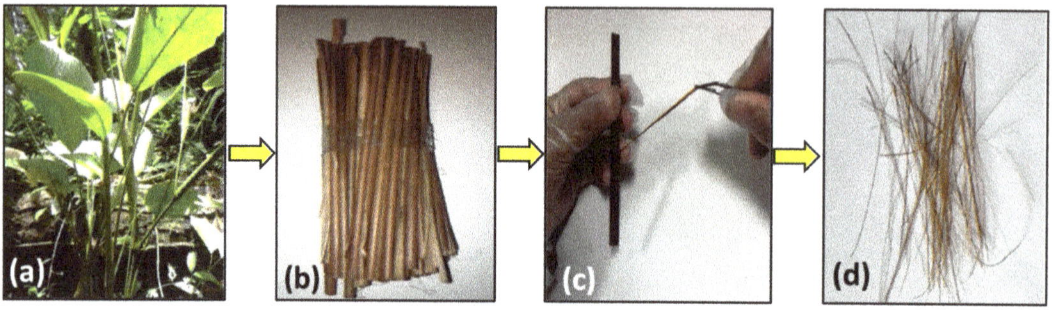

Figure 15. Guaruman (*Ischnoshiphon Koern*): (**a**) guaruman plant; (**b**) as-received, mechanically divided splints from the stalk; (**c**) manual separation of fibers from the splint; (**d**) bunch of the final isolated fibers (**d**). Adapted with permission from ref. [253]. Copyright 2020, Elsevier. Licensed under CC BY-NC-ND 4.0.

Guaruman fiber is underexplored concerning the assessment of its properties, encompassing both the individual characteristics and properties of the fiber, as well as its application as reinforcement in composites. The lack of dissemination, coupled with the potential difficulty in obtaining the plant, may constrain studies on this fiber. However, although rare, some works have been identified in the literature. Reis et al. [253] conducted an analysis of the characteristics and properties of guaruman fibers, comparing them with commonly studied natural fibers. The results, presented in Table 9, highlight the favorable properties of guaruman compared to other natural fibers, such as low density and a reduced microfibril angle (MFA).

Table 9. Comparison of the mechanical properties and microfibril angle of guaruman fiber with other fibers known in the literature [253].

Fiber	Density (g/cm^3)	Tensile Strength (MPa)	Young Modulus (GPa)	MFA (°)
Guaruman	0.57	614	21	7.8
Jute	1.45	597	20	8.0
Ramie	1.50	685	44	6.2
Hemp	1.45	539	35	7.5
Sisal	1.38	478	19	20.0
PALF	1.44	180	59	11.5
Coir	1.52	135	5	51.0

Continuing the study of guaruman fibers, Reis et al. [254] conducted research in which they employed guaruman fibers as reinforcement in epoxy matrix composites for use in multi-layer ballistic armor. To achieve this, they developed a prototype ballistic vest with a front layer made of aluminum oxide ceramic (Al_2O_3), the second layer composed of a composite using 30 vol.% guaruman fiber, and the final layer made of Kevlar. Ballistic tests were conducted using 7.62 × 51 mm caliber ammunition, and the prototype's performance was assessed by measuring the indentation in the clay positioned behind it. The indentation in the clay, known as Backface Signature (BFS), should be less than 44 mm. The results of the ballistic tests, depicted in Figure 16, demonstrated that the prototype met the standard, with BFS below 44 mm. To illustrate, the authors compared the results with composites produced under similar conditions but using PALF fibers, achieving similar outcomes, meeting ballistic test standards, and showcasing the potential of guaruman fiber for military applications.

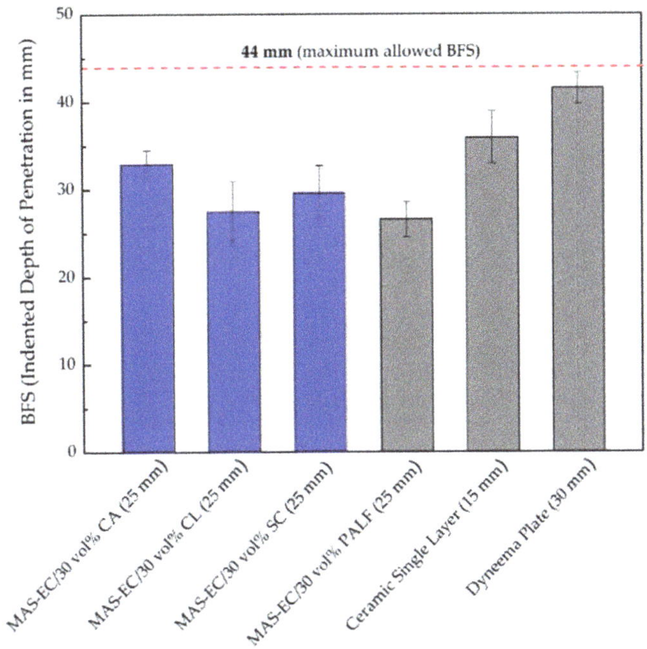

Figure 16. Results of BFS of composites produced with guaruman fibers after ballistic testing using 7.62 × 51 mm ammunition. Reprinted with permission from ref. [254].

A more recent example is the work of Azevedo et al. [86], in which the authors conducted a preliminary study on the use of guaruman fiber as reinforcement in cementitious matrix composites. They used a Portland cement matrix of type CPIII, widely used in Brazilian construction, and added guaruman fibers in fractions of 2.5, 5, and 7.5 wt.% Additionally, they investigated two different conditions: untreated fibers and fibers treated with alkali using NaOH. Through the characterization of the composites, the authors observed that the addition of fibers reduced the density of the cementitious composite as the fiber content increased. However, the alkali-treated fibers showed an even greater reduction in density and, consequently, in liquid absorption.

Guaruman fiber has good mechanical properties and great potential to be explored in composites and sustainable applications, both in the field of materials science and engineering and in other areas of knowledge. However, the lack of knowledge about this fiber compared to other natural fibers, such as jute and sisal, for example, limits its utilization.

3.7. Periquiteira

The Periquiteira (*Cochlospermum orinocense*), also known as tree cotton, Envira-Branca, or Buxixão, is a plant from the Bixaceae family. This plant is characterized by being a medium-sized tree, ranging from 12 to 28 m in height, with a straight cylindrical trunk that can measure 40 to 75 cm and remain unbranched for up to half of the tree's height. Its bark is whitish, with vertical fissures and fiber detachment, with 60 cm of cataphylls [255,256].

The wood has a coarse texture, a straight grain, is tasteless but slightly fragrant when fresh, lightweight, smooth, with low resistance to decay and attack by wood-eating insects, and it grows best in a sunny position. It is a fast-growing tree [256,257].

The occurrence of the Periquiteira is in Brazil, specifically in the Amazon Rainforest, but it also extends to countries in South America such as Peru, Colombia, Venezuela, and the Guianas. It mainly grows in more open areas of advanced secondary growth, in upland areas not subject to periodic flooding, plain areas, and highlands, usually on dry clay at altitudes of up to 450 m in Peru [256]. In Brazil, the Periquiteira is found in the states of Roraima, Rondônia, Amapá, Pará, Amazonas, Acre, Maranhão, and Mato Grosso [257].

Little is known about this plant, as there is still no precise information about its edible and/or medicinal uses. So far, it is known that the fiber is extracted from the inner bark of the fruit and is commonly used in rope manufacturing. The only studies found in the literature on the properties of periquiteira fibers are the works of [258,259]. The study conducted by Silva et al. [258] was the first to analyze the properties of periquiteira fibers, including the composition of lignocellulosic components and mechanical properties through tensile tests. The results of this study are presented in Table 10.

Table 10. Properties of periquiteira fiber [258].

Fiber	Hemicellulose (%)	Lignin (%)	Cellulose (%)	Tensile Strength (MPa)	Young Modulus (GPa)	Elongation (%)
Periquiteira	-	12.03	60.15	83.93–168.19	4.04–7.09	0.19–0.81

Based on the original results obtained by Silva et al. [258], the recently published paper by Pinheiro et al. [259] continued to map the knowledge about periquiteira fiber, bringing other fiber information such as density, crystallinity, microfibril angle, thermal stability, and tensile properties as a function of fiber diameter. As illustrated in Figure 17, the authors manually extracted the fiber from the plant's bark.

Figure 17. Periquiteira (*Cochlospermum orinocense*): (**a**) periquiteira plant; (**b**) bark extraction site; (**c**) bark of the plant; (**d**) periquiteira fibers. Adapted with permission from ref. [259]. Copyright 2023, MDPI AG. Licensed under CC BY 4.0.

Through the analysis, the authors identified the fiber diameter distribution, as well as characteristics such as high crystallinity (70.49%), low MFA (7.39°), thermal analysis results similar to other lignocellulosic fibers, and tensile strength values ranging from 100 to 255 MPa, where the smaller the fiber diameter range, the higher the mechanical strength.

In conclusion, the periquiteira plant holds great potential for application, both in the biomedical and botanical field due to its yet unknown properties, as well as in engineering and sustainability, thanks to the properties extracted from it and their possible applications. It is highly likely that in the coming years, new studies on this fiber will emerge, pushing the boundaries of knowledge about this plant and its fibers even further.

3.8. *Piassava*

Piaçava is a palm tree native to Brazil, belonging to the Arecaceae family. Its popular name comes from the indigenous Tupi language, meaning "Fibrous Plant". The different species of Piaçava are found mainly in the states of Acre (*Aphandria natalia*), Figure 18a and Bahia (*Attalea funifera*), and Amazonas (*Leopoldinia piassaba*). This palm is capable of growing in low-fertility soils that are unsuitable for many crops [260].

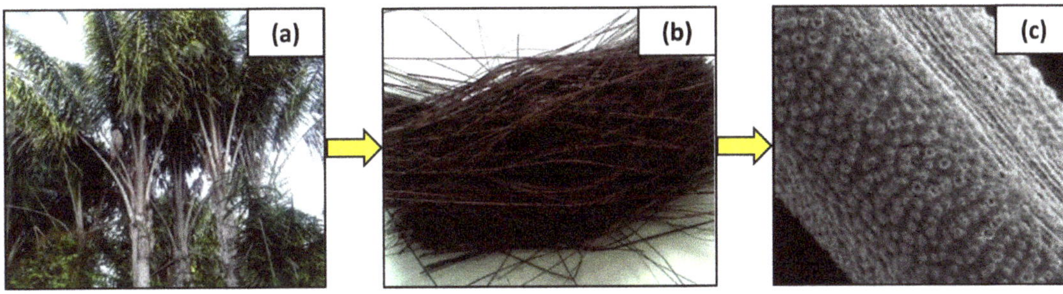

Figure 18. Piassava (*Attalea funifera*): (**a**) piassava palm tree; (**b**) piassava fibers extracted from leaf; (**c**) SEM micrograph of piassava fiber surface. Adapted with permission from ref. [261]. Copyright 2018, Scielo. Licensed under CC BY 4.0.

Among the species of Piassava, Bahia is the largest fiber producer in the country, representing 95% of the national production, followed by the Piassava from Amazonas and Acre. These species of Piassava differ in the characteristics of their fibers [119,262]. The fiber from Bahia is the most commercialized due to its long, rigid, and waterproof fibers that maintain their elasticity even when wet. On the other hand, the fibers from Amazonas are softer, more flexible, and elastic. These fibers are commonly used in the manufacturing of brushes, brooms, ropes, crafts, and also in the composition of rustic coverings [260].

The extraction of Piassava fibers is carried out through extractivism, with different systems in the producing states. In Bahia, there are associations of collectors in the communities, which generates income and, at the same time, preserves the ecosystem in the Atlantic Forest [262,263]. In Amazonas, on the other hand, extractivism occurs through aviation, where the boss provides advanced food and goods in exchange for the services of the collectors. This type of extractivism does not benefit the collectors, often leaving them in a situation similar to slavery [264].

The collection of Piassava plant fibers is conditioned by the water level, which means that during dry periods, collectors are isolated without the opportunity to receive supplies and goods since the rivers are not navigable. Piassavas can be classified in two ways: by height and by the presence or absence of previous cutting [265]. The cutting method varies according to the size of the palm tree. As for height, Piassavas are classified as follows: new Piassava (up to 2 m), garrote (2 to 4 m), garrotão (4 to 6 m), and giant or old (over 6 m). Regarding the type of cutting, they can be non-extractable (poor or fiberless), mamaipoca (already cut, ready for recutting), or virgin (never cut) [266]. Piassava fibers originate from the base of the palm leaves and are collected manually. In this process, they are untangled, arranged, cut, and then tied together for commercialization, generating the fibers as illustrated in Figure 18b,c [267–269].

Piassava fibers have been widely used due to the high production of this palm tree, which can yield around 8 to 10 kg of fibers per tree. These fibers have various applications, ranging from crafts, utensils, and ropes, to engineering, as reinforcement in composites. As a result, numerous research studies are conducted annually using this fiber.

An example of the application of piassava fibers as reinforcement in composites is a study conducted by Carvalho et al. [270], in which a preliminary investigation was carried out to assess the effectiveness of piassava fibers as a reinforcement agent in polyurethane matrix plates for flooring. In the study, the piassava fibers were crushed and added to the composite in powdered form, at volume fractions of 10, 20, and 30 vol.%. The authors conducted abrasion tests on the composites, following the NBR 14050 standard [271], and observed that the composite with 30 vol.% fiber content showed potential for application, although it did not fully meet the requirements of the standard. The authors mentioned the need to investigate other parameters to achieve the specified requirements.

Silva et al. [272] evaluated the performance of polyester matrix composites reinforced with different fibers in impact and bending tests. For comparison, they used raffia, jute, mallow, piassava, and sisal fibers as reinforcements in similar compositions. The results of the Charpy test are illustrated in Figure 19.

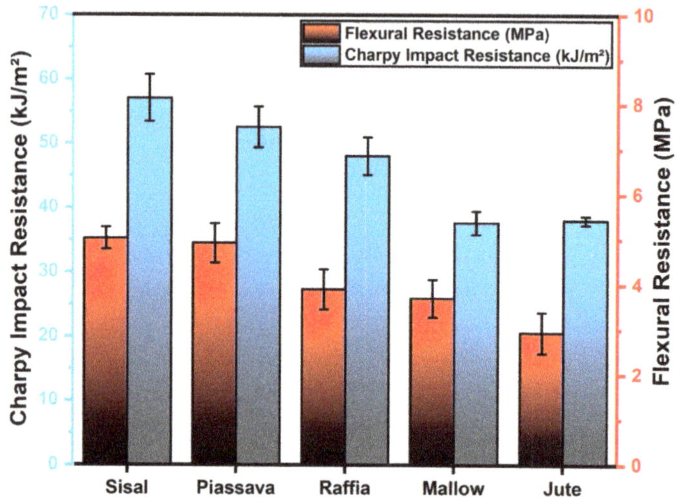

Figure 19. Comparison of impact resistance and flexural strength results as a function of the type of fiber used in polyester matrix composites. Graph produced with data results from ref. [272].

Based on the results obtained, the authors observed that piassava fibers have high resistance compared to other fibers such as mallow, raffia, and jute, although their performance was slightly lower than the composite with sisal fibers. Among all natural fibers, sisal is one of the most resistant, if not the most resistant natural fiber among the species [273,274].

As another example, Nunes et al. [275] produced composites from recycled polypropylene (PPr), using piassava fibers as a reinforcing agent. To improve the interfacial adhesion of these composites, the authors added MAPP to the material at a fraction of 10 wt.%. The content of piassava fibers in the composite varied between 0, 10, 20, and 30 wt.%. From the results of the characterization of the mechanical properties of the composites, the authors observed that the addition of piassava fibers in the fractions of 20 and 30 wt.% considerably increased the flexural strength of the composites, in which the group without the presence of fiber presented a value of 32.7 MPa, while the composites with 20 and 30 wt.% of piassava presented values of 41.7 and 45.0 MPa, respectively. This behavior was repeated in the tensile and Shore D hardness tests, in which the composites with 20 and 30% piassava fibers showed superior performance; however, in the impact test, the sample without fiber showed superior performance to all the composites. The piassava fiber allowed the distribution of stresses, increasing the mechanical strength of the PPr, but the presence of MAPP allowed an increase in interfacial adhesion, resulting in superior properties.

The application of piassava fiber has many benefits. Its mechanical strength and durability make it ideal for making utensils and even for reinforcing construction materials and engineering composites. In addition, its use contributes to the preservation of the environment, as piassava is a native plant and its sustainable extraction promotes the socioeconomic development of local communities. Given its unique properties and environmental benefits, piassava fiber certainly has a promising role to play in a wide range of industries.

3.9. Tucum

The tucum, scientifically named *Astrocaryum vulgare* (Figure 20a), is a palm tree typical of the Amazon region. This plant has several scientific names, such as *Astrocaryum chambira Burret* and *Astrocaryum aculeatum G.* The fruits of the tucum are called tucumã and are widely used in local cuisine. In addition, the palm has a thorny trunk and can be used as a living fence to protect crops in short-cycle forestry of pioneer species. The purpose of this fence is to protect the seedlings from herbivory by animals [276].

Figure 20. Tucum: (**a**) tucum palm tree; (**b**) tucum fruits; (**c**) tucum leaves exhibiting a thorny characteristic to ward off predators; (**d**) tucum fibers extracted from the leaf. Adapted from ref. [277].

Their fruits (Figure 20b) are up to 6 cm long and vary between 4 and 5 cm in diameter, with a rounded shape, a greenish color, and a sour taste. When ripe, the fruit is black in color and tastes sweet [278].

Tucum fibers are taken from the palm's leaves (Figure 20c), which are very resistant and have the following characteristics: sheath and petiole covered in flat, yellowish spines; sheath 1.1 m long; petiole 2.6 m long; rachis 4.8 m long; 160 spines per side, linear or linear-lanceolate, irregularly arranged and arranged in different planes; with small spines on the margins, midribs subterminal; midribs 1.51–1.63 m long and 4–4.5 cm wide [278,279].

Tucum fiber, illustrated in Figure 20d, is obtained after the palm leaves have been removed and dried. This fiber contains a considerable amount of cellulose, over 80%, and has been an important source of income for communities in the region, being widely used in handicrafts. For several indigenous communities in the northwestern Amazon region, tucum fiber is of significant value. The fibers obtained from the unexpanded leaves are used to make a wide variety of products, such as hammocks, bags, and fishing nets [203,280–283]. Harvesting and processing these fibers is part of Aboriginal traditions and represents important moments of social interaction [284].

In recent years, products made from tucum fibers have become very popular with tourists and in craft shops. The tucum palm has become an important cash crop for indigenous families. However, frequent extraction, often carried out in a destructive manner, has depleted the natural populations of tucum in some areas of the Amazon [282,285].

The application of tucum fiber in engineering composites has been attracting a great deal of interest from researchers in recent years, especially in Brazil. This interest has intensified due to the potential of this fiber to improve the properties of composites and open up new application possibilities in industry. Various studies have been carried out to better understand the characteristics and behavior of this fiber in composites, with the aim of optimizing its use and exploiting its full potential. Cunha et al. [192] compared the effect of tucum fibers (*Astrocaryum chambira Burret*) and mallow fibers in polyurethane (PU) matrix composites. The authors produced composites only reinforced with tucum fibers (TPU) and only with mallow fibers (MPU), in fractions of 30, 50, and 70 wt.%, and the properties of the fibers and composites were analyzed. XRD analysis showed that the crystallinity of the tucum fiber was higher than that of the mallow fiber, with values of 79.34 and 68.56%, respectively. The results of the thermogravimetric analysis (TGA) indicated that the tucum fiber has slightly higher thermal stability. In the mechanical analysis, using the flexural test, the composites reinforced with tucum fibers showed higher values than the composites reinforced with mallow fibers, in all compositions. Based on the water absorption results, the TPU composites showed much higher absorption values due to greater interfacial adhesion.

Oliveira et al. [286] investigated the mechanical and ballistic properties of epoxy matrix composites reinforced with tucum fibers from the genus *Astrocaryum vulgare*. The authors produced composites by the cold compression method, using 20 and 40 vol.% fractions of tucum fibers. The results obtained by the authors are shown in Figure 21.

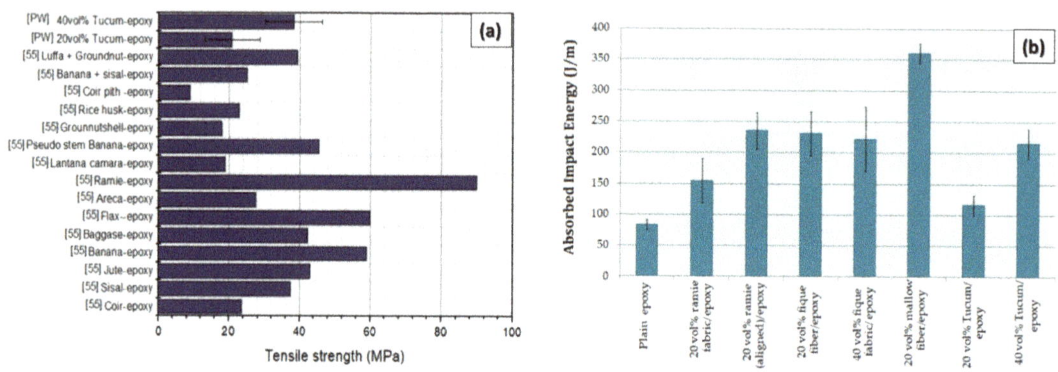

Figure 21. Results of the epoxy/tucum composites: (**a**) comparison of the tensile strength results of the epoxy/tucum composite with the literature; (**b**) comparison of the Izod impact resistance results. Adapted from ref. [286]. Copyright 2020, MDPI AG. Licensed under CC BY 4.0.

The results of the mechanical characterization of the epoxy/tucum composites were not good compared to some composites. As illustrated in Figure 21a, the tucum-reinforced composites showed superior performance to composites reported by other authors, reinforced with coir, ground nut shell, and lantana camara fibers. However, performance was much lower than composites reinforced with ramie, banana, and flax fibers, showing intermediate properties. In the Izod impact results, shown in Figure 21b, the composite reinforced with 20 vol.% tucum showed similar performance to plain epoxy, with little improvement in impact resistance, making the mechanical performance inferior to that of the other fibers used in the comparison, present in the Figure. On the other hand, the composite with 40 vol.% performed similarly to composites reinforced with fique fibers, both in the form of fibers, fabric, and ramie fibers, although the result was much lower than a composite reinforced with 20 vol.% mallow. The results of the ballistic tests carried out with 0.22 ammunition are inferior to plain epoxy. In general, the addition of natural fibers to epoxy in fractions ranging from 10 to 40 vol.% promotes an increase in ballistic resistance, as observed in studies with similar tests and compositions, but with epoxy/hemp [287] and epoxy/kenaf [288] composites. Even though the ballistic performance is lower, the authors report the importance of using natural fibers in composites for ballistic applications, where cost is a fundamental factor. The cost of natural fibers is significantly lower than synthetic fibers, and certain natural fibers have the potential for ballistic applications, aggregating sustainability and low production costs.

Another interesting example is the study carried out by Kieling et al. [23], in which the researchers produced composites using tucum powder, extracted from the seed of the genus *Astrocaryum aculeatum*. Recycled polypropylene was used as the polymer matrix. The composites were processed with different fractions of tucum (0, 10, 20, 30, 40, and 50 wt.%) and their properties were analyzed. During the tensile, flexural, and impact tests, the addition of tucum as a reinforcing agent resulted in a reduction in the mechanical performance of the composites. However, in the compression test, the tucum-reinforced composites showed a considerable improvement as the load increased. In relation to the water absorption test, the composites with 40 and 50 wt.% tucum showed a greater capacity to absorb liquid. This also resulted in greater resistance to the flammability test, where these two groups showed a lower flame propagation speed. Therefore, even with lower mechanical properties, tucum proved to be effective as a flame retardant additive in polypropylene.

Tucum fiber has significant potential to be explored as a reinforcement in composites. Although this fiber does not have excellent mechanical properties like sisal, curauá, or piassava, it can be used in studies in which different surface treatments can be applied to the fiber to improve interfacial adhesion with the composites and, consequently, their final properties. With the growing popularity of fiber in the engineering field, it is expected that new studies will emerge to further explore its potential.

3.10. Ubim

Ubim is a palm from the Arecaceae family, also known by its scientific name *Geonoma baculífera* [289]. The word ubim comes from the indigenous language, specifically the Tupi u'bi. The palm is also known by other names, such as *Geonoma estevaniana Burret, Gynestum baculiferum Poit., Geonoma acutiflora Mart* [290].

The Arecaceae family includes the genus *geonoma*, which is made up of small palms that generally grow in the understory. This genus is one of the largest in the Americas and is home to 15 species that are widely distributed across the continent, especially in tropical regions [291]. The species of *Geonoma* are commonly found in areas with high levels of rainfall, and are one of the most prevalent plant species in these environments. Palms of the genus *Geonoma* have a preference for riparian forest vegetation that occurs along watercourses, as well as open vegetation [292].

Ubim, illustrated in Figure 22, is a small cespitose palm with multiple, smooth stems and elongated, unbranched fibers. Its height varies between 1 and 4 m, with a diameter of 1 to 3 cm. The stem can be erect or partially creeping, and the plant has seven to twelve leaves, sparsely branched inflorescences, and globose or ovoid fruits. This species is typically found in the understory of forests with high rainfall, riparian forests, floodplains and igapós. The ubim is adapted to humid environments, is considered shade-tolerant, and generally grows in places with low incidence of direct light [289].

Figure 22. Ubim (*Geonoma baculífera*): (**a**) ubim plant; (**b**) ubim stem; (**c**) ubim fibers extracted from the stem. Adapted from ref. [293].

The occurrence of the ubim covers Central and South America, with records in the Guianas, Peru, Bolivia, and Venezuela. In Brazil, this plant can be found in the states of Amazonas, Acre, Amapá, Pará, Maranhão, and Piauí [294,295]. This species is widely used in the Amazon by extractivist communities who depend on the sustainable exploitation of various native species to meet their needs for construction materials for rural buildings.

The leaves of the ubim, when intertwined along a stick, form ubim "cloths", which are used as a covering in the constructions of these communities. This is a traditional practice, especially among those who live close to the border between Brazil and Bolivia. In Bolivia, ubim is commercially exploited, which indicates that the species may also have commercial potential on the Brazilian side [296].

In addition to its use in construction, ubim also has ecological importance for some indigenous and riverside communities. Ubim fibers (Figure 22b,c) are used to make baskets, mats, and other handicrafts. In addition, this plant has potential for ornamental purposes in gardens and interiors.

Ubim is widely used in the Amazon and Acre regions; however, the plant's potential application depends on the development of the market, which currently lacks a regular and abundant supply of the product. It also faces the apparent lack of awareness of its existence on the part of consumers in the city of Rio Branco, in the state of Acre, which represents the largest potential market for this product in the region [296]. In addition to the potential of ubim daughters, ubim fiber has yet to be widely applied as a reinforcing material in industrialized products and engineering applications. Unlike better-known fibers such as sisal, bamboo, curaua, coconut, jute, and others, there are practically no scientific reports studying the properties of ubim fiber and its application.

Studies into the properties of ubim fiber and its applications are still limited. Recently, a pioneering study by Marchi et al. [297] evaluated the characteristics and properties of this fiber. The researchers identified important characteristics for evaluating the fiber, such as crystallinity, microfibril angle, diameter, density, and cellulose content. The results of this study are shown in Table 11.

Table 11. Physical properties of ubim fiber [297].

Fiber	Diameter (μm)	Density (g/cm^3)	Cellulose (%)	Crystallinity (%)	Microfibril Angle (°)
Ubim	510–620	0.44–0.97	66	63–83	7.46

The authors' results show that the fiber has good characteristics such as high crystallinity, low microfibril angle, and low density. These characteristics suggest that the fiber has potential for application in engineering composites, combining low weight and possibly good mechanical properties.

Carrying on from their previous work, Marchi et al. [298] investigated the thermal, chemical, and ballistic properties of epoxy matrix composites reinforced with ubim fibers. Composites were produced with fiber fractions of 10, 20, and 30 vol.%, in which the properties were compared with pure epoxy. Fourier transform infrared spectroscopy (FTIR), thermogravimetric analysis (TGA), differential scanning calorimetry (DSC), and ballistic tests were carried out. The FTIR analysis showed peaks characteristic of other natural lignocellulosic fibers. The thermal analyses showed that the fibers slightly improved the thermal stability of the composites, even though the onset degradation temperatures (T_{onset}) and maximum degradation temperatures (T_{max}) occurred slightly earlier in the case of the composites. The data obtained from the ballistic test are shown in Table 12.

Table 12. Ballistic parameters obtained in ballistic test with 7.62 mm ammunition [298].

Samples	v_i (m/s)	v_r (m/s)	E_{abs} (J)	v_L (m/s)
Epoxy	812.58 ± 3.84	786.28 ± 5.93	203.82 ± 18.92	204.82 ± 9.36
10 vol.%	833.75 ± 12.81	810.28 ± 15.39	187.03 ± 25.99	195.98 ± 13.57
20 vol.%	808.42 ± 9.84	786.56 ± 10.81	169.07 ± 33.73	185.98 ± 18.00
30 vol.%	815.28 ± 8.70	794.85 ± 10.54	159.42 ± 26.32	180.79 ± 14.93

Based on the results of the ballistic test, the authors pointed out that the energy absorbed (E_{abs}) was reduced as a result of the increase in ubim fibers, but the velocity limit

(v_L) was also reduced. In general, the composites with fibers absorbed less energy, but reduced the velocity of the projectile, compared to pure epoxy, which points to a potential application of composites reinforced with ubim fibers for ballistic applications.

In summary, ubim fibers have potential application in engineering composites. Based on the promising results observed by Marchi et al. [298], it would be possible to expand the applications of ubim fibers by investigating other fiber properties, in addition to their application in composites with different matrices.

4. Final Remarks and Conclusions

The abundance of biodiversity in the Amazon region is reflected in the wide variety of natural fibers found in this area. Additionally, the local communities possess ancestral knowledge of the sustainable use of these resources, further underscoring the importance of preserving and valuing these fibers. The preservation of not only the Amazon rainforest but also the economic and social development of the communities that rely on these resources are intrinsically linked to this preservation.

It is important to highlight that there are several plants in the Amazon region whose fibers were not addressed in this study but also have promising applications in engineering composites. Some examples of these fibers include ouricuri [299], embira [300], taboa [301], titica vine [302,303], tururi [304–306], inajá [307], ubuçu [308,309], jarina [310], and pataua [311,312], among others. These fibers have significant potential for use in various industrial applications, replacing synthetic reinforcements with natural materials and promoting sustainability.

The development of new research is crucial to expand our knowledge about the properties of these natural fibers and further encourage their use in the industry. Advanced research is being conducted daily, covering topics such as hybrid composites, laminates, new processing methods, and many others. These studies aim to push the boundaries of knowledge about composite materials and maximize the potential of these natural fibers in engineering.

This review article highlights the potential of natural fibers found in the Amazon Rainforest for use in engineering composites and sustainable actions. The region's natural fibers have a high cellulose content and exhibit mechanical properties that make them a promising alternative to synthetic fibers in various applications.

The use of natural fibers contributes to sustainable actions by reducing the environmental impact of production processes and promoting the use of renewable resources. However, it is important to note that there are challenges and limitations to be overcome. More research and development are needed to optimize the properties of natural fibers and improve processing methods.

The use of natural fibers drives industrial development, whether in the production of clothing and textiles or as reinforcement in composites to enhance the properties of plastic or cementitious materials. Sectors such as automotive and construction benefit by combining the cost-effectiveness of these materials with the excellent properties that reinforcements can impart to the matrix. In the well-established Brazilian market, cotton fiber stands out as one of the most utilized, particularly in clothing manufacturing. However, Amazonian natural fibers have the potential to further boost the country's economy in the production of composite fabrics for engineering applications.

The lack of knowledge about these fibers has contributed to their underutilization in the national industry. Nevertheless, there is an observable growth in research involving these materials, exploring their properties and applications. Therefore, these fibers, previously unknown to the general public, have the potential to be incorporated into the industry on a large scale, generating income in the regions where the plants are extracted and boosting the economy of the Amazon region.

In summary, the use of natural fibers from the Amazon has the potential to drive economic growth and promote sustainable development. Furthermore, this utilization contributes to the preservation of the unique biodiversity of this important region.

Author Contributions: Conceptualization, S.N.M.; methodology, P.H.P.M.d.S.; software, P.H.P.M.d.S.; validation, P.H.P.M.d.S.; formal analysis, P.H.P.M.d.S. and B.F.d.A.F.C.; investigation, B.F.d.A.F.C., B.Z.M. and P.H.P.M.d.S.; resources, P.H.P.M.d.S.; data curation, P.H.P.M.d.S.; writing—original draft preparation, P.H.P.M.d.S., B.F.d.A.F.C. and B.Z.M.; writing—review and editing, P.H.P.M.d.S. and S.N.M.; visualization, P.H.P.M.d.S.; supervision, S.N.M.; project administration, S.N.M. All authors have read and agreed to the published version of the manuscript.

Funding: This research received no external funding.

Institutional Review Board Statement: Not applicable.

Informed Consent Statement: Not applicable.

Data Availability Statement: The data presented in this study are available on request from the corresponding authors.

Conflicts of Interest: The authors declare no conflicts of interest.

Abbreviations

The following abbreviations are used in this manuscript:

°	Angle symbol
°C	Celsius
µm	micrometer
Al_2O_3	Aluminium Oxide
BC	Before Christ
BFS	Backface Signature
CC	Creative Commons
DNFI	Discover Natural Fibers Initiative
DSC	Differential Scanning Calorimetry
E_{abs}	Energy absorbed
g	Gram
g/cm^3	Gram per cubic centimeter
GJ	Gigajoule
GPa	Gigapascal
HDPE	High Density Polyethylene
HIPS	High-Impact Polystyrene
J	Joule
kg	Kilogram
m/s	Meters per second
MAS	Multilayered Armor System
MFA	Microfibril Angle
mm	milimeter
MPA	Megapascal
MWCNT	Multi-Walled Carbon Nanotube
N	Newton
N_2	Nitrogen
NaOH	Sodium Hydroxide
NIJ	National Institute of Justice
nm	nanometer
NFRPC	Natural Fiber-Reinforced Polymer Composite
PHB	Polyhydroxybutyrate
PP	Polypropylene
PPr	Recycled Polypropylene
PP-g-MA	Maleic Anhydride
SDGs	Sustainable Development Goals
SEM	Scanning Electron Microscopy
tex	Unit of textile measurement
TGA	Thermogravimetric Analysis
ton	Tonne
US$	Dollar

UV	Ultraviolet
v_i	Initial velocity
v_L	Limit velocity
v_r	Residual velocity
vol.%	Volume percent
wt.%	Weight percent

References

1. Fisher, C.H. History of natural fibers. *J. Macromol. Sci. Chem.* **1981**, *15*, 1345–1375. [CrossRef]
2. Viot, C.R.; Wendel, J.F. Evolution of the cotton genus, Gossypium, and its domestication in the Americas. *Crit. Rev. Plant Sci.* **2023**, *42*, 1–33. [CrossRef]
3. Kieling, A.C.; de Macedo Neto, J.C.; Del Pino, G.G.; da Silva Barboza, R.; Diáz, F.R.V.; Rivera, J.L.V.; Fernández, M.V.; Ketterer, C.G.; Ortega, A.G.; Abarzúa, R.I. Development of an Epoxy Matrix Hybrid Composite with Astrocaryum Aculeatum (Tucumã) Endocarp and Kaolin from the Amazonas State in Brazil. *Polymers* **2023**, *15*, 2532. [CrossRef] [PubMed]
4. Pezzolo, D.B. *Tecidos: História, Rramas, Tipos e Usos*; Editora Senac São Paulo: São Paulo, Brazil, 2021.
5. Baldia, C.M.; Armitage, R.A. Archaeological Textiles as Secondary Plant and Animal Products. *Handb. Archaeol. Sci.* **2023**, *2*, 797–811.
6. Xhauflair, H.; Jago-On, S.; Vitales, T.J.; Manipon, D.; Amano, N.; Callado, J.R.; Tandang, D.; Kerfant, C.; Choa, O.; Pawlik, A. The invisible plant technology of Prehistoric Southeast Asia: Indirect evidence for basket and rope making at Tabon Cave, Philippines, 39–33,000 years ago. *PLoS ONE* **2023**, *18*, e0281415. [CrossRef] [PubMed]
7. Bairagi, N.; Selvadhas, A.; Archarya, S. Innovative Collaboration and Co-designing with Santhal and Mohli Tribes of Dumka, India. 2023. Available online: https://dl.designresearchsociety.org/iasdr/iasdr2023/fullpapers/175/ (accessed on 15 November 2023).
8. Galán-Guevara, C. Understanding change in traditional sustainable livelihoods: A complex socio-ecological system in an indigenous community in Mexico. *Community Dev. J.* **2023**, bsad010. [CrossRef]
9. Balsalobre-Lorente, D.; Abbas, J.; He, C.; Pilař, L.; Shah, S.A.R. Tourism, urbanization and natural resources rents matter for environmental sustainability: The leading role of AI and ICT on sustainable development goals in the digital era. *Resour. Policy* **2023**, *82*, 103445. [CrossRef]
10. Adebayo, T.S.; Ullah, S.; Kartal, M.T.; Ali, K.; Pata, U.K.; Ağa, M. Endorsing sustainable development in BRICS: The role of technological innovation, renewable energy consumption, and natural resources in limiting carbon emission. *Sci. Total Environ.* **2023**, *859*, 160181. [CrossRef]
11. Cardoso, R.L.B.; da Silva Rodrigues, J.; Ramos, R.P.B.; de Castro Correa, A.; Leão Filha, E.M.; Monteiro, S.N.; da Silva, A.C.R.; Fujiyama, R.T.; Candido, V.S. Use of Yarn and Carded Jute as Epoxy Matrix Reinforcement for the Production of Composite Materials for Application in the Wind Sector: A Preliminary Analysis for the Manufacture of Blades for Low-Intensity Winds. *Polymers* **2023**, *15*, 3682. [CrossRef]
12. Ravindran, G.; Mahesh, V.; Bheel, N.; Chittimalla, S.; Srihitha, K.; Sushmasree, A. Usage of Natural Fibre Composites for Sustainable Material Development: Global Research Productivity Analysis. *Buildings* **2023**, *13*, 1260. [CrossRef]
13. Elfaleh, I.; Abbassi, F.; Habibi, M.; Ahmad, F.; Guedri, M.; Nasri, M.; Garnier, C. A comprehensive review of natural fibers and their composites: An eco-friendly alternative to conventional materials. *Results Eng.* **2023**, *19*, 101271. [CrossRef]
14. Hasan, K.F.; Horváth, P.G.; Alpár, T. Potential natural fiber polymeric nanobiocomposites: A review. *Polymers* **2020**, *12*, 1072. [CrossRef] [PubMed]
15. Zhang, Z.; Cai, S.; Li, Y.; Wang, Z.; Long, Y.; Yu, T.; Shen, Y. High performances of plant fiber reinforced composites—A new insight from hierarchical microstructures. *Compos. Sci. Technol.* **2020**, *194*, 108151. [CrossRef]
16. Silva, J.R.S.d.; Fonseca, J.C.P.; Santos, T.d.S.; Oliveira, J.B.d.; Maquiné, T.M.; Freitas, B.M.d.; Silva, R.N.A.; Nascimento, N.R.d.; Costa, J.M.d.; Bello, R.H.; et al. Polymer composite produced with Brazil nut residues and high impact polystyrene. *Polímeros* **2023**, *32*. [CrossRef]
17. Initiative, D.N.F. World Natural Fibre Update November 2023. Discover Natural Fibers Initiative. Technical Report. 2023. Available online: https://dnfi.org (accessed on 15 November 2023).
18. da Silveira, P.H.P.M.; Santos, M.C.C.d.; Chaves, Y.S.; Ribeiro, M.P.; Marchi, B.Z.; Monteiro, S.N.; Gomes, A.V.; Tapanes, N.d.L.C.O.; Pereira, P.S.d.C.; Bastos, D.C. Characterization of Thermo-Mechanical and Chemical Properties of Polypropylene/Hemp Fiber Biocomposites: Impact of Maleic Anhydride Compatibilizer and Fiber Content. *Polymers* **2023**, *15*, 3271. [CrossRef] [PubMed]
19. Das, S.C.; La Rosa, A.D.; Goutianos, S.; Grammatikos, S. Effect of accelerated weathering on the performance of natural fibre reinforced recyclable polymer composites and comparison with conventional composites. *Compos. Part C Open Access* **2023**, *12*, 100378. [CrossRef]
20. Murali, B.; Karthik, K.; Marotrao, S.S.; Laxmaiah, G.; Yadav, A.S.; Prasanth, I.; Abbas, M. Mechanical and dynamic mechanical properties of hybrid kevlar/natural fiber composites. *Mater. Res. Express* **2023**, *10*, 105305. [CrossRef]
21. Vinod, A.; Sanjay, M.; Siengchin, S. Recently explored natural cellulosic plant fibers 2018–2022: A potential raw material resource for lightweight composites. *Ind. Crops Prod.* **2023**, *192*, 116099. [CrossRef]

22. Neto, J.S.; de Queiroz, H.F.; Aguiar, R.A.; Banea, M.D. A review on the thermal characterisation of natural and hybrid fiber composites. *Polymers* **2021**, *13*, 4425. [CrossRef]
23. Kieling, A.C.; Santana, G.P.; Santos, M.C.D.; Neto, J.C.D.M.; Pino, G.G.D.; Santos, M.D.D.; Duvoisin Jr, S.; Panzera, T.H. Wood-plastic composite based on recycled polypropylene and Amazonian tucumã (*Astrocaryum aculeatum*) endocarp waste. *Fibers Polym.* **2021**, *22*, 2834–2845. [CrossRef]
24. Park, H. A study on structural design and analysis of small wind turbine blade with natural fibre (flax) composite. *Adv. Compos. Mater.* **2016**, *25*, 125–142. [CrossRef]
25. Klose, L.; Meyer-Heydecke, N.; Wongwattanarat, S.; Chow, J.; Pérez García, P.; Carré, C.; Streit, W.; Antranikian, G.; Romero, A.M.; Liese, A. Towards Sustainable Recycling of Epoxy-Based Polymers: Approaches and Challenges of Epoxy Biodegradation. *Polymers* **2023**, *15*, 2653. [CrossRef] [PubMed]
26. Panzera, T.H.; Jeannin, T.; Gabrion, X.; Placet, V.; Remillat, C.; Farrow, I.; Scarpa, F. Static, fatigue and impact behaviour of an autoclaved flax fibre reinforced composite for aerospace engineering. *Compos. Part B Eng.* **2020**, *197*, 108049. [CrossRef]
27. Rangappa, S.M.; Siengchin, S.; Parameswaranpillai, J.; Jawaid, M.; Ozbakkaloglu, T. Lignocellulosic fiber reinforced composites: Progress, performance, properties, applications, and future perspectives. *Polym. Compos.* **2022**, *43*, 645–691. [CrossRef]
28. United Nations. Sustainable Development Goals. 2015. Available online: https://www.un.org/sustainabledevelopment/inequality (accessed on 20 November 2023).
29. Maximo, Y.I.; Hassegawa, M.; Verkerk, P.J.; Missio, A.L. Forest bioeconomy in Brazil: Potential innovative products from the forest sector. *Land* **2022**, *11*, 1297. [CrossRef]
30. Kardung, M.; Cingiz, K.; Costenoble, O.; Delahaye, R.; Heijman, W.; Lovrić, M.; van Leeuwen, M.; M'barek, R.; van Meijl, H.; Piotrowski, S.; et al. Development of the circular bioeconomy: Drivers and indicators. *Sustainability* **2021**, *13*, 413. [CrossRef]
31. Barañano, L.; Garbisu, N.; Alkorta, I.; Araujo, A.; Garbisu, C. Contextualization of the bioeconomy concept through its links with related concepts and the challenges facing humanity. *Sustainability* **2021**, *13*, 7746. [CrossRef]
32. Kamarudin, S.H.; Mohd Basri, M.S.; Rayung, M.; Abu, F.; Ahmad, S.; Norizan, M.N.; Osman, S.; Sarifuddin, N.; Desa, M.S.Z.M.; Abdullah, U.H.; et al. A review on natural fiber reinforced polymer composites (NFRPC) for sustainable industrial applications. *Polymers* **2022**, *14*, 3698. [CrossRef]
33. Barbosa, B.M.; Vaz, S., Jr.; Colodette, J.L.; de Siqueira, H.F.; da Silva, C.M.S.; Cândido, W.L. Effects of Kraft Lignin and Corn Residue on the Production of Eucalyptus Pellets. *BioEnergy Res.* **2023**, *16*, 484–493. [CrossRef]
34. Barros, M.A.L.; Silva, C.R.C.D.; Lima, L.M.D.; Farias, F.J.C.; Ramos, G.A.; Santos, R.C.D. A review on evolution of cotton in Brazil: GM, white, and colored cultivars. *J. Nat. Fibers* **2022**, *19*, 209–221. [CrossRef]
35. Soares, V.L.; Faber, M.d.O.; Monteiro, A.F.; Cammarota, M.C.; Ferreira-Leitão, V.S. Potential use of sisal juice as raw material for sequential biological production of hydrogen and methane. *Braz. J. Chem. Eng.* **2023**, 1–12. [CrossRef]
36. Raya, F.T.; Carvalho, L.M.; José, J.; da Cruz, L.P.; Almeida, R.L.; Delevatti, H.A.d.A.; Silveira, N.; da Silva, S.F.; Pissolato, M.D.; Oliveira, A.B.; et al. Rescuing the Brazilian Agave breeding program: Morphophysiological and molecular characterization of a new germplasm. *Front. Chem. Eng.* **2023**, *5*, 1218668. [CrossRef]
37. de Souza, L.A.G. Biodiversity of Fabaceae in the Brazilian Amazon and Its Timber Potential for the Future. In *Tropical Forests-Ecology, Diversity and Conservation Status*; IntechOpen: London, UK, 2023.
38. Rodrigues, L.d.A.P.; Nunes, D.D.G.; Hodel, K.V.S.; Viana, J.D.; Silva, E.P.; Soares, M.B.P. Exotic fruits patents trends: An overview based on technological prospection with a focus on Amazonian. *Heliyon* **2023**, *9*, e22060. [CrossRef] [PubMed]
39. Neves, P.D.; Cabral, M.R.; Santos, V.; Mafra, M.R.P.; Junior, H.S. Technical assessment of leaf fibers from curaua: An amazonian bioresource. *J. Nat. Fibers* **2022**, *19*, 5900–5909. [CrossRef]
40. Akil, H.; Omar, M.; Mazuki, A.M.; Safiee, S.; Ishak, Z.M.; Bakar, A.A. Kenaf fiber reinforced composites: A review. *Mater. Des.* **2011**, *32*, 4107–4121. [CrossRef]
41. Sanjay, M.R.; Arpitha, G.; Yogesha, B. Study on mechanical properties of natural-glass fibre reinforced polymer hybrid composites: A review. *Mater. Today Proc.* **2015**, *2*, 2959–2967. [CrossRef]
42. Tonk, R. Natural fibers for sustainable additive manufacturing: A state of the art review. *Mater. Today Proc.* **2021**, *37*, 3087–3090. [CrossRef]
43. Crewther, W. The stress—Strain characteristics of animal fibers after reduction and alkylation. *Text. Res. J.* **1965**, *35*, 867–877. [CrossRef]
44. Liu, X.; Wang, L.; Wang, X. Evaluating the softness of animal fibers. *Text. Res. J.* **2004**, *74*, 535–538. [CrossRef]
45. Ammar, Z.; Ibrahim, H.; Adly, M.; Sarris, I.; Mehanny, S. Influence of natural fiber content on the frictional material of brake pads—A review. *J. Compos. Sci.* **2023**, *7*, 72. [CrossRef]
46. Al-Kharabsheh, B.N.; Arbili, M.M.; Majdi, A.; Alogla, S.M.; Hakamy, A.; Ahmad, J.; Deifalla, A.F. Basalt fiber reinforced concrete: A compressive review on durability aspects. *Materials* **2023**, *16*, 429. [CrossRef] [PubMed]
47. Tao, W.; Wang, B.; Wang, N.; Guo, Y.; Li, J.; Zhou, Z. Research progress on basalt fiber-based functionalized composites. *Rev. Adv. Mater. Sci.* **2023**, *62*, 20220300. [CrossRef]
48. Dvorkin, L.; Konkol, J.; Marchuk, V.; Huts, A. Efficient, Fine-Grained Fly Ash Concrete Based on Metal and Basalt Fibers. *Materials* **2023**, *16*, 3969. [CrossRef]
49. Peng, Z.; Zhang, H.; Feng, Q.; Zheng, Y. Improvement of basalt fiber dispersion and its effect on mechanical characteristics of oil well cement. *J. Build. Eng.* **2023**, *76*, 107244. [CrossRef]

50. Wang, N.; Liu, H.; Huang, F. Effects of Hybrid Rockwool–Wood Fiber on the Performance of Asbestos-Free Brake Friction Composites. *Lubricants* **2023**, *11*, 27. [CrossRef]
51. Yavuz, H. Effect of limestone usage on tribological properties in copper and asbestos-free brake friction materials. *Ind. Lubr. Tribol.* **2023**, *75*, 238–245. [CrossRef]
52. Suphamitmongkol, W.; Khanoonkon, N.; Rungruangkitkrai, N.; Boonyarit, J.; Changniam, C.; Sampoompuang, C.; Chollakup, R. Potential of Pineapple Leaf Fibers as Sound and Thermal Insulation Materials in Thailand. *Prog. Appl. Sci. Technol.* **2023**, *13*, 26–32.
53. Patel, R.V.; Yadav, A.; Winczek, J. Physical, Mechanical, and Thermal Properties of Natural Fiber-Reinforced Epoxy Composites for Construction and Automotive Applications. *Appl. Sci.* **2023**, *13*, 5126. [CrossRef]
54. Khatri, H.; Naveen, J.; Jawaid, M.; Jayakrishna, K.; Norrrahim, M.; Rashedi, A. Potential of natural fiber based polymeric composites for cleaner automotive component Production-A comprehensive review. *J. Mater. Res. Technol.* **2023**, *25*, 1086–1104.
55. El Hawary, O.; Boccarusso, L.; Ansell, M.P.; Durante, M.; Pinto, F. An overview of natural fiber composites for marine applications. *J. Mar. Sci. Eng.* **2023**, *11*, 1076. [CrossRef]
56. Haramina, T.; Hadžić, N.; Keran, Z. Epoxy Resin Biocomposites Reinforced with Flax and Hemp Fibers for Marine Applications. *J. Mar. Sci. Eng.* **2023**, *11*, 382. [CrossRef]
57. Sangmesh, B.; Patil, N.; Jaiswal, K.K.; Gowrishankar, T.; Selvakumar, K.K.; Jyothi, M.; Jyothilakshmi, R.; Kumar, S. Development of sustainable alternative materials for the construction of green buildings using agricultural residues: A review. *Constr. Build. Mater.* **2023**, *368*, 130457.
58. Özkılıç, Y.O.; Beskopylny, A.N.; Stel'makh, S.A.; Shcherban, E.M.; Mailyan, L.R.; Meskhi, B.; Chernil'nik, A.; Ananova, O.; Aksoylu, C.; Madenci, E. Lightweight expanded-clay fiber concrete with improved characteristics reinforced with short natural fibers. *Case Stud. Constr. Mater.* **2023**, *19*, e02367. [CrossRef]
59. Mansor, M.; Nurfaizey, A.; Tamaldin, N.; Nordin, M. Natural fiber polymer composites: Utilization in aerospace engineering. In *Biomass, Biopolymer-Based Materials, and Bioenergy*; Elsevier: Amsterdam, The Netherlands, 2019; pp. 203–224.
60. Asim, M.; Saba, N.; Jawaid, M.; Nasir, M. Potential of natural fiber/biomass filler-reinforced polymer composites in aerospace applications. In *Sustainable Composites for Aerospace Applications*; Elsevier: Amsterdam, The Netherlands, 2018; pp. 253–268.
61. Dhas, J.E.R.; Arun, M. A review on development of hybrid composites for aerospace applications. *Mater. Today Proc.* **2022**, *64*, 267–273. [CrossRef]
62. Ramamoorthy, S.K.; Skrifvars, M.; Persson, A. A review of natural fibers used in biocomposites: Plant, animal and regenerated cellulose fibers. *Polym. Rev.* **2015**, *55*, 107–162. [CrossRef]
63. Bismarck, A.; Mishra, S.; Lampke, T. Plant fibers as reinforcement for green composites. In *Natural Fibers, Biopolymers, and Biocomposites*; CRC Press: Boca Raton, FL, USA, 2005; pp. 52–128.
64. Anandjiwala, R.D.; Blouw, S. Composites from bast fibres-prospects and potential in the changing market environment. *J. Nat. Fibers* **2007**, *4*, 91–109. [CrossRef]
65. Ticoalu, A.; Aravinthan, T.; Cardona, F. A review of current development in natural fiber composites for structural and infrastructure applications. In Proceedings of the Southern Region Engineering Conference (SREC 2010), Toowoomba, Australia, 11–12 November 2010.
66. Bhattacharyya, D.; Subasinghe, A.; Kim, N.K. Natural fibers: Their composites and flammability characterizations. *Multifunct. Polym. Compos.* **2015**, *1*, 102–143.
67. John, M.J.; Thomas, S. Biofibres and biocomposites. *Carbohydr. Polym.* **2008**, *71*, 343–364. [CrossRef]
68. Wei, J.; Meyer, C. Degradation mechanisms of natural fiber in the matrix of cement composites. *Cem. Concr. Res.* **2015**, *73*, 1–16. [CrossRef]
69. Jayaraman, K. Manufacturing sisal–polypropylene composites with minimum fibre degradation. *Compos. Sci. Technol.* **2003**, *63*, 367–374. [CrossRef]
70. Tsoumis, G. *Science and Technology of Wood: Structure, Properties, Utilization*; Van Nostrand Reinhold New York: New York, NY, USA, 1991; Volume 115.
71. Satyanarayana, K.; Ravikumar, K.; Sukumaran, K.; Mukherjee, P.; Pillai, S.; Kulkarni, A. Structure and properties of some vegetable fibres: Part 3 Talipot and palmyrah fibres. *J. Mater. Sci.* **1986**, *21*, 57–63. [CrossRef]
72. Komuraiah, A.; Kumar, N.S.; Prasad, B.D. Chemical composition of natural fibers and its influence on their mechanical properties. *Mech. Compos. Mater.* **2014**, *50*, 359–376. [CrossRef]
73. Karimah, A.; Ridho, M.R.; Munawar, S.S.; Adi, D.S.; Damayanti, R.; Subiyanto, B.; Fatriasari, W.; Fudholi, A. A review on natural fibers for development of eco-friendly bio-composite: Characteristics, and utilizations. *J. Mater. Res. Technol.* **2021**, *13*, 2442–2458. [CrossRef]
74. Brink, M.; Escobin, R. *Plant Resources of South-East Asia*; Backhuys Publ.: Leiden, The Netherlands, 2003.
75. Stokke, D.D.; Wu, Q.; Han, G. *Introduction to Wood and Natural Fiber Composites*; John Wiley & Sons: Hoboken, NJ, USA, 2013; 314p, ISBN 978-0-470-71091-3.
76. Oksman, K.; Bengtsson, M. Wood fibre thermoplastic composites: Processing, properties and future developments. 2007.
77. Mwaikambo, L.Y.; Ansell, M.P. The effect of chemical treatment on the properties of hemp, sisal, jute and kapok for composite reinforcement. *Die Angew. Makromol. Chem.* **1999**, *272*, 108–116. [CrossRef]

78. Koohestani, B.; Darban, A.; Mokhtari, P.; Yilmaz, E.; Darezereshki, E. Comparison of different natural fiber treatments: A literature review. *Int. J. Environ. Sci. Technol.* **2019**, *16*, 629–642. [CrossRef]
79. Moudood, A.; Rahman, A.; Öchsner, A.; Islam, M.; Francucci, G. Flax fiber and its composites: An overview of water and moisture absorption impact on their performance. *J. Reinf. Plast. Compos.* **2019**, *38*, 323–339. [CrossRef]
80. Baltazar-y Jimenez, A.; Bismarck, A. Wetting behaviour, moisture up-take and electrokinetic properties of lignocellulosic fibres. *Cellulose* **2007**, *14*, 115–127. [CrossRef]
81. Kostic, M.; Pejic, B.; Skundric, P. Quality of chemically modified hemp fibers. *Bioresour. Technol.* **2008**, *99*, 94–99. [CrossRef]
82. Shahzad, A. Hemp fiber and its composites—A review. *J. Compos. Mater.* **2012**, *46*, 973–986. [CrossRef]
83. Li, Z.; Li, Z.; Ding, R.; Yu, C. Composition of ramie hemicelluloses and effect of polysaccharides on fiber properties. *Text. Res. J.* **2016**, *86*, 451–460. [CrossRef]
84. Zhou, J.; Li, Z.; Yu, C. Property of ramie fiber degummed with Fenton reagent. *Fibers Polym.* **2017**, *18*, 1891–1897. [CrossRef]
85. Wang, C.; Bai, S.; Yue, X.; Long, B.; Choo-Smith, L.P. Relationship between chemical composition, crystallinity, orientation and tensile strength of kenaf fiber. *Fibers Polym.* **2016**, *17*, 1757–1764. [CrossRef]
86. Azevedo, A.R.; Lima, T.E.; Reis, R.H.; Oliveira, M.S.; Candido, V.S.; Monteiro, S.N. Guaruman fiber: A promising reinforcement for cement-based mortars. *Case Stud. Constr. Mater.* **2022**, *16*, e01029. [CrossRef]
87. Wang, W.M.; Cai, Z.S.; Yu, J.Y.; Xia, Z.P. Changes in composition, structure, and properties of jute fibers after chemical treatments. *Fibers Polym.* **2009**, *10*, 776–780. [CrossRef]
88. Mishra, S.; Mohanty, A.K.; Drzal, L.T.; Misra, M.; Hinrichsen, G. A review on pineapple leaf fibers, sisal fibers and their biocomposites. *Macromol. Mater. Eng.* **2004**, *289*, 955–974. [CrossRef]
89. Zhu, Z.; Hao, M.; Zhang, N. Influence of contents of chemical compositions on the mechanical property of sisal fibers and sisal fibers reinforced PLA composites. *J. Nat. Fibers* **2018**, *17*, 101–112. [CrossRef]
90. Tenazoa, C.; Savastano, H.; Charca, S.; Quintana, M.; Flores, E. The effect of alkali treatment on chemical and physical properties of ichu and cabuya fibers. *J. Nat. Fibers* **2021**, *18*, 923–936. [CrossRef]
91. Chokshi, S.; Parmar, V.; Gohil, P.; Chaudhary, V. Chemical composition and mechanical properties of natural fibers. *J. Nat. Fibers* **2022**, *19*, 3942–3953. [CrossRef]
92. Saragih, S.W.; Lubis, R.; Wirjosentono, B.; Eddyanto, E. Characteristic of abaca (*Musa textilis*) fiber from Aceh Timur as bioplastic. In *AIP Conference Proceedings*; AIP Publishing: Melville, NY, USA, 2018; Volume 2049, p. 020058. [CrossRef]
93. Hassan, M.M.; Wagner, M.H.; Zaman, H.; Khan, M.A. Physico-mechanical performance of hybrid betel nut (*Areca catechu*) short fiber/seaweed polypropylene composite. *J. Nat. Fibers* **2010**, *7*, 165–177. [CrossRef]
94. Yusriah, L.; Sapuan, S.; Zainudin, E.; Mariatti, M. Exploring the potential of betel nut husk fiber as reinforcement in polymer composites: Effect of fiber maturity. *Procedia Chem.* **2012**, *4*, 87–94. [CrossRef]
95. Ibrahim, M.M.; Dufresne, A.; El-Zawawy, W.K.; Agblevor, F.A. Banana fibers and microfibrils as lignocellulosic reinforcements in polymer composites. *Carbohydr. Polym.* **2010**, *81*, 811–819. [CrossRef]
96. Jagadeesh, D.; Venkatachalam, R.; Nallakumarasamy, G. Characterisation of banana fiber-a review. *J. Environ. Nanotechnol.* **2015**, *4*, 23–26.
97. Chollakup, R.; Smitthipong, W.; Kongtud, W.; Tantatherdtam, R. Polyethylene green composites reinforced with cellulose fibers (coir and palm fibers): Effect of fiber surface treatment and fiber content. *J. Adhes. Sci. Technol.* **2013**, *27*, 1290–1300. [CrossRef]
98. Adeniyi, A.G.; Onifade, D.V.; Ighalo, J.O.; Adeoye, A.S. A review of coir fiber reinforced polymer composites. *Compos. Part B Eng.* **2019**, *176*, 107305. [CrossRef]
99. Yueping, W.; Ge, W.; Haitao, C.; Genlin, T.; Zheng, L.; Feng, X.Q.; Xiangqi, Z.; Xiaojun, H.; Xushan, G. Structures of bamboo fiber for textiles. *Text. Res. J.* **2010**, *80*, 334–343. [CrossRef]
100. Nahar, S.; Hasan, M. Effect of chemical composition, anatomy and cell wall structure on tensile properties of bamboo fiber. *Eng. J.* **2013**, *17*, 61–68. [CrossRef]
101. Satyanarayana, K.; Guimarães, J.; Wypych, F. Studies on lignocellulosic fibers of Brazil. Part I: Source, production, morphology, properties and applications. *Compos. Part A Appl. Sci. Manuf.* **2007**, *38*, 1694–1709. [CrossRef]
102. Guimaraes, J.L.; Frollini, E.; Da Silva, C.; Wypych, F.; Satyanarayana, K. Characterization of banana, sugarcane bagasse and sponge gourd fibers of Brazil. *Ind. Crops Prod.* **2009**, *30*, 407–415. [CrossRef]
103. Tanobe, V.O.; Sydenstricker, T.H.; Munaro, M.; Amico, S.C. A comprehensive characterization of chemically treated Brazilian sponge-gourds (*Luffa cylindrica*). *Polym. Test.* **2005**, *24*, 474–482. [CrossRef]
104. Johar, N.; Ahmad, I.; Dufresne, A. Extraction, preparation and characterization of cellulose fibres and nanocrystals from rice husk. *Ind. Crops Prod.* **2012**, *37*, 93–99. [CrossRef]
105. Panthapulakkal, S.; Zereshkian, A.; Sain, M. Preparation and characterization of wheat straw fibers for reinforcing application in injection molded thermoplastic composites. *Bioresour. Technol.* **2006**, *97*, 265–272. [CrossRef] [PubMed]
106. Reddy, K.O.; Maheswari, C.U.; Shukla, M.; Rajulu, A.V. Chemical composition and structural characterization of Napier grass fibers. *Mater. Lett.* **2012**, *67*, 35–38. [CrossRef]
107. Tomczak, F.; Satyanarayana, K.G.; Sydenstricker, T.H.D. Studies on lignocellulosic fibers of Brazil: Part III–Morphology and properties of Brazilian curauá fibers. *Compos. Part A Appl. Sci. Manuf.* **2007**, *38*, 2227–2236. [CrossRef]
108. Malkapuram, R.; Kumar, V.; Negi, Y.S. Recent development in natural fiber reinforced polypropylene composites. *J. Reinf. Plast. Compos.* **2009**, *28*, 1169–1189. [CrossRef]

109. Yasin, P.; Ramana, M.V.; Vamshi, C.K.; Pradeep, K. A study of continuous Henequen/Epoxy composites. *Mater. Today Proc.* **2019**, *18*, 3798–3811. [CrossRef]
110. Ioelovich, M.; Leykin, A. Structural investigations of various cotton fibers and cotton celluloses. *BioResources* **2008**, *3*, 170–177. [CrossRef]
111. Dorez, G.; Ferry, L.; Sonnier, R.; Taguet, A.; Lopez-Cuesta, J.M. Effect of cellulose, hemicellulose and lignin contents on pyrolysis and combustion of natural fibers. *J. Anal. Appl. Pyrolysis* **2014**, *107*, 323–331. [CrossRef]
112. Bacci, L.; Di Lonardo, S.; Albanese, L.; Mastromei, G.; Perito, B. Effect of different extraction methods on fiber quality of nettle (*Urtica dioica* L.). *Text. Res. J.* **2011**, *81*, 827–837. [CrossRef]
113. Viju, S.; Thilagavathi, G. Characterization of surface modified nettle fibers for composite reinforcement. *J. Nat. Fibers* **2022**, *19*, 1819–1827. [CrossRef]
114. Laftah, W.A.; Wan Abdul Rahman, W.A. Pulping process and the potential of using non-wood pineapple leaves fiber for pulp and paper production: A review. *J. Nat. Fibers* **2016**, *13*, 85–102. [CrossRef]
115. Neto, A.R.S.; Araujo, M.A.; Barboza, R.M.; Fonseca, A.S.; Tonoli, G.H.; Souza, F.V.; Mattoso, L.H.; Marconcini, J.M. Comparative study of 12 pineapple leaf fiber varieties for use as mechanical reinforcement in polymer composites. *Ind. Crops Prod.* **2015**, *64*, 68–78. [CrossRef]
116. Duarte, G.V.; Ramarao, B.V.; Amidon, T.E.; Ferreira, P.T. Effect of hot water extraction on hardwood kraft pulp fibers (*Acer saccharum*, sugar maple). *Ind. Eng. Chem. Res.* **2011**, *50*, 9949–9959. [CrossRef]
117. Neus Anglès, M.; Salvadó, J.; Dufresne, A. Steam-exploded residual softwood-filled polypropylene composites. *J. Appl. Polym. Sci.* **1999**, *74*, 1962–1977. [CrossRef]
118. Bledzki, A.; Franciszczak, P.; Osman, Z.; Elbadawi, M. Polypropylene biocomposites reinforced with softwood, abaca, jute, and kenaf fibers. *Ind. Crops Prod.* **2015**, *70*, 91–99. [CrossRef]
119. Elzubair, A.; Bonelli, C.M.C.; Suarez, J.C.M.; Mano, E.B. Morphological, structural, thermal and mechanical characterization of piassava fibers. *J. Nat. Fibers* **2007**, *4*, 13–31. [CrossRef]
120. Arce, C.; Garzon, E.; Sanchez-Soto, P.J. Phyllite clays as raw materials replacing cement in mortars: Properties of new impermeabilizing mortars. *Constr. Build. Mater.* **2019**, *224*, 348–358. [CrossRef]
121. Lertwattanaruk, P.; Suntijitto, A. Properties of natural fiber cement materials containing coconut coir and oil palm fibers for residential building applications. *Constr. Build. Mater.* **2015**, *94*, 664–669. [CrossRef]
122. Lilargem Rocha, D.; Tambara Júnior, L.U.D.; Marvila, M.T.; Pereira, E.C.; Souza, D.; de Azevedo, A.R.G. A review of the use of natural fibers in cement composites: Concepts, applications and Brazilian history. *Polymers* **2022**, *14*, 2043. [CrossRef]
123. Carr, D.J.; Cruthers, N.M.; Laing, R.M.; Niven, B.E. Fibers from Three Cultivars of New Zealand Flax (*Phormium tenax*). *Text. Res. J.* **2005**, *75*, 93–98. [CrossRef]
124. Sreenivasan, V.; Somasundaram, S.; Ravindran, D.; Manikandan, V.; Narayanasamy, R. Microstructural, physico-chemical and mechanical characterisation of Sansevieria cylindrica fibres—An exploratory investigation. *Mater. Des.* **2011**, *32*, 453–461. [CrossRef]
125. Syed, N.F.; Zakaria, M.H.; Bujang, J.S. Fiber Characteristics and Papermaking of Seagrass Using Hand-beaten and Blended Pulp. *BioResources* **2016**, *11*, 5358–5380. [CrossRef]
126. Mathew, L.; Joseph, K.; Joseph, R. Isora fibre: Morphology, chemical composition, surface modification, physical, mechanical and thermal properties—A potential natural reinforcement. *J. Nat. Fibers* **2007**, *3*, 13–27. [CrossRef]
127. Chirayil, C.J.; Joy, J.; Mathew, L.; Mozetic, M.; Koetz, J.; Thomas, S. Isolation and characterization of cellulose nanofibrils from Helicteres isora plant. *Ind. Crops Prod.* **2014**, *59*, 27–34. [CrossRef]
128. Sreekala, M.; Kumaran, M.; Thomas, S. Oil palm fibers: Morphology, chemical composition, surface modification, and mechanical properties. *J. Appl. Polym. Sci.* **1997**, *66*, 821–835. [CrossRef]
129. Ammar, H.; Abid, M.; Abid, S. Cellulose fibers obtained by organosolv process from date palm rachis (*Phoenix dactylifera* L.). In Proceedings of the Materiaux 2010, Mahdia, Tunisia, 4–7 January 2010; IOP Publishing: Bristol, UK, 2012; 28, p. 012002. [CrossRef]
130. Sivamurugan, P.; Selvam, R.; Pandian, M.; Ashraf, M.S.; Chakrapani, I.S.; Thanikasalam, A.; Roshith, P.; Ramesh, K.; Ramesh, B. Extraction of novel biosilica from finger millet husk and its coconut rachilla-reinforced epoxy biocomposite: Mechanical, thermal, and hydrophobic behaviour. *Biomass Convers. Biorefinery* **2022**, 1–9. [CrossRef]
131. da Silva Nicolau, G.; Weber, R.P.; Monteiro, S.N.; Loureiro, G.A.; Lavinsky, A.A.; da Fonseca, L.V.; da Silva, E.G.; dos Santos, P.L.B.; Abranches, R.P.; Machado, V.C. Influence of Mercerization Process on the Surface of Coconut Fiber for Composite Reinforcement. In Proceedings of the Materials Science Forum, Chennai, India, 16–17 September 2020; Trans Tech Publications: Wollerau, Switzerland, 2020; Volume 1012, pp. 37–42.
132. Brígida, A.; Calado, V.; Gonçalves, L.; Coelho, M. Effect of chemical treatments on properties of green coconut fiber. *Carbohydr. Polym.* **2010**, *79*, 832–838. [CrossRef]
133. Kulandaivel, N.; Muralikannan, R.; KalyanaSundaram, S. Extraction and characterization of novel natural cellulosic fibers from pigeon pea plant. *J. Nat. Fibers* **2018**, *17*, 769–779. [CrossRef]
134. Fiore, V.; Scalici, T.; Valenza, A. Characterization of a new natural fiber from *Arundo donax* L. as potential reinforcement of polymer composites. *Carbohydr. Polym.* **2014**, *106*, 77–83. [CrossRef]
135. Bledzki, A.K.; Mamun, A.A.; Volk, J. Physical, chemical and surface properties of wheat husk, rye husk and soft wood and their polypropylene composites. *Compos. Part A Appl. Sci. Manuf.* **2010**, *41*, 480–488. [CrossRef]

136. Belkhir, S.; Koubaa, A.; Khadhri, A.; Ksontini, M.; Nadji, H.; Smiti, S.; Stevanovic, T. Seasonal effect on the chemical composition of the leaves of *Stipa tenacissima* L. and implications for pulp properties. *Ind. Crops Prod.* **2013**, *44*, 56–61. [CrossRef]
137. Guna, V.; Ilangovan, M.; Adithya, K.; CV, A.K.; Srinivas, C.; Yogesh, S.; Nagananda, G.; Venkatesh, K.; Reddy, N. Biofibers and biocomposites from sabai grass: A unique renewable resource. *Carbohydr. Polym.* **2019**, *218*, 243–249. [CrossRef] [PubMed]
138. Nath Barman, D.; Haque, M.A.; Kang, T.H.; Kim, G.H.; Kim, T.Y.; Kim, M.K.; Yun, H.D. Effect of mild alkali pretreatment on structural changes of reed (Phragmites communis Trinius) straw. *Environ. Technol.* **2014**, *35*, 232–241. [CrossRef] [PubMed]
139. Loganathan, T.M.; Sultan, M.T.H.; Ahsan, Q.; Jawaid, M.; Naveen, J.; Shah, A.U.M.; Hua, L.S. Characterization of alkali treated new cellulosic fibre from Cyrtostachys renda. *J. Mater. Res. Technol.* **2020**, *9*, 3537–3546. [CrossRef]
140. Chow, P.; Nakayama, F.S.; Blahnik, B.; Youngquist, J.A.; Coffelt, T.A. Chemical constituents and physical properties of guayule wood and bark. *Ind. Crops Prod.* **2008**, *28*, 303–308. [CrossRef]
141. Song, Y.; Han, G.; Li, M.; Jiang, W.; Li, X.; Zhang, Y.; Yu, R. Performance analysis of kudzu fiber prepared by using combined steam explosion and chemical degumming. *J. Nat. Fibers* **2017**, *14*, 759–768. [CrossRef]
142. Kazama, E.H.; Ferreira, F.M.; Silva, A.R.B.d.; Fiorese, D.A. Influência do sistema de colheita nas características da fibra do algodão. *Rev. Ceres* **2016**, *63*, 631–638. [CrossRef]
143. Mohammed, L.; Ansari, M.N.; Pua, G.; Jawaid, M.; Islam, M.S. A review on natural fiber reinforced polymer composite and its applications. *Int. J. Polym. Sci.* **2015**, *2015*, 243947. [CrossRef]
144. Petroudy, S.D. Physical and mechanical properties of natural fibers. In *Advanced High Strength Natural Fibre Composites in Construction*; Elsevier: Amsterdam, The Netherlands, 2017; pp. 59–83.
145. Madsen, B.; Gamstedt, E.K. Wood versus plant fibers: Similarities and differences in composite applications. *Adv. Mater. Sci. Eng.* **2013**, *2013*, 564346. [CrossRef]
146. Joseph, K.; Thomas, S.; Pavithran, C. Effect of chemical treatment on the tensile properties of short sisal fibre-reinforced polyethylene composites. *Polymer* **1996**, *37*, 5139–5149. [CrossRef]
147. Khalid, M.Y.; Imran, R.; Arif, Z.U.; Akram, N.; Arshad, H.; Al Rashid, A.; Garcia Marquez, F.P. Developments in chemical treatments, manufacturing techniques and potential applications of natural-fibers-based biodegradable composites. *Coatings* **2021**, *11*, 293. [CrossRef]
148. Sonawane, G.H.; Patil, S.P.; Sonawane, S.H. Nanocomposites and its applications. In *Applications of Nanomaterials*; Elsevier: Amsterdam, The Netherlands, 2018; pp. 1–22.
149. Meneghetti, P.; Qutubuddin, S. Synthesis, thermal properties and applications of polymer-clay nanocomposites. *Thermochim. Acta* **2006**, *442*, 74–77. [CrossRef]
150. Vilaseca, F.; Corrales, F.; Llop, M.F.; Pelach, M.A.; Mutjé, P. Chemical treatment for improving wettability of biofibres into thermoplastic matrices. *Compos. Interfaces* **2005**, *12*, 725–738. [CrossRef]
151. Faruk, O.; Bledzki, A.K.; Fink, H.P.; Sain, M. Biocomposites reinforced with natural fibers: 2000–2010. *Prog. Polym. Sci.* **2012**, *37*, 1552–1596. [CrossRef]
152. Gholampour, A.; Ozbakkaloglu, T. A review of natural fiber composites: Properties, modification and processing techniques, characterization, applications. *J. Mater. Sci.* **2020**, *55*, 829–892. [CrossRef]
153. Rahman, A.S.; Mathur, V.; Asmatulu, R. Effect of nanoclay and graphene inclusions on the low-velocity impact resistance of Kevlar-epoxy laminated composites. *Compos. Struct.* **2018**, *187*, 481–488. [CrossRef]
154. Sanjay, M.; Madhu, P.; Jawaid, M.; Senthamaraikannan, P.; Senthil, S.; Pradeep, S. Characterization and properties of natural fiber polymer composites: A comprehensive review. *J. Clean. Prod.* **2018**, *172*, 566–581. [CrossRef]
155. Amiandamhen, S.; Meincken, M.; Tyhoda, L. Natural fibre modification and its influence on fibre-matrix interfacial properties in biocomposite materials. *Fibers Polym.* **2020**, *21*, 677–689. [CrossRef]
156. Nurazzi, N.; Asyraf, M.; Rayung, M.; Norrrahim, M.; Shazleen, S.; Rani, M.; Shafi, A.; Aisyah, H.; Radzi, M.; Sabaruddin, F.; et al. Thermogravimetric analysis properties of cellulosic natural fiber polymer composites: A review on influence of chemical treatments. *Polymers* **2021**, *13*, 2710. [CrossRef]
157. Mohammed, M.; Rahman, R.; Mohammed, A.M.; Adam, T.; Betar, B.O.; Osman, A.F.; Dahham, O.S. Surface treatment to improve water repellence and compatibility of natural fiber with polymer matrix: Recent advancement. *Polym. Test.* **2022**, *115*, 107707. [CrossRef]
158. Corrales, F.; Vilaseca, F.; Llop, M.; Girones, J.; Mendez, J.; Mutje, P. Chemical modification of jute fibers for the production of green-composites. *J. Hazard. Mater.* **2007**, *144*, 730–735. [CrossRef]
159. Gallego, R.; Piras, C.C.; Rutgeerts, L.A.; Fernandez-Prieto, S.; De Borggraeve, W.M.; Franco, J.M.; Smets, J. Green approach for the activation and functionalization of jute fibers through ball milling. *Cellulose* **2020**, *27*, 643–656. [CrossRef]
160. Sood, M.; Dwivedi, G. Effect of fiber treatment on flexural properties of natural fiber reinforced composites: A review. *Egypt. J. Pet.* **2018**, *27*, 775–783. [CrossRef]
161. Ku, H.; Wang, H.; Pattarachaiyakoop, N.; Trada, M. A review on the tensile properties of natural fiber reinforced polymer composites. *Compos. Part B Eng.* **2011**, *42*, 856–873. [CrossRef]
162. Benítez, A.N.; Monzón, M.D.; Angulo, I.; Ortega, Z.; Hernández, P.M.; Marrero, M.D. Treatment of banana fiber for use in the reinforcement of polymeric matrices. *Measurement* **2013**, *46*, 1065–1073. [CrossRef]
163. Huda, M.; Drzal, L.; Ray, D.; Mohanty, A.; Mishra, M. Natural-fiber composites in the automotive sector. In *Properties and Performance of Natural-Fibre Composites*; Elsevier: Amsterdam, The Netherlands, 2008; pp. 221–268.

164. da Costa Tavares, M.G. A Amazônia brasileira: Formação histórico-territorial e perspectivas para o século XXI. *GEOUSP Espaço E Tempo (Online)* **2011**, *15*, 107–121.
165. Rorato, A.C.; Dal'Asta, A.P.; Lana, R.M.; Dos Santos, R.B.; Escada, M.I.S.; Vogt, C.M.; Neves, T.C.; Barbosa, M.; Andreazzi, C.S.; Dos Reis, I.C.; et al. Trajetorias: A dataset of environmental, epidemiological, and economic indicators for the Brazilian Amazon. *Sci. Data* **2023**, *10*, 65. [CrossRef]
166. Buchdid Camargo Neves, L.T.; Dos Santos Campos, D.C.; Souza Mendes, J.K.; Urnhani, C.O.; Mattos De Araujo, K.G. Quality of Fruits Manually Processed of Acai (*Euterpe oleracea* MART.) and BACABA (*Oenocarpus bacaba* MART.). *Rev. Bras. De Frutic.* **2015**, *37*, 729–738. [CrossRef]
167. Yokomizo, G.K.I.; Mochiutti, S.; de Queiroz, J.A.L.; dos Santos, G.R.; Furtado, R.G.; Brandão, A.P.; Colares, I.B. Parameter estimates for genetic characters of assai palm trees fruits in Amapá State. *Ciência Florest.* **2016**, *26*, 985–993. [CrossRef]
168. Nascimento, R.J.S.d.; Couri, S.; Antoniassi, R.; Freitas, S.P. Composição em ácidos graxos do óleo da polpa de açaí extraído com enzimas e com hexano. *Rev. Bras. De Frutic.* **2008**, *30*, 498–502. [CrossRef]
169. Rogez, H. *Açai: Preparo, Composição e Melhoramento da Conservação*; EDUFPA: Belém, Brazil, 2000; 313p.
170. Rodrigues, R.B.; Lichtenthäler, R.; Zimmermann, B.F.; Papagiannopoulos, M.; Fabricius, H.; Marx, F.; Maia, J.G.; Almeida, O. Total oxidant scavenging capacity of Euterpe oleracea Mart.(açaí) seeds and identification of their polyphenolic compounds. *J. Agric. Food Chem.* **2006**, *54*, 4162–4167. [CrossRef]
171. Matias dos Santos, G.; Arraes Maia, G.; Machado de Sousa, P.H.; Correia da Costa, J.M.; Wilane de Figueiredo, R.; Matias do Prado, G. Correlação entre atividade antioxidante e compostos bioativos de polpas comerciais de açaí (*Euterpe oleracea* Mart). *Arch. Latinoam. De Nutr.* **2008**, *58*, 187–192.
172. Global Biodiversity Information Facility GBIF. Ocorrência Euterpe Oleracea Mart. 2023. Available online: https://www.gbif.org/pt/species/5293398 (accessed on 10 November 2023).
173. Costa, D.d.S.; El Banna, W.R.; Lima, L.S.; Almeida, D.F.; Santos, E.d.J.S.d.; Lopes, C.E.P.; Fujiyama, R.T. Compósito De Fibras De Açaí E Resina Poliéster. In Proceedings of the 68th abm International Annual Congress, Belo Horizonte, Brazil, 30 July–2 August 2013; pp. 1393–1401.
174. Bufalino, L.; de Sena Neto, A.R.; Tonoli, G.H.D.; de Souza Fonseca, A.; Costa, T.G.; Marconcini, J.M.; Colodette, J.L.; Labory, C.R.G.; Mendes, L.M. How the chemical nature of Brazilian hardwoods affects nanofibrillation of cellulose fibers and film optical quality. *Cellulose* **2015**, *22*, 3657–3672. [CrossRef]
175. Subramoniam, R.; Huisingh, D.; Chinnam, R.B.; Subramoniam, S. Remanufacturing Decision-Making Framework (RDMF): Research validation using the analytical hierarchical process. *J. Clean. Prod.* **2013**, *40*, 212–220. [CrossRef]
176. Al-Oqla, F.M.; Sapuan, M.S.; Ishak, M.R.; Aziz, N.A. Combined multi-criteria evaluation stage technique as an agro waste evaluation indicator for polymeric composites: Date palm fibers as a case study. *BioResources* **2014**, *9*, 4608–4621. [CrossRef]
177. Al-Oqla, F.M.; Sapuan, S. Natural fiber reinforced polymer composites in industrial applications: Feasibility of date palm fibers for sustainable automotive industry. *J. Clean. Prod.* **2014**, *66*, 347–354. [CrossRef]
178. Castro, C.D.P.d.C.; Dias, C.G.B.T.; Faria, J.d.A.F. Production and evaluation of recycled polymers from açaí fibers. *Mater. Res.* **2010**, *13*, 159–163. [CrossRef]
179. Bastos, L.P.; Lima, L.d.C.; Santos, G.B.; Melo, G.d.S.V.d.; Mesquita, A.L.A. Simulação com painéis de fibra de açaí para melhoria da inteligibilidade da fala em sala de aula. *Ambiente Construído* **2021**, *21*, 45–63. [CrossRef]
180. Martins, M.A.; Mattoso, L.H.C.; Pessoa, J.D.C. Comportamento térmico e caracterização morfológica das fibras de mesocarpo e caroço do açaí (*Euterpe oleracea* Mart.). *Rev. Bras. De Frutic.* **2009**, *31*, 1150–1157. [CrossRef]
181. de Carvalho, C.M.R.G.; Castro, G.C.; dos Santos, G.M.; Campelo, J.E.G.; Sarmento, J.L.R.; Carvalho, M.D.F. Aproveitamento integral do coco babaçu como estratégia de educação e desenvolvimento social sustentável. *Extensão Em Foco* **2020**, *21*, 156–164.
182. Lorenzine, H.; Sousa, H.; Coelho, L.; Medeiros, J.; Nikolaus, B. *Palmeiras No Brasil*; Plantarum: Nova Odessa, Brazil, 1996; p. 70.
183. Lorenzi, H.; Noblick, L.; Kahn, F.; Ferreira, E. *Flora Brasileira-Arecaceae (Palmeiras)*; Instituto Plantarum: Nova Odessa, Brazil, 2010.
184. González-Pérez, S.E.; Coelho-Ferreira, M.; Robert, P.d.; Garcés, C.L.L. Conhecimento e usos do babaçu (Attalea speciosa Mart. e Attalea eichleri (Drude) AJ Hend.) entre os Mebêngôkre-Kayapó da Terra Indígena Las Casas, estado do Pará, Brasil. *Acta Bot. Bras.* **2012**, *26*, 295–308. [CrossRef]
185. Silva, G.; Barroso, M.E.; Silva, D.L.; Conceição, G. A importância do coco babaçu para a comunidade escolar e extrativista no município de São João do Arraial/PI. *Agrar. Acad.* **2016**, *3*. [CrossRef]
186. Santana, S.A.; Vieira, A.P.; da Silva Filho, E.C.; Melo, J.C.; Airoldi, C. Immobilization of ethylenesulfide on babassu coconut epicarp and mesocarp for divalent cation sorption. *J. Hazard. Mater.* **2010**, *174*, 714–719. [CrossRef] [PubMed]
187. Global Biodiversity Information Facility GBIF. Ocorrência Attalea Speciosa Mart. ex Spreng. 2023. Available online: https://www.gbif.org/pt/species/8174773 (accessed on 10 November 2023).
188. Carrazza, L.R.; Ávila, J.C.C.; Silva, M.L.D. *Manual Tecnológico de Aproveitamento Integral do Fruto e da Folha do Babaçu (Attalea spp.)*, 2nd ed.; ISPN: Brasília, Brazil, 2012; ISBN 978-85-63288-06-6.
189. Soler, M.P.; Vitali, A.d.A.; Muto, E.F. Tecnologia de quebra do coco babaçu (*Orbignya speciosa*). *Food Sci. Technol.* **2007**, *27*, 717–722. [CrossRef]
190. de Morais, J.P.G.; Campana, M.; Del Valle, T.A.; Moreira, T.G.; da Silva, E.D.R.; do Prado, R.F.; de Oliveira, R.E. Inclusion of babassu bran produced in milk production in Amazonia. *Trop. Anim. Health Prod.* **2021**, *53*, 527. [CrossRef]

191. Costa, R.M.; Leite, M.R.L.; Matos, S.d.S.; Sousa, R.C.M.d.; Silva-Matos, R.R.S.d.; Bezerra, A.A.d.C. Quality of pre-sprouted sugarcane seedlings grown on substrates with decomposed babassu palm stem. *Pesqui. Agropecu. Trop.* **2021**, *51*. [CrossRef]
192. dos Anjos, S.S.N.; do Nascimento Neto, J.O. Avaliação do impacto de política de subvenção econômica na cadeia produtiva de biodiesel de Babaçu. *Rev. Em Agronegócio E Meio Ambiente* **2021**, *14*, 1–16. [CrossRef]
193. Chaves, Y.S.; da Silveira, P.H.P.; Neuba, L.d.M.; Junio, R.F.P.; Ribeiro, M.P.; Monteiro, S.N.; Nascimento, L.F.C. Evaluation of the density, mechanical, thermal and chemical properties of babassu fibers (*Attalea speciosa.*) for potential composite reinforcement. *J. Mater. Res. Technol.* **2023**, *23*, 2089–2100. [CrossRef]
194. De Lemos, A.L.; Mauss, C.J.; Santana, R.M.C. Characterization of natural fibers: Wood, sugarcane and babassu for use in biocomposites. *Cellul. Chem. Technol.* **2017**, *51*, 711–718.
195. Chaves, Y.S.; da Silveira, P.H.P.M.; Monteiro, S.N.; Nascimento, L.F.C. Babassu Coconut Fibers: Investigation of Chemical and Surface Properties (*Attalea speciosa*). *Polymers* **2023**, *15*, 3863. [CrossRef]
196. Furtado, J.B.d.M.; Furtado Filho, P.A.; Oliveira, T.P.; Caetano, M.R.d.S.; Araujo, I.M.d.S.; Figueiredo, F.C.; Santos Júnior, J.R.d. Enhancement of the photodegradative potential of polymer composites containing babassu fiber. *Mater. Res.* **2020**, *23*. [CrossRef]
197. de Oliveira Libório Dourado, J.B.; Alves, M.E.R.; de Oliveira Júnior, W.A.; de Oliveira, B.L.M.; de Jesus Bezerra da Silva, H.; López-Galindo, A.; Viseras, C.; Furtini, M.B.; dos Santos, V.B. Babassu Fibers as Green Mortar Additives. *J. Nat. Fibers* **2023**, *20*, 2256471. [CrossRef]
198. Marinho, V.A.D.; Almeida, T.; Carvalho, L.; Canedo, E.L. Aditivação e biodegradação de compósitos PHB/babaçu. *Rev. Eletrônica De Mater. E Process.* **2018**, *13*, 37–41.
199. Endress, B.A.; Horn, C.M.; Gilmore, M.P. Mauritia flexuosa palm swamps: Composition, structure and implications for conservation and management. *For. Ecol. Manag.* **2013**, *302*, 346–353. [CrossRef]
200. Peters, C.M.; Balick, M.J.; Kahn, F.; Anderson, A.B. Oligarchic forests of economic plants in Amazonia: Utilization and conservation of an important tropical resource. *Conserv. Biol.* **1989**, *3*, 341–349. [CrossRef] [PubMed]
201. Lawrence, A.; Phillips, O.L.; Ismodes, A.R.; Lopez, M.; Rose, S.; Wood, D.; Farfan, A.J. Local values for harvested forest plants in Madre de Dios, Peru: Towards a more contextualised interpretation of quantitative ethnobotanical data. *Biodivers. Conserv.* **2005**, *14*, 45–79. [CrossRef]
202. da Cruz Demosthenes, L.C.; Nascimento, L.F.C.; Monteiro, S.N.; Costa, U.O.; da Costa Garcia Filho, F.; da Luz, F.S.; Oliveira, M.S.; Ramos, F.J.H.T.V.; Pereira, A.C.; Braga, F.O. Thermal and structural characterization of buriti fibers and their relevance in fabric reinforced composites. *J. Mater. Res. Technol.* **2020**, *9*, 115–123. [CrossRef]
203. Mesa, L.; Galeano, G. Palms uses in the Colombian Amazon. *Caldasia* **2013**, *35*, 351–369.
204. Abreu, Y.; Ávila, R.; Gonçalves, T. The buriti agro extractivism: An alternative for the development of the Brazilian Amazon Region. *Int. J. Soc. Sci. Entrep.* **2014**, *1*, 189–197.
205. Virapongse, A. Social mechanisms and mobility: Buriti palm (*Mauritia flexuosa*) extractivism in Brazil. *Hum. Ecol.* **2017**, *45*, 119–129. [CrossRef]
206. Sosnowska, J.; Walanus, A.; Balslev, H. Asháninka palm management and domestication in the Peruvian Amazon. *Hum. Ecol.* **2015**, *43*, 451–466. [CrossRef] [PubMed]
207. Draper, F.C.; Roucoux, K.H.; Lawson, I.T.; Mitchard, E.T.; Coronado, E.N.H.; Lähteenoja, O.; Montenegro, L.T.; Sandoval, E.V.; Zaráte, R.; Baker, T.R. The distribution and amount of carbon in the largest peatland complex in Amazonia. *Environ. Res. Lett.* **2014**, *9*, 124017. [CrossRef]
208. Demosthenes, L.C.d.C.; Luz, F.S.d.; Nascimento, L.F.C.; Monteiro, S.N. Buriti Fabric Reinforced Epoxy Composites as a Novel Ballistic Component of a Multilayered Armor System. *Sustainability* **2022**, *14*, 10591. [CrossRef]
209. da Silva Santos, R.; de Souza, A.A.; De Paoli, M.A.; de Souza, C.M.L. Cardanol–formaldehyde thermoset composites reinforced with buriti fibers: Preparation and characterization. *Compos. Part A Appl. Sci. Manuf.* **2010**, *41*, 1123–1129. [CrossRef]
210. Júnior, H.L.O.; Moraes, A.; Poletto, M.; Zattera, A.J.; Amico, S.C. Chemical composition, tensile properties and structural characterization of buriti fiber. *Cellul. Chem. Technol.* **2016**, *50*, 15–22.
211. Pelegrini, K.; Donazzolo, I.; Brambilla, V.; Coulon Grisa, A.M.; Piazza, D.; Zattera, A.J.; Brandalise, R.N. Degradation of PLA and PLA in composites with triacetin and buriti fiber after 600 days in a simulated marine environment. *J. Appl. Polym. Sci.* **2016**, *133*. [CrossRef]
212. Monteiro, S.N.; Calado, V.; Rodriguez, R.J.; Margem, F.M. Thermogravimetric stability of polymer composites reinforced with less common lignocellulosic fibers—An Overview. *J. Mater. Res. Technol.* **2012**, *1*, 117–126. [CrossRef]
213. Neves Monteiro, S.; de Oliveira Braga, F.; Pereira Lima, E.; Henrique Leme Louro, L.; Wieslaw Drelich, J. Promising curaua fiber-reinforced polyester composite for high-impact ballistic multilayered armor. *Polym. Eng. Sci.* **2017**, *57*, 947–954. [CrossRef]
214. Monteiro, S.N.; Lima, É.P.; Louro, L.H.L.; Da Silva, L.C.; Drelich, J.W. Unlocking function of aramid fibers in multilayered ballistic armor. *Metall. Mater. Trans. A* **2015**, *46*, 37–40. [CrossRef]
215. *NIJ 0101.06*; Ballistic Resistance of Body Armor. National Institute of Justice: Washington, DC, USA, 2008.
216. Cattani, I.M.; Baruque-Ramos, J. Buriti palm fiber (*Mauritia flexuosa* MART.): Characterization and studies for its application in design products. *Key Eng. Mater.* **2016**, *668*, 63–74. [CrossRef]
217. Castro, M.F.d.O.; Xavier, G.S.M.; Dias, C.B.; Santos, K.d.S.d.; Leão, R. Resistência à flexão de compósitos reforçados com fibra de buriti tratada com NaOH. In Proceedings of the Congresso Técnico Científico da Engenharia e da Agronomia—CONTECC, Virtual, 15–17 September 2021; pp. 1–6.

218. Junio, R.F.P.; de Mendonça Neuba, L.; Souza, A.T.; Pereira, A.C.; Nascimento, L.F.C.; Monteiro, S.N. Thermochemical and structural characterization of promising carnauba novel leaf fiber (*Copernicia prunifera*). *J. Mater. Res. Technol.* **2022**, *18*, 4714–4723. [CrossRef]
219. Junio, R.F.P.; Nascimento, L.F.C.; Neuba, L.d.M.; Souza, A.T.; Moura, J.V.B.; Garcia Filho, F.d.C.; Monteiro, S.N. Copernicia prunifera leaf fiber: A promising new reinforcement for epoxy composites. *Polymers* **2020**, *12*, 2090. [CrossRef] [PubMed]
220. Melo, J.D.D.; Carvalho, L.F.M.; Medeiros, A.M.; Souto, C.R.; Paskocimas, C.A. A biodegradable composite material based on polyhydroxybutyrate (PHB) and carnauba fibers. *Compos. Part B Eng.* **2012**, *43*, 2827–2835. [CrossRef]
221. Monteiro, S.N.; Lopes, F.P.D.; Barbosa, A.P.; Bevitori, A.B.; Silva, I.L.A.D.; Costa, L.L.D. Natural lignocellulosic fibers as engineering materials—An overview. *Metall. Mater. Trans. A* **2011**, *42*, 2963–2974. [CrossRef]
222. Junio, R.F.P.; de Mendonça Neuba, L.; Pereira, A.C.; Souza, A.T.; Isidoro, L.M.M.; Nascimento, L.F.C.; Monteiro, S.N. Carnauba leaf fibers: Correlation among diametrical variation, physical and mechanical properties. *J. Mater. Res. Technol.* **2023**, *22*, 1888–1899. [CrossRef]
223. Tiwari, Y.M.; Sarangi, S.K. Characterization of raw and alkali treated cellulosic Grewia Flavescens natural fiber. *Int. J. Biol. Macromol.* **2022**, *209*, 1933–1942. [CrossRef]
224. Eduardo, A.F.; Nóbrega, M.V.d.; Ferreira, R.L.d.S. Evaluation of the behavior of mortars produced with fibers from the straw of carnauba: Effects of the content of addition and length used. *Matéria* **2021**, *26*. [CrossRef]
225. Onuaguluchi, O.; Banthia, N. Plant-based natural fibre reinforced cement composites: A review. *Cem. Concr. Compos.* **2016**, *68*, 96–108. [CrossRef]
226. Silva, E.J.d.; Silva, P.D.d.; Marques, M.L.; Fornari Junior, C.; Garcia, F.C.; Luzardo, F.H. Resistência à compressão de argamassas em função da adição de fibra de coco. *Rev. Bras. De Eng. Agríc. E Ambient.* **2014**, *18*, 1268–1273. [CrossRef]
227. De Pellegrin, M.; Acordi, J.; Montedo, O. Influence of the length and the content of cellulose fibers obtained from sugarcane bagasse on the mechanical properties of fiber-reinforced mortar composites. *J. Nat. Fibers* **2019**, *18*, 111–121. [CrossRef]
228. Castro, D.O.d.; Ruvolo-Filho, A.; Frollini, E. Materials prepared from biopolyethylene and curaua fibers: Composites from biomass. *Polym. Test.* **2012**, *31*, 880–888. [CrossRef]
229. Souza, S.; Ferreira, M.; Sain, M.; Ferreira, M.; Pupo, H.; Cherian, B.; Leão, A. The use of curaua fibers as reinforcements in composites. In *Biofiber Reinforcements in Composite Materials*; Elsevier: Amsterdam, The Netherlands, 2015; pp. 700–720.
230. Santos, P.A.; Spinacé, M.A.; Fermoselli, K.K.; De Paoli, M.A. Efeito da forma de processamento e do tratamento da fibra de curauá nas propriedades de compósitos com poliamida-6. *Polímeros* **2009**, *19*, 31–39. [CrossRef]
231. Leão, A.; Sartor, S.M.; Caraschi, J.C. Natural fibers based composites—Technical and social issues. *Mol. Cryst. Liq. Cryst.* **2006**, *448*, 161–763. [CrossRef]
232. de Freitas, A.E.; Padilha, F.d.J.; Barros, S.d.S.; Khan, T.M.; Pereira, B.; Barbosa, W.T.; Barbosa, J.D.; Calderaro, F.L.; da Silva, S.; Quirino, M.G.; et al. Curauá fiber from plants produced by tissue culture: Thermal, mechanical, and morphological characterizations. *Cellulose* **2023**, *30*, 2841–2858. [CrossRef]
233. Garcés, C.L.L.; Pérez, S.E.G.; Silva, J.A.d.; Araújo, M.O.d.; Coelho-Ferreira, M. Objetos indígenas para o mercado: Produção, intercâmbio, comércio e suas transformações. Experiências Ka'apor e Mebêngôkre-Kayapó. *Bol. Do Mus. Para. Emilio Goeldi. Cienc. Humanas* **2015**, *10*, 659–680. [CrossRef]
234. Costa, U.O.; Nascimento, L.F.C.; Garcia, J.M.; Bezerra, W.B.A.; da Luz, F.S.; Pinheiro, W.A.; Monteiro, S.N. Mechanical properties of composites with graphene oxide functionalization of either epoxy matrix or curaua fiber reinforcement. *J. Mater. Res. Technol.* **2020**, *9*, 13390–13401. [CrossRef]
235. Zah, R.; Hischier, R.; Leão, A.L.; Braun, I. Curauá fibers in the automobile industry—A sustainability assessment. *J. Clean. Prod.* **2007**, *15*, 1032–1040. [CrossRef]
236. Gomes, A.; Matsuo, T.; Goda, K.; Ohgi, J. Development and effect of alkali treatment on tensile properties of curaua fiber green composites. *Compos. Part A Appl. Sci. Manuf.* **2007**, *38*, 1811–1820. [CrossRef]
237. Gomes, A.; Goda, K.; Ohgi, J. Effects of alkali treatment to reinforcement on tensile properties of curaua fiber green composites. *JSME Int. J. Ser. A Solid Mech. Mater. Eng.* **2004**, *47*, 541–546. [CrossRef]
238. Ferreira, S.R.; de Andrade Silva, F.; Lima, P.R.L.; Toledo Filho, R.D. Effect of hornification on the structure, tensile behavior and fiber matrix bond of sisal, jute and curauá fiber cement based composite systems. *Constr. Build. Mater.* **2017**, *139*, 551–561. [CrossRef]
239. Teixeira, F.P.; de Andrade Silva, F. On the use of natural curauá reinforced cement based composites for structural applications. *Cem. Concr. Compos.* **2020**, *114*, 103775. [CrossRef]
240. Teixeira, R.S.; Santos, S.F.d.; Christoforo, A.L.; Savastano, H., Jr.; Lahr, F.A.R. Extrudability of cement-based composites reinforced with curauá (*Ananas erectifolius*) or polypropylene fibers. *Constr. Build. Mater.* **2019**, *205*, 97–110. [CrossRef]
241. Meliande, N.M.; Oliveira, M.S.; Silveira, P.H.P.M.d.; Dias, R.R.; Marçal, R.L.S.B.; Monteiro, S.N.; Nascimento, L.F.C. Curaua—Aramid Hybrid Laminated Composites for Impact Applications: Flexural, Charpy Impact and Elastic Properties. *Polymers* **2022**, *14*, 3749. [CrossRef]
242. Meliande, N.M.; Silveira, P.H.P.M.d.; Monteiro, S.N.; Nascimento, L.F.C. Tensile Properties of Curaua—Aramid Hybrid Laminated Composites for Ballistic Helmet. *Polymers* **2022**, *14*, 2588. [CrossRef] [PubMed]

243. Meliande, N.M.; Oliveira, M.S.; Pereira, A.C.; Balbino, F.D.P.; da Silva Figueiredo, A.B.H.; Monteiro, S.N.; Nascimento, L.F.C. Ballistic properties of curaua-aramid laminated hybrid composites for military helmet. *J. Mater. Res. Technol.* **2023**, *25*, 3943–3956. [CrossRef]
244. da Silva, A.O.; de Castro Monsores, K.G.; Oliveira, S.d.S.; Weber, R.P.; Monteiro, S.N. Ballistic behavior of a hybrid composite reinforced with curaua and aramid fabric subjected to ultraviolet radiation. *J. Mater. Res. Technol.* **2018**, *7*, 584–591. [CrossRef]
245. de Campos, A.; Claro, P.C.; Luchesi, B.R.; Miranda, M.; Souza, F.V.; Ferreira, M.D.; Marconcini, J.M. Curaua cellulose sheets dip coated with micro and nano carnauba wax emulsions. *Cellulose* **2019**, *26*, 7983–7993. [CrossRef]
246. Frollini, E.; Bartolucci, N.; Sisti, L.; Celli, A. Biocomposites based on poly (butylene succinate) and curaua: Mechanical and morphological properties. *Polym. Test.* **2015**, *45*, 168–173. [CrossRef]
247. Borsoi, C.; Scienza, L.C.; Zattera, A.J. Characterization of composites based on recycled expanded polystyrene reinforced with curaua fibers. *J. Appl. Polym. Sci.* **2013**, *128*, 653–659. [CrossRef]
248. Barbalho, G.H.d.A.; Nascimento, J.J.d.S.; Silva, L.B.d.; Gomez, R.S.; Farias, D.O.d.; Diniz, D.D.S.; Santos, R.S.; Figueiredo, M.J.d.; Lima, A.G.B.d. Bio-Polyethylene Composites Based on Sugar Cane and Curauá Fiber: An Experimental Study. *Polymers* **2023**, *15*, 1369. [CrossRef] [PubMed]
249. Neto, J.S.; Cavalcanti, D.K.; da Cunha Ferro, L.E.; de Queiroz, H.F.; Aguiar, R.A.; Banea, M.D. Effect of Multi-Walled Carbon Nanotubes on the Mechanical and Thermal Properties of Curauá Natural-Fiber-Reinforced Composites. *C* **2023**, *9*, 102. [CrossRef]
250. Santos, N.S.S.; Carneiro, S.C.P.M.H.d.S. Utilização do trançado de palha como estratégia para o desenvolvimento sustentável do setor moveleiro. *Simpósio De Eng. De Produção* **2005**, *13*, 2005.
251. Costa , M.O.; Souza, F.S.; Andrade, S.M.C. Ensaios de tração e de absorção de água em compósitos de resina poliester e fibra da tala de Guarumã. In Proceedings of the 21° CBECIMAT - Congresso Brasileiro de Engenharia e Ciência dos Materiais, Cuiabá, Brazil, 9–13 November 2014; Volume 21, pp. 3736–3741.
252. Pinheiro, M.A.; Gomes, L.G.; Silva, A.C.R.d.; Candido, V.S.; Reis, R.H.M.; Monteiro, S.N. Guaruman: A natural Amazonian fiber with potential for Polymer composite reinforcement. *Mater. Res.* **2019**, *22*, e20190092. [CrossRef]
253. Reis, R.H.M.; Nunes, L.F.; Oliveira, M.S.; de Veiga Junior, V.F.; Filho, F.D.C.G.; Pinheiro, M.A.; Candido, V.S.; Monteiro, S.N. Guaruman fiber: Another possible reinforcement in composites. *J. Mater. Res. Technol.* **2020**, *9*, 622–628. [CrossRef]
254. Reis, R.H.M.; Nunes, L.F.; da Luz, F.S.; Candido, V.S.; da Silva, A.C.R.; Monteiro, S.N. Ballistic Performance of Guaruman Fiber Composites in Multilayered Armor System and as Single Target. *Polymers* **2021**, *13*, 1203. [CrossRef] [PubMed]
255. Batista, J.S.; de Moraes, M.F.; da Silva, R.S.D.; da Cruz, W.J.A.; Carniello, M.A. Cochlospermum orinocense (bixaceae) ocorrentes na bacia do guaporé, mato grosso, Brasil. In Proceedings of the 64° Congresso Nacional de Botânica, Belo Horizonte, Brazil, 10–15 November 2013; p. 1.
256. Figueiredo, P.S.d.; Ribeiro, É.K.M.D.; Lacerda, D.M.A.; Girnos, E.C. Estratégia reprodutiva de *Cochlospermum orinocense* (Kunth) Steud.: Fenologia, biologia floral e sistema de cruzamento em uma espécie pioneira de florestas na Amazônia. *Braz. J. Bot.* **2009**, *32*, 781–792. [CrossRef]
257. Reflora. *Cochlospermum Orinocense* (Kunth) Steud. 2023. Available online: https://floradobrasil.jbrj.gov.br/reflora/listaBrasil/ConsultaPublicaUC/BemVindoConsultaPublicaConsultar.do?invalidatePageControlCounter=&idsFilhosAlgas=&idsFilhosFungos=&lingua=&grupo=&familia=null&genero=&especie=&autor=&nomeVernaculo=&nomeCompleto=Cochlospermum+orinocense&formaVida=null&substrato=null&ocorreBrasil=QUALQUER&ocorrencia=OCORRE&endemismo=TODOS&origem=TODOS&regiao=QUALQUER&estado=QUALQUER&ilhaOceanica=32767&domFitogeograficos=QUALQUER&bacia=QUALQUER&vegetacao=TODOS&mostrarAte=SUBESP_VAR&opcoesBusca=TODOS_OS_NOMES&loginUsuario=Visitante&senhaUsuario=&contexto=consulta-publica (accessed on 10 November 2023).
258. da Silva, R.N.A.; de Macedo Neto, J.C.; Kimura, S.P.R. Natural fiber for reinforcement in matrix polymeric. *Indep. J. Manag. Prod.* **2022**, *13*, 154–167. [CrossRef]
259. Pinheiro, M.A.; Ribeiro, M.M.; Rosa, D.L.S.; Nascimento, D.d.C.B.; da Silva, A.C.R.; Dos Reis, M.A.L.; Monteiro, S.N.; Candido, V.S. Periquiteira (*Cochlospermum orinocense*): A Promising Amazon Fiber for Application in Composite Materials. *Polymers* **2023**, *15*, 2120. [CrossRef]
260. Monteiro, S.N. Properties and structure of Attalea funifera piassava fibers for composite reinforcement—A critical discussion. *J. Nat. Fibers* **2009**, *6*, 191–203. [CrossRef]
261. Ferreira, D.C.d.O.N.; Ferreira, A.d.S.; Monteiro, S.N. Weibull analysis of tensile tested piassava fibers with different diameters. *Matéria* **2018**, *23*. [CrossRef]
262. Pimentel, N.M.; Del Menezzi, C. Rendimento do processamento dos produtos oriundos da fibra vegetal da piaçava (*Attalea funifera*). *Nativa* **2020**, *8*, 137–144. [CrossRef]
263. Bonelli, C.; Elzubair, A.; Suarez, J.C.M.; Mano, E.B. Comportamento térmico, mecânico e morfológico de compósitos de polietileno de alta densidade reciclado com fibra de piaçava. *Polímeros* **2005**, *15*, 256–260. [CrossRef]
264. Turini, E. Proposta de preços mínimos safra 2013/2014. Produtos da sociobiodiversidade. *Propos. De Preços Mínimos* **2014**, *3*, 5–19.
265. Fornari, C.I.; Fornari, J.; Celso, C. Avaliação da força de cisalhamento de fibras longas de piaçava em poliéster insaturado. *Estud. Tecnol. Em Eng.* **2013**, *9*. [CrossRef]
266. Guimarães, C.A.L.; Silva, L.A.M. *Piaçava da Bahia (Attalea funifera Martius): Do Extrativismo à Cultura Agrícola*; Editus: Ilhéus, Brazil, 2012.

267. Agrize, P.L.; da Silva, B.D.L.; Carvalho Veiga, B.; Rocha, C.A.A.; Garcia Filho, F.d.C.; Braga, F.d.O. Structure and mechanical behavior of lignosulfonate-treated piassava (*Attalea funifera*) fibers. *Matéria* **2023**, *28*, e20230041. [CrossRef]
268. dos Santos, F.M.; de Souza, T.F.; Barquete, D.M.; Amado, F.D. Comparative analysis of the sisal and piassava fibers as reinforcements in lightweight cementitious composites with EVA waste. *Constr. Build. Mater.* **2016**, *128*, 315–323. [CrossRef]
269. Miranda, C.S.; Fiuza, R.P.; Carvalho, R.F.; José, N.M. Efeito dos tratamentos superficiais nas propriedades do bagaço da fibra de piaçava Attalea funifera martius. *Quím. Nova* **2015**, *38*, 161–165.
270. Carvalho, J.P.R.G.d.; Simonassi, N.T.; Lopes, F.P.D.; Monteiro, S.N.; Vieira, C.M.F. Novel sustainable castor oil-based polyurethane biocomposites reinforced with piassava fiber powder waste for high-performance coating floor. *Sustainability* **2022**, *14*, 5082. [CrossRef]
271. *NBR 14050*; Sistemas de Revestimentos de alto Desempenho, à base de Resinas Epoxídicas e Agregados Minerais-Projeto, Execução e Avaliação do Desempenho-Procedimento. Associação Brasileira de Normas Técnicas: São Paulo, Brazil, 1998.
272. Silva, D.S.; Ribeiro, M.M.; da Silva Rodrigues, J.; de Castro Corrêa, A.; Costa, D.C.L.; de Oliveira Costa, H.A.; da Silva, F.J.A.R.; dos Santos, A.J.G.; da Silva, M.H.P.; Fujiyama, R.T. Properties of flexural and impact of matrix composites polyester reinforced with short lignocellulosic fibers. *Res. Soc. Dev.* **2022**, *11*, e32511326612. [CrossRef]
273. Thomas, B.C.; Jose, Y.S. A study on characteristics of sisal fiber and its performance in fiber reinforced concrete. *Mater. Today Proc.* **2022**, *51*, 1238–1242. [CrossRef]
274. Fernandes, R.A.P.; Silveira, P.H.P.M.d.; Bastos, B.C.; Pereira, P.S.d.C.; Melo, V.A.d.; Monteiro, S.N.; Tapanes, N.d.L.C.O.; Bastos, D.C. Bio-based composites for light automotive parts: Statistical analysis of mechanical properties; effect of matrix and alkali treatment in sisal fibers. *Polymers* **2022**, *14*, 3566. [CrossRef] [PubMed]
275. Nunesa, S.G.; da Silvab, L.V.; Amicob, S.C.; Dantas, J.; Vianac, F.D.R.A. Study of Composites Produced with Recovered Polypropylene and Piassava Fiber. *Mater. Res.* **2017**, *20*, 144–150. [CrossRef]
276. Marinelli, A.L.; Monteiro, M.R.; Ambrósio, J.D.; Branciforti, M.C.; Kobayashi, M.; Nobre, A.D. Desenvolvimento de compósitos poliméricos com fibras vegetais naturais da biodiversidade: Uma contribuição para a sustentabilidade amazônica. *Polímeros* **2008**, *18*, 92–99. [CrossRef]
277. Vasques, J.C. Nanocompósito de Poli(óxido de etileno) Reforçado Com Nanocristais de Celulose Extraídos da Fibra de Tucum. Master's Thesis, Universidade de São Paulo, São Paulo, Brazil, 2018.
278. Ferreira, E.L. *Manual das Palmeiras do Acre, Brasil*; Instituto Nacional de Pesquisas/Universidade Federal do Acre: Rio Branco, Brazil, 2005.
279. Abreu, R.; Nunes, N.L. Tecendo a tradição e valorizando o conhecimento tradicional na Amazônia: O caso da "linha do tucum". *Horiz. Antropol.* **2012**, *18*, 15–43. [CrossRef]
280. Jensen, O.H.; Balslev, H. Etnobotánica de la palma de fibra *Astrocaryum chambira* (Arecaceae) en la Amazonía ecuatoriana. *Econ. Bot.* **1995**, *49*, 309–319. [CrossRef]
281. Vormisto, J. Making and marketing chambira hammocks and bags in the village of Brillo Nuevo, northeastern Peru. *Econ. Bot.* **2002**, *56*, 27–40. [CrossRef]
282. Coomes, O.T. Rain forest 'conservation-through-use'? Chambira palm fibre extraction and handicraft production in a land-constrained community, Peruvian Amazon. *Biodivers. Conserv.* **2004**, *13*, 351–360. [CrossRef]
283. Bernal, R.; Torres, C.; García, N.; Isaza, C.; Navarro, J.; Vallejo, M.I.; Galeano, G.; Balslev, H. Palm management in south america. *Bot. Rev.* **2011**, *77*, 607–646. [CrossRef]
284. Acevedo, L.M.G. El tejido en chambira, una actividad que une más que sogas. *Bol. De Antropol. Univ. De Antioq.* **2005**, *19*, 164–185. [CrossRef]
285. Castaño-Arboleda, N.; Cárdenas, D.; Rodriguez, E.O. *Ecología, Aprovechamiento y Manejo Sostenible de Nueve Especies de Plantas del Departamento del Amazonas, Generadoras de Productos Maderables y No Maderables*; Instituto Amazónico de Investigaciones Científicas "SINCHI": Bogotá, Colombia, 2006; ISBN 978-958-8317-21-2.
286. Oliveira, M.S.; Luz, F.S.d.; Teixeira Souza, A.; Demosthenes, L.C.d.C.; Pereira, A.C.; Filho, F.d.C.G.; Braga, F.d.O.; Figueiredo, A.B.H.d.S.; Monteiro, S.N. Tucum fiber from Amazon Astrocaryum vulgare palm tree: Novel reinforcement for polymer composites. *Polymers* **2020**, *12*, 2259. [CrossRef] [PubMed]
287. Ribeiro, M.P.; de Mendonça Neuba, L.; da Silveira, P.H.P.M.; da Luz, F.S.; da Silva Figueiredo, A.B.H.; Monteiro, S.N.; Moreira, M.O. Mechanical, thermal and ballistic performance of epoxy composites reinforced with Cannabis sativa hemp fabric. *J. Mater. Res. Technol.* **2021**, *12*, 221–233. [CrossRef]
288. da Silva, T.T.; Silveira, P.H.P.M.d.; Figueiredo, A.B.H.d.S.; Monteiro, S.N.; Ribeiro, M.P.; Neuba, L.d.M.; Simonassi, N.T.; Garcia Filho, F.d.C.; Nascimento, L.F.C. Dynamic mechanical analysis and ballistic performance of kenaf fiber-reinforced epoxy composites. *Polymers* **2022**, *14*, 3629. [CrossRef] [PubMed]
289. Henderson, A.; Galeano, G.; Bernal, R. *Field Guide to the Palms of the Americas*; Princeton University Press: Princeton, NJ, USA, 2019; Volume 5390.
290. Anderson, A.B. Os nomes e usos de palmeiras entre uma tribo de índios Yanomama. *Acta Amaz.* **1977**, *7*, 5–13. [CrossRef]
291. Jardim, M.A.G.; Santos, G.d.; Medeiros, T.D.S.; Francez, D.d.C. Diversidade e estrutura de palmeiras em floresta de várzea do estuário amazônico. *Amazôn. Ciênc. Desenvolv.* **2007**, *2*, 67–84.
292. Punt, W.; Wessels Boer, J. A palynological study in geonomoid palms. *Acta Bot. Neerl.* **1966**, *15*, 266–275. [CrossRef]

293. Marchi, B.Z. Caracterização e Avaliação Balística de Compósitos Epóxi Reforçados Com Fibras de Ubim. Ph.D. Thesis, Instituto Militar de Engenharia, Rio de Janeiro, Brazil, 2023.
294. de Granville, J.J. La distribución de las palmas en Guyana Francesa. *Acta Amaz.* **1989**, *19*, 115–138. [CrossRef]
295. Rocha, A.E.S.d.; Silva, M.F.F.d. Aspectos fitossociológicos, florísticos e etnobotânicos das palmeiras (Arecaceae) de floresta secundária no município de Bragança, PA, Brasil. *Acta Bot. Bras.* **2005**, *19*, 657–667. [CrossRef]
296. Siviero, A.; Lin, C.M.; Silveira, M.; Daly, D.C.; Wallace, R.H. *Etnobotânica e Botânica Econômica do Acre*; Edufac: Rio Branco, Brazil, 2016; 410p, ISBN 978-85-8236-027-9.
297. Marchi, B.Z.; Oliveira, M.S.; Bezerra, W.B.A.; de Sousa, T.G.; Candido, V.S.; da Silva, A.C.R.; Monteiro, S.N. Ubim fiber (*Geonoma baculífera*): A less known Brazilian amazon natural fiber for engineering applications. *Sustainability* **2021**, *14*, 421. [CrossRef]
298. Marchi, B.Z.; Silveira, P.H.P.M.d.; Bezerra, W.B.A.; Nascimento, L.F.C.; Lopes, F.P.D.; Candido, V.S.; Silva, A.C.R.d.; Monteiro, S.N. Ballistic Performance, Thermal and Chemical Characterization of Ubim Fiber (*Geonoma baculifera*) Reinforced Epoxy Matrix Composites. *Polymers* **2023**, *15*, 3220. [CrossRef] [PubMed]
299. Silva, T.S.; Henrique, D.C.; Meili, L.; SOLETTI, J.I.; Carvalho, S. Utilização da fibra do Ouricuri (*Syagrus coronata*) na remoção do corante azul de metileno: Variáveis de processo e isoterma de adsorção. In Proceedings of the XXXVII Congresso Brasileiro de Sistemas Particulados, São Carlos, Brazil, 18–21 October 2015; pp. 1–10.
300. Silvestrim, R.G.; Santana, R.M.C.; Moura, A.d.S. Avaliação comparativa da degradação por intemperismo natural e por compostagem de compósitos de PLA reforçado com fibra vegetal da árvore Embira. In Proceedings of the Congresso Brasileiro de Polímeros, Bento Gonçalves, Brazil, 27–31 January 2019; pp. 1–5.
301. Holanda, F.S.R.; Castro, J.D.S.d.; Santos, L.D.V.; Andrade, C.E.C.d.; Griza, S.; Rodrigues Junior, J.J.; Pedrotti, A. Degradação de geotêxteis de Typha latifolia Linn utilizados em técnicas de bioengenharia de solos. *Ciênc. Florest.* **2020**, *30*, 1147–1160. [CrossRef]
302. da Cunha, J.d.S.C.; Nascimento, L.F.C.; da Luz, F.S.; Monteiro, S.N.; Lemos, M.F.; da Silva, C.G.; Simonassi, N.T. Physical and Mechanical Characterization of Titica Vine (*Heteropsis flexuosa*) Incorporated Epoxy Matrix Composites. *Polymers* **2021**, *13*, 4079. [CrossRef] [PubMed]
303. Cunha, J.d.S.C.d.; Nascimento, L.F.C.; Luz, F.S.d.; Garcia Filho, F.d.C.; Oliveira, M.S.; Monteiro, S.N. Titica Vine Fiber (*Heteropsis flexuosa*): A Hidden Amazon Fiber with Potential Applications as Reinforcement in Polymer Matrix Composites. *J. Compos. Sci.* **2022**, *6*, 251. [CrossRef]
304. Melo, D.Q.; Vidal, C.B.; da Silva, A.L.; Teixeira, R.N.; Raulino, G.S.C.; Medeiros, T.C.; Fechine, P.B.; Mazzeto, S.E.; De Keukeleire, D.; Nascimento, R.F. Removal of Cd^{2+}, Cu^{2+}, Ni^{2+}, and Pb^{2+} ions from aqueous solutions using tururi fibers as an adsorbent. *J. Appl. Polym. Sci.* **2014**, *131*. [CrossRef]
305. Vidal, C.B.; Melo, D.Q.; Raulino, G.S.; da Luz, A.D.; da Luz, C.; Nascimento, R.F. Multielement adsorption of metal ions using Tururi fibers (*Manicaria Saccifera*): Experiments, mathematical modeling and numerical simulation. *Desalination Water Treat.* **2016**, *57*, 9001–9008. [CrossRef]
306. Monteiro, A.S.; Baruque-Ramos, J. Amazonian Tururi palm fiber material (*Manicaria saccifera* Gaertn). In Proceedings of the 2nd International Conference on Natural Fibers, São Miguel, Portugal, 27–29 April 2015; Springer: Berlin/Heidelberg, Germany, 2016; pp. 127–137, ISBN 978-94-017-7515-1.
307. Viana, Á.L.; Mady, F.T.M.; do Carmo, M.A.; da Silva Guimarães, D.F. Pecíolo de inajá (Maximiliana maripa [aubl.] Drud) como fonte de matéria prima para produção de papel na Amazônia. *Rev. Eletrônica Em Gestão Educ. E Tecnol. Ambient.* **2014**, 1512–1520.
308. Oliveira, A.; d'Almeida, J. Characterization of ubuçu (*Manicaria saccifera*) natural fiber mat. *Polym. Renew. Resour.* **2014**, *5*, 13–28. [CrossRef]
309. Baruque-Ramos, J.; Monteiro, A.; Midani, M.; Seyam, A.F.M.; Leonardi, B.; Pennas, L.; Cattani, I. Textile Palm Fibers from Amazon Biome. *Mater. Res. Proc.* **2019**, *11*, 262–274.
310. Chu, Y.; Meyers A, M.A.; Wang, B.; Yang, W.; Jung, J.Y.; Coimbra, C.F. A sustainable substitute for ivory: The Jarina seed from the Amazon. *Sci. Rep.* **2015**, *5*, 14387. [CrossRef] [PubMed]
311. Cruz, O.M.; de Souza, R.L.F.; de Freitas, R.R.; de Souza, L.M.; Mafra, E.R.M.L.; da Silva e Silva, B.M.; Bufalino, L. The properties of the mesocarp fibers of patauá, a multiple-use palm from the Amazonia forest. *SN Appl. Sci.* **2019**, *1*, 1317. [CrossRef]
312. Cruz, O.M.; Dias, M.C.; Oliveira, D.N.P.S.D.; Silva, M.G.D.; Souza, T.M.D.; Mendes, L.M.; Matos, L.C.; Bufalino, L. Characterization of raw and thermochemically-treated mesocarp fibers of Oenocarpus bataua, an Amazon palm. *CERNE* **2023**, *29*, e103219. [CrossRef]

MDPI
St. Alban-Anlage 66
4052 Basel
Switzerland
www.mdpi.com

Eng Editorial Office
E-mail: eng@mdpi.com
www.mdpi.com/journal/eng

www.ingramcontent.com/pod-product-compliance
Lightning Source LLC
Chambersburg PA
CBHW061448220926
51850CB00004B/41